TM 9-2320-272-24-3
5 Ton M939 Series Truck
Direct and General Support Maintenance Manual

Vol 3 of 4

June 1998

This manual contains maintenance information for the 5 ton M939 US Military Trucks. This is volume 3 of 4 in the Direct / General Support Manual Series. M939 series trucks are a 5 ton heavy duty 6x6 truck. Cargo versions were designed to transport 10,000 pounds of cargo in all terrain and all weather conditions. Originally designed in the 1970's to replace the M39 and M809 series of vehicles. 44,590 units were produced. This manual is printed to help private owners in the maintenance of their vehicles.

Should you have suggestions or feedback on ways to improve this book please send email to Books@OcotilloPress.com

---

Edited 2021 Ocotillo Press
ISBN 978-1-954285-65-1

Ocotillo Press
Houston, TX 77017
Books@OcotilloPress.com

***ARMY TM 9-2320-272-24-3**
AIR FORCE TO 36A12-1C-1155-2-3

This publication supersedes TM 9-2320-272-20-1, October 1985, and changes 1 through 4; TM 9-2320-272-20-2, October 1985, and changes 1 through 3; TM 9-2320-272-34-1, June 1986, and changes 1 through 2 TM 9-2320-272-34-2, June 1986, and changes 1 and 2; and TM 9-2320-358-24&P, October 1992

TECHNICAL MANUAL
VOLUME 3 OF 4

UNIT, DIRECT SUPPORT, AND GENERAL SUPPORT MAINTENANCE MANUAL FOR

TRUCK, 5-TON, 6X6, M939, M939A1, M939A2 SERIES TRUCKS (DIESEL)

TRUCK, CARGO: 5-TON, 6X6, DROPSIDE,
M923 (2320-01-050-2084) (EIC: BRY); M923A1 (2320-01-206-4087) (EIC: BSS); M923A2 (2320-01-230-0307) (EIC: BS7);
M925 (2320-01-047-8769) (EIC: BRT); M925A1 (2320-01-206-4088) (EIC: BST); M925A2 (2320-01-230-0308) (EIC: BS8);

TRUCK, CARGO: 5-TON, 6X6 XLWB,
M927 (2320-01-047-8771) (EIC: BRV); M927A1 (2320-01-206-4089) (EIC: BSW); M927A2 (2320-01-230-0309) (EIC: BS9);
M928 (2320-01-047-8770) (EIC: BRU); M928A1 (2320-01-206-4090) (EIC: BSX); M928A2 (2320-01-230-0310) (EIC: BTM);

TRUCK, DUMP: 5-TON, 6X6,
M929 (2320-01-047-8756) (EIC: BTH); M929A1 (2320-01-206-4079) (EIC: BSY); M929A2 (2320-01-230-0305) (EIC: BTN);
M930 (2320-01-047-8755) (EIC: BTG); M930A1 (2320-01-206-4080) (EIC: BSZ); M930A2 (2320-01-230-0306) (EIC: BTO);

TRUCK, TRACTOR: 5-TON, 6X6,
M931 (2320-01-047-8753) (EIC: BTE); M931A1 (2320-01-206-4077) (EIC: BS2); M931A2 (2320-01-230-0302) (EIC: BTP);
M932 (2320-01-047-8752) (EIC: BTD); M932A1 (2320-01-205-2684) (EIC: BS5); M932A2 (2320-01-230-0303) (EIC: BTQ);

TRUCK, VAN, EXPANSIBLE: 5-TON, 6X6,
M934 (2320-01-047-8750) (EIC: BTB); M934A1 (2320-01-205-2682) (EIC: BS4); M934A2 (2320-01-230-0300) (EIC: BTR);

TRUCK, MEDIUM WRECKER: 5-TON, 6X6,
M936 (2320-01-047-8754) (EIC: BTF); M936A1 (2320-01-206-4078) (EIC: BS6); M936A2 (2320-01-230-0304) (EIC: BTT).

DEPARTMENTS OF THE ARMY AND THE AIR FORCE

JUNE 1998

WARNING

EXHAUST GASES CAN KILL

1. DO NOT operate vehicle engine in enclosed area.
2. DO NOT idle vehicle engine with windows closed.
3. DO NOT drive vehicle with inspection plates or cover plates removed.
4. BE ALERT at all times for odors.
5. BE ALERT for exhaust poisoning symptoms. They are:
   - Headache
   - Dizziness
   - Sleepiness
   - Loss of muscular control
6. IF YOU SEE another person with exhaust poisoning symptoms:
   - Remove person from area
   - Expose to open air
   - Keep person warm
   - Do not permit person to move
   - Administer artificial respiration or CPR, if necessary*

   * For artificial respiration, refer to FM 21-11.
7. BE AWARE: The field protective mask for Nuclear, Biological, or Chemical (NBC) protection will not protect you from carbon monoxide poisoning.

   THE BEST DEFENSE AGAINST EXHAUST POISONING IS ADEQUATE VENTILATION.

WARNING SUMMARY

- Hearing protection is required for the driver and passenger. Hearing protection is also required for all personnel working in and around this vehicle while the engine is running (AR-40-5 and TB MED 501).
- If required to remain inside vehicle during extreme heat, occupants should follow the water intake, work/rest cycle, and other stress preventive measures (FM 21-10, Field Hygiene and Sanitation).
- If NBC exposure is suspected, all air filter media should be handled by personnel wearing protective equipment. Consult with your unit NBC officer or NBC NCO for appropriate handling or disposal instructions.
- This vehicle has been designed to operate safely and efficiently within the limits specified in this TM. Operation beyond these limits is prohibited by IAW AR 70-1 without written approval from the commander, U.S. Army Tank-automotive and Armaments Command, ATTN: AMCPEO-CM-S, Warren, MI 48397-5000.
- Never work under dump body unless safety braces are properly positioned. Failure to do this will result in injury to personnel.
- During winching operation, never stand between vehicles. Assistant must remain in secondary vehicle to engage service brake if cable snaps or automatic brake fails while towing vehicle. Failure to do this may result in injury to personnel.
- Accidental or intentional introduction of liquid contaminants into the environment is in violation of state, federal, and military regulations. Refer to Army POL (para. 1-7) for information concerning storage, use, and disposal of these liquids. Failure to do so may result in injury or death.
- Cleaning solvents are flammable and toxic. Do not use near open flame and always have a fire extinguisher nearby when solvents are used. Use only in well-ventilated places, wear protective clothing, and dispose of cleaning rags in approved container. Failure to do this will result in injury to personnel and/or damage to equipment.
- Eyeshields must be worn when cleaning with compressed air. Compressed air source will not exceed 30 psi (207 kPa). Failure to do so may result in injury to personnel.
- Extreme care should be taken when removing surge tank filler cap if temperature gauge reads above 175°F (79°C). Steam or hot coolant under pressure will cause injury.
- Alcohol used in the alcohol evaporator is flammable, poisonous, and explosive. Do not smoke when removing alcohol evaporator or adding fluid, and do not drink fluid. Failure to do this will result in injury or death.
- Do not perform electrical circuit testing fuel tank with fill cap or sending unit removed. Fuel may ignite, causing injury to personnel.
- When performing battery maintenance, ensure batteries are seated and clamped down, all rubber boots are installed, clamps are well down on battery posts, and all battery cables lie flat against the top of the batteries. Failure to do this may result in injury to personnel and/or damage to equipment.
- Ensure companion seatbelts are not caught inside battery box. This will cause belts to rot which may lead to injury of personnel.
- On M936/A1/A2 model vehicles, remove spare tire prior to changing tire and install tire in spare tire carrier after tire change is complete. Operation of crane and/or vehicle engine while vehicle is on jacks may result in injury to personnel or damage to equipment.
- Never assemble or disassemble tire and rim assembly while inflated, use inflation to seat lockring on split rim or tire on two-piece rim, or inflate a tire without a tire inflation cage. Injury to personnel may result.
- Do not disconnect air lines or hoses, remove safety valves or CTIS components, or perform brake chamber repairs before draining air reservoirs. Small parts under pressure may shoot out with high velocity, causing injury to personnel.

WARNING SUMMARY (Contd)

- Remove all jewelry when working on electrical circuits. Jewelry coming in contact with electrical circuits may produce a short circuit, causing extreme heat, explosions, and fling particles of metal. Failure to do so will result in injury or death and damage to equipment.
- Use eyeshields and follow instructions carefully when performing assembling, disassembling, or maintenance on this device. Components of this device are under spring tension and may shoot out at a high velocity. Failure to do so will result in injury to personnel.
- Do not remove hoses with engine running or start engine with hoses removed. High-pressure fluids may cause hoses to whip violently and spray randomly Failure to do so may result in injury to personnel.
- Keep hands out from between metal surfaces when removing heavy components. Failure to do so may result in injury to personnel.
- Keep personnel out from under equipment and components of equipment when supported by only a lifting device. Sudden loss of lifting power or shift in load may result in injury or death.
- Do not drain engine, transmission, or radiator fluids, or remove lines containing these fluids, when hot. Doing so may result in injury to personnel.
- Vehicle will become charged with electricity if it contacts or breaks high-voltage wires. Do not leave vehicle while high-voltage lines are in contact with vehicle. Failure to do so may result in injury to personnel.
- Wear hand protection when handling lifting and winching cables, hot exhaust components, and parts with sharp edges. Failure to do so may result in injury to personnel.
- Do not perform fuel system procedures while smoking or within 50 ft (15.2 m) of sparks or open flame. Diesel fuel is highly flammable and can explode easily, causing injury or death to personnel and/or damage to equipment.
- Ensure drainvalve on aftercooler is open when filling cooling system. Failure to do so may result in injury to personnel.
- Turbocharger intake fins are extremely sharp and turn at very high rpm. Keep hands and loose items away from intake openings. Failure to do so may result in injury to personnel.
- Do not place hands between frame and radiator when removing screws from trunnion or lifting radiator. Sudden changes in support may cause the radiator to shift, causing injury to personnel.
- Air pressure may create airborne debris. Use eye protection or injury to personnel may result.
- Air system components are subject to high pressure. Always relieve pressure before loosening or removing air system components.
- Wear safety goggles when using a hammer.
- Ether is extremely flammable. Do not perform ether start system procedures near fire. Injury to personnel may result.

TECHNICAL MANUAL
NO. 9-2320-272-24-3

TECHNICAL ORDER
NO. 36A12-1C-1155-2-3

HEADQUARTERS
DEPARTMENTS OF THE ARMY AND THE AIR FORCE
Washington, D.C., 30 JUNE 1998

TECHNICAL MANUAL
VOLUME 3 OF 4
UNIT, DIRECT SUPPORT, AND GENERAL SUPPORT MAINTENANCE MANUAL
FOR
TRUCK, 5-TON, 6X6, M939, M939A1, M939A2 SERIES TRUCKS (DIESEL)

TRUCK	MODEL	EIC	NSN WITHOUT WINCH	NSN WITH WINCH
Cargo, Dropside	M923	BRY	2320-01-050-2084	
Cargo, Dropside	M923A1	BSS	2320-01-206-4087	
Cargo, Dropside	M923A2	BS7	2320-01-230-0307	
Cargo, Dropside	M925	BRT		2320-01-047-8769
Cargo, Dropside	M925A1	BST		2320-01-206-4088
Cargo, Dropside	M925A2	BS8		2320-01-230-0308
Cargo	M927	BRV	2320-01-047-8771	
Cargo	M927A1	BSW	2320-01-206-4089	
Cargo	M927A2	BS9	2320-01-230-0309	
Cargo	M928	BRU		2320-01-047-8770
Cargo	M928A1	BSX		2320-01-206-4090
Cargo	M928A2	BTM		2320-01-230-0310
Dump	M929	BTH	2320-01-047-8756	
Dump	M929A1	BSY	2320-01-206-4079	
Dump	M929A2	BTN	2320-01-230-0305	
Dump	M930	BTG		2320-01-047-8755
Dump	M930A1	BSZ		2320-01-206-4080
Dump	M930A2	BTO		2320-01-230-0306
Tractor	M931	BTE	2320-01-047-8753	
Tractor	M931A1	BS2	2320-01-206-4077	
Tractor	M931A2	BTP	2320-01-230-0302	
Tractor	M932	BTD		2320-01-047-8752
Tractor	M932A1	BS5		2320-01-205-2684
Tractor	M932A2	BTQ		2320-01-230-0303
Van, Expansible	M934	BTB	2320-01-047-8750	
Van, Expansible	M934A1	BS4	2320-01-205-2682	
Van, Expansible	M934A2	BTR	2320-01-230-0300	
Medium Wrecker	M936	BTF		2320-01-047-8754
Medium Wrecker	M936A1	BS6		2320-01-206-4078
Medium Wrecker	M936A2	BTT		2320-01-230-0304

REPORTING OF ERRORS AND RECOMMENDING IMPROVEMENTS

You can help improve this manual. If you find any mistakes or if you know of a way to improve the procedures, please let us know. Mail your letter or DA Form 2028 (Recommended Changes to Publications and Blank Forms), or DA Form 2028-2 located in back of this manual, directly to: Director, Armament and Chemical Acquisition and Logistics Activity, ATTN: AMSTA-AC-NML, Rock Island, IL 61299-7630. A reply will be furnished to you. You may also provide DA Form 2028-2 information via datafax or e-mail:

- E-mail: amsta-ac-nml.@ria-emh2.army.mil
- Fax: DSN 783-0726 or commercial (309) 782-0726

DISTRIBUTION STATEMENT A Approved for public release; distribution is unlimited.

*This publication supersedes TM 9-2320-272-20-1, 24 October 1985, and changes 1 through 4; TM 9-2320-272-20-2, 25 October 1985, and changes 1 through 3; TM 9-2320-272-34-1, 10 June 1986, and changes 1 through 2; TM 9-2320-272-34-2, 10 June 1986, and changes 1 and 2; and TM 9-2320-358-24&P, 21 October 1992.

This publication is published in four volumes. TM 9-2320-272-24-1 contains chapters 1, 2, and 3 (through section IX). TM 9-2320-272-24-2 contains chapters 3 (sections X through XVI) and 4 (sections I through III). TM 9-2320-272-24-3 contains chapter 4 (sections IV through XVI). TM 9-2320-272-24-4 contains chapters 5 and 6 and appendices A through H. Volume 1 contains a table of contents for the entire manual. Volumes 1, 2, and 3 contain an alphabetical index covering tasks found in their respective volume. Volume 4 contains an alphabetical index covering all tasks found in the entire manual.

TABLE OF CONTENTS

VOLUME 3 OF 4

Section IV. ENGINE (M939A2) MAINTENANCE

4-38. ENGINE MAINTENANCE INDEX

4-39. ENGINE MOUNTING BRACKETS AND ISOLATORS REPLACEMENT

THIS TASK COVERS:

a. Removal **b. Installation**

INITIAL SETUP:

APLICABLE MODELS
M939A2

TOOLS
General mechanic's tool kit (Appendix E, Item 1)
Torque wrench (Appendix E, Item 144)

MATERIALS/PARTS
Nine locknuts (Appendix D, Item 301)
Tiedown strap (Appendix D, Item 693)
Five locknuts (Appendix D, Item 302)

PERSONNEL REQUIRED
Two

REFERENCES (TM)
TM 9-2320-272-10
TM 9-2320-272-24P

EQUIPMENT CONDITION
- Parking brake set (TM 9-2320-272-10).
- Surge tank removed (para. 3-62).
- Vibration damper removed (para. 4-42).
- Air cleaner hose removed (para. 3-13).

NOTE
Assistant will help with this procedure.

a. Removal

1. Remove tiedown strap (3) from hose (5). Discard tiedown strap (3).
2. Loosen two clamps (1) and remove hose (5) from thermostat housing connector (4) and radiator inlet tube (2).
3. Remove two locknuts (16), washers (15), and backing plate (17) from two screws (7) and cross-member (18). Discard locknuts (16).
4. Install chain and lifting device on two engine lifting brackets (6). Place tension on chain.
5. Remove two screws (7), washers (8), isolators (9), washers (8), and isolators (9) from front bracket (14).
6. Remove four screws (13), washers (12), and front bracket (14) from engine (10) and radiator bracket (11).

4-39. ENGINE MOUNTING BRACKETS AND ISOLATORS REPLACEMENT (Contd)

1
2
5
3
4

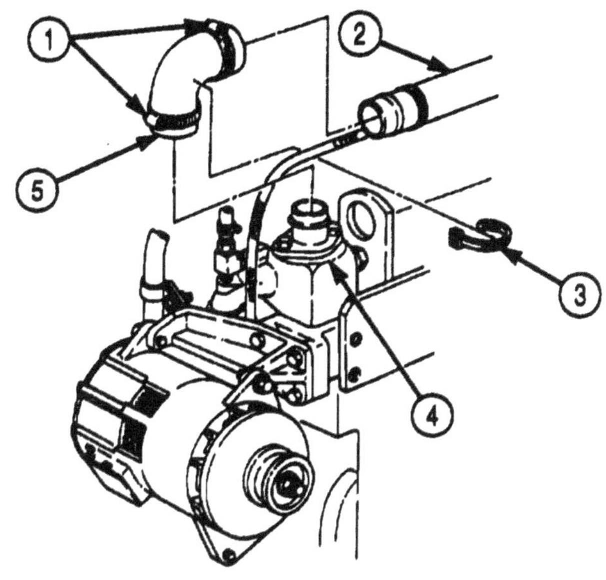

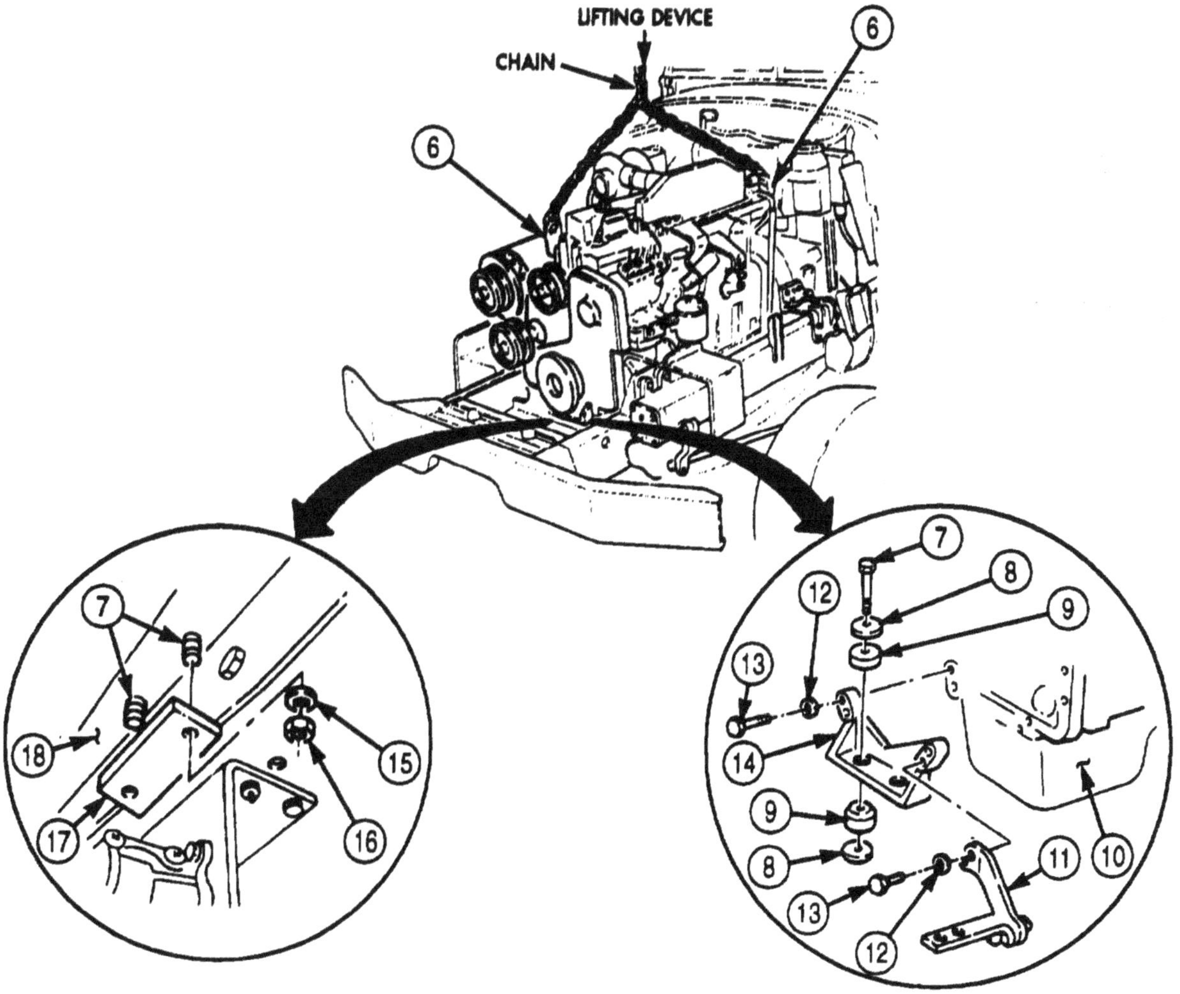

LIFTING DEVICE
CHAIN
6
6
7
18
15
17
16
7
8
12
9
13
14
9
8
11
10
13
12

4-39. ENGINE MOUNTING BRACKETS AND ISOLATORS REPLACEMENT (Contd)

NOTE

- Tag mounting brackets for installation.
- Perform steps 7 through 9 for removing right engine mounting brackets.

7. Remove two screws (1), washers (2), and four isolators (3) from upper mounting bracket (9) and lower mounting bracket (8).
8. Remove four locknuts (7), washers (6), screws (4), and lower mounting bracket (8) from frame rail (5). Discard locknuts (7).
9. Remove four screws (10), washers (11), and upper mounting bracket (9) from flywheel housing (12).

NOTE

Perform steps 10 through 13 for removing left engine mounting brackets.

10. Remove two locknuts (13), washers (14), isolators (15), screws (21), and isolators (15) from upper mounting bracket (30) and lower mounting bracket (18). Discard locknuts (13).
11. Remove locknut (24), washer (25), screw (26), washer (25), locknut (27), washer (17), screw (16), washer (17), shim (29), and engine support (28) from frame rail (22) and lower mounting bracket (18). Discard locknuts (24) and (27).
12. Remove four locknuts (20), washers (19), screws (23), and lower mounting bracket (18) from frame rail (22). Discard locknuts (20).
13. Remove four screws (32), washers (31), and upper mounting bracket (30) from flywheel housing (12).

b. Installation

NOTE

Perform steps 1 through 4 for installing left engine mounting brackets.

1. Install upper mounting bracket (30) on flywheel housing (12) with four washers (31) and screws (32). Tighten screws (32) 70-90-lb-ft (95-122 N·m).
2. Install lower mounting bracket (18) on frame rail (22) with four screws (23), washers (19), and new locknuts (20). Tighten top two locknuts (20) 80-95 lb-ft (109-129 N·m). Tighten bottom two locknuts (20) 55-70 lb-ft (75-95 N·m).
3. Install shim (29) and engine support (28) on frame rail (22) and lower mounting bracket (18) with washer (17), screw (16), washer (17), new locknut (27), washer (25), screw (26), washer (25), and new locknut (24). Tighten locknuts (24) and (27) 85 lb-ft (115 N·m).
4. Install four isolators (15) on upper (30) and lower (18) mounting brackets with two screws (21), washers (14) and new locknuts (13). Finger-tighten screws (21).

NOTE

Perform steps 5 through 7 for installing right engine mounting brackets.

5. Install upper mounting brackets (9) on flywheel housing (12) with four washers (11) and screws (10). Tighten screws (10) 70-90 lb-ft (95-122 N·m).
6. Install lower mounting bracket (8) on frame rail (5) with four screws (4), washers (6), and new locknuts (7).
7. Install four isolators (3) on upper mounting bracket (9) and lower mounting bracket (8) with two washers (2) and screws (1). Finger-tighten screws (1).

4-39. ENGINE MOUNTING BRACKETS AND ISOLATORS REPLACEMENT (Contd)

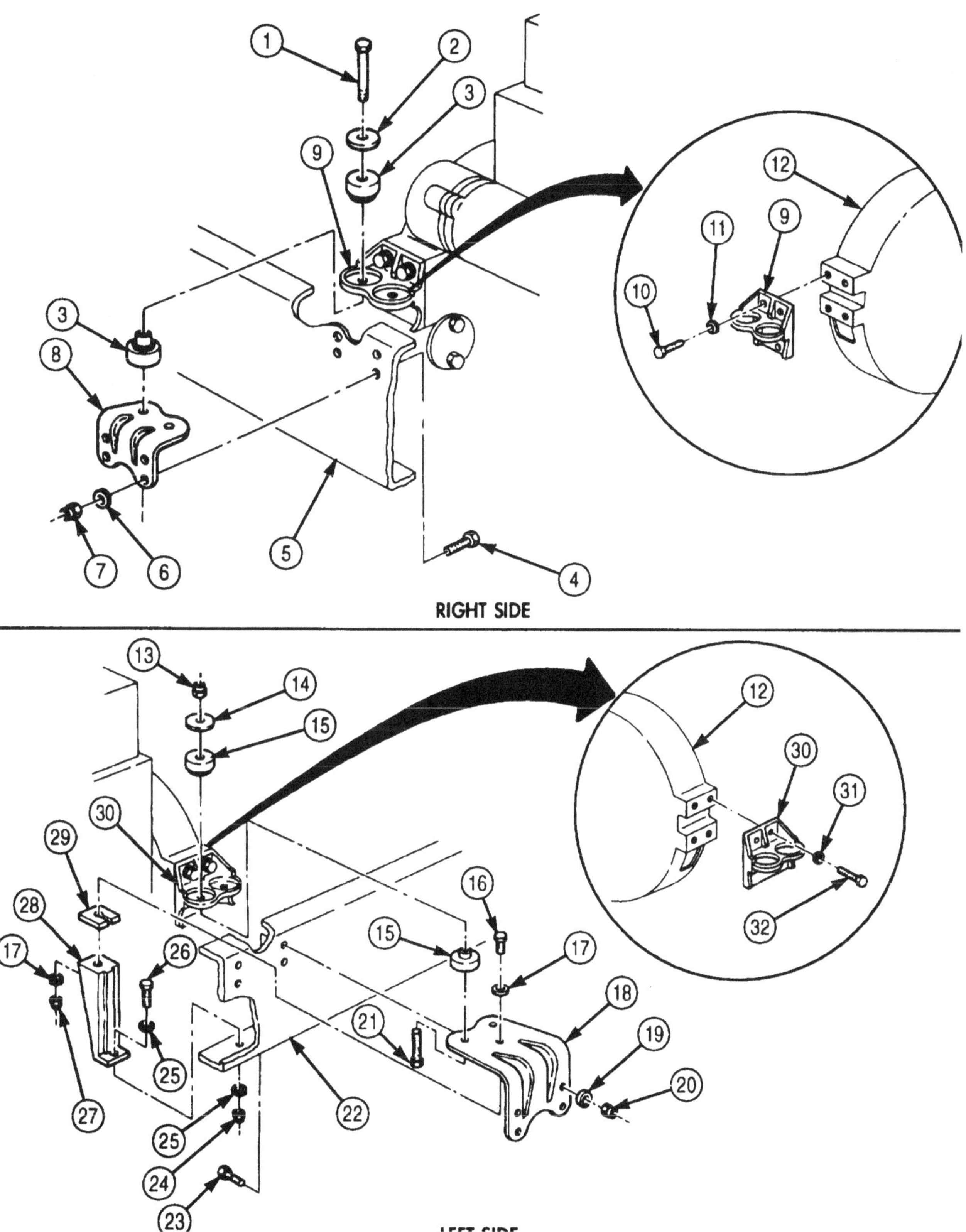

4-39. ENGINE MOUNTING BRACKETS AND ISOLATORS REPLACEMENT (Contd)

8. Position front bracket (9), four isolators (4), washers (3), and two screws (2) on crossmember (13) and install backing plate (12) with two washers (10) and new locknuts (11). Tighten locknuts (11) 75-85 lb-ft (102-115 N·m).
9. Slowly release tension on chain and lower engine (5) while aligning holes in front bracket (9) and radiator bracket (6) with holes in engine (5).
10. Install front bracket (9) and radiator bracket (6) on engine (5) with four washers (7) and screws (8). Tighten screws (8) 80 lb-ft (109 N·m).
11. Tighten two screws (14) and locknuts (15) 120-140 lb-ft (163-190 N·m).
12. Remove lifting device and chain from two engine lifting brackets (1).
13. Install hose (20) on radiator inlet tube (17) and thermostat housing connector (19) with two clamps (16) and new tiedown strap (18).

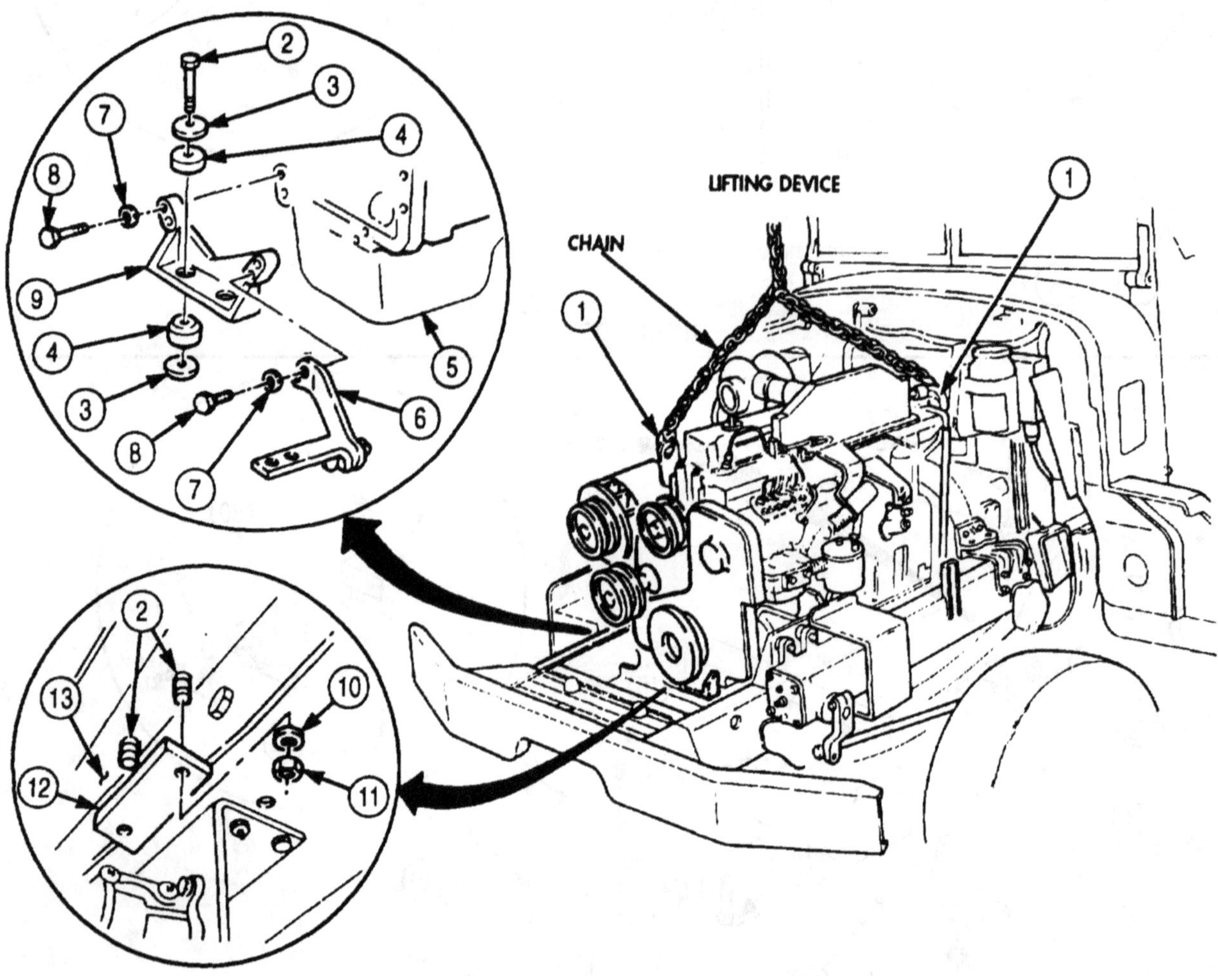

4-39. ENGINE MOUNTING BRACKETS AND ISOLATORS REPLACEMENT (Contd)

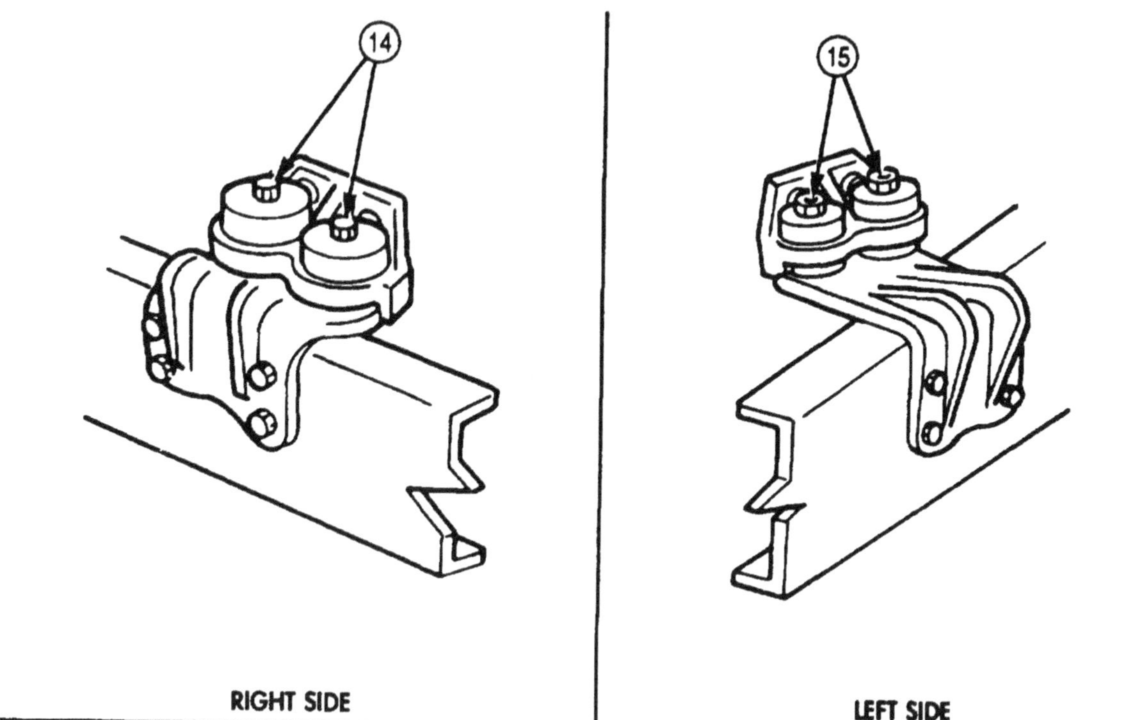

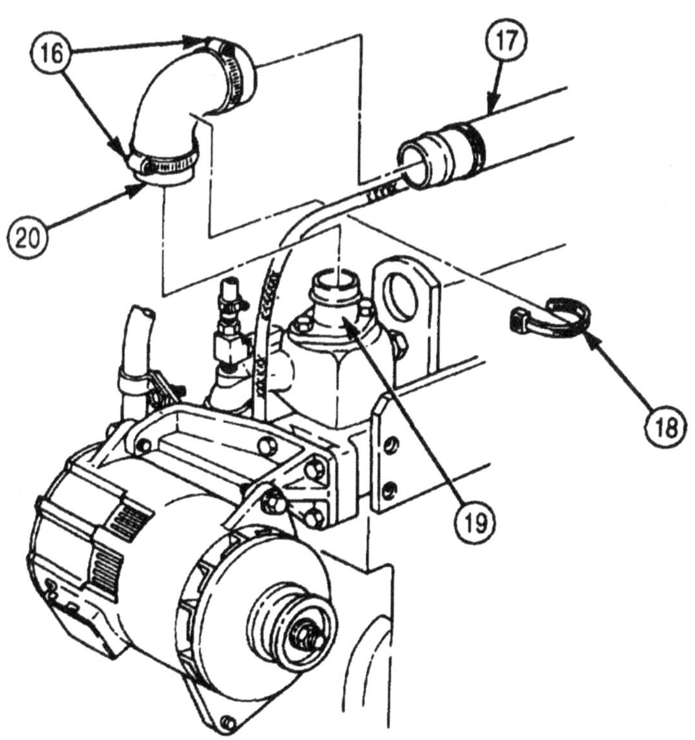

FOLLOW-ON TASKS:
- Install air cleaner hose (para. 3-13).
- Install vibration damper (para. 4-42).
- Install surge tank (para. 3-62).

4-40. ENGINE LIFTING BRACKETS REPLACEMENT

THIS TASK COVERS:

a. Removal

b. Installation

INITIAL SETUP:

APPLICABLE MODELS
M939A2

TOOLS
General mechanic's tool kit (Appendix E, Item 1)

MATERIALS/PARTS
Tiedown strap (Appendix D, Item 690)

REFERENCES (TM)
TM 9-2320-272-10
TM 9-2320-272-24P

EQUIPMENT CONDITION
- Parking brake set (TM 9-2320-272-10).
- Hood raised and secured (TM 9-2320-272-10).
- Thermostat and thermostat housing removed (para. 3-66).

a. Removal

1. Remove tiedown strap (2) and overflow hose (1) from lifting bracket (6). Discard tiedown strap (2).
2. Remove two screws (7) and lifting bracket (6) from engine (5).
3. Remove two screws (4) and lifting bracket (3) from engine (5).

b. Installation

1. Install lifting bracket (3) on engine (5) with two screws (4).
2. Install lifting bracket (6) on engine (5) with two screws (7).
3. Install overflow hose (1) on lifting bracket (6) with new tiedown strap (2).

4-40. ENGINE LIFTING BRACKETS REPLACEMENT (Contd)

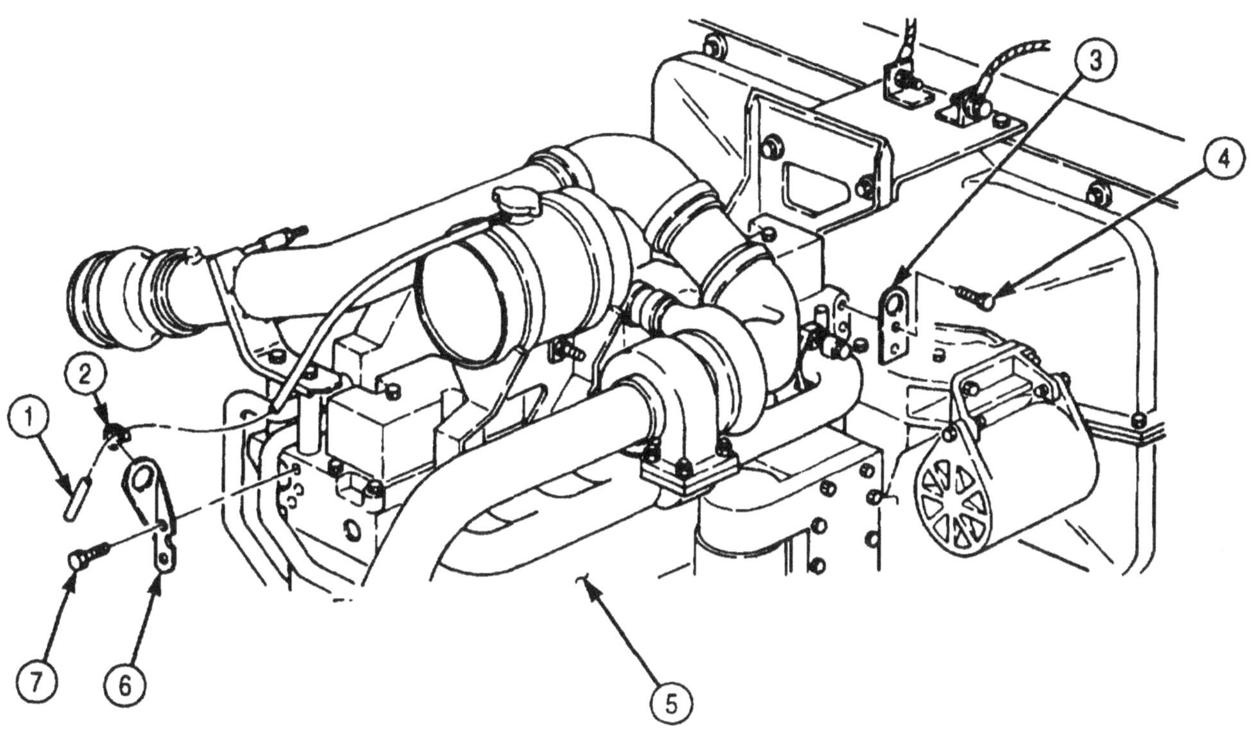

FOLLOW-ON TASK: Install thermostat and thermostat housing (para. 3-66).

4-41. ROCKER LEVERS, PUSH RODS, AND CYLINDER HEAD REPLACEMENT

THIS TASK COVERS:

a. Rocker Levers and Push Rods Removal
b. Cylinder Head Removal
c. Cylinder Head Installation
d. Rocker Levers and Push Rods Installation

INITIAL SETUP:

APPLICABLE MODELS

M939A2

SPECIAL TOOLS

Torque angle gauge (Appendix E, Item 141)

TOOLS

General mechanic's tool kit (Appendix E, Item 1)
Torque wrench (Appendix E, Item 144)

MATERIALS/PARTS

Seal (Appendix D, Item 598)
Fuel filter (Appendix D, Item 135)
Lockwasher (Appendix D, Item 364)
Gasket (Appendix D, Item 188)
Two washers (Appendix D, Item 645)
Two washers (Appendix D, Item 644)
Two washers (Appendix D, Item 714)
Washer (Appendix D, Item 711)
Lubrication oil (Appendix C, Item 48)
Diesel fuel Appendix C, Item 42)

REFERENCES (TM)

TM 9-2320-272-24P

EQUIPMENT CONDITION

- Parking brake set (TM 9-2320-272-10).
- Aftercooler removed (para. 3-76).
- Exhaust manifold removed (para. 4-50).
- Fan clutch and hose removed (para. 3-74).
- Fuel injector tubes removed (para. 3-19).
- Valve covers removed (para. 3-6).

a. Rocker Levers and Push Rods Removal

1. Loosen twelve locknuts (3) and turn adjusting screws (2) two full turns counterclockwise.
2. Remove twelve screws (1), six retaining clamps (9), rocker lever assemblies (8), and rocker lever supports (7) from cylinder head (5).
3. Remove oil manifold (4) from cylinder head (5).
4. Remove twelve push rods (6) from cylinder head (5).

4-41. ROCKER LEVERS, PUSH RODS, AND CYLINDER HEAD REPLACEMENT (Contd)

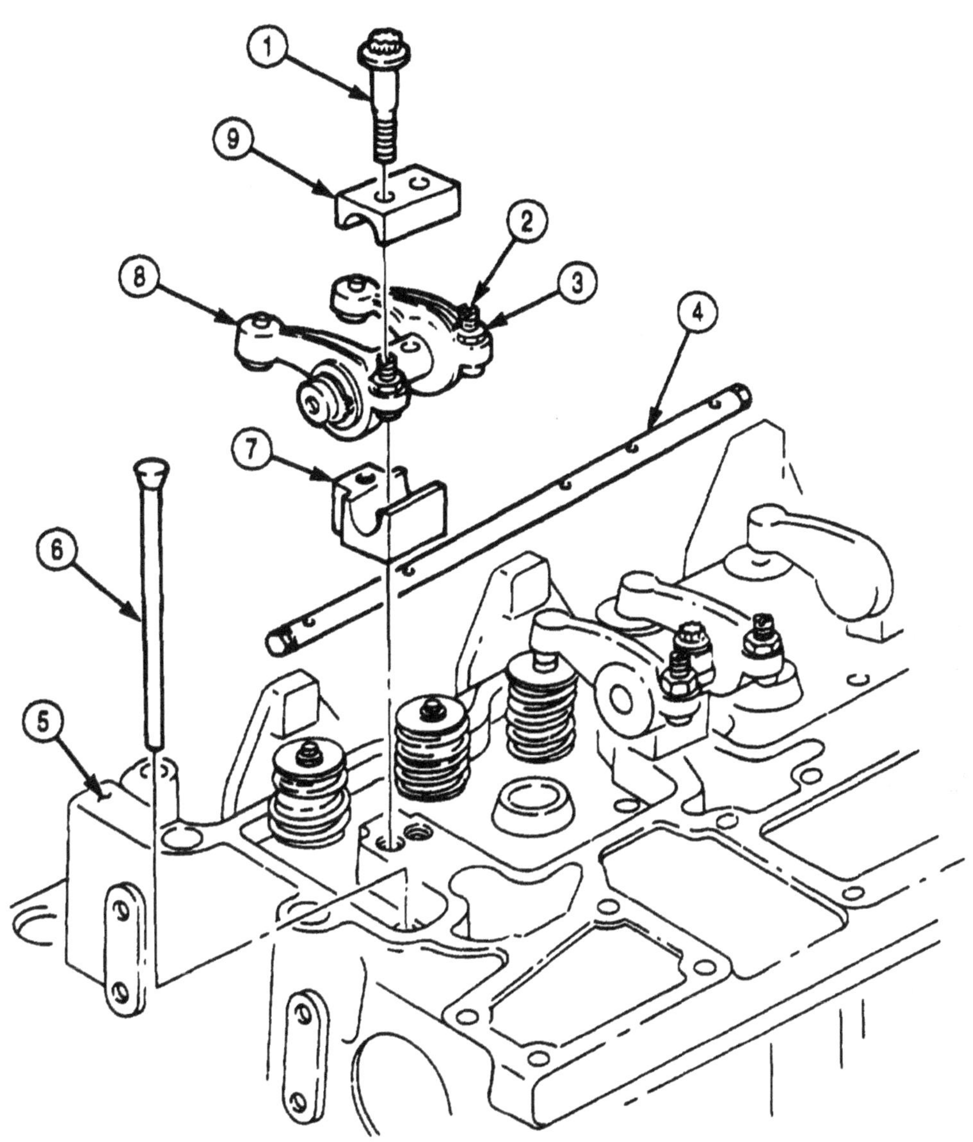

4-41. ROCKER LEVERS, PUSH RODS, AND CYLINDER HEAD REPLACEMENT (Contd)

b. Cylinder Head Removal

1. Disconnect ether supply tube (1) from atomizer nozzle (20).
2. Remove two screws (23), washers (21), and bracket (22) from cylinder head (9). Discard washers (21).
3. Remove screw (14), washer (13), ground strap (12), lockwasher (11), and temperature sensor (10) from cylinder head (9). Discard lockwasher (11).
4. Remove screw (15), washer (8), fuel supply tube (7), and washer (6) from cylinder head (9). Discard washers (6) and (8).
5. Remove screw (17), adapter (18), washer (19), fuel supply line (4), and washer (5) from cylinder head (9). Discard washers (5) and (19).
6. Remove fuel filter (16) from cylinder head (9). Discard fuel filter (16).

NOTE

Perform step 7 for engines equipped with external bypass system.

7. Disconnect tube (3) from adapter (2).

4-41. ROCKER LEVERS, PUSH RODS, AND CYLINDER HEAD REPLACEMENT (Contd)

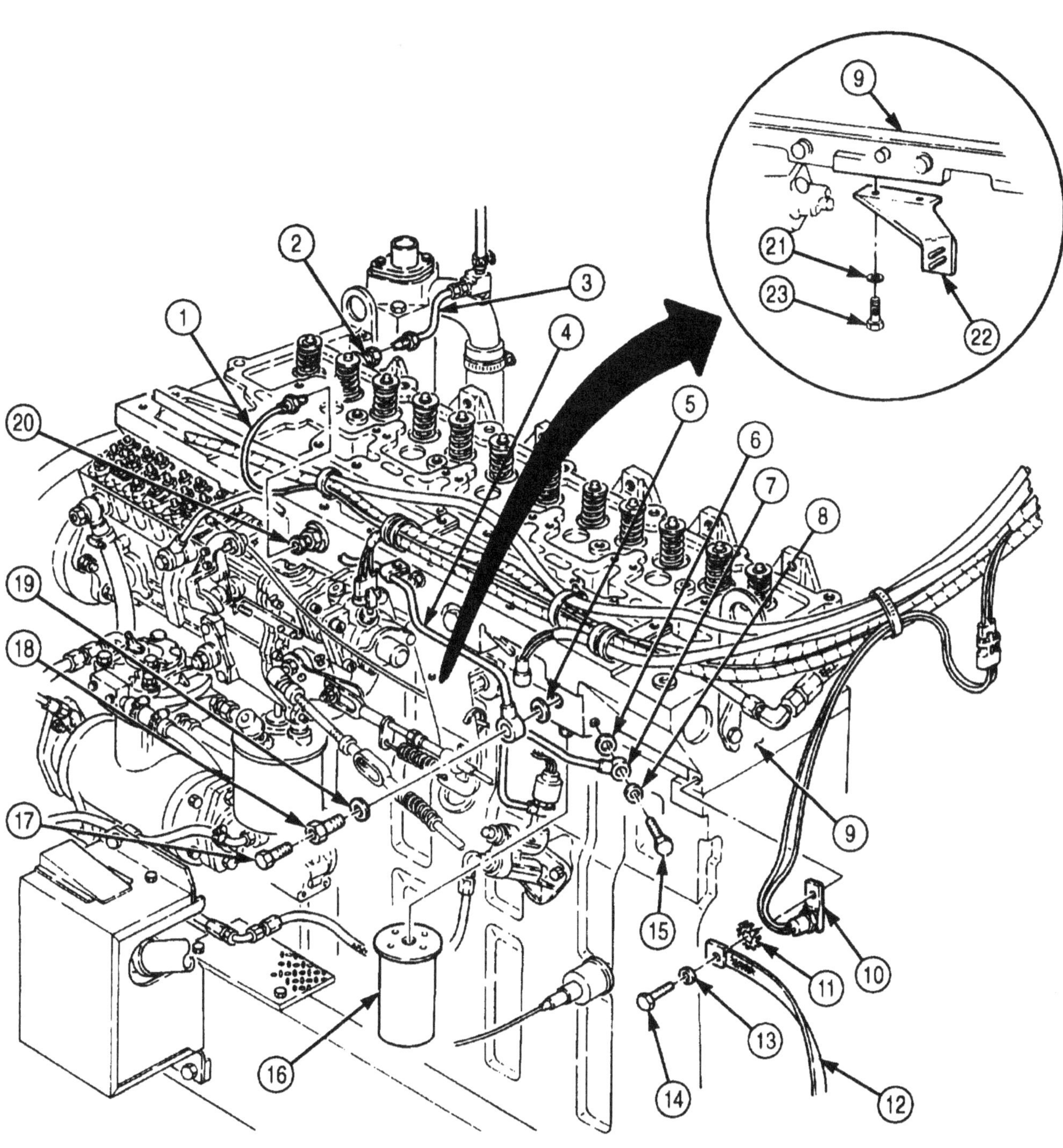

4-41. ROCKER LEVERS, PUSH RODS, AND CYLINDER HEAD REPLACEMENT (Contd)

8. Remove water inlet tube (5) and seal (4) from elbow (3). Discard seal (4).
9. Remove screw (7) from air supply tube (6).
10. Loosen clamp (1) and remove air supply tube (6) from coupling (2).

NOTE

- Tag all screws for installation.
- Assistant will help with step 11.

11. Remove fourteen screws (11) and twelve screws (12) from cylinder head (13).
12. Remove cylinder head (13) and gasket (22) from engine block (23). Discard gasket (22).
13. Remove adapter (24), atomizer (25), adapters (10), (20), and (21), elbow (19), plug (18), and washer (17) from cylinder head (13). Discard washer (17).
14. Remove nipple (16) from cylinder head (13).
15. Remove two screws (14) and (8) and lifting brackets (15) and (9) from cylinder head (13).

NOTE

Perform step 16 if coupling is damaged.

16. Remove coupling (26) from cylinder head (13).

4-41. ROCKER LEVERS, PUSH RODS, AND CYLINDER HEAD REPLACEMENT (Contd)

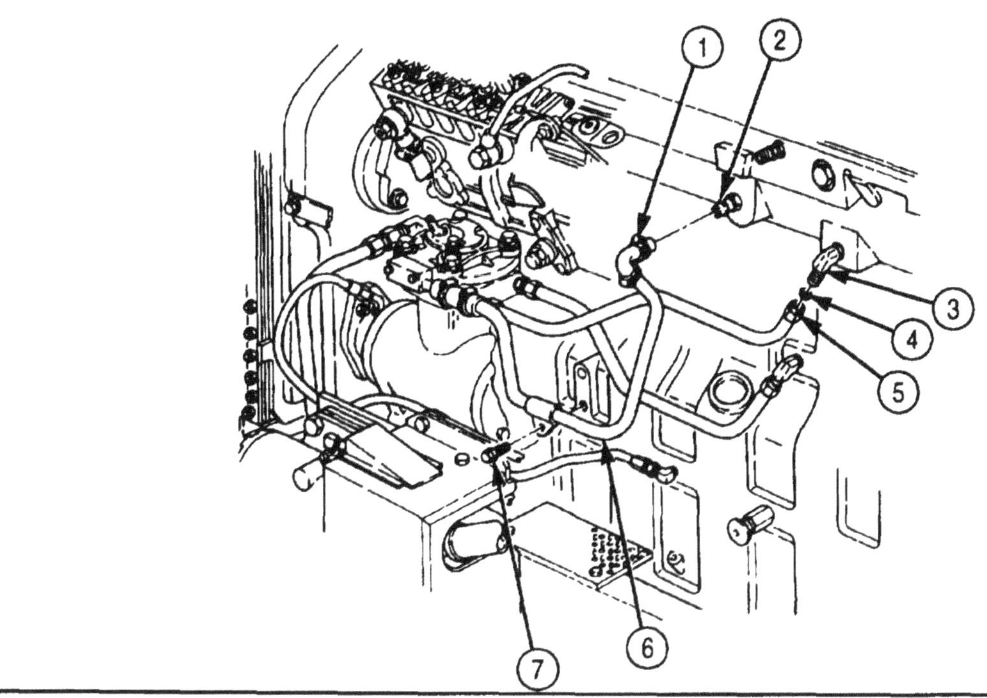

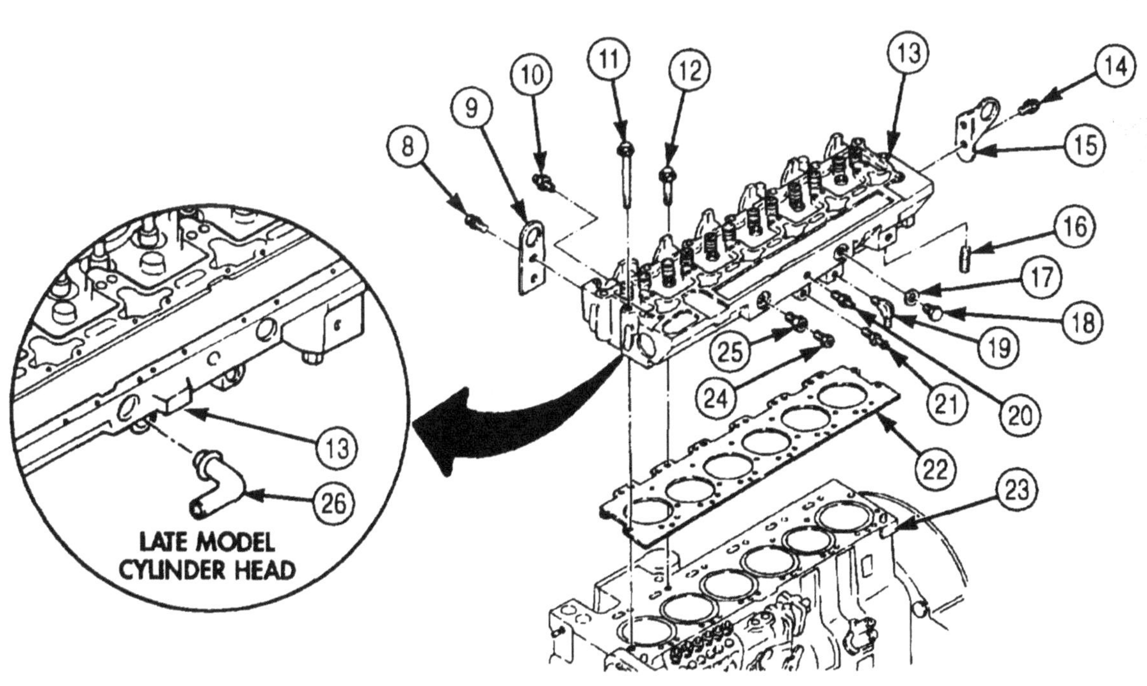

4-41. ROCKER LEVERS, PUSH RODS, AND CYLINDER HEAD REPLACEMENT (Contd)

c. Cylinder Head Installation

NOTE

- Cylinder head assemblies are interchangeable between early model engines, 44487830 and before, and late model engines, after 44487830. When interchanging cylinder head assemblies, 3913111, early model engines, and 3917767, late model engines, ensure the corresponding cylinder head screws are used and air compressor plumbing matches the applicable engine model configuration.
- Perform step 1 if installing new late model cylinder head.

1. Install coupling (19) on cylinder head (6).
2. Install lifting brackets (2) and (8) on cylinder head (6) with two screws (1) and (7).
3. Install new washer (10), plug (11), elbow (12), adapters (3), (13), and (14), atomizer (18), adapter (17), and nipple (9) on cylinder head (6).

NOTE

Assistant will help with step 4.

4. Position new gasket (15) and cylinder (6) on engine block (16) and install with fourteen screws (4) and twelve screws (5).

NOTE

- Perform step 5 for late model engines.
- Perform step 6 for early model engines.

5. Tighten screws (4) and (5) using the following steps:
   a. Tighten screws (4) and (5) in sequence shown.
   b. Loosen screws (4) and (5) in sequence shown.
   c. Tighten screws (4) and (5) 50 lb-ft (68 N·m) in sequence shown.
   d. Tighten long screws (4) 105 lb-ft (142 N·m) in sequence shown.
   e. Using torque angle gauge, rotate screws (4) and (5) an additional 75-105 degrees in sequence shown.
6. Tighten screws (4) and (5) using the following steps:
   a. Tighten screws (4) and (5) 30 lb-ft (41 N·m) is sequence shown.
   b. Tighten screws (4) and (5) 110 lb-ft (149 N·m) in sequence shown.
   c. Tighten screws (4) and (5) 160 lb-ft (217 N·m) in sequence shown.
7. Connect air supply tube (25) to coupling (21) with clamp (20).
8. Install air supply tube (25) on engine block (16) with screw (26).
9. Install new seal (23) and water inlet tube (24) on elbow (22).

4-41. ROCKER LEVERS, PUSH RODS, AND CYLINDER HEAD REPLACEMENT (Contd)

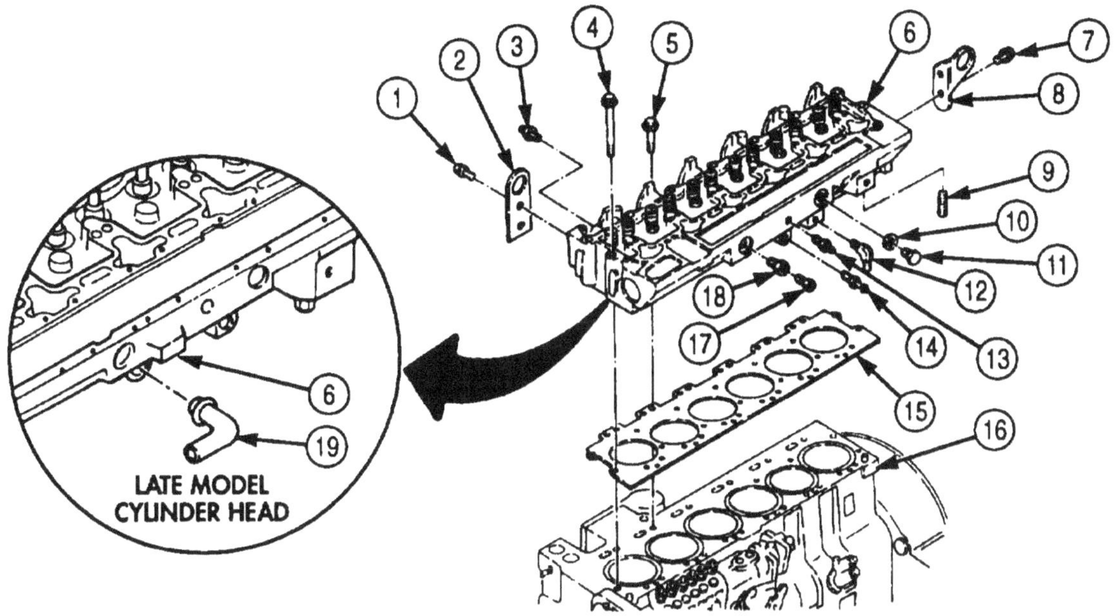

LATE MODEL CYLINDER HEAD

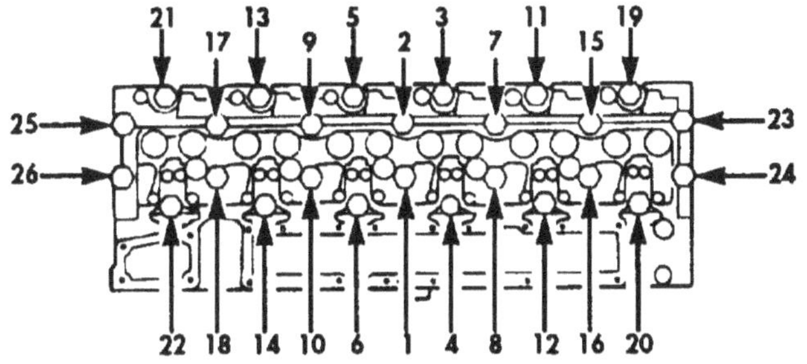

TORQUE SEQUENCE

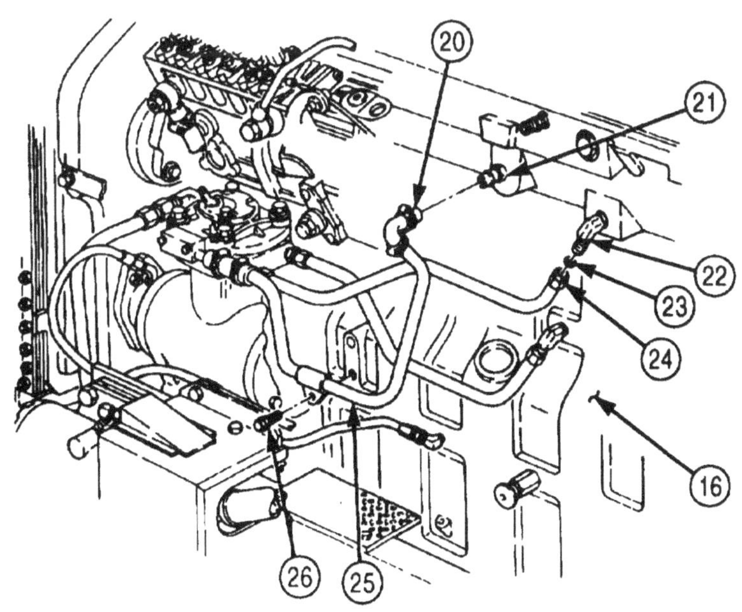

4-41. ROCKER LEVERS, PUSH RODS, AND CYLINDER HEAD REPLACEMENT (Contd)

NOTE

Perform step 10 for engines equipped with external bypass system.

10. Connect tube (3) to adapter (2).
11. Install bracket (22) on cylinder head (8) with two new washers (24) and screws (23).

NOTE

Fill fuel filter and coat outer seal with diesel fuel before installation.

12. Install new fuel filter (17) on cylinder head (8) and engine block (16). Tighten until snug.
13. Install new lockwasher (11), temperature sensor (10), and ground strap (12) on cylinder head (8) with washer (13) and screw (14).

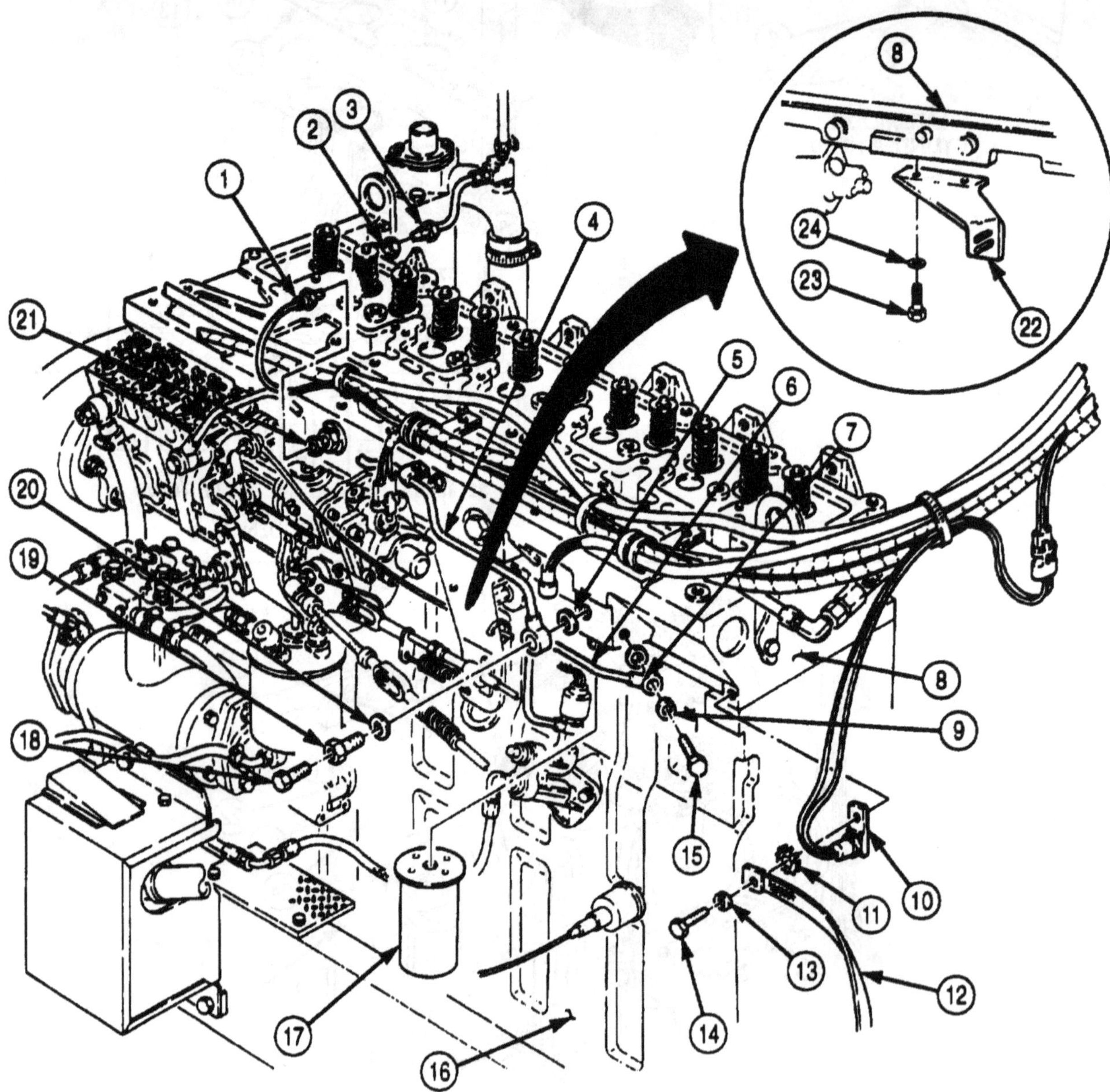

4-41. ROCKER LEVERS, PUSH RODS, AND CYLINDER HEAD REPLACEMENT (Contd)

14. Connect ether supply tube (1) on atomizer nozzle (21).
15. Install fuel supply tube (4) on cylinder head (8) with new washers (5) and (20), adapter (19), and screw (18).
16. Install fuel supply tube (6) on cylinder head (8) with new washers (7) and (9) and screw (15).

d. Rocker Levers and Push Rods Installation

NOTE

Apply light coat of oil to all parts before installation.

1. Install twelve push rods (25) in cylinder head (8).
2. Position oil manifold (30) on cylinder head (8) with flat surface facing downward.
3. Install six rocker lever supports (26) and rocker lever assemblies (27) on cylinder head (8) and oil manifold (30) with six retaining clamps (28) and twelve screws (29). Tighten screws (29) 55 lb-ft (75 N·m).

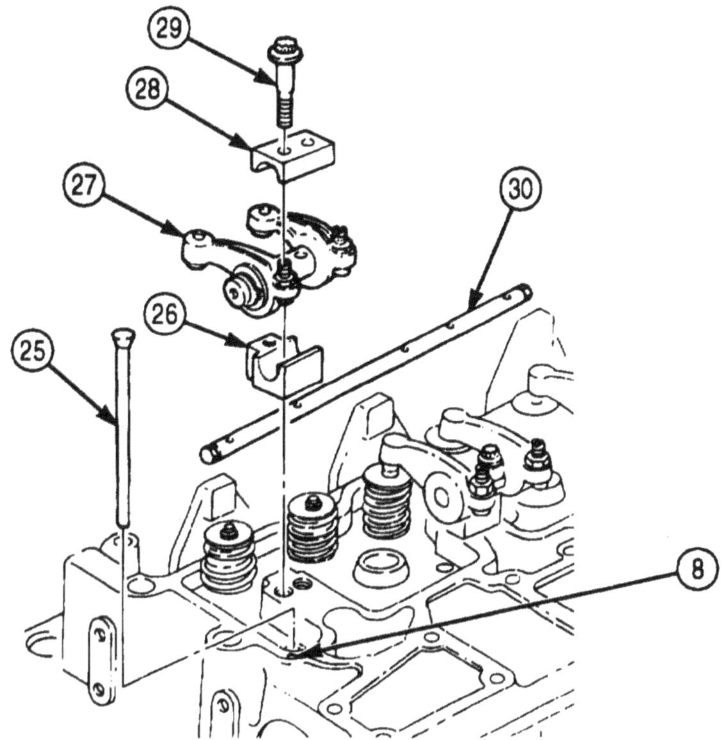

FOLLOW-ON TASKS:
- Install fan clutch and hose (para. 3-74).
- Install exhaust manifold (para. 4-50).
- Install aftercooler (para. 3-76).
- Install fuel injector tubes (para. 3-19).
- Install valve cover (para. 3-6).

4-42. VIBRATION DAMPER MAINTENANCE

THIS TASK COVERS:

a. Removal
b. Inspection
c. Installation

INITIAL SETUP:

APPLICABLE MODELS
M939A2

SPECIAL TOOLS
Engine barring tool (Appendix E, Item 43)

TOOLS
General mechanic's tool kit (Appendix E, Item 1)
Torque wrench (Appendix E, Item 144)

MATERIALS/PARTS
O-ring (Appendix D, Item 444)

PERSONNEL REQUIRED
Two

REFERENCES (TM)
TM 9-2320-272-10
TM 9-2320-272-24P

EQUIPMENT CONDITION
- Parking brake set (TM 9-2320-272-10).
- Engine drivebelt removed (para. 3-71).
- Fan and fan shroud removed (para. 3-64).

NOTE
Assistant will help with this procedure.

a. Removal

1. Remove plug (4) and O-ring (3) from flywheel housing (2). Discard O-ring (3).
2. Using engine barring tool, prevent crankshaft (1) from turning.
3. Remove four screws (6) and vibration damper (5) from crankshaft (1).

b. Inspection

1. Inspect vibration damper (5) for bends, breaks, and damaged pulley grooves. Replace vibration damper (5) if bent, broken, or grooves are damaged.
2. Inspect vibration damper alignment marks (8). Replace vibration damper (5) if alignment marks (8) are 0.063 in. (1.60 mm) out of alignment.
3. Inspect rubber member (7) for wear and damage. Replace vibration damper (5) if damaged or worn more than 0.125 in. (3.18 mm) deeper than face of vibration damper (5).

c. Installation

1. Using engine barring tool, prevent crankshaft (1) from turning.
2. Install vibration damper (5) on crankshaft (1) with four screws (6). Tighten screws (6) 150 lb-ft (203 N·m).
3. Install new O-ring (3) and plug (4) in flywheel housing (2).

4-42. VIBRATION DAMPER MAINTENANCE (Contd)

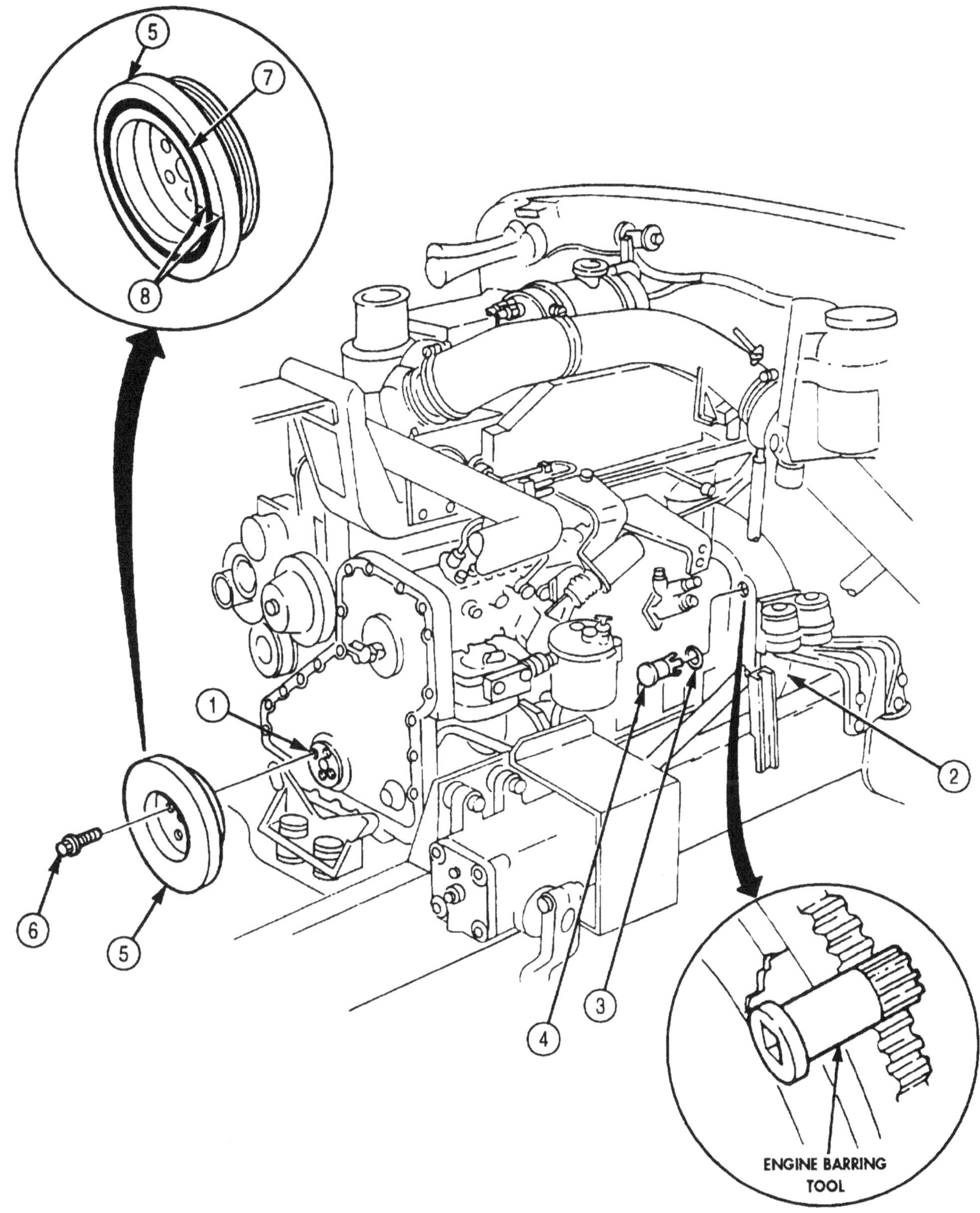

FOLLOW-ON TASKS:
- Install fan and fan shroud (para. 3-64).
- Install engine drivebelt (para. 3-71).

4-43. FLEXPLATE AND FLYWHEEL HOUSING MAINTENANCE

THIS TASK COVERS:

a. Removal
b. Disassembly
c. Cleaning and Inspection
d. Assembly
e. Installation

INITIAL SETUP:

APPLICABLE MODELS
M939A2

SPECIAL TOOLS
Engine barring tool (Appendix E, Item 43)

TOOLS
General mechanic's tool kit (Appendix E, Item 1)
Torque wrench (Appendix E, Item 144)

MATERIALS/PARTS
Seal (Appendix D, Item 602)
Gasket (Appendix D, Item 189)
Antiseize tape (Appendix C, Item 72)
Gasket sealant (Appendix C, Item 30)
Lubricating oil (Appendix C, Item 50)

PERSONNEL REQUIRED
Two

REFERENCES (TM)
TM 9-2320-272-24P

EQUIPMENT CONDITION
Transmission (out-of-vehicle) removed (para. 4-72).

CAUTION

Do not use timing pin to prevent engine from turning. Doing so may cause damage to timing pin.

NOTE

Assistant will help with this procedure.

a. Removal

1. Using engine barring tool, prevent crankshaft (13) from turning.
2. Remove eight screws (7), clamping ring (8), flexplate (9), and crankshaft adapter (6) from crankshaft (13).
3. Remove twelve screws (5), flywheel housing (4), and seal (3) from rear cover (14). Discard seal (3).

NOTE

Perform step 4 if dowel pins are damaged.

4. Remove two dowel pins (2) from engine block (1).

b. Disassembly

1. Remove two pipe plugs (18) from flywheel housing (4).
2. Remove two screws (15), cover plate (16), and gasket (17) from flywheel housing (4). Discard gasket (17).

NOTE

Tag mounting brackets for installation. One on each side of flywheel housing. One side shown.

3. Remove eight screws (12), washers (11), and two mounting brackets (10) from flywheel housing (4).

4-43. FLEXPLATE AND FLYWHEEL HOUSING MAINTENANCE (Contd)

c. Cleaning and Inspection

1. For general cleaning instructions, refer to para. 2-14.
2. Inspect flexplate (9) for cracks, breaks, elongated holes, and damaged teeth. Replace flexplate (9) if cracked, broken, holes are elongated, or teeth are damaged.
3. Inspect flywheel housing (4) for cracks, breaks, and stripped threads. Replace or repair flywheel housing (4) if cracked, broken, or threads are stripped.

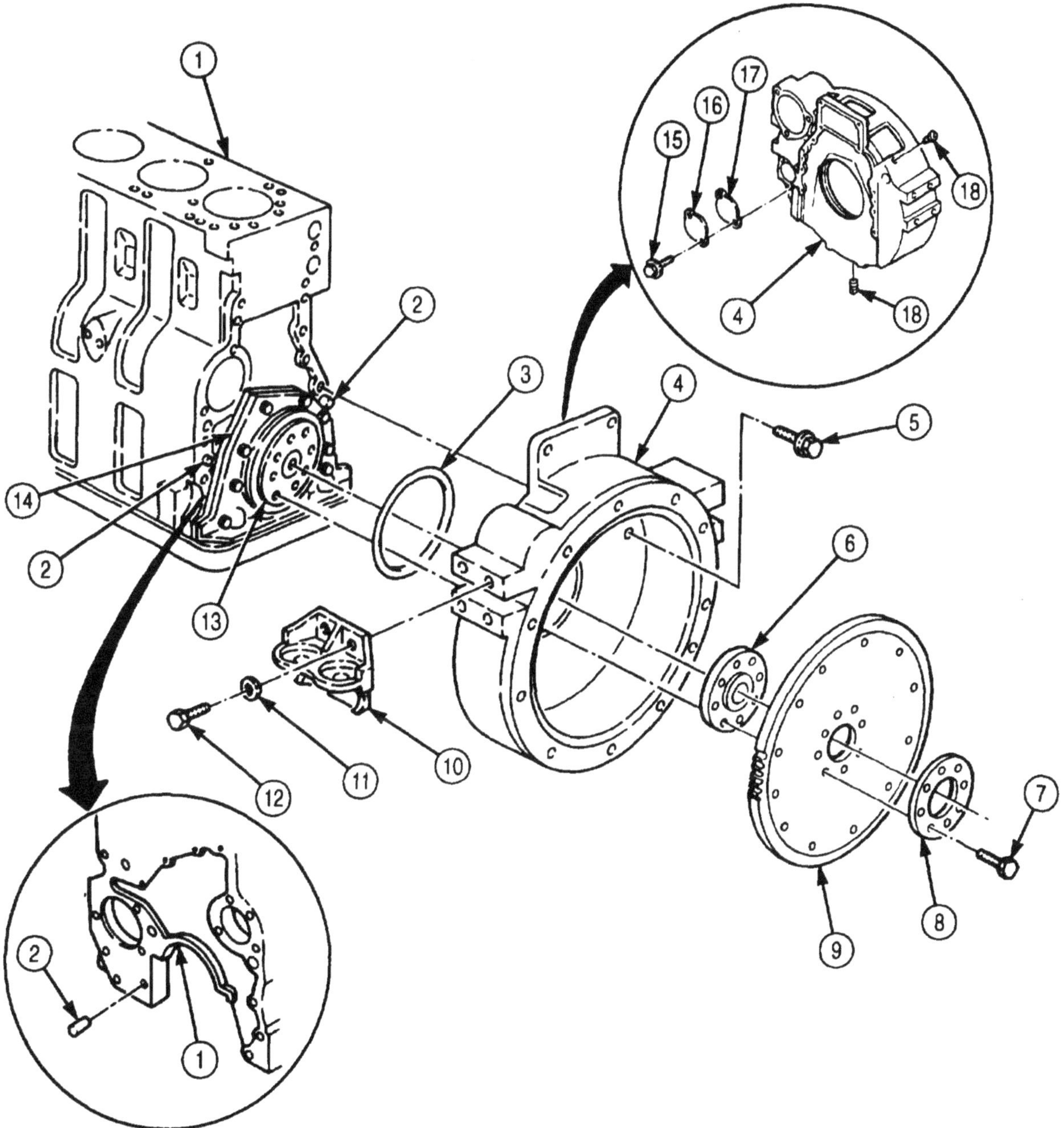

4-43. FLEXPLATE AND FLYWHEEL HOUSING MAINTENANCE (Contd)

d. Assembly

1. Install two mounting brackets (10) on flywheel housing (4) with eight washers (11) and screws (12). Tighten screws (12) 70-90 lb-ft (95-122 N·m).
2. Install new gasket (17) and cover plate (16) on flywheel housing (4) with two screws (1.5).

NOTE

Wrap male pipe threads with antiseize tape before installation.

3. Install two pipe plugs (18) on flywheel housing (4).

e. Installation

NOTE

Perform step 1 if dowel pins were removed.

1. Install new dowel pins (2) on engine block (1).
2. Install new seal (3) on rear cover (14).
3. Apply gasket sealant to threads of twelve screws (5) and mating surfaces of flywheel housing (4) and engine block (1).
4. Position flywheel housing (4) on engine block (1). Ensure flywheel housing (4) is properly aligned on dowel pins (2) and seal (3) is not damaged.
5. Install flywheel housing (4) on engine block (1) with twelve screws (5). Tighten screws (5) 45 lb-ft (61 N·m) in sequence shown.
6. Lubricate eight screws (9) with clean engine oil.

CAUTION

Do not use timing pin to prevent engine from turning. Doing so may cause damage to timing pin.

7. Install crankshaft adapter (6), flexplate (7), and clamping ring (8) on crankshaft (13) with eight screws (9). Tighten screws (9) 100 lb-ft (136 N·m) in sequence shown.

4-43. FLEXPLATE AND FLYWHEEL HOUSING MAINTENANCE (Contd)

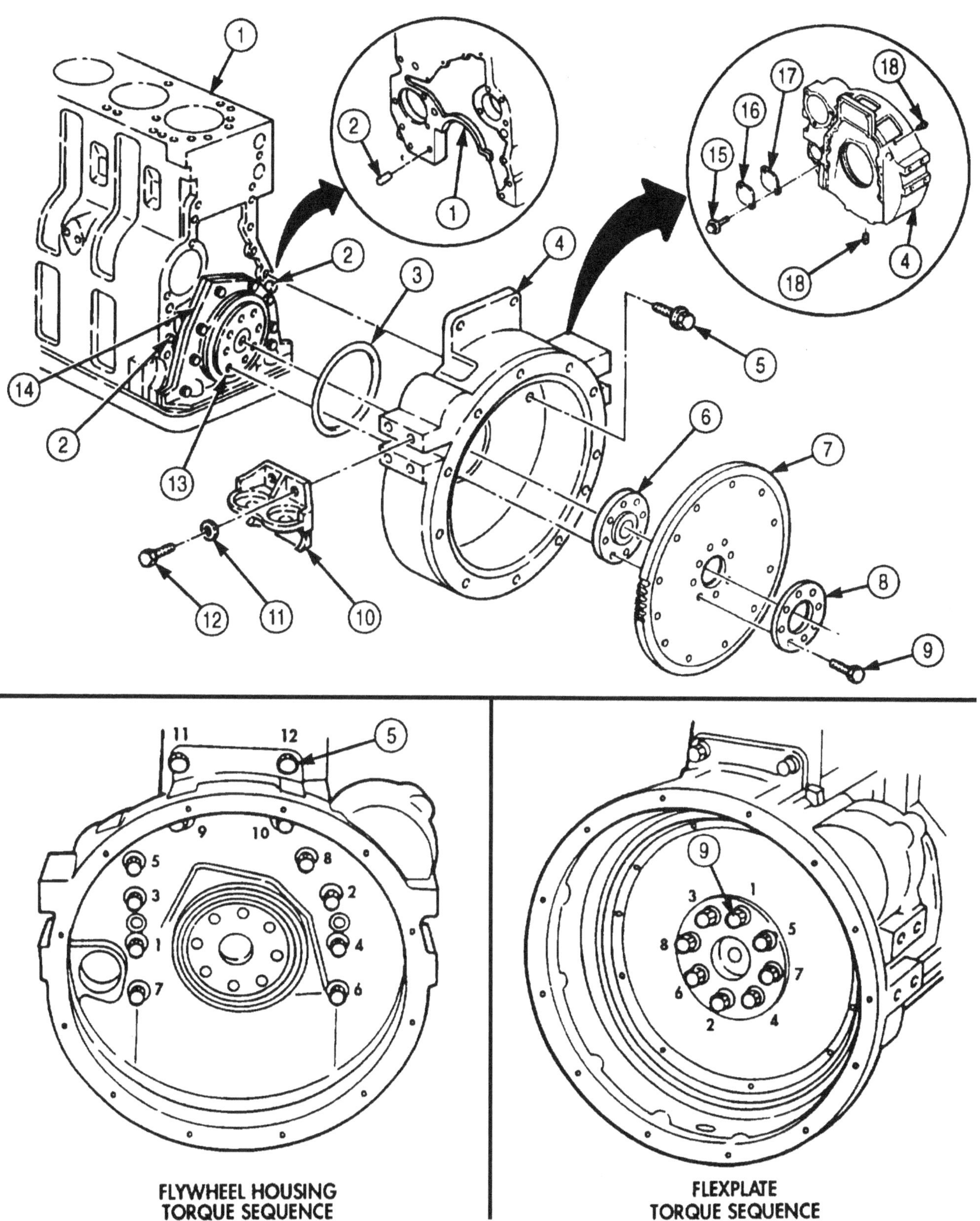

FOLLOW-ON TASK: Install transmission (out-of-vehicle) (para. 4-72).

4-44. REAR COVER AND OIL SEAL MAINTENANCE

THIS TASK COVERS:

a. Removal
b. Cleaning and Inspection
c. Installation

INITIAL SETUP:

APPLICABLE MODELS
M939A2

TOOLS
General mechanic's tool kit (Appendix E, Item 1)
Torque wrench (Appendix E, Item 146)

MATERIALS/PARTS
Oil seal kit (Appendix D, Item 505)
Four lockwashers (Appendix D, Item 408)
Gasket (Appendix D, Item 190)
Drycleaning solvent (Appendix C, Item 71)
Rags (Appendix C, Item 58)

REFERENCES (TM)
TM 9-2320-272-24P

EQUIPMENT CONDITION
Flexplate and flywheel housing removed (para. 4-43).

GENERAL SAFETY INSTRUCTIONS
- Drycleaning solvent is flammable and toxic. Do not use near an open flame.
- Keep fire extinguisher nearby when using drycleaning solvent.

a. Removal

1. Remove four screws (8), washers (9), and lockwashers (10) from rear cover (4) and oil pan (11). Discard lockwashers (10).
2. Remove eight screws (5), rear cover (4), and gasket (3) from engine block (1). Discard gasket (3).
3. Remove rear oil seal (7) from rear cover (4). Discard rear oil seal (7).

b. Cleaning and Inspection

WARNING

Drycleaning solvent is flammable and toxic. Do not use near open flame and always have a fire extinguisher nearby when solvents are used. Use only in well-ventilated places, wear protective clothing, and dispose of cleaning rags in approved container. Failure to do this may result in injury or death to personnel and/or damage to equipment.

1. Clean sealing surface of crankshaft (2) with drycleaning solvent and dry with clean rag.
2. Inspect crankshaft (2) for nicks, gouges, and stripped threads. Replace crankshaft (2) if nicked, gouged, or threads are stripped.
3. Inspect rear cover (4) for nicks and gouges. Replace rear cover (4) if nicked or gouged.
4. Inspect exposed area of oil pan gasket (12) for damage. Replace oil pan gasket (12) (para. 4-47), if damaged.

c. Installation

1. Install new gasket (3) and rear cover (4) on engine block (1) with eight screws (5). Finger-tighten screws (5).
2. Install alignment tool on crankshaft (2) and rear cover (4). Ensure rear cover (4) is centered and level with rails of oil pan (11).
3. Tighten screws (5) 7 lb-ft (9 N·m) and remove alignment tool.

4-44. REAR COVER AND OIL SEAL MAINTENANCE (Contd)

4. Install pilot (6) in new rear oil seal (7). Ensure tapered end of pilot (6) and rear oil seal (7) face the same direction.
5. Install pilot (6) and rear oil seal (7) on crankshaft (2).
6. Remove pilot (6) from crankshaft (2) and rear oil seal (7). Discard pilot (6).
7. Using alignment tool, alternately tap rear oil seal (7) until seated in rear cover (4).
8. Install four new lockwashers (10), washers (9), and screws (8) on rear cover (4) and oil pan (11).

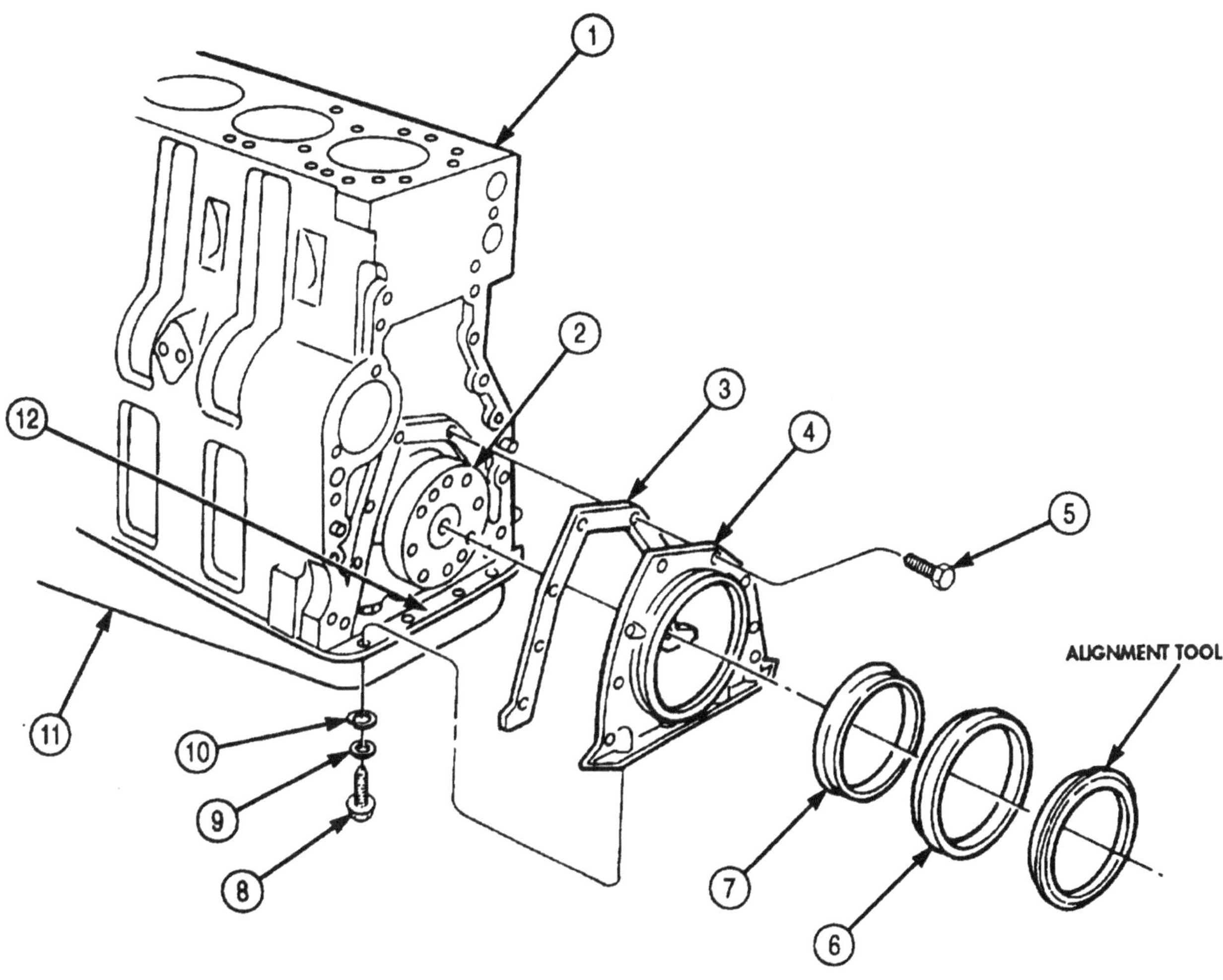

FOLLOW-ON TASK: Install flexplate and flywheel housing (para. 4-43).

4-45. REAR OIL SEAL (IN VEHICLE) MAINTENANCE

THIS TASK COVERS:

a. Removal
b. Cleaning and Inspection
c. Installation

INITIAL SETUP:

APPLICABLE MODELS
M939A2

SPECIAL TOOLS
Engine barring tool (Appendix E, Item 43)

TOOLS
General mechanic's tool kit (Appendix E, Item 1)
Slide hammer (Appendix E, Item 121)
Torque wrench (Appendix E, Item 144)

MATERIALS/PARTS
Rear oil seal kit (Appendix D, Item 505)
Rags (Appendix C, Item 58)
Drycleaning solvent (Appendix C, Item 71)

PERSONNEL REQUIRED
Two

REFERENCES (TM)
TM 9-2320-272-24P

EQUIPMENT CONDITION
Transmission (in-vehicle) removed (para. 4-71).

GENERAL SAFETY INSTRUCTIONS
- Drycleaning solvent is flammable and toxic. Do not use near an open flame.
- Keep Are extinguisher nearby when using drycleaning solvent.

a. Removal

1. Using barring tool, prevent crankshaft (4) from turning.
2. Remove eight screws (6), clamping ring (1), flexplate (2), and crankshaft adapter (3) from crankshaft (4) and flywheel housing (5).
3. Drill two 0.109-in. (2.77-mm) holes (7) in rear oil seal (8) on opposite sides.
4. Using slide hammer and screw, and alternating from one hole (7) to the other, remove rear oil seal (8) from crankshaft (4) and flywheel housing (5). Discard rear oil seal (8).

b. Cleaning and Inspection

WARNING

Drycleaning solvent is flammable and toxic. Do not use near open flame and always have a fire extinguisher nearby when solvents are used. Use only in well-ventilated places, wear protective clothing, and dispose of cleaning rags in approved container. Failure to do this may result in injury or death to personnel and/or damage to equipment.

1. Clean sealing surface of crankshaft (4) with drycleaning solvent and dry with clean rag.
2. Inspect crankshaft (4) for nicks, burrs, and gouges. If excessively nicked, burred, or gouged, report problem to General Support maintenance.

c. Installation

CAUTION

Ensure crankshaft and rear cover sealing surfaces are clean and dry. Failure to do so may prevent proper sealing.

1. Install pilot (9) in new rear oil seal (8) ensuring tapered end of pilot (9) and rear oil seal (8) face the same direction.

4-45. REAR OIL SEAL (IN VEHICLE) MAINTENANCE (Contd)

2. Install pilot (9) and rear oil seal (8) on crankshaft (4).
3. Using alignment tool, drive rear oil seal (8) on crankshaft (4).
4. Remove alignment tool from crankshaft (4).
5. Remove pilot (9) from crankshaft (4) and rear oil seal (8).
6. Install crankshaft adapter (3), flexplate (2), and clamping ring (1) on crankshaft (4) with eight screws (6). Tighten screws (6) 100 lb-ft (136 N·m) in sequence shown.

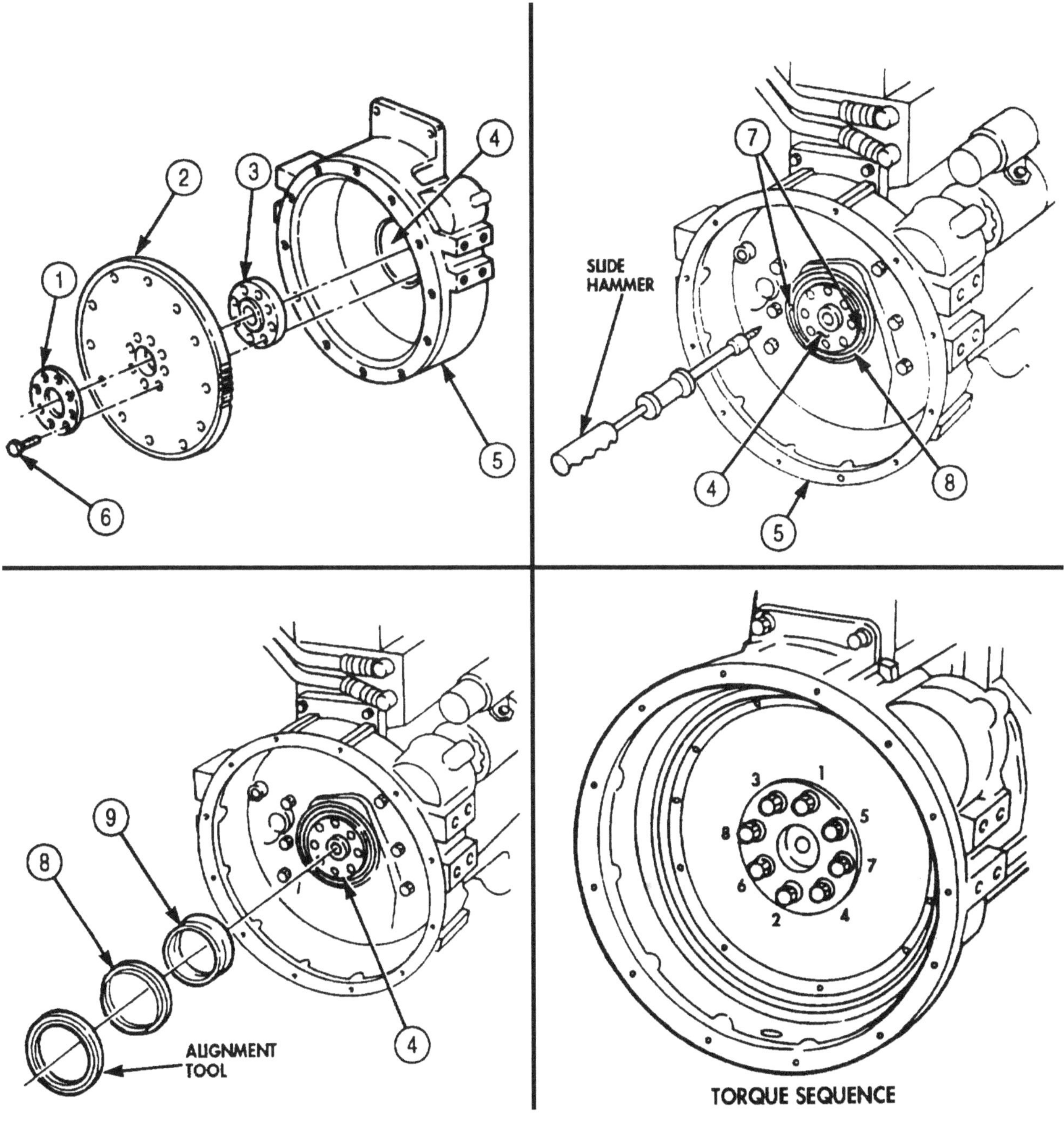

FOLLOW-ON TASK: Install transmission (in-vehicle) (para. 4-71).

4-46. FRONT GEARCASE HOUSING AND TIMING PIN MAINTENANCE

THIS TASK COVERS:

a. Front Gearcase Housing Removal
b. Timing Pin Removal
c. Cleaning and Inspection
d. Timing Pin Installation
e. Front Gearcase Housing Installation

INITIAL SETUP:

APPLICABLE MODELS
M939A2

SPECIAL TOOLS
Engine barring tool (Appendix E, Item 43)

TOOLS
General mechanic's tool kit (Appendix E, Item 1)
Torque wrench (Appendix E, Item 144)

MATERIALS/PARTS
O-ring (Appendix D, Item 445)
Gasket (Appendix D, Item 191)
Gasket (Appendix D, Item 192)
Sealing compound (Appendix C, Item 68)
Antiseize compound (Appendix C, Item 10)
Gasket compound (Appendix C, Item 29)

REFERENCES (TM)
TM 9-2320-272-24P

EQUIPMENT CONDITION
- Engine mounted on repair stand (para. 4-9).
- Rocker levers and pushrods removed (para. 4-41).
- Vibration damper removed (para. 4-42).
- Front gearcase cover removed (para. 4-48).
- Fuel injection pump removed (para. 4-57).

a. Front Gearcase Housing Removal

NOTE

Assistant will help with step 1.

1. Locate Top Dead Center (TDC) for cylinder No. 1. While turning engine, push down on timing pin (2) until timing pin (2) enters hole (3) in camshaft (1).
2. Remove four screws (13), washers (14), and front engine mount (15) from gearcase housing (5).
3. Remove four front oil pan screws (11), washers (10), and washers (9) from gearcase housing (5) and oil pan (12).
4. Remove twenty-four screws (4), gear housing (5), and gasket (6) from engine block (7). Discard gasket (6).

b. Timing Pin Removal

1. Remove ring (19), timing pin (2), and O-ring (18) from timing pin housing (17). Discard O-ring (18).
2. Remove two torx screws (20), timing pin housing (17), and gasket (16) from gearcase housing (5). Discard gasket (16).

c. Cleaning and Inspection

1. For general cleaning instructions, refer to para. 2-14.
2. For general inspection instructions, refer to para 2-15.
3. Repair or replace all parts failing inspection.
4. Inspect oil pan gasket (8) for tears. If torn, replace oil pan gasket (8) (para. 4-47).

4-46. FRONT GEARCASE HOUSING AND TIMING PIN MAINTENANCE (Contd)

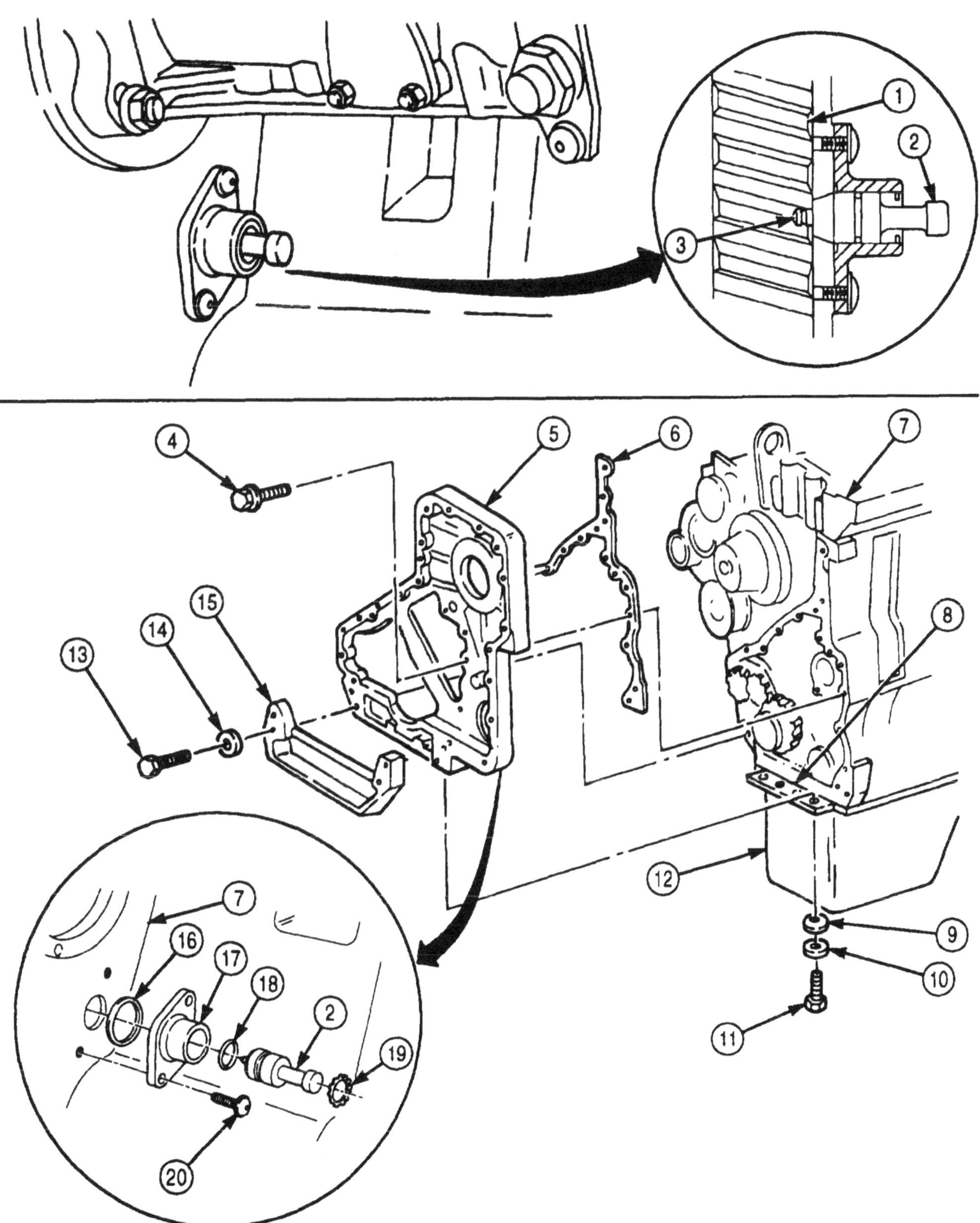

4-46. FRONT GEARCASE HOUSING AND TIMING PIN MAINTENANCE (Contd)

d. Timing Pin Installation

1. Apply antiseize compound to threads of two torx screws (18).
2. Install new gasket (13) and timing pin housing (14) on gear housing (2) with two torx screws (18). Tighten torx screws (18) 4 lb-ft (5 N·m).
3. Install new O-ring (15), timing pin (16), and ring (17) in timing pin housing (14).

e. Front Gearcase Housing Installation

1. Trim 0.0625 in, (1.59 mm) off bottom of new gasket (3) and apply gasket compound to both sides.
2. Apply sealing compound to front of new oil pan gasket (5) (if replaced).
3. Position new gasket (3) on engine block (4).
4. Install gear housing (2) on engine block (4) with twenty-four screws (1). Tighten screws (1) 18 lb-ft (24 N·m).
5. Install four washers (8), washers (6), and front oil pan screws (7) on oil pan (9) and gear housing (2). Tighten screws (7) 18 lb-ft (24 N·m).
6. Install front engine mount (12) on gear housing (2) with four washers (11) and screws (10). Tighten screws (10) 82 lb-ft (111 N·m).

4-46. FRONT GEARCASE HOUSING AND TIMING PIN MAINTENANCE (Contd)

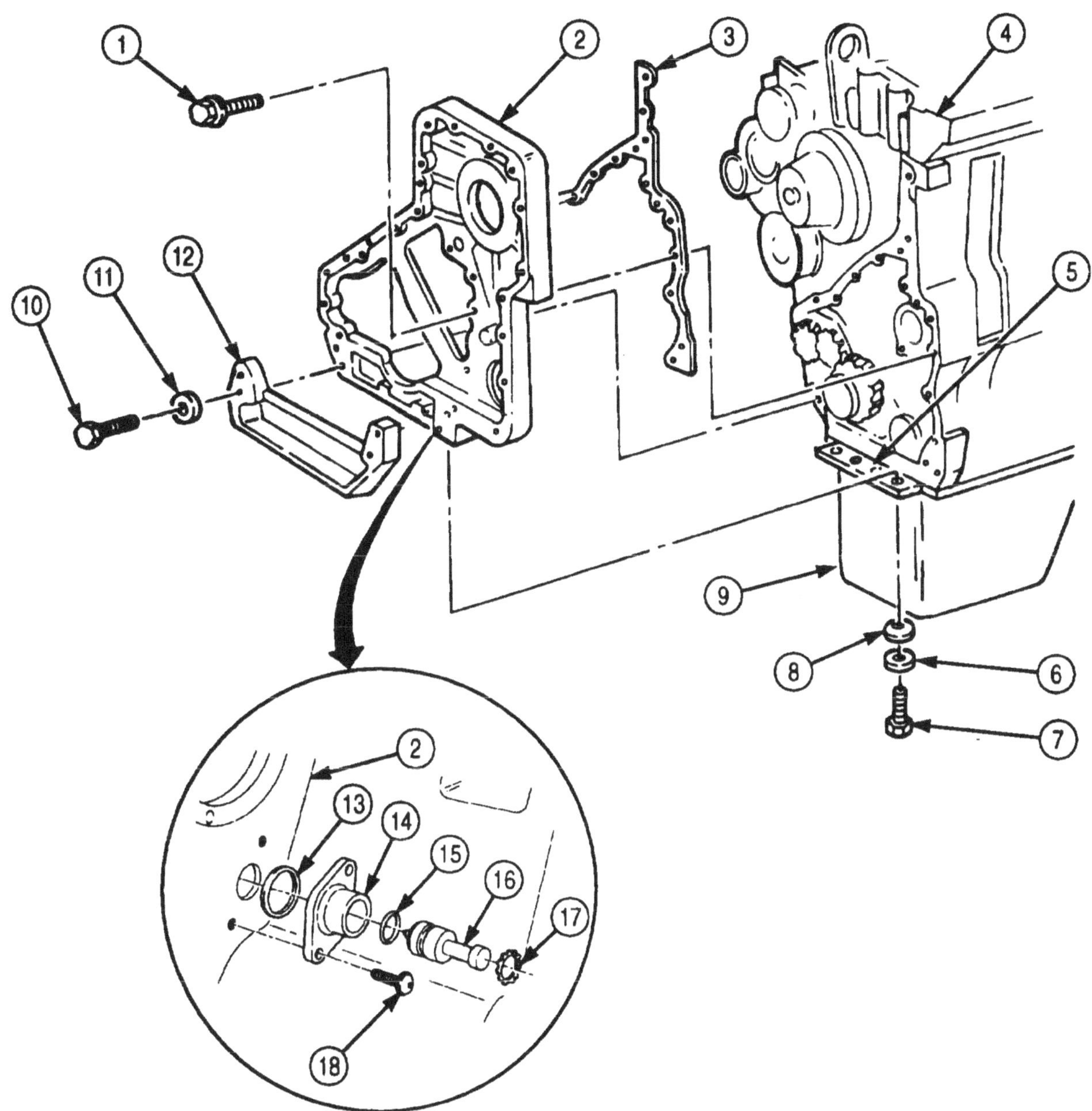

FOLLOW-ON TASKS:
- Install fuel injection pump (para. 4-57).
- Install front gearcase cover (para. 4-48).
- Install vibration damper (para. 4-42).
- Install rocker levers and pushrods (para. 4-41).
- Remove engine from repair stand (para. 4-9).

4-47. OIL PAN AND OIL SUCTION TUBE MAINTENANCE

THIS TASK COVERS:

a. Oil Pan Removal
b. Oil Suction Tube Removal
c. Cleaning and Inspection
d. Oil Suction Tube Installation
e. Oil Pan Installation

INITIAL SETUP:

APPLICABLE MODELS
M939A2

TOOLS
General mechanic's tool kit (Appendix E, Item 1)

MATERIALS/PARTS
Thirty-two lockwashers (Appendix D, Item 380)
Seal (Appendix D, Item 595)
Washer (Appendix D, Item 711)
Gasket (Appendix D, Item 194)
Gasket (Appendix D, Item 193)
Locknut (Appendix D, Item 303)
Sealing compound (Appendix C, Item 64)
Sealing compound (Appendix C, Item 65)

REFERENCES (TM)
LO 9-2320-272-12
TM 9-2320-272-10
TM 9-2320-272-24P

EQUIPMENT CONDITION
- Parking brake set (TM 9-2320-272-24-10).
- Engine oil drained (LO 9-2320-272-12).

a. Oil Pan Removal

1. Remove thirty-two screws (5), washers (4), lockwashers (3), oil pan (6), and gasket (2) from engine block (1). Discard gasket (2) and lockwashers (3).
2. Remove two plugs (8), seal (7), and washer (9) from oil pan (6). Discard seal (7) and washer (9).

b. Oil Suction Tube Removal

1. Remove two screws (12) and bracket (11) from engine block (1).
2. Remove two screws (14). spacers (15), oil suction tube (13), and gasket (10) from engine block (1). Discard gasket (10).
3. Remove locknut (18), washer (17), screw (16), and bracket (11) from oil suction tube (13). Discard locknut (18).

c. Cleaning and Inspection

1. For general cleaning instructions, refer to para. 2-14.
2. For general inspections instructions, refer to para. 2-15.
3. Replace all parts failing inspection.

d. Oil Suction Tube Installation

1. Install bracket (11) on oil suction tube (13) with screw (16), washer (17), and new locknut (18).
2. Install new gasket (10) and oil suction tube (13) on engine block (1) with two spacers (15) and screws (14).
3. Install bracket (11) on engine block (1) with two screws (12).

4-47. OIL PAN AND OIL SUCTION TUBE MAINTENANCE (Contd)

e. Oil Pan Installation

NOTE

Prior to installation, fill joint between oil pan rail, front gear case cover, and rear cover with sealing compound.

1. Apply sealing compound to mating surfaces of engine block (1) and oil pan (6).
2. Position new gasket (2) on oil pan (6) with raised bead facing oil pan (6).
3. Install oil pan (6) on engine block (1) with thirty-two new lockwashers (3), washers (4), and screws (5).
4. Install new seal (7), new washer (9), and two plugs (8) on oil pan (6).

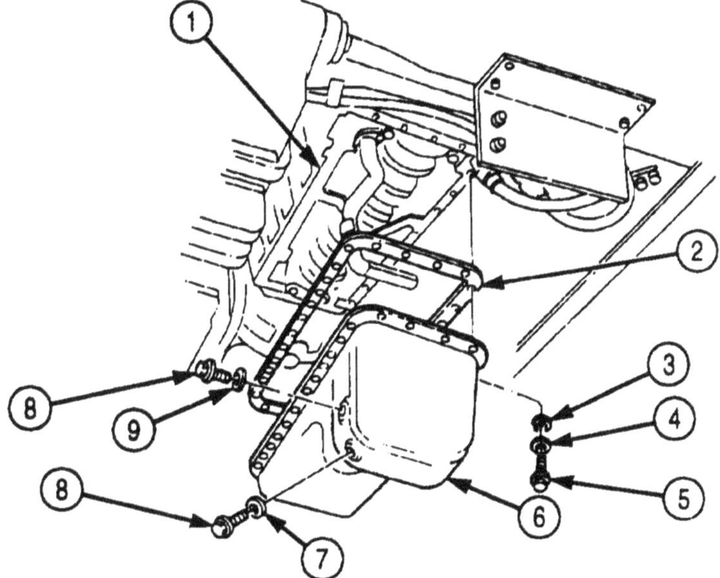

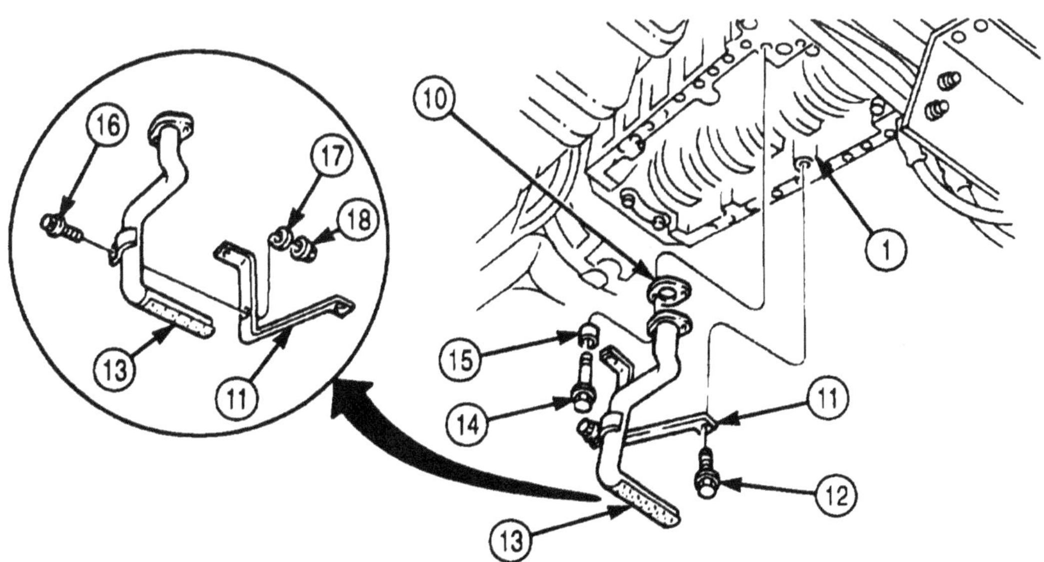

FOLLOW-ON TASKS: • Fill engine crankcase to proper level (LO 9-2320-272-12).
• Check oil pan for leaks.

4-48. FRONT GEARCASE COVER AND OIL SEAL MAINTENANCE

THIS TASK COVERS:

a. Removal
b. Cleaning and Inspection
c. Installation

INITIAL SETUP:

APPLICABLE MODELS
M939A2

TOOLS
General mechanic's tool kit (Appendix E, Item 1)
Torque wrench (Appendix E, Item 146)

MATERIALS/PARTS
Oil seal kit (Appendix D, Item 506)
Gasket (Appendix D, Item 195)
Drycleaning solvent (Appendix C, Item 71)
Rags (Appendix C, Item 58)
Sealing compound (Appendix C, Item 68)

REFERENCES (TM)
TM 9-2320-272-24P

EQUIPMENT CONDITION
- Tachometer drive removed (para. 4-51).
- Vibration damper removed (para. 4-42).

GENERAL SAFETY INSTRUCTIONS
- Drycleaning solvent is flammable and toxic. Do not use near an open flame.
- Keep fire extinguisher nearby when using drycleaning solvent.

a. Removal

NOTE

Gearcase mounting screws are different lengths. Tag locations of screws for installation.

1. Remove twenty-two screws (1) from gearcase cover (5) and gear housing (2).
2. Remove gearcase cover (5) and gasket (4) from gear housing (2). Discard gasket (4).
3. Remove oil seal (6) from gearcase cover (5). Discard oil seal (6).

b. Cleaning and Inspection

1. For general cleaning instructions, refer to para. 2-14.
2. For general inspection instructions, refer to para. 2-15.

WARNING

Drycleaning solvent is flammable and toxic. Do not use near open flame and always have a fire extinguisher nearby when solvents are used. Use only in well-ventilated places, wear protective clothing, and dispose of cleaning rags in approved container. Failure to do this may result in injury or death to personnel and/or damage to equipment.

3. Clean sealing surface of crankshaft (3) with drycleaning solvent. Dry with clean rag.

c. Installation

1. Apply sealing compound to mating surfaces of gearcase cover (5) and gear housing (2).
2. Install new gasket (4) and gearcase cover (5) on gear housing (2) with twenty-two screws (1). Finger-tighten screws (1).
3. Install alignment tool on crankshaft (3) to align gearcase cover (5).
4. Tighten twenty-two screws (1) 20 lb-ft (27 N·m).
5. Remove alignment tool from crankshaft (3).

4-48. FRONT GEARCASE COVER AND OIL SEAL MAINTENANCE (Contd)

6. Apply sealing compound to outside diameter of new oil seal (6).
7. Place oil seal (6) on pilot (7).
8. Install oil seal (6) with pilot (7) on crankshaft (3).
9. Remove pilot (7) from crankshaft (3).
10. Using alignment tool, drive oil seal (6) into gearcase cover (5).

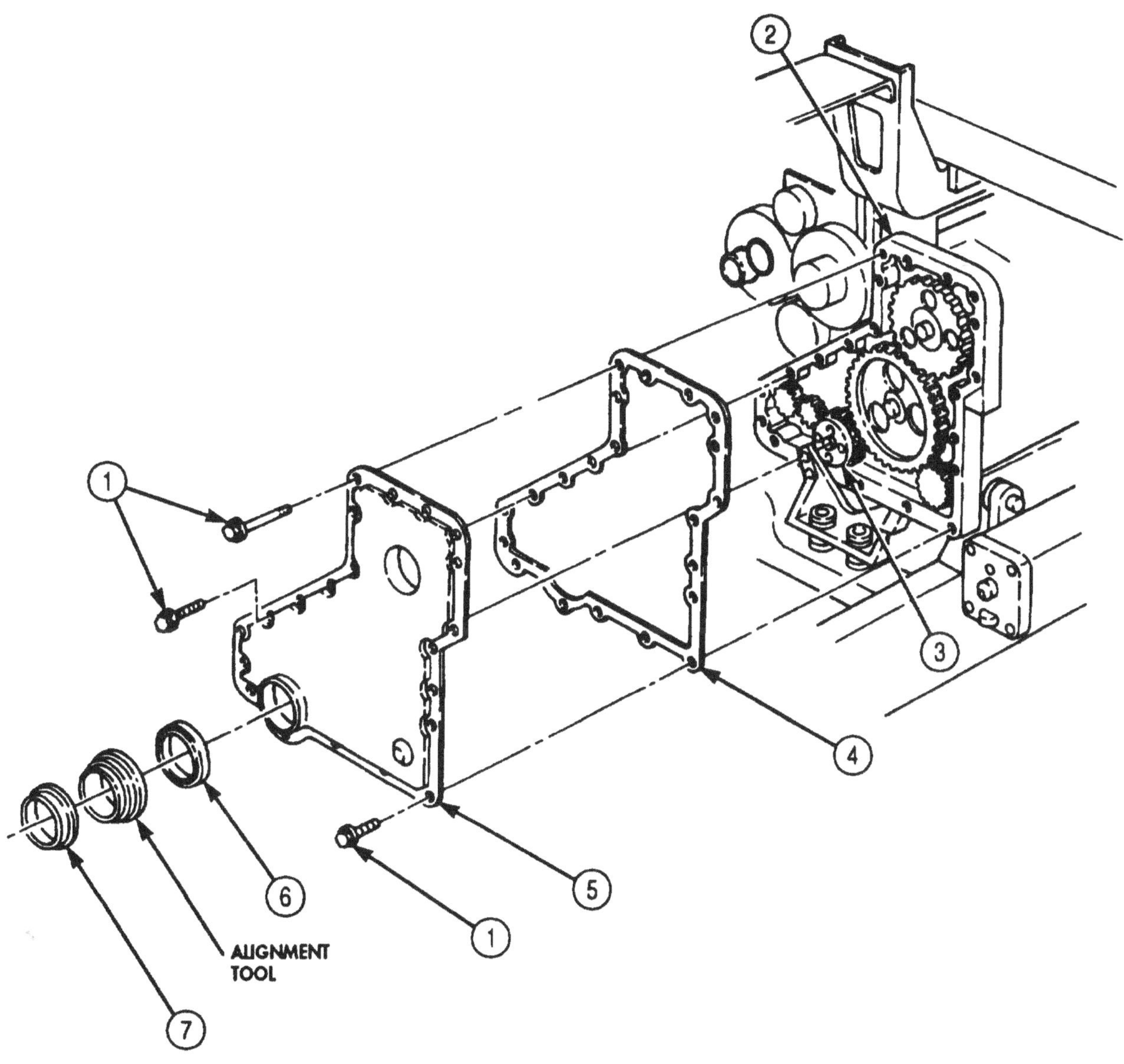

FOLLOW-ON TASKS: • Install vibration damper (para. 4-42).
• Install tachometer drive (para. 4-51).

4-49. OIL PUMP MAINTENANCE

THIS TASK COVERS:

a. Gear Backlash Test
b. Removal
c. Cleaning and Inspection
d. Installation

INITIAL SETUP:

APPLICABLE MODELS
M939A2

TOOLS
General mechanic's tool kit (Appendix E, Item 1)
Dial indicator (Appendix E, Item 36)
Torque Wrench (Appendix E, Item 144)

MATERIALS/PARTS
Oil pump kit (Appendix D, Item 494)
Pressure regulator plunger (Appendix D, Item 524)
Washer (Appendix D, Item 716)
Engine oil (Appendix C, Item 50)

REFERENCES (TM)
TM 9-2320-272-10
TM 9-2320-272-24P

EQUIPMENT CONDITION
Front gearcase cover removed (para. 4-48).

a. Gear Backlash Test

1. Install dial indicator on gear housing flange (1). Ensure anvil is positioned against tooth of drive gear (4).
2. Holding idler gear (3) stationary, turn drive gear (4) against teeth of idler gear (3).
3. Set dial indicator to zero.
4. Turn drive gear (4) in opposite direction and take backlash reading.
5. Backlash reading must be between 0.003-0.013 in. (0.08-0.330 mm). If backlash reading is not within limits, replace oil pump (6).
6. Position anvil of dial indicator on tooth of idler gear (3).
7. Holding crankshaft gear (2) stationary, turn idler gear (3) against teeth of crankshaft gear (2).
8. Set dial indicator to zero.
9. Turn idler gear (3) in opposite direction and take backlash reading.
10. Backlash reading must be between 0.003-0.013 in. (0.08-0.330 mm). If backlash reading is not within limits, replace oil pump (6).

b. Removal

Remove four screws (5) and oil pump (6) from engine block (7).

4-49. OIL PUMP MAINTENANCE (Contd)

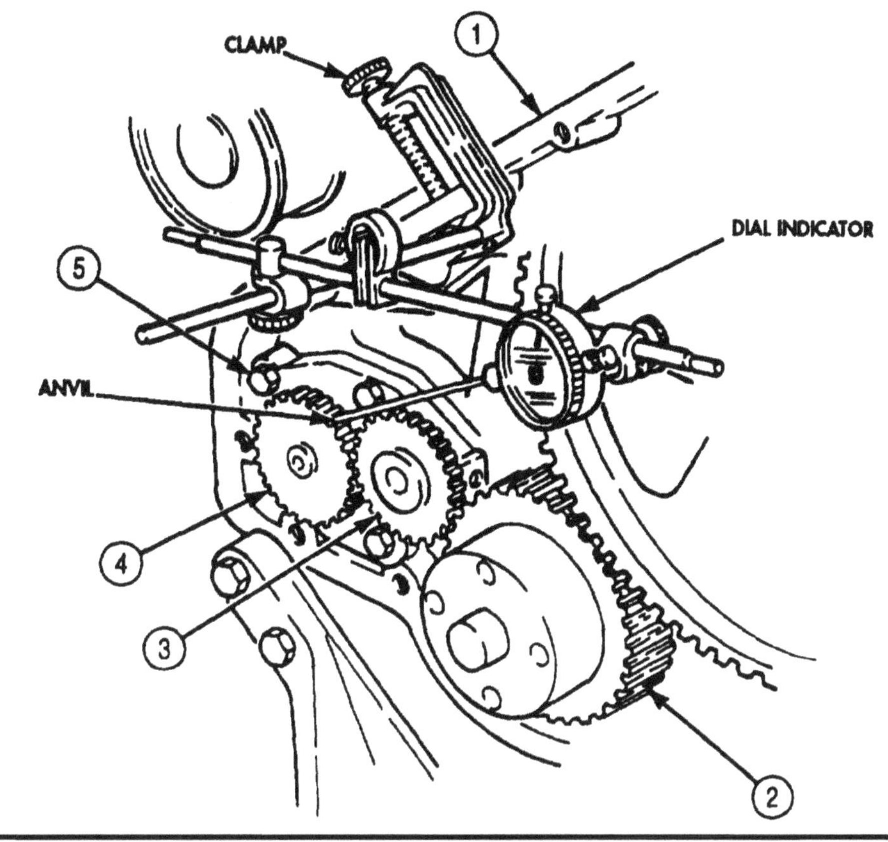

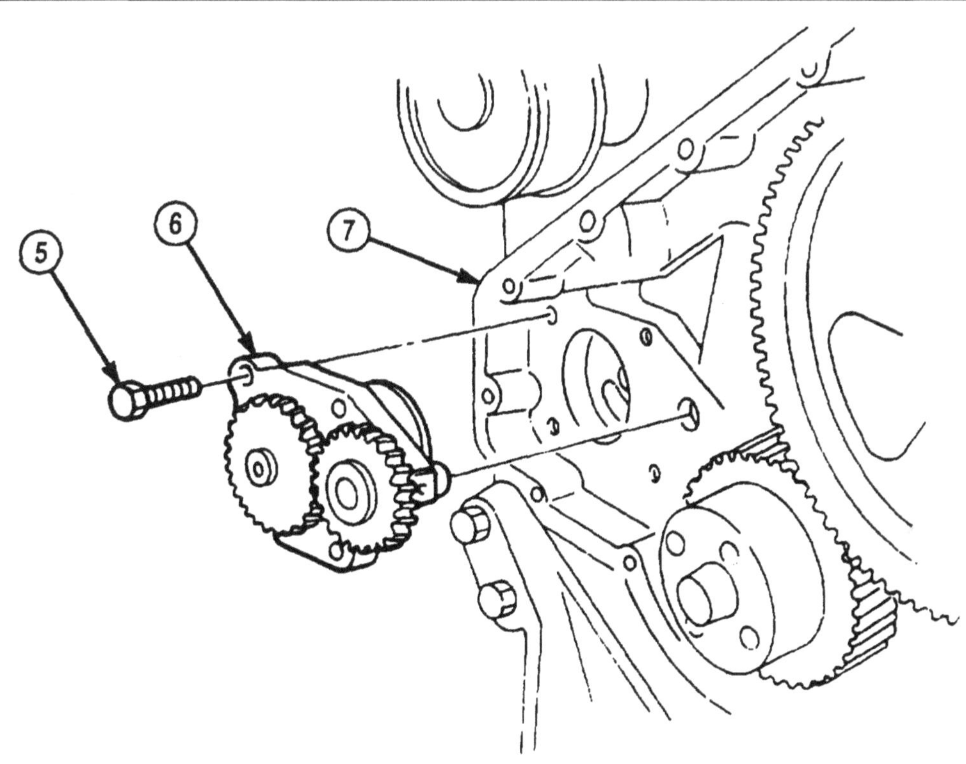

4-49. OIL PUMP MAINTENANCE (Contd)

c. Cleaning and Inspection

1. For general cleaning instructions, refer to para. 2-14.
2. Remove backing plate (1) from oil pump housing (4).

NOTE

Mark gerotor planetary position for installation.

3. Mark word TOP on gerotor planetary (3) and remove from oil pump housing (4)
4. Inspect gerotor planetary (3) and oil pump housing (4) for damage. If damaged, replace oil pump (6).
5. Install gerotor planetary (3) in oil pump housing (4). Ensure word TOP faces upwards.
6. Using feeler gauge, measure distance between tip of gerotor drive gear (2) and high spot of gerotor planetary (3). If distance exceeds 0.007 in. (0.1778 mm), replace oil pump (6).
7. Place straight edge across oil pump housing (4).
8. Using feeler gauge, measure distance between straight edge and gerotor planetary (3) and drive gear (2). If distance exceeds 0.005 in. (0. 0.127 mm), replace oil pump (6).
9. Using feeler gauge, measure distance between gerotor planetary (3) and oil pump housing (4). If distance exceeds 0.015 in (0.381 mm), replace oil pump (6).

d. Installation

NOTE

- Installation of oil pump kit 3802278 is required for early model engine if inspection required replacement of oil pump.
- Perform steps 1 and 2 for early model engine.

1. Remove plug (13), washer (12), spring (14), and pressure regulator plunger (11) from oil filter head (10). Discard washer (12) and pressure regulator plunger (11).
2. Install new pressure regulator plunger (11) and spring (14) in oil filter head (10) with new washer (12) and plug (13). Tighten plug (13) 60 lb-ft (81 N·m).

CAUTION

Lubricate oil pump with clean engine oil prior to installation. Failure to do so may cause damage to equipment.

3. Align idler pin (9) with gear housing bore (8).
4. Install oil pump (6) on gear housing (7) with four screws (5).
5. Tighten four screws (5) in the following sequence:
   a. Tighten screws (5) 4 lb-ft (5 N·m).
   b. Tighten screws (5) 18 lb-ft (24 N·m).

4-49. OIL PUMP MAINTENANCE (Contd)

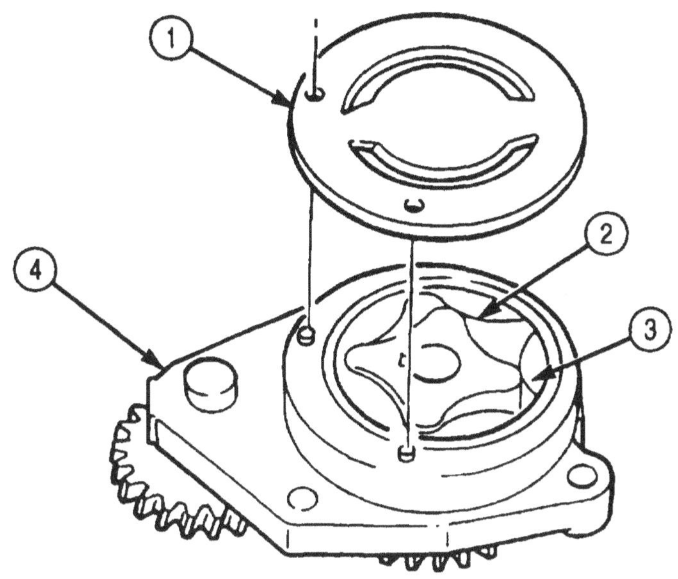

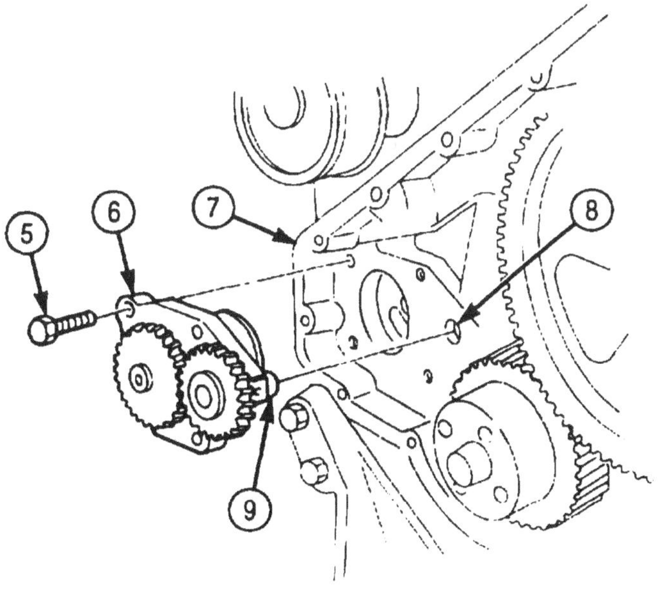

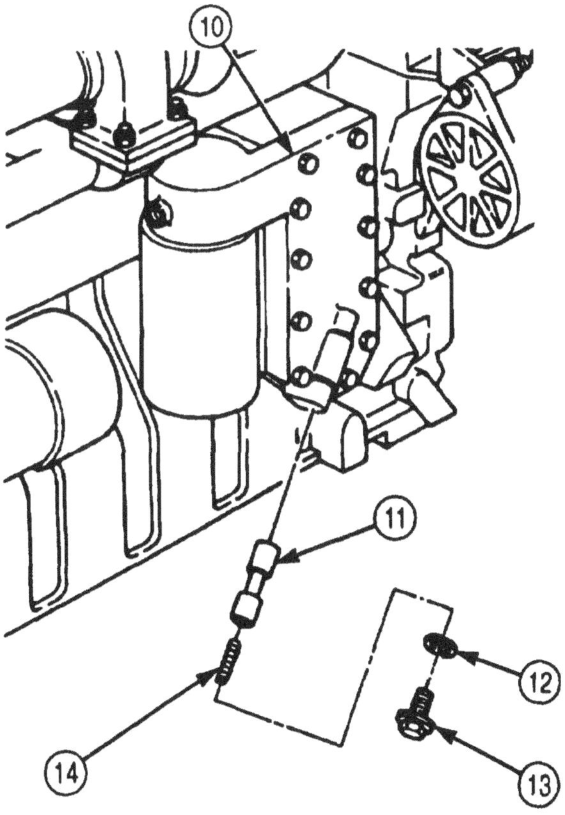

FOLLOW-ON TASK: Install front gearcase cover (para. 4-48).

4-50. EXHAUST MANIFOLD REPLACEMENT

THIS TASK COVERS:

a. Removal **b. Installation**

INITIAL SETUP:

APPLICABLE MODELS
M939A2

TOOLS
General mechanic's tool kit (Appendix E, Item 1)
Torque wrench (Appendix E, Item 146)

MATERIALS/PARTS
Six gaskets (Appendix D, Item 196)
Six lockplates (Appendix D, Item 340)
Antiseize compound (Appendix C, Item 10)

REFERENCES (TM)
TM 9-2320-272-24P

EQUIPMENT CONDITION
Turbocharger and coolant lines removed (para. 3-21).

a. Removal

1. Remove screw (1) from bracket (2) and exhaust manifold (3).
2. Bend tabs of lockplates (5) away from screws (4).
3. Remove twelve screws (4), six lockplates (5), exhaust manifold (3), and six gaskets (6) from cylinder head (7). Discard lockplates (5) and gaskets (6).

b. Installation

1. Position six new gaskets (6) and exhaust manifold (3) on cylinder (7) and install with six new lockplates (5) and twelve screws (4). Tighten screws (4) 30 lb-ft (41 N·m).
2. Bend tabs of lockplates (5) against head of screws (4).
3. Apply antiseize compound to threads of screw (1) and install on bracket (2) and exhaust manifold (3). Tighten screw (1) 30 lb-ft (41 N·m).

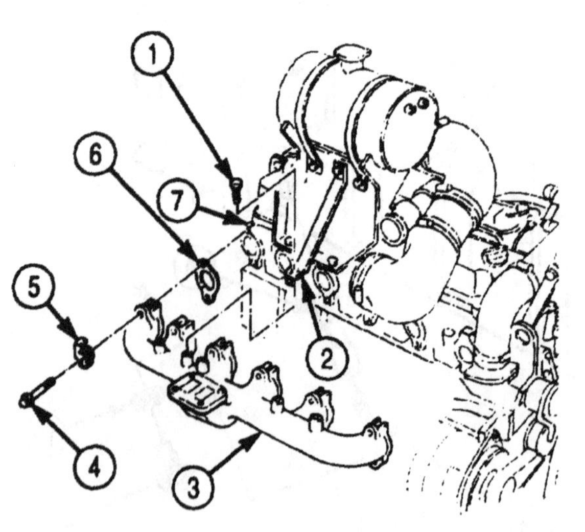

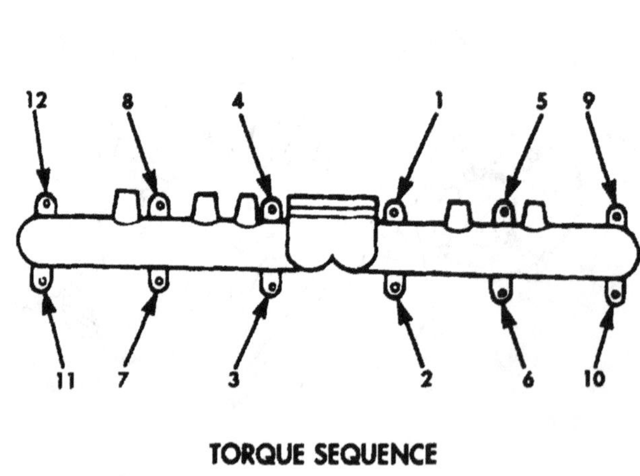

TORQUE SEQUENCE

FOLLOW-ON TASK: Install turbocharger and coolant lines (para. 3-21).

4-51. TACHOMETER DRIVE MAINTENANCE

THIS TASK COVERS:

a. Removal
b. Cleaning and inspection
c. Installation

INITIAL SETUP:

APPLICABLE MODELS
M939A2

TOOLS
General mechanic's tool kit (Appendix E, Item 1)

MATERIALS/PARTS
Gasket (Appendix D, Item 197)
Seal (Appendix D, Item 603)

REFERENCES (TM)
TM 9-2320-272-10
TM 9-2320-272-24P

EQUIPMENT CONDITION
• Parking brake set (TM 9-2320-272-10).
• Hood raised and secured (TM 9-2320-272-10).

a. Removal

1. Disconnect tachometer drive cable (8) from tachometer drive (9).
2. Remove cover (3) and seal (4) from front gearcase cover (5) by turning cover (3) and seal (4) counterclockwise. Discard seal (4).
3. Remove drive hub (6) from tachometer drive (9).
4. Remove two screws (1), washers (2), tachometer drive (9), and gasket (7) from cover (3). Discard gasket (7).

b. Cleaning and Inspection

1. For general cleaning instructions, refer to para. 2-14.
2. For general inspection instructions, refer to para. 2-15.
3. Replace all parts failing inspection.

c. Installation

1. Install new gasket (7) and tachometer drive (9) on cover (3) with two washers (2) and screws (1).
2. Install drive hub (6) on tachometer drive (9).
3. Install new seal (4) and cover (3) on front gearcase cover (5).
4. Connect tachometer drive cable (8) to tachometer drive (9).

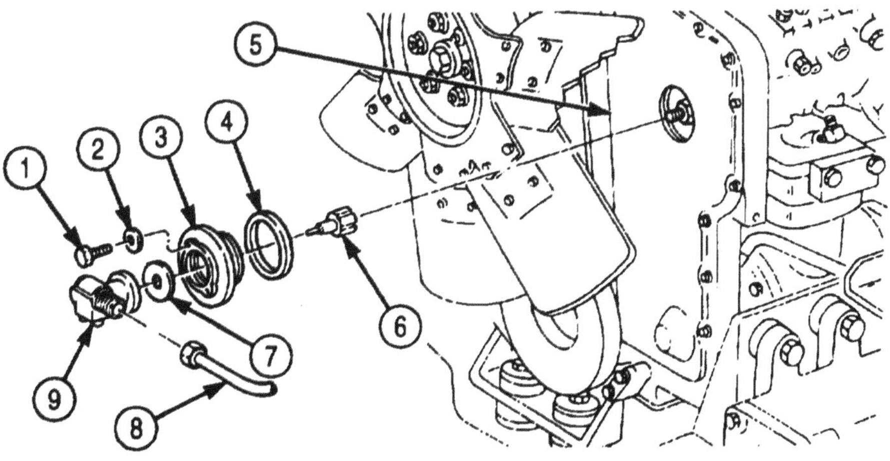

4-52. AIR COMPRESSOR MAINTENANCE

THIS TASK COVERS:

a. Disassembly
b. Cleaning and Inspection
c. Assembly

INITIAL SETUP:

APPLICABLE MODELS
M939A2

TOOLS
General mechanic's tool kit (Appendix E, Item 1)
Gear puller (Appendix E, Item 102)
Torque wrench (Appendix E, Item 144)
Ring compressor (Appendix E, Item 32)

MATERIALS/PARTS
Air compressor repair kit (Appendix D, Item 4)
Two screw-assembled lockwashers (Appendix D, Item 579)
Two screw-assembled lockwashers (Appendix D, Item 580)
Four lockwashers (Appendix D, Item 364)
Compression ring (Appendix D, Item 38)
Gasket (Appendix D, Item 199)
Gasket (Appendix D, Item 186)

MATERIAL/PARTS (Contd)
Gasket (Appendix D, Item 200)
Gasket (Appendix D, Item 198)
Gasket (Appendix D, Item 138)
Four lockwashers (Appendix D, Item 382)
Compression ring (Appendix D, Item 39)
Three oil rings (Appendix D, Item 517)
Engine oil (Appendix C, Item 49)
Antiseize tape (Appendix C, Item 72)

REFERENCES (TM)
TM 9-2320-272-24P

EQUIPMENT CONDITION
Air compressor removed (para. 3-206).

a. Disassembly

1. Remove elbow (2) from unloader valve body (1).
2. Remove elbow (17) and two adapters (7) from compressor head cover (18).
3. Remove four screws (10), lockwashers (9), washers (8), adapter (11), and gasket (12) from compressor crankcase (16). Discard lockwashers (9) and gasket (12).
4. Remove two screws (13), air inlet block (14), and gasket (15) from compressor head cover (18). Discard gasket (15).
5. Remove two screw-assembled lockwashers (20), washers (19), unloader valve body (1), unloader valve spring (3), intake valve (4), and intake valve spring (6) from compressor head cover (18). Discard screw-assembled lockwashers (20), unloader valve spring (3), and intake valve spring (6).
6. Remove intake valve (4) from valve seat (5). Discard intake valve (4).
7. Remove unloader valve cap (23), packing seal (22), and O-ring (21) from unloader valve body (1). Discard unloader valve cap (23), packing seal (22), O-ring (21), and unloader valve body (1).

4-52. AIR COMPRESSOR MAINTENANCE (Contd)

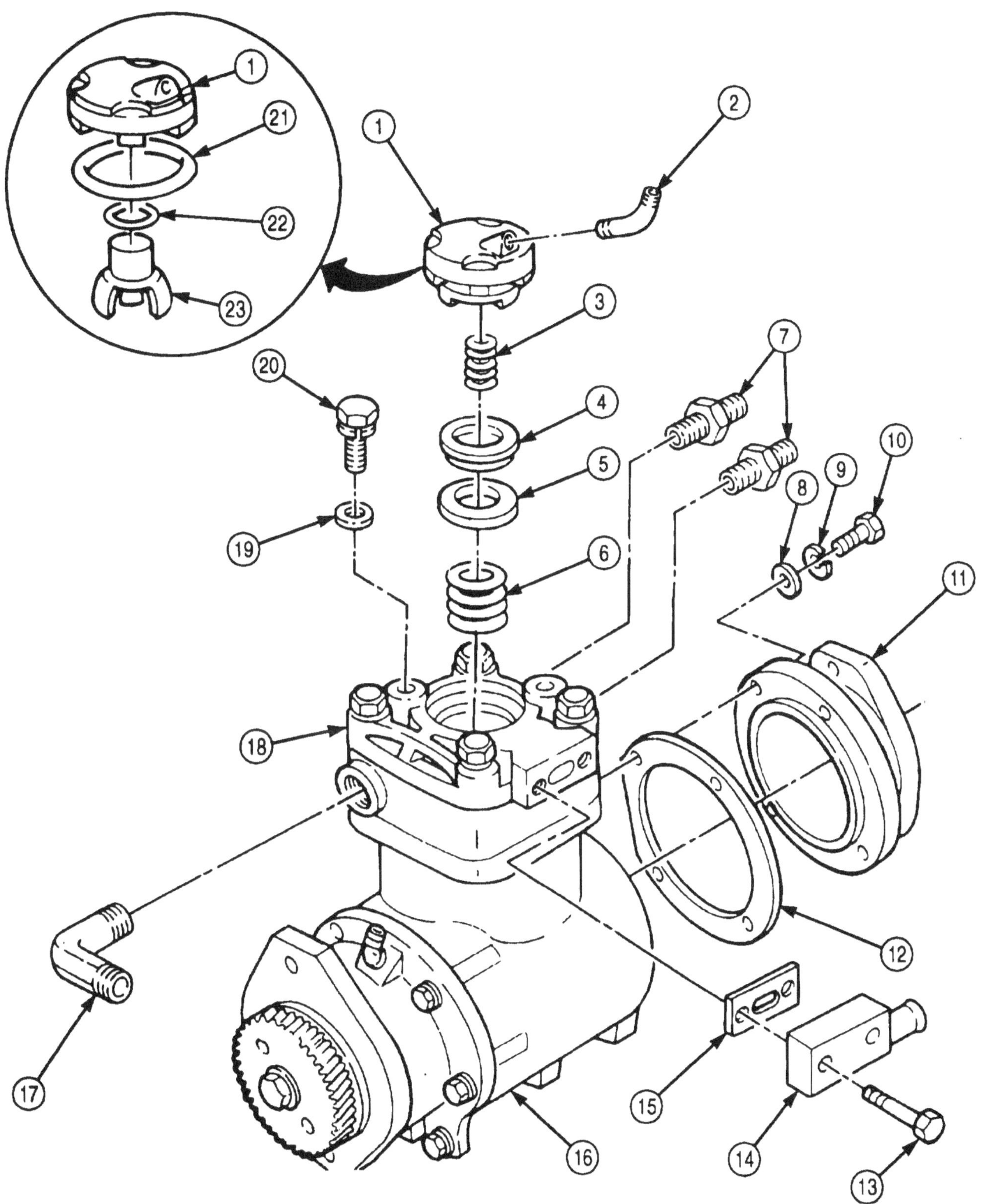

4-52. AIR COMPRESSOR MAINTENANCE (Contd)

8. Remove four screws (l), lockwashers (2), washers (3) compressor head cover (4) gasket (5), compressor head (6), and gasket (7) from compressor crankcase (16). Discard lockwashers (2) and gaskets (5) and (7).
9. Press exhaust valve seat (33) from hole (34) on bottom of compressor head (6).
10. Remove O-ring (32), exhaust valve seat (33), seal (31). exhaust valve (30), exhaust valve spring (29), and wear plate (28) from compressor head (6). Discard O-ring (32), exhaust valve seat (33), seal (31), exhaust valve (30), exhaust valve spring (29), and wear plate (28).
11. Remove two screw-assembled lockwashers (17), four screws (19), and washers (18) from compressor crankcase (16) and support (25). Discard screw-assembled lockwashers (17).

CAUTION

Use care when removing crankshaft from connecting rod to avoid damage.

12. Rotate crankshaft (27) 90 degrees before or after top dead center of piston (14) for ease of removal.
13. Remove support (25), crankshaft (27), and gasket (26) from compressor crankcase (16). Discard gasket (26).
14. Install crankshaft (27) and support (25) in soft-jawed vise.
15. Remove screw (24) and washer (23) from drive gear (22) and crankshaft (27).

NOTE

Compressor drive gear is drilled and tapped for metric threads.

16. Using gear puller, remove drive gear (22) from crankshaft (27).
17. Using arbor press, remove spacer (21), thrust washer (20), and crankshaft (27) from support (25).
18. Remove support (25) from vise.
19. Remove piston (14) and connecting rod (15) from compressor crankcase (16).
20. Remove top (9) and intermediate (10) compression rings from piston (14). Discard compression rings (9) and (10).

CAUTION

Do not drive piston pin from bore. Damage to piston may result.

NOTE

It may be necessary to heat piston in hot water to remove piston pin.

21. Remove two oil rings (11) and oil ring expander (12) from piston (14). Discard oil rings (11) and oil ring expander (12).
22. Remove two retaining rings (8), piston pin (13), and connecting rod (15) from piston (14).

4-52. AIR COMPRESSOR MAINTENANCE (Contd)

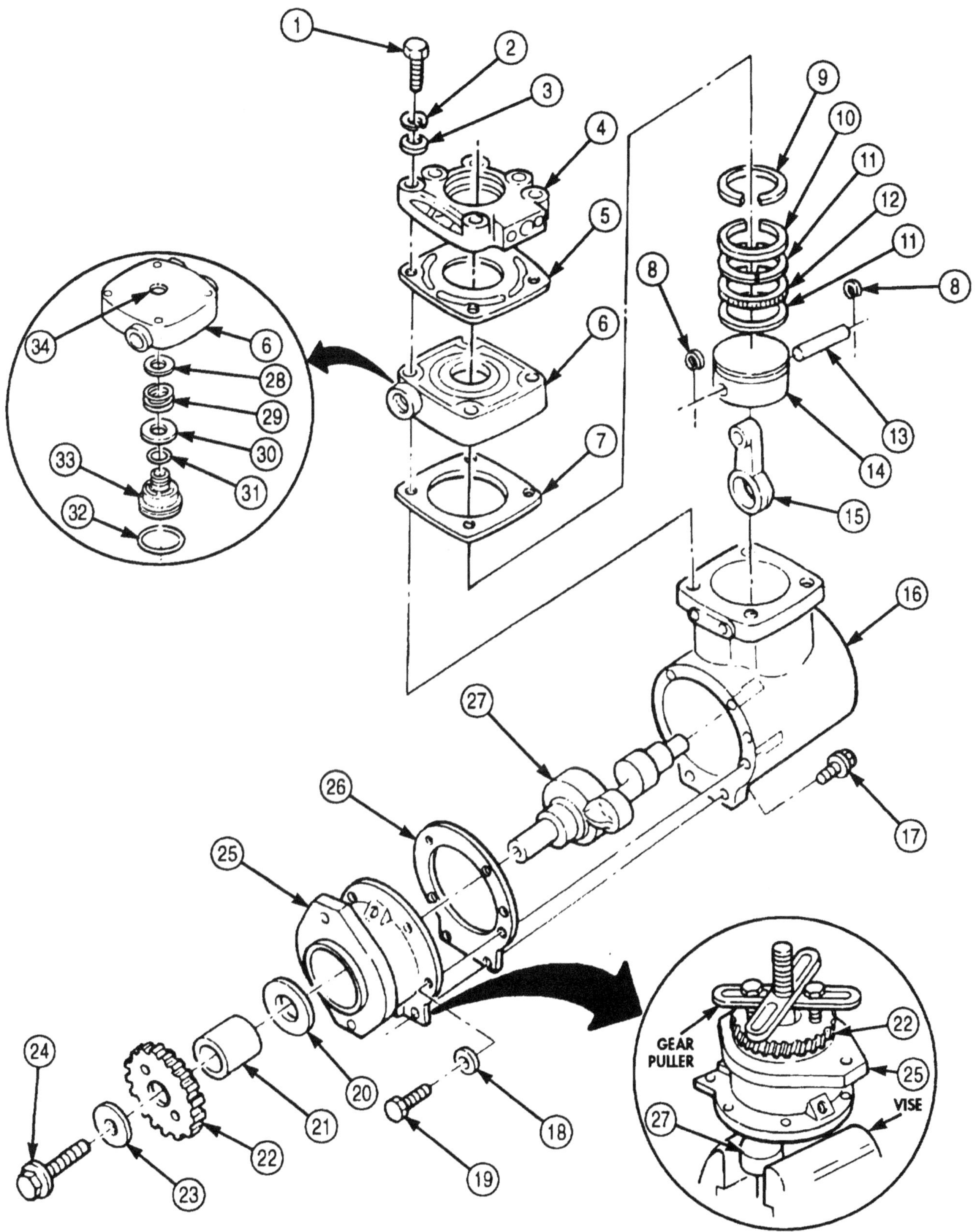

4-52. AIR COMPRESSOR MAINTENANCE (Contd)

b. Cleaning and Inspection

1. For general cleaning instructions, refer to para. 2-14.
2. For general inspection instructions, refer to para. 2-15.
3. Inspect compressor head (2) and compressor head cover (1) for cracks and breaks. If cracked or broken, replace compressor head (2) or compressor head cover (1).
4. Inspect crankshaft (6) for scratches, scoring, and wear. Refer to table 4-6, Air Compressor Wear Limits, for measurements. If damaged or worn, replace crankshaft (5).
5. Inspect bore (4) of support (3) for scratches, scoring, and wear. Refer to table 4-6, Air Compressor Wear Limits, for measurements. If damaged or won, replace support (3).
6. Inspect thickness of thrust washer (10) for wear. Refer to table 4-6, Air compressor Wear Limits, for measurements. If damaged or worn, replace thrust washer (10).
7. Inspect connecting rod (6), for breaks, bends, twists, scoring, and wear. If bent, broken, twisted, scored, or worn, replace connecting rod (6).
8. Measure inside diameter of piston pin bore (7) and crankshaft journal (8). Refer to table 4-6, Air Compressor Wear Limits, for measurements. If damaged or worn, replace connecting rod (6).
9. Measure outside diameter of piston pin (9). Refer to table 4-6, Air Compressor Wear Limits, for measurements. If damaged or worn, replace piston pin (9).
10. Inspect piston (11) for breaks, cracks, scoring, and wear. If broken, cracked, scored, or worn, replace piston (11).
11. Measure outside diameter of piston skirt (12). Refer to table 4-6, Air Compressor Wear Limits, for measurements. If damaged or worn, replace piston (11).
12. Measure inside diameter of piston pin bore (13). Refer to table 4-6, Air Compressor Wear Limits. for measurements. If damaged or worn, replace piston (11).

Table 4-6. Air Compressor Wear Limits.

ITEM NO.	ITEM/POINT OF MEASUREMENT	WEAR LIMITS/TOLERANCES	
		INCHES	MILLIMETERS
4	Support bore/inside diameter	1.877	47.68
5	Crankshaft/outside diameter A	1.877	47.68
	outside diameter B	1.933	49.10
	outside diameter C	1.871	47.52
7	Piston pin bore/inside diameter	0.689	17.50
8	Crankshaft journal bore/inside diameter	1.935	49.15
9	Piston pin/outside diameter	0.687	17.45
10	Thrust washer thickness	0.240	6.10
12	Piston skirt/outside diameter	3.617	91.87
13	Piston pin bore/inside diameter	0.689	17.50

4-52. AIR COMPRESSOR MAINTENANCE(Contd)

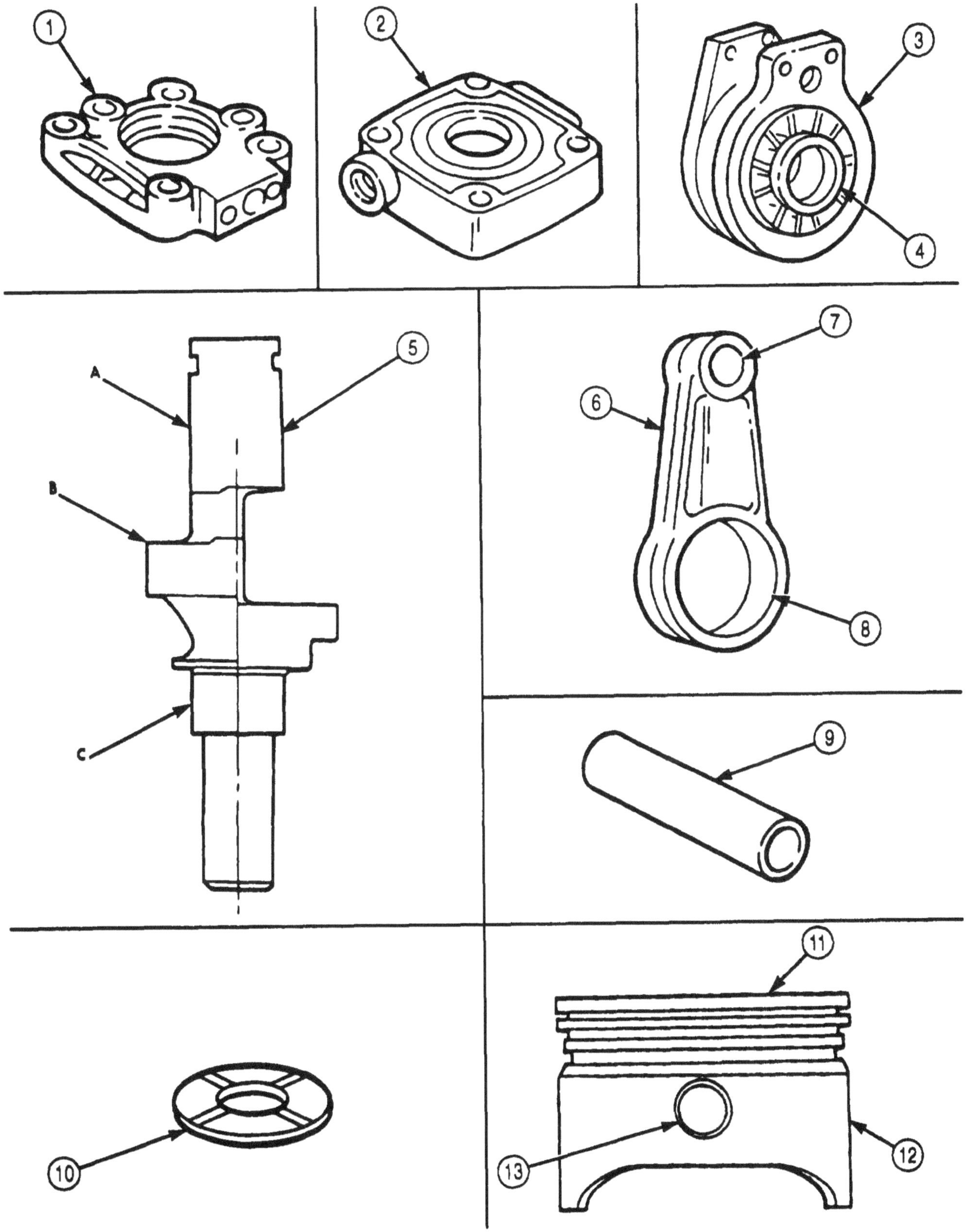

1
2
3
4
A
5
B
C
7
6
8
9
11
10
13
12

4-52. AIR COMPRESSOR MAINTENANCE (Contd)

13. Place new intermediate compression ring (3) in groove of piston (1). Using feeler gauge, measure distance between groove wall and intermediate compression ring (3). If distance exceeds 0.0045 in. (0.114 mm), replace piston (1).
14. Place new top compression ring (2) in groove of piston (1). Using feeler gauge, measure distance between groove wall and top compression ring (2). If distance exceeds 0.0045 in. (0.114 mm), replace piston (1).
15. Inspect compressor crankcase (4) for scoring, breaks, and wear. If scored, broken, or worn, replace compressor crankcase (4).
16. Measure inside diameter of piston bore (5) at two places, 1 in. (25.4 mm) below compressor head surface (6) of compressor crankcase (4). If inside diameter exceeds 3.629 in. (92.18 mm), replace compressor crankcase (4).
17. Subtract two bore diameter measurements. If difference of diameters exceed 0.002 in. (0.05 mm), replace compressor crankcase (4).
18. Inspect compressor crankcase bushing (8) for nicks, scoring, and wear. Measure inside diameter of bushing (8). If nicked, scored or inside diameter exceeds 1.878 in. (47.70 mm), replace compressor crankcase bushing (8).

NOTE

Perform steps 19 through 22 bushing requires replacement.

19. Support compressor crankcase (4) on arbor press with bushing bore (7) facing up.
20. Press bushing (8) from bushing bore (7). Discard bushing (8).
21. Lubricate new bushing (8) with clean engine oil.
22. Using arbor press, install new bushing (8) in bushing bore (7) until flush with bore surface.

4-52. AIR COMPRESSOR MAINTENANCE (Contd)

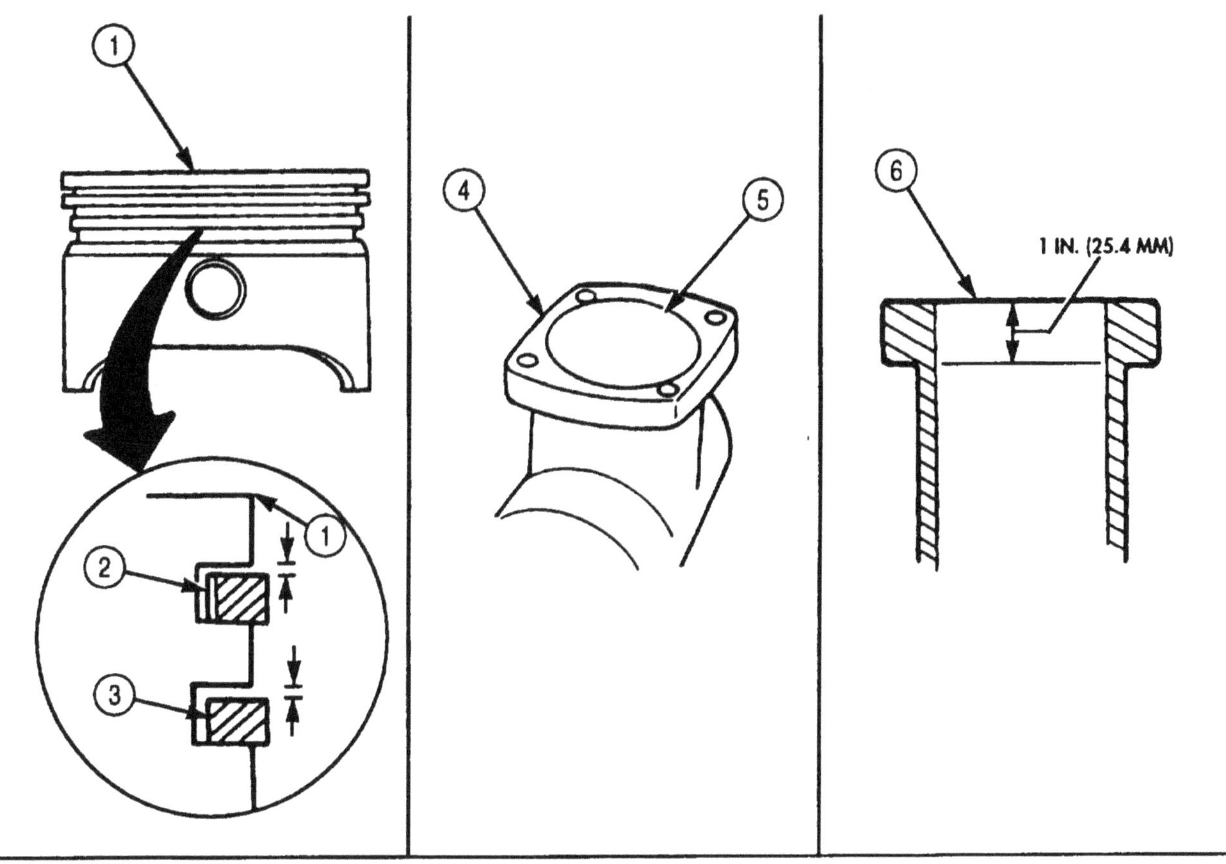

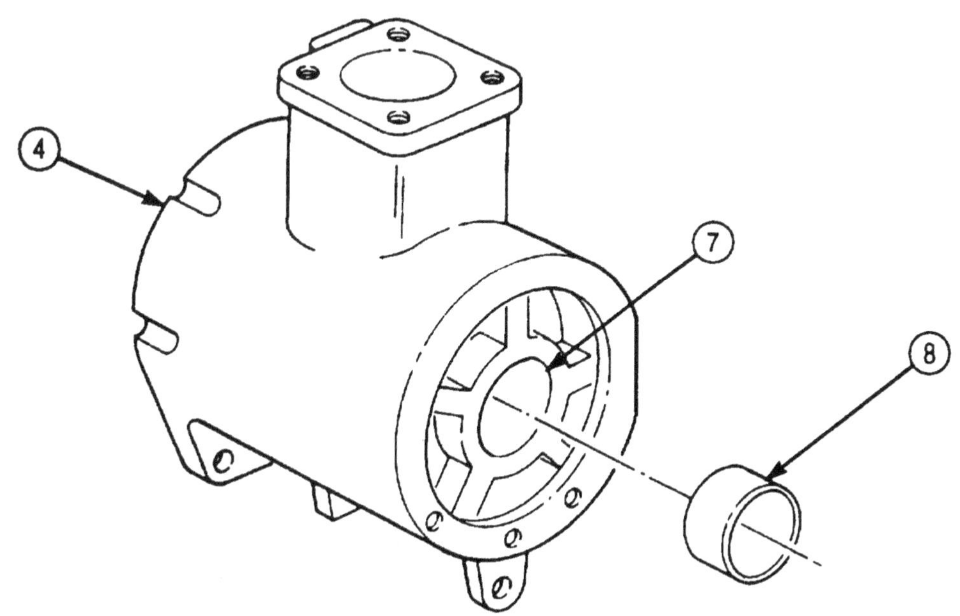

4-52. AIR COMPRESSOR MAINTENANCE (Contd)

c. Assembly

1. Install retaining ring (17) in groove of piston pin bore (19).

CAUTION

Do not drive piston pin into bore. Damage to piston may result.

NOTE

It may be necessary to heat piston in hot water to install piston pin.

2. Lubricate piston pin bore (19) with clean engine oil.
3. Install connecting rod (2) and piston pin (18) on piston (1) with retaining ring (17).
4. Install new oil ring expander (16) and two new oil rings (15) on bottom groove of piston (1). Ensure gaps of oil ring expander (16) and oil rings (15) are offset 180° from each other and 80° from piston pin bore (19).
5. Install new intermediate (14) and new top (13) compression rings on grooves of piston (1). Ensure compression rings (13) and (14) are installed with word TOP facing up, so gaps are 180° from each other and 80° from piston pin bore (19).
6. Lubricate piston (1) and piston bore (3) of compressor crankcase (4) with clean engine oil.
7. Using ring compressor, install piston (1) and connecting rod (2) in compressor crankcase (4).
8. Lubricate crankshaft (10) with clean engine oil and install in soft-jawed vise.
9. Install support (9), thrust washer (2), and spacer (22) on crankshaft (10).
10. Position drive gear (8) on crankshaft (10) and install with washer (23) and screw (20). Tighten screw (20) 95 lb-ft (129 Nžm).
11. Remove crankshaft (10) from soft-jawed vise.
12. Position new gasket (11) on support (9).

CAUTION

Use care when installing crankshaft into connecting rod to avoid damage.

13. Lubricate crankcase bushing (24) with clean engine oil.
14. Position piston (1) at 80° before or after top dead center.
15. Insert crankshaft (10) and support (9) in compressor crankcase (4) and connecting rod (2).
16. Turn drive gear (8) to rotate crankshaft (10) until seated in connecting rod journal bore (12) and crankcase bushing (20).
17. Install new gasket (11) and support (9) on compressor crankcase (4) with four washers (6), screws (7), and two new screw-assembled lockwashers (5). Tighten screws (7) and screw-assembled lockwashers (5) 30-35 lb-ft (41-47 N•m).

4-52. AIR COMPRESSOR MAINTENANCE (Contd)

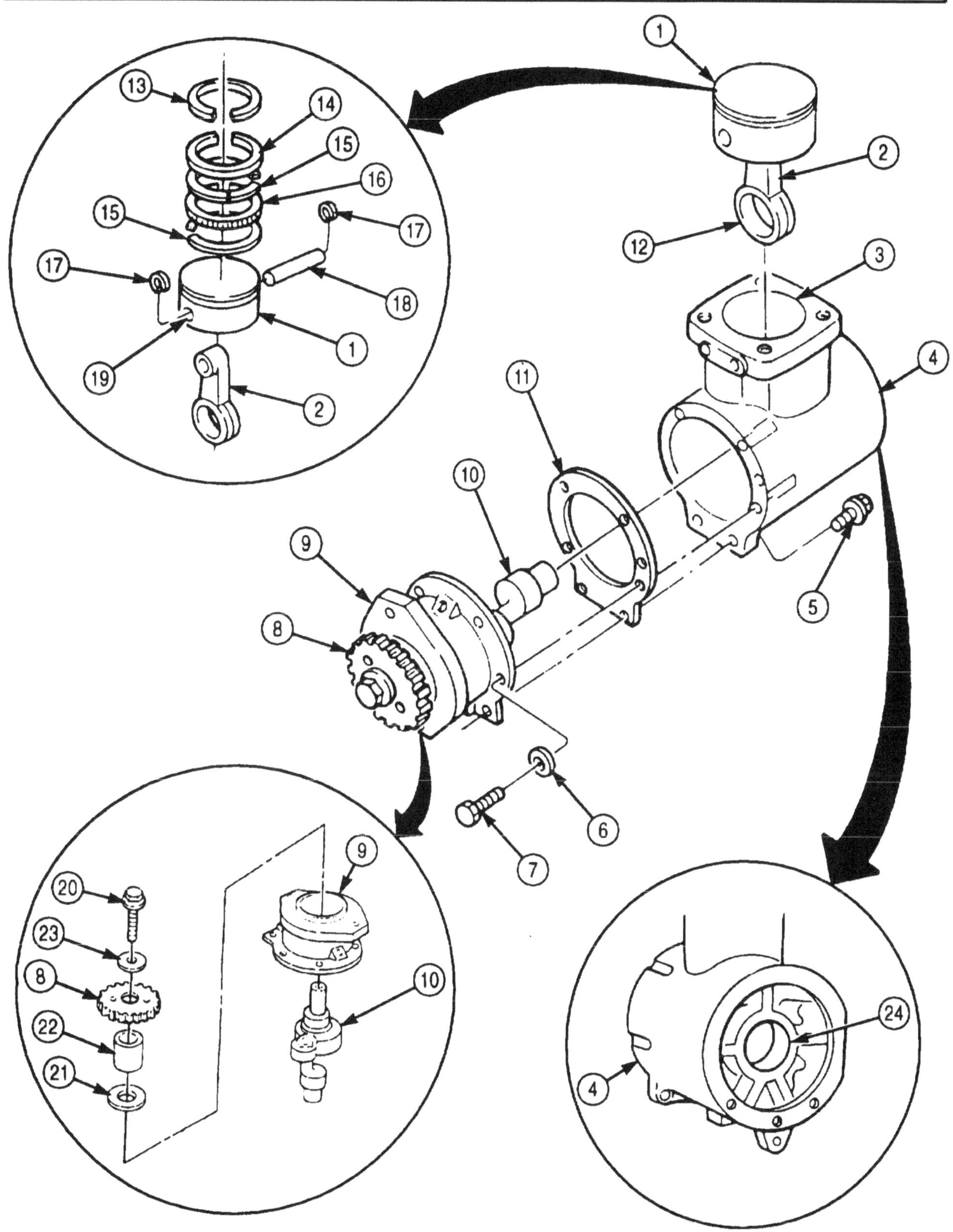

4-52. AIR COMPRESSOR MAINTENANCE(Contd)

18. Install new wear plate (14), new exhaust valve spring (13), new exhaust valve (12), new seal (11), new exhaust valve seat (10), and new O-ring (9) in compressor head (6).
19. Press exhaust valve seat (10) in hole in bottom side of cylinder head (6).
20. Coat new gaskets (5) and (7) with clean engine oil.
21. Install new gasket (7), compressor head (6), new gasket (5), and compressor head cover (4) on compressor crankcase (8) with four washers (3), new lockwashers (2), and screws (1).
22. Install new intake valve (18) on valve seat (19).
23. Install new O-ring (33), new packing seal (34), and new unloader valve cap (35) in new unloader valve body (15).
24. Install new intake valve spring (20), intake valve (18), valve seat (19), new unloader valve spring (17), and new unloader valve body (15) on compressor head cover (4) with two washers (31) and new screw-assembled washers (32).
25. Install new gasket (29) and air inlet block (28) on compressor head cover (4) with two screws (27).
26. Install new gasket (26) and adapter (25) on compressor crankcase (8) with four washers (22), new lockwashers (23), and screws (24).

NOTE

Wrap male threads with antiseize tape prior to installation.

27. Install two adapters (21) and elbow (30) on compressor head cover (4).
28. Install elbow (16) on unloader valve body (15).

4-52. AIR COMPRESSOR MAINTENANCE (Contd)

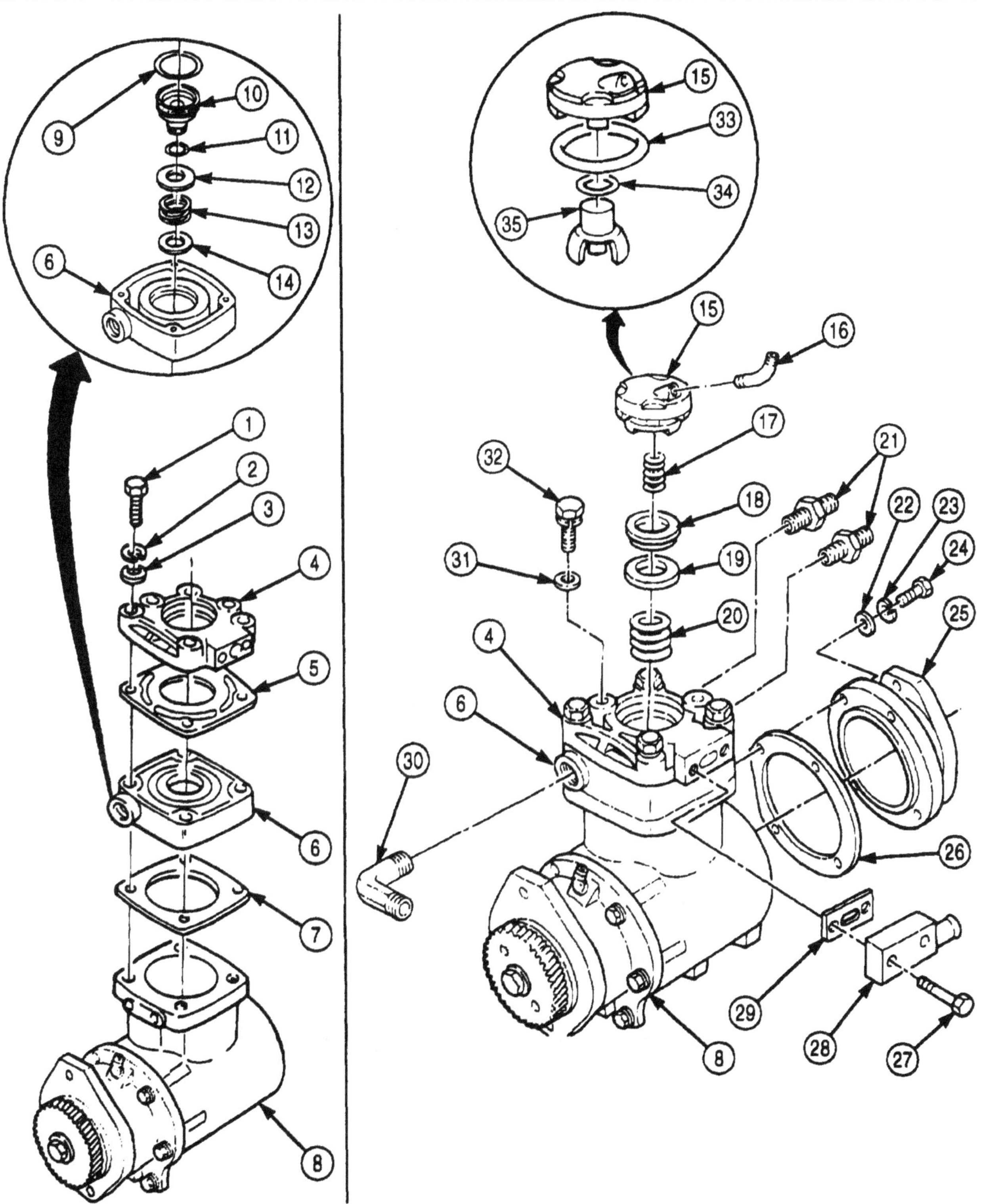

FOLLOW-ON TASK: Install air compressor (para. 3-206).

4-53. VALVE ADJUSTMENT

THIS TASK COVERS:
Adjustment

INITIAL SETUP:

APPLICABLE MODELS
M939A2

SPECIAL TOOLS
Engine barring tool (Appendix E, Item 43)

TOOLS
General mechanic's tool kit (Appendix E, Item 1)

MATERIALS/PARTS
O-ring (Appendix D, Item 444)
Chalk (Appendix C, Item 17)

PERSONNEL REQUIRED
TWO

REFERENCES (TM)
TM 9-2320-272-24P

EQUIPMENT CONDITION
Valve cover removed (para. 3-6).

NOTE

Assistant will help with this procedure.

Adjustment

1. Remove plug (8) and O-ring (7) from flywheel housing (6). Discard O-ring (7).

CAUTION

Ensure timing pin is pulled out after top dead center is reached. Failure to do so may damage timing pin.

2. Using engine barring tool, rotate crankshaft (9) clockwise and engage timing pin (10).
3. Make aligning chalk marks (12) on vibration damper (13) and front gearcase cover (11).
4. Loosen twelve locknuts (1), and back off adjusting screws (2) counterclockwise from push rods (5).

NOTE

Step 5 through 8 are performed for each cylinder.

5. Place feeler gauge, between rocker lever head (3) and valve stem (4).
6. Adjust valve clearance to proper limits. Refer to table 4-7, Valve Clearance Adjustment Limits, step A
7. Tighten adjusting screw (2) until a slight drag is felt on feeler gauge.
8. Tighten locknut (1) 18 lb-ft (24 N-m).
9. Rotate crankshaft (9) clockwise one full turn and realign chalk marks (12).
10. Adjust valve clearance to proper limits using table 4-7
11. Install new O-ring (7) and plug (8) in flywheel housing (6).

Table 4-7. Valve Clearance Adjustment Limits.

CYLINDER	STEP A	STEP B
1	I and E	
2	I	E
3	E	I
4	I	E
5	E	I
6		I and E

Valve clearance setting:
I = Intake valve gap 0.012 in. (0.30 mm)
E = Exhaust valve gap 0.024 in. (0.61 mm)

4-53. VALVE ADJUSTMENT (Contd)

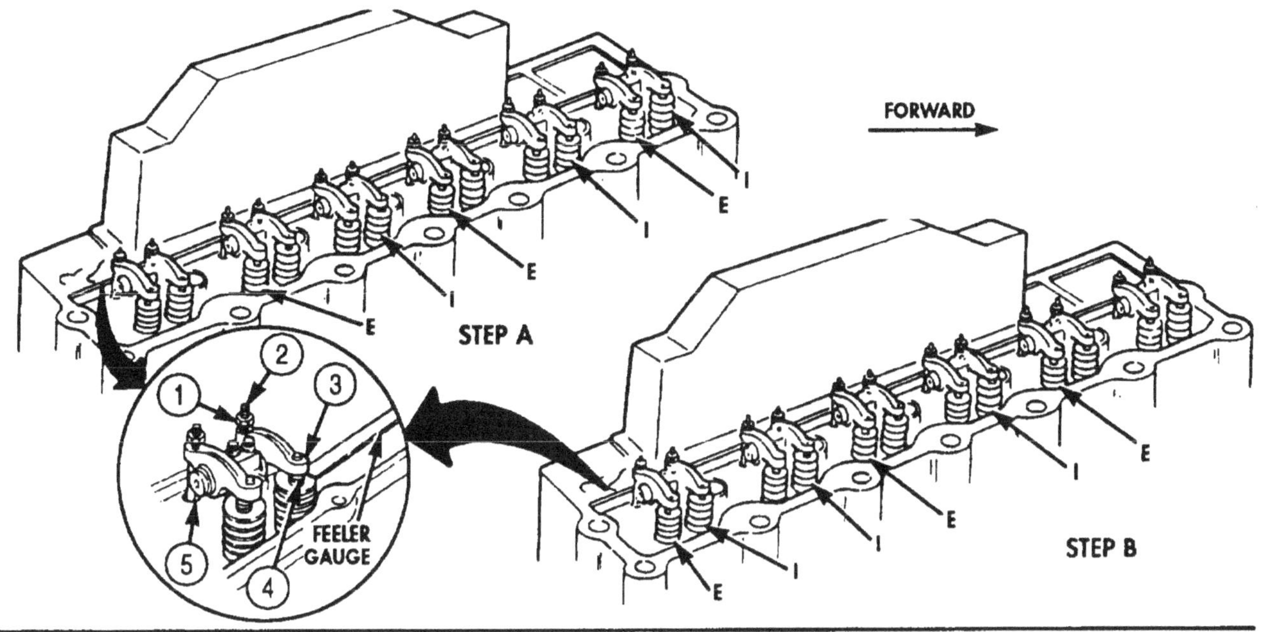

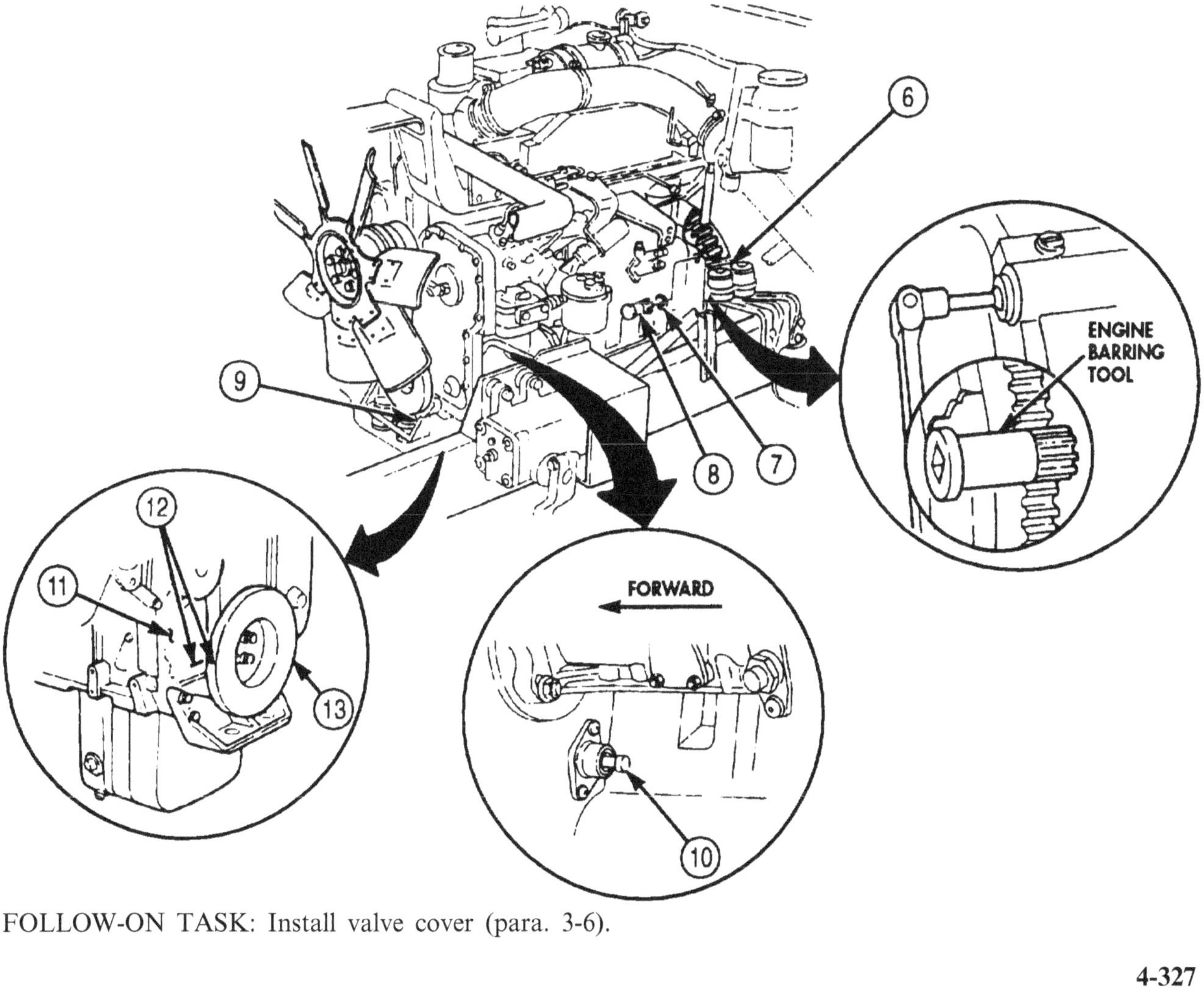

FOLLOW-ON TASK: Install valve cover (para. 3-6).

4-54. ENGINE RUN-IN TEST AND IDLE SPEED ADJUSTMENT

THIS TASK COVERS:

a. Engine Run-in Test **b. Idle Speed Adjustment**

INITIAL SETUP:

APPLICABLE MODELS
M939A2

TOOLS
General mechanic's tool kit (Appendix E, Item 1)

PERSONNEL REQUIRED
TWO

REFERENCES (TM)
LO 9-2320-272-12
TM 9-2320-272-10
TM 9-2320-272-24P

EQUIPMENT CONDITION
- Parking brake set (TM 9-2320-272-10).
- Hood raised and secured (TM 9-2320-272-10)

a. Engine Run-in Test

1. Pull emergency engine stop control (4) out all the way.

CAUTION

Do not operate starter continuously for more than 30 seconds at a time. Wait two minutes between periods of starter operation. Failure to do so may result in damage to equipment.

2. Crank engine until oil pressure registers on oil pressure gauge (3).

NOTE

Prime fuel system if engine fails to start (para. 3-22).

3. Push emergency engine stop control (4) all the way in.
4. Start engine (TM 9-2320-272-10) and run at 1,000-2,000 rpm for 30 minutes.
5. Check oil pressure gauge (3) and water temperature gauge (2).
   a. If oil pressure gauge (3) does not register or suddenly drops to less than 10 psi (69 kPa), stop engine and troubleshoot lubrication system (para. 4-2).
   b. If water temperature gauge (2) increases above 205°F (96°C), stop engine and allow to cool. If engine overheats again, troubleshoot cooling system (para. 2-21).
6. Stop engine and inspect for leaks.

b. Idle Speed Adjustment

NOTE

Assistant will help with this procedure.

Start engine (TM 9-2320-272-10) and check tachometer (1) for proper idle speed, 550-650 rpm. If idle speed is not within limits, adjust accelerator linkage (5).

a. Loosen locknut (7).

b. Turn throttle rod (6) clockwise to increase idle speed; counterclockwise to decrease idle speed.

c. Tighten locknut (7).

4-54. ENGINE RUN-IN TEST AND IDLE SPEED ADJUSTMENT (Contd)

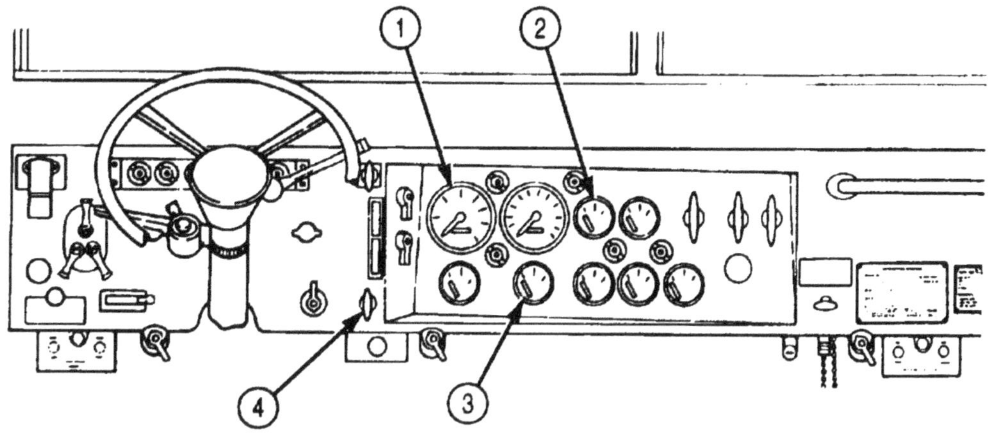

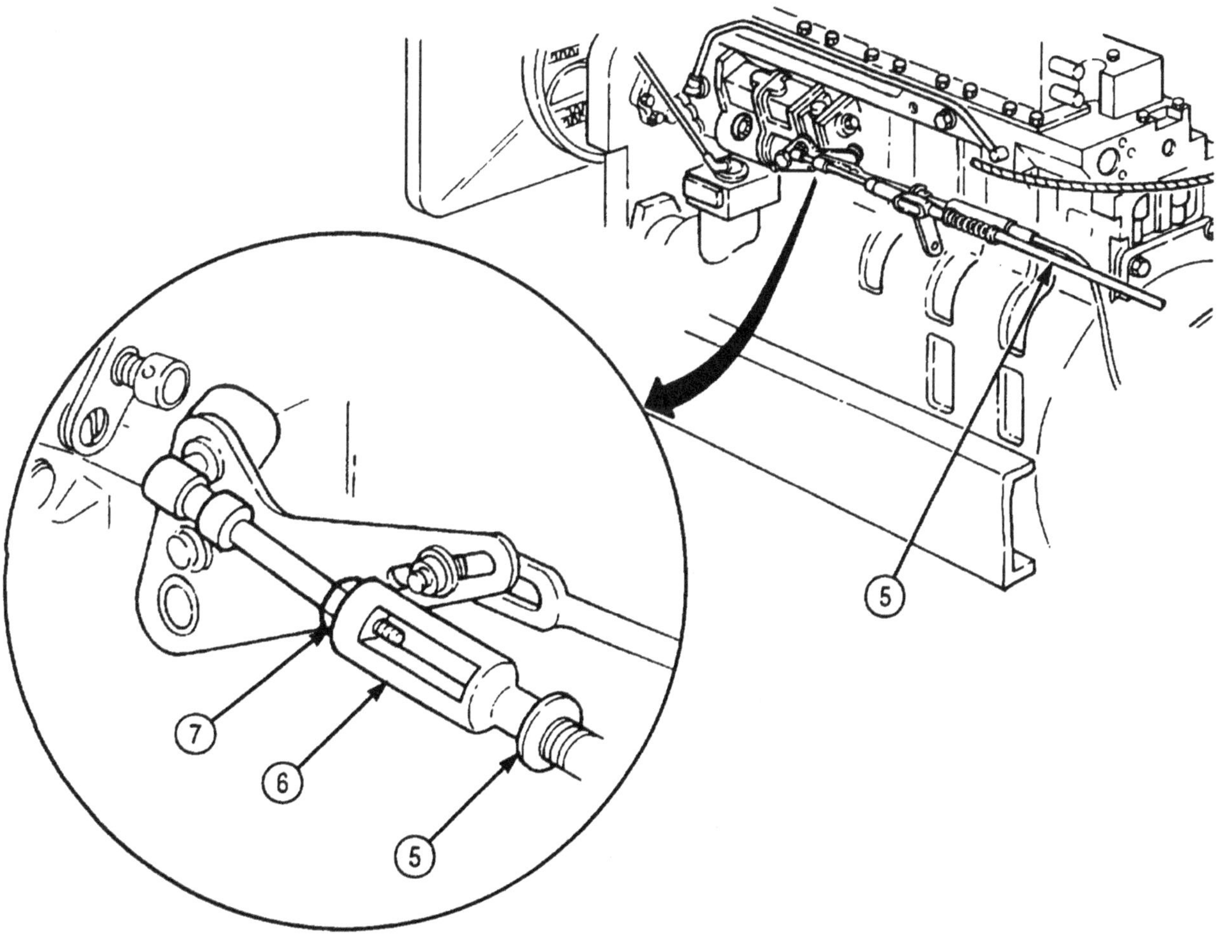

FOLLOW-ON TASK: Change engine oil after first 500 mi. (805 km) (LO 9-2320-272-12).

4-55. FUEL INJECTOR MAINTENANCE

THIS TASK COVERS:

a. Removal
b. Disassembly
c. Cleaning and Inspection
d. Assembly
e. Installation

INITIAL SETUP:

APPLICABLE MODELS
M939A2

SPECIAL TOOLS
Nozzle cleaningkit (Appendix E, Item 87)

TOOLS
General mechanic's tool kit (Appendix E, Item 1)
Torque wrench (Appendix E, Item 146)

MATERIALS/PARTS
Six washers (Appendix D, Item 712)
Six injector sleeves (Appendix D, Item 256)
Cleaning cloth (Appendix C, Item 21)
Diesel fuel (Appendix C, Item 42)
Drycleaning solvent (Appendix C, Item 71)

REFERENCES (TM)
TM 9-2320-272-10
TM 9-2320-272-24P

EQUIPMENT CONDITION
Fuel injector tubes removed (para. 3-19).

SPECIAL ENVIRONMENTAL CONDITIONS
Perform procedure in dust-free area.

GENERAL SAFETY INSTRUCTIONS
- Diesel fuel is flammable. Do not perform this task near open flame.
- Keep fire extinguisher nearby when using drycleaning solvent.
- Drycleaning solvent is flammable and toxic. Do not use near an open flame.

WARNING

Diesel fuel is flammable. Do not perform fuel system procedures near open flame. Injury to personnel or damage to equipment may result.

a. Removal

1. Remove screw (1), injector clamp (2), and injector (3) from cylinder head (5).
2. Cover injector bore (4) with clean cloth.

b. Disassembly

WARNING

Drycleaning solvent is flammable and toxic. Do nuse near open flame and always have a fire extinguisher nearby when solvents are used. Use only in well-ventilated places, wear protective clothing, and dispose of cleaning rags in approved container. Failure to do this may result in injury or death to personnel and/or damage to equipment.

CAUTION

Improper cleaning methods and use of unauthorized cleaning solvents can damage fuel injector.

1. Clean exterior of injector (3) with drycleaning solvent.

NOTE

Use injector clamp to support injector for disassembly.

2. Secure injector clamp (2) in soft-jawed vise.
3. Install injector (3) into injector clamp (2), with injector nozzle (7) facing upward.

4-55. FUEL INJECTOR MAINTENANCE(Contd)

4. Remove washer (6) from nozzle nut (9). Discard washer (6).
5. Remove nozzle nut (9) from injector sleeve (8).

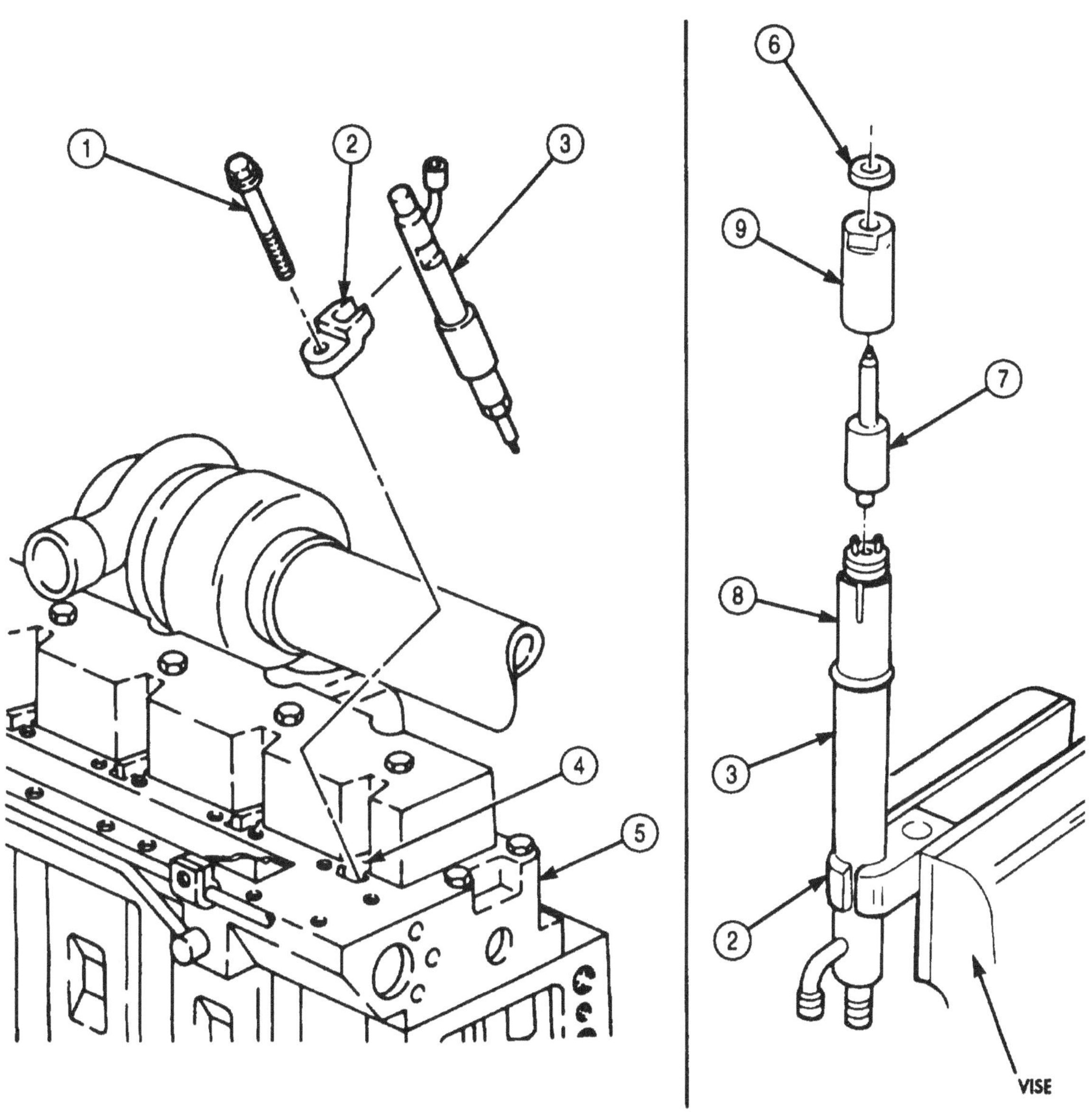

4-55. FUEL INJECTOR MAINTENANCE (Contd)

CAUTION

Injector needle valve and nozzle are a matched pair. Do not interchange. Do not touch internal parts unless hands are moistened with diesel fuel or damage to equipment may result.

6. Remove injector nozzle (2) and needle valve (7) from intermediate plate (6).
7. Place injector nozzle (2) and needle valve (7) in clean diesel fuel.
8. Remove intermediate plate (6) from injector sleeve (3).
9. Place intermediate plate (6) in clean diesel fuel.

NOTE

When removing injector, keep injector in upward position. This will keep pressure spindle, pressure spring, and shim from falling out of injector.

10. Remove injector (4) from injector clamp (5).
11. Tip injector (4) into hand and remove pressure spindle (8), pressure spring (10) and shim (9) from injector (4).
12. Remove injector sleeve (3) from injector (4). Discard injector sleeve (3).

c. Cleaning and Inspection

1. Inspect nozzle nut (1) for damaged threads. Replace nozzle nut (1) if damaged.
2. Using 0.0001.in. (0.0025-mm) cleaning needle from cleaning kit, clean injector nozzle spray holes (11) on injector nozzle (2).

WARNING

Drycleaning solvent is flammable and toxic. Do not use near open flame and always have a fire extinguisher nearby when solvents are used. Use only in well-ventilated places, wear protective clothing, and dispose of cleaning rags in approved container. Failure to do this may result in injury or death to personnel and/or damage to equipment.

3. Clean residue from injector nozzle (2) with drycleaning solvent and dip in clean diesel fuel.

NOTE

For cleaning purposes, dip scraper and hardwood in clean diesel fuel.

4. Using scraper from cleaning kit, clean nozzle seat (13) on injector nozzle (2).
5. Using hardwood from cleaning kit, polish exterior of nozzle seat (13) on injector nozzle (2).
6. Using scraper from cleaning kit, clean interior ring groove (12) on injector nozzle (2).
7. Using brass brush from cleaning kit, clean needle valve (7).

NOTE

Pressure shoulder of needle valve will normally have a machined appearance.

8. Inspect tip of needle valve (7) for excessive wear and erosion. If worn or eroded, replace injector (4).

NOTE

Perform steps 8 through 10 to check needle valve and injector nozzle alignment,

9. Dip needle valve (7) in clean diesel fuel and insert into injector nozzle (2) until seated.
10. Pull needle valve (7) one third of the way out of injector nozzle (2).
11. Needle valve (7) must slide back into injector nozzle (2) under its own weight. If needle valve (7) does not slide back into injector nozzle (2), replace injector (4).

4-55. FUEL INJECTOR MAINTENANCE (Contd)

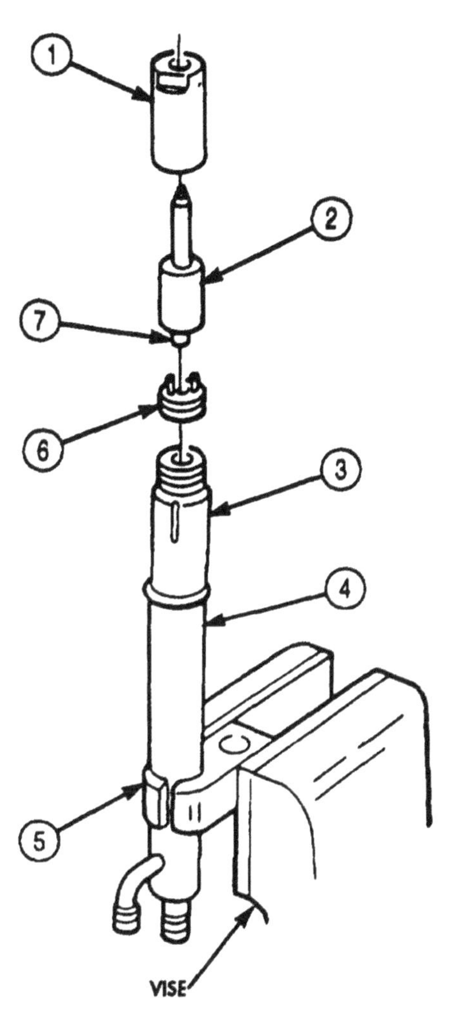

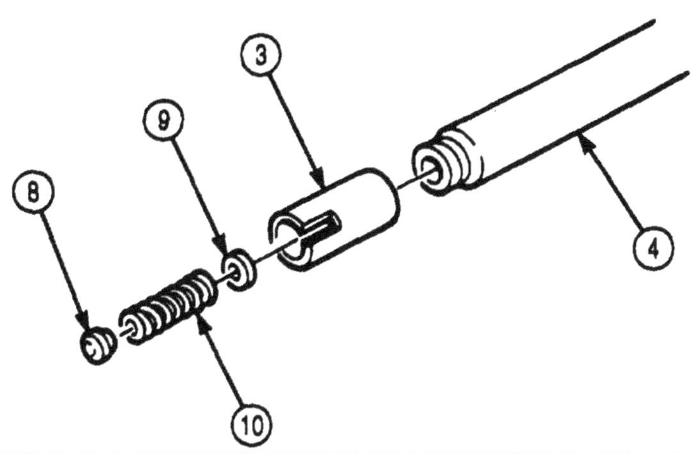

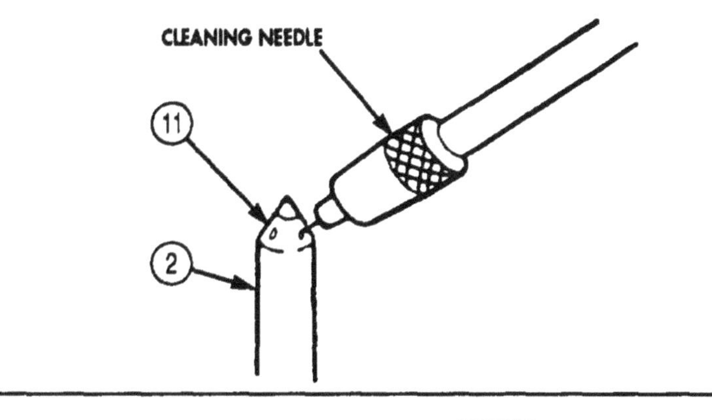

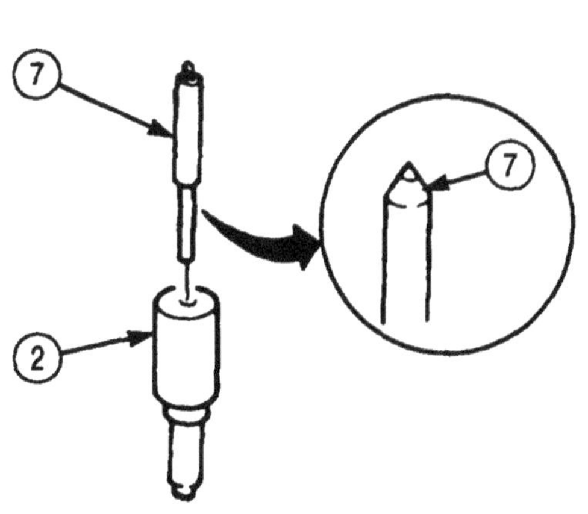

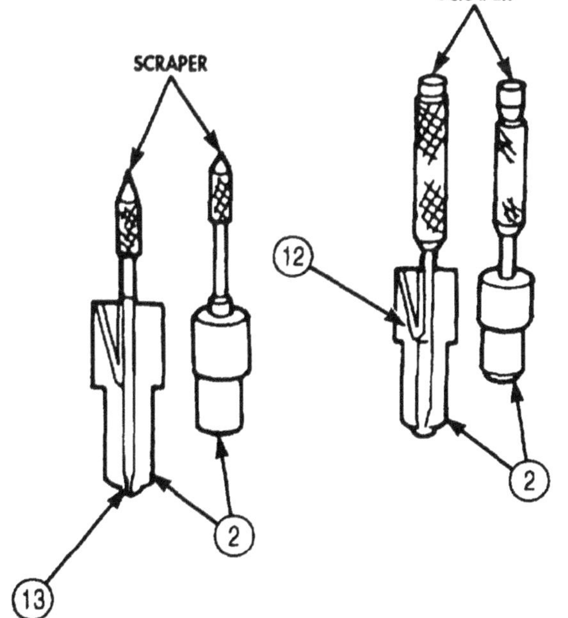

4-55. FUEL INJECTOR MAINTENANCE (Contd)

d. Assembly

NOTE

Fuel injector must be calibrated before installation (para. 4-56).

1. Secure injector clamp (5) in soft-jawed vise.

CAUTION

Lubricate parts with clean diesel fuel before assembly. Do not use lubricating oil. Oil can crystallize under excessive heat causing damage to injector.

2. Install injector (41 in injector clamp (5) with injector (4) facing upward.
3. Install new injector sleeve (3) on injector (4).
4. Install shim (9), pressure spring (l0), and pressure spindle (8) in injector (4).
5. Install intermediate plate (6), needle valve (7), and injector nozzle (2) on injector (4) with nozzle nut (1). Tighten nozzle nut (1) 20 lb-ft (27 N•m).
6. Remove injector (4) from injector clamp (5).
7. Remove injector clamp (5) from soft-jawed vice.

e. Installation

1. Install new washer (13) on assembled injector (12).
2. Remove cleaning cloth from injector bore (14).

NOTE

Coat injector and injector bore with clean diesel fuel.

3. Position injector clamp (16) into injector bore (14) of cylinder head (15).
4. Install assembled injector (12) into injector bore (14) of cylinder head (15).
5. Install injector clamp (16) and assembled injector (12) on cylinder head (15) with screw (11). Tighten screw (11) 20 lb-ft (27 N•m).

4-55. FUEL INJECTOR MAINTENANCE (Contd)

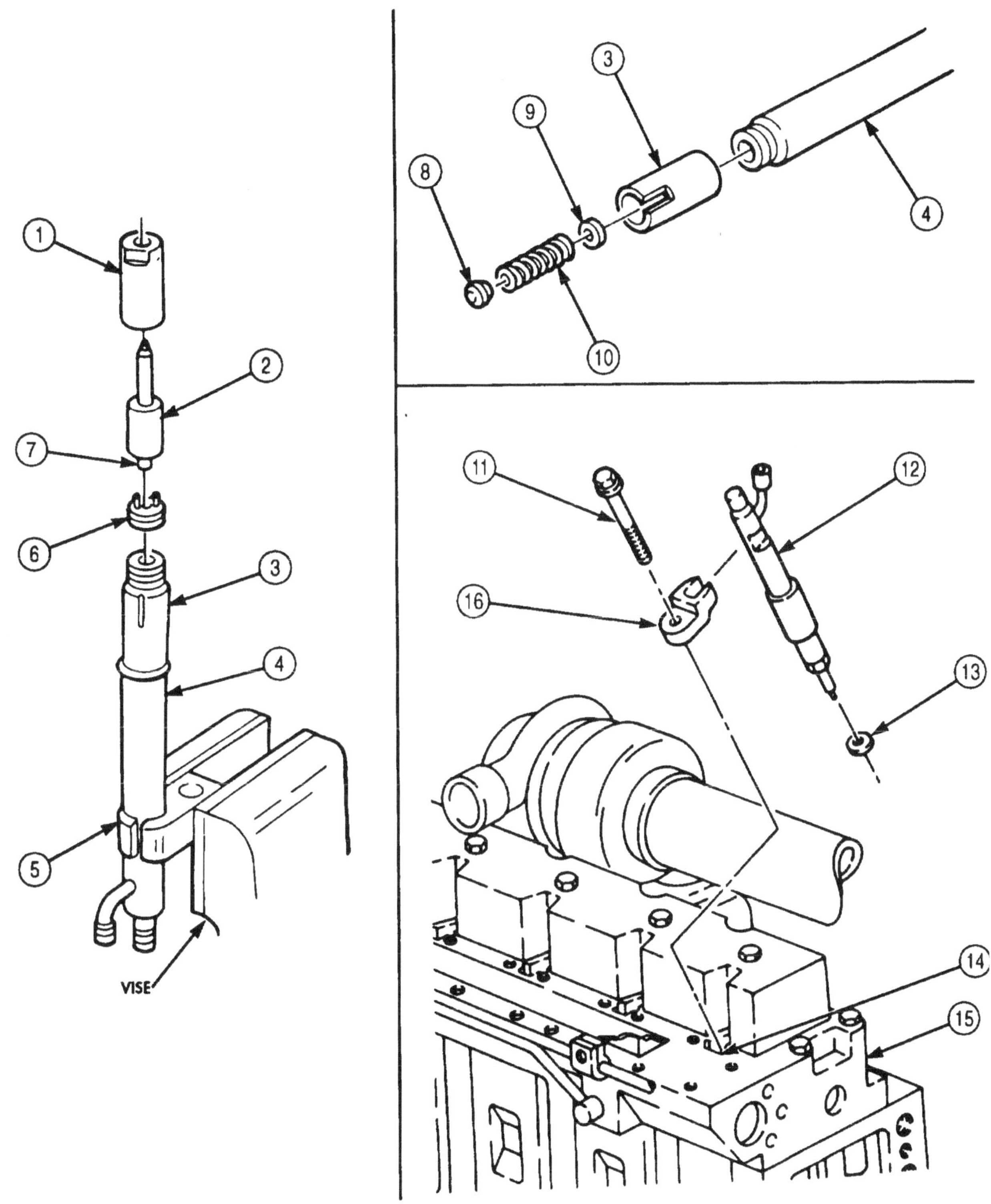

FOLLOW-ON TASK: Install fuel injector tubes (para. 3-19).

4-56. FUEL INJECTOR CALIBRATION

THIS TASK COVERS:

a. Opening Pressure Test
b. Leakage Test
c. Spray Pattern Test

INITIAL SETUP:

APPLICABLE MODELS
M939A2

SPECIAL TOOLS
Tube reducer (Appendix E, Item 151)
Injector nozzle tester (Appendix E, Item 71)

TOOLS
General mechanic's tool kit (Appendix E, Item 1)

MATERIALS/PARTS
Diesel fuel (Appendix C, Item 42)

REFERENCES (TM)
TM 9-2320-272-24P

EQUIPMENT CONDITION
Fuel injector removed (para. 4-55).

GENERAL SAFETY INSTRUCTIONS
- All personnel must keep hands clear of test spray.
- Diesel fuel is flammable. Do not perform this task near open flame.

a. Opening Pressure Test

WARNING

- Keep hands and body clear of test spray. Pressure of diesel fuel is sufficient enough to penetrate skin, causing injury to personnel.
- Diesel fuel is flammable. Do not perform fuel system procedures near open flame. Injury to personnel or damage to equipment may result.

NOTE

- Use clean diesel fuel when calibrating injector.
- Fuel injectors are calibrated the same. This procedure covers calibration of one injector.

1. Install tube reducer (3) on injector (2).
2. Install test line (11) to adapter (3).
3. Open valve on injector nozzle tester.

NOTE

Operate lever on injector nozzle tester one stroke per second.

4. Operate lever on injector nozzle tester until injector (2) ejects test spray.
5. Observe pressure indicator on pressure gauge when spray begins.
6. Correct pressure reading should be 2,973 psi (20,499 kpa).

NOTE

- Shim is located below pressure spring.
- Increasing shim thickness will increase opening pressure. Decreasing shim thickness will decrease opening pressure.

7. If opening pressure deviates from limits, change shim inside of injector (2), as required (para. 4-55).

b. Leakage Test

1. Open valve on injector nozzle tester.

4-56. FUEL INJECTOR CALIBRATION (Contd)

2. Operate lever on injector nozzle tester to obtain a constant pressure of 290 psi (1,999 kpa).
3. Check fuel injector (2) for diesel fuel leakage for approximately 10 seconds.
4. If diesel fuel leakage occurs, replace fuel injector (2).

c. Spray Pattern Test

NOTE

- Chatter/spray pattern test indicates ability of needle valve to move freely and to correctly atomize the fuel. Chatter is caused by the needle valve opening and closing under specific circumstances.
- Used injectors should not be evaluated for chatter at lower speeds. A used injector can be used if it passes the leak test, opening pressure test, and has a uniform spray pattern.

Operate lever at approximately three strokes per second. Fuel injector (2) must eject a uniform and well-atomized spray from four nozzle spray holes (4). If proper spray pattern is not obtained, replace fuel injector (2).

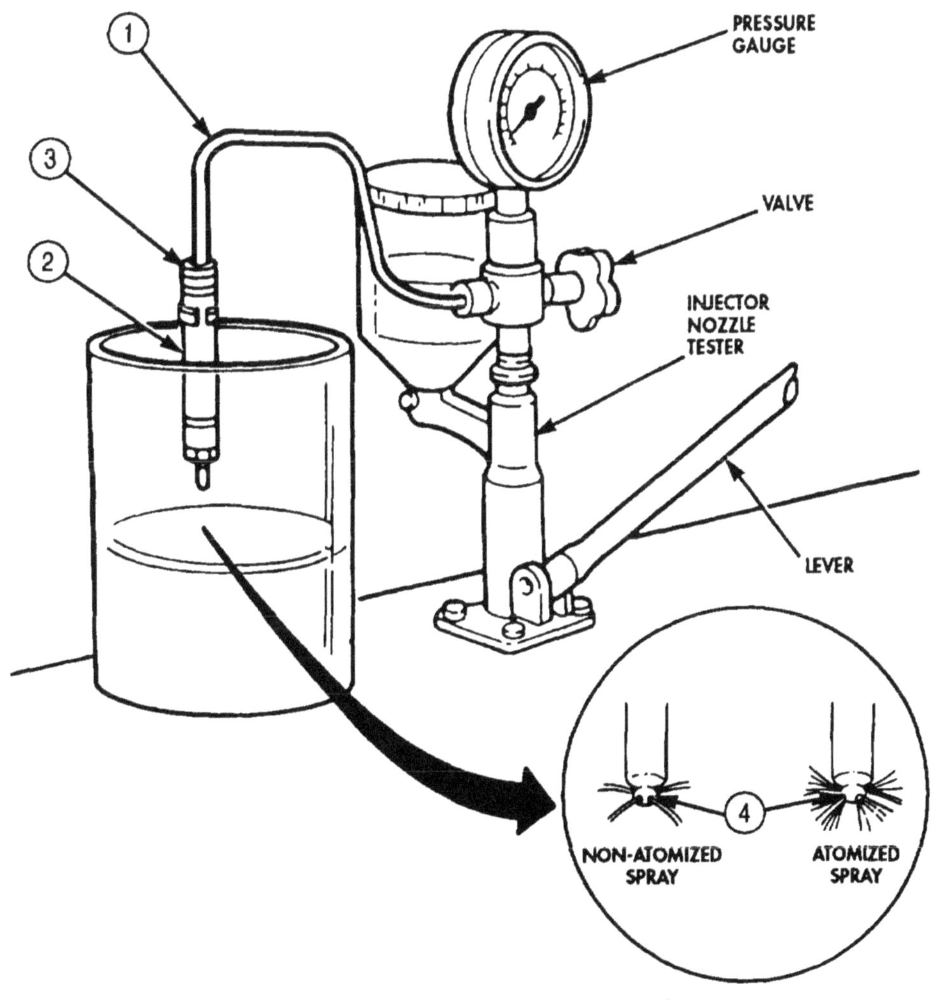

FOLLOW-ON TASK: Install fuel injector (para. 4-55)

4-57. FUEL INJECTION PUMP MAINTENANCE

THIS TASK COVERS:

a. Removal
b. Installation
c. Adjustment

INITIAL SETUP:

APPLICABLE MODELS
M939A2

SPECIAL TOOLS
Engine barring tool (Appendix E, Item 43)
Puller kit (Appendix E, Item 103)
Box wrench (Appendix E, Item 161)

TOOLS
General mechanic's tool kit (Appendix E, Item 1)
Torque wrench (Appendix E, Item 144)

MATERIALS/PARTS
Two cotter pins (Appendix D, Item 46)
Two copper washers (Appendix D, Item 42)
Copper washer (Appendix D,Item 43)
Lockwasher (Appendix D,Item 383)
Lockwasher (Appendix D, Item 345)
O-ring (Appendix D, Item 447)
O-ring (Appendix D, Item 444)
O-ring (Appendix D, Item 446)
Two lockwashers (Appendix D, Item 398)
Break-off screw (Appendix D, Item 18)
Cap and plug set (Appendix C, Item 14)
Lubrication oil (Appendix C, Item 49)
Safety wire (Appendix C, Item 79)

REFERENCES (TM)
TM 9-2320-272-24P

EQUIPMENT CONDITION
- Throttle control solenoid removed (para. 3-46).
- Air fuel control (AFC) tube removed (para. 3-18).
- Tachometer drive removed (para. 4-51).

GENERAL SAFETY INSTRUCTIONS
Diesel fuel is flammable. Do not perform this task near open flames.

a. Removal

WARNING

Diesel fuel is flammable. Do not perform fuel system procedures near open flame. Injury to personnel may result.

1. Remove cotter pin (18), washer (19) and cable pivot (15) from shutoff valve lever (12). Discard cotter pin (18).
2. Remove screw (17) clamp (1) and control cable (2) from fuel pump bracket (16).
3. Remove screw (13) throttle connector (14) and control cable (2) from cable pivot (15).
4. Remove cotter pin (3) and washer (4) from link pin (6). Discard cotter pin (3).
5. Remove modulator control cable (5) from throttle lever (7).
6. Compress socket (10) and remove accelerator linkage (9) from ball joint (8).
7. Remove ball joint (8) and lockwasher (11) from throttle lever (7). Discard lockwasher (11).

4-57. FUEL INJECTION PUMP MAINTENANCE (Contd)

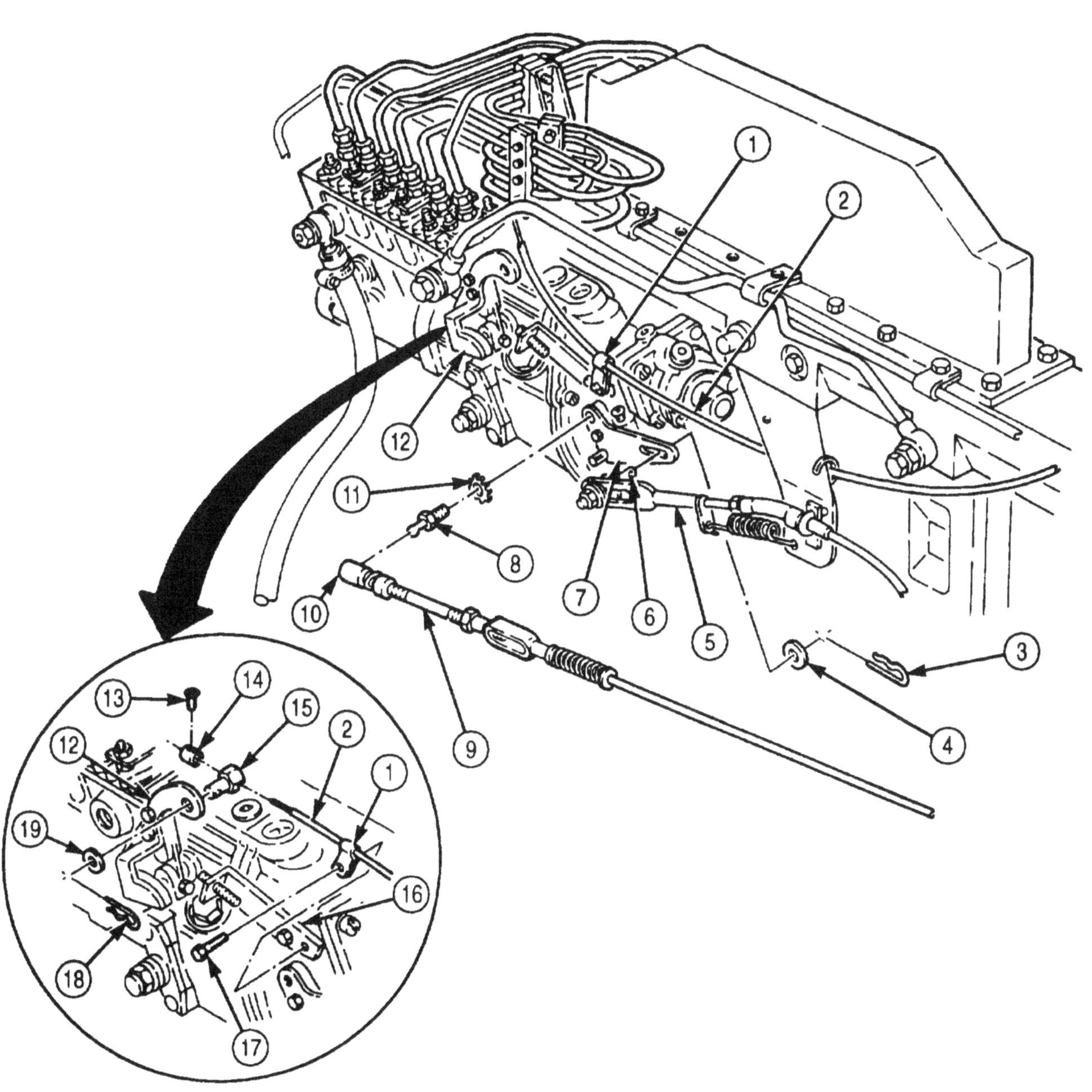

4-57. FUEL INJECTION PUMP MAINTENANCE (Contd)

CAUTION

- Clean area around connections before disconnecting lines and hoses. Failure to do so may result in damage to fuel pump.
- Cap or plug all openings immediately after disconnecting lines and hoses to prevent contamination, Failure to do so may result in damage to fuel pump.

NOTE

Have drainage container ready to catch excess fuel.

8. Loosen clamp (10) and remove fuel return hose (9) from fuel injection pump return nipple (11).
9. Disconnect six injector fuel lines (12) from fuel injection pump (8).
10. Remove screw (1), two connection washers (2), and fuel supply tube (3) from fuel injection pump (8).
11. Remove screw (6), bushing (6), two connector washers (4), and fuel supply tube (3) from cylinder head (7).
12. Remove plug (16) and O-ring (15) from flywheel housing (14). Discard O-ring (15).

NOTE

Assistant will help with step 13.

13. Using barring tool, secure engine flywheel (13) and remove nut (17) and lockwasher (24) from fuel pump shaft (20). Discard lockwasher (24).
14. Using barring tool, rotate engine until timing pin (21) enters hole in camshaft gear (22). Piston No. 1 is now at top dead center.

NOTE

Do not remove fuel injector pump gear from gear housing.

15. Using puller, remove fuel pump gear (23) from fuel pump shaft (20).
16. Remove four nuts (19) and fuel injection pump (8) from gear housing (18).
17. Remove O-rings (25) and (34) from fuel injection pump (8). Discard O-rings (25) and (34).

NOTE

Perform steps 18 through 20 if installing new fuel injection pump.

18. Remove two screws (29) and fuel shutoff lever (30) from fuel injection pump (8).
19. Remove two screws (28) and throttle lever (27) from shaft (26).
20. Remove plug (32), two copper washers (33), and fitting (31) from fuel injection pump (8). Discard copper washers (33).

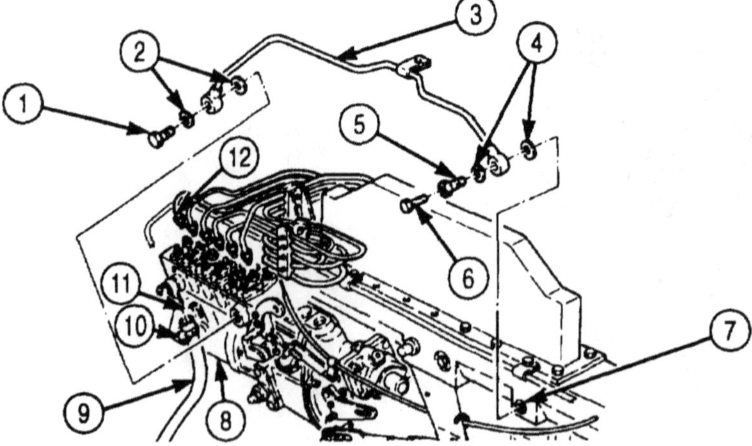

4-57. FUEL INJECTION PUMP MAINTENANCE (Contd)

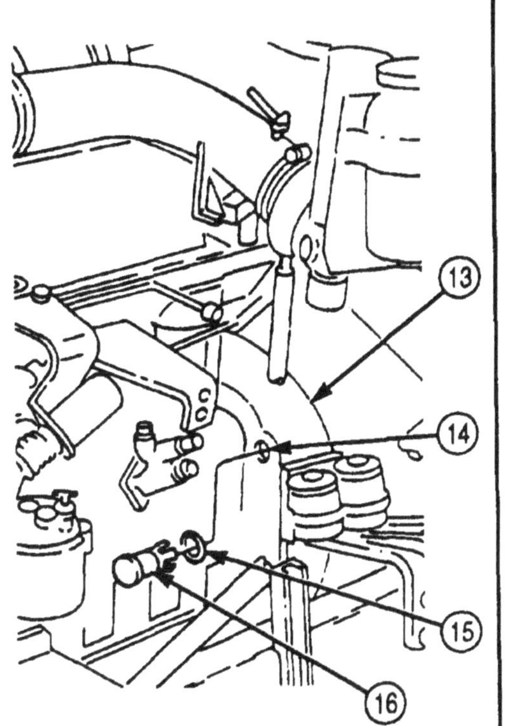

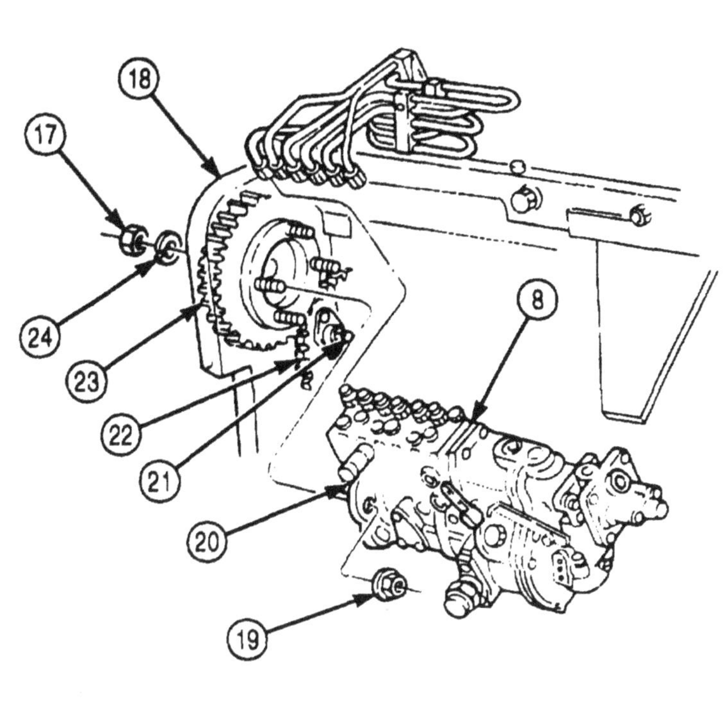

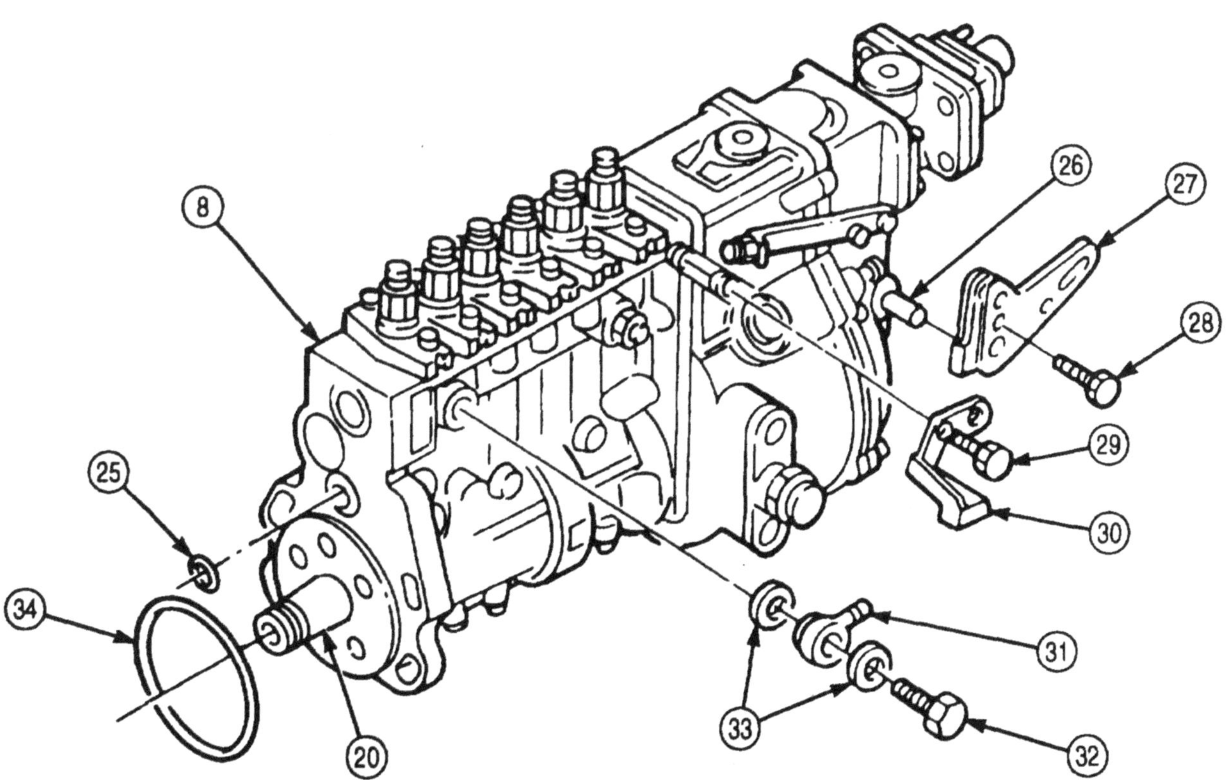

4-57. FUEL INJECTION PUMP MAINTENANCE (Contd)

b. Installation

NOTE

Perform steps 1 through 4 to ensure fuel injection pump timing is correct.

1. Remove plug (6), copper washer (5), and timing pin (4) from fuel injection pump (3).
2. Install nut (1) on shaft (8).

NOTE

Assistant will help with step 3.

3. If timing tooth (14) is not aligned with timing pin hole (13), rotate fuel pump shaft (8) until timing tooth (14) aligns with timing pin hole (13).
4. Remove nut (1) from shaft (8).
5. Reverse timing pin (4) and install slot of timing pin (4) over timing tooth (14).
6. Install copper washer (5) and plug (6) on fuel pump (3).

NOTE

Perform steps 7 through 9 if new injection pump was installed.

7. Install two new copper washers (22) and fitting (20) on fuel pump (3) with screw (21).
8. Install control lever (16) on shaft (15) with two screws (17).
9. Install fuel shutoff lever (19) on fuel injection pump (3) with two screws (18).
10. Install new O-rings (24) and (23) on fuel injection pump (3).
11. Install fuel injection pump (3) on four mounting studs (2) with nuts (7).
12. Install fuel pump gear (11) on fuel injection pump (3) with new lockwasher (12) and nut (1).
13. Disengage engine timing pin (9) from camshaft gear (10).
14. Remove plug (6), copper washer (5), and timing pin (4) from fuel injection pump (3). Discard copper washer (5).
15. Reverse timing pin (4) so slot of timing pin (4) faces outward from fuel injection pump (3).
16. Install new copper washer (5) and plug (6) on fuel injection pump (3).

NOTE

Assistant will help with step 17.

17. Using engine barring tool to prevent flywheel (25) from turning, tighten nut (1) 60 lb-ft (81 N-m).
18. Remove engine barring tool and install new O-ring (27) and plug (28) in flywhcel housing (26).

4-57. FUEL INJECTION PUMP MAINTENANCE (Contd)

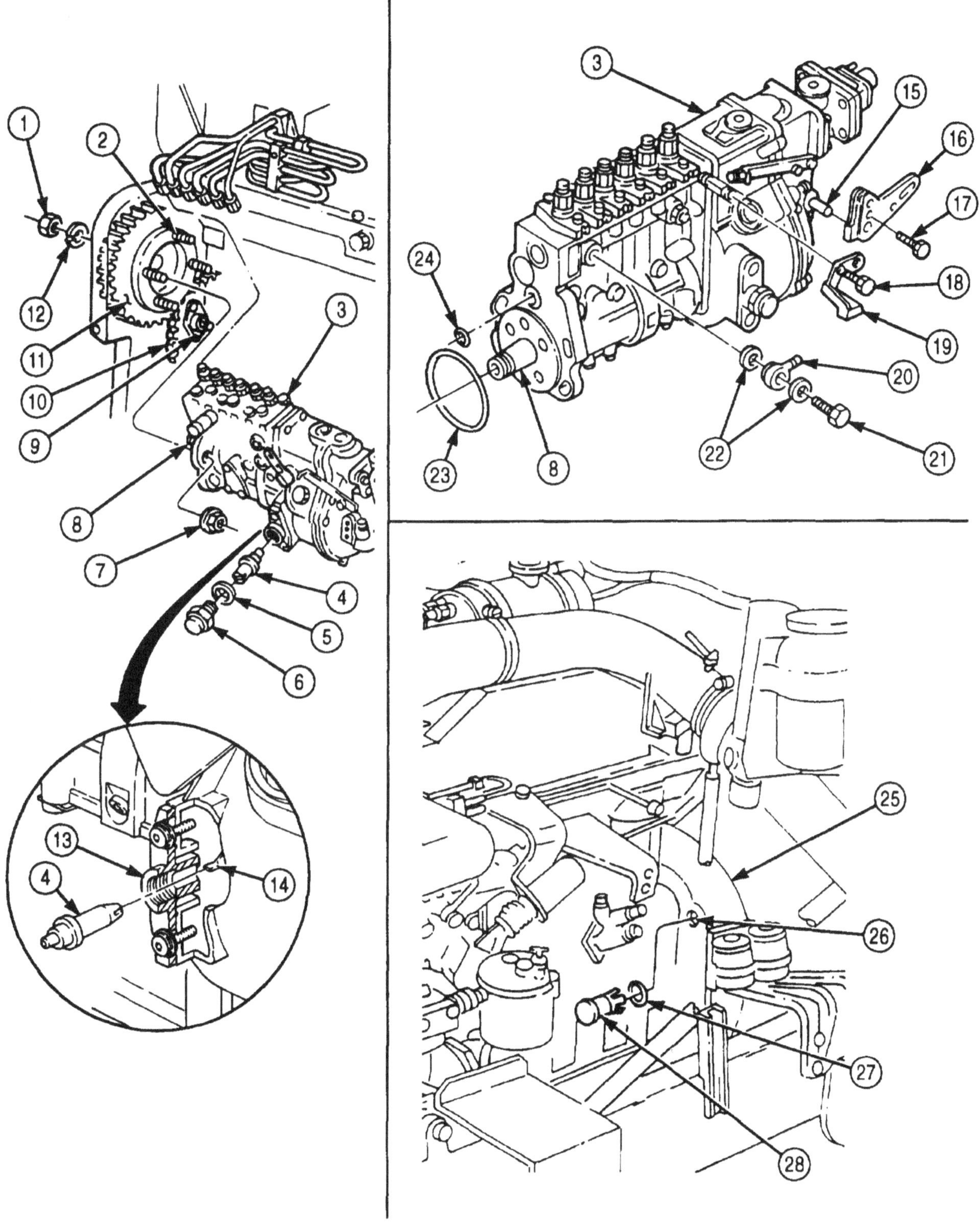

4-57. FUEL INJECTION PUMP MAINTENANCE (Contd)

19. Install fuel supply tube (3) on cylinder head (9) with two connector washers (4), bushing (5), and screw (6).
20. Install fuel return tube (3) on fuel injection pump (23) with two connector washers (2) and screw (1).
21. Connect six fuel injector fuel lines (24) to fuel injection pump (23).
22. Install fuel return hose (20) on fuel injection pump fuel return nipple (22) with clamp (21).
23. Install new lockwasher (16) and ball joint (17) on throttle lever (15).
24. Compress socket (18) and install accelerator linkage (12) on ball joint (17).
25. Place modulator control cable (13) on throttle lever (15).
26. Install washer (11) and new cotter pin (10) on link pin (14).
27. Install throttle connector (26) in cable pivot (27).
28. Insert control cable (8) in throttle connector (26) and install screw (25).
29. Install cable pivot (27) on shutoff valve lever (19) with washer (31) and new cotter pin (30).
30. Install control cable (8) on fuel pump bracket (28) with clamp (7) and screw (29).

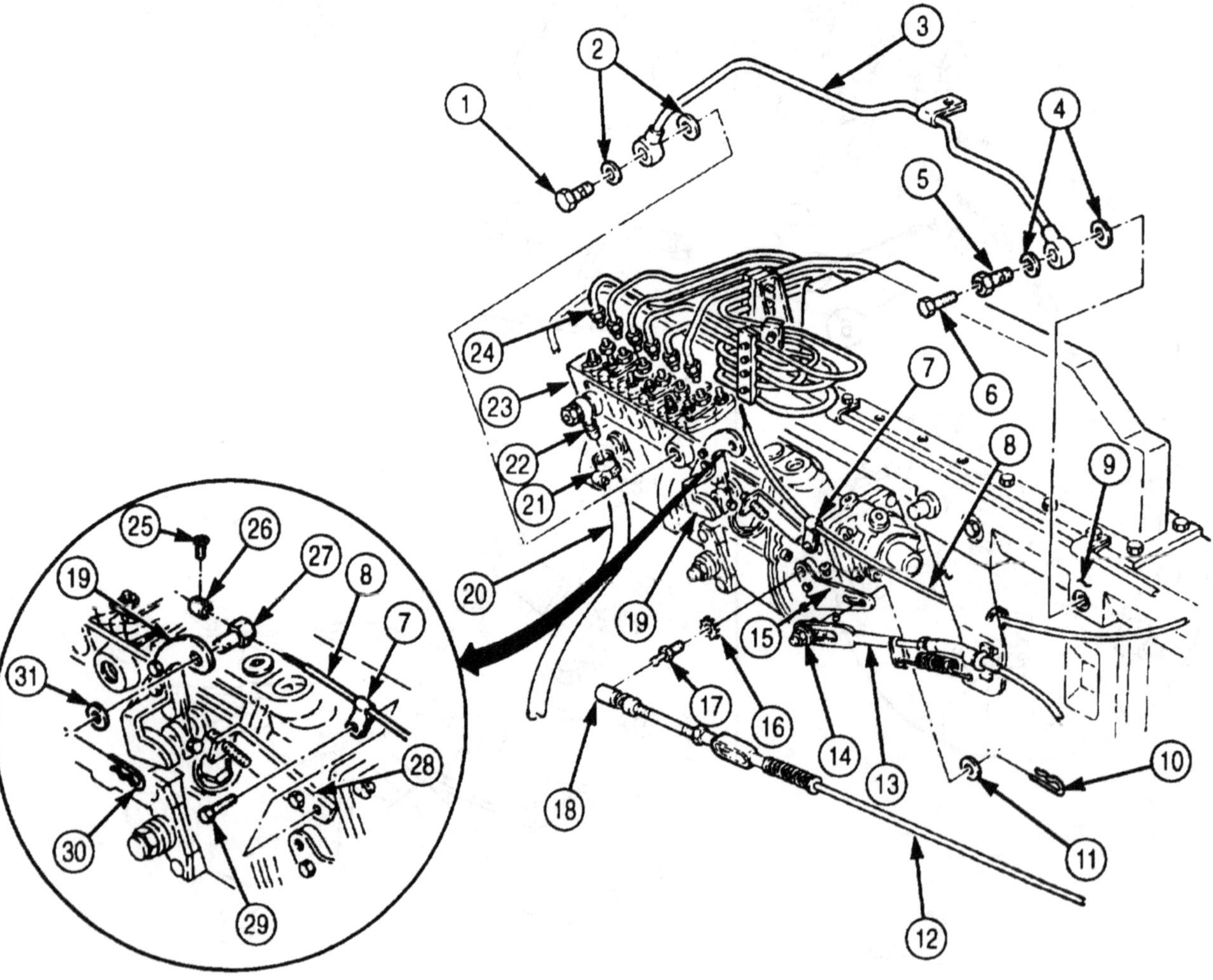

4-57. FUEL INJECTION PUMP MAINTENANCE (Contd)

c. Adjustment

NOTE

- Perform adjustments procedure for engine models serial number 44629589 and before.
- It may be necessary to cut on break-off screw for removal.

1. Remove safety wire (41) from protective cap (37). Discard safety wire (41).
2. Remove break-off screw (40), screw (39), two lockwashers (38), and protective cap (37) from manifold pressure compensator (34). Discard lockwashers (38).
3. Install break-off screw (40) and screw (39) on manifold pressure compensator (34) to hold stop (42) in position during adjustment.

NOTE

Do not turn adjustment screw more than one-half turn.

4. Hold adjusting screw (36) and loosen locknut (35). Turn adjusting screw (36) one-half turn clockwise and tighten locknut (35).
5. Remove break-off screw (40) and screw (39) from manifold pressure compensator (34). Discard break-off screw (40).
6. Install protective cap (37) on manifold pressure compensator (34) with two new lockwashers (38), new break-off screw (40), and screw (39).
7. Install new safety wire (41) on protective cap (37).
8. Remove plug (33) from governor housing (32).

NOTE

Do not turn star wheel more than six clicks during adjustment.

9. Using screwdriver, turn star wheel (44) clockwise toward engine (43) until six clicks are heard.
10. Install plug (33) on governor housing (32).
11. Stamp governor housing (32) and engine (43) data plate with field fix FF152 to identify fuel injection pump adjustment.

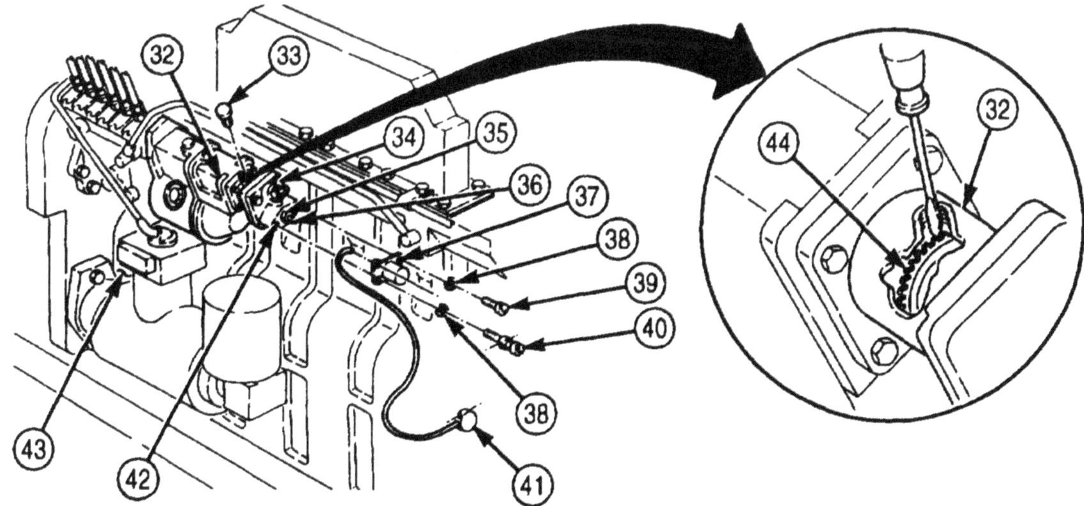

FOLLOW-ON TASKS:
- Install tachometer drive (para. 4-51).
- Install air fuel control (AFC) tube (para. 3-18).
- Install throttle control solenoid (para. 3-46).
- Prime fuel system (para. 3-22).

4-58. TURBOCHARGER MAINTENANCE

THIS TASK COVERS:

a. Disassembly
b. Cleaning and Inspection
c. Assembly

INITIAL SETUP:

APPLICABLE MODELS
M939A2

TOOLS
General mechanic's tool kit (Appendix E, Item 1)
Torque wrench (Appendix E, Item 146)
Dial indicator (Appendix E, Item 36)

MATERIALS/PARTS
Turbocharger repair kit (Appendix D, Item 700)
Abrasive cloth (Appendix C, Item 20)
Antiseize compound (Appendix C, Item 10)
Lubricating oil (Appendix C, Item 49)

REFERENCES (TM)
TM 9-2320-272-24P

EQUIPMENT CONDITION
Turbocharger removed (para. 3-21).

a. Disassembly

1. Place turbine housing (26) in soft-jawed vise.

NOTE

Perform step 2 to aid alignment during assembly.

2. Using scriber, scribe a line across compressor housing (1), diffuser (6), bearing housing (18), and) turbine housing (26).

CAUTION

Use care when removing compressor and turbine housing. Failure to do so may result in damage to turbine or compressor blades.

3. Remove nut (2), clamp (3), compressor housing (1), and O-ring (4) from diffuser (6). Discard O-ring (4).
4. Straighten tabs on two lockplates (25) and remove four screws (24) and two lockplates (25) from bearing housing (18). Discard lockplates (25) and screws (24).

NOTE

- Locknut has left-hand thread.
- It may be necessary to secure turbine shaft for removal of locknut.

5. Remove locknut (14), impeller (5), diffuser (6), O-ring (7), oil slinger (12), split ring (13), and oil baffle (11) from turbine shaft (2). Discard O-ring (7), split ring (13), and locknut (14).
6. Remove three screws (8), thrust bearing (10), and thrust collar (9) from bearing housing (18). Discard thrust bearing (10).
7. Straighten tabs on two lockplates (16) and remove four screws (15), two lockplates (16), and clamping plates (17) from turbine housing (26). Discard lockplates (16) and screws (15).
8. Remove bearing housing (18) from turbine housing (26).
9. Remove turbine shaft (21) and heat shield (22) from bearing housing (18).
10. Remove split ring (20) from turbine shaft (21). Discard split ring (20).
11. Remove four retaining rings (23) and two bearings (19) from bearing housing (18). Discard retaining rings (23) and bearings (19).

4-58. TURBOCHARGER MAINTENANCE (Contd)

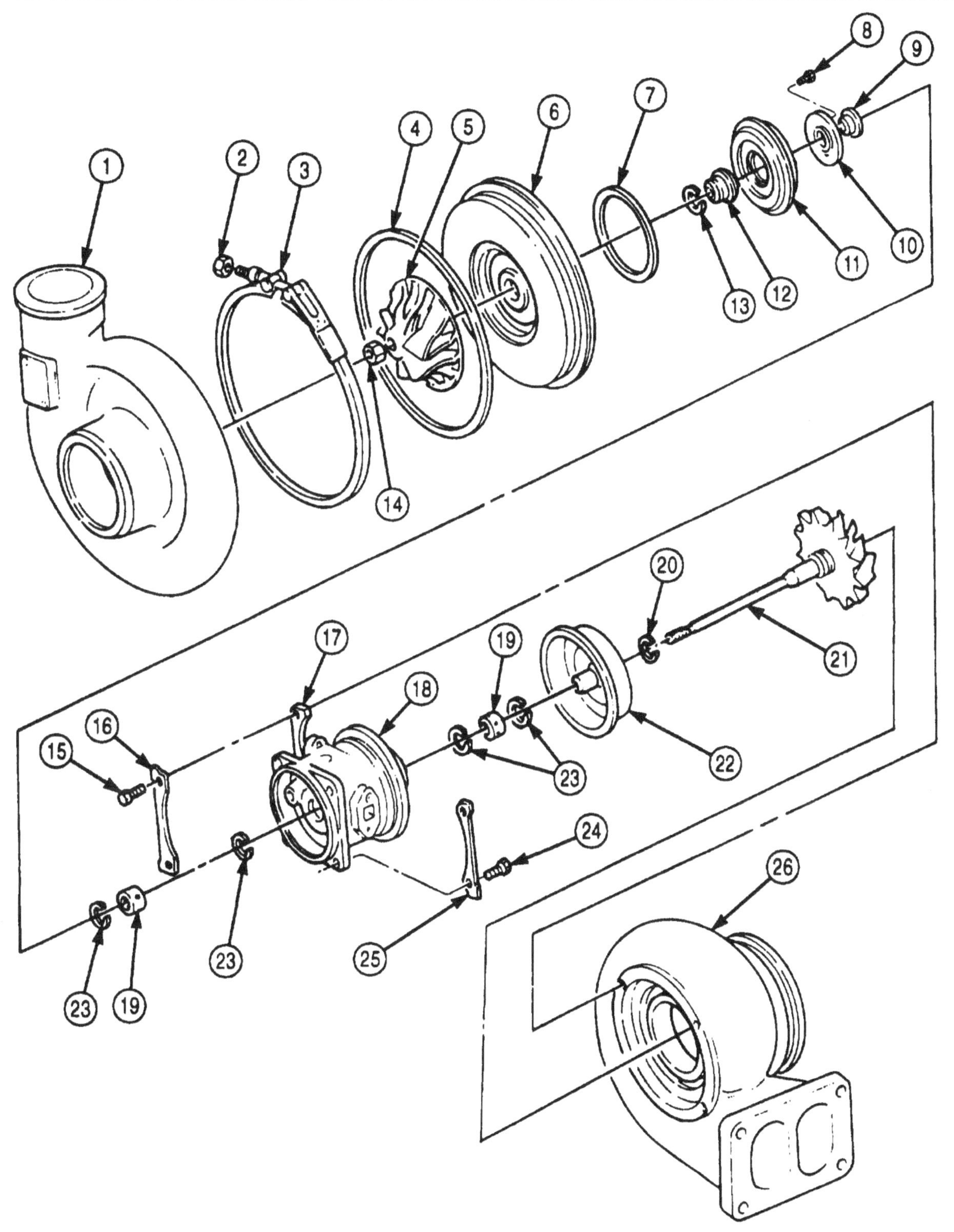

4-58. TURBOCHARGER MAINTENANCE (Contd)

b. Cleaning and Inspection

1. For general cleaning instructions, refer to para. 2-14.
2. For general inspection instructions, refer to para. 2-15.
3. Inspect thrust collar (3) for nicks and scratches. If nicked or scratched, replace thrust collar (3).
4. Using abrasive cloth, remove carbon build up from turbine housing (12) and polish bearing surface on turbine shaft (1).
5. Inspect turbine shaft (1) for wear. Measure outside diameter of bearing surfaces. If outside diameter is less than 0.432 in. (10.97 mm), replace turbine shaft (1).
6. Inspect compressor housing (4) for cracks and excessive wear. If cracked or worn, replace compressor housing (4).

NOTE

Coat new split ring with clean lubricating oil

1. Install new split ring (13) in groove on turbine shaft (1).
2. Install heat shield (14) on turbine shaft (1).

NOTE

Lubricate new bearing with clean lubricating oil.

3. Install two new bearings (11) and four new retaining rings (9) in bearing housing (10). Ensure beveled side of retaining rings (9) face bearings (11).
4. Install bearing housing (10) on turbine shaft (1). Rotate bearing housing (10) and press downward to seat split ring (13) on turbine shaft (1).

CAUTION

Turbocharger balance is very important. Check and recheck balance marks during assembly. An unbalanced turbocharger can result in damage to engine components.

5. Align scribed mark on thrust collar (3) with balance mark on turbine shaft (1).
6. Install thrust collar (3) on turbine shaft (1).

NOTE

Lubricate new thrust bearing with clean lubricating oil.

7. Install new thrust bearing (15) into bearing housing (10) with three screws (8). Tighten screws (8) 40 lb-in. (4.5 N•m).
8. Install oil baffle (16) into bearing housing (10).

NOTE

Perform step 9 so scribed line on oil slinger is visible when oil slinger is installed on turbine shaft.

9. Using scriber, scribe a line on top of oil slinger (2).
10. Install new split ring (7) on oil slinger (2).

4-58. TURBOCHARGER MAINTENANCE (Contd)

NOTE

Coat oil slinger with clean lubricating oil.

11. Install oil slinger (2) into diffuser (5) so that split ring (7) is seated in diffuser (5).
12. Install new O-ring (6) into diffuser (5).
13. Align scribed mark on oil slinger (2) with balance mark on turbine shaft (1).
14. Install diffuser (5) on turbine shaft (1).

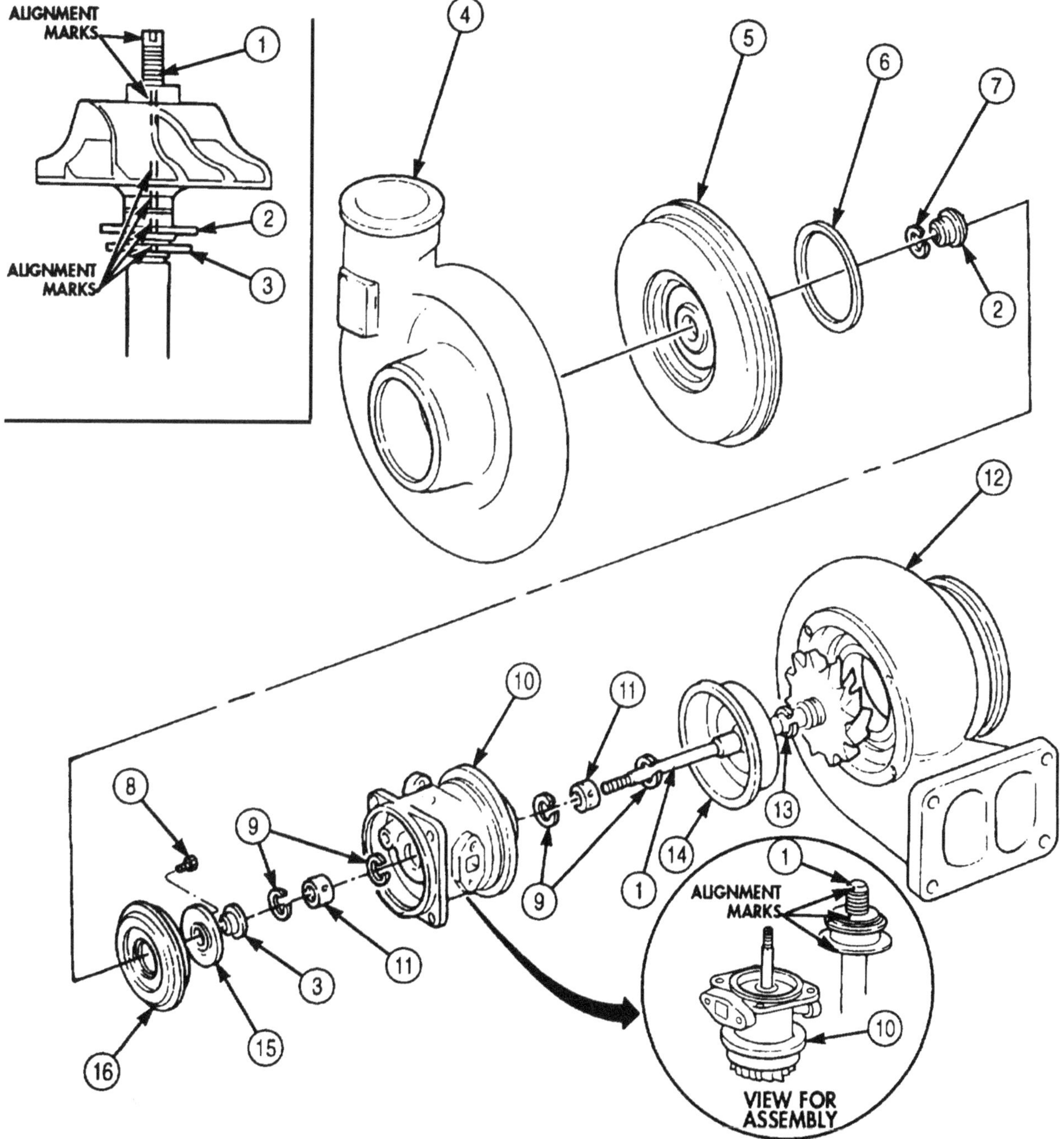

4-58. TURBOCHARGER MAINTENANCE (Contd)

CAUTION

Ensure impeller and turbine shaft balance marks are aligned. Failure to do so may result in damage to engine.

15. Install impeller (6) on turbine shaft (2) as follows:
    a. Align balance mark on impeller (6) with balance mark on turbine shaft (1).
    b. Install impeller (6) on turbine shaft (1) with new left-hand thread locknut (9). Tighten locknut (9) 12 lb-in. (1.4 N•m).
16. Install bearing housing (2) into turbine housing (13) and align scribe marks.

NOTE

Coat threads of four new screws with antiseize compound.

17. Install two new lockplates (11) and clamping plates (12) on turbine housing (13) with four new screws (10). Bend four tabs of two lockplates (11) against four screws (10).
18. Align scribe marks on diffuser (8) to scribe marks on bearing housing (2).
19. Install two new lockplates (15) into diffuser (8) with four new screws (14). Bend four tabs of two lockplates (15) against four screws (14).

CAUTION

Failure to obtain proper end play or radial clearance may result in damage to engine.

20. Check turbine shaft (1) for end play:
    a. Install turbine housing (13) in soft-jawed vise.
    b. Install dial indicator on bearing housing (2) so dial indicator plunger is against turbine shaft (1).
    c. Push turbine shaft (1) into bearing housing (2) and set dial indicator to read zero.
    d. Pull turbine shaft (1) away from bearing housing (2) and read dial indicator.
    e. End play must measure between 0.001-0.003 in. (0.03-0.08 mm).
    f. If end play is not within normal range, disassemble turbocharger and inspect turbine shaft (tasks a and b).
21. Check radial clearance:
    a. Place dial indicator plunger against impeller (6), just below locknut (9).
    b. Push impeller (6) away from dial indicator and set dial indicator to read zero.
    c. Pull impeller (6) toward dial indicator and take reading.
    d. Radial clearance must measure between 0.012-0.018 in. (0.305-0.457 mm).
    e. If radial clearance is not within normal range, disassemble bearing housing (2) and inspect thrust collar (tasks a and b).
    f. Remove dial indicator from bearing housing (2).
22. Coat threads of clamp (5) with antiseize compound and install nut (4) on clamp (5). Do not tighten nut (4).
23. Install new O-ring (7), clamp (5), and compressor housing (3) and align scribe marks on compressor housing to turbine housing (13).
24. Tighten nut (4) 120 lb-in. (13.6 N•m).

4-58. TURBOCHARGER MAINTENANCE (Contd)

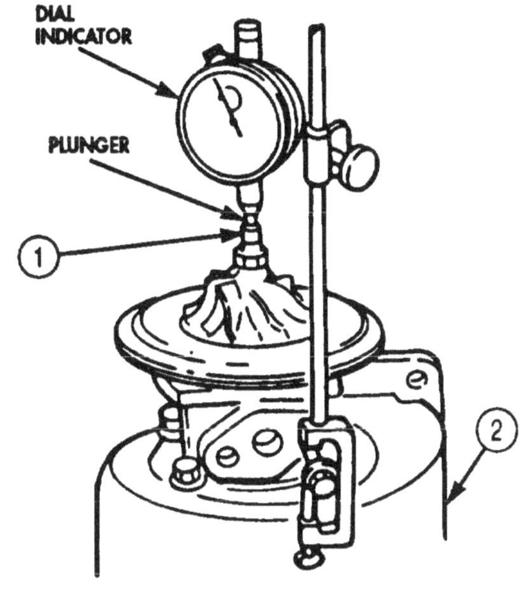

DIAL INDICATOR
PLUNGER
1
2

TURBINE SHAFT END PLAY

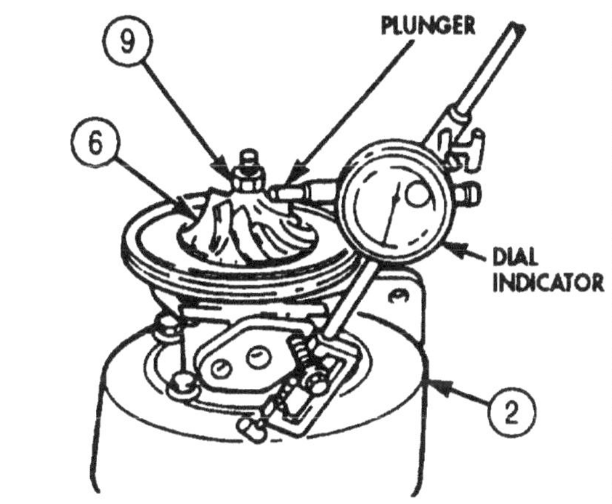

9
PLUNGER
6
DIAL INDICATOR
2

TURBINE SHAFT RADIAL CLEARANCE

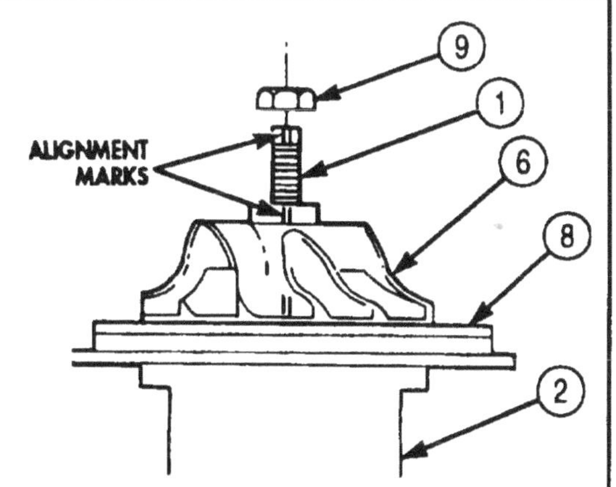

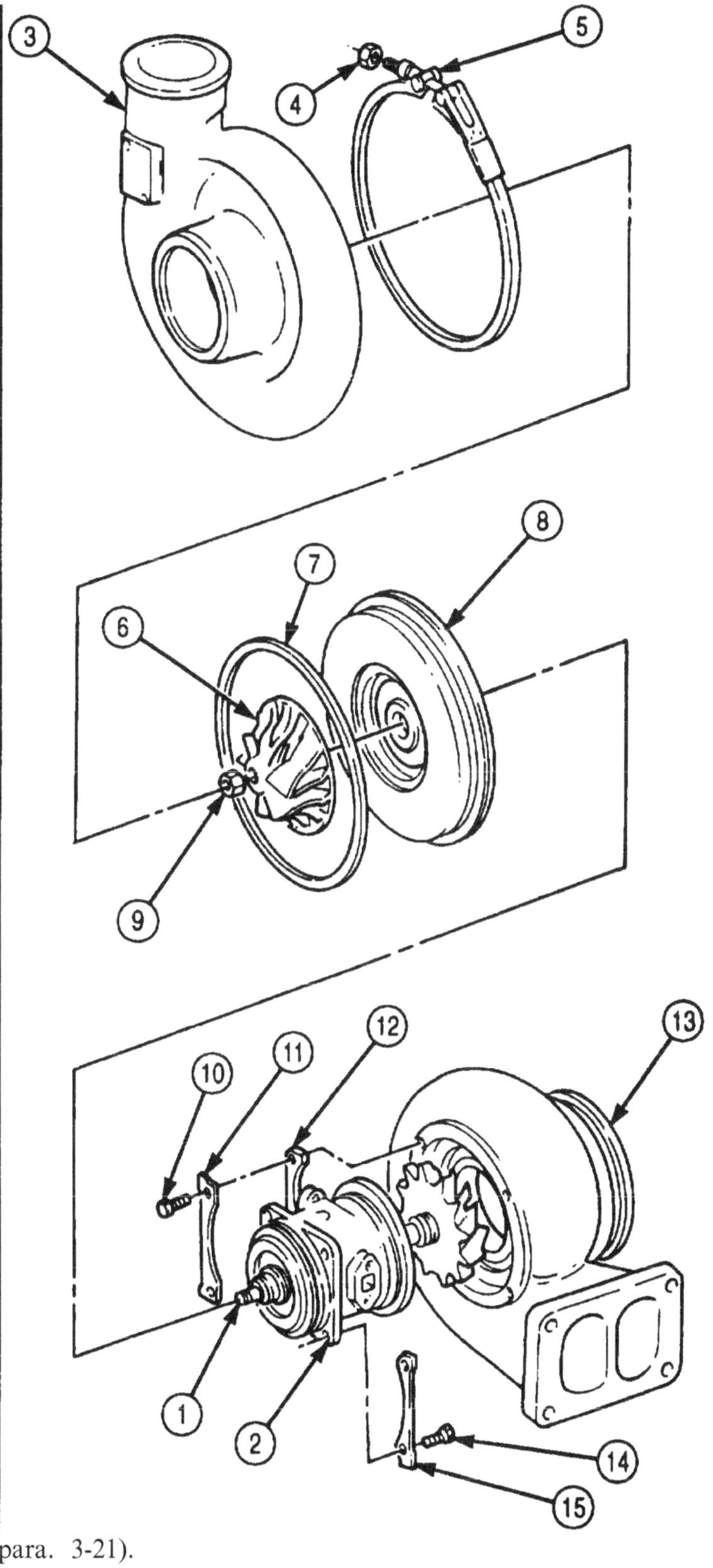

9
1
ALIGNMENT MARKS
6
8
2

3 4 5 6 7 8 9 10 11 12 13 1 2 14 15

FOLLOW-ONTASK: Install turbocharger (para. 3-21).

Section V. COOLING SYSTEM MAINTENANCE

4-59. COOLING SYSTEM MAINTENANCE INDEX

4-60. RADIATOR MAINTENANCE

THIS TASK COVERS:

a. Disassembly
b. Cleaning, Inspection, and Repair
c. Assembly

INITIAL SETUP:

APPLICABLE MODELS
All

TOOLS
General mechanic's tool kit (Appendix E, Item 1)
Torque wrench (Appendix E, Item 146)

MATERIALS/PARTS
Two gaskets (Appendix D, Item 201)
Eighty locknuts (Appendix D, Item 304)
GAA grease (Appendix C, Item 28)

REFERENCES (TM)
TM 9-2320-272-10
TM 9-2320-272-24P
TM 750-254

EQUIPMENT CONDITION
- Radiator removed (para. 3-59 or para. 3-60).
- Radiator fan shroud removed (para. 3-63 or para. 3-64).

a. Disassembly

NOTE

- Tag screws for installation.
- Mark positions of two shroud mounting brackets for installation.

1. Remove forty locknuts (l), washers (2), thirty-six screws (7), two clamping strips (6), four screws (12), and two shroud mounting brackets (11) from inlet tank (3) and radiator core (5). Discard locknuts (1).
2. Remove inlet tank (3) and gasket (4) from radiator core (5). Discard gasket (4).

NOTE

- Tag screws and radiator side braces for installation.
- Mark position of shroud mounting brackets for installation,

3. Remove forty locknuts (16), washers (15), thirty-six screws (13), four screws (18), two shroud mounting brackets (19), two clamping strips (14), upper radiator side brace (8), lower radiator side brace (9), and four clamping strips (10) from outlet tank (17) and radiator core (5). Discard locknuts (16).
4. Remove outlet tank (17) and gasket (4) from radiator core (5). Discard gasket (4).

4-60. RADIATOR MAINTENANCE (Contd)

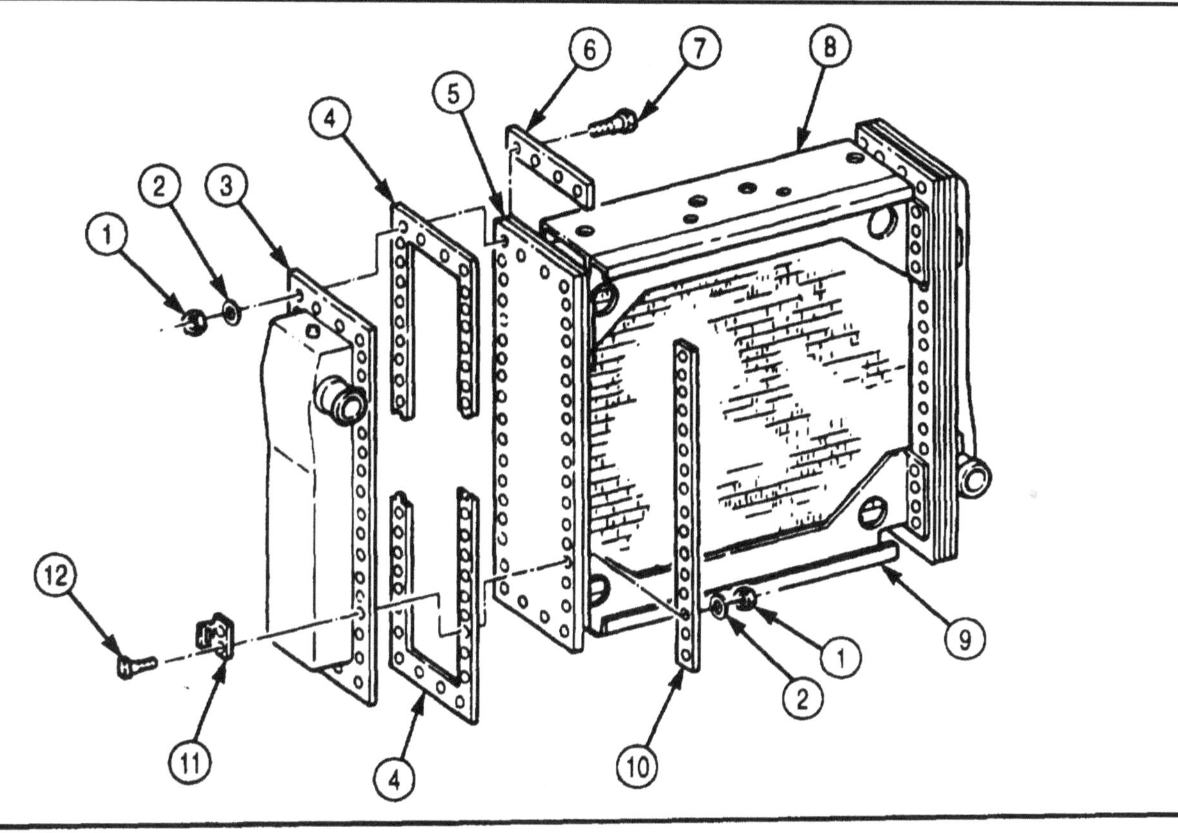

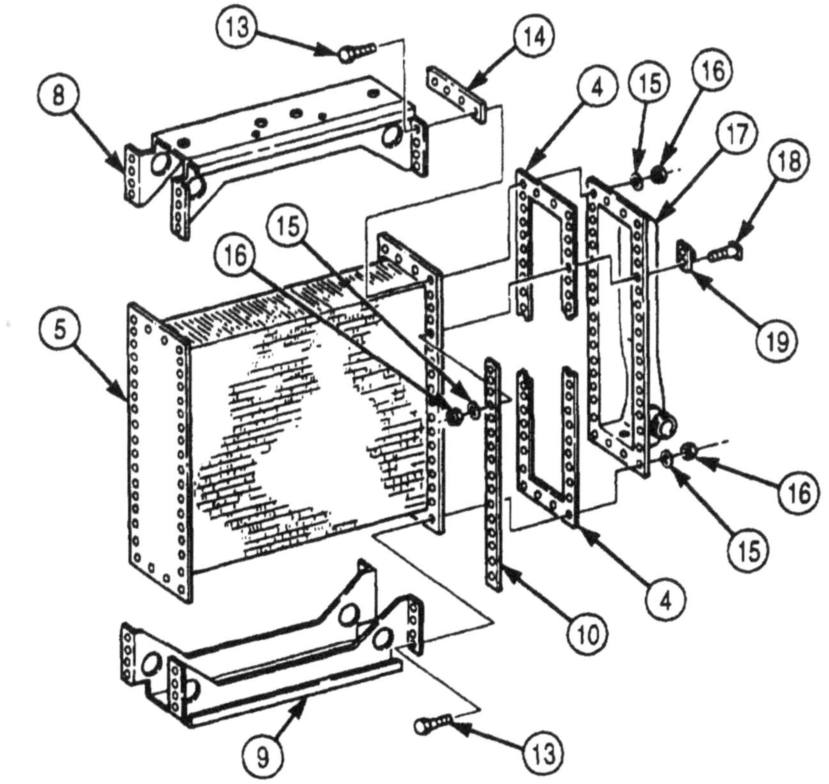

4-60. RADIATOR MAINTENANCE (Contd)

b. Cleaning, Inspection, and Repair

Clean, inspect, and repair radiator core (1) (TM 750-254).

c. Assembly

1. Position lower radiator side brace (12), upper radiator side brace (2), and two clamping strips (4) on radiator core (1).
2. Interlock new gasket (5) and align to screw holes on radiator core (1). Apply GAA grease to gasket (5) to hold it in place on radiator core (1).
3. Align holes in gasket (51, outlet tank (10), and two clamping strips (11) to radiator core (1).
4. Install outlet tank (10) on radiator core (11, upper radiator side brace (2), and lower radiator side brace (12) with thirty-six screws (3), two shroud mounting brackets (8), four screws (9), forty washers (6), and new locknuts (7). Tighten locknuts (7) 19 lb-ft (26 N•m).
5. Align new gasket (15) to screw holes in radiator core (1). Apply GAA grease on gasket (15) to hold it in place on radiator core (1).
6. Install two clamping strips (16), inlet tank (19), and two clamping strips (18) on radiator core (1) with two shroud mounting brackets (20), four screws (21), thirty-six screws (17), forty washers (14), and new locknuts (13). Tighten locknuts (13) 19 lb-ft (26 N-m).

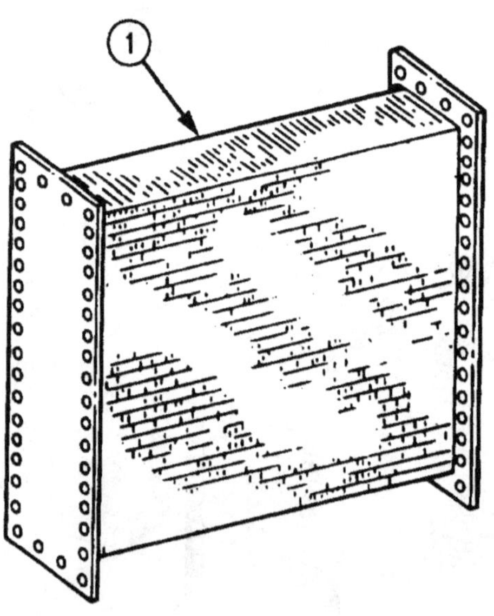

4-60. RADIATOR MAINTENANCE (Contd)

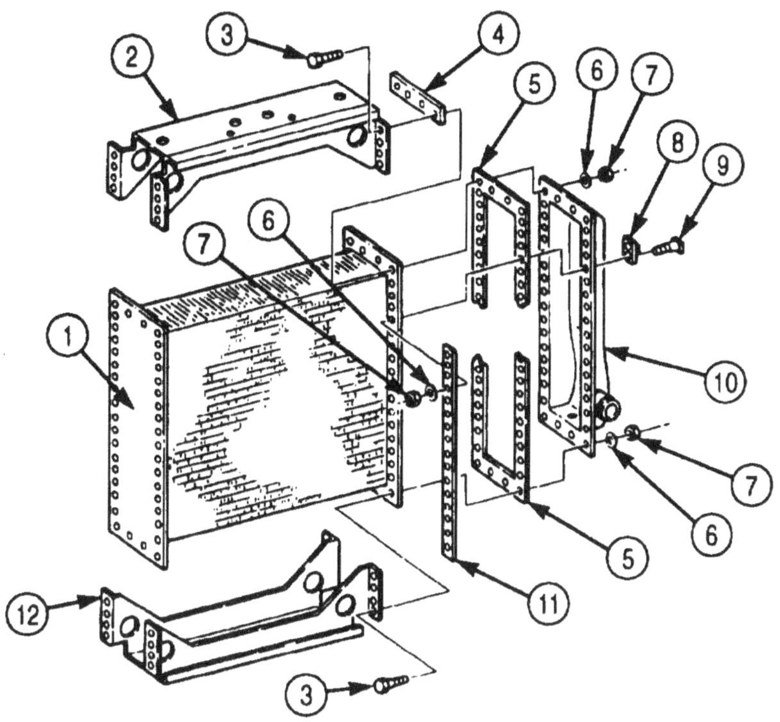

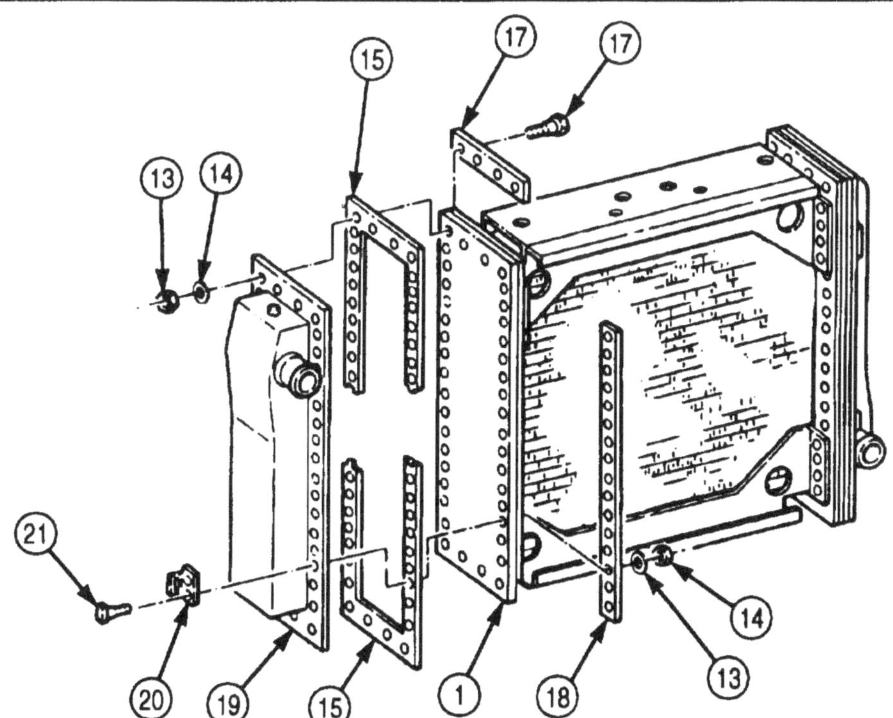

FOLLOW-ON TASKS: • Install radiator (para. 3-59 or para. 3-60).
• Install fan shroud (para. 3-63 or para. 3-64).
• Service cooling system (para. 3-53).
• Start engine (TM 9-2320-272-10) and check for coolant leaks.

4-61. FAN DRIVE CLUTCH (M939/A1) MAINTENANCE

THIS TASK COVERS:

a. Disassembly
b. Cleaning and Inspection
c. Assembly
d. Fan Drive Clutch Operation Check

INITIAl SETUP:

APPLICABLE MODELS
M939lA1

SPECIAL TOOLS
Mandrel (Appendix E, Item 116)
Bearing replacer (Appendix E, Item 13)

TOOLS
General mechanic's tool kit (Appendix E, Item 1)
Torque wrench (Appendix E, Item 146)
Vise
Arbor press

MATERIALS/PARTS
O-ring (Appendix D, Item 451)
Locknut (Appendix D, Item 305)
Snapring (Appendix D, Item 659)

MATERIALS/PARTS (Cod)
O-ring (Appendix D, Item 448)
O-ring (Appendix D, Item 449)
O-ring (Appendix D, Item 450)
Bearing (Appendix D, Item 11)
Two bearings (Appendix D, Item 12)
Rags (Appendix C, Item 58)
GAA grease (Appendix C, Item 28)

REFERENCES (TM)
TM 9-2320-272-10
TM 9-2320-272-24P

EQUIPMENT CONDITION
Fan drive clutch removed (para. 3-75).

GENERAL SAFETY INSTRUCTIONS
- Piston spring is under compression and may cause injury if not properly removed.
- When cleaning with compressed air, wear eyeshield.

a. Disassembly

1. Install clutch assembly (7) in soft-jawed vise.
2. Remove seal plug (1) from clutch assembly (7).
3. Remove O-rings (2) and (31 from seal plug (1). Discard O-rings (2) and (3).
4. Remove expansion locknut (4). washer (5), and spacer (6) from shaft (10). Discard locknut (4).
5. Remove clutch assembly bracket (9) from clutch assembly (7).
6. Remove spacer (8) from shaft (10).
7. Remove clutch assembly (7) from vise and set on workbench with thrust plate (11) facing up.

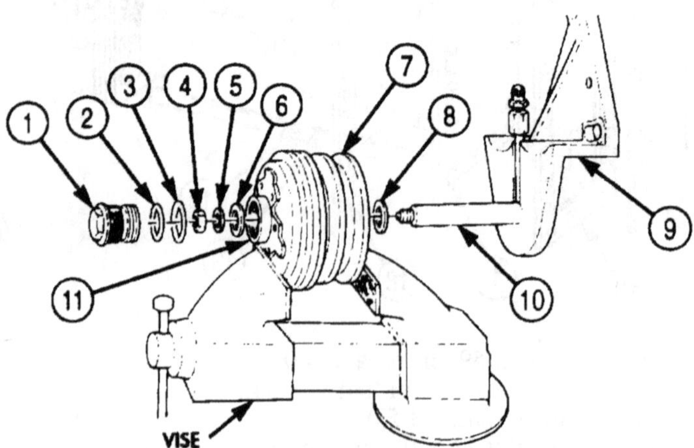

4-61. FAN DRIVE CLUTCH (M939/A1) MAINTENANCE (Contd)

8. Remove six screws (12) holding thrust cap backup plate (13) to clutch assembly (7), leaving two screws (12) installed opposite one another.

WARNING

The piston spring, located between thrust cap and clutch facing, is under compression and may cause injury to personnel if screws are not removed as directed below.

NOTE

Assistant will help with step 9.

9. While assistant holds down thrust cap backup plate (13), alternately loosen and remove two remaining screws (12) from thrust cap backup plate (13) and clutch assembly (7). Have assistant slowly release thrust cap backup plate (13) from clutch assembly (7).

NOTE

Step 10 is required only if thrust cap fails to release.

10. Install two screws (12) into opposite sides of thrust cap backup plate (13) and then loosen screws (12) evenly about halfway. Tap screws (12) gently until backup plate (13) releases, and repeat step 9.

NOTE

Clutch housing and clutch drive pulley alignment must be marked for assembly.

11. Remove thrust cap backup plate (13), spacer (14), clutch drive pulley (15), piston spring (16), and clutch facing (17) from clutch housing (18).

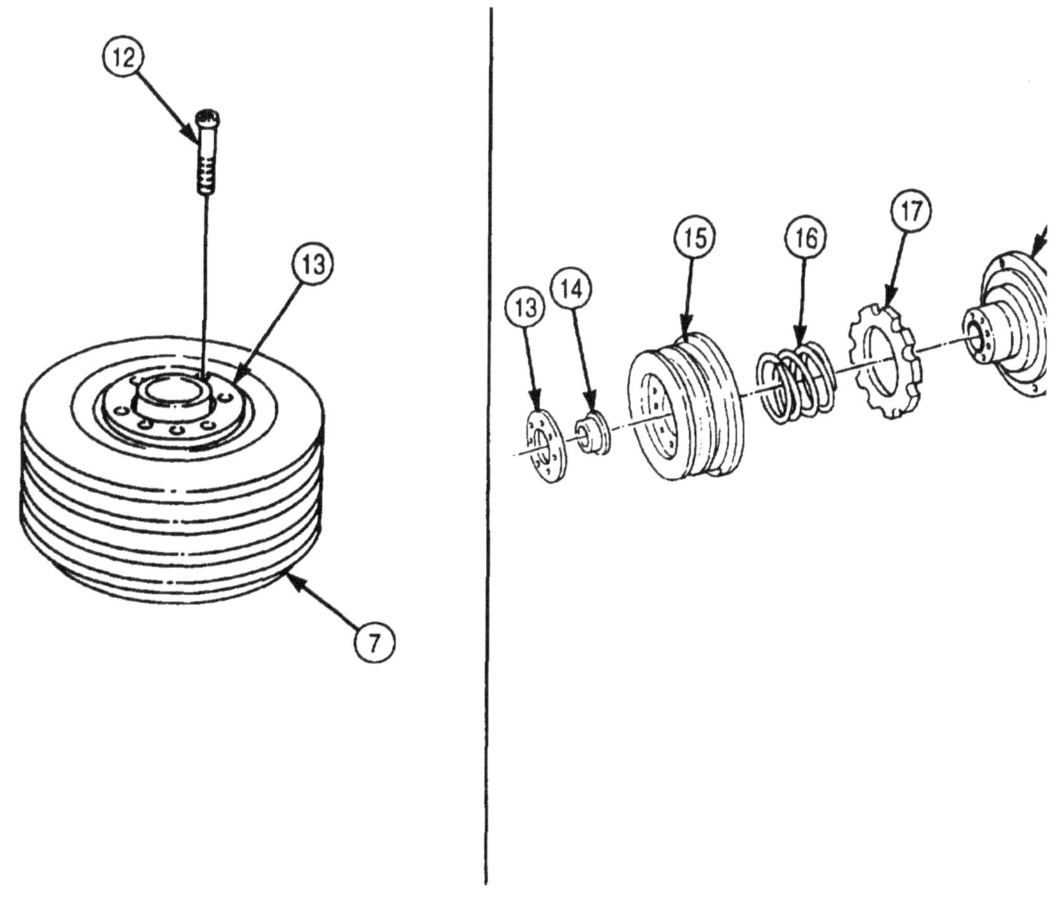

4-6 1. FAN DRIVE CLUTCH (M939/A1) MAINTENANCE(Contd)

12. Remove snapring (1) from drive pulley (2). Discard snapring (1).

CAUTION

Do not attempt to remove drive pins from drive pulley. These pins are set by manufacturer and must not be disturbed.

13. Place drive pulley (2) on arbor press with drive pins facing down. Use wood blocks to provide stand off on arbor press bed.
14. Using bearing replacer, press thrust cap (3) out of drive pulley (2).

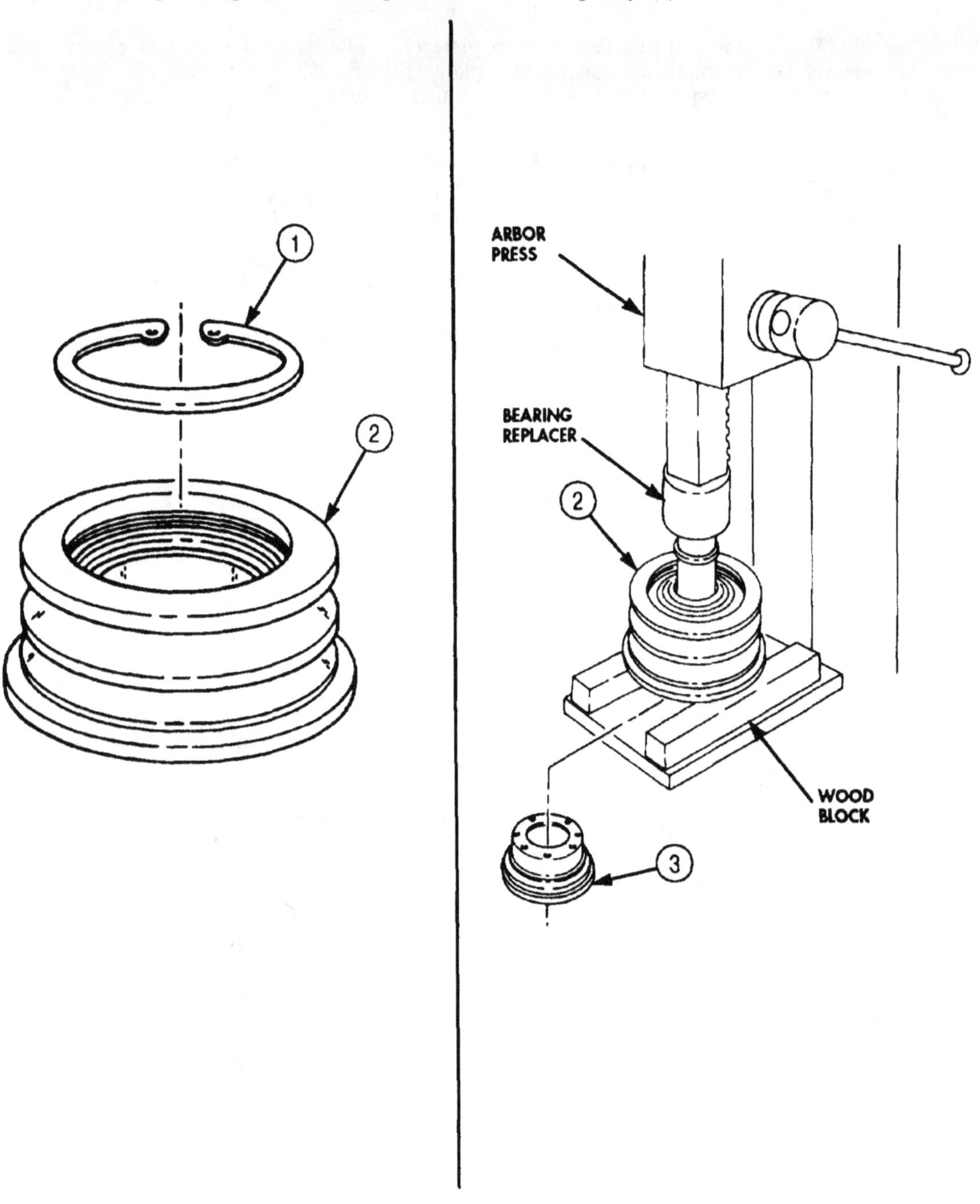

4-61. FAN DRIVE CLUTCH (M939/A1) MAINTENANCE(Contd)

15. Turn drive pulley (2) over, place on wood blocks on arbor press bed, and use a mandrel to press bearing (4) out of drive pulley (2). Discard bearing (4).

NOTE

Before removing piston, seal plug must be reinstalled in clutch housing and then apply compressed air to discharge piston.

16. Install seal plug (5) in clutch housing (6). Tighten seal plug (5) 40 lb-fi (54 N-m).

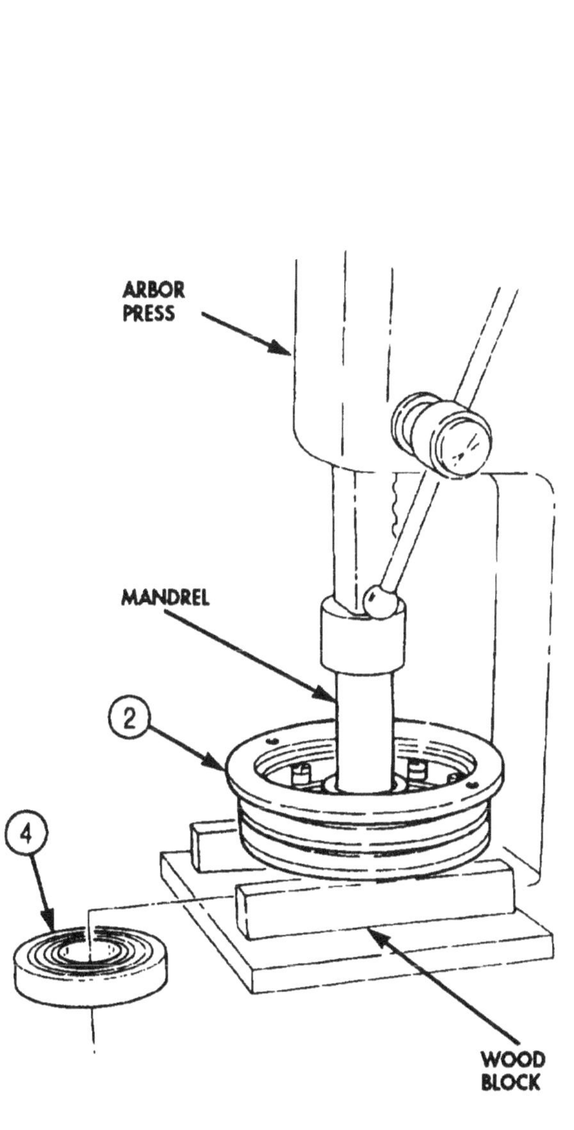

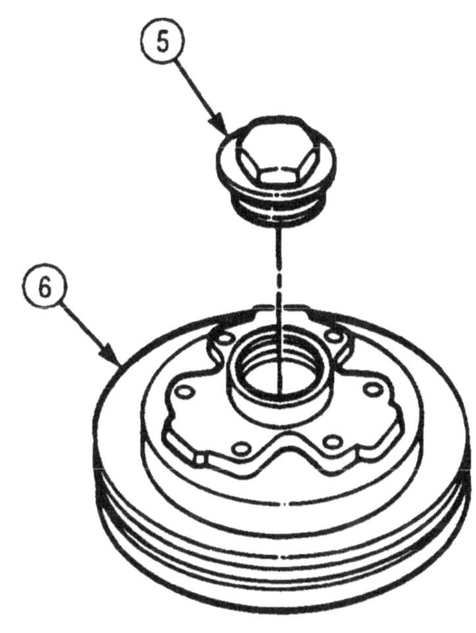

4-61. FAN DRIVE CLUTCH (M939/A1) MAINTENANCE(Contd)

17. Place clutch housing (1) on shaft (7) of clutch assembly bracket (6), install air adapter (4) on air adapter fitting (3), and connect shop air hose (5). Do not apply air pressure at this time.
18. Ensure clutch assembly bracket (6) and shaft (7) are firmly anchored to bench with clutch housing (1) over shaft (7).
19. Place shop towel or rag around shaft (7) to protect piston (2).

WARNING

When working with compressed air, eyeshields must be worn. Failure to wear eyeshields may result in injury to personnel.

NOTE

Assistant will help with step 20.

20. While assistant holds clutch housing (1) firmly in place, momentarily open shop air hose (5) and apply 60 psi (414 kPa) to partially discharge piston (2). Ensure piston (2) discharges evenly. If piston (2) discharges unevenly, press piston (2) back into clutch housing (1). Repeat discharge action.

CAUTION

Extreme care must be used when removing piston from clutch housing to avoid damaging piston. Do not use sharp tools, and do not pry**or use force.**

21. Place clutch housing (1) on workbench with piston (2) side up.
22. Insert two thin-bladed screwdrivers opposite each other and under lip above O-ring (9).

NOTE

Assistant will help with step 23.

23. Pulling evenly at several points around outside diameter of piston (2), remove piston (2) from clutch housing (1).
24. Remove O-rings (8) and (9) from piston (2). Discard O-rings (8) and (9).
25. Remove seal plug (10) from clutch housing (1),

4-61. FAN DRIVE CLUTCH (M939/A1) MAINTENANCE(Contd)

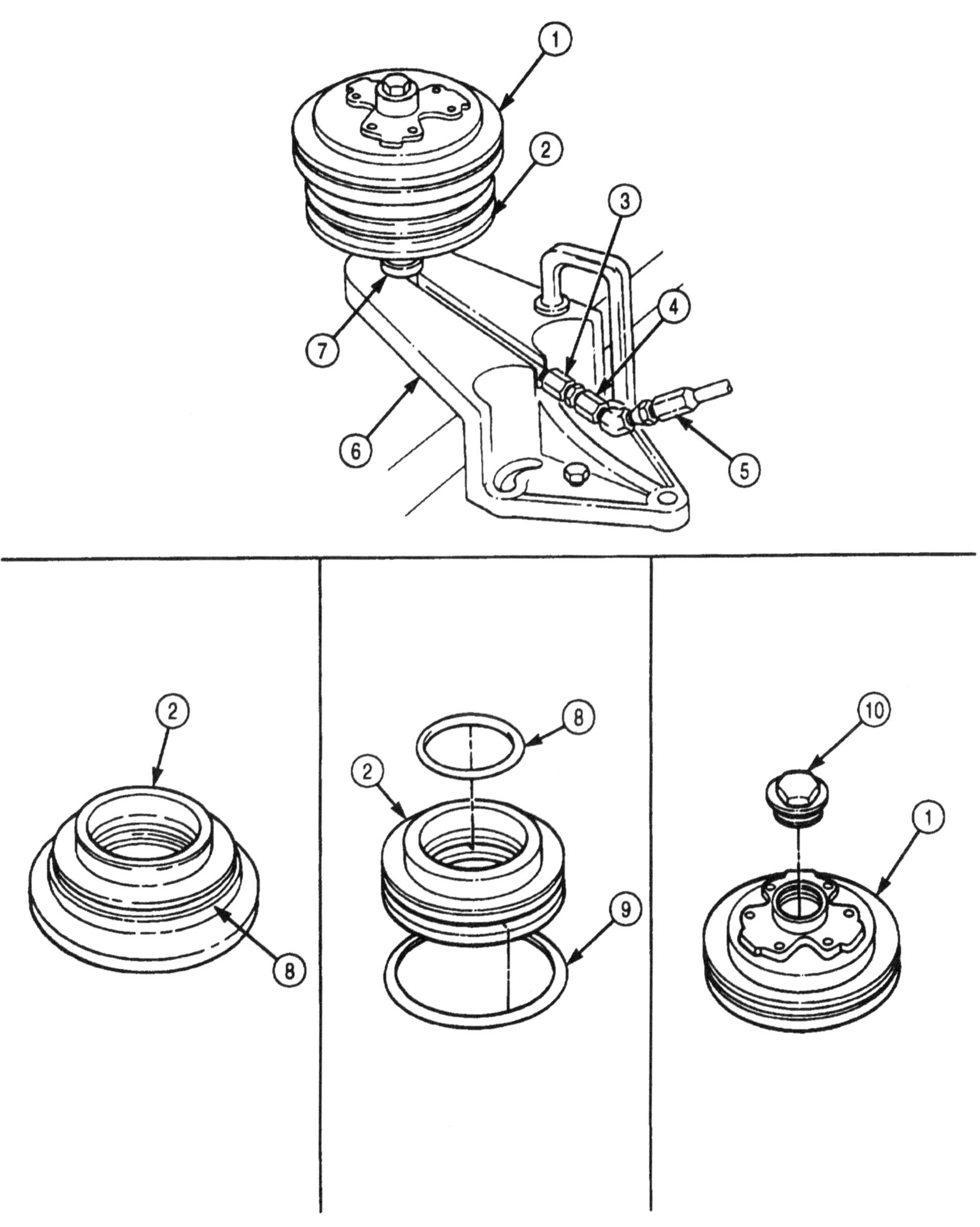

4-61. FAN DRIVE CLUTCH (M939/A1) MAINTENANCE (Contd)

26. Place clutch housing (2) on arbor press with fan hub (1) facing up. Support clutch housing (2) with wood blocks.
27. Using a mandrel, press spacer (4), spacer (5), and bearing (3) from clutch housing (2). Discard bearing (3).
28. Using wood blocks on bed of arbor press, remove fan clutch bearing (6) from spacer (4). Discard bearing (6).

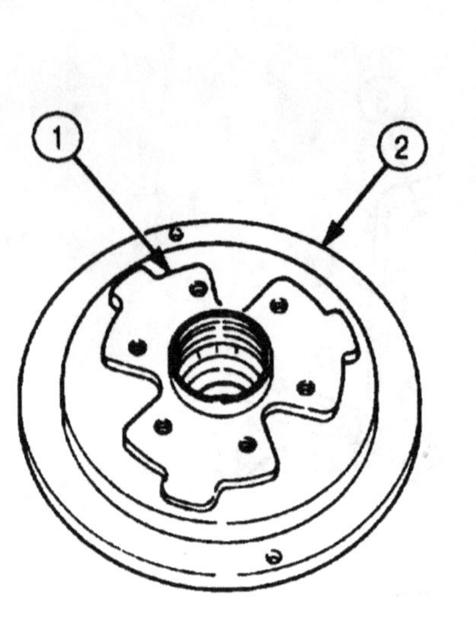

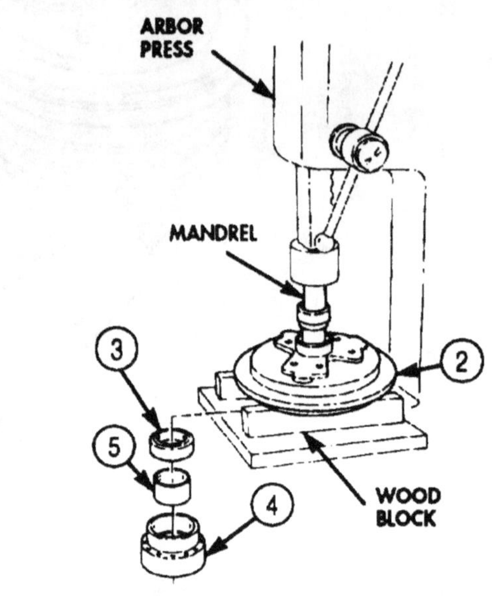

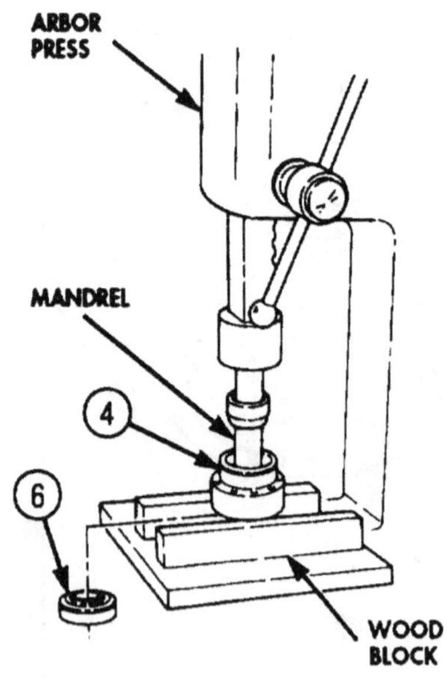

4-61. FAN DRIVE CLUTCH (M939/A1) MAINTENANCE (Contd)

b. Cleaning and Inspection

1. For general cleaning instructions, refer to para. 2-14.
2. For general inspection procedures, refer to para. 2-15.
3. Inspect clutch housing (2), drive pulley (11), and spacers (4), (5), and (7), bracket (13), clutch facing (8), and thrust cap (10) for chips, cracks, or breaks. Replace part(s) if chipped, cracked, or broken.
4. Inspect clutch housing (2) and shaft (12) for gouged or stripped threads. Replace clutch housing (2) or shaft (12) if threads are damaged.
5. Inspect return spring (9) for cracks and weak coils. Replace spring (9) if cracked or coils are weak.

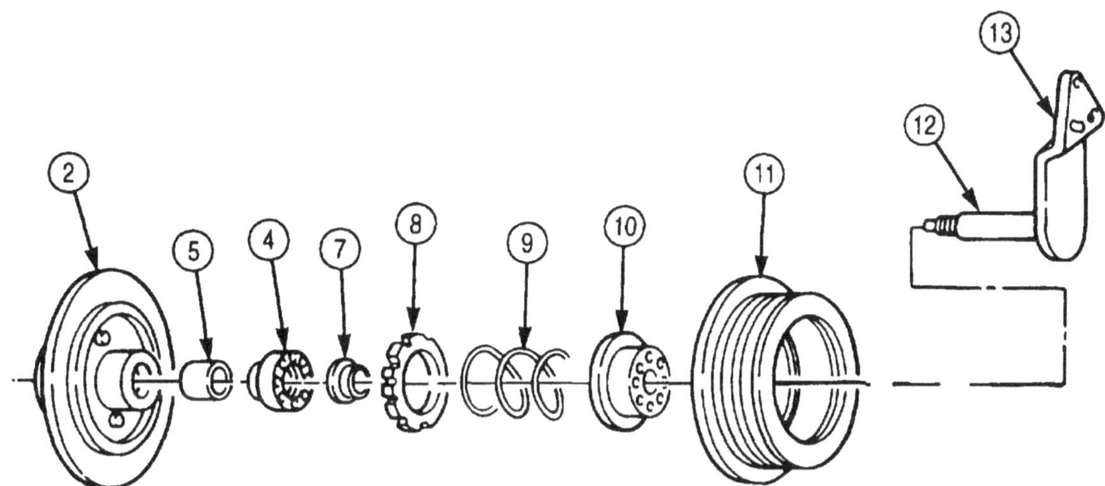

4-61. FAN DRIVE CLUTCH (M939/A1) MAINTENANCE (Contd)

c. Assembly

1. Lightly coat bore (4) of clutch housing (3) with GAA grease.
2. Place clutch housing (3) on arbor press supported by two wood blocks. Piston bore (2) must be facing up.
3. Using mandrel, install new bearing (5) in clutch housing (3).
4. Using mandrel, install bearing spacer (1) in clutch housing (3). Ensure holes in spacer (1) align with holes in clutch housing (3). If holes are not aligned, remove bearing spacer (1) and install again.
5. Using arbor press and mandrel, install spacer (7) and new bearing (6) into clutch housing (3).
6. Remove clutch housing (3) from arbor press and place on workbench with piston bore (2) side facing up.

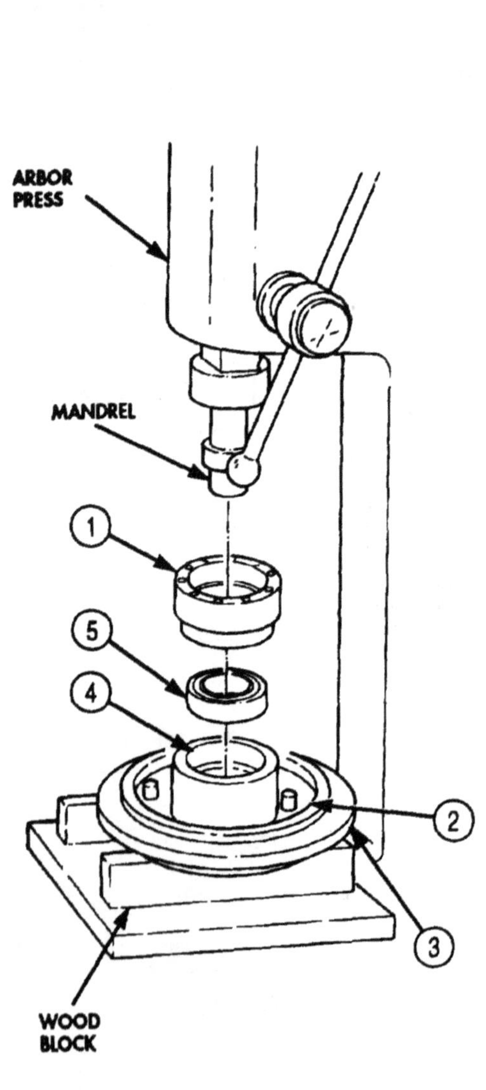

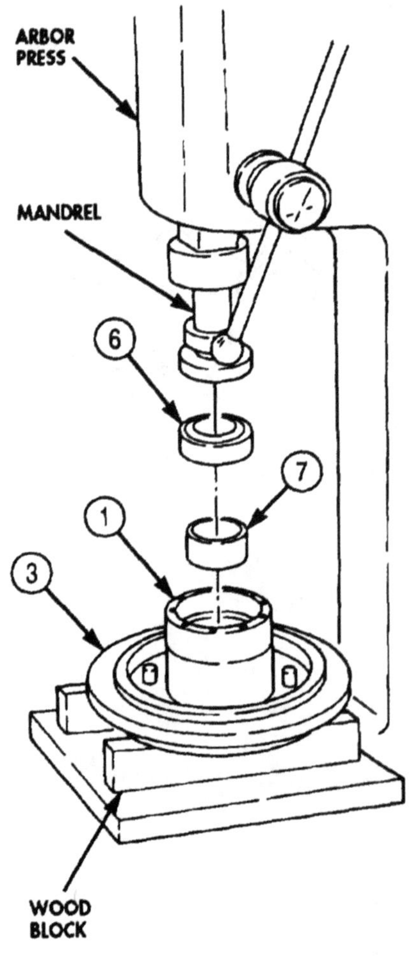

4-61. FAN DRIVE CLUTCH (M939/A1) MAINTENANCE (Contd)

7. Lightly coat new O-rings (8) and (10) with GAA grease.
8. Install O-ring (8) in outside groove of piston (9). Ensure O-ring (8) is not twisted.
9. Install O-ring (10) in bore of piston (9). Ensure O-ring (10) is not twisted.
10. Coat piston bore (2) of clutch housing (3) with GAA grease and set with hinge pins (11) facing up.

CAUTION

When installing piston, use care not to cut or damage O-rings.

11. Using both hands, press downward and rotate piston (9) until seated in clutch housing (3). Ensure piston (9) is seated over three hinge pins (11) in clutch housing (3).

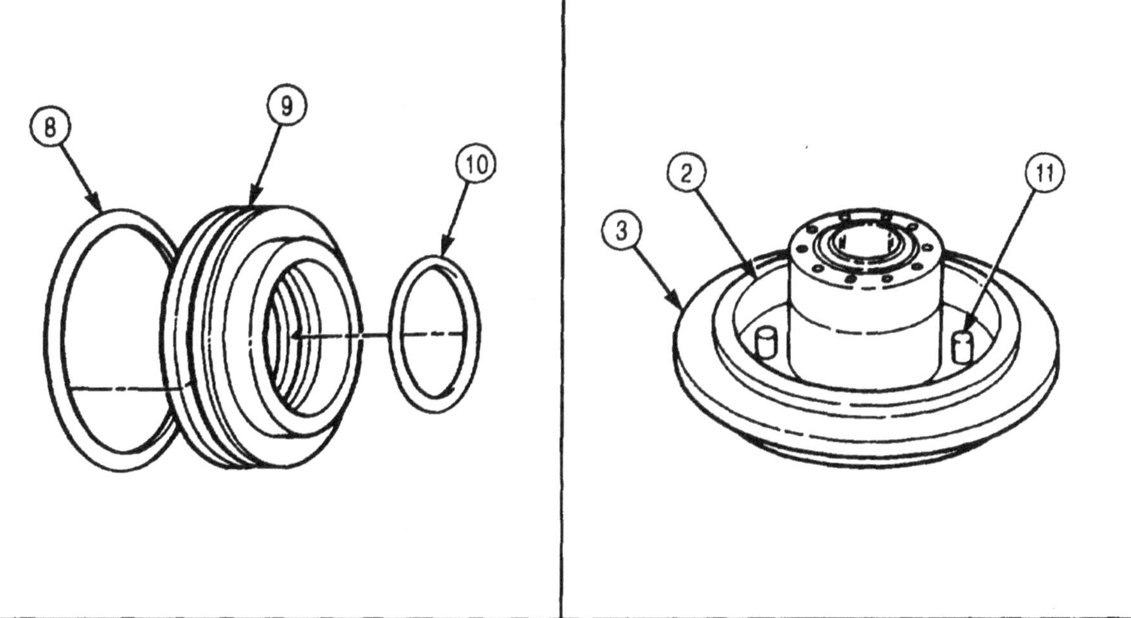

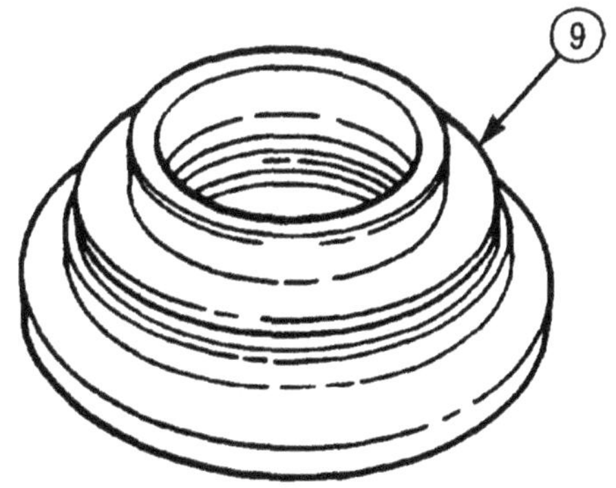

4-61. FAN DRIVE CLUTCH (M939/A1) MAINTENANCE (Contd)

12. Place drive pulley (3) on arbor press, with drive pins (5) facing downward. Use wood blocks on arbor press bed to support drive pulley (3).
13. Using bearing replacer and arbor press, press new bearing (1) into drive pulley (3).
14. Turn drive pulley (3) over so drive pins (5) face up.
15. Using bearing replacer and arbor press, install thrust cap (4).
16. Turn drive pulley (3) over to ensure snapring groove (2) is fully exposed. If groove (2) is not exposed, press bearing (1) until snapring groove (2) is exposed.
17. Using snapring pliers, install new snapring (6) on drive pulley (3).

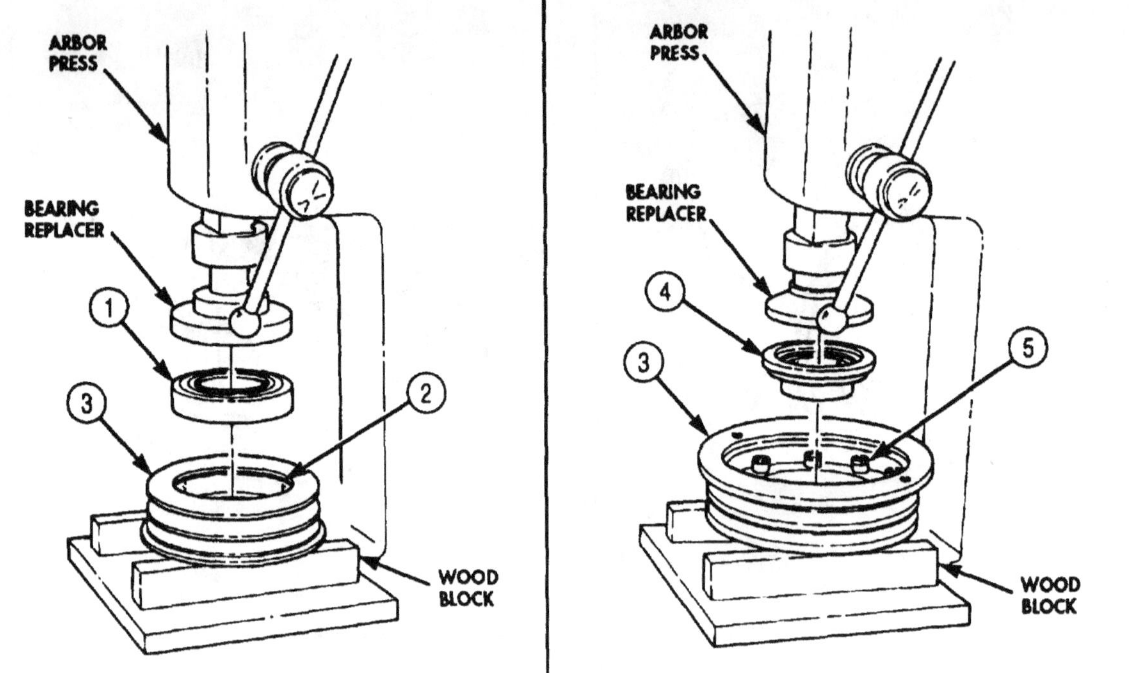

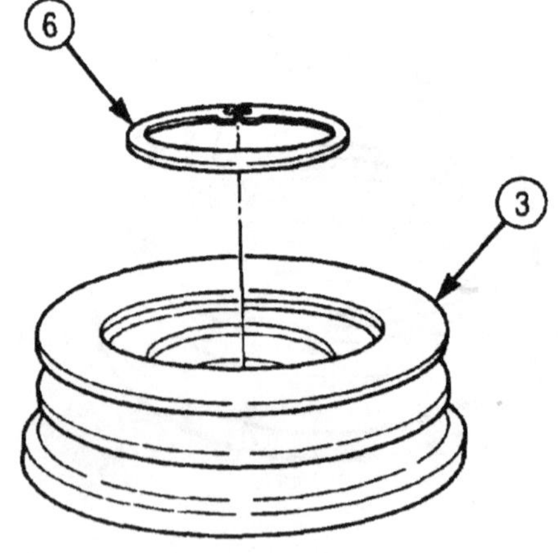

4-61. FAN DRIVE CLUTCH (M939/A1) MAINTENANCE (Contd)

18. Align scribe marks on flanges of drive pulley (3) and clutch housing (12), and install clutch facing (11), return spring (10), clutch housing (12), spacer (8), and thrust cap backup plate (7) on drive pulley (3).
19. Using two fan override lockup screws (13), stored in fan clutch assembly bracket (15), temporarily secure clutch housing (12) and drive pulley (3) together.

NOTE

Assistant will help with step 20.

20. Align screw holes in thrust cap backup plate (7) to holes in thrust cap (9) on clutch housing (12) and, while assistant holds down backup plate (7), install eight screws (14) on clutch housing (12). Tighten screws (14) 70 lb-in. (8 N•m).
21. Remove two fan override screws (13) from clutch housing (12) and drive pulley (3) and install screws (13) in holes (16) on fan clutch assembly bracket (15).

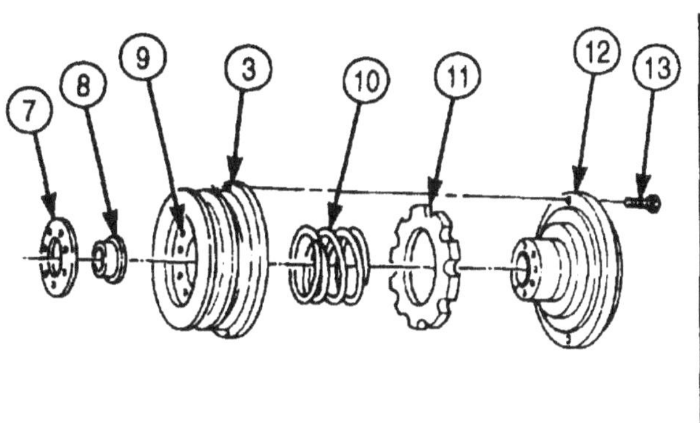

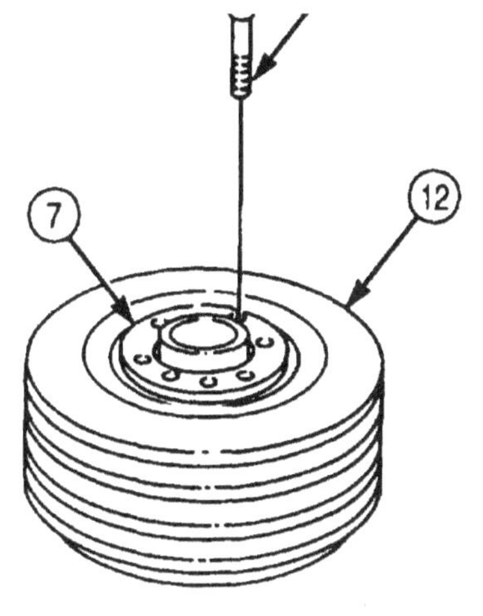

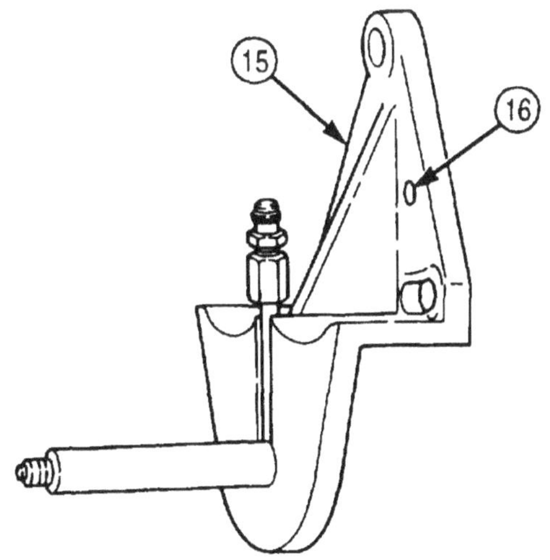

4-61. FAN DRIVE CLUTCH (M939/A1) MAINTENANCE (Contd)

22. Install clutch assembly (7) in soft-jawed vise.
23. Install spacer (8) on shaft (9).
24. Install shaft (9) in clutch assembly (7).
25. Position spacer (6), with cut out facing toward clutch assembly (7), and install in clutch assembly (7) with washer (5) and new expansion locknut (4). Tighten locknut (4) 65 lb-ft (88 N•m).

CAUTION

O-rings must not be twisted after installation.

26. Coat new O-rings (2) and (3) with GAA grease and install O-rings (2) and (3) on seal plug (1).

CAUTION

When installing seal plug, use care not to cut or damage O-rings.

27. Install seal plug (1) in clutch assembly (7). Tighten plug (1) 40 lb-ft (54 N•m).
28. Install fan drive clutch (para. 3-75).

d. Fan Drive Clutch Operation Check

1. Start engine (TM 9-2320-272-10) and run until coolant temperature reaches 175-195°F (79-91°C).
2. Place a piece of cardboard in front of radiator core until engine temperature reaches 200°F (93°C). If fan does not engage when temperature reaches 200°F (93°C), refer to mechanical troubleshooting.

4-61. FAN DRIVE CLUTCH (M939/A1) MAINTENANCE (Contd)

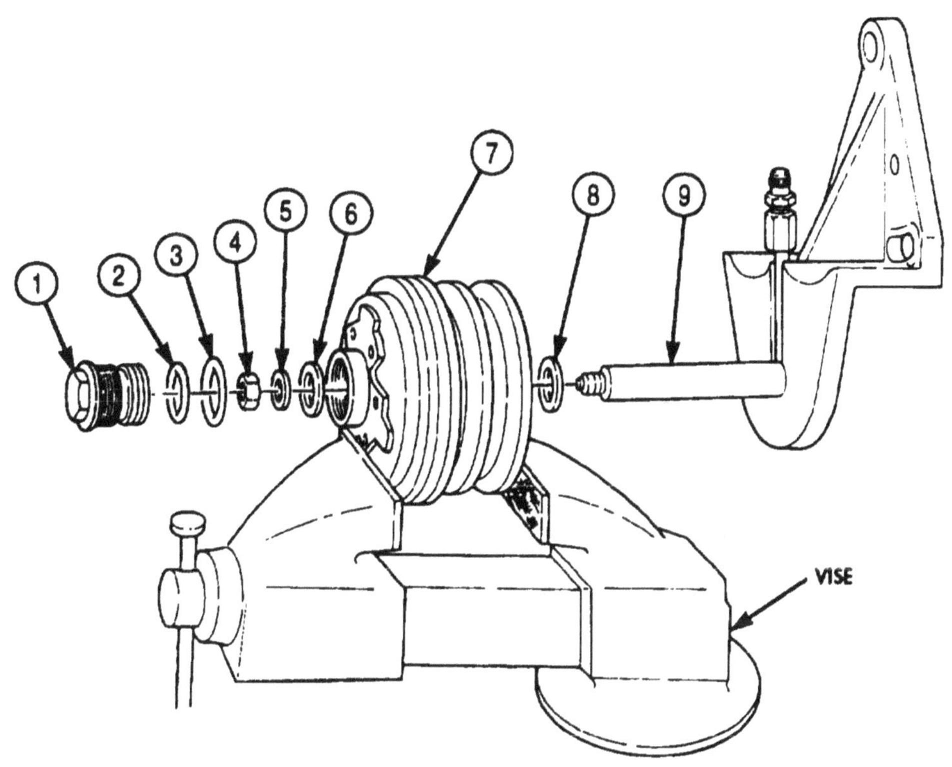

4-62. FAN DRIVE CLUTCH (M939A2) MAINTENANCE

THIS TASK COVERS:

a. Disassembly
b. Cleaning and Inspection
c. Assembly

INITIAL SETUP:

APPLICABLE MODELS
M939A2

TOOLS
General mechanic's tool kit (Appendix E, Item 1)
Micrometer (Appendix E, Item 80)
Snap gauge (Appendix E, Item 123)
Torque wench (Appendix E, Item 144)
Arbor press
Vise

MATERIALS/PARTS
Fan drive repair kit (Appendix D. Item 108)
Adhesive sealant (Appendix C, Item 4)
GAA grease (Appendix C, Item 28)
Rags (Appendix C, Item 58)
Antiseize tape (Appendix C, Item 72)

REFERENCES (TM)
TM 9-2320-272-10
TM 9-2320-272-24P

EQUIPMENT CONDITION
Fan clutch removed (para. 3-74).

GENERAL SAFETY INSTRUCTIONS
- Return spring is under compression and may cause injury if not properly removed.
- When cleaning with compressed air, wear eyeshield.

a. Disassembly

1. Install bracket (4) in soft-jawed vise, with housing (3) facing up.
2. Place prybar between studs (1) to keep housing (3) from turning.
3. Remove cap (2) from housing (3).
4. Remove locknut (5), spring washer (6), spacer (7), housing (3), pulley (8), and spacer (9) from shaft (10). Discard locknut (5), spring washer (6), and spacer (7).

WARNING

Return spring, located between thrust cap and piston, is under compression. Wear eyeshields and follow disassembly instructions carefully, or injury to personnel may result.

5. Install housing (3), facing up, in a soft-jawed vise.
6. Remove six screws (11) from housing (3), leaving two screws (11) installed opposite one another.

NOTE

Assistant will help with steps 7 and 8.

7. While assistant holds down thrust cap (12), alternately loosen and remove two remaining screws (11) from housing (3).
8. Slowly release and remove thrust cap (12), return spring (13), clutch facing (14), and spacer (15) from housing (3). Discard clutch facing (14) and spacer (15).
9. Remove piston (16) from housing (3) as follows:

   a. Install cap (2) on housing (3), and place housing (3) on bracket (4).

   b. Wrap rag around base of shaft (10).

4-62. FAN DRIVE CLUTCH (M939A2) MAINTENANCE (Contd)

WARNING

Eyeshields must be worn when working with compressed air. Failure to wear eyeshields may result in injury to personnel.

c. Firmly hold top of housing (3) and apply 50 psi (345 kPa) of compressed air to pipe (17).

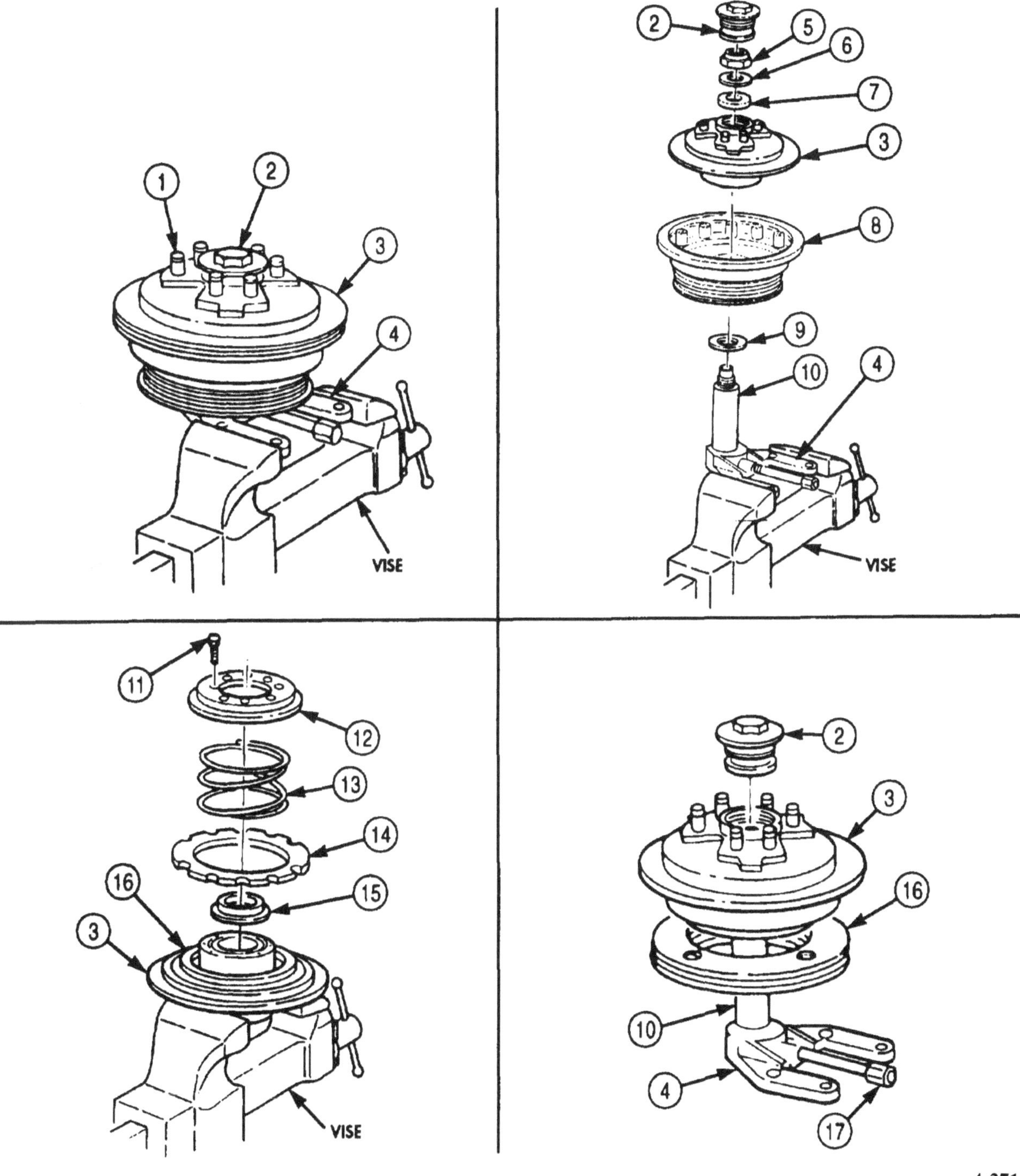

4-62. FAN DRIVE CLUTCH (M939A2) MAINTENANCE (Contd)

10. Remove cap (1), O-rings (2) and (11), housing (3), and piston (4) from shaft (5). Discard cap (1) and O-rings (2) and (11).
11. Remove exterior O-ring (10) and interior O-ring (9) from piston (4). Discard O-rings (9) and (10).

NOTE

Perform step 12 if pipe or fitting is damaged.

12. Remove fitting (7) and pipe (6) from bracket (8).
13. Remove two bearings (13) and spacer (15) from housing (3) as follows:
    a. Position housing (3) on arbor press with fan hub (12) facing up.
    b. Press two bearings (13) and spacer (15) out of housing bore (14). Discard bearings (13) and spacer (15).
14. Remove roller bearing (20) from pulley (17) as follows:
    a. Remove retaining ring (16) from front of pulley (17). Discard retaining ring (16).
    b. Remove snapring (18) from back of pulley (17).
    c. Position pulley (17) on arbor press with front facing up.
    d. Press roller bearing (20) and sleeve (19) through bore of pulley (17). Discard roller bearing (20) and sleeve (19).

b. Cleaning and Inspection

1. For general cleaning instructions, refer to para. 2-14.
2. Inspect housing (3), pulley (17), spacer (22), bracket (8), shaft (5), piston (4), and thrust cap (21) for chips, cracks, and bends. Replace damaged part(s).
3. Inspect housing (3), pulley (17), and shaft (5) for gouged or stripped threads. Repair or replace damaged part(s).

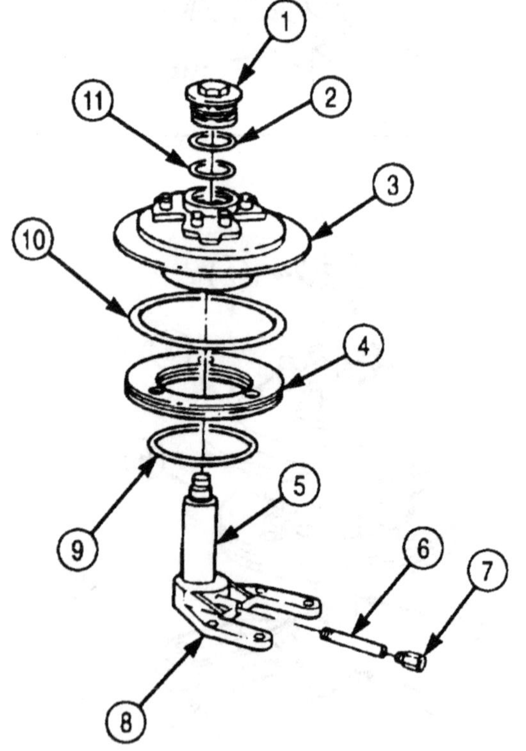

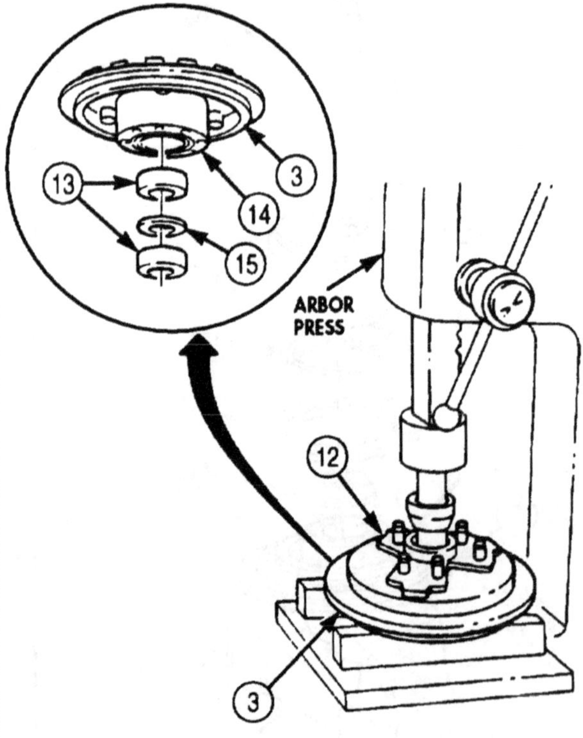

4-62. FAN DRIVE CLUTCH (M939A2) MAINTENANCE (Contd)

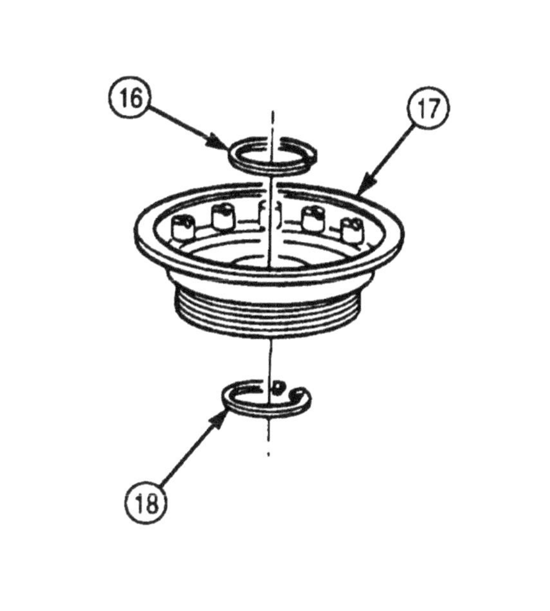

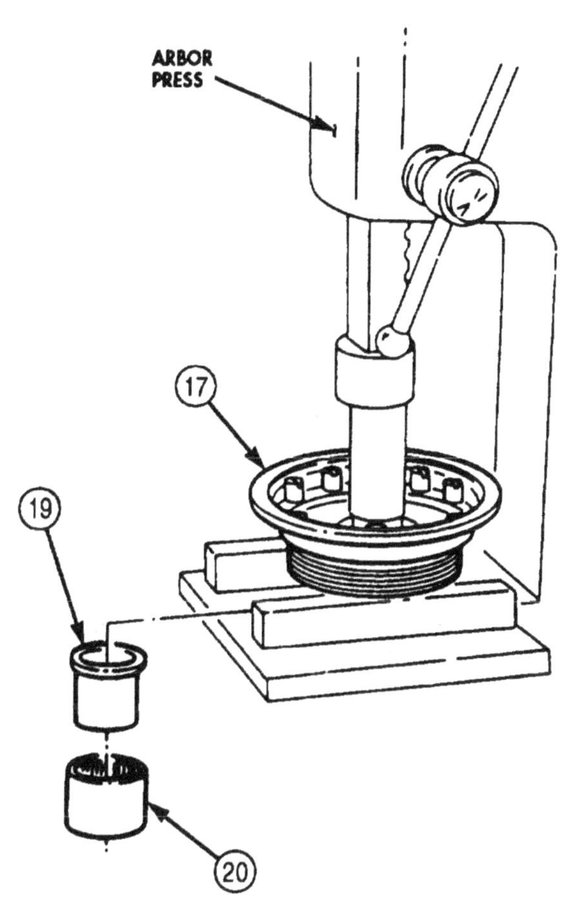

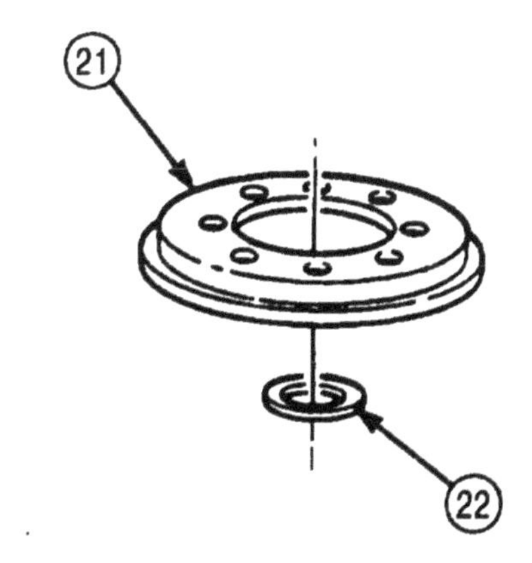

4-62. FAN DRIVE CLUTCH (M939A2) MAINTENANCE (Contd)

c. Assembly

1. Install new sleeve (2) on new roller bearing (1) as follows:
   a. Position sleeve (2) on arbor press, with flanged end of sleeve (2) facing down.
   b. Press roller bearing (1) on sleeve (2) until bearing (1) bottoms out on sleeve (2).
2. Install new roller bearing (1) on pulley (4) as follows:
   a. Install new retaining ring (6) on front side of pulley bore (3).
   b. Position pulley (4) on arbor press with front facing down.

NOTE

Coat outside surface of roller bearing with adhesive sealant before installation.

   c. With sleeve (2) facing pulley (4), press roller bearing (1) into pulley bore (3) until sleeve (2) contacts retaining ring (6).
   d. Install snapring (5) on back side of pulley bore (3).
3. Install two new bearings (7) and new spacer (8) on housing (10) as follows:
   a. Position housing (10) on arbor press with three guide pins (9) facing up.

NOTE

Coat outside surface of bearings with adhesive sealant before installation.

   b. Press one bearing (7) into housing bore (11) until bearing (7) is flush with bottom of housing (10).
   c. Insert spacer (8) in housing bore (11).
   d. Press second bearing (7) into housing bore (11) until second bearing (7) is flush with top of housing (10).

NOTE

Coat new O-rings with GAA grease before installation.

4. Install new interior O-ring (16) on inside of groove of piston (17).
5. Install new exterior O-ring (18) on outside groove of piston (17).

NOTE

Apply GAA grease to inside surface of piston before installation.

6. Align and install piston (17) on three guide pins (9).
7. Position housing (10) on flat surface with fan hub (19) facing down.
8. Align new spacer (15), new clutch facing (20), return spring (14), and thrust cap (13) on piston (17).

WARNING

Thrust cap is spring-loaded. Wear eyeshields and follow assembly instructions carefully, or injury to personnel may result.

NOTE

- Assistant will help with step 9.
- Coat eight thrust screws with adhesive sealant before installation.

9. Install eight screws (12) on thrust cap (13) and housing (10).
10. Holding housing (10), turn clutch facing (20) to ensure facing (20) spins freely between thrust cap (13) and piston (17).

NOTE

If clutch facing does not spin freely, thrust cap must be removed and steps 8 through 10 repeated.

4-62. FAN DRIVE CLUTCH (M939A2) MAINTENANCE (Contd)

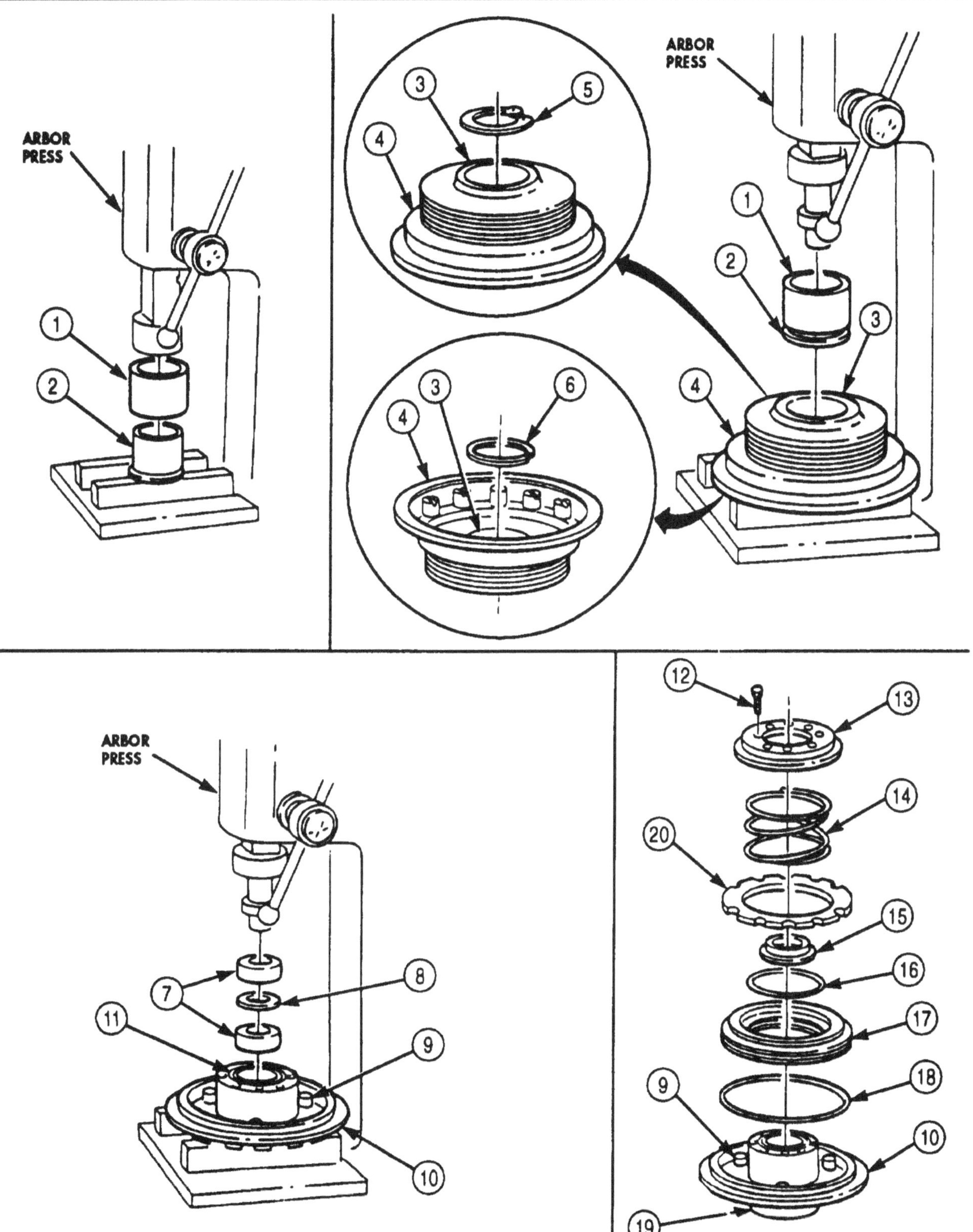

4-62. FAN DRIVE CLUTCH (M939A2) MAINTENANCE (Contd)

11. Install bracket (10) in soft-jawed vise facing up.

NOTE

- Perform step 12 if pipe or fitting was removed.
- Wrap all male pipe threads with antiseize tape before installation.

12. Install pipe (8) and fitting (9) on bracket (10).
13. Install spacer (11) on shaft (7).
14. Install pulley (6) on shaft (7) with drive pins (12) facing up.
15. Align clutch facing (5) with drive pins (12). and install housing (13) on pulley (6).
16. Install new top spacer (4) on shaft (7) with flat side facing up.
17. Install new spring washer (14) on shaft (7) with crown side of spring washer (14) facing up.
18. Install new locknut (3) on shaft (7). Tighten locknut (3) 65 lb-ft (88 N•m).

NOTE

Coat O-rings with GAA grease before installation.

19. Install new O-rings (2) and (15) on top and bottom grooves of new cap (1).
20. Place prybar between studs (16) to prevent housing (13) from turning, and install cap (1) on housing (13). Tighten cap (1) 40 lb-ft (54 N•m).
21. Install fan clutch (para. 3-74).
22. Start engine (TM 9-2320-272-10).
23. Run engine until coolant temperature reaches 175-195°F (79-91°C).
24. Place cardboard in front of radiator core until engine temperature reaches 200°F (93°C).
25. If fan does not engage when engine temperature reaches 200°F (93°C), refer to mechanical troubleshooting.

4-62. FAN DRIVE CLUTCH (M939A2) MAINTENANCE (Contd)

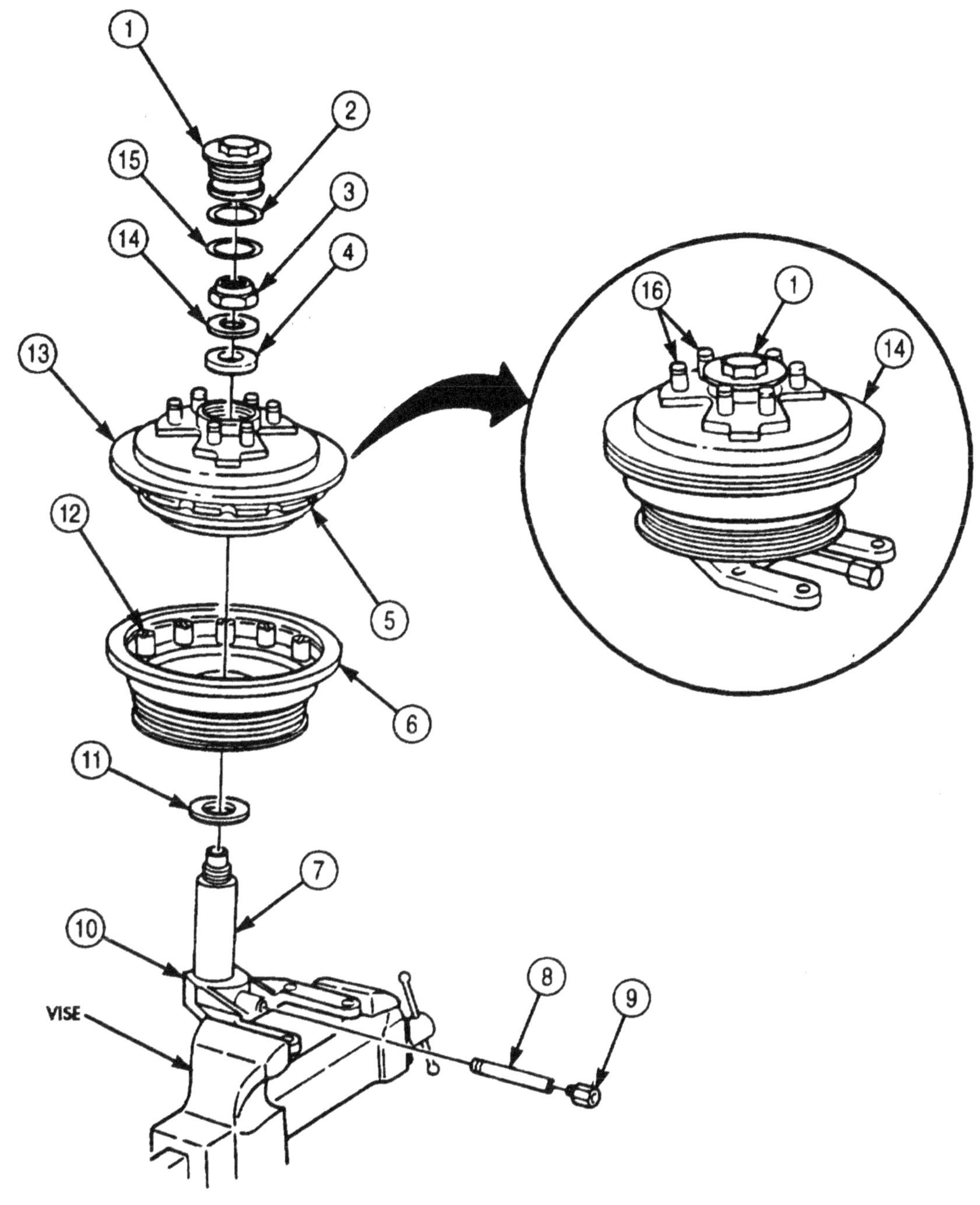

4-63. WATER PUMP (M939/A1) MAINTENANCE

THIS TASK COVERS:

a. Disassembly
b. Cleaning and Inspection
c. Assembly

INITIAL SETUP:

APPLICABLE MODELS
M939/A1

TOOLS
General mechanic's tool kit (Appendix E, Item 1)
Mandrel (Appendix E, Item 116)
Puller kit (Appendix E, Item 101)
Arbor press

MATERIALS/PARTS
Two seals (Appendix D, Item 635)
Snapring (Appendix D, Item 660)
Relief fitting (Appendix D, Item 532)
GM grease (Appendix C, Item 28)
Liquid soap (Appendix C, Item 27)

REFERENCES (TM)
TM 9-214
TM 9-2320-272-24P

EQUIPMENT CONDITION
Water pump removed (para. 3-68).

GENERAL SAFETY INSTRUCTIONS
When cleaning with compressed air, wear eyeshields and ensure source does not exceed 30 psi (207 kPa).

a. Disassembly

1. Using a gear puller, remove drive pulley (2) from shaft (1) on water pump body (3).
2. Using a gear puller, remove water pump impeller (4) from shaft (1) on water pump body (3).
3. Remove seal (5) from shaft (1). Discard seal (5).
4. Remove snapring (7) from ring groove (8) inside bore (6). Discard snapring (7).
5. Using two supporting wood blocks, place water pump body (3) on arbor press with bore (6) facing down.
6. Remove shaft. (1) by pressing on impeller end of shaft (1). Press shaft (1) out of bore (6).
7. Position and clamp a mechanical wedge-type puller between grease grooves (9) on spacer (10) and remove bearing (8) and spacer (10) from shaft (1).

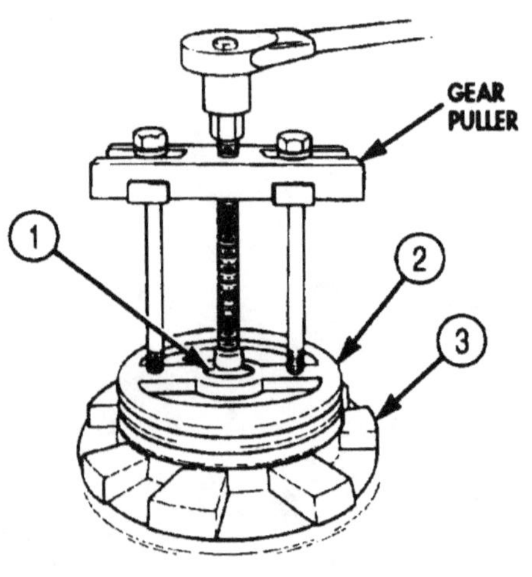

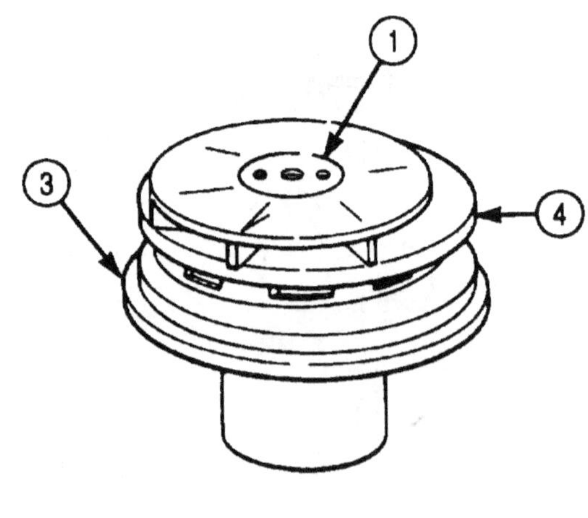

4-63. WATER PUMP (M939/A1) MAINTENANCE (Contd)

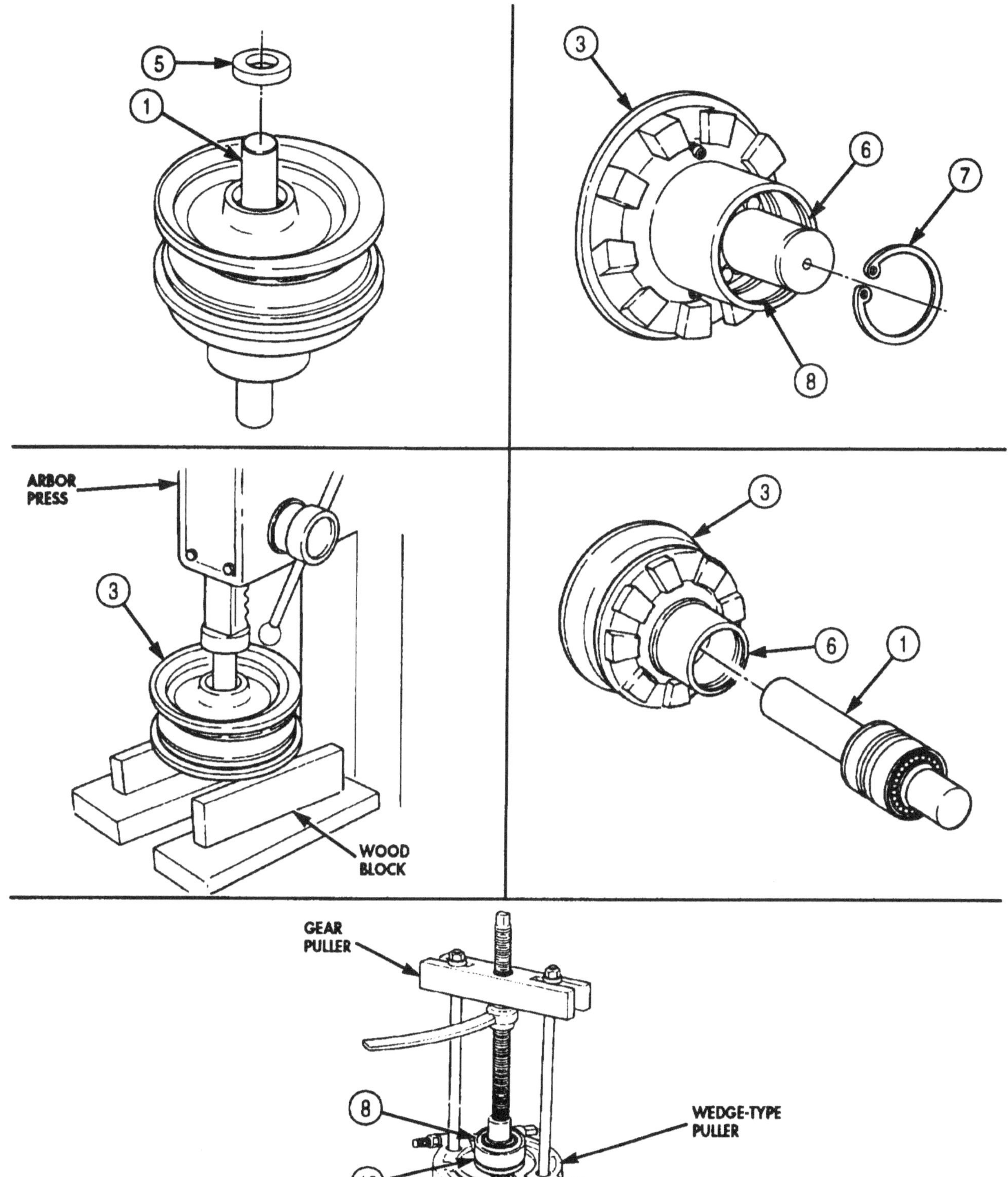

4-63. WATER PUMP (M939/A1) MAINTENANCE (Contd)

8. Remove snapring (1) from snapring groove (2) on shaft (3).
9. Using arbor press, remove shaft (3) by pressing through bearing (4).
10. Press water pump seal (6) out of impeller end of water pump body (5). Discard seal (6).
11. Remove plug (7) from water pump body (5).
12. Remove vent relief fitting (8) from water pump body (5). Discard relief fitting (8).

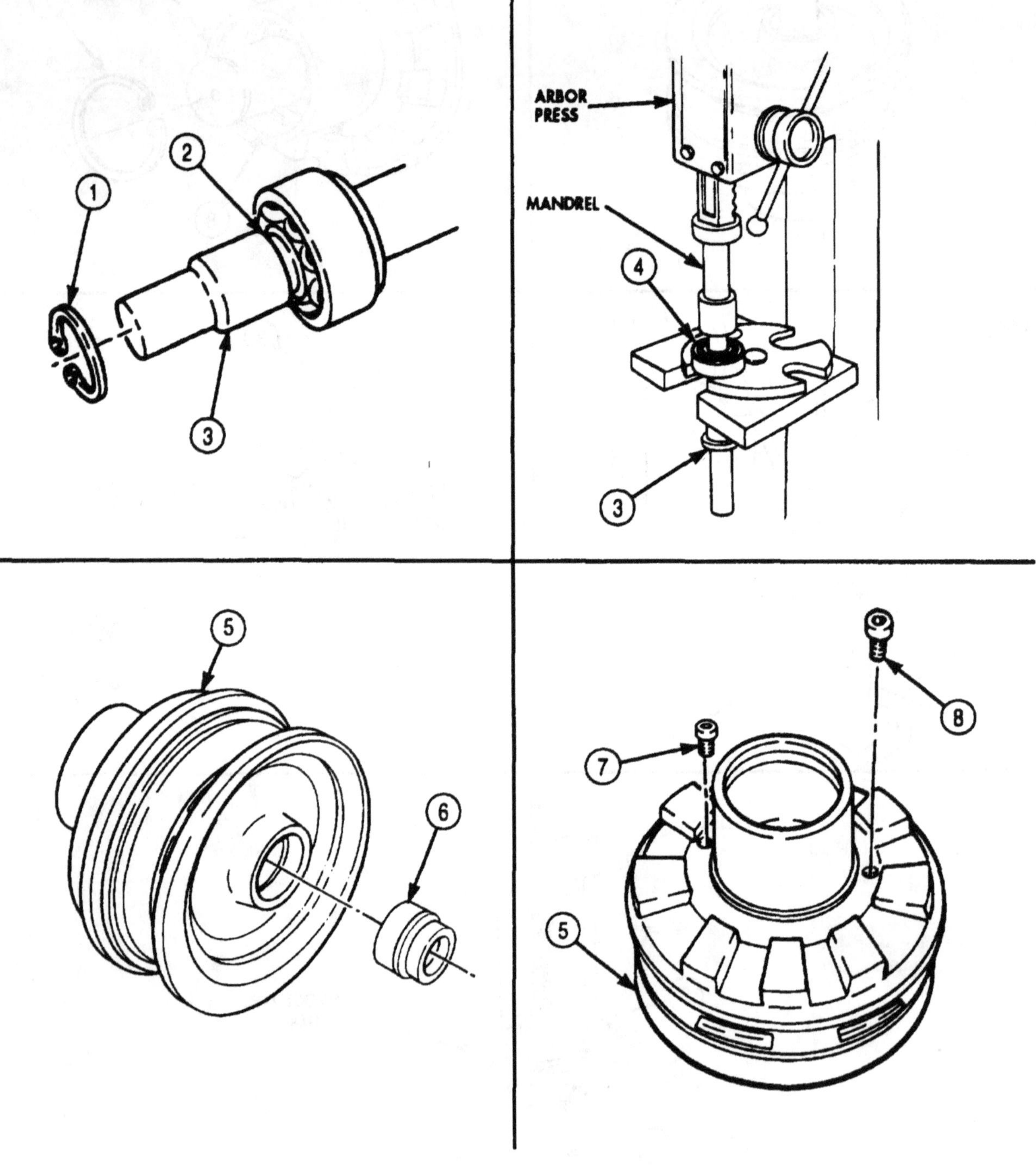

4-63. WATER PUMP (M939/A1) MAINTENANCE (Contd)

b. Cleaning and Inspection

WARNING

Eyeshields must be worn when cleaning with compressed air. Compressed air source will not exceed 30 psi (207 kPa). Failure to wear eyeshields may result in injury to personnel.

CAUTION

Rotate bearings very slowly while cleaning. Do not spin bearing races with compressed air when drying bearings. Serious damage to bearings could result.

NOTE

Service bearings in clean surroundings.

1. Clean bearings (4) and (10), water pump body (5), shaft (3), and water pump impeller (12) (para. 2-14 and TM 9-214).
2. Inspect bearings (4) and (10) (para. 2-15 and TM 9-214).
3. Inspect spacer (9) for cracks and galls. Replace spacer (9) if cracked or galled.
4. Inspect water pump impeller (12) for cracks and heavy corrosion. Replace impeller (12) if cracked. Remove corrosion with wire brush. Refer to table 4-8 for critical measurement of bore.
5. Inspect shaft (3) for cracks, scoring, and galling. Replace shaft (3) if cracked or galled. Using micrometer, check outside diameter (O.D.). Refer to table 4-8 for shaft (3) wear limits.
6. Inspect water pump body (5) for cracks, pits, and heavy corrosion. Replace water pump body (5) if cracked, pitted, or heavily corroded. Using a snap gauge, check inside diameter (I.D.) of bore (11) for wear. Refer to table 4-8 for bore (11) replacement wear limits.

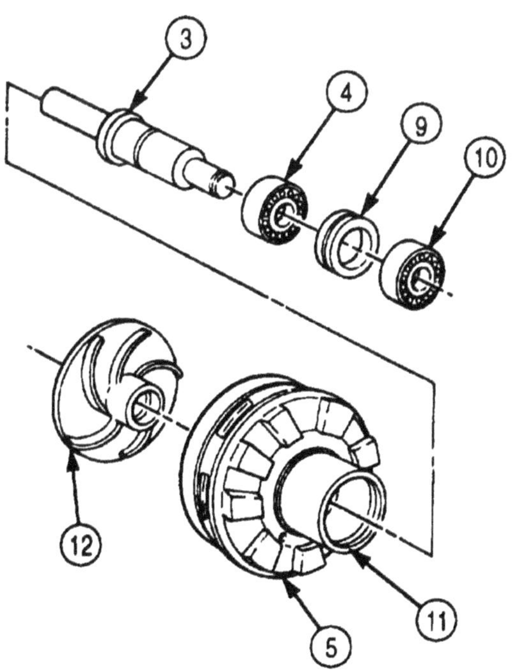

4-63. WATER PUMP (M939/A1) MAINTENANCE (Contd)

NOTE

- This table provides the minimum, maximum, and worn serviceability standards for the water pump assembly. The wear limits indicate the point to which a part or parts may be worn before replacement is required.
- An asterisk (*) in the wear limits column indicates the part or parts must be replaced or repaired when worn beyond the dimensions specified in the "size and fit of new parts" column.

Table 4-8. Repair and Replacement Standards for Water Pump.

REF. NO.	POINT OF MEASUREMENT	SIZE AND FIT OF NEW PARTS	WEAR LIMITS
WATER PUMP BODY			
1	Housing bearing bore	2.4408-2.4414 in. (61.996-62.012 mm)	2.4494 in. (62.215 mm)
2	Housing bore (impeller end) water pump body (I.D.)	1.5000-1.5200 in. (38.100-38.608 mm)	*
	Minimum press fit between shaft and impeller	0.001 in. (0.03 mm)	
3	Drive pulley bore inside diameter (I.D.)	0.663-0.6673 in. (16.84-16.949 mm)	*
	Minimum press fit between shaft and pulley	0.001 in. (0.03 mm)	
SHAFT			
4	Shaft diameter (O.D.) Impeller end	0.6262-0.6267 in. (15.905-15.918 mm)	*
5	Shaft diameter (O.D.) Seal seat location	0.6262-0.6267 in. (15.905-15.918 mm)	*
6	Shaft diameter (O.D.) Inner bearing	0.9843-0.9847 in. (25.001-25.011 mm)	*
7	Shaft diameter (O.D.) Outer bearing	0.9843-0.9847 in. (25.001-25.011 mm)	*
8	Shaft diameter (O.D.) Pulley end	0.6693-0.6696 in. (17.002-17.007 mm)	*
IMPELLER			
9	Impeller bore inside diameter (I.D.)	0.624-0.625 in. (15.85-15.88 mm)	*
10	Impeller vane to pump body clearance (cast iron)	0.020-0.040 in. (0.51-1.02 mm)	*

4-63. WATER PUMP (M939/A1) MAINTENANCE (Contd)

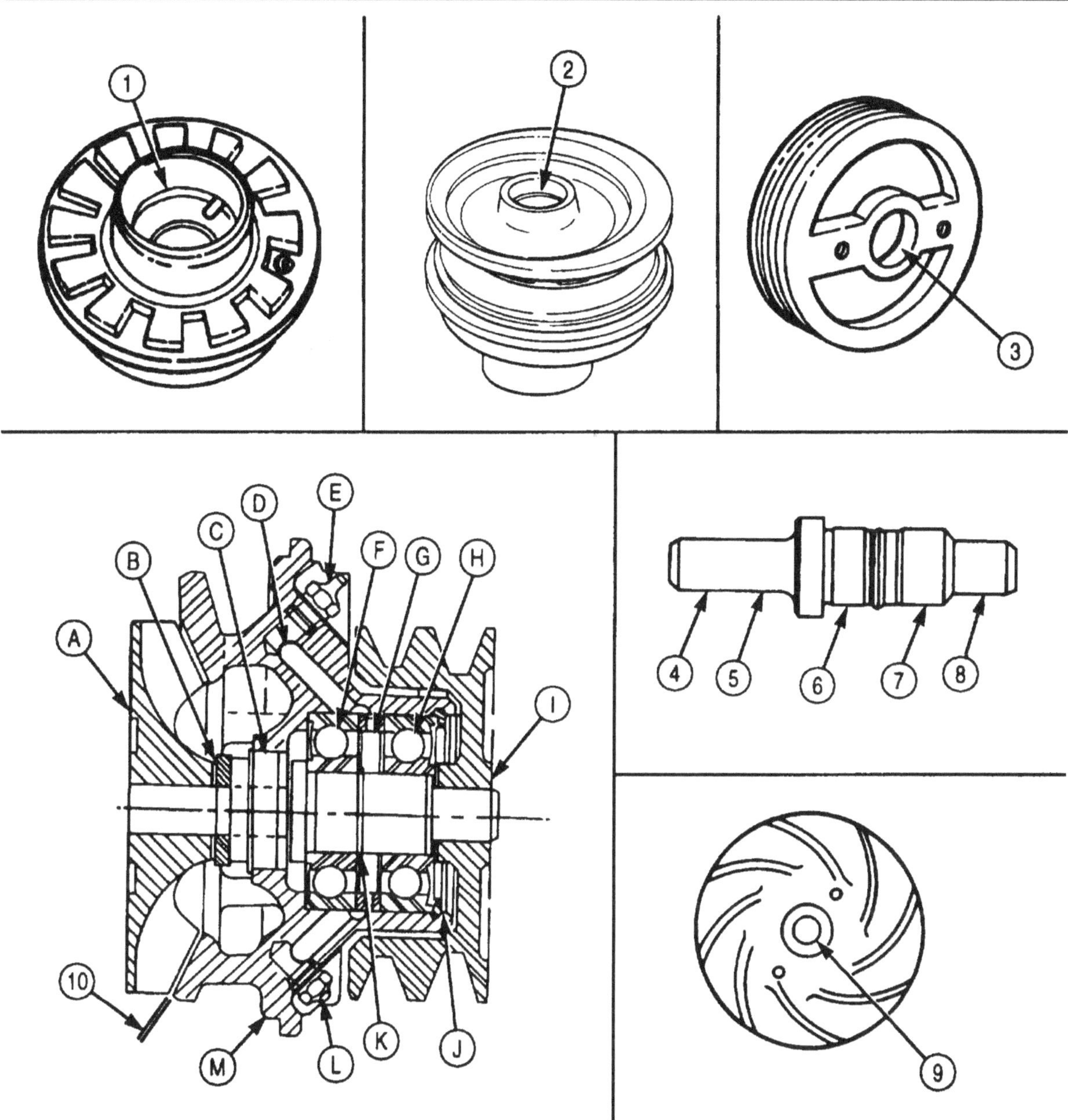

A	IMPELLER	H	OUTER BEARING
B	SEAT - WATER PUMP SEAL	I	DRIVE PULLEY
C	SEAL - WATER PUMP	J	SNAPRING
D	LUBRICANT PASSAGE	K	SNAPRING
E	GREASE FITTING	L	RELIEF FITTING
F	INNER BEARING	M	WATER PUMP BODY
G	SPACER		

4-63. WATER PUMP (M939/A1) MAINTENANCE (Contd)

c. Assembly

NOTE

New replacement bearings are installed as they are removed from packages, without cleaning or repacking. Original bearings, approved by inspection for reuse, must be packed after cleaning and inspection.

1. Pack bearing (3) with GM grease and coat shaft (1) with light coat of grease (TM 9-214).
2. Using arbor press and mandrel, press shaft (1) through bearing (3) until shoulder (2) of shaft (1) bottoms on bearing (3).
3. Install snapring (4) in snapring groove (5) on shaft (1).
4. Install spacer (7) on shaft (1) and seat against bearing (3).
5. Using arbor press and mandrel, press bearing (6) over shaft (1) until seated against bearing (3).

NOTE

Carbon face of seal must be kept free of grease.

6. Coat new seal (9) with liquid soap to ease installation and, using mandrel, install seal (9) at impeller end in water pump body (8).
7. Position water pump body (8) on arbor press with bore (10) facing up.
8. Align shaft assembly (1) straight with bore (10) and, with impeller end down, press shaft (1) into water pump body (8).
9. Install new snapring (11) into snapring groove (12) in bore (10).

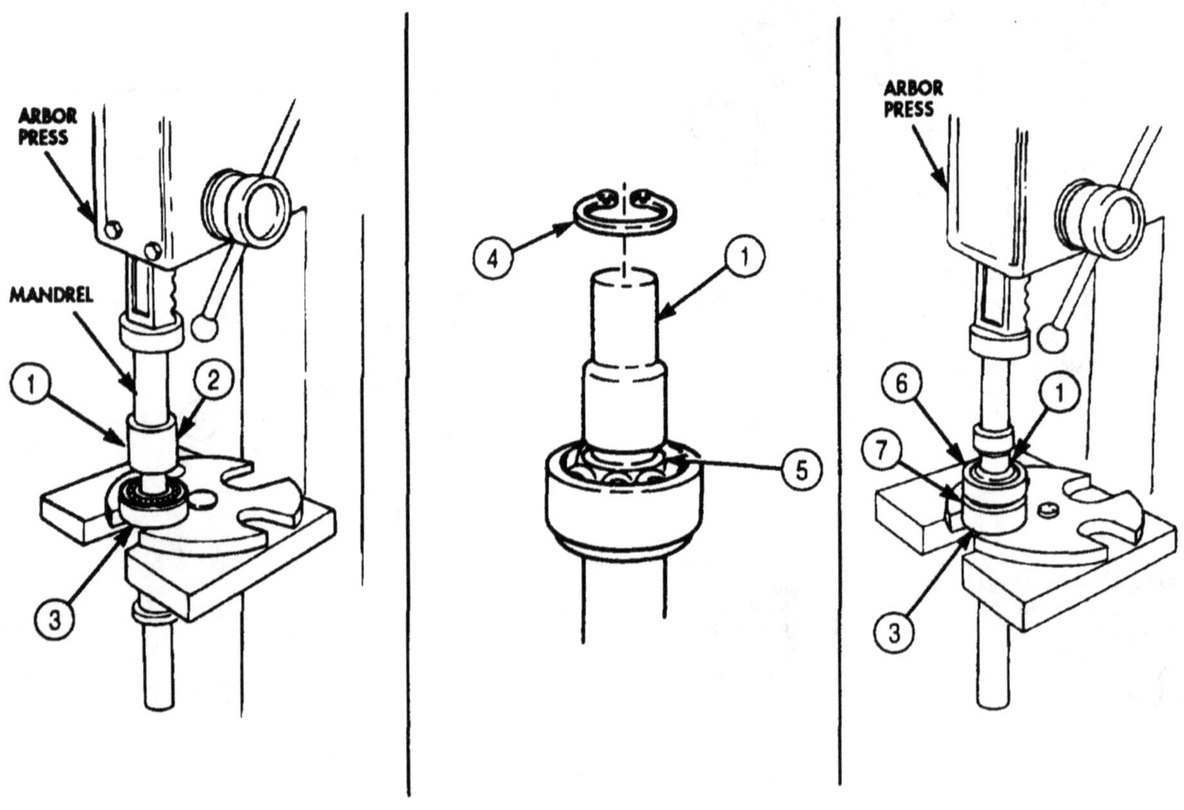

4-63. WATER PUMP (M939/A1) MAINTENANCE (Contd)

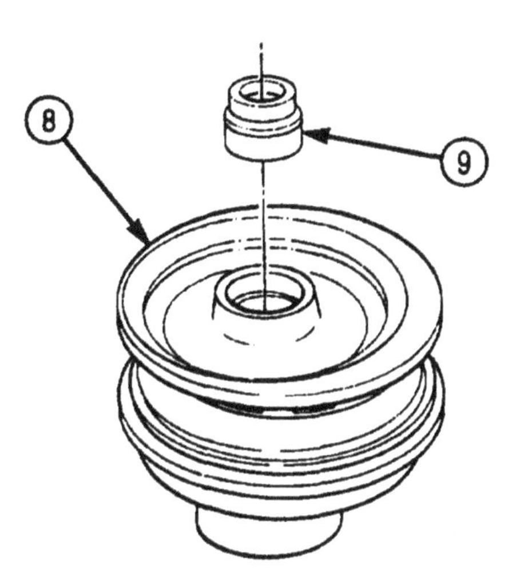

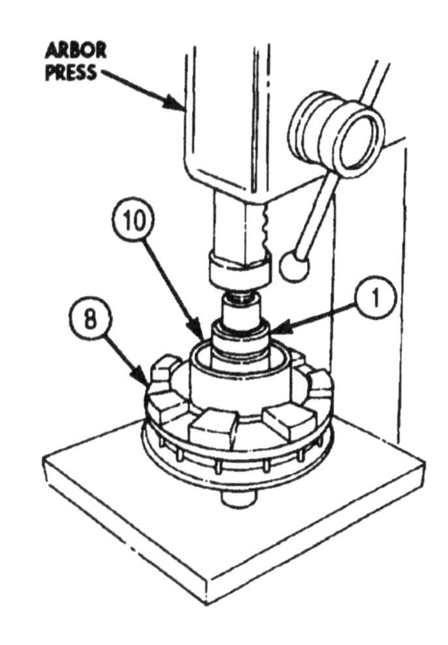

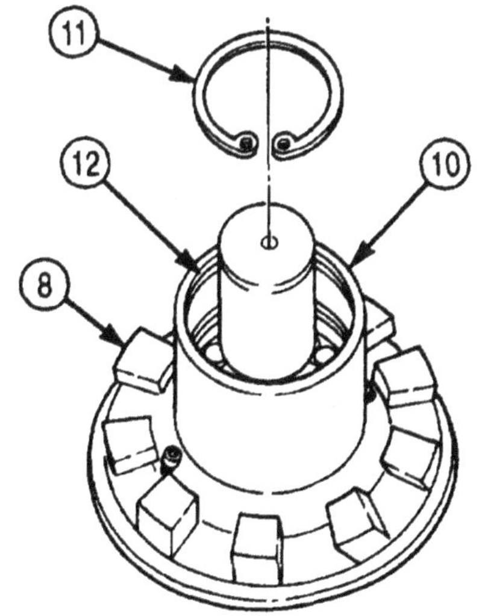

4-63. WATER PUMP (M939/A1) MAINTENANCE (Contd)

10. Install new relief fitting (4) in hole of water pump body (5).

CAUTION

Water pumps are lubricated only after rebuild. The lubrication passages are normally kept plugged to prevent overgreasing which can cause seal blowout and damage to water pump.

NOTE

Steps 11 and 12 must be performed to prevent water pump damage.

11. Install grease fitting (2) in passage (1). Grease fitting (2) is installed temporarily to fill water pump cavity with lubricant.

NOTE

Use care when lubricating water pump. Do not allow grease to come into contact with shaft.

12. Protecting shaft (3) from grease, fill water pump body (5) with 0.60-0.70 cu in. or 0.31-0.37 oz of GAA grease. Remove and discard grease fitting (2).

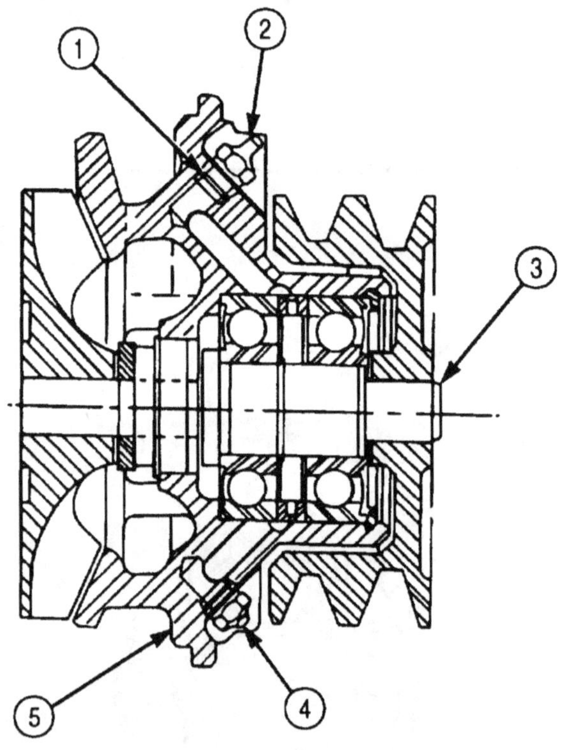

4-63. WATER PUMP (M939/A1) MAINTENANCE (Contd)

13. Install plug (6) in passage (1).
14. Place water pump body (5) on arbor press directly on impeller end of shaft (3).
15. Position drive pulley (7) on shaft (3), and press pulley (7) until seated on shaft (3).
16. Place water pump body (5) on arbor press with drive pulley (7) face down.
17. Apply coating of liquid soap on new water pump seal (8) and position seal (8) over shaft (3) with stainless steel surface facing up. Install seal (8) over shaft (3).
18. Using mandrel, press water pump impeller (9) on shaft (3). Refer to table 4-8 for impeller (9) vane clearance to pump body (5).

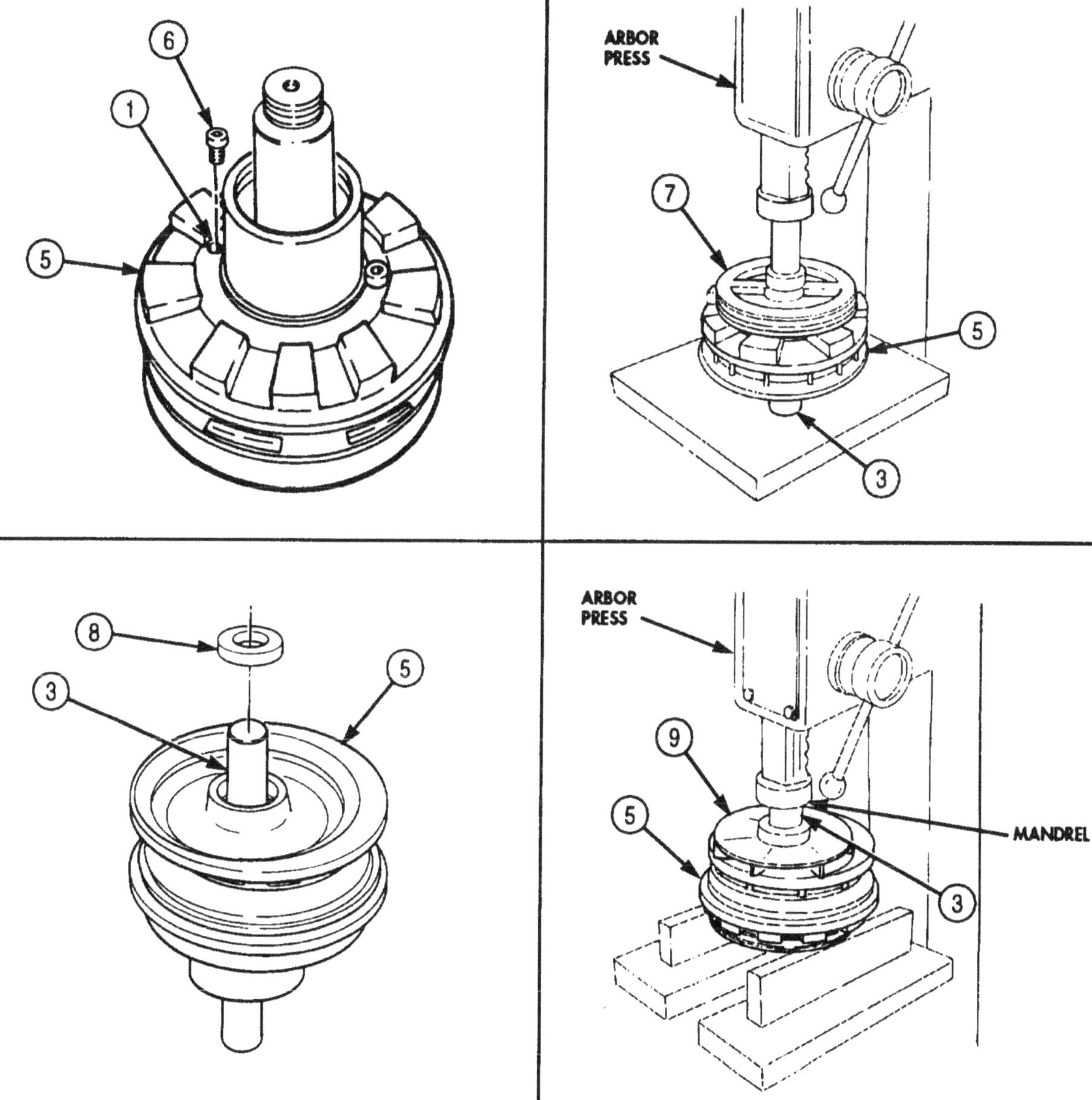

FOLLOW-ON TASK: Install water pump (para. 3-68)

Section VI. ELECTRICAL SYSTEMS MAINTENANCE

4-64. ELECTRICAL SYSTEMS MAINTENANCE INDEX

4-65. FRONT LIGHTS CABLE ASSEMBLY (M939/A1) REPLACEMENT

THIS TASK COVERS:

a. Removal **b. Installation**

INITIAL SETUP:

APPLICABLE MODELS
M939/A1

TOOLS
General mechanic's tool kit (Appendix E, Item 1)

MATERIALS/PARTS
Four lockwashers (Appendix D, Item 369)
Eight locknuts (Appendix D, Item 299)
Lockwasher (Appendix D, Item 379)
Ten locknuts (Appendix D, Item 291)

REFERENCES (TM)
TM 9-2320-272-10
TM 9-2320-272-24P

EQUIPMENT CONDITION
- Hood raised and secured (TM 9-2320-272-10).
- Parking brake set (TM 9-2320-272-10).
- Battery ground cables disconnected (para. 3-126).
- Splash shields removed (TM 9-2320-272-10).

a. Removal

NOTE

Tag wires and connectors for installation.

1. Disconnect front light cable connector (24) from wiring harness connector (21).
2. Remove four nuts (26), lockwashers (25), screws (22), and wiring harness connector (21) from left fender (23). Discard lockwashers (25).
3. Remove four locknuts (19) and screws (14) from right fender (13) and wiring cover (11). Discard locknuts (19).
4. Remove grommet (18) from wiring cover (11).
5. Disconnect wires (12), (15), (16), and (17) from cable assembly (20).
6. Remove wiring cover (11) from cable assembly (20).
7. Remove four locknuts (5) and screws (1) from left fender (2) and wiring cover (4). Discard locknuts (5).
8. Remove grommet (7) from wiring cover (4).
9. Disconnect wires (8), (9), (10), and (3) from cable assembly (6).
10. Remove wiring cover (4) from cable assembly (6).

4-65. FRONT LIGHTS CABLE ASSEMBLY (M939/A1) REPLACEMENT (Contd)

4-65. FRONT LIGHTS CABLE ASSEMBLY (M939/A1) REPLACEMENT (Contd)

NOTE

Left and right headlamps are removed the same. Steps 11 through 13 cover the right headlamp assembly.

11. Disconnect connectors (11) and (12) from right headlamp (13).
12. Disconnect right headlamp ground connector (14) from headlamp (13).
13. Remove screw (7), washer (8), ground wire (9), and lockwasher (10) from panel (15). Discard lockwasher (10).
14. Disconnect connector (16) from blackout lamp (17).

NOTE

M936/A1/A2 vehicles have only eight screws and clamps to be removed in step 15.

15. Remove ten locknuts (5), screws (6), and clamps (4) from vehicle (2) and cable assembly (1). Discard locknuts (5).

b. Installation

CAUTION

Do not install wiring harness to hood retaining rod mounting bracket screw. Movement will cause clamp to cut harness.

NOTE

M936 vehicles use only eight screws and clamps for step 1.

1. Route cable (1) along perimeter of grille (3), and install ten clamps (4) on vehicle (2) with screws (6) and locknuts (5).
2. Connect connector (16) to blackout lamp (17).

NOTE

Left and right headlamps are installed the same. Steps 3 and 4 cover the right headlamps.

3. Connect connectors (11) and (12) on right headlamp (13).
4. Connect connector (14) on right headlamp (13) and secure eye end of ground wire (9) to panel (15) with new lockwasher (10), washer (8), and screw (7).

4-65. FRONT LIGHTS CABLE ASSEMBLY (M939/A1) REPLACEMENT (Contd)

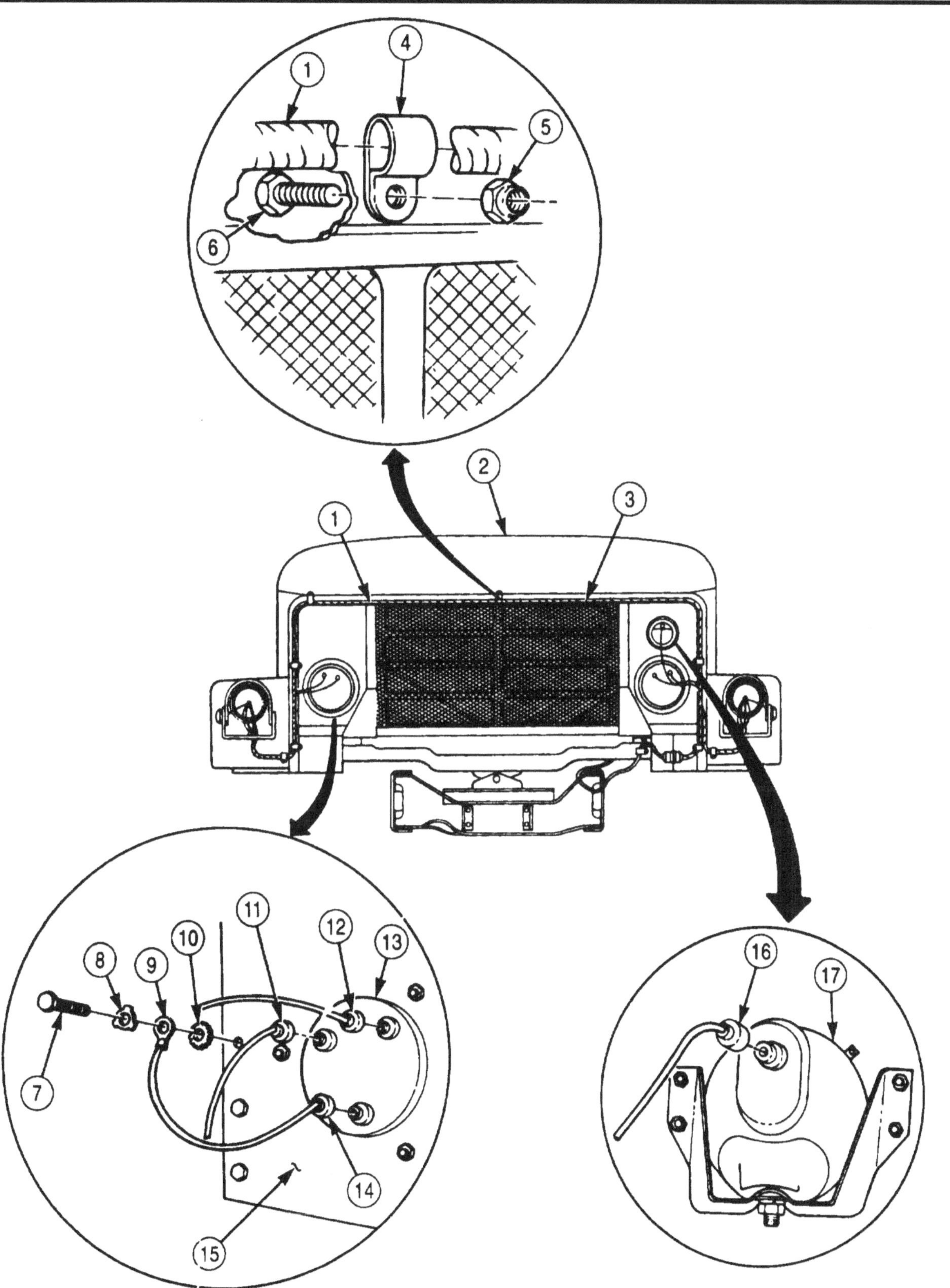

4-65. FRONT LIGHTS CABLE ASSEMBLY (M939/A1) REPLACEMENT (Cod)

5. Insert ends of wires (13), (18), (19), and (20) through hole in wiring cover (14) and connect to mating connectors of cable (15).
6. Install grommet (17) in wiring cover (14).
7. Install wiring cover (14) on fender (12) with four screws (11) and new locknuts (16).
8. Insert ends of wires (7), (8), (9), and (10) through hole in wiring cover (6) and connect to mating connectors of cable (5).
9. Install grommet (3) in wiring cover (6).
10. Install wiring cover (6) on fender (1) with four screws (2) and new locknuts (4).
11. Install front wiring harness connector (23) on fender (26) with four screws (24), new lockwashers (22), and nuts (21).
12. Connect front lights cable connector (25) to front wiring harness connector (23).

4-65. FRONT LIGHTS CABLE ASSEMBLY (M939/A1) REPLACEMENT (Contd)

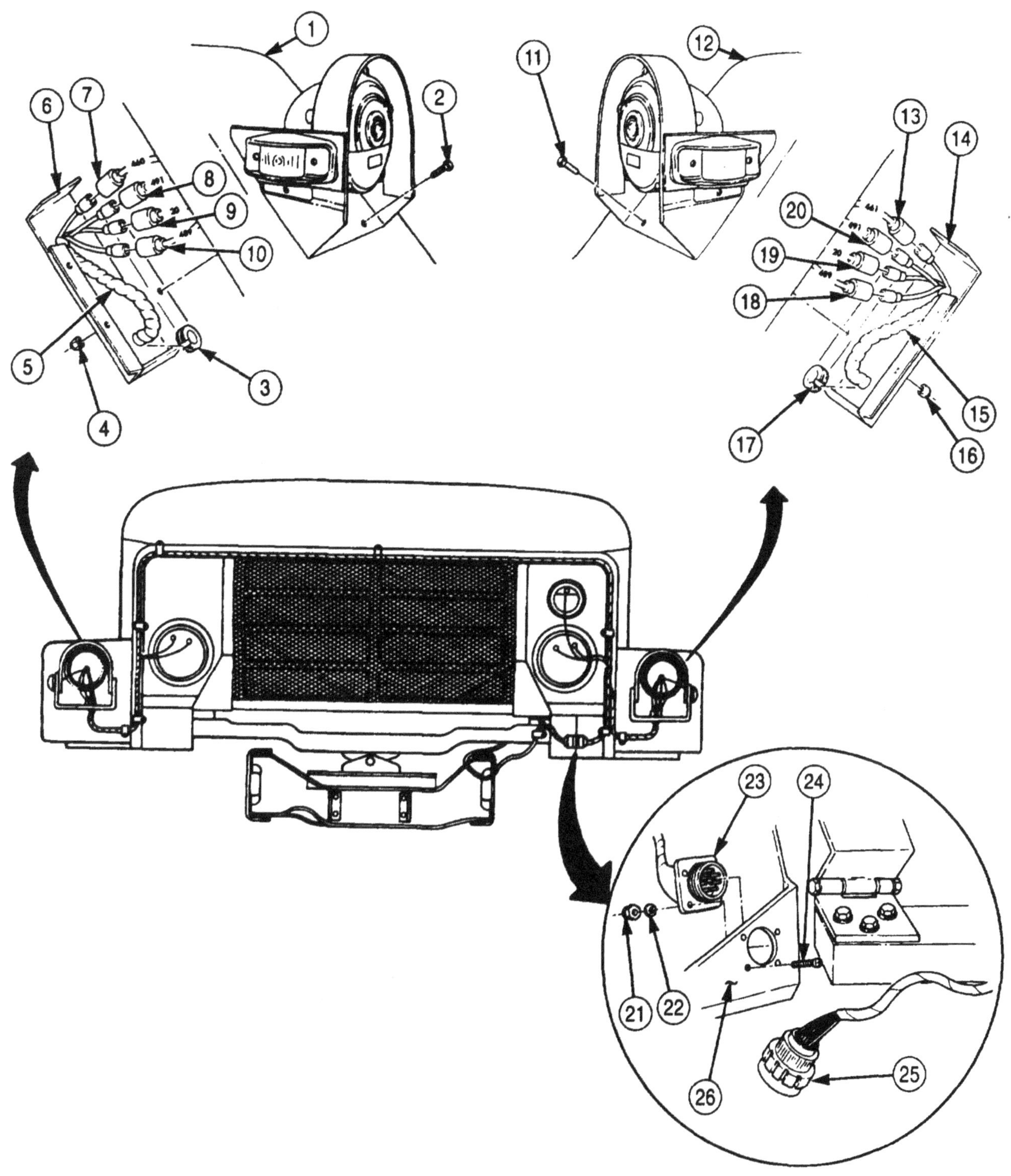

FOLLOW-ON TASKS:
- Connect battery ground cables (para. 3-126).
- Check operation of all front lights (TM 9-2320-272-10).
- Install splash shields (TM 9-2320-272-10).

4-46. FRONT WIRING HARNESS (M939/A1) REPLACEMENT

THIS TASK COVERS:

a. Removal **b. Installation**

INITIAL SETUP:

APPLICABLE MODELS
M939/A1

TOOLS
General mechanic's tool kit (Appendix E, Item 1)

MATERIALS/PARTS
Cotter pin (Appendix D, Item 53)
Four tiedown straps (Appendix D, Item 694)
Gasket sealant, (Appendix C, Item 30)
Three tiedowns straps (Appendix D, Item 684)
Lockwasher (Appendix D, Item 364)
Two lockwashers (Appendix D, Item 416)
Two lockwashers (Appendix D, Item 379)
Three lockwashers (Appendix D, Item 376)
Eight lockwashers (Appendix D, Item 371)
Four lockwashers (Appendix D, Item 345)
Four locknuts (Appendix D, Item 291)
Spring nut (Appendix D, Item 673)
Lockwasher (Appendix D, Item 403)
Two screw-assembled lockwashers (Appendix D, Item 587)

PERSONNEL REQUIRED
TWO

REFERENCES
TM 9-2320-272-10
TM 9-2320-272-24P

EQUIPMENT CONDITION
- Parking brake sot (TM 9-2320-272-10).
- Splash shields removed (TM 9-2320-272-10).
- Air intake pipe removed (para. 3-14).
- Battery ground cables disconnected (para. 3-126).
- Failsafe warning module removed (para. 3-106).
- Main light switch removed (para. 3-108).
- Protective control box removed (para. 3-115).
- Turn signal flasher removed (para. 3-114).

a. Removal

NOTE

Tag wires, connectors, and cables for installation.

1. Disconnect front wiring harness connector (8) from front lights cable assembly receptacle (1).
2. Remove nut (5), washer (3), screw (2), washer (3), cable clamp (4), and wiring harness (7) from radiator (6).
3. Disconnect tachometer pulse sender connector (19) from fuel pump connector (18).
4. Disconnect fuel pressure transducer connector (17) on fuel pump (16) from front wiring harness connector (20).
5. Remove screw (11), washer (12), and cable clamp (10) from engine intake manifold (9).
6. Remove nut (14) and two wires (13) from fuel pump shutoff solenoid (15).
7. Disconnect wire (23) from engine temperature sending unit (22) and remove wiring harness from behind air compressor (21) and intake manifold (9).
8. Disconnect wires (25) and (26) from ether start fuel pump pressure switch (24).

4-66. FRONT WIRING HARNESS (M939/A1) REPLACEMENT (Contd)

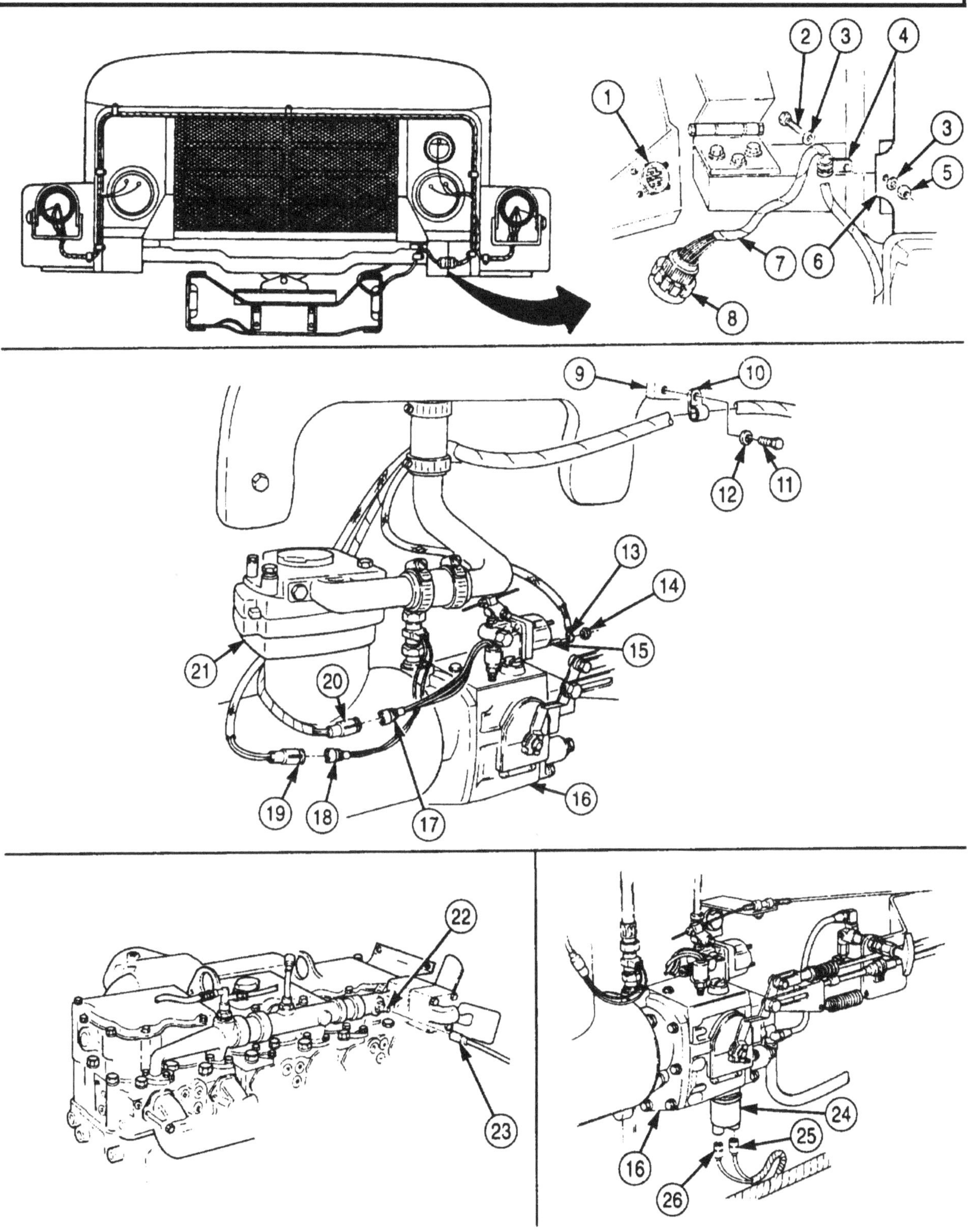

4-66. FRONT WIRING HARNESS (M939/A1) REPLACEMENT (Contd)

9. Disconnect wire (2) from oil pressure sending unit (1).
10. Remove screw-assembled washer (3), washer (4), and wire (5) from starter solenoid (6).

NOTE

M936 vehicles may not have lockwasher.

11. Remove nut (15), lockwasher (16), wire (17), and wire (14) from solenoid post (7). Discard lockwasher (16).
12. Place wire (17) on post (7) and secure with nut (15).
13. Remove nut (12), lockwasher (13), wire (11), and wire (10) from starter motor terminal (8). Discard lockwasher (13).
14. Place wire (11) and grounding sleeve (9) on post (8) and secure with nut (12).
15. Remove screw-assembled washer (20), wire (21), ground strap (22), and lockwasher (23) from intake manifold (24). Discard lockwasher (23).
16. Disconnect connectors (18) and (19).

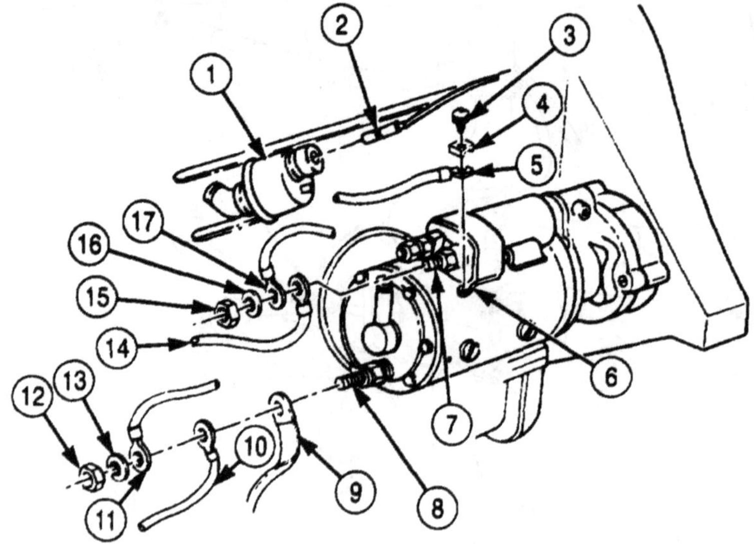

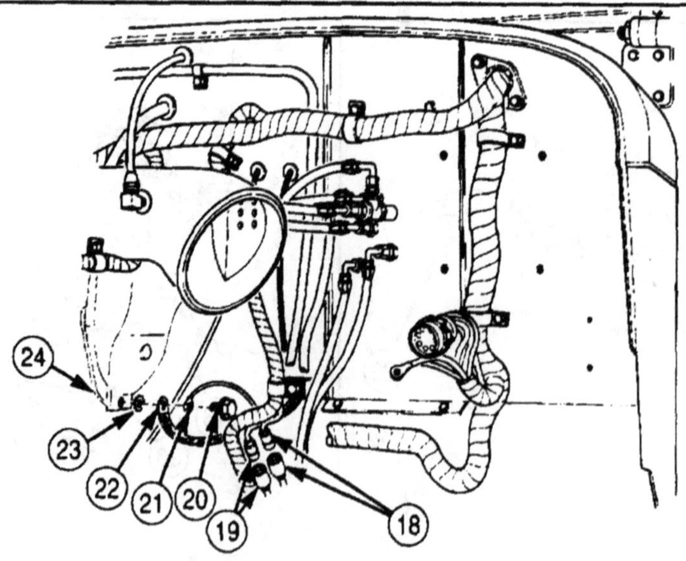

4-66. FRONT WIRING HARNESS (M939/A1) REPLACEMENT (contd)

17. Disconnect wires (25) and (32) from horn circuit breaker (26).
18. Disconnect wires (27) and (30) from transmission control and spring brake circuit breaker (28).
19. Remove and discard two tiedown straps (29) from wiring harness (31).
20. Disconnect wire (34) from engine temperature switch wire (33).
21. Disconnect wire (36) from personnel hot water heater (36).
22. Disconnect wires (43) and (44) from horn solenoid (46).
23. Disconnect wire (41) from transorb diode coupling assembly wire (42).
24. Remove four tiedown straps (40) from front wiring harness cable (31). Discard tiedown straps (40).
25. Remove three screws (39) and cable clamps (38) from firewall (37). Tag cable clamps (38) for installation.

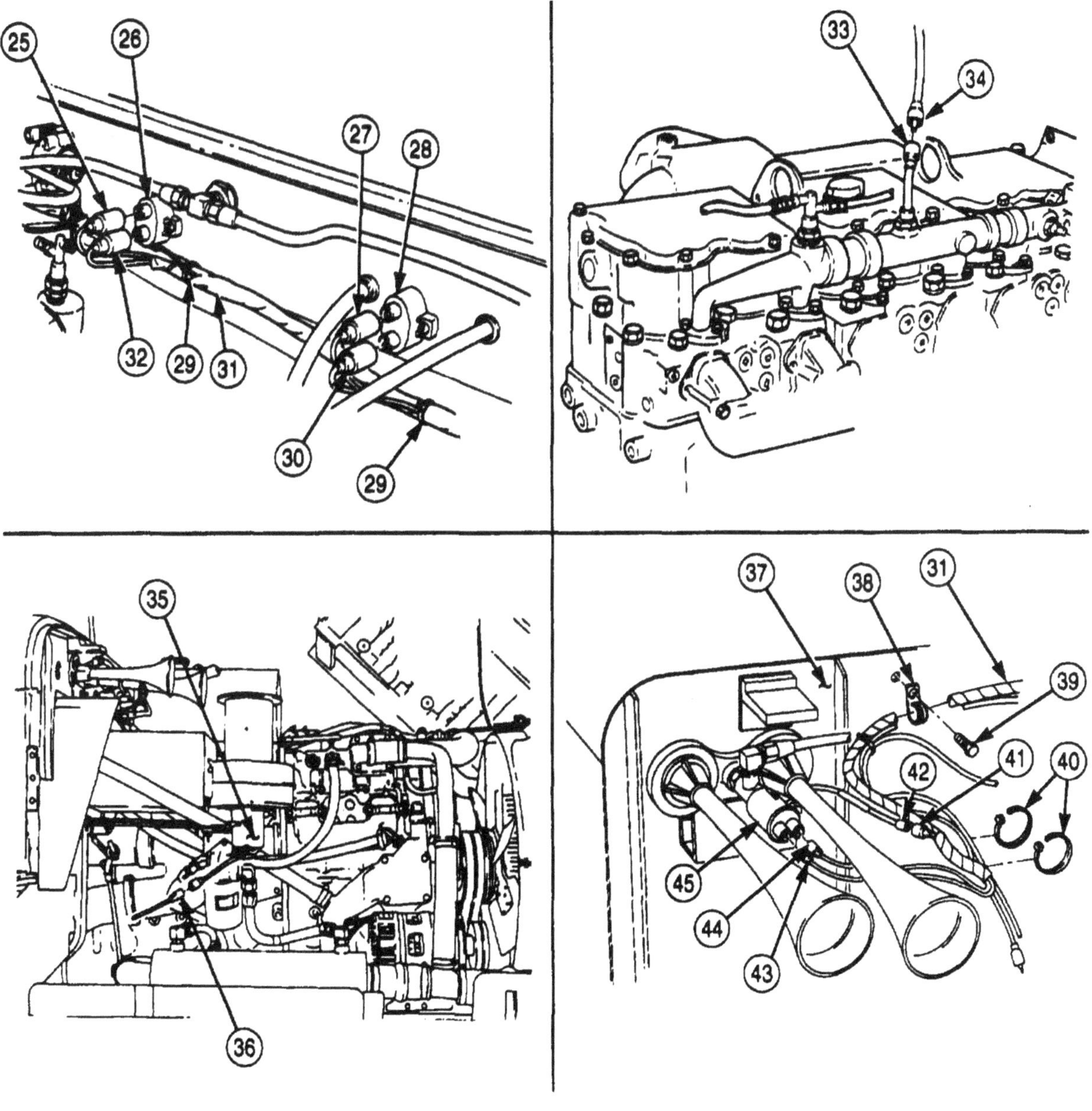

4-66. FRONT WIRING HARNESS (M939/A1) REPLACEMENT (Contd)

26. Remove two screw-assembled lockwashers (5) and terminal cover (4) from alternator (10). Discard screw-assembled lockwashers (5).
27. Remove two screws (3), lockwashers (2), and wire retaining strap (1) from alternator (10). Discard lockwashers (2).
28. Remove screw (15), lockwasher (14), and wire (13) from alternator (10). Discard lockwasher (14).

NOTE

Sealant must be removed before removing wires.

29. Remove nut (19), lockwasher (18), washer (17), and wire (16) from alternator (10). Discard lockwasher (18).
30. Remove nut (6), lockwasher (7), washer (8), and wire (9) from alternator (10). Discard lockwasher (7).
31. Disconnect connectors (11).

NOTE

One tiedown strap is near alternator. Other two are located along inside of right frame rail.

32. Remove and discard three tiedown straps (12).

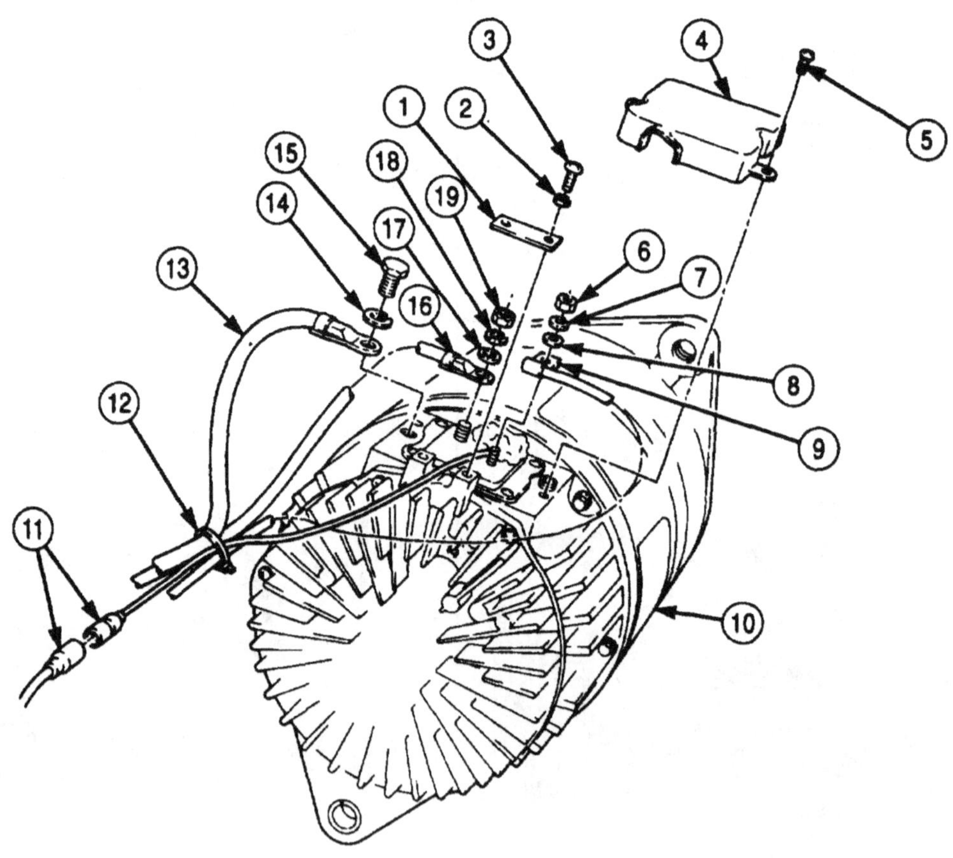

4-66. FRONT WIRING HARNESS (M939/A1) REPLACEMENT (Contd)

33. Disconnect wires (22), (23), and (24) from headlight beam selector switch (25).
34. Remove four nuts (26), lockwashers (27), screws (21), and connector (20) from rear wiring harness receptacle (28). Discard lockwashers (27).
35. Disconnect wire (30) from wire (29).
36. Disconnect wire (32) from transfer case switch capacitor wire (31).

NOTE

Perform step 37 for M936 wrecker only.

37. Disconnect wire (34) from 5th gear lock-up capacitor wire (33).

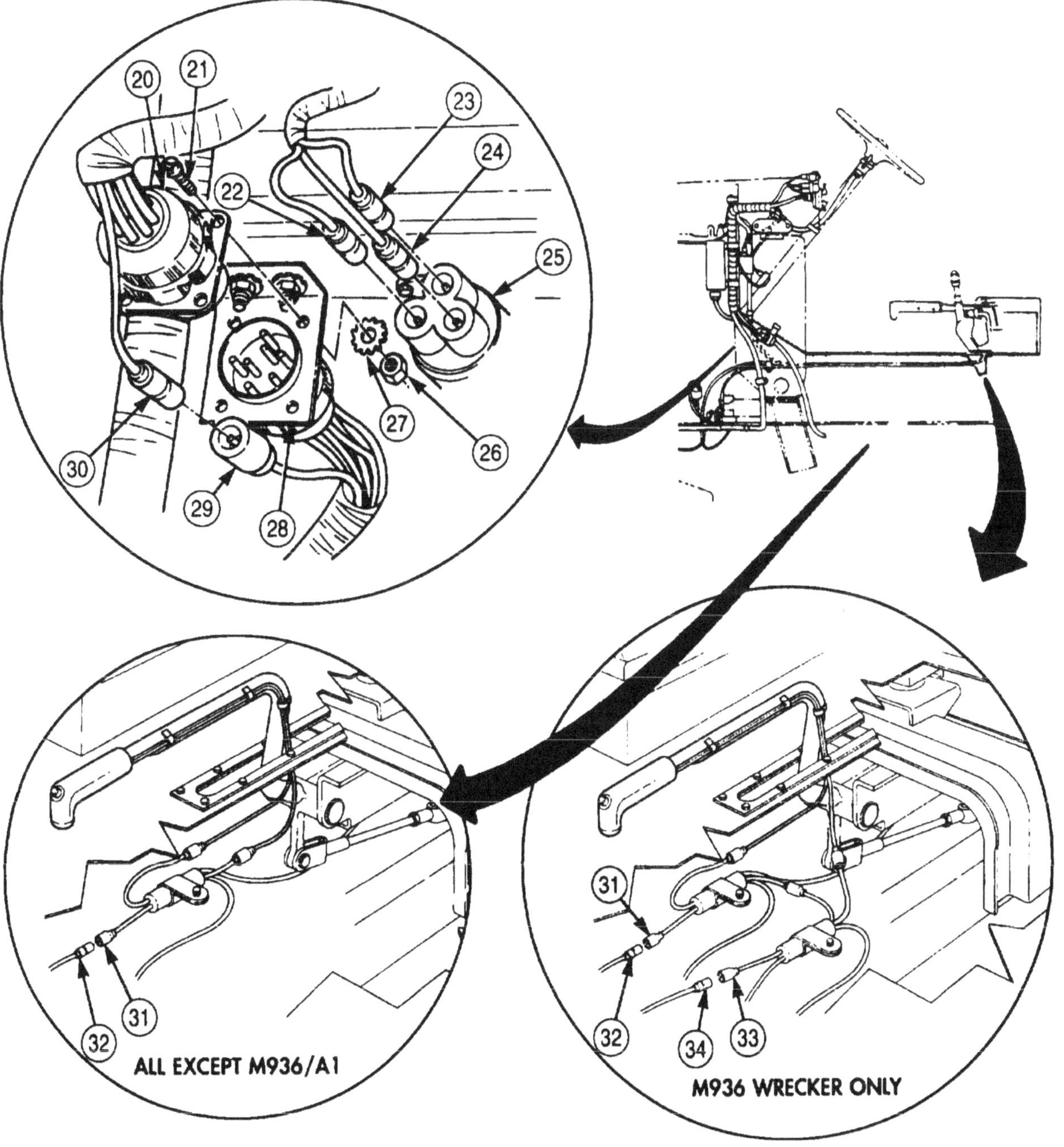

4-66. FRONT WIRING HARNESS (M939/A1) REPLACEMENT (Contd)

38. Remove locknut (7), screw (1), lockwasher (9), and alternator ground wire (8) from frame crossmember (3) and air line bracket (5). Discard lockwasher (9) and locknut (7).
39. Remove locknut (6), screw (2), and air line clamp (4) from air line bracket (5).
40. Disconnect wire (10) from parking brake switch wire (11).
41. Disconnect front harness connector (13) from turn signal control (12).
42. Disconnect wire (15) from horn switch (14).
43. Disconnect wires (17) and (18) from electrical circuit breaker (16).
44. Disconnect wire (19) from heater blower motor circuit breaker (20).

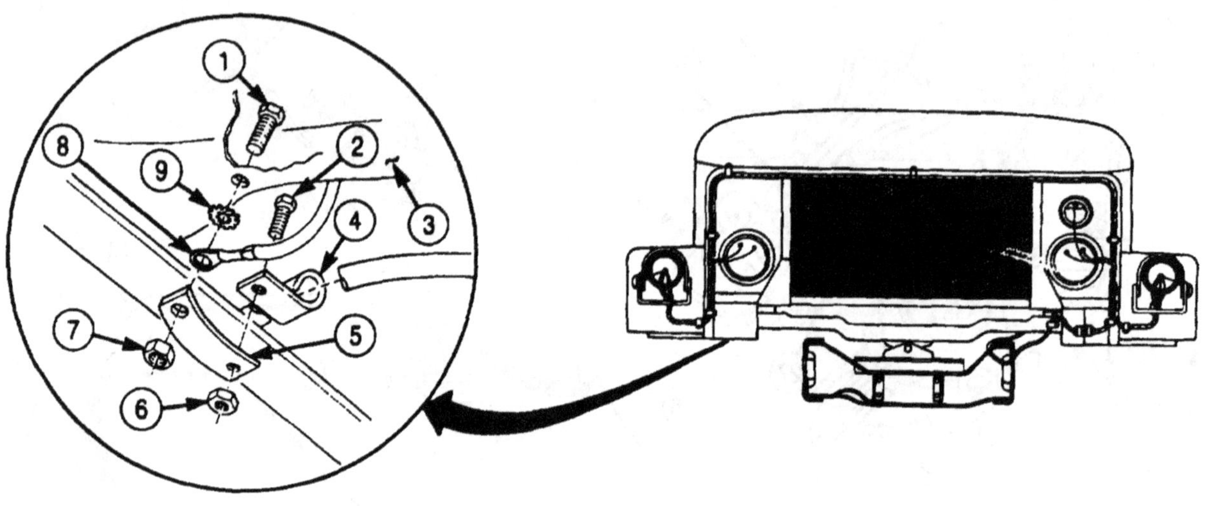

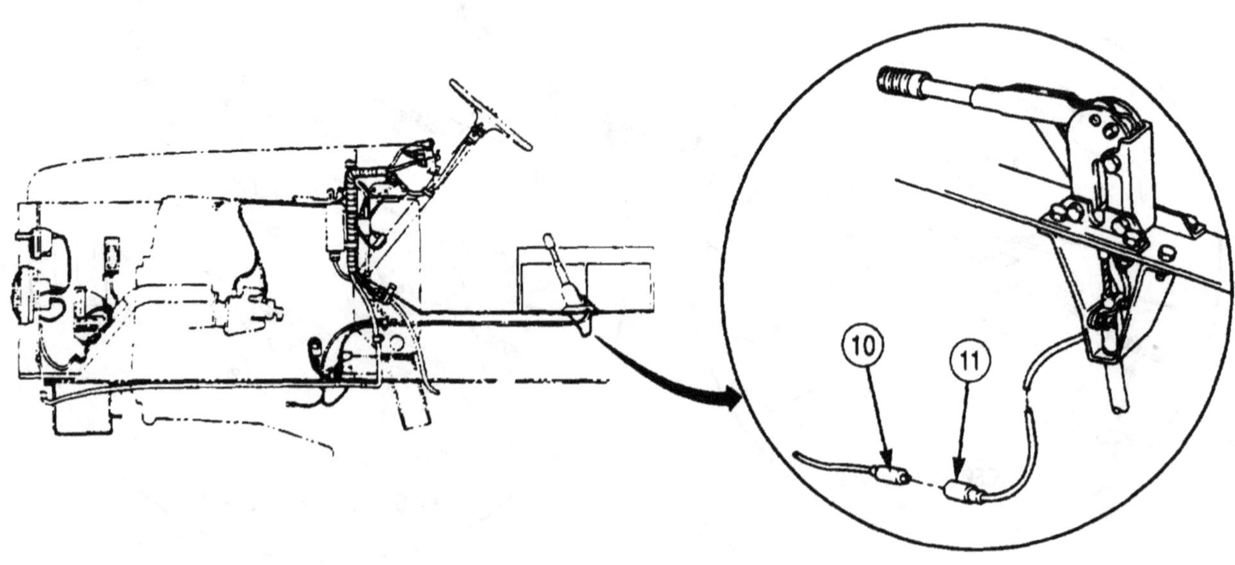

4-66. FRONT WIRING HARNESS (M939/A1) REPLACEMENT (Contd)

NOTE

Perform step 45 for M929, M930, M931, M932, and M936 vehicles only.

45. Disconnect wires (23), (24), and (25) from fuel selector switch (22).
46. Disconnect connector (26) from front wheel drive lock-in switch (21).

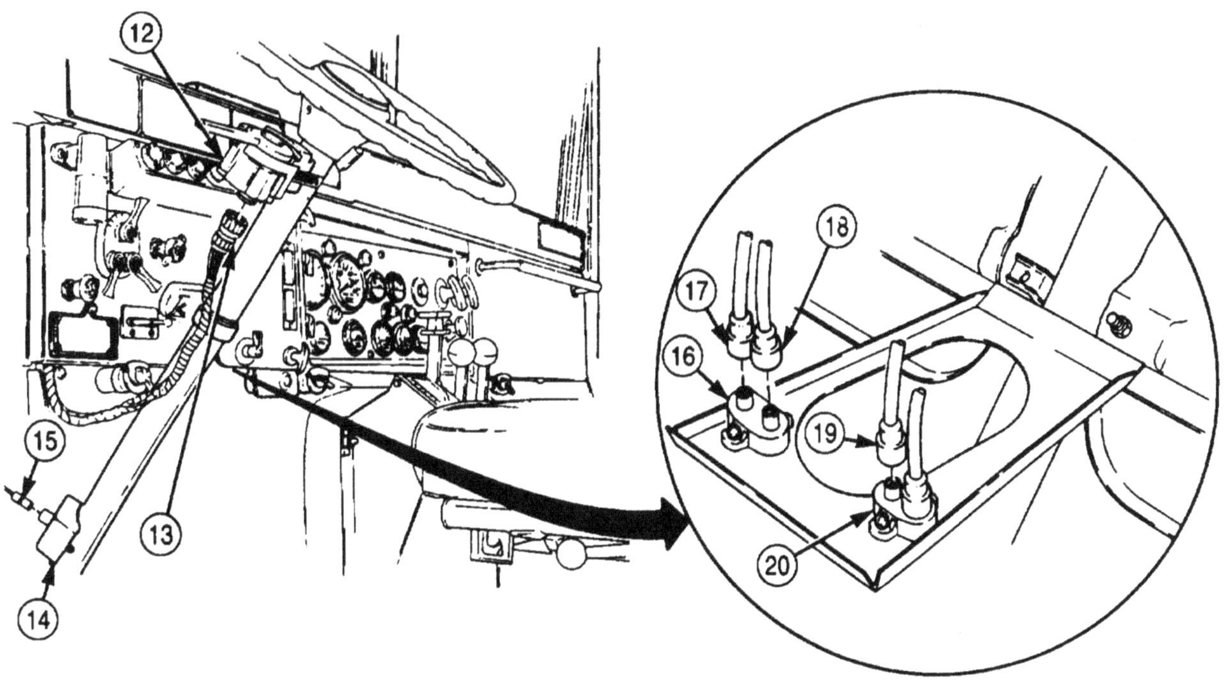

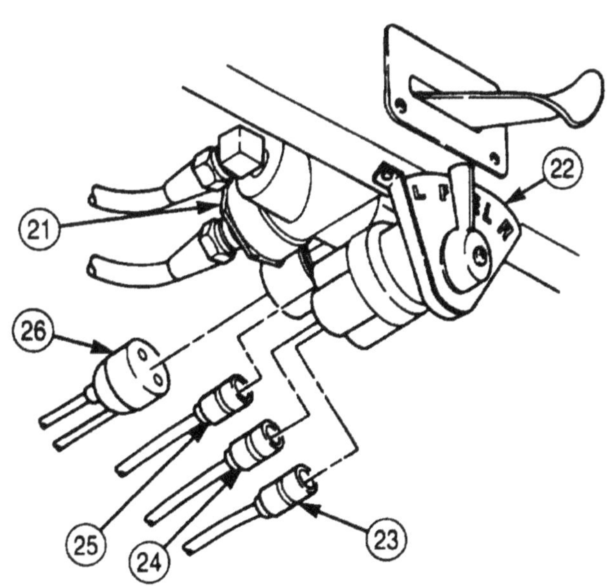

M929, M930, M931, M932, AND M936 VEHICLES

4-66. FRONT WIRING HARNESS (M939/A1) REPLACEMENT (Contd)

47. Disconnect wires (2) and (3) from ether start switch (1).
48. Disconnect wires (5), (6), and (7) from heater blower motor switch (4).
49. Disconnect wire (9) from warning signal lamp switch (8).

NOTE

After harness wires have been removed in steps 50 through 52, cables and hardware should be installed on terminal adapters for installation.

50. Remove nut (10), screw (14), battery cables (11) and (12), and wire (13) from terminal adapter (15).
51. Remove nut (16), screw (19), wire (20), and battery cable (18) from terminal adapter (17).
52. Remove nut (21), screw (23), battery cables (22) and (26), and wire (24) from terminal adapter (25).
53. Push wires (13), (20), and (24) through hole in cab floor.
54. Remove four screws (36) from warning light panel (28) and pull panel (28) away from instrument panel (29).
55. Disconnect wire (30) from parking brake indicator light (27).
56. Disconnect wire (31) from low air pressure indicator light (41).
57. Disconnect wire (32) from spring brake override indicator light (40).
58. Disconnect wire (33) from engine hot indicator light (39).
59. Disconnect wire (34) from axle lock-in indicator light (38).
60. Disconnect wire (35) from high-beam indicator light (37).

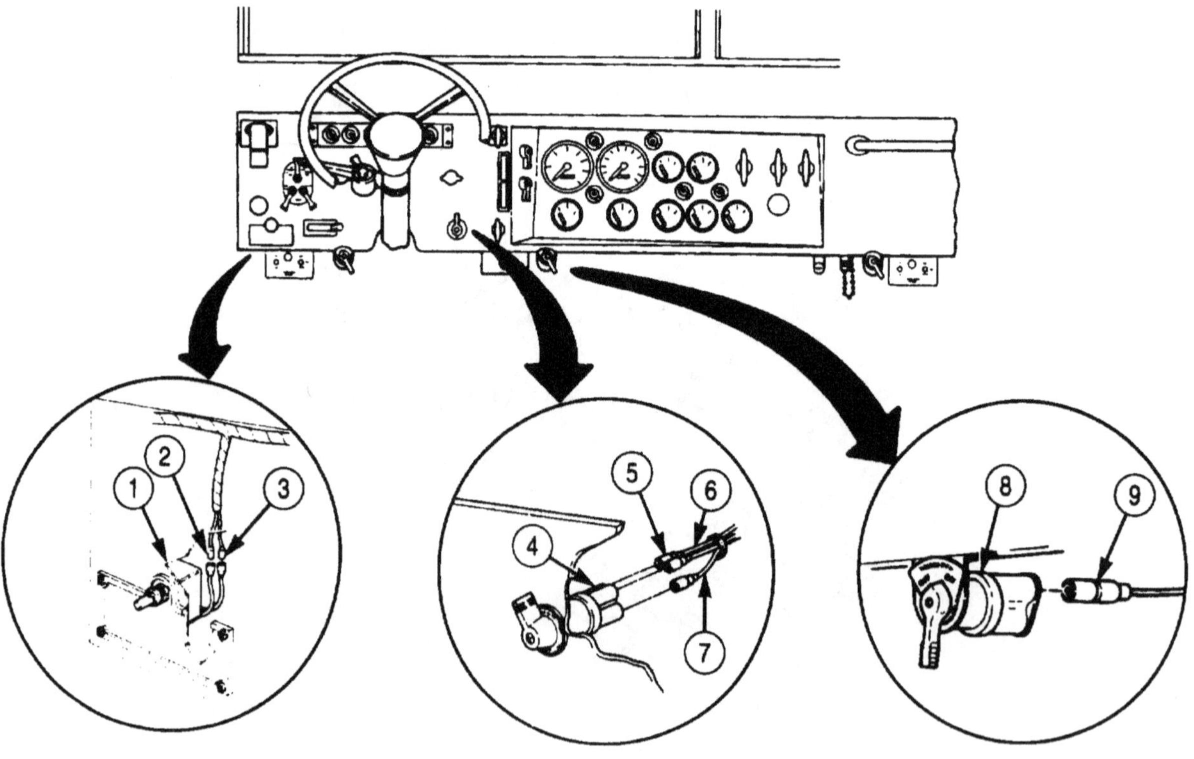

4-66. FRONT WIRING HARNESS (M939/A1) REPLACEMENT (Contd)

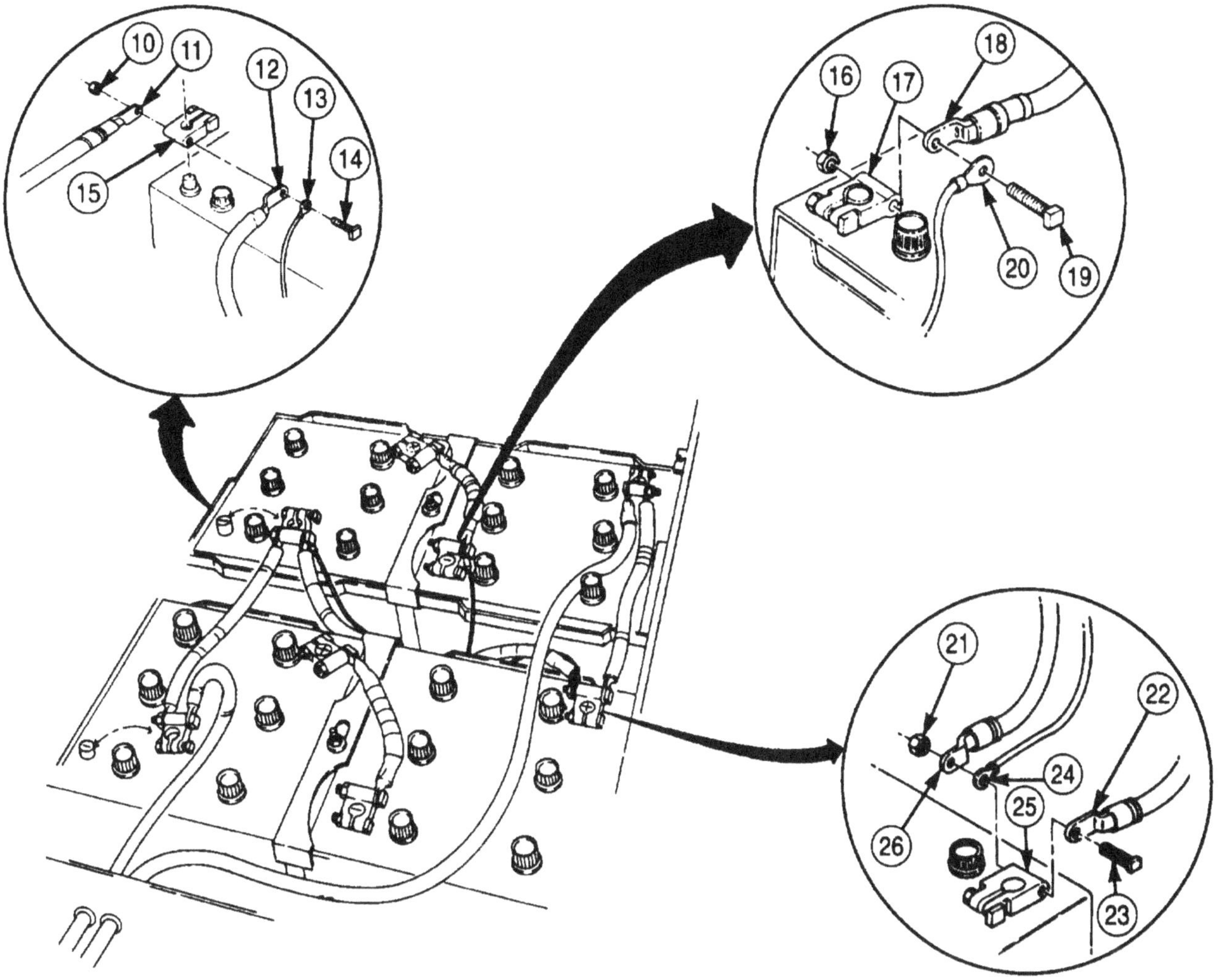

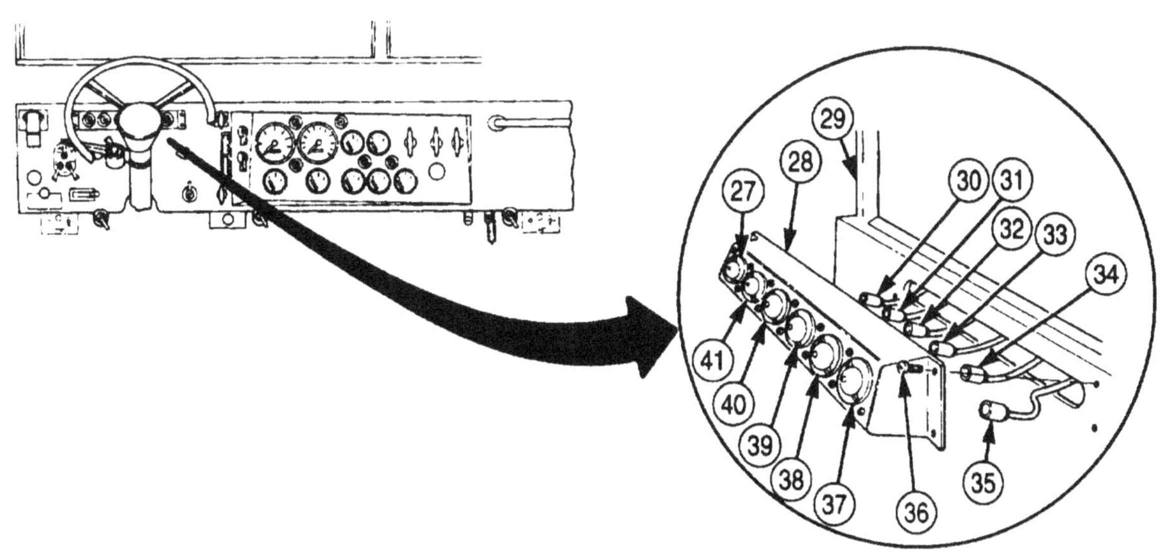

4-66. FRONT WIRING HARNESS (M939/A1) REPLACEMENT (Contd)

61. Remove eight screws (3) from instrument cluster (1) and pull cluster (1) away from instrument panel (2).
62. Remove screw (7), retainer nut (9), and clamp (8) from heater assembly (4) in right side of engine compartment.
63. Remove cotter pin (10) from fresh air control cable (6). Discard cotter pin (10).
64. Remove fresh air control cable (6) and spring nut (5) from heater (4). Discard spring nut (5).
65. From behind instrument cluster (1), disconnect tachometer driveshaft (18).
66. Disconnect speedometer driveshaft (17) from instrument cluster (1).
67. Disconnect air tube (11) from instrument cluster (1).
68. Disconnect wire (20), two wires (21), and wire (22) from battery switch (19).
69. Disconnect wires (23), (24), and (25) from starter switch (26).
70. Disconnect wire (31) from wire (32) on instrument cluster wire assembly (14).
71. Disconnect five wires (16) from five instrument cluster lights (15).
72. Disconnect wire (27) from fuel gauge (28).
73. Disconnect wire (30) from oil pressure gauge (29).
74. Disconnect wire (33) from transmission oil temperature gauge (36).
75. Disconnect wire (13) from engine temperature gauge (12).
76. Disconnect two wires (34) from spring brake pressure switch (36).

4-66. FRONT WIRING HARNESS (M939/A1) REPLACEMENT (Contd)

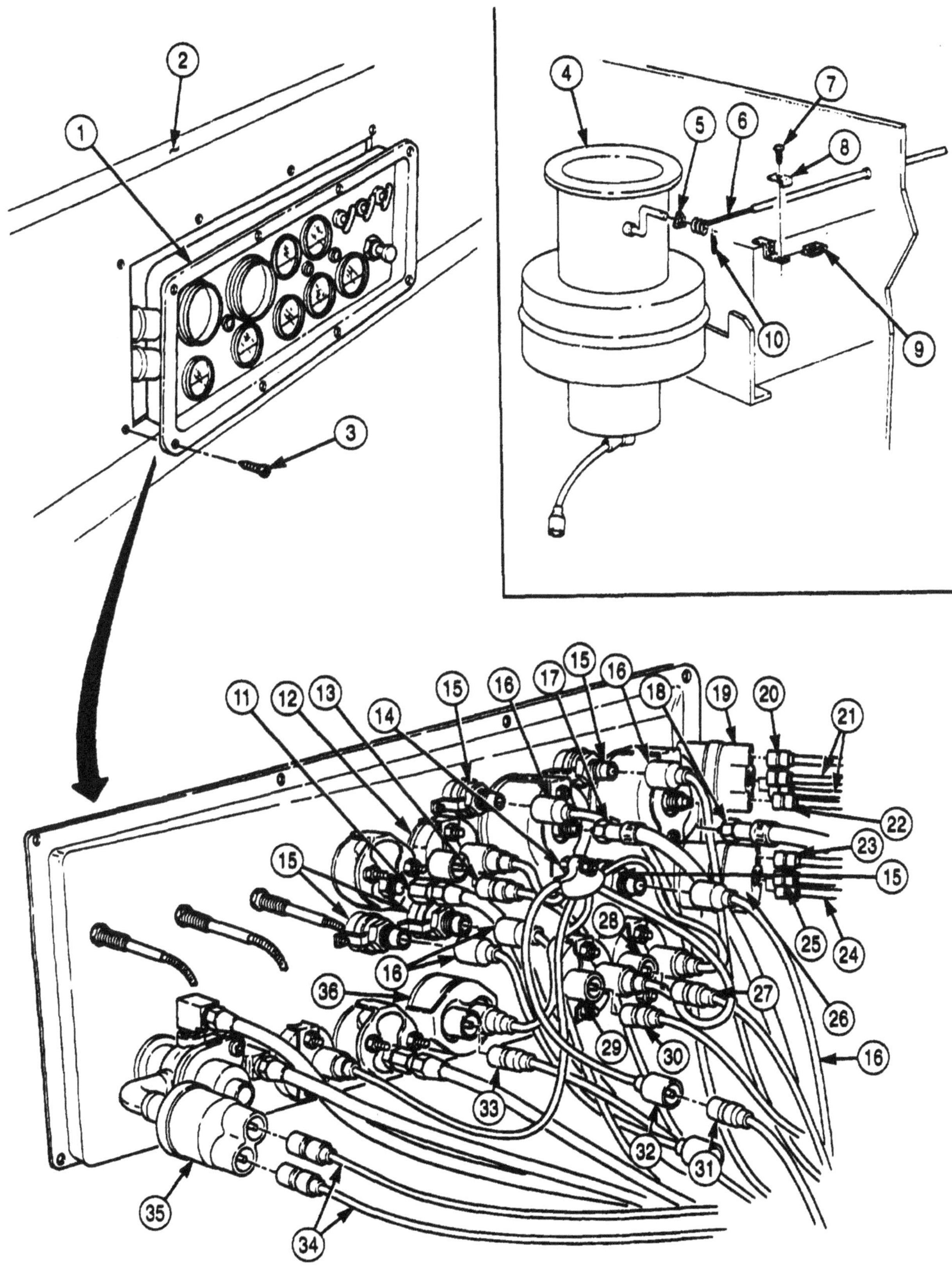

4-66. FRONT WIRING HARNESS (M939/A1) REPLACEMENT (Contd)

77. Place transmission selector assembly (1) in N (neutral) position.
78. Remove four screws (3), lockwashers (4), and transmission selector assembly (1) from console (2). Discard lockwashers (4).
79. Disconnect transmission selector wires (5), (6), and (7) from three harness leads (8).

NOTE

Perform step 80 for M936 vehicles only.

80. Disconnect connector (9) from connector (10) and from floodlight switch (25) and auxiliary receptacle (26).
81. Remove cap (17) from connector (22).
82. Remove nut (13), lockwasher (14), screw (19), and ground wire (15) from mounting bracket (21). Discard lockwasher (14).
83. Remove nut (11), lockwasher (12), screw (18), and cap chain (16) from mounting bracket (21). Discard lockwasher (12).
84. Remove two nuts (24), lockwashers (23), screws (20), and connector (22) from three mounting brackets (21). Discard lockwashers (23).

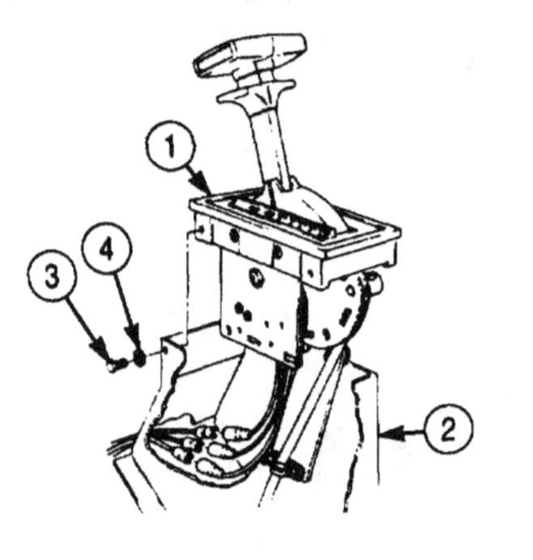

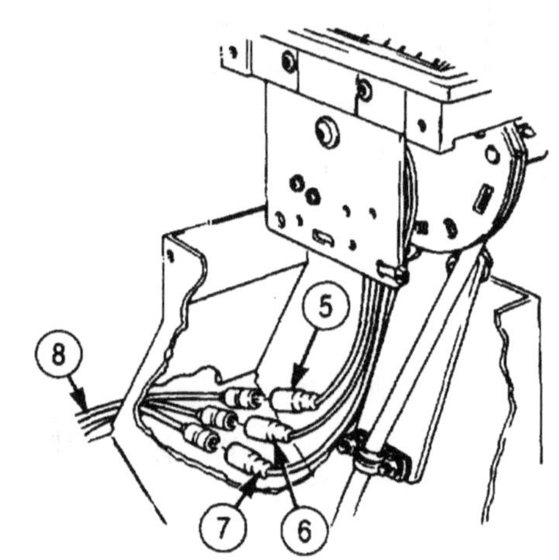

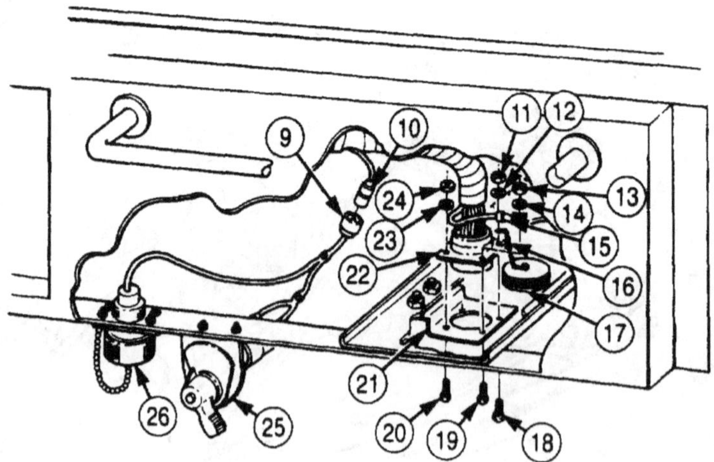

4-66. FRONT WIRING HARNESS (M939/A1) REPLACEMENT (Contd)

85. Remove two screws (27), washers (28), and grommets (29) from firewall (30).
86. Remove two locknuts (34), seven screws (32), and harness clamps (33) from front harness (31) and vehicle. Discard locknuts (34).

NOTE

Tag clamps for installation.

87. Remove front wiring harness (31) from vehicle.

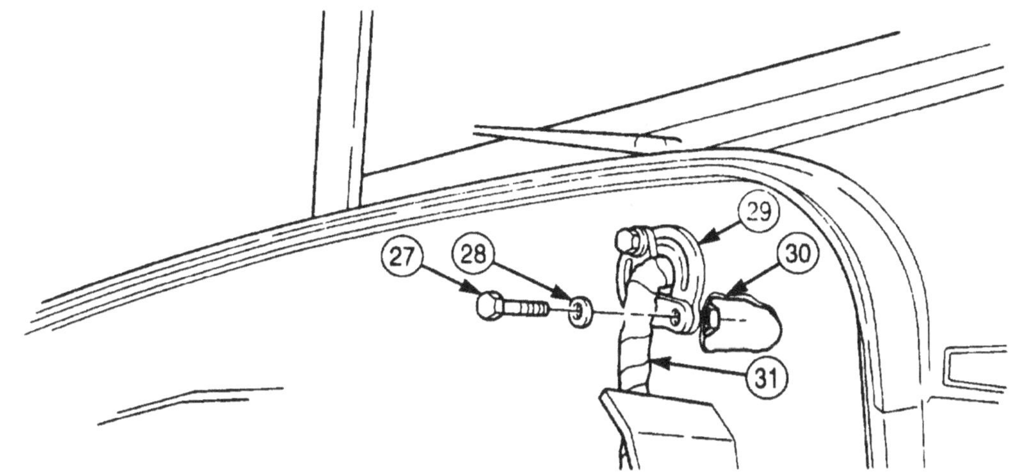

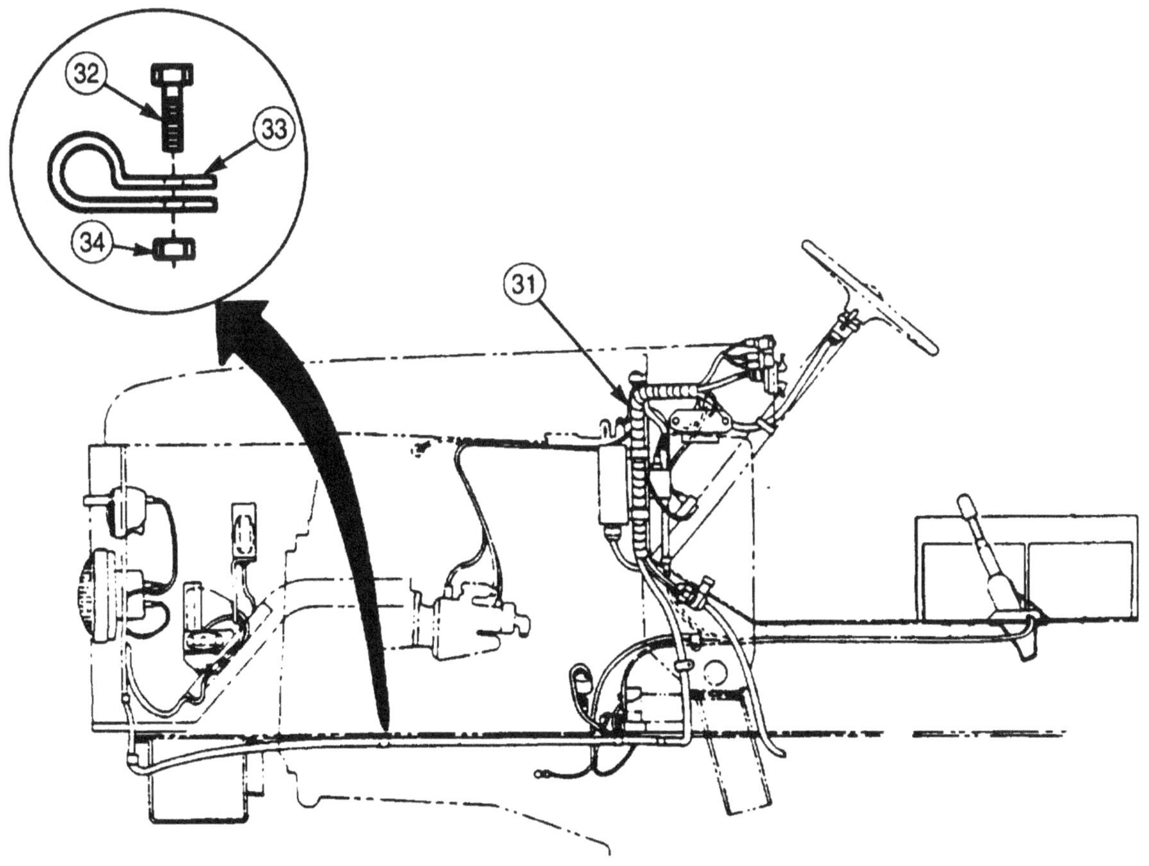

4-66. FRONT WIRING HARNESS (M939/A1) REPLACEMENT (Contd)

b. Installation

CAUTION

Use care when routing harness. Snagging may result, and forceful pulling will cause damage to harness.

NOTE

Assistant will help with step 1.

1. Position new front wiring harness (5) on vehicle.
2. Insert front wiring harness (5) through hole in firewall (4) and route as high as possible behind instrument panel (6).
3. Install two grommets (3) on tirewall (4) with two washers (2) and screws (1).
4. Install seven harness clamps (8) on front wiring harness (5), and install with screws (7) and new locknuts (9).

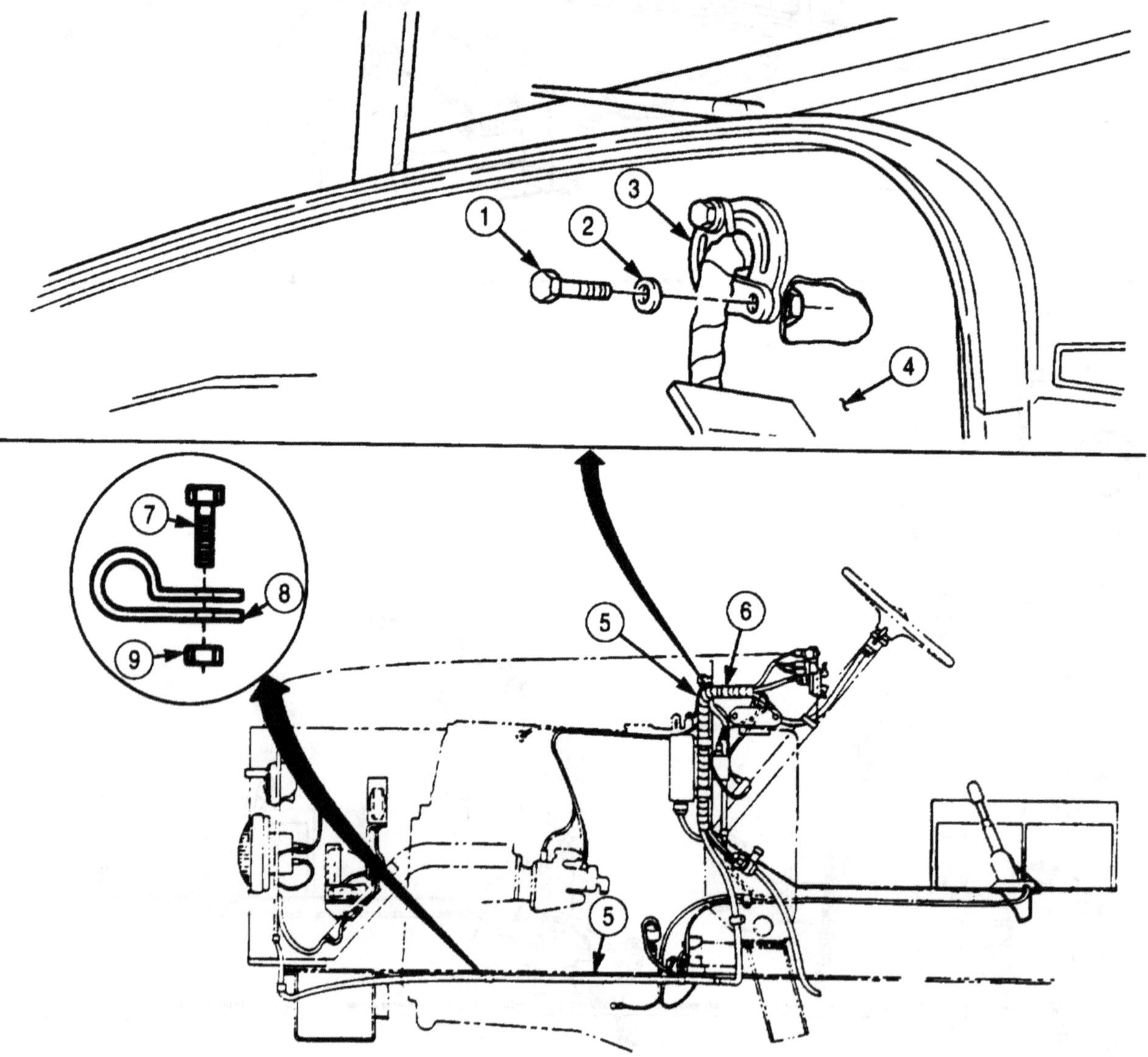

4-66. FRONT WIRING HARNESS (M939/A1) REPLACEMENT (Contd)

5. Install connector (26) on top of mounting bracket (22) under right side of instrument panel (23) with two screws (21), new lockwashers (27), and nuts (28).
6. Install cap chain (17) on mounting bracket (22) with screw (19), new lockwasher (13), and nut (12).
7. Install ground wire (16) on mounting bracket (22) with screw (20), new lockwasher (15), and nut (14).
8. Install cap (18) on connector (26).
9. Connect connector (11) to connector (10) of floodlight switch (24) and auxiliary receptacle (25).
10. Place wires (33), (34), and (35) in console (30) and connect to selector assembly wires (32), (31), and (36).
11. Install transmission selector assembly (29) on console (30) with four new lockwashers (38) and screws (37).

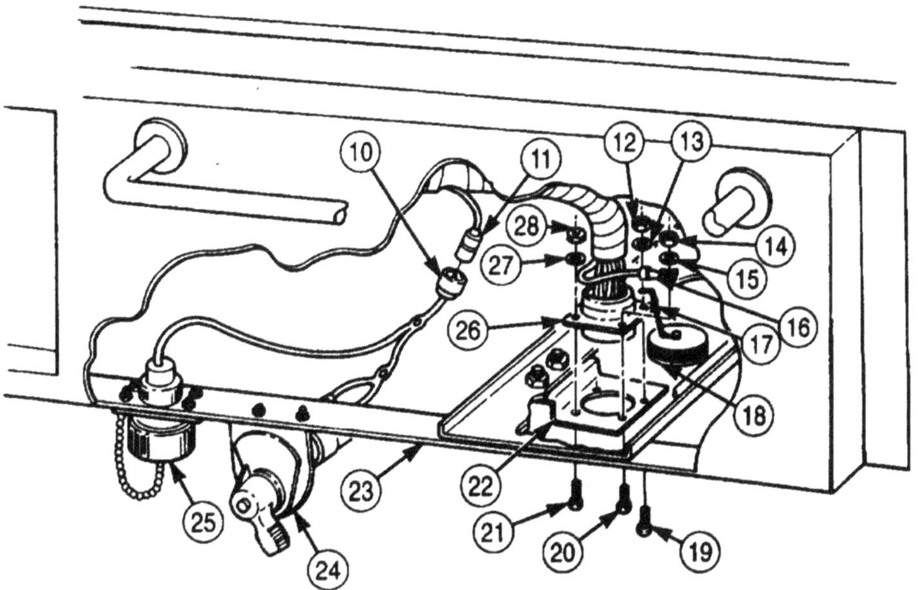

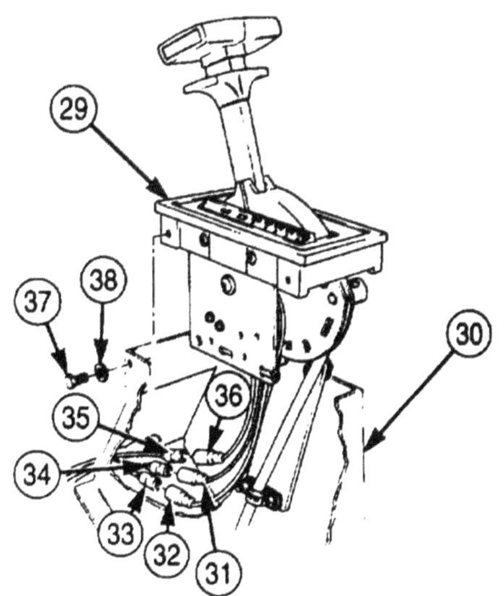

4-66. FRONT WIRING HARNESS (M939/A1) REPLACEMENT (Contd)

12. Connect two wires (24) to spring brake pressure switch (25).
13. Connect wire (3) to engine temperature gauge (2).
14. Connect wire (23) to transmission oil temperature gauge (27).
15. Connect wire (19) to engine oil pressure gauge (20).
16. Connect wire (17) to fuel gauge (18).
17. Connect five wires (6) to instrument cluster lights (5).
18. Connect wire (21) to wire (22) on instrument cluster wire assembly (4).
19. Connect wires (13), (14), and (15) to starter switch (16).
20. Connect wire (10), two wires (11), and wire (12) to battery switch (9).
21. Connect tachometer driveshaft (8) to tachometer on instrument panel (26).
22. Connect speedometer driveshaft (7) to speedometer on panel (26).
23. Install air tube (1) to air pressure gauge (28) on panel (26).

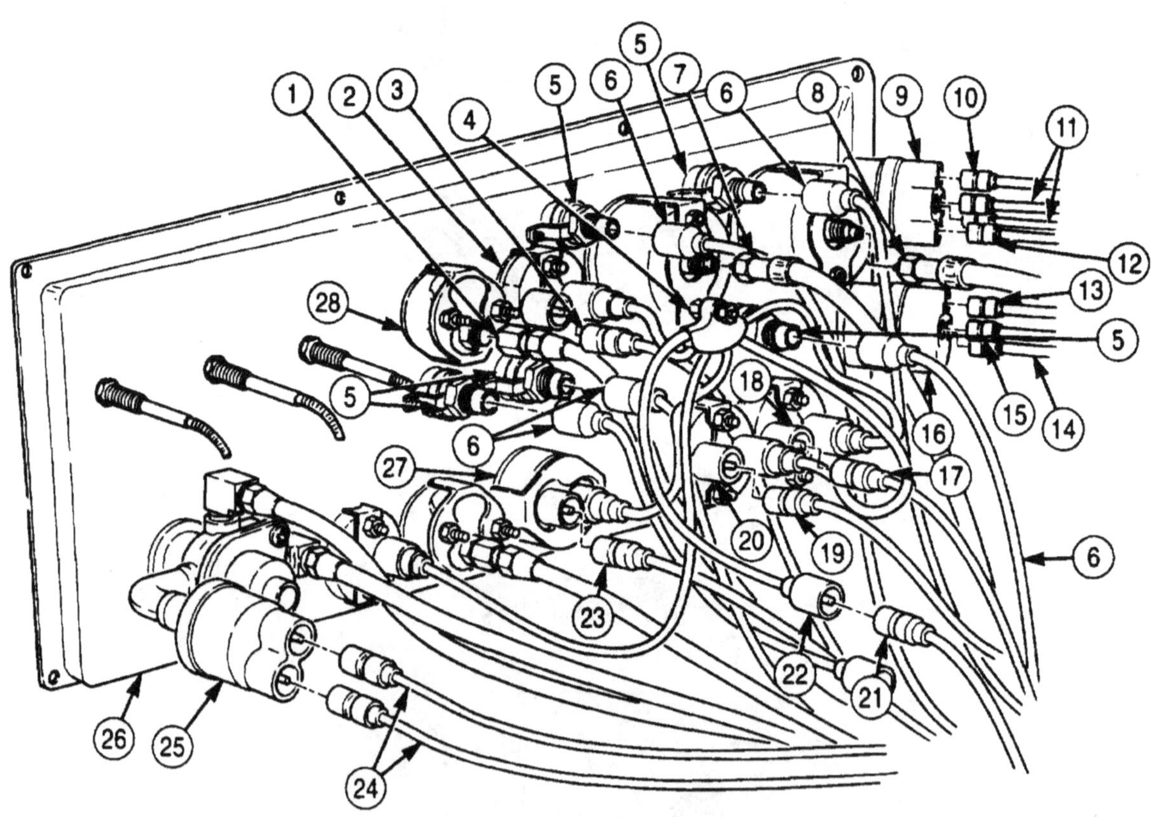

4-66. FRONT WIRING HARNESS (M939/A1) REPLACEMENT (Contd)

24. Install fresh air control cable (31) on heater assembly (29) in right side of engine compartment with new spring nut (30) and new cotter pin (35).
25. Secure cable (31) with screw (32), clamp (33), and retainer nut (34).
26. Install instrument cluster (36) on instrument panel (26) with eight screws (37).

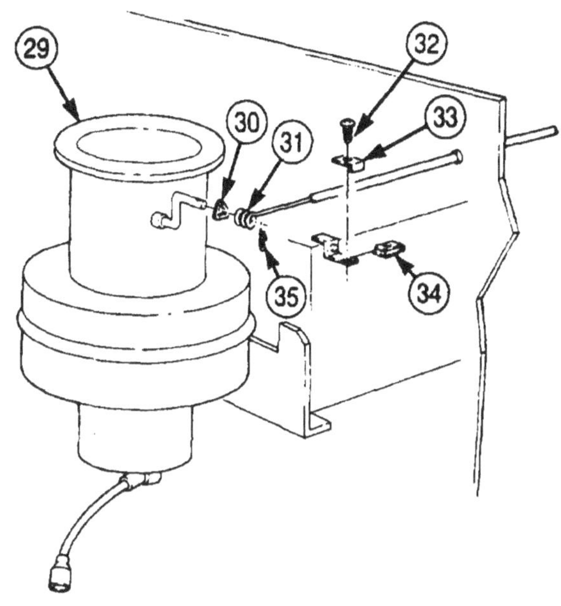

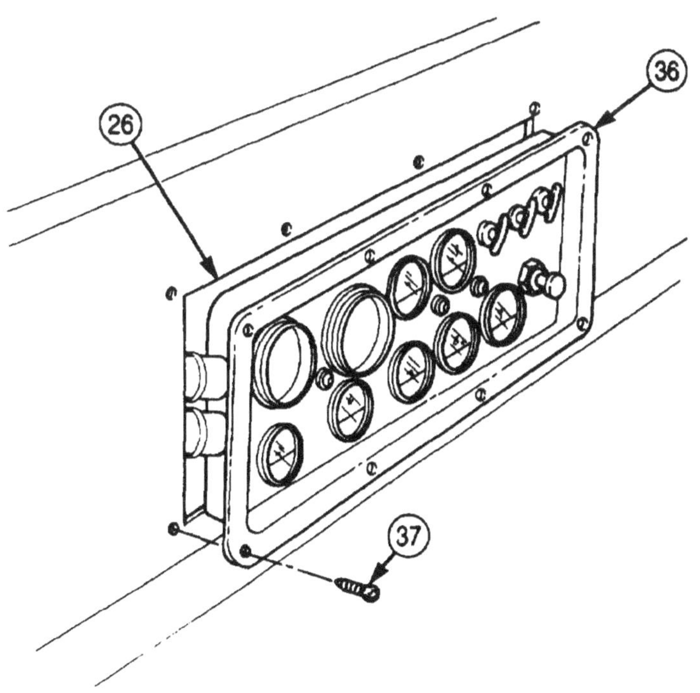

4-66. FRONT WIRING HARNESS (M939/A1) REPLACEMENT (Contd)

27. Connect wire (8) to high-beam indicator light (10) on warning light panel (1).
28. Connect wire (7) to axle lock-in indicator light (11).
29. Connect wire (6) to engine hot indicator light (12).
30. Connect wire (5) to spring brake override indicator light (13).
31. Connect wire (4) to low air pressure indicator light (14).
32. Connect wire (3) to parking brake indicator light (16).
33. Install warning light panel (1) on instrument panel (2) with four screws (9).

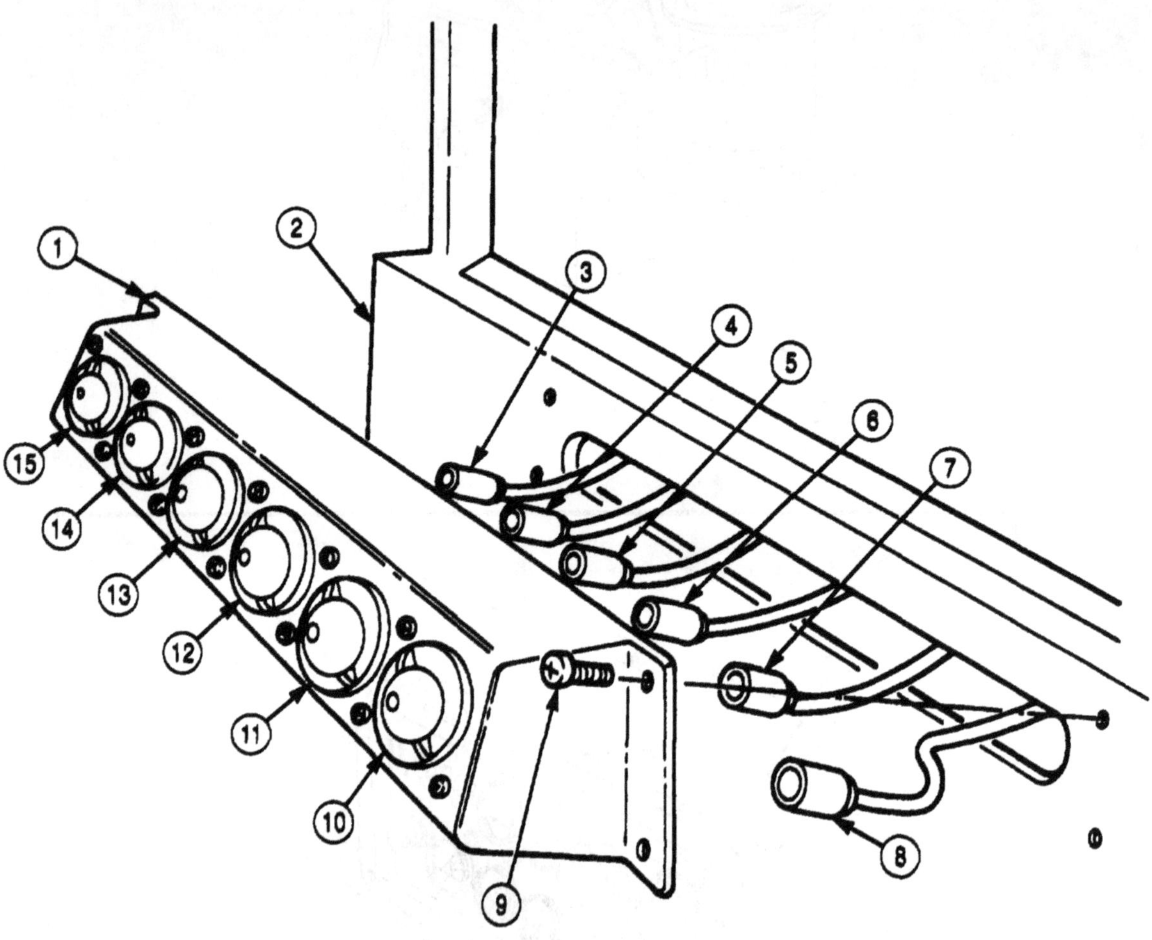

4-66. FRONT WIRING HARNESS (M939/A1) REPLACEMENT (Contd)

34. Push wires (22), (27), and (31) through hole in cab floor.
35. Install cables (20) and (21) and wire (22) on terminal adapter (16) with screw (23) and nut (19).
36. Install cable (25) and wire (27) on terminal adapter (17) with screw (26) and nut (24).
37. Install wire (31) and cables (29) and (32) on terminal adapter (18) with screw (30) and nut (28).

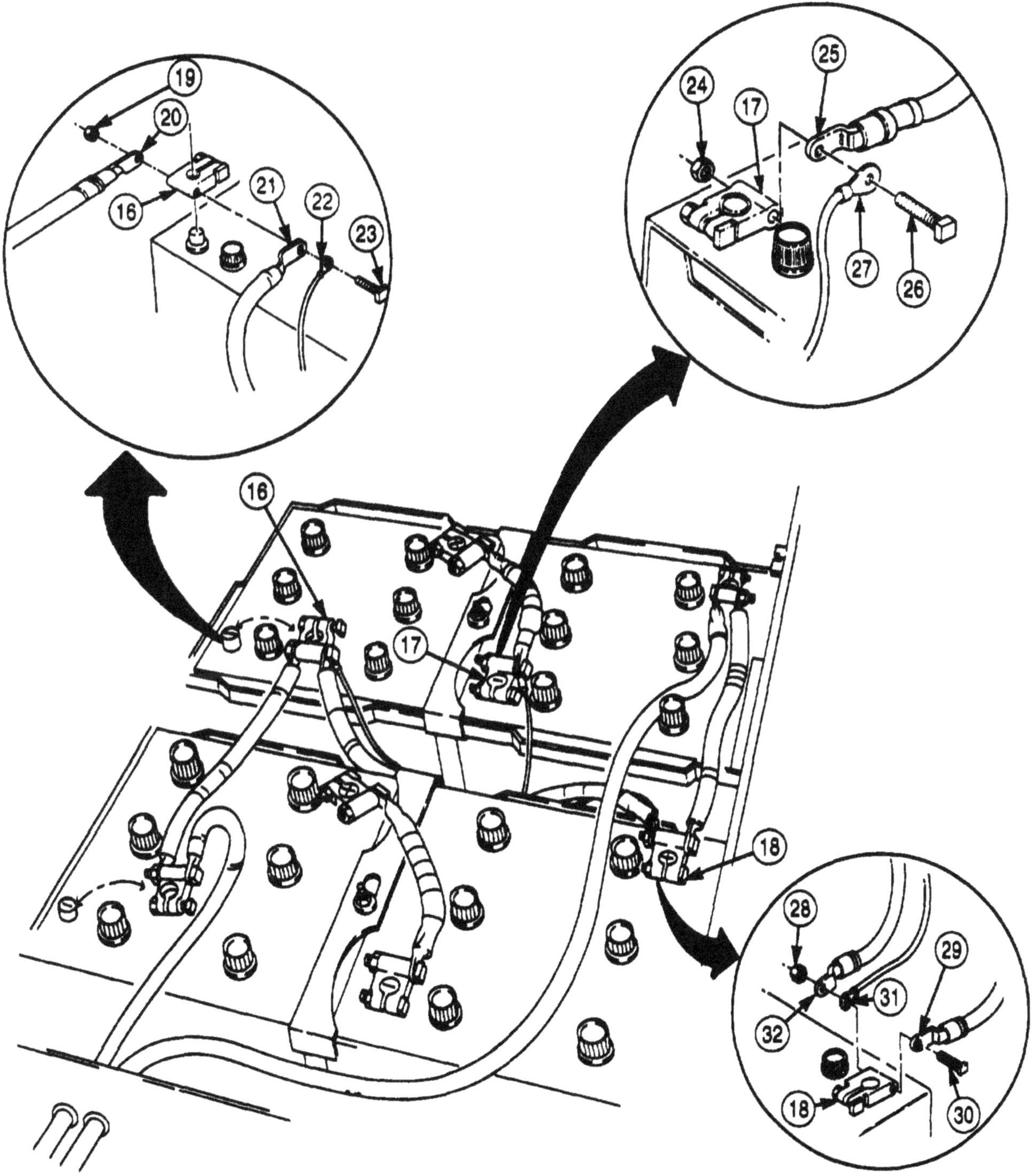

4-66. FRONT WIRING HARNESS (M939/A1) REPLACEMENT (Contd)

NOTE

Perform step 38 only for M936/A1 vehicles.

38. Connect wire (9) to warning signal lamp switch (8).
39. Connect wires (5), (6), and (7) to heater blower motor switch (4).
40. Connect wires (2) and (3) to ether start switch (1).
41. Connect connector (15) to front wheel drive lock-in switch (10).

NOTE

Perform step 42 for M929, M930, M931, M932, and M936 vehicles only.

42. Connect wires (12),(13), and (14) to fuel selector switch (11).

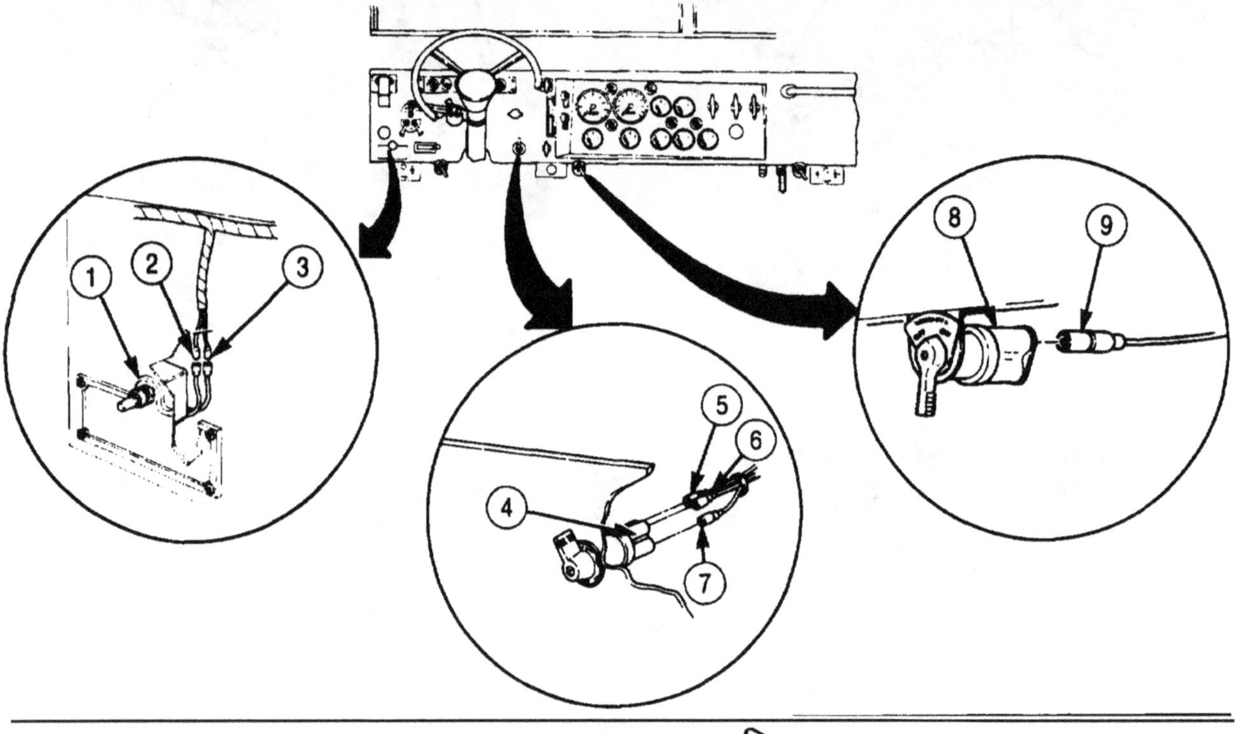

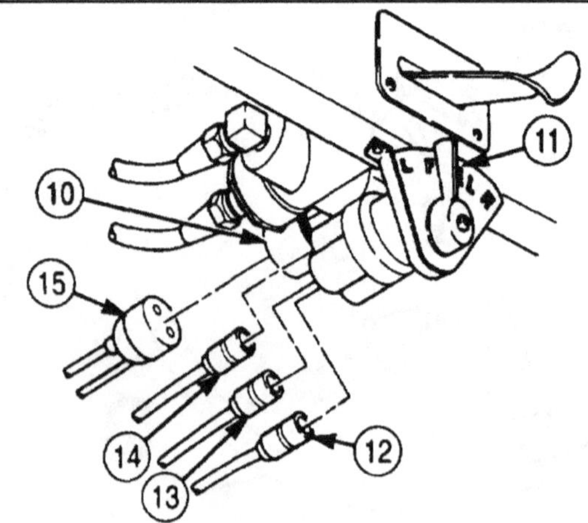

M929, M930, M931, M932, AND M936 VEHICLES

4-66. FRONT WIRING HARNESS (M939/A1) REPLACEMENT (Contd)

43. Connect wire (22) to heater blower circuit breaker (23).
44. Connect wires (20) and (21) to electrical gauge circuit breaker (24).
45. Connect wire (19) to horn switch (18).
46. Connect front harness connector (17) to turn signal control (16).
47. Install air line clamp (28) on air line bracket (29) with screw (26) and locknut (30).
48. Install alternator ground wire (32) on frame crossmember (27) with new lockwasher (33), air line bracket (29), screw (25), and new locknut (31).

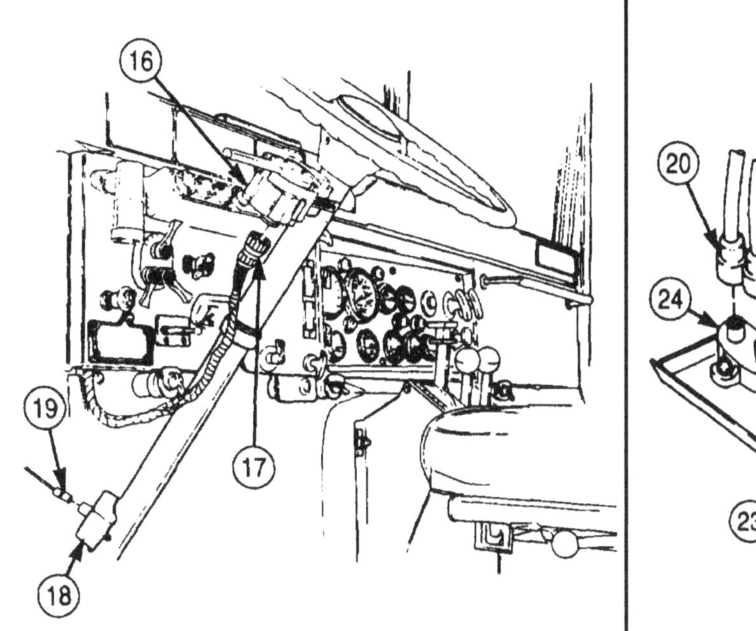

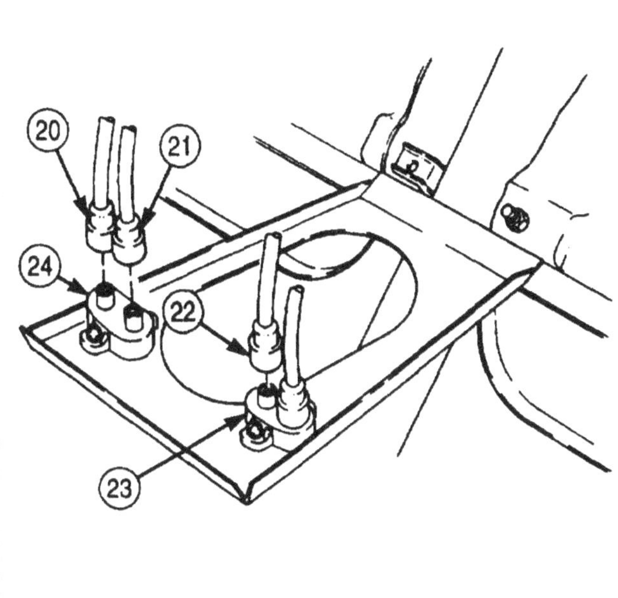

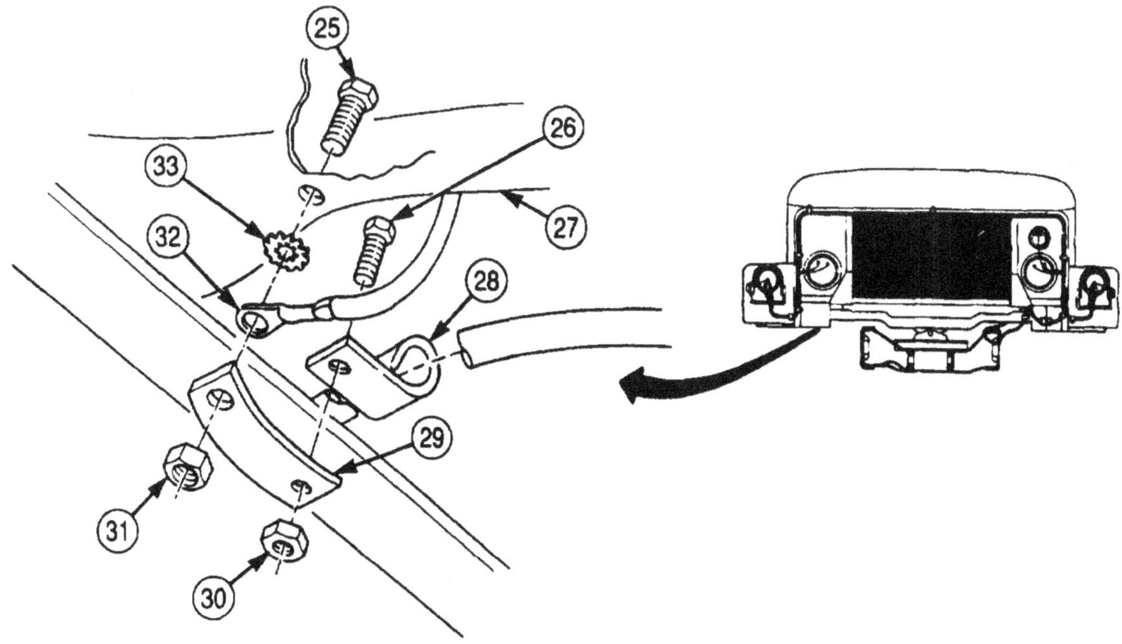

4-66. FRONT WIRING HARNESS (M939/A1) REPLACEMENT (Contd)

49. Connect wire (1) to parking brake switch wire (2).
50. Connect wire (17) to 5th gear lock-up capacitor wire (16).
51. Connect wire (15) to transfer case switch capacitor wire (14).
52. Connect wire (12) to wire (11).
53. Install connector (13) on receptacle of rear wiring harness receptacle (10) with four screws (3), new lockwashers (9), and nuts (8).

NOTE

Perform step 53 for M929/A1, M930/A1, M931/A1, M932/A1, and M936/A1 vehicles only.

54. Connect wires (4), (5), and (6) to headlight beam selector switch (7).

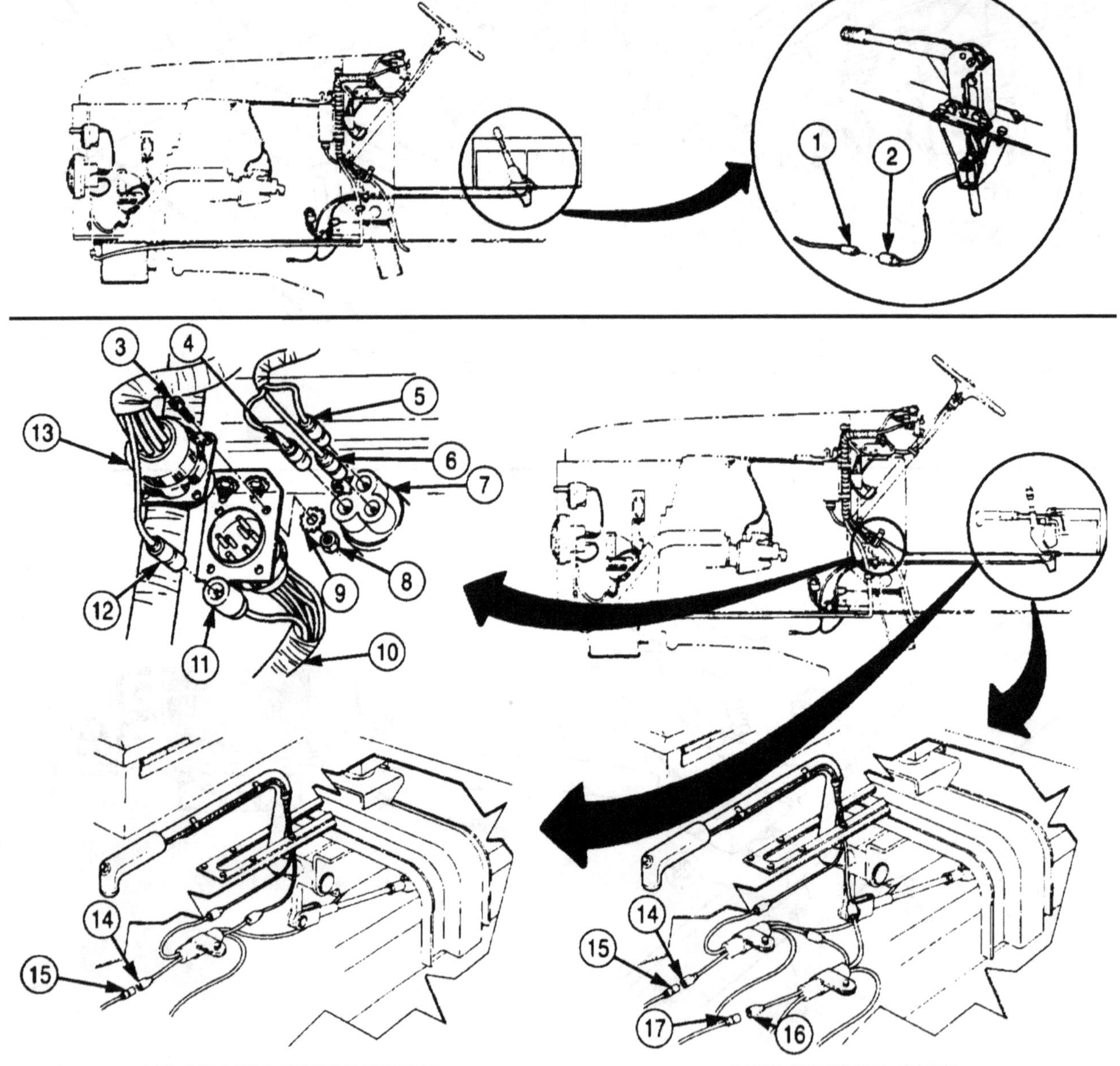

ALL MODELS EXCEPT M936/A1

M936 WRECKER ONLY

4-66. FRONT WIRING HARNESS (M939/A1) REPLACEMENT (Contd)

55. Connect two connectors (31) of alternator (30) wiring.
56. Install wire (29) on alternator (30) with washer (28), new lockwasher (27), and nut (26). Tighten nut (26) 20-25 lb-in. (2-3 N•m).
57. Install wire (33) on alternator (30) with washer (18), new lockwasher (19), and nut (20). Tighten nut (20) 45-55 lb-in. (5-6 N-m).
58. Install wire retaining strap (21) over wire (31) and on alternator (30) with two new lockwashers (22) and screws (23).
59. Install wire (34) on alternator (30) with new lockwasher (35) and screw (36). Tighten screw (36) 82-102 lb-in. (9-12 N•m).
60. Apply sealant on wires (29) and (33) and install cover (24) on alternator (30) with two new screw-assembled lockwashers (25).

NOTE

Two tiedown straps go over wires on inside right frame rail.

61. Install three tiedown straps (32).
62. Connect wire (41) to transorb diode coupling assembly wire (42).
63. Connect wires (43) and (44) to horn solenoid (45).
64. Install four tiedown straps (40) on rear of harness.
65. Install three cable clamps (38) on firewall (37) with three screws (39).

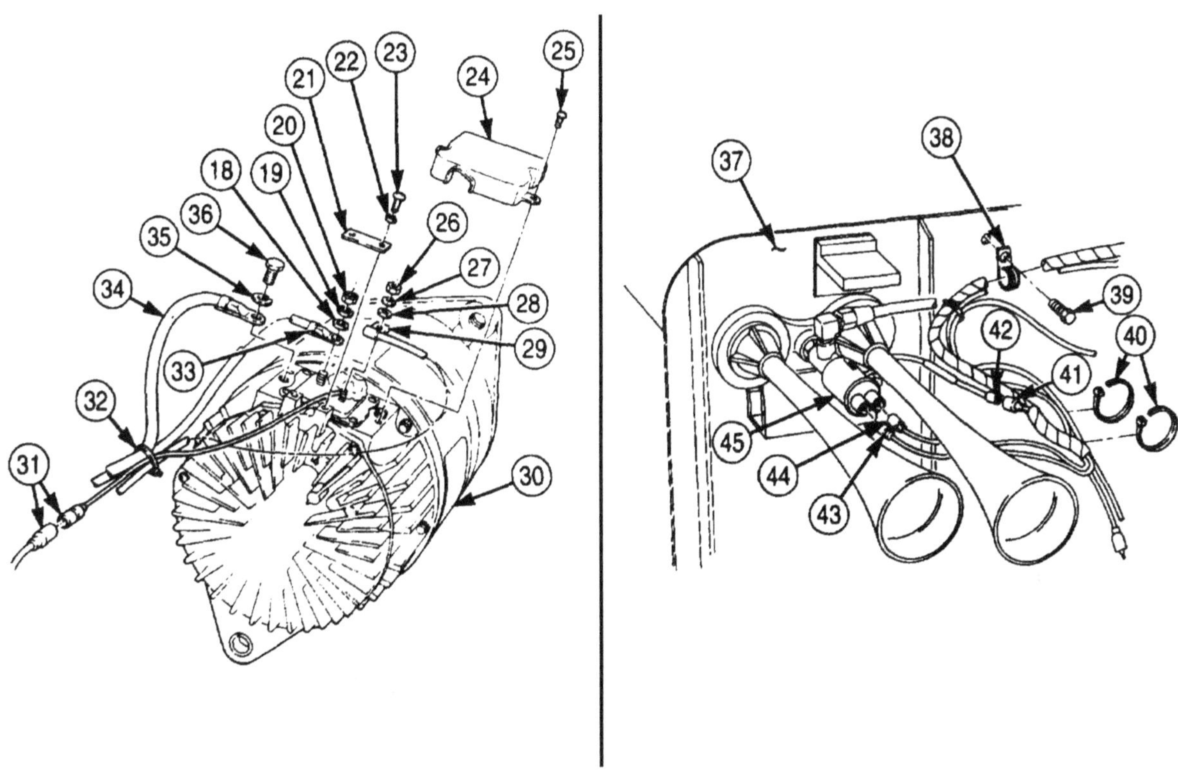

4-66. FRONT WIRING HARNESS (M939/A1) REPLACEMENT (Contd)

66. Connect wire (2) to engine temperature switch wire (1).
67. Connect wire (5) to wire (4) on personnel heater (3).
68. Connect wires (9) and (10) to transmission control and spring break circuit breaker (7).
69. Connect wires (12) and (13) to horn circuit breaker (6).
70. Install two tiedown straps (8) on wiring harness (11).

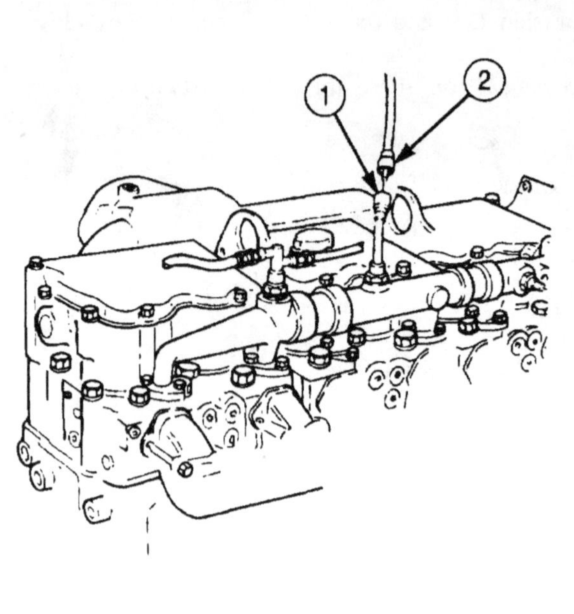

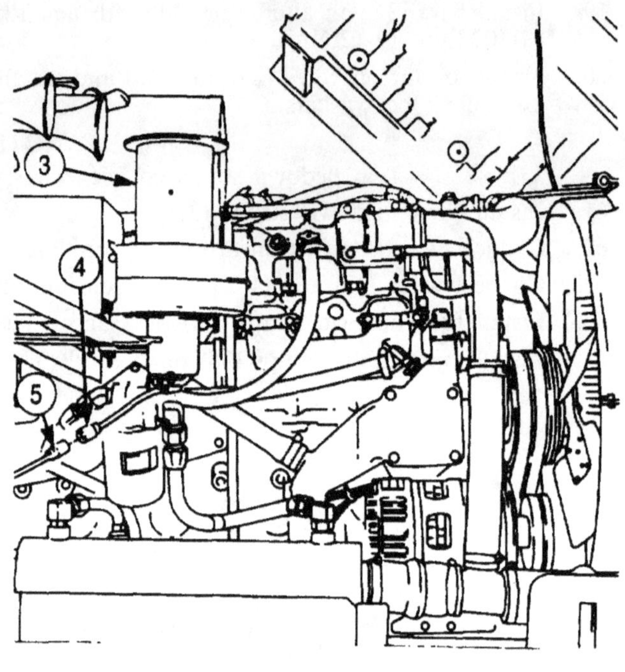

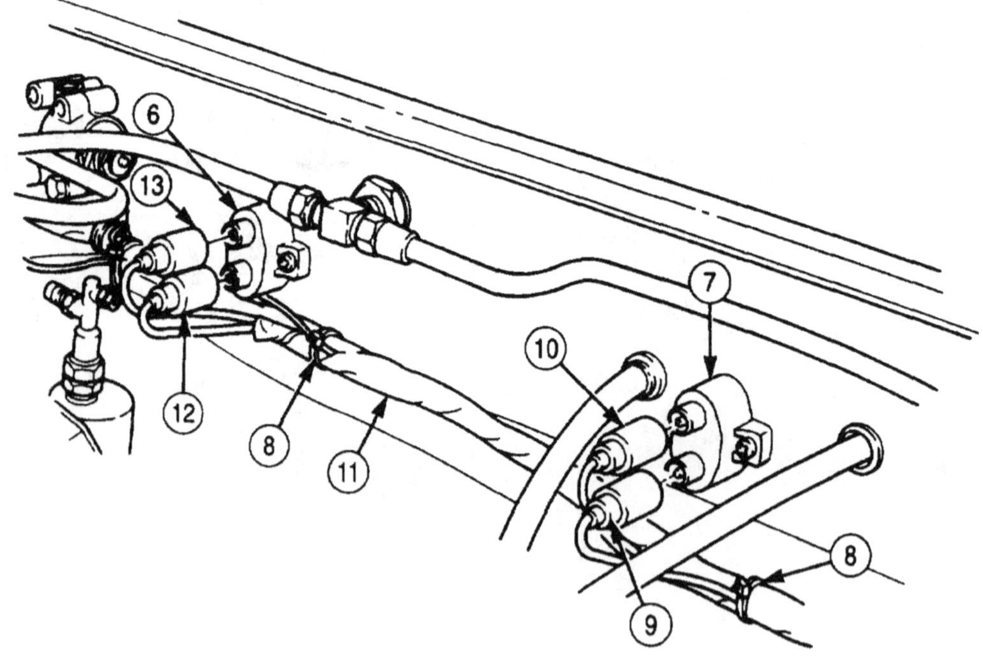

4-66. FRONT WIRING HARNESS (M939/A1) REPLACEMENT (Contd)

71. Install ground strap (22) and wires (23) and (24) on starter post (21) with new lockwasher (26) and nut (25).
72. Install wires (27) and (30) on starter post (20) with new lockwasher (29) and nut (28).
73. Install wire (18) on solenoid terminal (19) with washer (17) and screw-assembled washer (16).
74. Connect wire (15) to oil pressure sending unit (14).
75. Connect two wires (31) and (32).
76. Install ground wire (34) on intake manifold (37) with new lockwasher (36), ground strap (35), and screw-assembled washer (33).

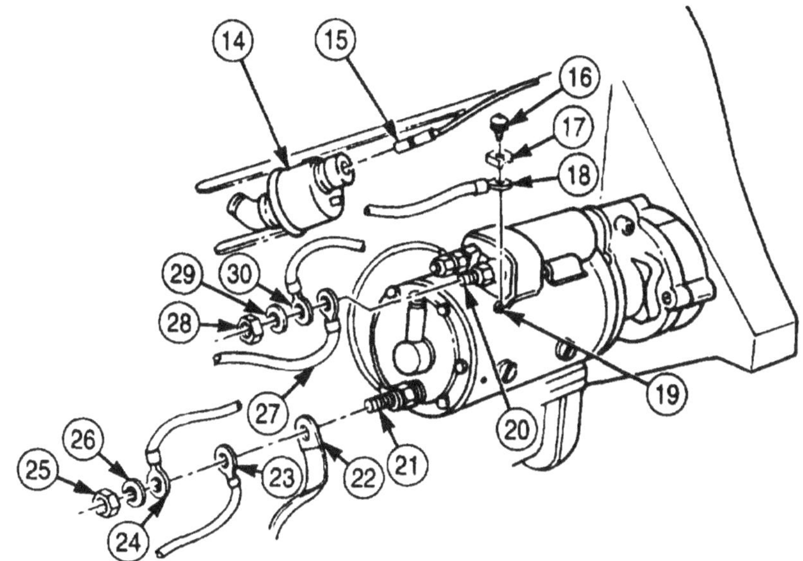

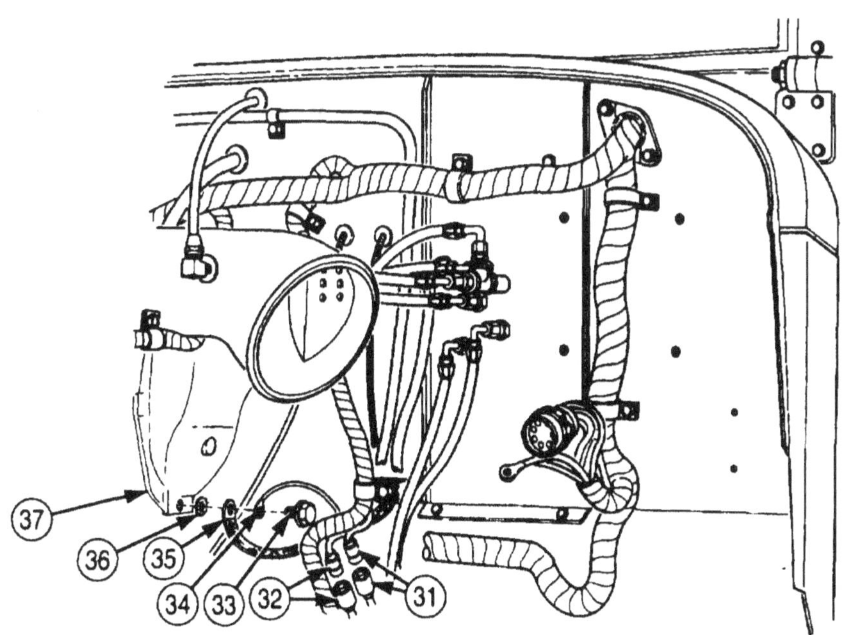

4-66. FRONT WIRING HARNESS (M939/A1) REPLACEMENT (Contd)

77. Connect wires (3) and (2) on ether start fuel pressure switch (1)
78. Connect wire (5) to engine temperature sending unit (4).
79. Install wire (11) on fuel pump solenoid terminal (13) with nut (12).

NOTE

Perform step 80 for all vehicles except M936/A1.

80. Connect fuel pressure transducer wires (14) and (17).
81. Connect tachometer pulse sender wires (15) and (16).
82. Install clamp (9) on harness (6) and install on intake manifold (10) with washer (8) and screw (7).

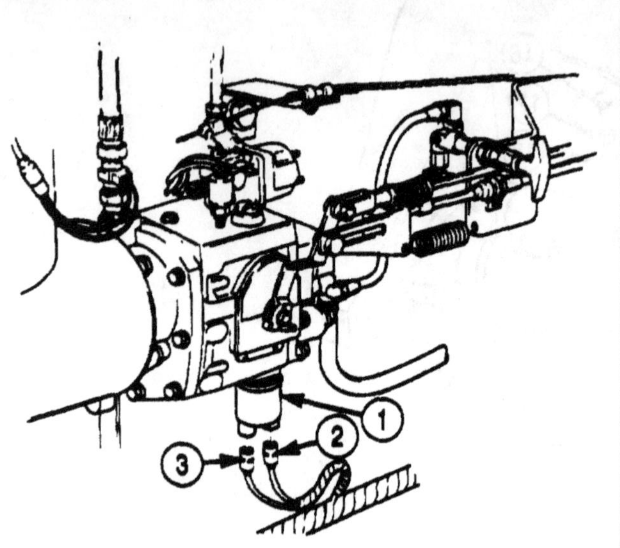

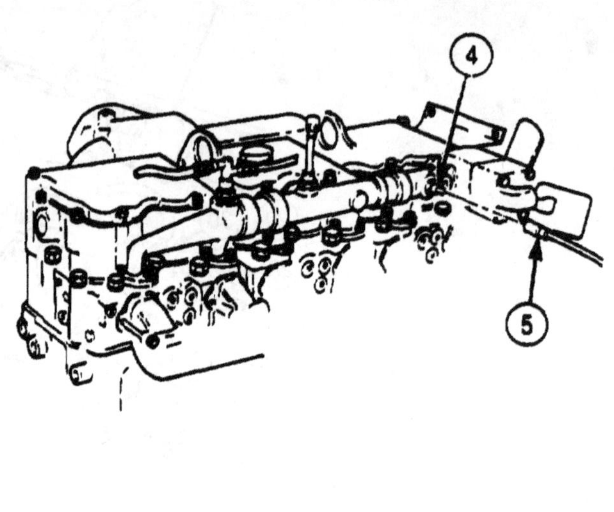

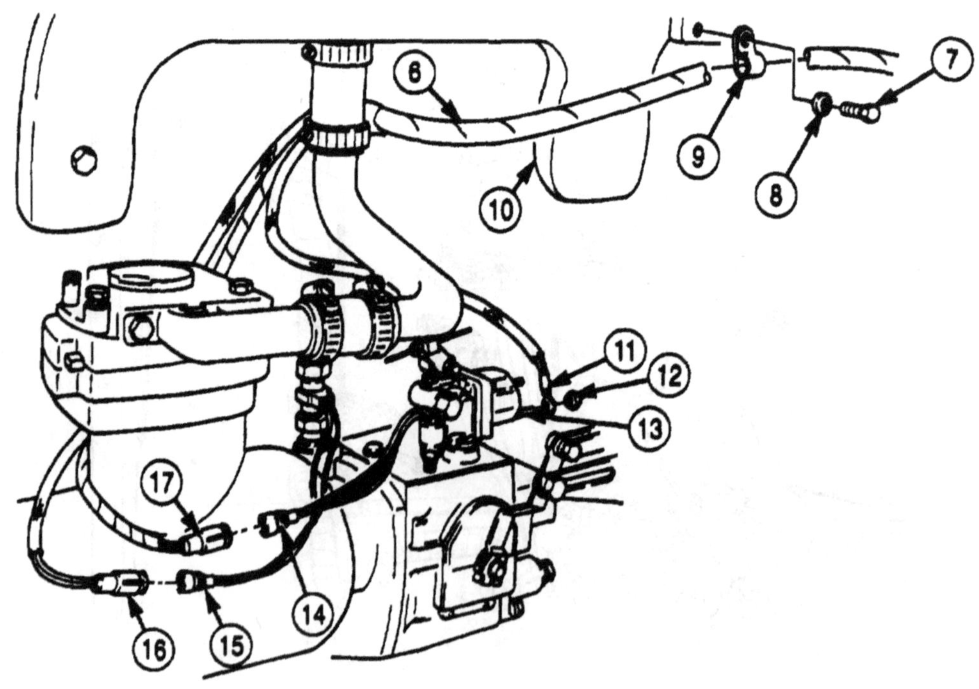

4-66. FRONT WIRING HARNESS (M939/A1) REPLACEMENT (Contd)

83. Connect front wiring harness connector (26) to front lights cable assembly receptacle (18).
84. Install cable clamp (21) on harness (24) and install on radiator (23) with washer (20), screw (19), washer (20), and nut (22).
85. Install seven harness clamps (27) on front wiring harness (6) with seven screws (26) and two nuts (28).

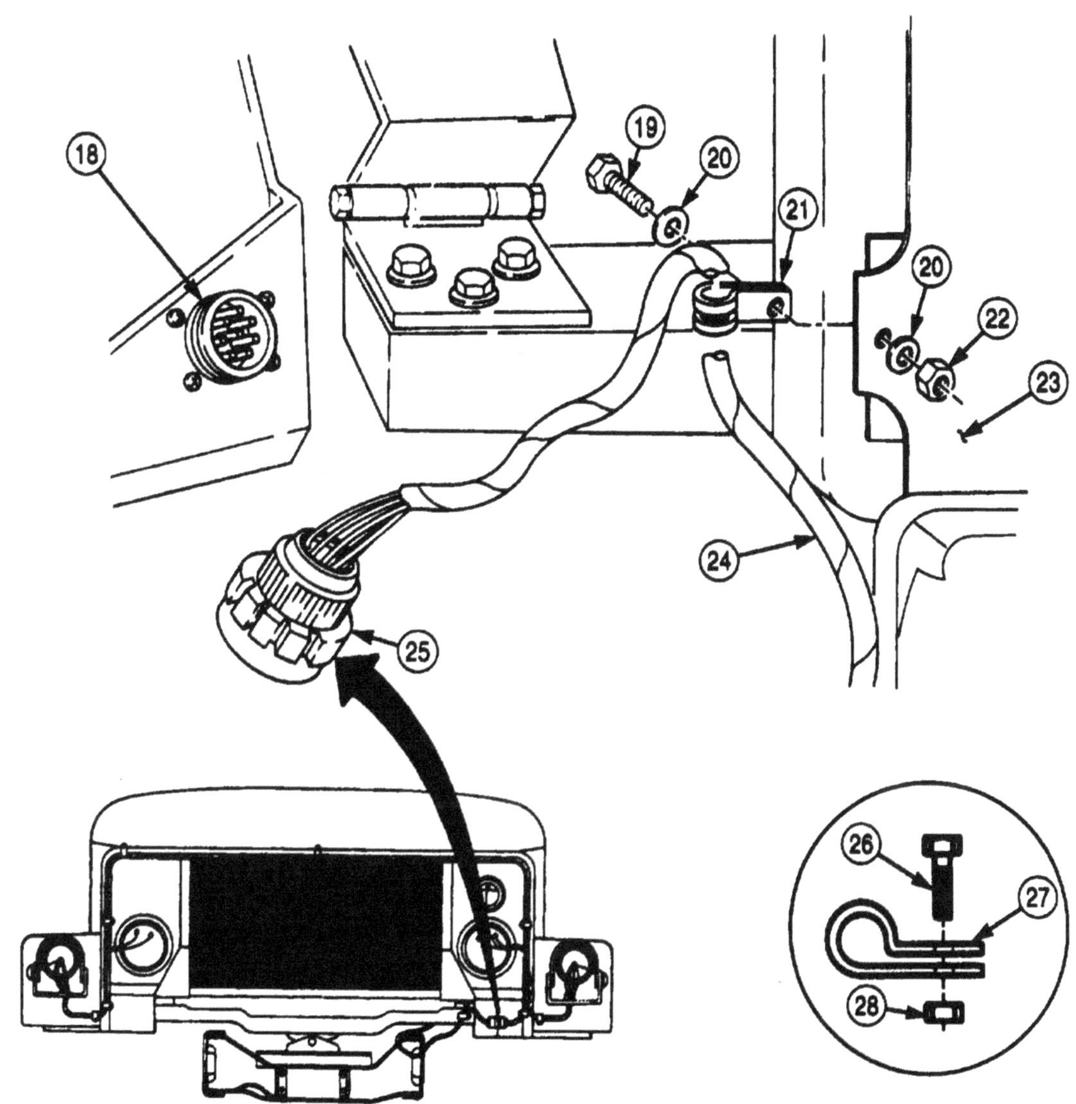

FOLLOW-ON TASKS:
- Install main light switch (para. 3-108).
- Install failsafe warning module (para. 3-106).
- Install turn signal flasher (para. 3-114).
- Install protective control box (para. 3-115).
- Install battery ground cables (para. 3-126).
- Install splash shields (TM 9-2320-272-10).
- Install air intake pipe (para. 3-14).

4-67. FRONT WIRING HARNESS (M939A2) REPLACEMENT

THIS TASK COVERS:

a. Removal **b. Installation**

INITIAL SETUP:

APPLICABLE MODELS
M939A2

TOOLS
General mechanic's tool kit (Appendix E, Item 1)

MATERIALS/PARTS
Cotter pin (Appendix D, Item 53)
Locknut (Appendix D, Item 291)
Spring nut (Appendix D, Item 673)
Tiedown straps (Appendix D, Item 690)
Four lockwashers (Appendix D, Item 416)
Eight lockwashers (Appendix D, Item 371)
Four lockwashers (Appendix D, Item 345)
Two screw-assembled lockwashers (Appendix D, Item 587)
Lockwasher (Appendix D, Item 401)
Lockwasher (Appendix D, Item 403)
Lockwasher (Appendix D, Item 364)
Locknut (Appendix D, Item 282)
Sealing compound (Appendix C, Item 62)

PERSONNEL REQUIRED
TWO

REFERENCES (TM)
TM 9-2320-272-10
TM 9-2320-272-24P

EQUIPMENT CONDITION
- Splash shields removed (TM 9-2320-272-10).
- Protective control box removed (para. 3-115).
- Turn signal flasher removed (para. 3-114).
- Failsafe wiring module disconnected (para. 3-106).
- Main headlight switch disconnected (para. 3-108).
- Air intake pipe removed (para. 3-14).

---

a. Removal

NOTE

- For inspection and repair of front wiring harness, refer to para. 3-131.
- Tag wires and connectors for installation.

1. Remove nut (4), lockwasher (3), and wire (6) from solenoid (1). Discard lockwasher (3).
2. Remove nut (8), lockwasher (7), battery cable (10), and wire (9) from solenoid (1). Discard lockwasher (7).
3. Remove nut (5), lockwasher (2), and wire (12) from starter (16). Discard lockwasher (2).
4. Remove nut (13), lockwasher (ll), and wires (14) and (15) from starter (16).
5. Remove two screw-assembled lockwashers (32) and terminal cover (31) from alternator (26). Discard screw-assembled lockwashers (32).

NOTE

Sealing compound must be removed from wiring before removing wires.

6. Disconnect wire (34) from lead (35).
7. Remove screw (36), lockwasher (37), and wire (38) from alternator (26). Discard lockwasher (37).
8. Remove nut (28), lockwasher (27), and wire (39) from alternator (26). Discard lockwasher (27).
9. Remove nut (30), lockwasher (29), and wire (33) from alternator (26). Discard lockwasher (29).
10. Remove screw (51), clamp (52), and front wiring harness (50) from engine block (40).
11. Remove screw (46), washer (45), ground strap (44), and ground wires (43), (42), and (41) from engine block (40).
12. Remove screw (48), washer (47), clamp (49), and wiring harnesses (50) and (53) from engine block (40).
13. Disconnect wire (25) from coolant temperature sending unit (24).

4-67. FRONT WIRING HARNESS (M939A2) REPLACEMENT (Contd)

14. Remove locknut (20), washer (21), screw (23), washer (21), clamp (22), and front wiring harness (18) from radiator (19). Discard locknut (20).
15. Disconnect wiring harness (18) from front lights cable receptacle (17).

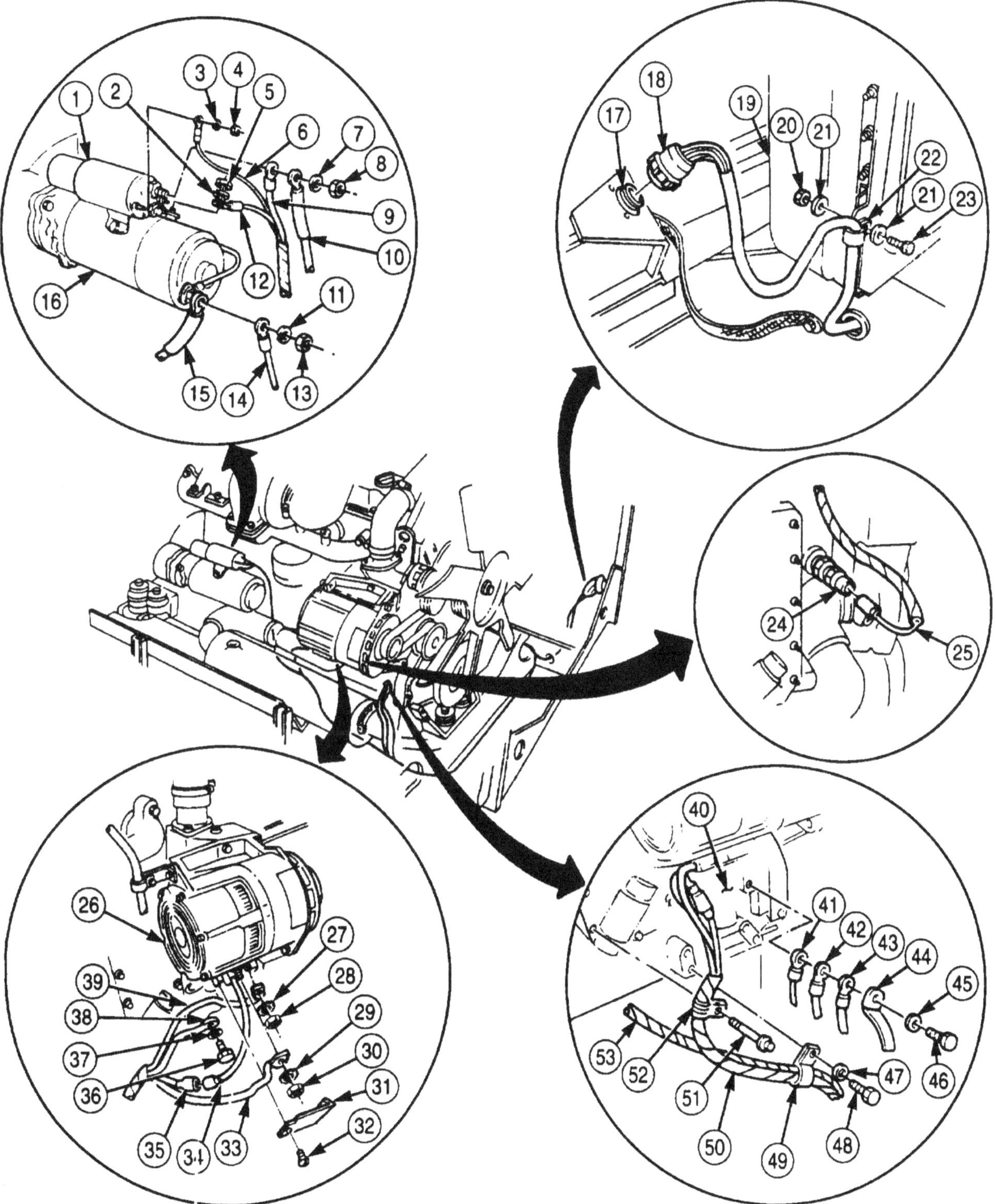

4-67. FRONT WIRING HARNESS (M939A2) REPLACEMENT (Contd)

16. Remove two screws (7) and clamps (6) from firewall (5).
17. Remove two nuts (10), washers (9), screws (12), and clamps (13) from frame rail (8).
18. Disconnect wire (11) from oil pressure sending unit (14).
19. Remove two screws (1), washers (2), lockwashers (4), and grommets (3) from firewall (5).
20. Disconnect wires (16) and (17) from horn solenoid (15).
21. Disconnect wire (29) from transorb diode coupling (30).
22. Disconnect wires (20) and (21) from circuit breaker (19).
23. Disconnect wires (23) and (24) from circuit breaker (22).
24. Disconnect wire (31) from personnel hot water heater wire (32).
25. Remove three screws (28) and clamps (18) from firewall (5).
26. Remove screw (26), washer (27), and clamp (25) from firewall (5).

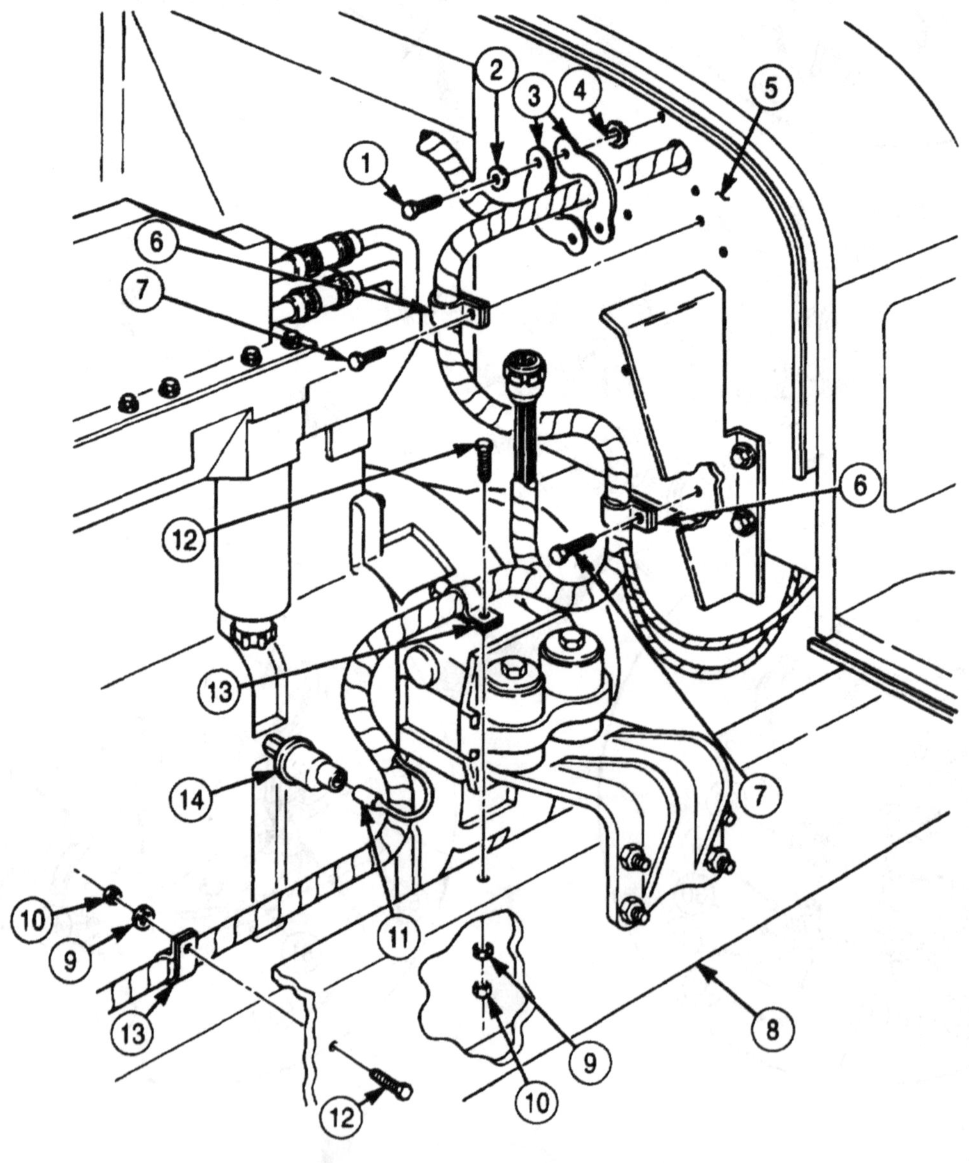

4-67. FRONT WIRING HARNESS (M939A2) REPLACEMENT (Contd)

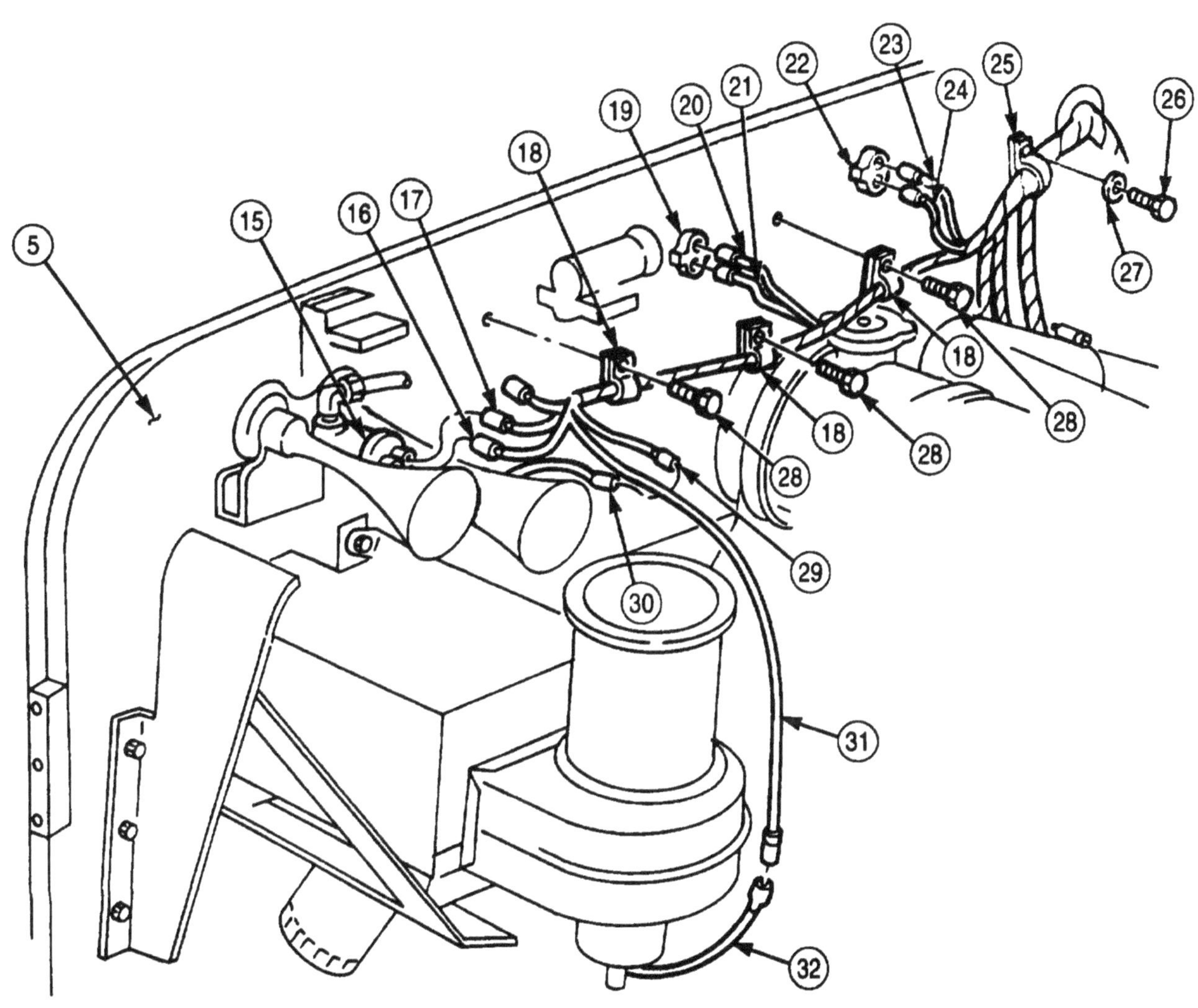

4-67. FRONT WIRING HARNESS (M939A2) REPLACEMENT (Contd)

27. Disconnect connector (17) from throttle control solenoid (20).
28. Disconnect connector (18) from fuel pressure transducer (19).
29. Disconnect wires (10) and (12) from two connectors (11).
30. Remove screw (5), washer (4), and ground wire (6) from flywheel housing (16).
31. Remove screw (15), washer (14), clamp (13), bracket (7), and ground strap (9) from firewall (8).
32. Remove screw (1), washer (2), and clamp (3) from firewall (8).
33. Remove tiedown straps (21) from dryer heater wire (22). Discard tiedown straps (21).
34. Disconnect wire (23) from dryer heater wire (22).

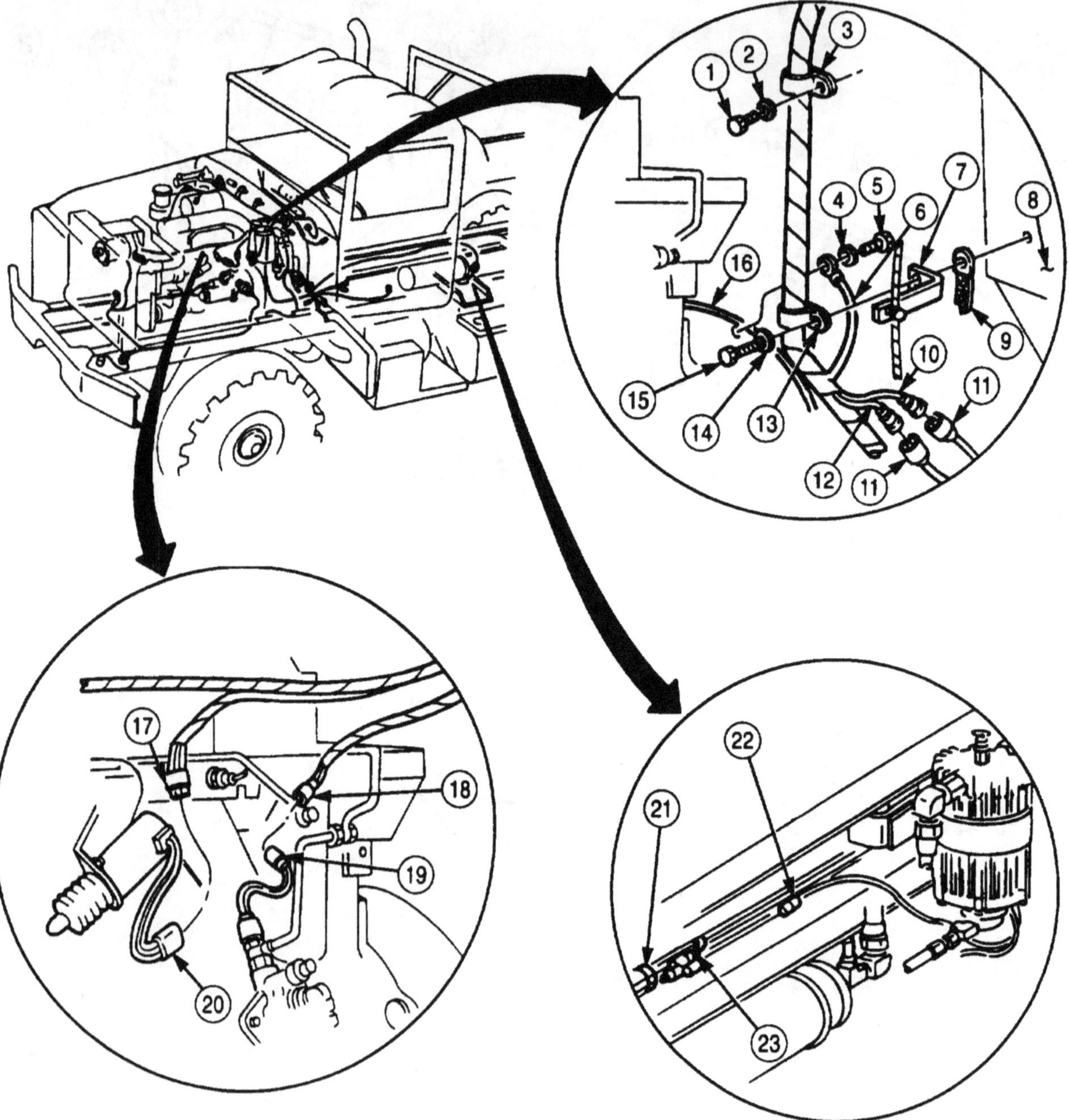

4-67. FRONT WIRING HARNESS (M939A2) REPLACEMENT (Contd)

35. Disconnect three wires (25) from high-beam selector switch (26).

NOTE

Perform step 36 for M929A2, M930A2, M931A2, M932A2, and M936A2 vehicles only.

36. Disconnect wire (30) from wire (31).
37. Remove four nuts (27), lockwashers (28), and screws (24) from front wiring harness connector (29) and rear wiring harness connector (32). Discard lockwashers (28).
38. Disconnect front wiring harness connector (29) from rear wiring harness connector (32).
39. Disconnect wire (33) from parking brake switch wire (34).
40. Disconnect wire (35) from transfer case switch capacitor lead (36).

NOTE

Perform step 41 for M936A2 vehicles only.

41. Disconnect wire (38) from 5th gear lock-up capacitor lead (37).

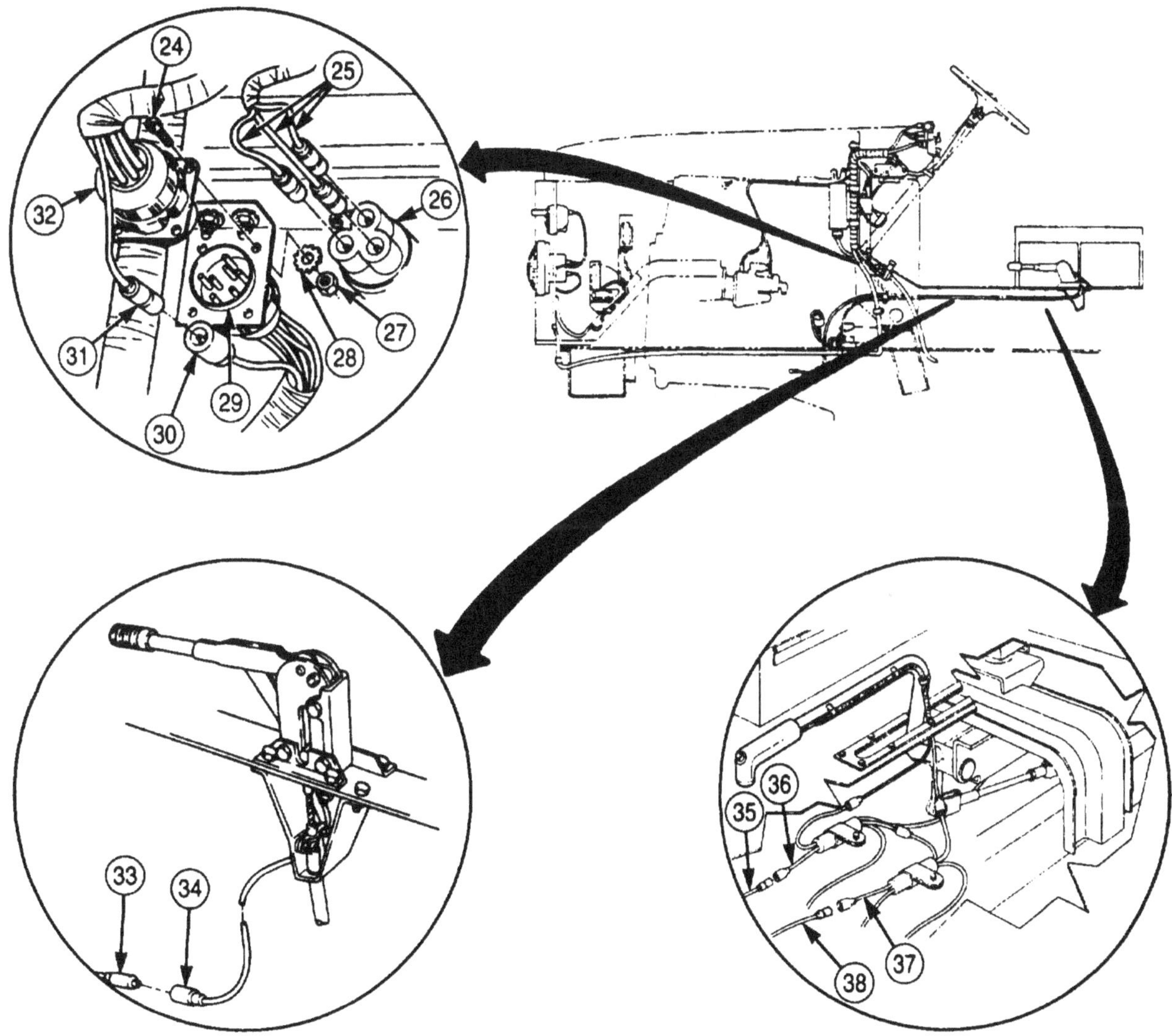

4-67. FRONT WIRING HARNESS (M939A2) REPLACEMENT (Contd)

42. Disconnect front harness connector (2) from turn signal control (1).
43. Disconnect wire (4) from horn switch (3).
44. Disconnect two wires (5) from electrical gauge circuit breaker (8).
45. Disconnect two wires (6) from heater blower motor circuit breaker (7).
46. Disconnect connector (10) from tachometer pulse sensor (9).

NOTE

Perform step 47 for M936A2 vehicles only.

47. Disconnect wire (11) from warning signal lamp switch (12).
48. Disconnect three wires (17) from heater blower motor switch (18).
49. Disconnect two wires (19) from ether start switch wires (20).
50. Disconnect wire (13) from CTIS blackout connector (14).
51. Disconnect CTIS power supply connector (16) from CTIS wiring harness connecter (15).

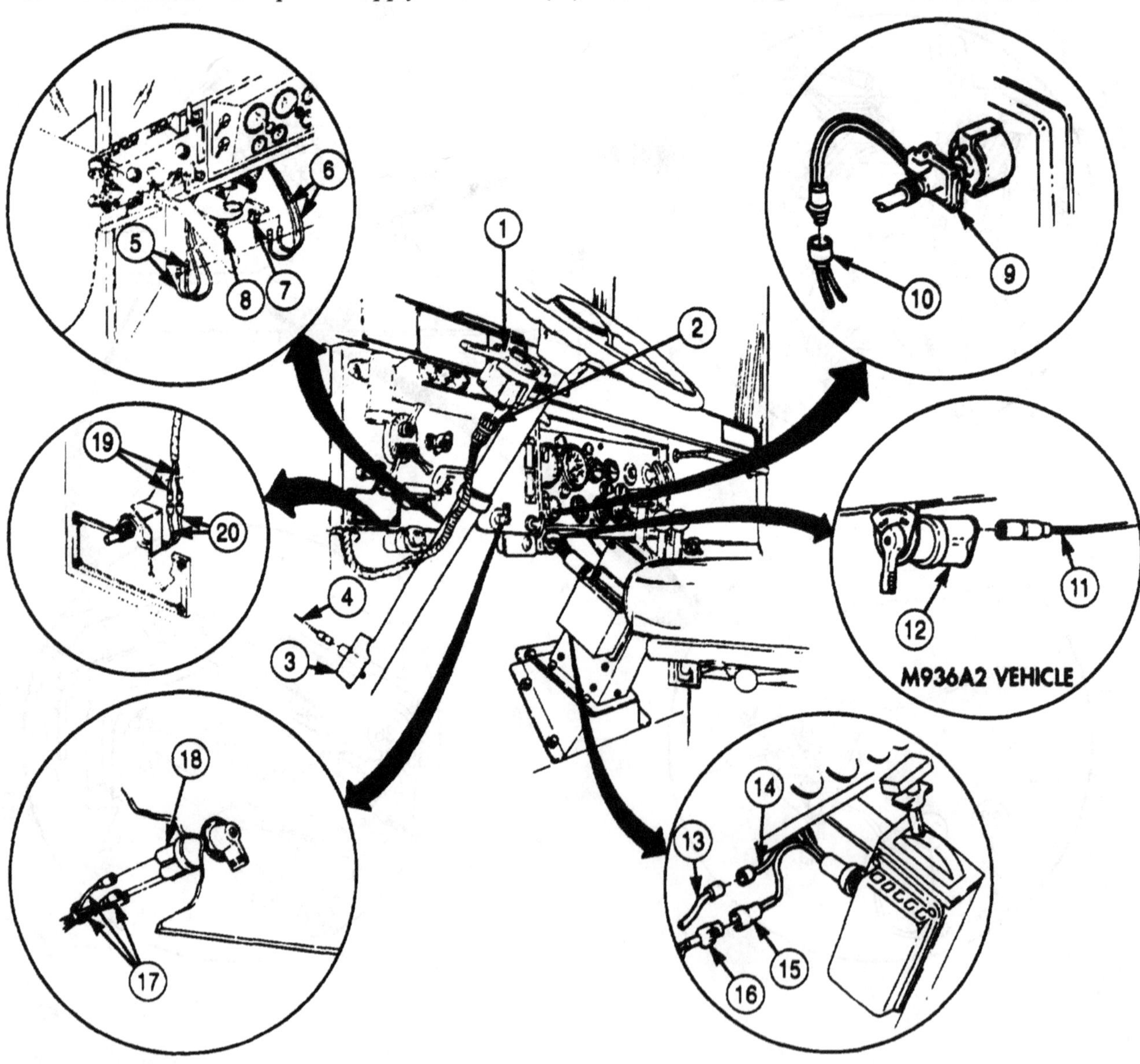

4-67. FRONT WIRING HARNESS (M939A2) REPLACEMENT (Contd)

52. Disconnect connector (24) from front wheel drive lock-in switch (21).

NOTE

Perform step 53 for M929A2, M930A2, M931A2, M932A2, and M936A2 vehicles only.

53. Disconnect three wires (23) from fuel selector switch (22).
54. Remove four screws (32) and warning light panel (25) from instrument panel (26).
55. Disconnect wire (27) from parking brake indicator light (37).
56. Disconnect wire (28) from low air pressure indicator light (36).
57. Disconnect wire (29) from spring brake override indicator light (35).
58. Disconnect wire (30) from axle lock-in indicator light (34).
59. Disconnect wire (31) from high-beam indicator light (33).
60. With selector (39) in N (neutral), remove four screws (43), lockwashers (44), and transmission selector (38) from tower (40). Discard lockwashers (44).
61. Disconnect three wires (42) from wires (41).

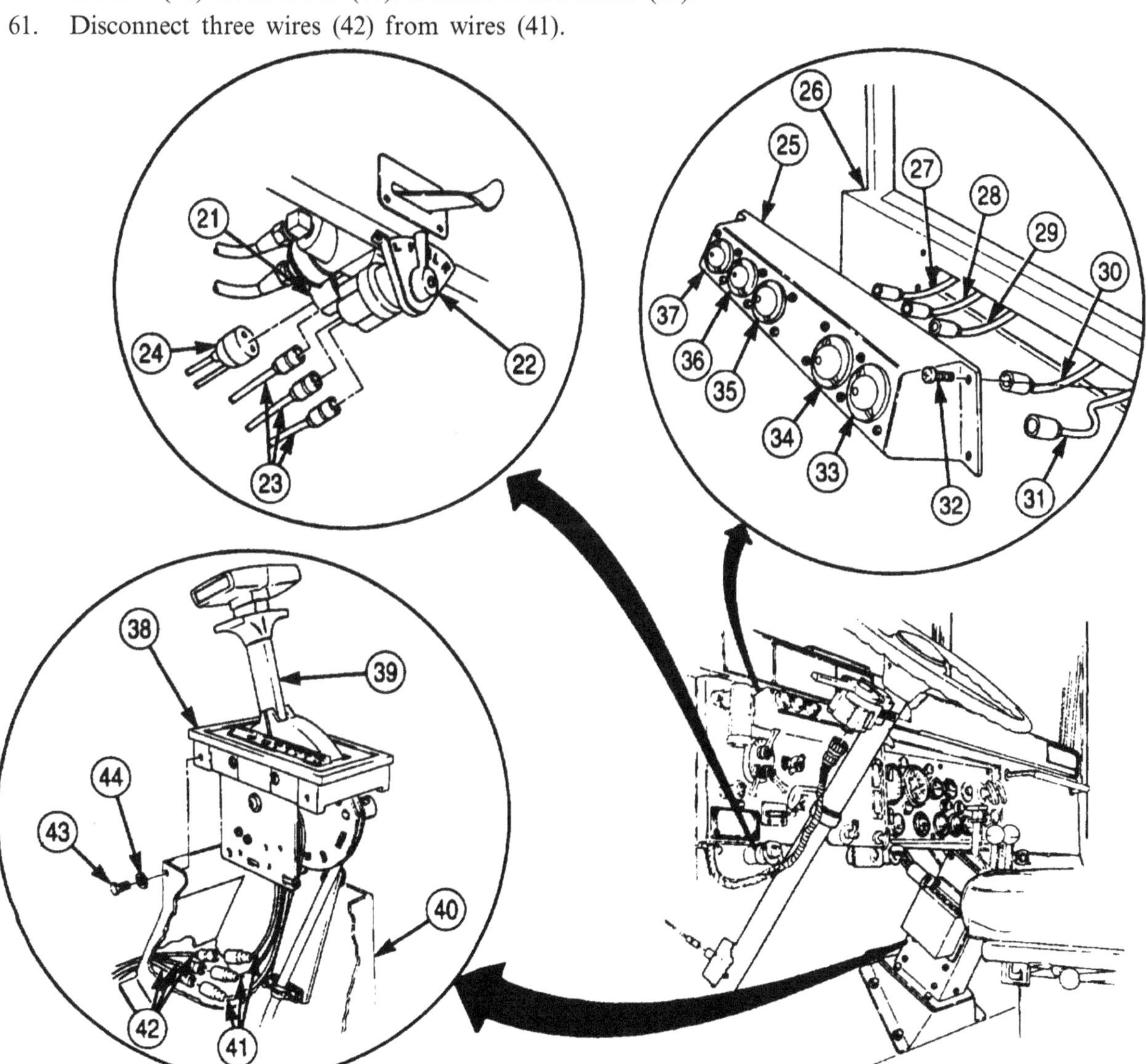

4-67. FRONT WIRING HARNESS (M939A2) REPLACEMENT (Contd)

62. Remove eight screws (3) and instrument cluster (1) from instrument panel (2).
63. Remove screw (7), clamp (8), and retainer nut (9) from fresh air control cable (6) and bracket (10).
64. Remove cotter pin (11), fresh air control cable (6), and spring nut (5) from heater (4). Discard cotter pin (11) and spring nut (5).
65. Disconnect three wires (16) from battery switch (15).
66. Disconnect three wires (17) from starter switch (14).
67. Disconnect wire (28) from instrument cluster wiring harness (27).
68. Disconnect five wires (18), (19), (22), (24), and (31) from five instrument cluster lights (12).
69. Disconnect wire (20) from fuel gauge (21).
70. Disconnect wire (25) from oil pressure gauge (26).
71. Disconnect wire (29) from transmission oil temperature gauge (30).
72. Disconnect wire (23) from engine temperature gauge (13).
73. Disconnect two wires (32) from spring brake pressure switch (33).

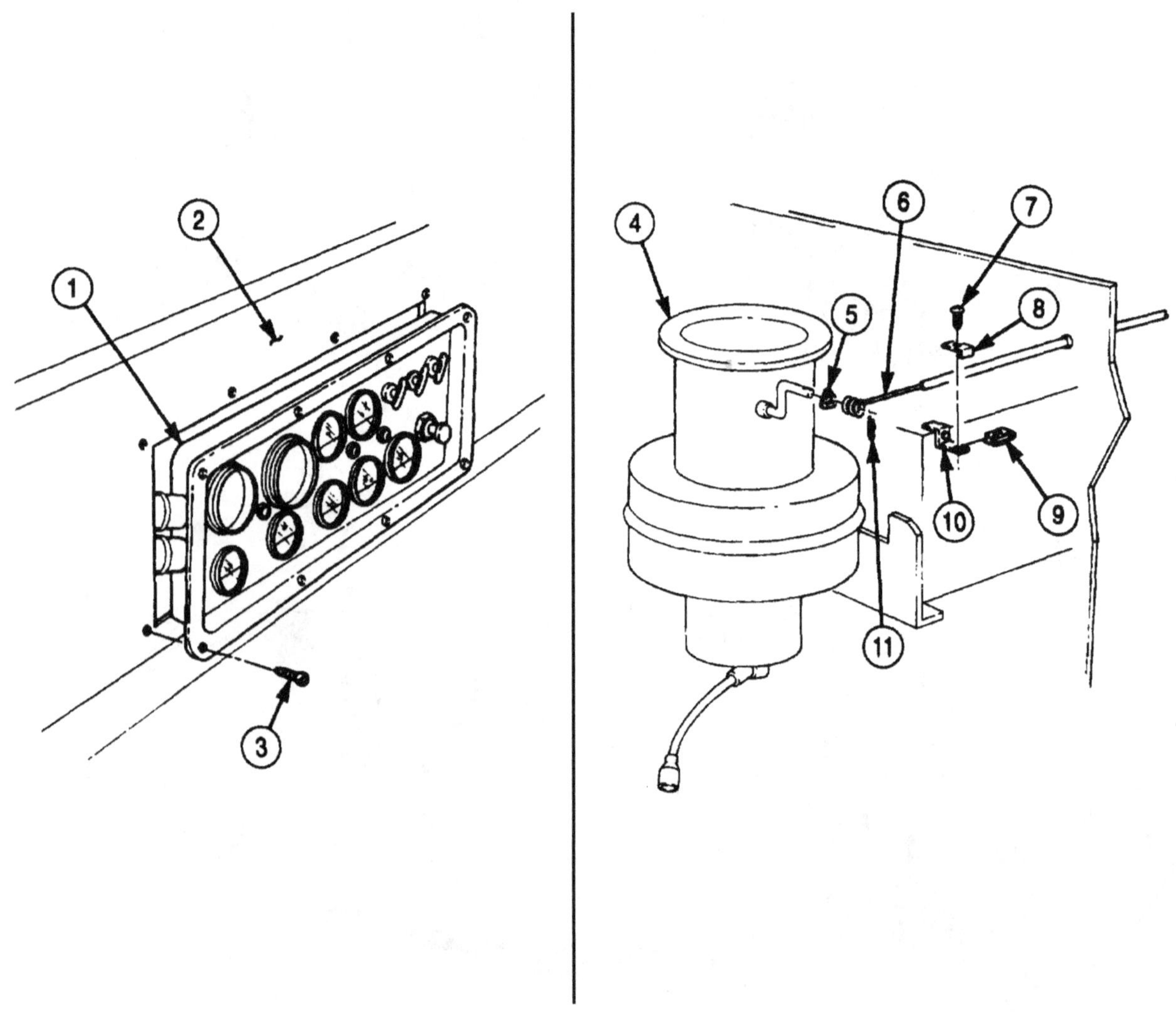

4-67. FRONT WIRING HARNESS (M939A2) REPLACEMENT (Contd)

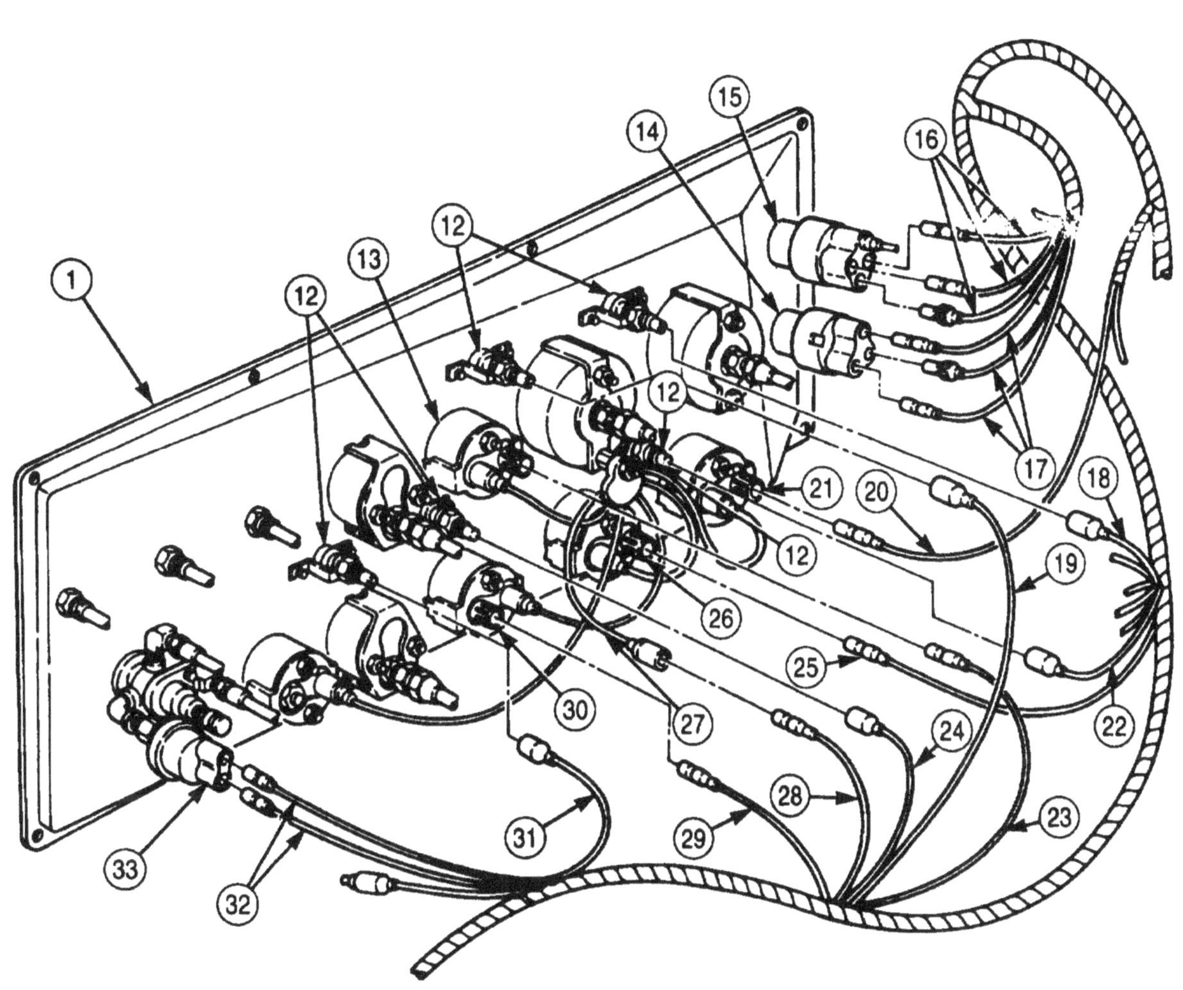

4-67. FRONT WIRING HARNESS (M939A2) REPLACEMENT (Contd)

NOTE

Perform step 74 for M936A2 vehicles only.

74. Disconnect wire (2) from floodlight switch and auxiliary receptacle wire (1).
75. Remove cap (11) from diagnostic connector (16).
76. Remove nut (7), lockwasher (8), ground wire (9), and screw (13) from mounting bracket (15). Discard lockwasher (8).
77. Remove nut (5), lockwasher (6), cap chain (10), and screw (12) from mounting bracket (15). Discard lockwasher (6).
78. Remove two nuts (3), lockwashers (4), screws (14), and diagnostic connector (16) from mounting bracket (15). Discard lockwashers (4).

NOTE

After harness wires have been removed in steps 79 through 82, cables and hardware should be installed on terminal adapters for installation.

79. Remove nut (22), screw (23), battery cables (26) and (25), and wire (24) from terminal clamp (19).
80. Remove nut (27), screw (29), battery cable (28), and wire (30) from terminal clamp (17).
81. Remove nut (31), screw (35), battery cables (32) and (34), and wire (33) from terminal clamp (18).

NOTE

Assistant will help with step 82.

82. Carefully remove front wiring harness (21) from vehicle (20).

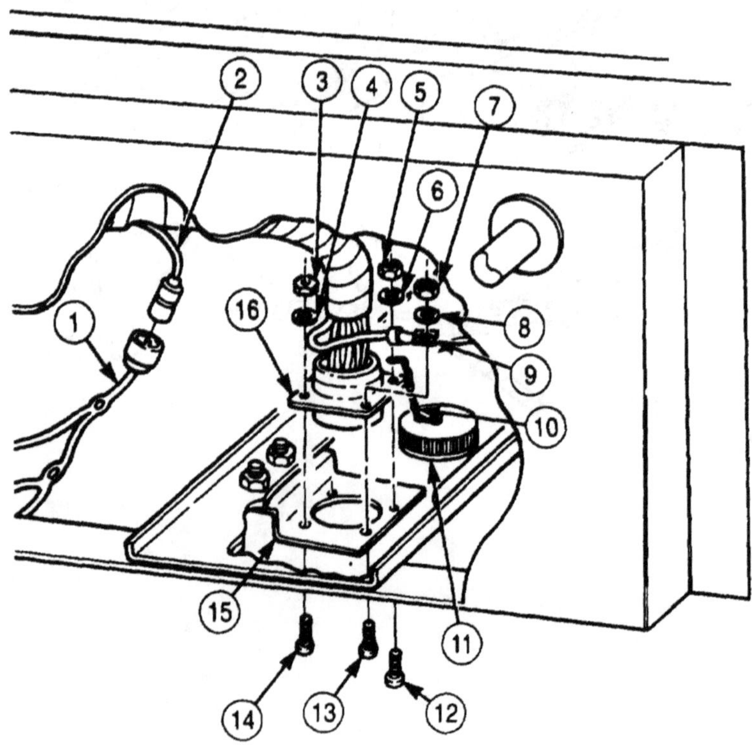

4-67. FRONT WIRING HARNESS (M939A2) REPLACEMENT (Contd)

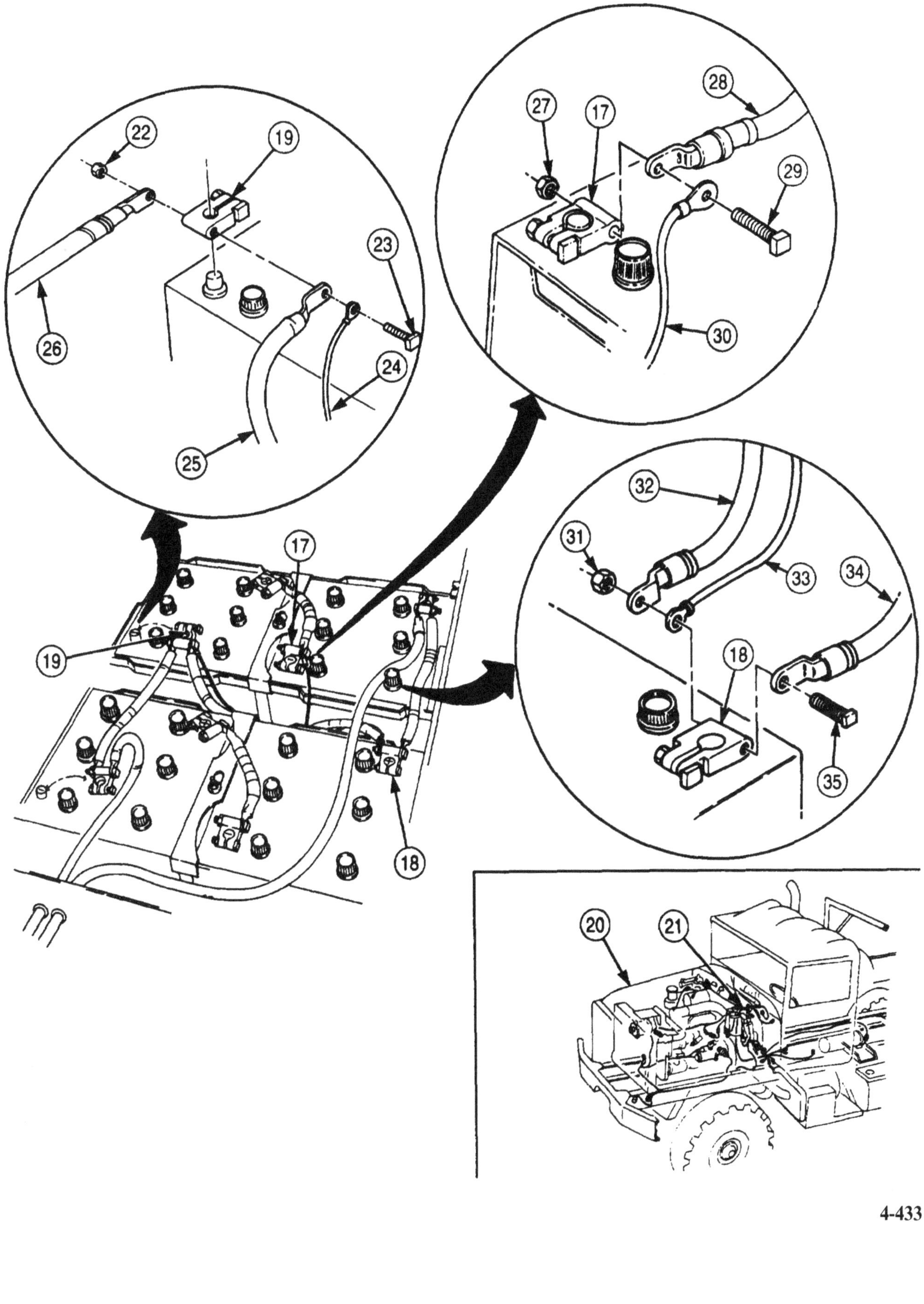

22
19
23
26
24
25
27
17
28
29
30
32
31
33
34
18
35
17
19
18
20
21

4-67. FRONT WIRING HARNESS (M939A2) REPLACEMENT (Contd)

b. Installation

CAUTION

Use care when routing harness. Snagging may result, and forceful pulling will cause damage to harness.

NOTE

Assistant will help with step 1.

1. Position front wiring harness (2) in the vehicle (1), routing branches in approximate positions and through firewall (3).
2. Connect two wires (24) to spring brake pressure switch (25).
3. Connect wire (15) to engine temperature gauge (5).
4. Connect wire (21) to transmission oil temperature gauge (22).
5. Connect wire (17) to oil pressure gauge (18).
6. Connect wire (12) to fuel gauge (13).
7. Connect five wires (10), (11), (14), (16), and (23) to five instrument cluster lights (4).
8. Connect wire (20) to instrument cluster wiring harness (19).
9. Connect three wires (9) to starter switch (6).
10. Connect three wires (8) to battery switch (7).
11. Install fresh air control cable (28) on heater (26) with new spring nut (27) and new cotter pin (33).
12. Install clamp (30) on bracket (32) with clamp (30), nut (31), and screw (29).
13. Install instrument cluster assembly (34) on instrument panel (35) with eight screws (36).

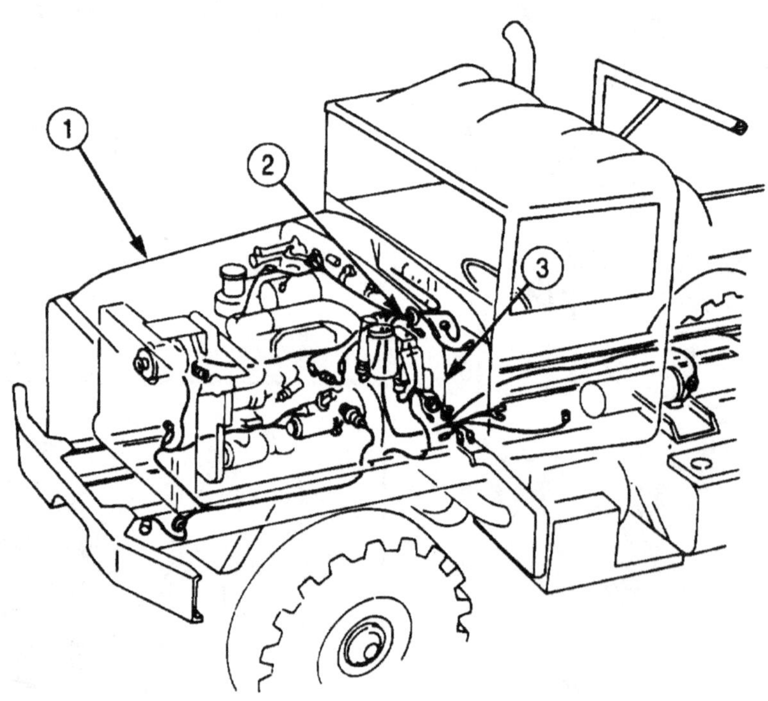

4-67. FRONT WIRING HARNESS (M939A2) REPLACEMENT (Contd)

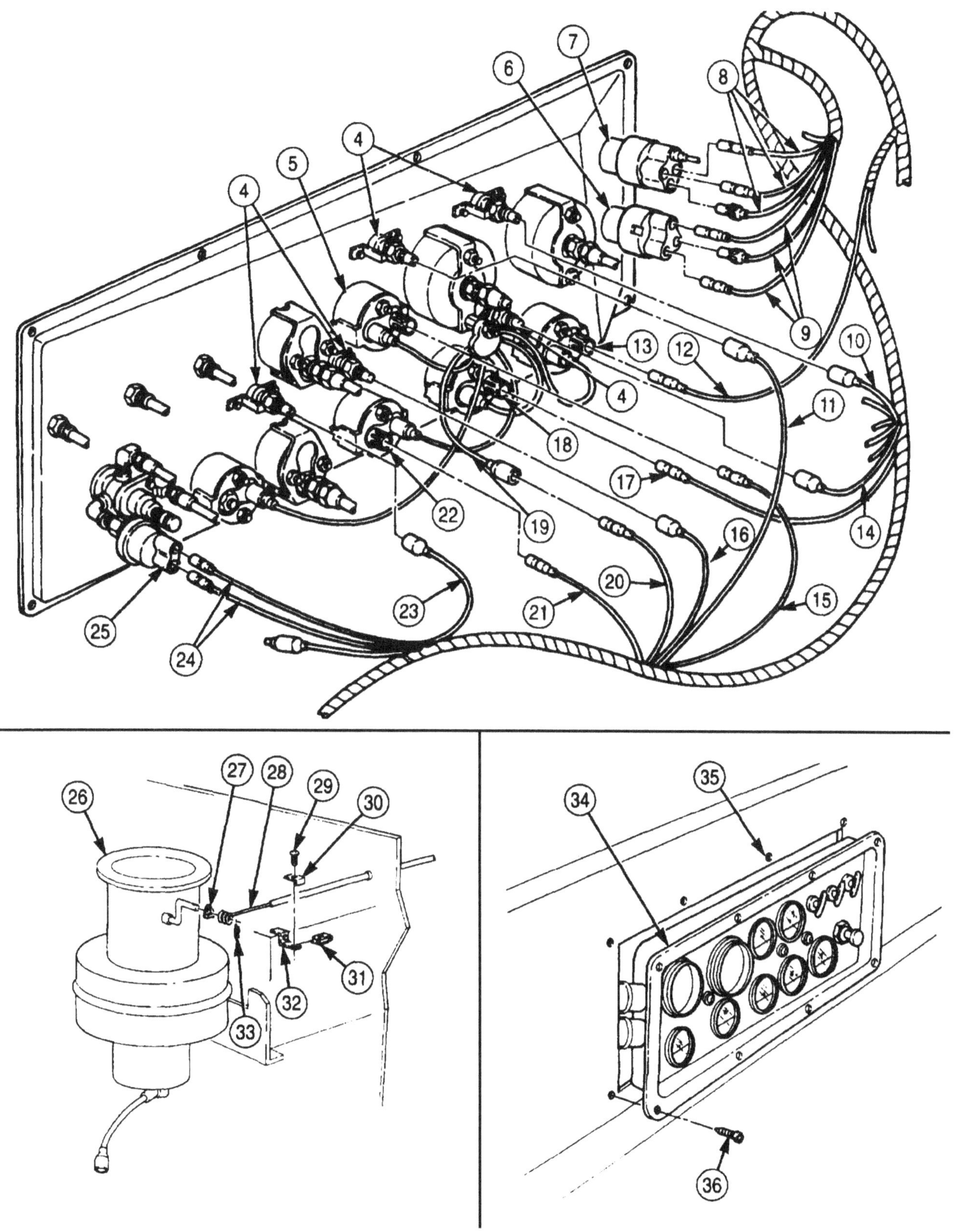

4-67. FRONT WIRING HARNESS (M939A2) REPLACEMENT (Contd)

14. Install wire (15) and battery cables (16) and (14) on terminal clamp (3) with screw (17) and nut (13).
15. Install wire (12) and battery cable (10) on terminal clamp (2) with screw (11) and nut (9).
16. Install wire (6) and battery cables (7) and (8) on terminal clamp (1) with screw (5) and nut (4).
17. Install diagnostic connector (20) on mounting bracket (33) with two screws (32), new lockwashers (22), and nuts (23).
18. Install cap chain (28) on mounting bracket (33) with screw (30), new lockwasher (25), and nut (24).
19. Install ground wire (21) on mounting bracket (33) with screw (31), new lockwasher (27), and nut (26).
20. Install cap (29) on diagnostic connector (20).

NOTE

Perform step 21 for M936A2 vehicles only.

21. Connect wire (19) to floodlight switch and auxiliary receptacle wire (18).

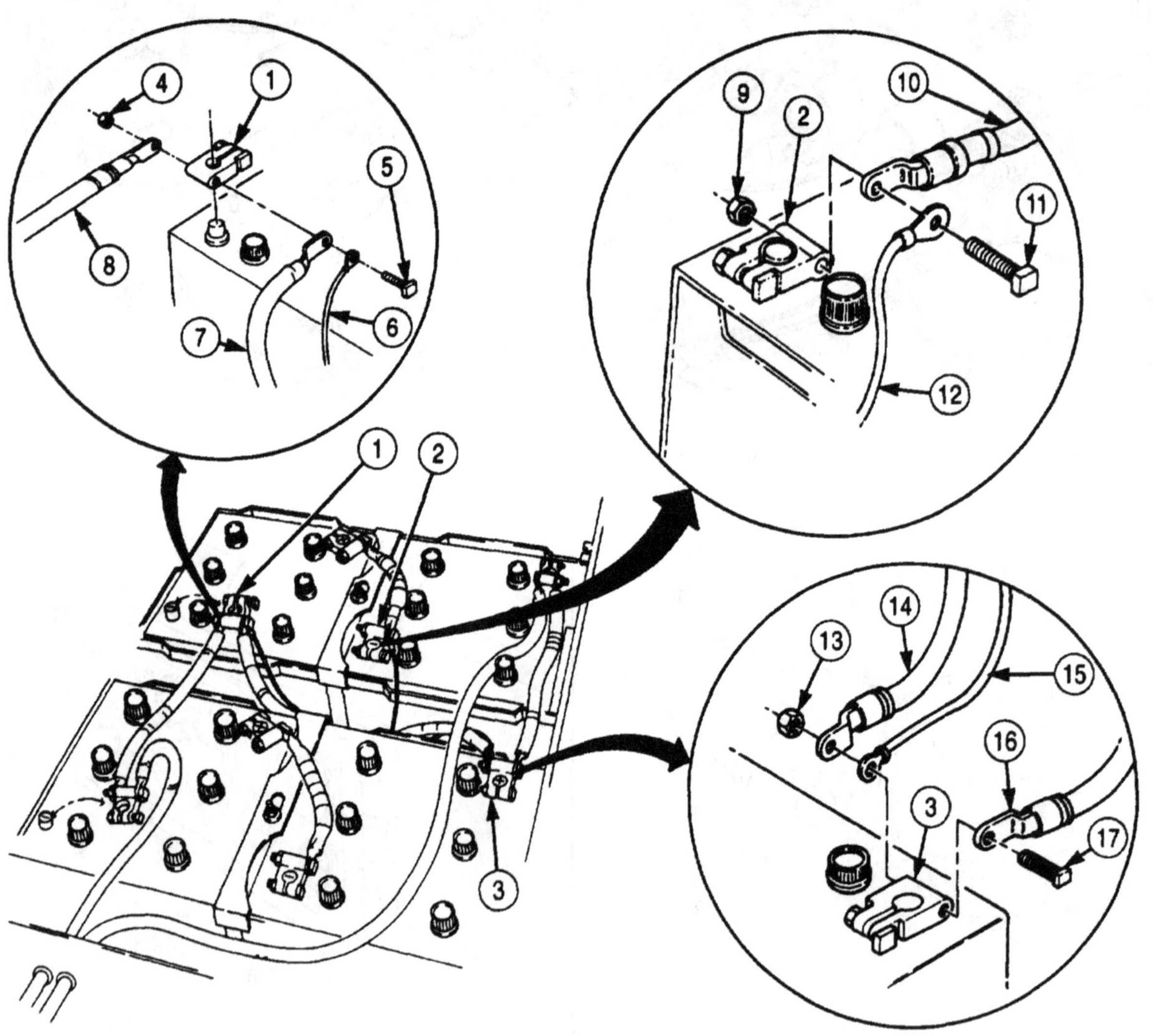

4-67. FRONT WIRING HARNESS (M939A2) REPLACEMENT (Contd)

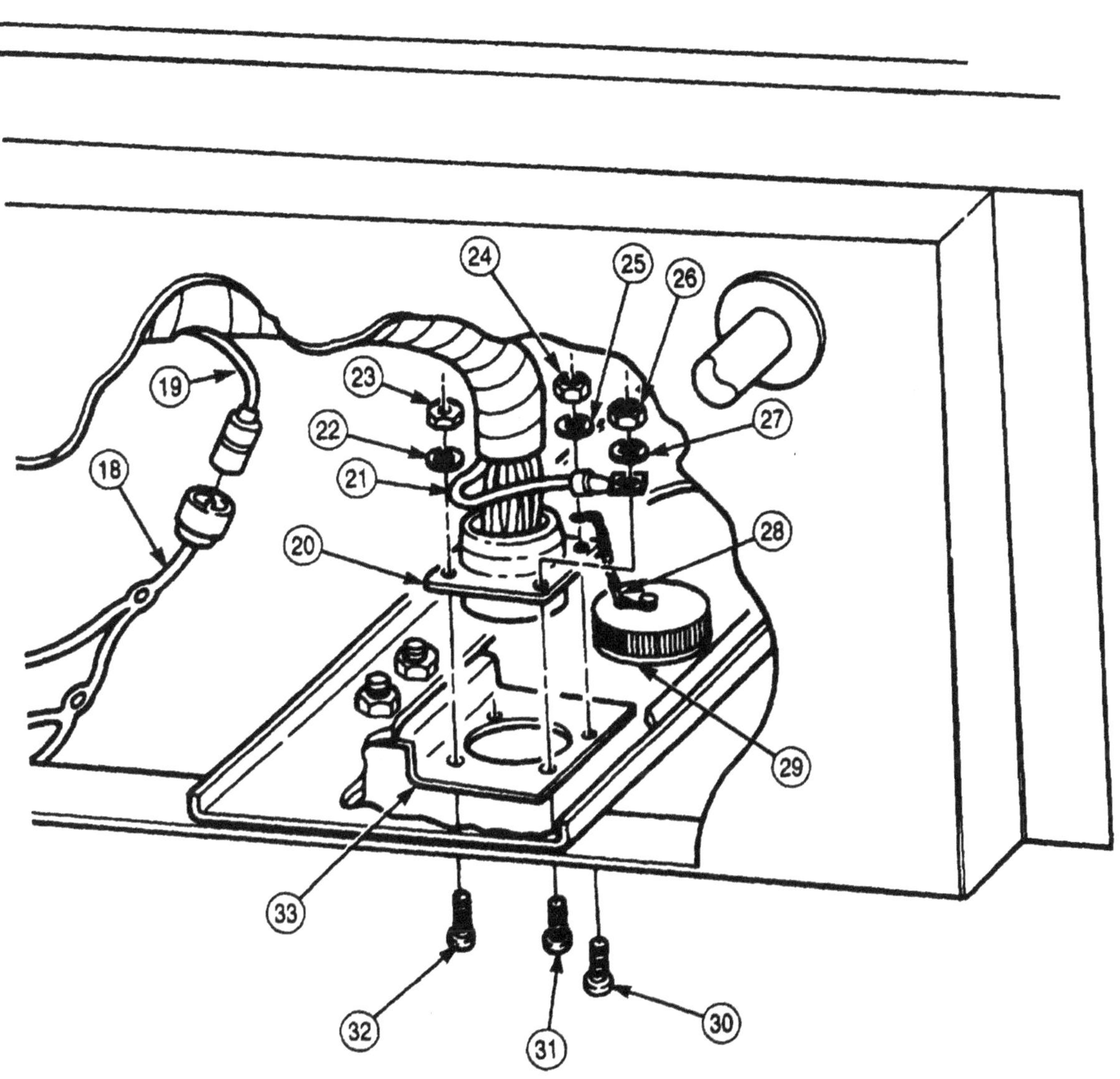

4-67. FRONT WIRING HARNESS (M939A2) REPLACEMENT (Contd)

22. Connect three wires (5) to wires (4).
23. With selector (2) in N (neutral), install transmission selector (1) on tower (3) with four new lockwashers (7) and screws (6).
24. Connect wire (14) to high-beam indicator light (16).
25. Connect wire (13) to axle lock-in indicator light (17).
26. Connect wire (12) to spring brake override indicator light (18).
27. Connect wire (11) to low pressure indicator light (19).
28. Connect wire (10) to parking brake indicator light (20).
29. Install warning light panel (8) on instrument panel (9) with four screws (15).

NOTE

Perform step 30 for M929A2, M930A2, M931A2, M932A2, and M936A2 vehicles only.

30. Connect three wires (22) to fuel selector switch (21).
31. Connect connector (23) to front wheel drive lock-in switch (24).

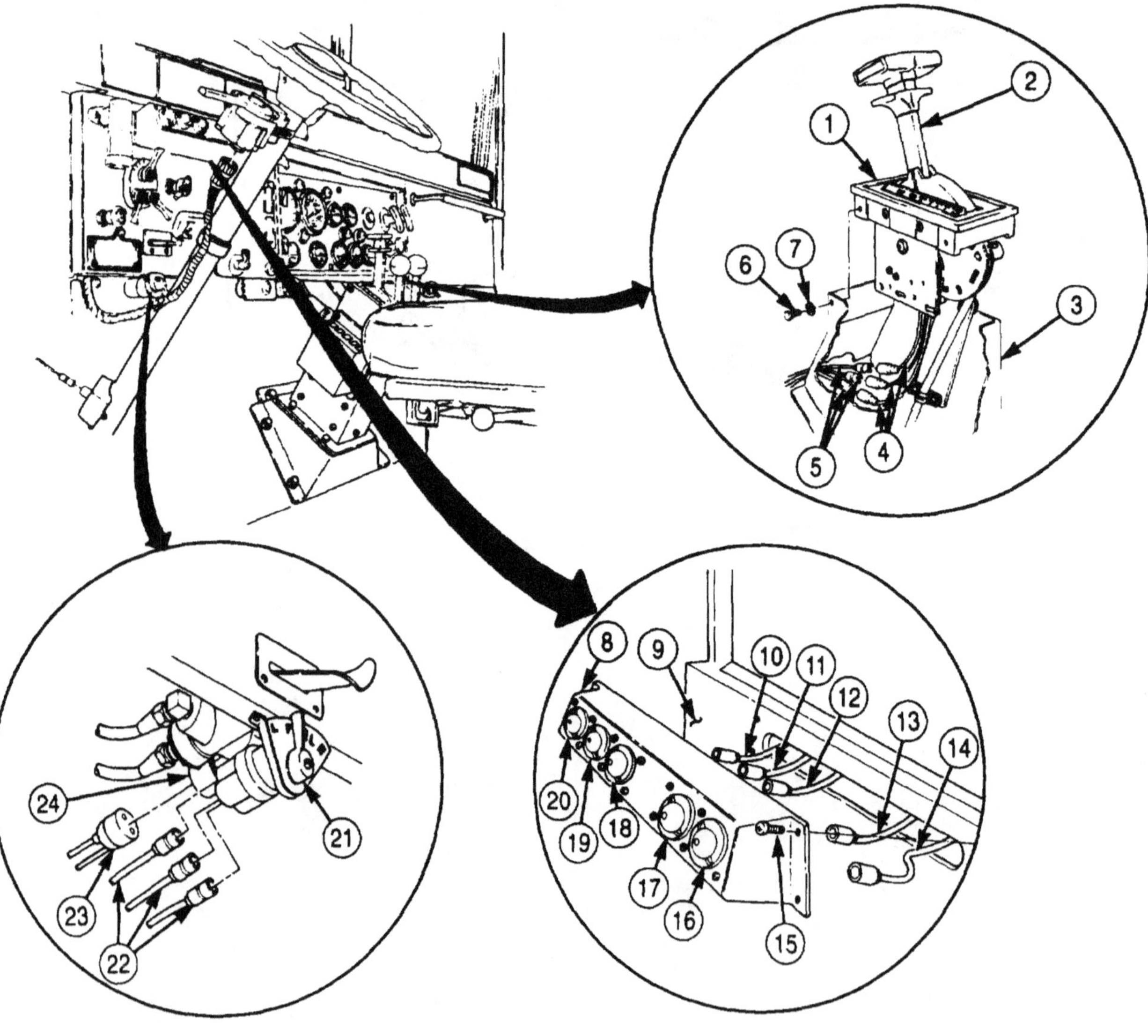

4-67. FRONT WIRING HARNESS (M939A2) REPLACEMENT (Contd)

32. Connect two wires (29) to ether start switch wires (30).
33. Connect three wires (32) to heater blower motor switch (31).

NOTE

Perform step 34 for M936A2 vehicles only.

34. Connect wire (34) to warning signal lamp switch (33).
35. Connect wire (36) to CTIS and blackout connector (37).
36. Connect CTIS power supply connector (35) to CTIS wiring harness (38).
37. Connect wire (28) to horn switch (27).
38. Connect front harness connector (25) to turn signal control (26).
39. Connect connector (44) to tachometer pulse sensor (43).
40. Connect two wires (40) to electrical gauge circuit breaker (39).
41. Connect two wires (42) to heater blower motor circuit breaker (41).

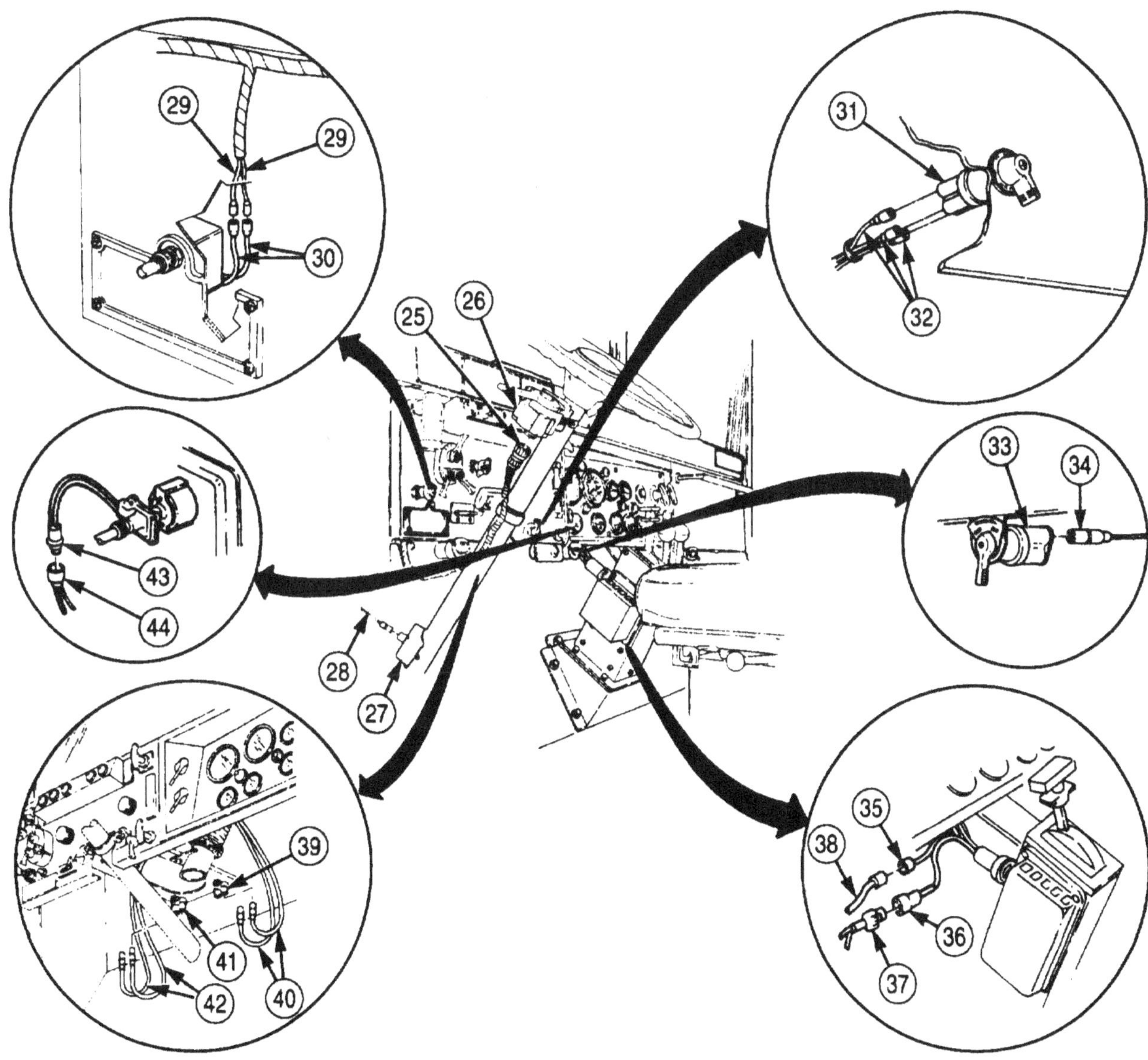

4-67. FRONT WIRING HARNESS (M939A2) REPLACEMENT (Contd)

42. Connect wire (1) to transfer case switch capacitor lead (2).

NOTE

Perform step 43 for M936A2 vehicles.

43. Connect wire (4) to 5th gear lock-up capacitor lead (3).
44. Connect wire (5) to parking brake switch wire (6).
45. Connect front wiring harness connector (15) to rear wiring harness connector (12).
46. Install four screws (7), new lockwashers (11), and nuts (10) on front wiring harness connector (15) and rear wiring harness connector (12).

NOTE

Perform step 47 for M929A2, M930A2, M931A2, M932A2, and M936A2 vehicles only.

47. Connect wire (14) to wire (13).
48. Connect three wires (8) to high-beam selector switch (9).

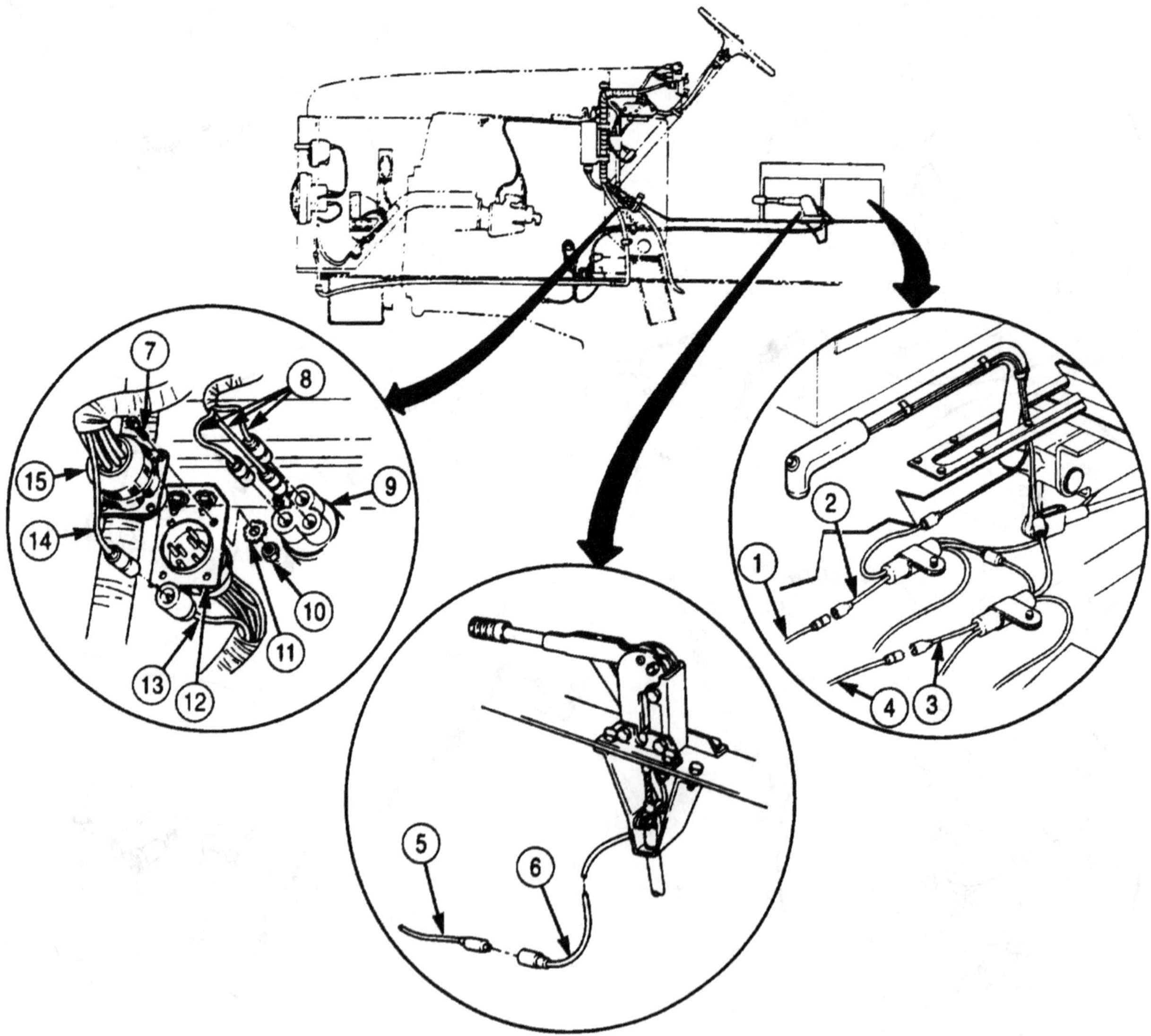

4-67. FRONT WIRING HARNESS (M939A2) REPLACEMENT (Contd)

49. Connect wire (18) to air dryer heater wire (17).
50. Install new tiedown straps (16) on wire (18).
51. Install clamp (19) on firewall (24) with washer (33) and screw (32).
52. Install ground strap (23), bracket (25), and clamp (28) on firewall (24) with washer (29) and screw (30).
53. Install ground wire (20) on flywheel housing (31) with washer (21) and screw (22).
54. Connect two wires (26) to connectors (27).
55. Connect connector (35) to fuel pressure transducer (36).
56. Connect connector (34) to throttle control solenoid (37).

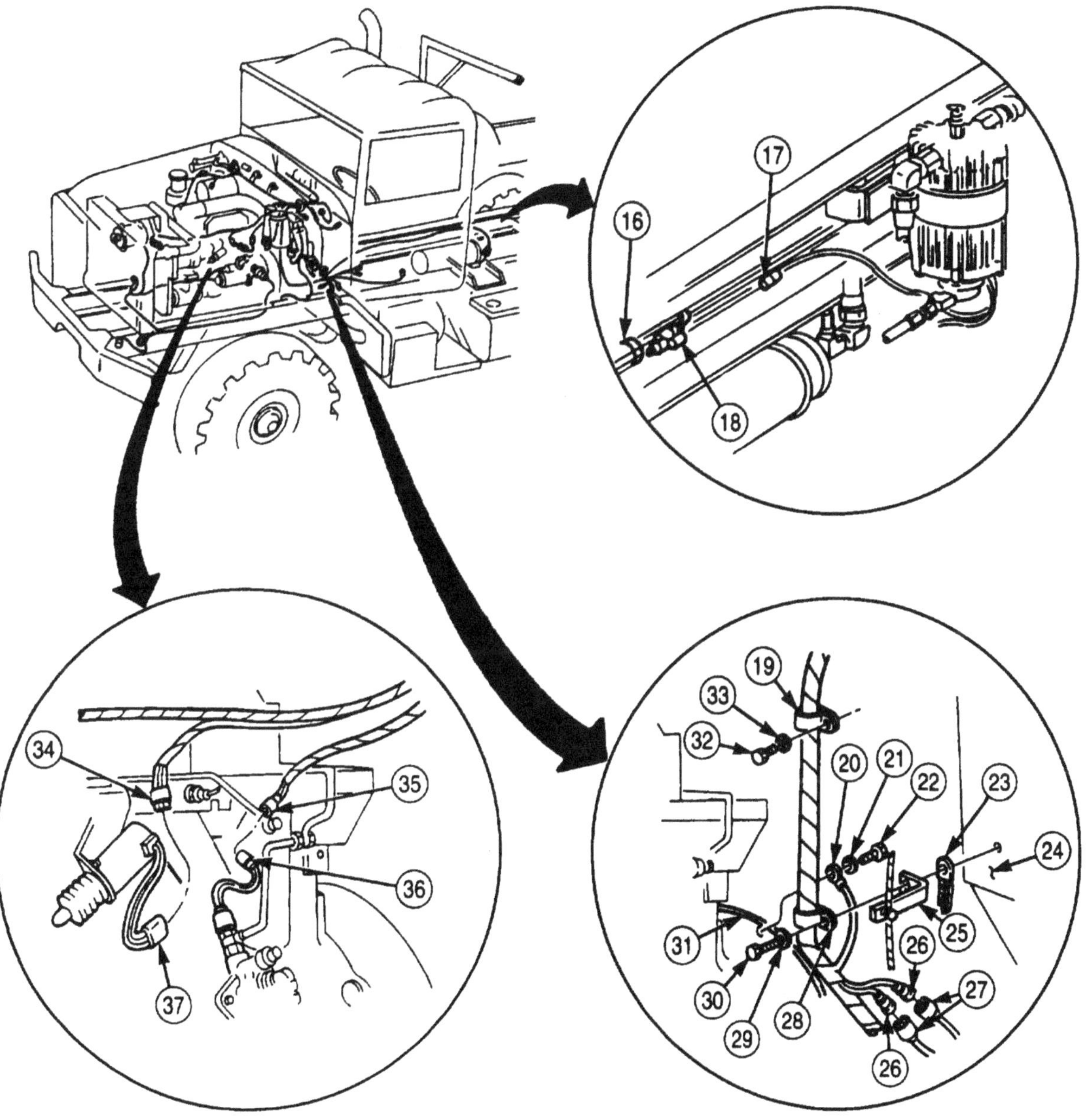

4-67. FRONT WIRING HARNESS (M939A2) REPLACEMENT (Contd)

57. Install clamp (12) on right firewall (1) with washer (14) and screw (13).
58. Install three clamps (5) on right firewall (1) with three screws (15).
59. Connect wire (18) to personnel hot water heater wire (19).
60. Connect wires (10) and (11) to circuit breaker (9).
61. Connect wires (7) and (8) to circuit breaker (6).
62. Connect wire (16) to transorb diode coupling (17).
63. Connect wires (3) and (4) to horn solenoid (2).
64. Install grommets (22) on left firewall (24) with two lockwashers (23), washers (21), and screws (20).
65. Connect wire (30) to oil sending unit (33).
66. Install two clamps (32) on frame rail (27) with screws (31), washers (28), and nuts (29).
67. Install two clamps (25) on left firewall (24) with two screws (26).

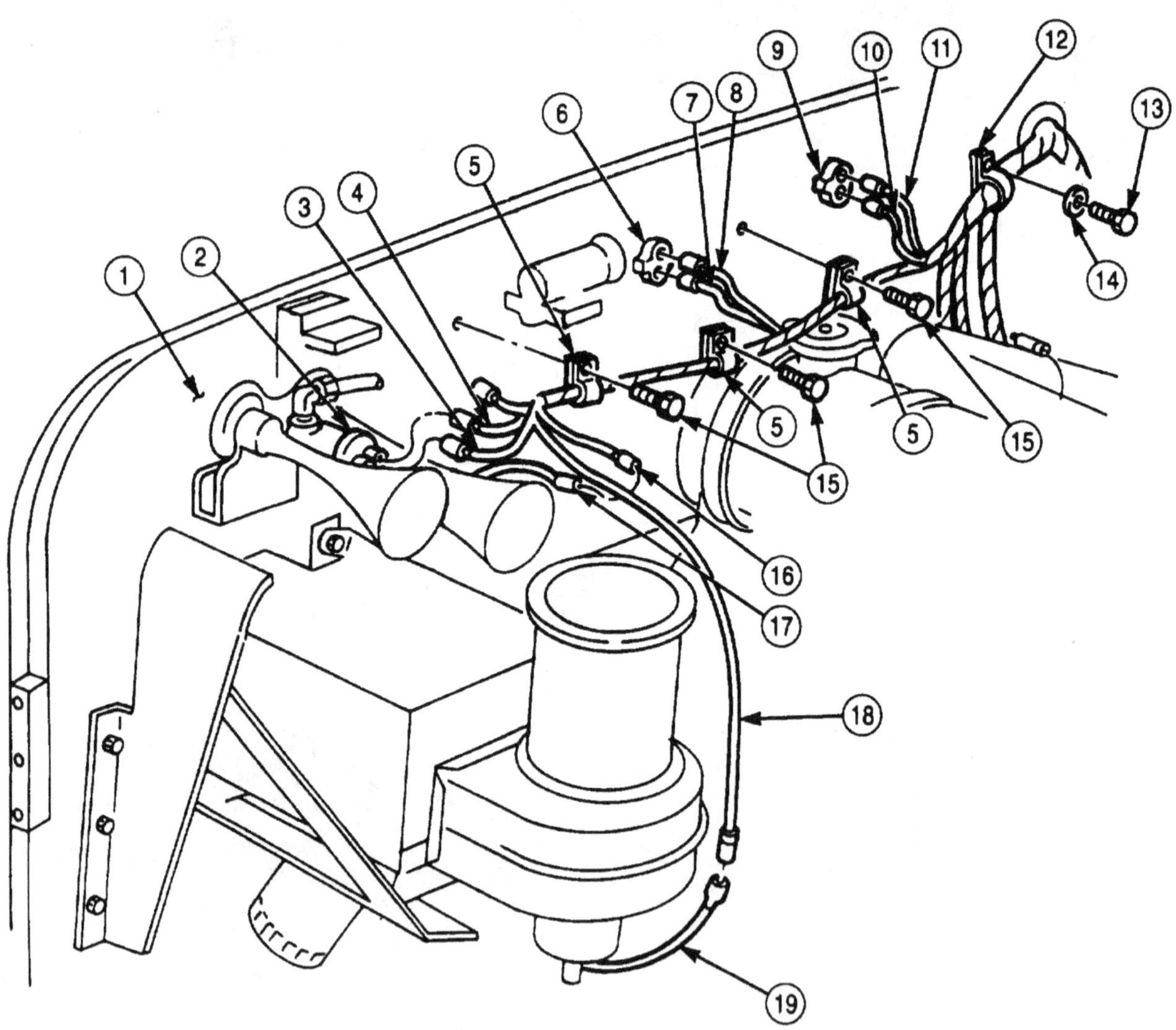

4-67. FRONT WIRING HARNESS (M939A2) REPLACEMENT (Contd)

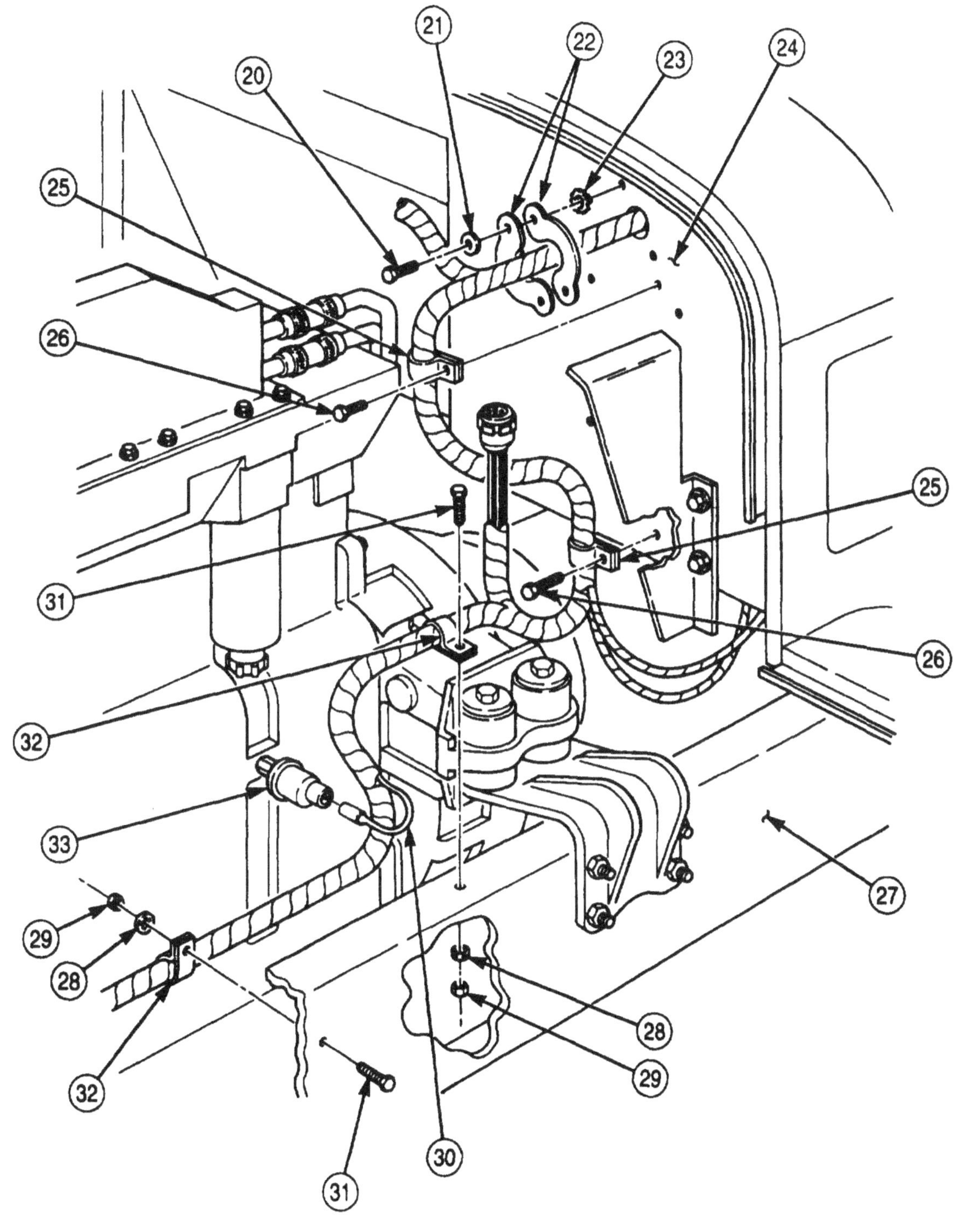

4-67. FRONT WIRING HARNESS (M939A2) REPLACEMENT (Contd)

68. Install front wiring harness (1) on front crossmember (2).
69. Connect connector (4) to front lights cable receptacle (3).
70. Install front wiring harness (1) and clamp (8) on radiator (5) with washer (7), screw (9), washer (7), and new locknut (6)
71. Connect wire (24) to coolant temperature sending unit (25).
72. Install wiring harnesses (1) and (20) and clamp (19) on engine block (10) with washer (17) and screw (18).
73. Install ground wires (11), (12), and (13) and ground strap (14) on engine block (10) with washer (15) and screw (16).
74. Install front wiring harness branch (20) and clamp (22) on oil cooler (23) with screw (21).

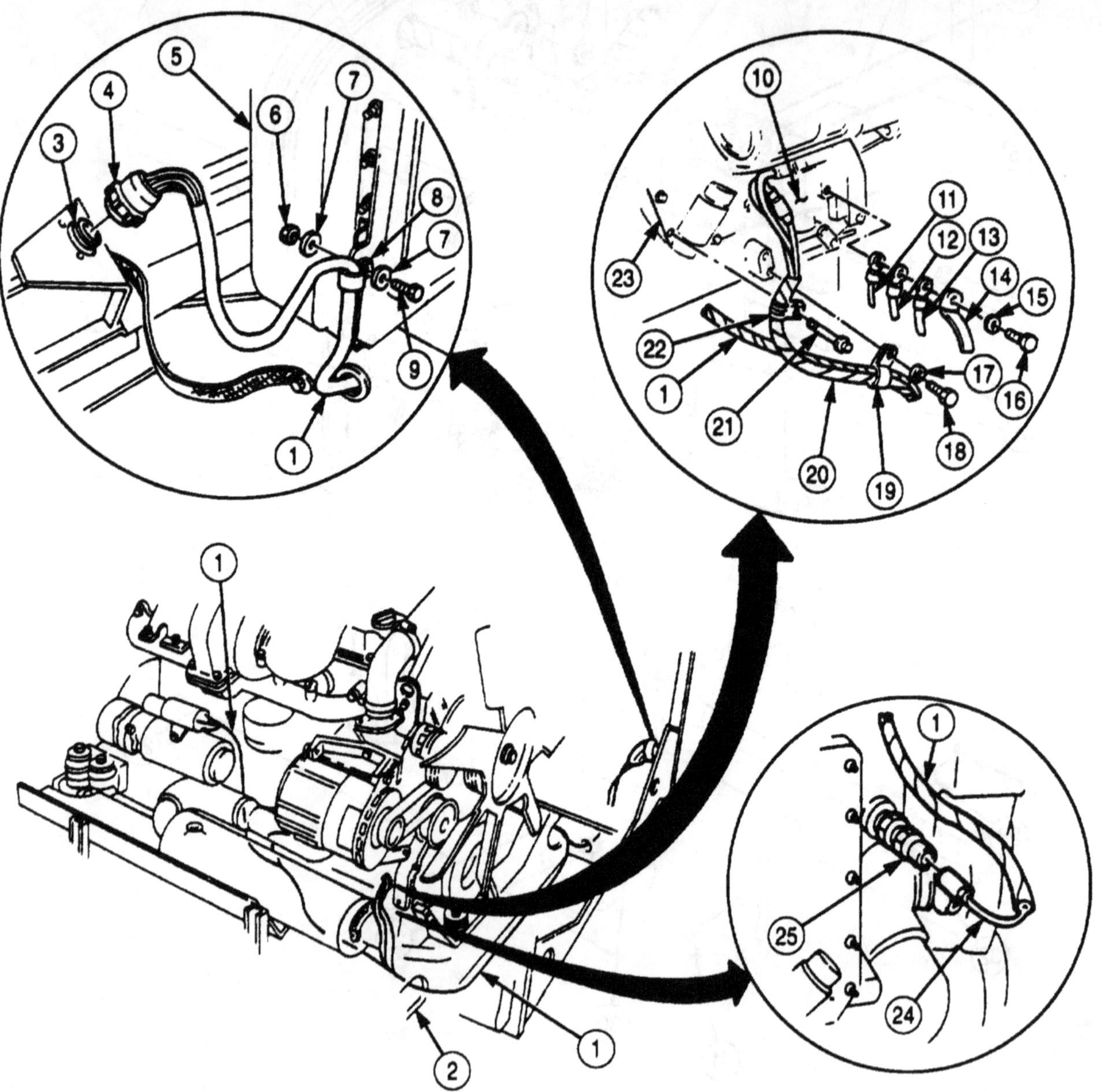

4-67. FRONT WIRING HARNESS (M939A2) REPLACEMENT (Contd)

75. Install wire (50) on alternator (42) with new lockwasher (45) and nut (46).
76. Install wire (55) on alternator (42) with new lockwasher (43) and nut (44).
77. Install wire (54) on alternator (42) with new lockwasher (53) and screw (52).
78. Connect wire (51) to connector (49).

NOTE

Apply sealing compound to terminal cover and wires for installation.

79. Install terminal cover (47) on alternator (42) with two new screw-assembled lockwashers (48).
80. Install wires (37) and (38) on starter (39) with new lockwasher (36) and nut (35).
81. Install wire (34) on starter (39) with new lockwasher (40) and nut (41).
82. Install wire (30) and battery cable (33) on solenoid (26) with new lockwasher (31) and nut (32).
83. Install wire (27) on solenoid (26) with new lockwasher (26) and nut (29).

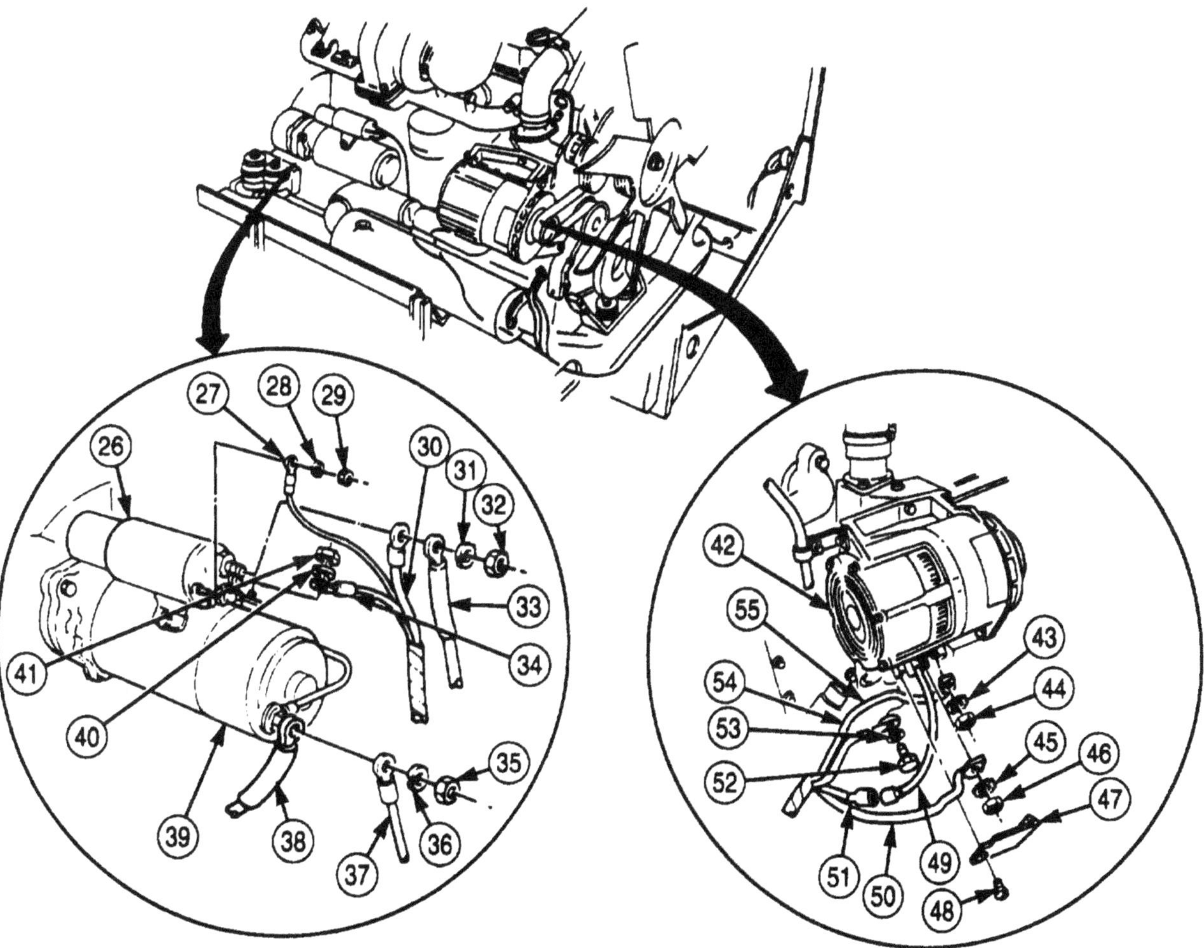

FOLLOW-ON TASKS:
- Install air intake pipe (para. 3-14).
- Connect main light switch (para. 3-108).
- Connect failsafe wiring module (para. 3-106).
- Install turn signal flasher (para. 3-114).
- Install protective control box (para. 3-115).
- Install splash shields (TM 9-2320-272-10).

4-68. REAR WIRING HARNESS REPLACEMENT

THIS TASK COVERS:

a. Removal **b. Installation**

INITIAL SETUP:

APPLICABLE MODELS
All

TOOLS
General mechanic's tool kit (Appendix E, Item 1)

MATERIALS/PARTS
Locknut (Appendix D, Item 288)
Locknut (M936/A1/A2) (Appendix D, Item 323)
Lockwasher (Appendix D, Item 360)
Two lockwashers (M931/A1/A2, M932/A1/A2) (Appendix D, Item 402)
Locknut (M931/A1/A2, M932/A1/A2) (Item 286)

MANUAL REFERENCES (TM)
TM 9-2320-272-10
TM 9-2320-272-24P

EQUIPMENT CONDITION
- Parking brake set (TM 9-2320-272-10).
- Hood raised and secured (TM 9-2320-272-10).
- Battery ground cables disconnected (para. 3-126).

a. Removal

NOTE

Tag wires, cables, and connectors for installation.

1. Disconnect wire (1) from transmission temperature transmitter (3) and remove from spring tension tab (2).
2. Disconnect rear wiring harness connector (5) from front wiring harness plug (4).

NOTE

Perform step 3 for M929/A1/A2, M9301/A1/A2, M931/A1/A2, M932/A1/A2, and M936/A1/A2 vehicles only.

3. Disconnect wire (6) from wire (7).
4. Disconnect wires (8) and (12) from wires (9) and (11) of neutral start switch (10).

4-68. REAR WIRING HARNESS REPLACEMENT (Contd)

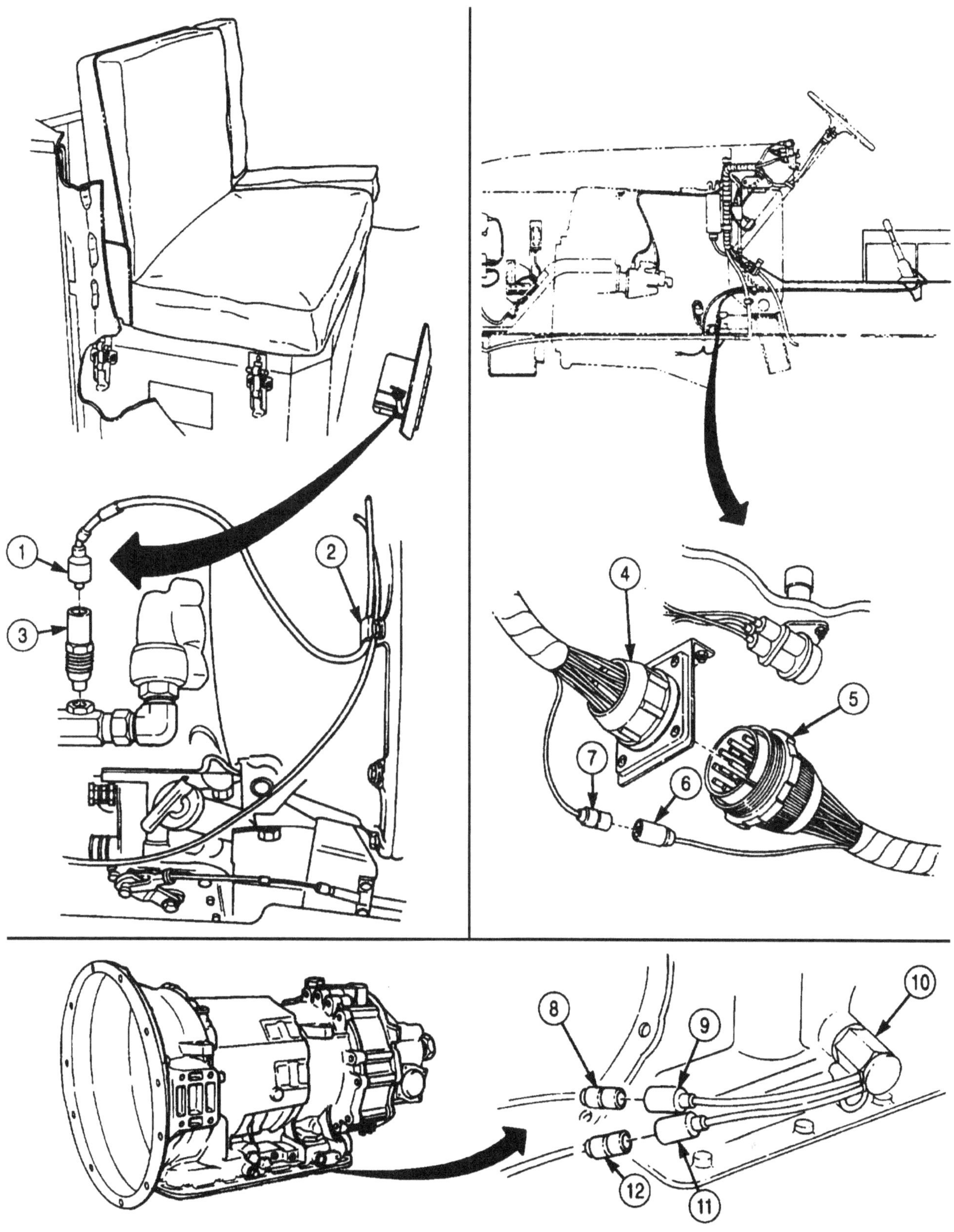

4-68. REAR WIRING HARNESS REPLACEMENT (Contd)

5. Disconnect wire (1) from fuel level sending unit (2) on left-side fuel tank (3).

NOTE

Perform step 6 for M929/A1/A2, M930/A1/A2, M931/A1/A2, M932/A1/A2, and M936/A1/A2 vehicles only.

6. Disconnect wire (6) for left- and right-side marker lights (4) from connector (5).

NOTE

Perform step 7 on all models except M931/A1/A2 and M932/A1/A2 vehicles.

7. Disconnect two wires (8) from stoplight switch (7).

NOTE

Perform step 8 for M931/A1/A2 and 932/A1/A2 vehicles only.

8. Disconnect two wires (10) from stoplight switch (9).
9. Disconnect wire (11) from primary low air pressure switch (12).

NOTE

Perform step 10 for models M929/A1/A2, M930/A1/A2, M931/A1/A2, M932/A1/A2, and M936/A1/A2 vehicles only.

10. Disconnect wire (13) from fuel level sending unit (14) on right-side fuel tank (15).

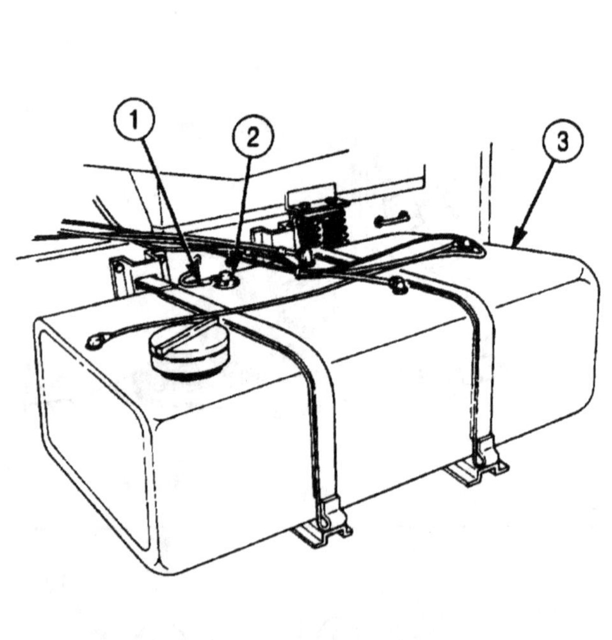

ALL VEHICLES

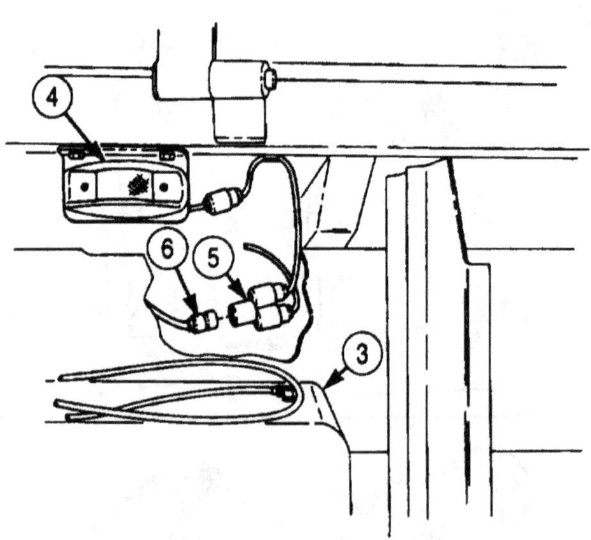

M923/A1/A2, M925/A1/A2, M927/A1/A2, AND M928/A1/A2 VEHICLES

4-68. REAR WIRING HARNESS REPLACEMENT (Contd)

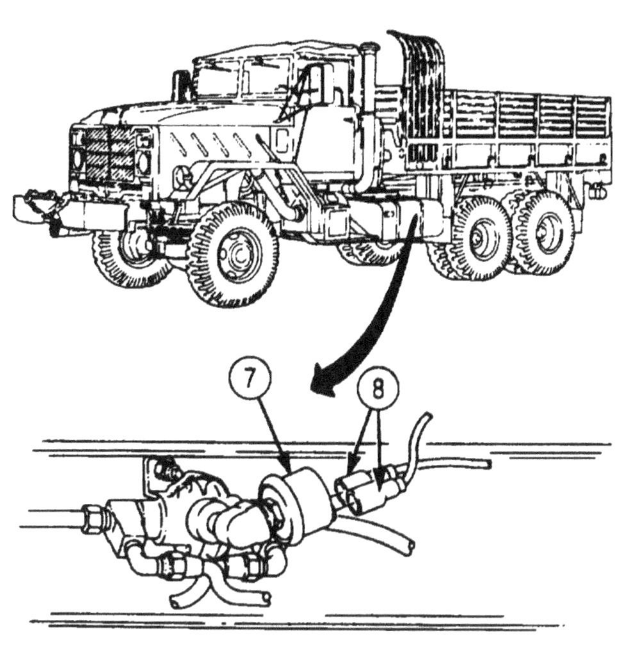

ALL EXCEPT M931/A1/A2 AND M939/A1/A2 VEHICLES

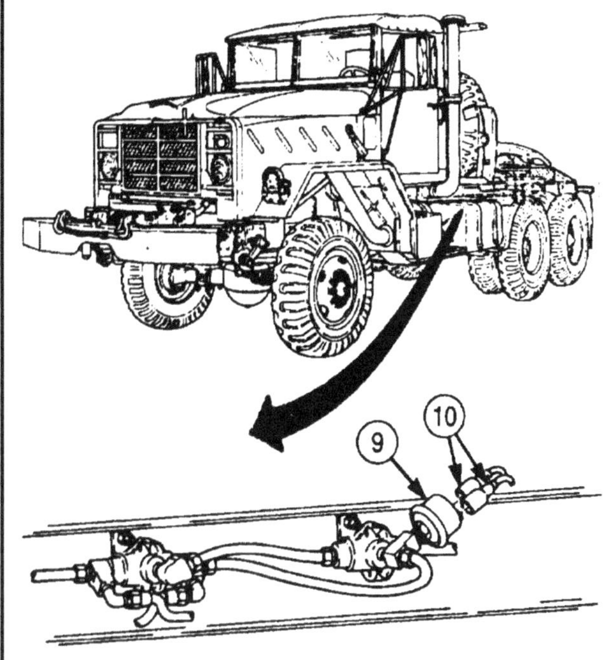

M9321/A1/A2 AND M932/A1/A2 VEHICLES

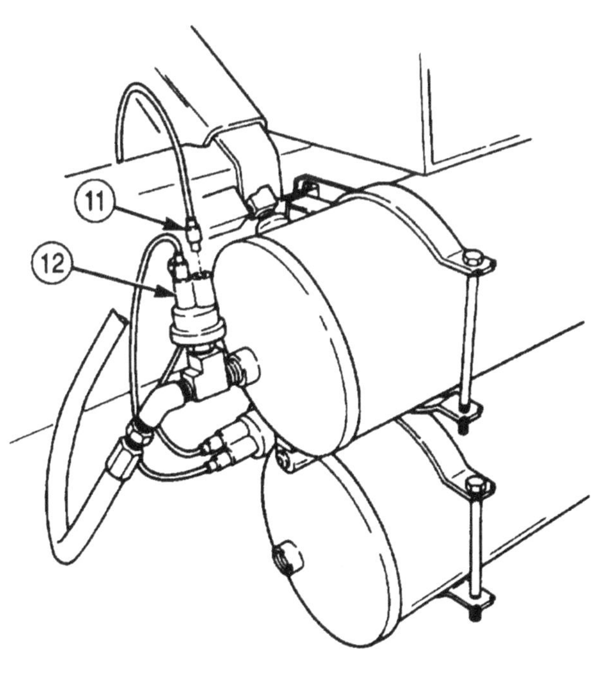

ALL VEHICLES

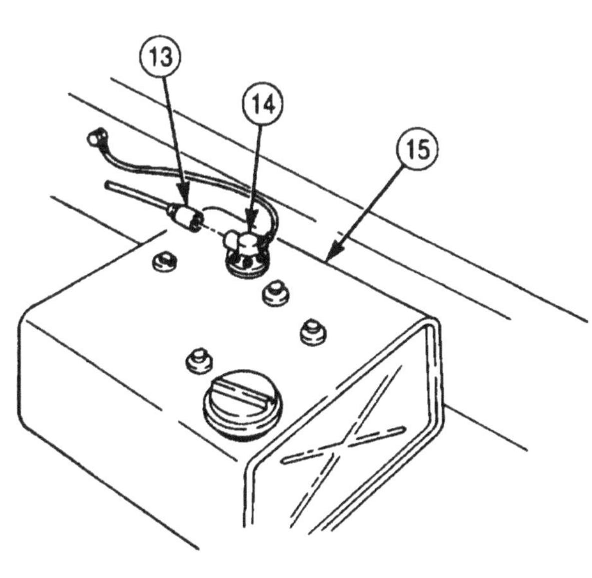

M929/A1/A2, M930/A1/A2, M931/A1/A2, M932/A1/A2, AND M936/A1/A2 VEHICLES

4-68. REAR WIRING HARNESS REPLACEMENT (Contd)

NOTE

Perform step 11 for M923/A1/A2, M925/A1/A2, M927/A1/A2, and M928/A1/A2 vehicles.

11. Disconnect wire (4) for right rear and left rear side marker light (3) from wire (5).

NOTE

Perform step 12 for M923/A1/A2 and M925/A1/A2 vehicles.

12. Open tension tab (13) and remove wires (6), (7), (8), and (9) from right rear composite light (10) and left rear composite light (22).
13. Disconnect right rear composite light wires (6), (7), (8), and (9) from wires (2), (1), (11), and (12).
14. Disconnect left rear composite light wires (21), (23), (24), and (25) from wires (20), (28), (27), and (26).

NOTE

Perform step 15 for M929/A1/A2, M930/A1/A2, and M936/A1/A2 vehicles.

15. Disconnect wire (29) for left side and right side marker lights (18) and (19) from connector (30).

NOTE

Perform step 16 for M936/A1/A2 vehicles.

16. Remove locknut (17), screw (15), and wood block (16) from left-side frame rail (14). Discard locknut (17).

NOTE

Steps 17 through 19 apply to M934/A1/A2 vehicles.

17. Disconnect wires (36) for left rear composite light (33) from plug (35).
18. Disconnect wire (34) for left rear clearance light (32) from wire (37).
19. Disconnect wire (39) of left rear blackout clearance light (31) from wire (38).

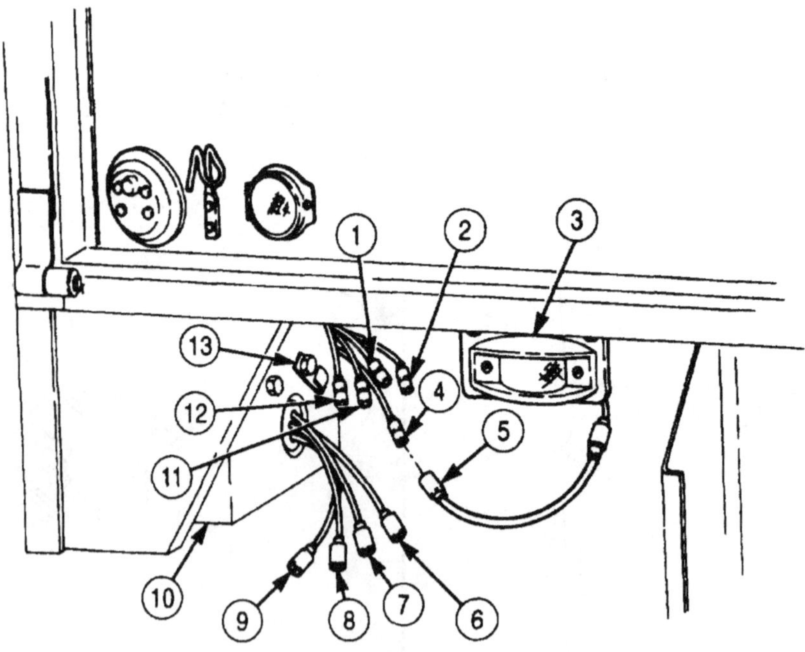

4-68. REAR WIRING HARNESS REPLACEMENT (Contd)

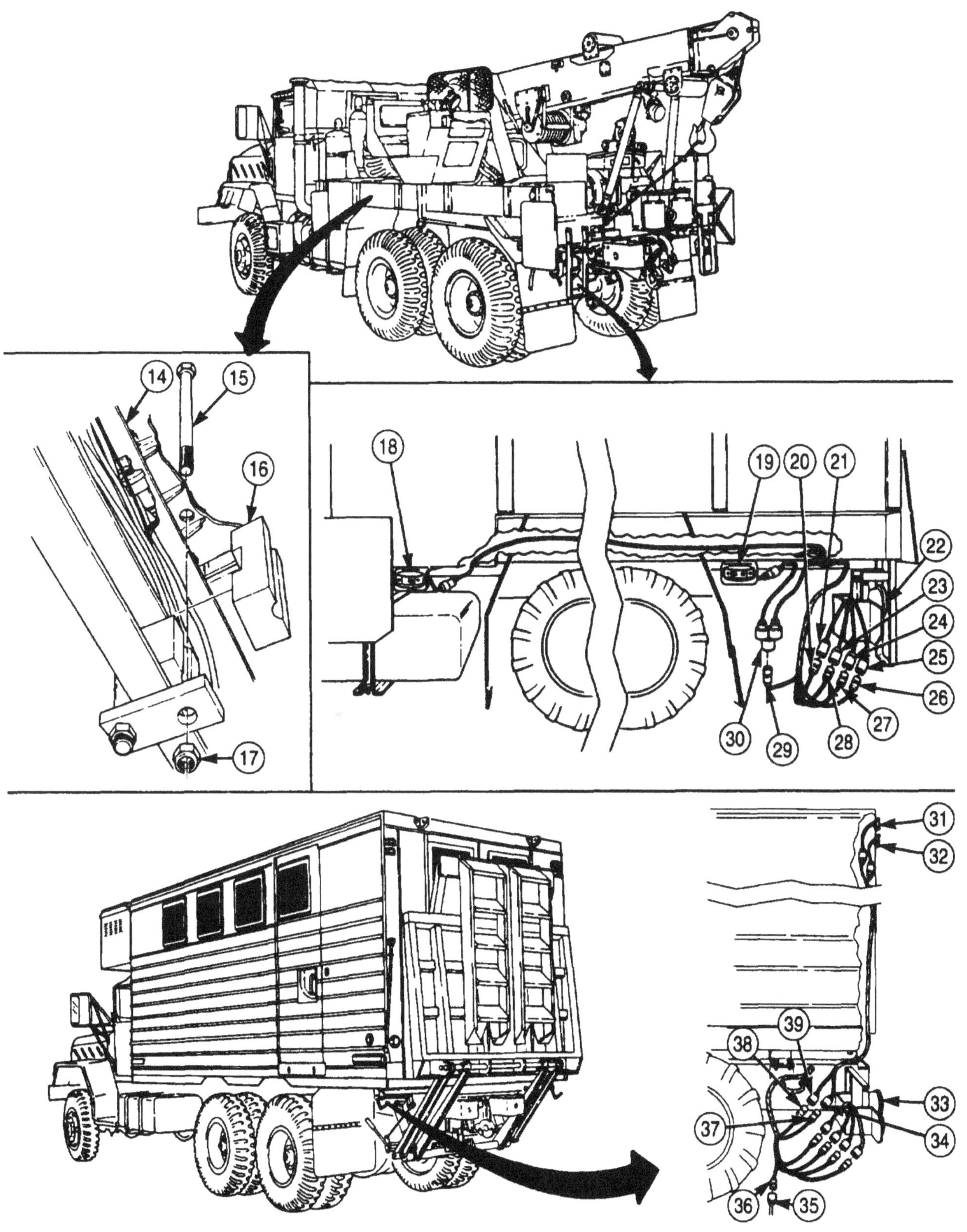

4-68. REAR WIRING HARNESS REPLACEMENT (Contd)

NOTE

Steps 20 through 22 apply to M931/A1/A2 and M932/A1/A2 vehicles only.

20. Remove two nuts (10), screws (1), and clamps (11) from frame rail (2).
21. Disconnect harness connector (8) from semitrailer receptacle (6).
22. Remove two nuts (9), ground wires (7) and (3), two lockwashers (4), and screws (5) from semitrailer receptacle (6). Discard lockwashers (4).

NOTE

- Tag clamps for installation.
- Refer to table 4-9 for number of harness clamps installed on each vehicle.

23. Remove nut (15), screw (13), and harness clamp (14) from wiring harness (12)

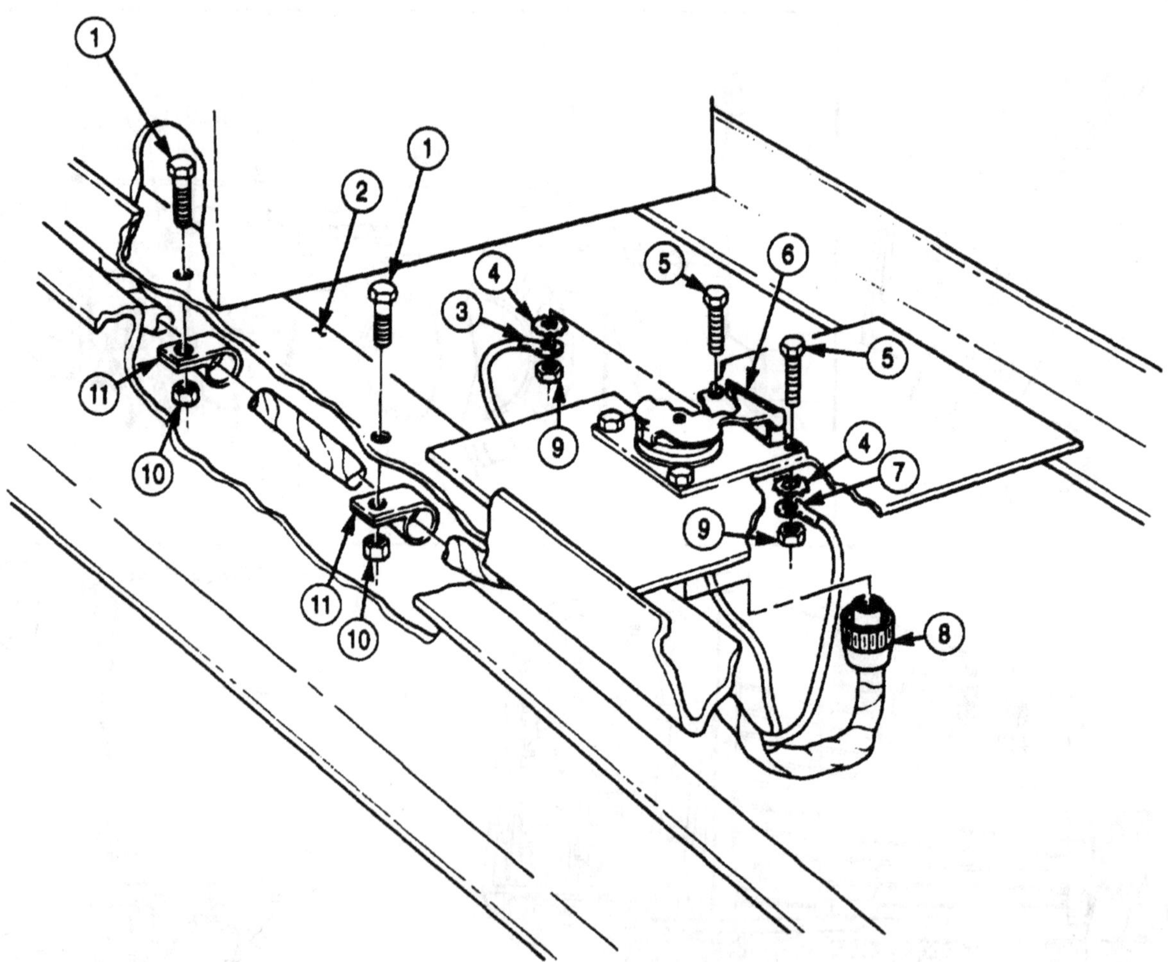

M931/A1/A2 AND M932/A1/A2 VEHICLES

4-68. REAR WIRING HARNESS REPLACEMENT (Contd)

Table 4-9. Clamps Required for Rear Wiring Harness.

VEHICLE	NUMBER OF CLAMPS
M923/A1/A2, M925/A1/A2	18
M927/A1/A2, M9281/A1/A2	10
M929/A1/A2, M9301/A1/A2	11
M931/A1/A2, M932/A1/A2	12
M936/A1/A2	10

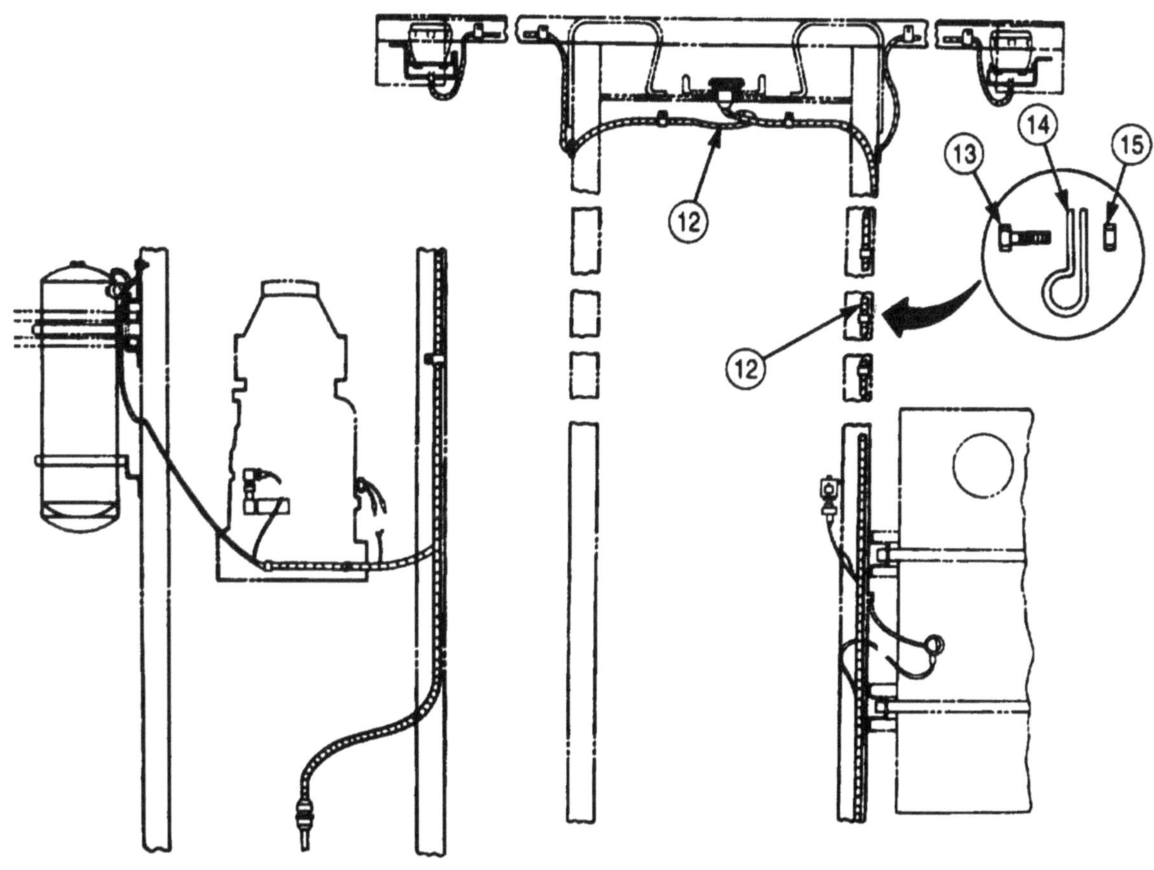

4-68. REAR WIRING HARNESS REPLACEMENT (Contd)

NOTE

Perform step 24 for M931/A1/A2 and M932/A1/A2 vehicles only.

24. Remove locknut (4), screw (2), clamp (3), two ground wires (6), and lockwasher (5) from rear crossmember frame rail (1). Discard locknut (4) and lockwasher (5).

NOTE

Perform step 25 for all vehicles except M931/A1/A2 and M932/A1/A2 vehicles.

25. Remove locknut (11), screw (7), two ground wires (10), and lockwasher (9) from rear frame rail (8). Discard locknut (11) and lockwasher (9).

NOTE

Receptacle cover must be lifted and held open to remove top two screws.

26. Remove four locknuts (15), screws (17), receptacle cover (12), and trailer cable receptacle (13) from rear crossmember frame (8). Discard locknuts (15).
27. Pull rear portions of rear wiring harness (16) through hole (14) in rear crossmember frame (8).

b. Installation

CAUTION

Use care when routing harness. Snagging may occur, and forceful pulling will result in damage to harness.

1. Route rear wiring harness (16) through hole (14) in rear crossmember frame (8) and lay out in approximate position.

NOTE

Receptacle cover must be lifted and held open to install top two screws.

2. Install trailer cable receptacle (13) and receptacle cover (12) with four screws (17) and new locknuts (15).

4-68. REAR WIRING HARNESS REPLACEMENT (Contd)

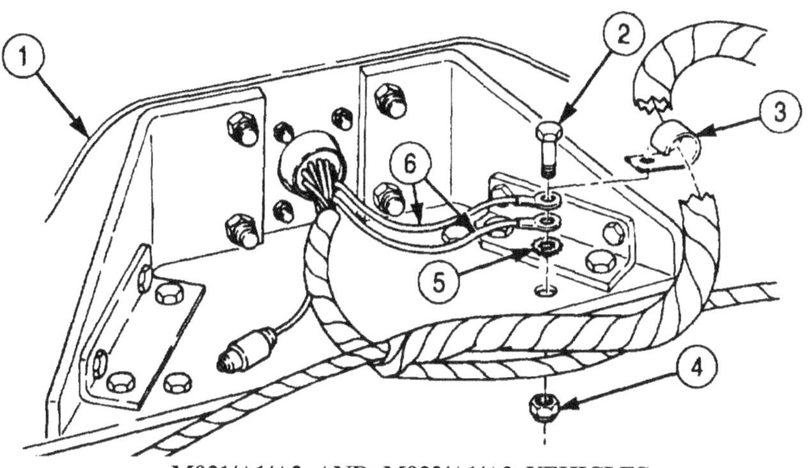

M931/A1/A2 AND M932/A1/A2 VEHICLES

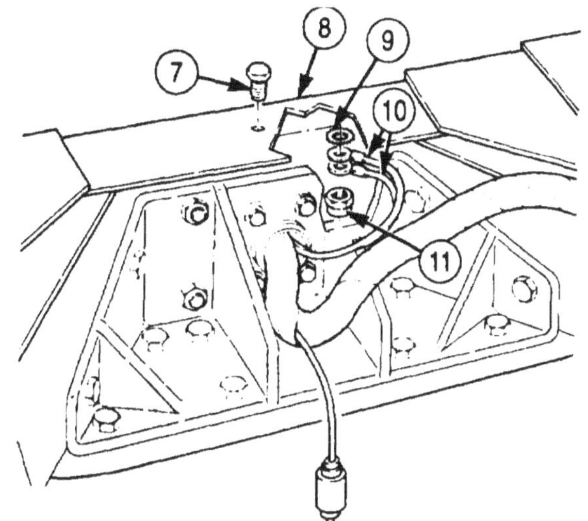

ALL EXCEPT M931/A1/A2 AND M932/A1/A2 VEHICLES

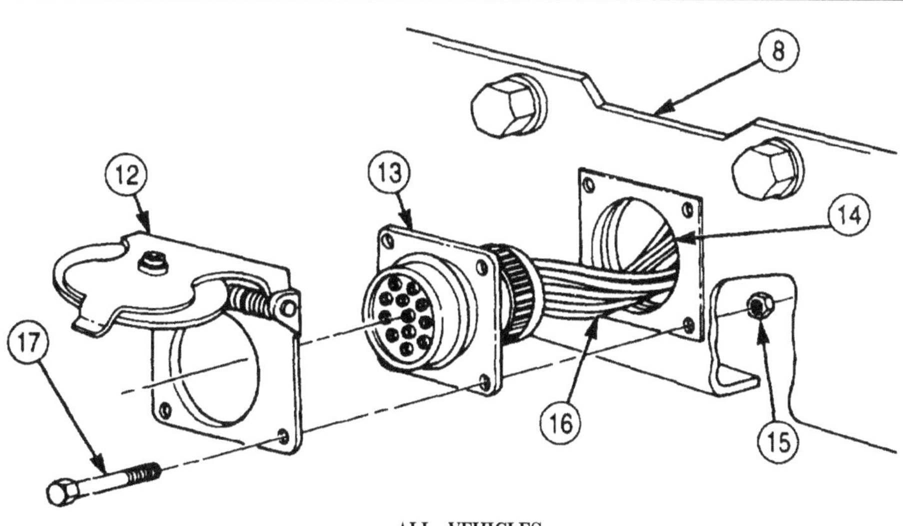

ALL VEHICLES

4-68. REAR WIRING HARNESS REPLACEMENT (Contd)

NOTE

Perform step 3 for all vehicles except M931/A1/A2 and M932/A1/A2.

3. Install two ground wires (4) on frame crossmember (2) with screw (1), new lockwasher (3), and new locknut (5).

NOTE

Perform step 4 for M931/A1/A2 and M932/A1/A2 vehicles.

4. Install two ground wires (11) on frame crossmember (6) with screw (7), clamp (8), new lockwasher (10), and locknut (9).

NOTE

Perform step 5 for M923/A1/A2, M925/A1/A2, M927/A1/A2, and M928/A1/A2 vehicles only.

5. Connect wire (16) for right- and left-side marker lights (14) to wire (15).
6. Connect wires (13), (12), (22), and (23) for right rear composite light (21) to wires (17), (18), (19), and (20).

NOTE

Perform step 7 for M923/A1/A2 and M925/A1/A2 vehicles only.

7. Close spring tension tab (24) around wires (17), (18), (19), and (20).

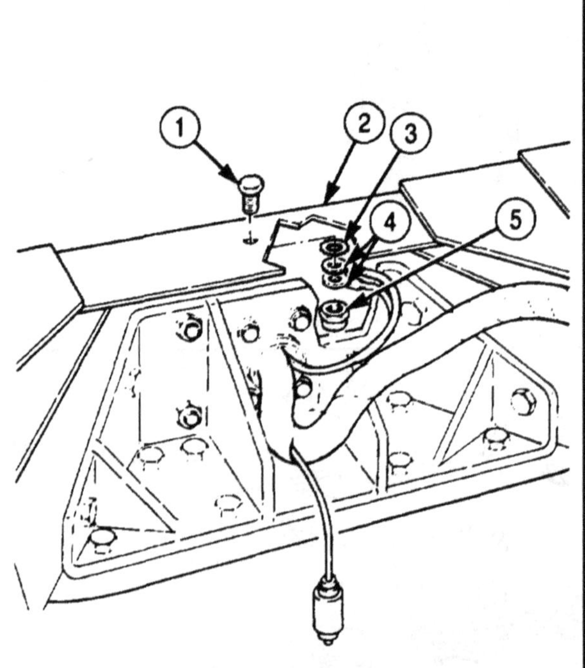

ALL EXCEPT M931/A1/A2 AND M932/A1/A2 VEHICLES

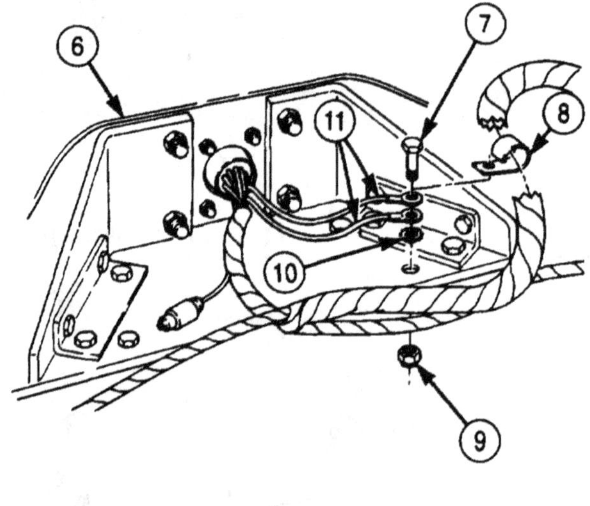

M931/A1/A2 AND M932/A1/A2 VEHICLES

4-68. REAR WIRING HARNESS REPLACEMENT (Contd)

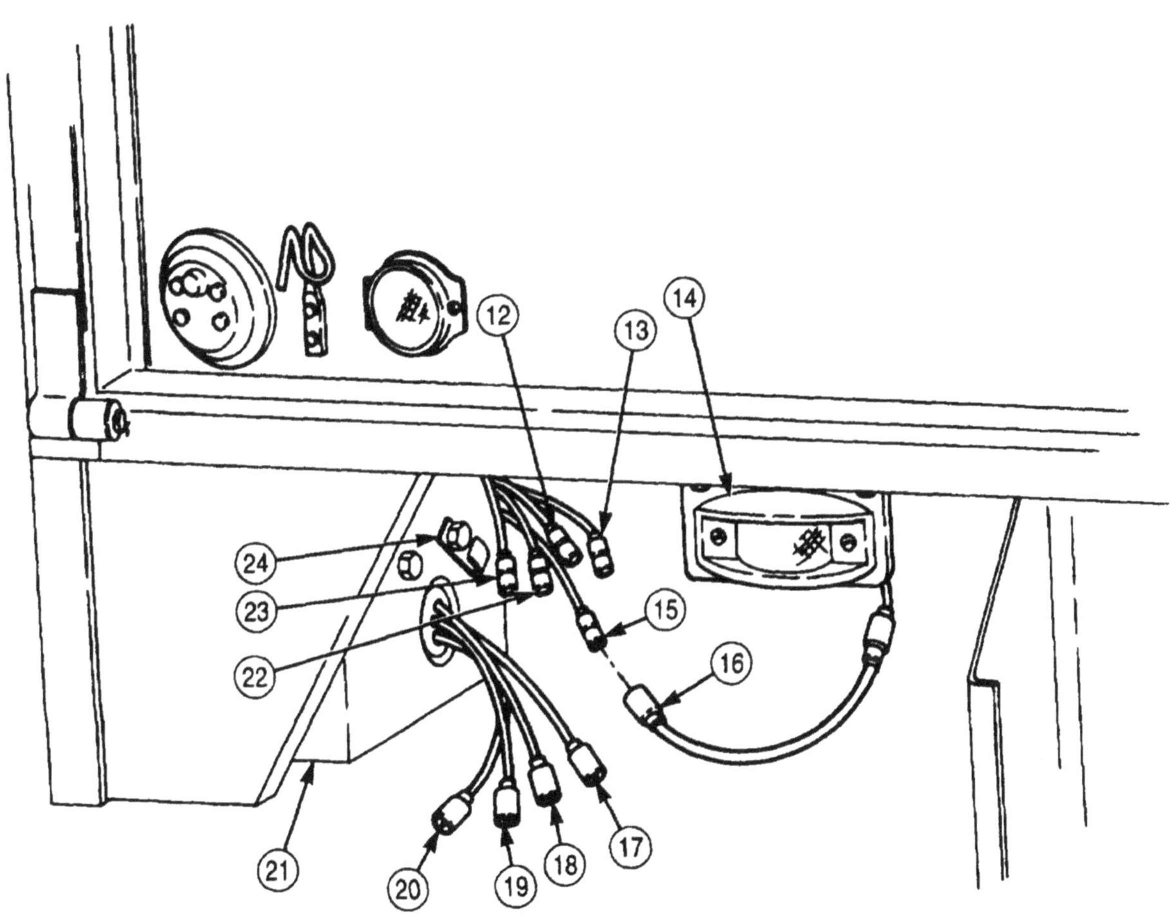

4-68. REAR WIRING HARNESS REPLACEMENT (Contd)

NOTE

Perform step 8 for M936/A1/A2 vehicles only.

8. Install wood block (3), screw (2), and new locknut (4) on left-side frame rail (1).
9. Connect wires (8), (10), (11), and (12) for left rear composite light (9) to wires (7), (15), (14), and (13).

NOTE

Perform step 10 for M929/A1/A2, M930/A1/A2, and M936/A1/A2 vehicles only.

10. Connect wire (16) for left-side and right-side marker lights (5) and (6) to connector (17).

NOTE

- Perform steps 11 through 15 for M934/A1/A2 vehicles only.
- Perform these steps in front of left rear composite light only.

11. Connect blackout clearance light (18) wires (25) and (24).
12. Connect clearance light (19) wires (20) and (23).
13. Install wire (22) in wire plug (21).

4-68. REAR WIRING HARNESS REPLACEMENT (Contd)

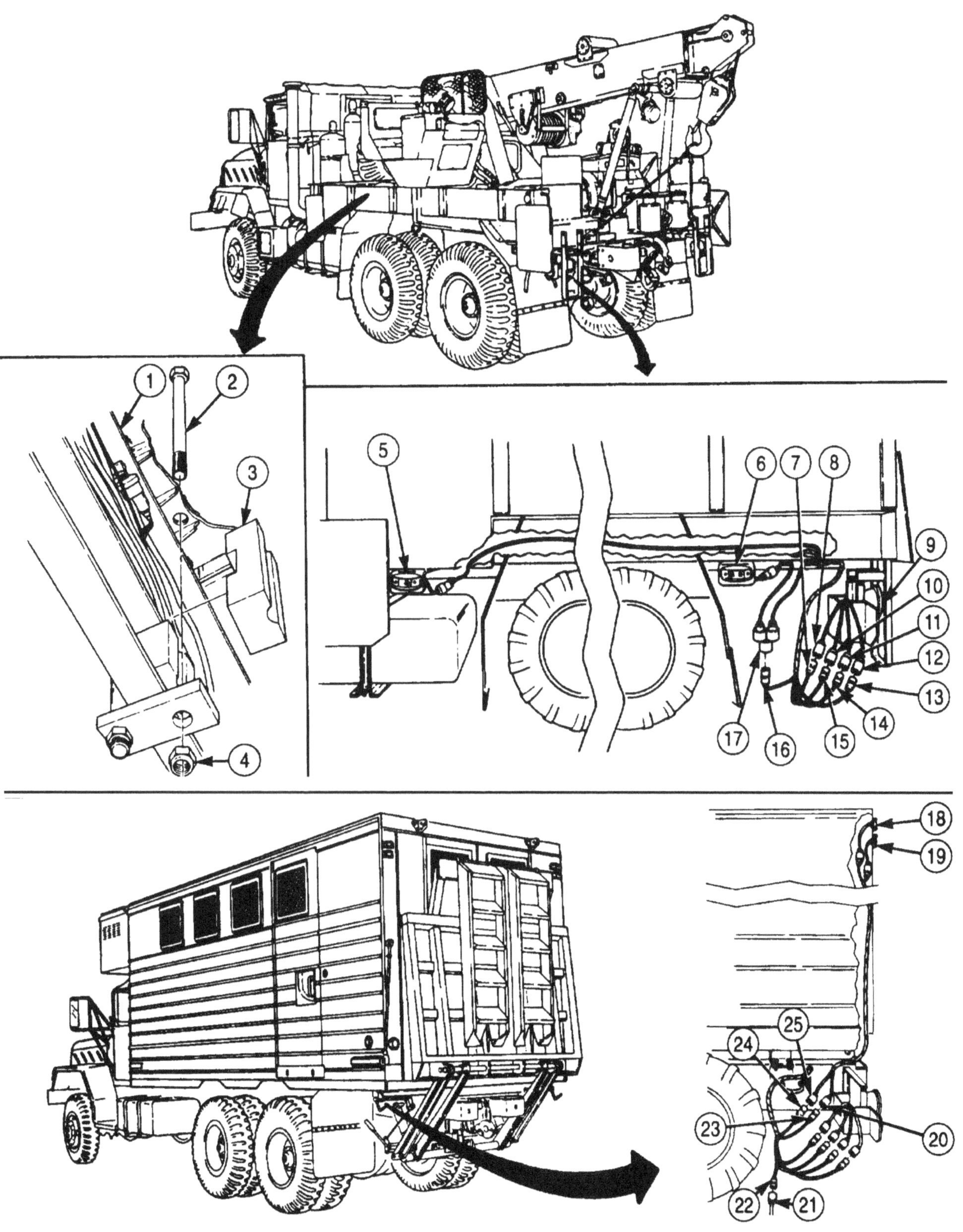

4-68. REAR WIRING HARNESS REPLACEMENT (Contd)

14. Connect wire (1) to fuel level sending unit (2) on right-side fuel tank (3).
15. Connect wire (4) to primary low air pressure switch (5).

NOTE

Perform step 16 for all except M931/A1/A2 and M932/A1/A2 vehicles.

16. Connect two wires (7) to stoplight switch (6).

NOTE

Perform step 17 for M931/A1/A2 and M932/A1/A2 vehicles.

17. Connect two wires (9) to stoplight switch (8).

NOTE

Perform step 18 for M923/A1/A2, M925/A1/A2, M927/A1/A2, and M928/A1/A2 vehicles only.

18. Connect wire (13) to connecter (11) for left- and right-side marker lights (10).
19. Connect wire (14) to fuel level sending unit (15) on left-side fuel tank (12).

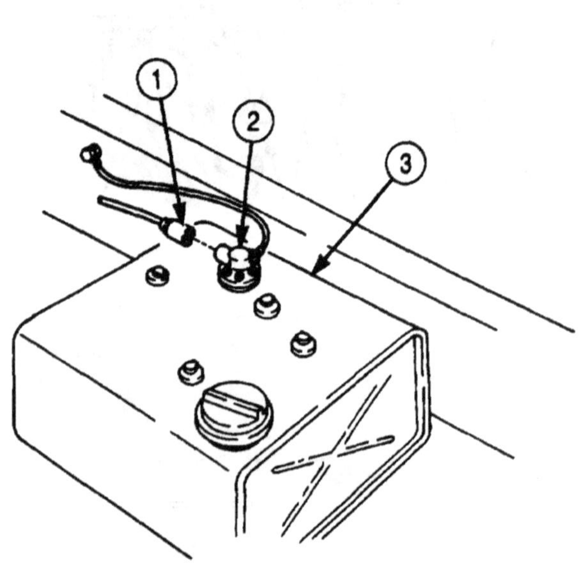

M929/A1/A2, M930/A1/A2, M931/A1/A2, M932/A1/A2, AND M936/A1/A2 VEHICLES

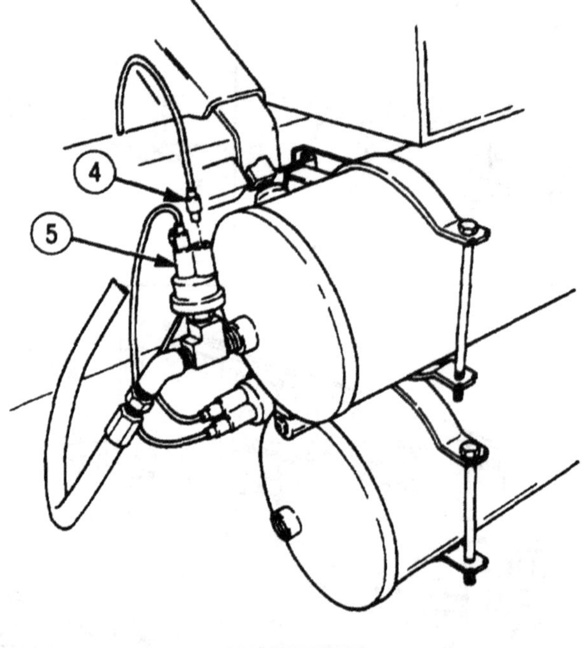

ALL VEHICLES

4-68. REAR WIRING HARNESS REPLACEMENT (Contd)

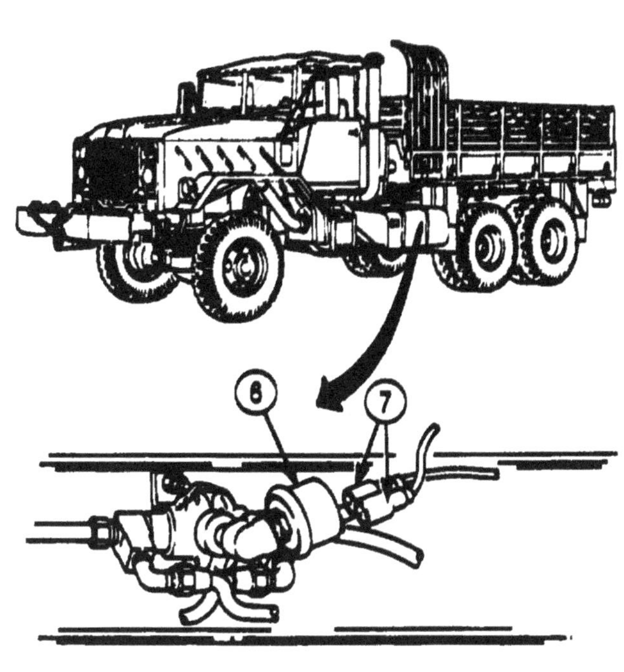

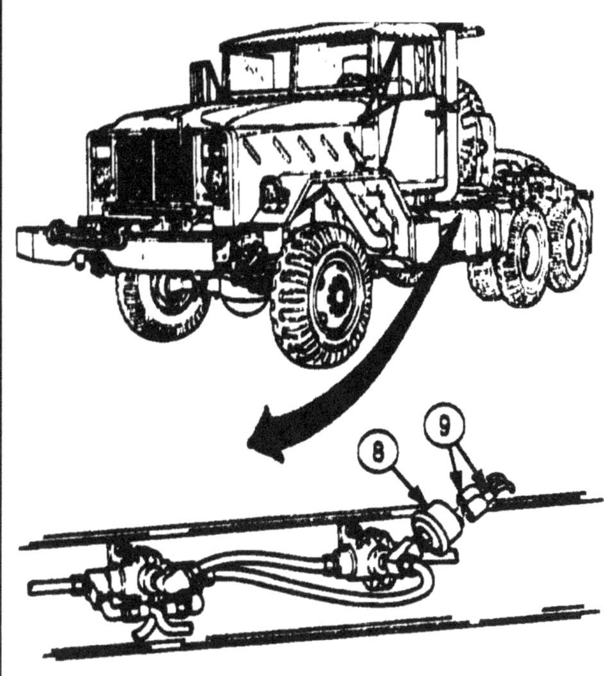

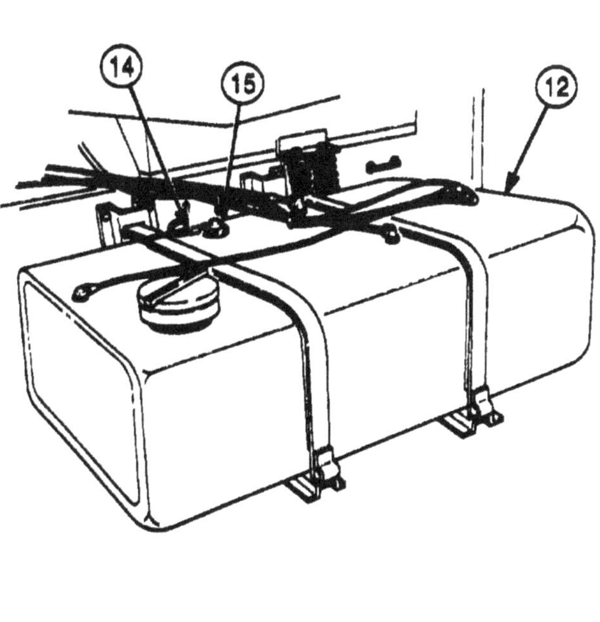

4-68. REAR WIRING HARNESS REPLACEMENT (Cod)

20. Connect wires (1) and (4) to wires (2) and (3).
21. Connect rear wiring harness connector (6) to front wiring harness receptacle (5).

NOTE

Perform step 22 for M929/A1/A2, M930/A1/A2, M931/A1/A2, M932/A1/A2, and M936/A1/A2 vehicles only.

22. Connect wires (7) and (8).
23. Connect wire (9) to transmission temperature sending unit (12).
24. Close spring tension tab (10) around wires (9) and (11).

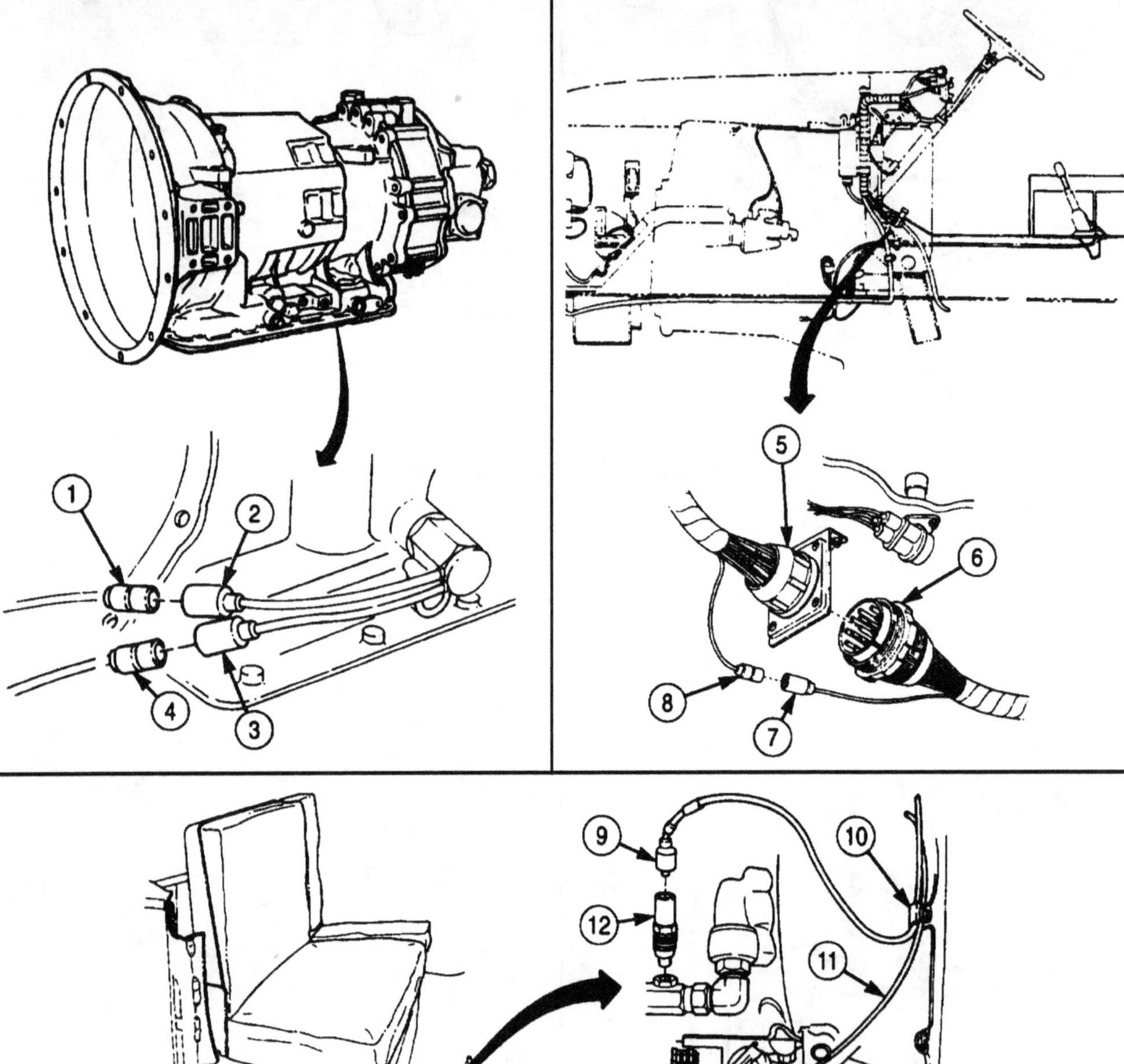

4-68. REAR WIRING HARNESS REPLACEMENT (Contd)

NOTE

Perform steps 25 through 27 for M931/A1/A2 and M932/A1/A2 vehicles only.

25. Connect harness connector (19) to semitrailer receptacle (17).
26. Install two ground wires (15) and (18) on receptacle (17) and plate (21) with two screws (16), new lockwashers (14), and nuts (20).
27. Install two clamps (24) and wiring harness (25) on frame rail (22) with two screws (13) and nuts (23).

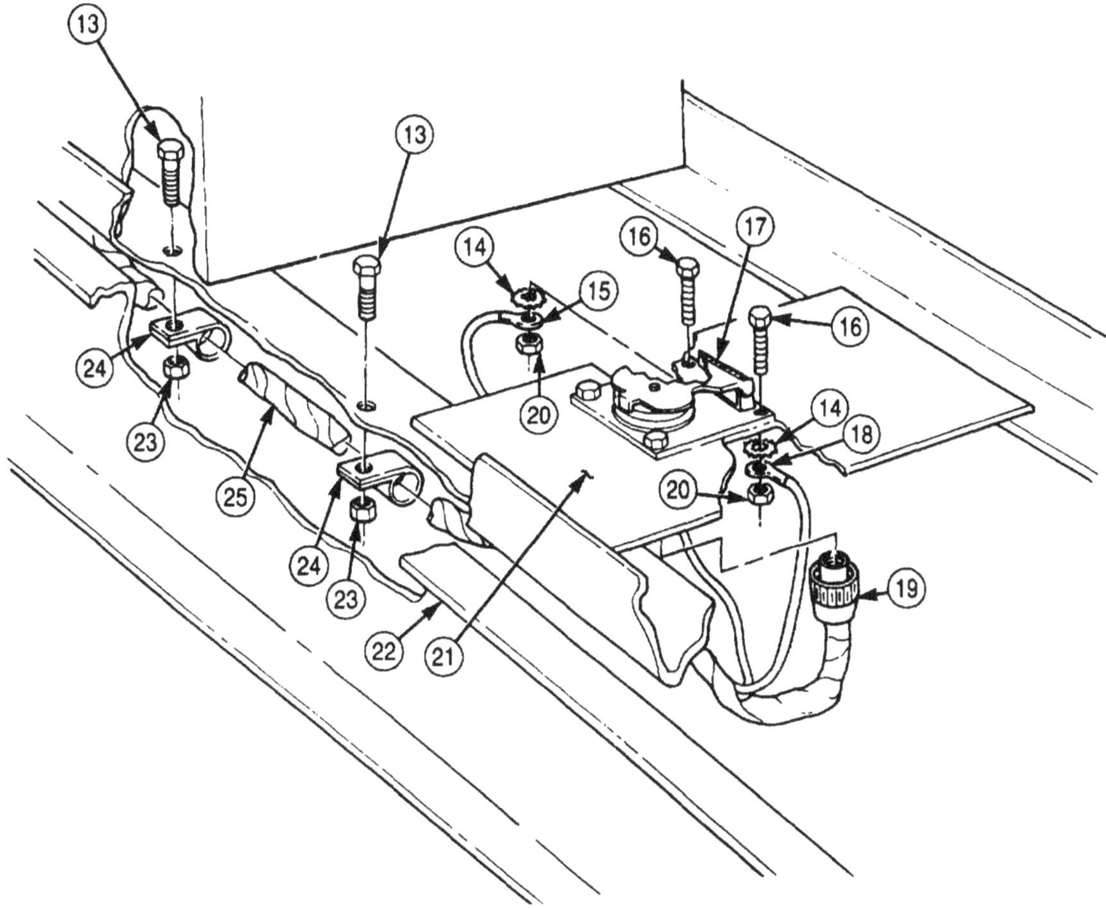

4-68. REAR WIRING HARNESS REPLACEMENT (Contd)

NOTE

Refer to table 4-9 and accompanying diagram for number of harness clamps installed on each vehicle.

28. Install correct number of harness clamps (14) on rear wiring harness (12) with corresponding number of screws (13) and nuts (15).

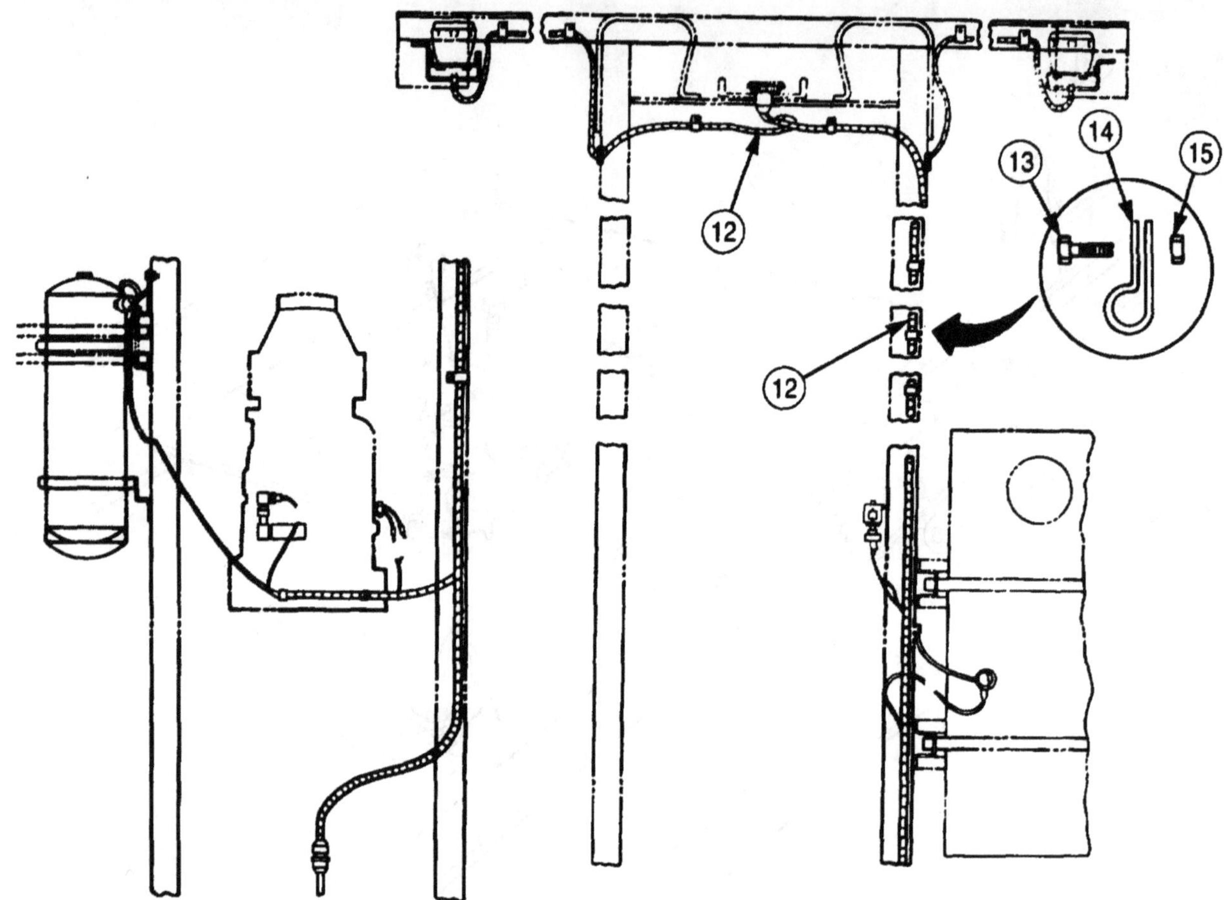

FOLLOW-ON TASKS

- Connect battery ground cables (para. 3-126).
- Start vehicle (TM 9-2320-272-10) and check operation of all rear lights, fuel selector switch, fuel gauge, low air pressure warning lights and gauges, and transmission temperature gauge.

Section VII. TRANSMISSION MAINTENANCE

4-69. TRANSMISSION MAINTENANCE INDEX

4-70. GENERAL

a. M939, M939A1, and M939A2 series vehicles utilize one of two Allison MT654CR model automatic transmissions. Early model (part number 6885292) and later model (part number 23040127) transmissions are easily identified by the part number on the equipment data plate located on the lower right rear of the transmission housing. The major differences between models is that later models use a larger oil pump and a different converter hub. Both models are interchangeable, replaced, and maintained the same way for all series vehicles. Some internal components are not interchangeable and are noted in the procedures.

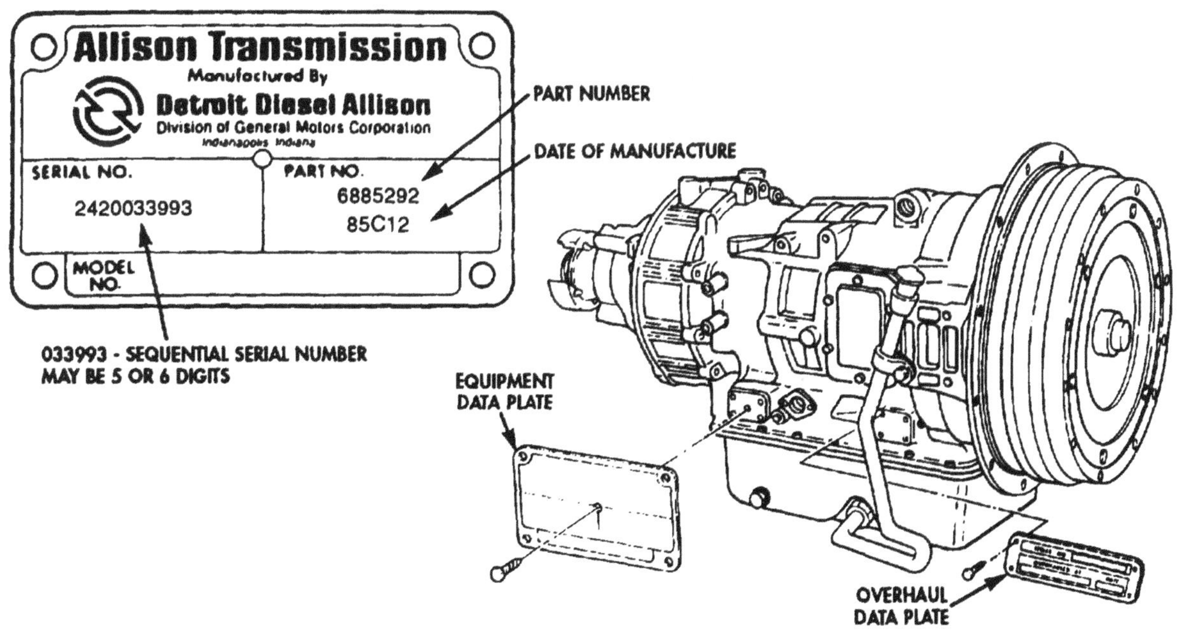

4-71. TRANSMISSION REPLACEMENT (IN-VEHICLE)

THIS TASK COVERS:

a. Removal **b. Installation**

INITIAL SETUP:

APPLICABLE MODELS

All

SPECIAL TOOLS

Engine barring tool (Appendix E, Item 8)

TOOLS

General mechanic's tool kit (Appendix E, Item 1)
Transmission jack (Appendix E, Item 147)
Torque wrench (Appendix E, Item 144)

MATERIALS/PARTS

Gasket (Appendix D, Item 202)
Thirteen lockwashers (Appendix D, Item 354)
Eight locknuts (Appendix D, Item 291)
Cotter pin (Appendix D, Item 66)
Two lockwashers (Appendix D, Item 377)
Gasket (Appendix D, Item 203)
Cap and plug set (Appendix C, Item 14)

PERSONNEL REQUIRED

TWO

REFERENCES (TM)

LO 9-2320-272-12
TM 9-2320-272-10
TM 9-2320-272-24P

EQUIPMENT CONDITION

- Transmission oil drained (para. 3-133).
- Transmission-to-transfer case propeller shaft removed (para. 3-148).
- Transmission oil dipstick removed (para. 3-134).
- Transmission breather removed (para. 3-136).
- Transmission modulator removed (para. 3-145).
- Transmission fifth gear lock-in solenoid valve and bracket removed (para. 4-80).
- Transmission neutral start switch removed (para. 3-98).
- Winch control valve removed (if so equipped) (para. 4-179 or para. 4-178).
- Transmission power takeoff removed (if so equipped) (para. 4-211).

GENERAL SAFETY INSTRUCTIONS

- Torque converter must be removed with transmission.
- Keep transmission tilted slightly downward.

a. Removal

CAUTION

Plug all openings to prevent dirt from entering transmission. Damage may occur if dirt or dust enters transmission.

1. Disconnect wire (1) from transmission oil temperature transmitter (8).
2. Remove screw (2), lockwasher (3), and clamp (4) with wire (7) from transmission flange (5). Discard lockwasher (3).
3. Tie wires (1) and (7) clear of transmission (6).

NOTE

Perform steps 3 and 4 for vehicles equipped with transmission Power Takeoff (PTO).

4. Remove two nuts (10), screws (13), retainer strap (14), spacer plate (12), and PTO cable (15) from bracket (11).
5. Remove cotter pin (9), washer (18), and pin (16) on PTO cable (15) from select lever (17) . Discard cotter pin (9).

4-71. TRANSMISSION REPLACEMENT (IN-VEHICLE) (Contd)

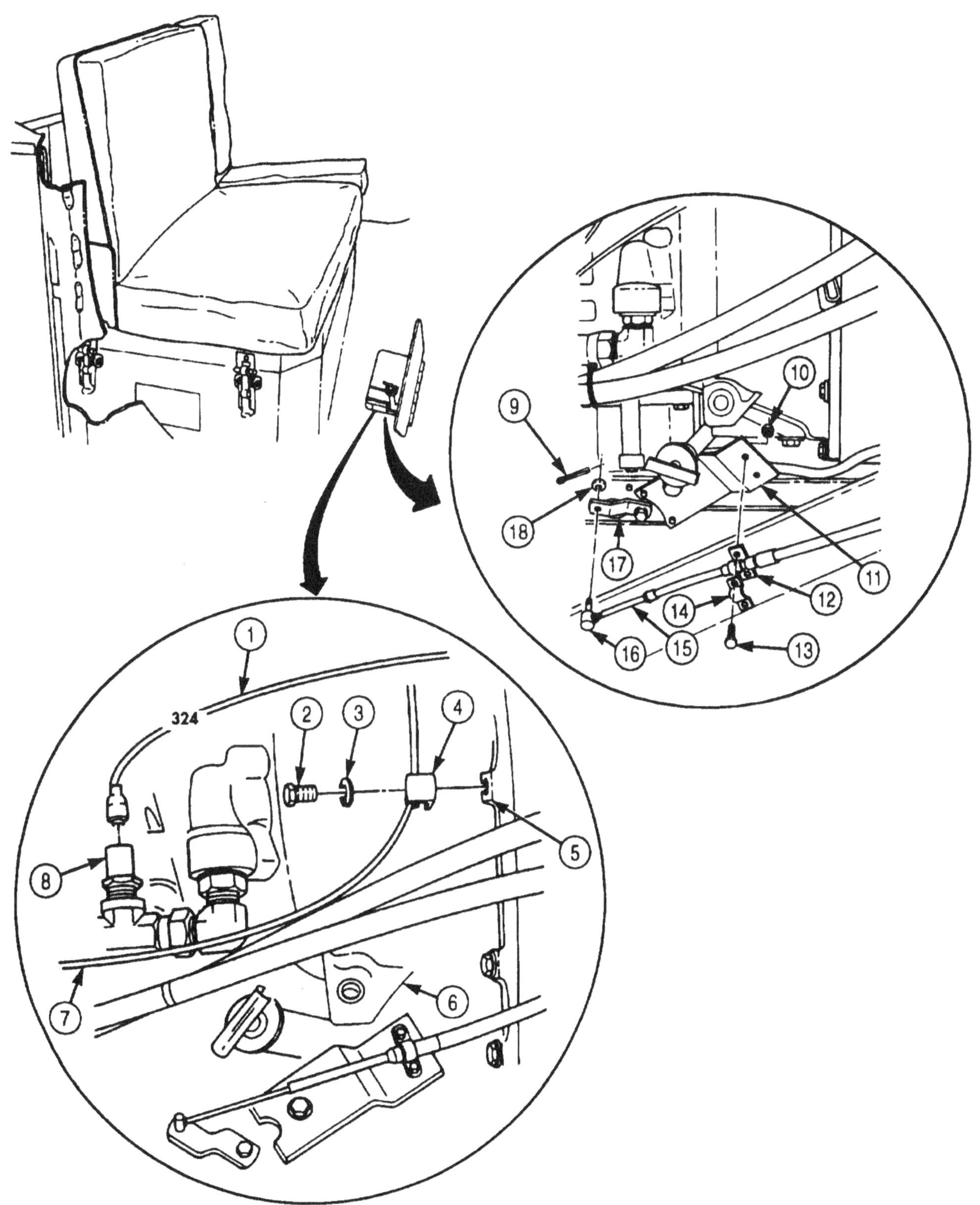

4-71. TRANSMISSION REPLACEMENT (IN-VEHICLE) (Contd)

6. Disconnect oil cooler filter-to-transmission supply hose (1) from lubrication valves adapter (4) on transmission (2).

NOTE

Perform step 6 if vehicle is equipped with transmission PTO.

7. Disconnect PTO oil return hose (8) from lubrication valve adapter (4).
8. Disconnect transmission-to-oil cooler return hose (5) from temperature transmitter adapter (6) on transmission (2).
9. Remove two screws (8) and lockwashers (7) from rear support bracket (9) and transmission (2). Discard lockwashers (7).

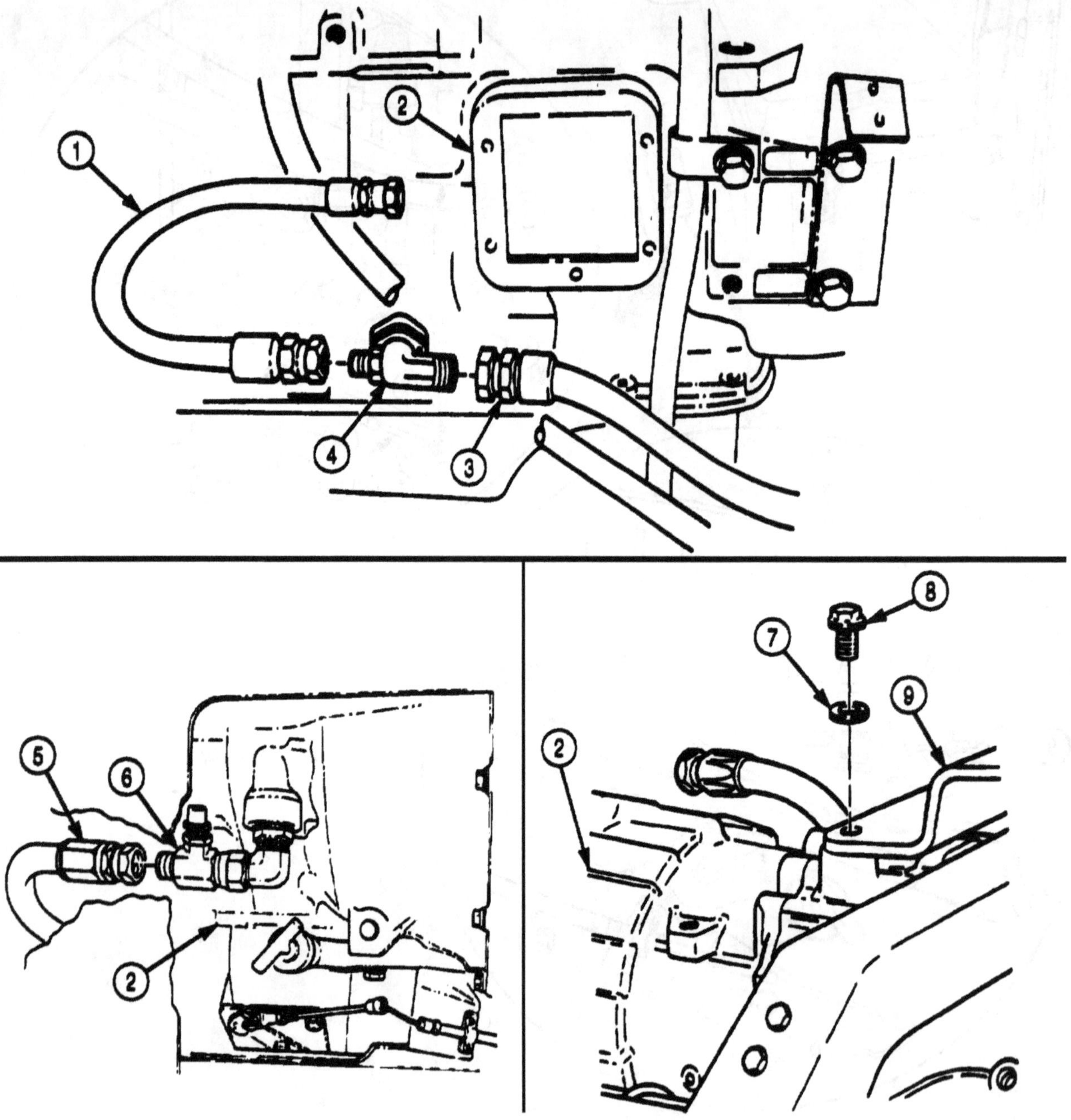

4-71. TRANSMISSION REPLACEMENT (IN-VEHICLE) (Contd)

10. Remove eight locknuts (12), screws (11), and front axle propeller shaft (13) from transfer case flange (10). Tie propeller shaft (13) aside to provide clearance for transmission (2) removal from vehicle. Discard locknuts (12).

NOTE

- The use of shim stock tube in torque converter access hole prevents converter screw from falling down behind flywheel. A loose screw may cause a lockup condition, preventing crankshaft rotation required to remove remaining screws.
- Engine crankshaft may be turned using barring tool,

11. Remove two screws (18), access cover (17), and gasket (16) from flywheel housing (14). Discard gasket (16).
12. Using engine barring tool, rotate crankshaft until screw (19) is visible in access hole (21).
13. Roll shim stock (20) into tube form and size to fit diameter of access hole (21). Insert shim stock (20) in access hole (21) and position over screw (19) against flywheel (15).
14. Remove remaining twelve screws (19) from flywheel (15) and converter (22) in the same manner.
15. Remove four screws (26), lockwashers (25), and clamp (24) from 9, 11, 1, and 3 o'clock positions on flywheel housing (23) and transmission (2), Discard lockwashers (25).

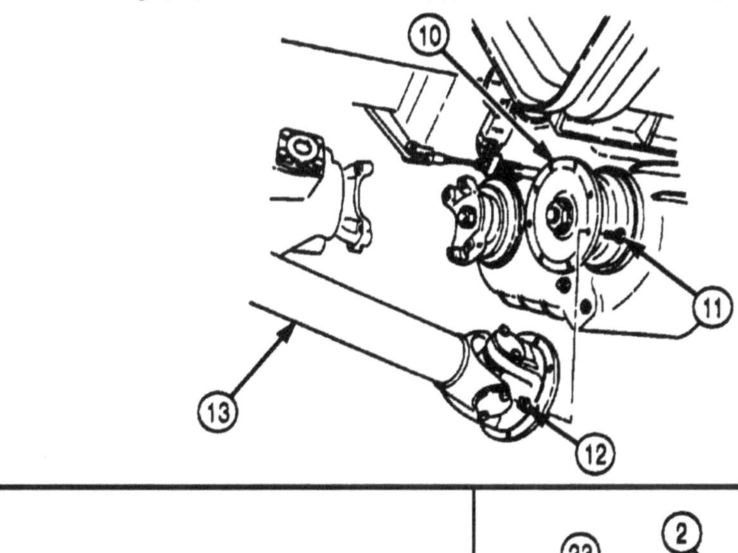

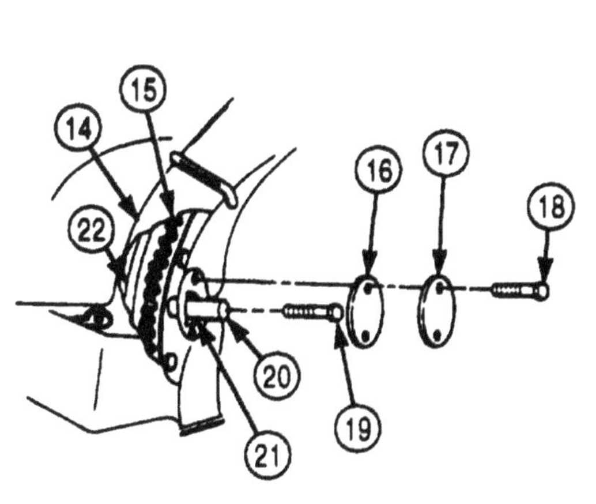

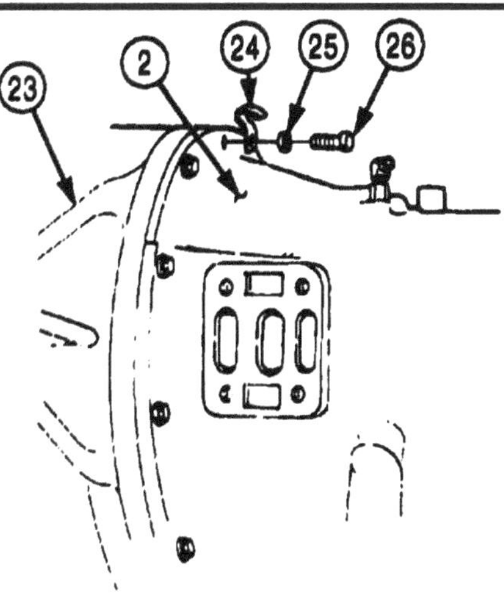

4-71. TRANSMISSION REPLACEMENT (IN-VEHICLE) (Contd)

NOTE

Guide or alignment screws can be made from extra screws that are longer and of the same thread size (appendix F, item 2).

16. Install four guide screws (5) at 9, 11, 1, and 3 o'clock positions on flywheel housing (1).
17. Position transmission jack under transmission (2) and raise until weight of transmission (2) is supported by jack.
18. Remove eight remaining screws (4) and lockwashers (3) from flywheel housing (1) and transmission (2). Discard lockwashers (3).

WARNING

- Keep rear of transmission tilted slightly downward at all times to prevent converter from sliding off and causing injury to personnel or damage to equipment.
- Torque converter must be removed with transmission as an assembly to prevent injury to personnel and damage to converter.

19. Separate transmission (2) from flywheel housing (1), keep level until clear of guide screws (5), then tilt rear of transmission (2) downward slightly to prevent converter (11) from separating from transmission (2).
20. Remove gasket (6) from transmission (2) or flywheel housing (1). Discard gasket (6).
21. Install retaining strap (9) on converter (11) with two screws (12).
22. Install retaining strap (9) on transmission (2) with four screws (10), washers (8), and nuts (7).
23. Raise front of vehicle until clearance is provided for removal of transmission (2) from under vehicle.
24. Remove transmission (2) and transmission jack from vehicle.
25. Remove four guide screws (5) from flywheel housing (1).

b. Installation

WARNING

- Keep rear of transmission tilted slightly downward at all times to prevent converter from sliding off and causing injury to personnel or damage to equipment.
- Torque converter and transmission must be installed as an assembly to prevent injury to personnel and damage to converter.

1. Install four guide screws (5) at 9, 11, 1, and 3 o'clock positions on flywheel housing (1).
2. Place transmission (2) on transmission jack.
3. Remove four nuts (7), washers (8), and screws (10) from retaining strap (9) and transmission (2).
4. Remove two screws (12) and retaining strap (9) from converter (11).
5. Position new gasket (6) on guide screws (5) and flywheel housing (1).
6. Raise front of vehicle until clearance is provided for installation of transmission (2) under vehicle.

CAUTION

Maintain transmission alignment to flywheel housing during installation to prevent damage to converter.

7. Position transmission (2) and transmission jack under vehicle.
8. Raise transmission (2), align with guide screws (5), and seat against flywheel housing (1).
9. Install transmission (2) on flywheel housing (1) with clamp (13), eight new lockwashers (14), and screws (15). Tighten screws (15) 36-40 lb-ft (49-54 N•m).
10. Remove four guide screws (5) from flywheel housing (1) and install four remaining new lockwashers (14) and screws (15). Tighten screws (15) 36-40 lb-ft (49-54 N•m).

4-71. TRANSMISSION REPLACEMENT (IN-VEHICLE) (Contd)

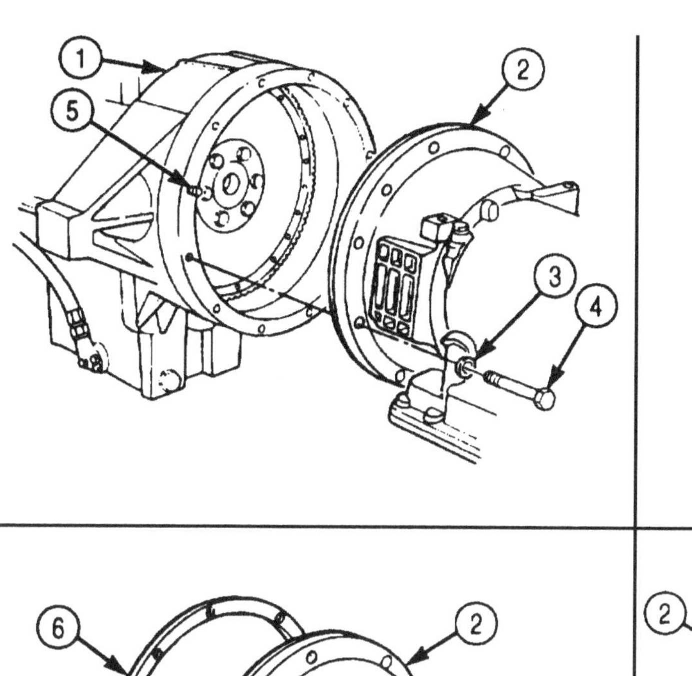

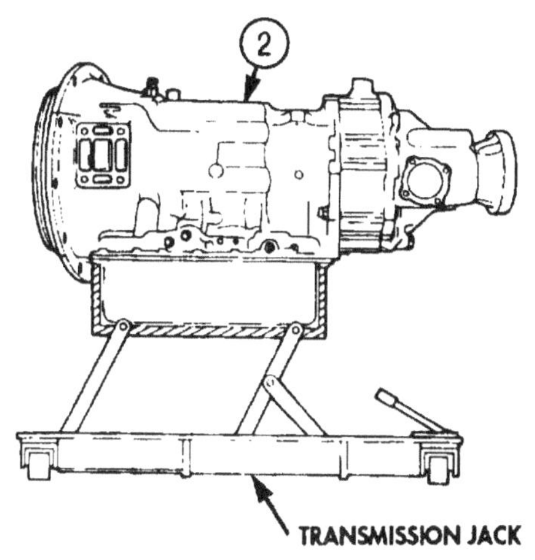

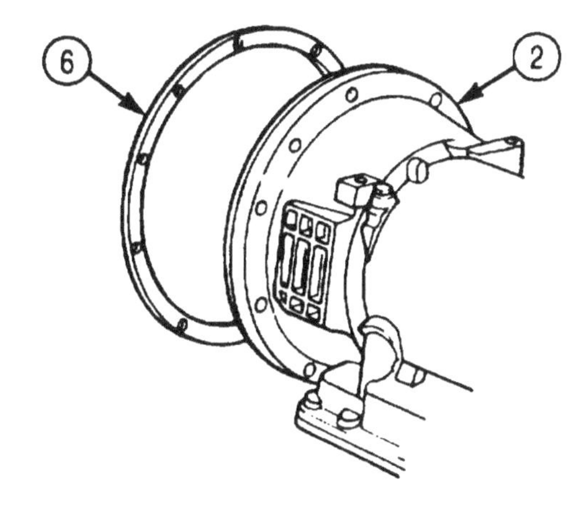

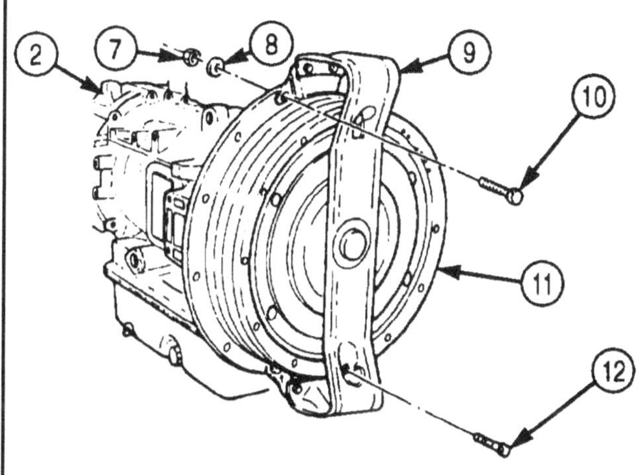

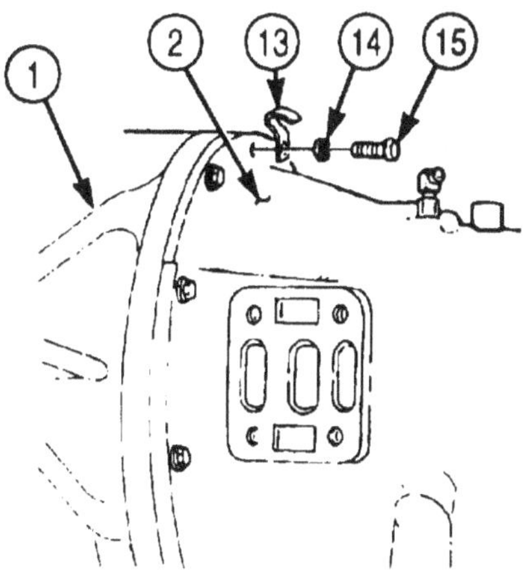

4-71. TRANSMISSION REPLACEMENT (IN-VEHICLE) (Contd)

NOTE

- Use engine barring tool to rotate engine crankshaft when aligning flywheel to converter screw holes.
- The use of shim stock tube in torque converter access hole prevents converter screw from falling down behind flywheel. A loose screw may cause a lockup condition, preventing crankshaft rotation required to install remaining screws.

11. Using engine barring tool, rotate crankshaft until screw holes in converter (8) and flywheel (1) align with access hole (7).
12. Roll shim stock (6) into tube form and size to fit diameter of access hole (7). Insert shim stock (6) in access hole (7) and position over screw hole against flywheel (1).
13. Install twelve screws (5) on flywheel (1) and converter (8) in the same manner. Tighten screws (5) 41-49 lb-ft (56-66 Nm).
14. Remove shim stock (6) from access hole (7).
15. Install new gasket (2) and cover plate (3) on flywheel housing (9) with two screws (4). Tighten screws (4) 5-8 lb-ft (7-11 Nm).
16. Connect oil cooler filter-to-transmission supply hose (10) to lubrication valve adapter (13) on transmission (11).

NOTE

Perform step 17 for vehicles equipped with transmission PTO.

17. Connect PTO oil return hose (12) to lubrication valve adapter (13) on transmission (11).
18. Connect transmission-to-oil cooler return hose (14) to temperature transmitter adapter (15) on transmission (11).
19. Install front axle propeller shaft (19) on transfer case flange (16) with eight screws (17) and new locknuts (18). Tighten locknuts (18) 32-40 lb-ft (43-54 Nm).

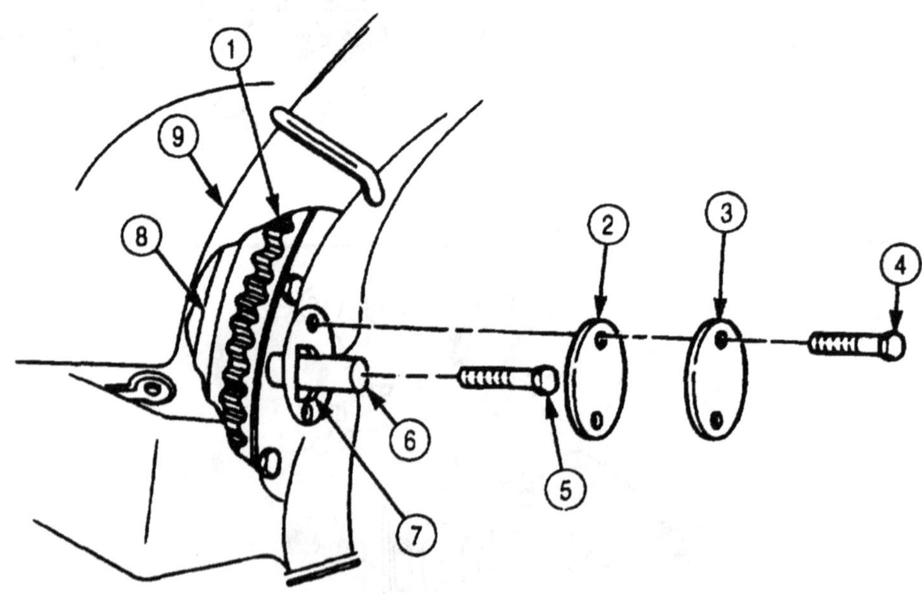

4-71. TRANSMISSION REPLACEMENT (IN-VEHICLE) (Contd)

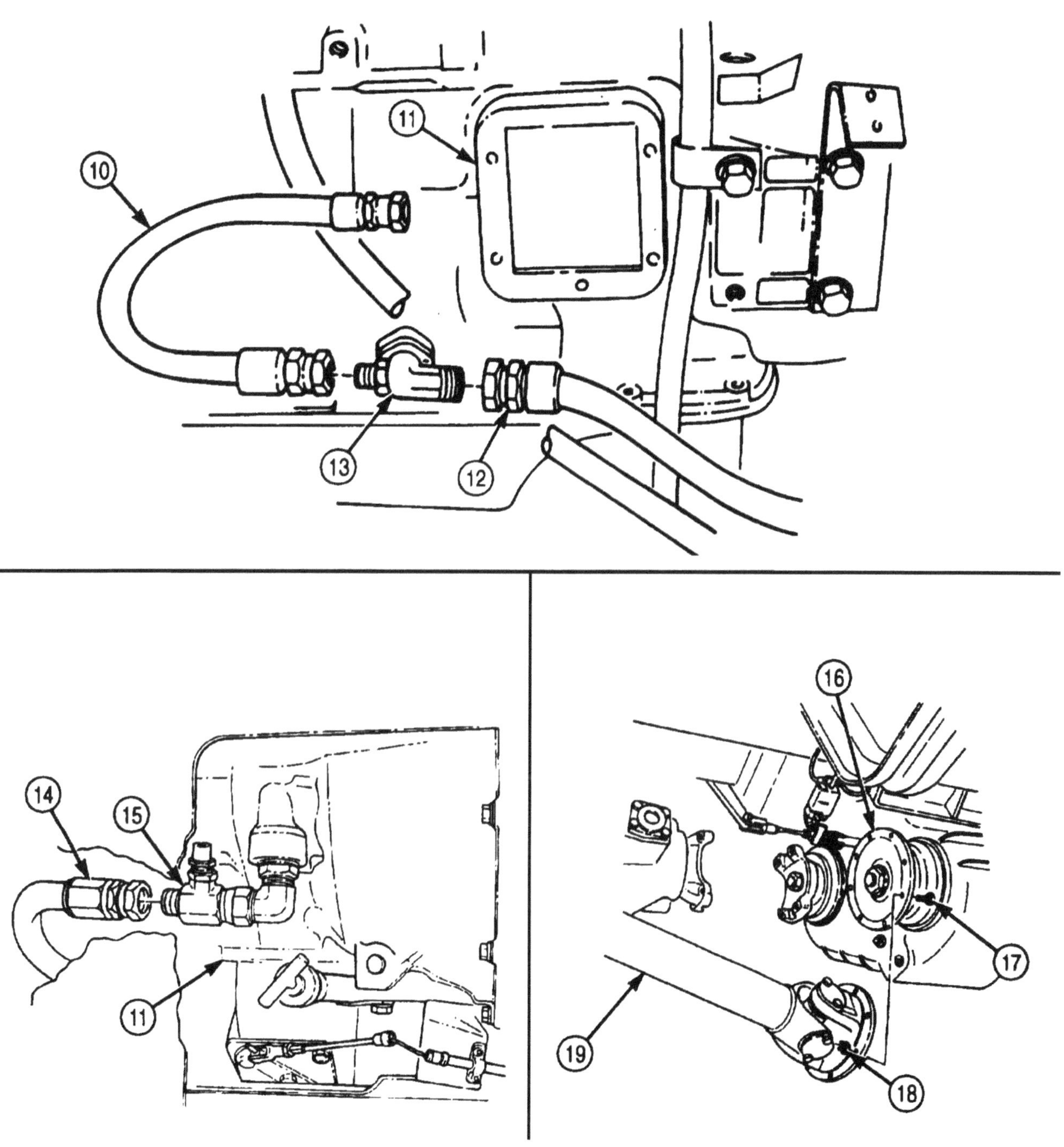

4-71. TRANSMISSION REPLACEMENT (IN-VEHICLE) (Contd)

20. Install two new lockwashers (3) and screws (2) on rear support bracket (4) and transmission (1). Tighten screws (2) 65-85 lb-ft (88-115 N•m).

NOTE

Perform steps 21 and 22 for vehicles equipped with transmission PTO.

21. Install pin (12) of PTO cable (11) on select lever (13) with washer (14) and new cotter pin (5).
22. Install PTO cable (11) on bracket (7) with spacer plate (8), retainer strap (10), two screws (9), and nuts (6).
23. Connect wire (15) to transmission oil temperature transmitter (22).
24. Install wire (20) on transmission flange (19) with clamp (18), new lockwasher (17), and screw (16).

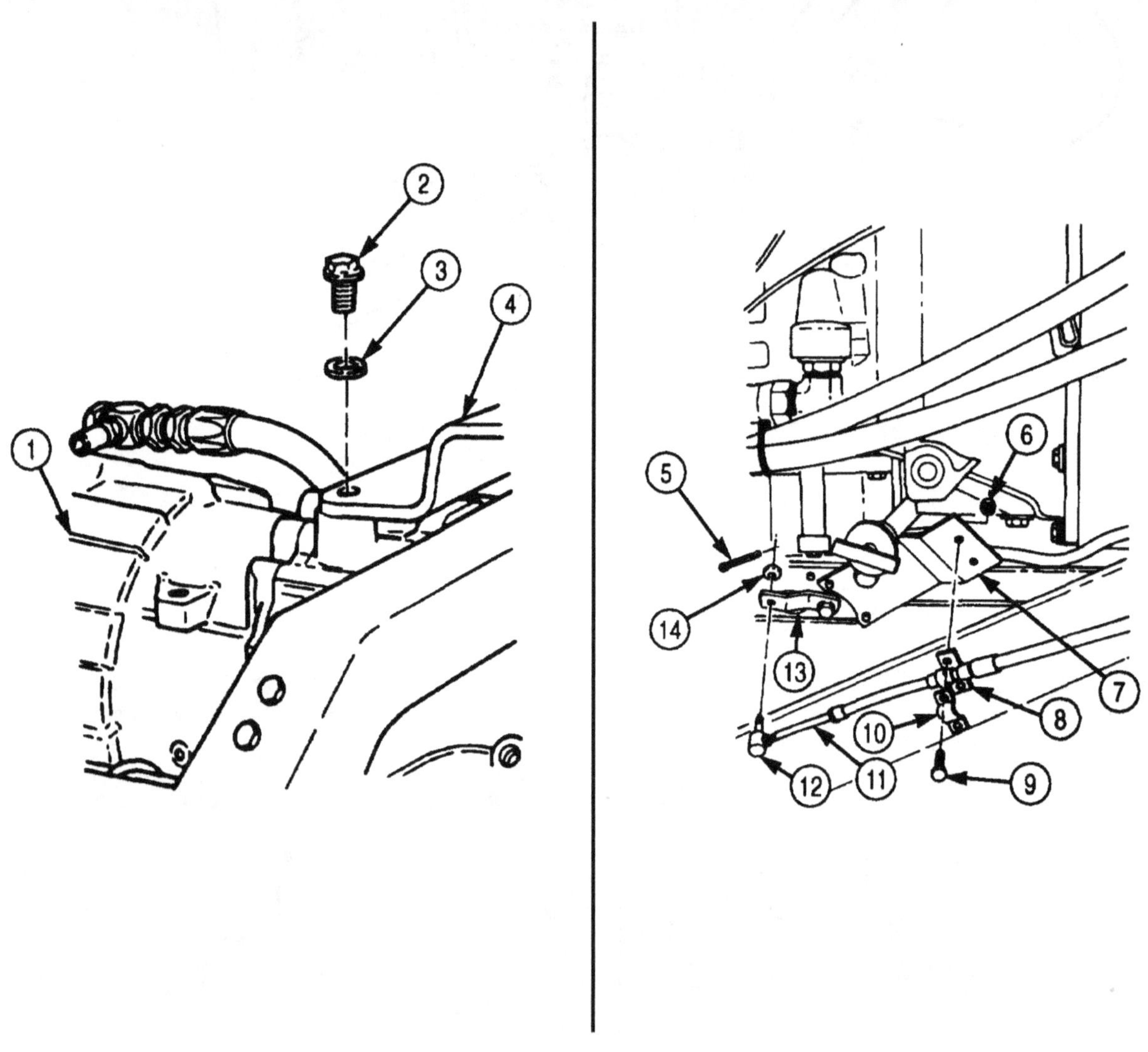

4-71. TRANSMISSION REPLACEMENT (IN-VEHICLE) (Contd)

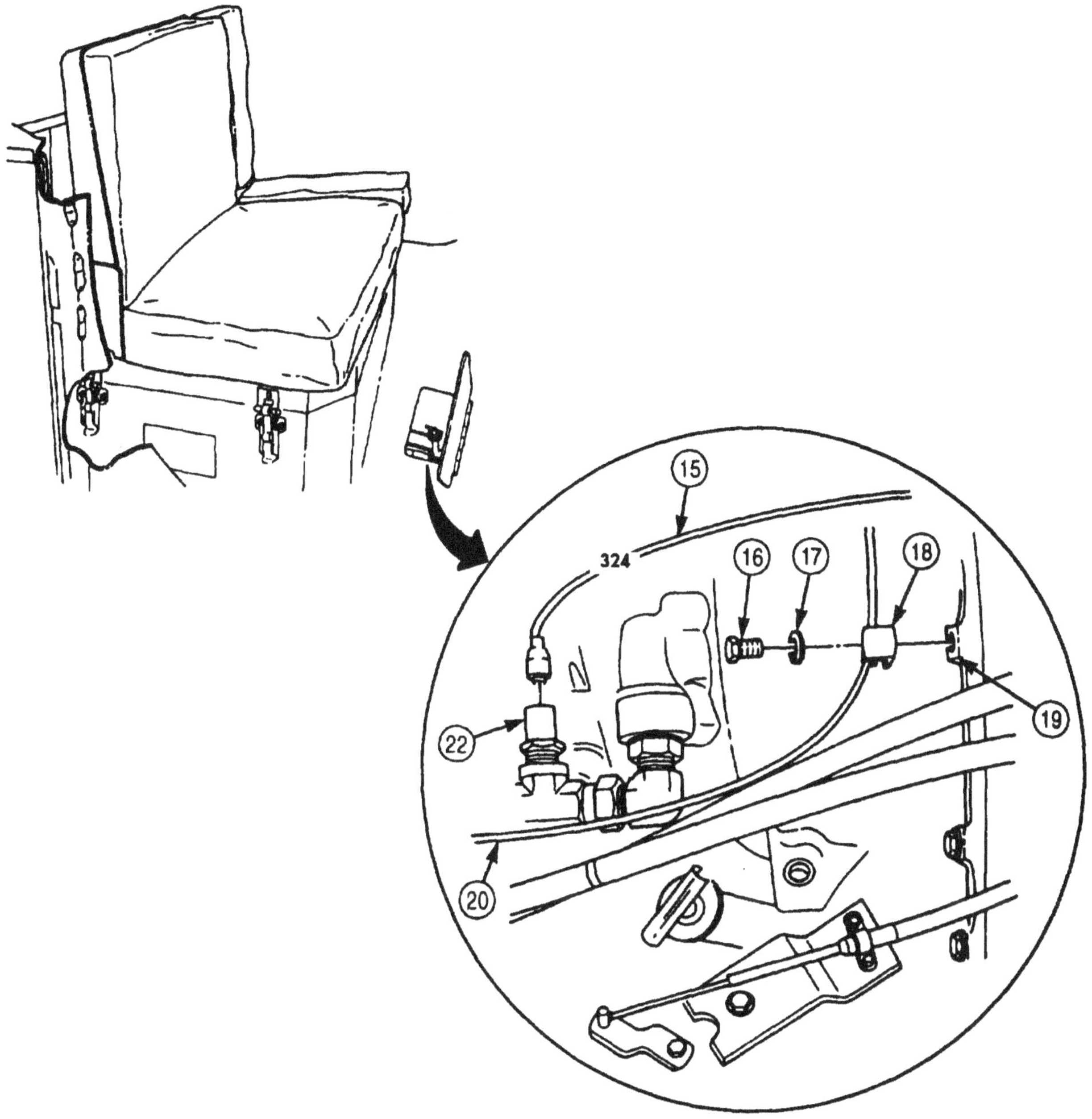

FOLLOW-ON TASKS:
- Install transmission power takeoff (if so equipped) (para. 4-211).
- Install winch control valve (if so equipped) (para. 4-179 or para. 4-178).
- Install transmission neutral start switch (para. 3-98).
- Install transmission fifth gear lock-in solenoid valve and bracket (para. 4-80).
- Install transmission modulator (para. 3-145).
- Install transmission breather (para. 3-136).
- Install transmission oil dipstick (para. 3-134).
- Install transmission-to-transfer case propeller shaft (para. 3-148).
- Fill transmission to proper oil level (LO 9-2320-272-12).
- Start engine (TM 9-2320-272-10) and road test vehicle.

4-72. TRANSMISSION REPLACEMENT (OUT-OF-VEHICLE)

THIS TASK COVERS:

a. Removal **b. Installation**

INITIAL SETUP:

APPLICABLE MODELS
All

SPECIAL TOOLS
Engine barring tool (Appendix E, Item 8)

TOOLS
General mechanic's tool kit (Appendix E, Item 1)
Torque wrench (Appendix E, Item 144)
Transmission jack (Appendix E, Item 147)

MATERIALS/PARTS
Gasket (Appendix D, Item 202)
Gasket (Appendix D, Item 203)
Twelve lockwashers (Appendix D, Item 354)

PERSONNEL REQUIRED
TWO

REFERENCES (TM)
TM 9-2320-272-24P

EQUIPMENT CONDITION
Power plant removed (para. 4-4 or 4-5).

GENERAL SAFETY INSTRUCTIONS
- Torque converter must be removed with transmission.
- Keep transmission tilted slightly downward.

a. Removal

NOTE

- The use of shim stock tube in torque converter access hole prevents converter screw from falling down behind flywheel. A loose screw may cause a lockup condition, preventing crankshaft rotation required to remove remaining screws.
- Engine crankshaft may be turned using barring tool.

1. Remove two screws (4), access cover (3), and gasket (2) from flywheel housing (1). Discard gasket (2).
2. Using engine barring tool, rotate crankshaft until screw (5) is visible in access hole (7).
3. Roll shim stock (6) into tube form and size to fit diameter of access hole (7). Insert shim stock (6) in access hole (7) and position over screw (5) against flywheel (9).
4. Remove remaining twelve screws (5) from flywheel (9) and converter (8) in the same manner,
5. Remove four screws (13), lockwashers (12), and clamp (11) from 9, 11, 1, and 3 o'clock positions on flywheel housing (1) and transmission (10). Discard lockwashers (12).

4-72. TRANSMISSION REPLACEMENT (OUT-OF-VEHICLE) (Contd)

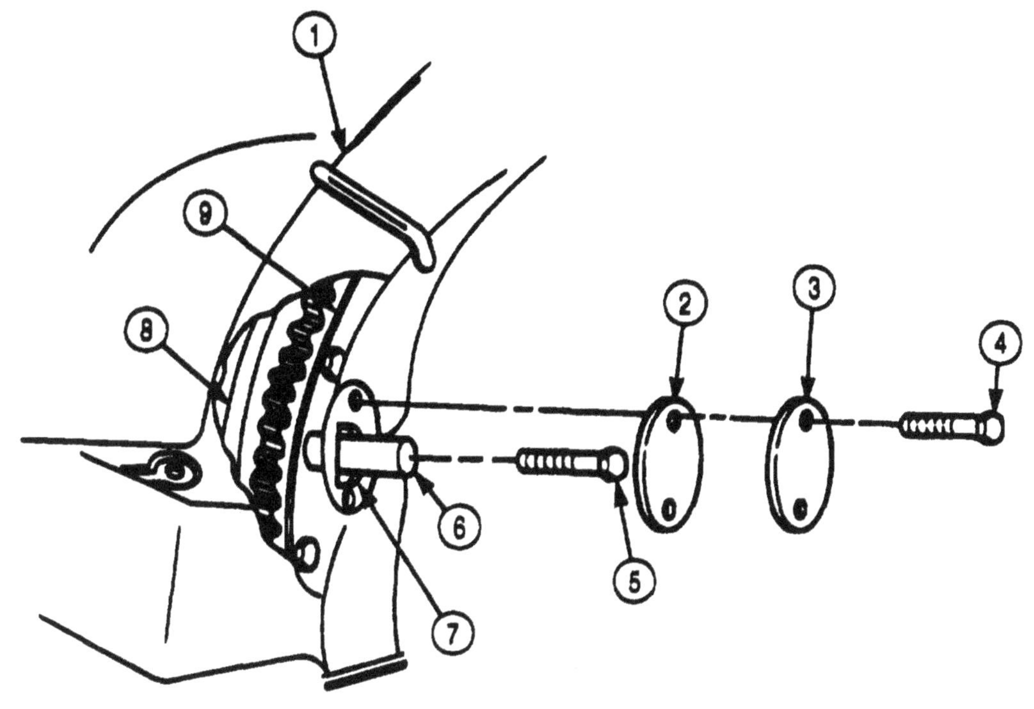

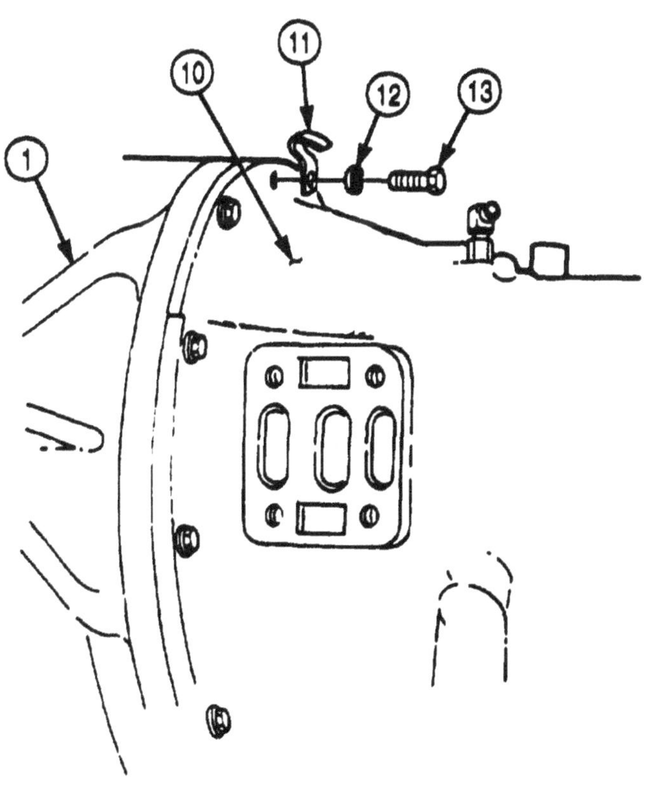

4-72. TRANSMISSION REPLACEMENT (OUT-OF-VEHICLE) (Contd)

NOTE

Guide or alignment screws can be made from extra screws that are longer and of the same thread size (appendix F, item 2).

6. Install four guide screws (1) at 9, 11, 1, and 3 o'clock positions on flywheel housing (1).
7. Position transmission jack under transmission (4) and raise until weight of transmission (4) is supported by jack.
8. Remove eight remaining screws (6) and lockwashers (5) from flywheel housing (2) and transmission (4). Discard lockwashers (5).

WARNING

- Keep rear of transmission tilted slightly downward at all times to prevent converter from sliding off and causing injury to personnel or damage to equipment.
- Torque converter and transmission must be removed as an assembly to prevent injury to personnel and damage to converter.

9. Separate transmission (4) from flywheel housing (2), keep level until clear of guide screws (1), then tilt rear of transmission (4) downward slightly to prevent converter (12) from separating from transmission (4).
10. Remove gasket (3) from transmission (4) or flywheel housing (2). Discard gasket (3).
11. Install retaining strap (10) on converter (12) with two screws (13).
12. Install retaining strap (10) on transmission (4) with four screws (11), washers (9), and nuts (8).
13. Remove four guide screws (1) from flywheel housing (2).

b. Installation

WARNING

- Keep rear of transmission tilted slightly downward at all times to prevent converter from sliding off and causing injury to personnel or damage to equipment.
- Torque converter must be installed with transmission as an assembly to prevent injury to personnel and damage to converter.

1. Install four guide screws (1) at 9, 11, 1, and 3 o'clock positions on flywheel housing (2).
2. Place transmission (4) on transmission jack.
3. Remove four nuts (8), washers (9), and screws (11) from retaining strap (10) and transmission (4).
4. Remove two screws (13) and retaining strap (10) from converter (12).
5. Position new gasket (3) on guide screws (1) and flywheel housing (2).

4-72. TRANSMISSION REPLACEMENT (OUT-OF-VEHICLE) (Contd)

CAUTION

Maintain transmission alignment to flywheel housing during installation to prevent damage to converter.

6. Raise transmission (4), align with guide screws (1), and seat against flywheel housing (2).
7. Install transmission (4) on flywheel housing (2) with clamp (7), eight new lockwashers (5), and screws (6). Tighten screws (6) 36-40 lb-ft (49-54 N•m).
8. Remove four guide screws (1) and install remaining four new lockwashers (5) and screws (6). Tighten screws (6) 36-40 lb-ft (49-54 N•m).

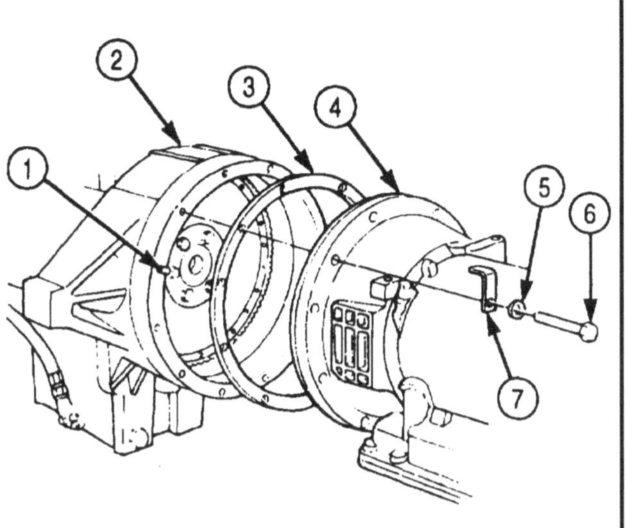

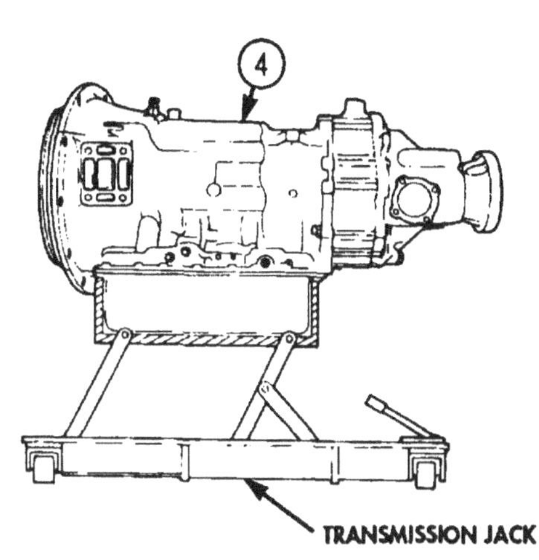

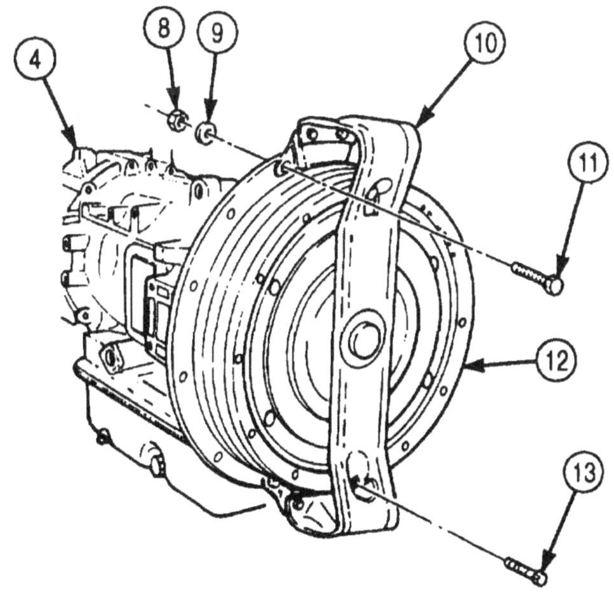

4-72. TRANSMISSION REPLACEMENT (OUT-OF-VEHICLE) (Contd)

NOTE

- Use engine barring tool to rotate engine crankshaft when aligning flywheel to converter screw holes.
- The use of shim stock tube in torque converter access hole prevents converter screw from falling down behind flywheel. A loose screw may cause a lockup condition, preventing crankshaft rotation required to install remaining screws.

9. Using engine barring tool, rotate crankshaft until screw holes in converter (9) and flywheel (1) align with access hole (8).
10. Roll shim stock (3) into tube form and size to tit diameter of access hole (8). Insert shim stock (3) in access hole (8) and position over screw hole and against flywheel (1).
11. Install twelve screws (7) on flywheel (1) and converter (9) in the same manner. Tighten screws (7) 41-49 lb-ft (56-66 N•m).
12. Remove shim stock (3) from access hole (8).
13. Install new gasket (4) and cover plate (5) on flywheel housing (2) with two screws (6). Tighten screws (6) 5-8 lb-ft (7-11 N•m).

4-72. TRANSMISSION REPLACEMENT (OUT-OF VEHICLE) (Contd)

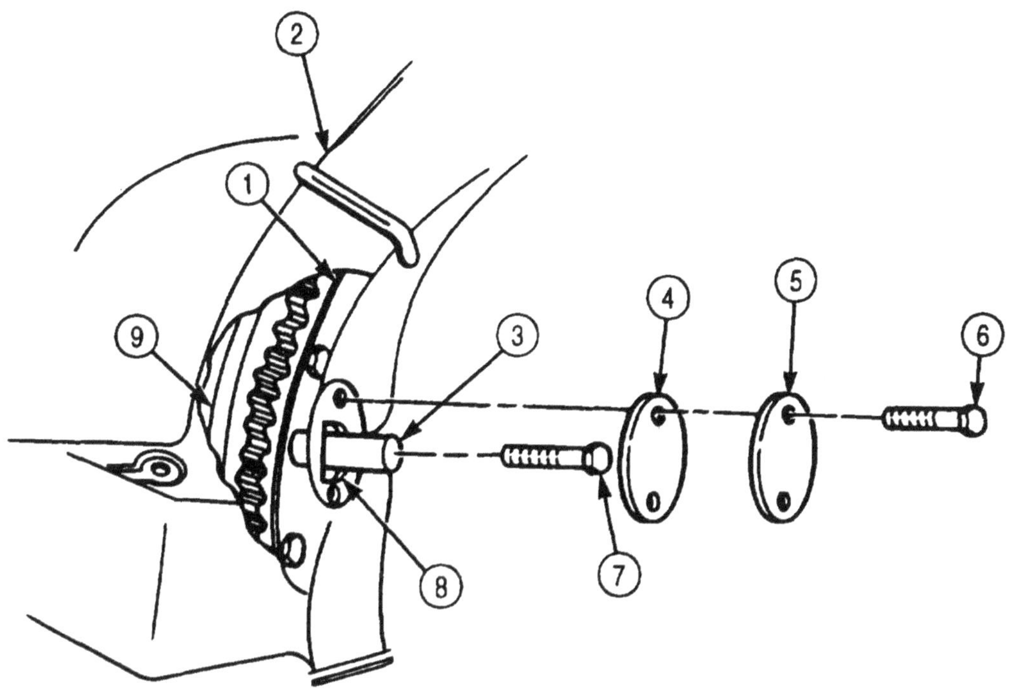

FOLLOW-ON TASK: Install power plant (para. 4-4 or 4-5).

4-73. TRANSMISSION SELECT LEVER REPLACEMENT

THIS TASK COVERS:

a. Removal **b. Installation**

INITIAL SETUP:

APPLICABLE MODELS
All

TOOLS
General mechanic's tool kit (Appendix E, Item 1)

MATERIALS/PARTS
Four lockwashers (Appendix D, Item 345)

REFERENCES (TM)
TM 9-2320-272-10
TM 9-2320-272-24P

EQUIPMENT CONDITION
- Parking brake set (TM 9-2320-272-10).
- Battery ground cables disconnected (para. 3-126).

a. Removal

1. Remove four screws (15) and lockwashers (16) from control lever console (17). Discard lockwashers (16).
2. Pull select lever assembly (18) away from control lever console (17).

NOTE

Tag leads for installation.

3. Disconnect three leads (14) from harness leads (13).
4. Remove two nuts (12), cable clamp (11), shim (9), two screws (7), and washers (8) from hanger plate (6).
5. Remove spring clip (2) and pull trunnion (3) and shift cable (10) from selector lever plate (5).
6. Remove selector lever assembly (18) from vehicle.

b. Installation

NOTE

Vehicle must be started in N (neutral) to check selector lever assembly. Vehicle will start if installation is correct. Remove and reinstall selector lever assembly if vehicle fails to start in N (neutral).

1. Place selector lever (1) in neutral position.
2. Place manual control linkage arm (19) in neutral position. Linkage arm (19) will be one detent down from full up position.
3. Install shim (9), shift cable (10), and cable clamp (11) on hanger plate (6) with two washers (8), screws (7), and nuts (12). Ensure cable clamp (11) seats in groove of shift cable (10) housing.
4. Loosen jamnut (4) and align cable trunnion (3) with first hole above elongated slot in selector lever plate (5). Cable trunnion (3) is turned clockwise to shorten and counterclockwise to lengthen. Tighten jamnut (4) and install in plate (5) with spring clip (2).
5. Connect three electrical leads (14) to harness leads (13).
6. Install selector lever assembly (18) in control lever console (17) with four new lockwashers (16) and screws (15).

4-73. TRANSMISSION SELECT LEVER REPLACEMENT (Contd)

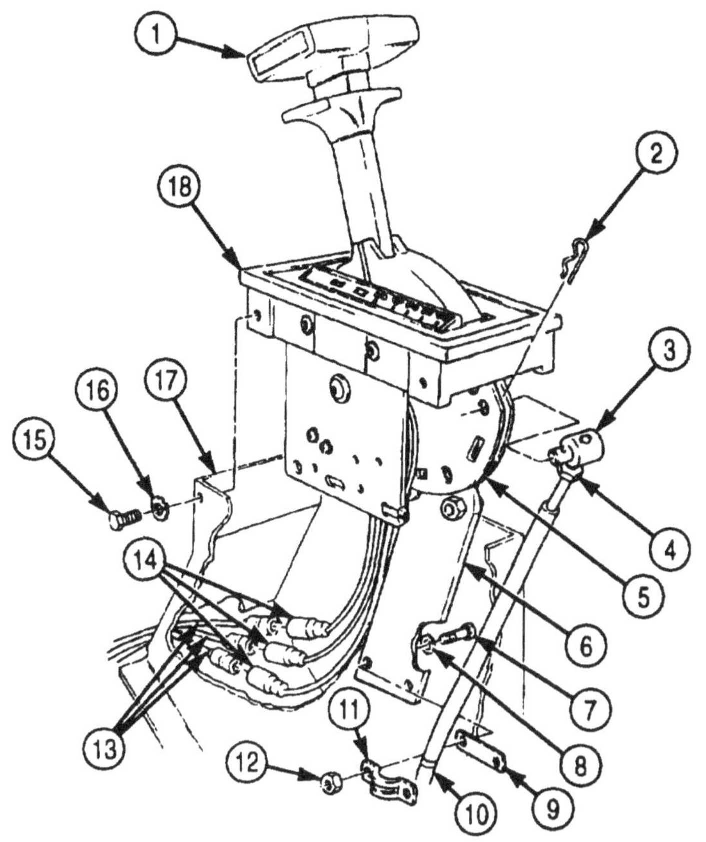

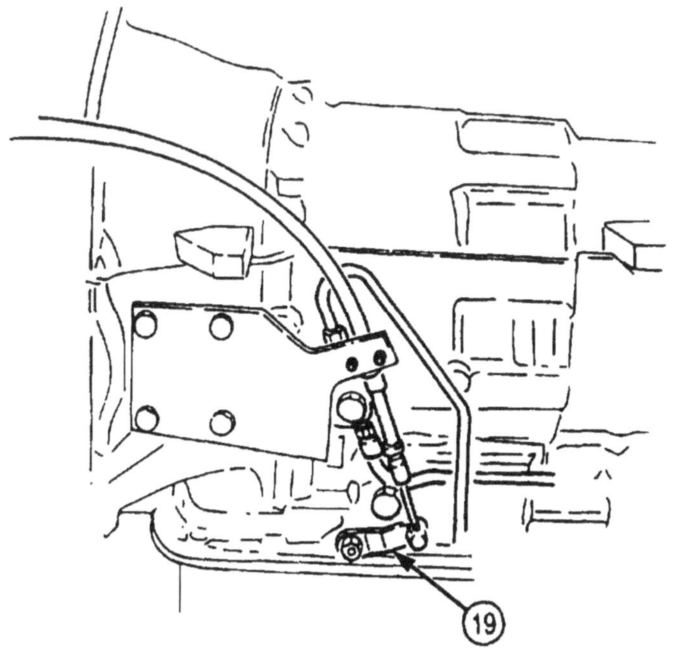

FOLLOW-ON TASKS:
- Connect battery ground cables (para. 3-126).
- Start engine (TM 9-2320-272-10) and road test vehicle.

4-74. TRANSMISSION SHIFT TOWER REPLACEMENT

THIS TASK COVERS:

a. Removal **b. Installation**

INITIAL SETUP:

APPLICABLE MODELS
All

TOOLS
General mechanic's tool kit (Appendix E, Item 1)

MATERIALS/PARTS
Eleven lockwashers (M939A2) (Appendix D, Item 345)
Gasket (M939A2) (Appendix D, Item 204)

REFERENCES (TM)
TM 9-2320-272-24P

EQUIPMENT CONDITION
- CTIS ECU removed (M939A2) (para. 3-468).
- Transmission selector lever removed (para. 4.73).

a. Removal

NOTE

Vehicles equipped with winch will have winch control tower attached to shift tower. Two screws and washers in step 1 will be removed from winch control tower.

1. Remove four screws (4), washers (3), and shift tower (2) from access cover (1).

NOTE

Perform steps 2 through 7 for M939A2 series vehicles,

2. Remove two screws (5) and washers (6) from bracket (7) and front mount bracket (12).
3. Remove screw (15) and rear mount bracket (14) from shift tower base (13).
4. Remove screw (10) and front mount bracket (12) from shift tower base (13).
5. Remove grommet (11) from front mount bracket (12).
6. Remove four screws (9), lockwashers (8), shift tower base (13), and gasket (16) from mounting bracket (17). Discard lockwashers (8) and gasket (16).
7. Remove seven screws (18), lockwashers (19), and mounting bracket (17) from floor (20). Discard lockwashers (19).

b. Installation

NOTE

- Perform steps 1 through 7 for M939A2 series vehicles.
- Do not tighten screws until transmission shift tower is fully installed.

1. Install mounting bracket (17) on floor (20) with seven new lockwashers (19) and screws (18).
2. Install new gasket (16) and shift tower base (13) on mounting bracket (17) with four new lockwashers (8) and screws (9).
3. Install grommet (11) on front mount bracket (12).
4. Install front mount bracket (12) on shift tower base (13) with screw (10). Finger-tighten screw (10).
5. Install rear mount bracket (14) on shift tower base (13) with screw (16). Finger-tighten screw (16).
6. Install front mount bracket (12) on bracket (7) with two washers (6) and screws (5). Finger-tighten screws (5).
7. Tighten screws (5), (10), and (15).

NOTE

Vehicles equipped with winch will have winch control tower installed attached to shift tower. Two screws and washers in step 8 will be installed in winch control tower.

8. Install shift tower (2) on access cover (1) with four washers (3) and screws (4).

4-74. TRANSMISSION SHIFT TOWER REPLACEMENT (Contd)

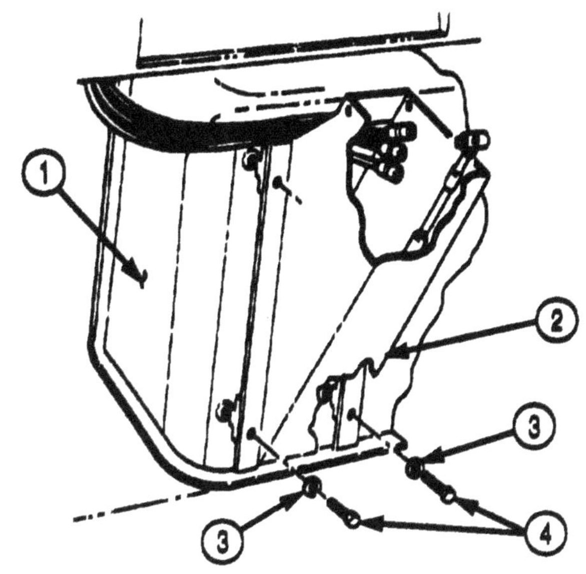

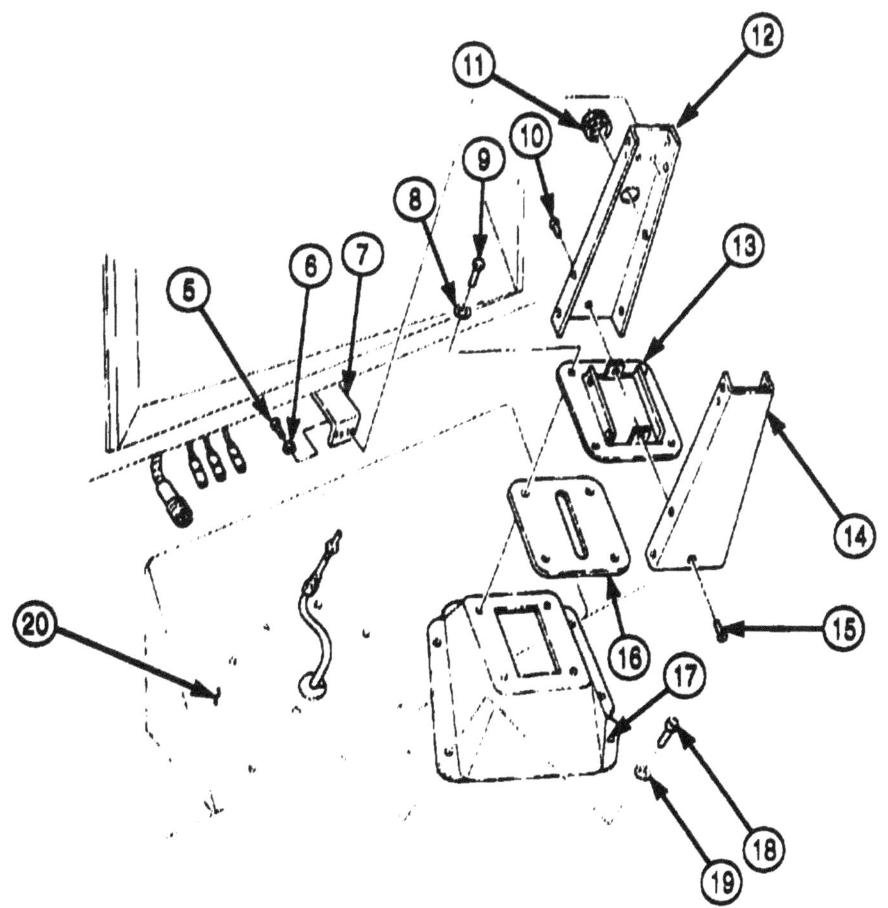

FOLLOW-ON TASKS:
- Install transmission selector lever (para. 4-73).
- Install CTIS ECU (M939A2) (para. 3-468).

4-75. TRANSMISSION SHIFT CABLE MAINTENANCE

THIS TASK COVERS:

a. Removal
b. Installation
c. Adjustment

INITIAL SETUP:

APPLICABLE MODELS
All

TOOLS
General mechanic's tool kit (Appendix E, Item 1)

MATERIALS/PARTS
Two locknuts (M939/A1) (Appendix D, Item 283)
Two locknuts (M939A2) (Appendix D, Item 334)

REFERENCES (TM)
TM 9-2320-272-10
TM 9-2320-272-24P

EQUIPMENT CONDITION
- Parking brake set (TM 9-2320-272-10).
- Left splash shield removed (TM 9-2320-272-10).
- Transmission selector lever removed (para. 4-73).

a. Removal

1. Remove two locknuts (1), U-bolt (4), clamp (3), and shift cable (2) from transmission lock-in solenoid bracket (10). Discard locknuts (1).

NOTE

Perform step 2 for M929/A1/A2 and M934/A1/A2 model vehicles.

2. Remove two locknuts (1), screws (11), clamp (3), and shift cable (2) from transmission lock-in solenoid bracket (10). Discard locknuts (1).
3. Remove spring clip (7) and shift cable (2) from manual control linkage (8).
4. Loosen jamnut (5) and remove trunnion (6) and jamnut (5) from shift cable (2).
5. Loosen jamnut (14) and remove trunnion (13) and jamnut (14) from shift cable (2) at shift tower (12).
6. Remove shift cable (2) from vehicle.
7. Remove grommet (15) from engine cover (16).

b. Installation

1. Install grommet (15) on engine cover (16).

NOTE

Assistant will help with step 2.

2. Route shift cable (2) from shift tower (12), through grommet (15), and to left side of transmission (9).
3. Install jamnut (14) and trunnion (13) on shift cable (2) at shift tower (12). Thread trunnion (13) on shift cable (2) until shift cable (2) is flush with trunnion (13). Tighten jamnut (14).
4. Install jamnut (5) and trunnion (6) on shift cable (2) at transmission (9). Do not tighten jamnut (5).
5. Install shift cable (2) on manual control linkage (8) with spring clip (7).

4-75. TRANSMISSION SHIFT CABLE MAINTENANCE (Contd)

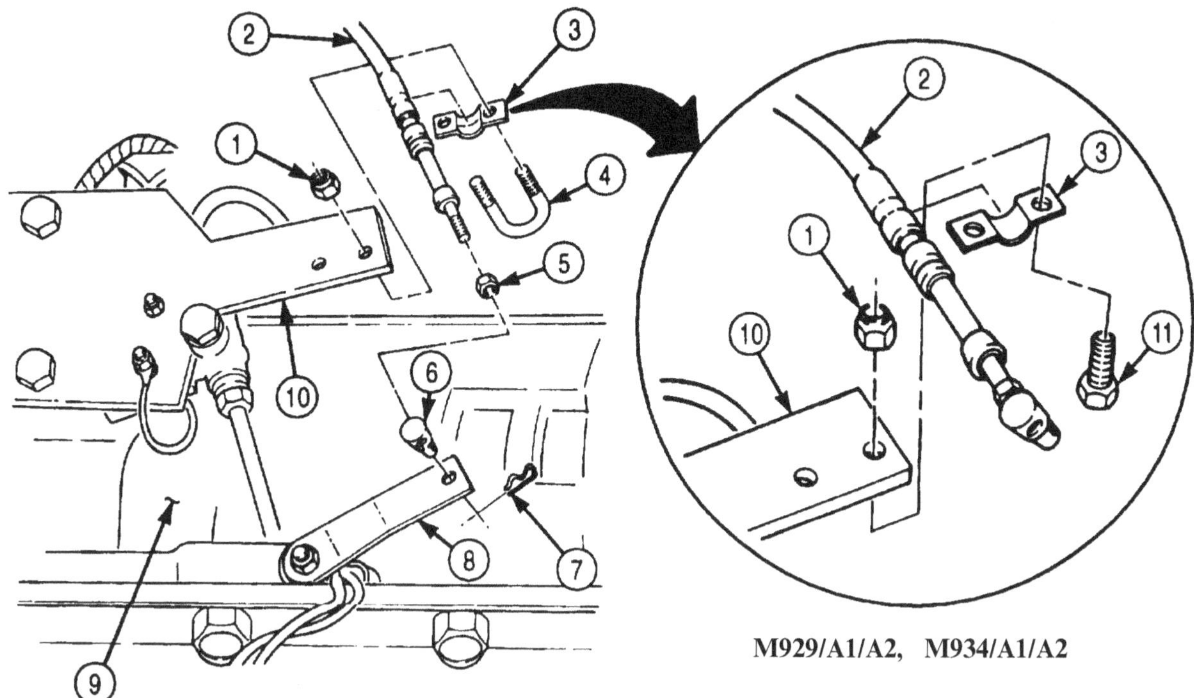

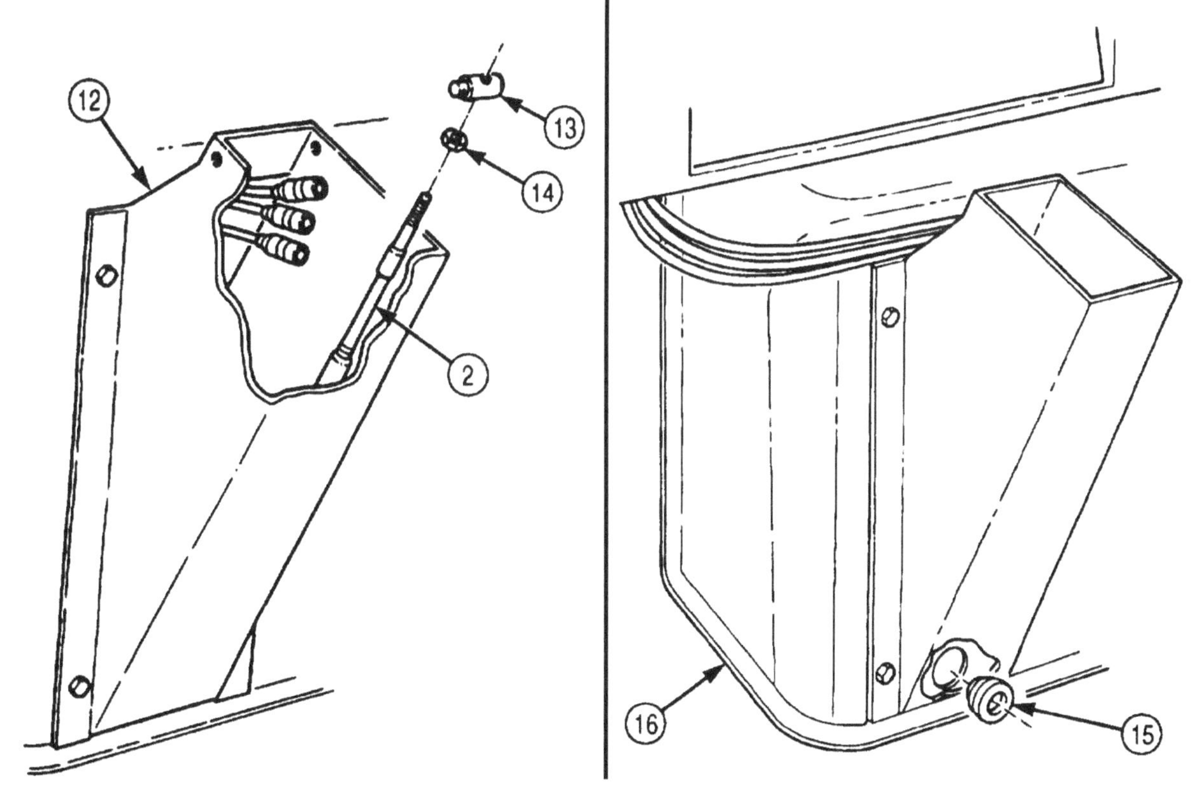

4-75. TRANSMISSION SHIFT CABLE MAINTENANCE (Contd)

NOTE

Perform step 6 for M929/A1/A2 and M934/A1/A2 model vehicles.

6. Install shift cable (1) on transmission lock-in solenoid bracket (8) with clamp (2), two screws (10), and new locknuts (9), ensuring clamp (2) seats in groove of shift cable (1).
7. Install shift cable (1) on transmission lock-in solenoid bracket (8) with clamp (2), U-bolt (3), and two new locknuts (9), ensuring clamp (2) seats in groove of shift cable (1).
8. Install transmission selector lever (para. 4-73).

c. Adjustment

CAUTION

Ensure manual control linkage arm is in proper detent for each transmission selector position. Failure to do so may cause damage to transmission.

1. Remove spring clip (6) and shift cable (1) from manual control linkage (7), if necessary.
2. Place manual control linkage (7) and transmission selector lever in first gear position.
3. Align hole of manual control linkage (7) with trunnion (5). If trunnion (5) does not align, loosen jamnut (4), and turn trunnion (5) clockwise to shorten or counterclockwise to lengthen. Tighten jamnut (4).
4. Install shift cable (1) on manual control linkage (7) with spring clip (6).

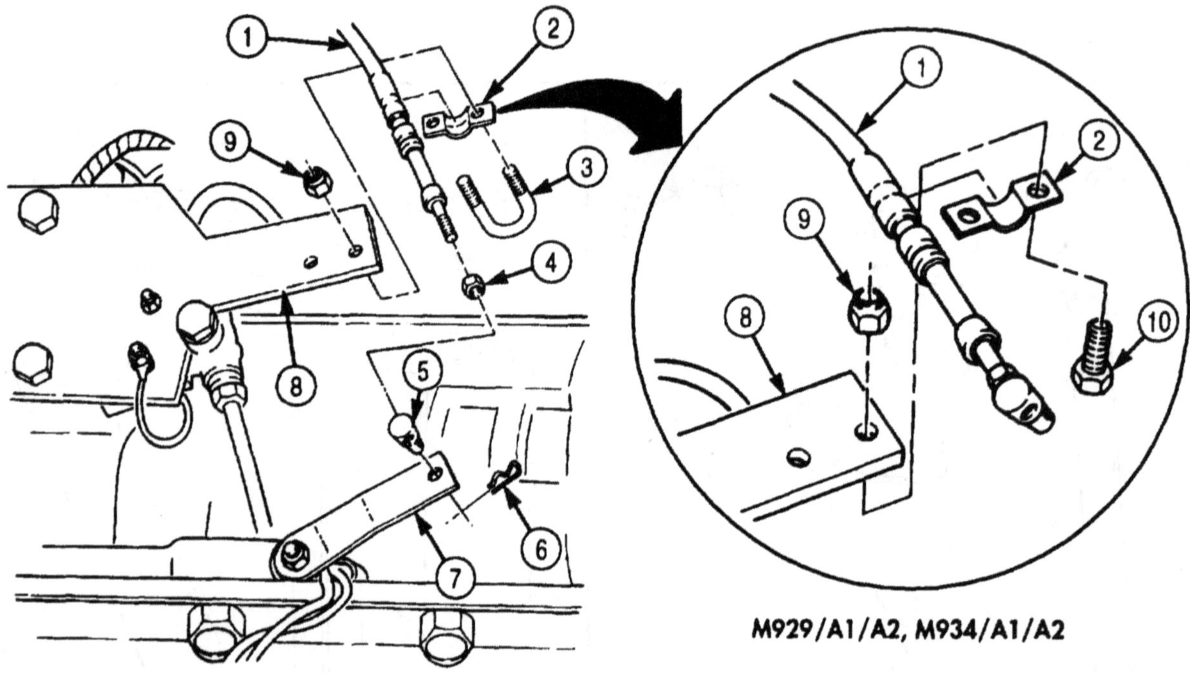

FOLLOW-ON TASKS:
- Install left splash shield (TM 9-2320-272-10).
- Start engine (TM 9-2320-272-10) and road test vehicle.

4-76. TRANSMISSION SELECTOR SHAFT OIL SEAL REPLACEMENT

THIS TASK COVERS:

a. Removal **b. Installation**

INITIAL SETUP:

APPLICABLE MODELS
All

SPECIAL TOOLS
Seal remover (Appendix E, Item 149)
Seal installer (Appendix E, Item 148)

TOOLS
General mechanic's tool kit (Appendix E, Item 1)
Torque wrench (Appendix E, Item 146)

MATERIALS/PARTS
Oil seal (Appendix D, Item 626)
Locknut (Appendix D, Item 324)
Cleaning cloth (Appendix C, Item 21)
Lubricating oil (Appendix C, Item 48)
Sealing compound (Appendix C, Item 65)

REFERENCES (TM)
LO 9-2320-272-12
TM 9-2320-272-10
TM 9-2320-272-24P

EQUIPMENT CONDITION
Parking brake set (TM 9-2320-272-10).

a. Removal

CAUTION

Clean around seal area to prevent entry of dirt. Damage to transmission may occur if dirt or dust enters transmission.

NOTE

Manual selector shaft locknut has metric threads.

1. Remove locknut (2) and manual control linkage (3) from selector shaft (1). Discard locknut (2).

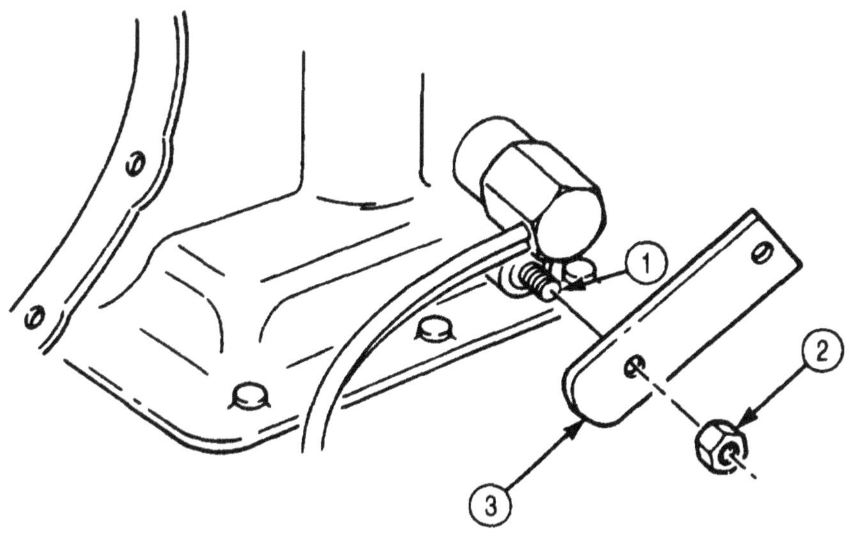

4-76. TRANSMISSION SELECTOR SHAFT OIL SEAL REPLACEMENT (Contd)

2. Position tapered end (4) of seal remover over selector shaft (6). Turn seal remover clockwise to thread into seal (3).
3. Finger-tighten bolt (5) on seal remover until it cannot be turned by hand. At this point, bolt (5) has contacted selector shaft (6).

NOTE

Have drainage container ready to catch oil.

4. Turn square head of bolt (5) until seal (3) slides from bore (2) of transmission housing (1).
5. Remove seal (3) from seal remover. Discard seal (3).

b. Installation

1. Apply lubricating oil to bore (2) of transmission housing (1) and selector shaft (6).
2. Apply a thin coat of sealing compound to outer edge of new seal (3).
3. Position seal (3) on seal installer with lip of seal (3) facing away from tool.

CAUTION

Use care when installing seal. Lip of seal may split on selector shaft.

4. Drive seal (3) into bore (2) of transmission housing (1) until seal (3) seats at counterbore position of bore (2).

CAUTION

When installing manual control linkage, a metric locknut must be used. Failure to do so will damage selector shaft.

5. Install manual control linkage (7) on selector shaft (6) with new locknut (8). Tighten locknut (8) 22-30 lb-ft (30-41 N•m).

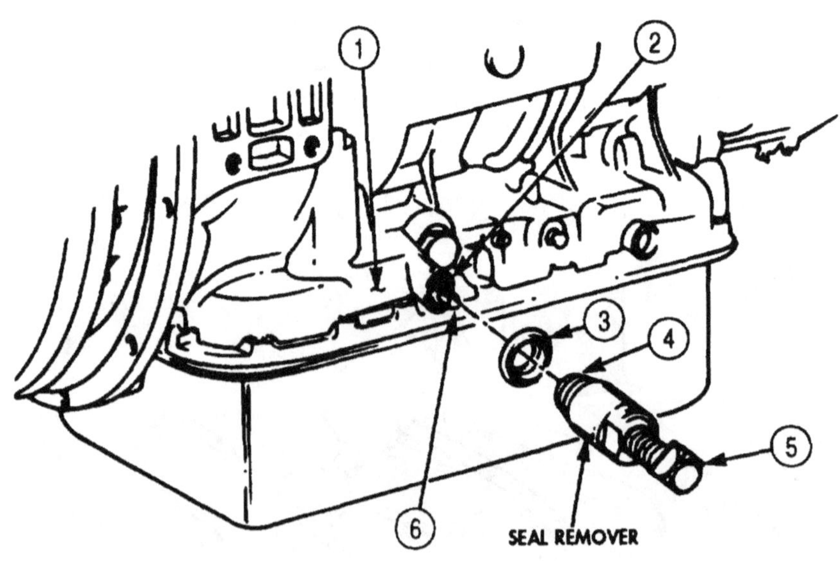

4-76. TRANSMISSION SELECTOR SHAFT OIL SEAL REPLACEMENT (Contd)

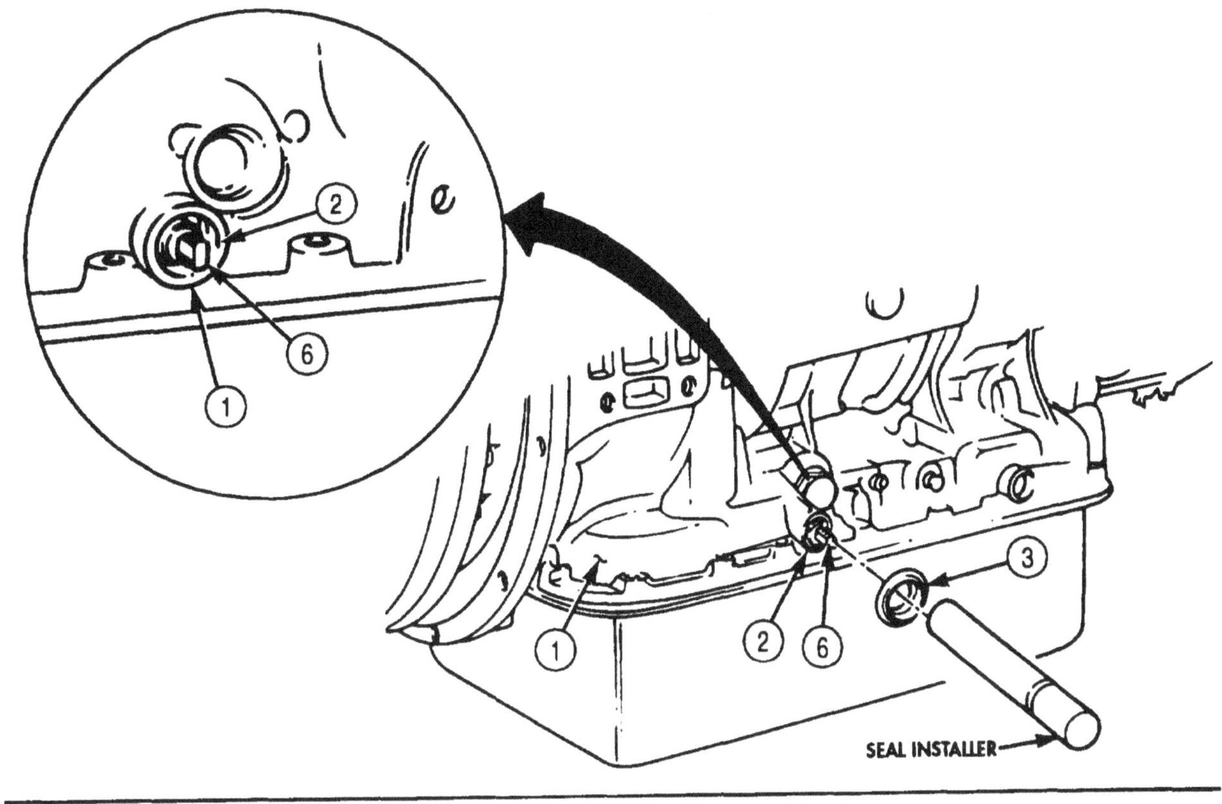

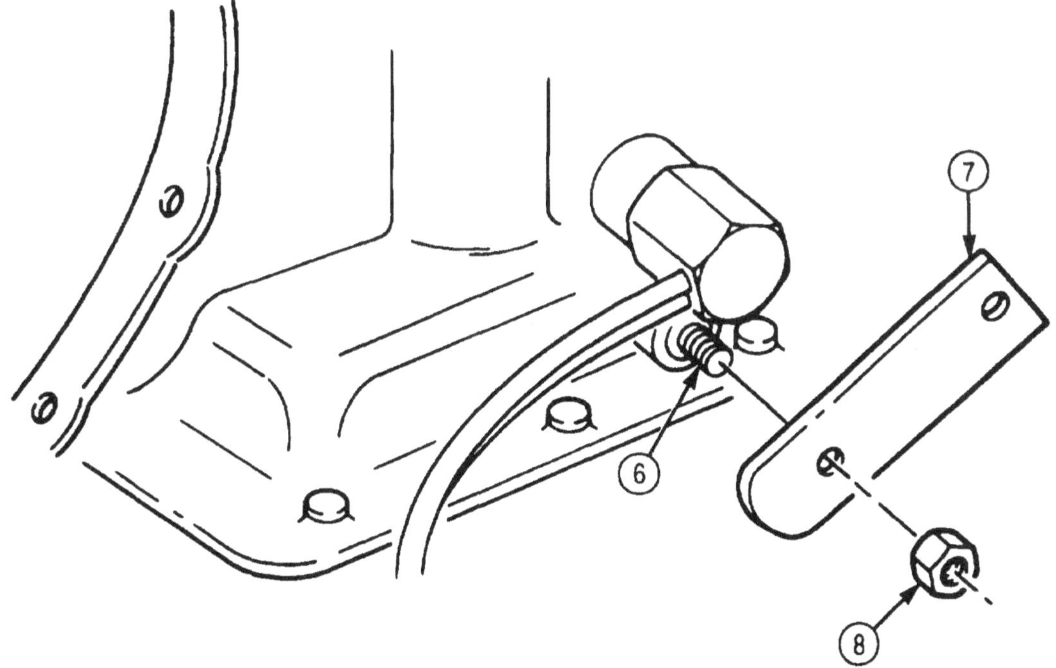

FOLLOW-ON TASK: Fill transmission to proper level (LO 9-2320-272-12).

4-77. TRANSMISSION MOUNT BUSHING REPLACEMENT

THIS TASK COVERS:

a. Removal **b. Installation**

INITIAL SETUP:

APPLICABLE MODELS
All

TOOLS
General mechanic's tool kit (Appendix E, Item 1)
Torque wrench (Appendix E, Item 144)

MATERIALS/PARTS
Two lockwashers (Appendix D, Item 350)
Two lockwashers (Appendix D, Item 392)

REFERENCES (TM)
TM 9-2320-272-10
TM 9-2320-272-24P

EQUIPMENT CONDITION
Parking brake ret (TM 9-2320-272-10).

a. Removal

1. Remove screw (2), lockwasher (1), two screws (3), lockwashers (4), and bracket (5) from transmission (7) and bushing (10). Discard lockwashers (1) and (4).
2. Remove screw (8), lockwasher (9), and bushing (10) from crossmember (6). Discard lockwasher (9).

b. Installation

1. Install bushing (10) on bracket (5) with new lockwasher (1) and screw (2), Tighten screw (2) 75-83 lb-ft (102-113 N•m).
2. Install bracket (5) and bushing (10) on crossmember (6) with new lockwasher (9) and screw (8). Tighten screw (8) 75-83 lb-ft (102-113 N•m).
3. Install bracket (5) on transmission (7) with two new lockwashers (4) and screws (3). Tighten screws (3) 75-83 lb-ft (102-113 N•m).

4-77. TRANSMISSION MOUNT BUSHING REPLACEMENT (Contd)

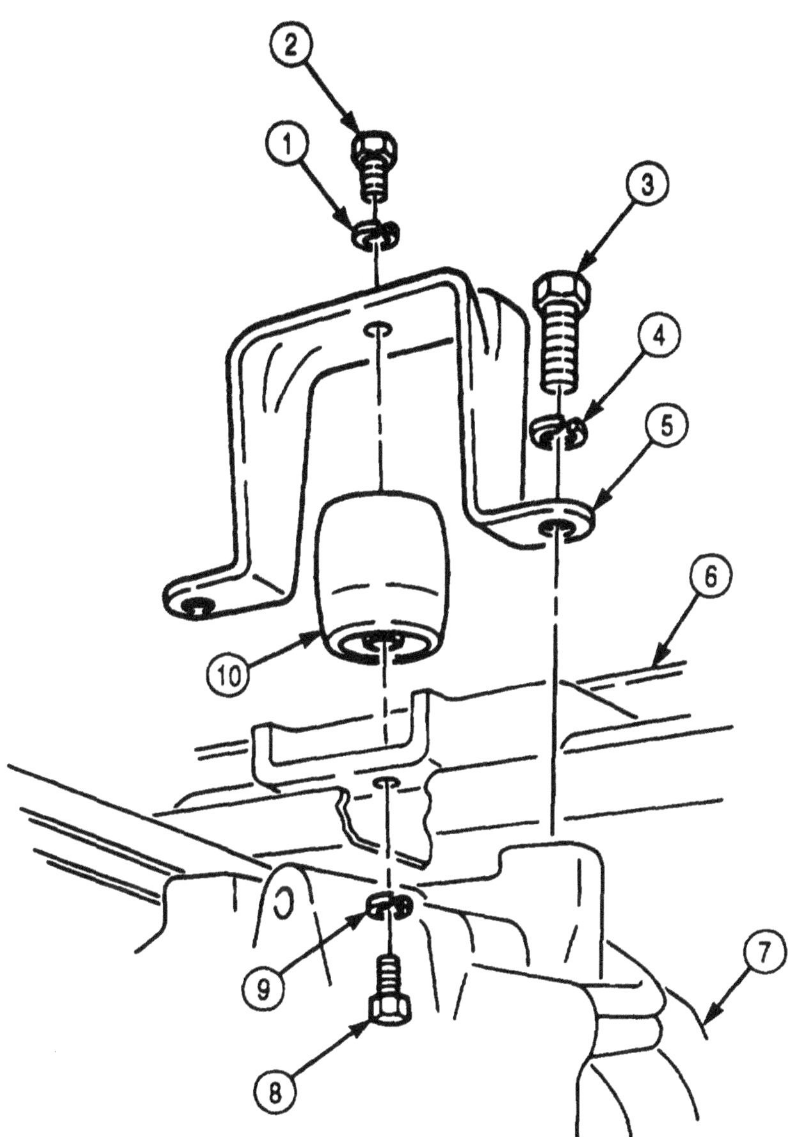

4-78. TRANSMISSION OUTPUT SHAFT YOKE AND OIL SEAL REPLACEMENT

THIS TASK COVERS:

a. Removal **b. Installation**

INITIAL SETUP:

APPLICABLE MODELS
All

SPECIAL TOOLS
Output shaft seal and dust shield remover (Appendix E, Item 92)
Dust shield installer (Appendix E, Item 39)
Oil seal installer (Appendix E, Item 91)
Driver handle (Appendix E, Item 37)

TOOLS
General mechanic's tool kit (Appendix E, Item 1)
Mechanical puller (Appendix E, Item 102)
Torque multiplier

MATERIALS/PARTS
Locknut (Appendix D, Item 325)
Oil seal (Appendix D, Item 503)
Lubricating oil (Appendix C, Item 48)

REFERENCES (TM)
LO 9-2320-272-12
TM 9-2320-272-10
TM 9-2320-272-24P

EQUIPMENT CONDITION
- Parking brake set (TM 9-2320-272-10).
- Propeller shaft removed (para. 3-148).

SPECIAL ENVIRONMENTAL CONDITIONS
Work area must be clean and free from blowing dust and dirt.

a. Removal

1. Remove locknut (3) from output shaft (1). Discard locknut (3).
2. Using mechanical puller, remove companion flange (2) from output shaft (1).

NOTE

Have drainage container ready to catch oil.

3. Using output shaft seal and dust shield remover, remove dust shield (4) and oil seal (5) from bore of transmission housing (6). Discard oil seal (5).

b. Installation

1. Using oil seal installer and driver handle, install new oil seal (5) in bore of transmission housing (6). Ensure rubber lip faces inside of rear cover (7).
2. Apply lubricating oil to inside diameter of oil seal (5).
3. Using dust shield installer and driver handle, install dust shield (4) in bore of transmission housing (6) with cupped side facing out until installer tool seats on face of rear cover (7).
4. Slide companion flange (2) on output shaft (1) and install with new locknut (3). Using torque multiplier, tighten locknut (3) 600-800 lb-ft (814-1,085 N·m).

4-78. TRANSMISSION OUTPUT SHAFT YOKE AND OIL SEAL REPLACEMENT (Contd)

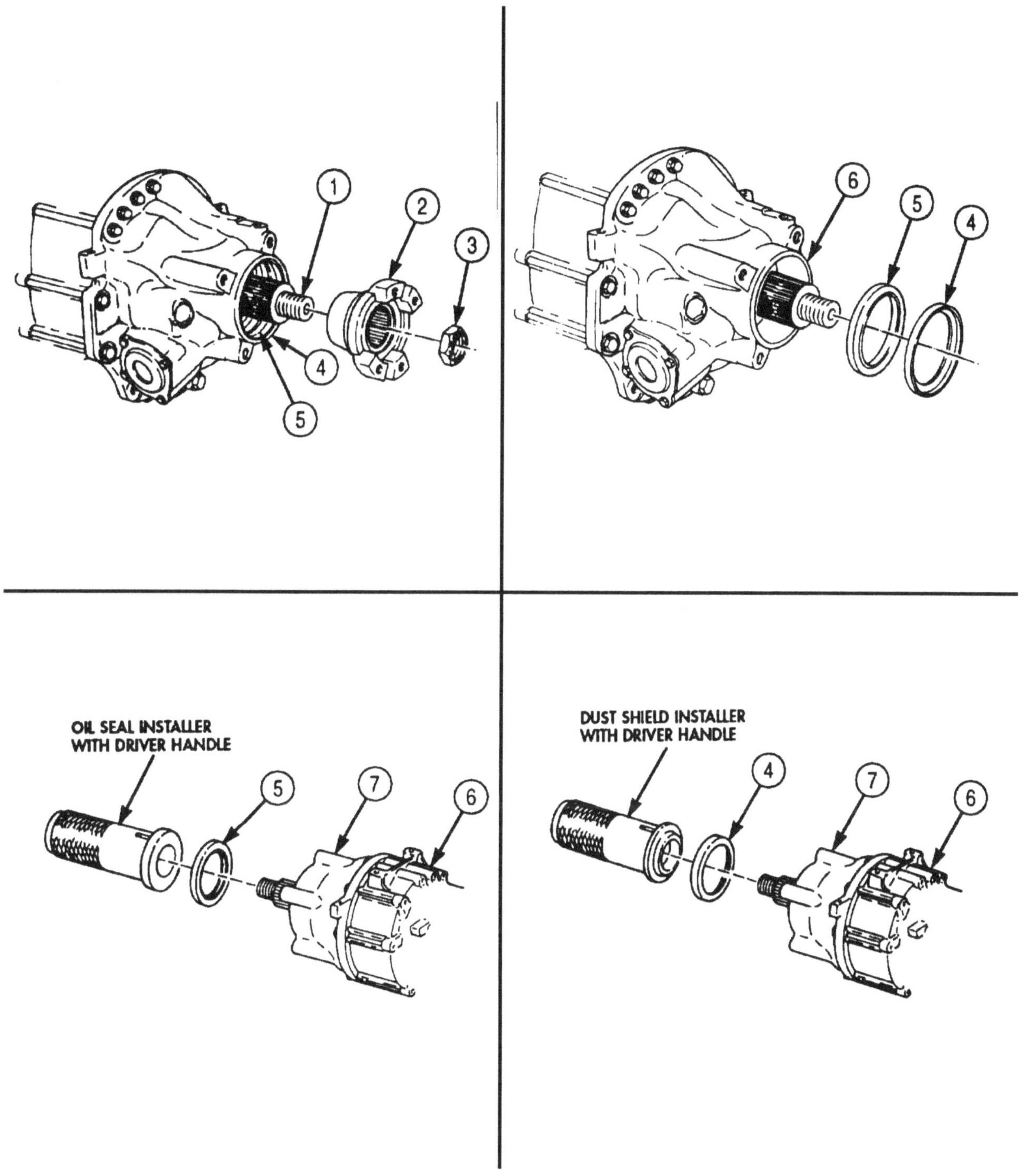

FOLLOW-ON TASKS:
- Fill transmission to proper fluid level (LO 9-2320-272-12).
- Install propeller shaft (para. 3-148).
- Start engine (TM 9-2320-272-10) and road test vehicle.

4-79. TRANSMISSION LUBRICATION VALVE REPLACEMENT

THIS TASK COVERS:

a. Removal **b. Installation**

INITIAL SETUP:

APPLICABLE MODELS
All

TOOLS
General mechanic's tool kit (Appendix E, Item 1)
Torque wrench (Appendix E, Item 146)

MATERIALS/PARTS
Locknut (Appendix D, Item 306)
O-ring (Appendix D, Item 470)
Gasket (Appendix D, Item 205)
Lubrication valve (Appendix D, Item 422)
Spring guide (Appendix D, Item 671)
Valve spring (Appendix D, Item 709)
Antiseize tape (Appendix C, Item 72)

REFERENCES (TM)
LO 9-2320-272-12
TM 9-2320-272-10
TM 9-2320-272-24P

EQUIPMENT CONDITION
Parking brake set (TM 9-2320-272-10).

SPECIAL ENVIRONMENTAL CONDITIONS
Work area must be clean and free from blowing dust and dirt.

CAUTION

Clean around lubrication valve area to prevent entry of dirt. Damage to transmission may occur if dirt or dust enters transmission.

a. Removal

NOTE

- Perform step 1 for vehicles equipped with front winch.
- Perform step 2 for vehicles equipped with transmission power takeoff (PTO).
- Have container ready to catch oil.

1. Remove locknut (1), screw (6), two clamps (2), and hoses (3) and (5) from hanger strap (4). Discard locknut (1).
2. Remove transmission-to-PTO supply hose (7) and adapter (8) from elbow (10).
3. Disconnect transmission oil filter-to-transmission supply hose (11) from elbow (10).
4. Remove elbow (10) and O-ring (12) from lubrication valve housing (13). Discard O-ring (12).
5. Remove two screws (9), lubrication valve housing (13), gasket (14), spring guide (15), lubrication valve (16), and valve spring (17) from transmission (18). Discard gasket (14), spring guide (15), lubrication valve (16), and valve spring (17).

4-79. TRANSMISSION LUBRICATION VALVE REPLACEMENT (Contd)

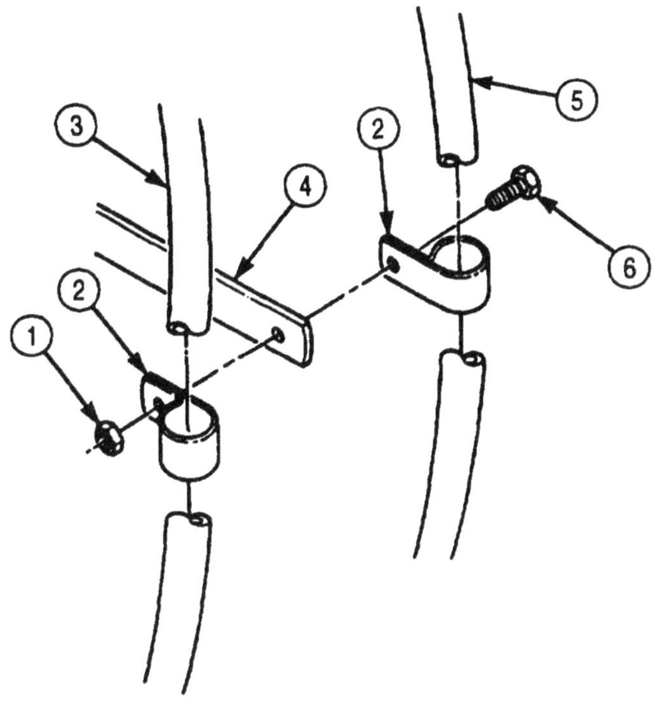

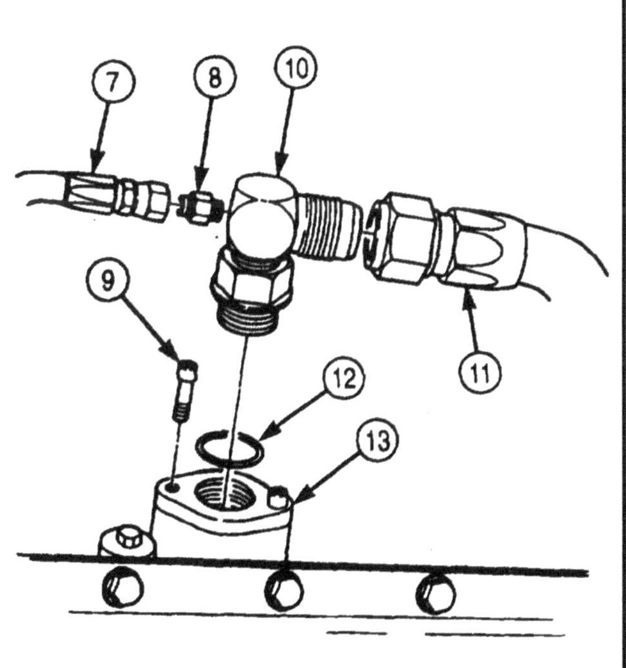

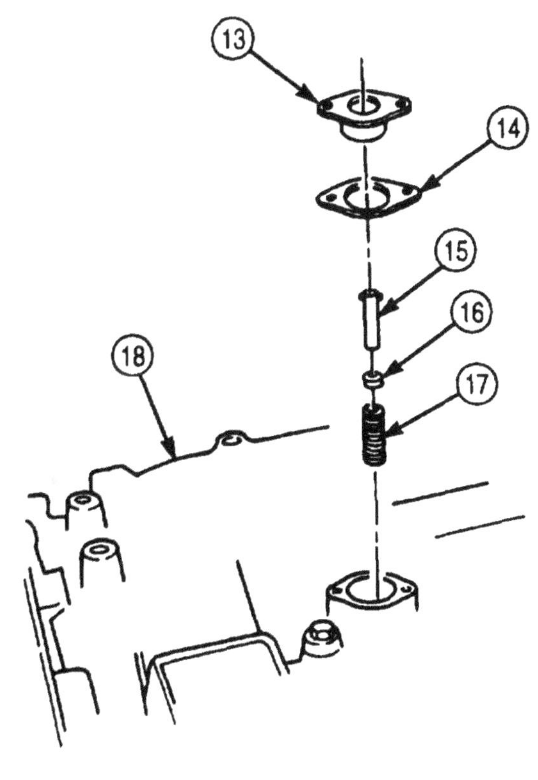

4-79. TRANSMISSION LUBRICATION VALVE REPLACEMENT (Contd)

b. Installation

1. Install new valve spring (5), new lubrication valve (4), and new spring guide (3) in bore (6) of transmission (7). Ensure raised side of lubrication valve (4) faces outward.
2. Install new gasket (2) and lubrication valve housing (1) on transmission (7) with two screws (13). Tighten screws (13) 9-11 lb-ft (12-15 Nm).
3. Place new O-ring (12) on elbow (10) and install elbow (10) on lubrication valve housing (1). Tighten elbow (10) until aligned and jamnut (14) seats, then tighten jamnut (14) until O-ring (12) seats.
4. Connect transmission oil filter-to-transmission supply hose (11) on elbow (10).

NOTE

- Perform step 5 for vehicles equipped with transmission PTO.
- Perform step 6 for vehicles equipped with front winch.
- Wrap all male threads with antiseize tape before installation,

5. Install adapter (9) and transmission-to-PTO supply hose (8) on elbow (10).
6. Install hoses (17) and (19) on hanger bracket (18) with two clamps (16), screw (20), and new locknut (15).

4-79. TRANSMISSION LUBRICATION VALVE REPLACEMENT (Contd)

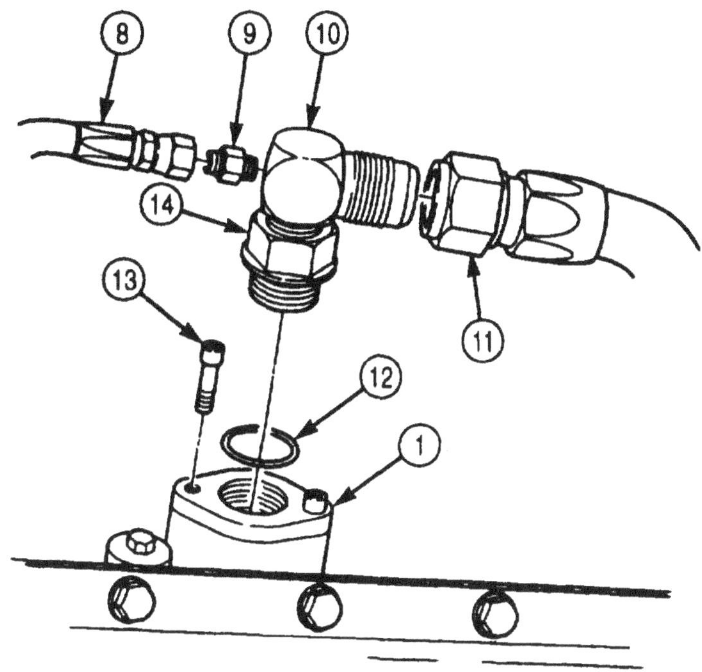

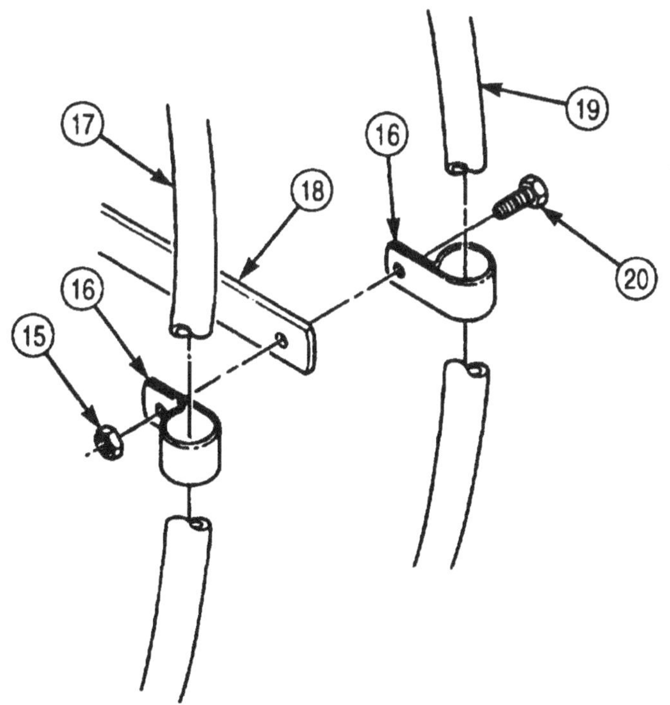

FOLLOW-ON TASKS:
- Fill transmission to proper fluid level (LO 9-2320-272-12).
- Start engine (TM 9-2320-272-10) and road test vehicle.

4-80. TRANSMISSION FIFTH GEAR LOCK-IN SOLENOID REPLACEMENT

THIS TASK COVERS:

a. Removal **b. Installation**

INITIAL SETUP:

APPLICABLE MODELS
All

TOOLS
General mechanic's tool kit (Appendix E, Item 1)

MATERIALS/PARTS
Lockwasher (Appendix D, Item 357)
Two locknuts (Appendix D, Item 274)
Tiedown strap (Appendix D, Item 696)
Cap and plug set (Appendix C, Item 14)
Antiseize tape (Appendix C, Item 72)

REFERENCES (TM)
LO 9-2320-272-12
TM 9-2320-272-10
TM 9-2320-272-24P

EQUIPMENT CONDITION
- Parking brake set (TM 9-2320-272-10).
- Transmission shift cable removed (para. 4-75).

SPECIAL ENVIRONMENTAL CONDITIONS
Work area must be clean and free from blowing dust and dirt.

a. Removal

CAUTION

Clean surrounding surface area and plug all open ports to prevent the entry of dirt. Damage to transmission may occur if dirt or dust enters transmission.

NOTE

Have drainage container ready to catch oil.

1. Disconnect main pressure line (4) from adapters (5) and (17).
2. Disconnect governor pressure line (8) from adapters (7) and (21).
3. Disconnect lead (1) from connector (16).
4. Remove tiedown strap (12) from harness (13) and ground lead (10). Discard tiedown strap (12).
5. Remove two locknuts (14), screws (9), ground lead (10), lockwasher (11), and fifth gear lock-in solenoid (6) from support bracket (3). Discard locknuts (14) and lockwasher (11).
6. Remove adapters (5) and (7) from fifth gear lock-in solenoid (6).
7. Remove four screws (2) and support bracket (3) from transmission (15).
8. Remove adapter (17) from transmission main pressure port (18).

NOTE

Only M936/A1/A2 series vehicles are equipped with check valve.

9. Remove adapter (21) and check valve (20) from transmission governor pressure port (19).

4-80. TRANSMISSION FIFTH GEAR LOCK-IN SOLENOID REPLACEMENT (Contd)

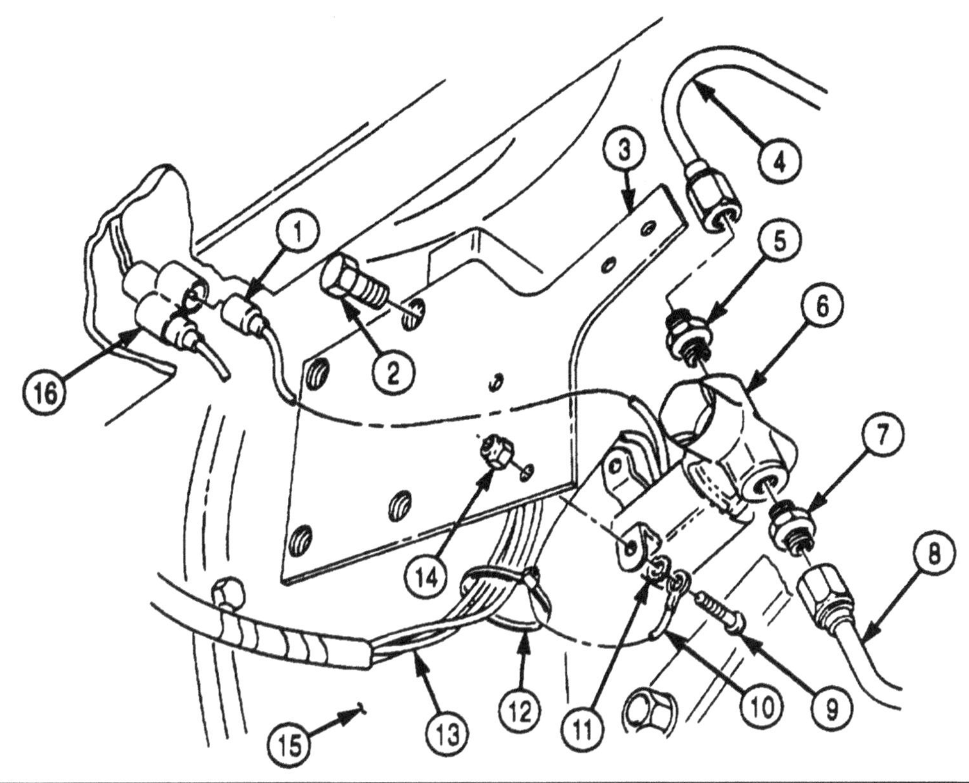

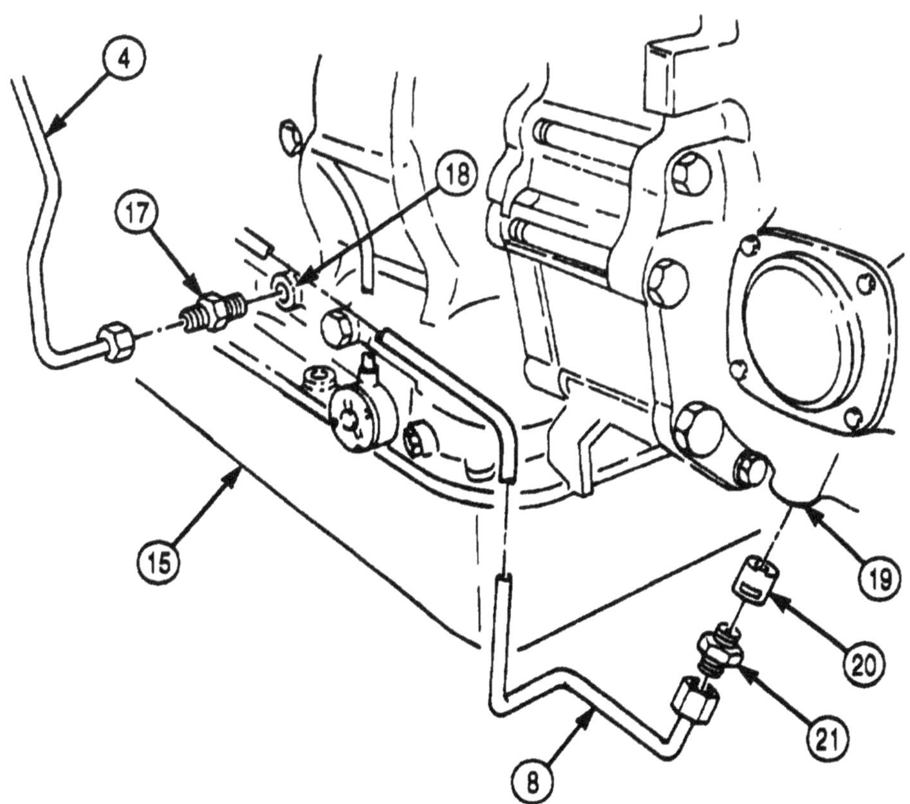

4-80. TRANSMISSION FIFTH GEAR LOCK-IN SOLENOID REPLACEMENT (Contd)

b. Installation

NOTE

- Wrap all male threads with antiseize tape before installation.
- Only M936/A1/A2 series vehicles are equipped with check valve.
- When installing check valve, ensure end with small bleed holes is facing away from adapter.

1. Install check valve (5) and adapter (6) on transmission governor pressure port (4).
2. Install adapter (2) on transmission main pressure port (3).
3. Install support bracket (11) on transmission (8) with four screws (10).
4. Install adapters (12) and (14) on fifth gear lock-in solenoid (13).
5. Install fifth gear lock-in solenoid (13) and ground lead (16) on support bracket (11) with two screws (15), new lockwasher (17), and two new locknuts (20).
6. Install new tiedown strap (18) on harness (19) and ground lead (16).
7. Connect lead (9) to connector (21).
8. Connect governor pressure line (7) to adapters (14) and (6).
9. Connect main pressure line (1) to adapters (12) and (2).

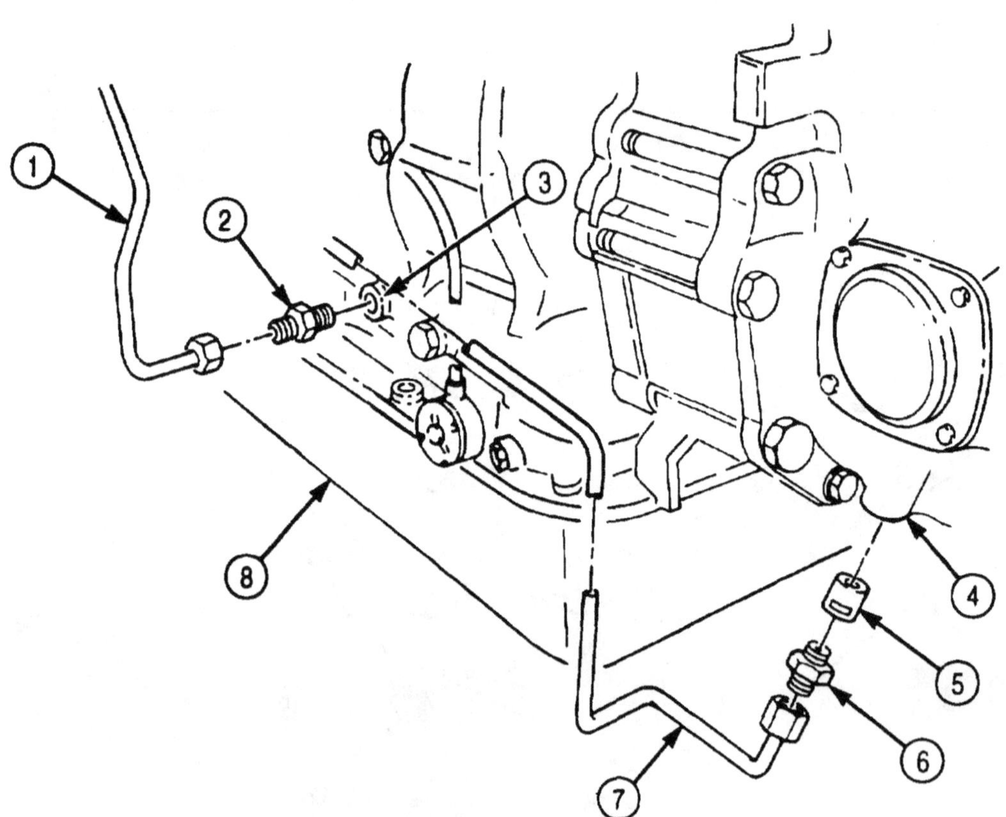

4-80. TRANSMISSION FIFTH GEAR LOCK-IN SOLENOID REPLACEMENT (Contd)

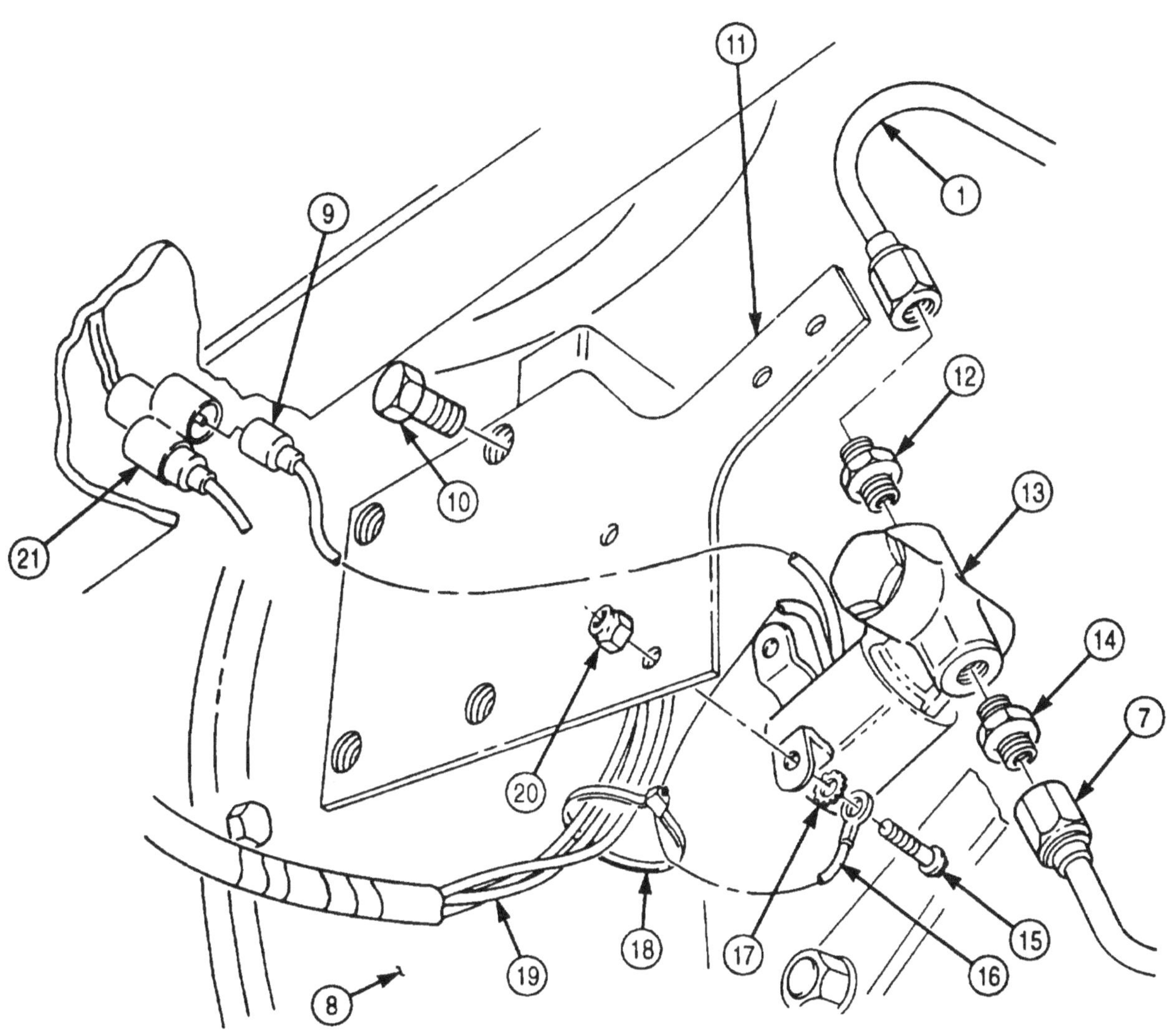

FOLLOW-ON TASKS:
- Install transmission shift cable (para. 4-75).
- Fill transmission to proper fluid level (LO 9-2320-272-12).
- Start engine (TM 9-2320-272-10) and road test vehicle.

Section VIII. TRANSFER CASE MAINTENANCE

4-81. TRANSFER CASE MAINTENANCE INDEX

4-82. TRANSFER CASE INTERLOCK VALVE (M936/A1/A2) REPLACEMENT

THIS TASK COVERS:

a. Removal **b. Installation**

INITIAL SETUP:

APPLICABLE MODELS
M936/A1/A2

TOOLS
General mechanic's tool kit (Appendix E, Item 1)

MATERIALS/PARTS
Antiseize tape (Appendix C, Item 72)

REFERENCES (TM)
TM 9-2320-272-24P

EQUIPMENT CONDITION
Transfer case removed (M936/A1/A2) (para. 4-95).

---

a. Removal

1. Remove two screws (1) and interlock valve (3) from bracket (5).
2. Remove three elbows (2) and adapter (4) from interlock valve (3).

b. Installation

NOTE

Wrap all male threads with antiseize tape before installation,

1. Install adapter (4) and three elbows (2) on interlock valve (3).
2. Install interlock valve (3) on bracket (5) with two screws (1).

4-82. TRANSFER CASE INTERLOCK VALVE (M936/A1/A2) REPLACEMENT (Contd)

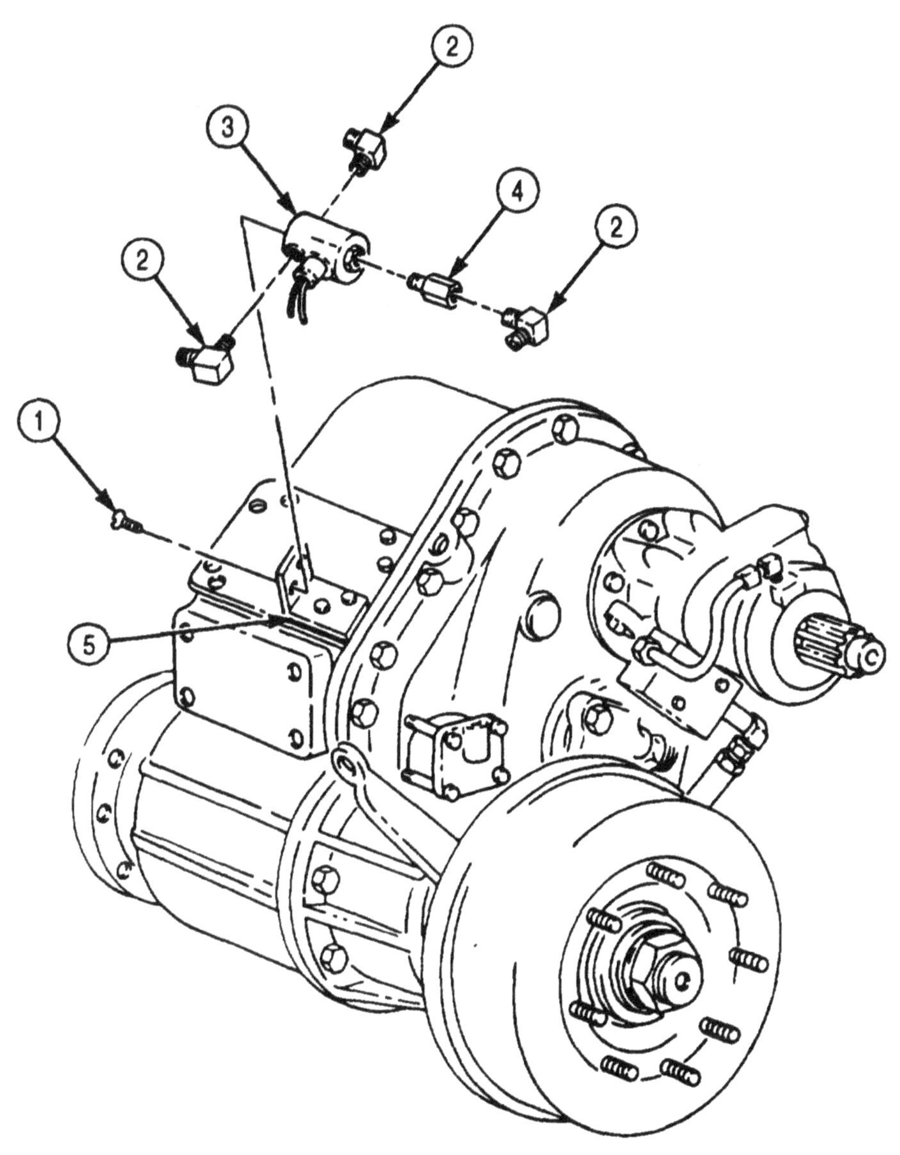

FOLLOW-ON TASK: Install transfer case (M936/A1/A2) (para. 4-95).

4-83. TRANSFER CASE INTERLOCK AIR CYLINDER REPLACEMENT

THIS TASK COVERS:

a. Removal **b. Installation**

INITIAL SETUP:

APPLICABLE MODELS
All

TOOLS
General mechanic's tool kit (Appendix E, Item 1)

MATERIALS/PARTS
Two lockwashers (Appendix D, Item 382)
Antiseize tape (Appendix C, Item 72)

REFERENCES (TM)
TM 9-2320-272-10
TM 9-2320-272-24P

EQUIPMENT CONDITION
- Parking brake set (TM 9-2320-272-10).
- Air reservoirs drained (TM 9-2320-272-10).
- Spare tire removed (M923/A1/A2, M929/A1/A2) (TM 9-2320-272-10).
- Dump body raised (TM 9-2320-272-10).
- Transfer case removed (M936/A1/A2) (para. 4-95).

GENERAL SAFETY INSTRUCTIONS
Do not disconnect air lines before draining air reservoirs.

WARNING

Do not disconnect air lines before draining air reservoirs. Small parts under pressure may shoot out with high velocity, causing injury to personnel.

a. Removal

1. Disconnect supply line (3) from elbow (2).

NOTE

- Perform step 2 for all vehicles except M936/A1/A2.
- Interlock air cylinder is threaded in transfer case housing.

2. Remove two screws (7), lockwashers (6), and parking brake cable bracket (8) from transfer case (5). Discard lockwashers (6).
3. Remove interlock air cylinder (1) and push rod (4) from transfer case (5).
4. Remove elbow (2) from interlock air cylinder (1).

b. Installation

NOTE

Wrap all male threads with antiseize tape before installation.

1. Install elbow (2) on interlock air cylinder (1).
2. Install push rod (4) and interlock air cylinder (1) on transfer case (5).

NOTE

Perform step 3 for all vehicles except M936/A1/A2.

3. Install parking brake cable bracket (8) on transfer case (5) with two new lockwashers (6) and screws (7).
4. Connect supply line (3) to elbow (2).

4-83. TRANSFER CASE INTERLOCK AIR CYLINDER REPLACEMENT (Contd)

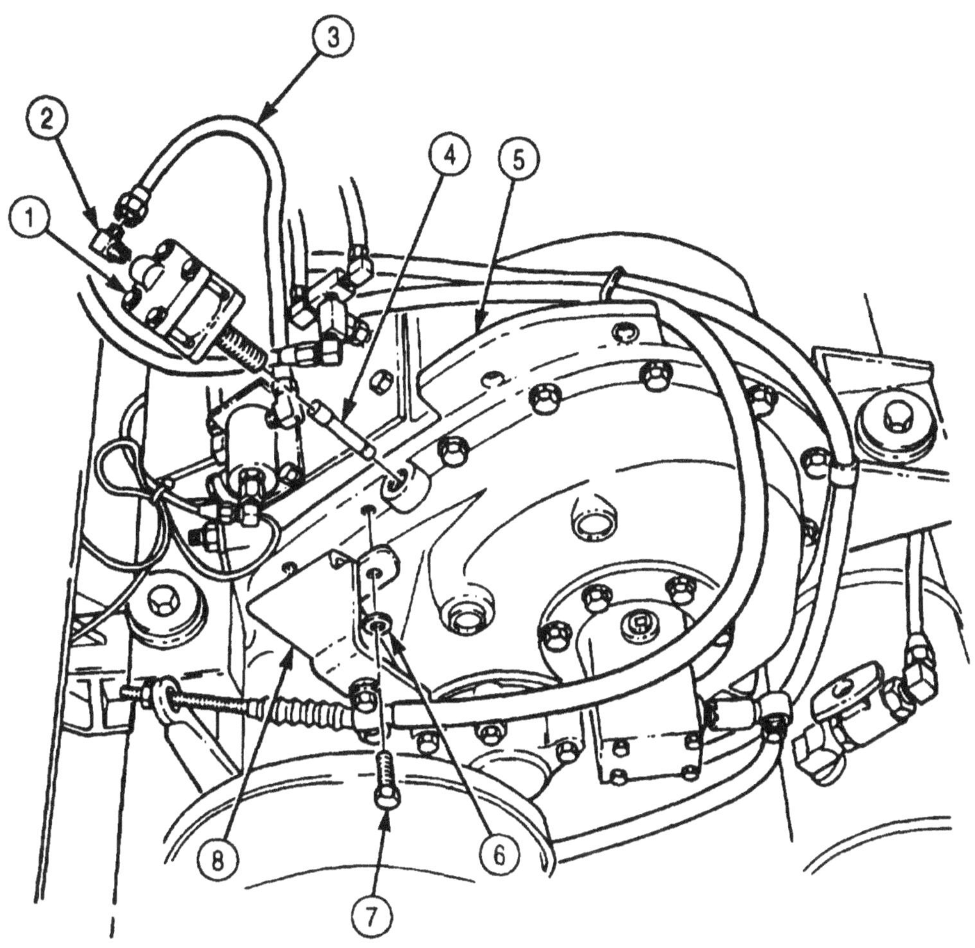

FOLLOW-ON TASKS:
- Install transfer case (M939/A1/A2) (para. 4-95).
- Install spare tire (M923/A1/A2, M929/A1/A2) (TM 9-2320-272-10).
- Lower dump body (TM 9-2320-272-10).
- Start engine (TM 9-2320-272-10), allow air pressure to build to normal operating pressure, and check for air leaks. Road test vehicle.

4-84. TRANSFER CASE FRONT AXLE ENGAGEMENT AIR CYLINDER REPLACEMENT

THIS TASK COVERS:

a. Removal

b. Installation

INITIAL SETUP:

APPLICABLE MODELS
All

TOOLS
General mechanic's tool kit (Appendix E, Item 1)
Torque wrench (Appendix E, Item 146)

MATERIALS/PARTS
Antiseize tape (Appendix C, Item 72)

REFERENCES (TM)
TM 9-2320-272-10
TM 9-2320-272-24P

EQUIPMENT CONDITION
- Parking brake set (TM 9-2320-272-10).
- Air reservoirs drained (TM 9-2320-272-10).
- Transfer case-to-forward rear axle propeller shaft removed (M936/A1/A2) (para. 3-150).

GENERAL SAFETY INSTRUCTIONS
Do not disconnect air lines before draining air reservoirs.

WARNING

Do not disconnect air lines before draining air reservoirs. Small parts under pressure may shoot out with high velocity, causing injury to personnel.

a. Removal

1. Remove air line (2) and elbow (1) from air cylinder cover (3).
2. Bend tabs of locking washers (11) away from heads of screws (10).
3. Remove four screws (10), locking washers (11), air cylinder cover (3), air cylinder (8), and seal (6) from transfer case (4).
4. Remove washers (5) and (9) from transfer case (4) and air cylinder cover (3).
5. Remove plunger (7) from air cylinder (8).

b. Installation

1. Install plunger (7) in air cylinder (8).
2. Install washers (5) and (9) on transfer case (4) and air cylinder cover (3).
3. Install seal (6), air cylinder (8), and air cylinder cover (3) on transfer case (4) with four locking washers (11) and screws (10). Tighten screws (10) 6-10 lb-ft (8-14•N).
4. Bend tabs of locking washers (11) over heads of screws (10).

NOTE

Wrap all male threads with antiseize tape before installation.

5. Install elbow (1) and air line (2) on air cylinder cover (3).

4-84. TRANSFER CASE FRONT AXLE ENGAGEMENT AIR CYLINDER REPLACEMENT (Contd)

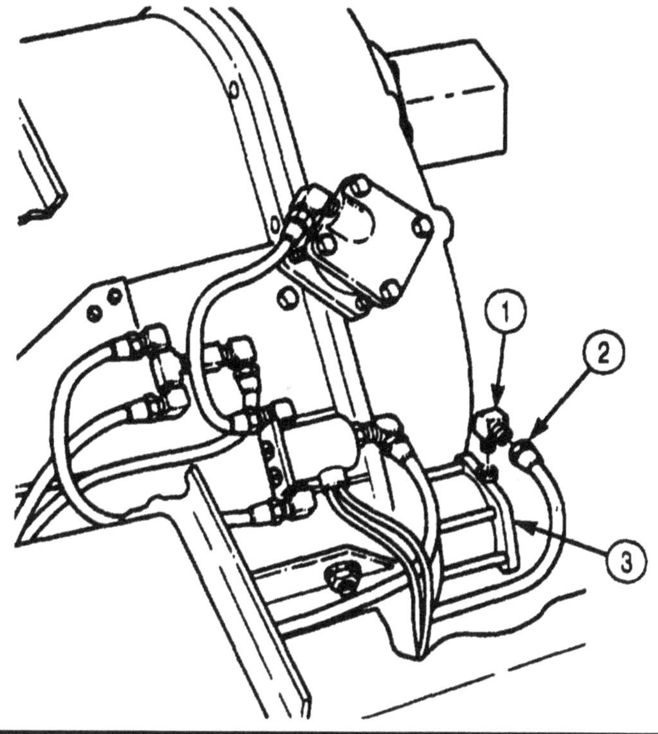

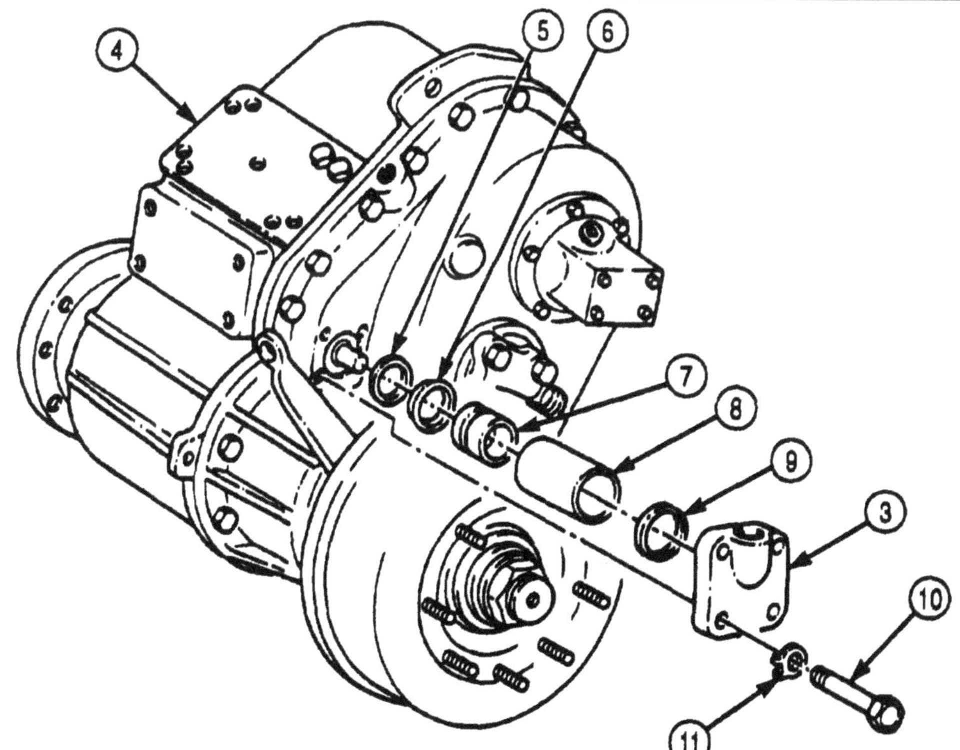

FOLLOW-ON TASKS: Install transfer case-to-forward rear axle propeller shaft (para. 3-150).
- Start engine (TM 9-2320-272-10), allow air pressure to build to normal operating pressure, and check for air leaks. Road test vehicle.

4-85. TRANSFER CASE SHIFT LEVER MAINTENANCE

THIS TASK COVERS:

a. Removal
b. Cleaning, Inspection, and Repair
c. Installation

INITIAL SETUP:

APPLICABLE MODELS
All

TOOLS
General mechanic's tool kit (Appendix E, Item 1)
Arbor press

MATERIALS/PARTS
Two cotter pins (Appendix D, Item 60)
Crocus cloth (Appendix C, Item 20)
Drycleaning solvent (Appendix C, Item 71)

REFERENCES (TM)
TM 9-2320-272-10
TM 9-2320-272-24P

EQUIPMENT CONDITION
- Parking brake set (TM 9-2320-272-10).
- Transfer case switch removed (para. 4-90).

GENERAL SAFETY INSTRUCTIONS
- Drycleaning solvent is flammable and toxic. Do not use near open flame.
- Keep fire extinguisher nearby when using drycleaning solvent.

---

a. Removal

1. Remove cotter pin (5), pin (7), and shift rod (6) from shift lever (2). Discard cotter pin (5).
2. Remove cotter pin (4), washer (3), pin (8), and shift lever (2) from shift lever bracket (9). Discard cotter pin (4).

b. Cleaning, Inspection, and Repair

WARNING

Drycleaning solvent is flammable and toxic. Do not use near open flame and always have a fire extinguisher nearby when solvents are used. Use only in well-ventilated places, wear protective clothing, and dispose of cleaning rags in approved container. Failure to do this may result in injury to personnel and/or damage to equipment.

1. Clean shift lever (2) with drycleaning solvent.
2. Inspect shift lever (2) for cracks and breaks. Replace shift lever (2) if cracked or broken.
3. Inspect shift lever bushing (1) for cracks, breaks, and pits. Remove pits with crocus cloth. Replace shift lever bushing (1) if cracked or broken.

NOTE

Perform steps 4 and 5 if bushing requires replacement.

4. Using arbor press, remove bushing (1) from shift lever (2). Discard bushing (1).
5. Using arbor press, install new bushing (1) on shift lever (2).

c. Installation

1. Install shift lever (2) on shift lever bracket (9) with pin (8), washer (3), and new cotter pin (4).
2. Install shift rod (6) on shift lever (2) with pin (7) and new cotter pin (5).

4-85. TRANSFER CASE SHIFT LEVER MAINTENANCE (Contd)

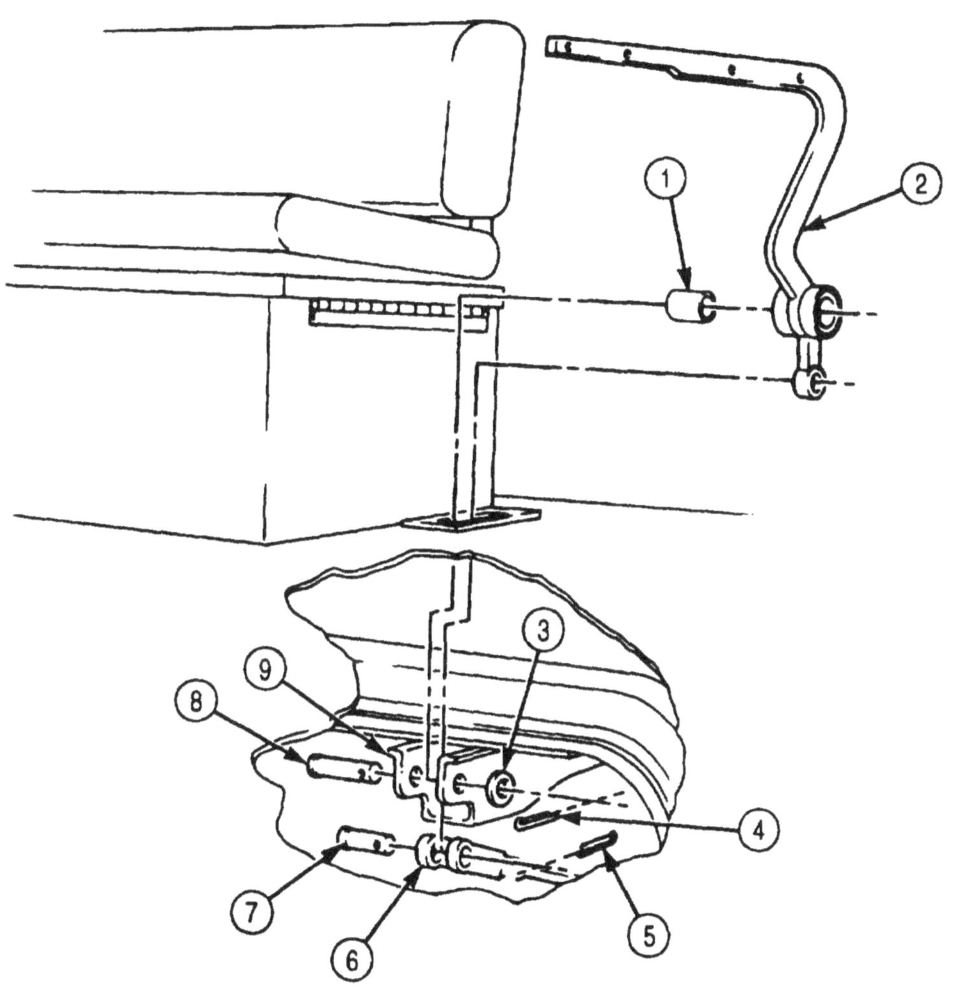

FOLLOW-ON TASK: Install transfer case switch (para. 4-90).

4-86. TRANSFER CASE SHIFT LEVER SHIFT ROD MAINTENANCE

THIS TASK COVERS:

a. Removal
b. Adjustment
c. Installation

INITIAL SETUP:

APPLICABLE MODELS
All

TOOLS
General mechanic's tool kit (Appendix E, Item 1)

MATERIALS/PARTS
Two cotter pins (Appendix D, Item 60)

REFERENCES (TM)
TM 9-2320-272-10
TM 9-2320-272-24P

EQUIPMENT CONDITION
Parking brake set (TM 9-2320-272-10).

a. Removal

Remove two cotter pins (7), pins (2), and shift lever shift rod (4) from shift lever (1) and cross shaft lever (6). Discard cotter pins (7).

b. Adjustment

1. Loosen jamnut (3) on shift lever shift rod (4) and adjust clevis (8) until distance between centers of holes in clevises (5) and (8) is 9-5/8 in. (24.4 cm).
2. Tighten jamnut (3).

c. Installation

Install shift lever shift rod (4) on shift lever (1) and cross shaft lever (6) with two pins (2) and new cotter pins (7).

4-86. TRANSFER CASE SHIFT LEVER SHIFT ROD MAINTENANCE (Contd)

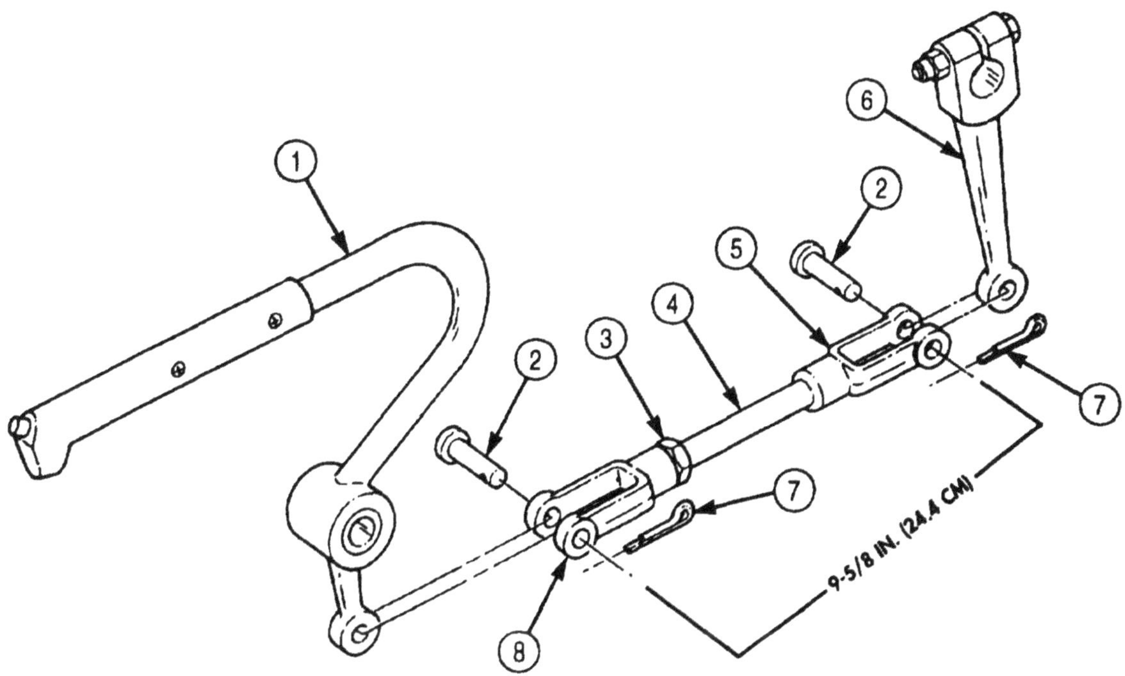

FOLLOW-ON TASK: Check transfer case shift lever for proper operation (TM 9-2320-272-10). Road test vehicle.

4-87. TRANSFER CASE CROSS SHAFT REPLACEMENT

THIS TASK COVERS:

a. Removal **b. Installation**

INITIAL SETUP:

APPLICABLE MODELS
All

TOOLS
General mechanic's tool kit (Appendix E, Item 1)

MATERIALS/PARTS
Two cotter pins (Appendix D, Item 60)
Two woodruff keys (Appendix D, Item 729)

REFERENCES (TM)
TM 9-2320-272-10
TM 9-2320-272-24P

EQUIPMENT CONDITION
Parking brake set (TM 9-2320-272-10).

a. Removal

1. Remove cotter pin (15), pin (13), and shift lever rod (16) from cross shaft lever (14). Discard cotter pin (15).
2. Remove cotter pin (10), pin (8), and transfer case shift rod (9) from cross shaft lever (7). Discard cotter pin (10).
3. Loosen nut (1) and screw (2) and remove cross shaft lever (14) and woodruff key (12) from cross shaft (5). Discard woodruff key (12).
4. Remove cross shaft (5) and cross shaft lever (7) from shift lever bracket (3).
5. Loosen nut (4) and screw (6) and remove cross shaft lever (7) and woodruff key (11) from cross shaft (5). Discard woodruff key (11).

b. Installation

1. Install new woodruff key (11) and cross shaft lever (7) on cross shaft (5). Do not tighten nut (4).
2. Install cross shaft (5) with cross shaft lever (7) on shift lever bracket (3).
3. Install new woodruff key (12) and cross shaft lever (14) on cross shaft (5) and tighten nut (1) and screw (2).
4. Tighten nut (4) and screw (6).
5. Install transfer case shift rod (9) on cross shaft lever (7) with pin (8) and new cotter pin (10).
6. Install shift lever rod (16) on cross shaft lever (14) with pin (13) and new cotter pin (15).

4-87. TRANSFER CASE CROSS SHAFT REPLACEMENT (Contd)

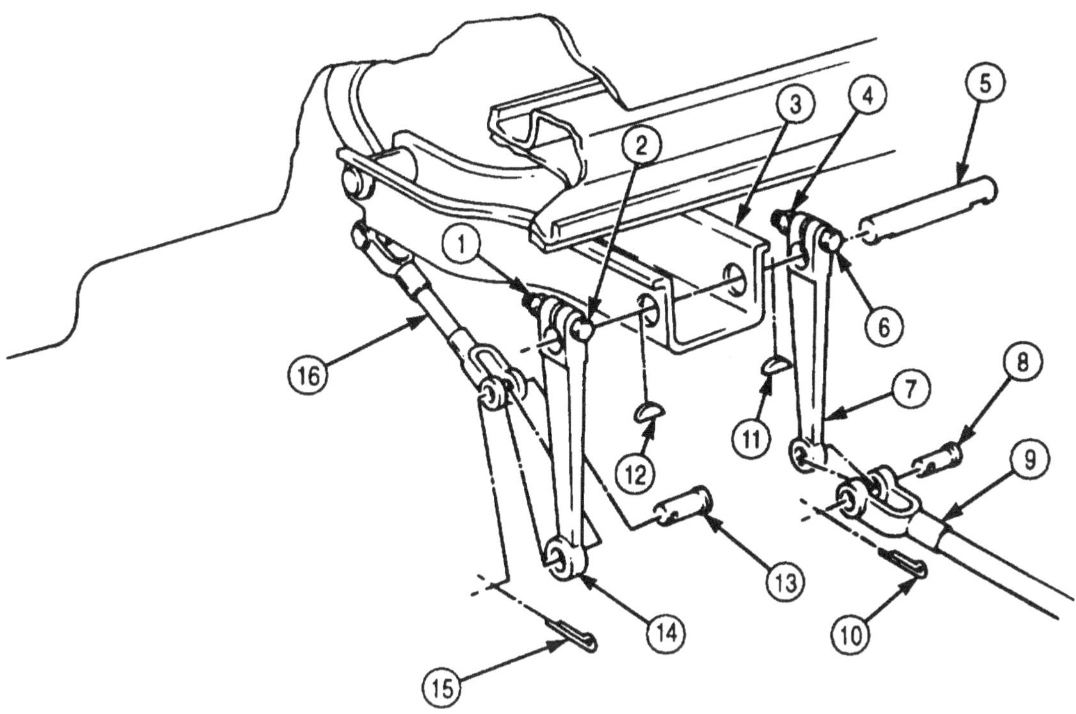

FOLLOW-ON TASK: Check transfer case shift lever for proper operation (TM 9-2320-272-10). Road test vehicle.

4-88. TRANSFER CASE SHIFT ROD MAINTENANCE

THIS TASK COVERS:

a. Removal
b. Cleaning, Inspection, and Repair
c. Adjustment
d. Installation

INITIAL SETUP:

APPLICABLE MODELS
All

TOOLS
General mechanic's tool kit (Appendix E, Item 1)
Vise

MATERIALS/PARTS
Two cotter pins (Appendix D, Item 60)

REFERENCES (TM)
TM 9-2320-272-10
TM 9-2320-272-24P

EQUIPMENT CONDITION
Parking brake set (TM 9-2320-272-10).

GENERAL SAFETY INSTRUCTIONS
- Drycleaning solvent is flammable and toxic. Do not use near open flame.
- Keep fire extinguisher nearby when using drycleaning solvent.

---

a. Removal

1. Remove cotter pin (6), pin (5), and transfer case shift rod (7) from shift shaft (4). Discard cotter pin (6).
2. Remove cotter pin (1), pin (3), and transfer case shift rod (7) from cross shaft lever (2). Discard cotter pin (1).

b. Cleaning, Inspection, and Repair

WARNING

Drycleaning solvent is flammable and toxic. Do not use near open flame and always have a fire extinguisher nearby when solvents are used. Use only in well-ventilated places, wear protective clothing, and dispose of cleaning rags in approved container. Failure to do this may result in injury to personnel and/or damage to equipment.

1. Clean transfer case shift rod (7) with drycleaning solvent.
2. Inspect transfer case shift rod (7) for cracks and breaks. Replace shift rod (7) if cracked or broken.
3. Inspect valve cam (8) for cracks, breaks, and bends Replace valve cam (8) if cracked, broken, or bent.

NOTE

Perform steps 4 through 7 if valve cam requires replacement.

4. Place transfer case shift rod (7) in soft-jawed vise.
5. Remove clevis (9), valve cam (8), washer (10), and jamnut (11) from shift rod (7).
6. Install jamnut (11), washer (10), valve cam (8), and clevis (9) on shift rod (7). Do not tighten jamnut (11).
7. Remove transfer case shift rod (7) from soft-jawed vise.

4-88. TRANSFER CASE SHIFT ROD MAINTENANCE (Contd)

c. Adjustment

1. Loosen jamnut (11) if necessary,
2. Adjust clevis (12) until distance between centers of holes is 9-17/32 in. (24.2 cm).
3. Tighten jamnut (11).

d. Installation

1. Install transfer case shift rod (7) on cross shaft lever (2) with pin (3) and new cotter pin (1).
2. Install transfer case shift rod (7) on shift shaft (4) with pin (5) and new cotter pin (6).

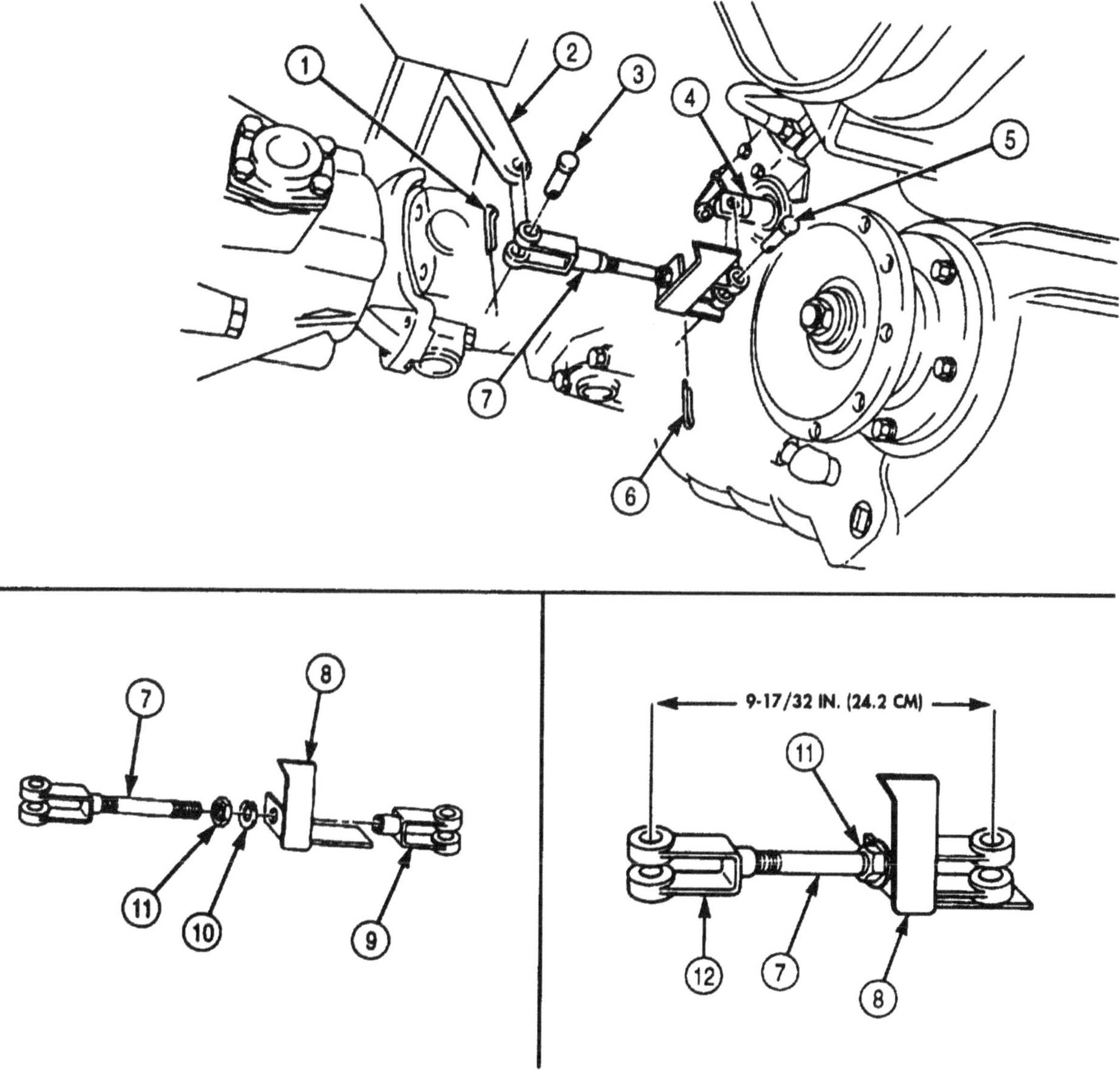

FOLLOW-ON TASK: Check transfer case shift lever for proper operation (TM 9-2320-272-10). Road test vehicle.

4-89. TRANSFER CASE SPEEDOMETER DRIVE GEAR AND DRIVEN SHAFT REPLACEMENT

THIS TASK COVERS:

a. Removal **b. Installation**

INITIAL SETUP:

APPLICABLE MODELS
All

TOOLS
General mechanic's tool kit (Appendix B, Item 1)
Vise

MATERIALS/PARTS
Woodruff key (Appendix D, Item 727)
Snapring (Appendix D, Item 663)
Gasket sealant (Appendix C, Item 30)

REFERENCES (TM)
LO 9-2320-272-12
TM 9-2320-272-10
TM 9-2320-272-24P

EQUIPMENT CONDITION
- Parking brake set (TM 9-2320-272-10).
- Transfer case oil drained (LO 9-2320-272-12).
- Parking brakeshoes and dustcovers removed (paras. 3-176 and 3-177).

a. Removal

NOTE

Speedometer drive gear cover screws are different lengths. Note location and position for installation.

1. Disconnect speedometer driveshaft (6) from drive adapter (5).
2. Remove two screws (2), washers (1), four screws (7), washers (8), and speedometer drive gear cover (9) from transfer case (13).
3. Place speedometer drive gear cover (9) in soft-jawed vise.
4. Remove drive adapter (5), driven shaft (4), and sleeve bushing (3) from speedometer drive gear cover (9).
5. Remove snapring (10), speedometer drive gear (11), and woodruff key (12) from intermediate shaft (14). Discard snapring (10) and woodruff key (12).

b. Installation

1. Install new woodruff key (12) and speedometer drive gear (11) on intermediate shaft (14) with new snapring (10).
2. Install sleeve bushing (3), driven shaft (4), and drive adapter (5) on speedometer drive gear cover (9).
3. Apply gasket sealant on contact surface of speedometer drive gear cover (9) and install on transfer case (13) with four washers (8), screws (7), two washers (1), and screws (2).
4. Connect speedometer driveshaft (6) on drive adapter (5).

4-89. TRANSFER CASE SPEEDOMETER DRIVE GEAR AND DRIVEN SHAFT REPLACEMENT (Contd)

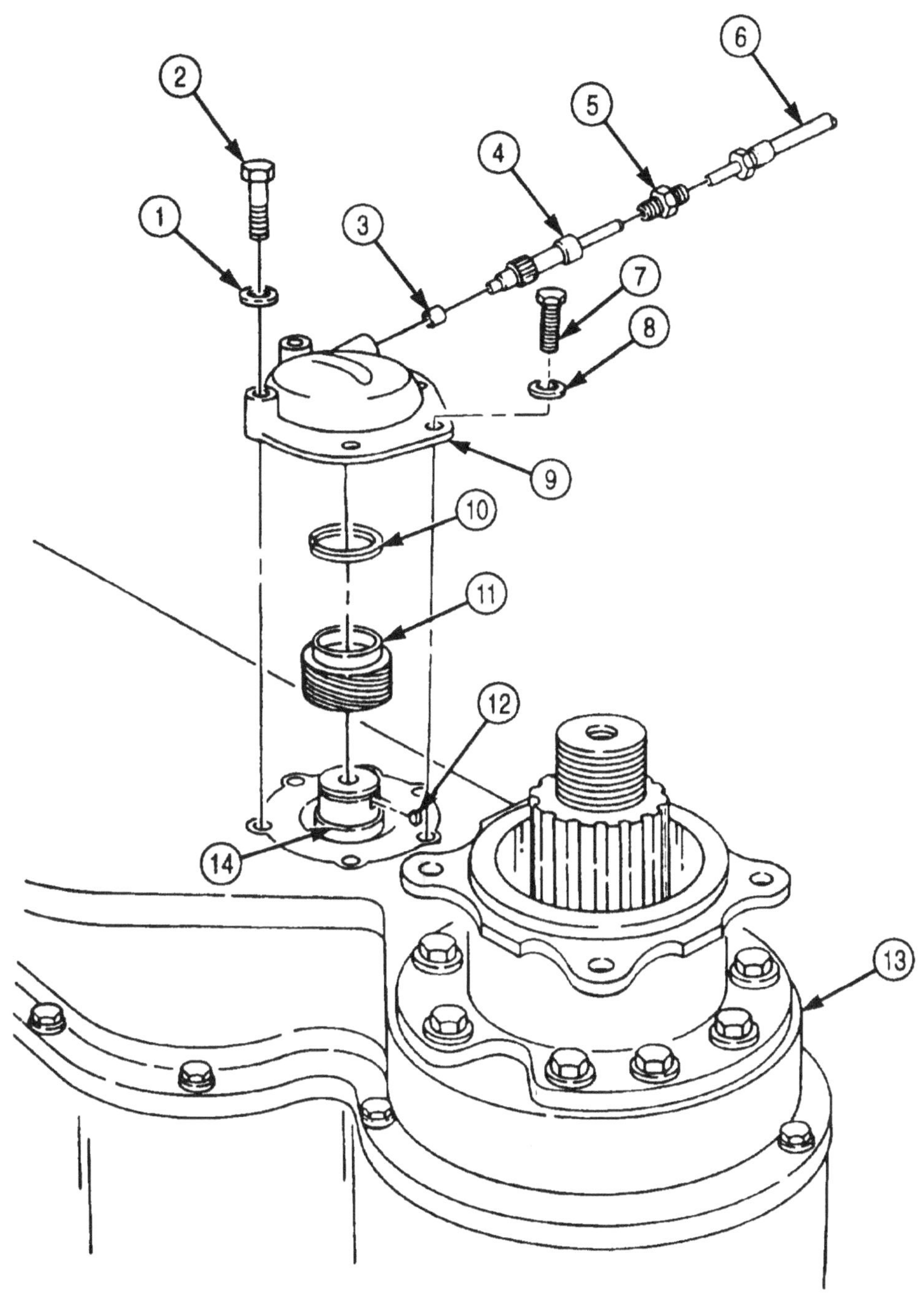

FOLLOW-ON TASKS:
- Install parking brakeshoes and dustcovers (paras. 3-176 and 3-177).
- Fill transfer case to proper fluid level (LO 9-2320-272-12).
- Start engine (TM 9-2320-272-10) and check for oil leaks. Road test vehicle.

4-90. TRANSFER CASE SWITCH REPLACEMENT

THIS TASK COVERS:

a. Removal **b. Installation**

INITIAL SETUP:

APPLICABLE MODELS
All

TOOLS
General mechanic's tool kit (Appendix E, Item 1)

MATERIALS/PARTS
Two lockwashers (Appendix D, Item 352)
Four lockwashers (Appendix D, Item 384)

REFERENCES (TM)
TM 9-2320-272-10
TM 9-2320-272-24P

EQUIPMENT CONDITION
Parking brake set (TM 9-2320-272-10).

a. Removal

NOTE

Tag all leads for installation.

1. Disconnect two switch leads (10) from connectors (11).
2. Remove grommet (8) and switch leads (10) from cab floor (9).
3. Remove two screws (7), lockwashers (6), clamps (5), and switch leads (10) from transfer case shift lever (4). Discard lockwashers (6).
4. Remove four screws (3), lockwashers (2), and switch (1) from transfer case shift lever (4). Discard lockwashers (2).

b. Installation

1. Install switch (1) on transfer case shift lever (4) with four new lockwashers (2) and screws (3).
2. Install two switch leads (10) on transfer case shift lever (4) with two clamps (5), new lockwashers (6), and screws (7).
3. Insert two switch leads (10) through cab floor (9) and connect to connectors (11).
4. Install grommet (8) around switch leads (10) and install on cab floor (9).

4-90. TRANSFER CASE SWITCH REPLACEMENT (Contd)

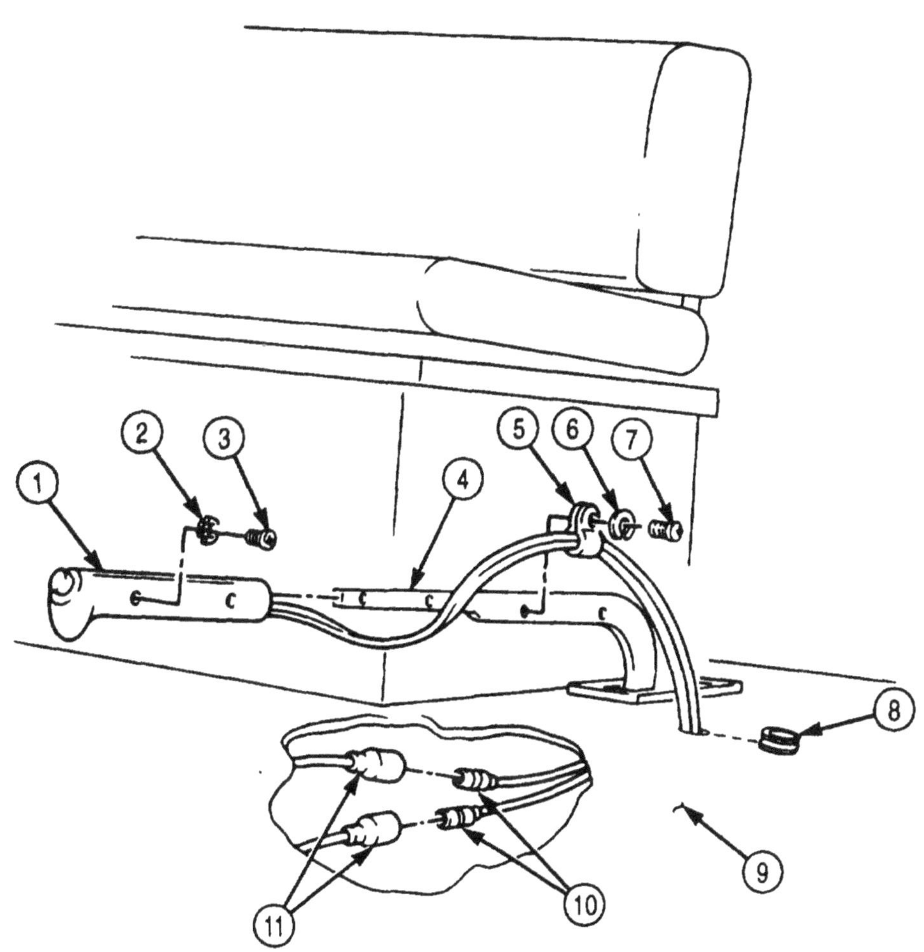

FOLLOW-ON TASK: Check transfer case shift lever for proper operation (TM 9-2320-272-10). Road test vehicle.

4-91. TRANSFER CASE CAPACITOR REPLACEMENT

THIS TASK COVERS:

a. Removal **b. Installation**

INITIAL SETUP:

APPLICABLE MODELS
All

TOOLS
General mechanic's tool kit (Appendix E, Item 1)

REFERENCES (TM)
TM 9-2320-272-10
TM 9-2320-272-24P

EQUIPMENT CONDITION
- Parking brake set (TM 9-2320-272-10).
- Transfer case-to-front axle propeller shaft removed (para. 3-149).

a. Removal

NOTE

Tag all leads for installation.

1. Disconnect capacitor leads (5), (6), and (9) from transfer case switch connector (4), transmission solenoid adapter (7), and interlock valve adapter (8).
2. Disconnect capacitor leads (1) and (14) from transfer case switch connector (3) and front wiring harness connector (13).
3. Remove screw (12), front wiring harness clamp (11), and capacitor (10) from crossmember (2).

b. Installation

1. Install capacitor (10) and front wiring harness clamp (11) on crossmember (2) with screw (12).
2. Connect. capacitor leads (5), (6), and (9) to transfer case switch connector (4), transmission solenoid adapter (7), and interlock valve adapter (8).
3. Connect capacitor leads (1) and (14) to transfer case switch connector (3) and front wiring harness connector (13).

4-91. TRANSFER CASE CAPACITOR REPLACEMENT (Contd)

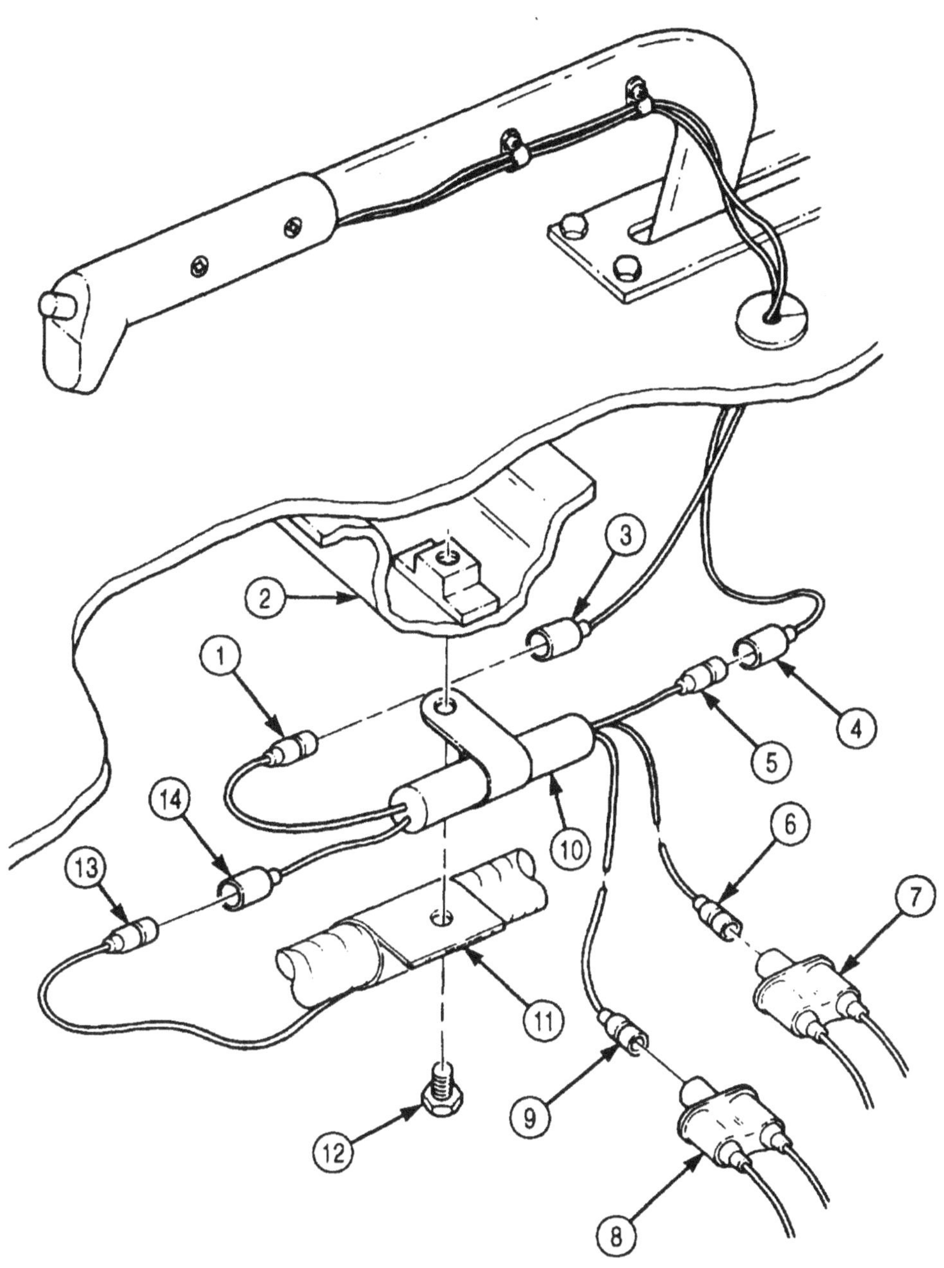

FOLLOW-ON TASKS:
- Install transfer case-to-front axle propeller shaft (para. 3-149).
- Check transfer case shift lever for proper operation (TM 9-2320-272-10). Road test vehicle.

4-92. TRANSFER CASE TRANSORB DIODE REPLACEMENT

THIS TASK COVERS:

a. Removal **b. Installation**

INITIAL SETUP:

APPLICABLE MODELS
All

TOOLS
General mechanic's tool kit (Appendix E, Item 1)

MATERIALS/PARTS
Two locknuts (Appendix D, Item 313)
Lockwasher (Appendix D, Item 371)

REFERENCES (TM)
TM 9-2320-272-10
TM 9-2320-272-24P

EQUIPMENT CONDITION
Parking brake set (TM 9-2320-272-10).

---

a. Removal

NOTE

Assistant will help with step 1.

1. Remove locknut (9), washer (10), transorb diode ground wire (8), locknut (11), ground wire (7), lockwasher (12), cable clamp (6), and screw (5) from frame (4). Discard locknuts (9) and (11) and lockwasher (12).
2. Disconnect transorb diode wire (2) from front wiring harness connector (1) and remove transorb diode (3) from frame (4).

b. Installation

1. Connect transorb diode wire (2) and transorb diode (3) to front wiring harness connector (1) between frame (4).
2. Install cable clamp (6), new lockwasher (12), and ground wire (7) on frame (4) with screw (5) and new locknut (11).
3. Install transorb diode ground wire (8) on screw (5) with washer (10) and new locknut (9).

4-92. TRANSFER CASE TRANSORB DIODE REPLACEMENT (Contd)

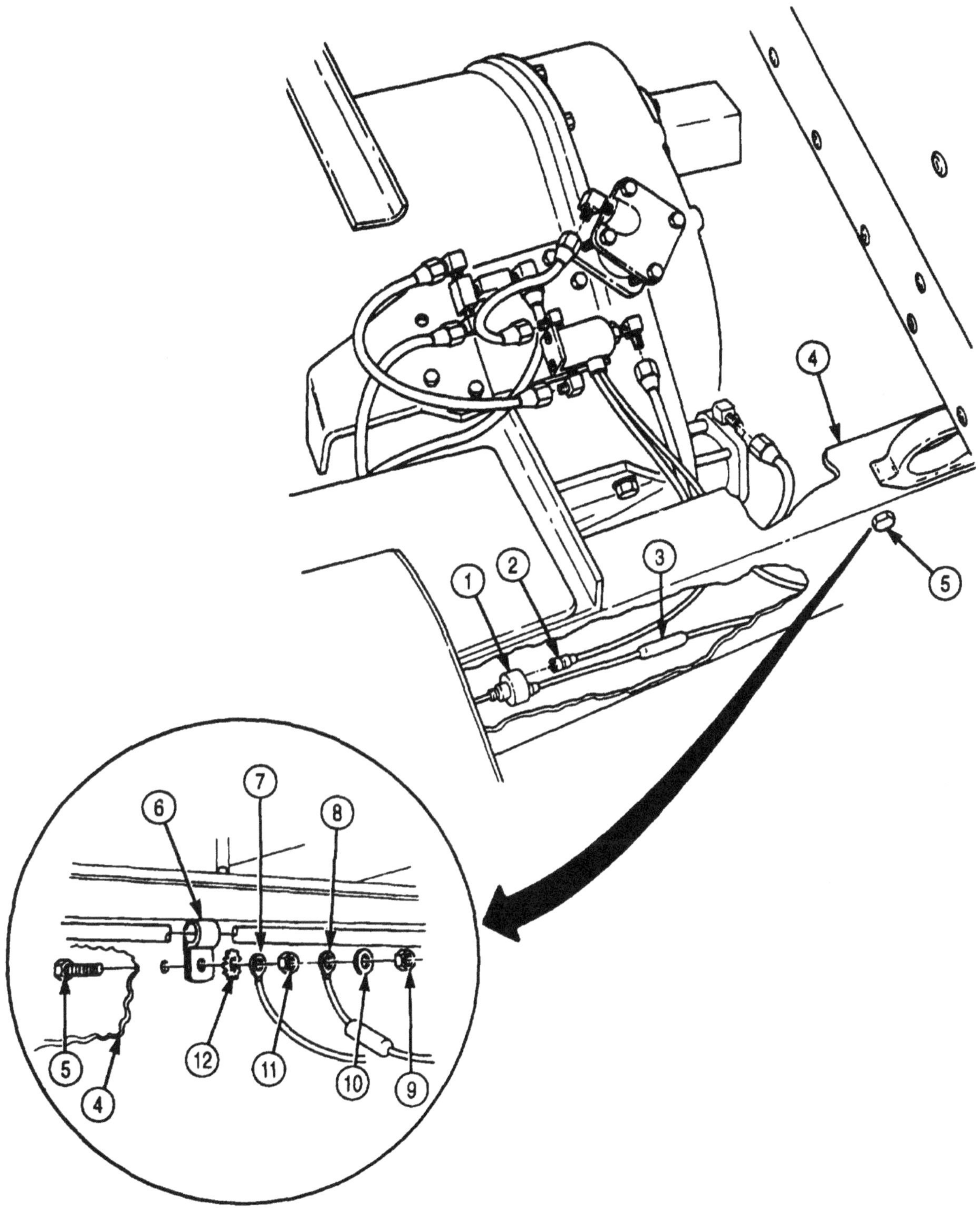

FOLLOW-ON TASK: Check transfer case shift lever for proper operation (TM 9-2320-272-10). Road test vehicle.

4-93. TRANSFER CASE OIL PUMP REPLACEMENT

THIS TASK COVERS:

a. Removal **b. Installation**

INITIAL SETUP:

APPLICABLE MODELS
All

TOOLS
General mechanic's tool kit (Appendix E, Item 1)
Torque wrench (Appendix E, Item 144)

MATERIALS/PARTS
Gasket sealant (Appendix C, Item 30)
Antiseize tape (Appendix C, Item 72)

REFERENCES (TM)
LO 9-2320-272-12
TM 9-2320-272-10
TM 9-2320-272-24P

EQUIPMENT CONDITION
- Parking brake set (TM 9-2320-272-10).
- Transfer case oil drained (LO 9-2320-272-12).

NOTE

Have drainage container ready to catch oil

a. Removal

1. Disconnect hose (4) from elbow (3).
2. Remove elbow (3) from oil pump (1).

NOTE

Mark position of oil pump for installation.

3. Remove six screws (5), washers (6), and oil pump (1) from transfer case (2).

b. Installation

1. Apply gasket sealant to mating surfaces of oil pump (1) and transfer case (2).
2. Install oil pump (1) on transfer case (2) with six washers (6) and screws (5). Tighten screws (5) 40-65 lb-ft (54-88 N•m).
3. Apply antiseize tape to male threads of elbow (3) and install on oil pump (1).
4. Connect hose (4) to elbow (3).

4-93. TRANSFER CASE OIL PUMP REPLACEMENT (Contd)

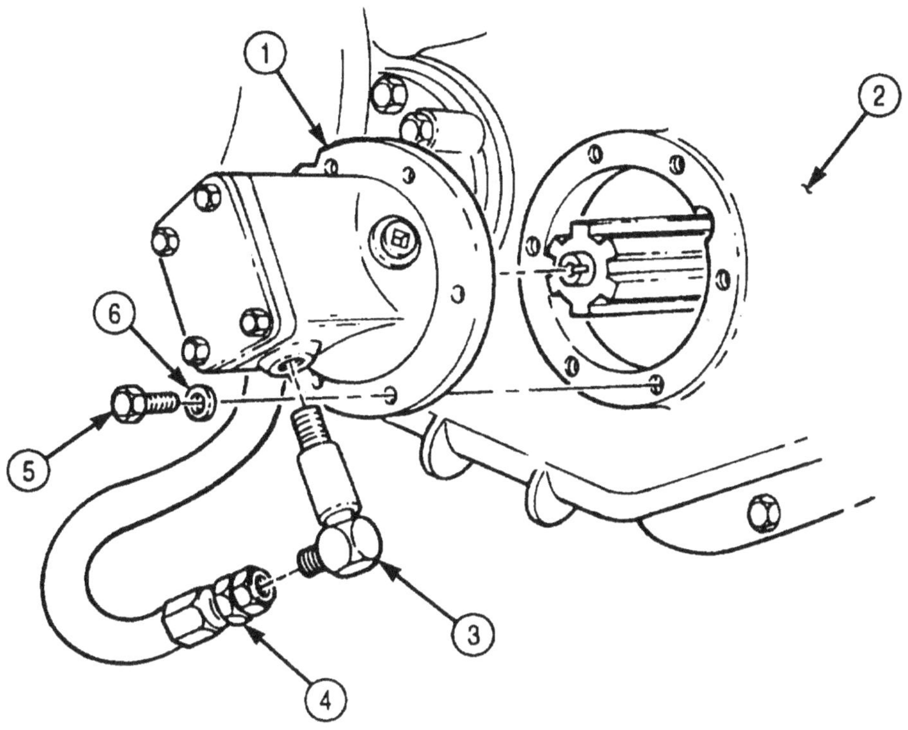

FOLLOW-ON TASKS:
- Fill transfer case to proper fluid level (LO 9-2320-272-12).
- Start engine (TM 9-2320-272-10) and check for leaks. Road test vehicle.

4-94. TRANSFER CASE REPLACEMENT

THIS TASK COVERS:

a. Removal **b. Installation**

INITIAL SETUP:

APPLICABLE MODELS
All (except M936/A1/A2)

TOOLS
General mechanic's tool kit (Appendix E, Item 1)
Torque wrench (Appendix E, Item 144)
Transmission jack (Appendix E, Item 147)
Hydraulic jack
Lifting device

MATERIALS/PARTS
Ten locknuts (Appendix D, Item 321)
Four lockwashers (Appendix D, Item 361)
Locknut (Appendix D, Item 308)
Ten locknuts (Appendix D, Item 291)
Antiseize tape (Appendix C, Item 72)

PERSONNEL REQUIRED
Two

REFERENCES (TM)
LO 9-2320-272-12
TM 9-2320-272-10
TM 9-2320-272-24P

EQUIPMENT CONDITION
- Parking brake set (TM 9-2320-272-10).
- Transfer case oil drained (LO 9-2320-272-12).
- Wet reservoir removed (para. 3-185).
- Transfer case-to-forward rear axle propeller shaft removed (para. 3-149).
- Spare tire removed (M923/A1/A2, M925/A1/A2, M929/A1/A2, and M930/A1/A2) (TM 9-2320-272-10).
- Transfer case interlock valve removed (para. 3-143).
- Transfer case front axle engagement control valve removed (para. 3-144).
- Transfer case shift rod removed (para. 4-88).

a. Removal

NOTE

Ensure transfer case shift lever is in HIGH position to prevent propeller shaft from turning when loosening screws.

1. Remove four screws (8), lockwashers (7), and propeller shaft (1) from transfer case input flange (2). Discard lockwashers (7).
2. Remove eight locknuts (5), screws (4), and propeller shaft (6) from transfer case front output flange (3). Discard locknuts (5).
3. Remove two locknuts (18), screws (9), clamp (10), parking brake cable (11), and spacer (20) from parking brake cable bracket (19). Discard locknuts (18).
4. Remove locknut (21) and parking brake cable (11) from parking brake lever (22). Discard locknut (21).
5. Disconnect speedometer driveshaft (15) from speedometer drive adapter (17).
6. Remove screw (12), washer (13), clamp (14), and speedometer driveshaft (15) from transfer case (16).
7. Remove screw (25), clamp (24) with parking brake cable (11), and clamp (23) with speedometer driveshaft (15) from transfer case (16).

NOTE

Tag all air lines for installation.

8. Disconnect two vent lines (32) from elbows (27).
9. Disconnect interlock vent line (26) from elbow (27).
10. Disconnect supply line (31) from air cylinder elbow (30).
11. Disconnect supply line (29) from interlock air cylinder elbow (28).

4-94. TRANSFER CASE REPLACEMENT (Contd)

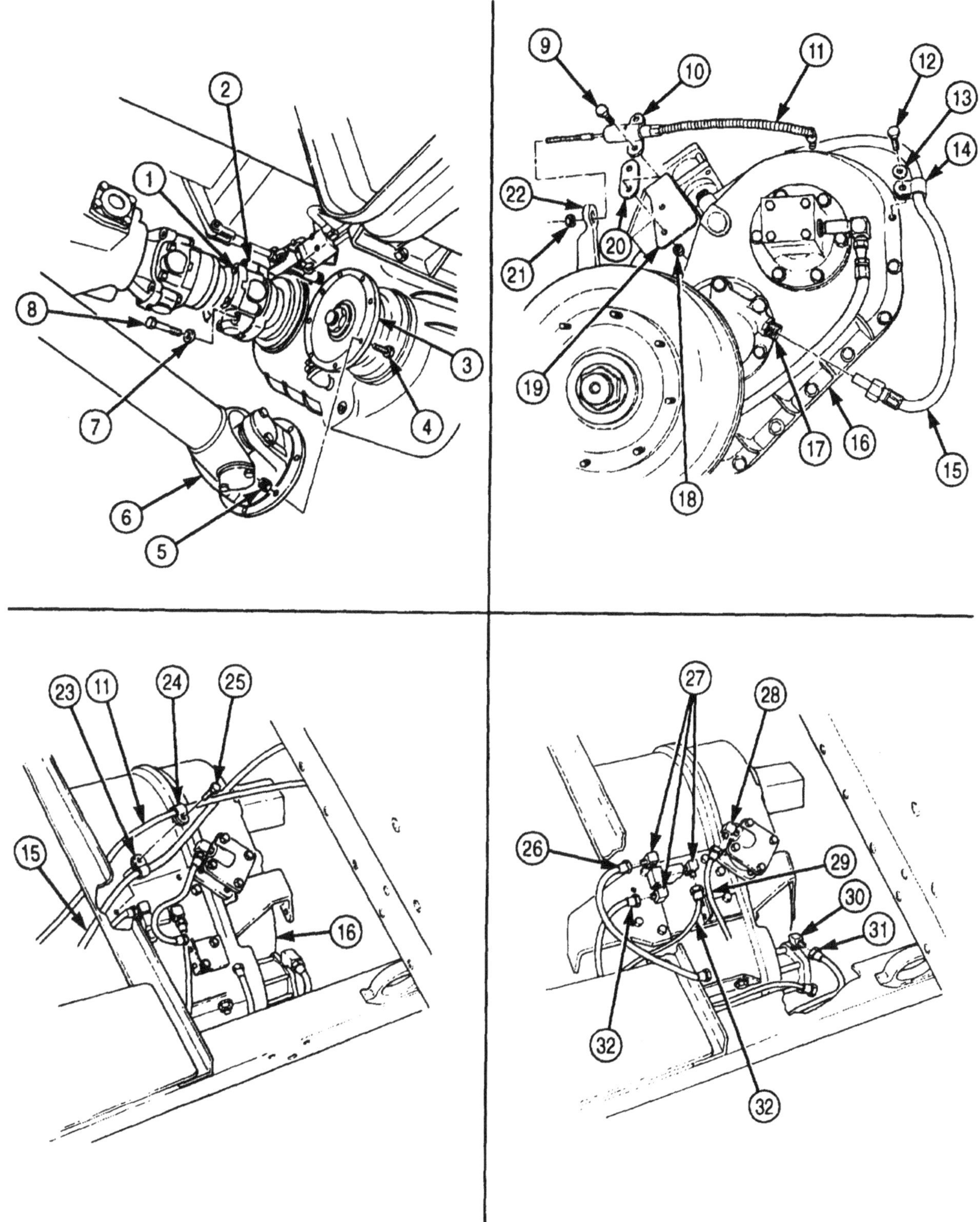

4-94. TRANSFER CASE REPLACEMENT (Contd)

12. Remove three elbows (1) from vent tees (2) and (9).
13. Remove vent tees (2) and (9) and fitting (8) from inspection cover (7).
14. Remove elbow (5) from air cylinder (6).
15. Remove elbow (3) from interlock air cylinder (4).
16. Remove two screws (12) and interlock valve bracket (13) from transfer case (10).
17. Remove three screws (11) and control valve bracket (14) from transfer case (10).
18. Remove two screws (16), washers (17), and parking brake cable bracket (15) from transfer case (10).
19. Position hydraulic jack under transfer case (10).
20. Remove three locknuts (24), screws (19), washers (18), and insulators (20) from mounting brackets (21) and frame (23). Discard locknuts (24).
21. Raise transfer case (10) high enough to remove three insulators (22) from mounting brackets (21) and frame (23).
22. Remove seven locknuts (25) and two mounting brackets (21) from transfer case (10). Discard locknuts (25).
23. Lower hydraulic jack and remove transfer case (10) from vehicle.
24. Using lifting device, remove transfer case (10) from hydraulic jack.
25. Remove seven studs (26) from transfer case (10).

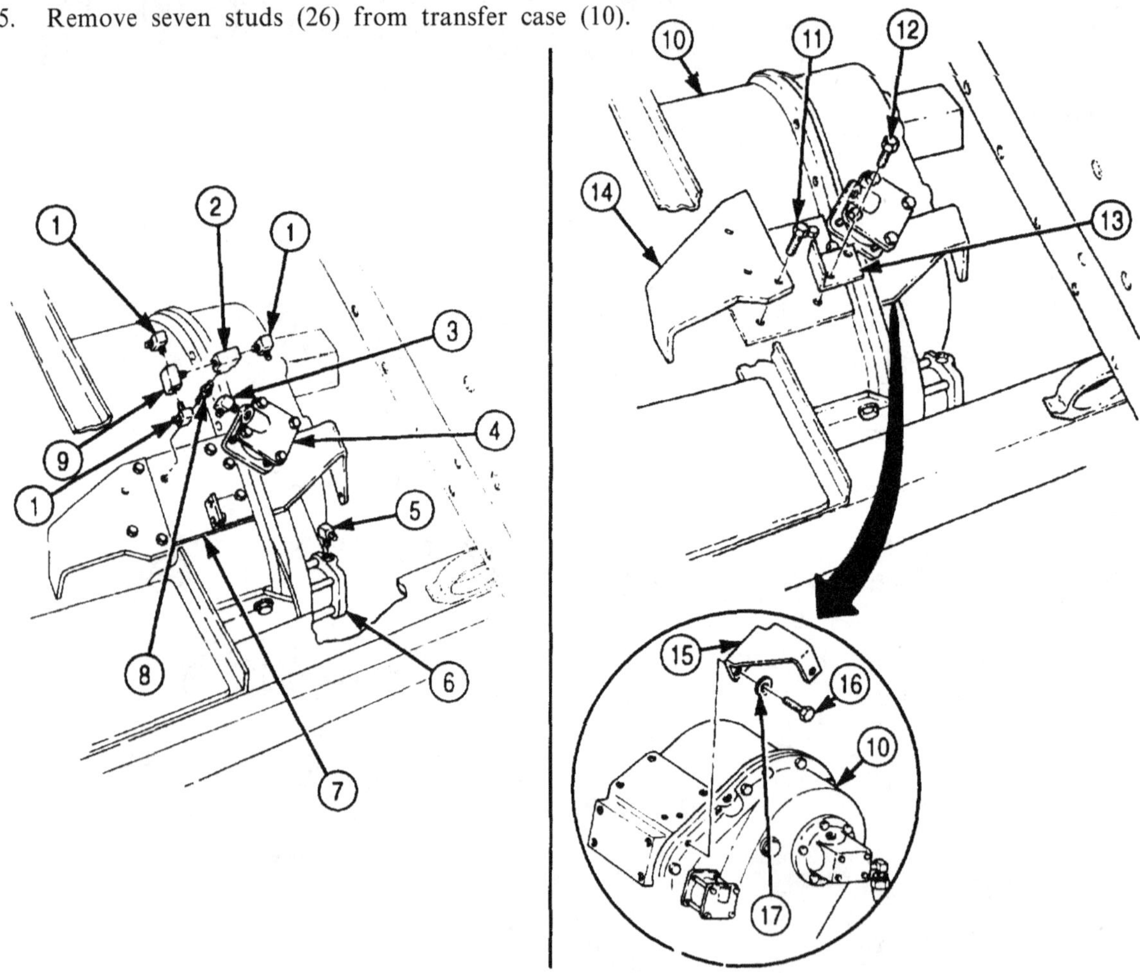

4-94. TRANSFER CASE REPLACEMENT (Contd)

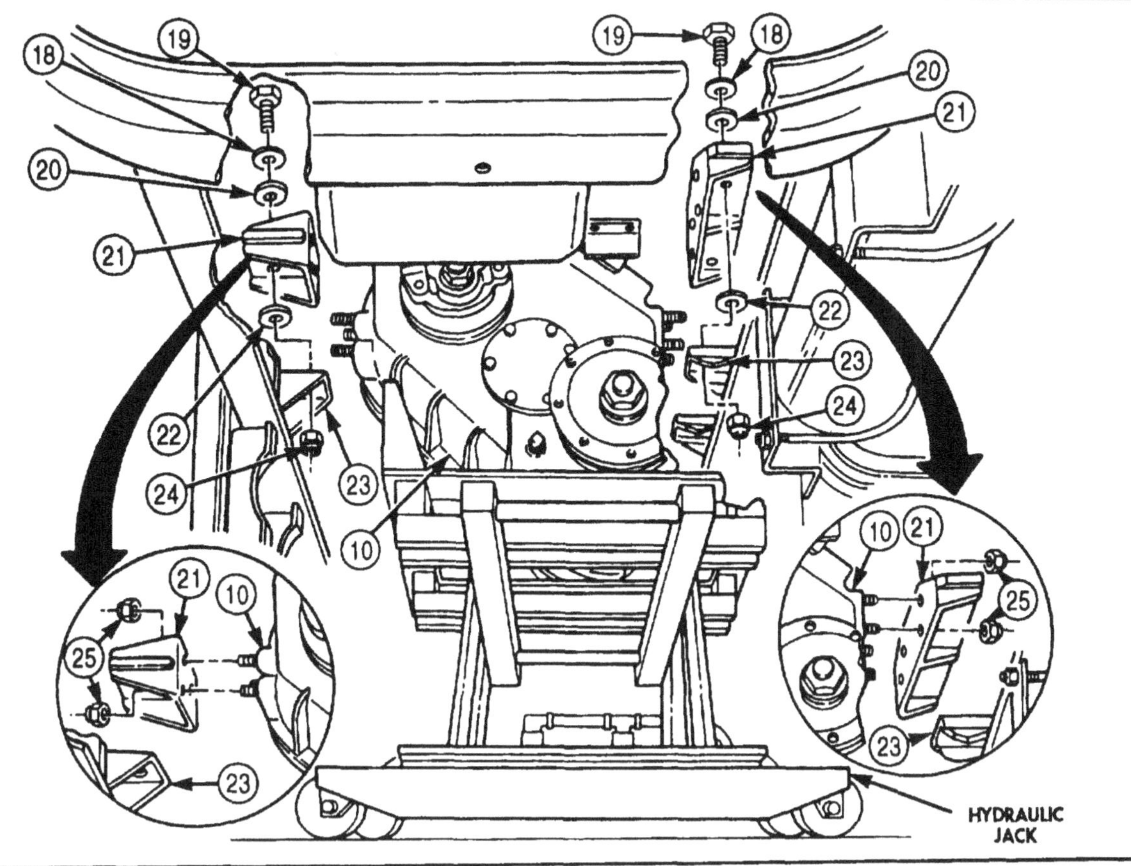

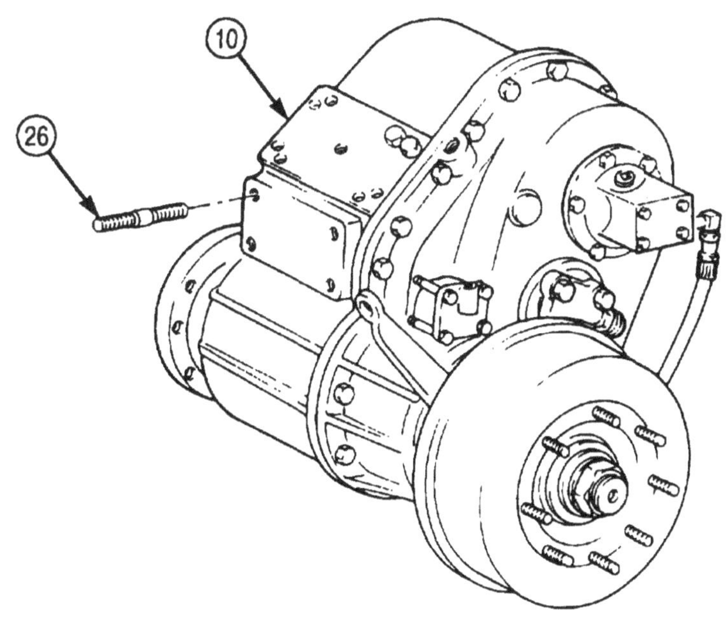

4-94. TRANSFER CASE REPLACEMENT (Contd)

b. Installation

1. Install seven etude (1) on transfer case (2).
2. Place transfer case (2) on hydraulic jack and position under vehicle.
3. Raise transfer case (2) in position high enough to allow for installation of mounting brackets (6).
4. Install two mounting brackets (6) on transfer case (2) with seven new locknuts (10). Tighten locknuts (10) 125-135 lb-ft (170-183 N•m).
5. Place three insulators (7) between mounting brackets (6) and frame (8).
6. Slowly lower transfer case (2) and align holes in mounting brackets (6), insulators (7), and frame (8).
7. Install three insulators (5), washers (3), screws (4), and new locknuts (9) on mounting brackets (6) and frame (8). Tighten locknuts (9) 50-60 lb-ft (68-81 N•m).
8. Remove hydraulic jack from transfer case (2).

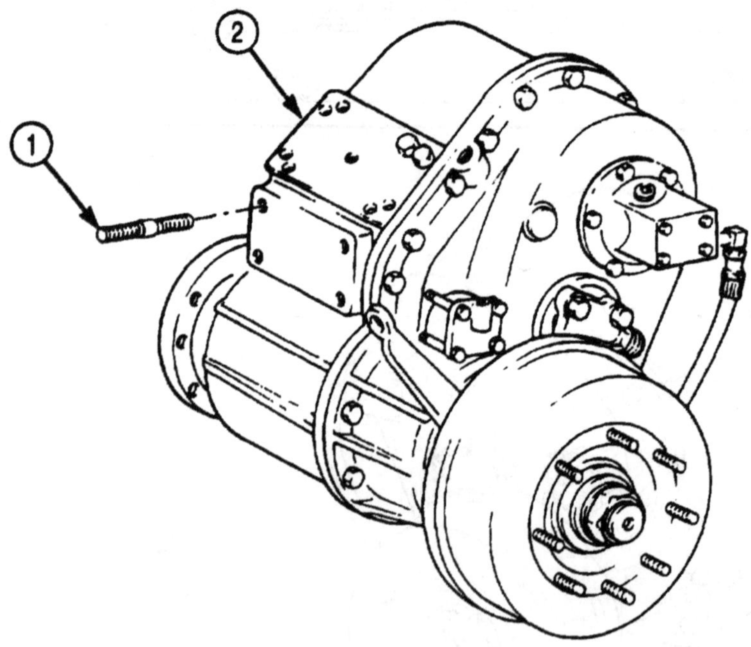

4-94. TRANSFER CASE REPLACEMENT (Contd)

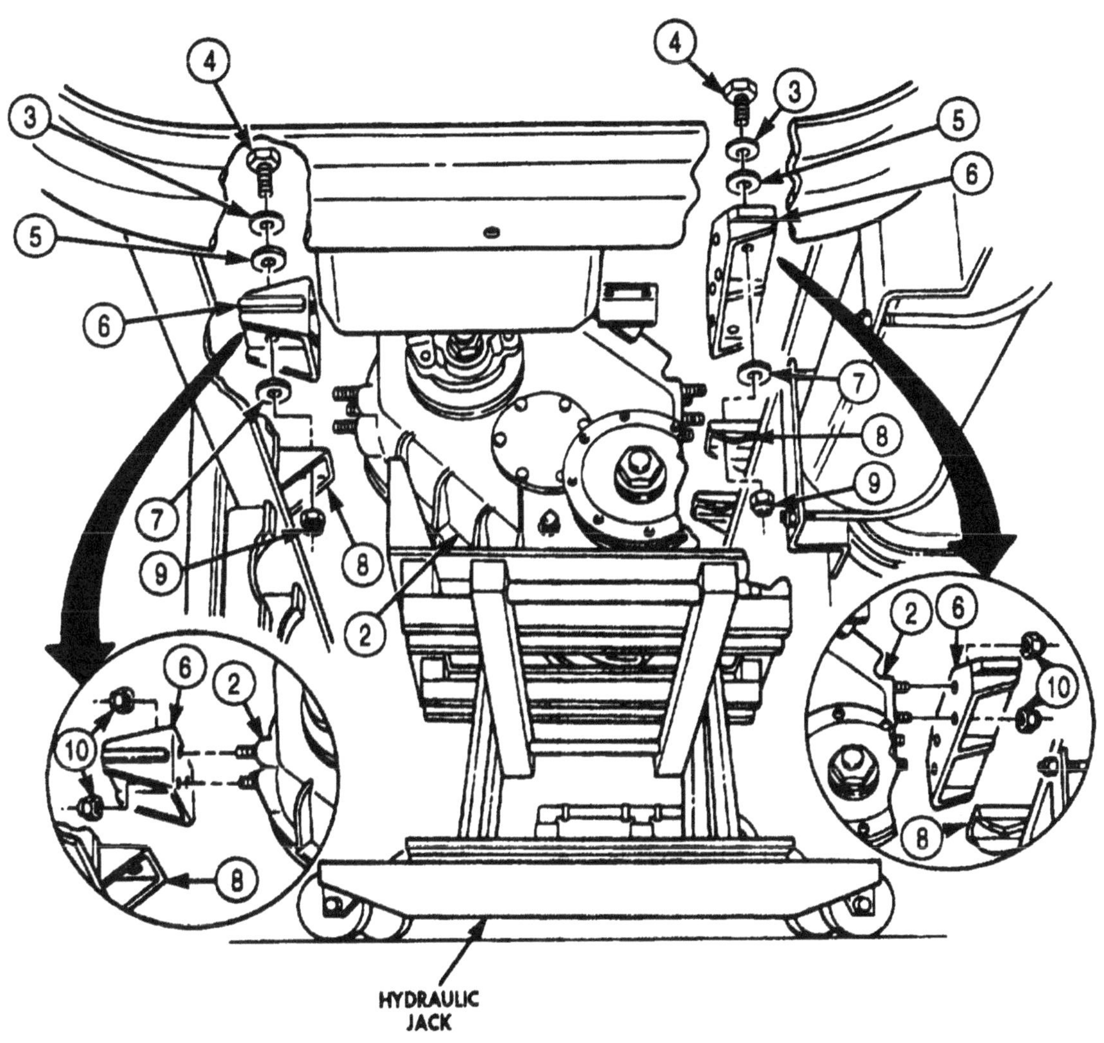

4-94. TRANSFER CASE REPLACEMENT (Contd)

9. Install parking brake cable bracket (6) on transfer case (1) with two washers (7) and screws (8). Tighten screws (8) 40-55 lb-ft (54-75 N•m).
10. Install control valve bracket (5) on transfer case (1) with three screws (2). Finger-tighten screws (2).
11. Install interlock valve bracket (4) on transfer case (1) with two screws (3). Finger-tighten screws (3).

NOTE

Wrap all male pipe threads with antiseize tape before installation.

12. Install elbow (11) on interlock air cylinder (12).
13. Install elbow (13) on air cylinder (14).
14. Install fitting (16) and vent tees (10) and (17) on inspection cover (15).
15. Install three elbows (9) on vent tees (10) and (17).
16. Connect supply line (19) to interlock air cylinder elbow (11).
17. Connect supply line (20) to air cylinder elbow (13).
18. Connect interlock vent line (18) to elbow (9).
19. Connect two vent lines (21) to elbows (9).
20. Install clamp (22) with parking brake cable (26), and clamp (24) with speedometer driveshaft (23), on transfer case (1) with screw (25).
21. Tighten screws (2) and (3) 30-40 lb-ft (41-54 N•m).

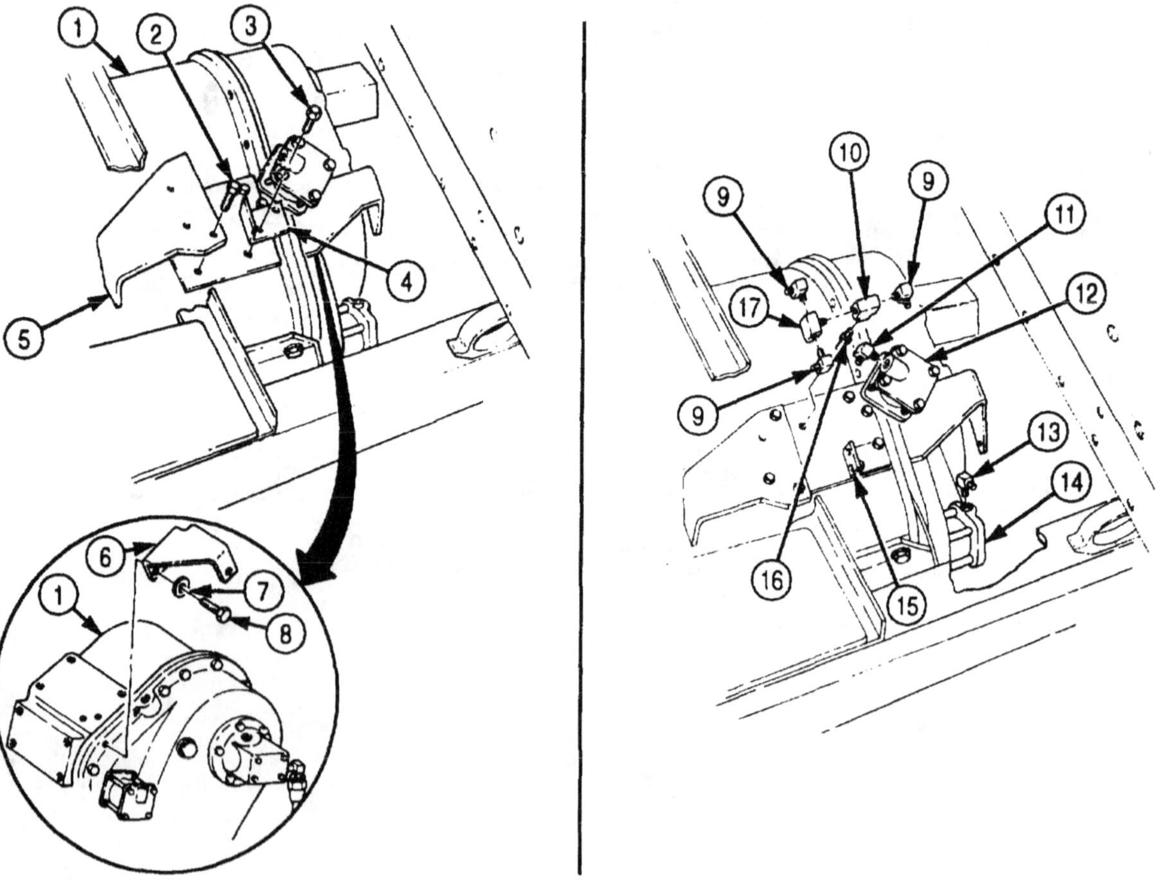

4-94. TRANSFER CASE REPLACEMENT (Contd)

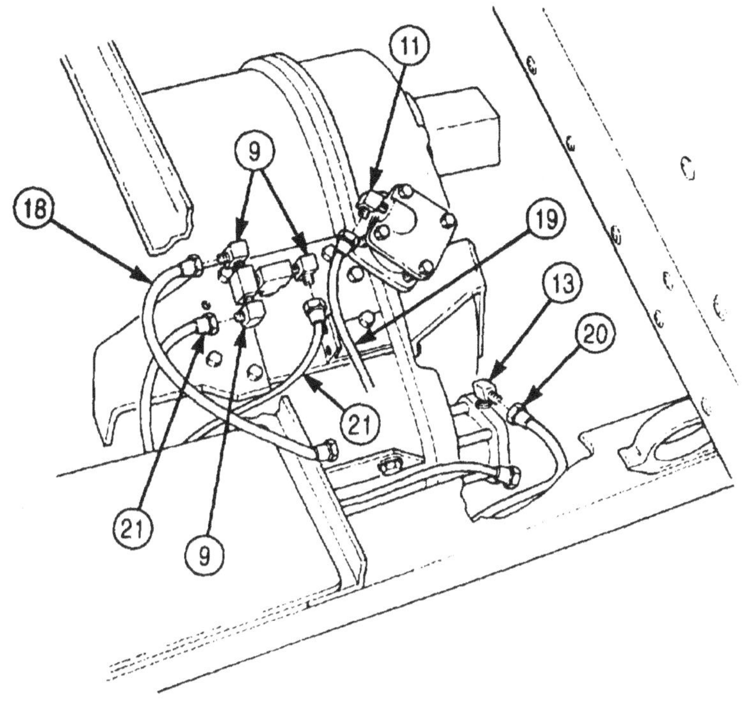

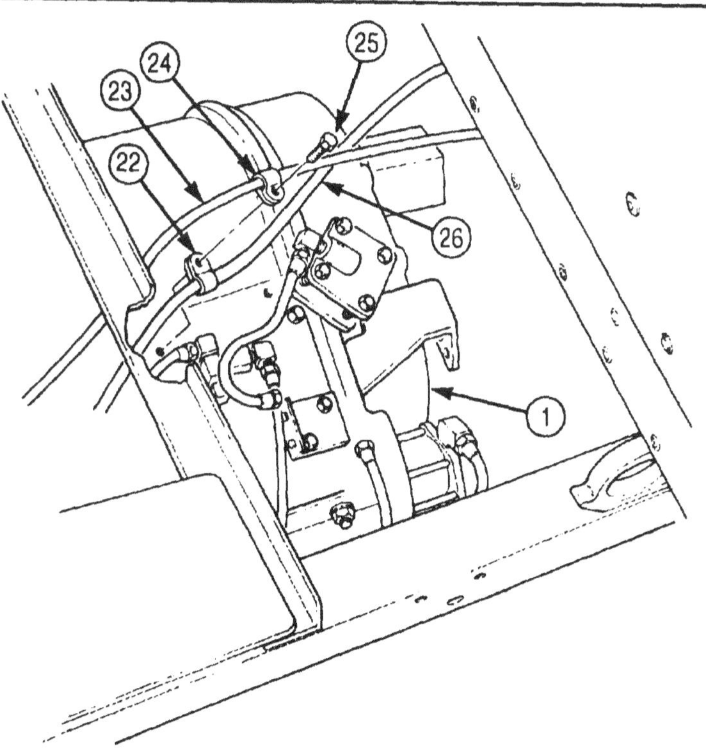

4-94. TRANSFER CASE REPLACEMENT (Contd)

22. Connect speedometer drive cable (7) to speedometer drive adapter (9).
23. Install clamp (6) and speedometer drive cable (7) on transfer case (8) with washer (5) and screw (4).
24. Install parking brake cable (3) on parking brake lever (13) with new locknut (12).
25. Install parking brake cable (3) and spacer (14) on parking brake cable bracket (11) with clamp (2), two screws (1), and new locknuts (10).
26. Install propeller shaft (20) on transfer case front output flange (17) with eight screws (18) and new locknuts (19). Tighten locknuts (19) 32-40 lb-ft (43-54 N•m).
27. Install propeller shaft (15) on transfer case input flange (16) with four new lockwashers (21) and screws (22). Tighten screws (22) 32-40 lb-ft (43-54 N•m).

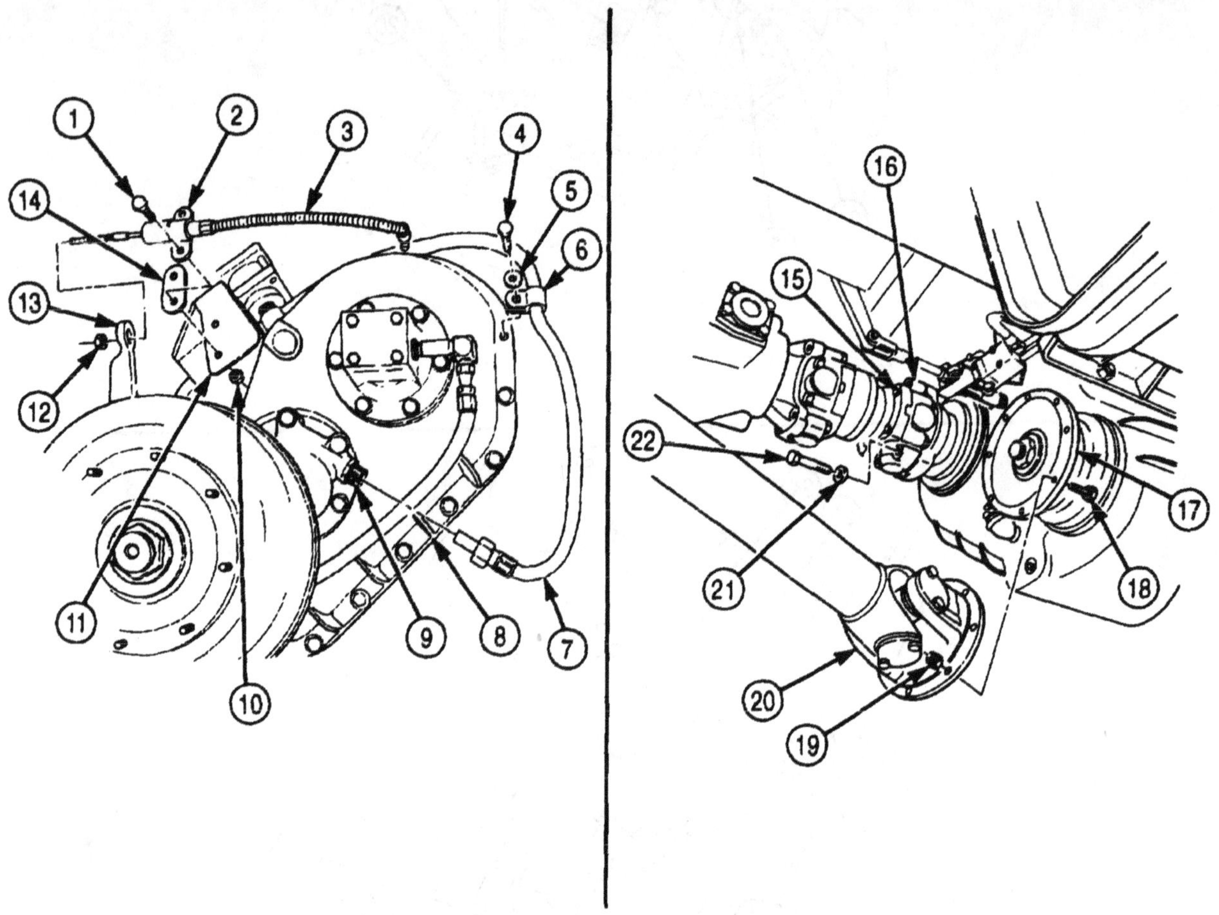

FOLLOW-ON TASKS:
- Install transfer case shift rod (para. 4-88).
- Install transfer case front axle engagement control valve (para. 3-144).
- Install transfer case interlock valve (para. 3-143).
- Install transfer case-to-forward rear axle propeller shaft (para. 3-149).
- Install wet reservoir (para. 3-185).
- Install spare tire (M923/A1/A2, M925/A1/A2, M929/A1/A2, and M930/A1/A2) (TM 9-2320-272-10).
- Adjust parking brake (TM 9-2320-272-10).
- Fill transfer case to proper fluid level (LO 9-2320-272-12).

4-95. TRANSFER CASE (M936/A1/A2) REPLACEMENT

THIS TASK COVERS:

a. Removal **b. Installation**

INITIAL SETUP:

APPLICABLE MODELS
M936/A1/A2

TOOLS
General mechanic's tool kit (Appendix E, Item 1)
Torque wrench (Appendix E, Item 144)
Transmission jack (Appendix E, Item 147)

MATERIALS/PARTS
Ten locknuts (Appendix D, Item 291)
Lockwasher (Appendix D, Item 371)
Cotter pin (Appendix D, Item 84)
Locknut (Appendix D, Item 313)
Locknut (Appendix D, Item 307)
Locknut (Appendix D, Item 308)
Locknut (Appendix D, Item 320)
Ten locknuts (Appendix D, Item 321)
Three locknuts (Appendix D, Item 294)
Antiseize tape (Appendix C, Item 72)

PERSONNEL REQUIRED
Two

REFERENCES (TM)
LO 9-2320-272-12
TM 9-2320-272-10
TM 9-2320-272-24P

EQUIPMENT CONDITION
- Parking brake set (TM 9-2320-272-10).
- Transfer case oil drained (LO 9-2320-272-12).
- Transfer case-to-forward rear axle propeller shaft removed (para. 3-149).
- Transmission-to-transfer case propeller shaft removed (para. 3-148).
- Transfer case front axle engagement control valve removed (para. 3-144).
- Transfer case shift rod removed (para. 4-88).

a. Removal

1. Remove four nuts (1), washers (2), screws (5), and propeller shaft (4) from Power Takeoff (PTO) flange (3).
2. Remove cotter pin (13), pin (6), and PTO cable (7) from PTO select lever (14). Discard cotter pin (13).
3. Remove two locknuts (12), screws (9), clamp (8), PTO cable (7), and spacer (10) from PTO cable bracket (11). Discard locknuts (12).

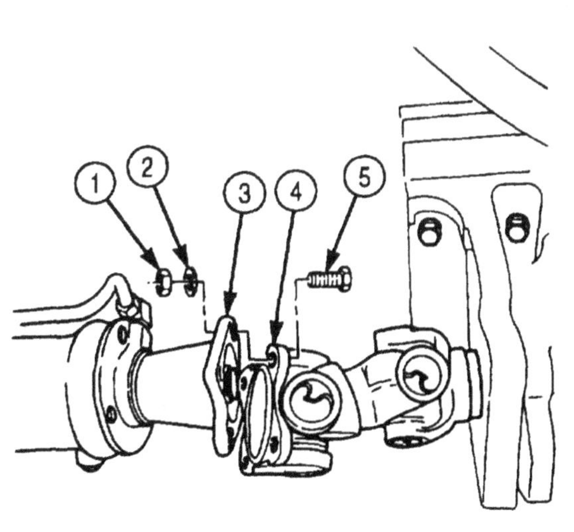

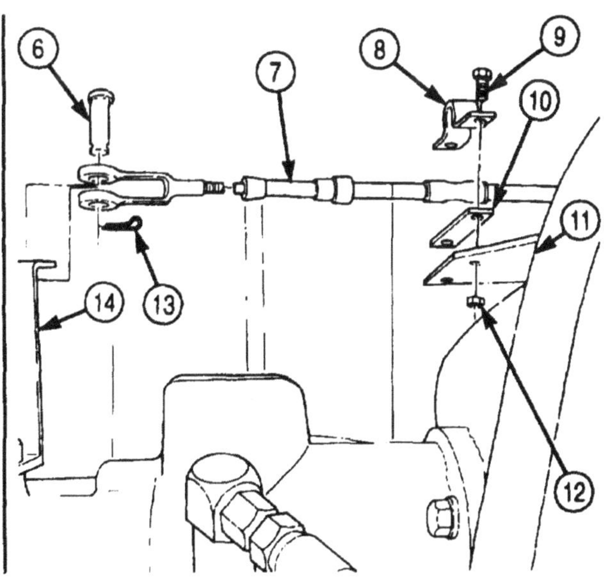

4-95. TRANSFER CASE (M936/A1/A2) REPLACEMENT (Contd)

4. Remove two screws (8), washers (9), and PTO cable bracket (10) from transfer case (12).
5. Remove locknut (16) and parking brake cable (4) from parking brake lever (17). Discard locknut (16).
6. Remove two locknuts (14), screws (2), clamp (3), parking brake cable (4), and spacer (1) from parking brake cable bracket (15). Discard locknuts (14).
7. Disconnect speedometer driveshaft (11) from speedometer drive adapter (13).
8. Remove screw (6), washer (5), clamp (7), and speedometer drive cable (11) from transfer case (12).
9. Remove eight locknuts (20), screws (19), and propeller shaft (21) from transfer case front output flange (18). Discard locknuts (20).
10. Remove two screws (26), washers (27), and parking brake cable bracket (25) from transfer case cover (28).
11. Remove screw (24), clamp (23) with parking brake cable (4), and clamp (22) with speedometer drive cable (11) from transfer case (12).

NOTE

Tag all air lines and wires for installation.

12. Disconnect supply line (33) from air cylinder elbow (32).
13. Disconnect supply line (39) from interlock air cylinder elbow (29) and interlock valve elbow (38).
14. Disconnect supply line (31) from interlock valve elbow (30).
15. Disconnect interlock vent line (37) from elbows (36) and (40).
16. Disconnect two vent lines (42) from elbows (41).
17. Disconnect interlock valve lead (35) from connector (34).
18. Remove locknut (45), washer (44), transorb diode ground wire (46), locknut (47), ground wire (48), lockwasher (49), cable clamp (50), and screw (43) from frame (51). Discard locknuts (45) and (47) and lockwasher (49).

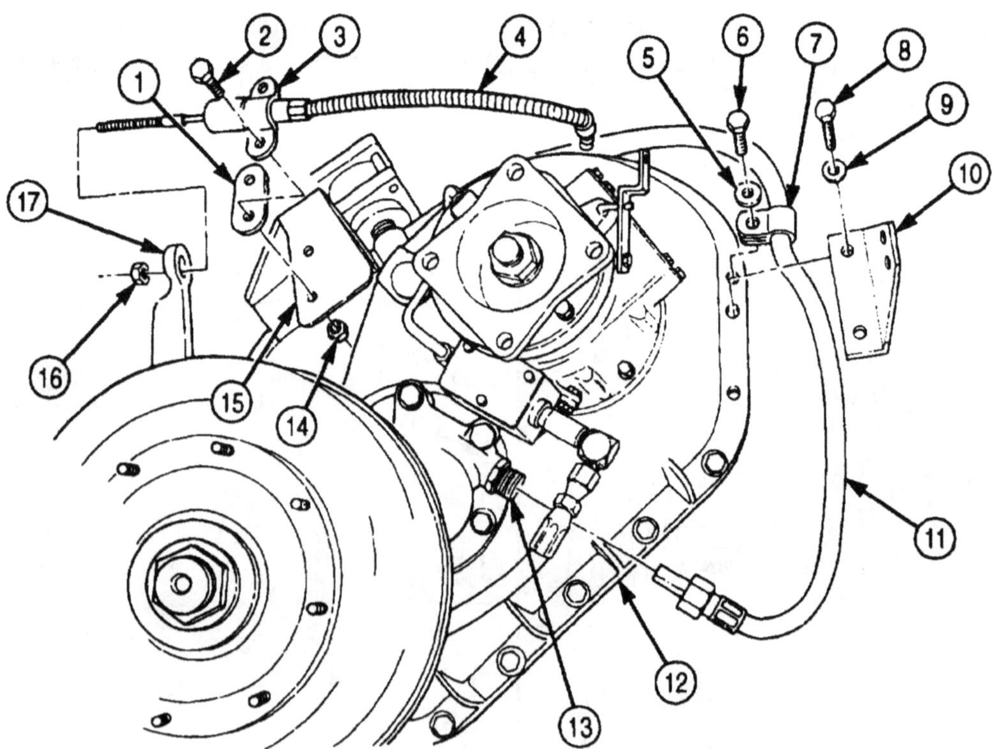

4-95. TRANSFER CASE (M936/A1/A2) REPLACEMENT (Contd)

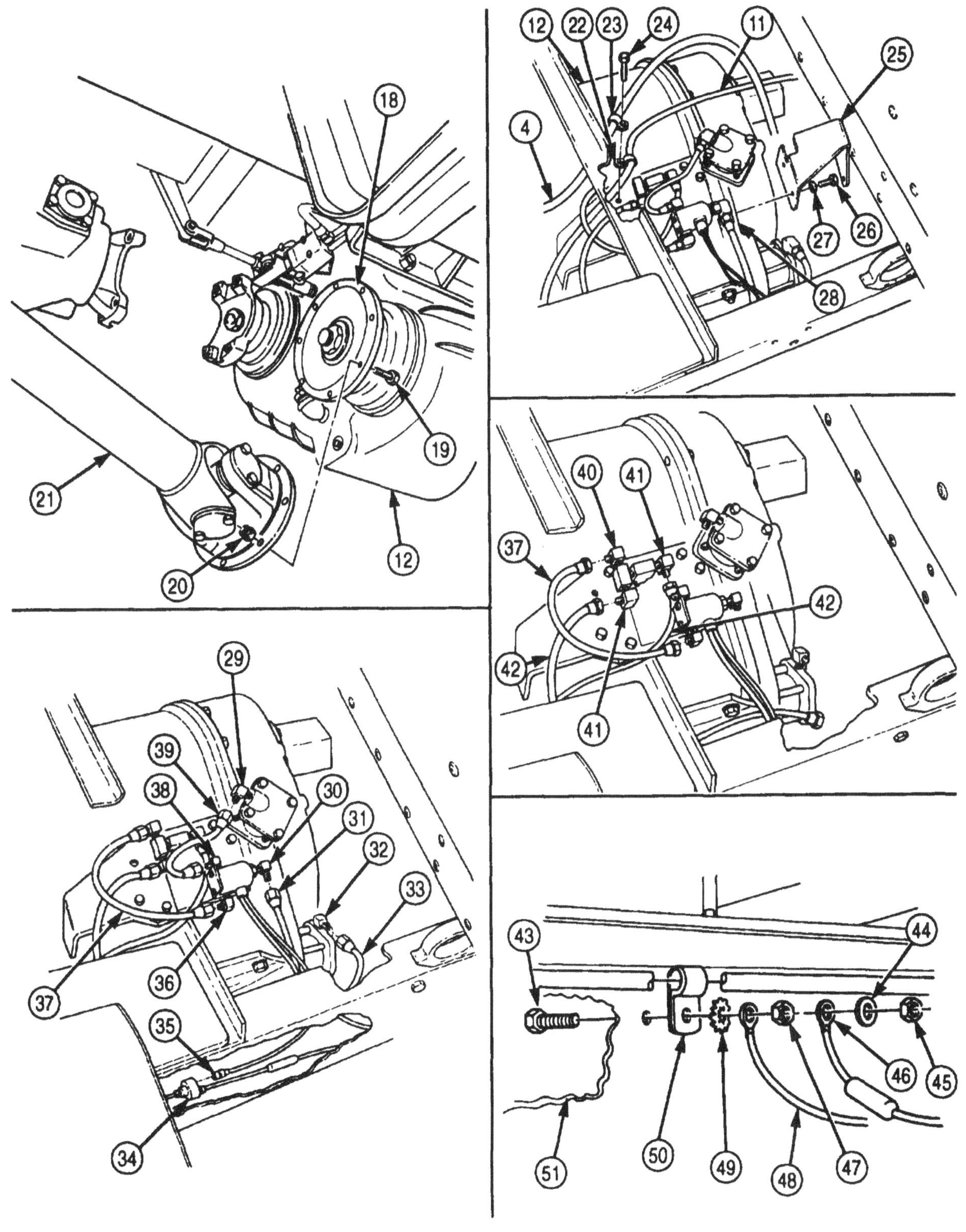

4-95. TRANSFER CASE (M936/A1/A2) REPLACEMENT (Contd)

19. Position transmission jack under transfer case (12).
20. Remove three locknuts (7), screws (1), washers (2), and insulators (3) from mounting brackets (4) and frame (6). Discard locknuts (7).
21. Raise transfer case (8) high enough to remove three insulators (5) from mounting brackets (4) and frame (6).
22. Remove three locknuts (9), screws (11), and right frame bracket (10) from frame (12). Discard locknuts (9).
23. Remove seven locknuts (13) and two mounting brackets (4) from transfer case (8). Discard locknuts (13).
24. Lower transmission jack and remove transfer case (8) from vehicle.

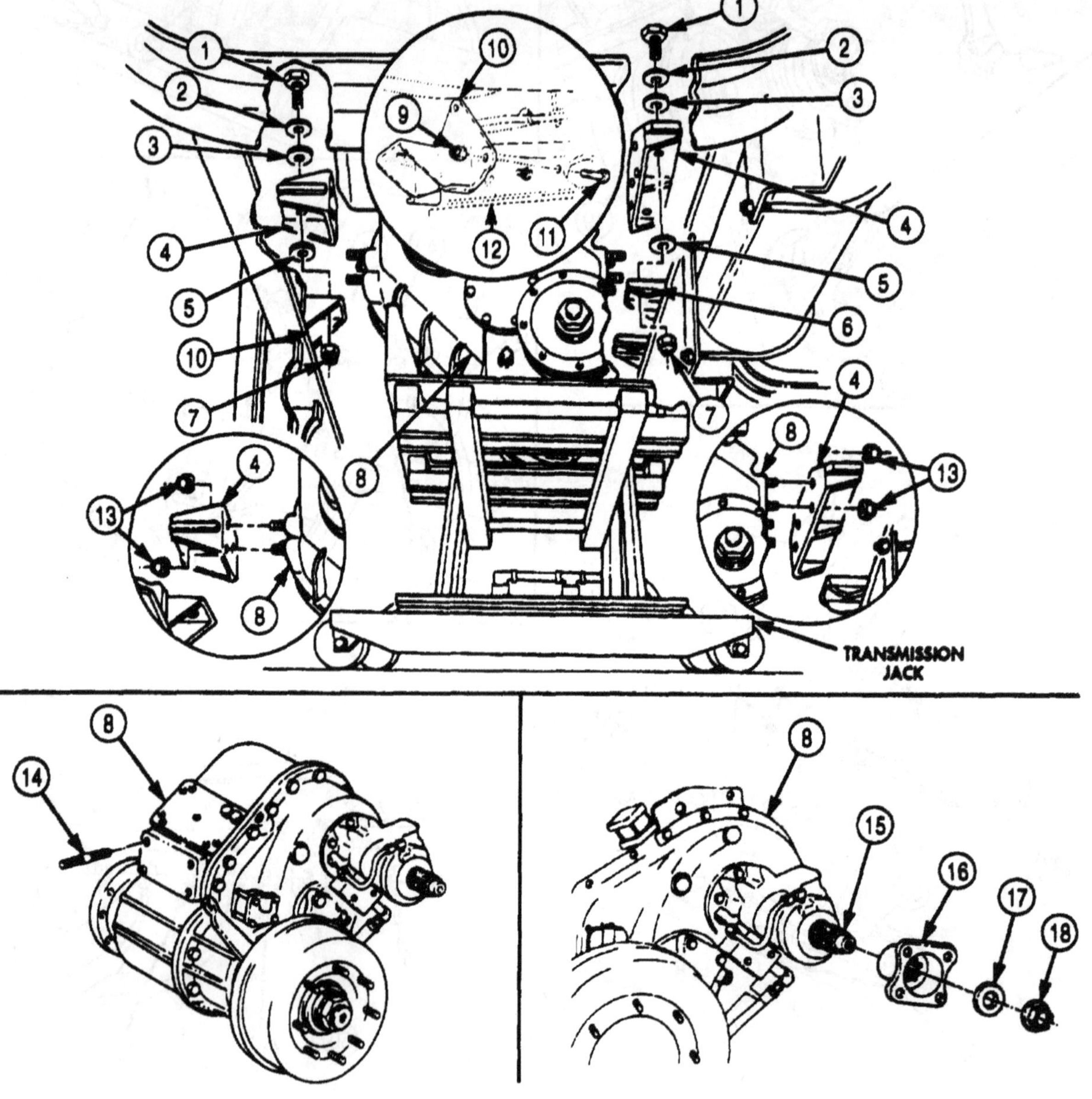

4-95. TRANSFER CASE (M936/A1/A2) REPLACEMENT (Contd)

25. Remove nut (18), washer (17), and flange (16) from PTO shaft (15).
26. Remove seven studs (14) from transfer case (8).
27. Using lifting device, remove transfer case (8) from hydraulic jack.
28. Remove three elbows (23) from vent tees (24) and (33).
29. Remove vent tees (24) and (33) and fitting (25) from inspection cover (31).
30. Remove elbow (26) from interlock air cylinder (27).
31. Remove elbow (29) from air cylinder cover (30).
32. Remove two screws (32) and interlock valve (28) from interlock valve bracket (21).
33. Remove two screws (22) and interlock valve bracket (21) from transfer case (8).
34. Remove three screws (20) and control valve bracket (19) from transfer case (8).

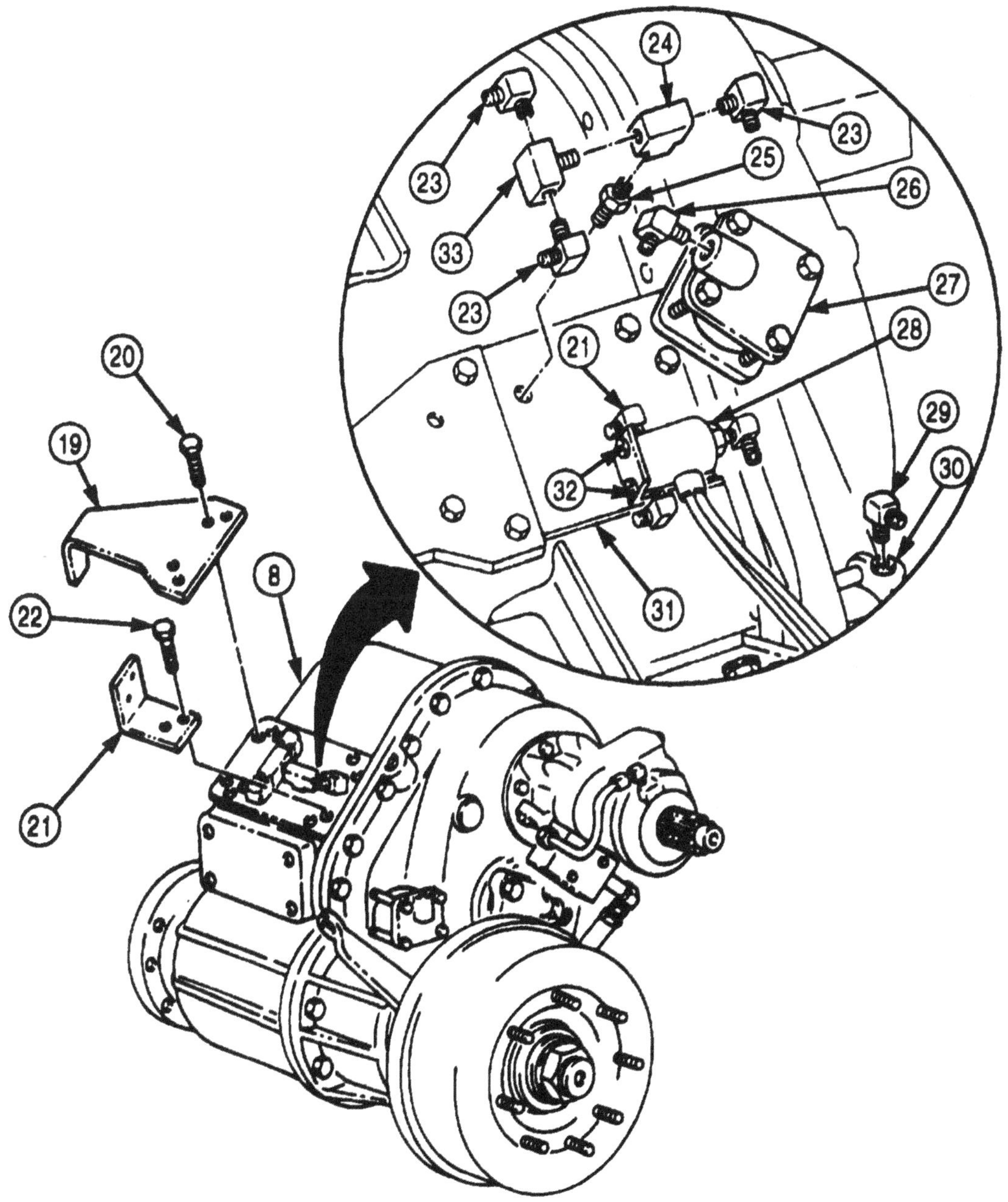

4-95. TRANSFER CASE (M936/A1/A2) REPLACEMENT (Contd)

b. Installation

1. Install control valve bracket (2) on transfer case (3) with three screws (1). Finger-tighten screws (1).
2. Install interlock valve bracket (4) on transfer case (3) with two screws (5). Tighten screws (1) and (5) 30-40 lb-ft (41-54 N•m).
3. Install interlock valve (11) on interlock valve bracket (4) with two screws (15).
4. Install elbow (12) on air cylinder cover (13).
5. Install elbow (9) on interlock air cylinder (10).
6. Install vent tees (7) and (16) and fitting (8) on inspection cover (14).
7. Install three elbows (6) on vent tees (7) and (16).
8. Using lifting device, position transfer case (3) on hydraulic jack.

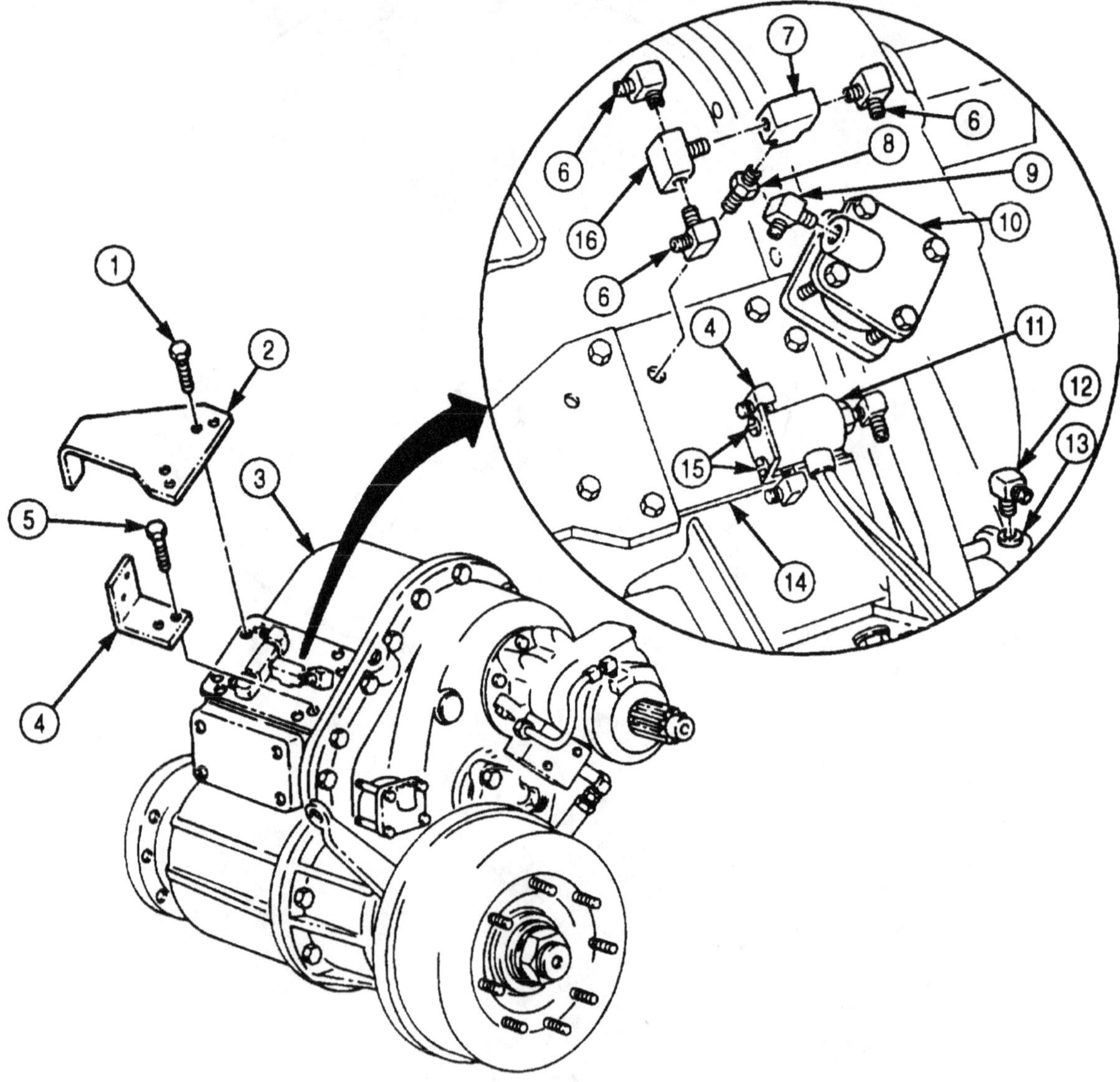

4-95. TRANSFER CASE (M936/A1/A2) REPLACEMENT (Contd)

9. Install flange (18) on PTO shaft (17) with washer (19) and nut (20).
10. Install seven studs (21) on transfer case (3).

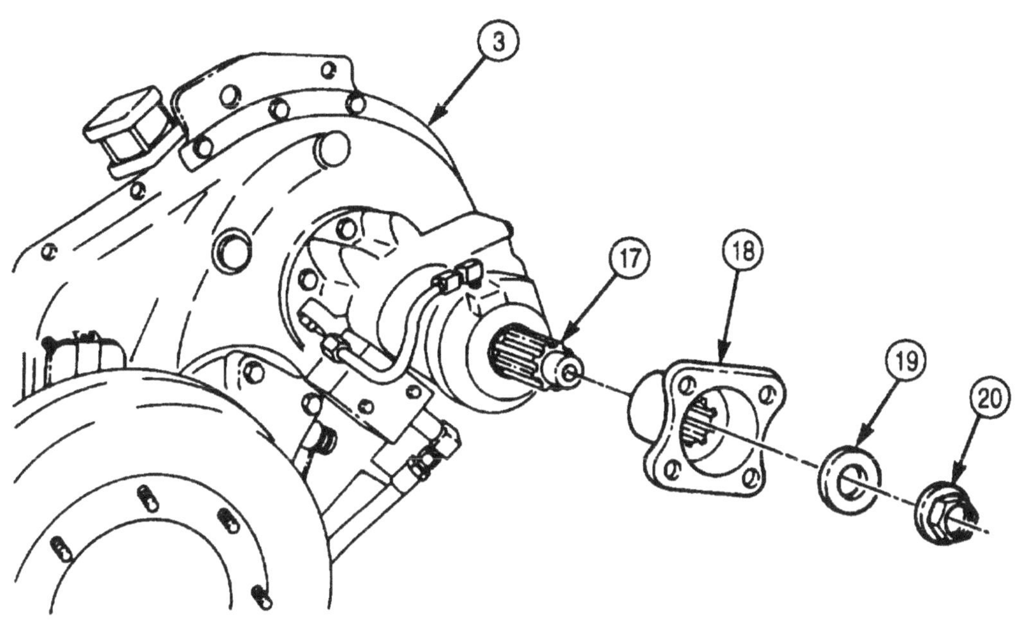

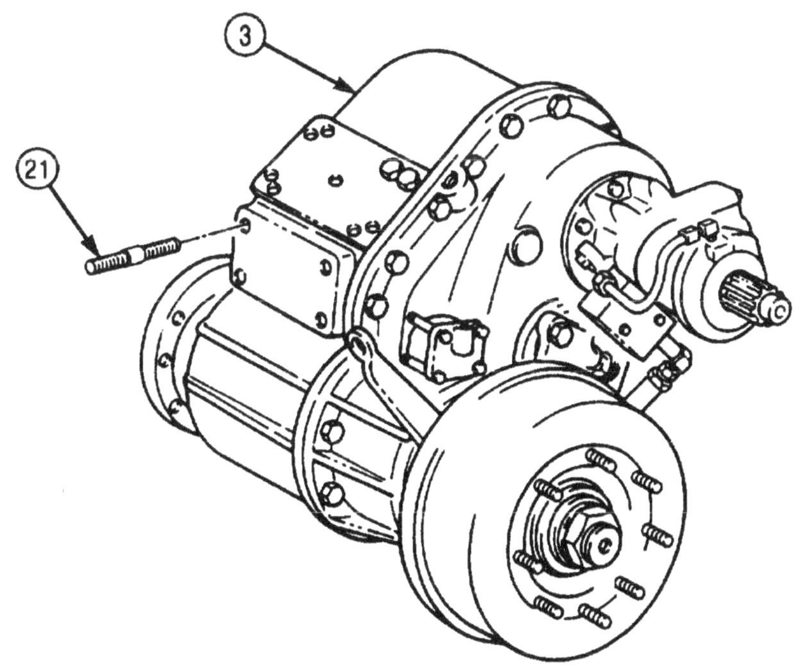

4-95. TRANSFER CASE (M936/A1/A2) REPLACEMENT (Contd)

11. Position hydraulic jack and transfer case (8) under vehicle.
12. Raise transfer case (8) in position high enough to allow for installation of mounting brackets (4).
13. Install two mounting brackets (4) on transfer case (8) with seven new locknuts (13). Tighten locknuts (13) 125-135 lb-ft (170-183 N•m).
14. Install right frame bracket (11) on frame (9) with three screws (12) and locknuts (10). Tighten locknuts (10) 120 lb-ft (163 N•m).
15. Place three insulators (5) between mounting brackets (4) and frame brackets (6).
16. Slowly lower transfer case (8) and align holes in mounting brackets (4), insulators (5), and frame brackets (6).
17. Install three insulators (3), washers (2), screws (1), and new locknuts (7) on mounting brackets (4) and frame brackets (6). Tighten locknuts (7) 50-60 lb-ft (68-81 N•m).
18. Remove hydraulic jack from transfer case (8) and vehicle.
19. Install cable clamp (21), new lockwasher (20). and ground wire (19) on frame (9) with screw (14) and new locknut (15).

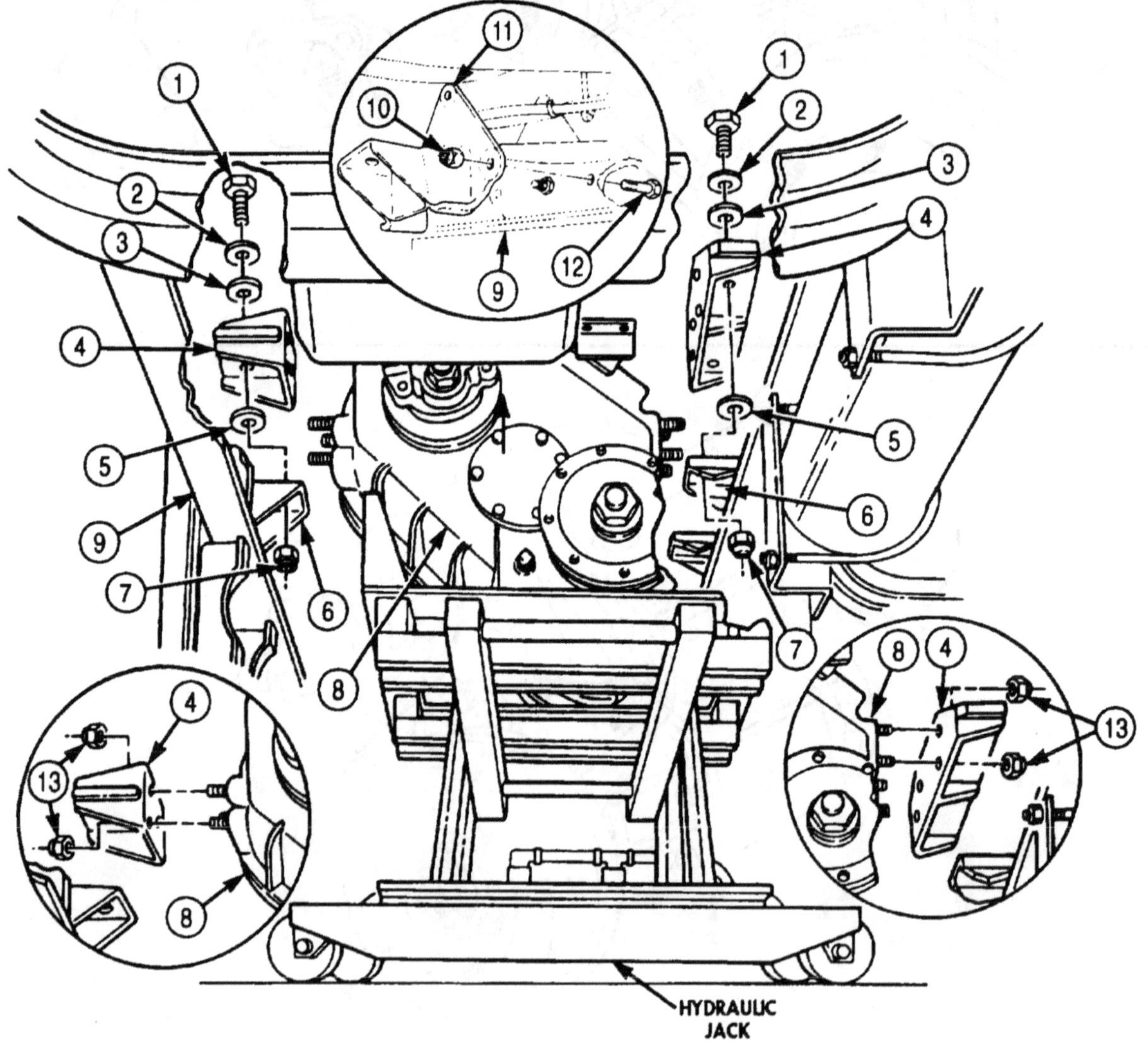

4-95. TRANSFER CASE (M936/A1/A2) REPLACEMENT (Contd)

20. Install transorb diode ground (18) on screw (14) with washer (17) and new locknut (16).
21. Connect two vent lines (25) to elbows (24).
22. Connect interlock vent line (22) to elbow (23) and elbow (35).
23. Connect supply line (30) to interlock valve elbow (29).
24. Connect supply line (27) to interlock air cylinder elbow (28) and interlock valve elbow (26).
25. Connect supply line (32) to air cylinder elbow (31).
26. Connect interlock valve lead (34) to connector (33).
27. Install clamp (37) with parking brake cable (40) and clamp (38) with speedometer drive cable (36) on transfer case (8) with screw (39). Tighten screw (39) 30-40 lb-ft (41-54 N•m).
28. Install parking brake cable bracket (41) on transfer case cover (44) with two washers (43) and screws (42). Tighten screws (42) 40-55 lb-ft (54-75 N•m).

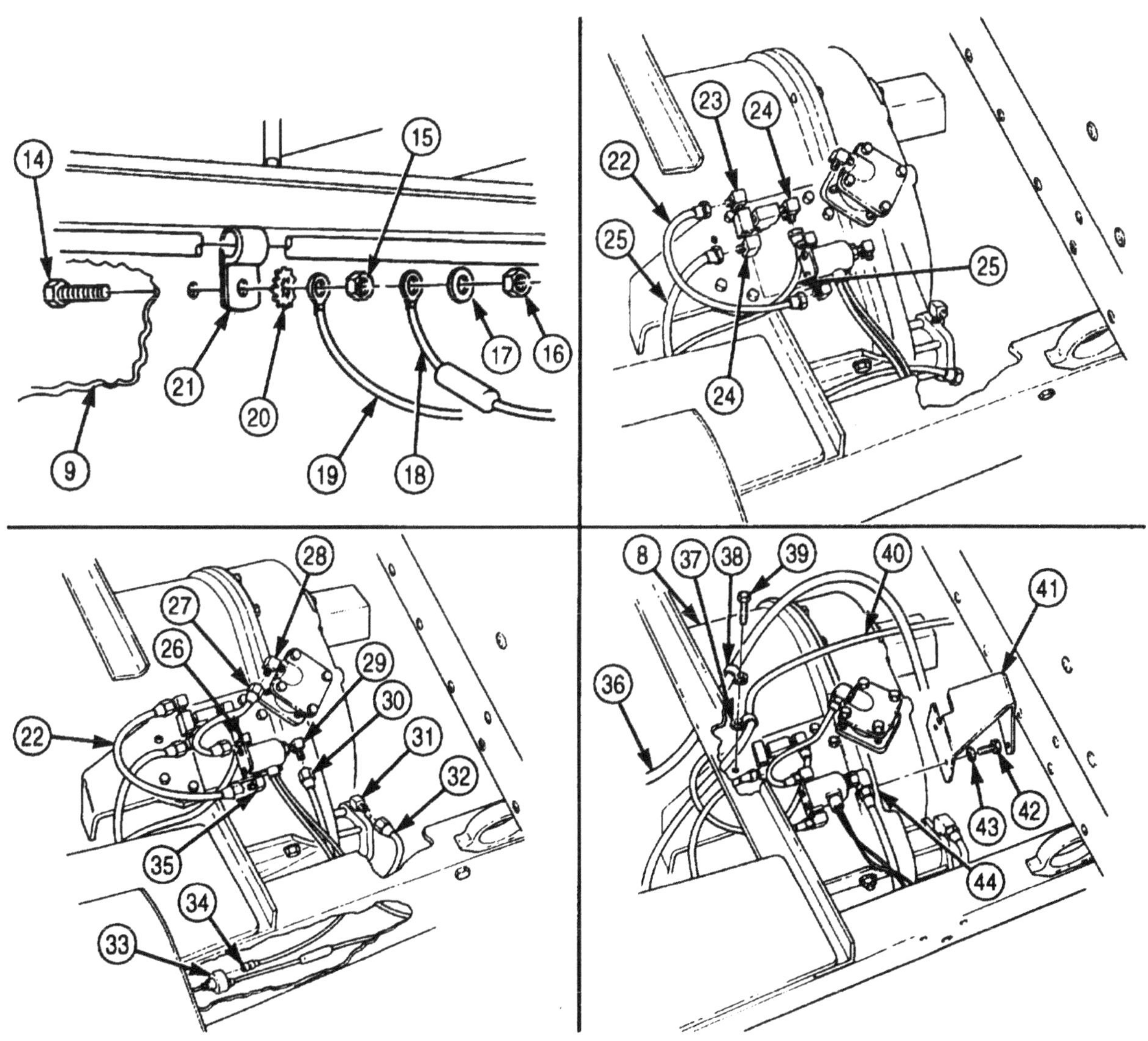

4-95. TRANSFER CASE (M936/A1/A2) REPLACEMENT (Contd)

29. Install propeller shaft (4) on transfer case front output flange (1) with eight screws (2) and new locknuts (3). Tighten locknuts (3) 32-40 lb-ft (43-54 N•m).
30. Connect speedometer driveshaft (15) on speedometer drive adapter (17).
31. Install clamp (11) and speedometer driveshaft (15) on transfer case (16) with washer (9) and screw (10).
32. Install parking brake cable (8) on parking brake lever (21) with new locknut (20).
33. Install parking brake cable (8) and spacer (5) on parking brake cable bracket (19) with clamp (7), two screws (6) and new locknuts (18).
34. Install PTO cable bracket (14) on transfer case (16) with two washers (13) and screws (12). Tighten screws (12) 40-55 lb-ft (54-75 N•m).
35. Install PTO cable (25) on PTO selector lever (22) with pin (23) and new cotter pin (24).
36. Install PTO cable (25) and spacer (28) on PTO cable bracket (14) with clamp (26), two screws (27), and new locknuts (29).
37. Install propeller shaft (33) on PTO flange (32) with four screws (34), new lockwashers (31), and locknuts (30). Tighten locknuts (30) 32-40 lb-ft (43-54 N•m).

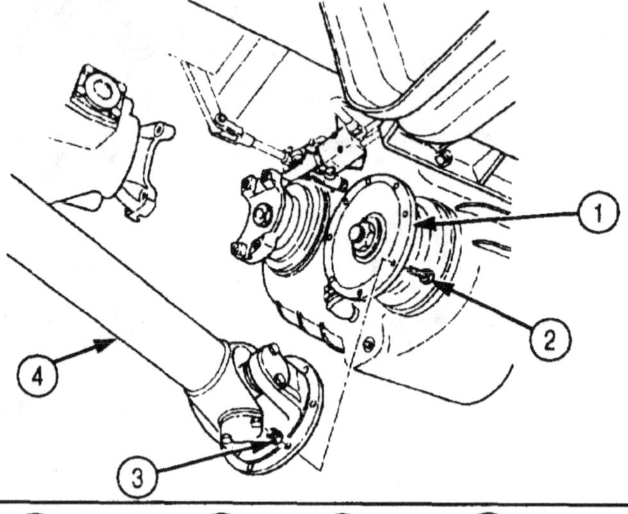

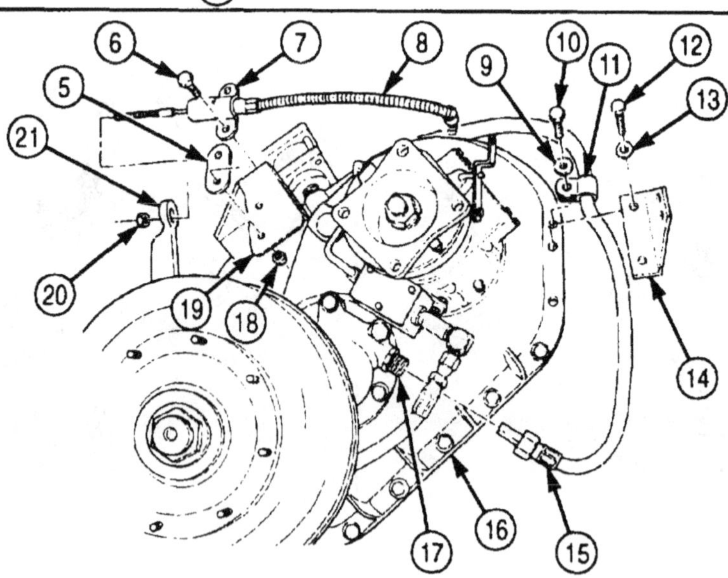

4-95. TRANSFER CASE (M936/A1/A2) REPLACEMENT (Contd)

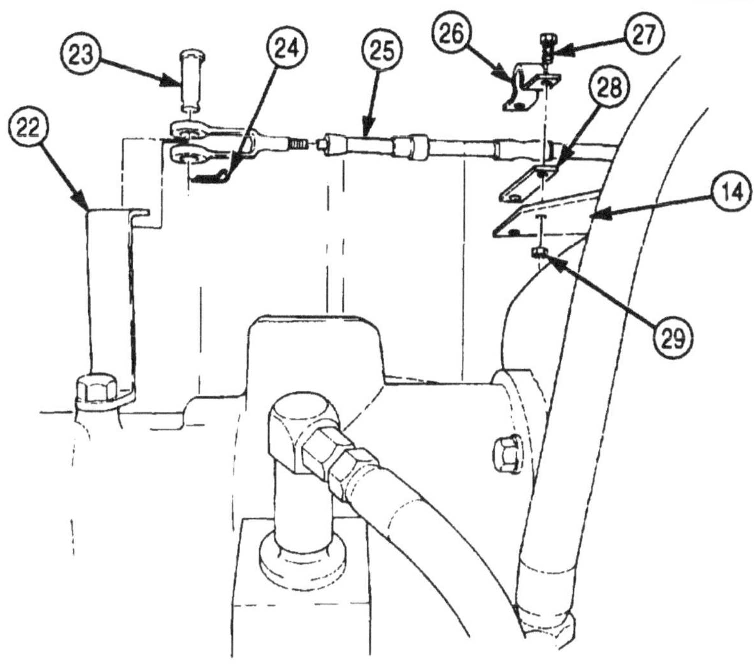

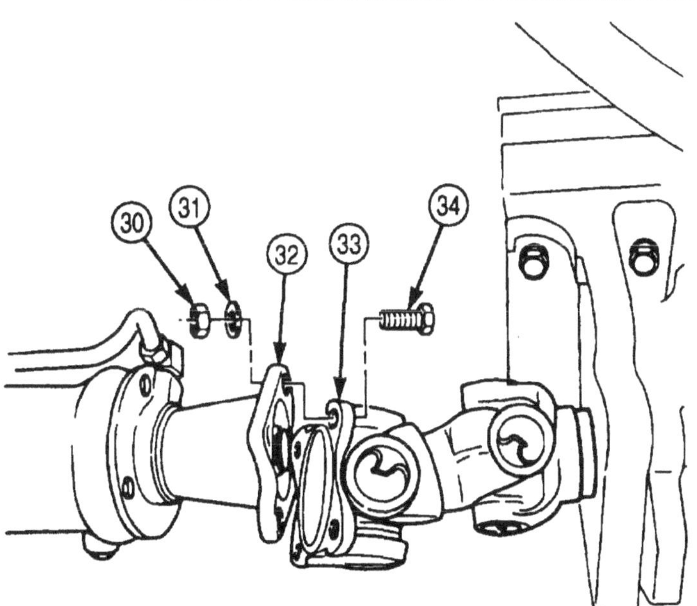

FOLLOW-ON TASKS:
- Install transfer case shift rod (para. 4-88).
- Install transfer case front axle engagement control valve (para. 3-144).
- Check parking brake adjustment (para. 3-179).
- Install transmission-to-transfer propeller shaft (para. 3-148).
- Install transfer case-to-forward rear axle propeller shaft (para. 3-149).
- Fill transfer case to proper oil level (LO 9-2320-272-12).

Section IX. FRONT AND REAR AXLE MAINTENANCE

4-96. FRONT AND REAR AXLE MAINTENANCE INDEX

4-97. SPINDLE BEARING SLEEVE REPLACEMENT

THIS TASK COVERS:

a. Removal

b. Installation

INITIAL SETUP:

APPLICABLE MODELS
All

SPECIAL TOOLS
Spindle bearing sleeve remover (Appendix E, Item 127)
Spindle bearing sleeve replacer (Appendix E, Item 128)

TOOLS
General mechanic's tool kit (Appendix E, Item 1)
Burnishing tool (Appendix E, Item 22)
Inside micrometer (Appendix E, Item 83)

REFERENCES (TM)
TM 9-2320-272-24P

EQUIPMENT CONDITION
Front axle shaft and universal joint removed (para. 3-154).

NOTE

Perform this procedure only if inspection of spindle bearing sleeve indicatea replacement is required.

a. Removal

Using spindle bearing sleeve remover, remove spindle bearing sleeve (1) from spindle (2). Discard spindle bearing sleeve (1).

4-97. SPINDLE BEARING SLEEVE REPLACEMENT (Contd)

b. Installation

1. Using spindle bearing sleeve replacer, install new spindle bearing sleeve (1) in spindle (2).
2. Using burnishing tool, machine inside diameter of spindle bearing sleeve (1) to 2.247-2.251 in. (57.074-57.176 mm).

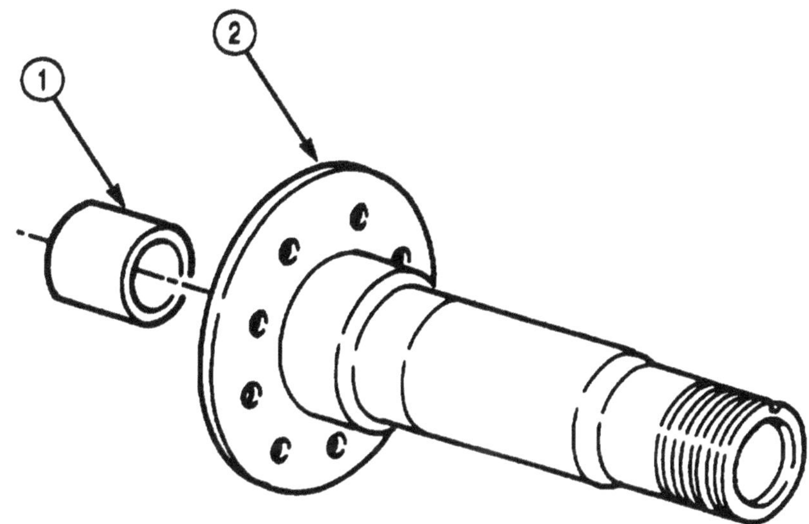

FOLLOW-ON TASK: Install Front axle shaft and universal joint (para. 3-154).

4-98. STEERING KNUCKLE MAINTENANCE

THIS TASK COVERS:

a. Removal
b. Cleaning and Inspection
c. Installation
d. End Play Test

INITIAL SETUP:

APPLICABLE MODELS
All

TOOLS
General mechanic's tool kit (Appendix E, Item 1)
Inside micrometer (Appendix E, Item 82)
Burnishing tool (Appendix E, Item 22)
Torque wrench (Appendix E, Item 145)
Dial indicator (Appendix E, Item 36)
Jack stand
Arbor press

MATERIALS/PARTS
Lockwire, 40 in. (Appendix D, Item 419)
Three cotter pins (Appendix D, Item 85)
Stop screw (Appendix D, Item 679)
Jamnut (Appendix D, Item 264)
Sleeve bearing (Appendix D, Item 657)
Four lockwashers (Appendix D, Item 399)
GAA grease (Appendix C, Item 28)
Sealing compound (Appendix C, Item 30)

REFERENCES (TM)
LO 9-2310-272-12
TM 9-2320-272-10
TM 9-2320-272-24P

EQUIPMENT CONDITION
Front axle shaft removed (para. 3-154).

a. Removal

NOTE

- Perform step 1 for left steering knuckle only.
- Perform step 2 for right steering knuckle only.

1. Remove cotter pin (10), nut (9), and drag link (8) from steering knuckle arm (3). Discard cotter pin (10).
2. Remove cotter pin (28), nut (27), and socket assembly (25) with ball stud (26) from steering knuckle arm (3). Discard cotter pin (28).
3. Remove cotter pin (13), nut (14), and tie rod (15) from steering knuckle (1). Discard cotter pin (13).
4. Remove four nuts (6), lockwashers (5), bushings (4), and steering knuckle arm (3) from steering knuckle (1). Discard lockwashers (5).
5. Remove lubrication fitting (7) from steering knuckle arm (3).
6. Insert two puller screws into screw holes (12) of upper sleeve (2), and remove from steering knuckle (1).
7. Remove spacer (11) from upper sleeve (2).
8. Remove two screws (24) and washers (23) from seal guard (22) and steering knuckle (1).
9. Remove grease fitting (19), four screws (20), washers (21), lower plate (18), and seal guard (22) from steering knuckle (1).
10. Insert two puller screws into screw holes (16) of lower sleeve (17), and remove from steering knuckle (1).
11. Remove lockwire (31) and twelve screws (29) from dust seal plate (30) and steering knuckle (1). Discard lockwire (31).

4-98. STEERING KNUCKLE MAINTENANCE (Contd)

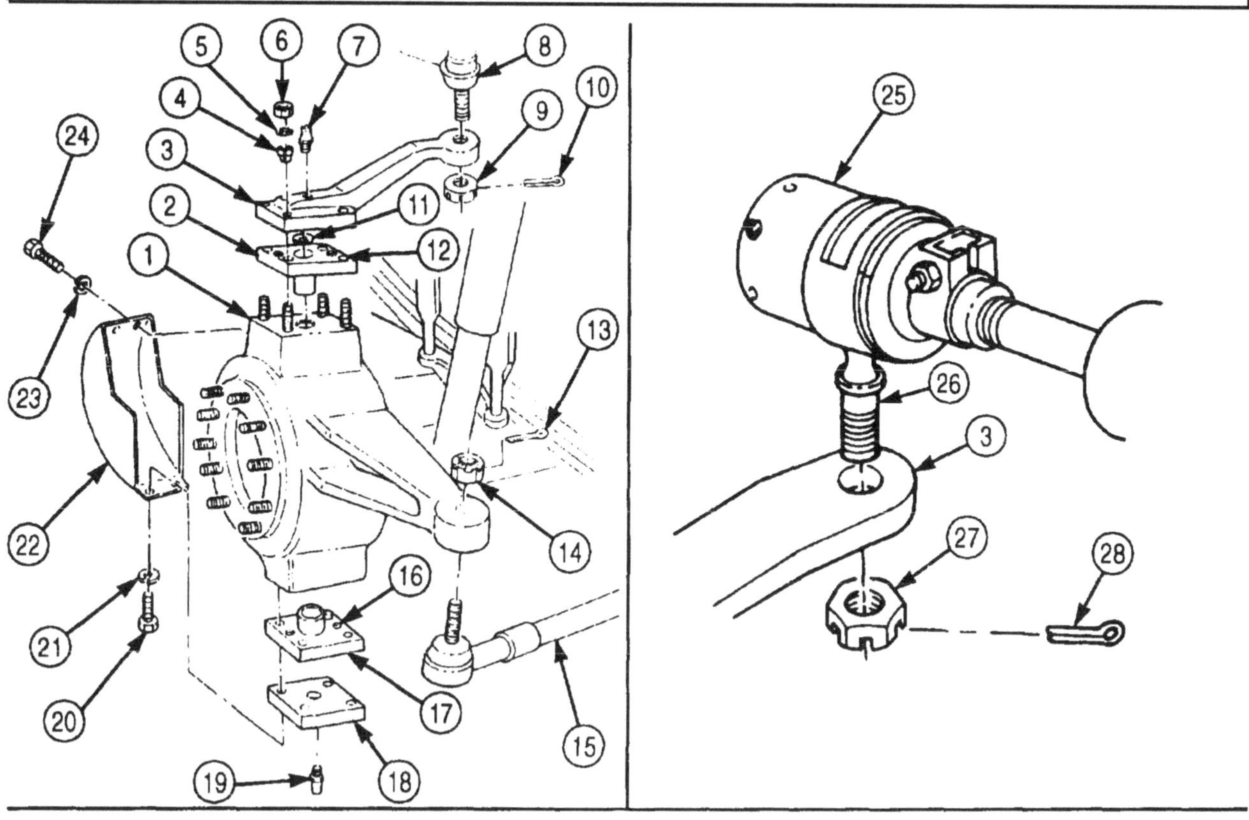

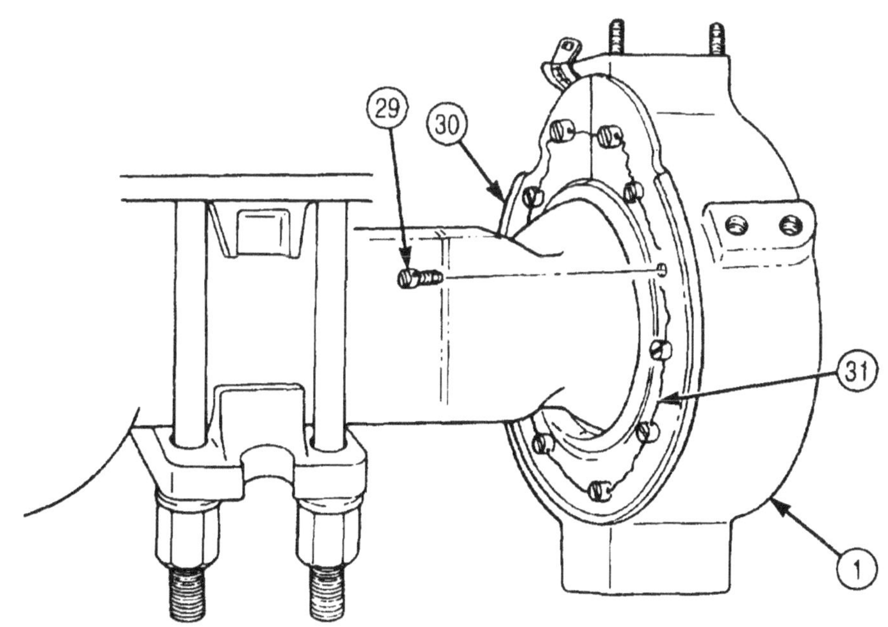

4-98. STEERING KNUCKLE MAINTENANCE (Contd)

12. Remove dust seal plate (4) and seal (3) from steering knuckle (7).
13. Remove steering knuckle (7) from axle kingpins (2) and (5).

b. Cleaning and Inspection

1. For general cleaning instructions, refer to para. 2-14.
2. For general cleaning instructions, refer to para. 2-15.
3. Inspect steering knuckle (7), lower plate (12), upper sleeve (10), and lower sleeve (11) for cracks and scores. Replace part (s) if cracked or scored.
4. Measure bearing inside diameter of upper and lower sleeves (10) and (11) for wear. If inside diameter is more than 1.505 in. (38.23 mm), replace sleeve (10) or (11).
5. Inspect studs (1) and (6) for stripped threads. Replace studs (1) or (6) if threads are stripped.
6. Inspect stop screw (8) for bends. Replace stop screw (8) if bent.

NOTE

- Perform steps 7 and 8 if replacing with new stop screw.
- Perform steps 9 through 11 if replacing with new sleeve bearings. Steps to replace sleeve bearings are the same for upper and lower sleeves.

7. Break weld and remove stop screw (8) and jamnut (9) from steering knuckle (7). Discard stop screw (8) and jamnut (9).
8. Install new stop screw (8) and new jamnut (9) on steering knuckle (7). Do not weld.
9. Using arbor press, remove sleeve bearing (14) from sleeve (13). Discard sleeve bearing (14).
10. Using arbor press, install new sleeve bearing (14) in sleeve (13).
11. Using burnisher, burnish inside diameter of sleeve bearing (14) to 1.500 in. (38.1 mm).

4-98. STEERING KNUCKLE MAINTENANCE (Contd)

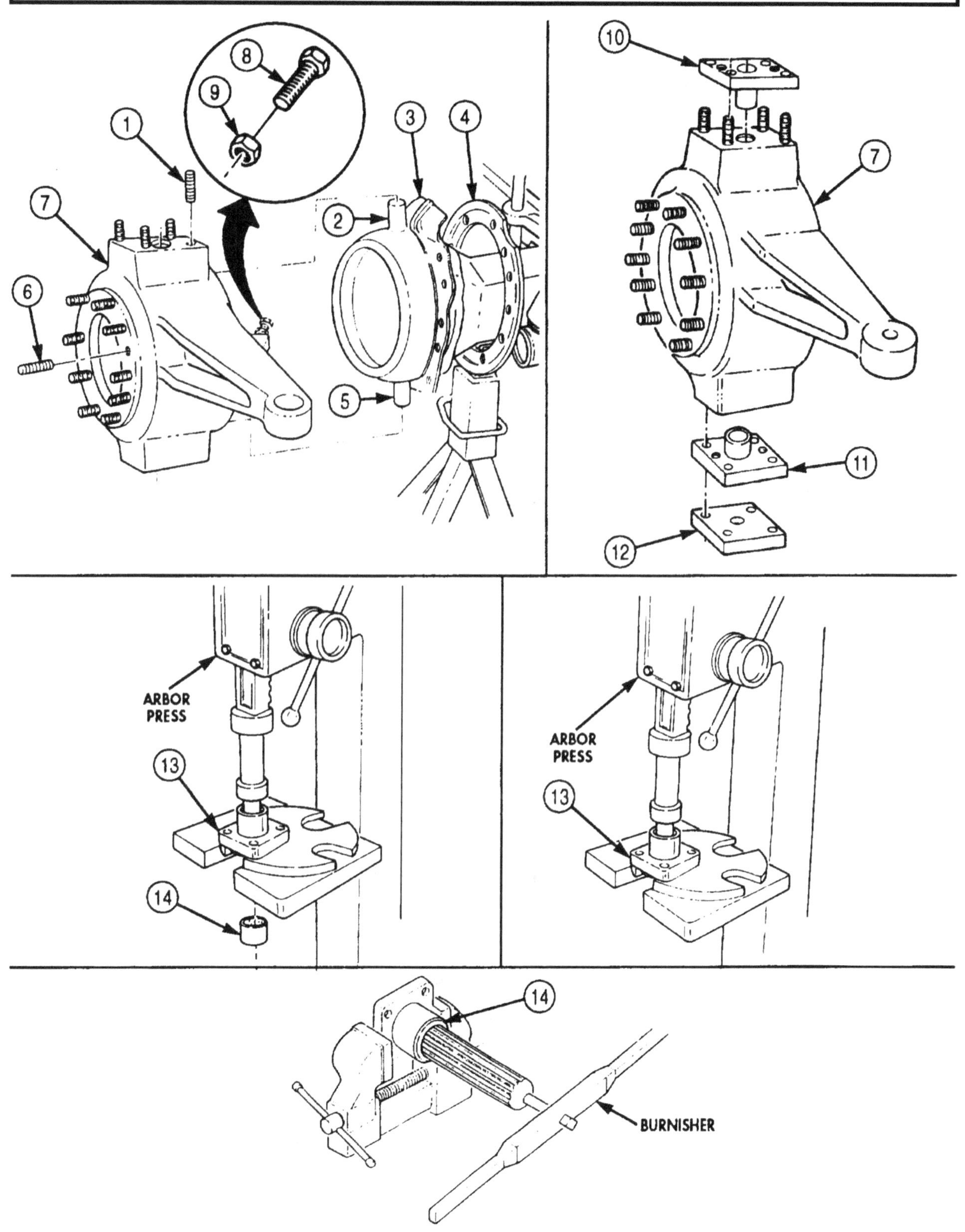

4-98. STEERING KNUCKLE MAINTENANCE (Contd)

c. Installation

1. Position steering knuckle (1) on kingpins (5) and (6) with studs (2) facing upward.
2. Apply GAA grease to bore of upper sleeve (3) and install on upper kingpin (5). Coat mounting flange with sealing compound.
3. Install spacer (4) on upper sleeve (3).
4. Apply GAA grease to bore of lower sleeve (7) and install on steering knuckle (1). Coat mounting flange with sealing compound.
5. Apply sealing compound to threads of studs (2).
6. Install steering knuckle arm (11) on studs (2) with four bushings (20), new lockwashers (8), and nuts (9). Tighten nuts (9) 155-170 lb-ft (210-231 N•m).
7. Install grease fitting (10) on steering knuckle arm (11).
8. Install seal (21) and dust seal plate (24) on steering knuckle (1) with twelve screws (23) and new lockwire (22).

NOTE

Lower sleeve must be installed with words WHEEL END facing down.

9. Apply sealing compound to contact surface of lower plate (12), and install lower plate (13) and seal guard (17) on steering knuckle (1) with four washers (16) and screws (15). Ensure seal guard (17) is installed in front two mounting holes. Tighten screws (15) 105-135 lb-ft (142-183 N•m)
10. Install grease fitting (14) on lower plate (12).
11. Install two washers (18) and screws (19) on seal guard (17) and steering knuckle (1). Tighten screws (19) 105-135 lb-ft (142-183 N•m).
12. Install tie rod (30) on steering knuckle (1) with nut (29). Tighten nut (29) 140-180 lb-ft (190-244 N•m).

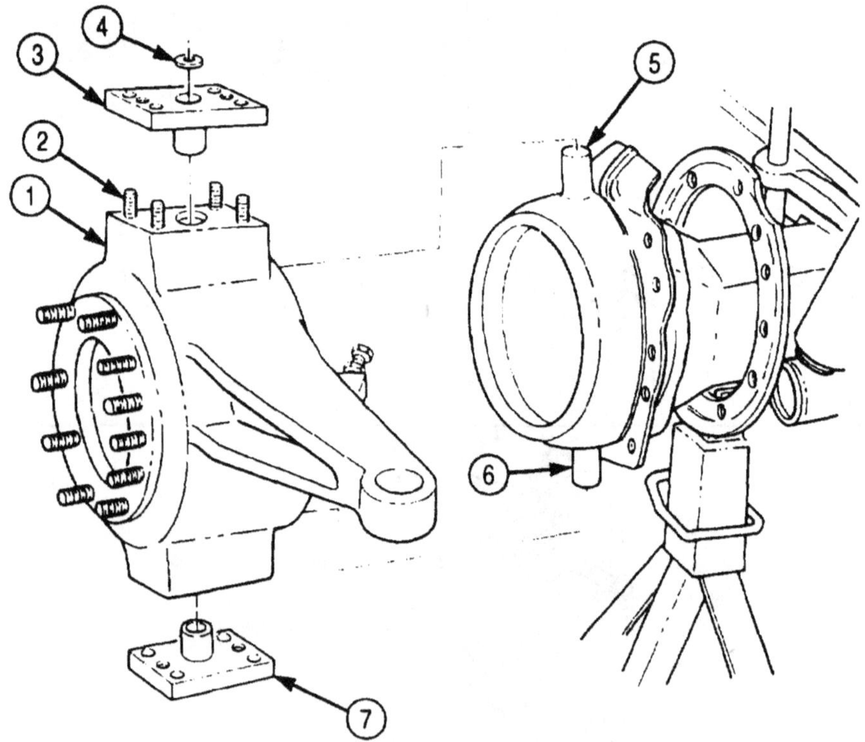

4-98. STEERING KNUCKLE MAINTENANCE (Contd)

13. Install new cotter pin (28) on nut (29) and tie rod (30).

NOTE

- Perform step 14 for left steering knuckle.
- Assistant will help with step 14.

14. Install drag link (25) on steering knuckle arm (11) with nut (27). Tighten nut (27) 140-180 lb-ft (190-244 N•m).
15. Install new cotter pin (26) on nut (27) and drag link (25).

NOTE

Perform step 16 for right steering knuckle arm.

16. Install socket assembly (31) with ball stud (34) on steering knuckle arm (11) with nut (33). Tighten nut (33) 140-180 lb-ft (190-244 N•m).
17. Install new cotter pin (32) on nut (33) and socket assembly (31).

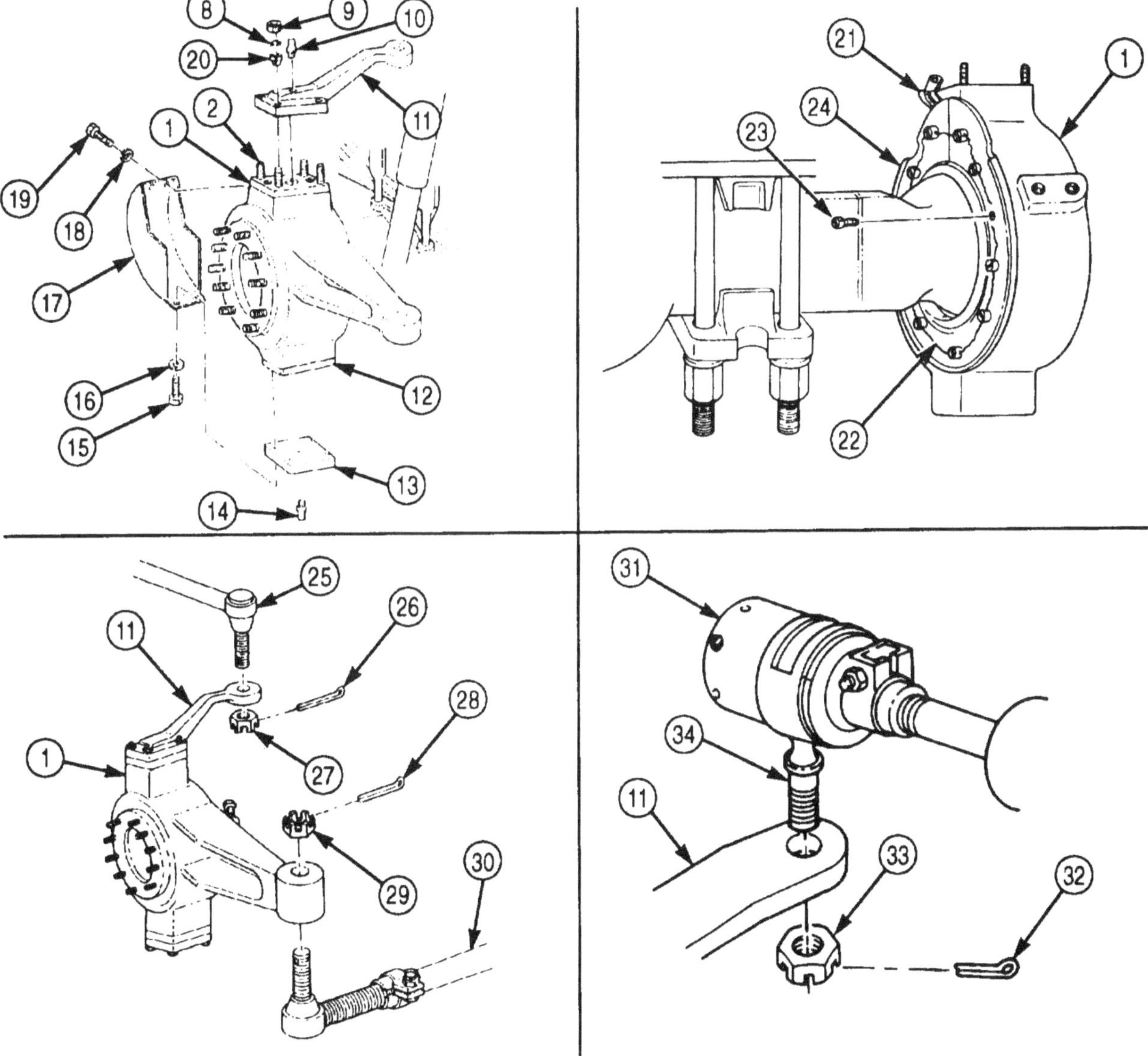

4-98. STEERING KNUCKLE MAINTENANCE (Contd)

d. End Play Test

1. Position dial indicator with magnetic stand on front axle housing (3) next to steering arm (2).
2. Adjust dial indicator until pointer touches center of steering arm mounting plate (1). Note reading on dial indicator pointer.
3. Position prybar and jack stand under steering knuckle (4).
4. Pry up on steering knuckle (4) and observe dial indicator reading.
   a. If dial indicator reading is 0.004-0.014 in. (0.1-0.36 mm), steering knuckle end play is within limits.
   b. If dial indicator reading is 0.014 in. (0.36 mm), or greater, remove upper sleeve and install a thicker spacer.
   c. If dial indicator reading is less than 0.004 in. (0.1 mm), remove upper sleeve and install a thinner spacer.
5. Remove dial indicator from front axle housing (3).
6. Remove prybar and jack stand from steering knuckle (4).

4-98. STEERING KNUCKLE MAINTENANCE (Contd)

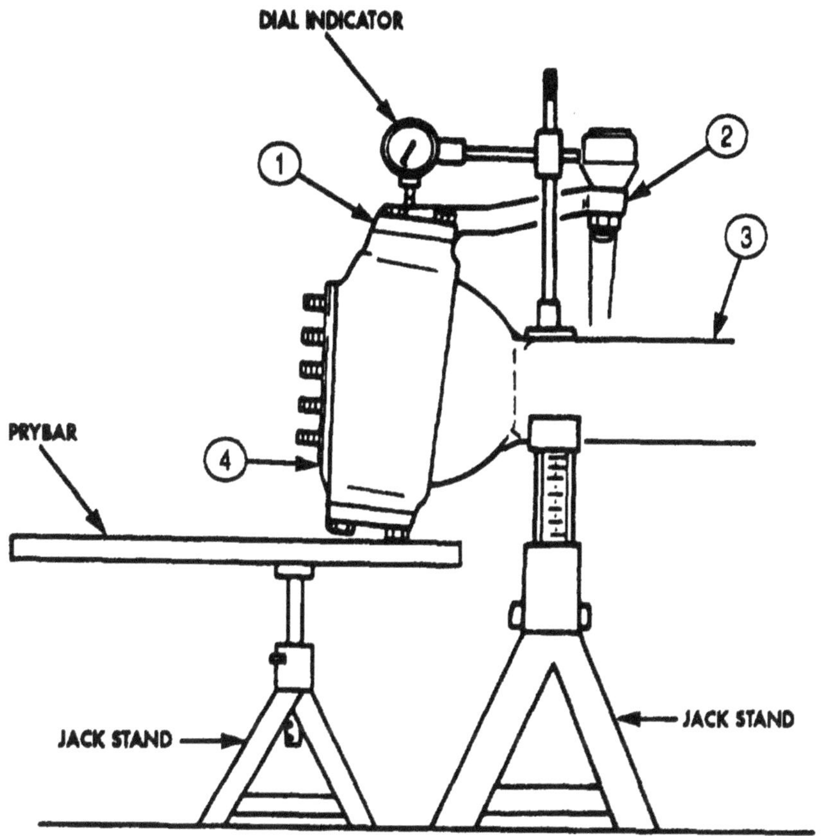

FOLLOW-ON TASKS:
- Install front axle shaft (para. 3-154).
- Lubricate steering knuckle (LO 9-2320-272-12).
- Adjust steering knuckle turn angle (para. 4-105).
- Start engine (TM 9-2320-272-10) and road test vehicle.

4-99. FRONT AXLE REPLACEMENT

THIS TASK COVERS:

a. Removal **b. Installation**

INITIAL SETUP:

APPLICABLE MODELS

All

TOOLS

General mechanic's tool kit (Appendix E, Item 1)
Torque wrench (Appendix E, Item 145)
Four jack stands
Hydraulic jack

MATERIALS/PARTS

Two locknuts (Appendix D, Item 286)
Cotter pin (Appendix D, Item 49)
Eight lockwashers (Appendix D, Item 362)
Eight locknuts (Appendix D, Item 291)
Cotter pin (Appendix D, Item 85)
Lubricating oil (Appendix C, Item 50)
Antiseize tape (Appendix C, Item 72)

REFERENCES (TM)

TM 9-2320-272-10
TM 9-2320-272-24P

EQUIPMENT CONDITION

- Parking brake set (TM 9-2320-272-10)
- Rear wheels chocked (TM 9-2320-272-10).
- Air reservoirs drained (TM 9-2320-272-10).
- Front wheels removed (para. 3-218 or 3-219).

GENERAL SAFETY INSTRUCTIONS

- Do not attempt to support weight of vehicle on hydraulic jack.
- Do not disconnect air lines before draining air reservoirs.

WARNING

Weight of vehicle must remain supported on jack stands at all times. Do not attempt to support weight of vehicle on hydraulic jack. Injury to personnel may result if jack fails.

a. Removal

1. Position hydraulic jack under front axle differential housing (3) and raise vehicle.
2. Position two jack stands under springs (2), ahead of left and right spring hangers (1).

NOTE

Do not fully lower hydraulic jack.

3. Lower hydraulic jack enough for springs (2) to rest on jack stands.

WARNING

Do not disconnect air lines before draining air reservoirs. Small parts under pressure may shoot out with high velocity, causing injury to personnel.

4. Disconnect primary line (4) and vent line (6) from left and right service brake chambers (8).
5. Remove adapter (5) and elbow (7) from left and right service brake chambers (8).

NOTE

- Perform step 6 for M939A2 series vehicles.
- Perform step 7 for left side of vehicle.
- Perform steps 8 through 11 for right side of vehicle.

6. Disconnect two air lines (13) from adapters (14) on relief safety valve (15).
7. Remove cotter pin (10), nut (9), and drag link (12) from steering knuckle arm (11). Discard cotter pin (10). Tie drag link (12) away from steering knuckle arm (11).
8. Remove cotter pin (16) from steering assist cylinder (17). Discard cotter pin (16).
9. Bend tabs of clip (18) away from felt pad (19) and steering knuckle ball (20).

4-99. FRONT AXLE REPLACEMENT (Contd)

10. Loosen adjustable plug (21) and tap steering assist cylinder (17) free from steering knuckle ball (20). Tie steering assist cylinder (17) away from steering knuckle ball (20).
11. Remove felt pad (19) and clip (18) from steering knuckle ball (20).

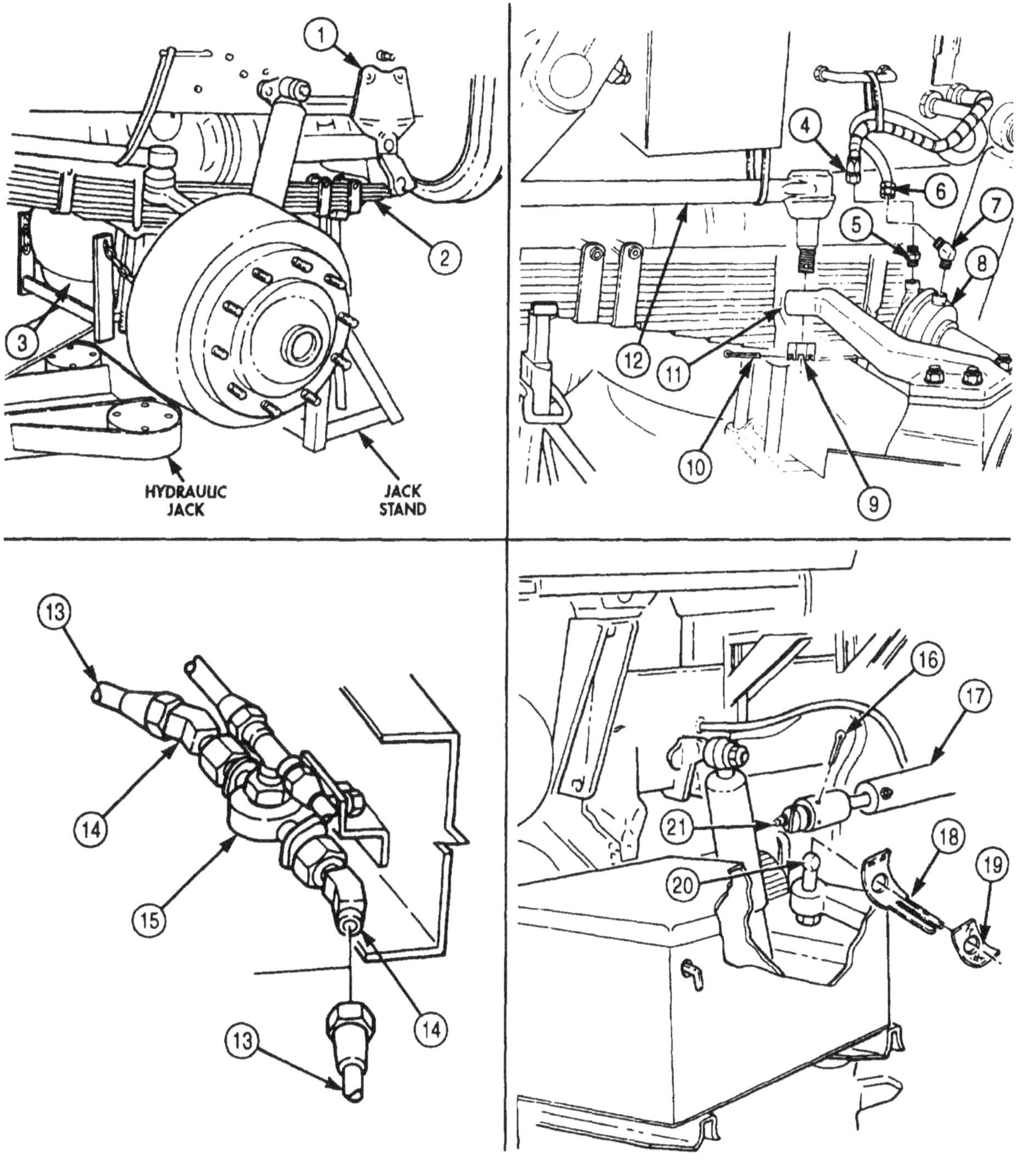

4-99. FRONT AXLE REPLACEMENT (Contd)

12. Remove locknut (12), washer (11), and shock absorber (2) from right and left lower spring seats (8). Discard locknuts (12).
13. Remove eight locknuts (3), screws (6), and propeller shaft flange (5) from differential flange (4). Discard locknuts (3). Tie propeller shaft (7) away from front axle (13).
14. Remove eight nuts (101, lockwashers (9), and right and left lower spring seats (8) from four spring U-bolts (1). Discard lockwashers (9).
15. Lower hydraulic jack until front axle (13) clears spring U-bolts (1).
16. Remove right and left upper spring seats (16) from front axle (13).

NOTE

Assistant will help with step 17.

17. Remove hydraulic jack and front axle (13) from under vehicle.

b. Installation

CAUTION

When positioning front axle, ensure spring U-bolts are properly aligned so spring center bolt heads enter alignment holes in front axle spring seats. Axle anchor pins must align to upper spring seat brackets or damage to equipment will result.

1. Place front axle (13) on hydraulic jack.
2. Position right and left upper spring seats (16) over anchor pins (15) on front axle (13).

NOTE

Assistant will help with step 3.

3. Raise hydraulic jack until spring U-bolts (1) are through upper spring seats (16) and springs (14) rest on upper spring seats (16).
4. Slide right and left lower spring seats (8) through four U-bolts (1) and install with eight new lockwashers (9) and nuts (10). Tighten nuts (10) 350-400 lb-ft (475-542 N•m).
5. Install propeller shaft (7) and propeller shaft flange (5) on differential flange (4) with eight screws (6) and new locknuts (3). Tighten locknuts (3) 32-40 lb-ft (43-54 N•m).
6. Install shock absorber (2) on right and left lower spring seats (8) with washer (11) and new locknut (12).

4-99. FRONT AXLE REPLACEMENT (Contd)

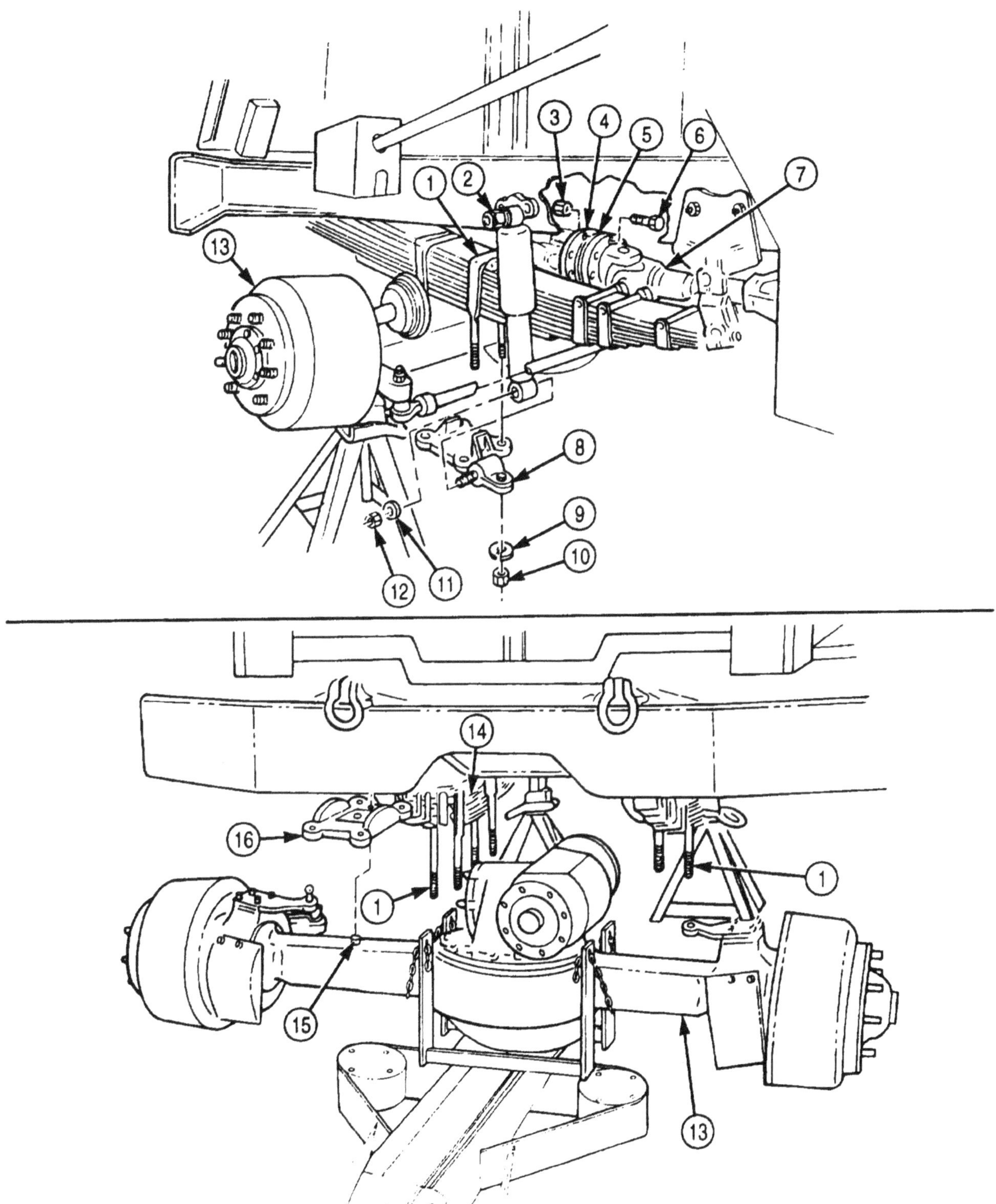

4-99. FRONT AXLE REPLACEMENT (Contd)

NOTE

- Perform steps 7 through 10 for right side of vehicle.
- Soak felt pad in lubricating oil before installation.

7. Install clip (3) and felt pad (4) on steering knuckle ball (5).
8. Install steering assist cylinder (2) on steering knuckle ball (5). Tighten adjustable plug (6), then back off until slots align with holes (7).
9. Install new cotter pin (1) on steering assist cylinder (2).
10. Bend tabs of clip (3) over felt pad (4) and steering knuckle ball (5).

NOTE

Perform steps 11 and 12 for left side of vehicle.

11. Install drag link (16) on steering knuckle arm (15) with nut (13). Tighten nut (13) 140-180 lb-ft (190-244 N•m).
12. Install new cotter pin (14) in nut (13) and drag link (16).

NOTE

- Wrap all male threads with antiseize tape before installation.
- Perform step 15 for M939A2 series vehicles.

15. Connect two air lines (17) to adapters (18) on relief safety valve (19).

13. Install elbow (11) and adapter (9) on left and right service brake chambers (12).
14. Connect two primary lines (8) and vent lines (10) to elbows (11) and adapters (9).

4-99. FRONT AXLE REPLACEMENT (Contd)

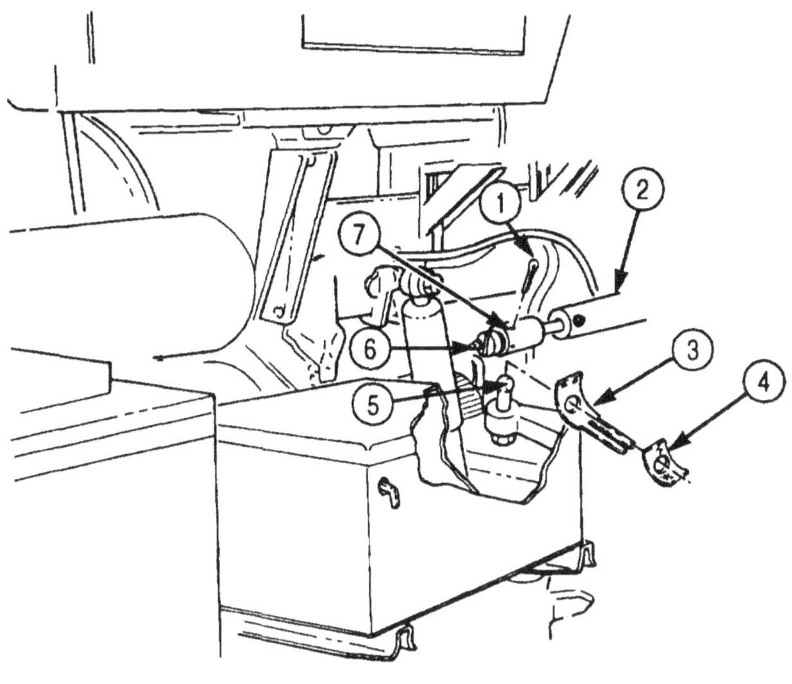

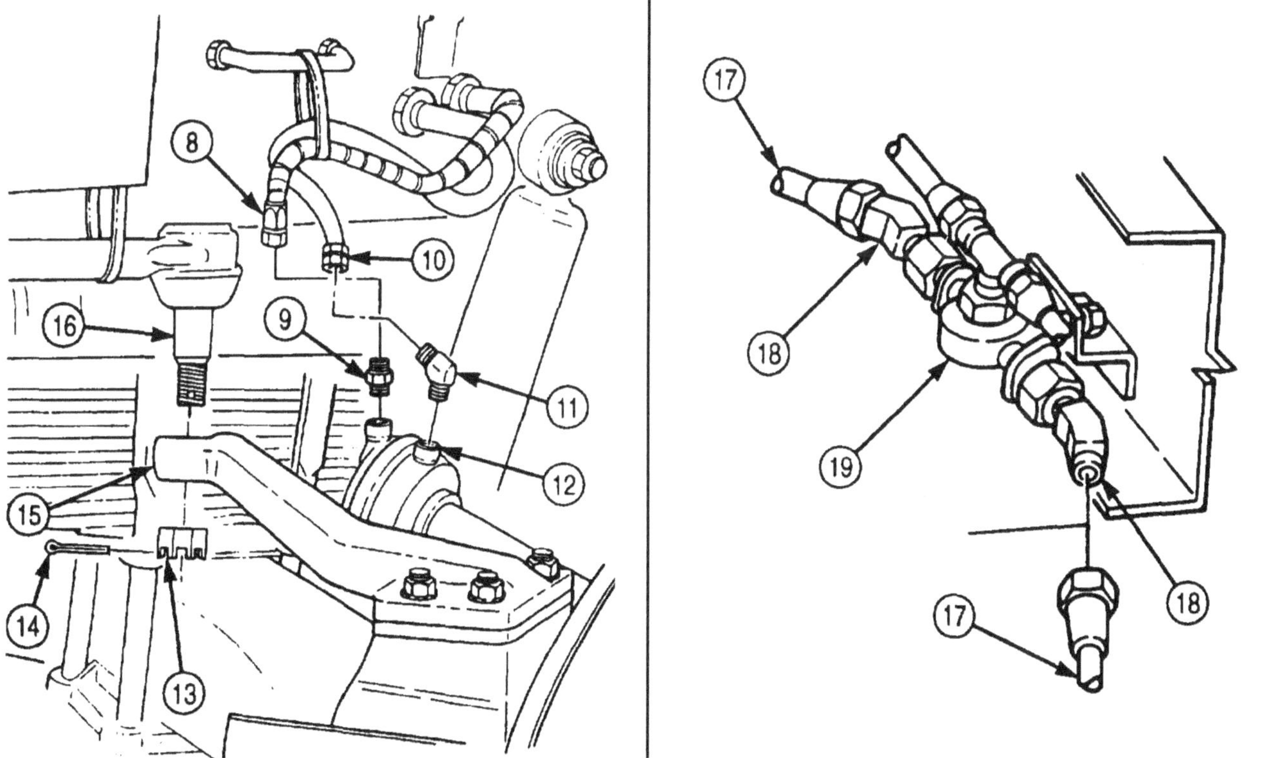

FOLLOW-ON TASKS:
- Install front wheels (para. 3-218 or 3-219).
- Lubricate front axle (LO 9-2320-272-12).
- Check and adjust steering assist cylinder travel (para. 3-233).
- Start engine (TM 9-2320-272-10), check for air leaks, and road test vehicle.

4-100. FRONT AND REAR AXLE CARRIER DIFFERENTIAL MAINTENANCE

THIS TASK COVERS:

a. Removal
b. Cleaning and Inspection
c. Installation

INITIAL SETUP:

APPLICABLE MODELS
All

TOOLS
General mechanic's tool kit (Appendix E, Item 1)
Torque wrench (Appendix E, Item 145)
Chain
Lifting device

MATERIALS/PARTS
Crocus cloth (Appendix C, Item 20)
Gasket sealing (Appendix C, Item 30)
Drycleaning solvent (Appendix C, Item 71)

PERSONNEL REQUIRED
TWO

REFERENCES
LO 9-2320-272-12
TM 9-2320-272-10
TM 9-2320-272-24P

EQUIPMENT CONDITION
- Front axle differential drained (LO 9-2320-272-12).
- Front axle removed (para. 4-99).
- Front axle shafts removed (para. 3-154).

GENERAL SAFETY INSTRUCTIONS
- Drycleaning solvent is flammable and toxic.
- Keep fire extinguisher nearby when using drycleaning solvent, Do not use near open flame.
- All personnel must stand clear during lifting operations.

NOTE

Front and rear axle carrier differentials are replaced the name way. This procedure covers replacement of front axle carrier differential,

a. Removal

1. Remove eighteen nuts (3) and washers (4) from carrier differential (2) and axle houring (5).
2. Attach chain and lifting device to carrier differential (2).

WARNING

All personnel must stand clear during lifting operations. A snapped cable, or swinging or shifting load, may result in injury to personnel.

CAUTION

When lifting carrier differential out of axle housing, use care not to damage mounting studs. Direct assistant to guide carrier differential out of axle housing.

3. Remove carrier differential (2) from axle housing (6).

b. Cleaning and Inspection

WARNING

- Drycleaning solvent is flammable and toxic. Do not use near an open flame and always have a fire extinguisher nearby when solvents are used. Use only in well-ventilated places, wear protective clothing, and dispose of cleaning rags in approved container. Failure to do this may result in injury to personnel and/or damage to equipment.
- Eyeshields must be worn when cleaning with a wire brush. Flying rust and metal particles may cause injury to personnel.

4-100. FRONT AND REAR AXLE CARRIER DIFFERENTIAL MAINTENANCE (Contd)

1. Using wire brush and drycleaning solvent, clean sealing compound from threads of mounting studs (6).
2. Inspect mating surfacer of differential flange (1) and axle housing (6) for nicks, burrs, and cracks, Remove any light burring with crocus cloth. Replace differential flange (1) or axle housing (6) if cracked, nicked, or heavily burred.
3. Inspect mounting studs (6) for bends or damaged threads. Replace mounting studs (6) if bent or threads are damaged.

c. Installation

NOTE

Differential flange and axle housing mating surfaces must be absolutely clean before applying sealing compound.

1. Apply generous but even amount of sealing compound around mating surfaces of differential flange (1), axle housing (5), and threads of mounting studs (6).
2. Attach chain and lifting device to carrier assembly (2),

WARNING

All personnel must stand clear during lifting operations. A snapped cable, or swinging or shifting load, may result in injury to personnel.

CAUTION

When installing carrier differential on axle housing, use care not to damage mounting studs. Direct assistant to guide carrier differential into axle housing.

3. Position carrier assembly (2) on axle housing (5) and install eighteen washers (4) and nuts (3). Tighten nuts (3) 160-205 lb-ft (217-278 N•m).
4. Remove lifting device and chain from carrier assembly (2).

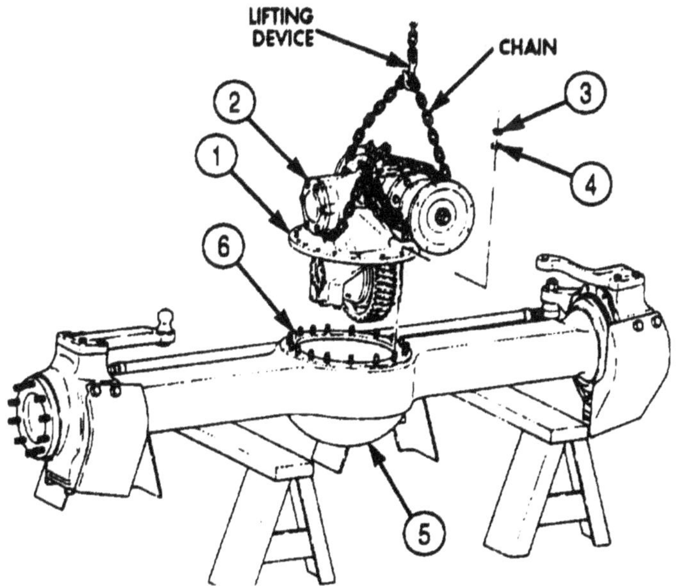

FOLLOW-ON TASKS: • Install front axle shafts (para. 3-154).
• Install front axle (para. 4-99).
• Fill differential carrier to proper oil level (LO 9-2320-272-12).
• Start engine (TM 9-2320-272-10) and road test vehicle.

4-101. FORWARD-REAR AND REAR-REAR AXLE REPLACEMENT

THIS TASK COVERS:

a. Removal b. Installation

INITIAL SETUP:

APPLICABLE MODELS

All

TOOLS

General mechanic's tool kit (Appendix E, Item 1)
Torque wrench (Appendix %%, Item 145)
Hydraulic jack
Jack stands

MATERIALS/PARTS

Eight locknuts (Appendix D, Item 291)
Two lockwashers (Appendix D, Item 354)
Six lockwashers (Appendix D, Item 362)
Lubricating oil (Appendix C, Item 48)
Antiseize tape (Appendix C. Item 72)

REFERENCES (TM)

LO 9-2320-272-12
TM 9-2320-272-10
TM 9-2320-272-24P

EQUIPMENT CONDITION

- Air reservoirs drained (TM 9-2320-272-10).
- Remove wheels (para. 3-218 or para. 3-219).

GENERAL SAFETY INSTRUCTIONS

- Do not attempt to support weight of vehicle on hydraulic jack.
- Do not disconnect air lines before draining air reservoirs.

WARNING

Weight of vehicle must remain supported on jack stands at all times. Do not attempt to support weight of vehicle on hydraulic jack. Injury to personnel may result if hydraulic jack fails.

NOTE

Forward-rear and rear-rear axles are replaced the same way. This procedure covers replacement of the rear-rear axle.

a. Removal

1. Position hydraulic jack under differential housing (7) and raise vehicle.
2. Position jack stands under spring seats (5) and rear axle (6), and lower vehicle until spring seats (5) and rear axle (6) rest on jack stands.
3. Remove eight locknuts (1), screws (4), and propeller shaft (3) from companion flange (2). Discard locknuts (1).

WARNING

Do not disconnect air lines before draining air reservoirs. Small parts under pressure may shoot out with high velocity, causing injury to personnel.

NOTE

- Perform step 4 for M939A2 series vehicles only.
- Tag all air lines for installation.

4. Disconnect two air lines (13) from elbows (14) on relief valve (15).
5. Disconnect primary line (9) and vent line (8) and remove elbows (10) and (12) from right and left service brake chambers (11).

4-101. FORWARD-REAR AND REAR-REAR AXLE REPLACEMENT (Contd)

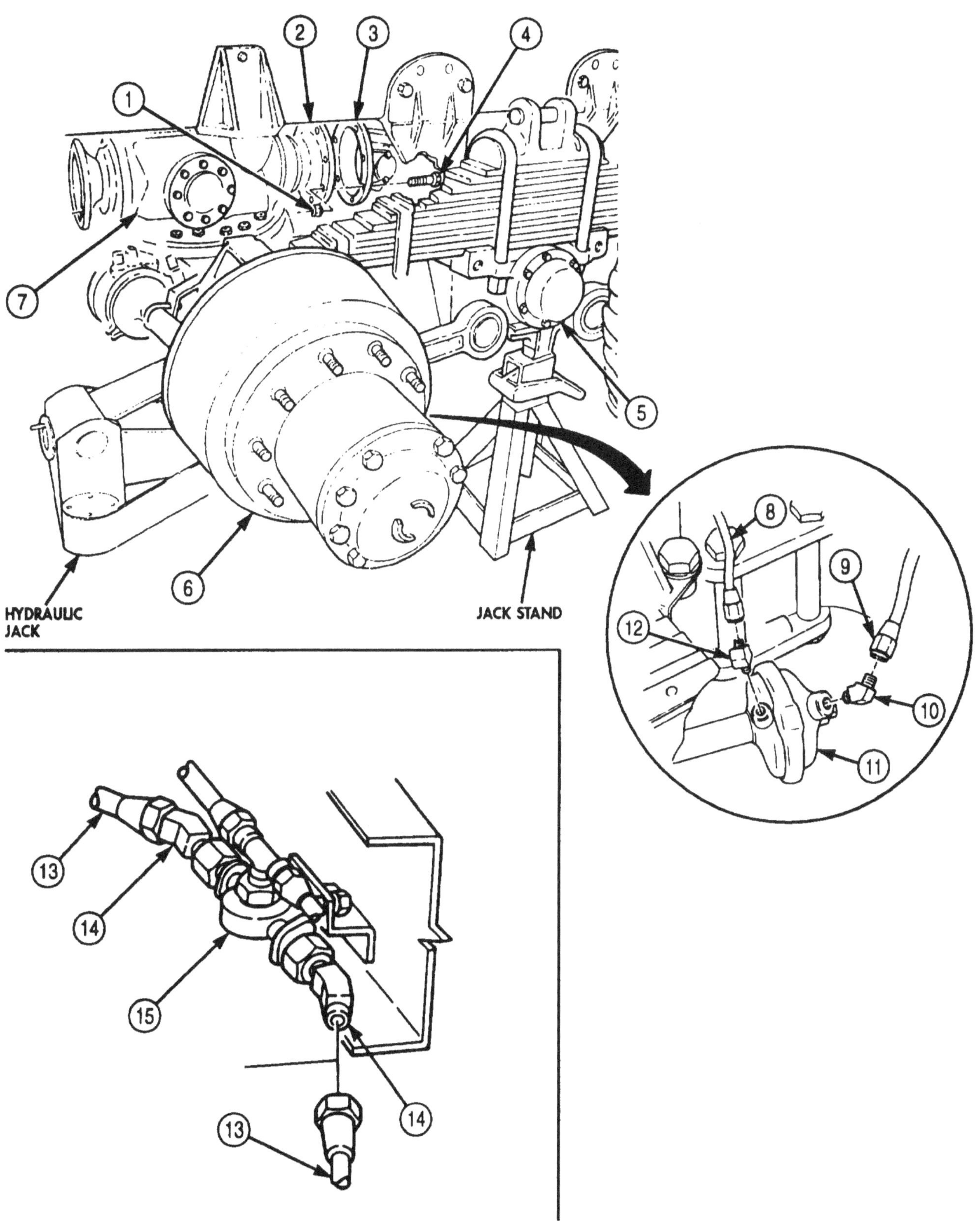

4-101. FORWARD-REAR AND REAR-REAR AXLE REPLACEMENT (Contd)

6. Disconnect vent lines (1) and (2), secondary line (11), and spring brake override line (10) and remove two adapter elbows (13), elbow (14), and tee fitting (15) from left and right spring brakes (12).
7. Remove primary line (6), secondary line (8), and two adapter fittings (5) from mounting plate (9).
8. Remove two screws (3), lockwashers (4), and mounting plate (9) from differential housing (7). Discard lockwashers (4).
9. Remove two nuts (16), washers (17), U-bolt (21), and U-bolt bracket (22) from right and left spring brakes (12).

NOTE

Assistant will help with step 10.

10. Remove nut (23), lockwasher (24), washer (25), screw (18), washer (19), and brake chamber bracket (26) from right and left upper spring brackets (20). Discard lockwashers (24).
11. Remove two nuts (36), washers (35), U-bolt (31), and U-bolt bracket (32) from right and left service brake chambers (30).
12. Remove nut (33), lockwasher (34), washer (28), screw (27), washer (28), and brake chamber bracket (29) from right and left upper spring brackets (30). Discard lockwashers (34).

NOTE

Assistant will help with steps 13 through 15.

13. Remove two nuts (41), lockwashers (40), and screws (39) from two upper torque rod brackets (37) and upper torque rod plates (42). Upper torque rod brackets (37) will remain attached to torque rods (38). Discard lockwashers (40).
14. Remove two nuts (46), lockwashers (45), screws (43), and right and left upper spring brackets (20) from left and right lead springs (47) and dowel pin (48) on rear axle (44). Discard lockwashers (45).
15. Remove spring seat wearpads (49) from right and left upper spring brackets (20).

NOTE

Assistant will help with steps 16 and 17.

16. Raise hydraulic jack and remove jack stands from rear axle (44).
17. Remove rear axle (44) from vehicle.

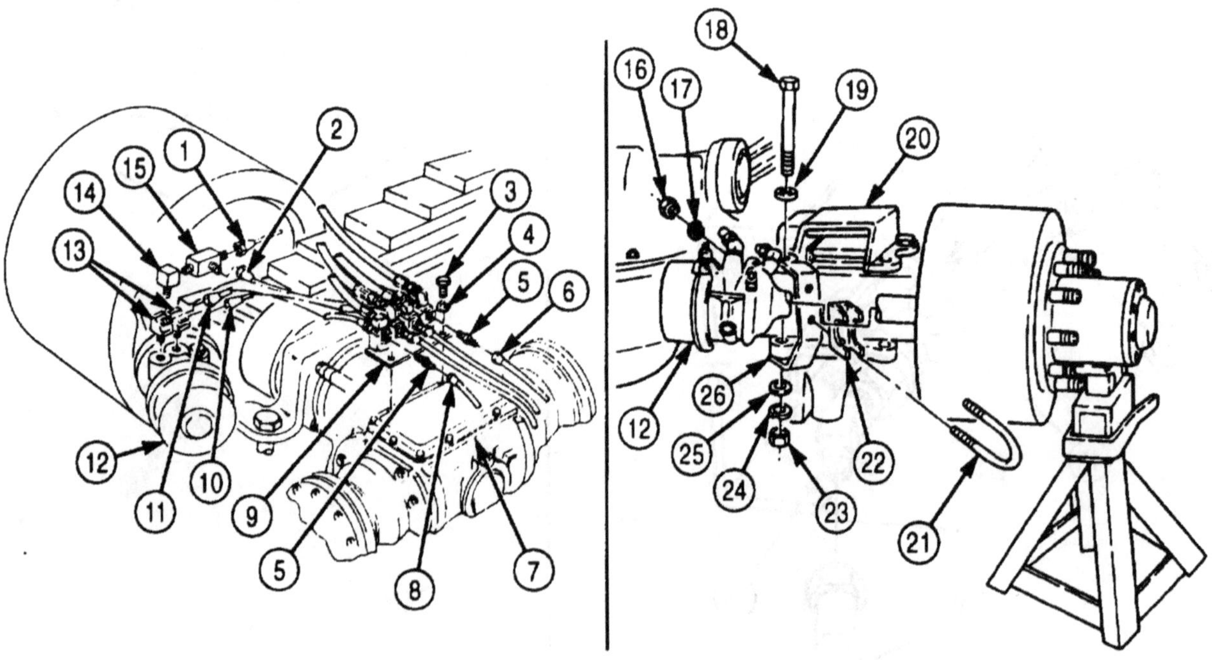

4-101. FORWARD-REAR AND REAR-REAR AXLE REPLACEMENT (Contd)

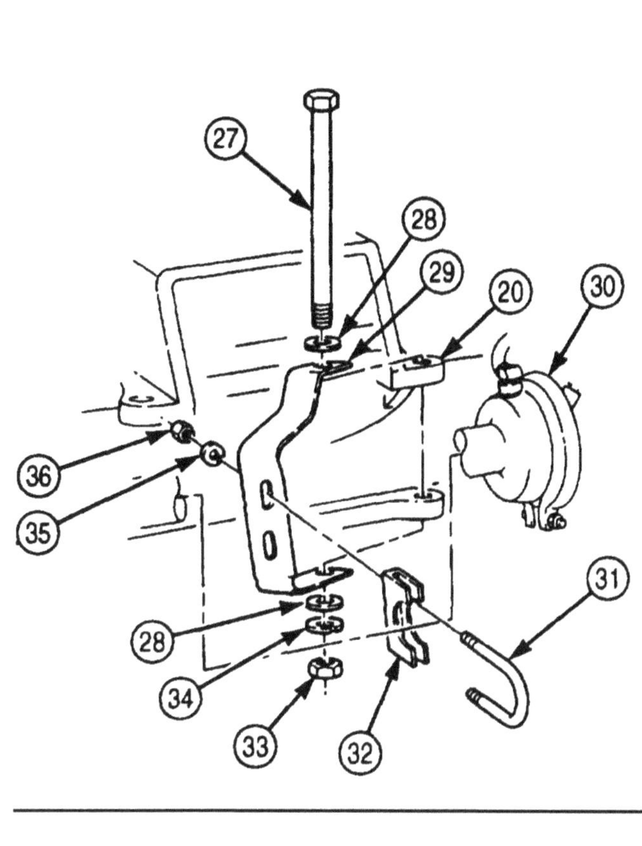

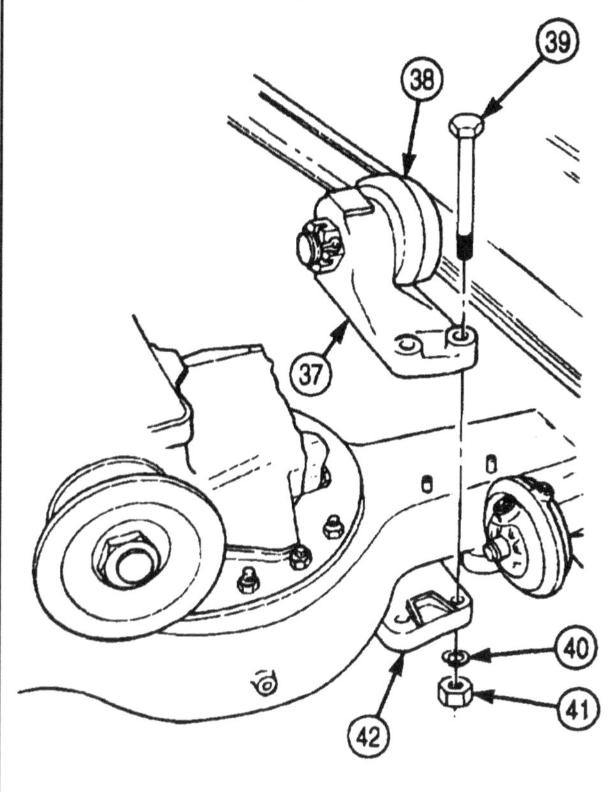

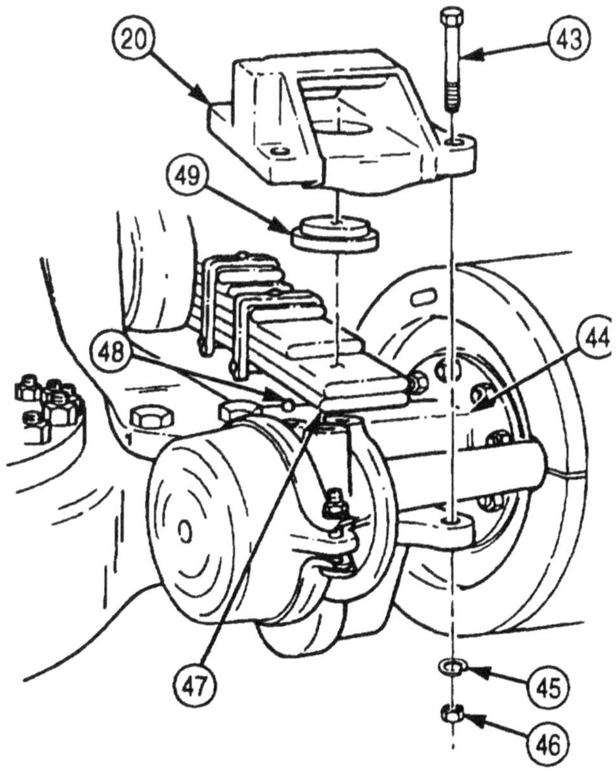

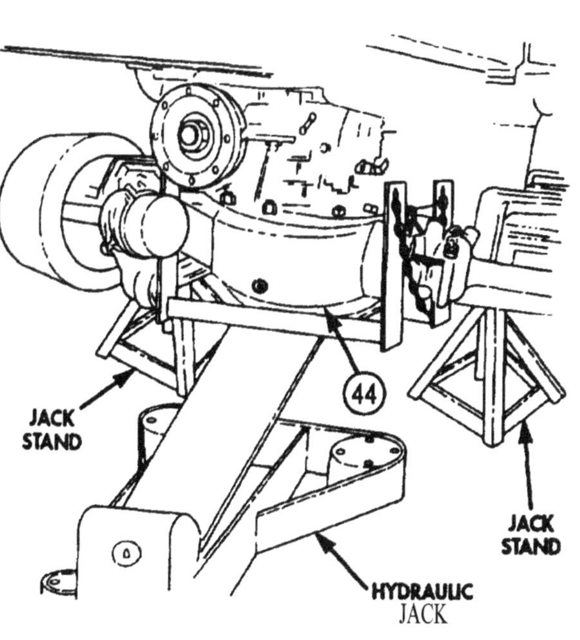

4-101. FORWARD-REAR AND REAR-REAR AXLE REPLACEMENT (Contd)

b. Installation

WARNING

Weight of vehicle must remain supported on jack stands at all times. Do not attempt to support weight of vehicle on hydraulic jack. Injury to personnel may result if hydraulic jack fails.

NOTE

Assistant will help with steps 1 through 3.

1. Position rear axle (1) on hydraulic jack.
2. Raise rear axle (1) and remove jack stands from under rear axle (1).
3. Using hydraulic jack, position rear axle (1) under springs (6).
4. Install spring seat wearpad (8) on left and right upper spring brackets (2).
5. Install upper spring bracket (2) on left and right leaf springs (6), position over rear axle (1) and dowel pin (7), and install two screws (3), new lockwashers (4), and nuts (5). Finger-tighten nuts (5).
6. Install brake chamber bracket (21) on left and right upper spring brackets (2) with washer (20), screw (19), washer (20), new lockwasher (26), and nut (25). Finger-tighten nuts (25).
7. Position two upper torque rod brackets (29) over dowel pins (34) on rear axle (l), and install on upper torque rod plates (31) with four screws (30), new lockwashers (32), and nuts (33).
8. Install U-bolt bracket (24) and U-bolt (23) on left and right service brake chamber brackets (21) with two washers (27) and nuts (28).

NOTE

Assistant will help with step 9.

9. Install brake chamber bracket (17) on left and right upper spring brackets (2) with washer (12), screw (11), washer (12), new lockwasher (16), and nut (15). Tighten nuts (5), (25), and (15) 350-375 lb-ft (475-509 N•m).
10. Install U-bolt bracket (14) and U-bolt (13) on left and right spring brakes (18) with two washers (10) and nuts (9).

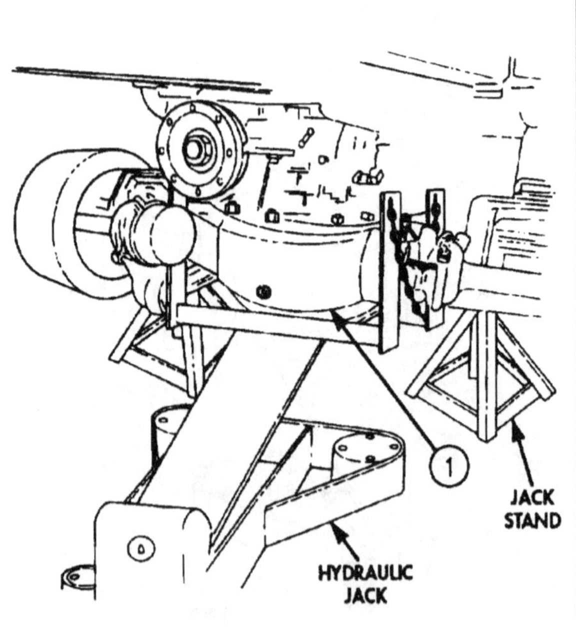

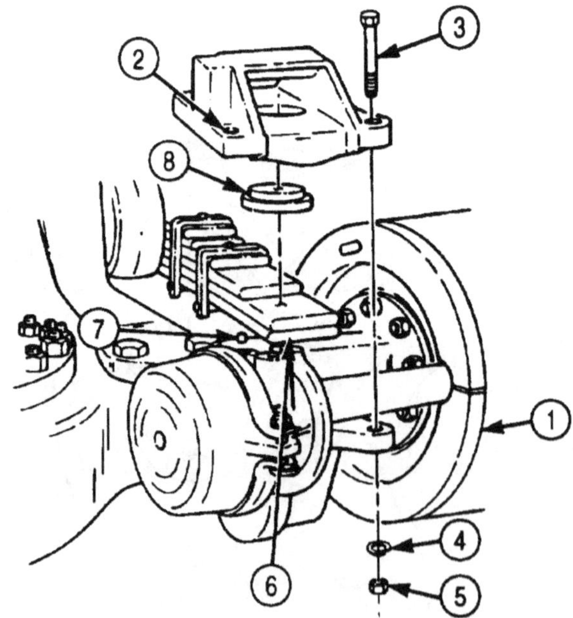

4-101. FORWARD-REAR AND REAR-REAR AXLE REPLACEMENT (Contd)

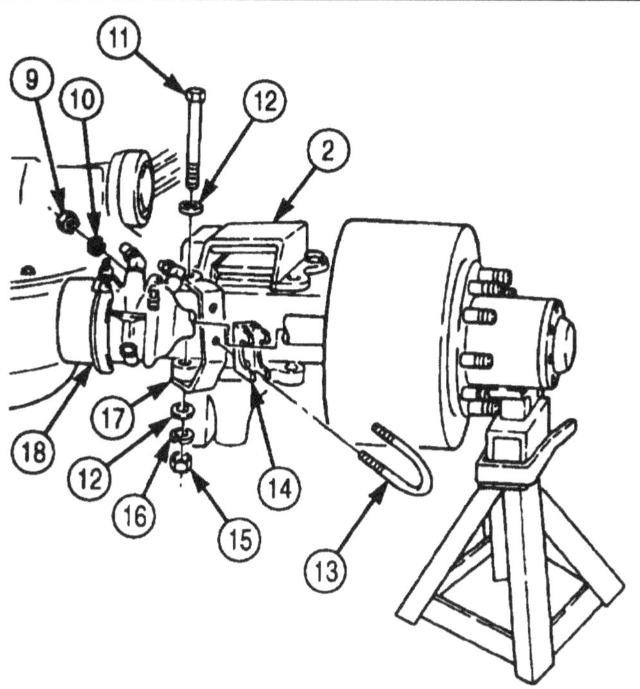

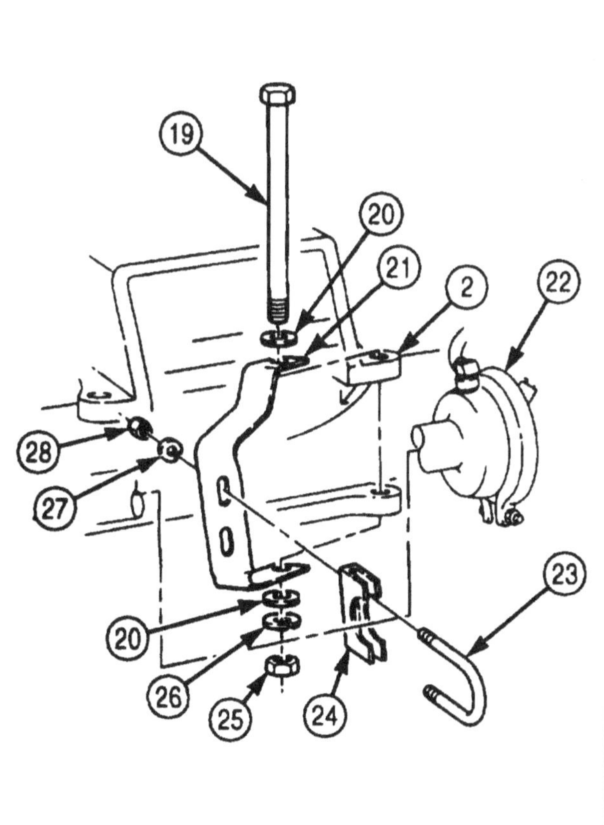

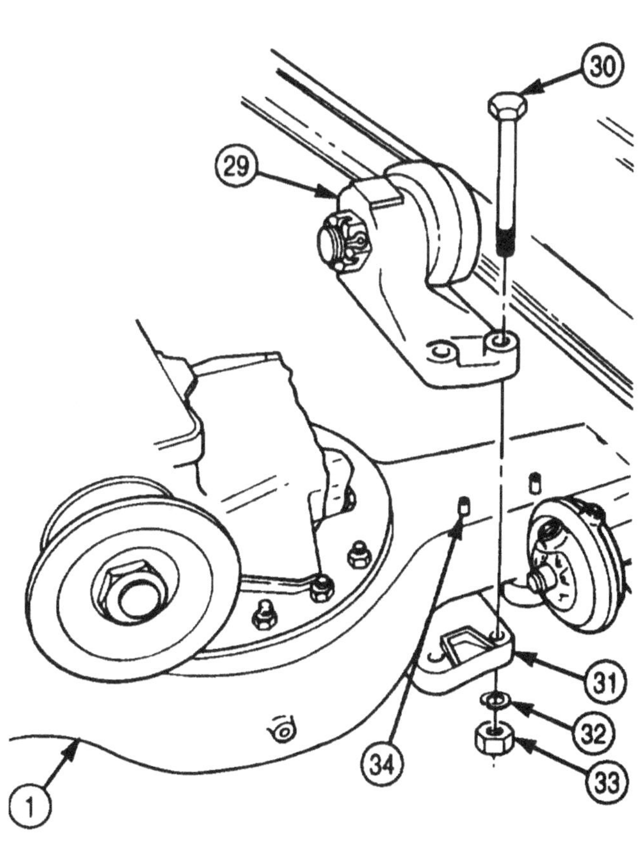

4-101. FORWARD-REAR AND REAR-REAR AXLE REPLACEMENT (Contd)

11. Install mounting plate (9) on differential housing (7) with two new lockwashers (4) and screws (3).

NOTE

Wrap all male threads with antiseize before installation.

12. Install two adapter fittings (5), secondary line (8), and primary line (6) on mounting plate (9).
13. Install tee fitting (16), elbow (14), two adapter elbows (13), spring brake override line (10), secondary line (11), and vent liner (1) and (2) on left and right spring brakes (12).

NOTE

Perform step 14 for M939A2 series vehicles,

14. Install elbows (27) and (26), vent line (24), and primary line (25) on right and left service brake chambers (12),
15. Connect two air liner (16) to elbows (17) on relief valve (18),
16. Install propeller shaft (20) on companion flange (10) with eight screws (21) and new locknuts (22). Tighten locknuts (22) 32-40 lb-ft (43-54 N•m).
17. Install wheels (para. 3-218 or para. 3-219).
18. Remove hydraulic jack and jack stands from rear axle (23).

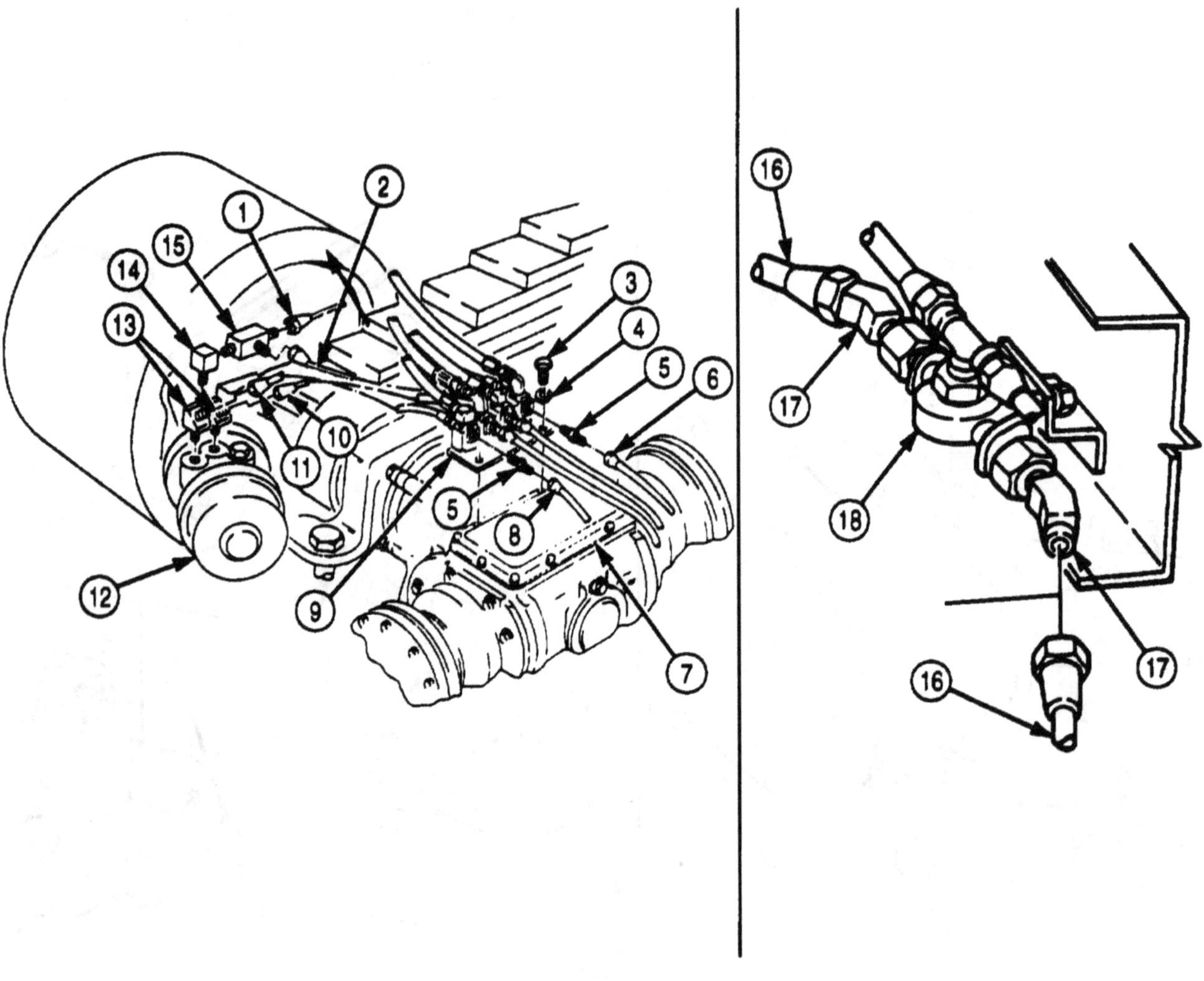

4-101. FORWARD-REAR AND REAR-REAR AXLE REPLACEMENT (Contd)

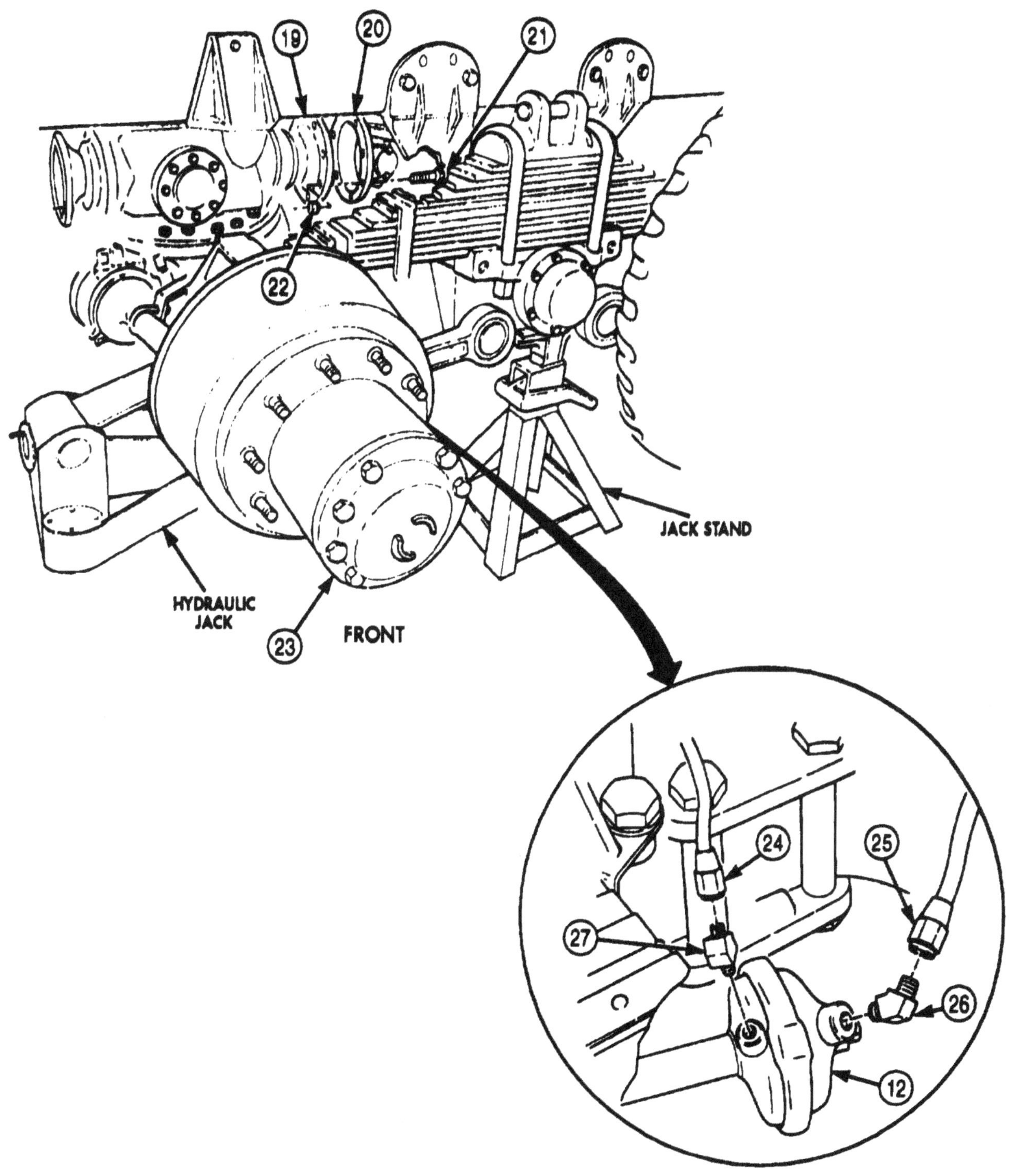

FOLLOW-ON TASKS: • Lubricate rear axle (LO 9-2320-272-12).
• Start engine (TM 9-2320-272-10) and allow air system to build to normal operating pressure. Check air lines for leaks. Road test vehicle.

4-102. FRONT DIFFERENTIAL 0IL SEAL REPLACEMENT

THIS TASK COVERS:

a. Removal

b. Installation

INITIAL SETUP:

APPLICABLE MODELS

All

TOOLS

General mechanic's tool kit (Appendix E, Item 1)
Torque wrench (Appendix E, Item 145)
Mechanical puller (Appendix E, Item 102)

MATERIALS/PARTS

Gasket (Appendix D, Item 206)
Oil seal (Appendix D, Item 500)
Cotter pin (Appendix D, Item 61)
Gasket (Appendix D, Item 207)
GAA grease (Appendix C, Item 28)
Sealing compound (Appendix C, Item 30)

REFERENCES (TM)

TM 9-2320-272-10
TM 9-2320-272-24P

EQUIPMENT CONDITION

- Parking brake set (TM 9-2320-272-10).
- Rear wheels chocked (TM 9-2320-272-10).
- Propeller shaft removed (para. 3-149).

a. Removal

1. Remove cotter pin (1) and nut (11) from driveshaft (8). Discard cotter pin (1).
2. Using mechanical puller, remove companion flange (2) from driveshaft (8).

CAUTION

Do not jam any tool between differential and pinion shaft retainer when removing front bearing cover. Shims will be damaged.

3. Remove eight screws (10), washers (9), front bearing cover (3), and gasket (6) from differential carrier (7). Discard gasket (6).
4. Remove oil seal (5) and gasket (4) from front bearing cover (3). Discard oil seal (5) and gasket (4).

b. Installation

1. Position new gasket (4) in front bearing cover (3).
2. Apply GAA grease to inside diameter of new oil seal (5). Apply sealing compound to outside diameter of new oil seal (5).
3. Install new oil seal (5) in front bearing cover (3).
4. Apply sealing compound to mating surfaces of new gasket (6).
5. Install new gasket (6) and front bearing cover (3) on differential carrier (7) with eight washers (9) and screws (10). Tighten screws (10) 93-120 lb-ft (126-163 N•m).
6. Install companion flange (2) on driveshaft (8) with nut (11). Tighten nut (11) 300-400 lb-ft (407-542 N•m).
7. Install new cotter pin (1) through nut (11) and driveshaft (8).

4-102. FRONT DIFFERENTIAL OIL SEAL REPLACEMENT (Contd)

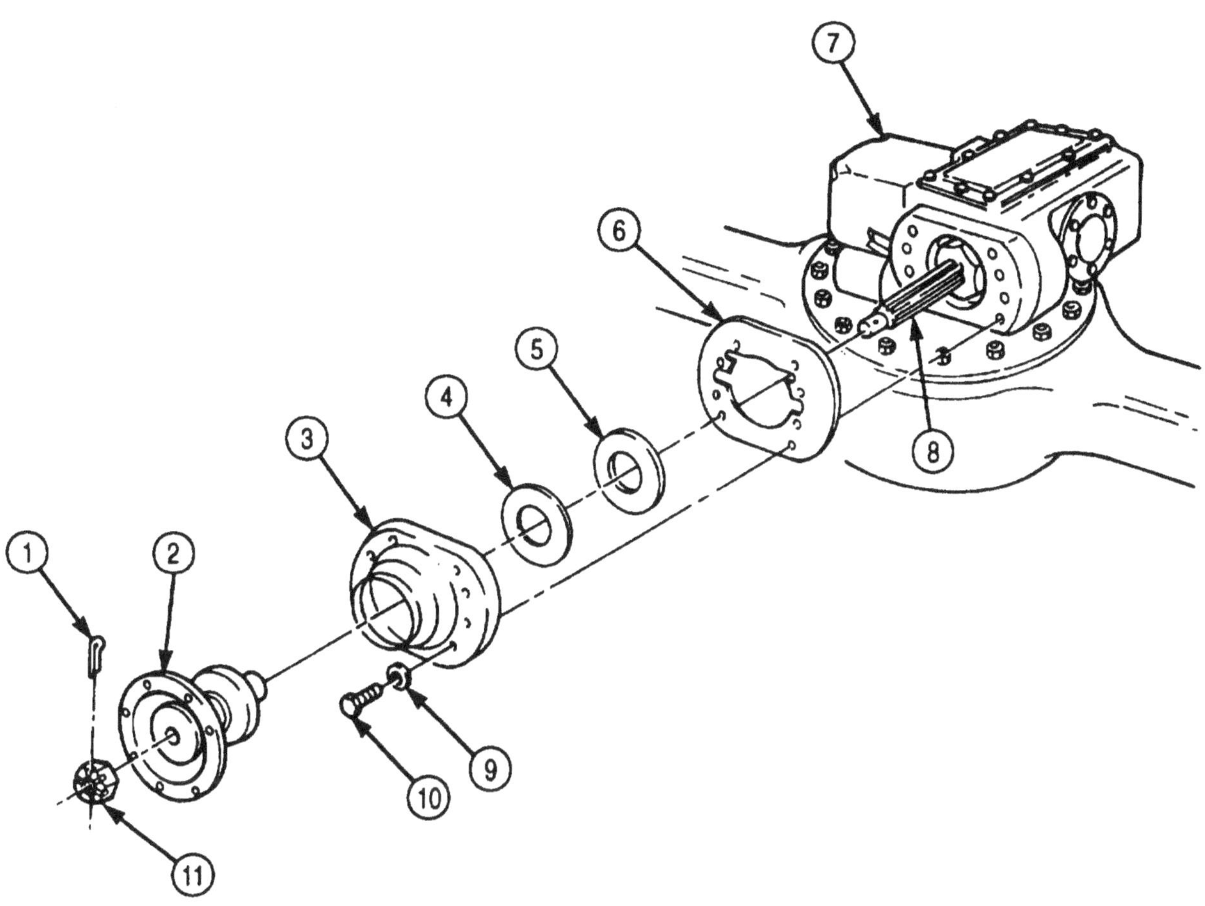

FOLLOW-ON TASKS:
- Install propeller shaft (para. 3-149).
- Start engine (TM 9-2320-272-10) and road test vehicle.
- Check front differential oil seal for leaks.

4-103. REAR DIFFERENTIAL OIL SEAL REPLACEMENT

THIS TASK COVERS:

a. Removal

b. Installation

INITIAL SETUP:

APPLICABLE MODELS

All

TOOLS

General mechanic's tool kit (Appendix E, Item 1)
Torque wrench (Appendix E, Item 145)
Mechanical puller (Appendix E, Item 102)

MATERIALS/PARTS

Oil seal (Appendix D, Item 500)
Cotter pin (Appendix D, Item 61)
Gasket (Appendix D, Item 208)
GAA grease (Appendix C, Item 28)
Sealing compound (Appendix C, Item 30)

REFERENCES (TM)

TM 9-2320-272-10
TM 9-2320-272-24P

EQUIPMENT CONDITION

- Parking brake set (TM 9-2320-272-10).
- Rear wheels chocked (TM 9-2320-272-10).
- Propeller shaft removed (para. 3-150).

a. Removal

1. Remove cotter pin (7) and nut (8) from driveshaft (2). Discard cotter pin (7).
2. Using mechanical puller, remove companion flange (6) from driveshaft (2).

CAUTION

Do not jam any tool between differential and pinion shaft retainer when removing front bearing cover. Shims will be damaged.

3. Remove six screws (9), washers (10), cap (4), and gasket (3) from differential carrier (1). Discard gasket (3).
4. Remove oil seal (5) from cap (4). Discard oil seal (5).

b. Installation

1. Apply GM grease to inside diameter of new oil seal (5). Apply sealing compound to outside diameter of oil seal (5).
2. Install oil seal (5) in cap (4).
3. Apply sealing compound to mating surfaces of new gasket (3).
4. Install gasket (3) and cap (4) on differential carrier (1) with six washers (10) and screws (9). Tighten screws (9) 24-40 lb-ft (33-54 N•m).
5. Install companion flange (6) on driveshaft (2) with nut (8). Tighten nut (8) 300-400 lb-ft (407-542 N•m).
6. Install new cotter pin (7) through nut (8) and driveshaft (2).

4-103. REAR DIFFERENTIAL OIL SEAL REPLACEMENT (Contd)

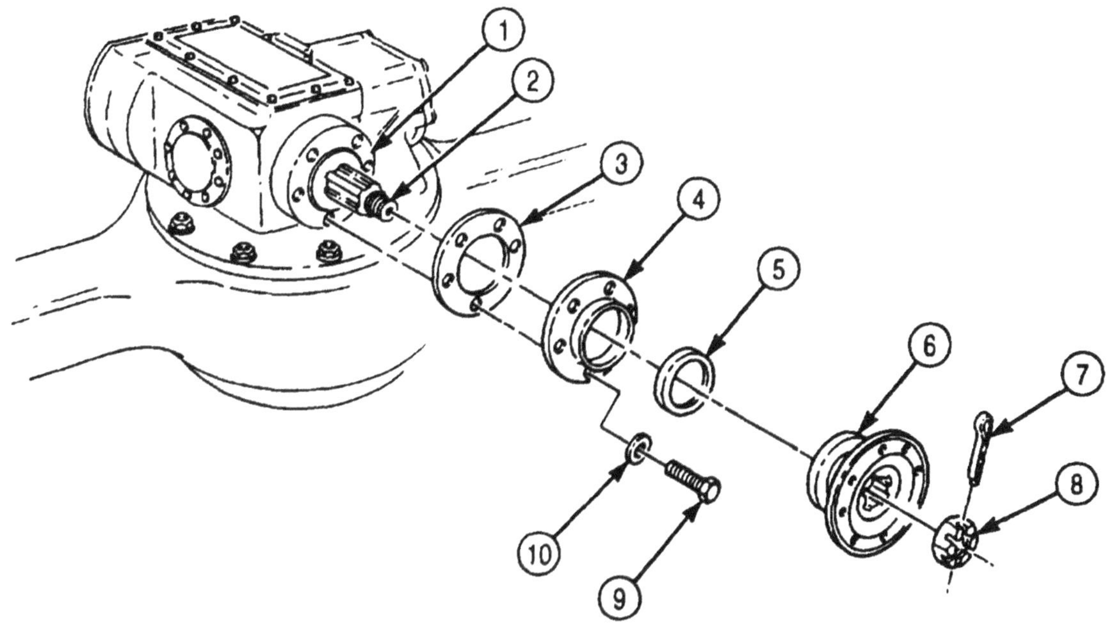

FOLLOW-ON TASKS:
- Install propeller shaft (para. 3-150).
- Start engine (TM 9-2320-272-10) and road test vehicle.
- Check front differential oil seal for leaks.

4-104. CROSS TUBE REPLACEMENT

THIS TASK COVERS:

a. Removal
b. Installation

INITIAL SETUP:

APPLICABLE MODELS
All

TOOLS
General mechanic's tool kit (Appendix E, Item 1)
Torque multiplier
Hydraulic jack
Jack stands

MATERIALS/PARTS
Eight locknuts (Appendix D, Item 331)

PERSONNEL REQUIRED
TWO

REFERENCES (TM)
TM 9-2320-272-24P

EQUIPMENT CONDITION
- Torque rods (lower) removed (para. 3-170).
- Rear springs removed (para. 3-167).
- Spring seat removed (para. 3-169).

GENERAL SAFETY INSTRUCTIONS
Weight of vehicle must be supported on jack stands at all times.

WARNING

Weight of vehicle must remain supported on jack stands at all times. Do not attempt to support weight of vehicle on hydraulic jack. Injury to personnel may result if hydraulic jack fails.

a. Removal

Remove eight locknuts (4), screws (2), and cross tube (3) from two rear axle bogie supports (1). Discard locknuts (4).

b. Installation

Install cross tube (3) on two rear axle bogie supports (1) with eight screws (2) and new locknuts (4). Tighten locknuts (4) 1,200-1,300 lb-ft (1,627-1,763 N•m).

4-104. CROSS TUBE REPLACEMENT (Contd)

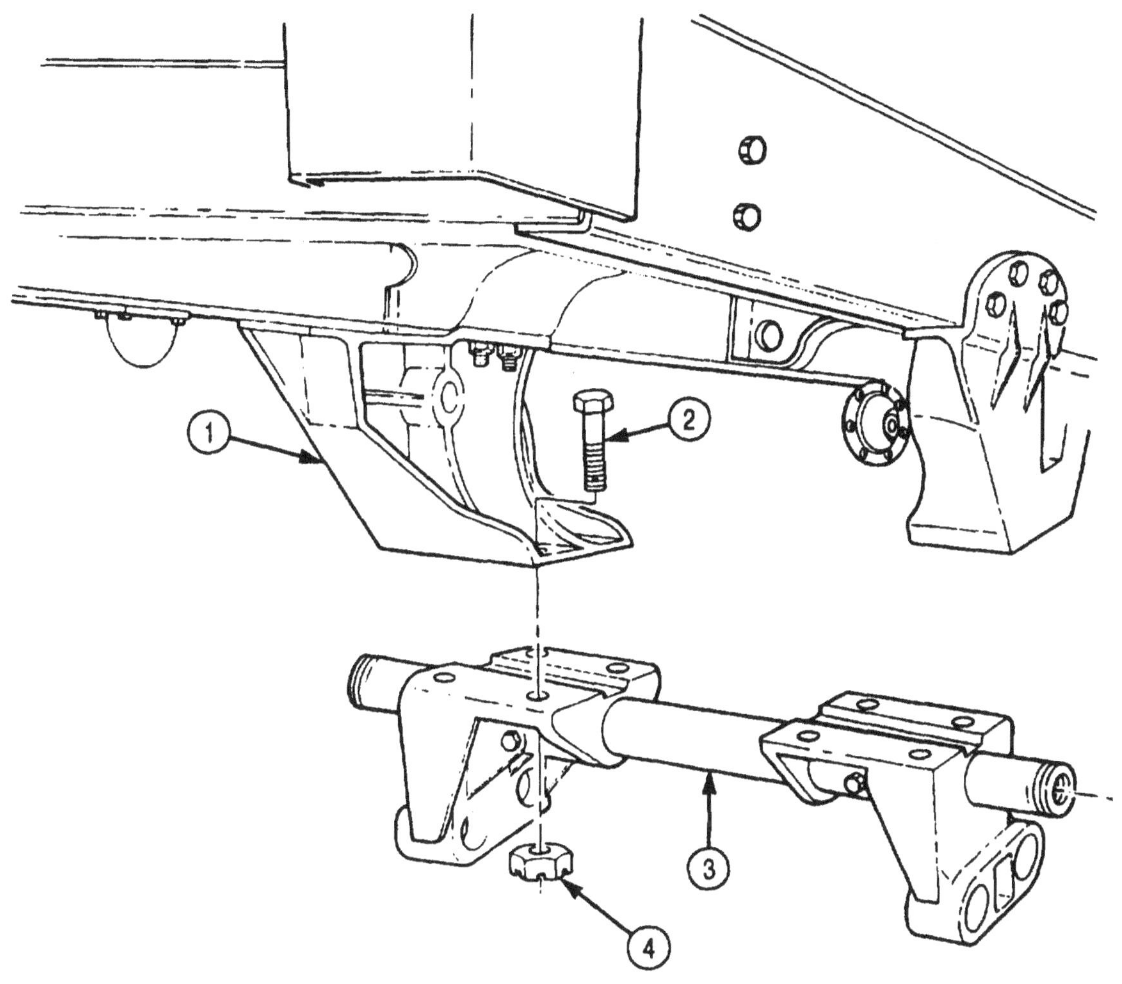

FOLLOW-ON TASKS:
- Install spring seat (para. 3-169).
- Install rear springs (para. 3-167).
- Install torque rods (lower) (para. 3-170).

4-105. STEERING STOP ADJUSTMENT

THIS TASK COVERS:

a. Check

b. Turning Angle Adjustment

INITIAL SETUP:

APPLICABLE MODELS

All

TOOLS

General mechanic's tool kit (Appendix E, Item 1)
Torque wrench (Appendix E, Item 144)
Wheel adjusting tools

MATERIALS/PARTS

Two adjustments screws (Appendix D, Item 571)
Two jamnuts (Appendix D, Item 264)

REFERENCES (TM)

TM 9-2820-272-10
TM 9-2820-272-24P

EQUIPMENT CONDITION

- Parking brake ret (TM 9-2920-272-10).
- Tires properly inflated (TM 9-2820-272-10).

1. Position front wheels straight ahead.
2. Place alignment device 3 in. (7.6 mm) from outer tire surface of left front wheel (1).
3. Place turning radius gauge against outer tire surface.
4. Center pointer with wheel (1) and align to 0 degree mark on bare scale.
5. Turn left front wheel (1) to outward stop and read degrees of travel on bare scale. Scale reading should be 28 degrees. If scale reading is not correct, perform turning angle adjustment (task b.).

b. Turning Angle Adjustment

1. Break welds and remove screw (4) and jamnut (6) from left steering knuckle (2). Discard screw (4) and jamnut (6).
2. Thread new nut (5) all the way on new screw (4).
3. Thread screw (4) all the way into tie rod arm (6), then all the way out, until screw (4) contacts axle housing (3). Screw (4) now becomes turn angle stop.
4. While holding screw (4), turn jamnut (5) until it contacts tie rod arm (6). Tighten nut (5) 66-86 lb-ft (88-116 N•m).
5. Tack-weld nut (6) to screw (4) and back of tie rod arm (6).

NOTE

Repeat task a. to check right front wheel turning angle adjustment.

4-105. STEERING STOP ADJUSTMENT (Contd)

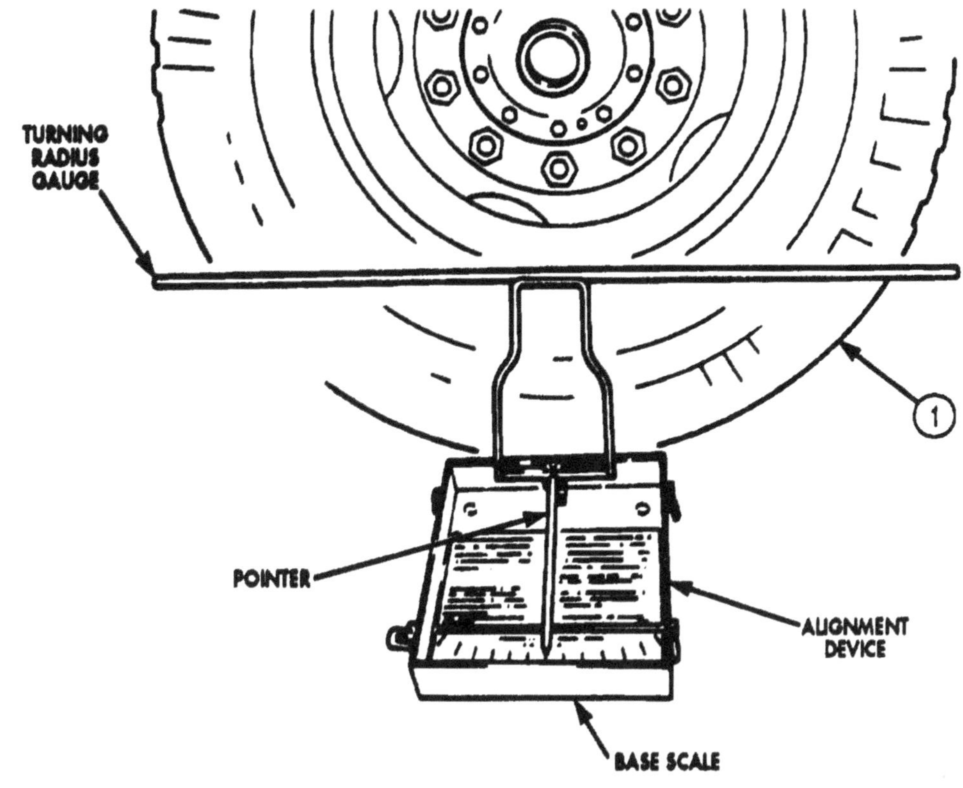

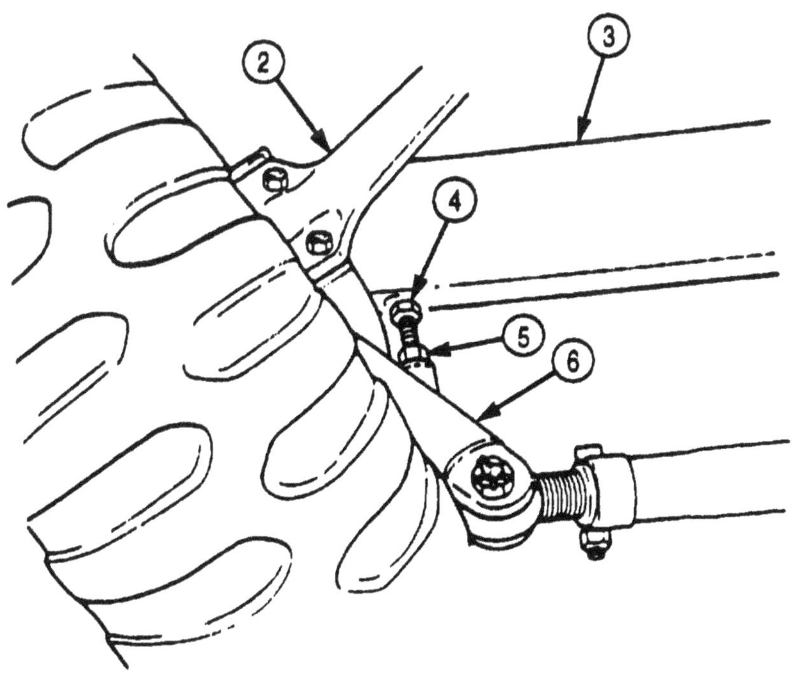

FOLLOW-ON TASK: Start engine (TM 9-2320-272-10) and road test vehicle.

4-106. AXLE LEAKAGE TEST

THIS TASK COVERS:

a. Test

b. Corrective Action

INITIAL SETUP:

TOOLS

General mechanic's tool kit (Appendix E, Item 1)
Air pressure gauge

REFERENCES (TM)

TM 9-2320-272-10
TM 9-2320-272-24P

EQUIPMENT CONDITION

- Parking brake set (TM 9-2320-272-10)
- Rear wheels chocked (TM 9-2320-272-10).

a. Test

1. Remove breather (1) from axle housing (4).
2. Install nipple (9) and tee (6) on axle housing (4).
3. Install air pressure gauge and quick-disconnect coupling (7) on tee (6).

CAUTION

Do not allow air pressure to exceed 15 psi (103 kPa). Seals will be damaged if air pressure exceeds 15 psi (103 kPa).

4. Connect air supply hose (8) to quick-disconnect coupling (7).
5. Turn on air supply and observe air gauge. When air pressure reaches 8 psi (55 kPa), disconnect air supply hose (8). Begin timing for a period of 45 seconds. After 45 seconds, check air gauge reading.
6. If air gauge reading indicates leakage of 5 psi (35 kPa) or faster, perform corrective action (task b.).

b. Corrective Action

1. Tighten all axle housing (4) and differential carrier (5) nuts and screws.
2. Reconnect air supply hose (8) to quick-disconnect coupling (7) and turn on air supply. When air pressure reaches 8 psi (55 kPa), disconnect air supply hose (8).
3. Apply soapsuds around front oil seal cover (2), rear oil seal cap (3), and base of differential carrier (5) and axle housing (4). Air bubbles indicate defective or missing sealing compound. Remove front oil seal cover (2), rear oil seal cap (3), or differential carrier (5) and reseal (para. 4-102, 4-103, or 3-158).
4. Remove quick-disconnect coupling (7), air pressure gauge, tee (6), and nipple (9) from axle housing (4).
5. Install breather (1) on axle housing (4).

4-106. AXLE LEAKAGE TEST (Contd)

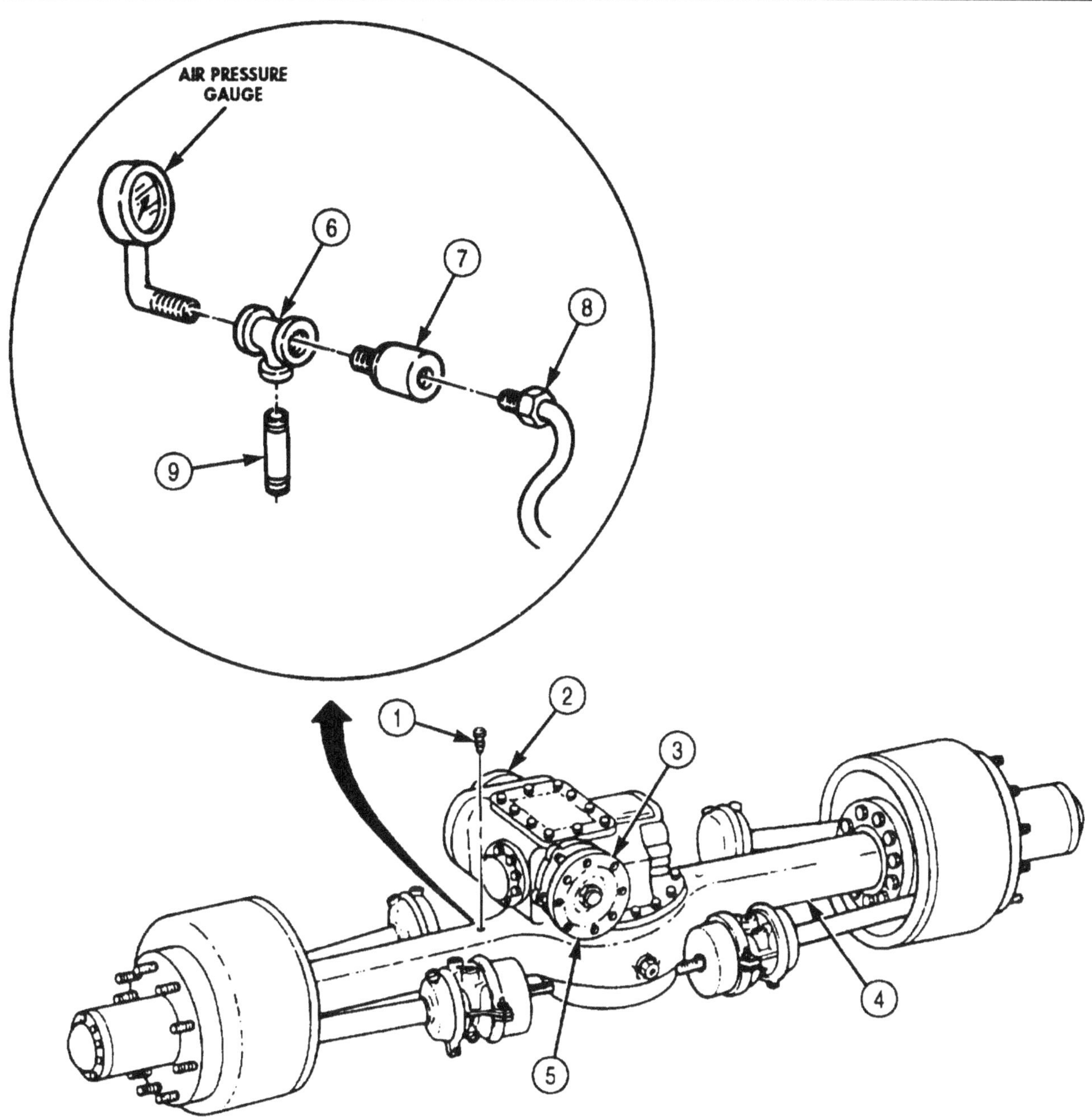

Section X. COMPRESSED AIR AND BRAKE SYSTEM MAINTENANCE

4-107. COMPRESSED AIR AND BRAKE SYSTEM MAINTENANCE INDEX

4-108. BRAKE PEDAL (TREADLE) VALVE REPAIR

THIS TASK COVERS:

a. Disassembly
b. Cleaning and Inspection
c. Assembly

INITIAL SETUP:

APPLICABLE MODELS
All

TOOLS
General mechanic's tool kit (Appendix E, Item 1)

MATERIALS/PARTS
Valve repair kit (Appendix D, Item 706)
O-ring (Appendix D, Item 452)
Pedal mounting plate gasket (Appendix D, Item 209)
Lint-free cloth (Appendix C, Item 21)

REFERENCES (TM)
TM 9-2320-272-24P

EQUIPMENT CONDITION
Brake pedal (treadle) valve removed (para. 3-197).

GENERAL SAFETY INSTRUCTIONS
Spring tension under valve retainer could cause flying parts.

NOTE

Do not perform this procedure unless valve repair kit is available.

a. Disassembly

1. Remove three screws (9), mounting plate (3), and gasket (2) from valve body (1). Discard gasket (2).
2. Remove breather (6), plunger (8), and O-ring (7) from mounting plate (3). Discard O-ring (7).
3. Loosen locknut (4) and remove stopscrew (5) from mounting plate (3).

4-108. BRAKE PEDAL (TREADLE) VALVE REPAIR (Contd)

4. Using pliers, grasp locknut (10) and remove lower piston assembly (11) from valve body (1). Discard lower piston assembly (11).

CAUTION

When performing step 5, be careful not to damage pedal valve body during upper piston assembly removal.

5. Position flat-tip punch on tabs (13) of upper assembly (12), rotate upper piston assembly (12) to loosen, then remove from valve body (1).

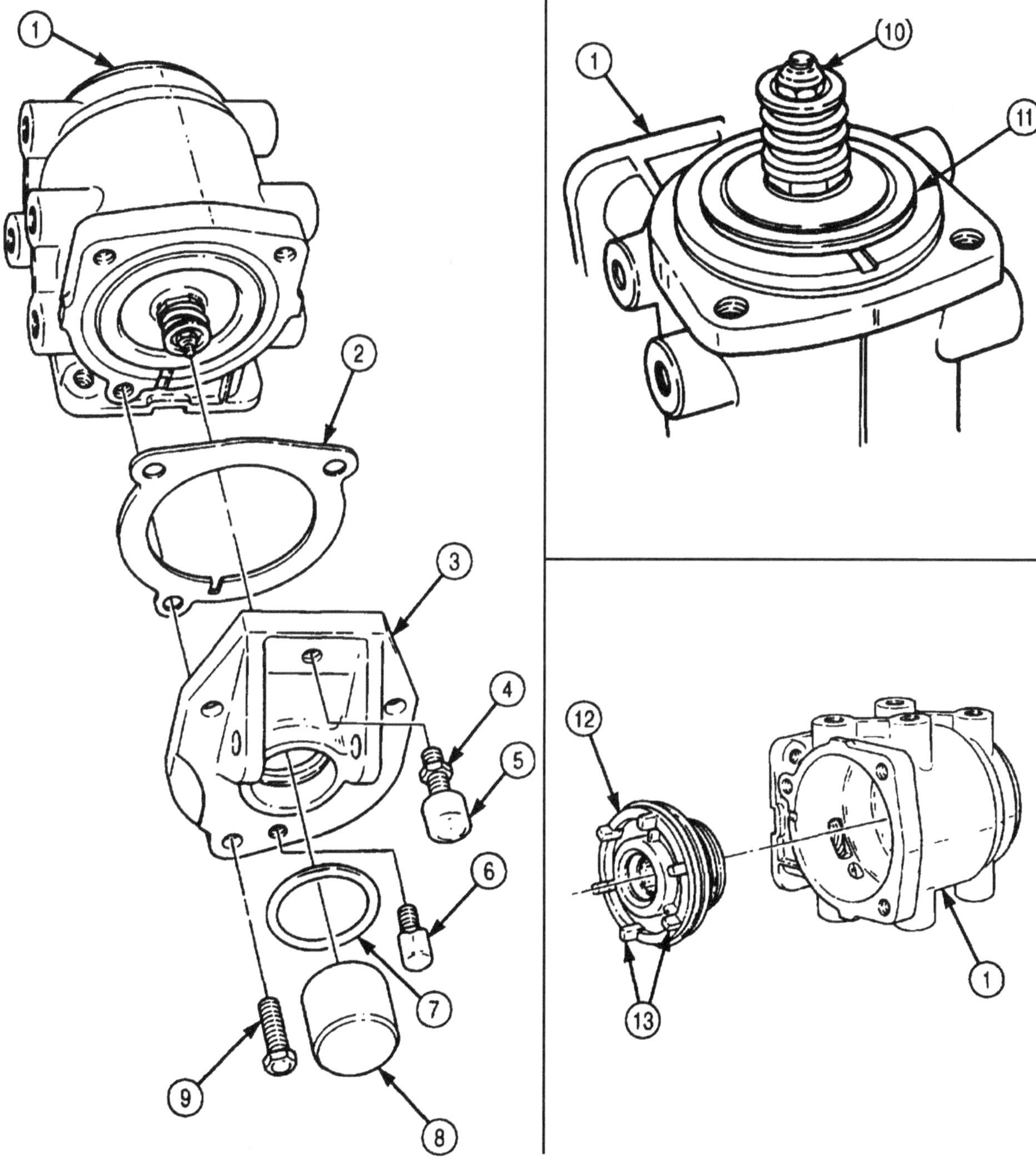

4-108. BRAKE PEDAL (TREADLE) VALVE REPAIR (Contd)

6. Remove O-rings (3) and (4) from grooves of piston (5). Discard O-rings (3) and (4).
7. Remove retaining ring (1) and exhaust valve assembly (2) from piston (5)

WARNING

Use caution when performing step 8. Spring tension under O-ring retainer will release when small retaining ring is removed. Flying parts may cause injury to personnel.

8. Remove retaining ring (6), washer (7), O-rings (8) and (9), O-ring retainer (10), spring (11), and valve retainer (12) from upper exhaust valve (13). Discard upper exhaust valve (13) and O-rings (8) and (9).

b. Cleaning and Inspection

1. Clean upper piston (5) and valve body (14) with lint-free cloth.
2. Inspect upper piston (5) and valve body (14) for cracks, breaks, chips, and pitting. Replace entire brake pedal (treadle) valve if either part is cracked, broken, chipped, or pitted.

c. Assembly

NOTE

Using grease supplied with valve repair kit, apply lubricant to all brake pedal (treadle) valve parts except rubber spring before assembly.

1. Position O-rings (9) and (8) on O-ring retainer (10).
2. Install valve retainer (12), spring (11), O-ring retainer (10), and washer (7) on new upper exhaust valve (13) with retaining ring (6).
3. Install exhaust valve assembly (2) in piston (5) with retaining ring (1).
4. Install O-rings (4) and (3) on piston (5).

4-108. BRAKE PEDAL (TREADLE) VALVE REPAIR (Contd)

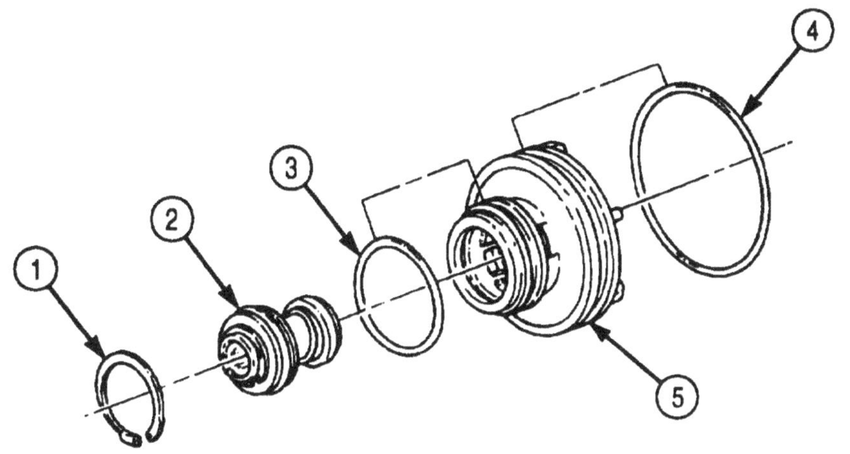

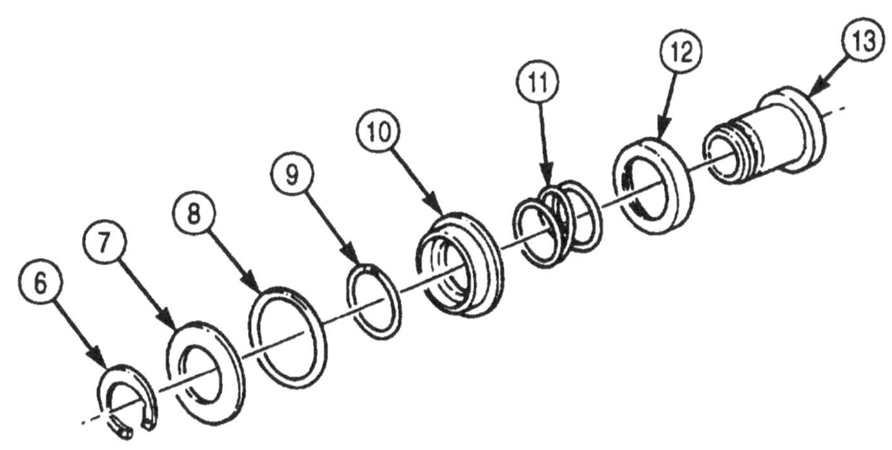

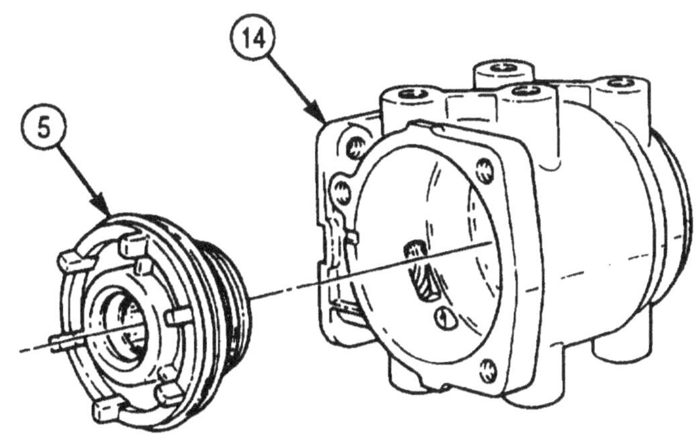

4-108. BRAKE PEDAL (TREADLE) VALVE REPAIR (Contd)

5. Install 0-rings (1), (4), and (8) on lower piston assembly (2).
6. Install upper piston assembly (5) in valve body (8). Use thumb pressure to seat properly,
7. Install lower piston assembly (2) in valve body (6). Use thumb pressure to seat properly,
8. Install stopscrew (10) on mounting plate (8). Do not tighten locknut (9).
9. Install breather (11), O-ring (12), and plunger (18) on mounting plate (8). Use hand pressure to seat plunger (18) properly.

NOTE

Discard retainer plats supplied with valve repair kit,

10. Install gasket (7) and mounting plate (8) on valve body (6) with three screws (14).

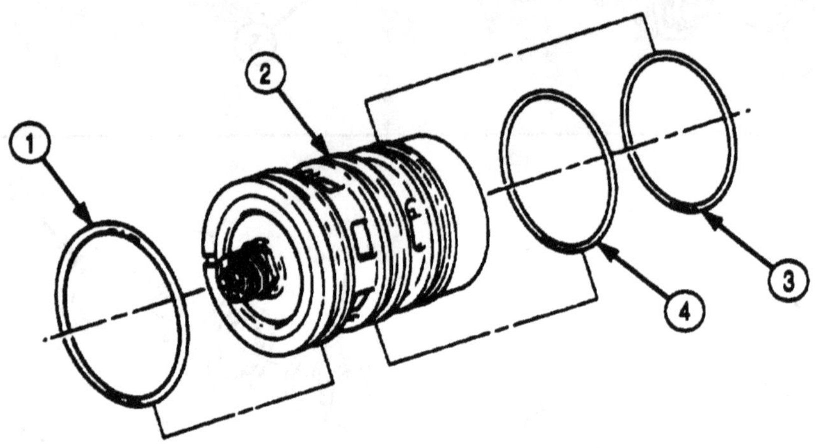

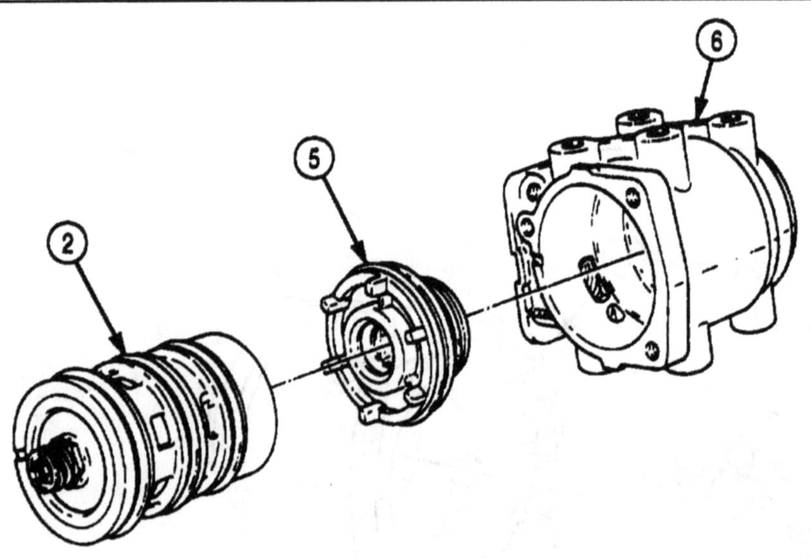

4-108. BRAKE PEDAL (TREADLE) VALVE REPAIR (Contd)

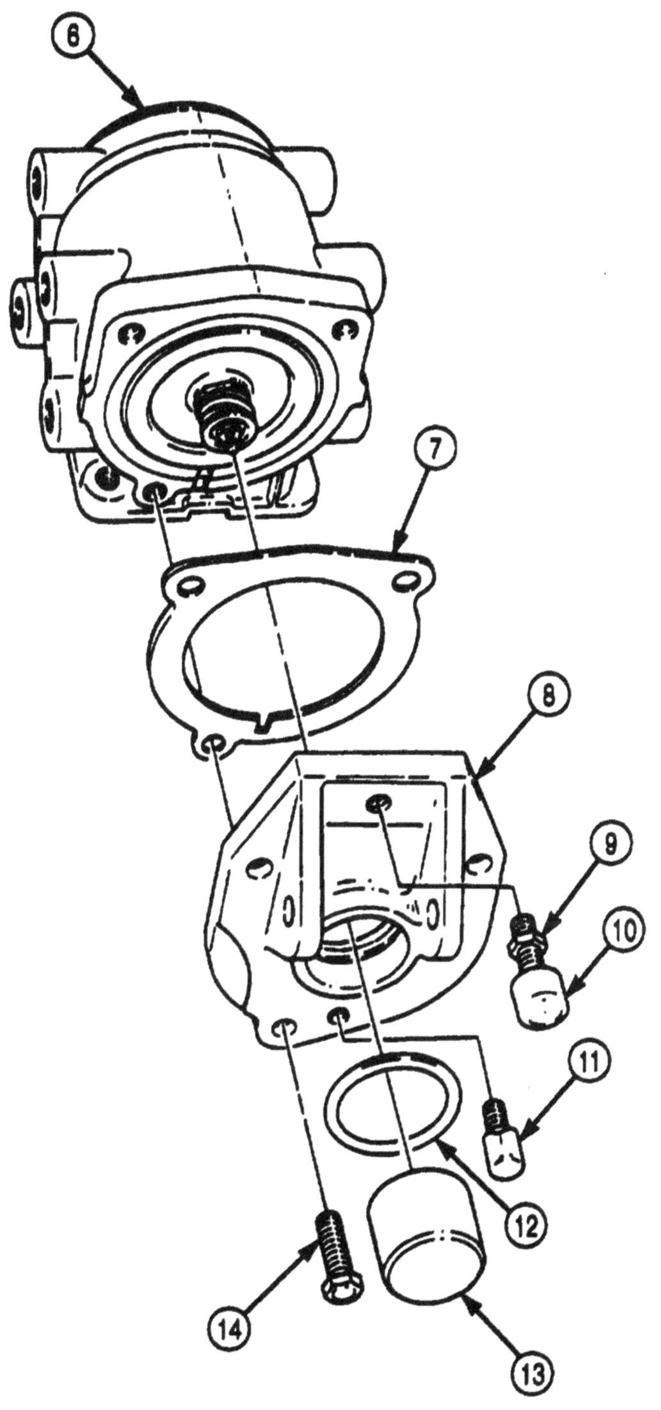

FOLLOW-ON TASK: Install brake pedal (treadle) valve (para. 3-197).

4-109. RELAY VALVE REPAIR

THIS TASK COVERS:

a. Disassembly
b. Cleaning and Inspection
c. Assembly

INITIAL SETUP

APPLICABLE MODELS

ALL

REFERENCES (TM)

TM 9-2320-272-24P

TOOLS

General mechanic's tool kit (Appendix E, Item 1)
Torque wrench (Appendix E, Item 146)

EQUIPMENT CONDITION

Relay valve removed (para. 3-187 or 3-188).

MATERIALS/PARTS

Valve repair kit (Appendix D, Item 707)
Four lockwashers (Appendix D, Item 386)
Four lockwashers (Appendix D, Item 387)
Lint-free cloth (Appendix C, Item 21)
Lubricating oil (Appendix C, Item 48)

NOTE

Do not perform this procedure unless valve repair kit is available.

a. Disassembly

1. Scribe an alignment mark across valve cover (2) and valve body (1).
2. Remove capnut (3), O-ring (4), and diaphragm (5) from valve cover (2). Discard O-ring (4) and diaphragm (5).
3. Remove four screws (6), lockwashers (7), and valve cover (2) from valve body (1). Discard lockwashers (7).
4. Remove seal ring (10) from valve body (1). Discard seal ring (10).
5. Remove relay piston (9) and O-ring (8) from valve cover (2). Discard O-ring (8).
6. Scribe an alignment mark across exhaust cover (11) and valve body (1).
7. Remove four screws (15), lockwashers (14), exhaust cover (11), gasket (12), and exhaust valve assembly (13) from valve body (1). Discard lockwashers (14), gasket (12), and exhaust valve assembly (13).

b. Cleaning and Inspection

1. Clean valve cover (2), valve body (1), exhaust cover (11), and relay piston (9) with lint-free cloth.
2. Inspect valve cover (2), valve body (1), exhaust cover (11), and relay piston (9) for cracks, breaks, chips, and pitting. Discard part(s) if cracked, broken, chipped, and pitted.

c. Assembly

1. Install new exhaust valve assembly (13) in valve body (1). Use thumb pressure to seat properly.
2. Align scribe marks on exhaust cover (11) and valve body (1) and install new gasket (12) and exhaust cover (11) on valve body (1) with four new lockwashers (14) and screws (15). Tighten screws (15) 2-3 lb-ft (3-4 N•m).
3. Apply lubricating oil on new O-ring (4) and install on capnut (3).
4. Install new diaphragm (5) and capnut (3) on valve cover (2). Tighten capnut (3) 12 lb-ft (16 N•m).
5. Apply lubricating oil to new O-ring (8) and install on relay piston (9).

4-109. RELAY VALVE REPAIR (Contd)

6. Install relay piston (9) in valve cover (2). Use thumb pressure to seat properly.
7. Apply lubricating oil on new seal ring (10) and install on valve body (1).
8. Align scribe marks on valve cover (2) and valve body (1) and install valve cover (2) on valve body (1) with four new lockwashers (7) and screws (6). Tighten screws (6) 12 lb-ft (16 N•m).

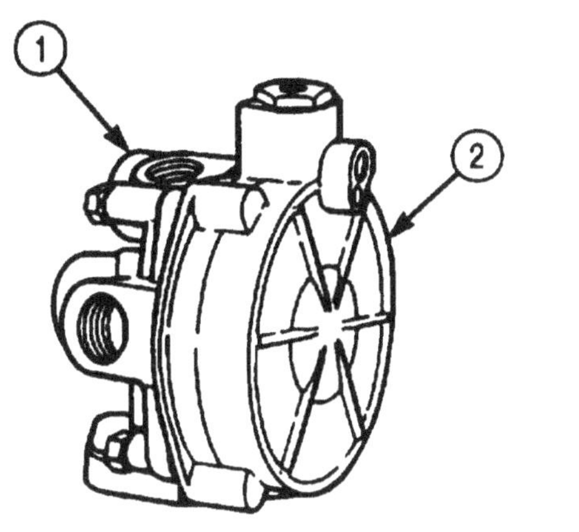

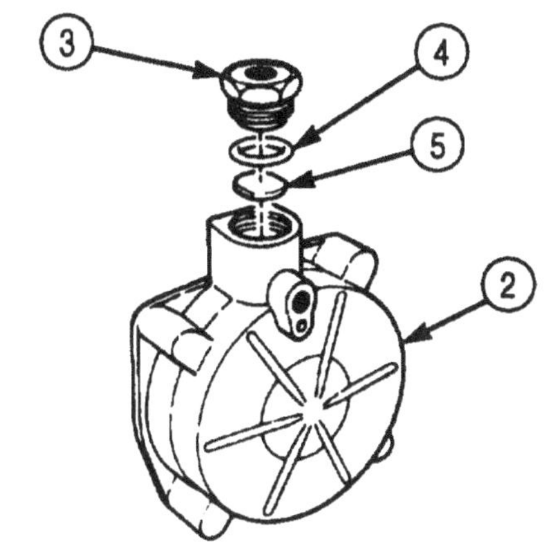

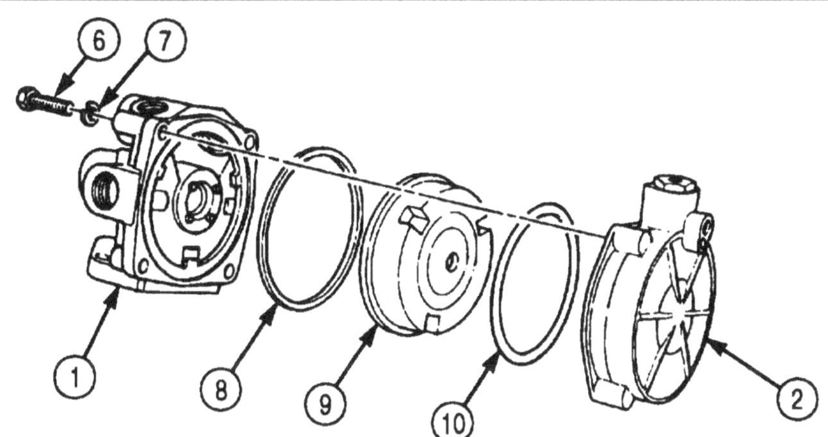

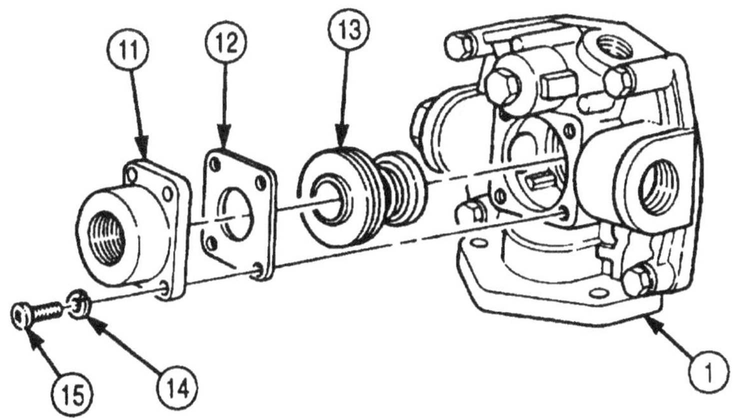

FOLLOW-ON TASK: Install relay valve (para. 3-187 or 3-188).

4-110. PARKING AND SERVICE BRAKEDRUM REPAIR

THIS TASK COVERS:

a. Cleaning
b. Inspection

INITIAL SETUP:

APPLICABLE MODELS
All

TOOLS
General mechanic's tool kit (Appendix E, Item 1)
Inside micrometer (Appendix E, Item 83)

MATERIALS/PARTS
Drycleaning solvent (Appendix C, Item 71)

REFERENCES (TM)
TM 9-2320-272-24P
TM 9-4910-482-10

EQUIPMENT CONDITION
Parking brakedrum removed (para. 3-223, 3-224, 3-460, or 3-461).

GENERAL SAFETY INSTRUCTIONS

- Drycleaning solvent is flammable and toxic. Do not use near open flame.
- Keep tire extinguisher nearby when using drycleaning solvent.
- Do not use compressed air or a dry brush when working in areas of asbestos dust.
- Wear a fiber mask or respirator when working around asbestos dust.

a. Cleaning

WARNING

- Drycleaning solvent is flammable and toxic. Do not use near an open flame and always have a fire extinguisher nearby when solvents are used. Use only in well-ventilated places, wear protective clothing, and dispose of cleaning rags in approved container. Failure to do this may result in injury to personnel and/or damage to equipment.
- Do not use compressed air or a dry brush for cleaning when working in areas where vehicle asbestos brake lining dust may accumulate. Remove asbestos dust and other residue from these areas using a soft bristle brush or cloth soaked with water. Breathing asbestos dust may cause injury to personnel.

Wash parking brakedrum (1) or service brakedrum (2) with water and soft bristle brush. Allow to air-dry. Clean with drycleaning solvent.

b. Inspection

1. Inspect parking brakedrum (1) or service brakedrum (2) for warps and cracks. Replace if warped or cracked.
2. Inspect parking brakedrum (1) or service brakedrum (2) for scores or pits. If scored or pitted, measure inside diameter at several positions. Replace parking brakedrum (1) if inside diameter is greater than 13.5 in. (342.9 mm). Replace service brakedrum (2) if inside diameter is greater than 16.625 in. (422.28 mm).
3. Resurface brakedrum to remove scores and/or pits (TM 9-4910-482-10).
4. If parking brakedrum (1) or service brakedrum (2) was resurfaced, measure inside diameter. Replace if inside diameter exceeds limits.

4-110. PARKING AND SERVICE BRAKEDRUM REPAIR (Contd)

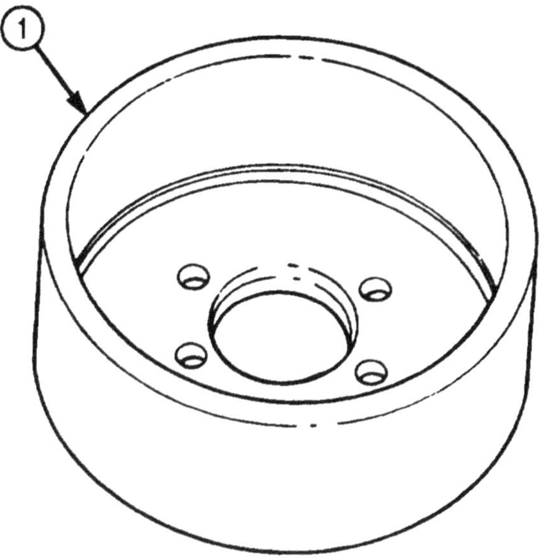

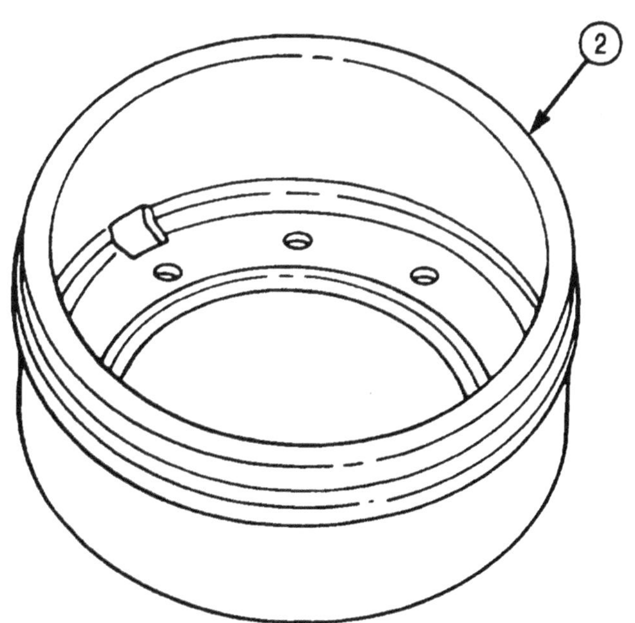

FOLLOW-ON TASK: Install brakedrum (para. 3-223, 3-224, 3-460, or 3-461).

4-111. PARKING AND SERVICE BRAKESHOE REPAIR

THIS TASK COVERS:

a. Parking Brakeshoe Disassembly
b. Service Brakeshoe Disassembly
c. Inspection
d. Service Brakeshoe Assembly
e. Parking Brakeshoe Assembly

INITIAL SETUP:

APPLICABLE MODELS

All

TOOLS

General mechanic's tool kit (Appendix E, Item 1)
Brake reliner (Appendix E, Item 19)

MATERIALS PARTS

Sixteen rivets (parking brake) (Appendix D, Item 544)
Thirty-two rivets (service brake) (Appendix D, Item 544)
Two brakeshoe linings (parking brake) (Appendix D, Item 16)
Two brakeshoe linings (service brake) (Appendix D, Item 16)

REFERENCES (TM)

TM 9-2320-272-24P
TM 9-4919-443-12

EQUIPMENT CONDITION

- Parking brakeshoes removed (para. 3-176).
- Service brakeshoes removed (para. 3-180).

CAUTION

- All parking brakeshoe linings must be replaced as matched sets. Failure to do so may result in damage to equipment.
- Service brakeshoe linings must be maintained in sets of four. Do not intermix brake linings from different kits. Damage to equipment may result.

a. Parking Brakeshoe Disassembly

1. Using brake reliner, remove twelve rivets (1) from each brakeshoe (3). Refer to TM 9-4919-443-12 for instructions on using brake reliner. Discard rivets (1).
2. Remove brake linings (2) from brakeshoes (3). Discard brake linings (2).

b. Service Brakeshoe Disassembly

1. Using brake reliner, remove twenty rivets (4) from each brakeshoe (6). Refer to TM 9-4919-443-12 for instructions on using brake reliner. Discard rivets (4).
2. Remove brake linings (5) from brakeshoes (6). Discard brake linings (5).

4-111. PARKING AND SERVICE BRAKESHOE REPAIR (Contd)

c. Inspection

Inspect parking brakeshoes (3) or service brakeshoes (6) for cracks, warpage, flat spots on faces, and out-of- round rivet holes. Replace brakeshoes (3) or (6) if cracked, warped, flat spots on face, or holes are out of round.

d. Service Brakeshoe Assembly

Using brake reliner, install new brake linings (5) on brakeshoes (6) with twenty new rivets (4). Refer to TM 9-4919-443-12 for instructions on using brake reliner.

e. Parking Brakeshoe Assembly

Using brake reliner, install new brake linings (2) on brakeshoes (3) with twelve new rivets (1). Refer to TM 9-4919-443-12 for instructions on using brake reliner.

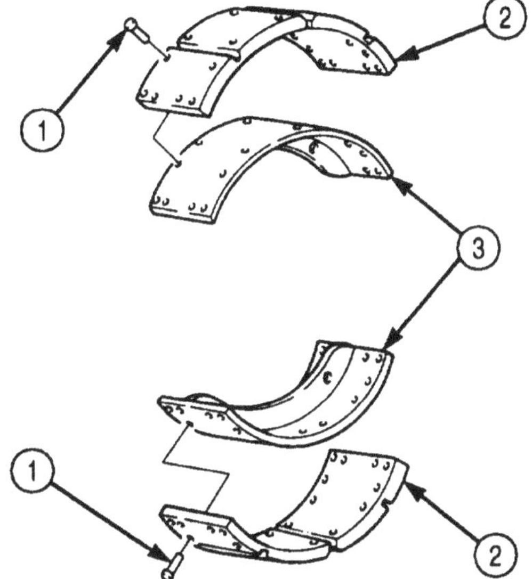

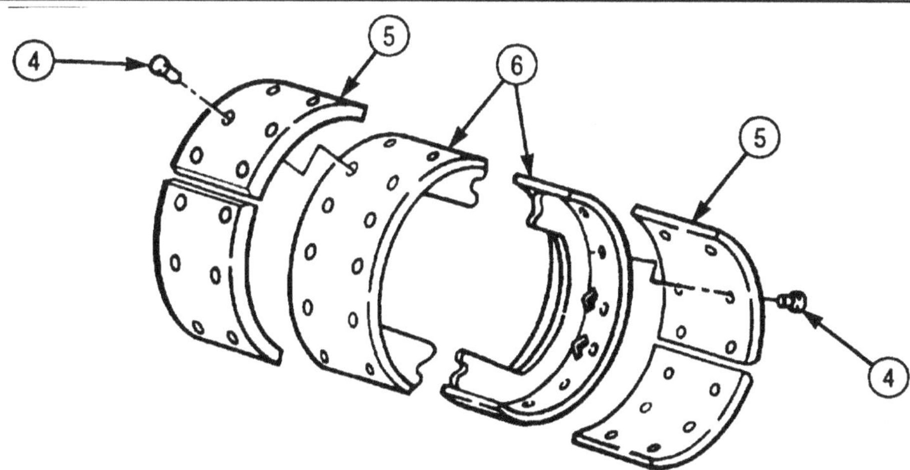

FOLLOW-ON TASKS:
- Install parking brakeshoes (para. 3-176).
- Install service brakeshoes (para. 3-180).

4-112. FRONT BRAKE ACTUATOR REPAIR

THIS TASK COVERS:

a. Adjustable Actuator Disassembly
b. Stationary Actuator Disassembly
c. Cleaning and Inspection
d. Stationary Actuator Assembly
e. Adjustable Actuator Assembly

INITIAL SETUP:

APPLICABLE MODELS
All

TOOLS
General mechanic's tool kit (Appendix E, Item 1)
Torque wrench (Appendix E, Item 146)

Anchor plunger kit (left) (Appendix D, Item 8)
Anchor plunger kit (right) (Appendix D, Item 7)
Adjuster parts kit (Appendix D, Item 3)
Lint-free cloth (Appendix C, Item 21)
Masking tape (Appendix C, Item 75)
GAA grease (Appendix C, Item 28)

REFERENCES (TM)
TM 9-2320-272-24P

EQUIPMENT CONDITION
Front brake spider removed (para. 3-189).

NOTE

- Twelve-degree wedge assembly should also be replaced when repairing front brake actuator. It is not offered in any of these Rockwell kits (TM 9-2320-272-24P).
- Do not perform this procedure unless repair kit is available.

a. Adjustable Actuator Disassembly

NOTE

Rockwell does not make a repair kit for the front spider wedge brake. The entire spider must be replaced. Individual parts are available under a separate NSN.

1. Remove wedge assembly (15) from back of adjusting plunger housing (11). Discard wedge assembly (15).
2. Remove two hollow screws (10), gaskets (9), springs (8), and adjusting pawls (7) from adjusting plunger housing (11). Discard hollow screws (10), gaskets (9), springs (8), and adjusting pawls (7).
3. Remove two adjusting bolts (3), plunger seals (41, adjusting sleeves (5), and adjusting plungers (6) Prom adjusting plunger housing (11). Discard adjusting bolts (3), plunger seals (4), adjusting sleeves (5), and adjusting plungers (6).

b. Stationary Actuator Disassembly

1. Remove two guide screws (12) and gaskets (13) from anchor plunger housing (14). Discard guide screws (12) and gaskets (13).
2. Remove two anchor plungers (1) and seals (2) from anchor plunger housing (14). Discard anchor plungers (1) and seals (2).

c. Cleaning and Inspection

Wipe adjusting plunger housing (11) and anchor plunger housing (14) with lint-free cloth and inspect for cracks, breaks, chips, and pitting. Replace entire spider assembly if adjusting plunger housing (11) or anchor plunger housing (14) is cracked, broken, chipped, or pitted.

4-112. FRONT BRAKE ACTUATOR REPAIR (Contd)

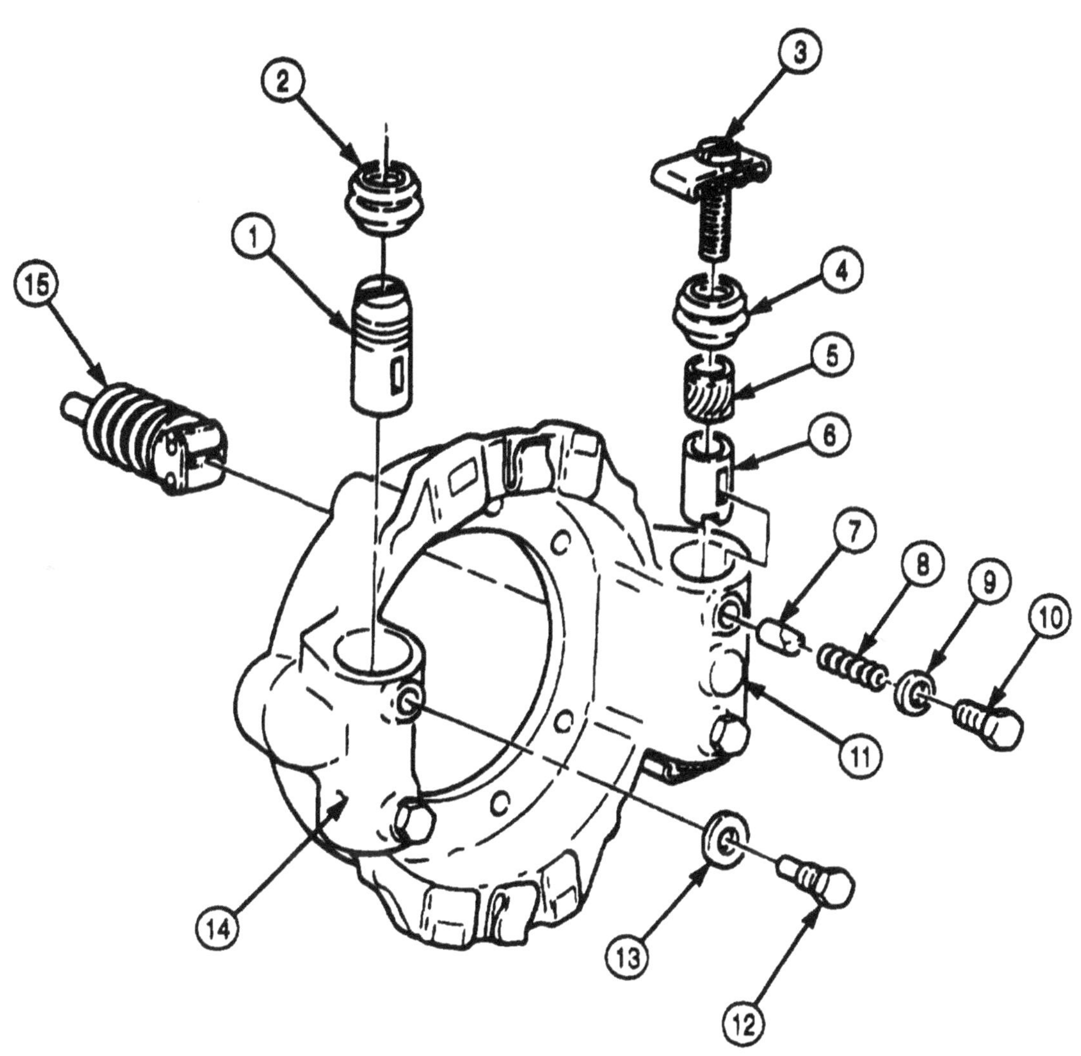

4-112. FRONT BRAKE ACTUATOR REPAIR (Contd)

d. Stationary Actuator Assembly

CAUTION

Keep all anchor parts together. Intermixing right and left plunger parts will cause brake system damage.

NOTE

Apply a light film of grease, included with new parts kit, to all metal parts, lips of seals, and plunger housing bores.

1. Place a piece of tape over slots of two new anchor plungers (2) to protect lips of seals (1).
2. Install two new seals (1) in double grooves of anchor plungers (2).

NOTE

Ensure correct anchor plunger, L or R, is installed in bore of anchor housing.

3. Align guide grooves (19) of anchor plungers (2) with guide screw holes (17) of anchor plunger housing (16).
4. Install anchor plungers (2) in anchor plunger housing (16). Using hammer and 1-3/4-in. (44.45-mm) wrench socket, seat seals (1) in anchor plunger housing (16).
5. Remove tape from anchor plungers (2).
6. Install two new gaskets (15) and new guide screws (14) in anchor plunger housing (16) and anchor plungers (2). Tighten guide screws (14) 15-20 lb-ft (20-27 N•m).

e. Adjustable Actuator Assembly

NOTE

Apply a light film of GAA grease to all parts prior to assembly.

1. Align guide groove (3) of two new adjusting plungers (7) with adjusting pawl holes (8) of adjusting plunger housing (13).
2. Install adjusting plungers (7) in adjusting plunger housing (13).
3. Position two new adjusting sleeves (6) on adjusting plunger (7).

NOTE

Adjusting pawls have teeth and flats on one end and chamfered edge on the other end. When performing step 4, ensure adjusting pawl is positioned with chamfer toward sleeve to align pawl and sleeve teeth.

4. Install two new adjusting pawls (9) and new springs (10) in adjusting plunger housing (13) with two new gaskets (11) and new hollow screws (12). Tighten hollow screws (12) 15-20 lb-ft (20-27 N•m).
5. Using adjusting bolt (4), rotate adjusting sleeve (6) and check for proper teeth meshing. If teeth are properly meshing, a clicking sound and ratchet feel will be indicated. Remove adjusting bolt (4).
6. Position inner lip of two new plunger seals (5) over adjusting sleeves (6). Using hammer and 1-3/4-in. (44.45-mm) wrench socket, seat outer ring of plunger seals (5) in adjusting plunger housing (13).

CAUTION

Do not bottom adjusting bolt against seal. Seal will be damaged.

7. Install two new adjusting bolts (4) in adjusting sleeve (6). Tighten adjusting bolts (4) until heads are showing above plunger seal (5).
8. Install new wedge assembly (18) in back of adjusting plunger housing (13). Check for proper operation of wedge assembly (18) and adjusting plungers (7).

4-112. FRONT BRAKE ACTUATOR REPAIR (Contd)

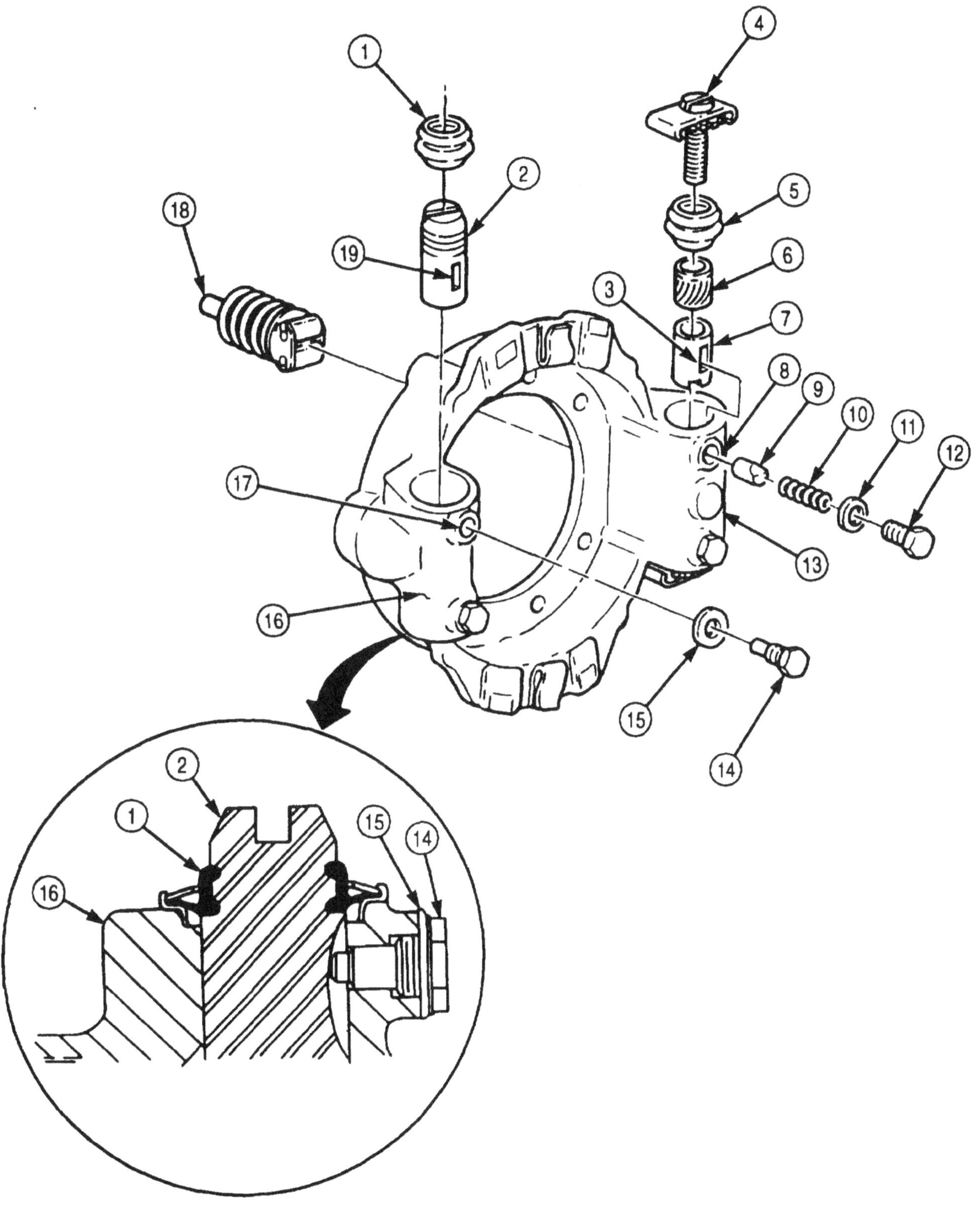

FOLLOW-ON TASK: Install front brake spider (para. 3-183).

4-113. REAR BRAKE ACTUATOR REPAIR

THIS TASK COVERS:

a. Disassembly
b. Cleaning and Inspection
c. Assembly

INITIAL SETUP:

APPLICABLE MODELS
All

TOOLS
General mechanic's tool kit (Appendix E, Item 1)
Torque wrench (Appendix E, Item 146)

MATERIALS/PARTS
Adjuster parts kit (Appendix D, Item 3)
Lint-free cloth (Appendix C, Item 21)
GAA grease (Appendix C, Item 28)
Masking tape (Appendix C, Item 75)

REFERENCES (TM)
TM 9-2320-272-24P

EQUIPMENT CONDITION
Rear brake spider removed (para. 3-184).

NOTE
Do not perform this procedure unless repair kit is available.

a. Disassembly

1. Remove wedge assembly (1) from back of plunger housing (6). Discard wedge assembly (1).
2. Remove guide screw (11) and gasket (12) from plunger housing (6). Discard guide screw (11) and gasket (12).
3. Remove hollow screw (10), gasket (9), spring (8), and adjusting pawl (7) from plunger housing (6). Discard hollow screw (10), gasket (9), spring (8), and adjusting pawl (7).
4. Remove anchor plunger (13) and seal (14) from plunger housing (6). Discard anchor plunger (13) and seal (14).
5. Remove adjusting bolt (2) from plunger housing (6). Discard adjusting bolt (2).
6. Remove adjusting plunger seal (3), adjusting sleeve (4), and adjusting plunger (5) from plunger housing (6). Discard adjusting plunger seal (3), adjusting sleeve (4), and adjusting plunger (5).

b. Cleaning and Inspection

Clean plunger housing (6) with lint-free cloth and inspect for cracks, breaks, chips, and pitting. Replace entire spider if cracked, broken, chipped, or pitted.

4-113. REAR BRAKE ACTUATOR REPAIR (Contd)

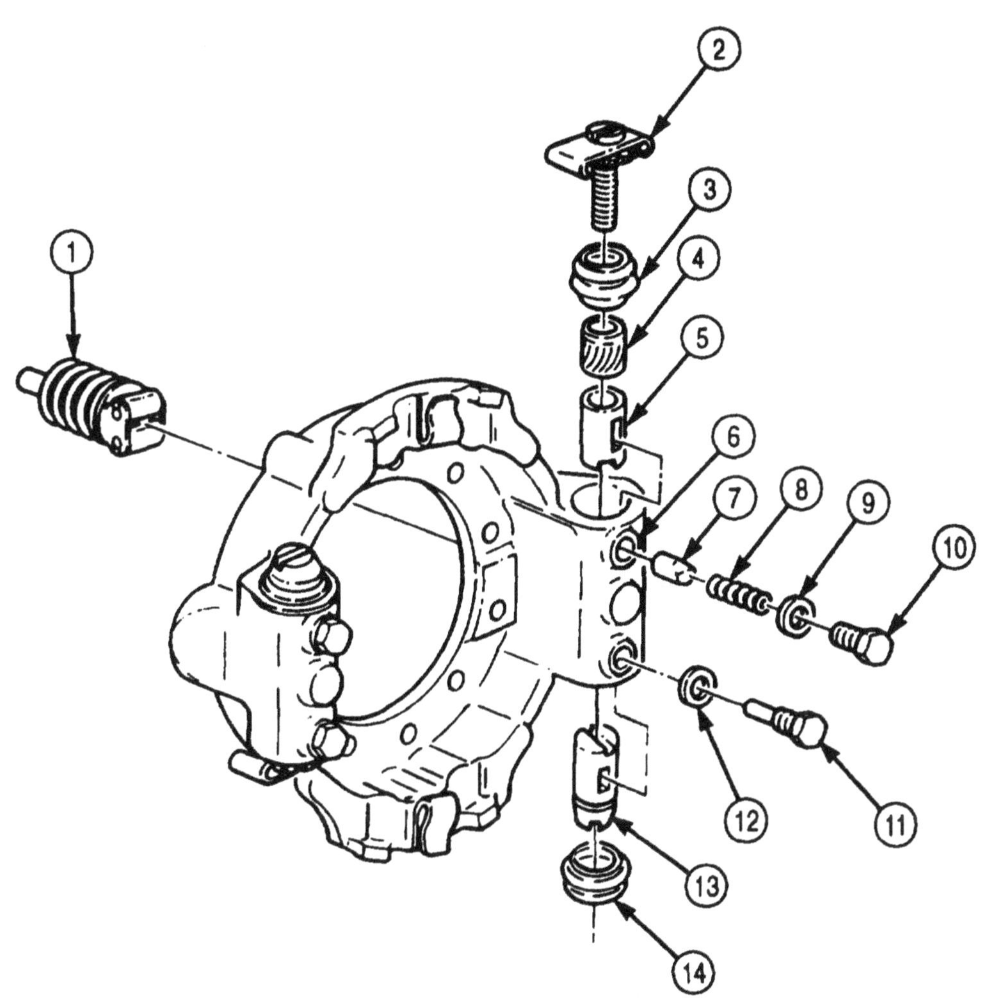

4-113. REAR BRAKE ACTUATOR REPAIR (Contd)

c. Assembly

NOTE

Apply a light film of GAA grease to all metal parts, lips of seals, and plunger housing bores prior to assembly.

1. Slide new seal (3) on new anchor plunger (1) until inner seal lip is in second plunger groove (4) and outer seal lip is in first plunger groove (2) of anchor plunger (1).
2. Place a piece of tape (5) over slot of anchor plunger (1) to protect lips of seal (3).

CAUTION

Anchor plungers and adjusting plungers are located vertically, opposite one another. Incorrect location will prevent automatic adjuster from working properly.

3. Remove tape from anchor plungers (1).
4. Align guide groove (10) of anchor plunger (1) with guide screw hole (9) of plunger housing (6).
5. Install anchor plunger (1) in plunger housing (6). Use hammer and 1-3/4-in. (44.45-mm) wrench socket to seat seal (3).
6. Install new gasket (7) and new guide screw (8) on plunger housing (6). Tighten guide screw (3) 15-20 lb-ft (20-27 N•m).
7. Align guide groove (12) of adjusting plunger (16) with adjusting pawl hole (17) of plunger housing (6).
8. Install adjusting plunger (16) in plunger housing (6).

NOTE

Adjusting pawls have teeth and flats on one end and chamfered edge on the other end. When performing step 9, ensure adjusting pawl is positioned with chamfer toward sleeve to align pawl and sleeve teeth.

9. Install new adjusting pawl (18) and new spring (19) in plunger housing (6) with new gasket (20) and new hollow screw (21). Tighten hollow screw (21) 15-20 lb-ft (20-27 N•m).
10. Using adjusting bolt (13), rotate adjusting sleeve (15) and check for proper teeth meshing. If teeth are properly meshing, a clicking sound and ratchet feel will be indicated.
11. Remove adjusting bolt (13).
12. Position inner lip of new adjusting plunger seal (14) in plunger housing (6). Using hammer and 1-3/4-in. (44.45-mm) wrench socket, seat adjusting plunger seal (14).

CAUTION

Do not bottom adjusting bolt against seal. Seal will be damaged.

13. Install new adjusting bolt (13) in adjusting sleeve (15). Tighten adjusting bolt (13) until head is showing above adjusting plunger seal (14).
14. Install new wedge assembly (11) in back of adjusting plunger housing (6). Check for proper operation of wedge assembly (11) and adjusting plungers (16).

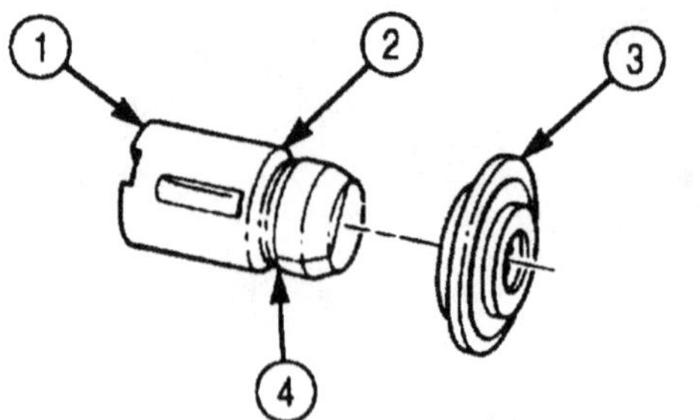

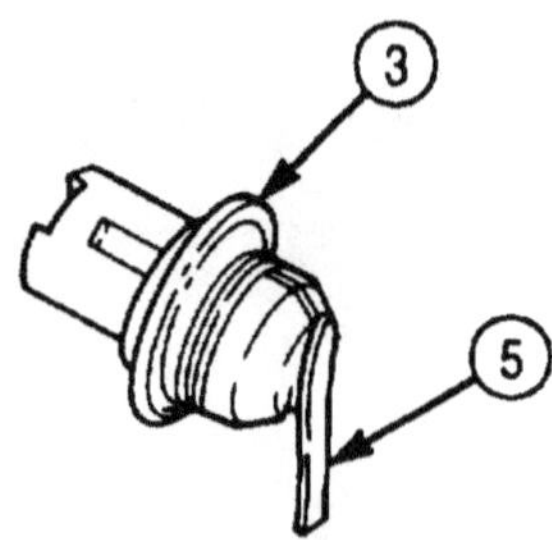

4-113. REAR BRAKE ACTUATOR REPAIR (Contd)

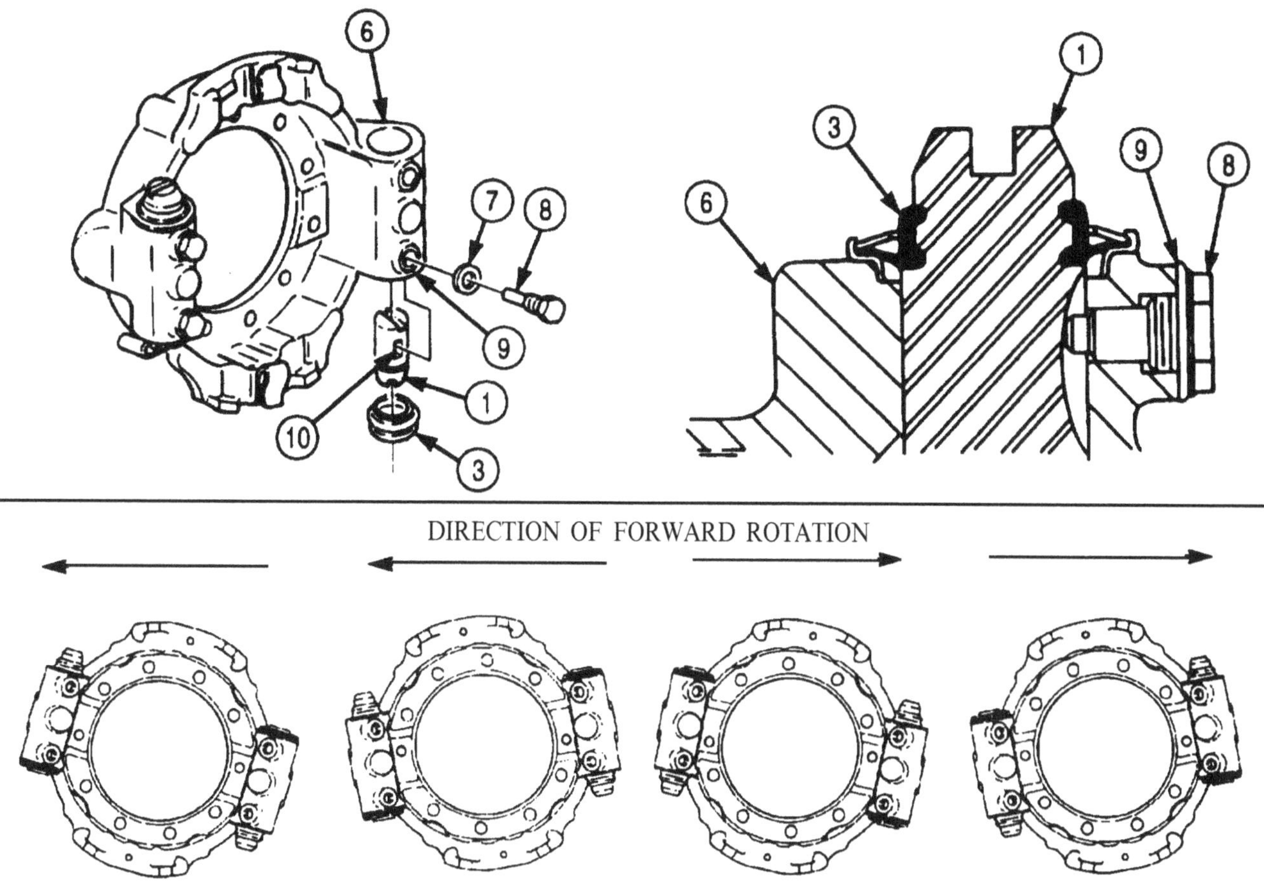

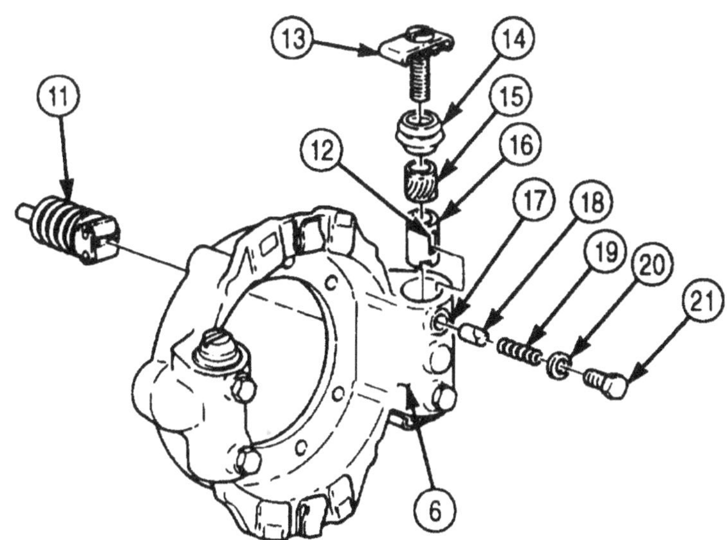

FOLLOW-ON TASK: Install rear brake spider (para. 3-184).

Section XI. POWER STEERING SYSTEM MAINTENANCE

4-114. POWER STEERING SYSTEM MAINTENANCE INDEX

4-115. GENERAL INFORMATION

The M939/A1/A2 series vehicles use one of two different models of integral power steering gears. Oil flow from the power steering pump supplies oil pressure for both models of steering gears. Both models direct oil flow and pressure to the same power assist cylinder mounted on the right aide of the vehicle. The two steering gears use different pitman arms. Both types of steering gears will work manually in the event of oil pressure failure.

ROSS STEERING GEAR. This is a recirculating ball-type steering gear with a rotary valve as the control element. It has both a worm gear and sector shaft adjustments in addition to two adjustable poppet valves.

SHEPPARD STEERING GEAR. This is a shuttle valve and piston-type steering gear with adjustable plungers controlling power-assisted steering angles. There are no other external mechanical adjustments.

4-116. UPPER AND LOWER STEERING COLUMN MAINTENANCE

THIS TASK COVERS:

a. Removal
b. Disassembly
c. Cleaning, Inspection, and Repair
d. Reassembly
e. Installation

INITIAL SETUP:

APPLICABLE MODELS
All

TOOLS
General mechanic's tool kit (Appendix E, Item 1)
Torque wrench (Appendix E, Item 146)

MATERIALS/PARTS
Lockwasher (Appendix D, Item 354)
Two locknuts (Appendix D, Item 291)

PERSONNEL REQUIRED
TWO

REFERENCES (TM)
TM 9-2320-272-10
TM 9-2320-272-24P

EQUIPMENT DESCRIPTION
- Parking brake set (TM 9-2320-272-10).
- Steering wheel removed (para. 3-226).
- Horn wire disconnected (para. 3-104).
- Turn signal indicator switch removed (para. 3-113).
- Trailer airbrake hand control valve removed (para. 3-210).
- Horn contact brush removed (para. 3-102).

a. Removal

NOTE

Before removing steering wheel columns, make sure front wheels are straight ahead for proper steering wheel alignment.

1. Remove nut (9), lockwasher (10), and screw (12) from universal joint (11). Discard lockwasher (10).
2. Remove two locknuts (14), screws (7), washers (8), and lower mounting clamp (6) from upper steering column (3) and firewall (13). Discard locknuts (14).

NOTE

Mark upper support clamp location on steering column for installation.

3. Remove two screws (5) and upper support clamp (4) from upper steering column (3) and instrument panel (1).
4. Remove upper steering column (3) from universal joint (11) and lift out through floorboard (15).
5. Slide two clamp bushings (2) off upper steering column (3).
6. Remove nut (16), lockwasher (17), and screw (19) and U-joint (18) from steering gear input shaft (20). Discard lockwasher (17).

4-116. UPPER AND LOWER STEERING COLUMN MAINTENANCE (Contd)

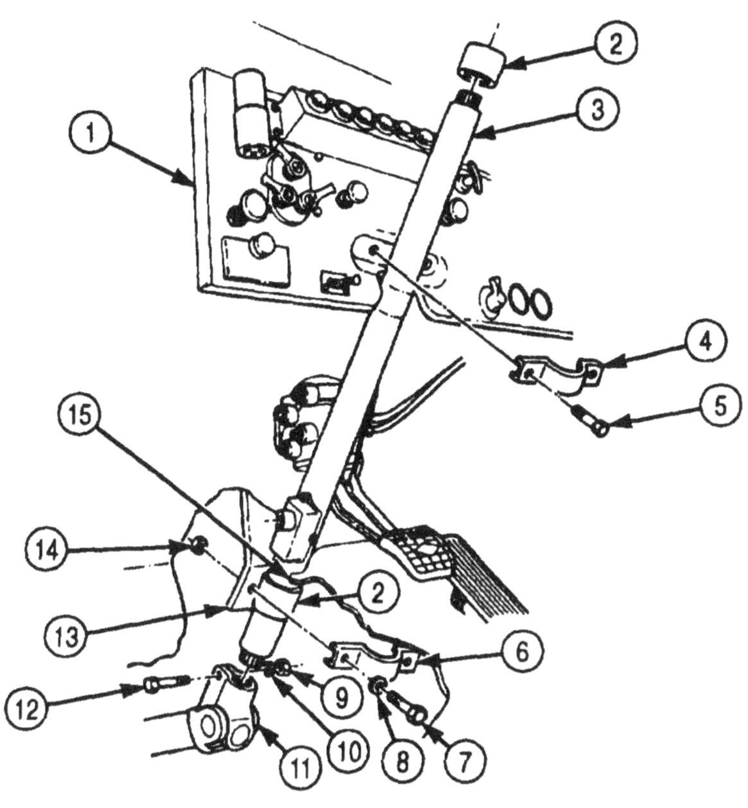

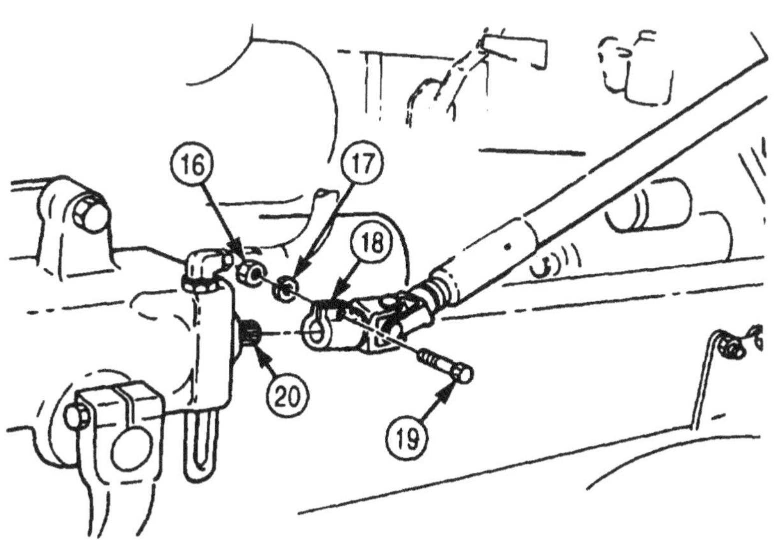

4-116. UPPER AND LOWER STEERING COLUMN MAINTENANCE (Contd)

b. Disassembly

1. Bend tabs (2) and slide retainer (1) and felt seal (3) back on lower section (7).
2. Remove lower steering column lower section (7) from lower steering column upper section (4).
3. Remove felt seal (3) and retainer (1) from lower steering column lower section (7). Discard retainer (1) and felt seal (3).
4. Remove grease fitting (6) from lower steering column upper section (4).
5. Remove two grease fittings (6) from two universal joints (5).

NOTE

Procedures to dissemble both universal joints are the same. Steps 6 through 8 are for the lower universal joint.

6. Remove four snaprings (9) from universal joint bearing caps (12).
7. Remove two universal joint bearing caps (10) and universal joint (11) from lower steering column end (8).
8. Remove two universal joint bearing caps (12) and universal joint (11) from lower steering column lower section (7).

c. Cleaning, Inspection, and Repair

WARNING

- Drycleaning solvent is flammable and toxic. Do not use near an open flame and always have a fire extinguisher nearby when solvents used. Use only in well-ventilated places, wear protective clothing, and dispose of cleaning rags in approved container. Failure to do this may result in injury to personnel and/or damage to equipment.
- Compressed air source will not exceed 30 psi (207 kPa). When cleaning with compressed air, eyeshields must be worn. Failure to wear eyeshields may result in injury to personnel.

1. Clean all steering column components with drycleaning solvent and dry with compressed air.
2. Inspect for cracks and burrs on splines. Repair small nicks or burrs. Replace lower section (7) and upper section (4) if cracked or minor repairs cannot be made.
3. Inspect bearing caps (12) for damage. Replace bearing caps (12) if damaged.
4. Inspect universal joint (11) for rough or uneven bearing surfaces. Replace universal joint (11) if surfaces are rough or uneven.

4-116. UPPER AND LOWER STEERING COLUMN MAINTENANCE (Contd)

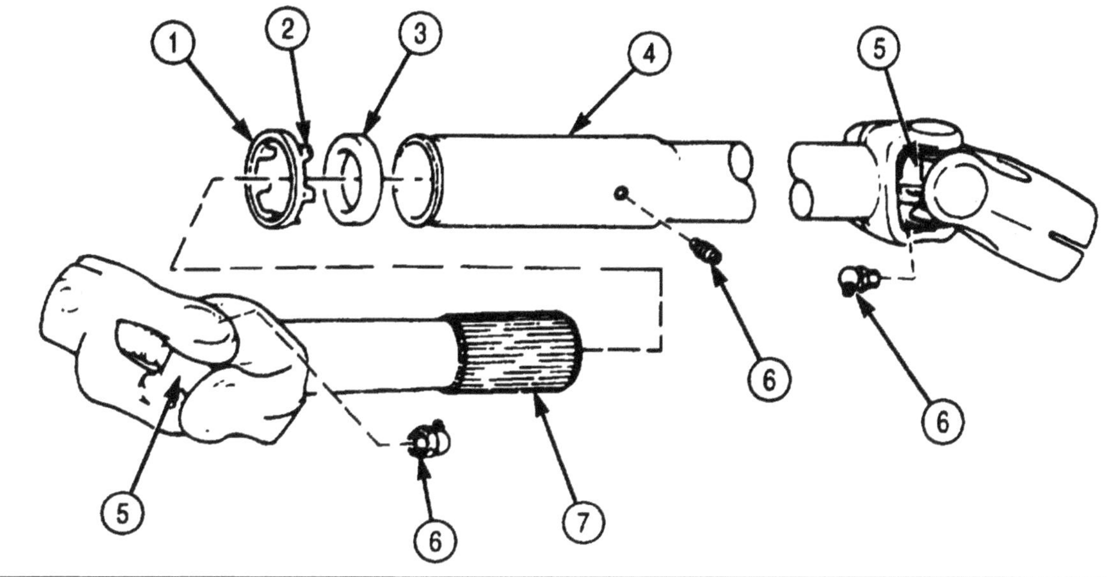

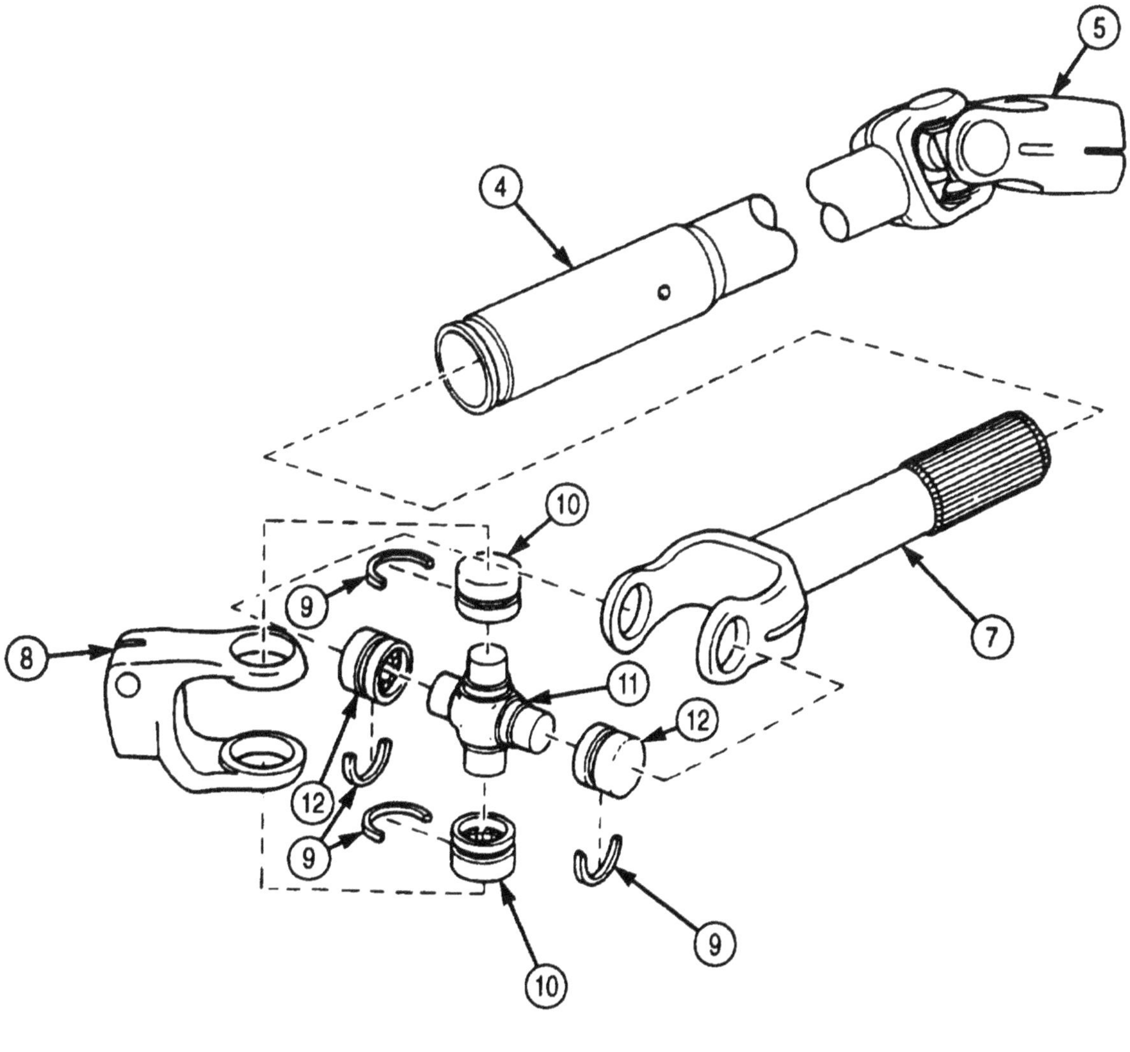

4-116. UPPER AND LOWER STEERING COLUMN MAINTENANCE (Contd)

d. Reassembly

1. Install two universal joint bearing caps (7) and universal joint (6) on lower steering column lower section (5).
2. Install two universal joint bearing caps (4) and universal joint (6) on lower steering column end (1).
3. Install four snaprings (2) on four universal joint bearing caps (4) and (7).
4. Install two grease fittings (12) on universal joints (11).
5. Install grease fitting (13) on lower steering column upper section (3).
6. Install new felt seal (10) and new retainer (8) on lower steering column lower section (5).
7. Install lower steering column lower section (5) on lower steering column upper section (3).
8. Slide felt seal (10) and retainer (8) forward to steering column upper section (3) and bend tabs (9) down.

4-116. UPPER AND LOWER STEERING COLUMN MAINTENANCE (Contd)

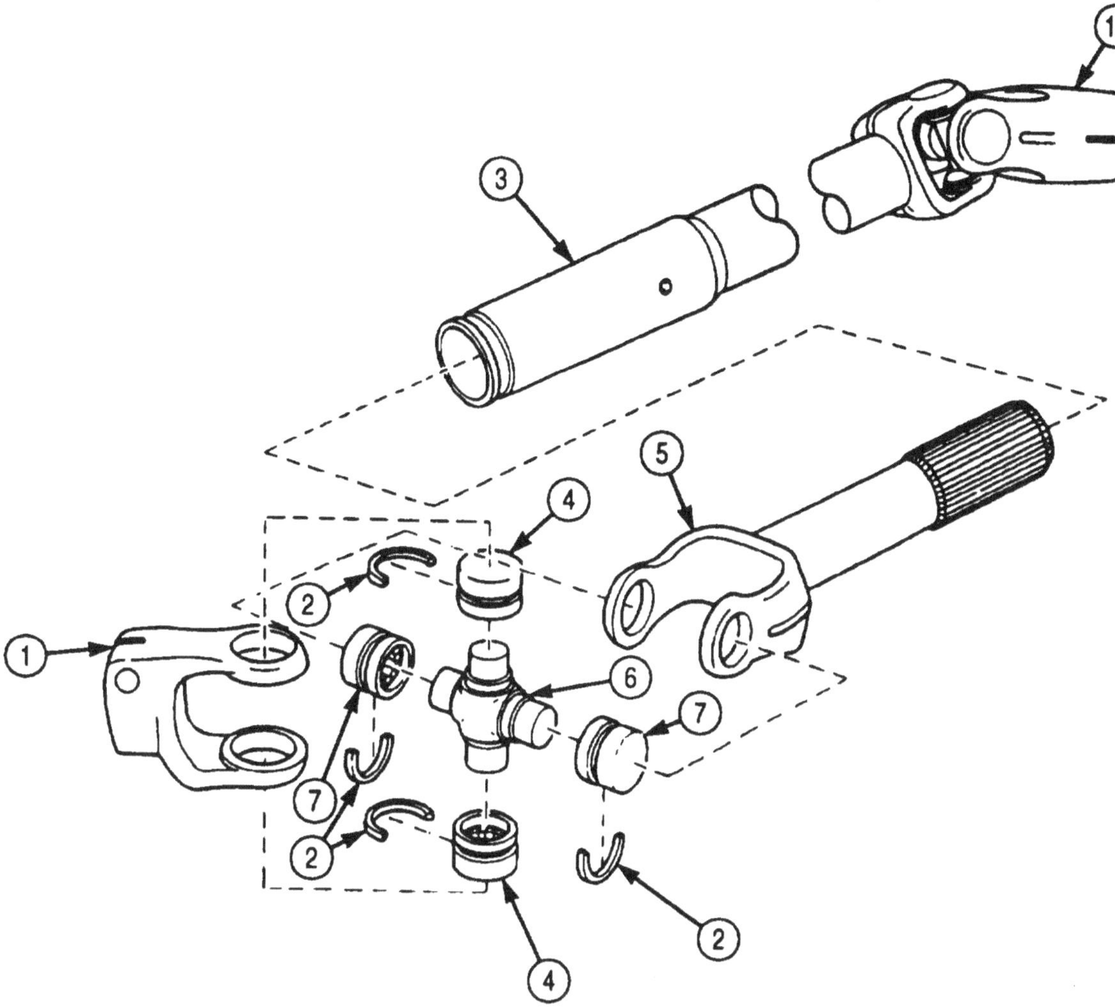

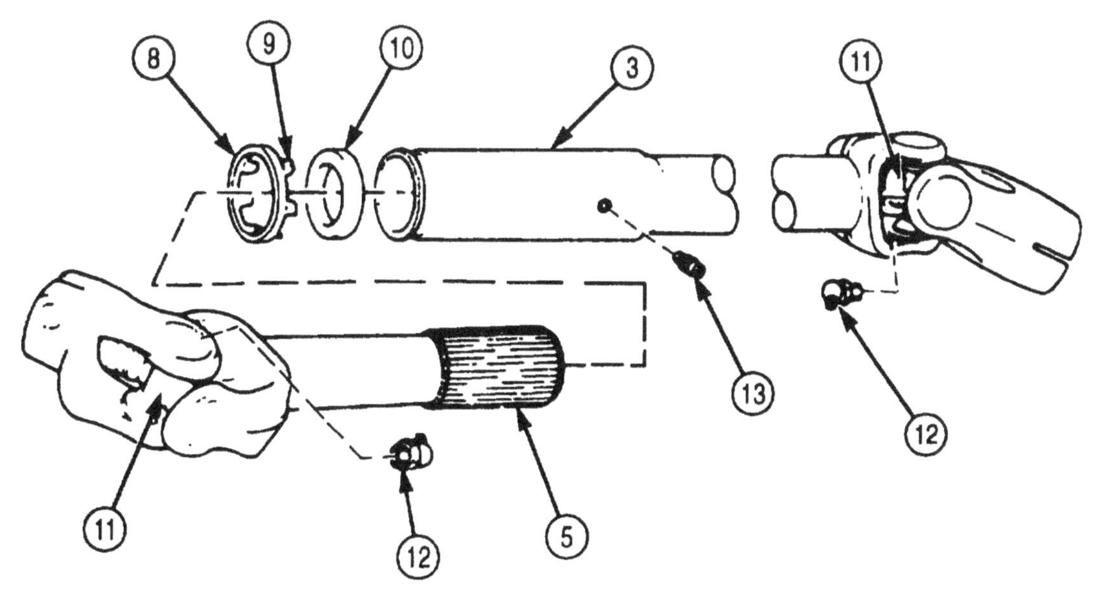

4-116. UPPER AND LOWER STEERING COLUMN MAINTENANCE (Contd)

e. Installation

1. Install U-joint (8) on steering gear input shaft (5) with screw (4), new lockwasher (2), and nut (1).
2. Install two clamp bushings (7) on upper steering column (8).

NOTE

Position horn contact brush next to firewall.

3. Lower upper steering column (8) through hole in floorboard (20) and insert splined end of upper steering column (8) into universal joint (16).
4. Install upper steering column (8) on universal joint (10) with screw (17), new lockwasher (15), and nut (14). Do not tighten nut (14).

NOTE

If installing a new steering column, mark clamp position on new column the same an on on old column.

5. Install upper steering column (8) on instrument panel (6) with upper support clamp (9) and two screws (10). Ensure clamp (9) is positioned over clamp bushing (7) at marked position.
6. Attach upper steering column (8) to firewall (18) with lower mounting clamp (11), two washers (18), screws (12), and new locknuts (19).
7. Tighten nut (14) 28-34 lb-ft (38-46 N•m),

4-116. UPPER AND LOWER STEERING COLUMN MAINTENANCE (Contd)

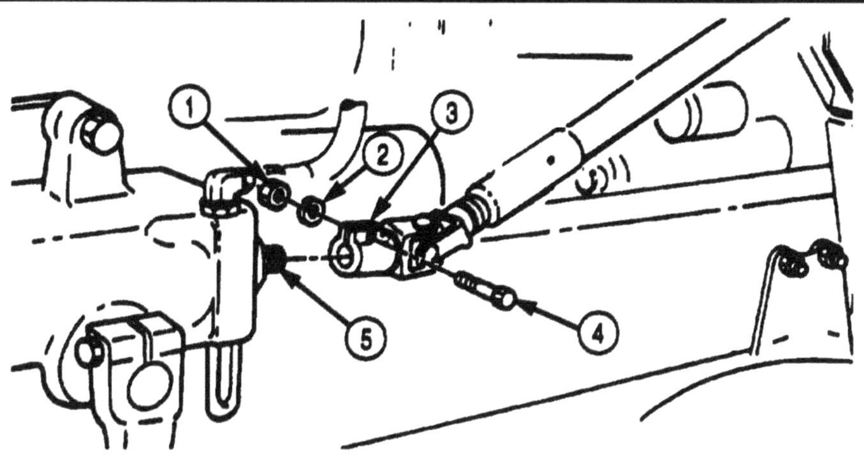

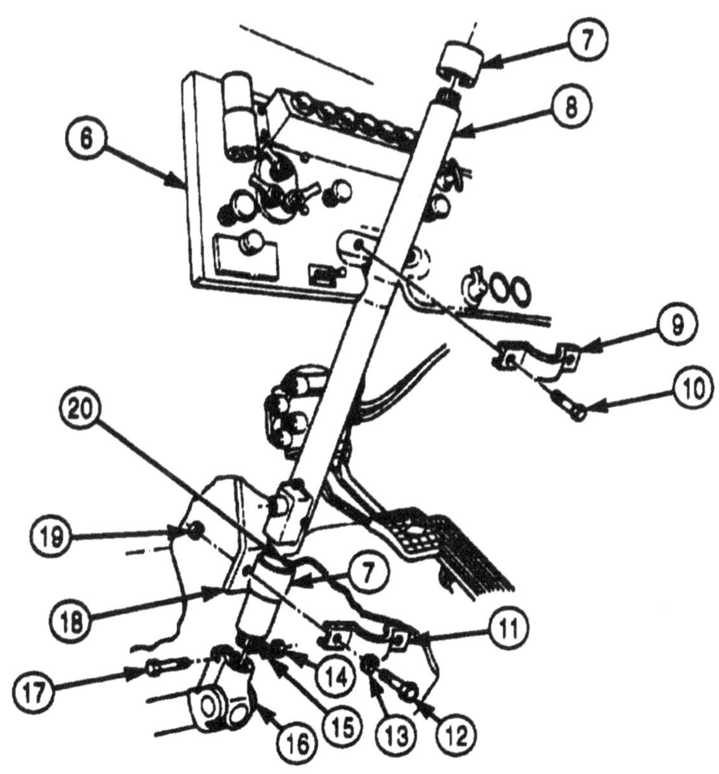

FOLLOW-ON TASKS:
- Install trailer airbrake hand control valve (para. 3-210).
- Install turn signal indicator switch (para. 3-113).
- Connect horn wire (para. 3-104).
- Install horn contact brush (para. 3-102).
- Install steering wheel (para. 3-226).
- Start engine (TM 9-2320-272-10) and road test vehicle.

4-117. STEERING GEAR (ROSS) AND MOUNTING BRACKET REPLACEMENT

THIS TASK COVERS:

a. Removal b. Installation

INITIAL SETUP:

APPLICATION MODELS

M939/A1

TOOLS

General mechanic's tool kit (Appendix E, Item 1)
Torque wrench (Appendix E, Item 145)
Mechanical puller (Appendix E, Item 110)

MATERIALS/PARTS

Four locknuts (Appendix D, Item 294)
Two O-rings (Appendix D, Item 437)
O-ring (Appendix D, Item 477)
Lockwasher (Appendix D, Item 382)
Locknut (Appendix D, Item 273)
Cap and plug set (Appendix C, Item 14)
Antiseize tape (Appendix C, Item 72)

REFERENCES (TM)

LO 9-2320-272-12
TM 9-2320-272-10
TM 9-2320-272-24P

EQUIPMENT CONDITION

- Parking brake set (TM 9-2320-272-10).
- Left splash shield removed (para. 3-301).
- Steering gear stone shield removed (para. 3-238).
- Drag link removed (para. 3-229).

CAUTION

Cap or plug all openings immediately after disconnecting lines and hoses to prevent contamination. Failure to do so may result in power steering gear damage.

NOTE

- Identify type of power steering gear. Removal and installation are different for the two steering gears. Refer to section II, chapter 1 and general information (para. 4-115).
- Ensure front wheels are straight ahead.

a. Removal

1. Remove nut (1), lockwasher (2), and screw (4) from U-joint (5) and steering gear input shaft (6). Discard lockwasher (2).
2. Open slot (3) in U-joint (5) and separate steering gear input shaft (6) and U-joint (5).
3. Remove locknut (9) and screw (7) from pitman arm (10). Discard locknut (9).

NOTE

Punch alignment marks on pitman arm and sector shaft.

4. Using a chisel, spread slot in top of pitman arm (10).
5. Using a puller, remove pitman arm (10) from sector shaft (8).

NOTE

- Have container ready to catch oil from disconnected lines.
- Tag lines and fittings for installation

6. Disconnect oil pressure line (13) and oil return line (16) from adapter elbow (12) and adapter (15).
7. Disconnect assist cylinder pressure lines (19) and (21) from adapter (18) and adapter elbow (20).

4-117. STEERING GEAR (ROSS) AND MOUNTING BRACKET REPLACEMENT (Contd)

8. Remove adapter elbows (12) and (20), adapters (15) and (18), and O-rings (14), (17), and (22) from steering gear housing (11). Discard O-rings (14), (17), and (22).

NOTE

Assistant will help with step 9.

9. Remove four nuts (23), washers (24), screws (27), steering gear (11), and mounting plate (26) from left frame rail (29).
10. Remove four locknuts (30), screws (28), and bracket (25) from left frame rail (29). Discard locknuts (30).

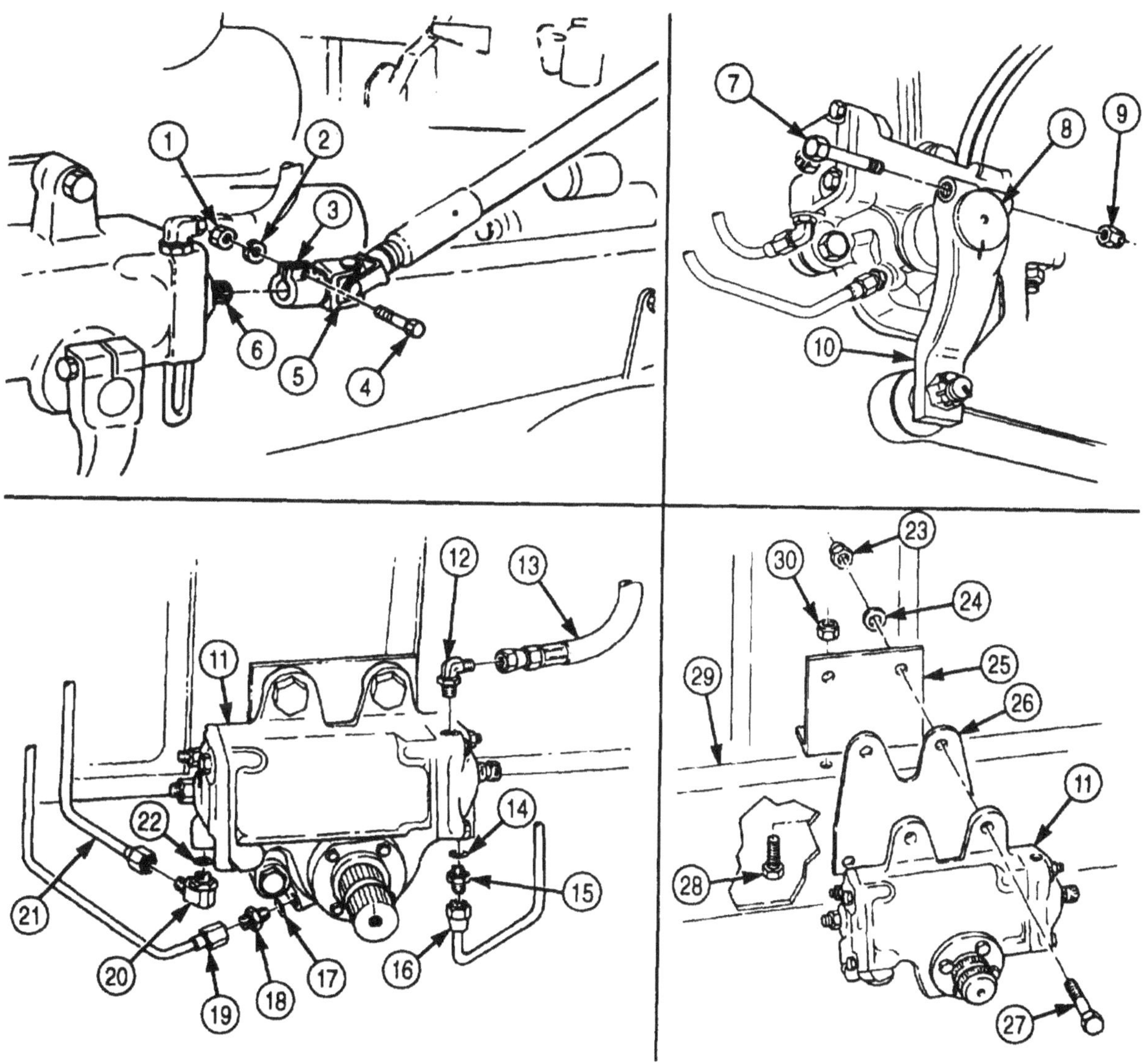

4-117. STEERING GEAR (ROSS) AND MOUNTING BRACKET REPLACEMENT (Contd)

b. Installation

NOTE

Assistant will help with steps 1 and 2.

1. Install mounting bracket (3) on left frame rail (3) with four screws (7) and new locknuts (9). Do not tighten locknuts (9).
2. Install mounting plate (4) and steering gear (5) on left frame rail (8) with four screws (6), washers (2), and nuts (1). Tighten nuts (1) 260-280 lb-ft (353-380 N•m).
3. Tighten locknuts (9) 69-70 lb-ft (81-95 N•m).

NOTE

Wrap male pipe threads with antiseize tape before installation.

4. Install new O-rings (12), (15), and (20), adapter elbows (10) and (18), and adapters (13) and (16) on steering gear housing (5).
5. Connect assist cylinder pressure lines (17) and (19) to adapter elbow (18) and adapter (16).
6. Connect oil return line (14) and oil pressure line (11) to steering gear housing adapter (13) and adapter elbow (10).

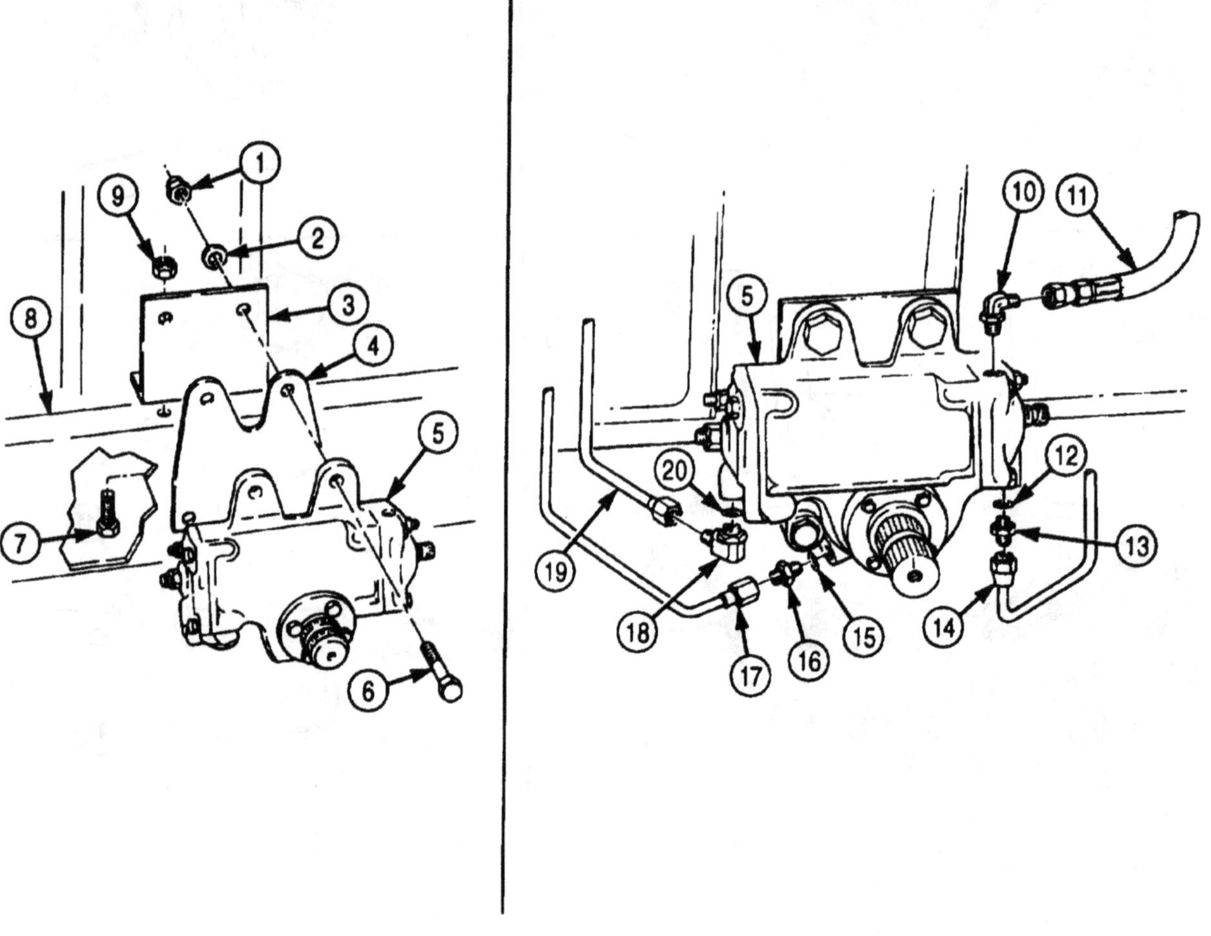

4-117. STEERING GEAR (ROSS) AND MOUNTING BRACKET REPLACEMENT (Contd)

7. Using alignment marks, install pitman arm (24) on sector shaft (22) with screw (21) and new locknut (23). Tighten locknut (23) 330-370 lb-ft (447-502 N•m).

NOTE

Before connecting lower steering column, ensure steering wheel spokes form a Y. Assistant will steady steering wheel during installation.

8. Position lower steering column U-joint (27) on steering gear input shaft (29), ensuring that screw holes in U-joint (27) align with groove in shaft (29).
9. Install screw (28) in sleeve (26) and shaft (29) and secure sleeve (26) on shaft (29) with new lockwasher (30) and nut (25). Tighten nut (25) 28-34 lb-ft (38-46 N•m).

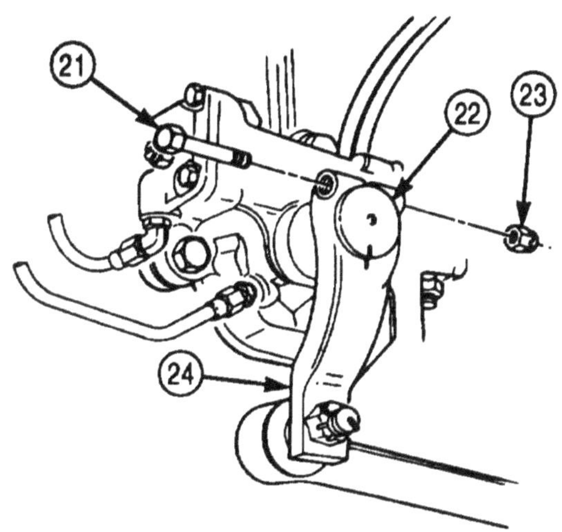

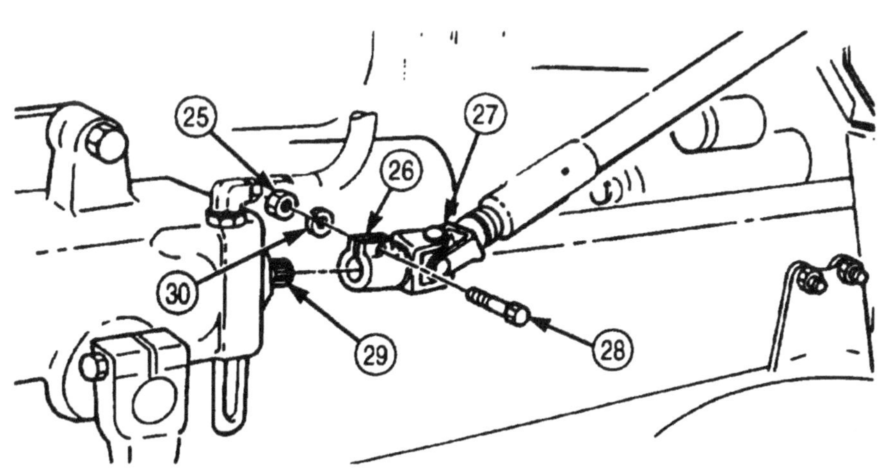

FOLLOW-ON TASKS:
- Install steering gear stone shield (para. 3-238).
- Install left splash shield (para. 3-301).
- Install drag link (para. 3-229).
- Fill steering gear to proper oil level (LO 9-2320-272-12).
- Start engine (TM 9-2320-272-10) and road test vehicle.

4-118. STEERING GEAR (SHEPPARD) AND MOUNTING BRACKET REPLACEMENT

THIS TASK COVERS

a. Removal

b. Installation

INITIAL SETUP

APPLICABLE MODELS

M939A2

TOOLS

General mechanic's tool kit (Appendix E, Item 1)
Torque wrench (Appendix E, Item 145)

MATERIALS/PARTS

Lockwasher (Appendix D, Item 382)
Four locknuts (Appendix D, Item 316)
Steering parts kit (Appendix D, Item 677)
Gasket and seal kit (Appendix D, Item 247)
Four locknuts (Appendix D, Item 294)
Four O-rings (Appendix D, Item 437)
Cap and plug set (Appendix C, Item 14)
Antiseize tape (Appendix C, Item 72)

REFERENCES (TM)

LO 9-2320-272-12
TM 9-2320-272-10
TM 9-2320-272-24P

EQUIPMENT DESCRIPTION

- Parking brake set (TM 9-2320-272-10).
- Left splash shield removed (para. 3-301).
- Steering gear stone shield removed (para. 3-238).
- Drag link disconnected (para. 3-229).

NOTE

- Identify type of power steering gear. Removal and installation is different for the two steering gears. Refer to chapter 1, section II for identification.
- Ensure front wheels are aligned straight ahead.

a. Removal

1. Remove nut (l), lockwasher (2), and screw (6) from sleeve (7) of universal joint (5). Discard lockwasher (2).
2. Using a chisel, open slot (3) in sleeve (7) of universal joint (5) and pull lower steering column (4) off steering gear input shaft (8).
3. Bend two long tabs (12) of retainer (11) out of notches in pitman arm (10).
4. Bend two short tabs (13) out of retainer (11) and remove retainer (11) from sector shaft (15). Discard retainer (11).

NOTE

Punch alignment marks on sector shaft and pitman arm.

5. Using puller, remove pitman arm (10) and seal (14) from sector shaft (15) of steering gear (9). Discard seal (14).

4-118. STEERING GEAR (SHEPPARD) AND MOUNTING BRACKET REPLACEMENT (Contd)

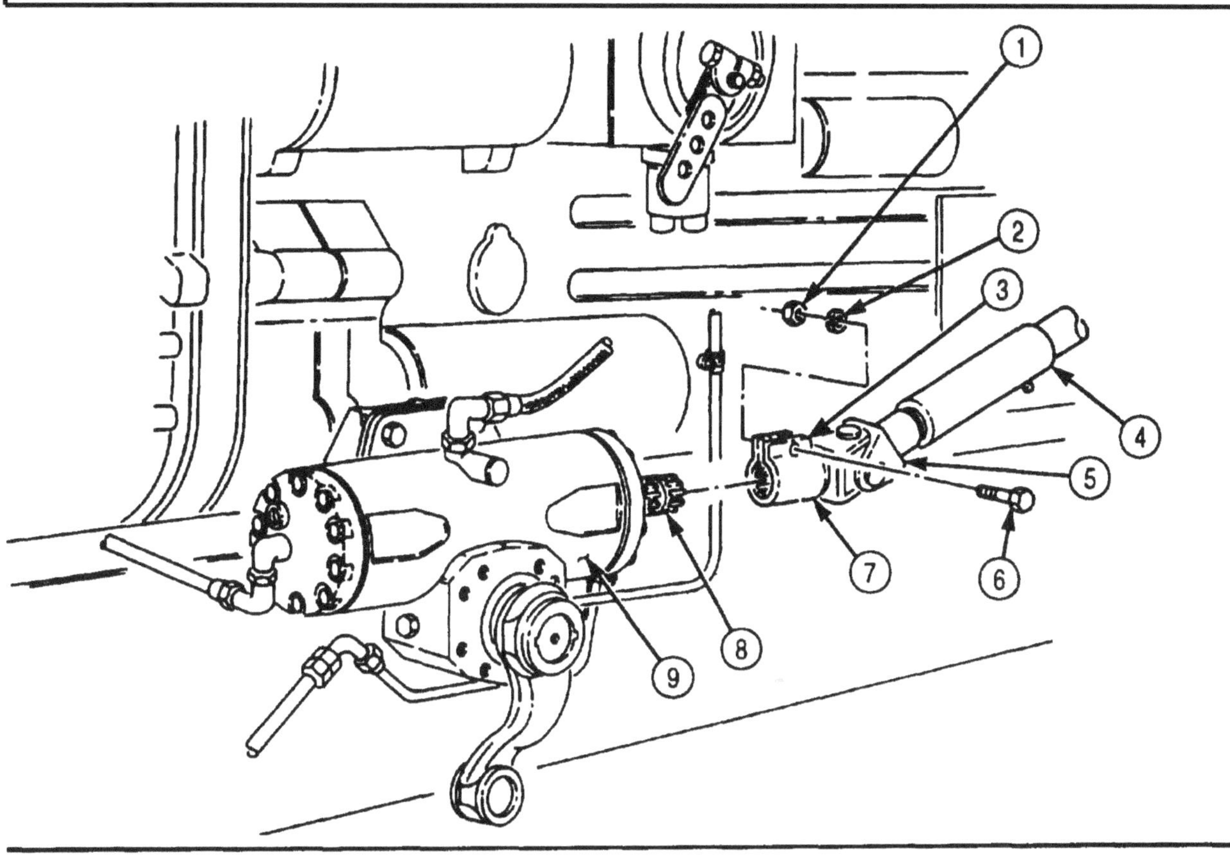

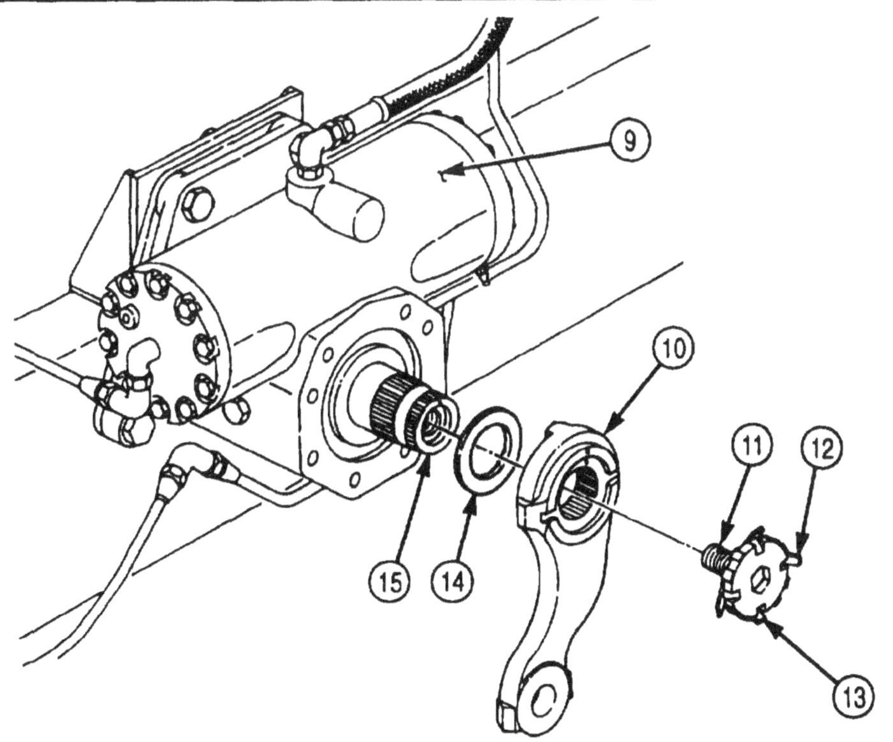

4-118. STEERING GEAR (SHEPPARD) AND MOUNTING BRACKET REPLACEMENT (Contd)

CAUTION

Cap or plug all openings immediately after disconnecting lines and hoses to prevent contamination. Failure to do so may result in steering system damage.

NOTE

- Have container ready to catch oil from disconnected lines.
- Tag all hydraulic lines and fittings for installation.

6. Disconnect oil pressure line (4) and oil return line (9) from adapter elbow (3) and adapter (8).
7. Disconnect assist cylinder pressure lines (14) and (16) from adapter elbow (15) and elbow (13).
8. Remove adapter elbow (3) and adapter (8) from steering gear (6).
9. Remove O-rings (5) and (7) from adapter elbow (3) and adapter (8). Discard O-rings (5) and (7).

NOTE

Perform steps 10 through 12 if old steering gear is to be reused.

10. Remove elbow (1) and adapter elbow (15) from front of steering gear (6).
11. Remove elbow (13), line (12), and adapter (11) from bottom rear of steering gear (6).
12. Remove O-rings (2) and (10) from elbow (1) and adapter (11). Discard O-rings (2) and (10).

NOTE

Assistant will help with stop 13.

13. Remove four locknuts (17), washers (18), screws (21), steering gear (6), and mounting plate (20) from left frame rail (23). Discard locknuts (17).
14. Remove four locknuts (24), screws (22), and mounting bracket (19) from left frame rail (23). Discard locknuts (24).

b. Installation

1. Install mounting bracket (19) on left frame rail (23) with four screws (22) and new locknuts (24). Do not tighten locknuts (24).
2. Install mounting plate (20) and steering gear (6) housing on left frame rail (23) and mounting bracket (19) with four screws (21), washers (18), and new locknuts (17). Tighten locknuts (17) 260-280 lb-ft (353-380 N•m).
3. Tighten locknuts (24) 60-70 lb-ft (81-85 N•m).

NOTE

Perform steps 4,5, and 6 if old steering gear is to be installed.

4. Install new O-rings (2) and (10) on elbow (1) and adapter (11).
5. Install elbow (1) and adapter elbow (16) on steering gear (6).
6. Install adapter (11), line (12), and elbow (13) on steering gear (6).
7. Install new O-rings (6) and (7) on adapter elbow (3) and adapter (8).
8. Install adapter elbow (3) and adapter (8) on steering gear (6).

NOTE

Wrap male pipe threads with antiseize tape before installation.

9. Connect assist cylinder pressure lines (14) and (16) to elbow (13) and adapter elbow (15).
10. Connect oil return line (9) and oil pressure line (4) to adapter (8) and adapter elbow (3).

4-118. STEERING GEAR (SHEPPARD) AND MOUNTING BRACKET REPLACEMENT (Contd)

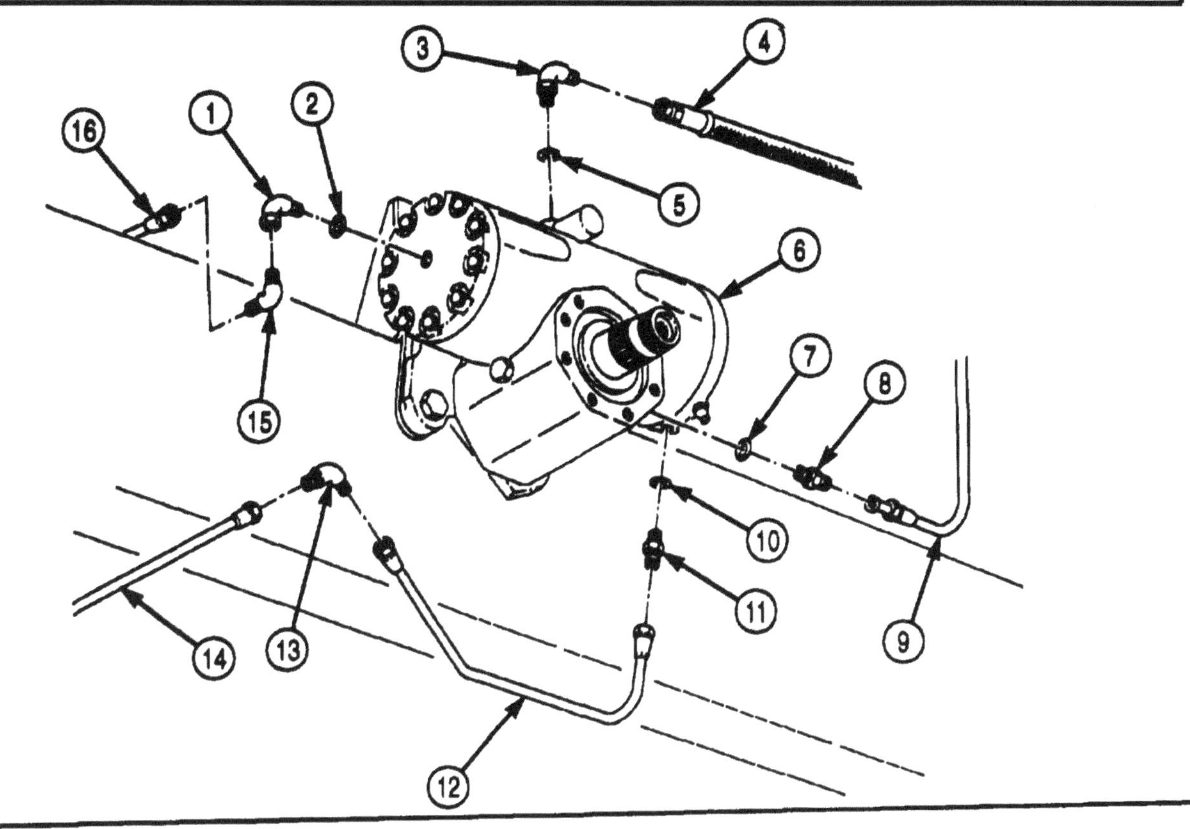

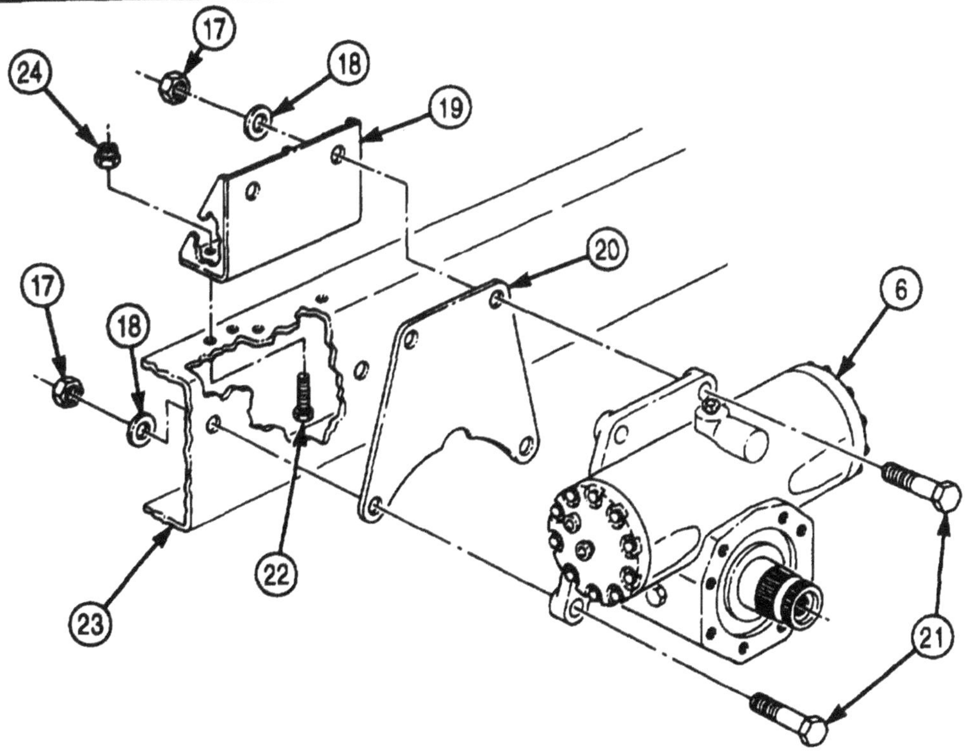

4-118. STEERING GEAR (SHEPPARD) AND MOUNTING BRACKET REPLACEMENT (Contd)

11. Press new friction washer (9) and new lockwasher (11) into slot on new retainer (4).
12. Press three new nylon balls (10) into indentations on retainer (4).
13. Aligning timing marks (1) and (2), position new seal (7) and pitman arm (3) on output shaft (8).
14. Install retainer (4) into output shaft (8) until drag of friction washer (9) is felt.
15. Align long tabs (5) to notches in pitman arm (3) and bend tabs (5) in notches. Back out retainer (4) up to 1/2 turn as necessary for tabs (5) alignment.
16. Tighten retainer (4) 225 lb-ft (305 N•m).
17. Bend two short tabs (6) into notches on retainer (4). Tighten retainer (4) if necessary to align tabs (6) with notches of retainer (4).

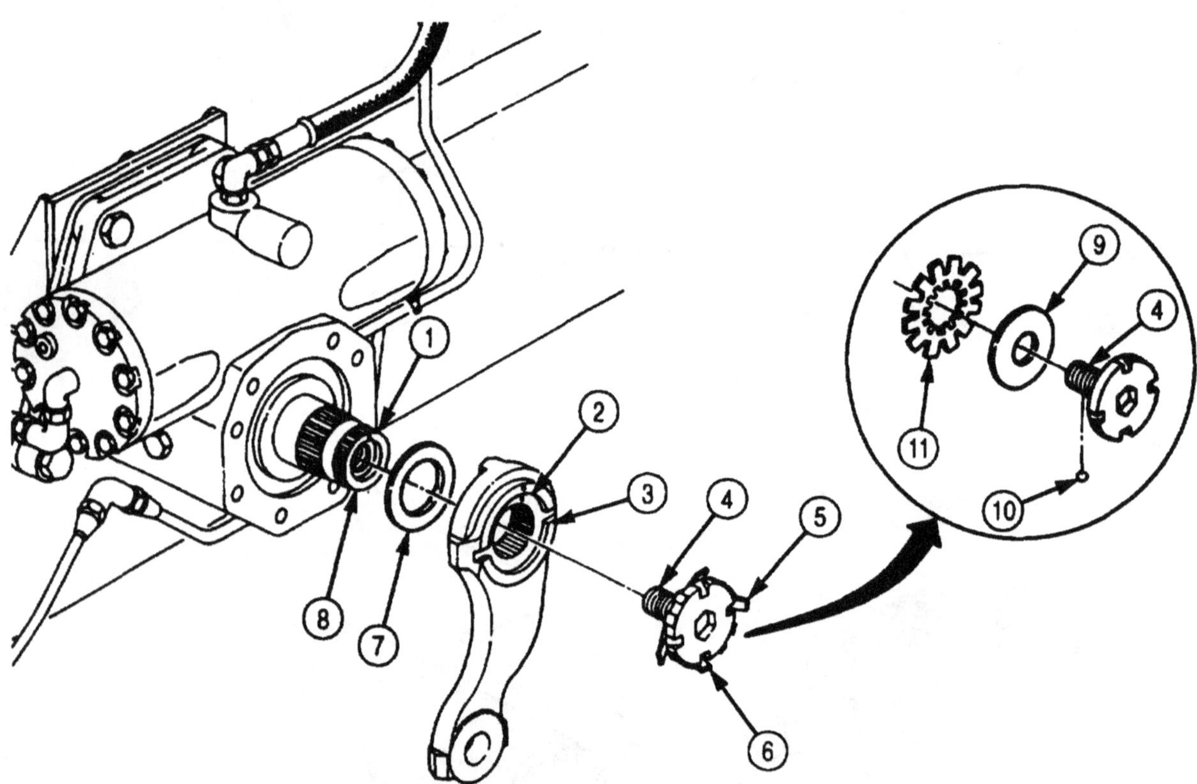

4-118. STEERING GEAR (SHEPPARD) AND MOUNTING BRACKET REPLACEMENT (Contd)

NOTE

- Before installing lower steering column, ensure steering wheel spokes form a Y.
- Assistant will steady steering wheel while universal joint is connected to steering gear.

18. Position universal joint (16) onto steering gear input shaft (17). Ensure screw holes in sleeve of universal joint (16) align with groove in shaft (17).
19. Install universal joint (16) on input shaft (17) with screw (15), new lockwasher (14), and nut (13). Tighten nut (13) 28-34 lb-ft (38-46 N•m).

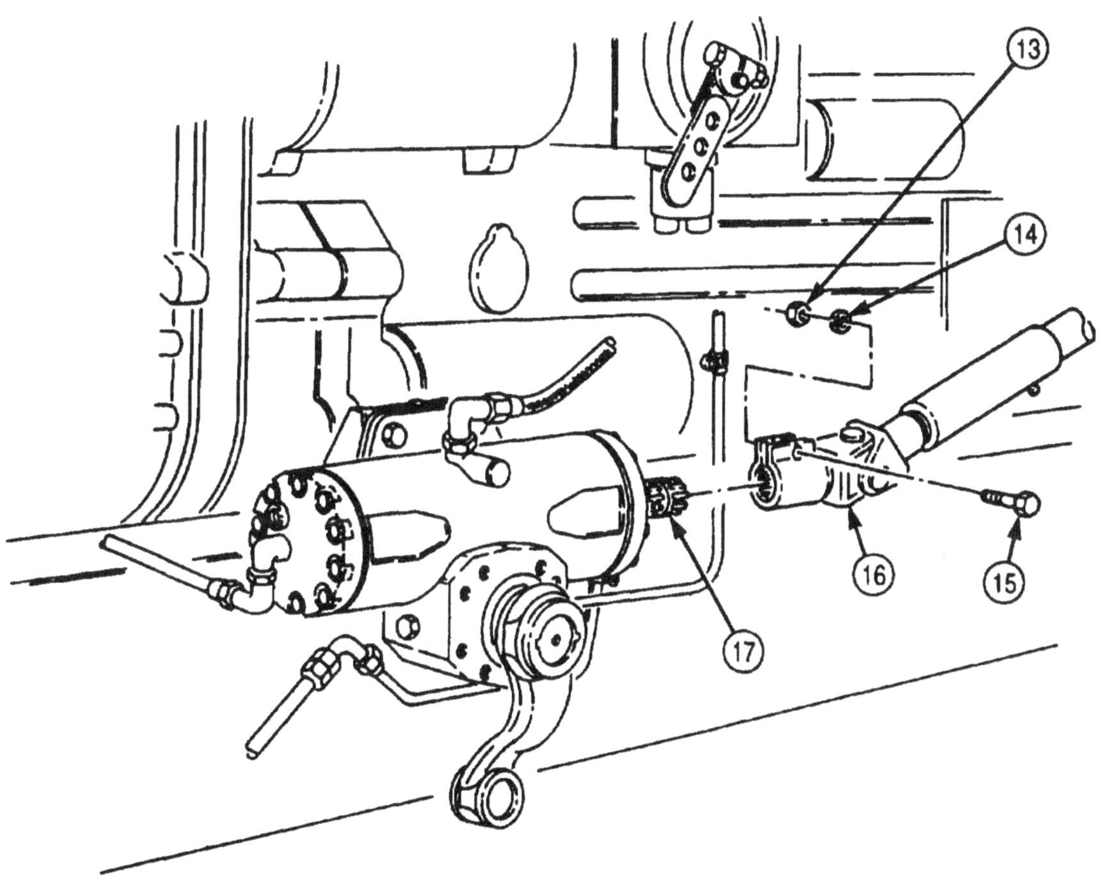

FOLLOW-ON TASKS:
- Install steering gear stone shield (para. 3-238).
- Install left splash shield (para. 3-301).
- Install drag link (para. 3-229).
- Fill steering gear to proper oil level (LO 9-2320-272-12).
- Start engine (TM 9-2320-272-10) and road test vehicle.

4-119. STEERING GEAR (ROSS) REPAIR

THIS TASK COVERS:

a. Disassembly
b. Cleaning, Inspection, and Repair
c. Assembly
d. Final Adjustment

INITIAL SETUP:

APPLICABLE MODELS

All

TOOLS

General mechanic's tool kit (Appendix E, Item 1)
Torque wrench (Appendix E, Item 145)
Arbor press
Mandrel
Vise

MATERIALS/PARTS

Gasket (Appendix D, Item 244)
Three O-rings (Appendix D, Item 483)
Two seal rings (Appendix D, Item 642)
Two lock tabs (Appendix D, Item 343)
Retaining ring (Appendix D, Item 540)
Two seals (Appendix D, Item 604)
Backing ring (Appendix D, Item 8)
Seal ring (Appendix D, Item 643)
Packing (Appendix D, Item 512)
O-ring (Appendix D, Item 482)
O-ring (Appendix D, Item 481)
Plastic backup washer (Appendix D, Item 521)
Two-piece seal (Appendix D, Item 701)

MATERIALS/PARTS (Cod)

Two O-rings (Appendix D, Item 484)
O-ring (Appendix D, Item 485)
GAA grease (Appendix C, Item 28)
Lubricating oil (Appendix C, Item 48)
Drycleaning solvent (Appendix C, Item 71)

REFERENCES (TM)

LO 9-2320-272-12
TM 9-2320-272-24P

EQUIPMENT CONDITION

- Steering gear removed (para. 4-117).
- Steering gear drained (para. LO 9-2320-272-12).

GENERAL SAFETY INSTRUCTIONS

- Keep fire extinguisher nearby when using drycleaning solvent.
- Drycleaning solvent is flammable and toxic. Do not use near open flame.
- Compressed air source will not exceed 30 psi (207 kPa).
- Eyeshields must be worn when cleaning with compressed air.

a. Disassembly

1. Position steering gear (3) in vise with sector shaft (2) in horizontal position.
2. Remove all paint or corrosion and position serrated end of sector shaft (2) with alignment mark (1) up. Adjust position of mark (1) by turning wormshaft (4).

4-119. STEERING GEAR (ROSS) REPAIR (Contd)

3. Loosen jamnut (5) on sector shaft adjusting screw (6).
4. Remove outer seal (13) from trunnion cover (8). Discard seal (13).
5. Remove four screws (7) from trunnion cover (8) and separate cover (8) from gear housing (3).
6. Remove packing (12), O-ring (11), O-ring (10), and plastic backup washer (9) from cover (8). Discard packing (12), O-rings (11) and (10), and plastic backup washer (9).
7. Remove six screws (20) from steering gear side cover (18) and steering gear housing (3).
8. Remove side cover (18), sector shaft (2), and gasket (19) from gear housing (3). Discard gasket (19).
9. Remove jamnut (5) from sector shaft adjusting screw (6).

NOTE

Sector shaft adjusting screw is part of sector shaft.

10. Turn sector shaft adjusting screw (6) clockwise until sector shaft (2) separates from side cover (18).
11. Remove seal retainer (21), two-piece seal (14), retaining ring (15), steel backup washer (16), and forty-four roller bearings (17) from side cover (18). Discard two-piece seal (14) and retaining ring (15).

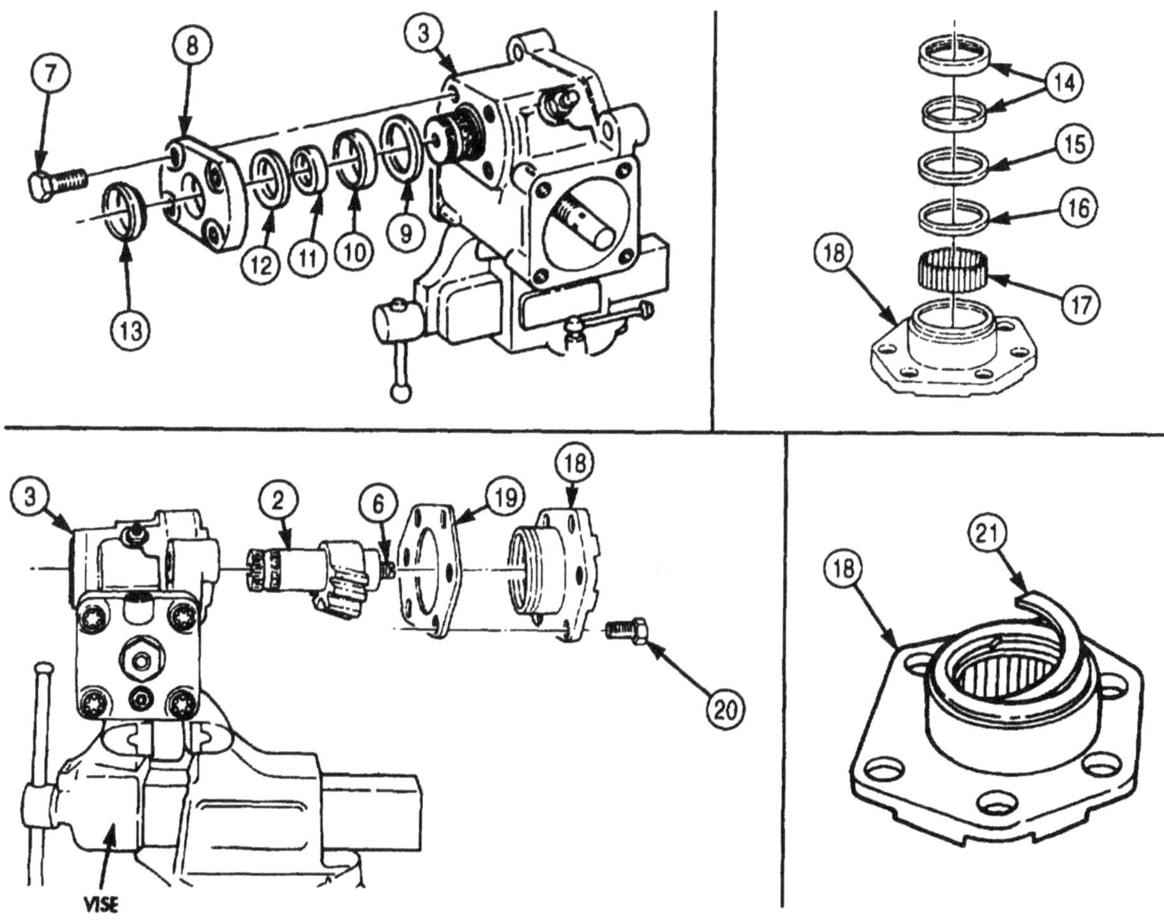

4-119. STEERING GEAR (ROSS) REPAIR (Contd)

12. Remove jamnut (5) and wormshaft adjusting screw (6) from end cover (2).
13. Remove jamnut (7) and poppet adjusting screw (8) from end cover (2).
14. Remove four screws (4) and washers (3) from end cover (2).
15. Remove end cover (2) and O-ring (9) from gear housing (1). Discard O-ring (9).
16. Remove four screws (11) from valve housing (10) and gear housing (1).

CAUTION

Alignment marks are located on valve sleeve and wormshaft so they can be assembled in their original position. Failure to align marks may cause damage to steering gear.

17. Remove valve housing (10) from gear housing (1), observing alignment marks (13) in valve housing (10) and on wormshaft (12).

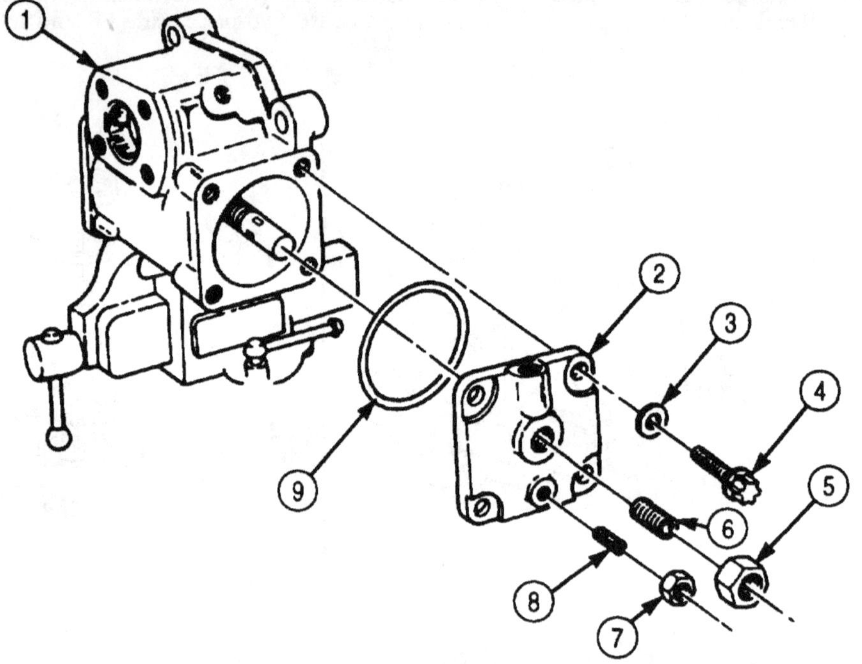

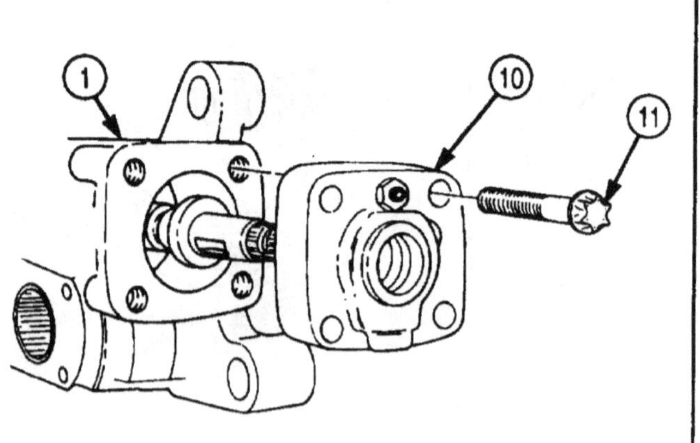

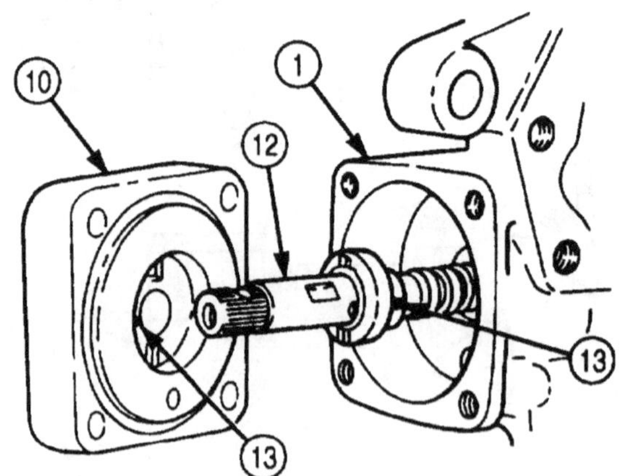

4-119. STEERING GEAR (ROSS) REPAIR (Contd)

18. Remove valve sleeve (18). inner thrust washer (17), thrust bearing (16), outer thrust washer (15), and O-ring (14) from mating side of valve housing (10). Discard O-ring (14).
19. Remove seal (19), retaining ring (20), steel backup washer (21), seal cup (25), and O-ring (24) from outside of bore of valve housing (10). Discard O-ring (24) and seal (19).
20. Remove jamnut (22) and poppet adjusting screw (23) from valve housing (10).
21. Remove two seal rings (26) and two O-rings (27) from valve sleeve (18). Discard O-rings (27) and seal rings (26).
22. Remove rack piston (29) and worm shaft (28) from end cover (2) side of gear housing (3).

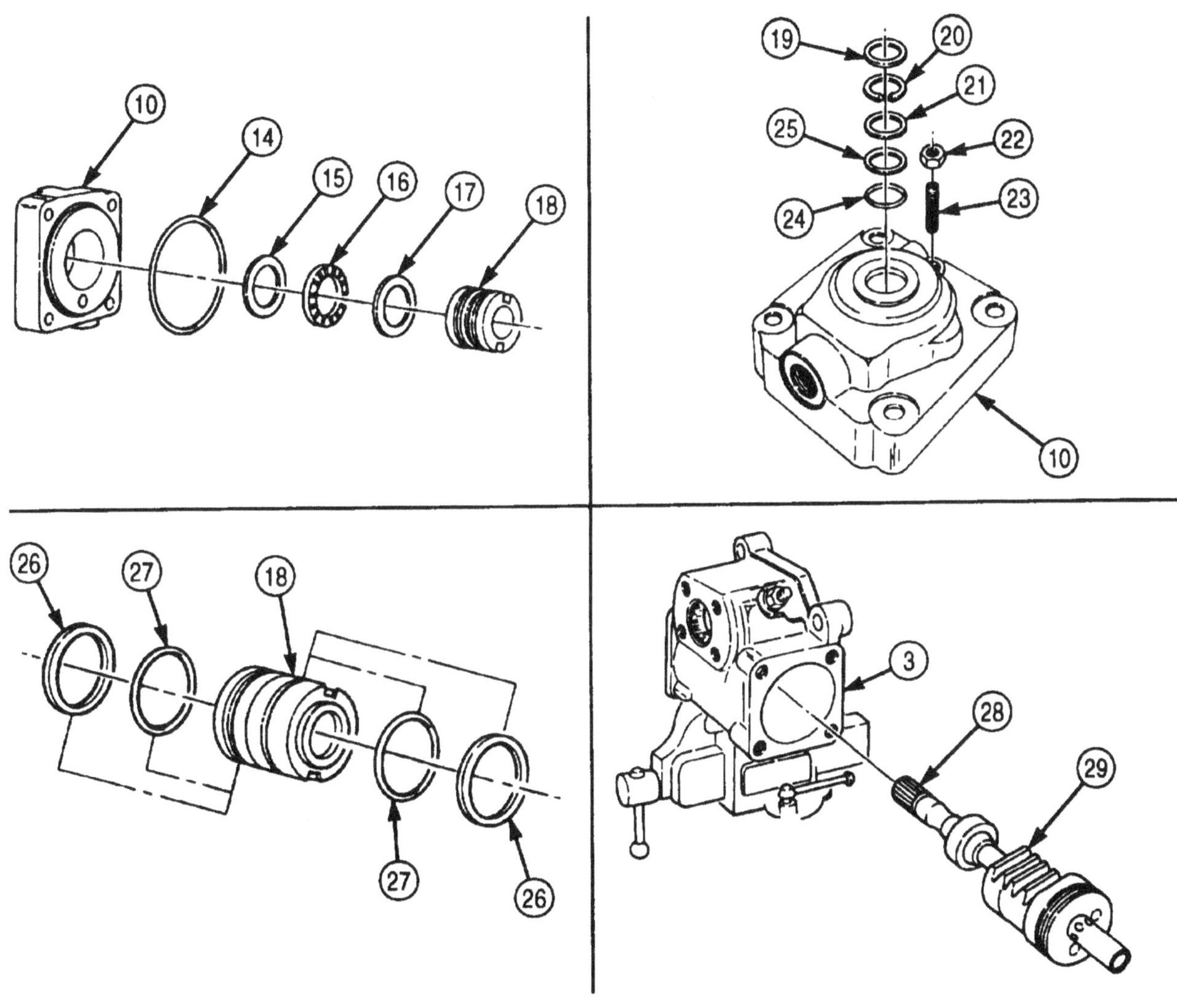

4-119. STEERING GEAR (ROSS) REPAIR (Contd)

23. Bend two locktabs (6) down and remove two screw (6), locktabs (5), and ball bearing return guide retaining clip (4) turn rack piston (2). Discard locktabs (6).
24. Remove two ball bearing return guides (3) from rack piston (2).
25. Turn rack piston (2) overso ball bearings (7) can roll out as wormshaft (1) is turned in each direction. Remove twenty-seven ball bearings (7).
26. Remove wormshaft (1) from rack piston (2).
27. Place rack piston (2) in a soft-jawed vise.
28. Remove poppet seat snapring (8) from each end of bore (13) in rack piston (2).
29. Push poppets (10) inward until poppet seats (9) can be removed from rack piston (2). Poppet spring (12) and spring rod (11) may partially come out of piston (2) when first poppet seat (9) is removed.
30. Remove two poppets (10), spring (12), and spring rod (11) from poppet bore (13).

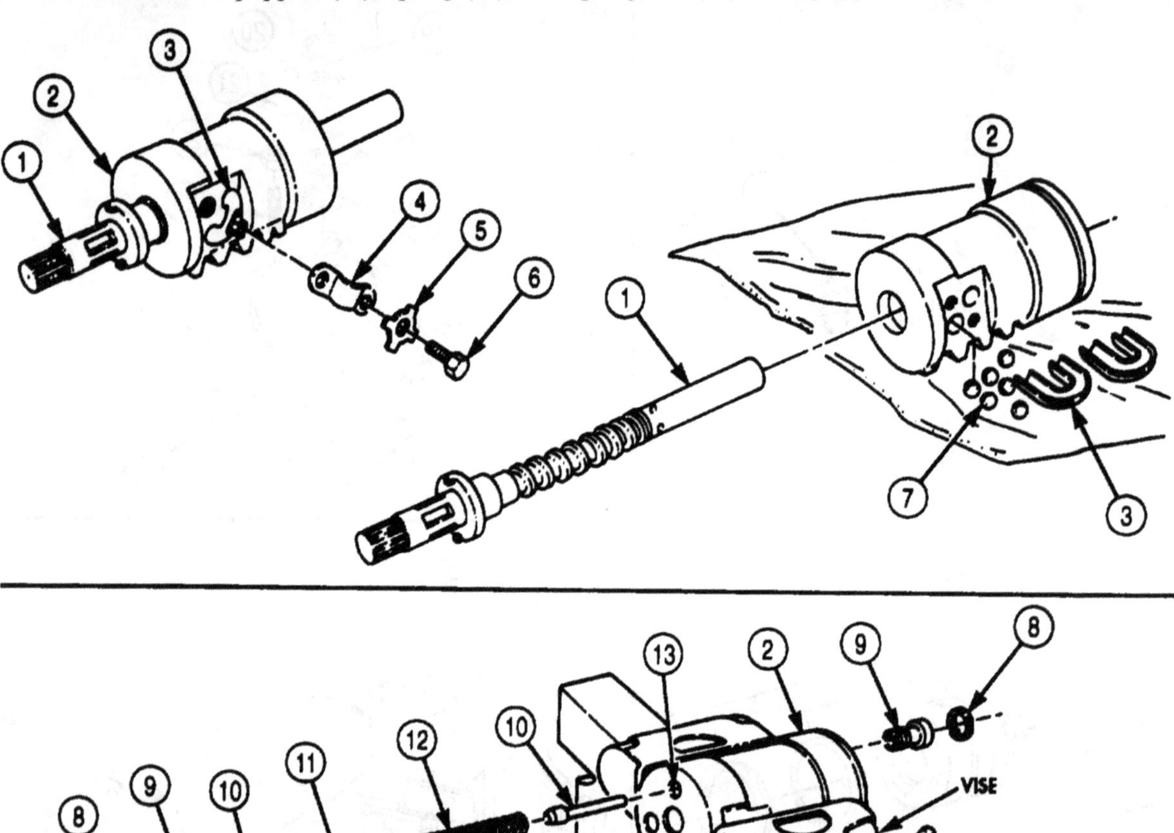

4-119. STEERING GEAR (ROSS) REPAIR (Contd)

31. Remove O-ring (14) and backing ring (16) from rack piston (2). Discard O-ring (14) and backing ring (16).
32. Remove seal ring (17) and retainer (16) from wormshaft (1). Discard seal ring (17).

NOTE

Perform step 33 only if sector shaft roller bearing requires replacement.

33. Using mandrel and arbor press, press roller bearing (19) out of gear housing (18). Discard roller bearing (19).

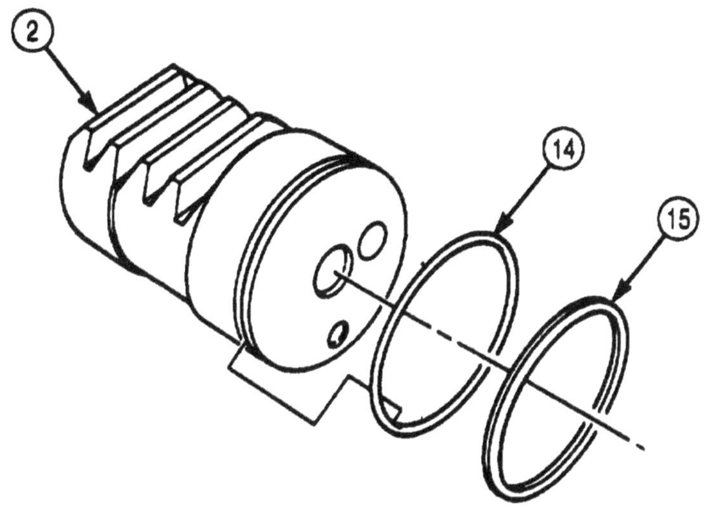

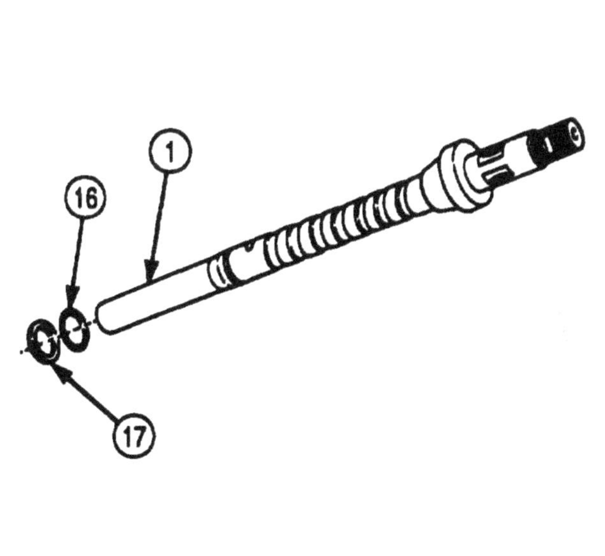

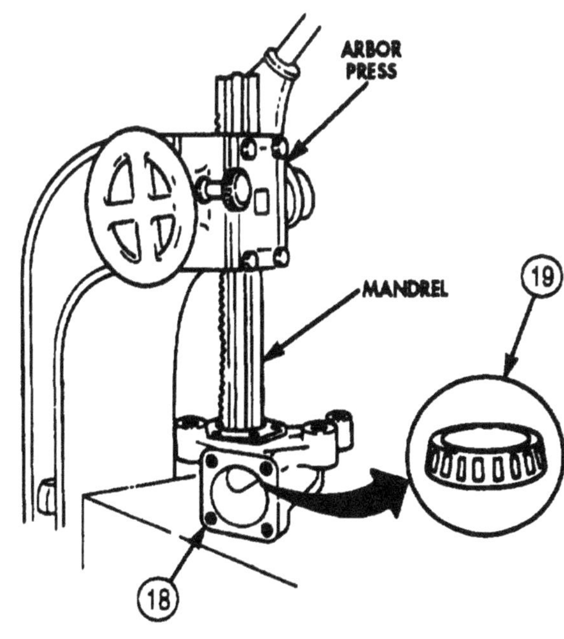

4-119. STEERING GEAR (ROSS) REPAIR (Contd)

b. Cleaning, Inspection, and Repair

WARNING

- Drycleaning solvent is flammable and toxic. Do not use near an open flame and always have a fire extinguisher nearby when solvents used. Use only in well-ventilated places, wear protective clothing, and dispose of cleaning rags in approved container. Failure to do this may result in injury to personnel and/or damage to equipment.
- Compressed air source will not exceed 30 psi (207 kPa). When cleaning with compressed air, eyeshields must be worn. Failure to wear eyeshields may result in injury to personnel.

1. Clean all steering gear components with drycleaning solvent and dry with compressed air.

NOTE

Light polished areas of steering gear components indicate normal wear.

2. Inspect bearing areas and tooth surfaces of sector shaft (3) and side cover (1) for pitting. Replace sector shaft (3) and side cover (1) if pitted.
3. Inspect roller bearings (2) in side cover (1) for pitting and scoring. Replace side cover (1) if roller bearings (2) are pitted or scored.
4. Inspect valve housing thrust bearing (6) and two thrust washers (5) for pitting and scoring. Replace thrust bearing (6) if pitted or scored
5. Inspect rack piston (9) for broken or missing teeth (10). Replace rack piston (9) if teeth (10) are broken or missing.
6. Inspect wormshaft (8) for dents and chipping. Replace wormshaft (8) if dented or chipped.
7. Inspect groove (7) in wormshaft (8) for chipping. Replace wormshaft (8) if grooves (7) are chipped.
8. Inspect gear housing (4) for cracks, scored bore, or damage. Replace gear housing (4) if cracked, bore is scored, or gear housing (4) is otherwise damaged.

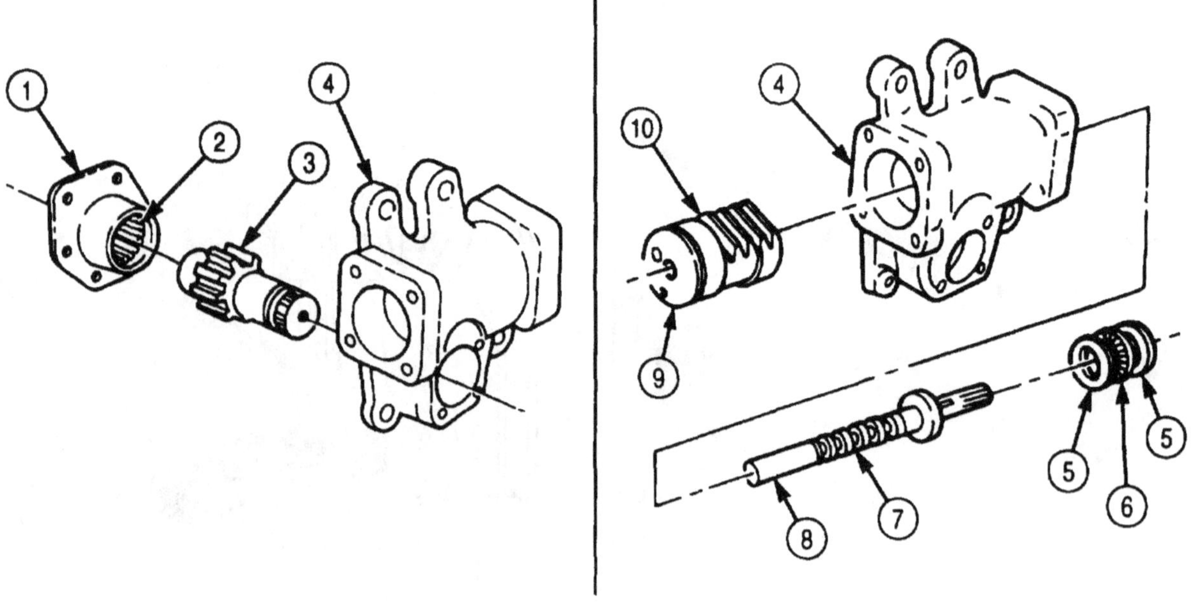

4-119. STEERING GEAR (ROSS) REPAIR (Contd)

c. Assembly

NOTE

- Coat all steering gear parts with clean lubricating oil before assembly.
- Step 1 is performed only if new roller bearing is installed.

1. Using arbor press and mandrel, press new roller bearing (2) into gear housing (4).
2. Install new O-ring (11) on rack piston (9).
3. Position rack piston (9) in soft-jawed vise.
4. Place two poppets (14), spring (16), and spring rod (15) in bore (17) of rack piston (9).
5. Install two poppet seats (13) on rack piston (9) with two snaprings (12). Tighten seats (13) 20-25 lb-ft (27-34 N•m).

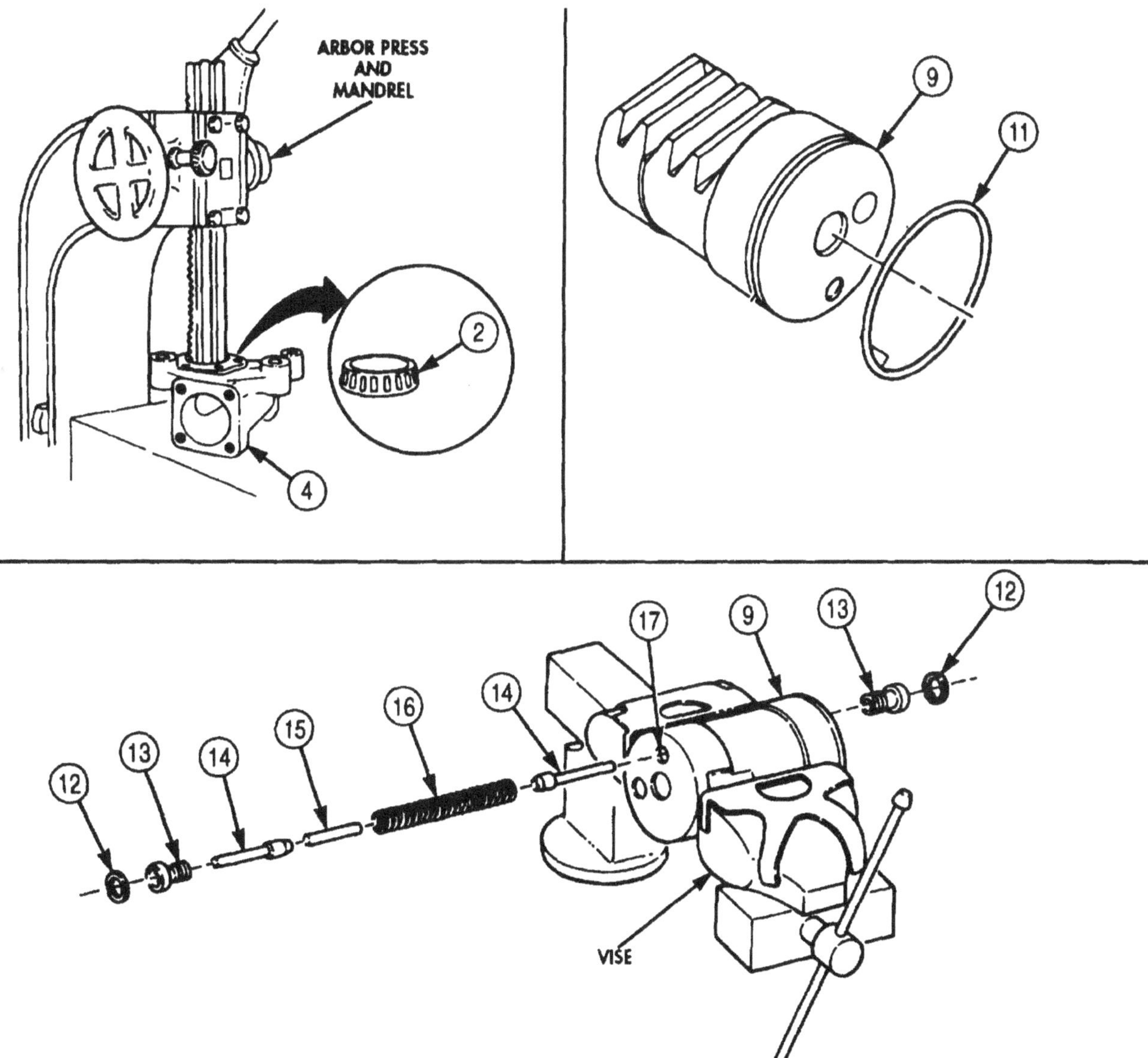

4-119. STEERING GEAR (ROSS) REPAIR (Contd)

6. Install new seal ring (1) on wormshaft (2).
7. Lightly coat wormshaft (2) with GM grease and install wormshaft (2) with end opposite seal ring (1) installed first into rack piston (3).
8. Position two ball bearing return guides (4) in rack piston (3).

NOTE

Ensure ball bearing return guides stay in place while installing ball bearings.

9. Drop twenty-seven ball bearings (5) through holes in ball bearing return guides (4) as worm shaft (2) is rotated in each direction.
10. Secure two ball bearing return guides (4) on rack piston (3) with retaining clip (6), two new locktabs (7), and screws (8). Tighten screws (8) 14-17 lb-ft (18-23 N•m) and bend locktabs (7) against flats on heads of screws (8).

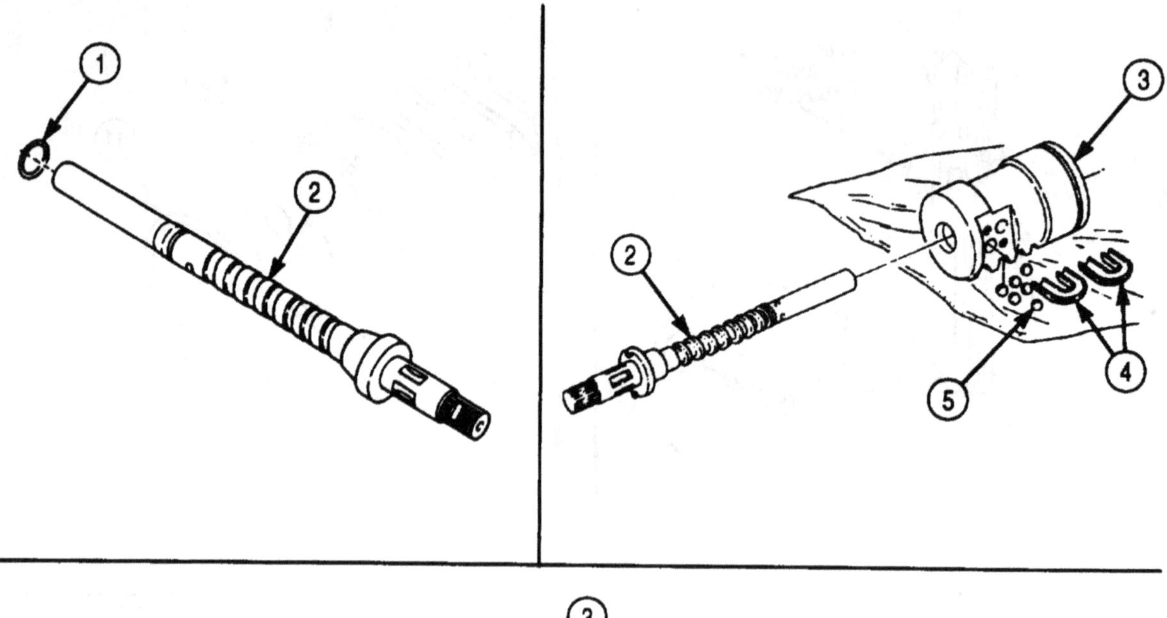

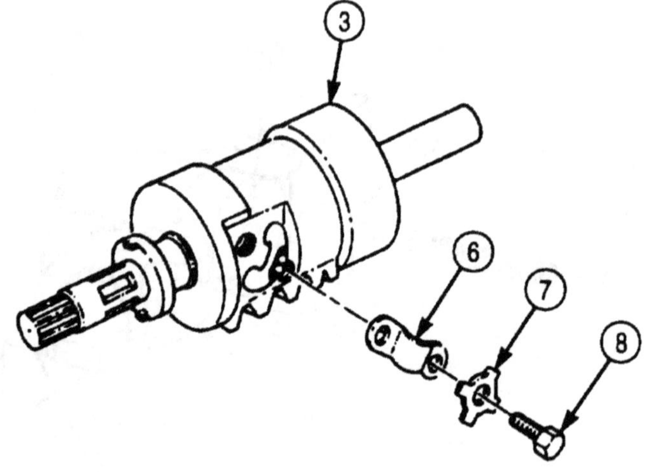

4-119. STEERING GEAR (ROSS) REPAIR (Contd)

11. Coat new backing ring (10) with clean GAA grease and install on rack piston (3).
12. Install rack piston (3) and wormshaft (2) through end cover side of gear housing (9).
13. Install wormshaft adjusting screw (14) and jamnut (15) a few turns in end cover (11).
14. Install poppet adjusting screw (17) and jamnut (16) a few turns in end cover (11).
15. Install new O-ring (18) and end cover (11) on housing (9) with four washers (12) and screws (13). Tighten screws (13) 105-115 lb-ft (142-156 N•m).
16. Install poppet adjusting screw (21) and jamnut (20) a few turns in valve housing (19).
17. Lubricate new valve housing O-ring (22) and install in groove (25) on valve housing (19).
18. Install thrust washer (23) and thrust bearing (24) in bore of valve housing (19).

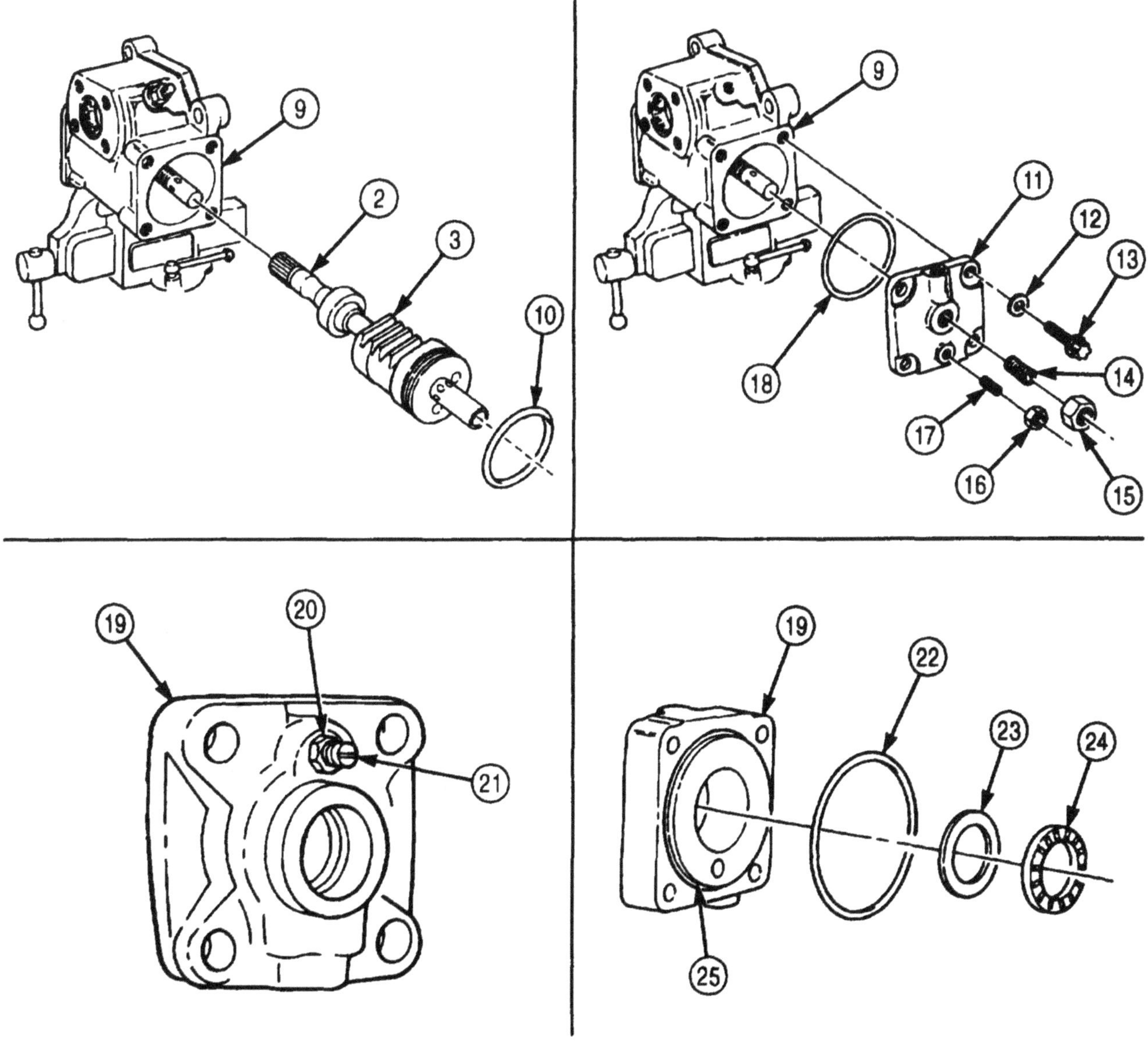

4-119. STEERING GEAR (ROSS) REPAIR (Contd)

19. Install two new O-rings (2) and new seal rings (1) on valve sleeve (3).
20. Coat end of valve sleeve (3) with GM grease and install thrust washer (5) on sleeve (3).
21. Install thrust washer (5) and sleeve (3) into valve housing (4). Valve sleeve (3) should be approximately 0.40 in. (10 mm) in from face of valve housing (4).
22. Position valve sleeve (3) in valve housing (4) so alignment marks (6) are aligned.

CAUTION

Alignment marks are located on valve sleeve, wormshaft, and valve housing so they can be assembled in their original position. Failure to align marks may cause damage to steering gear.

23. Position valve housing (4) on wormshaft (7).
24. Using a 12-point box-end wrench at an angle, rotate wormshaft (7) as needed to align marks (8) on wormshaft (7) and valve housing (4).
25. Install valve housing (4) on gear housing (9) with four screws (10). Tighten screws (10) 105-115 lb-ft (142-156 N•m).

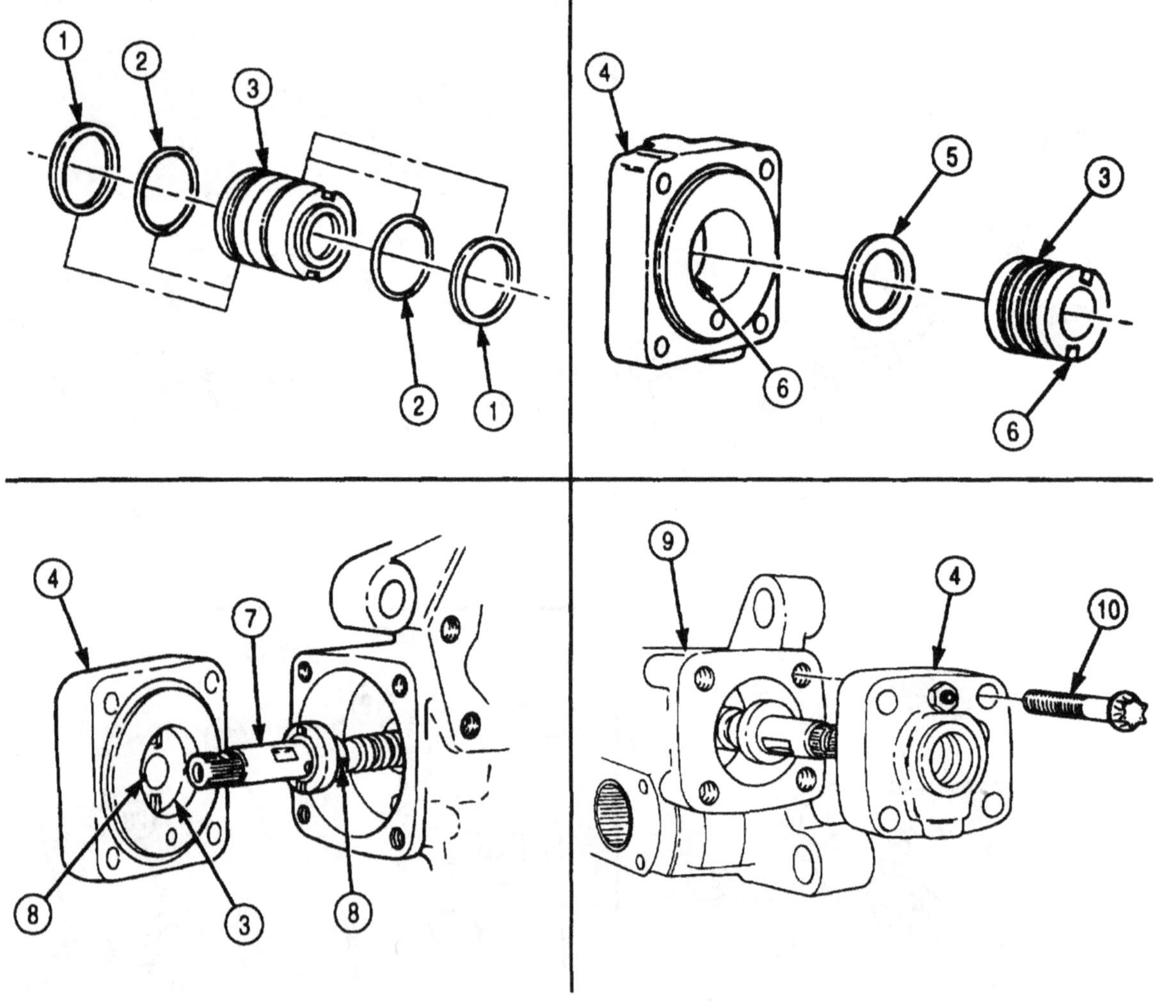

4-119. STEERING GEAR (ROSS) REPAIR (Contd)

26. Apply a generous amount of GAA grease to bearing race inside of cover (15) and install forty-four bearing rollers (14) in side cover (15).
27. Install steel backup washer (13), new retaining ring (12), and new two-piece seal (11) in steering gear side cover (15) with seal retainer (16).
28. Apply clean GAA grease to short bearing end of sector shaft (17) and place in side cover (15).

NOTE

Adjusting screw has left-hand threads.

29. Install sector shaft adjusting screw (18) counterclockwise through side cover (15) and into sector shaft (17).
30. Loosen adjusting screw (18) one turn clockwise to permit sector shaft (17) to turn freely in side cover (15).
31. Thread jamnut (19) a few turns onto adjusting screw (18).

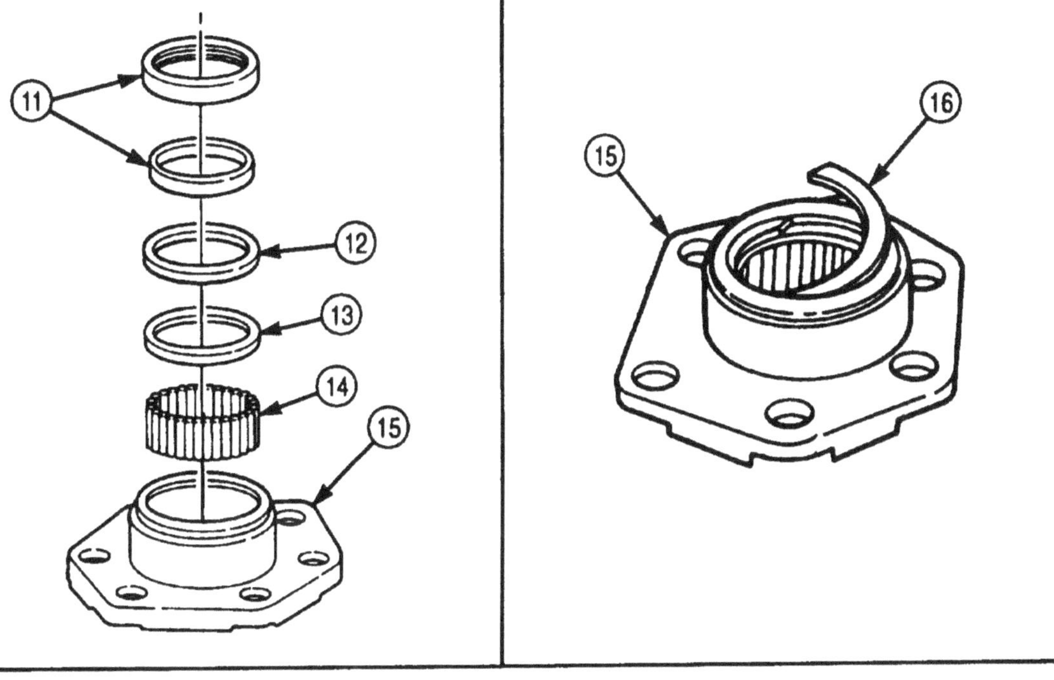

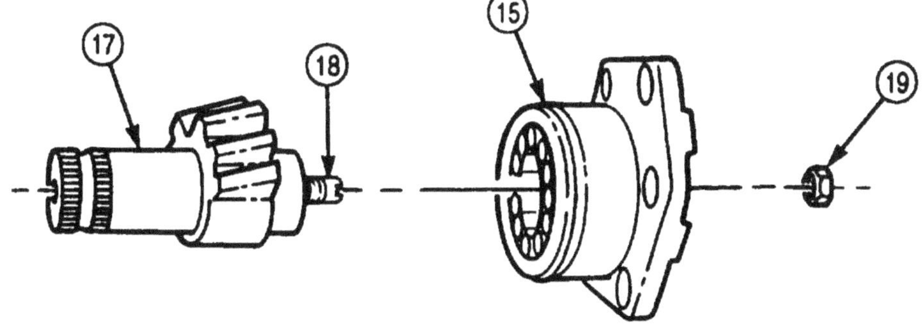

4-119. STEERING GEAR (ROSS) REPAIR (Contd)

32. Rotate wormshaft (2) until rack piston (1) is at center of steering gear travel.

CAUTION

When installing sector shaft, do not pinch new side cover gasket.

33. Align center tooth of sector shaft (7) to third notch from seal ring end of rack piston (1) and slide sector shaft (7) into gear housing (3).
34. Install new gasket (4) and side cover (6) on gear housing (3) with six screws (5). Using a criss-cross pattern, tighten screws (5) 150 lb-ft (203 N•m), followed by tightening to 210-240 lb-ft. (298-326 N•m).

CAUTION

The words OIL SIDE must be visible once the two-piece seal is assembled in trunnion cover. Otherwise, seal will not function, which could result in a loss of power steering assist.

35. Install new plastic backup washer (12), new two-piece seal (11), and new O-ring (10) in trunnion cover (9).
36. Ensuring that OIL SIDE of two-piece seal (11) faces housing (3), slide trunnion cover (9) carefully over serrated end of sector shaft (7) and install on gear housing (3) with four screws (8). Tighten screws 15-22 lb-ft (20-30 N•m).

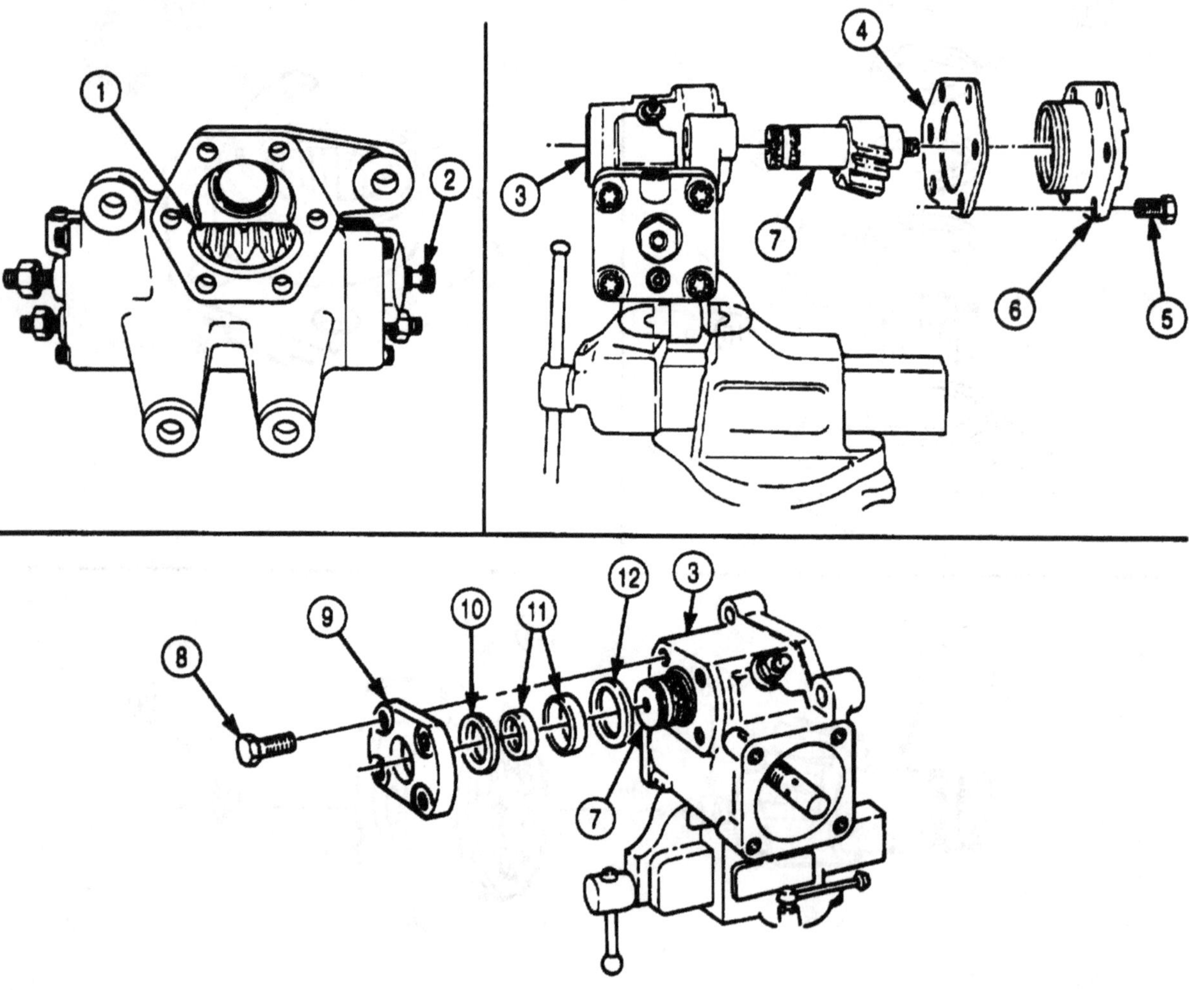

4-119. STEERING GEAR (ROSS) REPAIR (Contd)

37. Install new O-ring (17) into groove of seal cup (16).
38. Insert real cup (16) and steel backup washer (15) in valve housing (3) and seat using a real driver.
39. Using a real driver, install new seal (13) into valve housing (3) and then install retaining ring (14).
40. Using a real driver, install new outer seal (18) in trunnion cover (9).

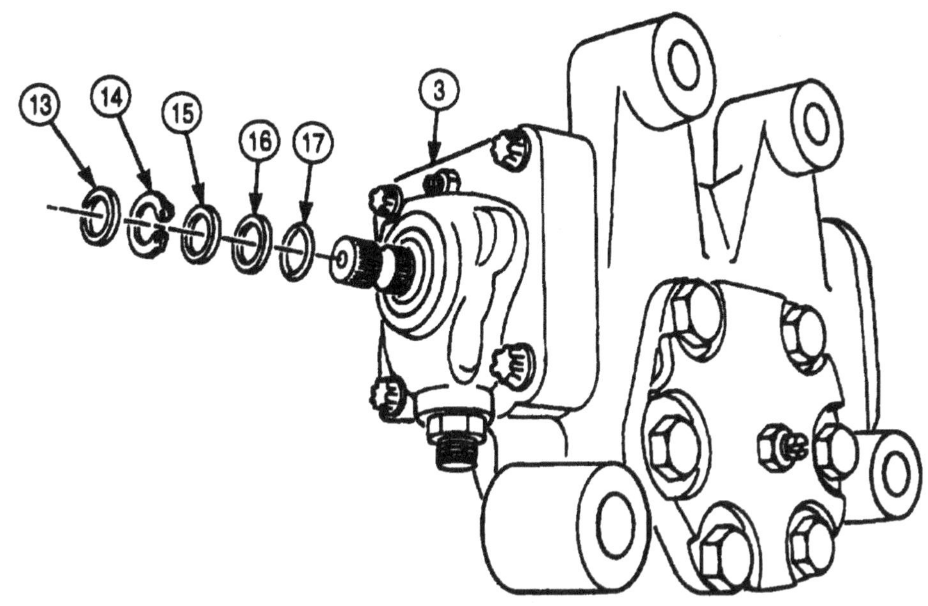

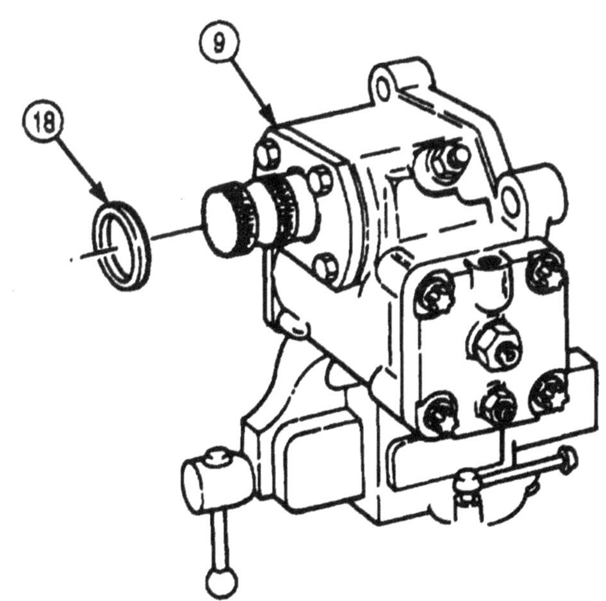

4-119. STEERING GEAR (ROSS) REPAIR (Contd)

d. Final Adjustment

1. Wormshaft (8) adjustment:
   a. Position steering gear (1) in vise with sector shaft (4) up and pointed away from bench.
   b. Set sector shaft (4) with alignment mark (5) up. This places sector shaft (4) in center of steering gear travel.
   c. Turn wormshaft adjusting screw (3) until seated against wormshaft (8).
   d. Tighten adjusting screw (3) 5-10 lb-ft (7-14 N•m).
   e. Back out adjusting screw (3) one turn.
2. Sector shaft (4) adjustment:
   a. Position torque wrench and 12-point socket on wormshaft (8) at an angle.
   b. While moving wormshaft (8) 1/4 turn each side of center, tighten sector shaft adjusting screw (7) until torque on wormshaft (8) is 25-30 lb-in. (2.8-3.4 N•m).
   c. Back out sector shaft adjusting screw (7) one turn. Note torque required to rotate wormshaft (8) 1/4 turn on each side of center of wormshaft (8). Record torque required.
   d. Move sector shaft adjusting screw (7) to provide a rise of 2-6 lb-in. (0.2-0.7 N•m) at a point 45 degrees each side of center of wormshaft (8).
   e. Tighten sector shaft adjusting screw jamnut (6) 45-50 lb-ft (62-68 N•m).
   f. Check torque required to rotate wormshaft (8). The torque must not exceed 20 lb-in. (2.3 N•m) to turn wormshaft (8) after jamnut (6) has been tightened. If torque exceeds 20 lb-in. (2.3 N•m), repeat step 2.
3. Wormshaft adjusting screw (3) adjustment:
   a. Using torque wrench and socket on wormshaft (8) and moving it 45 degrees each side of center, tighten wormshaft adjusting screw (3) until torque is 10-15 lb-in (1.1-1.7 N•m) higher than torque in step 2.
   b. Tighten jamnut (2) 70-80 lb-ft (95-108 N•m).
4. Rotate wormshaft (8) 90 degrees each side of center and note torque required. If torque exceeds 35 lb-in. (4 N•m), repeat step 3.

4-119. STEERING GEAR (ROSS) REPAIR (Contd)

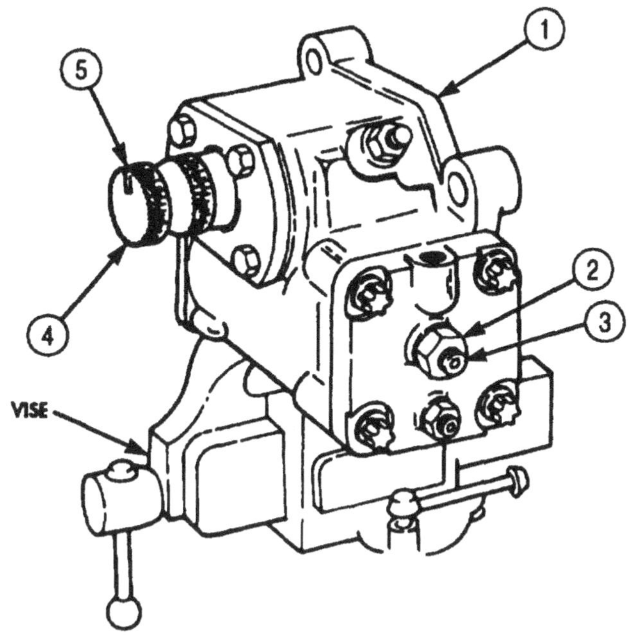

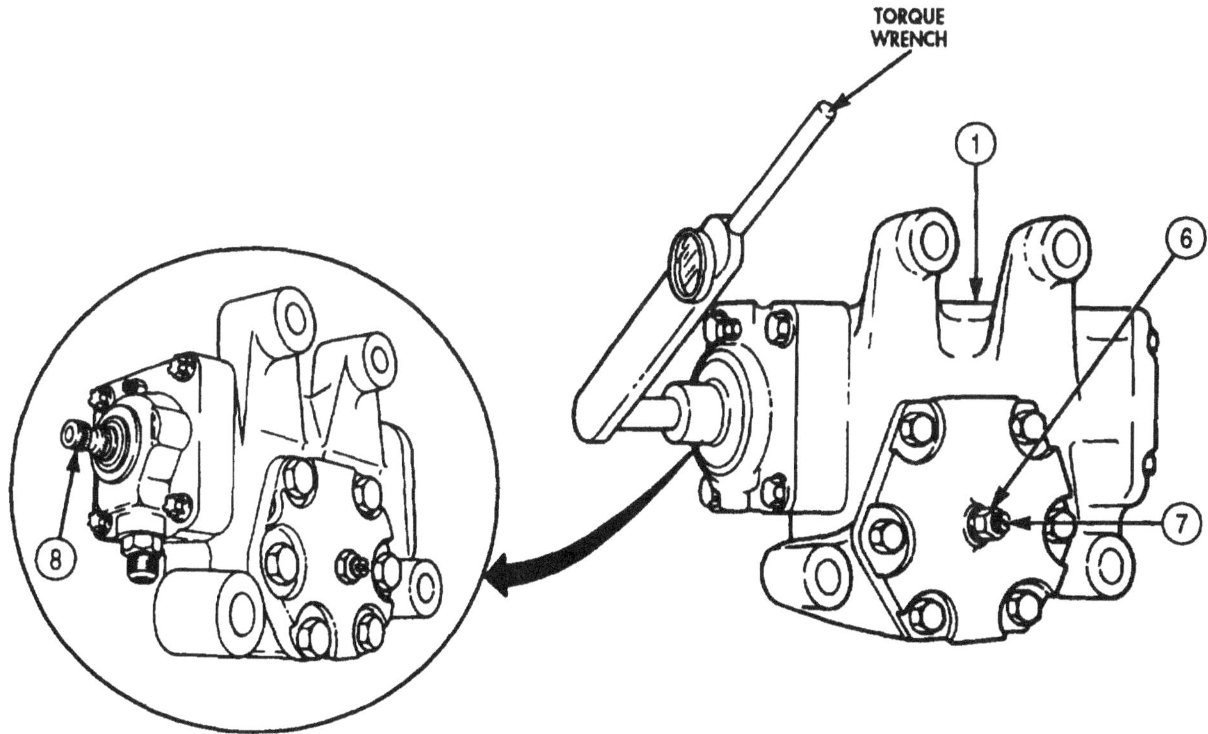

FOLLOW-ON TASKS:
- Install steering gear (para. 4-117).
- Fill steering gear to proper oil level (LO 9-2320-272-12).
- Perform final poppet valve adjustments (para. 4-123).

4-120. STEERING GEAR (SHEPPARD) REPAIR

THIS TASK COVERS:

a. Disassembly
b. Cleaning, Inspection, and Repair
c. Assembly

INITIAL SETUP

APPLICABLE MODELS
M939A2

TOOLS
General mechanic's tool kit (Appendix E, Item 1)
Torque wrench (Appendix E, Item 144)
Mandrel

MATERIALS/PARTS
Bearing retainer (Appendix D, Item 13)
Twenty lockwashers (Appendix D, Item 364)
Gasket and seal set (Appendix D, Item 247)
Steering parts kit (Appendix D, Item 678)
Eight lockwashers (Appendix D, Item 417)
Lubricating oil (Appendix C, Item 48)
Crocus cloth (Appendix C, Item 20)
GAA grease (Appendix C, Item 28)
Drycleaning solvent (Appendix C, Item 71)

REFERENCES (TM)
LO 9-2320-272-12
TM 9-214
TM 9-2320-272-24P

EQUIPMENT CONDITION
- Steering gear drained (LO 9-2320-272-12).
- Steering gear (Sheppard) removed (para. 4-118).

GENERAL SAFETY INSTRUCTIONS
- Keep fire extinguisher nearby when using drycleaning solvent.
- Drycleaning solvent is flammable and toxic. Do not use near open flame.
- Compressed air source will not exceed 30 psi (207 kPa).
- Eyeshields must be worn when cleaning with compressed air.

NOTE

- Use a clean, protected work surface for steering gear disassembly, repair, and assembly.
- High-pressure seal and washer are obsolete.
- Ensure high-pressure seal kit, P/N 2370461, is available before performing procedure.

a. Disassembly

1. Remove two relief valve plungers (10) from cylinder head (3) and bearing cap (7).
2. Remove two O-rings (11) from relief valve plungers (10). Discard O-rings (11).
3. Remove eight screws (13) and lockwashers (12) from output shaft cover (14) and steering gear housing (5). Discard lockwashers (12).
4. Using crocus cloth, remove any rust or scale from exposed shaft, except splines, on output shaft and gear (16).

NOTE

Mark output shaft cover and housing for installation.

5. Remove output shaft cover (14) from steering gear housing (5).
6. Remove O-ring (15) from output shaft cover (14). Discard O-ring (15).

4-120. STEERING GEAR (SHEPPARD) REPAIR (Contd)

7. Remove output shaft and gear (16) from steering gear housing (5).
8. Remove ten screws (1), lockwashers (2), and cylinder head (3) from steering gear housing (5). Discard lockwashers (2).
9. Remove O-ring (4) from cylinder head (3). Discard O-ring (4)
10. Remove ten screws (9), lockwashers (8), and bearing cap (7) from steering gear housing (5). Discard lockwashers (8).
11. Remove O-ring (6) from bearing cap (7). Discard O-ring (6).

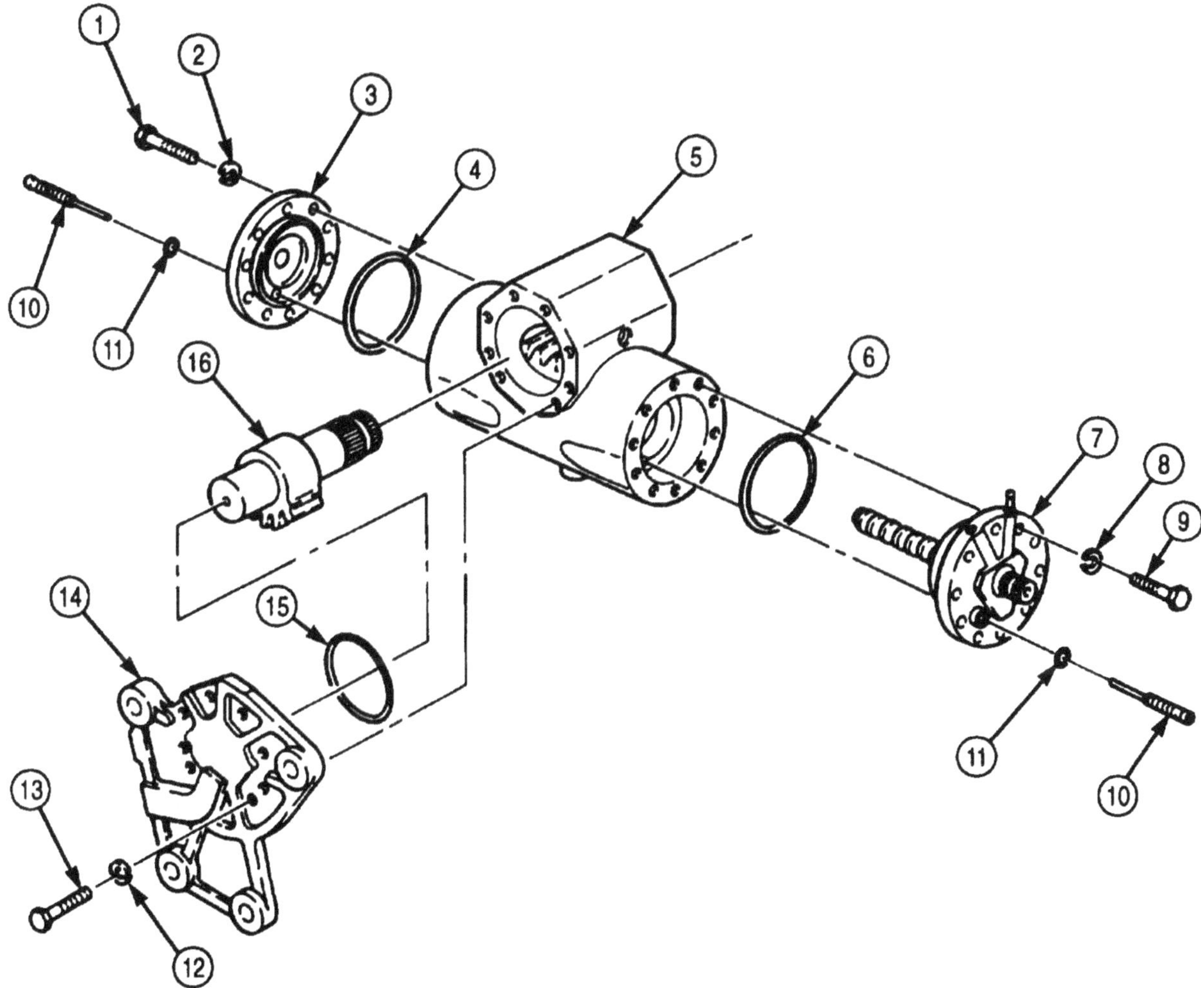

4-120. STEERING GEAR (SHEPPARD) REPAIR (Contd)

CAUTION

Support piston during removal. Failure to do so may result in damage to piston.

12. Remove piston (2) from steering gear housing (1).
13. Remove slipper seal (14), O-ring (13), and screen (18) from side of piston (2). Discard slipper seal (14), O-ring (13), and screen (18).
14. Remove two backup rings (11) and O-rings (12) from side of piston (2). Discard backup rings (11) and O-rings (12).
15. Using 0.078152 in. (2.0 mm) drill, drill out lockpin (22) from piston (2). Discard lockpin (22) remains.
16. Remove piston plug (3) from piston (2).
17. Remove backup ring (4) and O-ring (5) from piston plug (3). Discard backup ring (4), O-ring (5), and piston plug (3).
18. Using 0.078152 in. (2.0 mm) drill, drill out lockpin (21) from piston (2). Discard lockpin (21) remains.

NOTE

- Mark adjusting nut and piston for installation.
- Do not remove nylon ball inserts.

19. Remove adjusting nut (7) from piston (2).
20. Remove three nylon balls (6) from adjusting nut (7).
21. Remove two reversing springs (8) and valve (9) from piston (2). Discard reversing springs (7).
22. Remove valve positioning pin (20) from side of piston (2). Discard valve positioning pin (20).
23. Remove O-ring (19) from valve positioning pin (20). Discard O-ring (19).

CAUTION

Do not disturb adjusting nut in bottom of piston after removing reversing springs.

24. Remove two reversing springs (10) from piston (2). Discard reversing springs (10).
25. Remove valve seat (15), two check balls (16), spring (17), and valve seat (23) from piston (2). Discard check balls (16).

4-120. STEERING GEAR (SHEPPARD) REPAIR (Contd)

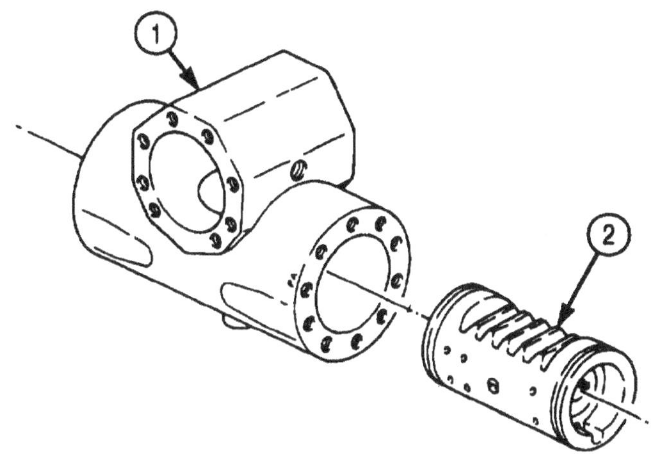

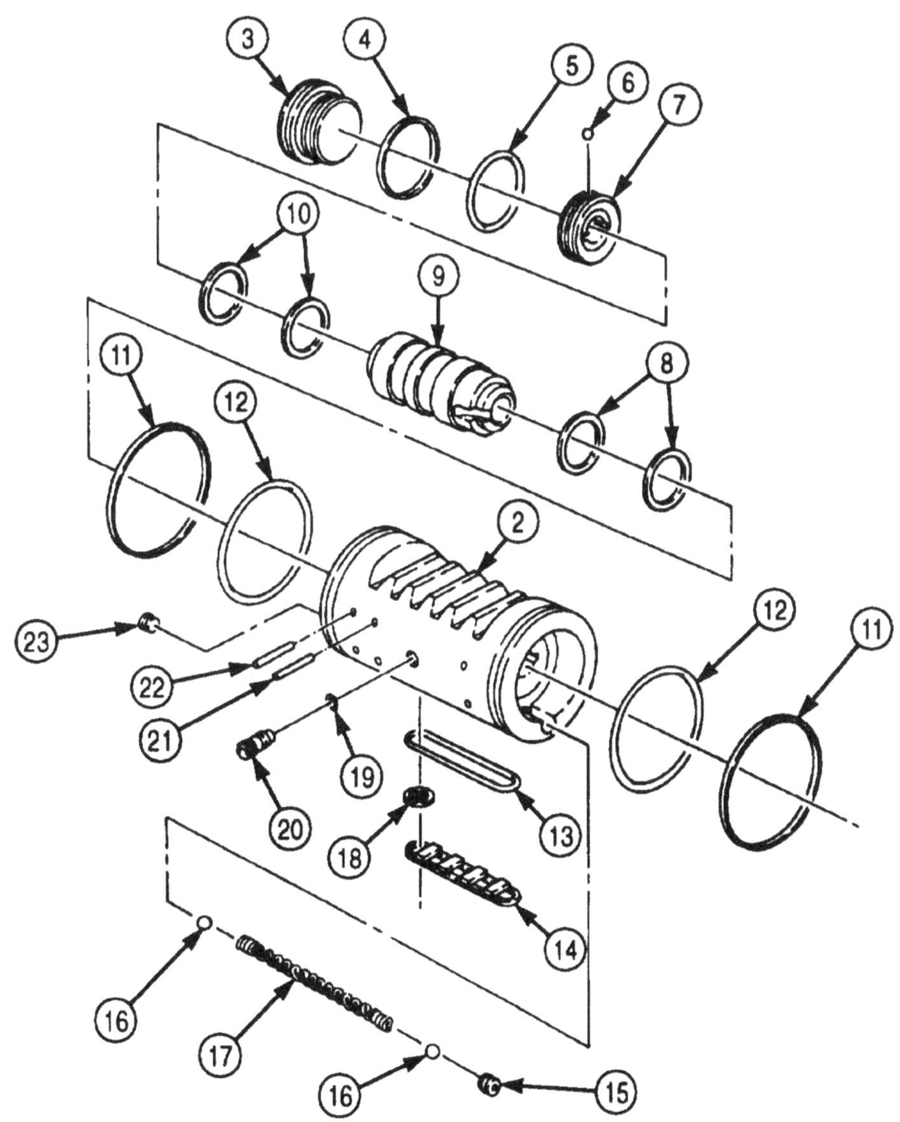

4-120. STEERING GEAR (SHEPPARD) REPAIR (Contd)

26. Remove oil real (14) from steering gear housing (15). Discard oil seal (14).
27. Remove bearing (13) from steering gear housing (15).
28. Using 0.078152 in. (2.0 mm) drill, drill lockpin (6) out of bearing cap (10). Discard lockpin (6) remains.
29. Remove bearing retainer (1) from bearing cap (10). Discard bearing retainer (1).
30. Press input shaft (2) and bearing (3) out of bearing cap (10).
31. Remove seals (9) and (8), high-pressure seal (4), and washer (6) from bearing cap (10). Discard high-presrure seal (4), washer (6), and seals (8) and (9).
32. Remove grease fitting (7) from bearing cap (10).
33. Remove bearing (11) from output shaft cover (12).

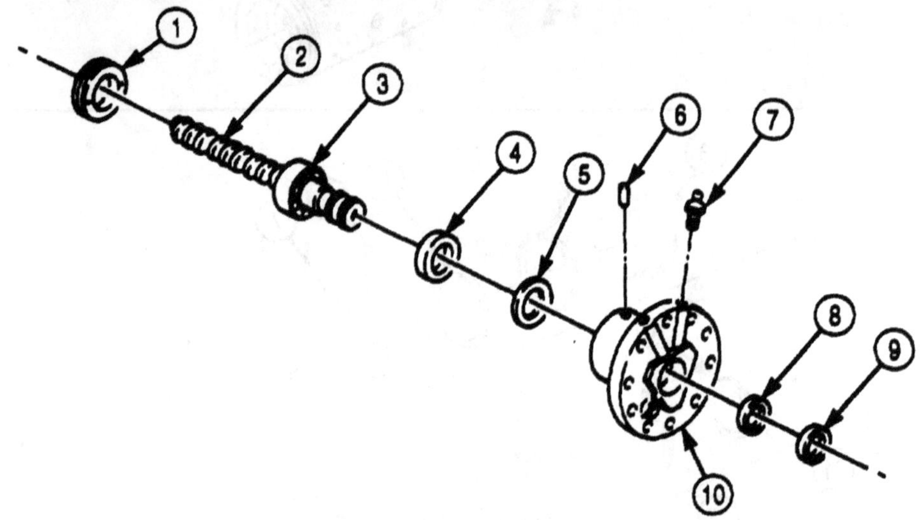

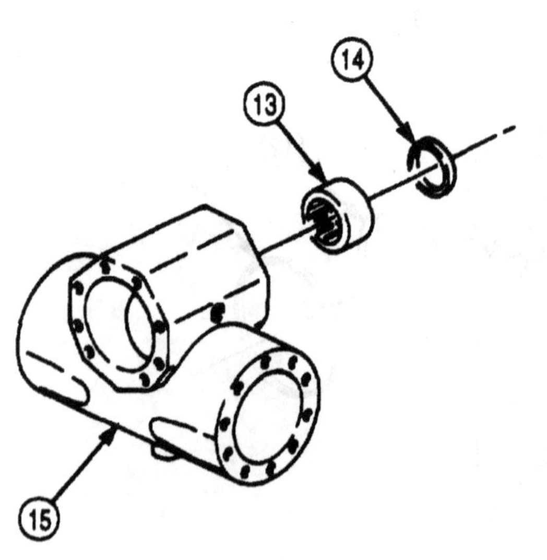

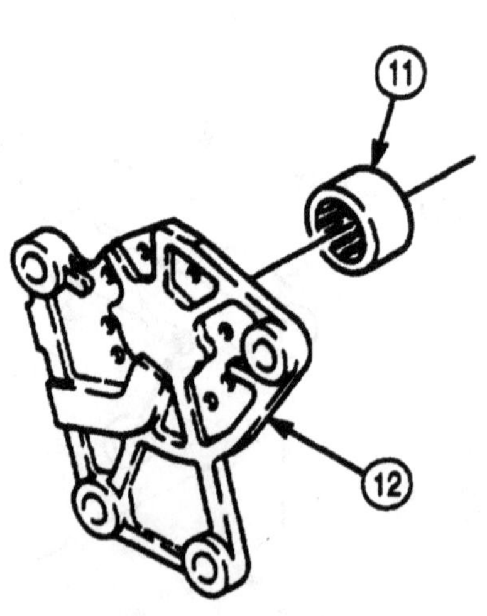

4-120. STEERING GEAR (SHEPPARD) REPAIR (Contd)

WARNING

- Drycleaning solvent is flammable and toxic, Do not use near an open flame and always have a fire extinguisher nearby when solvents are used. Use only in well-ventilated places, wear protective clothing, and dispose of cleaning rags in approved container. Failure to do this may result in injury to personnel and/or damage to equipment.
- Compressed air source will not exceed 30 psi (207 kPa). When cleaning with compressed air, eyeshields must be worn. Failure to wear eyeshields may result in injury to personnel.

1. Clean all steering gear components with drycleaning solvent and dry with compressed air.

NOTE

- Light polished areas on steering gear components indicate normal wear.
- Blue discoloration indicates overheating. Inspect carefully. Replace all discolored parts.

2. Inspect steering gear housing (15) for cracks, nicks, and pitted and scored bore (16). Replace steering gear housing (15) if cracked, nicked, or bore (16) is pitted or excessively scored.
3. Inspect input shaft (2) and bearing (3) for rough movement, discoloration, cracks, breaks, chipped threads (17), and damaged splines (18). Replace input shaft (2) as one assembly if discolored, cracked, broken, threads are chipped, or splines are damaged.
4. Inspect output shaft (19) for cracks, breaks, damaged splines (20), and damaged gear teeth (21). Replace output shaft (19) if damaged.

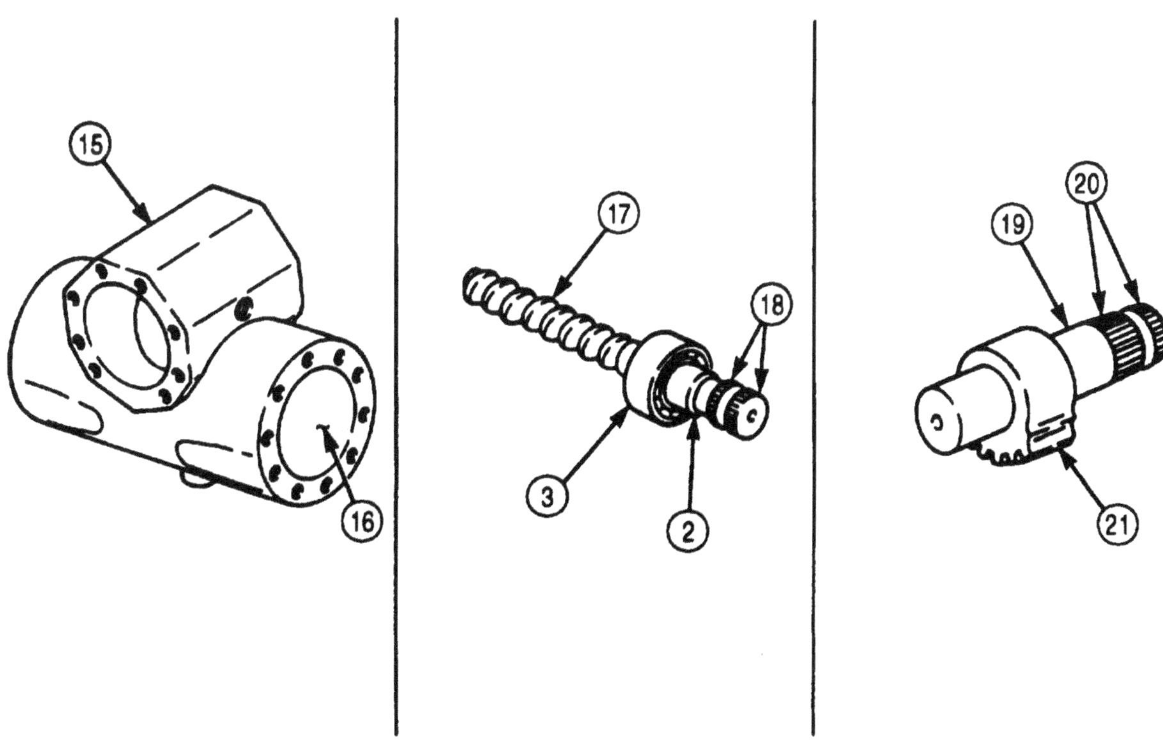

4-120. STEERING GEAR (SHEPPARD) REPAIR (Contd)

5. Inspect piston (3) for cracks, breaks, chipped seal grooves (2), and damaged threads. Replace piston (3) if damaged.

CAUTION

Do not disturb adjusting nut remaining in piston. Centering ability of valve will be lost.

6. Inspect valve bore (1) in piston (3) for pitting and scoring. Replace piston (3) if valve bore (1) is pitted or scored.

WARNING

Compressed air source will not exceed 30 psi (207 kPa). When cleaning with compressed air, eyeshields must be worn. Failure to wear eyeshields may result in injury to personnel.

7. Inspect oil passages in valve bore (1) for blockage. Clean oil passages with compressed air or soft wire if blocked.
8. Inspect valve (4) for cracks, breaks, burrs, chipped lands, and chipped or damaged threads. Replace valve (4) if lands are broken or chipped, or threads are damage.
9. Inspect two bearings (5) (TM 9-214). Replace bearings (5) if bearings fail inspection.
10. Inspect cylinder head (8), bearing cap (6), and output shaft cover (7) for cracks, breaks, and damage threads. Replace cylinder head (8), bearing cap (6), and output shaft cover (7) if damage is more than minor nicks and burrs.
11. Inspect spring (9) for bent or kinked coils, broken check ball ends, and discoloration. Replace spring (9) if damaged.

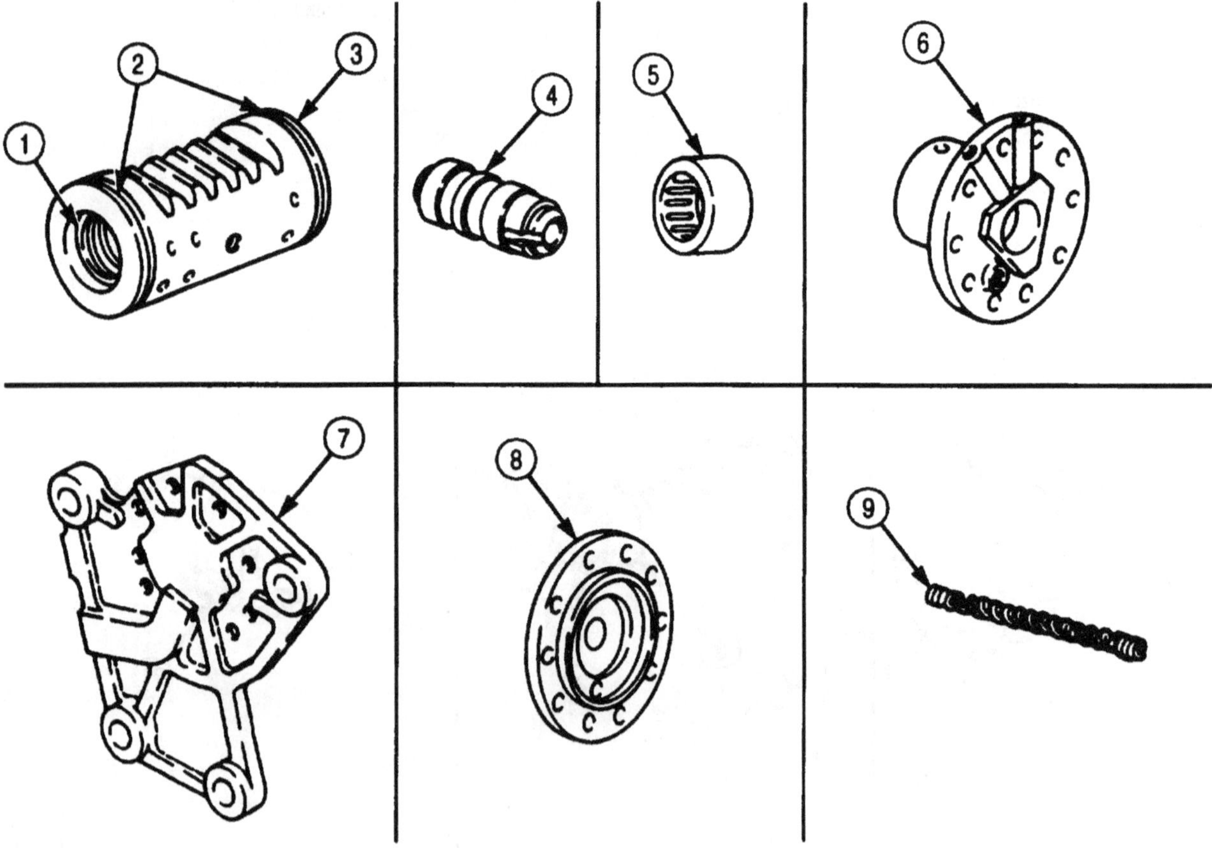

4-120. STEERING GEAR (SHEPPARD) REPAIR (Contd)

12. Inspect bearing retainer (10) for cracks and damaged threads. Replace retainer (10) if cracked or threads are damaged
13. Inspect adjusting nut (11) for cracks and damaged threads. Replace adjust nut (11) if cracked or teeth are damaged.
14. Inspect piston plug (13) for cracks, chipped or broken O-ring lips, and damaged or spanned holes. Replace piston plug (13) if cracked or damaged.
15. Inspect check valve seats (12) for nicks and burrs on seat surface and damaged slots or threads. Remove minor nicks and burrs. Replace check valve seats (12) is otherwise damaged.
16. Perform the following for all parts (except as noted):
    a. Remove minor nicks, burrs, and scoring with soft stone or crocus cloth.
    b. Replace all cracked, broken, or discolored parts.
    c. Chipped seal or O-ring grooves require part replacement.
    d. No repair to piston housing can be made. Replace piston housing if damaged.

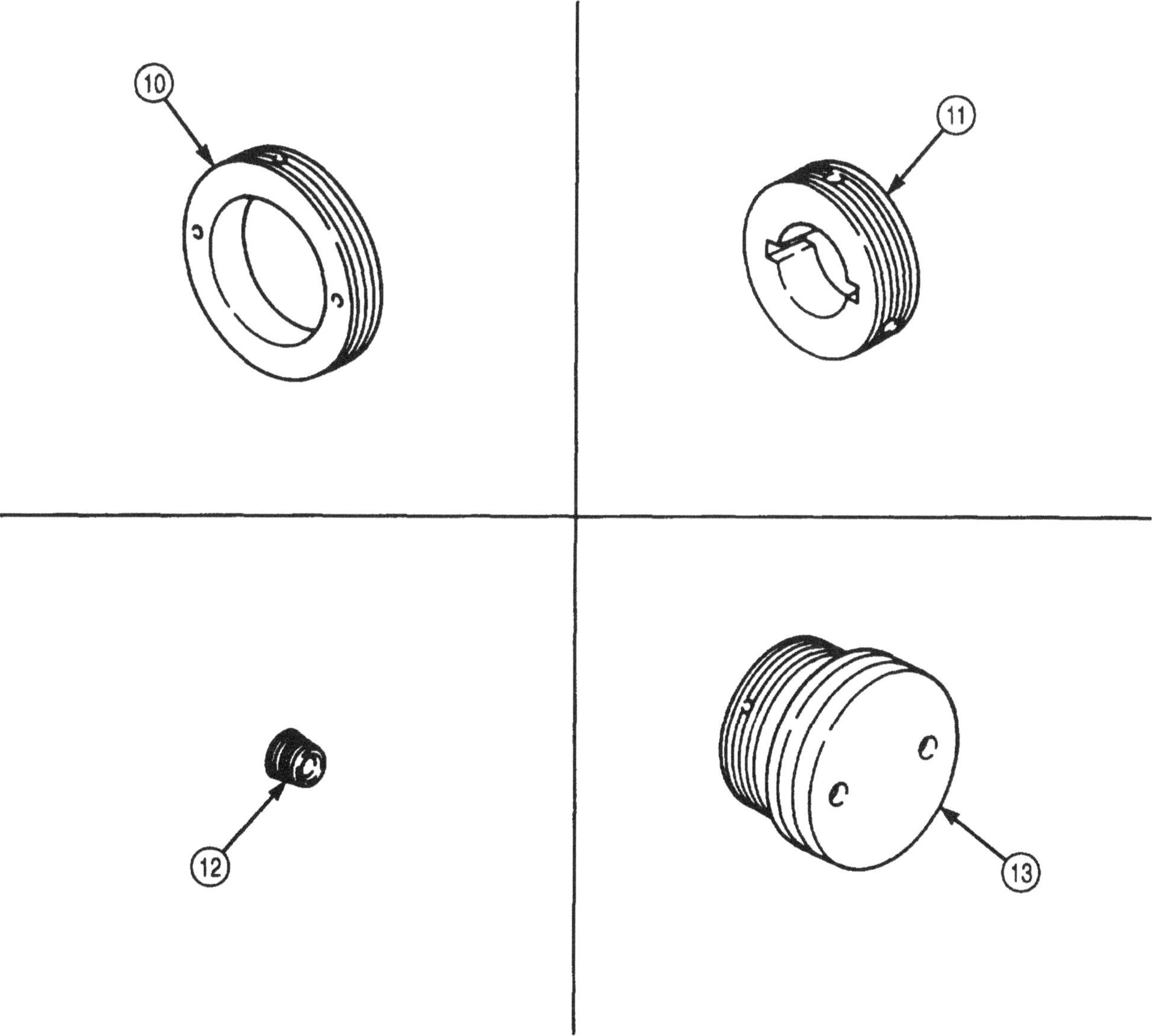

4-120. STEERING GEAR (SHEPPARD) REPAIR (Contd)

c. Assembly

NOTE

Coat all seals and O-rings with GAA grease before assembly.

1. Press bearing (2) into bore of steering gear housing (5).
2. Install new oil seal (3) in groove (4) of housing (5).
3. Install bearing (2) in output shaft cover (1).
4. Install valve (22) and two new reversing springs (13) in piston (18) as follows:
   a. Place springs (13) on long nose (12) of valve (22).
   b. Hold piston (18) with large opening down.
   c. Align slot (21) in valve (22) with positioning pin hole (15) in piston (18) and insert valve (22) up into piston (18) until reversing springs (13) contact internal adjusting nut (14).
   d. Turn piston (18) over.
5. Install new O-ring (16) on new positioning pin (17).
6. Install positioning pin (17) in piston (18) with nose of positioning pin (17) in slot (21) of valve (22).

NOTE

- Only move valve a small distance when checking end free play.
- Do not allow reversing springs to come off nose of valve.

7. Adjust positioning pin (17) until there is no radial free play, but valve (22) can move in bore of piston (18).

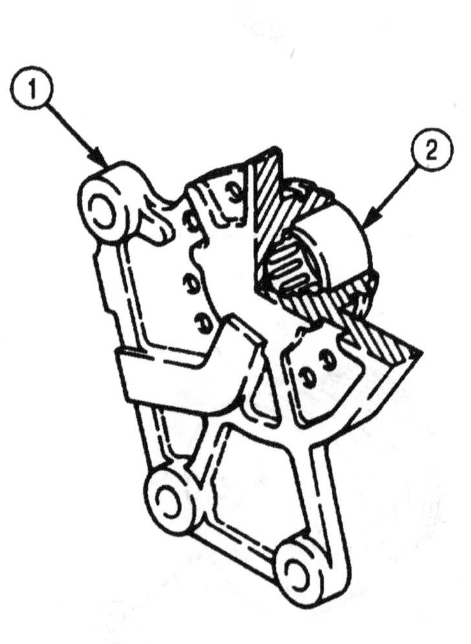

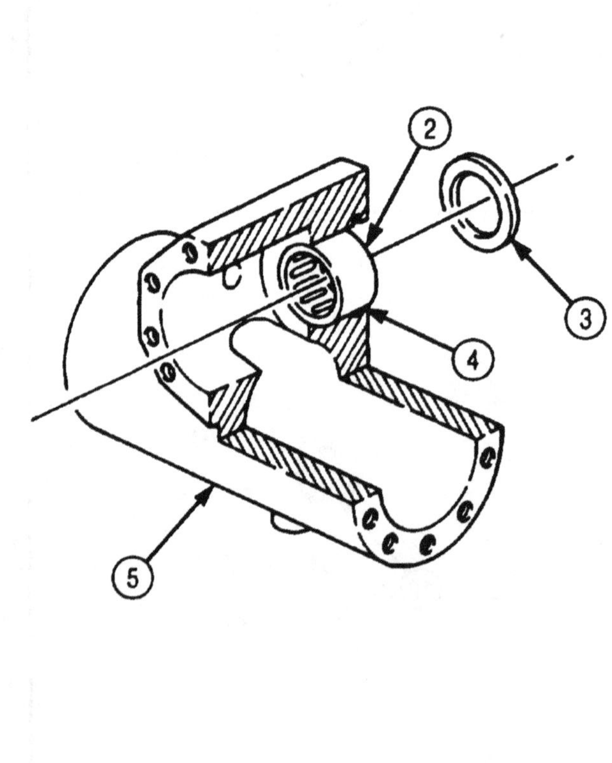

4-120. STEERING GEAR (SHEPPARD) REPAIR (Contd)

NOTE

Perform step 8 if new adjusting nut is to be installed.

8. Install two new reversing springs (23) over short nose of valve (22).
9. Press three new nylon balls (9) into recesses (10) of adjusting nut (11).
10. Install adjusting nut (11), shoulder side down, in piston (18) to recorded position.
11. Install new lockpin (19) in piston (18) and adjusting nut (11) until tip of lockpin (19) is below surface of piston (18).

NOTE

If old hole in piston and adjusting nut or plug do not line up, go to bottom pin hole in piston and drill into adjacent nut or plug and repeat step 11.

12. Install new O-ring (8) and new backup ring (7) on new piston plug (6).
13. Install piston plug (6) in piston (18).
14. Install new lockpin (20) in piston (18) and piston plug (6), and ensure tip of lockpin (20) is below surface of piston (18).

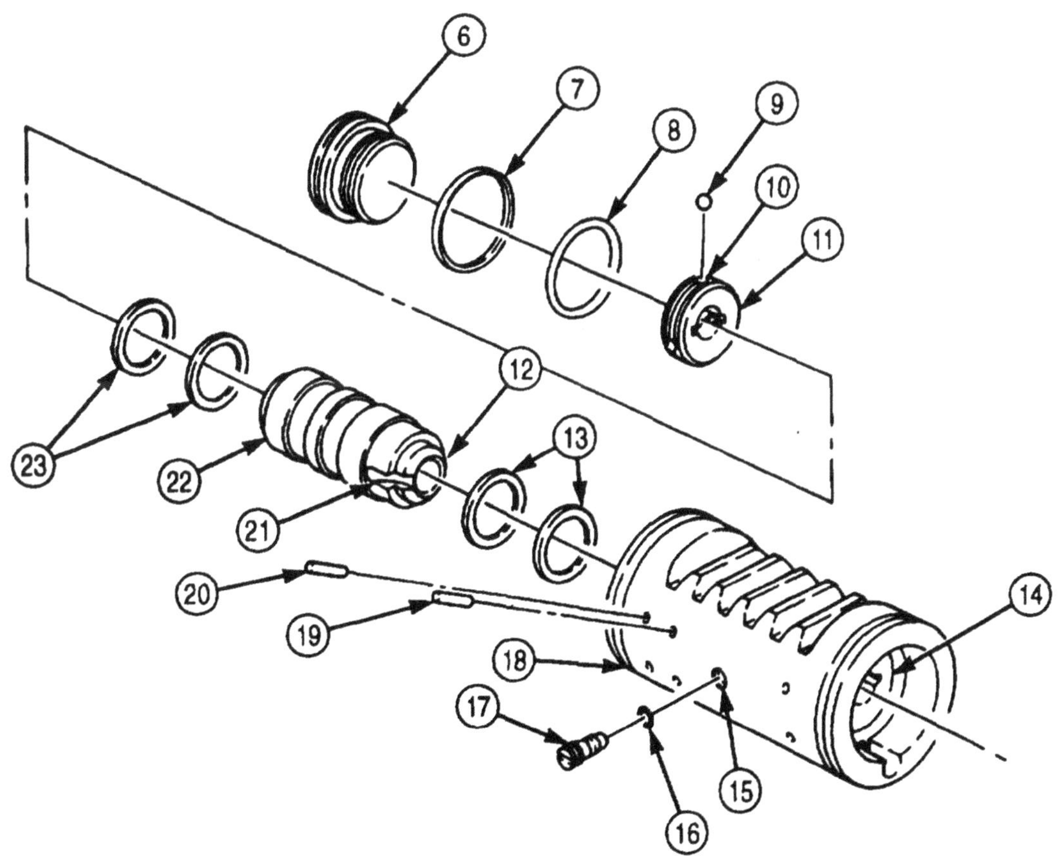

4-120. STEERING GEAR (SHEPPARD) REPAIR (Contd)

15. Install valve seat (11) in piston (3).
16. Install new check ball (7), spring (6), new check ball (5), and valve seat (4) in piston (3).
17. Install two new O-rings (2) and new backup rings (1) on piston (3).
18. Install new screen (9), new slipper seal (8). and new O-ring (10) on piston (3).
19. Install piston (3) in steering gear housing (12) starting from end of housing (12) marked PA As inner end of piston (3) approaches rear of output shaft bore (13), guide O-ring (2) and backup ring (1) of piston (3) to avoid pinching them on edge of bore (13).

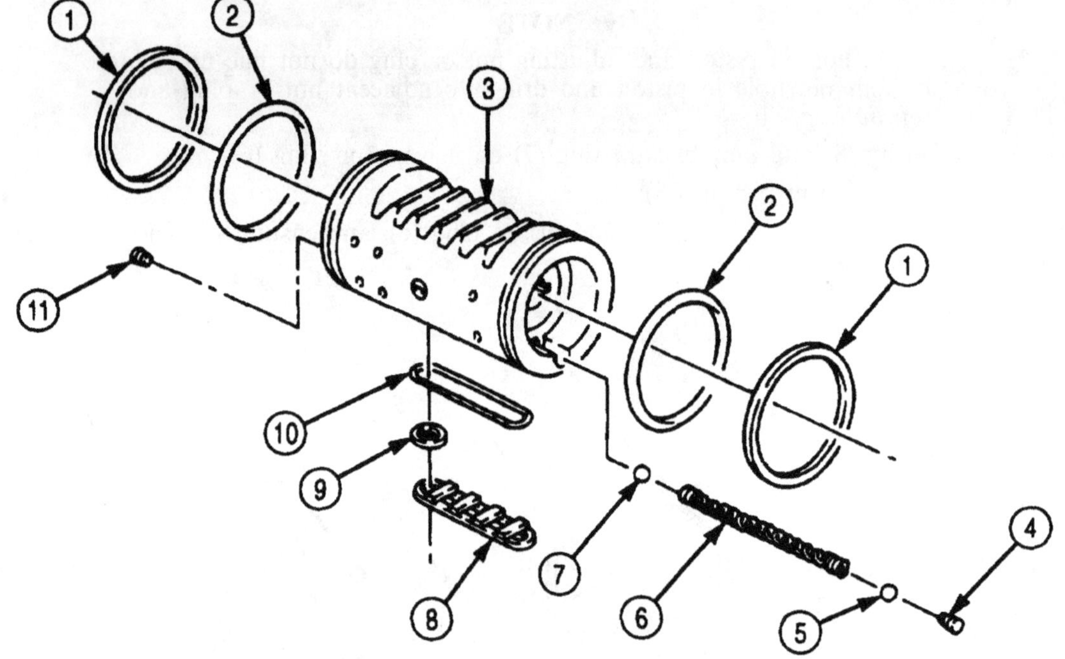

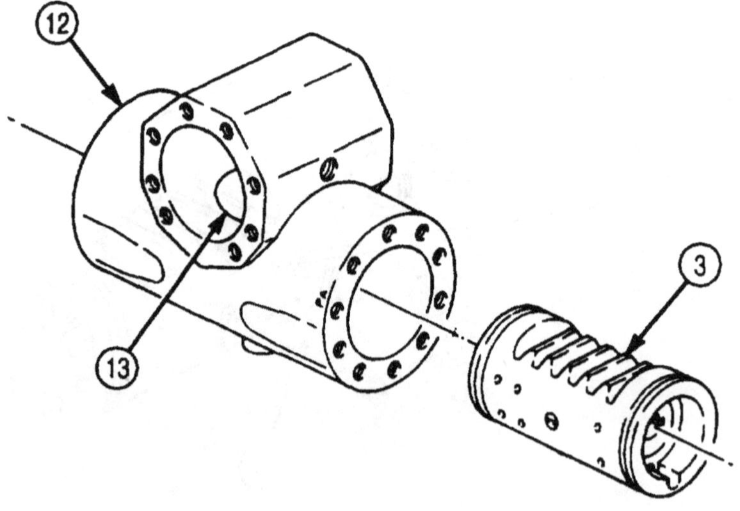

4-120. STEERING GEAR (SHEPPARD) REPAIR (Contd)

20. Place mandrel on firm surface and place bearing cap (17) so mandrel is seated on shoulder (23) of bearing cap (17).
21. Press new seal (15) into bore (16) of bearing cap (17) and seat against mandrel with lip of seal (15) up.
22. Press new seal (14) into bore (16) of bearing cap (17) and seat against seal (15) with lip of seal (14) up-

NOTE

Ensure high-pressure seal kit, P/N 2370461, is used to replace original seal.

23. Turn bearing cap (17) over, remove mandrel, and place bearing cap (17) on press bed with large end down.
24. Press new high-pressure seal (21) into bore (22) and seat against shoulder (23) with washer side of high-pressure seal (21) down.
25. Position new O-rings (18) and (20) on new retainer (19).
26. Install retainer (19) in bearing cap (17) and seat against high-pressure seal (21).

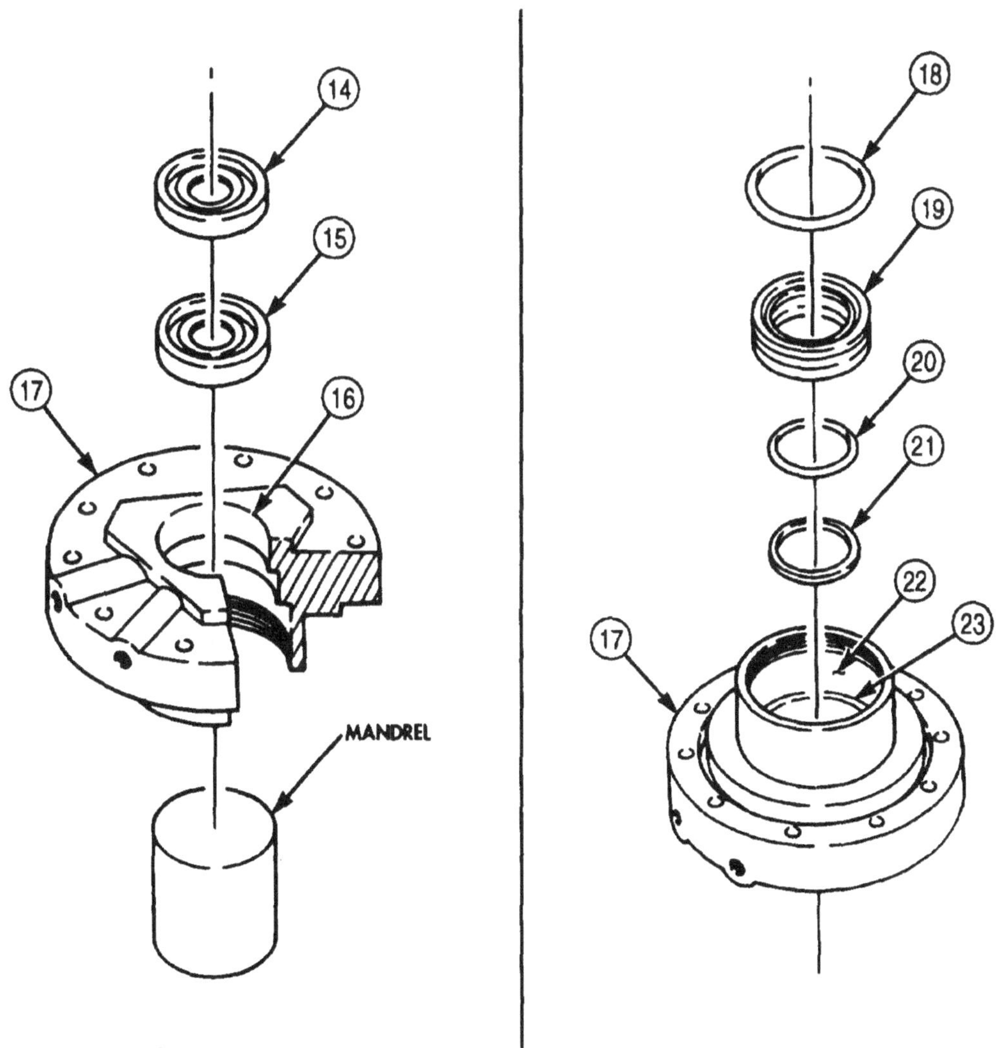

4-120. STEERING GEAR (SHEPPARD) REPAIR (Contd)

NOTE

Coat splines of input shaft with GAA grease to protect seals during assembly.

27. Press input shaft (2) and bearing (3) into bearing cap (6) until seated on shoulder (7) of bearing cap (6).
28. Install new bearing retainer (1) in hub of bearing cap (6).
29. Align new lockpin (4) holes on hub of bearing cap (6) with retainer (1).
30. Install lockpin (4) in bearing cap (6) and retainer (1) until tip of lockpin (4) is below surface of hub of bearing cap (6).
31. Install grease fitting (5) in bearing cap (6).
32. Install new O-ring (24) on cylinder head (10).
33. With marks aligned, install cylinder head (10) on housing (11) with ten new lockwashers (9) and screws (8). Tighten screws (8) 20-30 lb-ft (27-41 N•m).
34. Install new O-ring (26) on plunger (26).
35. Install plunger (26) in cylinder head (10).
36. Install new O-ring (13) on bearing cap (6).
37. Install bearing cap (6) on housing (11) with ten new lockwashers (14) and screws (15), turning input shaft (2) as necessary. Tighten screws (16) 20-30 lb-ft (27-41 N•m).

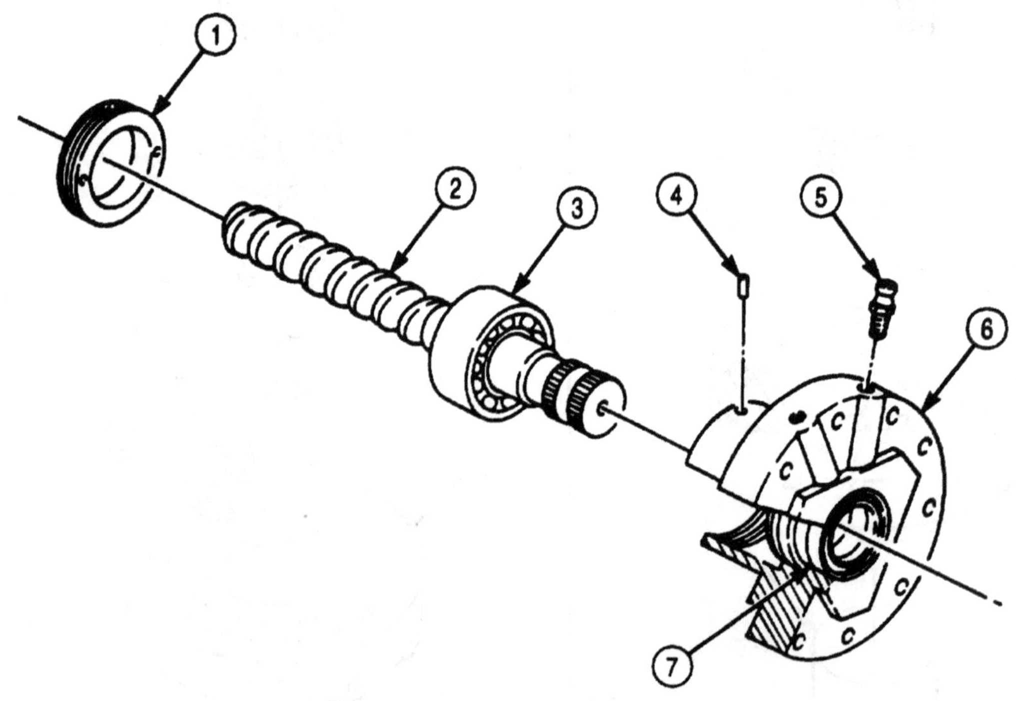

4-120. STEERING GEAR (SHEPPARD) REPAIR (Contd)

38. Install new O-ring (17) on relief valve plunger (16).
39. Install relief valve plunger (16) in bearing cap (6).

CAUTION

Wrap splines on output shaft with paper or tape. Failure to do so may result in damage to seal during assembly.

40. Install output shaft and gear (22) in steering gear housing (11). Ensure alignment marks (28) of output shaft and gear (22) are aligned with mark (27) on rack gear (23) of piston (12).
41. Install new O-ring (21) on output shaft cover (20).
42. Align marks on output shaft cover (20) and steering gear housing (11) and install shaft cover (20) on steering gear housing (11) with eight new lockwashers (18) and screws (19). Tighten screws (19) 85-95 lb-ft (116-129 N•m).

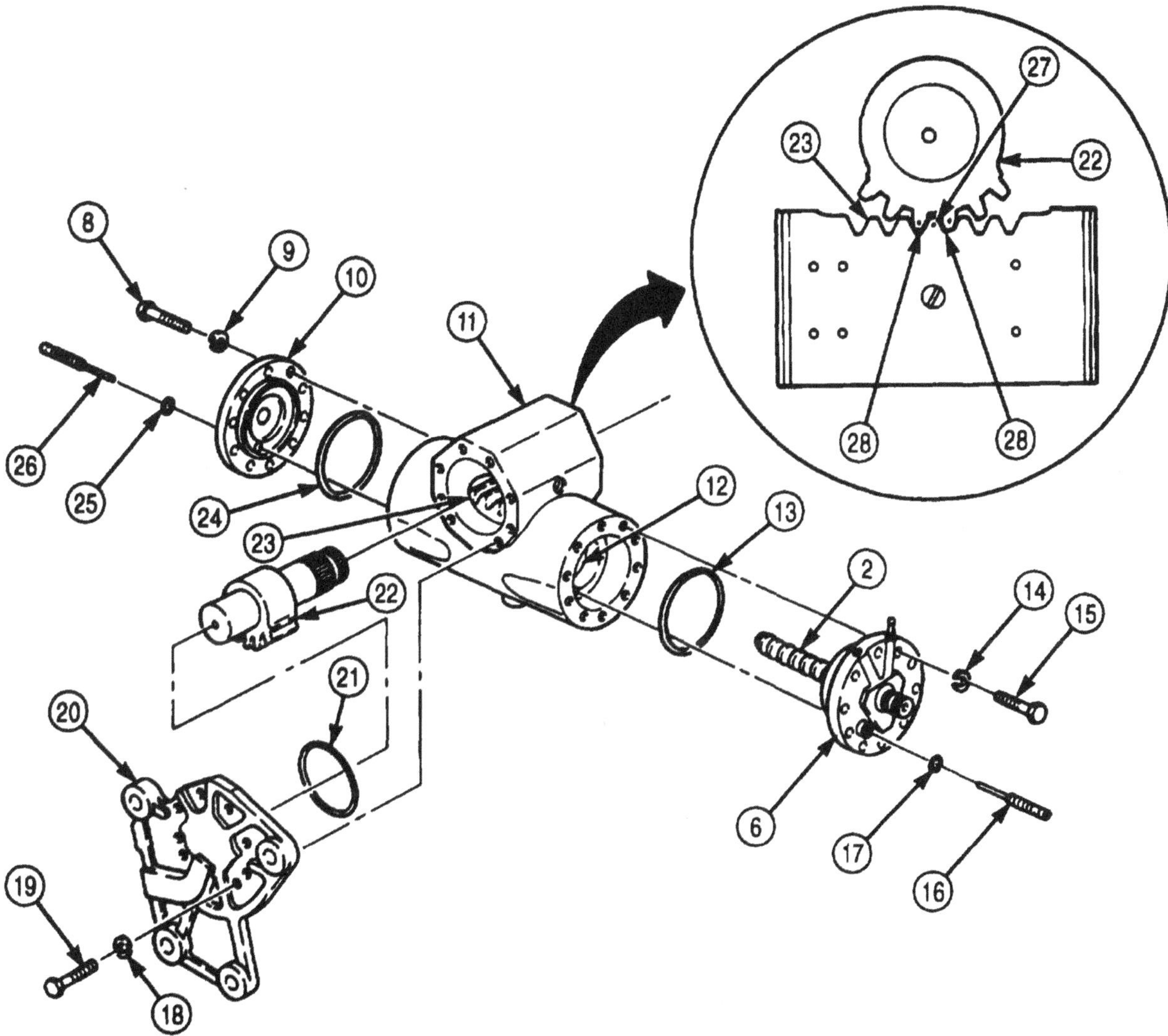

FOLLOW-ON TASKS:
- Install steering gear (para. 4-118).
- Fill steering gear to proper oil level (LO 9-2320-272-12).
- Perform final plunger adjustments (para. 4-124).

4-121. POWER STEERING ASSIST CYLINDER REPAIR

THIS TASK COVERS:

a. Disassembly
b. Cleaning, Inspection, and Repair
c. Assembly

INITIAL SETUP

APPLICATION MODELS

All

TOOLS

General mechanic's tool kit (Appendix E, Item 1)
Vise

MATERIALS/PARTS

Seal (Appendix D, Item 628)
Locknut (Appendix D, Item 332)
Packing ring (Appendix D, Item 513)
Lubricating oil (Appendix C, Item 48)
Drycleaning solvent (Appendix C, Item 71)

REFERENCES (TM)

LO 9-2320-272-12
TM 9-2320-272-24P

EQUIPMENT CONDITION

Power steering assist cylinder removed (para. 3-233).

GENERAL SAFETY INSTRUCTIONS

Keep fire extinguisher nearby when using drycleaning solvent.

WARNING

Drycleaning solvent is flammable and toxic. Do not use near an open flame and always have a fire extinguisher nearby when solvents are used. Use only in well-ventilated places, wear protective clothing, and dispose of cleaning rags in approved container. Failure to do this may result in injury to personnel and/or damage to equipment.

a. Disassembly

1. Thoroughly clean power steering assist cylinder (7) with drycleaning solvent.

NOTE

Have drainage container ready to catch oil.

2. Holding oil ports down, push piston rod (15) in and out to remove oil from cylinder (7).
3. Clamp steering assist cylinder (7) in vise.
4. Remove three screw-assembled washers (4), retainer (3), seal (2), and end plate (1) from top of steering assist cylinder (7).
5. Pushing cylinder gland (6) in, remove backup ring (5) from cylinder (7).
6. Remove cylinder gland (6), piston (11), and piston rod (15) as an assembly from cylinder (7).
7. Remove locknut (14) from piston rod (15). Discard locknut (14).
8. Turn piston (11) counterclockwise and remove from rod (15).
9. Remove cylinder gland (6) from rod (15).
10. Remove seal (13) and packing ring (12) from piston (11). Discard seal (13) and packing ring (12).
11. Remove snapring (22), retaining ring (21), oil seal (20), retaining ring (19), retaining ring (18), gland backup ring (17), and gland O-ring (16) from gland (6). Discard oil seal (20), gland backup ring (17), and gland O-ring (16).
12. Remove O-ring (24) and backup ring (23) from cylinder gland (6). Discard O-ring (24) and backup ring (23).
13. Remove plug (10), two ball sockets (9), and spring (8) from steering assist cylinder (7).

4-121. POWER STEERING ASSIST CYLINDER REPAIR (Contd)

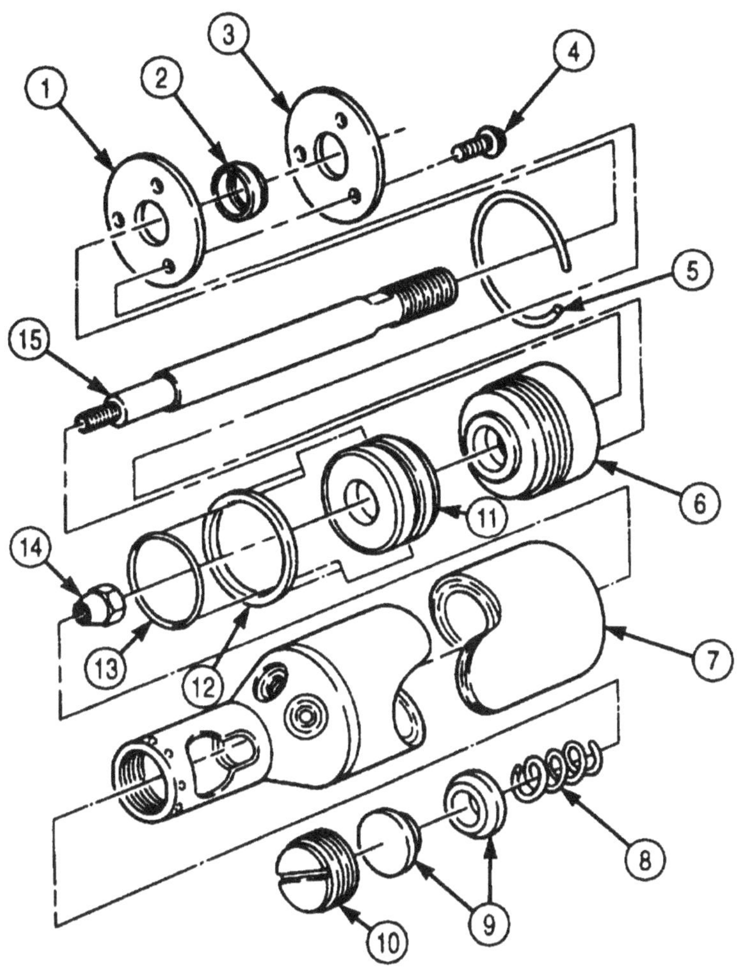

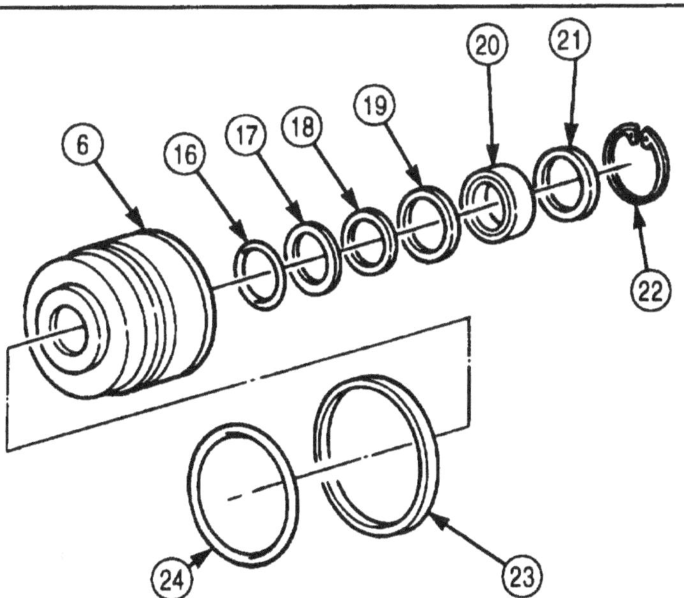

4-121. POWER STEERING ASSIST CYLINDER REPAIR (Contd)

b. Cleaning, Inspection, and Repair

WARNING

Drycleaning solvent is flammable and toxic. Do not use near an open flame and always have a fire extinguisher nearby when solvents are used. Use only in well-ventilated places, wear protective clothing, and dispose of cleaning rags in approved container. Failure to do this may result in injury to personnel and/or damage to equipment.

1. Clean steering assist cylinder (8) and related parts with drycleaning solvent and let air-dry.
2. Inspect steering assist cylinder (8), piston rod (15), piston (7), and cylinder gland (6) for scoring, cracks, nicks, or breaks. Replace steering assist cylinder (8) if any parts are scored, cracked, nicked, or broken.
3. Inspect two ball sockets (10) and spring (9) for cracks and breaks. Replace ball sockets (10) and spring (9) if broken or cracked.
4. Inspect plug (11) for cracking and damaged threads. Replace plug (11) if cracked or threads are damaged.

c. Assembly

NOTE

Coat all power steering assist cylinder components with clean lubricating oil before assembly

1. Install spring (9), two ball sockets (10), and plug (11) in steering assist cylinder (8).
2. Install new backup ring (23) and O-ring (24) on cylinder gland (6).
3. Install new gland O-ring (16), new gland backup ring (17), retaining rings (18) and (19), piston rod seal (20), and retaining ring (21) in gland (6) with snapring (22).
4. Install new packing ring (12) and new seal (13) on piston (7).
5. Install piston (7) on piston rod (15) with new locknut (14).
6. Install piston (7) and piston rod (15) in steering assist cylinder (8).
7. Using care not to damage gland (6) and internal seals and threads, slide cylinder gland (6) over piston rod (15) and into steering assist cylinder (8).
8. Install backup ring (5) in top of steering assist cylinder (8). Ensure backup ring (5) is seated in groove in cylinder (8).
9. Install end plate (1), new plate seal (2), and end plate retainer (3) on cylinder gland (6) with three screw-assembled washers (4).

4-121. POWER STEERING ASSIST CYLINDER REPAIR (Contd)

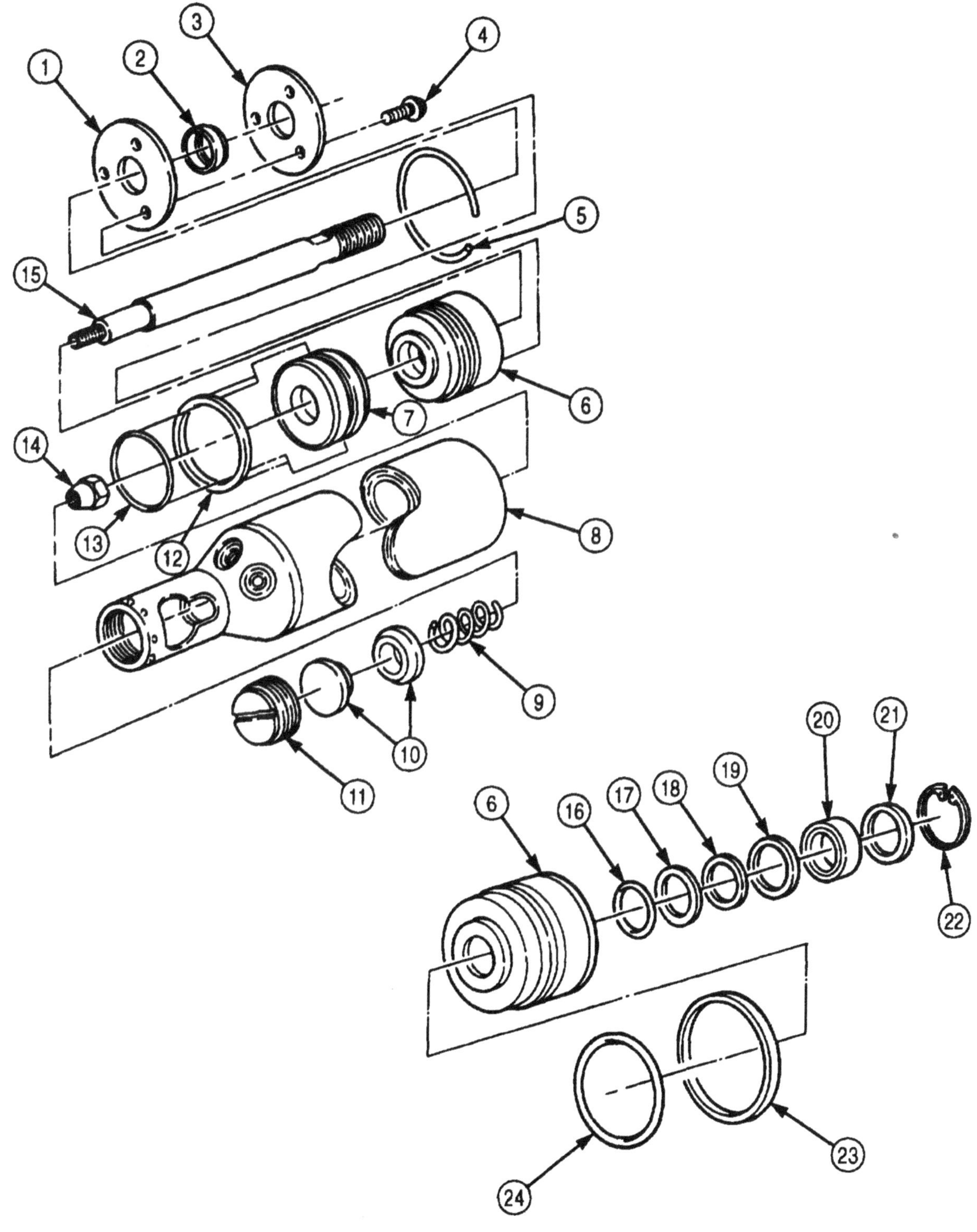

FOLLOW-ON TASKS:
- Install power steering assist cylinder (para. 3-233).
- Lubricate power assist cylinder (LO 9-2320-272-12).

4-122. POWER STEERING PUMP TEST EQUIPMENT SETUP

THIS TASK COVERS:

a. Installation b. Removal

INITIAL SETUP:

APPLICABLE MODELS
All

SPECIAL TOOLS
Power steering pressure gauge kit (Appendix E, Item 100)

TOOLS
General mechanic's tool kit (Appendix E, Item 1)

MATERIALS/PARTS
Lubricating oil (Appendix C, Item 48)

REFERENCES (TM)
LO 9-2320-272-12
TM 9-2320-272-10
TM 9-2320-272-24P

EQUIPMENT CONDITION
- Parking brake set (TM 9-2320-272-10).
- Left splash shield removed (para. 3-301).
- Steering gear stone shield removed (para. 3-238).

NOTE

Test equipment setup is the same for Ross and Sheppard steering gears. This procedure covers the Ross steering gear.

a. Installation

NOTE

Have drainage container ready to catch oil.

1. Disconnect hydraulic pressure line (5) from steering gear inlet elbow (8).
2. Connect load shutoff valve (7) to steering gear inlet elbow (8).
3. Connect tee adapter (6) to shutoff valve (7).
4. Connect pressure gauge to tee adapter (6).
5. Connect hydraulic pressure line (5) to tee adapter (6).
6. Disconnect hydraulic return line (9) from return tube (1).
7. Connect flow meter to return line (9).
8. Connect top end of flow meter to return tube (1) on steering pump reservoir (3).
9. Remove filler neck cover (4) from reservoir (3).
10. Install thermometer in pump reservoir filler neck (2).

b. Removal

1. Remove thermometer from steering pump reservoir (3).
2. Install filler neck cover (4) on pump reservoir filler neck (2).

NOTE

Have drainage container ready to catch oil.

3. Disconnect hydraulic return line (9) from flow meter.
4. Disconnect flow meter from return tube (1) on steering pump reservoir (3).
5. Connect hydraulic return line (9) to return tube (1) on steering pump reservoir (3).
6. Disconnect hydraulic pressure line (5) from tee adapter (6).

4-122. POWER STEERING PUMP TEST EQUIPMENT SETUP (Contd)

7. Disconnect pressure gauge from tee adapter (6).
8. Disconnect tee adapter (6) from load shutoff valve (7).
9. Disconnect load shutoff valve (7) from steering gear inlet elbow (8).
10. Connect hydraulic pressure line (5) to steering gear inlet elbow (8).

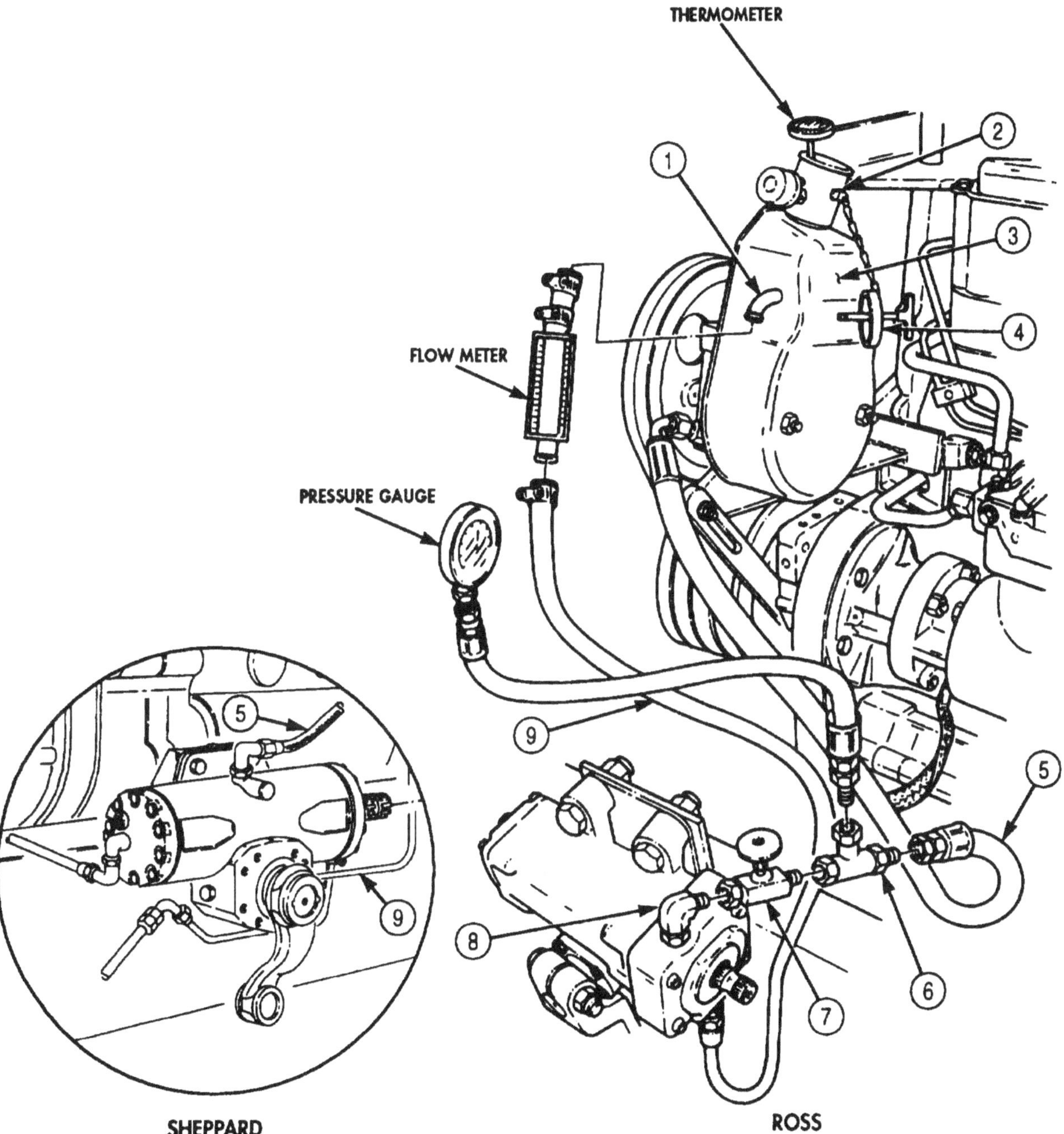

FOLLOW-ON TASKS: • Install steering gear stone shield (para. 3-238).
• Install left splash shield (para. 3-301).
• Fill power steering reservoir to proper level (LO 9-2320-272-12).

4-123. POWER STEERING TESTS AND ADJUSTMENT (ROSS)

THIS TASK COVERS:

a. Steering Pump Pressure Test
b. Steering Pump Flow Test
c. Steering Gear Internal Leakage Test
d. Poppet Adjustment

INITIAL SETUP:

APPLICABLE MODELS
All

TOOLS
General mechanic's tool kit (Appendix E, Item 1)

REFERENCES (TM)
LO 9-2320-272-12
TM 9-2320-272-10
TM 9-2320-272-24P

EQUIPMENT CONDITION
- Parking brake set (TM 9-2320-272-10).
- Power steering test equipment installed (para. 4-122).
- Power steering reservoir filled to proper oil level (LO 9-2320-272-121.

GENERAL SAFETY INSTRUCTIONS
Personnel must be clear of vehicle when vehicle engine is running.

WARNING

Personnel must be clear of vehicle when vehicle engine is running. Vehicle could suddenly move and cause injury to personnel and/or damage equipment.

a. Steering Pump Pressure Test

1. Start engine and warm up to operating temperature (TM 9-2320-272-10).

CAUTION

Never allow power steering oil temperature to exceed 180°F (82°C). Run all tests at the prescribed temperature range of 125-135°F (52-57°C).

2. Partially close load shutoff valve (2) until pressure gauge reads 1,000 psi (6,895 kPa).
3. Observe temperature reading on thermometer in power steering pump reservoir (1). Temperature must be between 125-135°F (52-57°C).

CAUTION

- Do not keep the load valve closed for longer than 5 seconds to avoid damaging pump.
- Closing load valve causes pump to operate at relief pressure, resulting in rapidly increasing power steering oil temperature.

4. Open load shutoff valve (2).
5. Close load shutoff valve (2) and observe pressure gauge reading. If gauge reads below 1,360-1,440 psi (9,377-9,929 kPa), repair or replace pump (3).
6. Open shutoff valve (2).

4-123. POWER STEERING TESTS AND ADJUSTMENT (ROSS) (Contd)

1. With thermometer in power steering pump reservoir (1), observe oil temperature. Temperature must be 125-135°F (52-57°C).
2. Observe oil flow rate indicated on flow meter. Flow rate should be 5.5-6.5 gpm (20.8-24.6 lpm).

CAUTION

Do not keep load valve closed for longer than 5 seconds to avoid damaging pump. Closing load valve causes pump to operate at relief pressure, resulting in rapidly increasing power steering oil temperature.

3. Close load shutoff valve (2) and observe pressure gauge and flow meter. Pressure should be 1,360-1,440 psi (9,377-9,929 kPa). Flow rate should be zero.
4. Immediately open load shutoff valve (2) and observe flow meter reading. Flow rate should return to original reading of 5.5-6.5 gpm (20.8-24.6 lpm). If not, replace pump (3).
5. Run engine at 1,200 rpm.
6. Fully close load shutoff valve (2) and observe pressure gauge and flow meter readings. Observe readings until steering pump (3) relief pressure is reached at 1,360-1,440 psi (9,377-9,939 kPa). Flow rate should be zero.
7. Open load shutoff valve (2) immediately. Flow rate should return to original reading of 5.5-6.5 gpm (20.8-24.6 lpm).

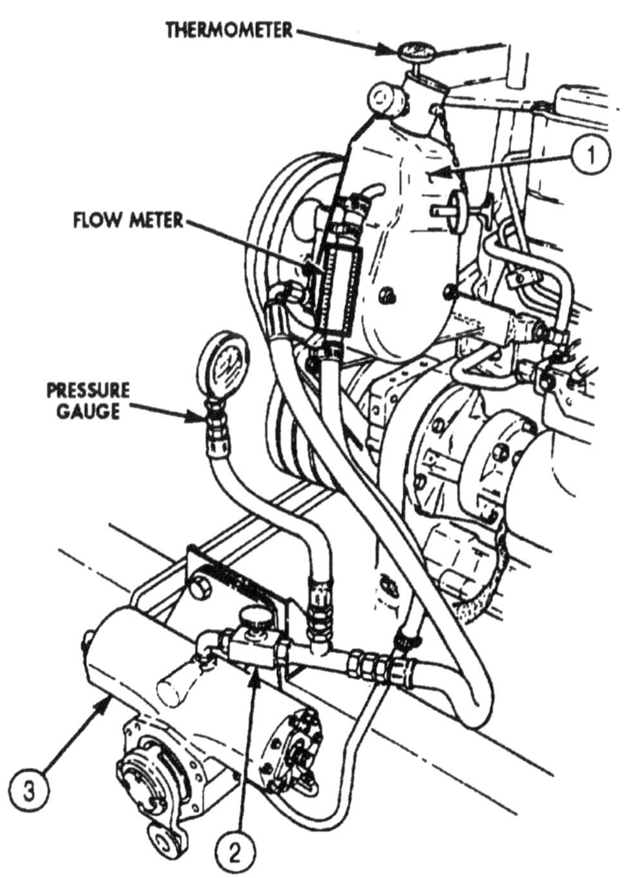

4-123. POWER STEERING TESTS AND ADJUSTMENT (ROSS) (Contd)

c. Steering Gear Internal Leakage Test

CAUTION

Damage could result to steering gear if steering gear does not allow wheels to reach axle stops. Poppets must be adjusted or steering gear will not bleed itself internally

1. Observe thermometer. Temperature must be 125-135°F (52-57°C).
2. Position steel spacer block (2) between axle (1) and axle stop (3) to prevent operation of poppets.

CAUTION

Do not hold steering wheel in full-turn position for more than 5-10 seconds. Pump damage could result.

NOTE

Assistant will rotate steering wheel.

3. Turn steering wheel (4) until axle stop (3) bottoms on steel spacer block (2).
4. Observe readings on pressure gauge and flow meter. Pressure should be 1,360-1,440 psi (9,277-9,529 kPa). Flow rate should be 0-1 gpm (0-4 lpm). If greater than 1 gpm (4 lpm), repair steering gear (9).

NOTE

Repeat steps 1 through 4 with wheels turned in opposite direction.

d. Poppet Adjustment

CAUTION

Do not hold steering wheel in full-turn position for more than 5-10 seconds. Pump damage could result.

NOTE

Assistant will rotate steering wheel.

1. Rotate steering wheel (4) to full left direction.
2. Observe sector shaft (7) clockwise rotation.
3. Loosen jamnut (8) and turn poppet adjusting screw (6) until pressure gauge reaches maximum pressure.
4. Tighten poppet adjusting screw (5) until pressure gauge shows a pressure drop of 200-400 psi (1,379-2,758 kPa).
5. Holding adjusting screw (5), tighten jamnut (8) 12-18 lb-ft (16-24 N•m).
6. Turn steering wheel (4) to full right direction and observe sector shaft (7) counterclockwise rotation.
7. Loosen jamnut (8) and turn poppet adjusting screw (6) until pressure gauge shows a pressure drop of 200-200 psi (1,379-1,758 kPa).
8. Hold screw (6) and tighten jamnut (8) 12-18 lb-ft (16-24 N•m).
9. With engine running, rotate steering wheel (4) from full left to full right several times to remove any air in power steering oil.

4-123. POWER STEERING TESTS AND ADJUSTMENT (ROSS) (Contd)

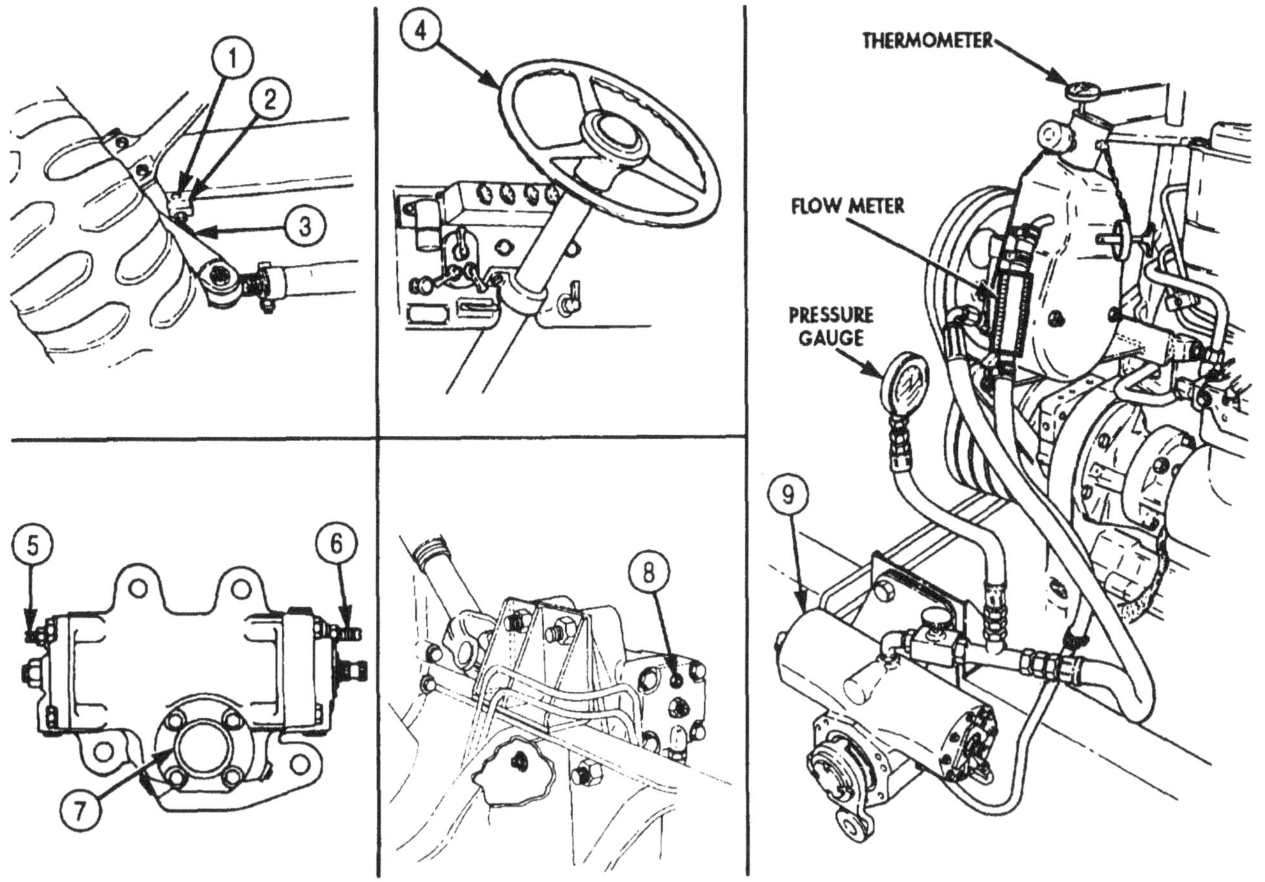

FOLLOW-ON TASKS:
- Remove test equipment (para. 4-122).
- Fill power steering reservoir to proper oil level (LO 9-2320-272-12).
- Check steering gear mechanical adjustments (para. 4-125).
- Start engine (TM 9-2320-272-10) and road test vehicle.

4-124. POWER STEERING TESTS AND ADJUSTMENT (SHEPPARD)

THIS TASK COVERS:

a. Steering Pump Pressure Test
b. Steering Pump Flow Test
c. Steering Gear Internal Leakage Test
d. Plunger Adjustment

INITIAL SETUP:

APPLICABLE MODELS
All

TOOLS
General mechanic's tool kit (Appendix E, Item 1)

REFERENCES (TM)
LO 9-2320-272-12
TM 9-2302-272-10
TM 9-2320-272-24P

EQUIPMENT CONDITION
- Parking brake set (TM 9-2320-272-10).
- Power steering test equipment installed (para. 4-122).
- Power steering reservoir filled to proper level (LO 9-2320-272-12).

GENERAL SAFETY INSTRUCTIONS
Personnel must be clear of vehicle when vehicle engine is running.

WARNING

Personnel must be clear of vehicle when vehicle engine is running. Vehicle could suddenly move and cause injury to personnel and/or damage to equipment.

a. Steering Pump Pressure Test

Refer to para. 4-123, task a.

b. Steering Pump Flow Test

Refer to para. 4-123, task b.

c. Steering Gear Internal Leakage Test

CAUTION

Damage could result to steering gear if steering gear does not allow wheels to reach axle stops. Plungers must be adjusted or steering gear will not bleed itself internally.

1. Observe temperature reading on thermometer. Temperature must be 125-135°F (52-57°C).
2. Turn two plungers (7), one on each end of steering gear (6), out to just below surface of boss (8).

NOTE

Spacers must be 0.125-0.187 in. (3.2-4.8 mm) thick.

3. Position steel spacer block (4) between axle (3) and axle stop (5).

4-124. POWER STEERING TESTS AND ADJUSTMENT (SHEPPARD) (Contd)

CAUTION

Do not hold steering wheel in full-turn position for more than 5-10 seconds. Pump damage could result.

NOTE

Assistant will rotate steering wheel.

6. Turn steering wheel (2) until axle stop (5) bottoms out on steel spacer block (4).
7. Observe readings on pressure gauge and flow meter. Pressure should be 1,360-1,440 psi (9,377-9,929 kPa). Flow rate should be 0-1 gpm (4 lpm). If flow rate is greater than 1 gpm (4 lpm), repair steering gear (6).

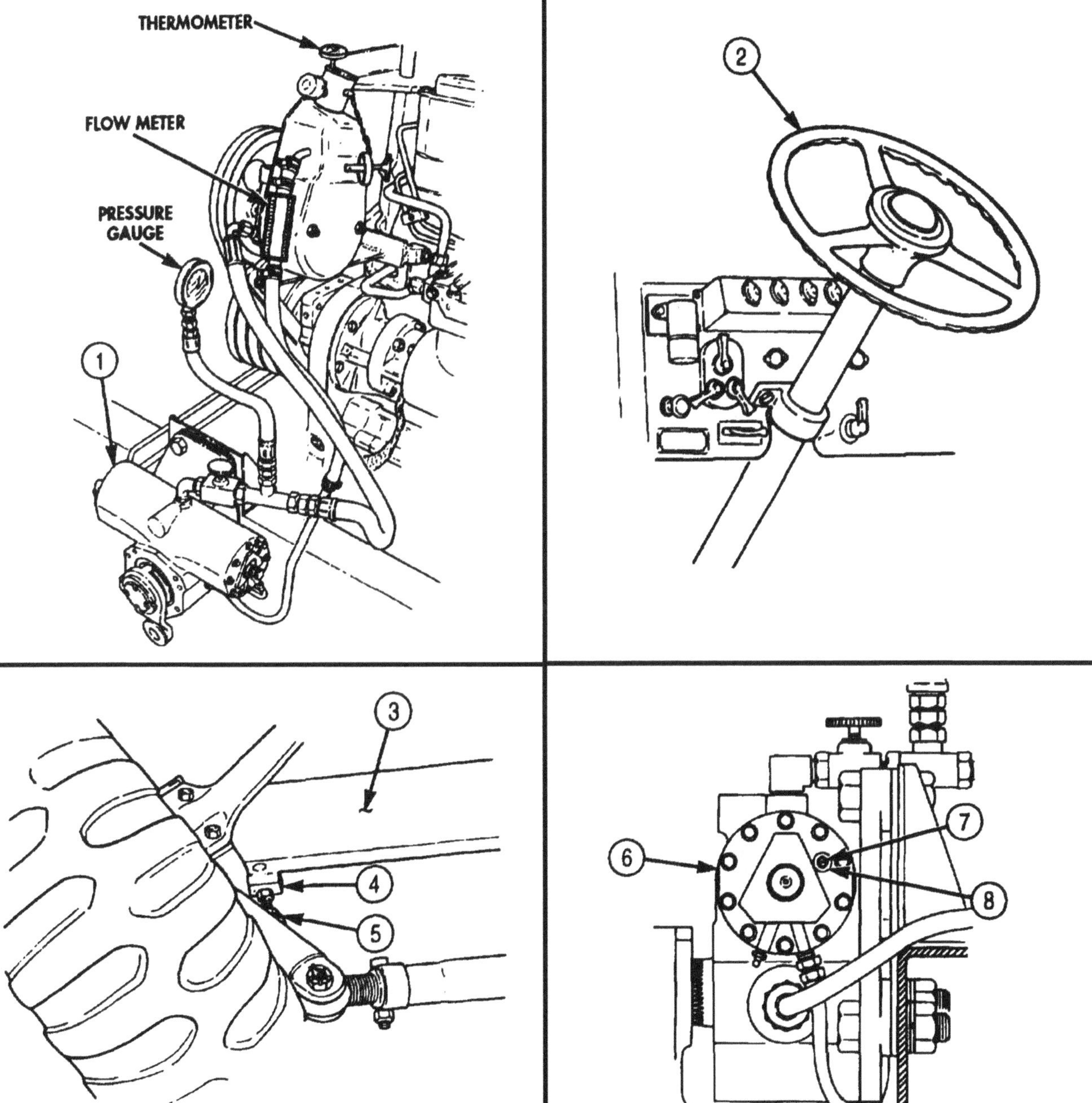

4-124. POWER STEERING TESTS AND ADJUSTMENT (SHEPPARD) (Contd)

d. Plunger Adjustment

CAUTION

Do not hold steering wheel in full-turn position for more than 5-10 seconds. Power steering pump damage could result.

NOTE

- Assistant will rotate steering wheel.
- Left side axle, axle stop and block shown. Right side is similar.

1. With steel spacer block (1) in place between axle (2) and axle stop (3), rotate steering wheel (4) to full left travel.
2. Observe that pitman arm (5) rotates clockwise.

NOTE

Turning plungers IN decreases power-assisted turning angle; turning plungers OUT increases power-assisted turning angle.

3. Turn rear plunger (8) of steering gear (7) clockwise until high-pressure hiss is heard. Pressure gauge must show 200-400 psi (1,379-2,785 kPa) pressure drop.
4. With steel spacer block (1) in place between axle (2) and axle stop (3), rotate steering wheel (4) to full right direction.
5. Observe that pitman arm (5) rotates counterclockwise.

NOTE

Turning plungers IN decreases power-assisted turning angle; turning plungers OUT increases power-assisted turning angle.

6. Turn front plunger (6) of steering gear (7) counterclockwise until high-pressure hiss is heard. Pressure gauge must show 200-400 psi (1,379-2,758 kPa) pressure drop.

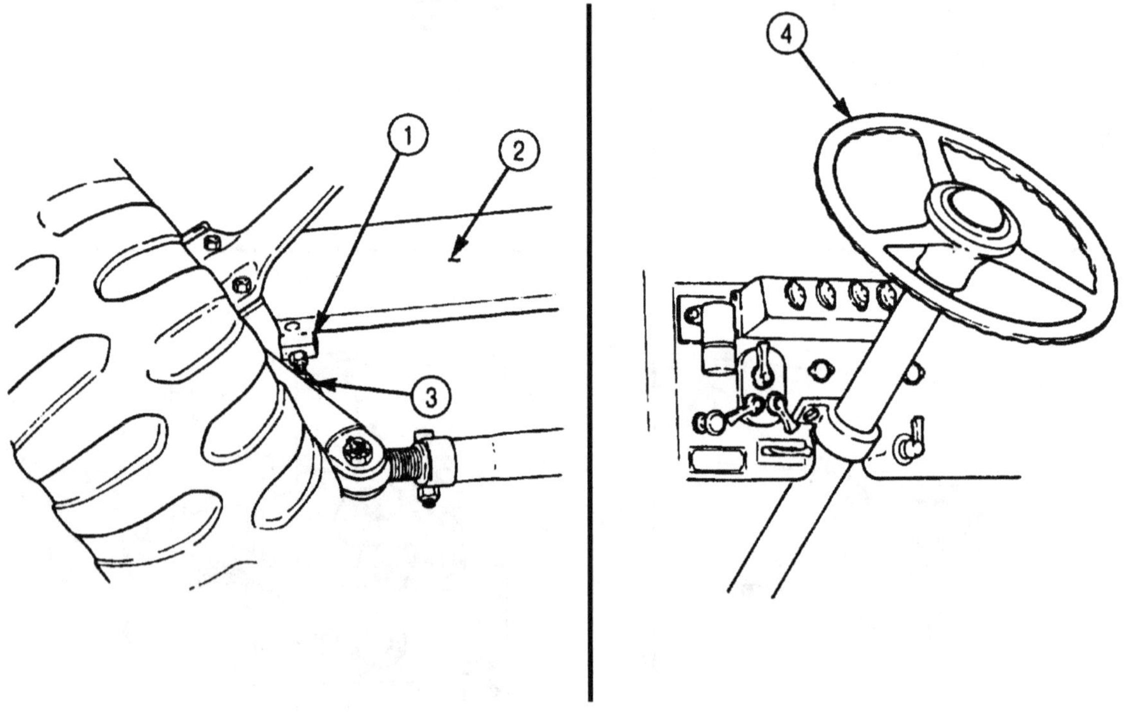

4-124. POWER STEERING TESTS AND ADJUSTMENT (SHEPPARD) (Contd)

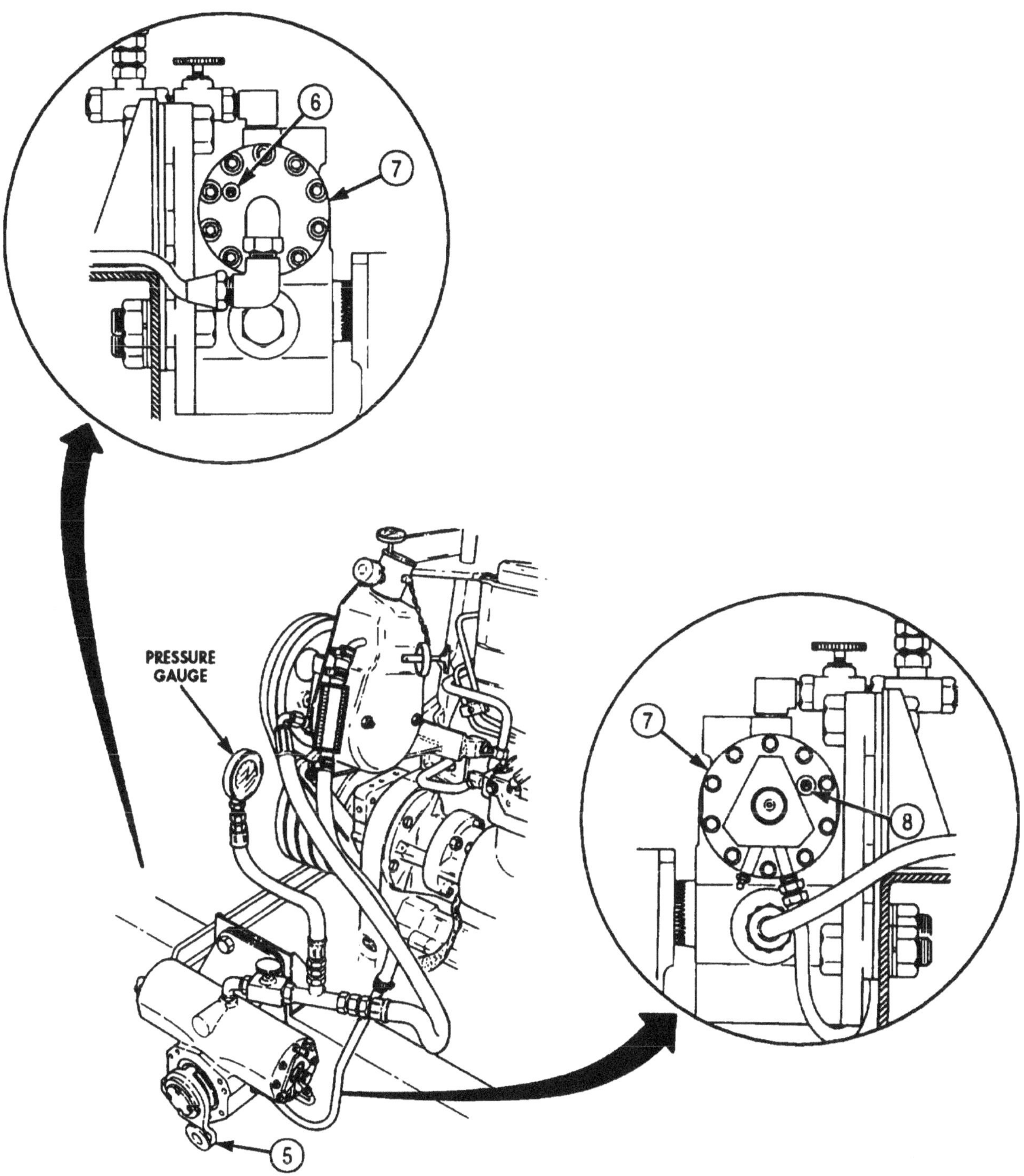

FOLLOW-ON TASKS:
- Remove test equipment (para. 4-122).
- Fill power steering reservoir to proper oil level (LO 9-2320-272-12).
- Start engine (TM 9-2320-272-10) and road test vehicle.

4-125. POWER STEERING GEAR ADJUSTMENT (ON-VEHICLE)

THIS TASK COVERS:

a. Sector Shaft Adjustment | **b. Wormshaft Adjustment**

INITIAL SETUP:

APPLICABLE MODELS
All

TOOLS
General mechanic's tool kit (Appendix E, Item 1)
Torque wrench (Appendix E, Item 144)

REFERENCES (TM)
TM 9-2320-272-10

PERSONNEL REQUIRED
Two

EQUIPMENT CONDITION
- Parking brake set (TM 9-2320-272-10).
- Left splash shield removed (para. 3-301).
- Drag link removed (para. 3-229).

a. Sector Shaft Adjustment

NOTE

Assistant will rotate steering wheel.

1. Rotate steering wheel (1) full travel in both directions and note halfway point.
2. Position steering wheel (1) at halfway point, and loosen jamnut (5) on right side of steering gear (2).
3. Tighten sector shaft adjusting screw (6) 10 lb-ft (14 N·m).
4. Turn sector shaft adjusting screw (6) counterclockwise one full turn.
5. Holding adjusting screw (6), tighten jamnut (5) 40-45 lb-ft (54-61 N·m).

b. Wormshaft Adjustment

1. Loosen jamnut (3) on wormshaft adjusting screw (4) on front of steering gear (2).
2. Loosen wormshaft adjusting screw (4) one full turn counterclockwise.
3. Tighten wormshaft adjusting screw (4) 5-6 lb-ft (7-8 N·m).
4. Holding adjusting screw (4), tighten jamnut (3) 70-80 lb-ft (95-109 N·m).

NOTE

Assistant will rotate steering wheel.

5. Rotate steering wheel (1) slightly in both directions and check for pulsations. If pulsations are felt, repeat steps 1 through 5.

4-125. POWER STEERING GEAR ADJUSTMENT (ON-VEHICLE) (Contd)

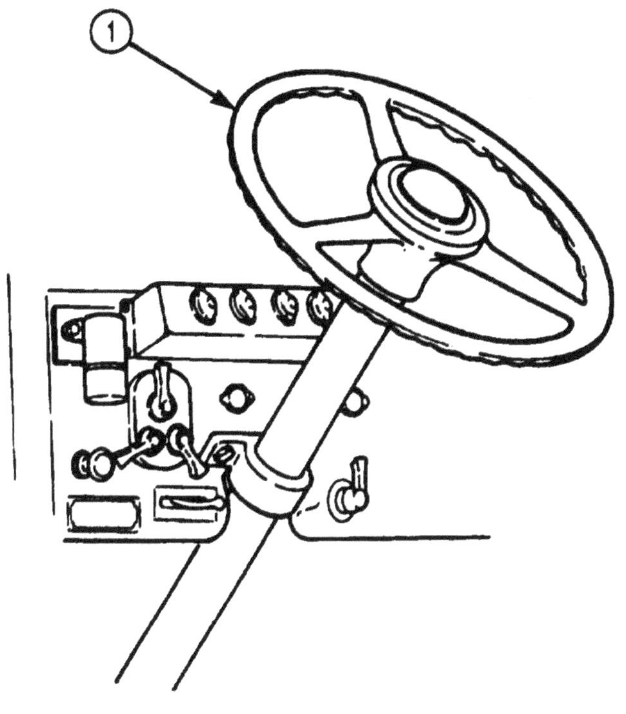

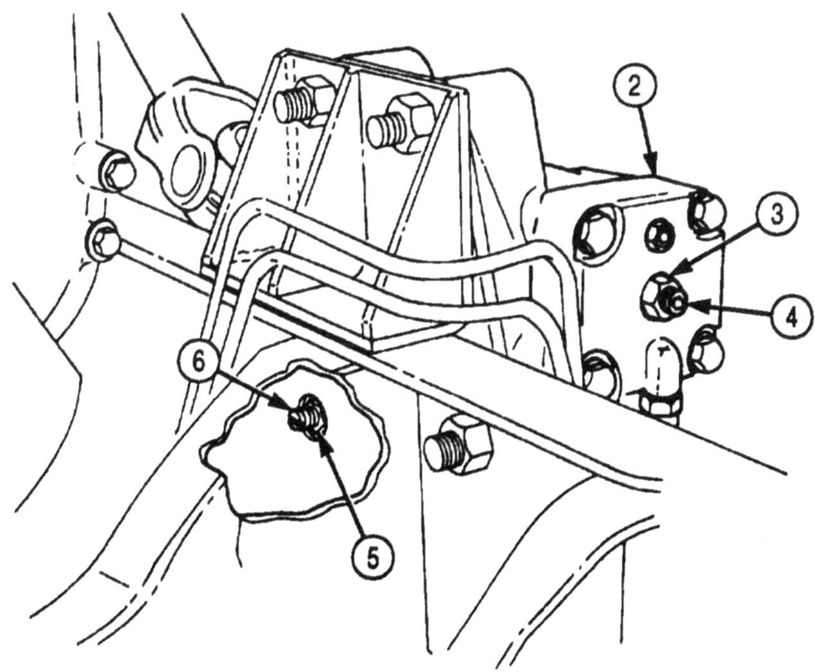

FOLLOW-ON TASKS:
- Install drag link (para. 3-229).
- Install left splash shield (para. 3-301).
- Start engine (TM 9-2320-272-10) and road test vehicle.

4-126. POWER STEERING PUMP AND RESERVOIR (M939/A1) REPAIR

THIS TASK COVERS:

a. Removal
b. Disassembly
c. Cleaning and Inspection
d. Assembly
e. Installation

INITIAL SETUP:

APPLICABLE MODELS

M939/A1

TOOLS

General mechanic's tool kit (Appendix E, Item 1)
Torque wrench (Appendix E, Item 144)
Mechanical puller kit (Appendix E, Item 102)
Vise

MATERIALS/PARTS

Three lockwashers (Appendix D, Item 354)
Power steering filter kit (Appendix D, Item 522)
Springtite assembly (Appendix D, Item 675)
Locknut (Appendix D, Item 327)
Adapter assembly (Appendix D, Item 2)
Woodruff key (Appendix D, Item 727)

MATERIALS/PARTS (Contd)

Drycleaning solvent (Appendix C, Item 71)
Cap and plug set (Appendix C, Item 14)
Rag (Appendix C, Item 58)

REFERENCES (TM)

TM 9-2320-272-10
TM 9-2320-272-24P

EQUIPMENT DESCRIPTION

- Parking brake set (TM 9-2320-272-10).
- Left splash shield removed (para. 3-301).

GENERAL SAFETY INSTRUCTIONS

Drycleaning solvent is flammable and toxic. Do not use near an open flame.

a. Removal

CAUTION

Cap or plug all openings after disconnecting lines and hoses to prevent contamination. Failure to do so may cause damage to equipment.

NOTE

- Have drainage container ready to catch hydraulic oil.
- Plug or cap all hoses.

1. Loosen clamp (16) and disconnect oil return hose (15) from inlet tube (1).
2. Disconnect power steering pressure hose (14) from outlet adapter (13).
3. Remove locknut (8), screw (12), two washers (9), and adjusting link (11) from power steering pump bracket (6). Discard locknut (8).
4. Remove two screws (3), lockwashers (4), and washers (5) from power steering pump bracket (6). Discard lockwashers (4).
5. Remove two drivebelts (10) from pulley (17).
6. Remove power steering pump bracket (6) and power steering pump (2) from engine bracket (7).

b. Disassembly

1. Place power steering pump bracket (6) and power steering pump (2) in vise.
2. Remove springtite screw (28) and washer (29) from shaft (32). Discard springtite screw (28).
3. Using mechanical puller, remove pump pulley (17) from shaft (32).
4. Remove woodruff key (27) from shaft (32). Discard woodruff key (27).
5. Remove reservoir cap and hook chain (18) from reservoir (20).

4-126. POWER STEERING PUMP AND RESERVOIR (M939/A1) REPAIR (Contd)

6. Remove two locknuts (36) from reservoir (20). Discard locknuts (36).

CAUTION

Tap reservoir with soft-headed hammer to loosen from pump base. Do not pry pump base with any tool. Doing so may damage pump base and reservoir.

NOTE

Mark index marks on pump base and reservoir for assembly.

7. Mark pump base (24) and reservoir (20) and remove pump reservoir (20) from pump base (24).
8. Remove two sealing washers (21) and mounting studs (22) from pump base (24). Discard sealing washers (21).
9. Remove pump base O-ring (23) from pump base (24). Discard O-ring (23).
10. Remove outlet adapter (26) and O-ring (25) from pump base (24). Discard O-ring (25) and outlet adapters (26).
11. Remove spring (34), oil filter (33), and washer (35) from reservoir (20). Discard spring (34), oil filter (33), and washer (35).
12. Remove filter breather (19) from reservoir (20).
13. Remove three screws (30), lockwashers (31), and pump base (24) from power steering pump bracket (6). Discard lockwashers (31).
14. Remove power steering pump bracket (6) from vise.

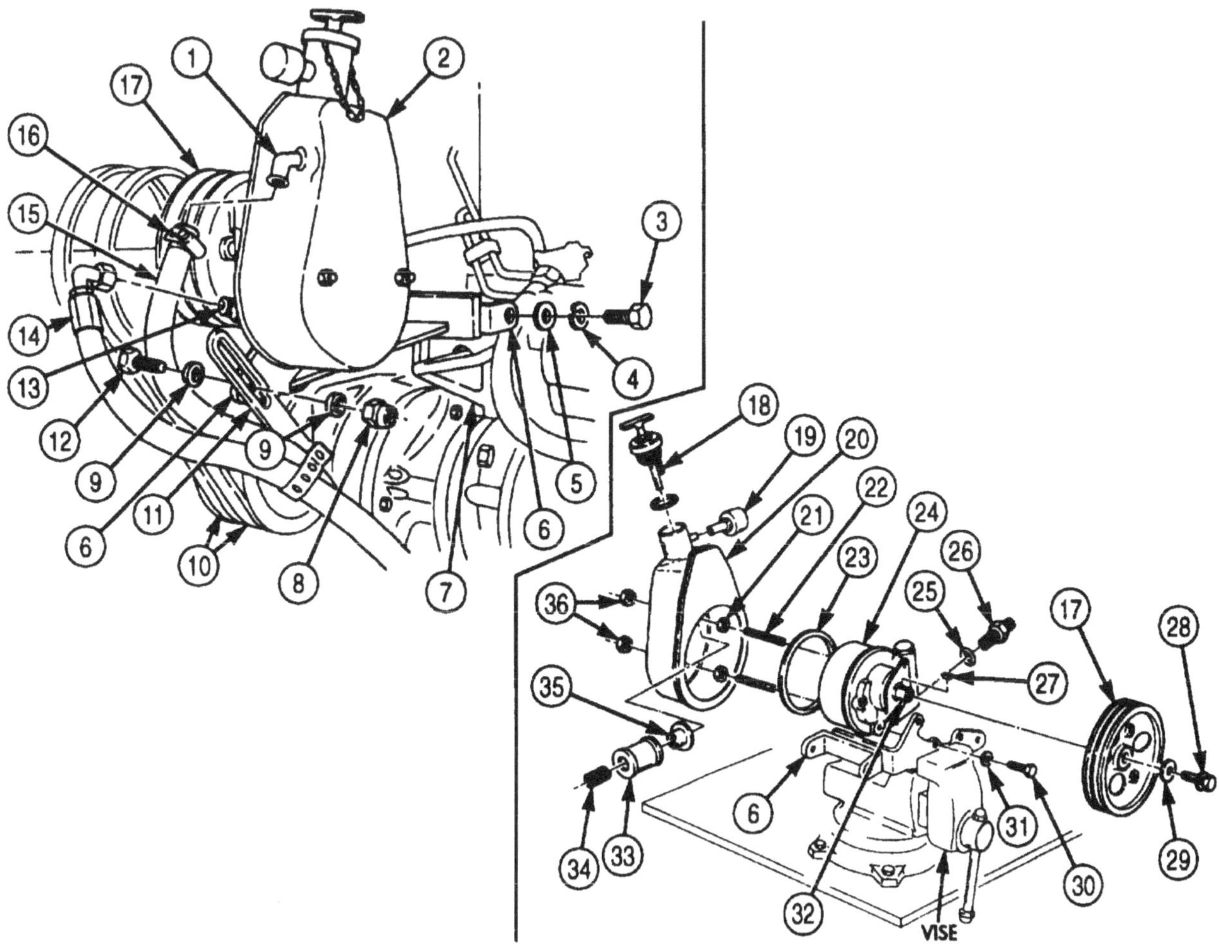

4-126. POWER STEERING PUMP AND RESERVOIR (M939/A1) REPAIR (Contd)

c. Cleaning and Inspection

WARNING

Drycleaning solvent is flammable and toxic. Do not use near an open flame and always have a fire extinguisher nearby when solvents are used. Use only in well-ventilated places, wear protective clothing, and dispose of cleaning rags in approved container. Failure to do this may result in injury to personnel and/or damage to equipment.

1. Clean pump base (7), reservoir (3), and filter breather (2) with drycleaning solvent. Dry with rag.
2. Inspect pump base (7) for cracks and shaft (16) for burned condition. Replace pump base (7) if cracked or if shaft (16) is burned.
3. Inspect mounting studs (5) for stripped threads. Replace mounting studs (5) if threads are stripped.
4. Inspect reservoir (3) for cracks. Replace reservoir (3) if cracked.

d. Assembly

1. Install filter breather (2) on reservoir (3).
2. Install new washer (20), new oil filter (19), and new spring (18) in reservoir (3). Position spring (18) to recess in reservoir (3).
3. Place power steering pump bracket (17) in vise.
4. Install pump base (7) on pump bracket (17) with three new lockwashers (15) and screws (14). Tighten screws (14) 25-31 lb-ft (34-42 N·m).
5. Install new O-ring (8) and new outlet adapter (9) in pump base (7). Tighten outlet adapter (9) 30-45 lb-ft (41-61 N·m).
6. Install new pump base O-ring (6) on pump base (7).
7. Install two mounting studs (5) in pump base (7). Tighten studs (5) 12-120 lb-in. (1-14 N·m).
8. Install two new sealing washers (4) on two mounting studs (5).
9. Install reservoir (3) over mounting studs (5) and on pump base (7) with two new locknuts (21).
10. Install hook chain and cap (1) on reservoir (3).
11. Place new woodruff key (10) in shaft (16) and install pulley (11) on shaft (16) with washer (13) and new springtite screw (12). Tighten screw (12) 18 lb-ft (24 N·m).
12. Remove power steering pump (23) and pump bracket (17) from vise.

e. Installation

1. Install power steering pump (23) and bracket (17) on engine bracket (27) with two washers (26), new lockwashers (25), and screws (24). Do not tighten screws (24).
2. Align pump bracket (17) on adjusting link (31) and install with washer (29), screw (32), washer (29), and new locknut (28). Do not tighten locknut (28).
3. Place two drivebelts (30) around pulley (11).
4. Connect power steering pressure hose (33) to adapter (9).
5. Connect hose (34) to inlet tube (22) and tighten clamp (35).

4-126. POWER STEERING PUMP AND RESERVOIR (M939/A1) REPAIR (Contd)

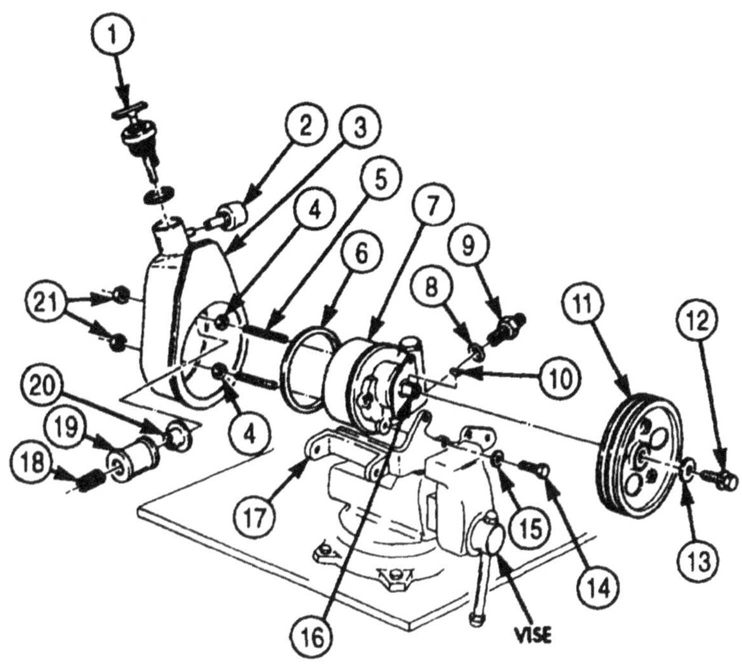

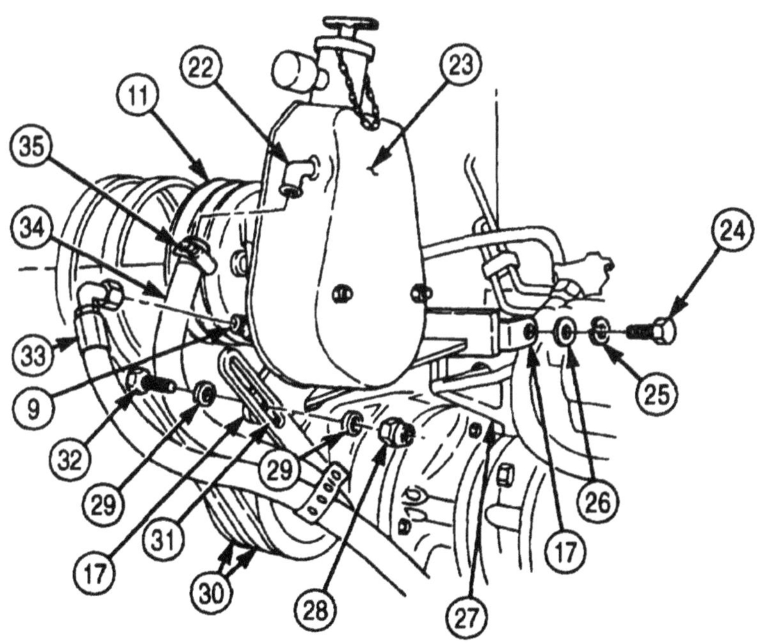

FOLLOW-ON TASKS:
- Adjust steering pump drivebelts (para. 3-230).
- Install left splash shield (para. 3-301).
- Fill power steering pump reservoir (LO 9-2320-272-12).
- Start engine (TM 9-2320-272-10).

4-127. POWER STEERING PUMP (M939A2) REPAIR

THIS TASK COVERS:

a. Disassembly
b. Cleaning and Inspection
c. Assembly

INITIAL SETUP:

APPLICABLE MODELS
M939A2

TOOLS
General mechanic's tool kit (Appendix E, Item 1)

MATERIALS/PARTS
Cover kit (Appendix D, Item 86)
Filter element kit (Appendix D, Item 123)
Flange gasket (Appendix D, Item 210)
Power steering parts kit (Appendix D, Item 523)
Pumping element kit (Appendix D. Item 525)
Lubricating oil (Appendix C, Item 49)
Petrolatum (Appendix C, Item 53)
Drycleaning solvent (Appendix C, Item 71)

REFERENCES (TM)
LO 9-2320-272-12
TM 9-2320-272-10
TM 9-2320-272-24P

EQUIPMENT CONDITION
Power steering pump removed (para. 3-236)

GENERAL SAFETY INSTRUCTIONS
- When cleaning with compressed air, wear eyeshields and ensure source pressure will not exceed 30 psi (207 kPa).
- Drycleaning solvent is flammable and toxic.
- Keep fire extinguisher nearby when using drycleaning solvent.
- Use eyeshields when removing steering pump reservoir cover.
- Use eyeshields when removing valve cap.

a. Disassembly

1. Remove screw (7), vent cover (6), and vent filter (5) from cover (4).

WARNING

Power steering pump reservoir cover is spring-compressed. Use eyeshields when removing. Failure to do so may result in injury to personnel.

2. Remove wingnut (1), washer (2), stud gasket (3), cover (4), and cover gasket (8). Discard stud gasket (3) and cover gasket (8).
3. Remove spring (9), filter cap (10), and filter (11) from reservoir (23). Discard filter cap (10) and filter (11).
4. Remove O-ring (20) and stud (12) from intake fitting (19). Discard O-ring (20).
5. Remove intake fitting (19) from reservoir plate (18).
6. Remove two screws (21), lockwashers (22), reservoir plate (18), reservoir (23), two reservoir screw gaskets (24), and intake gaskets (17) from pump body (16). Discard lockwashers (22) and gaskets (17) and (24).
7. Remove elbow (32) and O-ring (33) from pump body (16). Discard O-ring (33).
8. Remove outlet fitting (30) and O-ring (31) from pump cover (25). Discard O-ring (31).

WARNING

Valve cap is spring-compressed. Use eyeshields when removing. Failure to do so may result in injury to personnel.

9. Remove valve cap (29), O-ring (28), spring (27), and valve (26) from pump cover (25). Discard O-ring (28).
10. Remove three screws (13), flange (14), and flange gasket (15) from pump body (16). Discard flange gasket (15).

4-127. POWER STEERING PUMP (M939A2) REPAIR (Contd)

11. Remove five screws (48), pump cover (25), and O-ring (44) from pump body (16). Discard O-ring (44).
12. Remove shaft (42) and ten rollers (45) from pump body (16).
13. Remove snapring (47), carrier (46), and drive pin (43) from shaft (42).
14. Remove cam (40), port plate (39), end plate (36), and locating pin (41) from pump body (16).
15. Remove two O-rings (38) and seals (37) from end plate (36). Discard seals (37) and O-rings (38).
16. Remove O-ring (35) from back of pump body (16). Discard O-ring (35).
17. Remove two oil seals (34) through front of pump body (16). Discard oil seals (34).

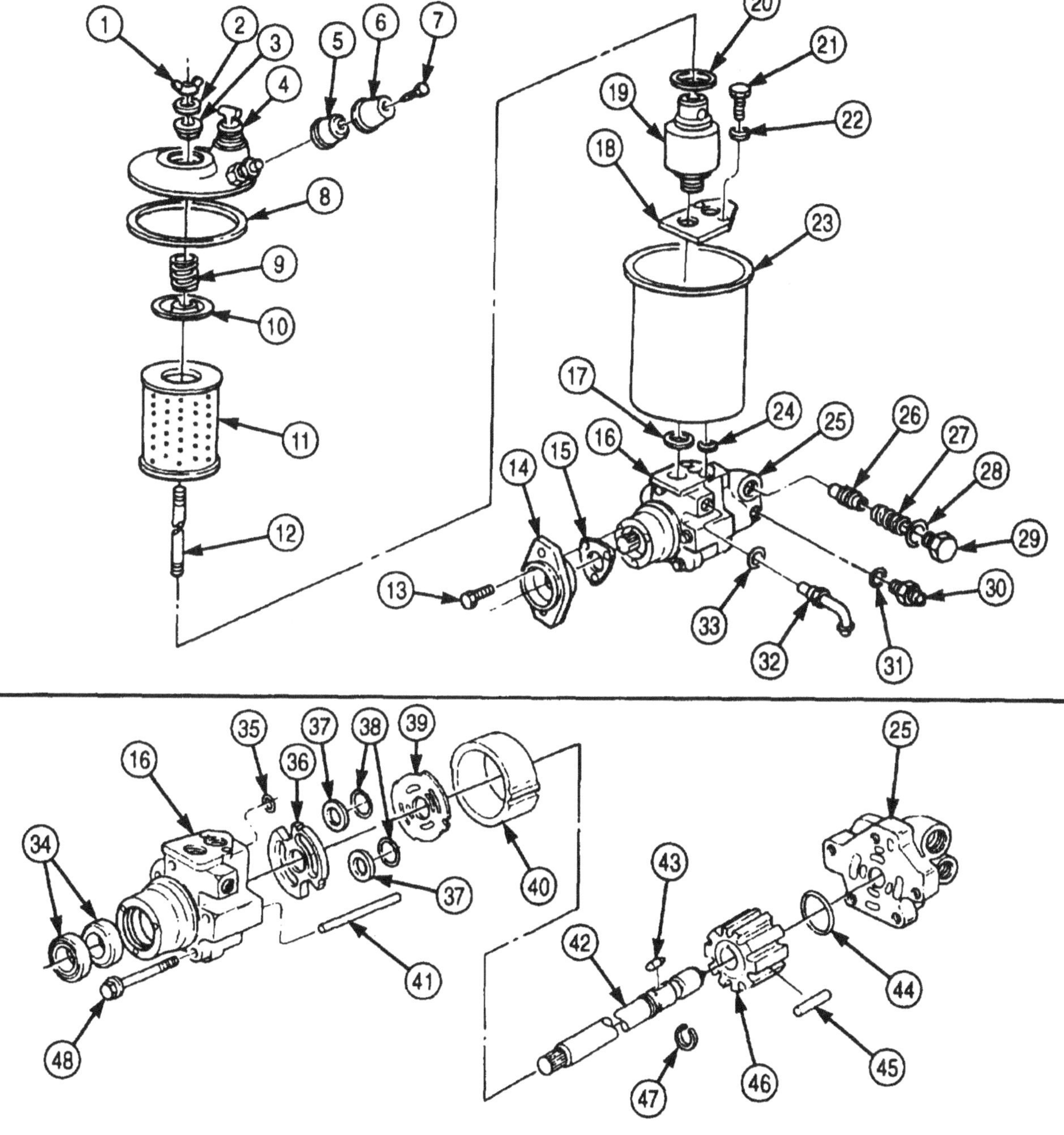

4-127. POWER STEERING PUMP (M939A2) REPAIR (Contd)

b. Cleaning and Inspection

1. For general cleaning instructions, refer to para. 2-14.

WARNING

- Drycleaning solvent is flammable and toxic. Do not use near on open flame and always have a fire extinguisher nearby when solvents are used. Use only in well-ventilated places, wear protective clothing, and dispose of cleaning rags in approved container. Failure to do this may result in injury to personnel and/or damage to equipment.
- Compressed air source will not exceed 30 psi (207 kPa). When cleaning with compressed air, eyeshields must be worn. Failure to do so may result in injury to personnel.

2. For general inspection instructions, refer to para. 2-15.
3. Replace all parts failing inspection.
4. Inspect bushing (22) of pump body (12) and pump cover (23) for excessive wear or gouging. Replace pump body (12) or pump cover (23) if bushing is worn or gouged.

c. Assembly

1. Install new oil seal (21) in pump body (12) with rubber side facing downward. Ensure oil seal (21) is seated against bottom of pump body (12).
2. Install new oil seal (20) in pump body (12) with rubber side facing upward. Ensure oil seal (20) is seated against oil seal (21).
3. Install locating pin (11) and new O-ring (13) on pump body (12).
4. Apply petrolatum to two new seals (8) and new O-rings (7) and install in end plate (9).
5. Align notch (10) of end plate (9) with locating pin (11) and install end plate (9) in pump body (12).
6. Align notch (14) of port plate (6) with locating pin (11) and install port plate (6) in pump body (12).

NOTE

- Ensure arrow of cam points clockwise when installing.
- Identification groove must face splines of shaft when installing carrier.

7. Align notch (15) of cam (5) with locating pin (11) and cam (5) in body (12). Ensure arrow (4) points clockwise.
8. Install drive pin (16), carrier (2), and snapring (18) on shaft (3). Ensure identification groove (17) faces splines (19) of shaft (3).
9. Lubricate shaft (3) and carrier (2) with clean engine oil and install in pump body (12).
10. Lubricate ten rollers (1) with clean engine oil and install in carrier (2).
11. Apply petrolatum to new O-ring (24) and install on pump body (12).
12. Install pump cover (23) on pump body (12) with five screws (25).

4-127. POWER STEERING PUMP (M939A2) REPAIR (Contd)

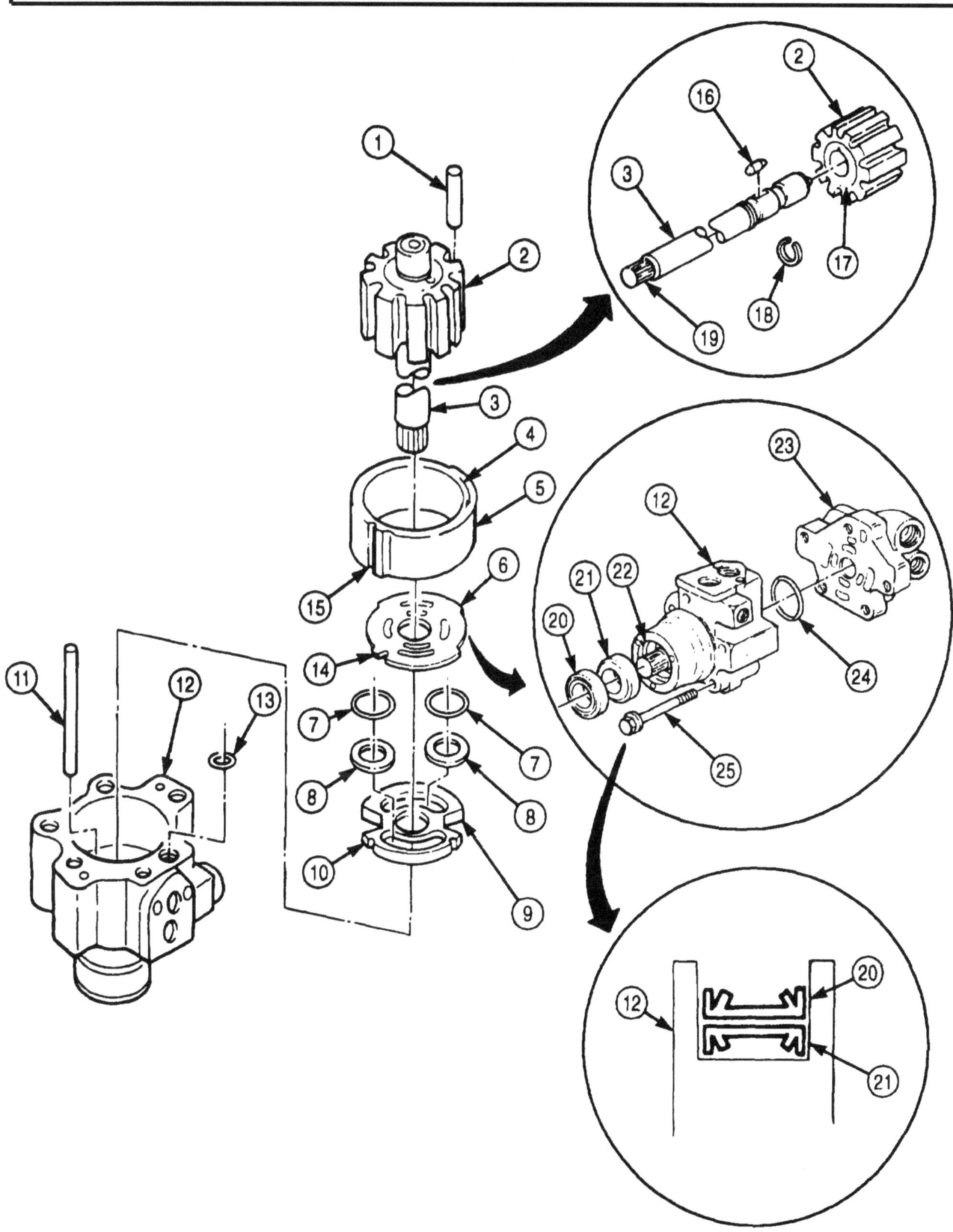

4-127. POWER STEERING PUMP (M939A2) REPAIR (Contd)

13. Install new flange gasket (30) and flange (29) on pump body (31) with three screws (28).
14. Coat new O-ring (22) with petrolatum and install on valve cap (23).
15. Install valve (20), spring (21), and valve cap (23) on pump cover (18).
16. Coat new O-ring (19) with petrolatum and install on outlet fitting (24).
17. Install outlet fitting (24) on pump cover (18).
18. Coat new O-ring (27) with petrolatum and install on elbow (25).
19. Install elbow (25) on pump body (31). Ensure flange (26) of elbow (25) seats on pump body (31).
20. Lubricate two new intake gaskets (32) and new reservoir screw gaskets (17) with clean engine oil and install in pump body (31).
21. Install reservoir (16) and reservoir plate (33) on pump body (31) with two new lockwashers (15) and screws (14). Finger-tighten screws (14).
22. Install intake fitting (34) on reservoir plate (33). Tighten screws (14).
23. Install new O-ring (12) and stud (13) on intake fitting (34).
24. Position new filter (11), new filter cap (10), and spring (9) on stud (13) and reservoir (16).
25. Install new cover gasket (8) and reservoir cover (4) on reservoir (16) with new stud gasket (1), washer (2), and wingnut (3).
26. Install vent filter (5) and vent cover (6) on reservoir cover (4) with screw (7).

4-127. POWER STEERING PUMP (M939A2) REPAIR (Contd)

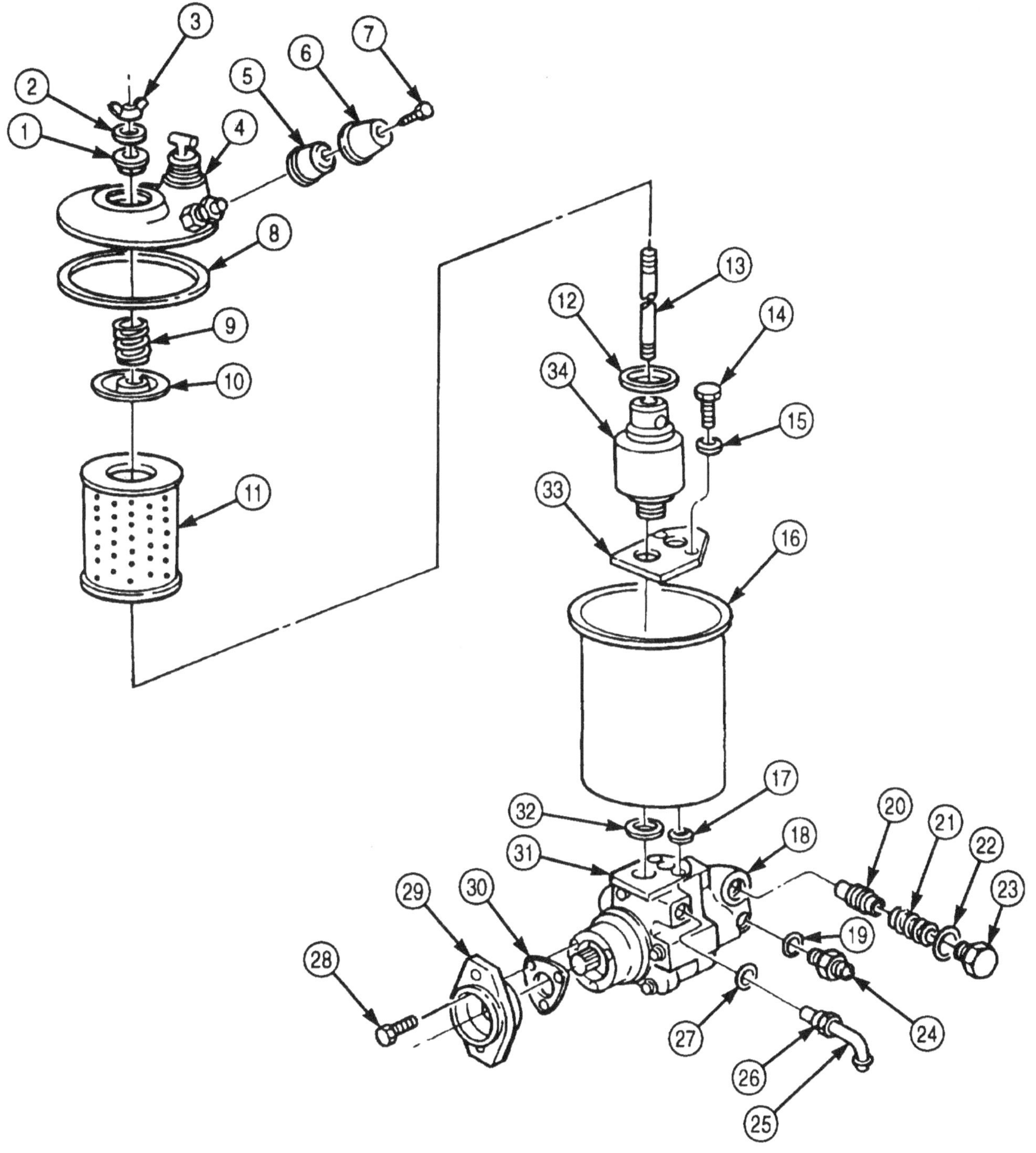

FOLLOW-ON TASKS:
- Install power steering pump (para. 3-236).
- Fill power steering pump reservoir (LO 9-2320-272-12).
- Start engine (TM 9-2320-272-10) and check for power steering leaks.

Section XII. FRAME MAINTENANCE

4-128. FRAME MAINTENANCE INDEX

4-129. FRAME REPAIR

Refer to TB 9-2300-247-40 for maintenance and repair of frames used on all M939/A1/A2 series vehicles. Refer to TM 9-2320-272-24P for authorized replacement parts used in frame repair.

4-130. FIFTH WHEEL MAINTENANCE

THIS TASK COVERS:

a. Disassembly
b. Cleaning and Inspection
c. Assembly

INITIAL SETUP:

APPLICABLE MODELS
M931/A1/A2, M932/A1/A2

TOOLS
General mechanic's tool kit (Appendix E, Item 1)
Torque wrench (Appendix E, Item 145)
Vernier calipers (Appendix E, Item 159)
Lifting device
Chains and attaching hardware

MATERIALS/PARTS
Cotter pin (Appendix D, Item 63)
Cotter pin (Appendix D, Item 64)
Three locknuts (Appendix D, Item 309)
Locknut (Appendix D, Item 296)
Two lockwashers (Appendix D, Item 388)
Lockwasher (Appendix D, Item 377)
Safety wire (Appendix D, Item 567)

PERSONNEL REQUIRED
Three

REFERENCES (TM)
LO 9-2320-272-12
TM 9-2320-272-24P

EQUIPMENT CONDITION
Fifth wheel removed (para. 3-248)

GENERAL SAFETY INSTRUCTIONS
- All personnel must stand clear during lifting operations.
- Eye protection must be worn when removing or installing springs under tension.
- Personnel must stand clear during release of plunger rack.
- Ensure lifting capacity is greater than weight of fifth wheel.

a Disassembly

1. Attach chain to fifth wheel housing (1) with two appropriately sized washers (2), screws (4), washers (2), and nuts (3).
2. Attach lifting device to chain. Apply tension to chain.
3. Remove two lubrication fittings (8) from fifth wheel housing (1).

CAUTION

Do not use heat to remove pivot pins from fifth wheel housing. Damage to equipment will result.

4. Remove two retaining pins (6) and pivot pins (7) from fifth wheel housing (1) and base (5).

WARNING

- All personnel must stand clear during lifting operations. A swinging or shifting load may cause injury to personnel.
- Ensure lifting capacity is greater than weight of fifth wheel (600 lb) (272 kg). Failure to do so may result in injury to personnel or damage to equipment.

5. Remove fifth wheel housing (1) from base (5).

4-130. FIFTH WHEEL MAINTENANCE (Contd)

NOTE

- Three personnel are required for step 6.
- Fifth wheel housing must be 6 in. (15 cm) above workbench for bushing removal.

6. Position fifth wheel housing (1) on workbench, with trailer mating surface (9) side down.

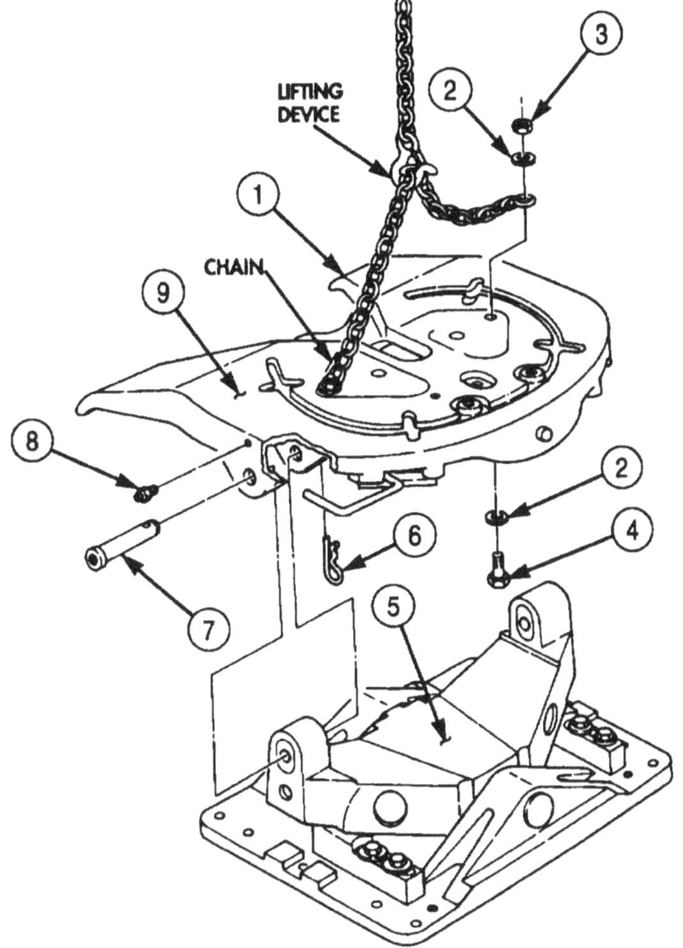

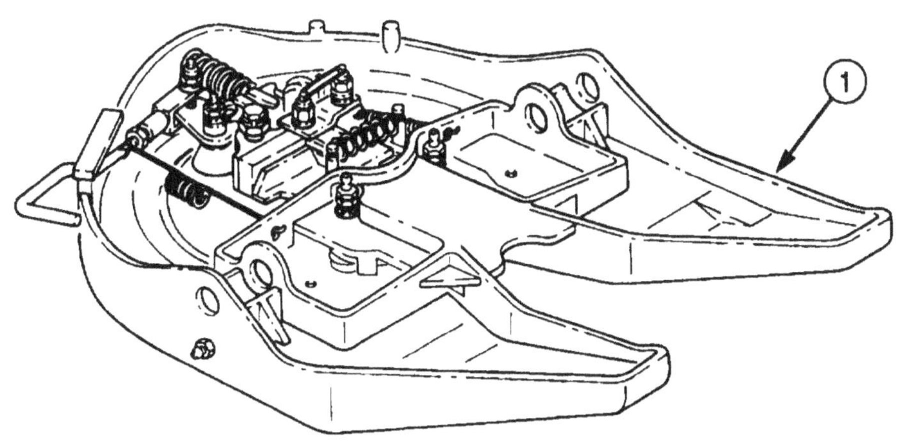

4-130. FIFTH WHEEL MAINTENANCE (Contd)

WARNING

- Personnel must stand clear during release of plunger rack. Failure to do so may result in injury to personnel.
- Eye protection must be worn when removing springs under tension. Failure to do so may result in injury to personnel.

7. Remove spring (29) and cotter pin (30) from fifth wheel housing (22) and handle (1). Discard cotter pin (30).
8. Remove plunger rack latch (17) and allow plunger rack (18) to slide to unload position.
9. Remove spring (16) from plunger rack latch (17) and cotter pin (15).
10. Remove cotter pin (15) from fifth wheel housing (22). Discard cotter pin (15).
11. Remove safety wire (9), two locknuts (12), washers (13), bracket (14), and plunger rack latch (17) from stud (21), screw (23), and fifth wheel housing (22). Discard safety wire (9) and locknuts (12).
12. Remove spring (7), plunger rack (18), screw (23), and spacer (20) from fifth wheel housing (22).
13. Remove screw (28), lockwasher (27), cam (26), and bushing (25) from fifth wheel housing (22). Discard lockwasher (27).
14. Remove locknut (6), washer (5), screw (3), and yoke (4) from pinion gear (19). Discard locknut (6).
15. Remove handle (1) from fifth wheel housing (22).
16. Loosen jamnut (2) and remove handle (1) from yoke (4) and jamnut (2) from handle (1).
17. Remove locknut (8), screw (24), washer (10), spacer (11), and pinion gear (19) from fifth wheel housing (22). Discard locknut (8).
18. Remove spring (32) from two coupler jaws (31).
19. Remove two lubrication fittings (34), screws (35), lockwashers (36), and washers (37) from coupler jaws (31). Discard lockwashers (36).
20. Remove two coupler jaws (31), compression arm (33), and two bushings (38) from fifth wheel housing (22).

4-130. FIFTH WHEEL MAINTENANCE (Contd)

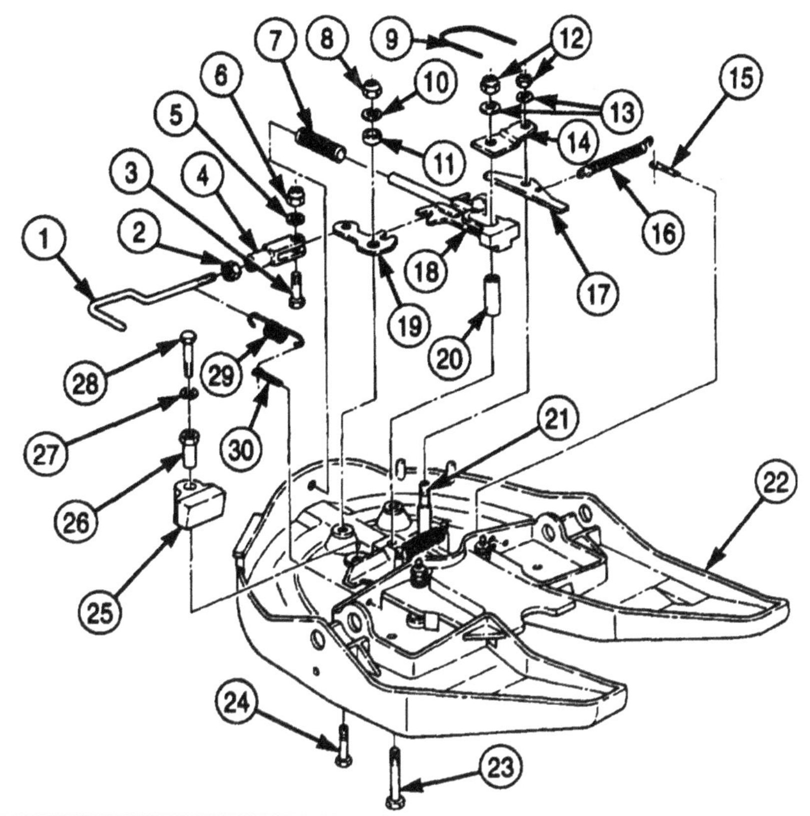

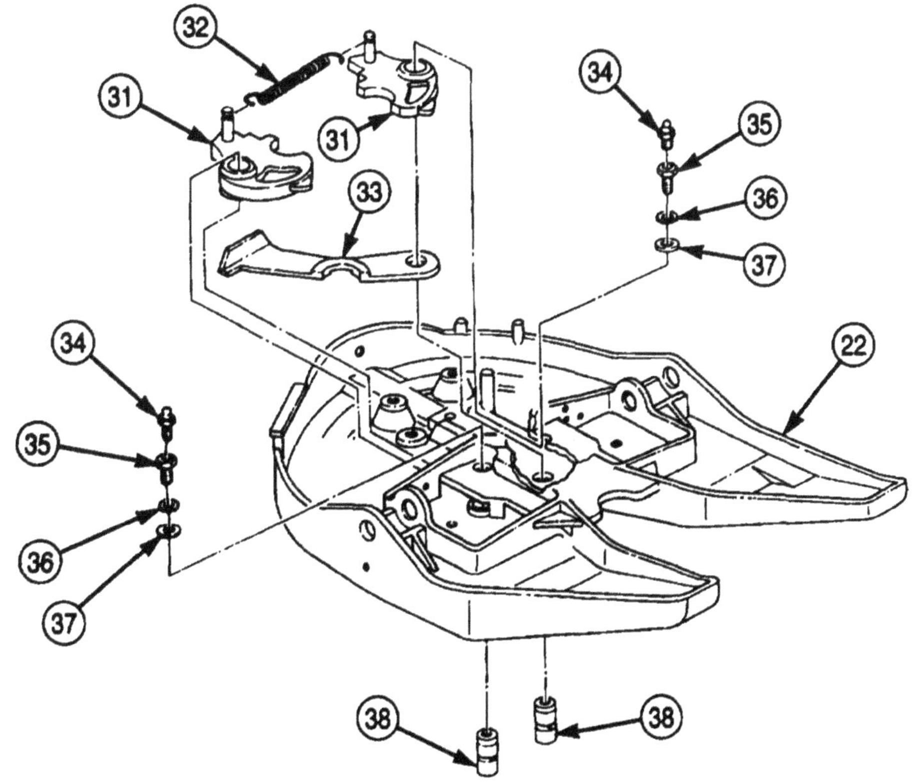

4-130. FIFTH WHEEL MAINTENANCE (Contd)

b. Cleaning and Inspection

1. For general cleaning instructions, refer to para. 2-14.
2. For general inspection instructions, refer to para. 2-15.
3. Inspect handle (1), plunger rack latch (3), pinion gear (6), stud (7), surface of fifth wheel housing (8), two coupler jaws (11), compression arm (14), and two bushings (15) for grooves, breaks, cracks, cuts, bends, stripped threads, and damaged gear teeth. Replace damaged part(s).
4. Inspect springs (2), (5), (12), and (13) for bends, breaks, collapsed coils, and correct spring lengths. Refer to table 4-10, Fifth Wheel Spring Free Length, for measurements. Replace damaged or worn part(s).
5. Inspect plunger rack (4), pivot pin (9), and cam (10) for wear. Refer to table 4-11, Fifth Wheel Wear Limits/Tolerances, for measurements. Replace damaged or worn part(s).

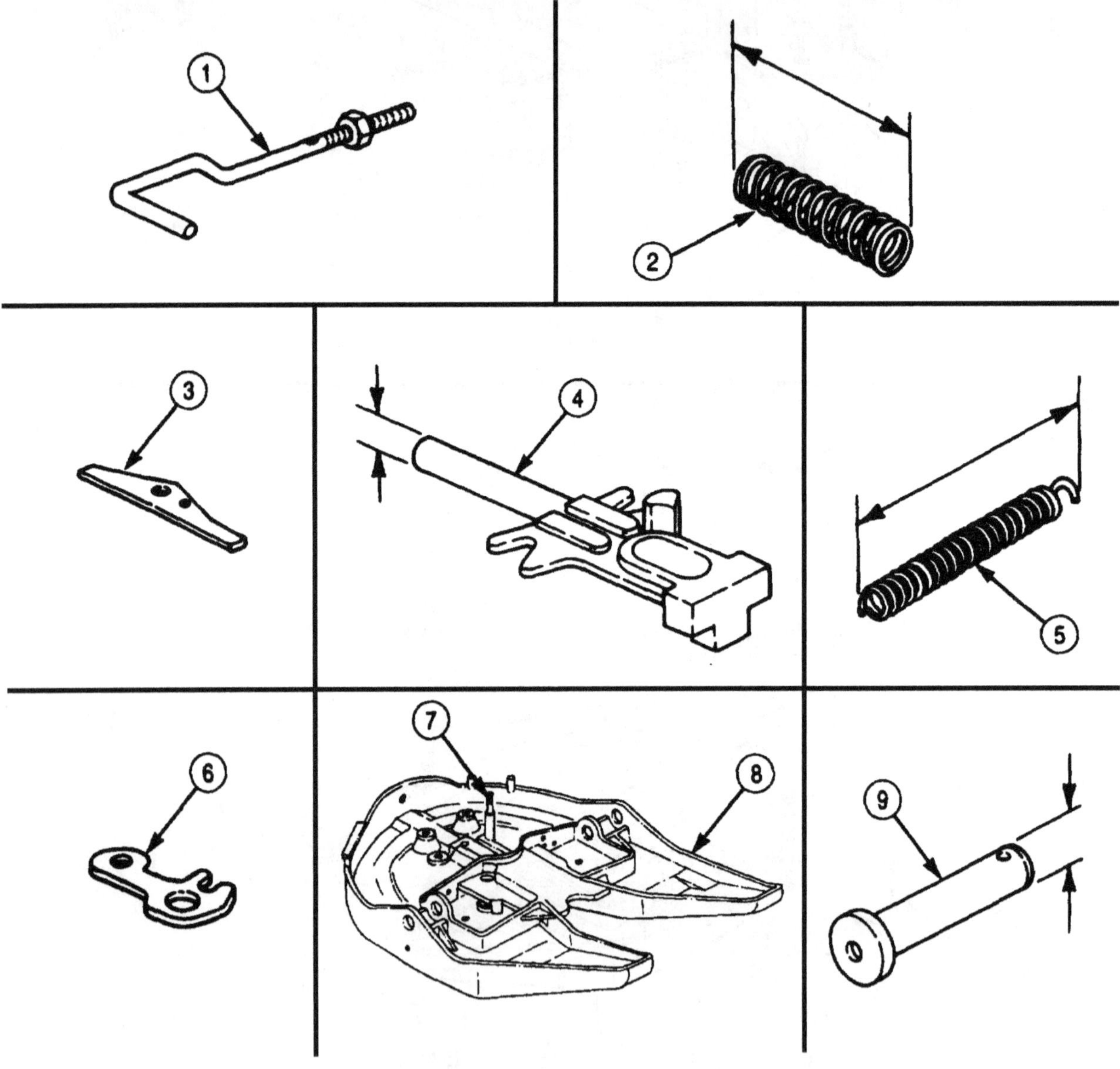

4-130. FIFTH WHEEL MAINTENANCE (Contd)

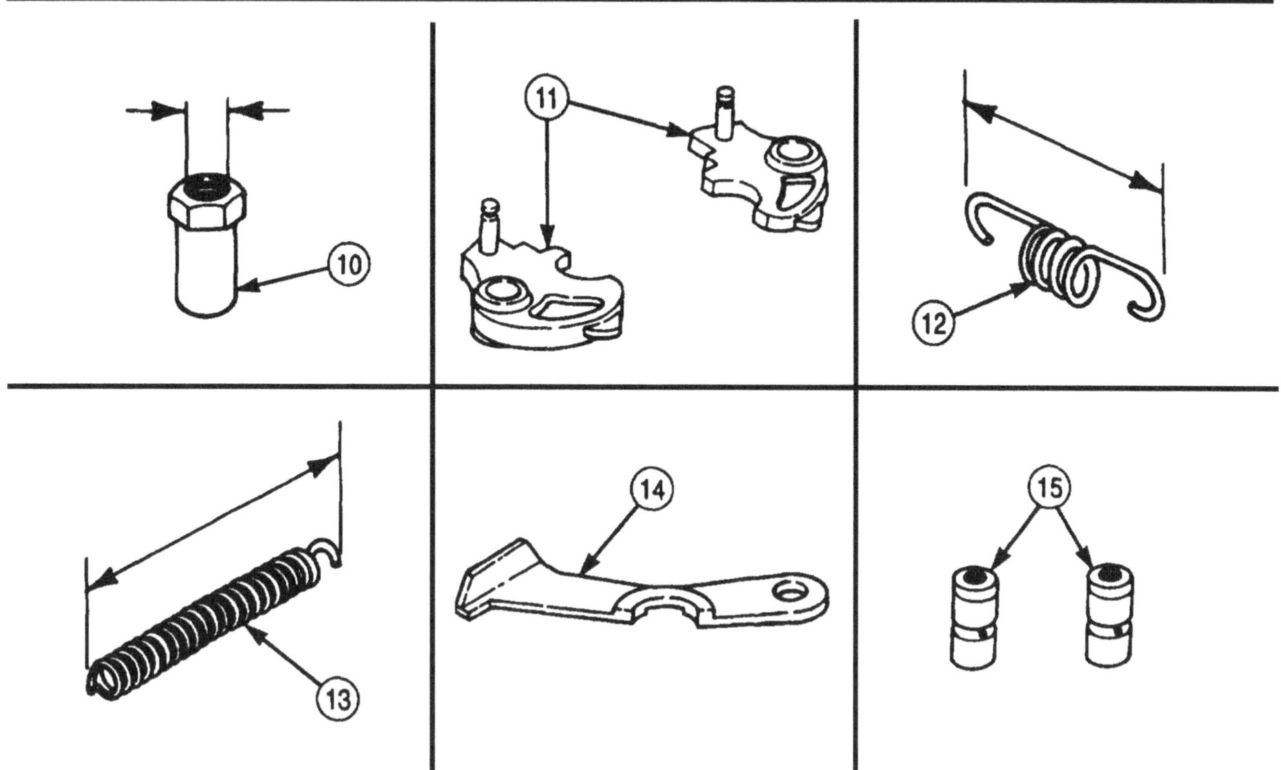

Table 4-10. Fifth Wheel Spring Free Length.

ITEM NO.	ITEM/POINT OF MEASUREMENT	SPRING FREE LENGTH	
		INCHES	MILLIMETERS
2	Locking plunger spring	7.25	184.2
5	Plunger rack latch spring	2.0625	52.388
12	Handle spring	6.00	152.4
13	Coupler jaw spring	2.31	58.674

Table 4-11. Fifth Wheel Wear Limits/Tolerances.

ITEM NO.	ITEM/POINT OF MEASUREMENT	WEAR LIMITS/TOLERANCES	
		INCHES	MILLIMETERS
4	Plunger rack - outer diameter	2.874-2.878	73.000-73.101
9	Pivot pin:		
	Outer diameter	1.47-1.53	37.338-38.862
	Length	6.444-6.59	163.678-167.386
10	Cam - inner diameter	1.093-1.099	27.762-27.915

4-130. FIFTH WHEEL MAINTENANCE (Cod)

c. Assembly

1. Install compression arm (3) and two coupler jaws (1) in fifth wheel housing (8) with two bushings (9).
2. Align hole in compression arm (3) with hole in fifth wheel housing (8).
3. Install two washers (7), new lockwashers (6), and screws (5) on two bushings (9). Tighten screws (5) 150-200 lb-ft (203-271 N·m).
4. Install two lubrication fittings (4) on screws (5).
5. Install spring (2) on two coupler jaws (1).
6. Install bushing (33) and cam (34) on fifth wheel housing (8) with new lockwasher (35) and screw (36).
7. Place spring (16) on plunger rack (27).
8. Install screw (31), spacer (29), and plunger rack (27) on fifth wheel housing (8).

NOTE

Add washers as needed to obtain proper height on plunger rack latch. Latch must set level with latch on plunger rack.

9. Install plunger rack latch (26), bracket (23), two washers (22), new locknuts (21), and safety wire (18) on screw (31) and stud (30).
10. Install new cotter pin (24) on fifth wheel housing (8), with eye of cotter pin (24) facing toward back of fifth wheel housing (8).

WARNING

Eye protection must be worn when installing springs under tension. Failure to do so may result in injury to personnel.

11. Install spring (25) on plunger rack latch (26) with cotter pin (24).
12. Install pinion gear (28) on fifth wheel housing (8) with screw (32), spacer (20), washer (19), and new locknut (17).
13. Install jamnut (11) on handle (10) and install handle (10) on yoke (13). Tighten jamnut (11) 80-95 lb-ft (108-129 N·m).
14. Install yoke (13) on pinion gear (28) with screw (12), washer (14), and new locknut (15).
15. Install new cotter pin (38) in fifth wheel housing (8).
16. Install spring (37) on handle (10) and cotter pin (38).

4-130. FIFTH WHEEL MAINTENANCE (Contd)

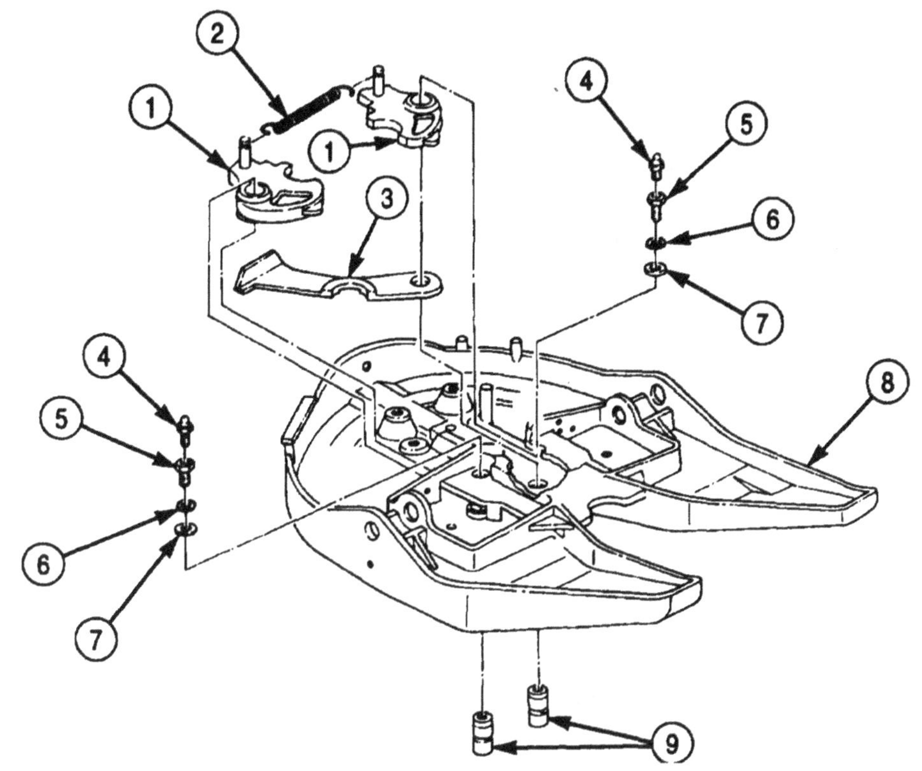

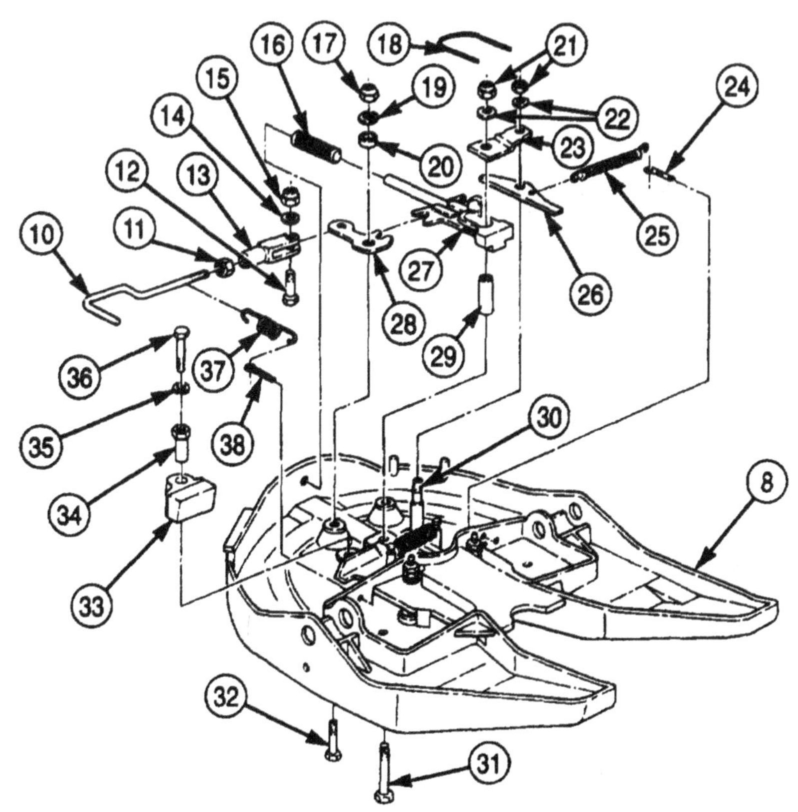

4-130. FIFTH WHEEL MAINTENANCE (Contd)

WARNING

- All personnel must stand clear during lifting operations. A swinging or shifting load may cause injury to personnel.
- Ensure lifting capacity is greater than weight of fifth wheel (600 lbs) (272 kg). Failure to do so may result in injury to personnel or damage to equipment.

17. Remove fifth wheel housing (1) from workbench.
18. Position fifth wheel housing (1) on base (5).
19. Insert two pivot pins (7) through fifth wheel housing (1) and base (5) and install with two retaining pins (6).
20. Install two lubrication fittings (8) on fifth wheel housing (1).
21. Remove lifting device from chain.
22. Remove two nuts (3), washers (2), screws (4), washers (2), and chain from fifth wheel housing (1).

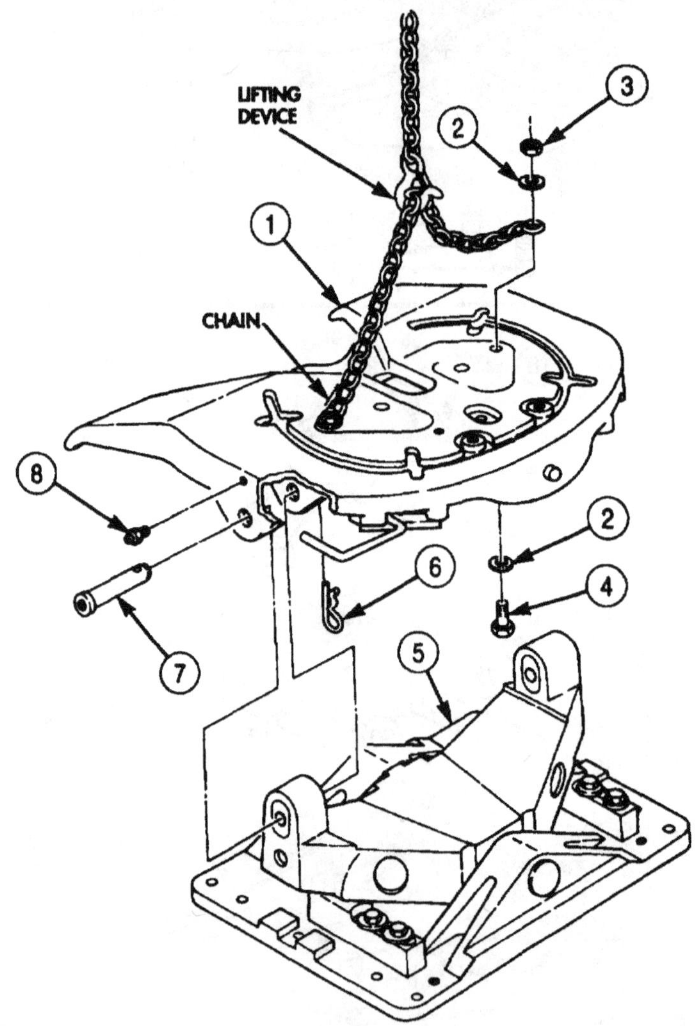

FOLLOW-ON TASKS:
- Install fifth wheel (para. 3-248).
- Lubricate fifth wheel (LO 9-2320-272-12).

Section XIII. BODY, CAB, AND HOOD MAINTENANCE

4-131. BODY, CAB, AND HOOD MAINTENANCE INDEX

4-132. HOOD CROSSMEMBER BRACKET MAINTENANCE

THIS TASK COVERS:

a. Removal
b. Cleaning and Inspection
c. Installation

INITIAL SETUP:

APPLICABLE MODELS
All

TOOLS
General mechanic's tool kit (Appendix E, Item 1)
Soft-faced hammer

MATERIALS/PARTS
Two lockwashers (Appendix D, Item 350)
Locknut (Appendix D, Item 294)
Insulator (Appendix D, Item 262)
Rags (Appendix C, Item 58)

REFERENCES (TM)
LO 9-2320-272-12
TM 9-2320-272-24P

EQUIPMENT CONDITION
Hood removed (para. 3-275).

a. Removal

1. Remove two screws (9), lockwashers (1), and trunnion bracket (2) from hood crossmember (3). Discard lockwashers (1).
2. Remove locknut (4), washer (5), through bolt (7), lower trunnion bracket (6), and insulator (8) from trunnion bracket (2). Use soft-faced hammer to remove through bolt (7) if necessary. Discard locknut (4) and insulator (8).

b. Cleaning and Inspection

1. Wipe trunnion bushings (10) of trunnion brackets (2) and (6) with clean rag.
2. Inspect trunnion bushings (10) for cracks, breaks, and chips. If cracked, broken, or chipped, replace bushings (10).

NOTE

Perform steps 3 and 4 if trunnion bushings are damaged.

3. Using steel drift and hammer, drive trunnion bushings (10) from trunnion bracket (6). Discard bushings (10).
4. Using steel drift and hammer, install new trunnion bushings (10) in trunnion bracket (6).

c. Installation

1. Install new insulator (8) and lower trunnion bracket (6) on trunnion bracket (2) with through bolt (7), washer (5), and new locknut (4). Tighten locknut (4) 96 lb-ft (130 N·m).
2. Install trunnion bracket (2) on hood crossmember (3) with two lockwashers (1) and screws (9).

4-132. HOOD CROSSMEMBER BRACKET MAINTENANCE (Contd)

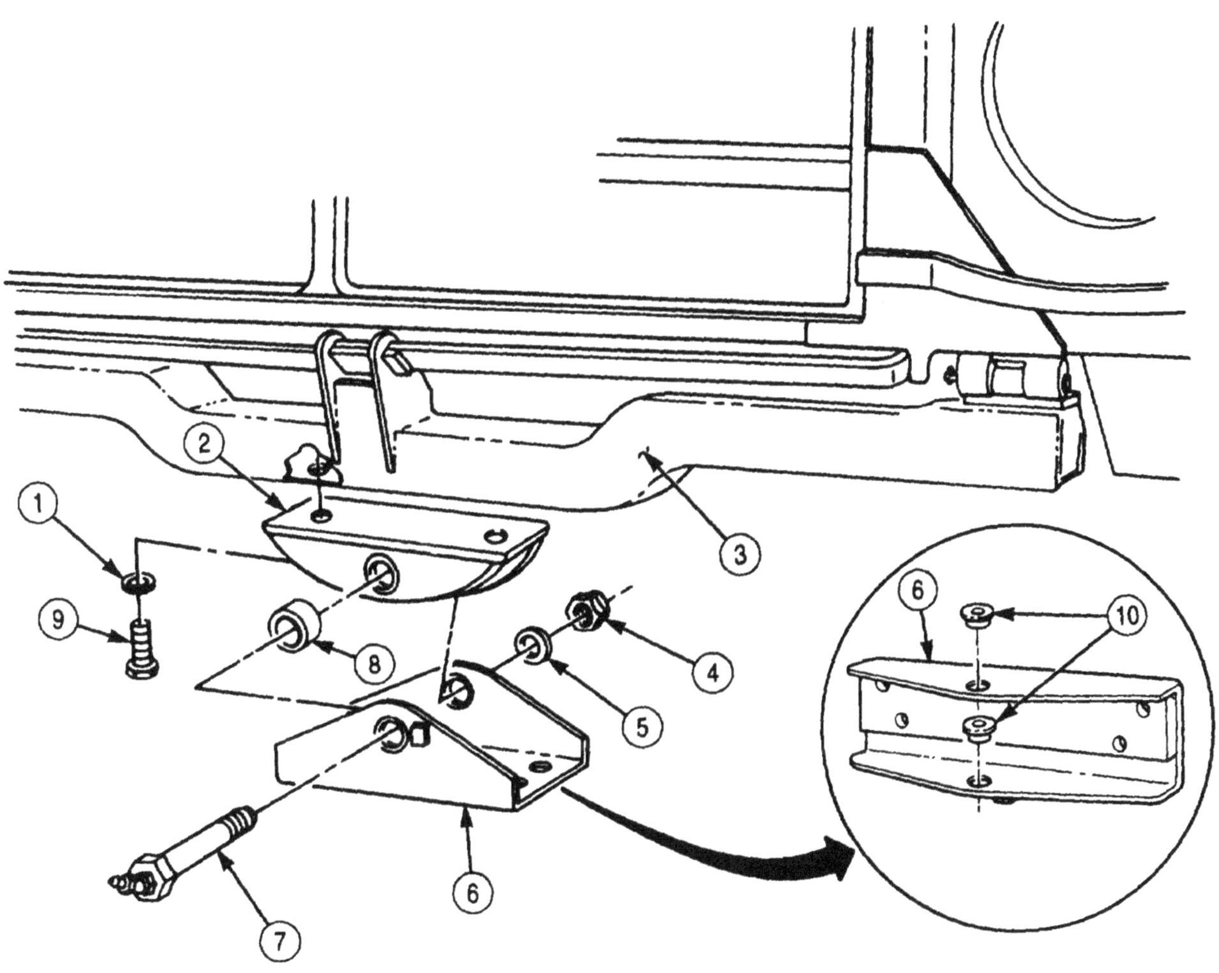

FOLLOW-ON TASKS:
- Lubricate through bolt (LO 9-2320-272-12).
- Install hood (para. 3-275).

4-133. HOOD HINGE REPLACEMENT

THIS TASK COVERS:

a. Removal **b. Installation**

INITIAL SETUP:

APPLICABLE MODELS
All

TOOLS
General mechanic's tool kit (Appendix E, Item 1)

MATERIALS/PARTS
Six screw-assembled lockwashers (Appendix D, Item 369
Cotter pin (Appendix D, Item 65)

REFERENCES (TM)
TM 9-2320-272-10
TM 9-2320-272-24P

EQUIPMENT CONDITION
Parking brake set (TM 9-2320-272-10).

NOTE

Left and right hood hinges are replaced the same. This procedure is for the left hood hinge.

a. Removal

1. Remove cotter pin (7) from hinge pin (3). Discard cotter pin (7).

NOTE

Ensure hood is supported when hinge pins are removed.

2. Remove hinge pin (3) from lower hinge (6) and hood (8).
3. Remove three screw-assembled lockwashers (2) and upper hinge half (1) from hood (8). Discard screw-assembled lockwashers (2).
4. Remove three screw-assembled lockwashers (4) and lower hinge half (6) from crossmember (5). Discard screw-assembled lockwashers (4).

b. Installation

NOTE

Coat hinges and hinge pins with GAA grease.

1. Install lower hinge half (6) on crossmember (5) with three new screw-assembled lockwashers (4).
2. Install upper hinge half (1) on hood (8) with three new screw-assembled lockwashers (2).
3. Install hood (8) on lower hinge half (6) with hinge pin (3) and new cotter pin (7).

4-133. HOOD HINGE REPLACEMENT (Contd)

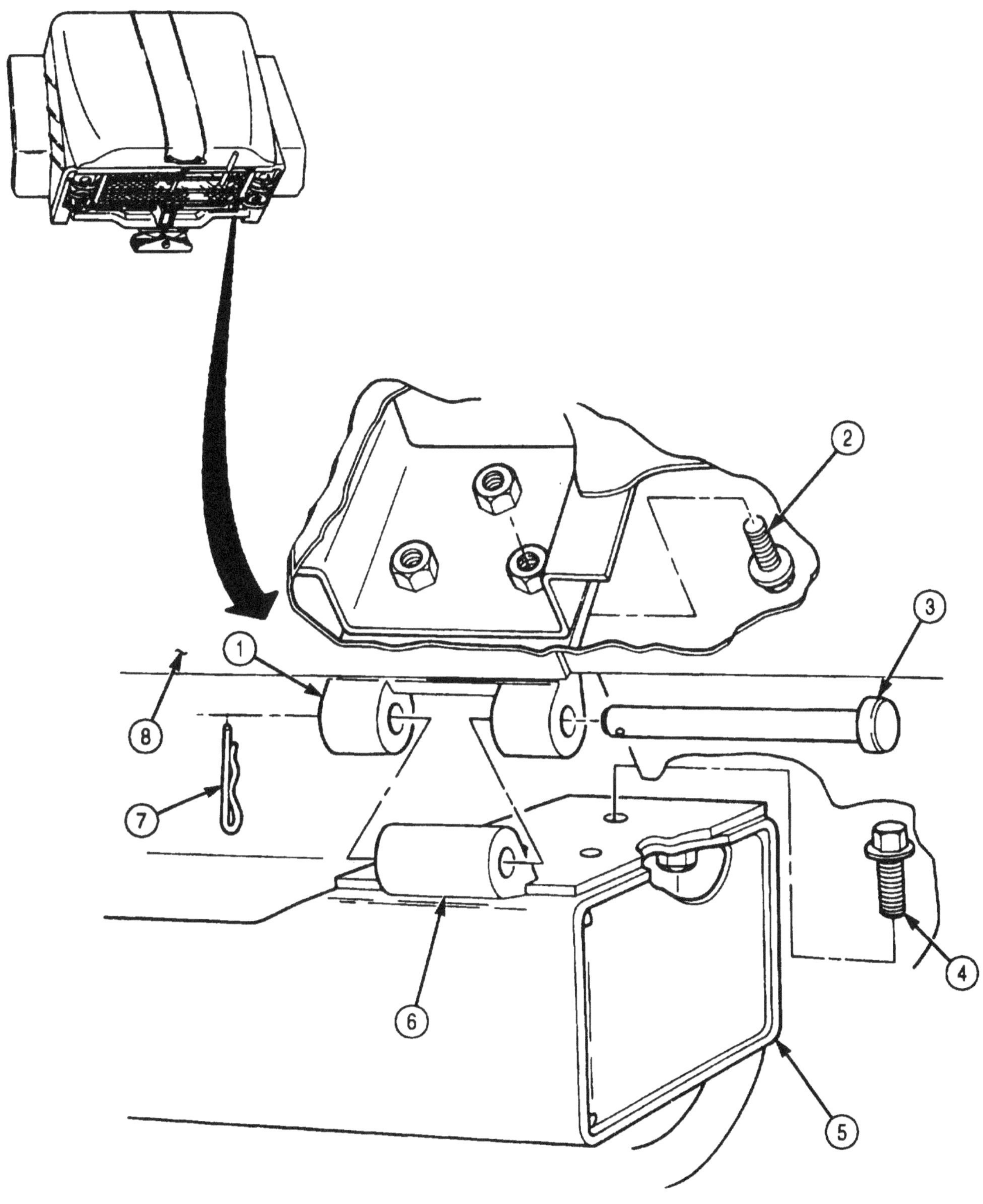

4-134. TORSION BARS AND CROSSMEMBERS REPLACEMENT

THIS TASK COVERS:

a. Removal **b. Installation**

INITIAL SETUP:

APPLICABLE MODELS
All

REFERENCES (TM)
TM 9-2320-272-24P

TOOLS
General mechanic's tool kit (Appendix E, Item 1)

EQUIPMENT CONDITION
Hood removed (para. 3-275).

MATERIALS/PARTS
Four locknuts (Appendix D, Item 291)
Two cotter pins (Appendix D, Item 65)
GAA grease (Appendix C, Item 28)

a. Removal

1. Remove two cotter pins (3) from two hinge pins (5). Discard cotter pins (3).
2. Remove four locknuts (7) and screws (2) from two torsion bars (1) and mounting brackets (8) and (10). Discard locknuts (7).

NOTE

Ensure hood and crossmember are supported when hinge pins are removed.

3. Remove two hinge pins (5) and bracket (9) from hood (4).
4. Remove two torsion bars (1) from mounting brackets (8) and (10).

b. Installation

NOTE

Coat hinges and hinge pins with GAA grease.

1. Position two torsion bars (1) in mounting brackets (10) and (8).
2. Install two torsion bars (1) on mounting brackets (8) and (10) with four screws (2) and locknuts (7).
3. Position hood (4) with bracket (9) and hinge (6) and install hinge pin (5) and cotter pin (3).

4-134. TORSION BARS AND CROSSMEMBERS REPLACEMENT (Contd)

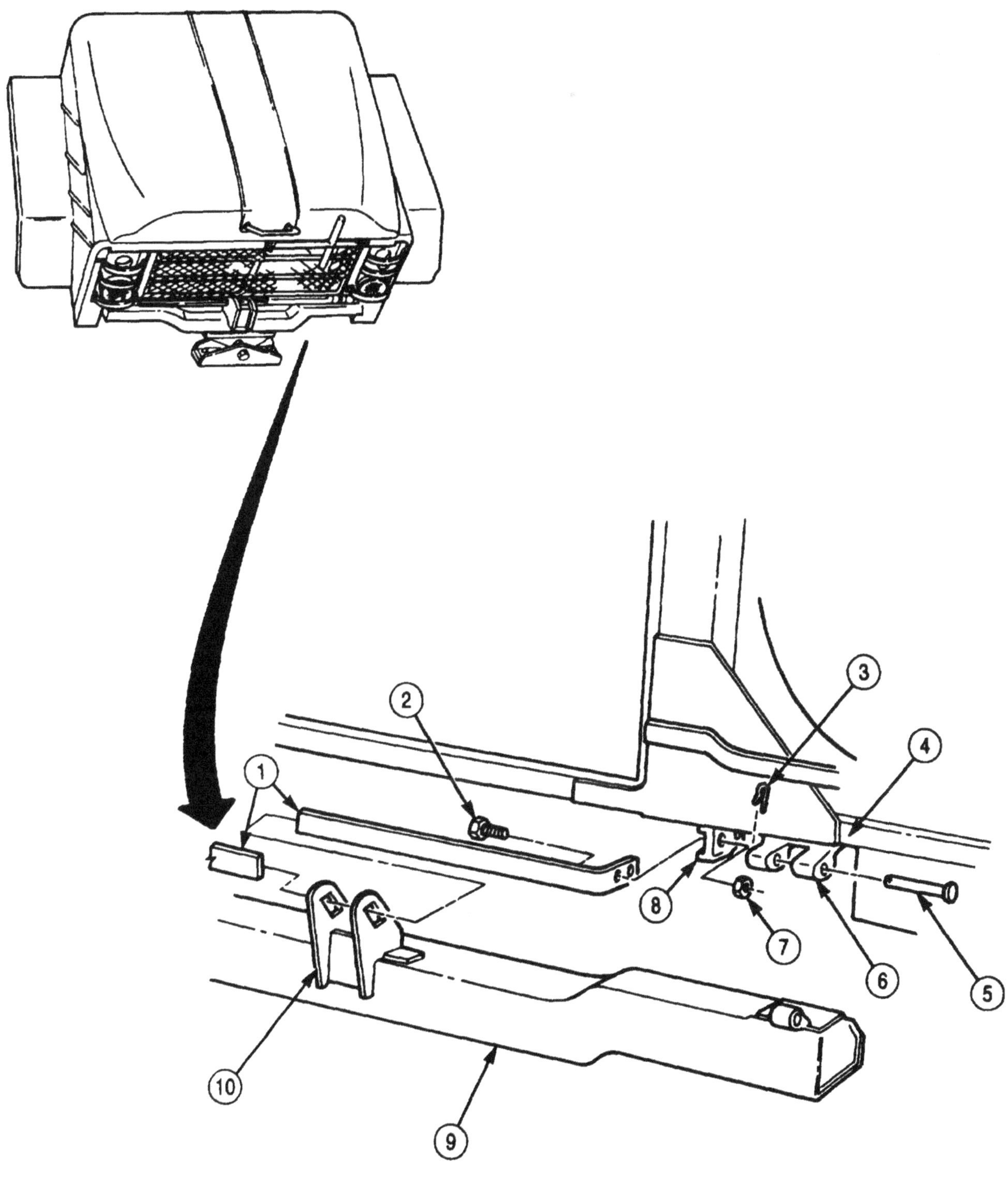

FOLLOW-ON TASK: Install hood (para. 3-275).

4-135. FRONT FENDER REPLACEMENT

THIS TASK COVERS:

a. Removal **b. Installation**

INITIAL SETUP:

APPLICABLE MODELS
All

TOOLS
General mechanic's tool kit (Appendix E, Item 1)

PERSONNEL REQUIRED
Two

REFERENCES (TM)
TM 9-2320-272-10
TM 9-2320-272-24P

EQUIPMENT CONDITION
- Parking brake set (TM 9-2320-272-10).
- Front composite light removed (para. 3-118).

a. Removal

1. Remove three screw-assembled washers (2) from fender reinforcement (3) and hood (1).
2. Remove two screw-assembled washers (6) from fender (4) and brushguard (7).
3. Remove eight screw-assembled washers (5) and fender (4) from hood (1).

b. Installation

1. Install fender (4) on hood (1) with eight screw-assembled washers (5).
2. Install fender (4) on brushguard (7) with two screw-assembled washers (6).
3. Install fender reinforcement (3) on hood (1) with three screw-assembled washers (2).

4-135. FRONT FENDER REPLACEMENT (Contd)

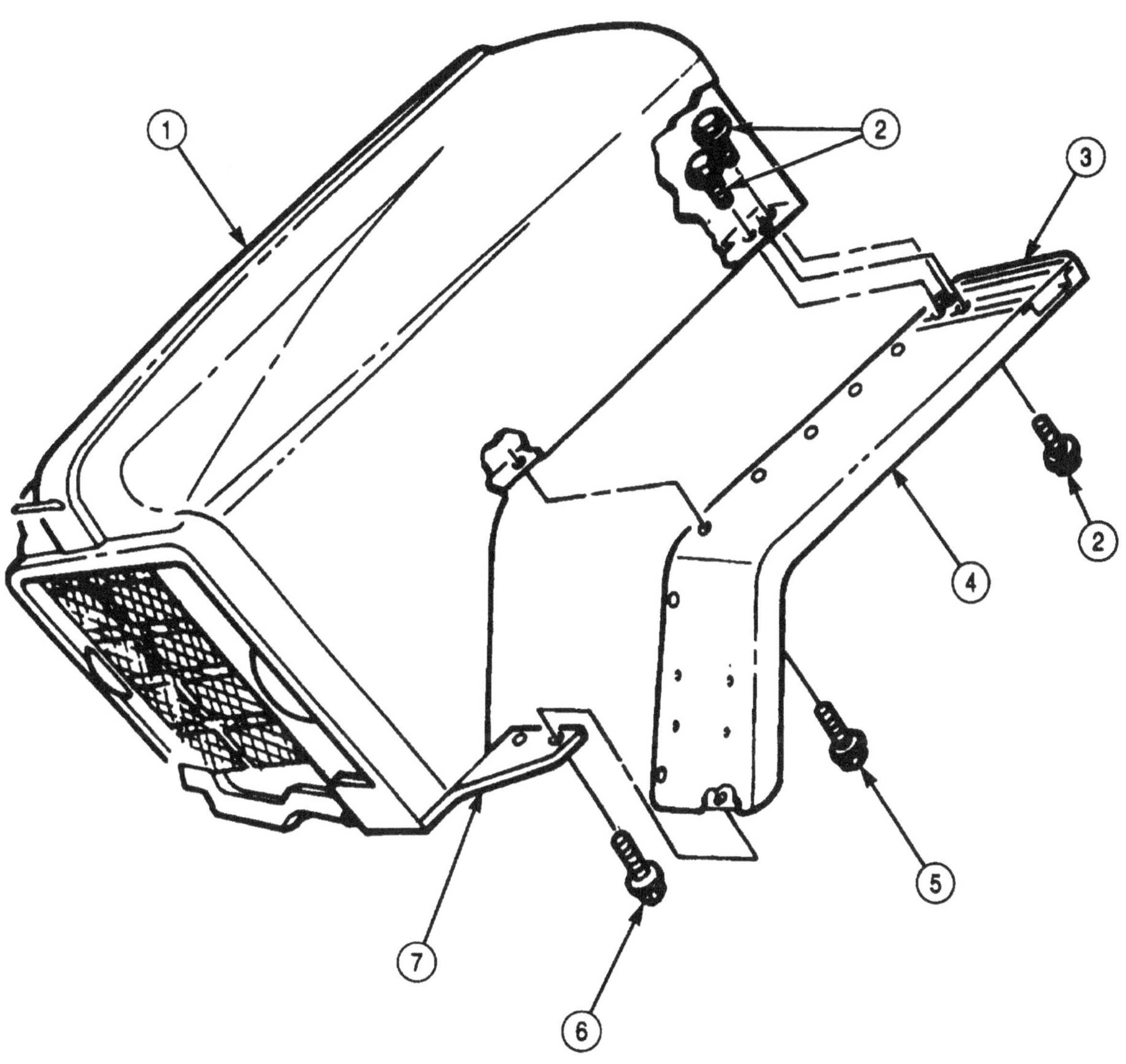

FOLLOW-ON TASK: Install front composite light (para. 3-118).

4-136. RADIATOR BRUSHGUARD MAINTENANCE

THIS TASK COVERS:

a. Removal
b. Inspection and Repair
c. Installation

INITIAL SETUP:

APPLICABLE MODELS
All

TOOLS
General mechanic's tool kit (Appendix E, Item 1)

REFERENCES (TM)
TM 9-237
TM 9-2320-272-24P

EQUIPMENT CONDITION
- Hood removed (para. 3-275).
- Blackout drive lamps removed (para. 3-117).
- Service headlamps removed (para. 3-116).
- Radiator baffles, seals, and plates removed (para. 3-63 or 3-64).
- Torsion bars and crossmember removed (para. 3-134).

a. Removal

1. Remove three screw-assembled washers (9) from left fender (10) and brushguard (3).
2. Remove four nuts (8), washers (7), screws (5), and wiring harness quick-disconnect (6) from brushguard (3).

NOTE

When performing steps 3 through 7, note location of clamps for installation.

3. Remove twelve screw-assembled washers (4) and harness clamps (1) from brushguard (3) and hood (2).
4. Remove six screw-assembled washers (12) and two upper hinge halves (11) from brushguard (3).

4-136. RADIATOR BRUSHGUARD MAINTENANCE (Contd)

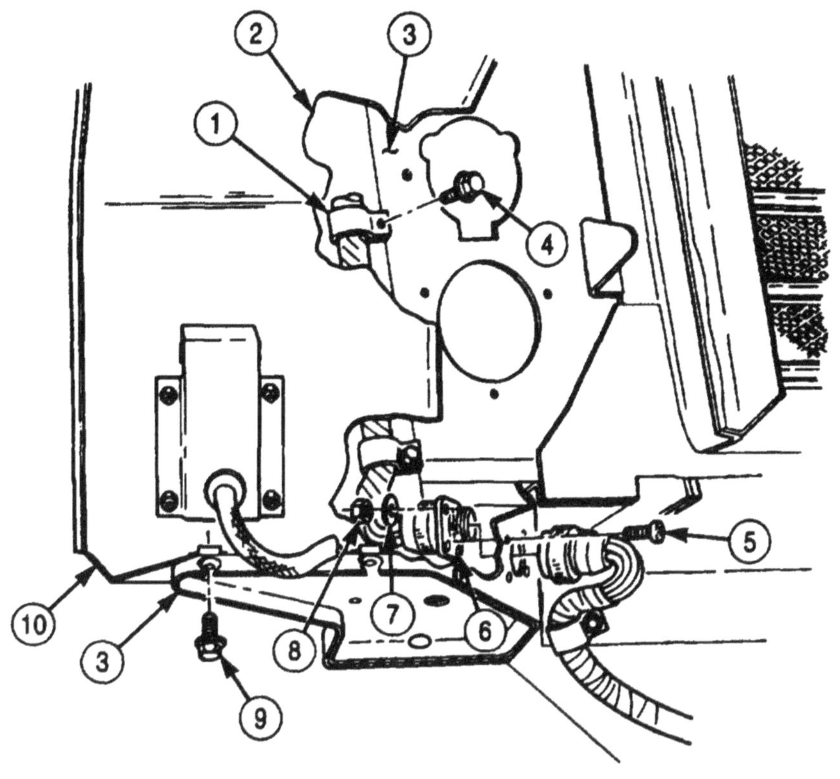

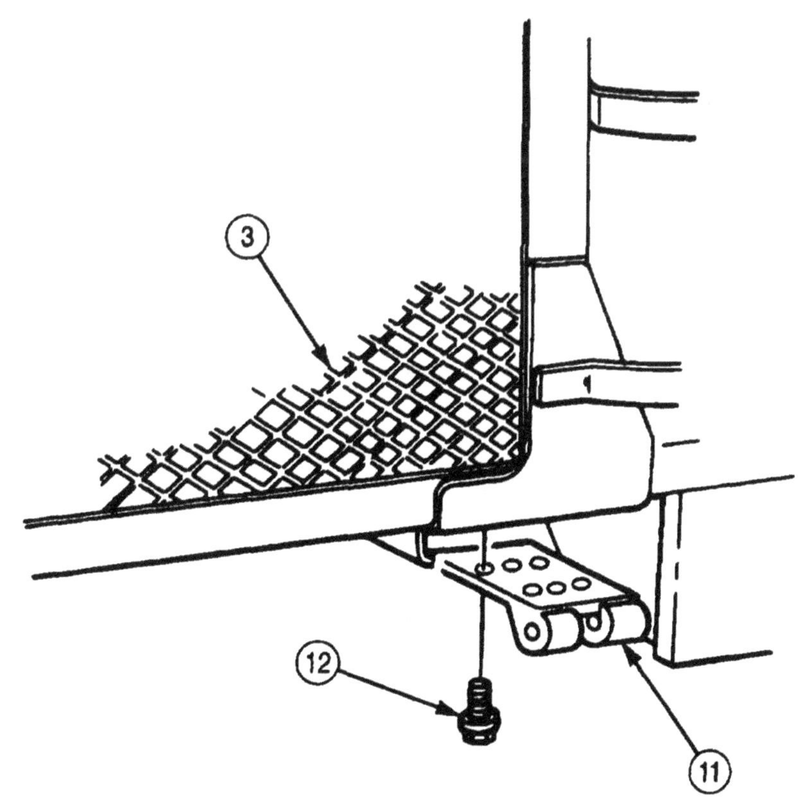

4-136. RADIATOR BRUSHGUARD MAINTENANCE (Contd)

5. Remove three screw-assembled washers (6) from right fender (5) and brushguard (7).
6. Remove clamp (9) and air duct (1) from elbow (8).
7. Remove four nuts (2), washers (3), screws (4), and elbow (8) from brushguard (7).
8. Remove ten nuts (11), screw-assembled washers (12), and brushguard (7) from hood (10).

b. Inspection and Repair

Insect brushguard (7) for cracks and breaks. If cracked or broken, replace or repair brushguard (7) (TM 9-237).

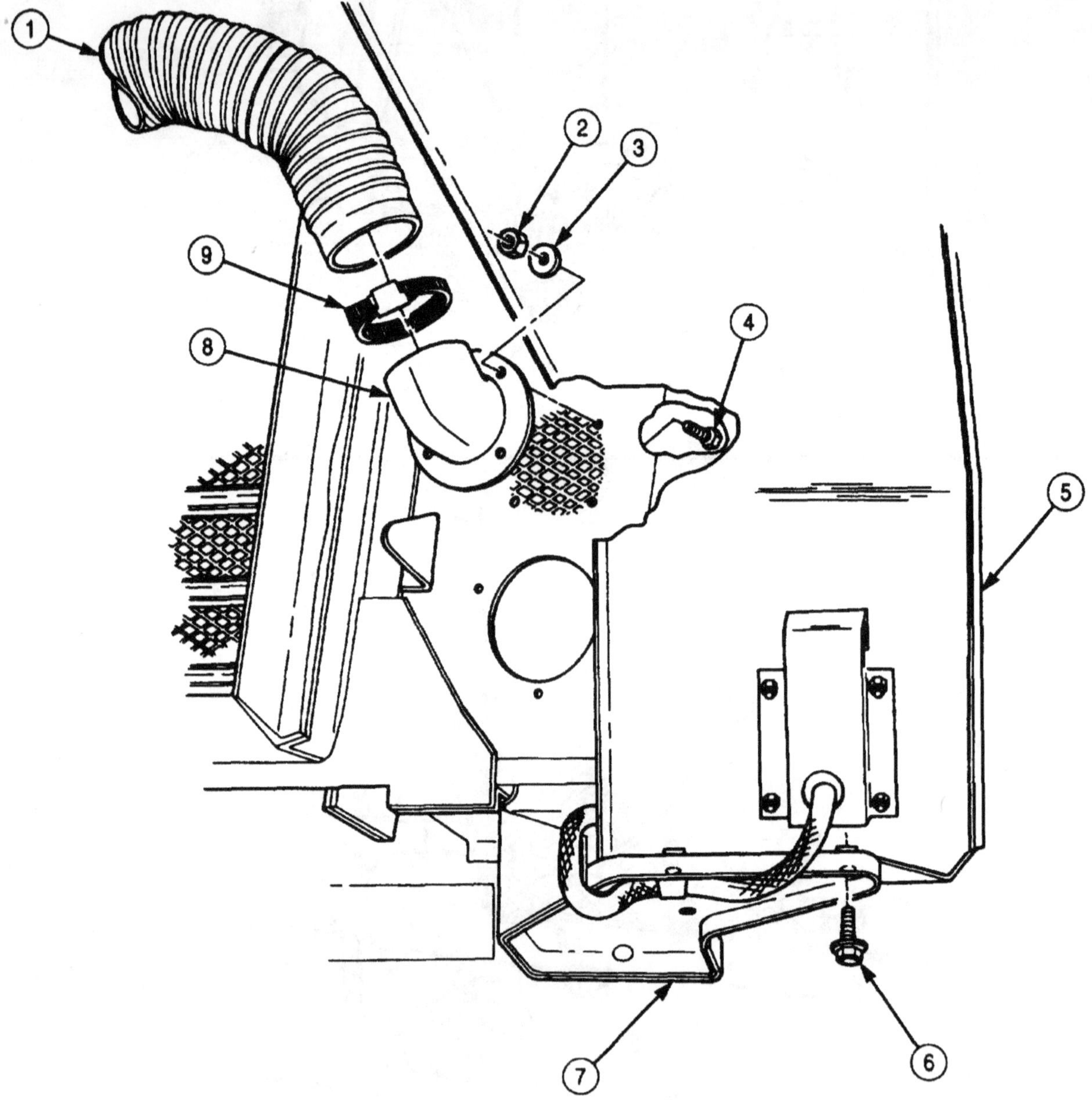

4-136. RADIATOR BRUSHGUARD MAINTENANCE (Contd)

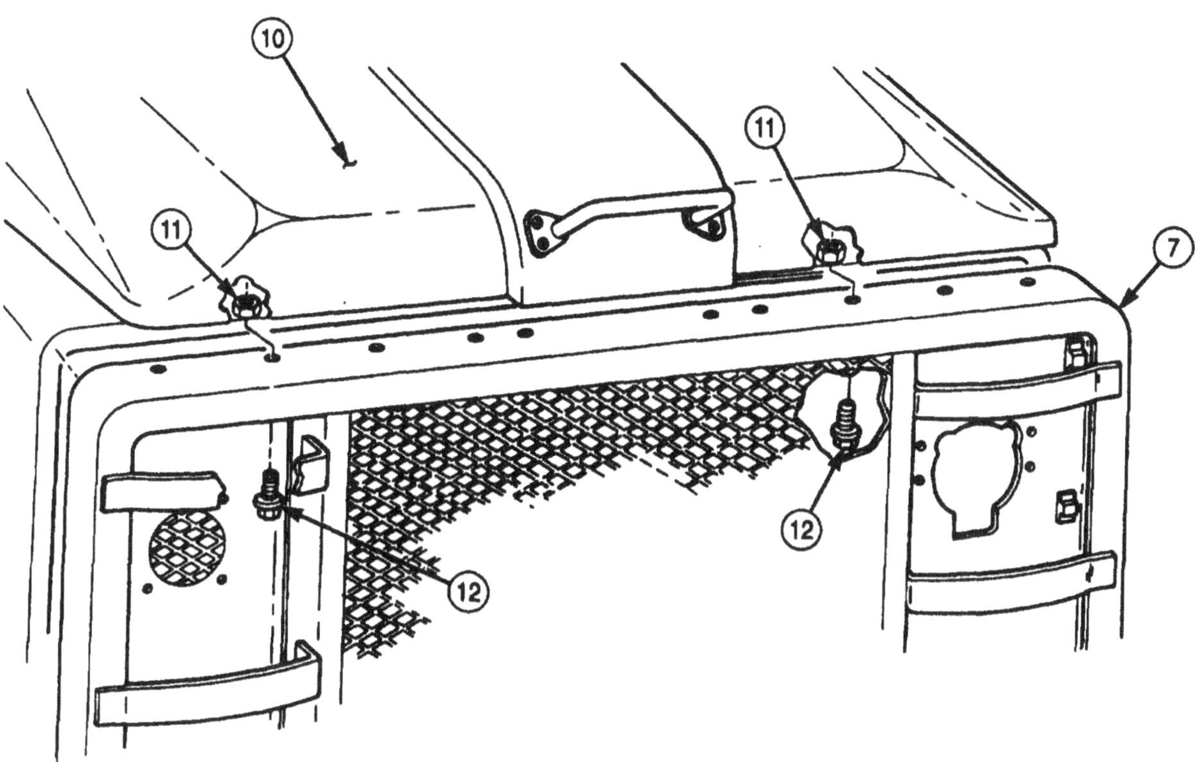

4-136. RADIATOR BRUSHGUARD MAINTENANCE (Contd)

c. Installation

1. Install brushguard (5) on hood (4), aligning clamps (3) on wiring harness (2) in proper position with ten screw-assembled washers (6) and nuts (1).
2. Install elbow (13) on brushguard (5) with four screws (10), washers (9), and nuts (8).
3. Install air duct (7) on elbow (13) with clamp (14).
4. Install brushguard (5) on right fender (11) with three screw-assembled washers (12).
5. Install two upper hinge halves (15) on brushguard (5) with three screw-assembled washers (16).
6. Install harness (26) and harness clamps (17) on hood (18) with twelve screw-assembled washers (19).
7. Install wiring harness quick-disconnect (21) on brushguard (5) with four screws (20), washers (22), and nuts (23).
8. Install three screw-assembled washers (24) on left fender (25) and brushguard (5).

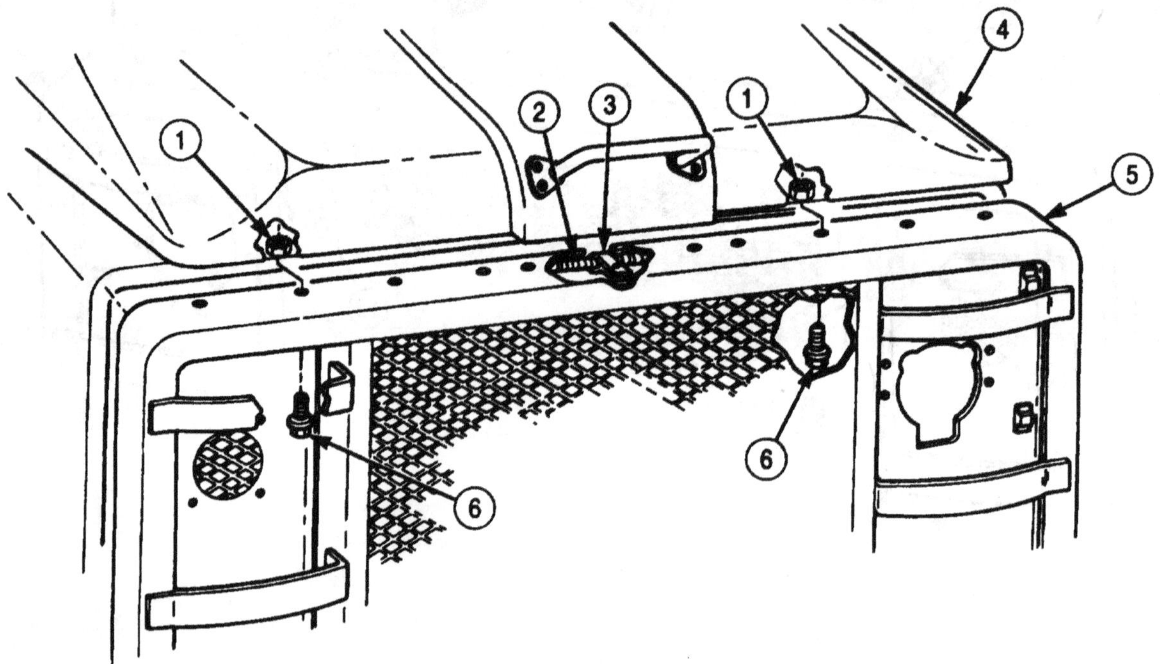

4-136. RADIATOR BRUSHGUARD MAINTENANCE (Contd)

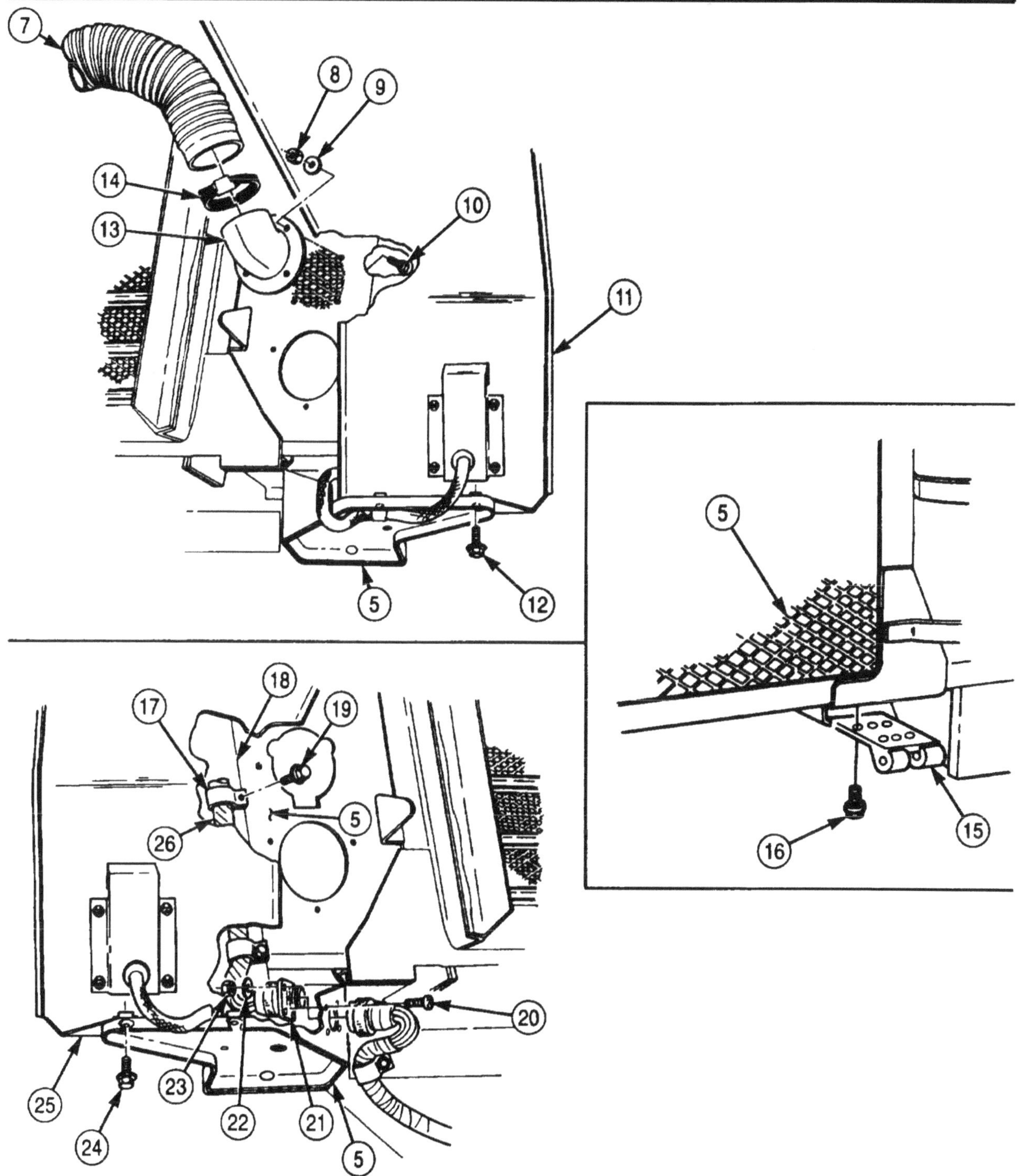

FOLLOW-ON TASKS:
- Install radiator baffles, seals, and plates (para. 3-64 or 3-63).
- Install torsion bars and crossmember (para. 4-134).
- Install service headlamps (para. 3-116).
- Install blackout drive lamps (para. 3-117).
- Install hood (para. 3-275).

4-137. CAB DOOR GLASS REPAIR

THIS TASK COVERS:

a. Disassembly **b. Assembly**

INITIAL SETUP:

APPLICABLE MODELS

All

TOOLS

General mechanic's tool kit (Appendix E, Item 1)

MATERIALS/PARTS

Regulator channel filler (Appendix D, Item 528)
Regulator channel seal (Appendix D, Item 529)
Frame seal (Appendix D, Item 131)

REFERENCES (TM)

TM 9-2320-272-10
TM 9-2320-272-24P

EQUIPMENT CONDITION

- Parking brake set (TM 9-2320-272-10).
- Cab door glass removed (para. 3-314).

GENERAL SAFETY INSTRUCTIONS

Eyeshields required when replacing cab door glass.

WARNING

Use eyeshields when replacing door glass. Glass could shatter, causing injury to personnel.

a Disassembly

1. Remove four screws (3), regulator channel (4), regulator channel filler (5), and regulator channel seal (6) from frame (2). Discard regulator channel filler (5) and regulator channel seal (6).
2. Carefully slide glass (7) from frame (2).
3. Remove frame seal (1) from frame (2). Discard frame seal (1).

b. Assembly

1. Position new frame seal (1) on top and sides of glass (7).
2. Position glass (7) with frame seal (1) in frame (2).
3. Install regulator channel seal (6), regulator channel filler (5), and regulator channel (4) on glass (7) and frame (2) with four screws (3).

4-137. CAB DOOR GLASS REPAIR (Contd)

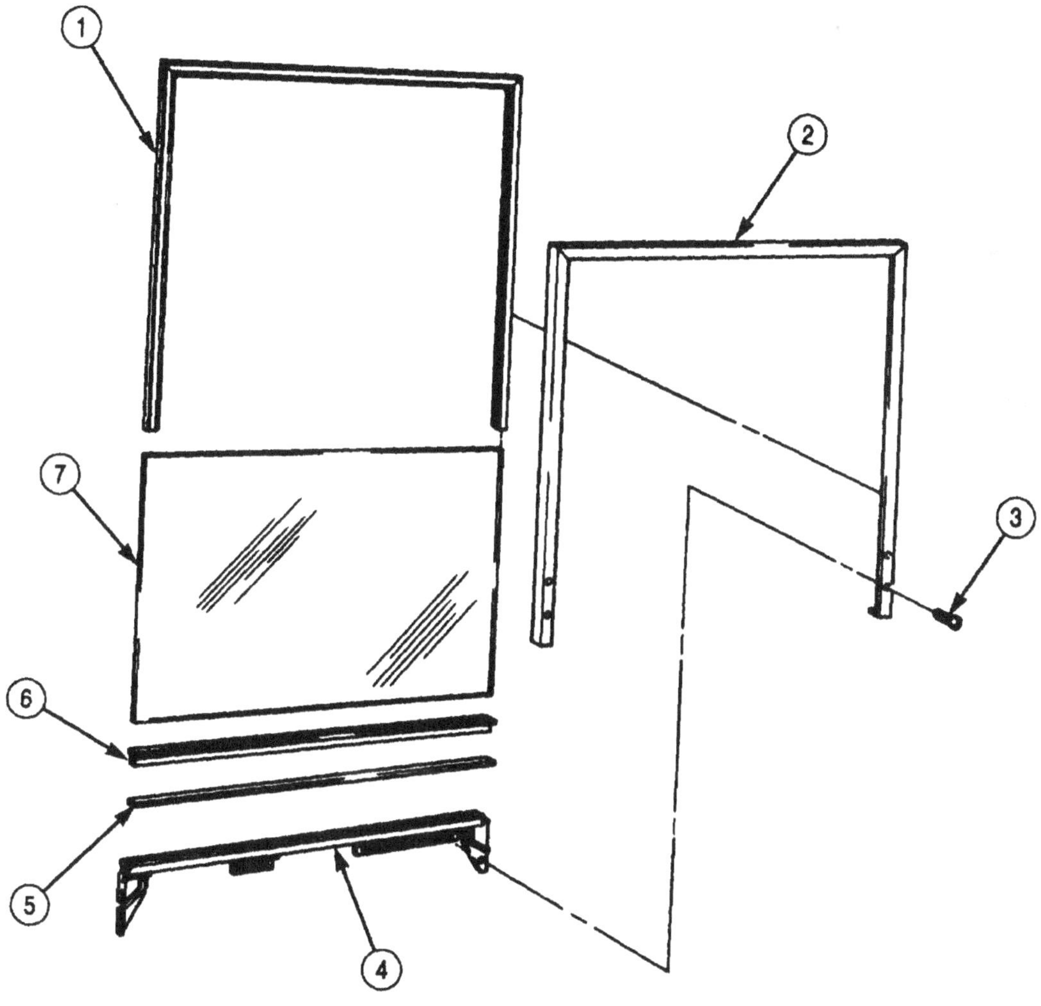

FOLLOW-ON TASK: Install cab door glass (para. 3-314).

4-138. CAB WINDSHIELD REPAIR

THIS TASK COVERS:

a. Disassembly **b. Assembly**

INITIAL SETUP:

APPLICABLE MODELS

All

TOOLS

General mechanic's tool kit (Appendix E, Item 1)

MATERIALS/PARTS

Four filler strips (Appendix D, Item 127)
Four lockwashers (Appendix D, Item 391)
Four lockwashers (Appendix D, Item 389)

REFERENCES (TM)

TM 9-2320-272-10
TM 9-2320-272-24P

EQUIPMENT CONDITION

- Parking brake set (TM 9-2320-272-10).
- Windshield removed (para. 3-278).

GENERAL SAFETY INSTRUCTIONS

Eyeshields required when replacing windshield glass.

WARNING

Use eyeshields when replacing windshield glass. Glass could shatter, causing injury to personnel.

a. Disassembly

1. Remove two capnuts (3), lockwashers (4), screws (5), and crosspiece (2) from frame (8). Discard lockwashers (4).
2. Remove filler strip (1) from crosspiece (2). Discard filler strip (1).
3. Remove four screws (7) and lockwashers (6) from frame (8). Discard lockwashers (6).
4. Remove two capnuts (14), lockwashers (13), screws (11), and locking latch (12) from frame (8). Discard lockwashers (13).
5. Remove glass (9) and three filler strips (10) from frame (8). Discard filler strips (10).

b. Assembly

1. Position new frame filler strip (10) around glass (9).
2. Install glass (9) with frame filler strip (10) in frame (8).
3. Install new crosspiece filler strip (1) over glass (9).
4. Install crosspiece (2) on glass (9) and tap into position until screw holes are aligned with holes in frame (8).

CAUTION

Do not overtighten crosspiece or windshield frame screws. Glass could be damaged.

5. Secure crosspiece (2) on frame (8) with two screws (5), new lockwashers (4), and capnuts (3).
6. Install four new lockwashers (6) and screws (7) on frame (8).
7. Install locking latch (12) on frame (8) with two screws (11), new lockwashers (13), and capnuts (14).
8. Trim crosspiece filler strip (1) and frame filler strip (10) even with frame (8) and crosspiece (2) edges.

4-138. CAB WINDSHIELD REPAIR (Contd)

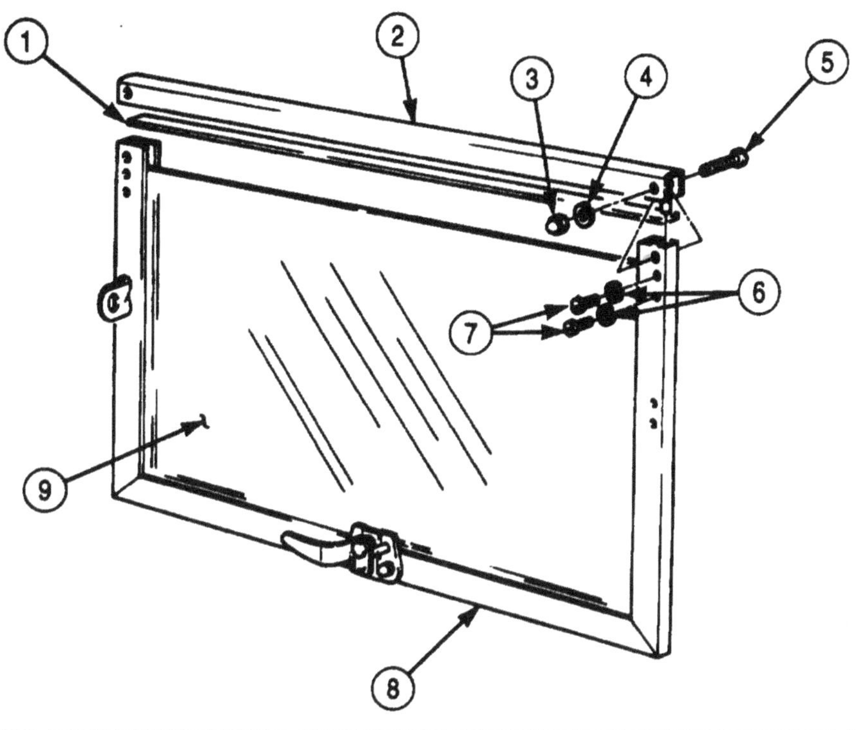

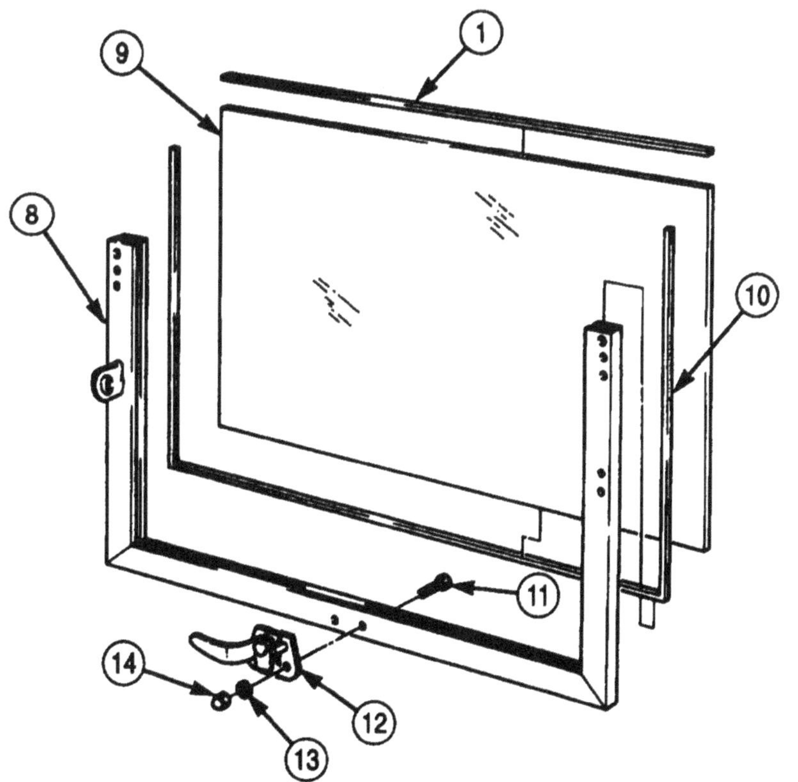

FOLLOW-ON TASK: Install windshield (para. 3-278).

4-139. RETRACTABLE WINDOW GLASS AND VAN DOOR WINDOW GLASS REPLACEMENT

THIS TASK COVERS:

a. Removal **b. Installation**

INITIAL SETUP:

APPLICABLE MODELS
All

TOOLS
General mechanic's tool kit (Appendix E, Item 1)

REFERENCES (TM)
TM 9-2320-272-10
TM 9-2320-272-24P

EQUIPMENT CONDITION
- Parking brake set (TM 9-2320-272-10).
- Retractable window removed (para. 3-351).
- Van rear door and side door window removed (para. 3-350).

a. Removal

1. Remove eighteen screws (3) and glass retainer (2) from window frame (1).
2. Remove glass unit (4) with seal (6) from window frame (1).
3. Remove seal (6) from two glass panes (5).

b. Installation

1. Install seal (6) on two glass panes (5).
2. Install glass unit (4) with seal (6) on window frame (1).
3. Install glass retainer (2) on window frame (1) with eighteen screws (3).

4-139. RETRACTABLE WINDOW GLASS AND VAN DOOR WINDOW GLASS REPLACEMENT (Contd)

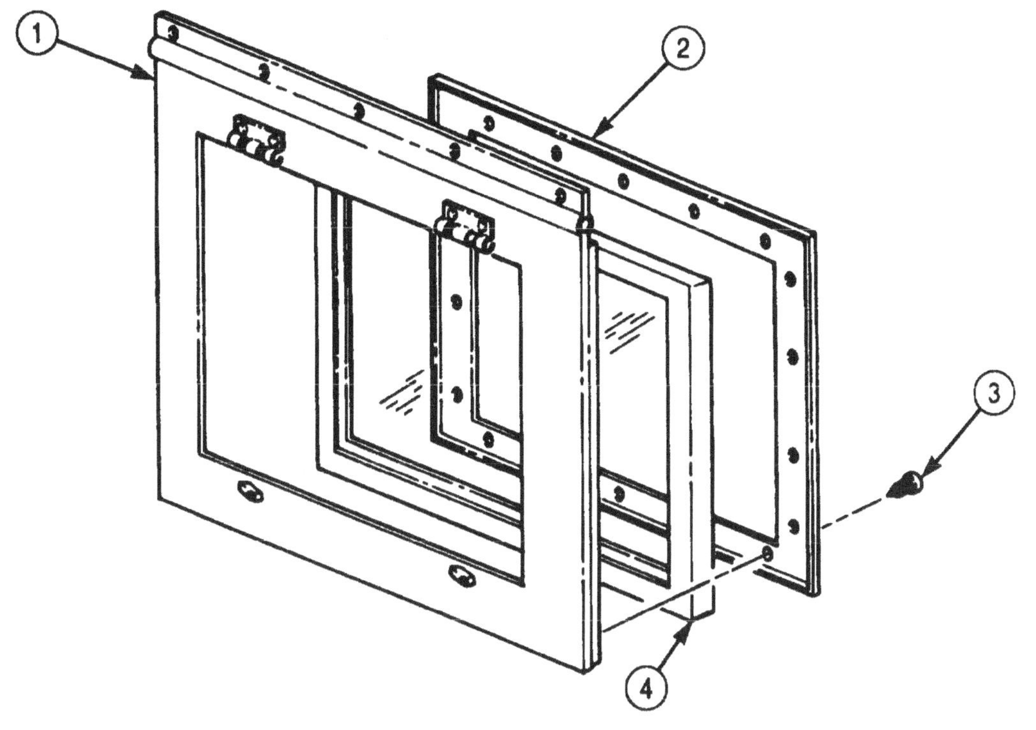

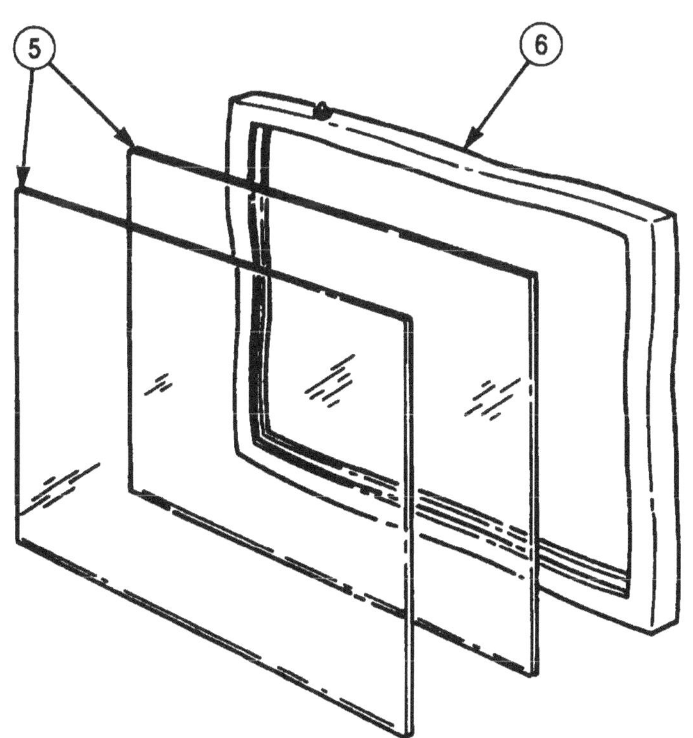

FOLLOW-ON TASKS:
- Install retractable window (para. 3-351).
- Install van rear door and side door window (para. 3-350).

4-140. ENGINE ACCESS COVER (IN-CAB) REPLACEMENT

THIS TASK COVERS:

a. Removal **b. Installation**

INITIAL SETUP:

APPLICABLE MODELS
All

TOOLS
General mechanic's tool kit (Appendix E, Item 1)

MATERIALS/PARTS
Adhesive (Appendix C, Item 7)

REFERENCES (TM)
TM 9-2320-272-10
TM 9-2320-272-24P

EQUIPMENT CONDITION
- Parking brake set (TM 9-2320-272-10).
- Transmission console removed (para. 474).
- Winch console removed (if equipped) (para. 4-74).
- Instrument cluster removed (para. 3-83).

a. Removal

1. Pull back cab floor insulation (3), firewall insulation (1), and access cover insulation (6) to expose screw-assembled washers (7).
2. Remove sixteen screw-assembled washers (7) and access cover (2) from firewall (5) and cab floor (4).

b. Installation

1. Install access cover (2) on firewall (5) and cab floor (4) with sixteen screw-assembled washers (7).
2. Apply adhesive to back side of cab floor insulation (3), firewall insulation (1), and access cover insulation (6) and press into place on access cover (2), firewall (5), and cab floor (4).

4-140. ENGINE ACCESS COVER (IN-CAB) REPLACEMENT (Contd)

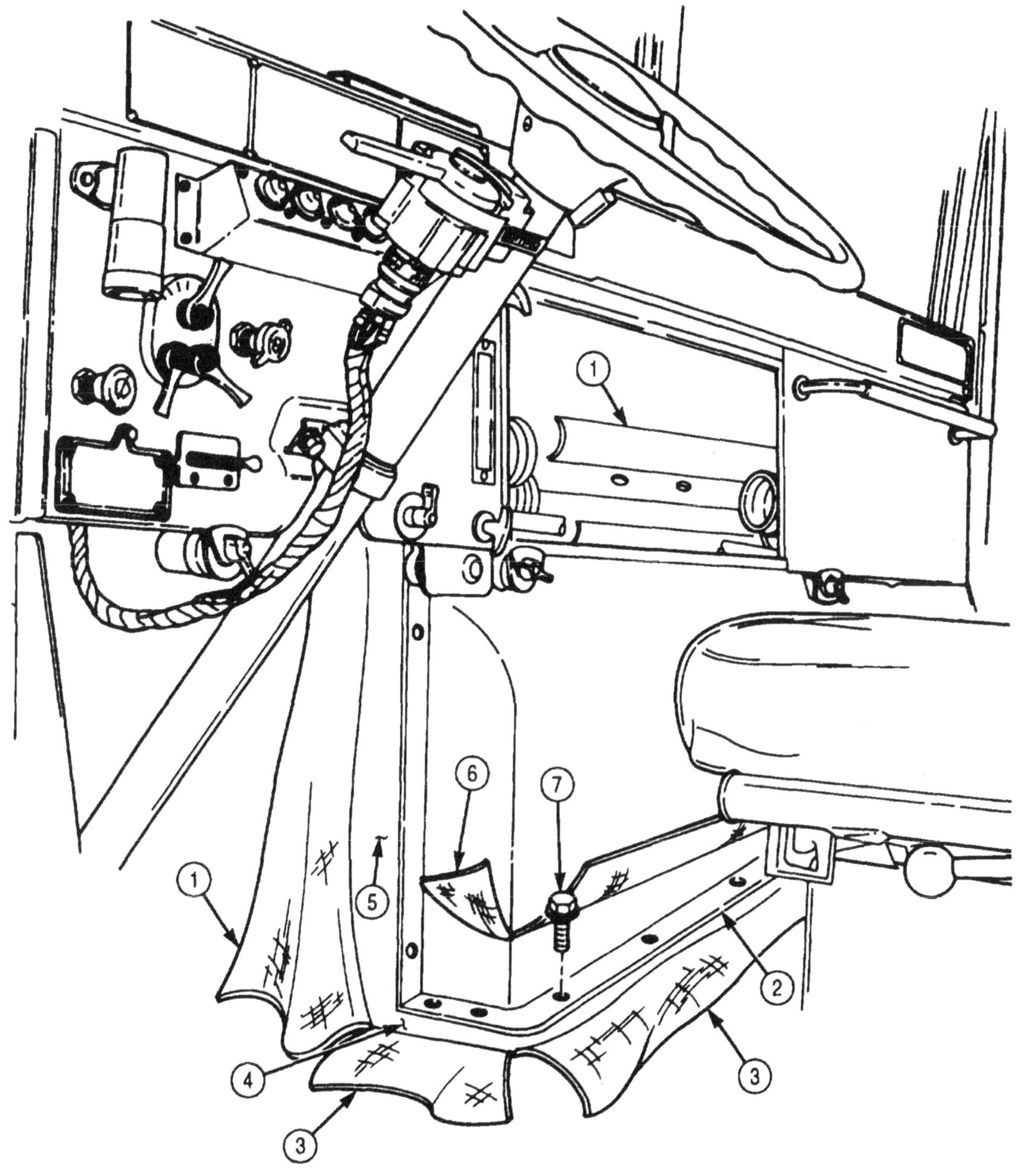

FOLLOW-ON TASKS:
- Install instrument cluster (para. 3-83).
- Install winch console, if equipped (para. 4-74).
- Install transmission console (para. 4-74).

Section XIV. SPECIAL PURPOSE BODY MAINTENANCE

4-141. SPECIAL PURPOSE BODY MAINTENANCE INDEX

4-142. CARGO BODY ASSEMBLY MAINTENANCE

THIS TASK COVERS:

a. Removal
b. Cleaning and Inspection
c. Repair
d. Installation

INITIAL SETUP:

APPLICABLE MODELS
M923/A1/A2, M925/A1/A2, M927/A1/A2, M928/A1/A2

TOOLS
General mechanic's tool kit (Appendix E, Item 1)
Two utility chains
Lifting device
Four shop stands

MATERIALS/PARTS
Eight locknuts (Appendix D. Item 321)
Ten locknuts (M923/A1/A2, M925/A1/A2) (Appendix D, Item 284)
Twenty locknuts (M927/A1/A2, M928/A1/A2) (Appendix D, Item 284)
Four locknuts (M923/A1/A2, M925/A1/A2) (Appendix D, Item 321)
Sixteen locknuts (M927/A1/A2, M928/A1/A2) (Appendix D, Item 321)
Two wood sills (M923/A1/A2, M925/A1/A2) (Appendix D, Item 725)
Four locknuts (M923/A1/A2, M925/A1/A2) (Appendix D, Item 273)
Wood sill (M927/A1/A2, M928/A1/A2) (Appendix D, Item 721)
Wood sill (M927/A1/A2, M928/A1/A2) (Appendix D, Item 722)
Wood sill (M927/A1/A2, M928/A1/A2) (Appendix D, Item 723)
Wood sill (M927/A1/A2, M928/A1/A2) (Appendix D, Item 724)

PERSONNEL REQUIRED
Three

MANUAL REFERENCES (TM)
TM 9-2320-272-10
TM 9-2320-272-24P
TM 9-237
TM 9-247
TM 43-0139

EQUIPMENT CONDITION
- Parking brake set (TM 9-2320-272-10).
- Wheels chocked (TM 9-2320-272-10).
- Front and side racks removed (TM 9-2320-272-10).
- Cargo body cover bows removed (para. 3-341).
- Wheel splash guards and brackets removed (para. 3-342).

GENERAL SAFETY INSTRUCTIONS
- Drycleaning solvent is flammable and toxic. Do not use near an open flame.
- Keep fire extinguisher nearby when using drycleaning solvent.
- Lifting device must have a capacity greater than the weight of the truck bed.
- All personnel must stand clear during lifting operations.

a. Removal

1. Remove four locknuts (5), washer (6), outer support springs (7), inner support springs (4), and screws (1) from two front upper holddown brackets (2) and lower holddown brackets (3). Discard locknuts (5).

NOTE

Extra-long wheelbase models (M927/A1/A2, M928/A1/A2) utilize six intermediate holddown brackets, twelve screws, and twelve locknuts.

2. Remove four locknuts (11) and screws (8) from four intermediate holddown brackets (9) and lower holddown brackets (10). Discard locknuts (11).

4-142. CARGO BODY ASSEMBLY MAINTENANCE (Contd)

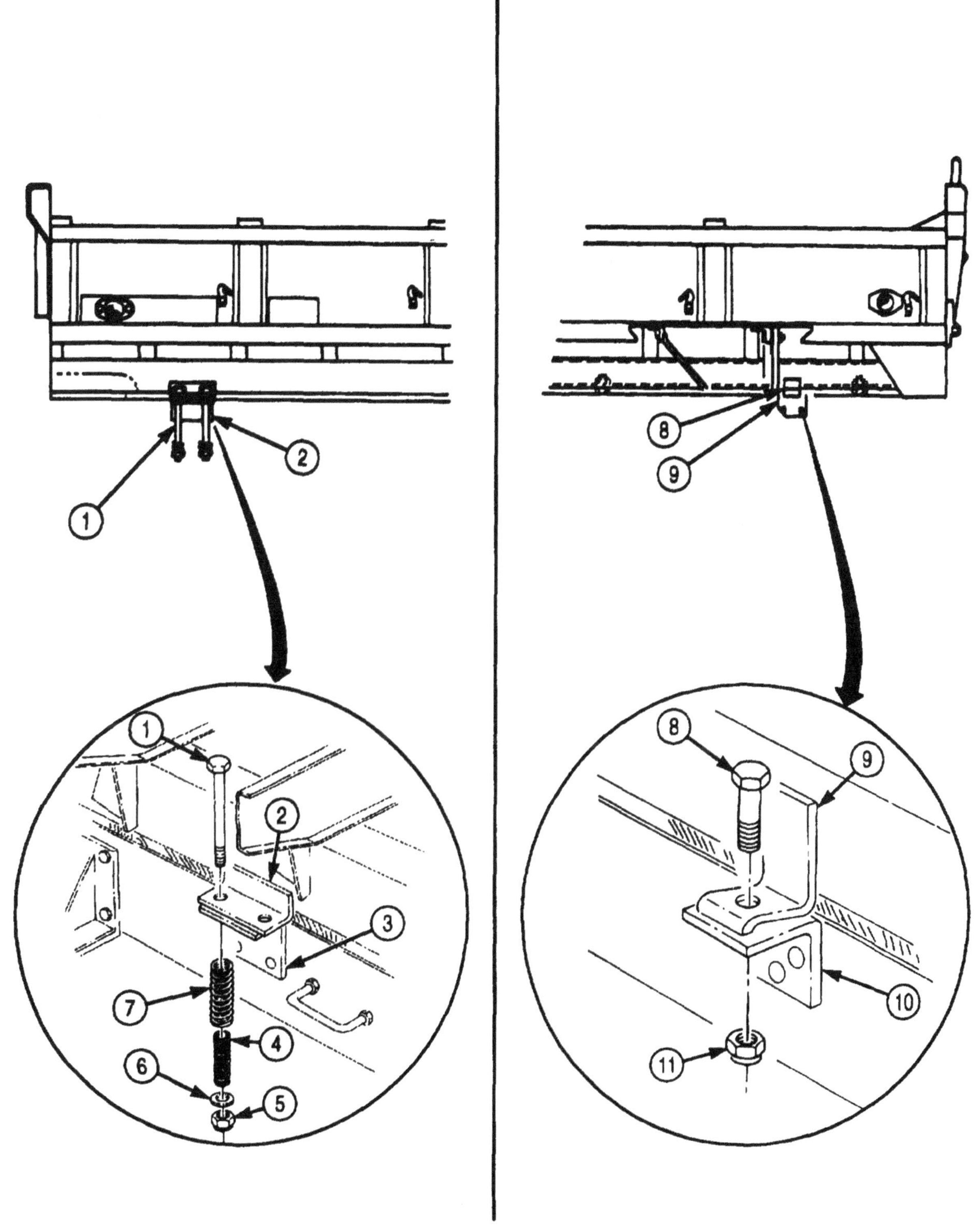

FRONT HOLDDOWN BRACKET

INTERMEDIATE HOLDDOWN BRACKET

4-142. CARGO BODY ASSEMBLY MAINTENANCE (Contd)

3. Disconnect four wires (13) from each of two rear composite lights (3). Tag wires (13) for installation.
4. Remove eight locknuts (10), screws (12), and wire clamps (11) from cargo body (2).
5. Wrap two utility chains around subframe beam (5) and fasten hooks over utility chains.
6. Attach lifting device to center of utility chains.

WARNING

- Lifting device must have a weight capacity greater than the weight of the truck bed to prevent injury or death to personnel and damage to equipment.
- All personnel must stand clear during lifting operations. A snapped cable, or swinging or shifting load, may result in injury or death to personnel.

NOTE

Mechanic will handle one guide line and direct hoisting operation. Assistant will handle other guide line. Second assistant will operate lifting device.

7. Connect two guide lines to front and rear lashing hooks (1).
8. Lift cargo body clear of frame (4) and place on four shop stands.
9. Disconnect lifting device, two utility chains, and guide lines from cargo body (2).

NOTE

Extra-long wheelbase models (M927/A1/A2, M928/A1/A2) utilize four wood sills, twenty washers, and twenty screws.

10. Remove ten locknuts (7), lockwashers (6), screws (9), and two wood sills (8) from cargo bed (5). Discard locknuts (7), lockwashers (6), and wood sills (8).

b. Cleaning and Inspection

WARNING

Drycleaning solvent is flammable and toxic. Do not use near open flame and always have a fire extinguisher nearby when solvents are used. Use only in well-ventilated places, wear protective clothing, and dispose of cleaning rags in approved container. Failure to do this may result in injury or death to personnel and/or damage to equipment.

1. Clean cargo body (2) with steam or wire brush and approved cleaning solvent (TM 9-247).
2. Inspect cargo body (2) for dents, cracks, breaks, and rust.

c. Repair

1. Weld or straighten as necessary (TM 9-237).
2. Strip and paint as necessary (TM 43-0139).

d. Installation

1. Wrap two utility chains around subframe beam (5) and fasten hooks over utility chains.
2. Connect two guide lines to front and rear lashing hooks (1).
3. Attach lifting device to center of utility chains.
4. Install two new wood sills (8) on cargo bed (2) with ten screws (9), new lockwashers (6), and new locknuts (7).

4-142. CARGO BODY ASSEMBLY MAINTENANCE (Contd)

WARNING

- Lifting device must have a weight capacity greater than the weight of the truck bed to prevent injury or death to personnel and damage to equipment.
- All personnel must stand clear during lifting operations. A snapped cable, or swinging or shifting load, may result in injury or death to personnel.

NOTE

Mechanic will handle one guide line and direct hoisting operation. Assistant will handle other guide line. Second assistant will operate lifting device.

5. Hoist cargo body (2) clear of shop stands.

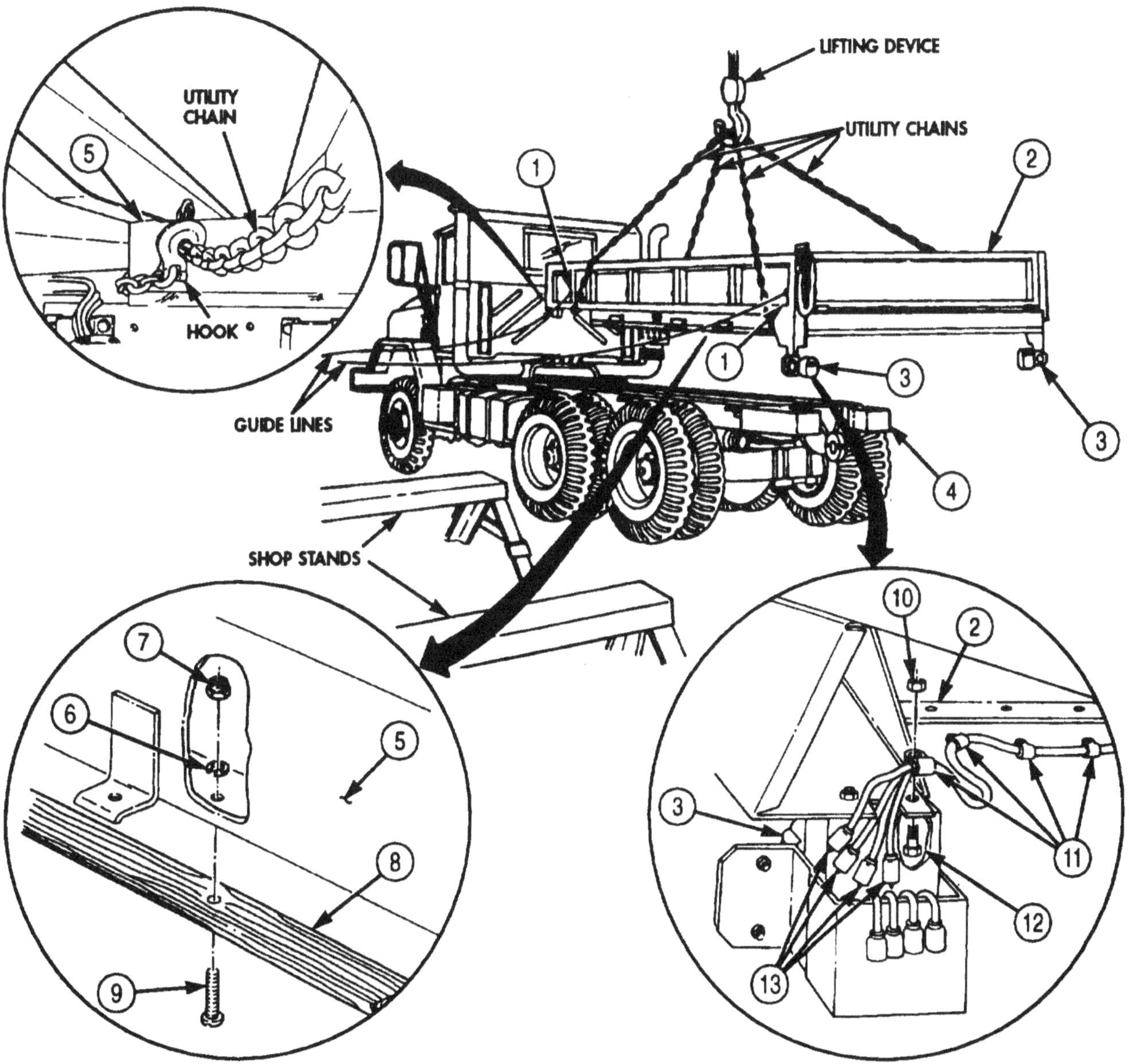

4-142. CARGO BODY ASSEMBLY MAINTENANCE (Contd)

6. Lower cargo body (12) onto frame (19) while using a drift pin to align two front upper holddown brackets (2) to lower holddown brackets (3), and four intermediate holddown brackets (9) to lower holddown brackets (10).
7. Place four outer support springs (7) over inner support springs (4), and position under lower holddown brackets (3), and install with four screws (1), washers (6), and locknuts (5). Tighten locknuts (5) until length of springs (7) and (4) are 6.75 in. ±0.06 in. (171.45 mm ± 1.5 mm).

NOTE

Extra-long wheelbase models (M927/A1/A2, M928/A1/A2) utilize six intermediate holddown brackets, twelve screws, and twelve locknuts.

8. Place four screws (8) through intermediate holddown brackets (9) and lower holddown brackets (10), and install four new locknuts (11).
9. Disconnect lifting device, utility chains, and guide lines from cargo body (12).
10. Connect four wires (16) to connectors (17) on each of two composite lights (18).
11. Install eight wire clamps (14) on cargo body (12) with eight screws (15) and locknuts (13).

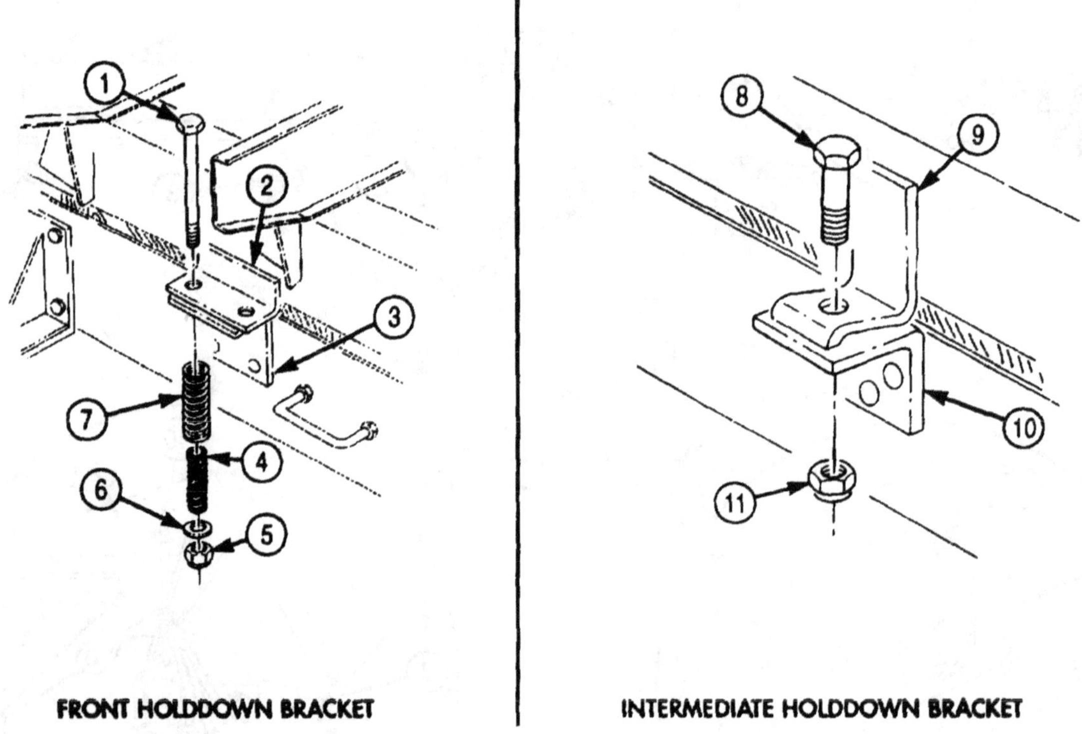

FRONT HOLDDOWN BRACKET

INTERMEDIATE HOLDDOWN BRACKET

4-142. CARGO BODY ASSEMBLY MAINTENANCE (Contd)

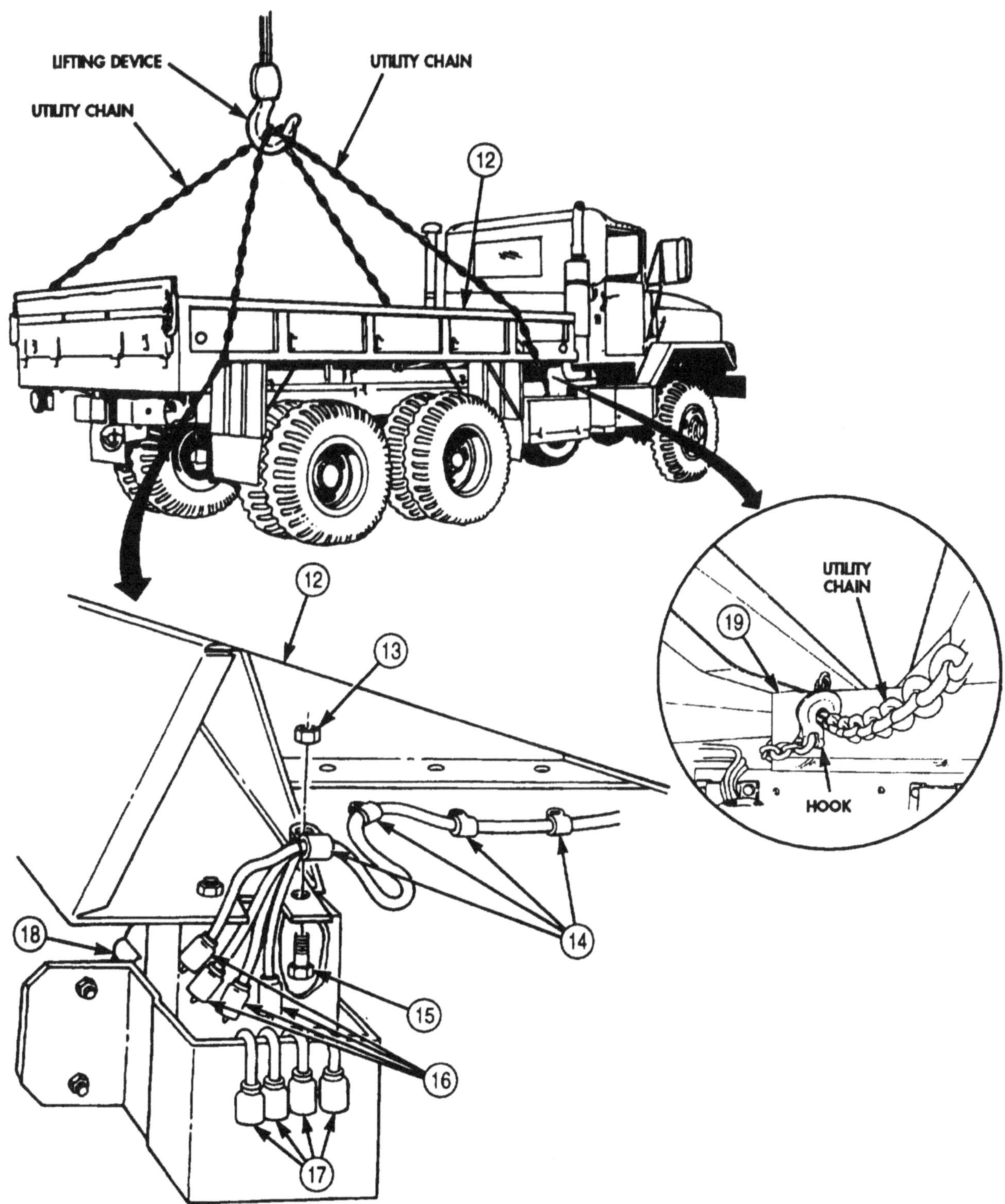

FOLLOW-ON TASKS:
- Install wheel splash guards and brackets (para. 3-342).
- Install cargo body cover bows (para. 3-341).
- Install front and side racks (TM 9-2320-272-10).

4-143. CARGO BODY FRAME RAIL SILL MAINTENANCE

THIS TASK COVERS:

a. **Removal**
b. **Inspection**
c. **Fabrication**
d. **Installation**

INITIAL SETUP:

APPLICABLE MODELS
M923/A1/A2, M925/A1/A2, M927/A1/A2. M928/A1/A2

TOOLS
General mechanic's tool kit (Appendix E, Item 1)
Torque wrench (Appendix E, Item 144)

MATERIALS/PARTS
Ten lockwashers (Appendix D, Item 715)
Green 383 coating (Appendix C, Item 22)
Wood (Appendix C, Item 82)
Wood preservative (Appendix C, Item 81)

MANUAL REFERENCES (TM)
LO 9-2320-272-12
TM 43-0137
TM 9-2320-272-24P

EQUIPMENT CONDITION
Cargo body assembly removed (para. 4-142).

a. Removal

1. Remove ten nuts (1), lockwashers (2), screws (3), and frame rail sill (4) from cargo body (5). Discard lockwashers (2).
2. Clean frame rail sill (4).

b. Inspection

Inspect frame rail sill (4) for rot, deterioration, cracks, or breaks. If defects are found, fabricate replacement.

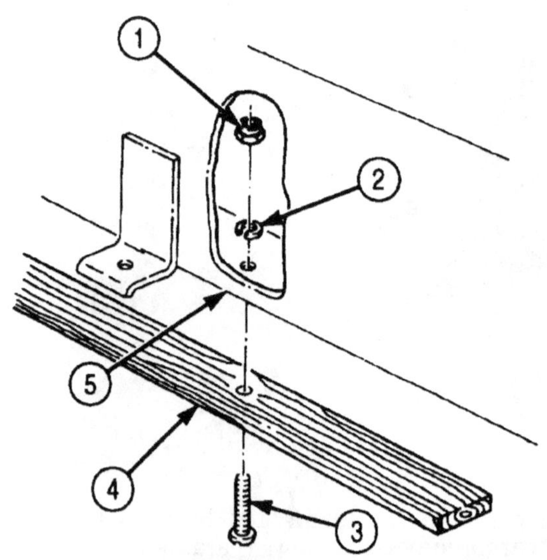

4-143. CARGO BODY FRAME RAIL SILL MAINTENANCE (Contd)

c. Fabrication

NOTE

Steps 1 and 2 apply to M927/A1/A2 and M928/A1/A2 model vehicles. Steps 3 and 4 apply to M923/A1/A2 and M925/A1/A2 model vehicles.

1. Cut frame rail sill (1) to dimensions and drill five counterbore holes (2) and three frame rest holes (3) as shown.
2. Seal and paint frame rail sill (1) (TM 43-0139).

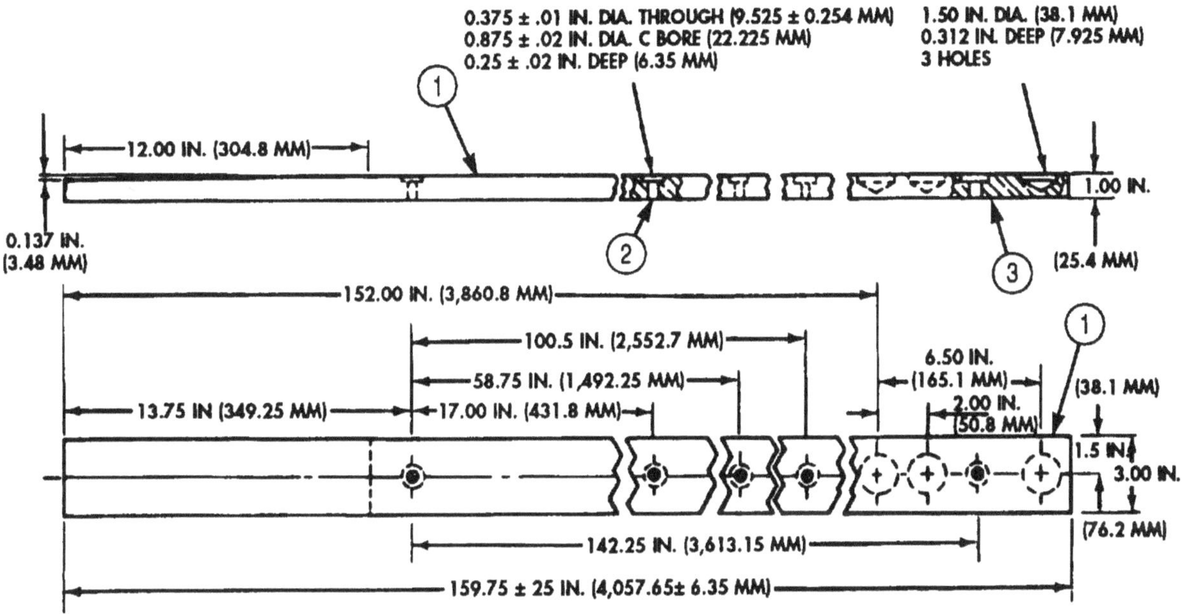

M923/A1/A2 AND M925/A1/A2 CARGO MODELS

4-143. CARGO BODY FRAME RAIL SILL MAINTENANCE

3. Cut frame rail sill (4) to dimensions and drill ten counterbore holes (5) and three frame root holes (6) as shown.
4. Seal and paint frame rail sill (4) (refer to TM 43-0139).

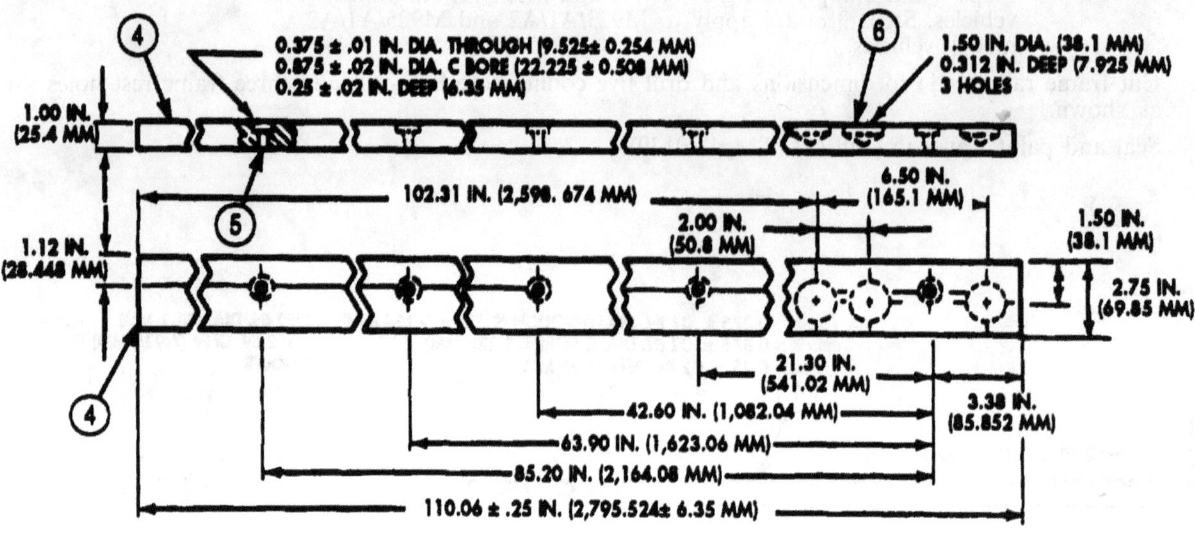

M927/A1/A2 AND M928/A1/A2 FRONT SECTION

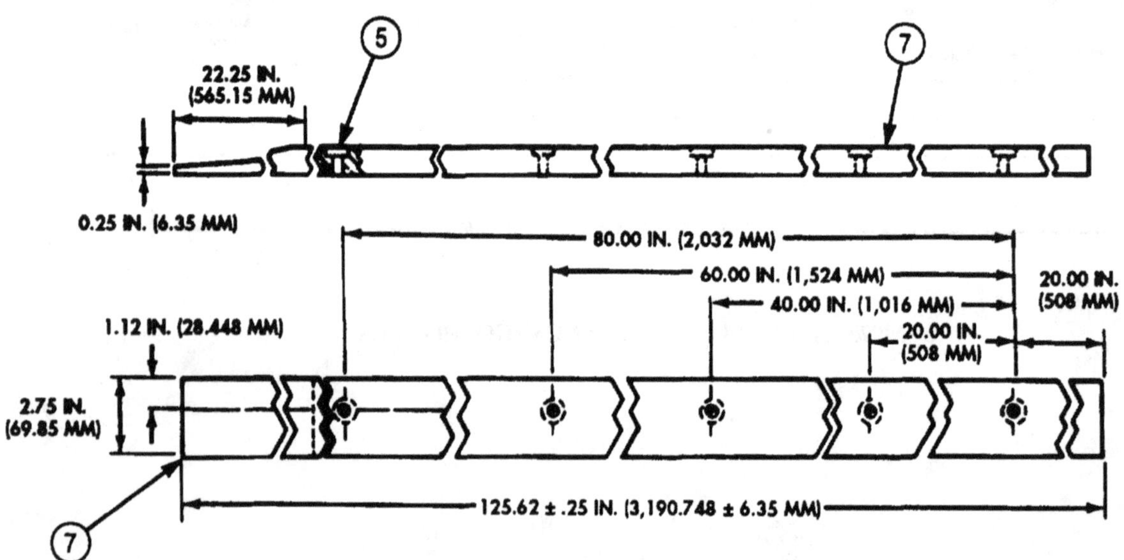

M927/A1/A2 AND M928/A1/A2 REAR SECTION

4-143. CARGO BODY FRAME RAIL SILL MAINTENANCE

d. Installation

1. Position frame rail sill (4) on cargo body (5).
2. Install frame rail sill (4) on cargo body (5) with ten screws (3), lockwashers (2), and nuts (1), ensuring heads of screws (3) rest in sill of counterbore. Tighten nuts (1) 100-110 lb-ft (11-12 N•m).

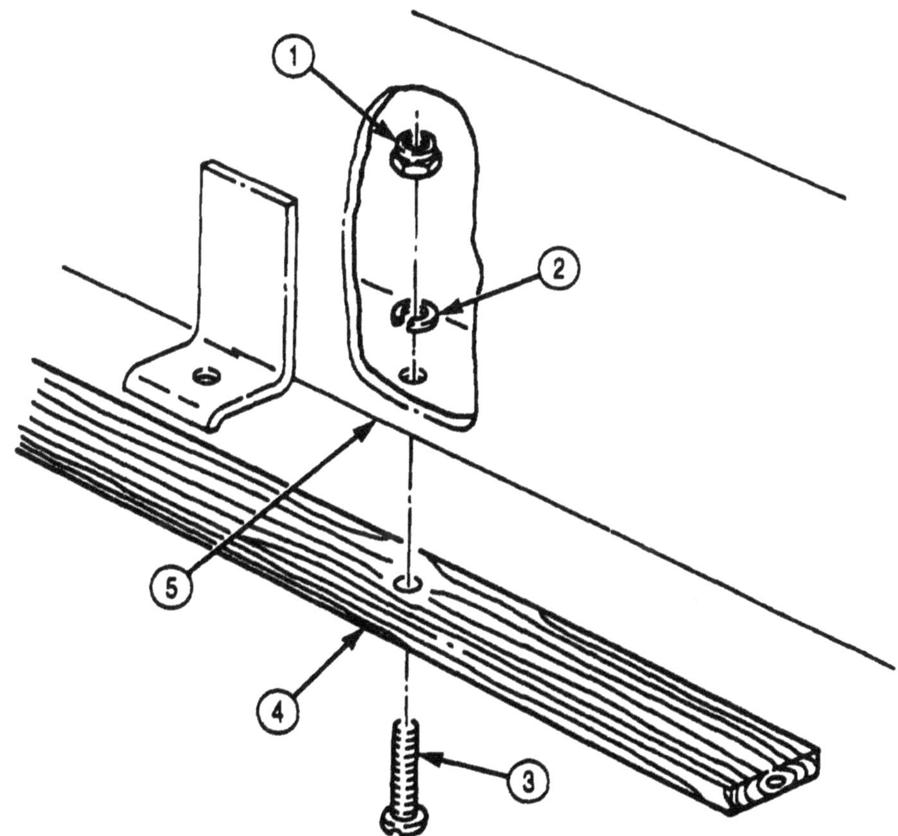

FOLLOW-ON TASK: Install cargo body assembly (para. 4-142).

4-144. DUMP BODY MAINTENANCE

THIS TASK COVERS:

a. Removal
b. Inspection and Repair
c. Installation

INITIAL SETUP:

APPLICABLE MODELS
M929/A1/A2, M930/A1/A2

TOOLS
General mechanic's tool kit (Appendix E, Item 1)
Torque wrench (Appendix E, Item 145)
Two utility chains
Lifting device
Shop stands

MATERIALS/PARTS
Locknut (Appendix D, Item 309)
Six locknuts (Appendix D, Item 321)
Locknut (Appendix D, Item 291)

PERSONNEL REQUIRED
Three

MANUAL REFERENCES (TM)
LO 9-2320-272-12
TM 43-0139
TM 9-2320-272-10
TM 9-2320-272-24P

EQUIPMENT CONDITION
- Parking brake set (TM 9-2320-272-10).
- Wheels chocked (TM 9-2320-272-10).

GENERAL SAFETY INSTRUCTIONS
- Position safety braces before working under raised dump body.
- Do not operate dump controls when dump body is removed.
- All personnel must stand clear during lifting operations.
- Lifting device must have a capacity greater than the weight of the dump body.

a. Removal

WARNING

- Lifting device must have a weight capacity greater than the combined weight of the dump body to prevent injury or death to personnel and damage to equipment.
- All personnel must stand clear during lifting operations. A snapped cable, or swinging or shifting load, may result in injury or death to personnel.
- Never work under dump body until safety braces are properly positioned. Injury to personnel may result if dump body suddenly lowers.

1. Place dump body (1) in raised position (TM 9-2320-272-10).
2. Place safety braces (6) in proper position (TM 9-2320-272-10).

NOTE

Step 3 applies to both left and right thrust plates.

3. Remove locknut (3) and screw (5) from thrust plate (2) and thrust plate pin (4). Discard locknut (3).
4. Remove safety braces (6) and place dump body (1) in lowest position (TM 9-2320-272-10).
5. Wrap two utility chains around subframe beam (9) and fasten hooks over utility chains.
6. Attach lifting device to center of utility chains.
7. Raise lifting device until slack is removed from chains.

4-144. DUMP BODY MAINTENANCE (Contd)

WARNING

Ensure dump control lever is in the NEUTRAL position and has not moved. Injury to personnel may result if lift cylinder is operated when not secured.

NOTE

Steps 8 through 11 apply to both left and right sides.

8. Remove thrust plate pin (4) from thrust plate (2) and roller arms (7).
9. Remove six locknuts (15) from hinge bracket (11). Discard locknuts (15).
10. Remove locknut (12), washer (13), and screw (14) from hinge bracket (11). Discard locknut (12).
11. Connect two guide lines to front and rear of body (1).

WARNING

All personnel must stand clear during lifting operations. A snapped cable, or swinging or shifting load, may result in injury or death to personnel.

NOTE

Mechanic will handle one guide line and direct hoisting operation. Assistant will handle other guide line. Second assistant will operate lifting device.

12. Lift dump body (1) clear of subframe (8) and place on shop stands.
13. Remove six screws (10) from dump body (1).
14. Remove guide lines, lifting device, and utility chains from dump body (1).

b. Inspection and Repair

1. Inspect dump body (1) for breaks, dents, cracks, and rust. If breaks, dents, cracks, and rusted-through areas are found, repair (TM 43-0139).
2. Clean and paint dump body (1) as necessary (TM 43-0139).

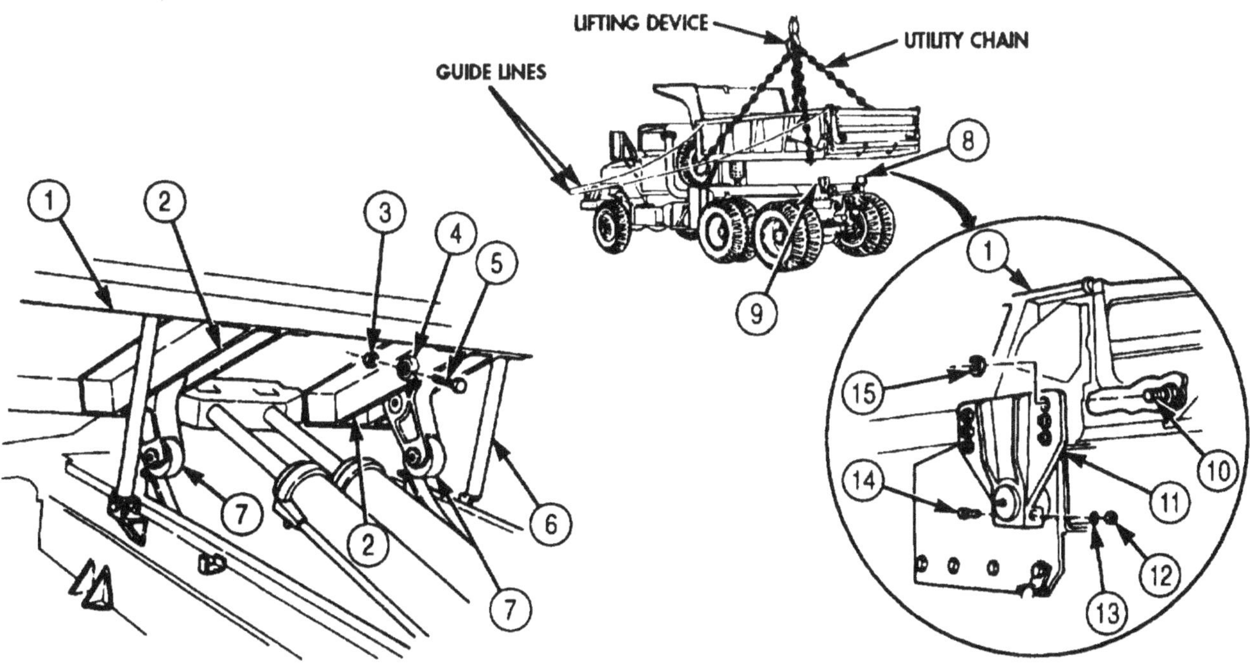

4-144. DUMP BODY MAINTENANCE (Contd)

c. Installation

WARNING

- Lifting device must have a weight capacity greater than the combined weight of the dump body to prevent injury or death to personnel and damage to equipment.
- Ensure dump control lever is in the NEUTRAL position and has not moved. Injury to personnel may result if lift cylinder is operated when not secured.

1. Wrap two utility chains around dump body (1) and fasten hooks over utility chains.
2. Attach lifting device to center of utility chains.
3. Connect two guide lines to front and rear of body (1).

WARNING

All personnel must stand clear during lifting operations. A snapped cable, or swinging or shifting load, may result in injury or death to personnel.

NOTE

- Mechanic will handle one guide line and direct hoisting operation. Assistant will handle other guide line. Second assistant will operate lifting device.
- Steps 4 and 7 through 9 apply to both left and right sides.

4. Install six screws (11) on left and right sides of dump body (1).
5. Hoist dump body (1) clear of shop stand.
6. Place dump body (1) on subframe (2).
7. Install two brackets (3) on six screws (11) with six locknuts (10). Tighten locknuts (10) 240 lb-ft (325 N•m).
8. Install screw (14), washer (13), and locknut (12) on each bracket (3). Tighten locknut (12) 35 lb-ft (48 N•m).
9. Align roller arms (9) and thrust plates (5) and install thrust plate pin (7).
10. Remove lifting device, utility chains, and guide lines from dump body (1).

WARNING

Never work under dump body until safety braces are properly positioned. Injury to personnel may result if dump body suddenly lowers.

11. Place dump body (1) in raised position (TM 9-2320-272-10).
12. Place safety braces (4) in proper position (TM 9-2320-272-10).

NOTE

Step 13 applies to both left and right thrust plates.

13. Install thrust plate pin (7) on thrust plate (5) with screw (8) and locknut (6). Tighten locknut (6) 35 lb-ft (48 N•m).

4-144. DUMP BODY MAINTENANCE (Contd)

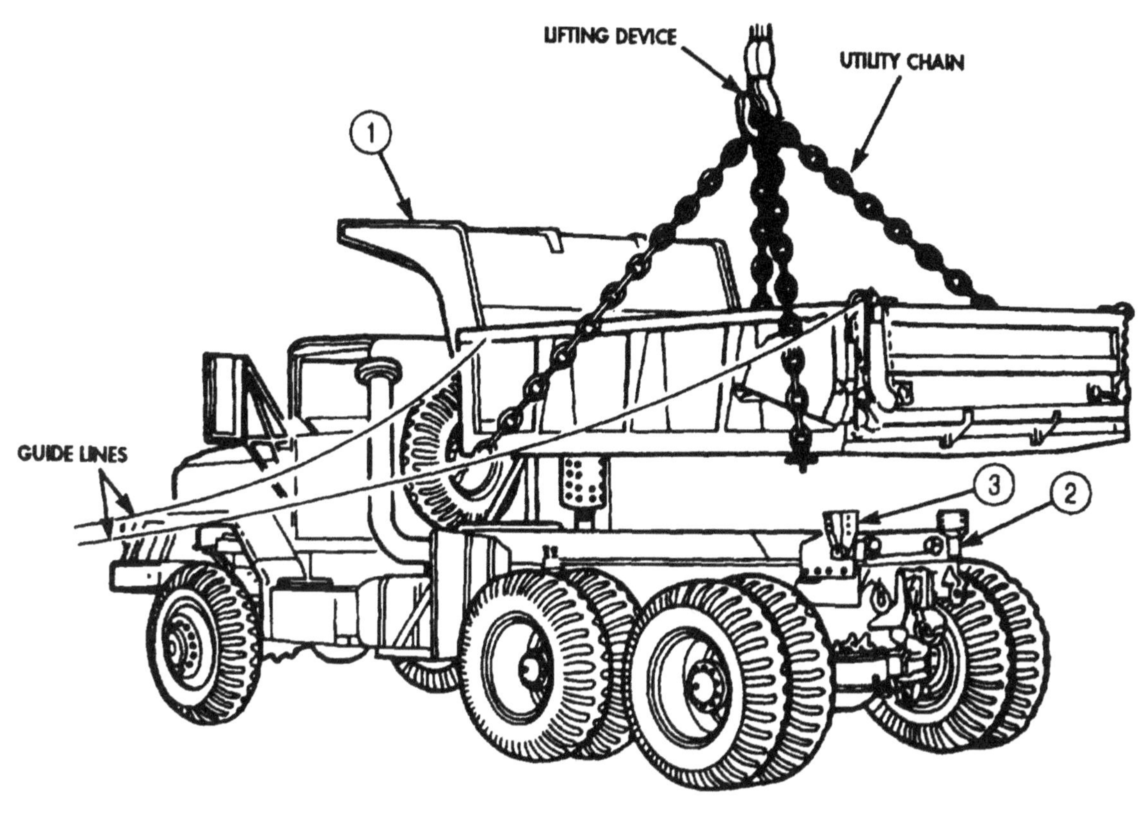

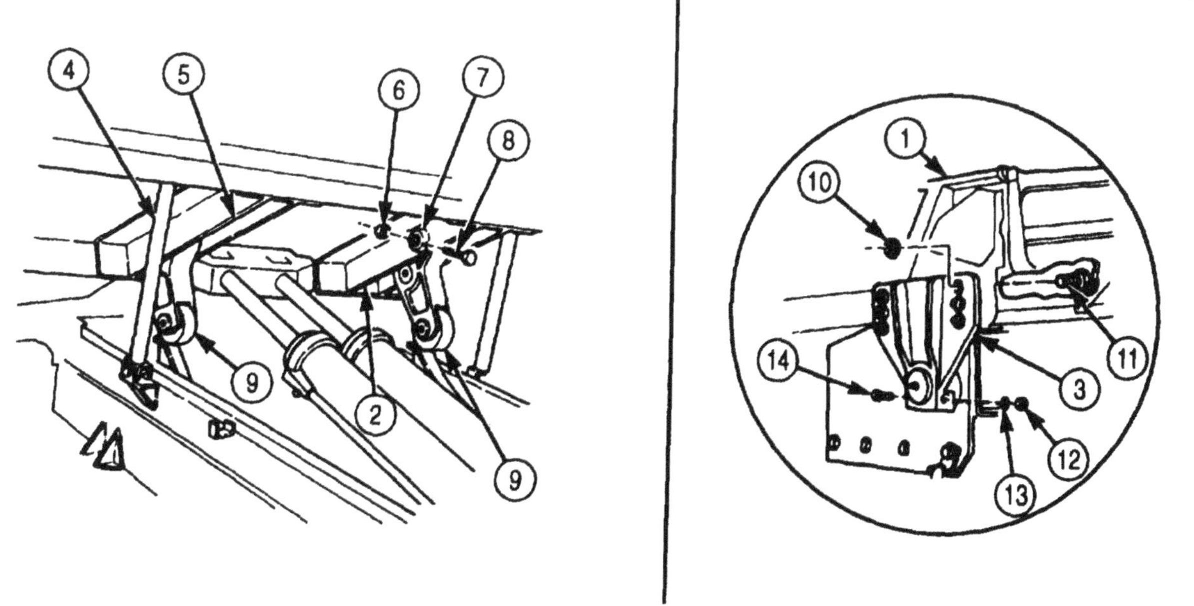

FOLLOW-ON TASK: Lubricate dump body (LO 9-2320-272-12).

4-145. DUMP SAFETY BRACE INSTALLATION

THIS TASK COVERS:

a. Removal **b. Installation**

INITIAL SETUP:

APPLICABLE MODELS
M929/A1/A2, M930/A1/A2

TOOLS
General mechanic's tool kit (Appendix E, Item 1)

MANUAL REFERENCES (TM)
TM 9-2320-272-10

EQUIPMENT CONDITION
- Parking brake set (TM 9-2320-272-10).
- Bump body in lowered position (TM 9-2320-272-10).

NOTE

Both left and right safety braces are removed in the same way. This procedure applies to left brace only.

a. Removal

1. Remove nut (5), washer (4), and screw (3) from safety brace hinge pin (1).
2. Using chisel and hammer, remove safety brace hinge pin (1) from bracket (2) and safety brace (6).
3. Remove safety brace (6) from bracket (2).

b. Installation

1. Install safety brace (6) on bracket (2).
2. Align holes in safety brace (6) and bracket (2), and install safety brace hinge pin (1) with screw (3), washer (4), and nut (5).

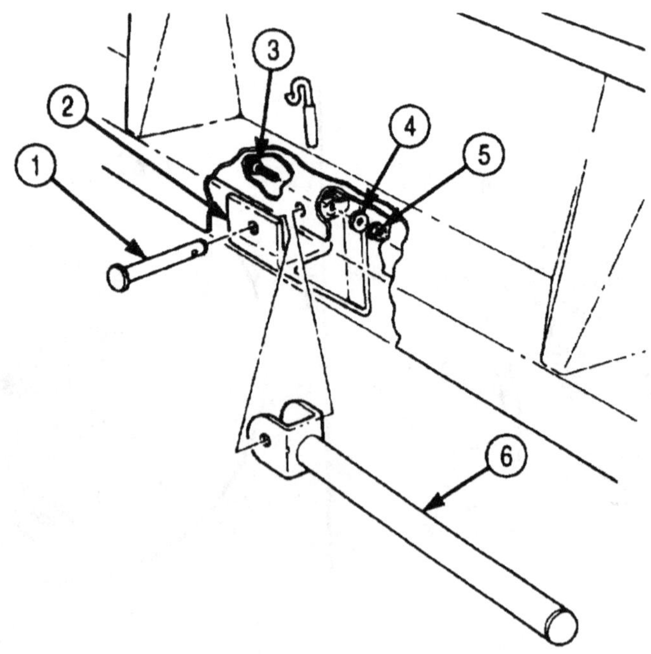

4-146. DUMP CAB PROTECTOR SHIELD MAINTENANCE

THIS TASK COVERS:

a. Removal
b. Inspection and Repair
c. Installation

INITIAL SETUP:

APPLICABLE MODELS
M929/A1/A2, M930/A1/A2

TOOLS
General mechanic's tool kit (Appendix E, Item 1)
Lifting device
Chains

MATERIALS/PARTS
Nineteen locknuts (Appendix D, Item 294)

PERSONNEL REQUIRED
Two

MANUAL REFERENCES (TM)
LO 9-2320-272-12
TM 9-2320-272-10
TM 9-2320-272-24P
TM 9-237
TM 43-0139

EQUIPMENT CONDITION
- Parking brake set (TM 9-2320-272-10).
- Wheels chocked (TM 9-2320-272-10).

GENERAL SAFETY INSTRUCTIONS
All personnel must stand clear during lifting operations.

a. Removal

1. Install two eyebolts (6) into rear holes (5) of cab protector (2) with two washers (4) and nuts (3).
2. Attach two utility chains to eyebolts (6) and opposite ends of cab protector front support channel (1).

NOTE

Mechanic will direct hoisting operation while assistant operates lifting device.

3. Attach lifting device to center of two utility chain and remove slack from utility chain.

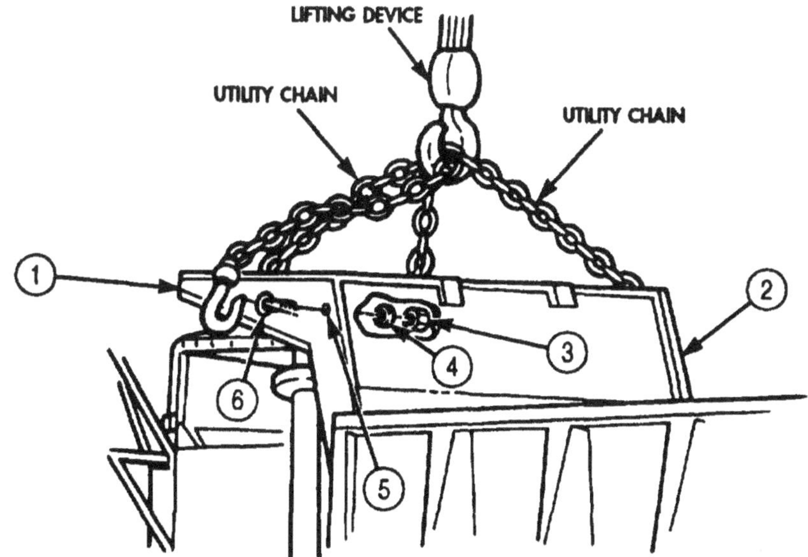

4-146. DUMP CAB PROTECTOR SHIELD MAINTENANCE (Contd)

4. Remove seven locknuts (13), washers (2), screws (1), and washers (2) from cab protector (3) and dump body (14). Discard locknuts (13).
5. Remove eight locknuts (7), washers (6), and screws (11) from cab protector (3) and two dump body extension brackets (12). Dircard locknuts (7).
6. Remove four locknuts (4), washers (5), and two screws (9) and (10) from cab protector (3) and dump body side support channel (8). Discard locknuts (4).

WARNING

All personnel must stand clear during lifting operations. A snapped cable, or swinging or shifting load, may result in injury or death to personnel.

7. Remove cab protector (3) from two dump body extension brackets (12) and position on four shop stands.
8. Remove lifting device and utility chains from cab protector (3).
9. Remove two nuts (16), washers (17), and eyebolts (19) from holes (18) in cab protector (3).

b. Inspection and Repair

1. Insect cab protector (3) for breaks, dents, and rust. If breaks, dents, and rusted-through areas are found, repair (TM 9-237).
2. Clean and paint as neceesary (TM 43-0139).

c. Installation

1. Install two eyebolts (19) into rear holes (18) of cab protector (3) with two washers (17) and nuts (16).
2. Attach two utility chains to eyebolts and opposite ends of cab protector front support channel (15).

NOTE

Mechanic will direct hoisting operation while assistant operates lifting device.

3. Attach lifting device to center of two utility chains and remove slack from utility chain.

WARNING

All personnel must stand clear during lifting operations. A snapped cable, or swinging or shifting load, may result in injury or death to personnel.

2. Remove cab protector (3) from four shop stands and position on two dump body extension brackets (12).
3. Using a drift pin, align mounting holes and install cab protector (3) on two dump body extension brackets (12) with eight screws (11), washers (6), and locknuts (7).
4. Install cab protector (3) on left and right dump body side support channel (8) with two screws (9) and (10), four washers (5), and locknuts (4).
5. Install cab protector (3) on dump body (14) with seven washers (2), screws (1), washers (2), and locknuta (13).
6. Remove lifting device and utility chains from cab protector (3).
7. Remove two nuts (16), washers (17), and eyebolts (19) from holes (18) in cab protector (3).

4-146. DUMP CAB PROTECTOR SHIELD MAINTENANCE (Contd)

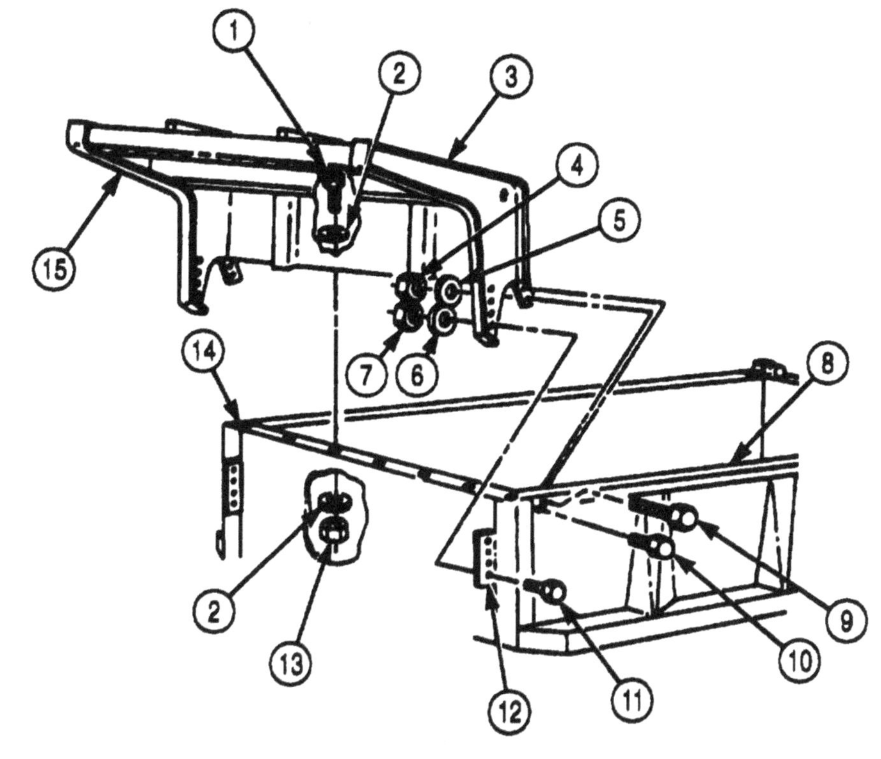

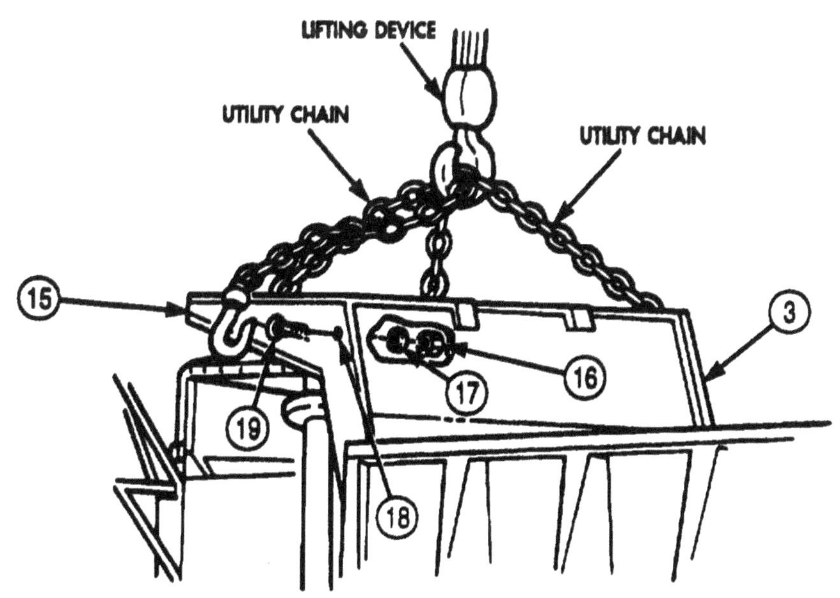

4-147. DUMP BODY HYDRAULIC HOSE REPLACEMENT

THIS TASK COVERS:

a. Removal **b. Installation**

INITIAL SETUP:

APPLICABLE MODELS
M929/A1/A2, M930/A1/A2

TOOLS
General mechanic's tool kit (Appendix E, Item 1)

MATERIALS/PARTS
Cap and plug set (Appendix C, Item 14)

MANUAL REFERENCES (TM)
LO 9-2320-272-12
TM 9-2320-272-10
TM 9-2320-272-24P

EQUIPMENT CONDITION
- Parking brake set (TM 9-2320-272-10).
- Dump body in lowered position (TM 9-2320-272-10).
- Hydraulic oil reservoir drained (LO 9-2320-272-12).

GENERAL SAFETY INSTRUCTIONS
- Position safety braces before working under raised dump body.
- Ensure dump control lever is in the NEUTRAL position and has not moved.
- Store or dispose of used oil properly.
- Do not remove lines when engine is running or start engine when lines are removed.

WARNING

- Never work under dump body until safety braces are properly positioned. Injury to personnel may result if dump body suddenly lowers.
- Ensure dump control lever is in the NEUTRAL position and has not moved. Injury to personnel may result if lift cylinder is operated when not secured.
- Accidental or intentional introduction of liquid contaminants into the environment is in violation of state, federal, and military regulations. Refer to Army POL (para. 1-8) for information concerning storage, use, and disposal of these liquids. Failure to do so may result in injury or death.
- Do not remove hoses with engine running or start engine with hoses removed. High-pressure fluids may cause hoses to whip violently and spray randomly. Doing so may result in injury to personnel.

NOTE

All hydraulic hoses are removed the same way This procedure covers dump hoist cylinder and safety lock hoses only.

a. Removal

1. Place dump body in raised position (TM 9-2320-272-10).
2. Place safety braces in proper position (TM 9-2320-272-10).

CAUTION

Plug all hydraulic lines or openings to prevent dirt from entering and damaging components.

4-147. DUMP BODY HYDRAULIC HOSE REPLACEMENT (Contd)

NOTE

- Cross fittings to hoist cylinder hoses must be disconnected from the cross fittings first. Then hoses can be removed from hoist cylinders.
- Have drainage container ready to catch oil.
- Drain hoses before plugging.
- Tag hoses for installation.

3. Disconnect four hydraulic hoses (6), (10), (12), and (14) from right cross fitting (11).
4. Disconnect four hydraulic hoses (5), (7), (9), and (13) from left cross fitting (8).
5. Disconnect two hydraulic hoses (6) and (12) from right and left cylinder ports (1) and (2).
6. Disconnect two hydraulic hoses (7) and (13) from right and left, cylinder ports (4) and (15).
7. Disconnect two hydraulic hoses (5) and (14) from safety lock cylinder (3).

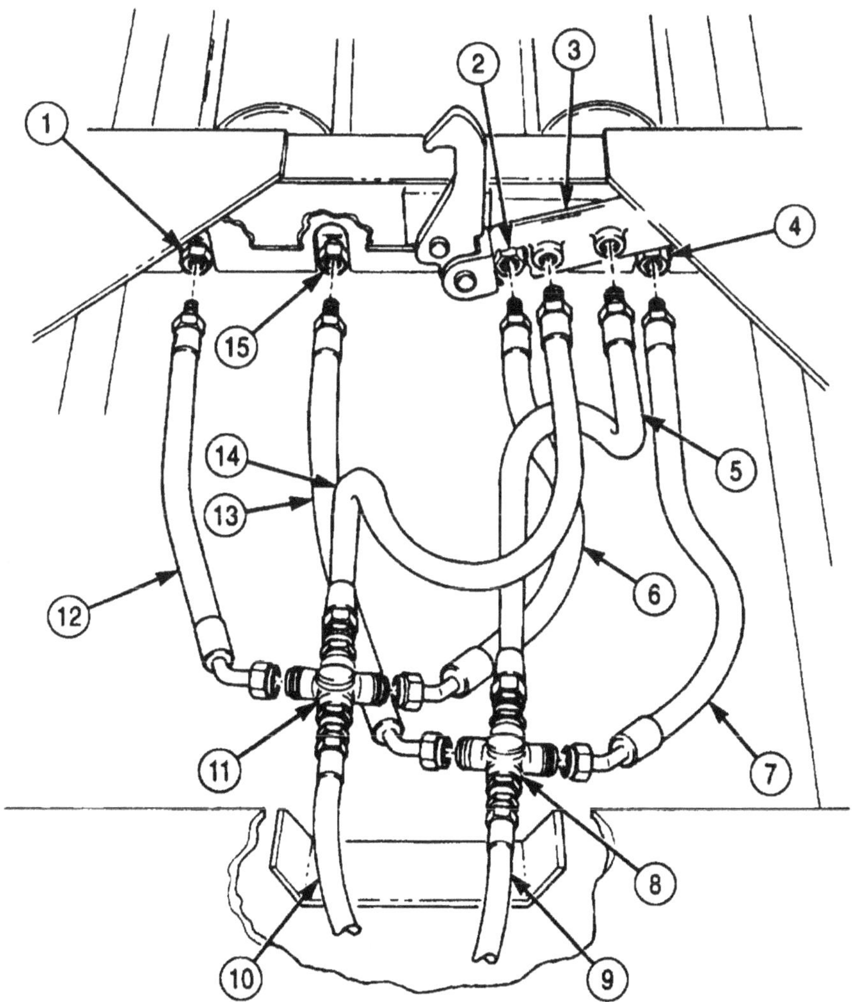

4-147. DUMP BODY HYDRAULIC HOSE REPLACEMENT (Contd)

8. Disconnect two hydraulic hoses (2) and (3) from control valve (1) and fittings (6) and (7).

b. Installation

NOTE

Ensure plugs are removed from all fittings before installation. Ensure no particles of plugging become trapped in dump body hydraulic system during installation of hoses and tubing.

1. Connect two hydraulic hoses (3) and (2) to fittings (4) and (5) on control valve (1).
2. Connect two hydraulic hoses (3) and (2) to cross fittings (6) and (7).
3. Connect two hydraulic hoses (16) to right and left cylinder ports (18) (with bypass tube (8) extending full length of cylinder) and right cross fitting (15).
4. Connect two hydraulic hoses (13) to right and left cylinder ports (11) (with bypass tube (9) extending midway of the cylinder) and left cross fitting (14).
5. Connect left cross fitting-to-safety lock cylinder hose (12) to left safety lock cylinder port (10) and left cross fitting (14).
6. Connect right cross fitting-to-safety lock cylinder hose (17) to right safety lock cylinder port (15) and right cross fitting (19).
7. Remove dump body safety braces and place in lowest position (LO 9-2320-272-12).

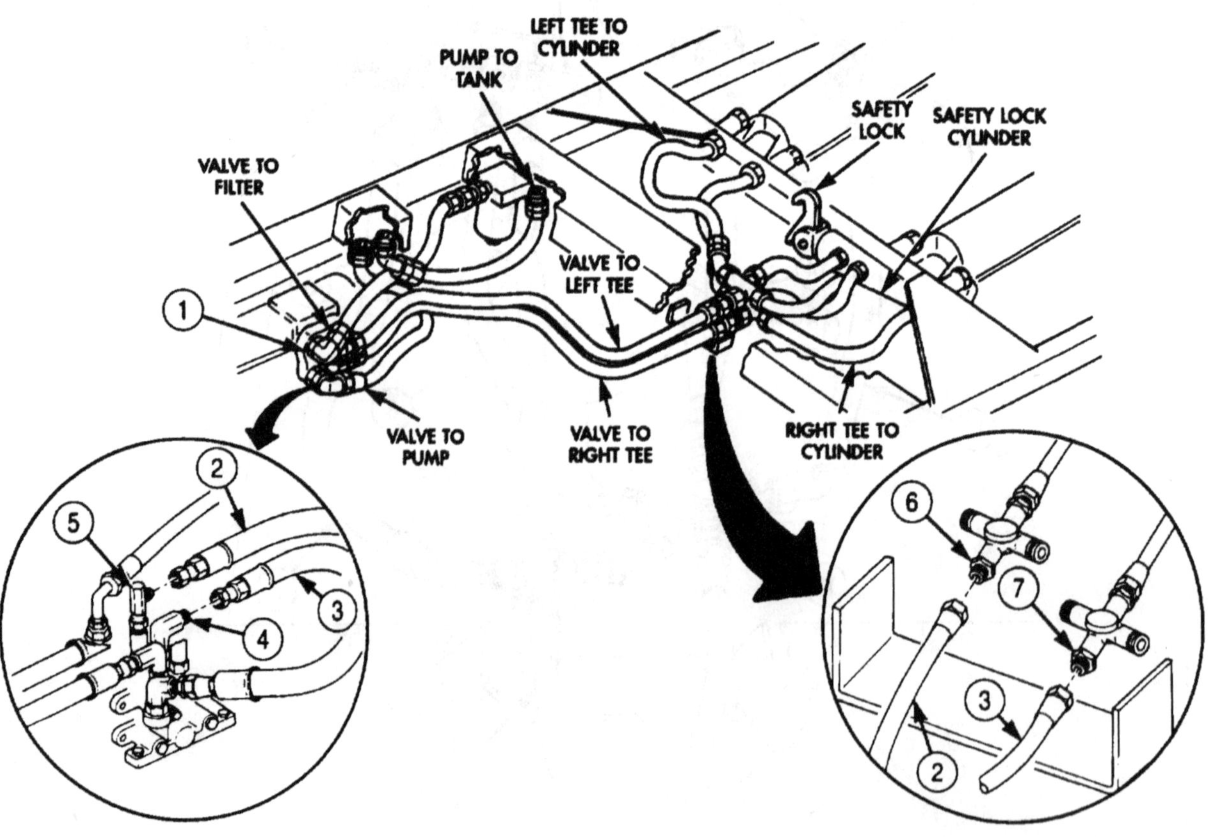

4-147. DUMP BODY HYDRAULIC HOSE REPLACEMENT (Contd)

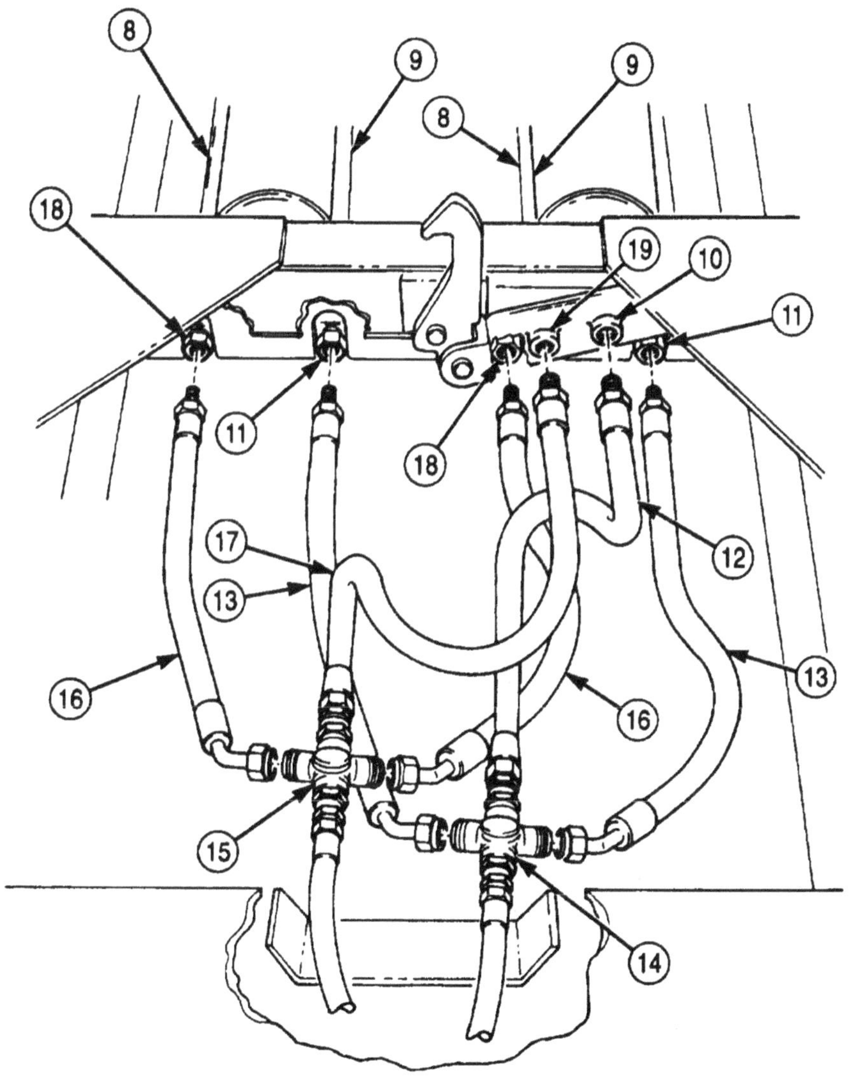

FOLLOW-ON TASKS:
- Fill hydraulic reservoir to proper oil level (LO 9-2320-272-12).
- Start engine (TM 9-2320-272-10) and operate through full range. Check for leaks and proper operation.

4-148. DUMP VALVE CABLE AND SHIFT LEVER REPLACEMENT

THIS TASK COVERS:

a. Removal **b. Installation and Adjustment**

INITIAL SETUP:

APPLICABLE MODELS
M929/A1/A2, M930/A1/A2

TOOLS
General mechanic's tool kit (Appendix E, Item 1)

MATERIALS/PARTS
Three locknuts (Appendix D, Item 313)
Six locknuts (Appendix D, Item 307)
Locknut (Appendix D, Item 294)
Fourteen locknuts (Appendix D, Item 291)
Two cotter pins (Appendix D, Item 62)
Tiedown strap (Appendix D, Item 690)

PERSONNEL REQUIRED
Two

MANUAL REFERENCES (TM)
LO 9-2320-272-12
TM 9-2320-272-10
TM 9-2320-272-24P

EQUIPMENT CONDITION
Parking brake set (TM 9-2320-272-10).

a. Removal

1. Remove cotter pin (9), washer (10), and clevis pin (13) from cable clevis (11) and shift lever arm (2). Discard cotter pin (9).
2. Remove four locknuts (8), screws (5), and shift lever (6) from two shift lever brackets (4), shims (7), and floor (15). Discard locknuts (8).
3. Remove woodruff key (19) from shift lever rod (6).
4. Remove locknut (3), screw (1), and shift lever arm (2) from cab floor (15). Discard locknut (3).
5. Remove two shift lever brackets (4) from shift lever rod (6).
6. Remove six locknuts (14), screws (18), seal (17), and retainer ring (16) from cab floor (15). Discard locknuts (14).
7. Loosen and remove cable clevis (11) and jamnut (29) from dump valve cable (35).
8. Remove two locknuts (26), screws (31), clamp (30), and shim (28) from cable conduit (12) and rear cab floor left bracket (27). Discard locknuts (26).
9. Remove locknut (25), screw (33), and clamp (32) from cable conduit (12) and rear cab floor right bracket (24). Discard locknut (25).
10. Remove tiedown strap (34) from cable conduit (12). Discard tiedown strap (34).
11. Remove cotter pin (22), washer (21), and clevis pin (20) from cable clevis (41) and control valve (23). Discard cotter pin (22).
12. Loosen and remove cable clevis (41) and jamnut (40) from dump valve cable (35).
13. Remove jamnut (40) and washer (39) from control valve (23) side of cable conduit (12) and lower bracket (38).
14. Pull cable conduit (12) through lower bracket (38). Remove washer (37) and nut (36) from cable conduit (12). Tag for installation.

4-148. DUMP VALVE CABLE AND SHIFT LEVER REPLACEMENT (Contd)

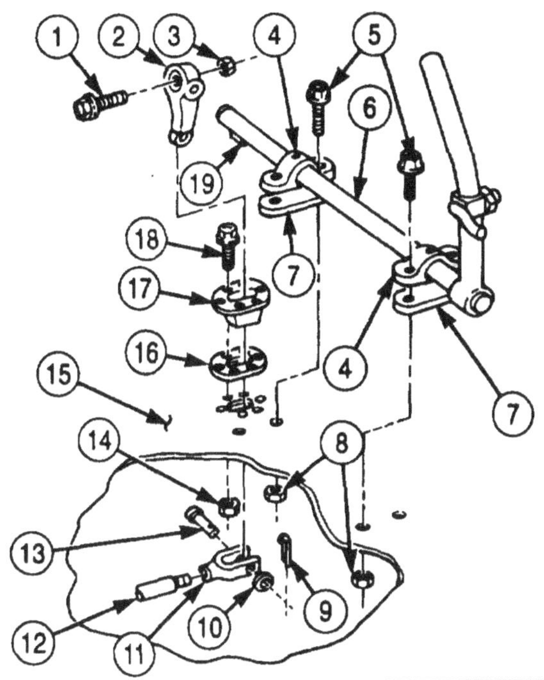

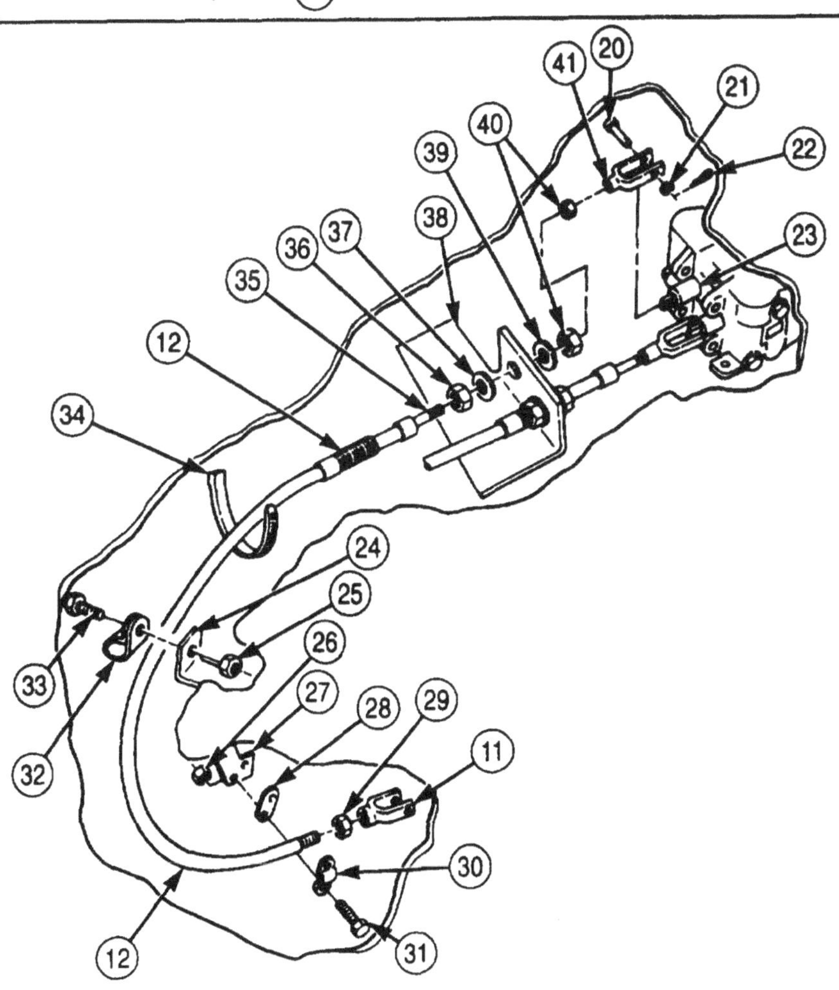

4-148. DUMP VALVE CABLE AND SHIFT LEVER REPLACEMENT (Cod)

b. Installation and Adjustment

1. Install nut (36) and washer (37) on cable conduit (12).
2. Place control valve (23) end of cable conduit (12) through lower bracket (38), and install cable conduit (12) on lower bracket (38) with washer (39) and jamnut (40).
3. Install cable clevis (41) on control valve (23) end of valve cable (35) with four turns.
4. Route cable conduit (12) under cab floor (15).
5. Install retainer ring (16) and seal (17) on cab floor (15) with six screws (18) and locknuts (14).
6. Slide two lever brackets (4) on lever rod (6).
7. Install woodruff key (19) on lever rod (6), and elide arm (2) over lever rod (6) and woodruff key (19).
8. Install screw (1) and locknut (3) on arm (2).
9. Position shift lever rod (6) on cab floor (15), and install lever rod (6), two brackets (4), and shims (7) with four screws (5) and locknuts (8).
10. Install cable clevis (11) on lever arm (2) with clevis pin (13), washer (10), and cotter pin (9).
11. Install cable conduit (12) and two shims (28) on rear cab bracket (27) with two clamps (30), screws (31), and locknuts (26).
13. Install cable conduit (12) on front cab bracket (24) with clamp (32), screw (33), and locknut (25).
14. Place shift lever in down position (TM 9-2320-272-10).
15. Pull valve shaft (23) out manually until seated. Valve shaft (23) will stop when seated.

NOTE

Cable throw length can be ajusted with jamnut.

16. Align hole of clevis pin (20) and hole of cable clevis (41) with hole in valve shaft (23). Use jamnuts (40) and (36).
17. Back off clevis (41) one turn and install on valve shaft (23) with clevis pin (20), washer (21), and cotter pin (22).
18. Install tiedown strap (34) on cable conduit (12).

4-148. DUMP VALVE CABLE AND SHIFT LEVER REPLACEMENT (Contd)

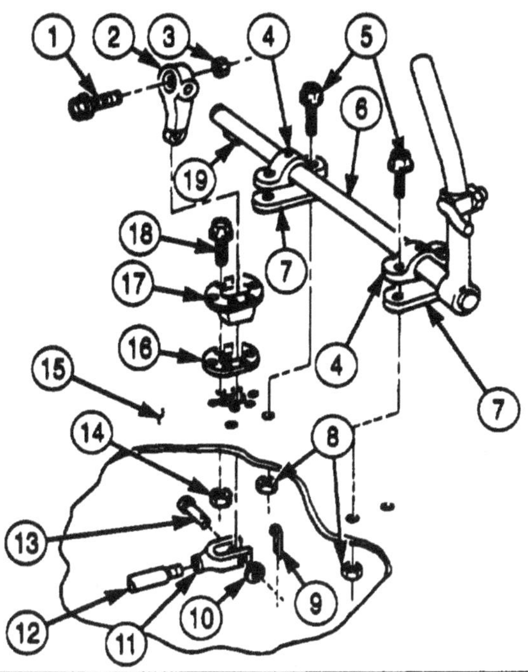

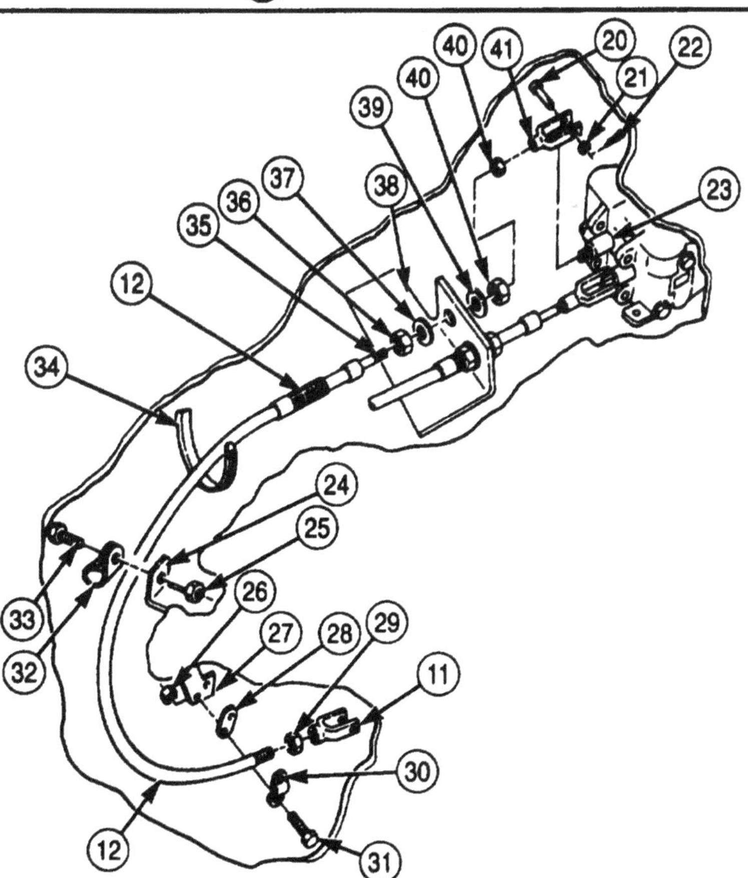

FOLLOW-ONTASKS: • Lubricate shift lever (LO 9-2320-272-12).
• Test dump cable operation (TM 9-2320-272-10).

4-149. DUMP SUBFRAME REPLACEMENT

THIS TASK COVERS:

a. Removal **b. Installation**

INITIAL SETUP:

APPLICABLE MODELS
M929/A1/A2, M930/A1/A2

TOOLS
General mechanic's tool kit (Appendix E, Item 1)

MATERIALS/PARTS
Eight locknuts (Appendix D, Item 321)
Four cotter pins (Appendix D, Item 69)
Cap and plug set (Appendix C, Item 14)

PERSONNEL REWIRED
Three

MANUAL REFERENCES (TM)
LO 9-2320-272-12
TM 9-2320-272-10
TM 9-2320-272-24P

EQUIPMENT CONDITION
- Parking brake set (TM 9-2320-272-10).
- Hydraulic oil reservoir drained (LO 9-2320-272-12).
- Dump body removed (para. 4-144).

GENERAL SAFETY INSTRUCTIONS
- Store or dispose of used oil properly.
- Do not remove lines when engine is running or start engine when lines are removed.
- All personnel must stand clear during lifting operations.

WARNING

- Accidental or intentional introduction of liquid contaminants into the environment is in violation of state, federal, and military regulations. Refer to Army POL (para. 1-8) for information concerning storage, use, and disposal of these liquids. Failure to do so may result in injury or death.
- Do not remove hoses with engine running or start engine with hoses removed. High-pressure fluids may cause hoses to whip violently and spray randomly. Doing so may result in injury to personnel.

CAUTION

Plug all hydraulic lines or openings to prevent dirt from entering and damaging components.

a. Removal

NOTE

- Have container ready to catch oil from hydraulic tubes, lines, and hoses.
- Drain hoses before plugging.
- Tag hoses for installation.

1. Disconnect pump-to-reservoir hydraulic tube (2) from pump adapter (3) and reservoir adapter (1).
2. Disconnect filter-to-reservoir hydraulic hose (5) from filter elbow (4) and reservoir adapter (6).

4-149. DUMP SUBFRAME REPLACEMENT (Contd)

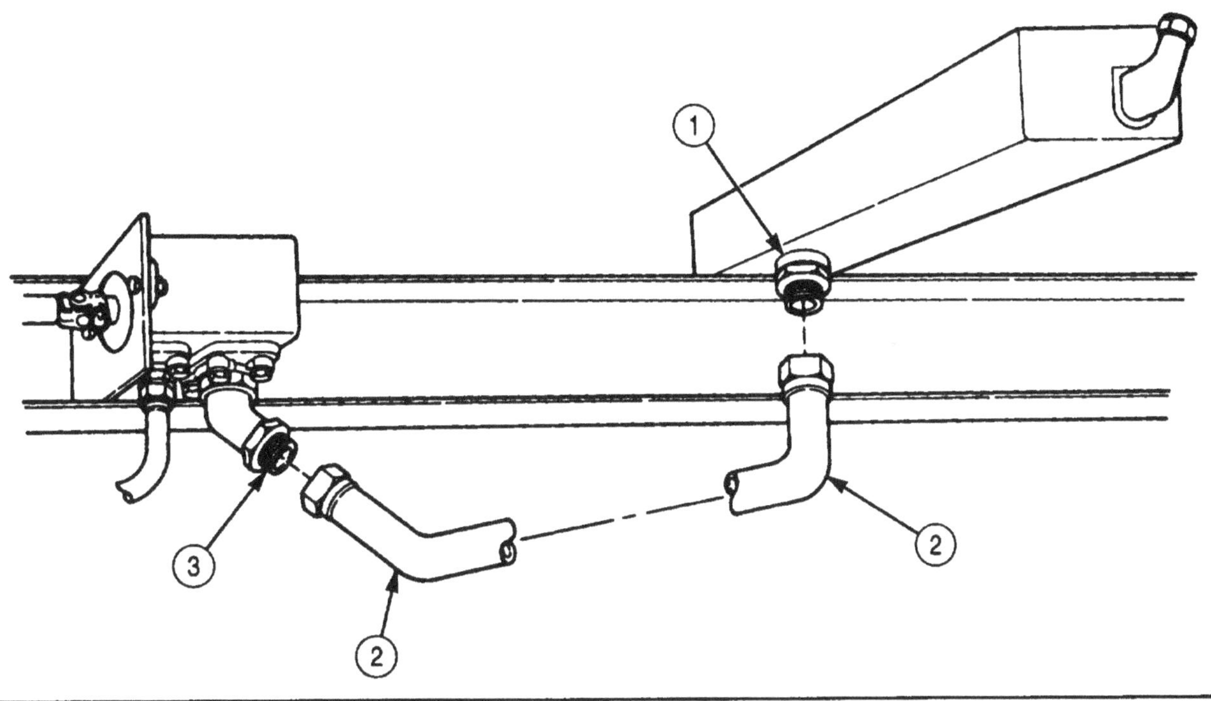

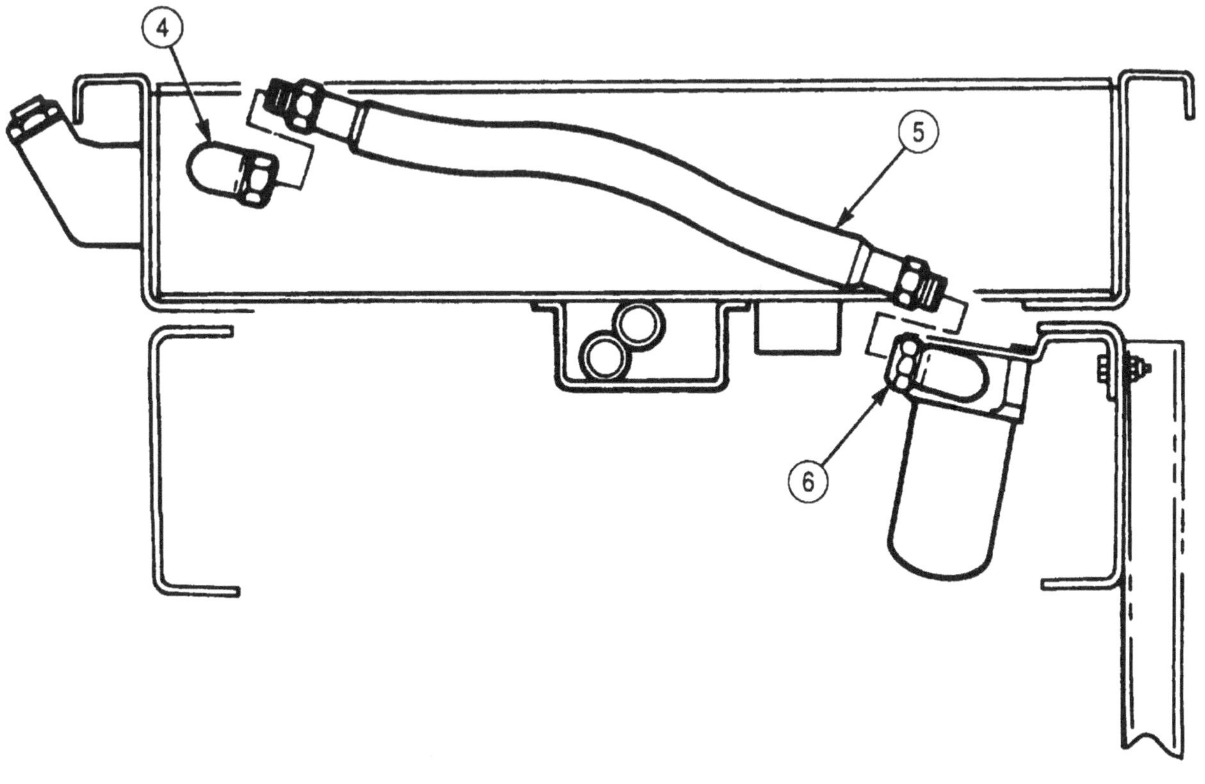

4-149. DUMP SUBFRAME REPLACEMENT (Contd)

NOTE

- Cross fittings to hoist cylinder hoses must be disconnected from the cross fittings first. Then hoses can be removed from hoist cylinders.
- Have drainage container ready to catch oil.
- Tag hoses for installation.

3. Disconnect three hydraulic hoses (6), (10), and (12) from right cross fitting (9).
4. Disconnect three hydraulic hoses (5), (7), and (11) from left cross fitting (8).
5. Disconnect two hydraulic hoses (6) and (10) from right and left cylinder ports (1) and (2).
6. Disconnect two hydraulic hoses (7) and (11) from right and left cylinder ports (4) and (13).
7. Disconnect two hydraulic hoses (5) and (12) from safety lock cylinder (3).

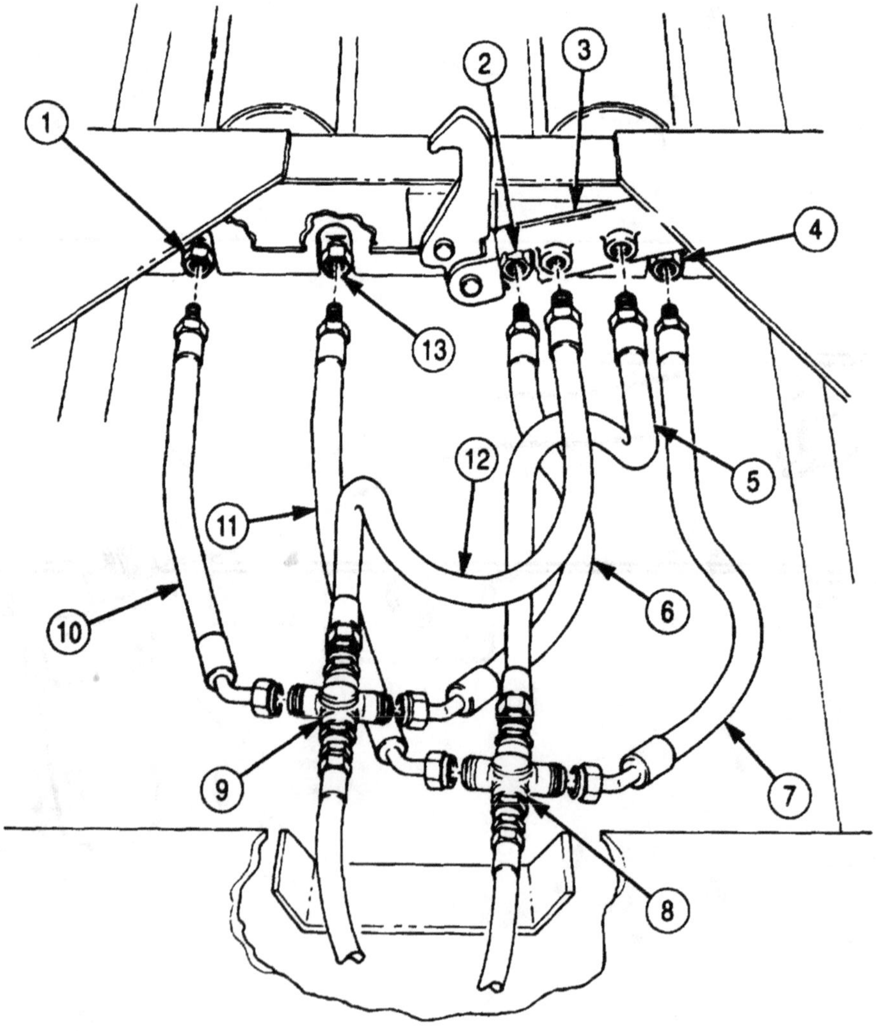

4-149. DUMP SUBFRAME REPLACEMENT (Contd)

8. Remove eight locknuts (17), screws (18), and two rear holddown brackets (16) from frame (19). Discard locknuts (17).
9. Attach two utility chains to frame channel (15).
10. Attach lifting device to center of two utility chains and remove slack from utility chain.
11. Remove two cotter pins (27), slotted nuts (26), washer (25), screws (20), upper half-keeper (21), spring (22), and lower half-keeper (23) from two front holddown brackets (24) and frame (19). Discard cotter pins (27).
12. Connect two guide lines to front and rear of subframe (14).

WARNING

All personnel must stand clear during lifting operations. A snapped cable, or swinging or shifting load, may result in injury or death to personnel.

NOTE

Mechanic will handle one guide line and direct hoisting operation. Assistant will handle other guide line. Second assistant will operate lifting device.

13. Lift subframe (14) clear of frame (19) and place on two shop stands.
14. Remove guide lines, lifting device, and utility chains from subframe (14).

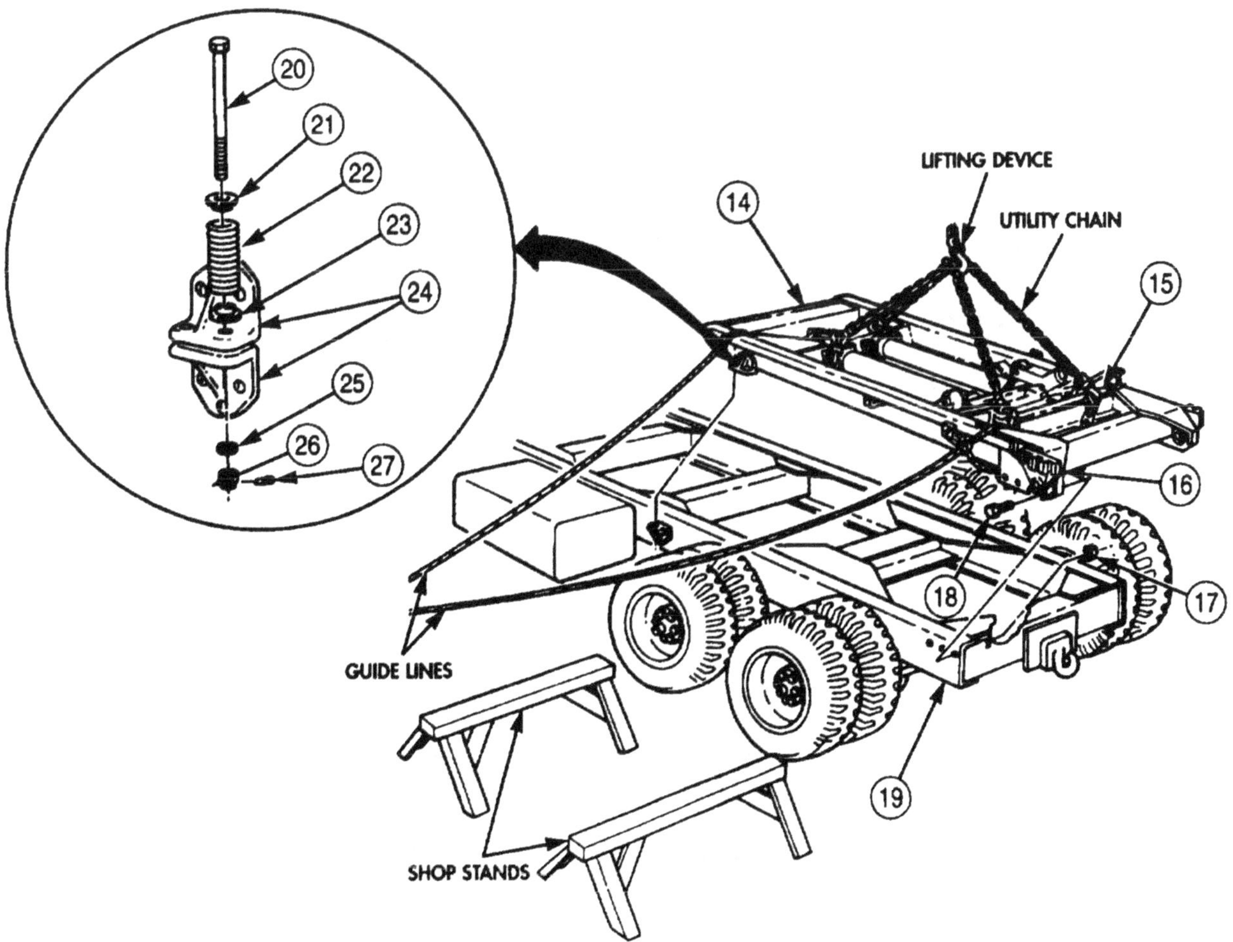

4-149. DUMP SUBFRAME REPLACEMENT (Contd)

b. Installation

1. Attach two utility chains to subframe (1).
2. Attach lifting device to center of two utility chains and remove slack from utility chain.
3. Connect two guide lines to front and rear of subframe (1).

WARNING

All personnel must stand clear during lifting operations. A snapped cable, or swinging or shifting load, may result in injury or death to personnel.

NOTE

Mechanic will handle one guide line and direct hoisting operation. Assistant will handle other guide line. Second assistant will operate lifting device.

4. Lift subframe (1) clear of two shop stands and lower onto frame (6).
5. Using drift pin, align holes in brackets (3) and (11).
6. Install two rear holddown brackets (3) on frame (6) with eight screws (5) and locknuts (4).
7. Remove guide lines, lifting device, and utility chains from subframe (1).
8. Install two front holddown brackets (11) on frame (6) with lower half-keeper (10), spring (9), upper half-keeper (8), screw (7), washer (12), slotted nut (13), and cotter pin (14).

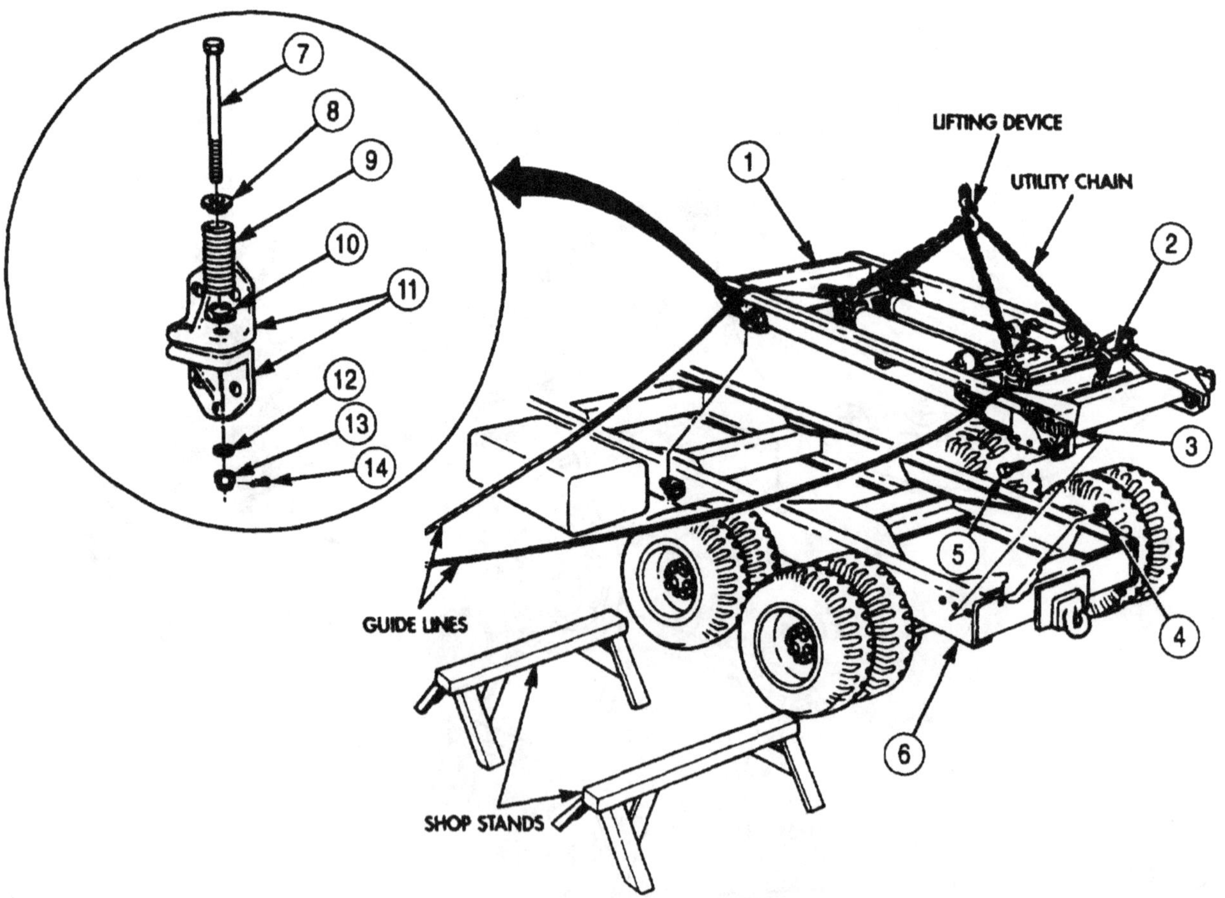

4-149. DUMP SUBFRAME REPLACEMENT (Contd)

9. Connect two hydraulic hoses (24) to left and right cylinder ports (26) (with bypass tube (15) which extends full length of cylinder) and right cross fitting (23).
10. Connect two hydraulic hoses (21) to left and right cylinder ports (19) (with bypass tube (16) which extends to middle of the cylinder) and left cross fitting (22).
11. Connect left cross fitting-to-safety lock cylinder hose (20) to left safety lock cylinder port (18) and left cross fitting (22).
12. Connect right cross fitting-to-safety lock cylinder hose (25) to right safety lock cylinder port (17), and right cross fitting (23).
13. Connect filter-to-reservoir hydraulic tube (28) to filter adapter (29) and reservoir elbow (27).
14. Connect pump-to-reservoir hydraulic tube (31) to pump adapter (32) and reservoir adapter (30).

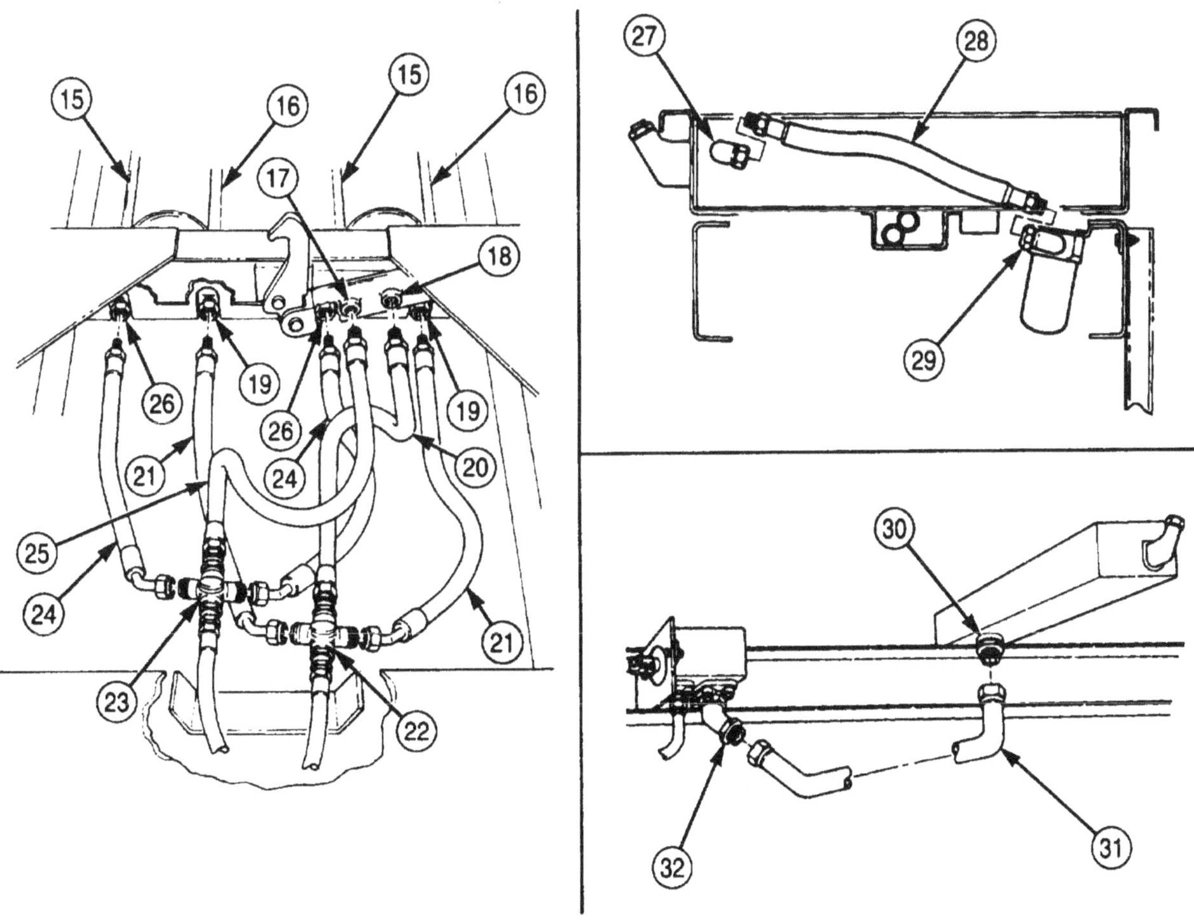

FOLLOW-ON TASKS:
- Install dump body (para. 4-144).
- Fill hydraulic reservoir to proper oil level (LO 9-2320-272-12).
- Start engine (TM 9-2320-272-10) and operate through full range. Check for leaks and proper operation.

4-150. DUMP HYDRAULIC PUMP REPLACEMENT

Procedure for replacement of the dump hydraulic pump can be found in pare. 4-176.

4-151. DUMP CONTROL VALVE REPLACEMENT

Procedure for replacement of the dump control valve can be found in para. 4-178.

4-152, WRECKER BODY REPLACEMENT

THIS TASK COVERS:

a. Removal **b. Installation**

INITIAL SETUP:

APPLICABLE MODELS
M936/A1/A2

TOOLS
General mechanic's tool kit (Appendix E, Item 1)
Lifting device
Two chains
Two guide lines

MATERIALS/PARTS
Sixteen locknuts (Appendix D, Item 323)
Sixteen locknuts (Appendix D, Item 310)

PERSONNEL REQUIRED
Three

MANUAL REFERENCES (TM)
TM 9-2320-272-10
TM 9-2320-272-24P

EQUIPMENT CONDITION
- Parking brake set (TM 9-2320-272-10).
- Crane turntable removed (para. 4-203).

GENERAL SAFETY INSTRUCTIONS
- Keep fire extinguisher nearby when using drycleaning solvent.
- All personnel must stand clear during lifting operations.
- Lifting device must have a weight capacity greater than the weight of the truck bed.

WARNING

- Lifting device must have a weight capacity greater than the combined weight of the truck bed to prevent injury or death to personnel and damage to equipment.
- All personnel must stand clear during lifting operations. A snapped cable, or swinging or shifting load, may result in injury or death to personnel.
- Drycleaning solvent is flammable and toxic. Do not use near open flame and always have a fire extinguisher nearby when solvents are used. Use only in well-ventilated places, wear protective clothing, and dispose of cleaning rags in approved container. Failure to do this may result in injury or death to personnel and/or damage to equipment.

NOTE

The left and right side body mounts are replaced the same. This procedure covers the left side body mounts.

a. Removal

1. Remove four locknuts (10), two U-bolts (11), and wood blocks from frame rail (3) and wrecker body (1). Discard locknuts (10).

4-152. WRECKER BODY REPLACEMENT (Contd)

2. Remove four locknuts (8) and two U-bolts (9) from frame rail (3) and wrecker body (1). Discard locknuts (8).
3. Remove four locknuts (6), two U-bolts (7), and four wood blocks from frame rail (3) and wrecker body (1). Discard locknuts (6).
4. Remove four locknuts (2), four mounting screws (5), two plates (4), and four wood blocks from frame rail (3) and wrecker body (1). Discard locknuts (2).
5. Remove sixteen locknuts (12), screws (13), and two brackets (14) from frame rail (3). Discard locknuts (12).
6. Attach two guide lines and utility chains to wrecker body (1) and lifting device to utility chains.

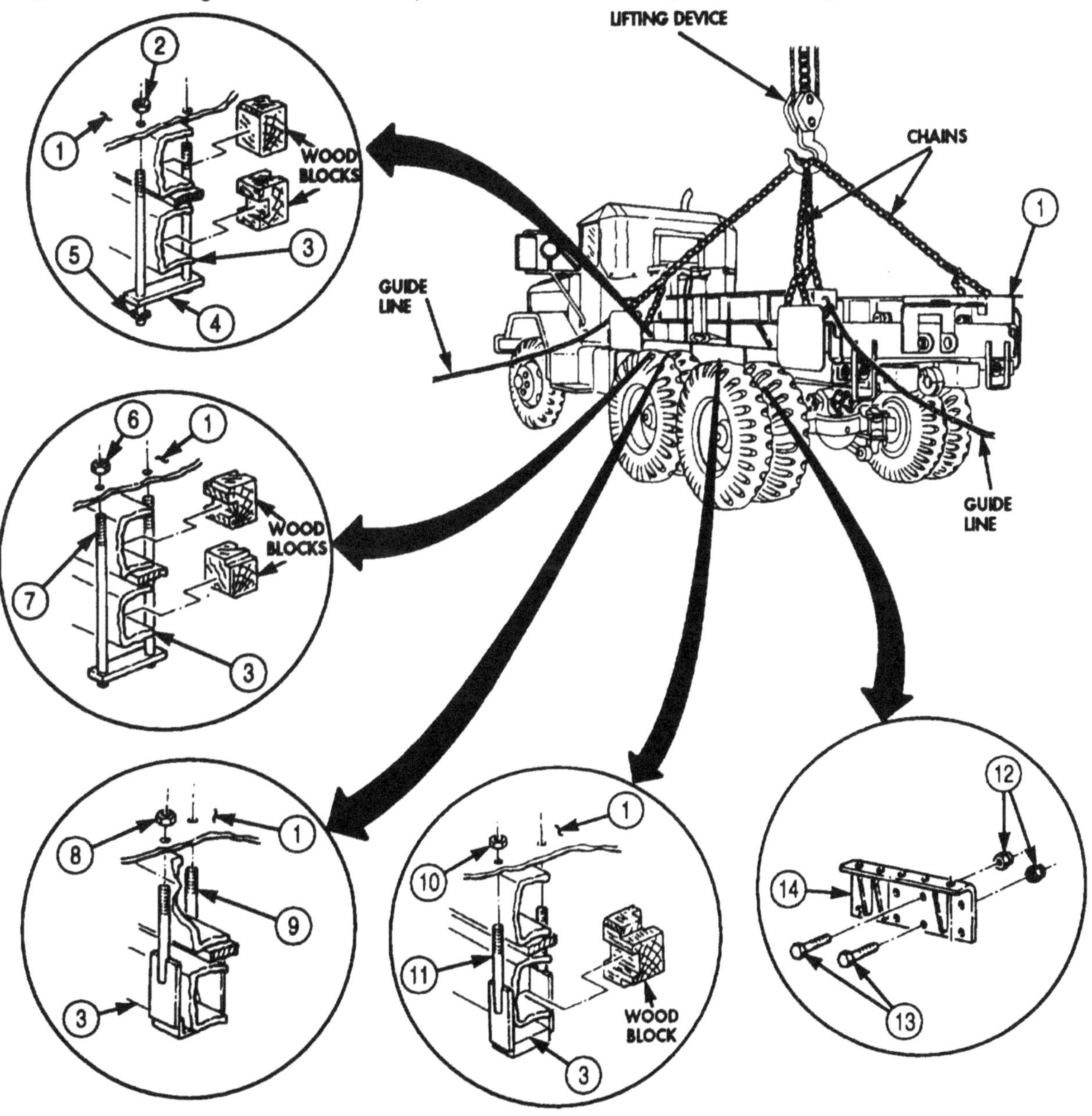

4-152. WRECKER BODY REPLACEMENT (Contd)

NOTE

Two assistants will help with steps 7 and 8.

7. Remove wrecker body (1) from frame rail (3) and lower onto support stands.
8. Remove lifting device, two chains, and guide lines from wrecker body (1).

b. Installation

1. Attach two guide lines and utility chains to wrecker body (1) and lifting device to utility chains.
2. Raise wrecker body (1) from support stands and position on frame rail (3). Ensure mounting holes of frame rail (3) and wrecker body (1) are aligned.
3. Install two wood blocks on frame rail (3) and wrecker body (1) with two U-bolts (10) and four locknuts (11).
4. Install two U-bolts (8) on frame rail (3) and wrecker body (1) with four locknuts (9).
5. Install four wood blocks on frame rail (3) and wrecker body (1) with two U-bolts (6) and four locknuts (7).
6. Install two plates (4) and four wood blocks on frame rail (3) and wrecker body (1) with four mounting screws (5) and locknuts (2).
7. Install two brackets (14) on frame rail (3) with sixteen screws (13) and ten locknuts (12).
10. Remove lifting device, two chains, and guide lines from wrecker body (1).

4-152. WRECKER BODY REPLACEMENT (Contd)

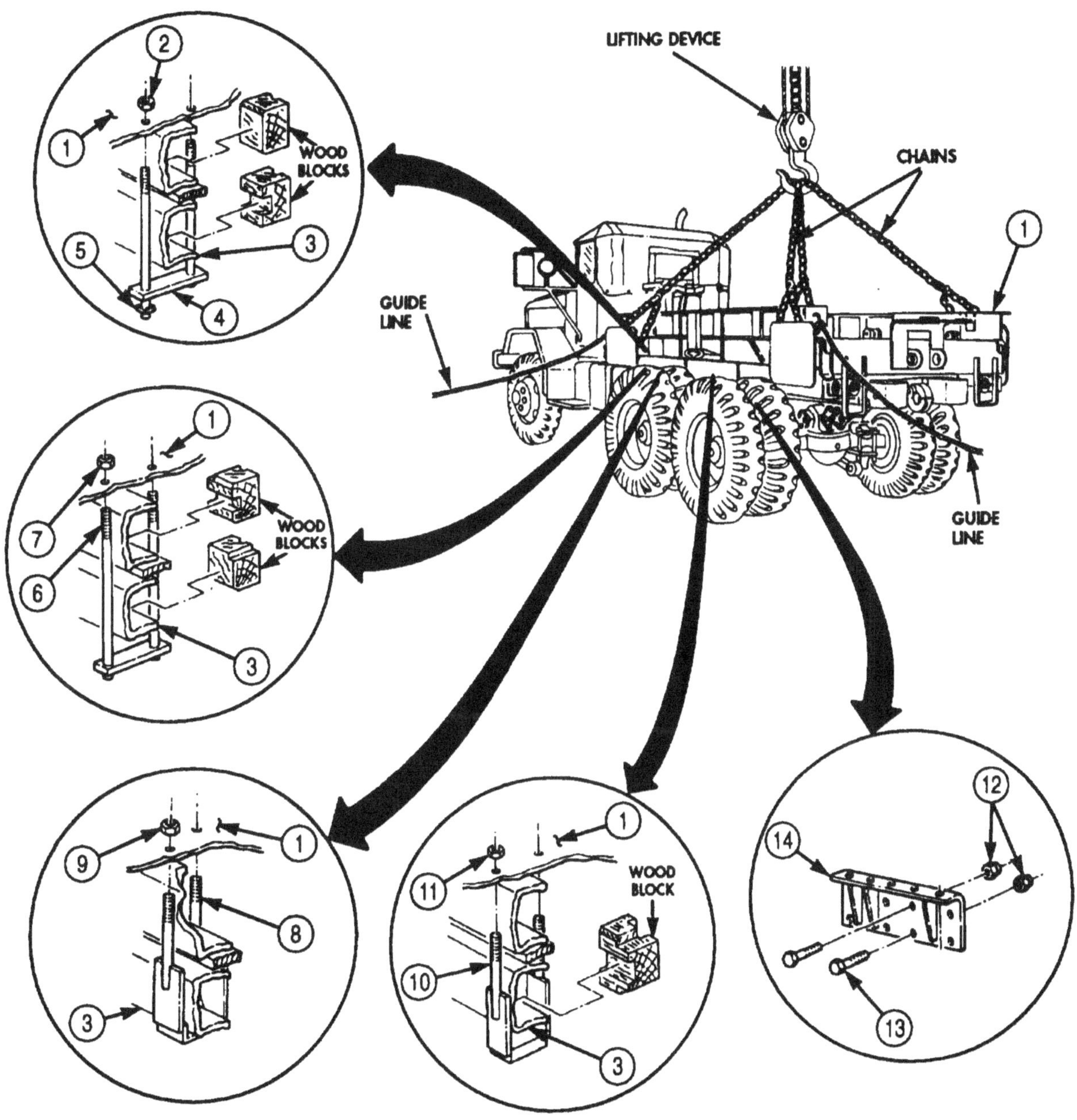

FOLLOW-ON TASK: Install crane turntable (para. 4-203).

4-153. REAR WINCH ROLLER REPAIR

THIS TASK COVERS:

a. Side Roller Disassembly
b. Top and Bottom Roller Disassembly
c. Cleaning and Inspection
d. Top and Bottom Roller Assembly
e. Side Roller Assembly

INITIAL SETUP:

APPLICABLE MODELS
All

TOOLS
General mechanic's tool kit (Appendix E, Item 1)

MATERIALS/PARTS
Four lockwashers (Appendix D, Item 354)
Two lockwashers (Appendix D, Item 377)
Four felt washers (Appendix D, Item 112)
GAA grease (Appendix C, Item 28)
Lubricating oil OE/HDO 30 (Appendix C, Item 50)
Drycleaning solvent (Appendix C, Item 71)

PERSONNEL REQUIRED
Two

MANUAL REFERENCES (TM)
LO 9-2320-272-12
TM 9-2320-272-10
TM 9-2320-272-24P

EQUIPMENT CONDITION
Parking brake set (TM 9-2320-272-10).

GENERAL SAFETY INSTRUCTIONS
- Drycleaning solvent is flammable and toxic. Do not use near an open flame.
- Keep fire extinguisher nearby when using drycleaning solvent.

---

a. Side Roller Disassembly

NOTE

Repair procedures for both side rollers are the same.
This procedure covers the left side roller.

1. Remove two screws (6), lockwashers (7), and lockplate (8) from bracket (3). Discard lockwashers (7).

NOTE

Assistant will help with steps 2 and 3.

2. Remove shaft (5) from bracket (3).
3. Remove roller (12), two thrust washers (9), and felt washers (10) from bracket (3). Discard felt washers (10).
4. Remove two bearings (11) from roller (12).
5. Remove two screws (1), lockwashers (14), plate (2), and spacer (13) from bracket (3). Discard lockwashers (14).
6. Remove two grease fittings (4) from shaft (5).

b. Top and Bottom Roller Disassembly

NOTE

Perform left side roller removal if bottom roller is to be repaired.

1. Remove two grease fittings (21) from shaft (22).
2. Remove two screws (24), lockwashers (25), and lockplate (23) from bracket (26). Discard lockwashers (25).

NOTE

Thrust washers may fall out.

3. Remove shaft (22) from roller (20).

4-153. REAR WINCH ROLLER REPAIR (Contd)

4. Remove two grease fittings (16) from tensioner sheave shaft (15).

NOTE

Assistant will help with step 5.

5. Remove roller (20) and two thrust washers (17) from bracket (26)
6. Remove two felt washers (18) and bearings (19) from roller (20). Discard felt washers (18).

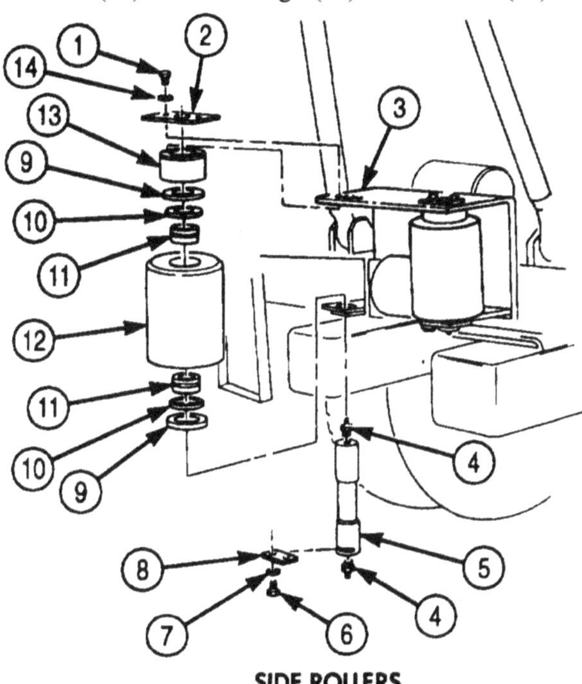

SIDE ROLLERS

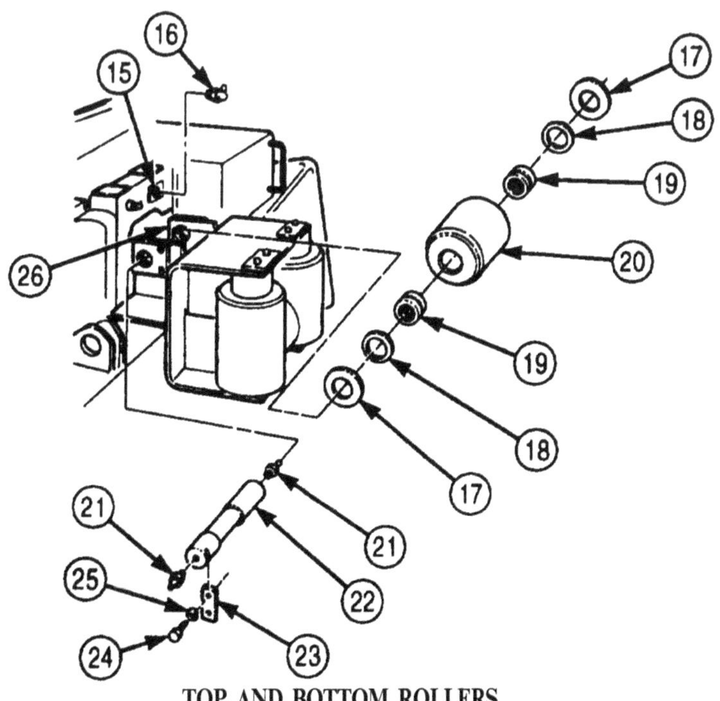

TOP AND BOTTOM ROLLERS

4-153. REAR WINCH ROLLER REPAIR (Contd)

c. Cleaning and Inspection

WARNING

Drycleaning solvent is flammable and toxic. Do not use near open flame and always have a fire extinguisher nearby when solvents are used. Use only in well-ventilated places, wear protective clothing, and dispose of cleaning rags in approved container. Failure to do this may result in injury or death to personnel and damage to equipment.

1. Clean roller (6) and shafts (8) and (17) assemblies with drycleaning solvent and allow to air-dry.
2. Inspect shafts (8) and (17) for cracks. Replace shafts (8) and (17), if cracked.
3. Measure bearing surfaces of shafts (8) and (17). If outer diameter of bearing surfaces of shafts (8) and (17) is less than 2.245 in. (57.02 mm), replace.
4. Inspect two thrust washers (3) and (21) for cracks. Replace thrust washers (3) and (21) if cracked.
5. Measure thickness of thrust washers (3) and (21). If thickness of thrust washers (3) and (21) is less than 0.055 in. (1.40 mm), replace.
6. Inspect two bearings (5) and (23) for damage (TM 9-214). If bearings (5) or (23) are damaged, replace.

d. Top and Bottom Roller Assembly

1. Pack two bearings (5) with grease and install in roller (6).
2. Soak two new felt washers (4) in lubricating oil prior to installation. Position two felt washers (4) and thrust washers (3) on each end of roller (6).

NOTE

Assistant will help with steps 3 and 4.

3. Slide roller (6), bearing (5), felt washers (4), and thrust washer (3) between brackets (12).
4. Install shaft (8) through brackets (12) and roller (6).
5. Align lockplate (9) with shaft (8) and install with two new lockwashers (11) and screws (10).
6. Install two grease fittings (7) on shaft (8).
7. Install two grease fittings (2) on tensioner sheave shaft (1).

NOTE

Install left side roller if removed (task e.).

e. Side Roller Assembly

1. Install two grease fittings (16) on shaft (17).
2. Install spacer (25) on bracket (15) with plate (14), two new lockwashers (26), and screws (13).
3. Pack two bearings (23) with grease and install in roller (24).
4. Soak two new felt washers (22) in lubricating oil prior to installation. Position two felt washers (22) and thrust washers (21) on each end of roller (24).

NOTE

Assistant will help with steps 5 and 6.

5. Slide roller (24), bearing (23), felt washers (22), and thrust washers (21) into bracket (15).
6. Install shaft (17) through bracket (15) and roller (24). Make sure locking slot is at bottom.
7. Align lockplate (20) with shaft (17) and install with two new lockwashers (19) and screws (18).

4-153. REAR WINCH ROLLER REPAIR (Contd)

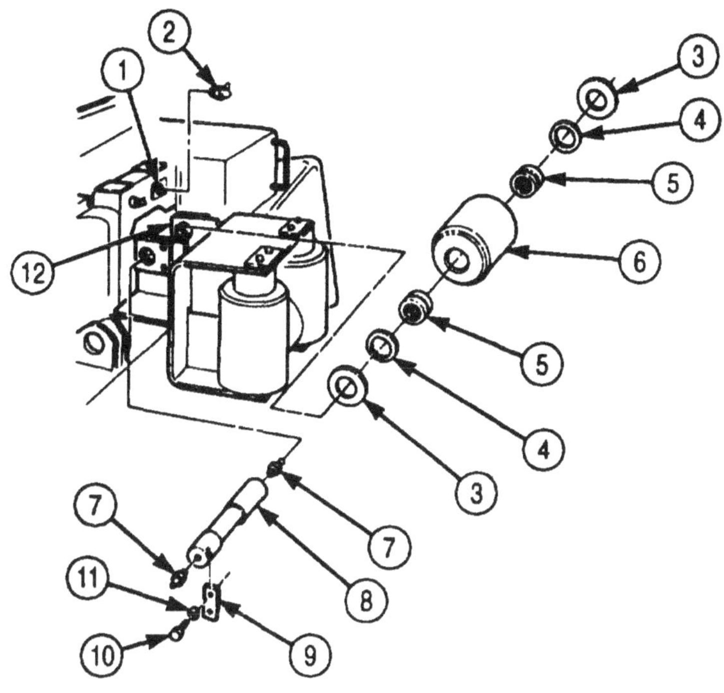

TOP AND BOTTOM ROLLERS

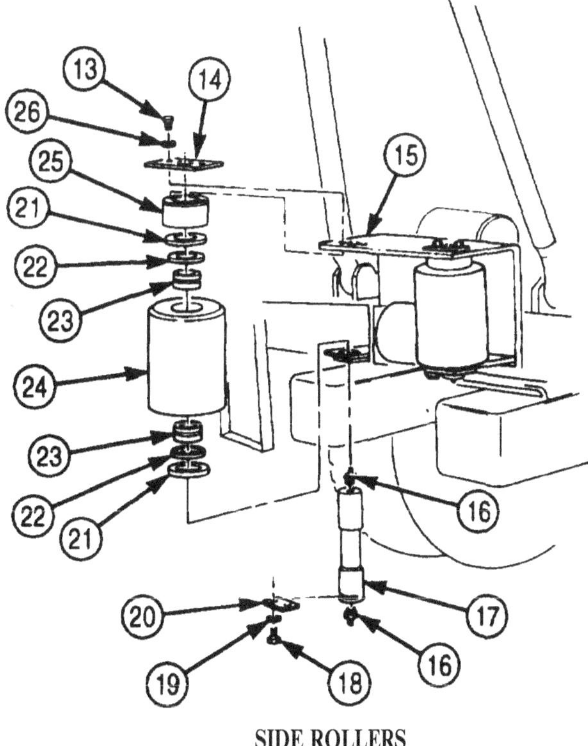

SIDE ROLLERS

FOLLOW-ON TASK: Lubricate roller assembly (LO 9-2320-272-12).

4-154. REAR DOOR MAINTENANCE

THIS TASK COVERS:

a. **Removal**
b. **Disassembly**
c. **Cleaning and Inspection**
d. **Assembly**
e. **Installation**

INITIAL SETUP:

APPLICABLE MODELS
M934/A1/A2

TOOLS
General mechanic's tool kit (Appendix E, Item 1)

MATERIALS/PARTS
Three cotter pins (Appendix D, Item 51)
Four locknuts (Appendix D, Item 272)
Six lockwashers (Appendix D, Item 400)
Gasket (Appendix D, Item 229)
Seven rivets (Appendix D, Item 542)
Weatherseal (Appendix D, Item 718)
Seal (Appendix D, Item 605)
Three rivets (Appendix D, Item 545)
Sealing compound (Appendix C, Item 15)
Adhesive (Appendix C, Item 7)

PERSONNEL REQUIRED
Two

MANUAL REFERENCES (TM)
LO 9-2320-272-12
TM 9-2320-272-10
TM 9-2320-272-24P
TM 43-0213

EQUIPMENT CONDITION
- Parking brake set (TM 9-2320-272-10).
- Ladder-removed (TM 9-2320-272-10).
- Rear door window removed (para. 3-350).

SPECIAL ENVIRONMENTAL CONDITIONS
Vehicle must be on a level surface.

NOTE

Left and right doors are maintained basically the same way. This procedure covers the left door except where otherwise indicated.

a. Removal

1. Remove two screws (5), lockwashers (6), strike (7), and door check (1) from door (4). Discard lockwashers (6).
2. Remove fifteen screws (3) and door (4) from van body (2).

b. Disassembly

1. Remove two screws (13) and angle bracket (12) from outer panel (11).
2. Remove four screws (14), lockwashers (15), and clamp (16) from outer panel (11). Discard lockwashers (15).
3. Remove four screws (18), two screws (8), washers (9), bushings (10), locknuts (17), and rack (19) from outer panel (11). Discard locknuts (17).

4-154. REAR DOOR MAINTENANCE (Contd)

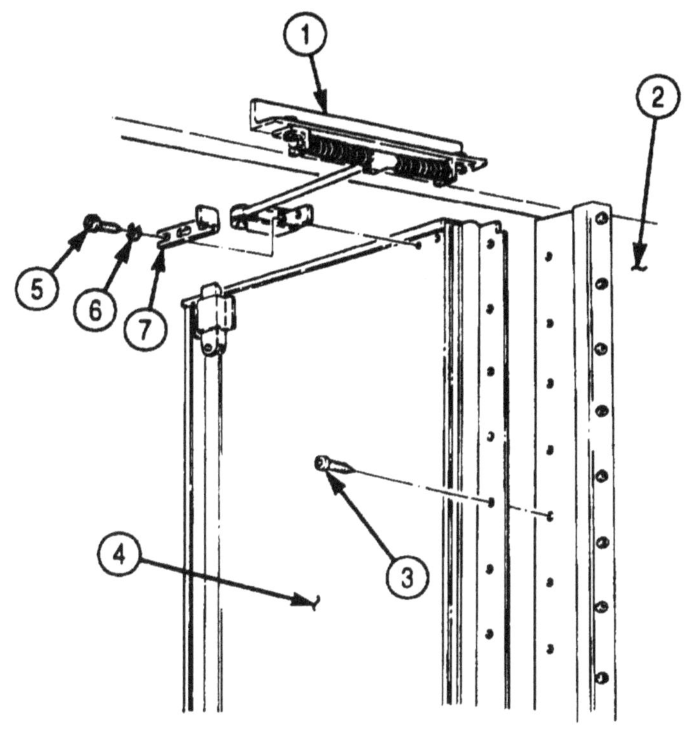

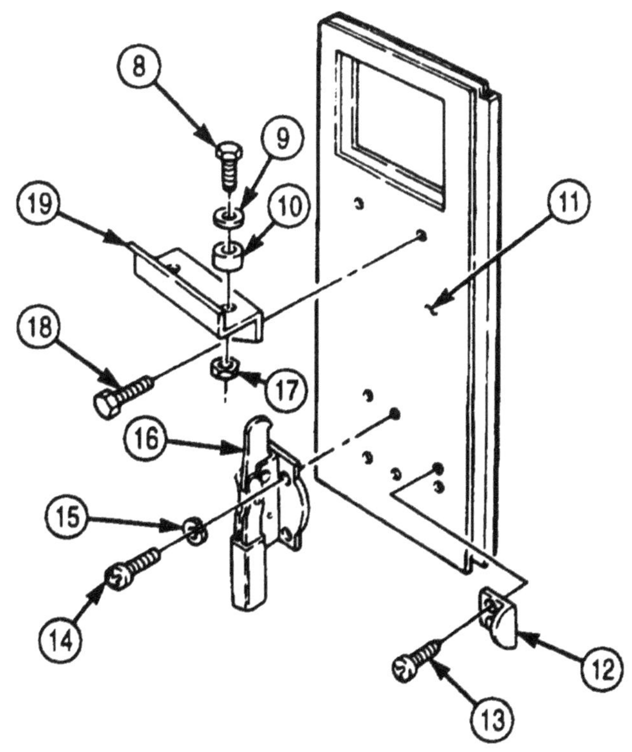

4-154. REAR DOOR MAINTENANCE (Contd)

4. Remove seven rivets (23) and seal (22) from door frame (32). Discard rivets (23).

NOTE

Step 5 applies to left rear door only.

5. Remove four screws (16), clip (17), six screws (19), and bracket (18) from inner panel (31).

NOTE

Step 6 applies to right rear door.

6. Remove three rivets (35), handle (34), and gasket (33) from outer panel (25). Discard rivets (35).
7. Remove two cotter pins (11), screws (7), lockwashers (8), and rod (9) from center case (15) and two bolt and case assemblies (5). Discard lockwashers (8) and cotter pins (11).
8. Remove pins (13) and (14) and handle (12) from center case (15).
9. Remove four screws (10) and center case (15) from inner panel (31).
10. Remove eight screws (6), two bolt and case assemblies (5), and spacer plates (4) from inner panel (3).
11. Remove sixteen screws (30), retainer (29), seal (28), hinge (27), and seal (26) from door frame (32). Discard seals (28) and (26).
12. Break adhesive seal and remove moldings (1), (2), and (3) from inner panel (31).

NOTE

Steps 13 and 14 apply to left rear door only.

13. Break adhesive seal and remove weatherseal (24) from door frame (32). Discard weatherseal (24).
14. Remove thirty-five screws (20) and molding (21) from door frame (32).

NOTE

Step 15 applies to right rear door only.

15. Remove thirty-four screws (20) and molding (21) from door frame (32).

c. Cleaning and Inspection

1. For general cleaning instructions, refer to para. 2-14.
2. Inspect all center case (15) movable parts for proper operation. Replace any parts if damaged.
3. Inspect rods (9) for breaks and bends. Replace rods (9) if damaged.
4. Inspect moldings (1), (2), (3), and (21) for cracks and bends. Replace moldings (1), (2), (3), or (21) if damaged.

d. Assembly

1. Rustproof all inside surfaces and boxed-in areas (TB 43-0213).

NOTE

- Apply sealing compound between exterior joints.
- Apply adhesive to rubber and metal surfaces for installation.
- Step 2 applies to right rear door only.

2. Install molding (21) on door frame (32) with thirty-four screws (20).

NOTE

Steps 3 and 4 apply to left rear door only.

3. Install molding (21) on door frame (32) with thirty-five screws (20).
4. Install weatherseal (24) on door frame (32) with adhesive.
5. Install moldings (1), (2), and (3) on inner panel (31) with adhesive.

4-154. REAR DOOR MAINTENANCE (Contd)

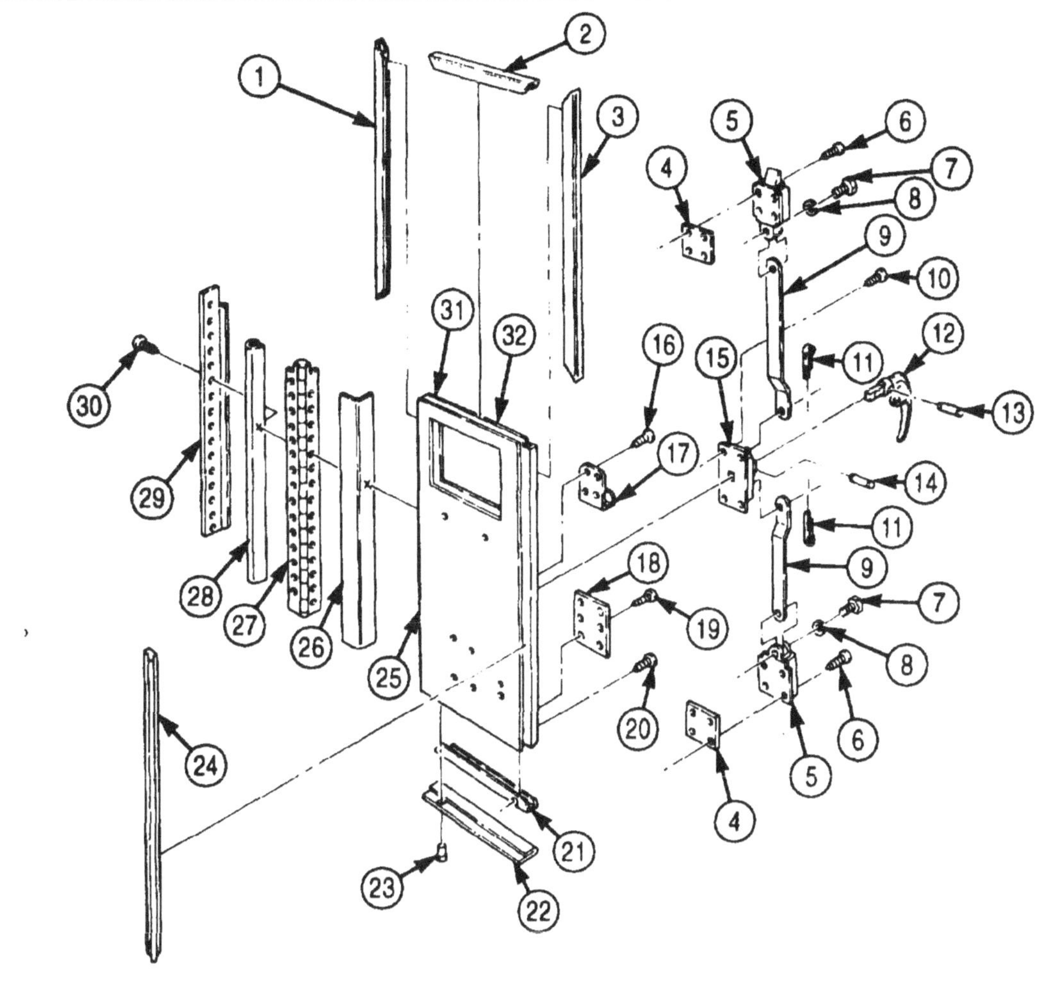

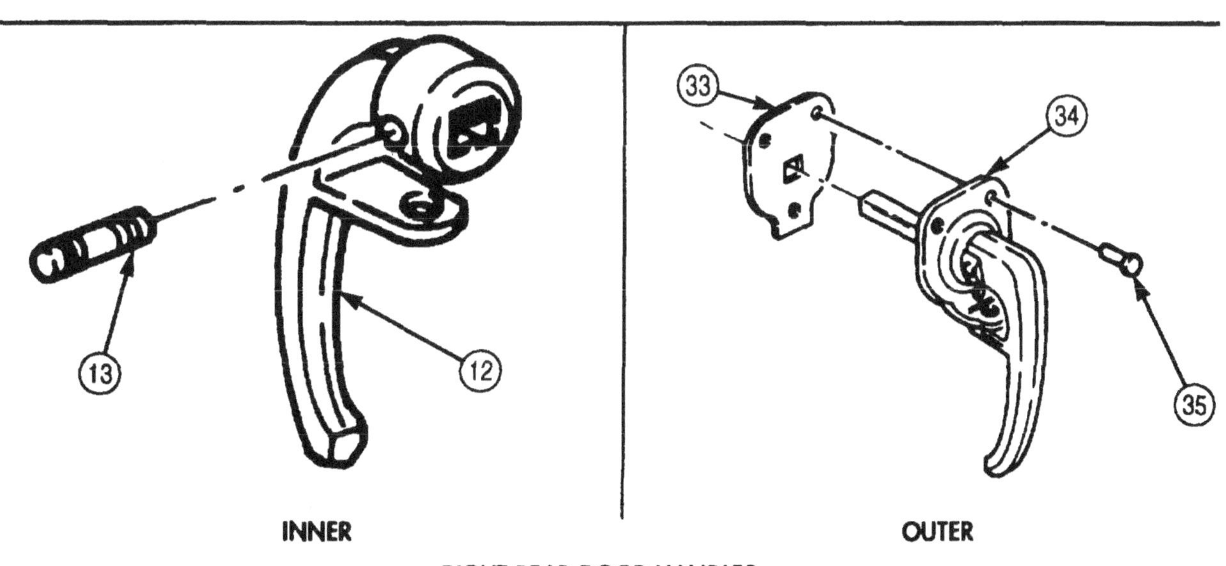

RIGHT REAR DOOR HANDLES

4-154. REAR DOOR MAINTENANCE (Contd)

6. Install new seal (10), hinge (9), new seal (8), and retainer (7) on door frame (12) with sixteen screws (6).
7. Install two spacer plates (16) and bolt and case assemblies (17) on inner panel (11) with eight screws (18).
8. Install center case (14) on inner panel (11) with four screws (21).
9. Install handle (2) on center case (14) with pins (1) and (23).
10. Install two rods (15) on center case (14) and two bolt and case assemblies (17) with two new cotter pins (22), new lockwashers (20), and screws (19).

NOTE

Step 11 applies to right rear door only.

11. Install new gasket (3) and handle (4) on outer panel (29) with three new rivets (5).
12. Install clip (24) and bracket (26) on outer panel (29) with four screws (13) and six screws (25).
13. Install seal (27) on door frame (12) with seven new rivets (28).
14. Install rack (40) on outer panel (29) with four screws (39), bushings (32), washers (31), two screws (30), and new locknuts (38).
15. Install clamp (37) on outer panel (29) with four screws (35) and new lockwashers (36).
16. Install two angle brackets (33) on outer panel (29) with two screws (34).

e. Installation

NOTE

- Align holes of door hinge with holes in van body for installation.
- Assistant will help with step 1.

1. Install door (29) on van body (43) with fifteen screws (44).
2. Install door (291 on door check (42) with strike (46), lockwasher (45), and screw (41).

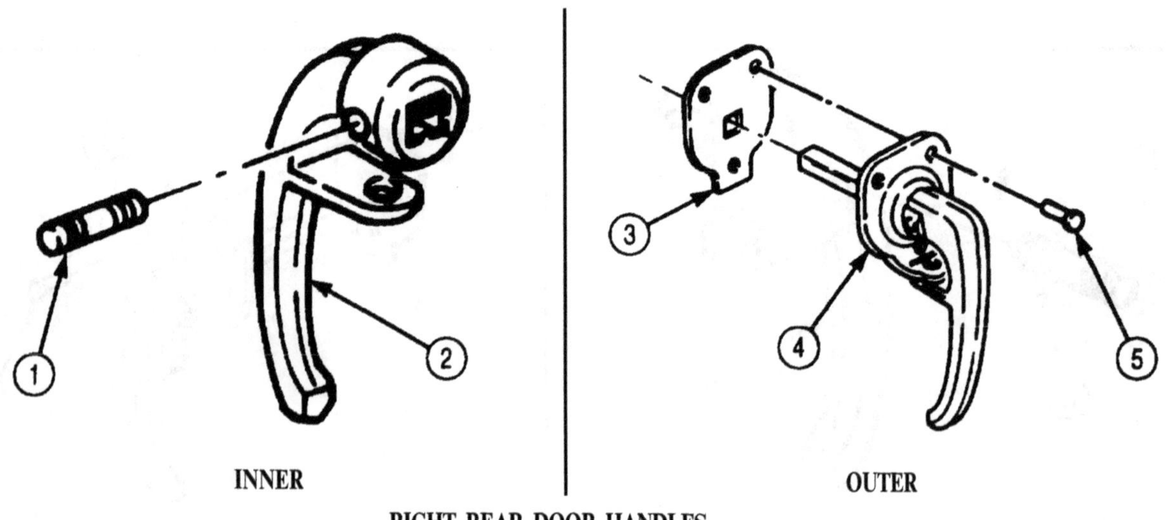

RIGHT REAR DOOR HANDLES

4-154. REAR DOOR MAINTENANCE (Contd)

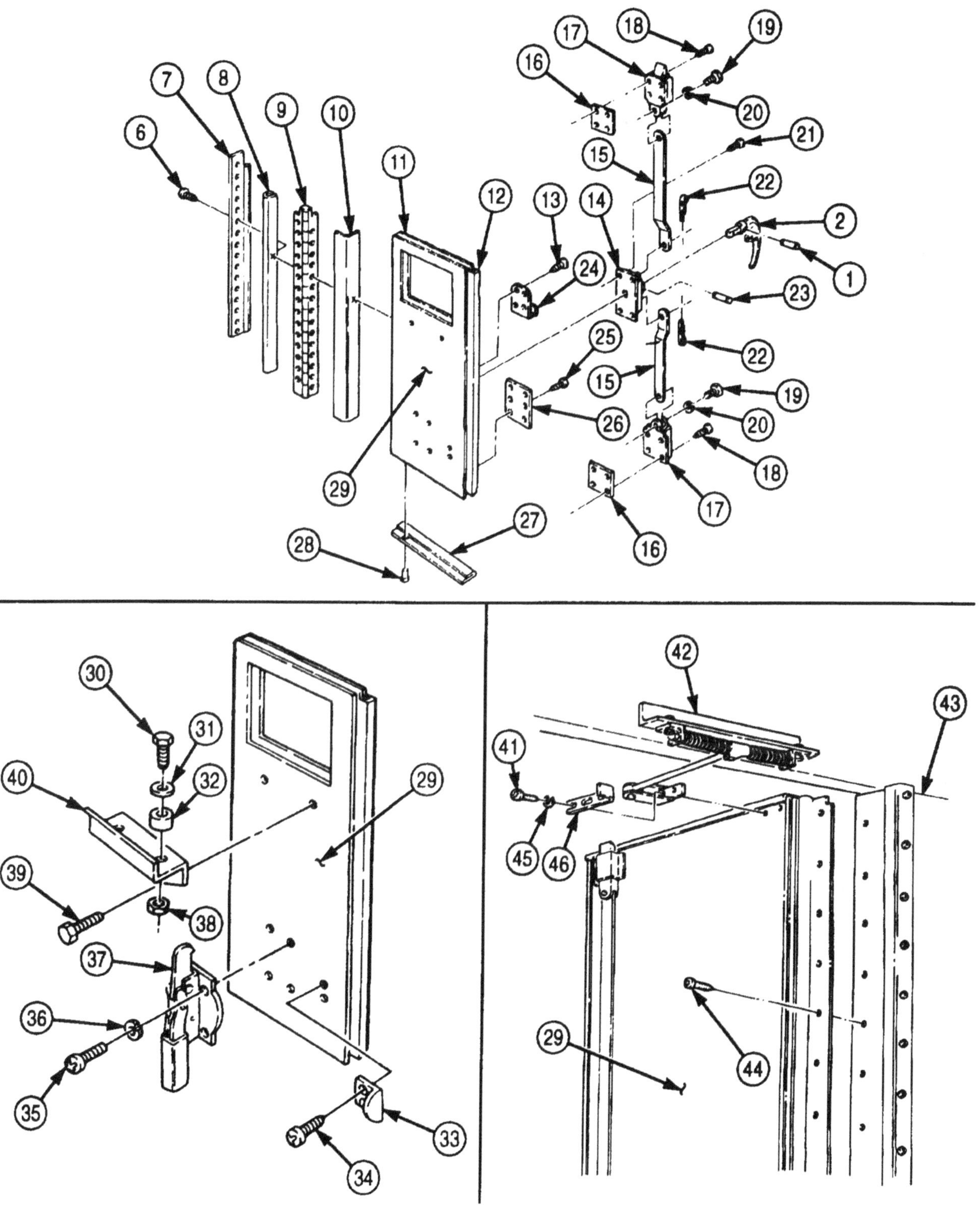

FOLLOW-ON TASKS:
- Install rear door window (para. 3-350).
- Install ladder (TM 9-2320-272-10).

4-155. HINGED END PANEL MAINTENANCE

THIS TASK COVERS:

a. **Removal**
b. **Disassembly**
c. **Cleaning and Inspection**
d. **Assembly**
e. **Installation**

INITIAL SETUP:

APPLICABLE MODELS
M934/A1/A2

TOOLS
General mechanic's tool kit (Appendix E, Item 1)

MATERIALS/PARTS
Seal (35 in.) (Appendix D, Item 632)
Seal (35 in.) (Appendix D, Item 633)
Seal (35 in.) (Appendix D, Item 606)
Twelve rivets (Appendix D, Item 546)
Eighteen rivets (Appendix D, Item 547)
Caulking compound (Appendix C, Item 15)

PERSONNEL REQUIRED
Two

MANUAL REFERENCES (TM)
TM 9-2320-272-10
TM 9-2320-272-24P

EQUIPMENT CONDITION
- Parking brake set (TM 9-2320-272-10).
- Van body sides fully extended and secured (TM 9-2320-272-10).

NOTE

All hinged end panels are removed the same way This procedure covers the left rear hinged end panel.

a. Removal

1. Remove sixteen screws (18) from end panel hinge (21) and side panel (20).

NOTE

Assistant will help with step 2.

2. Remove panel frame (7) as an assembly from side panel (20).

b. Disassembly

1. Remove sixteen screws (19), hinge (21), and seal (22) from panel frame (7). Discard seal (22).
2. Remove twenty-seven screws (8), retainer (10), and seal (9) from panel frame (7). Discard seal (9).
3. Remove ten screws (1), retainers (2) and (3), and seal (4) from channel (5). Discard seal (4).
4. Remove eighteen rivets (6), twenty screws (17), and channels (5) and (16) from outer skin (12) and inner skin (24). Discard rivets (6).
5. Remove two screws (15), strap (14), and spacer plate (13) from outer skin (12).

NOTE

- Perform steps 6 and 7 only if skins are to be replaced. See task c.
- Assistant will help with steps 6 and 7.

6. Remove twelve rivets (11) and outer skin (12) from panel frame (7). Discard rivets (11).
7. Remove thirty screws (25), eighteen screws (23), and inner skin (24) from panel frame (7).

4-155. HINGED END PANEL MAINTENANCE (Contd)

c. Cleaning and Inspection

1. Clean all parts (para 2-14).
2. Inspect channels (5) and (16) for cracks and bends. Replace channels (5) or (16) if cracked or broken.
3. Inspect retainers (2), (3), and (10), spacer plate (13), and strap (14) for bends and brakes. Replace retainers (2), (3), or (10), spacer plate (13), or strap (14) if bent or broken.
4. Inspect hinge (21) for cracks, breaks, corrosion, and proper operation.
5. Inspect inner skin (24) and outer skin (12) for tears and punctures. Replace inner skin (24) or outer skin (12) if tom and punctured.
6. Inspect panel frame (7) for cracks and bends. Replace panel frame (7) if cracked or broken.

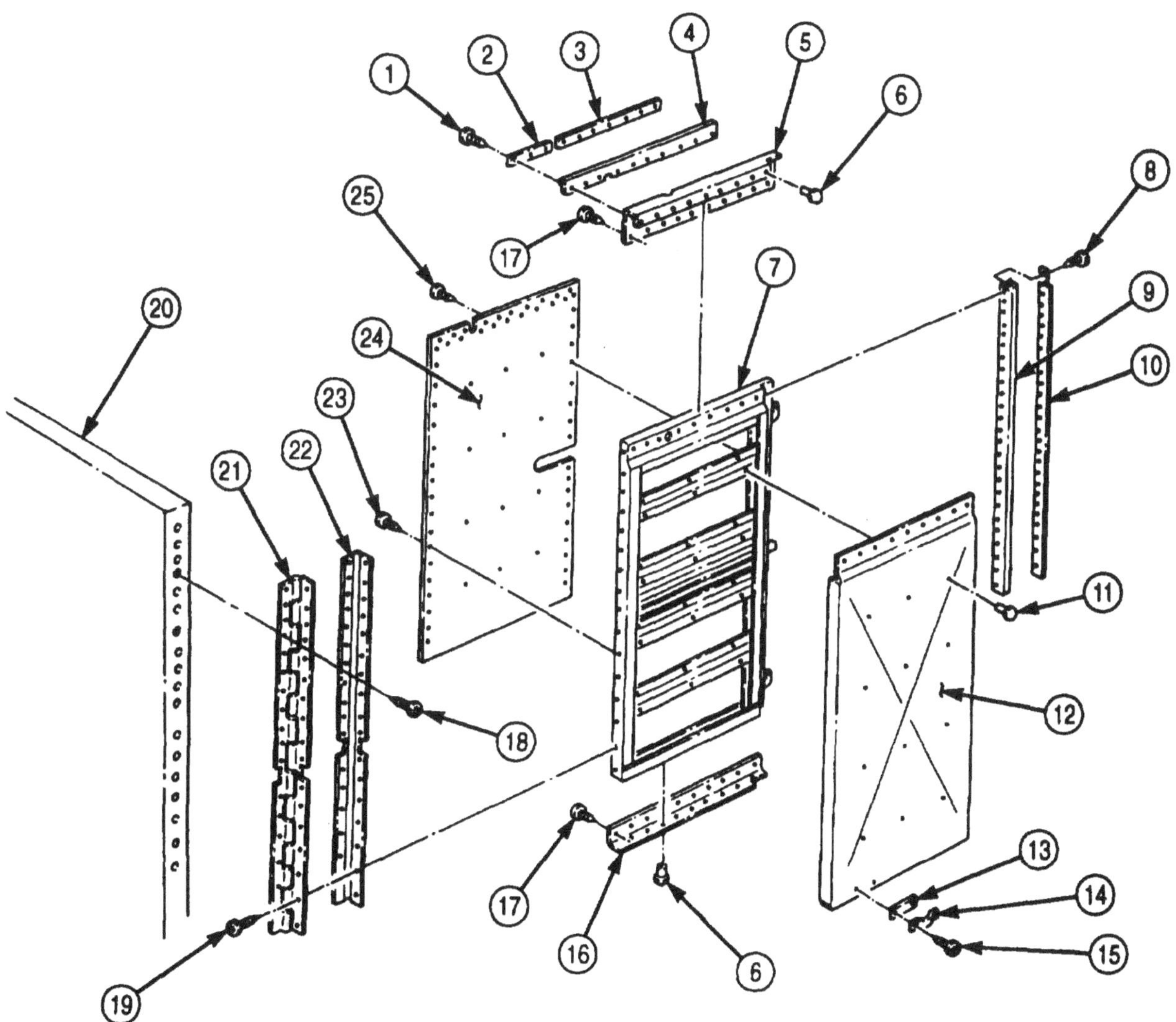

4-155. HINGED END PANEL MAINTENANCE (Contd)

d. Assembly

NOTE

- Seal all exterior joints with sealing compound.
- Perform steps 1 and 2 only if skins are to be replaced.
- Assistant will help with steps 1 and 2.

1. Install inner skin (24) on panel frame (7) with thirty screws (25) and eighteen screws (23).
2. Install outer skin (12) on panel frame (7) with twelve rivets (11).
3. Install spacer plate (13) and strap (14) on outer skin (12) with two screws (15).
4. Install channels (5) and (16) on outer skin (12) and inner skin (24) with eighteen new rivets (6) and twenty screws (17).
5. Install new seal (4) and retainers (2) and (3) on channel (5) with ten screws (1).
6. Install new seal (9) and retainer (10) on outer skin (12) with twenty-seven screws (8).
7. Install new seal (22) and hinge (21) on outer skin (12) with sixteen screws (19).

e. Installation

NOTE

Assistant will help you with step 1.

1. Position panel frame (7) as an assembly on side panel (20).
2. Install end panel hinge (21) on side panel (20) with sixteen screws (18).

4-155. HINGED END PANEL MAINTENANCE (Contd)

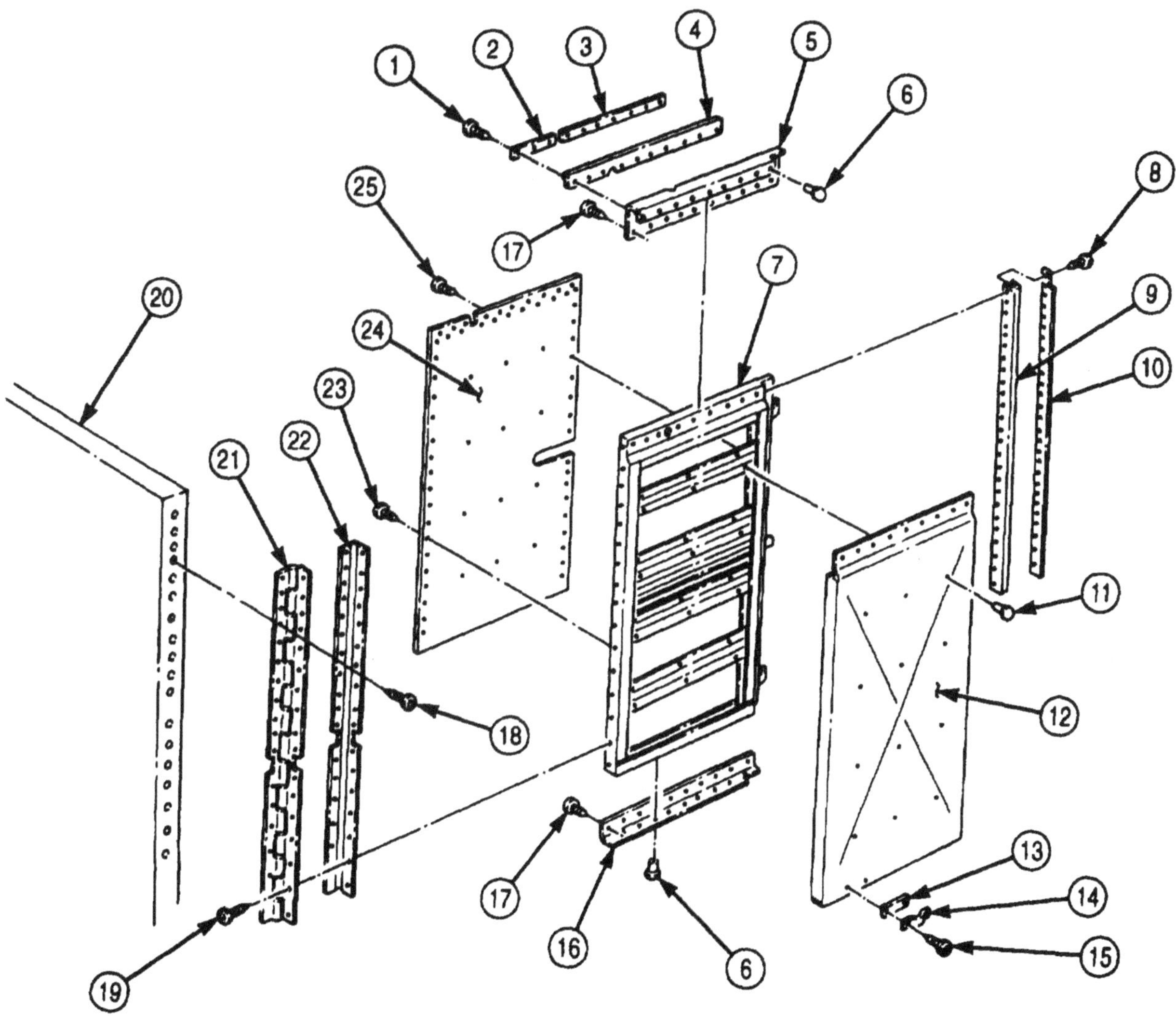

FOLLOW-ON TASK: Retract van body sides (TM 9-2320-272-10).

4-156. HEATER FUEL PUMP WIRING HARNESS MAINTENANCE

THIS TASK COVERS:

a. Removal
b. Repair
c. Installation

INITIAL SETUP:

APPLICABLE MODELS
M934/A1/A2

TOOLS
General mechanic's tool kit (Appendix E, Item 1)

MANUAL REFERENCES (TM)
TM 9-2320-272-10
TM 9-2320-272-24P

EQUIPMENT CONDITION
- Parking brake set (TM 9-2320-272-10).
- Battery ground cables disconnected (para. 3-126).

a. Removal

NOTE

Perform step 1 on M934 vehicles only.

Tag wires for installation.

1. Remove screw (1) and clamp (5) from fuel pump wiring harness (3).
2. Disconnect wiring harness (3) from fuel pump connector plug (2).
3. Disconnect wiring harness (3) from heater (6).
4. Remove wiring harness (3) from van body (4).

b. Repair

For van wiring harness repair, refer to para. 3-131.

c. Installation

1. Position wiring harness (3) on van body (4).
2. Connect wiring harness (3) to heater (6).
3. Connect wiring harness (3) to fuel pump connector plug (2).

NOTE

Perform step 4 on M934 vehicles only.

4. Install clamp (5) on wiring harness (3) with screw (1).

4-156. HEATER FUEL PUMP WIRING HARNESS MAINTENANCE (Contd)

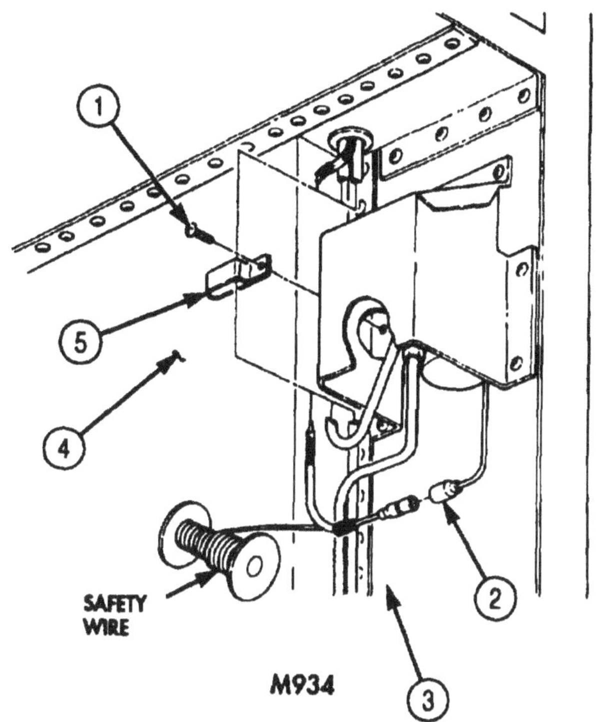

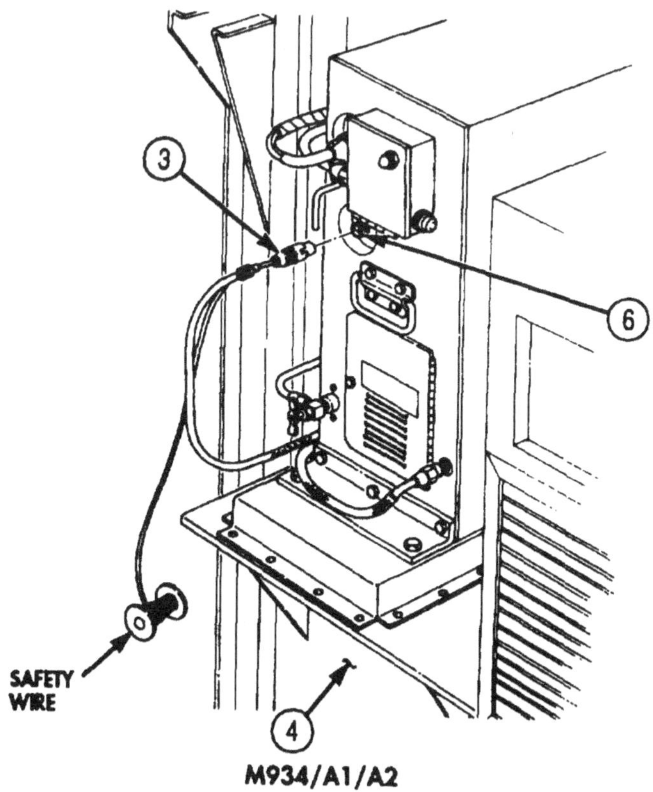

FOLLOW-ON TASK: Connect battery ground cables (para. 3-126).

4-157. RIGHT AND LEFT SIDE BLACKOUT HARNESS MAINTENANCE

THIS TASK COVERS:

a. Removal
b. Repair
c. Installation

INITIAL SETUP:

APPLICABLE MODELS
M934/A1/A2

TOOLS
General mechanic's tool kit (Appendix E, Item 1)

MANUAL REFERENCES (TM)
TM 9-2320-272-10
TM 9-2320-272-24P

EQUIPMENT CONDITION
- Parking brake set (TM 9-2320-272-10).
- Van body sides fully extended and secured (TM 9-2320-272-10).
- Hinged roof-operated blackout circuit plungers removed (para. 3-377).

a. Removal

NOTE

Tag wires for installation.

1. Remove six screws (21) and cover (20) from load center (15).
2. Remove screw (14) and wire (13) from neutral buss bar (16).
3. Remove screw (10) and wire (11) from relay (12).
4. Remove screws (5) and (9) and wires (7) and (8) from rear door blackout switch (6).

NOTE

Perform steps 5 through 8 for left and right side of van body.

5. Remove six screws (24) and cover (22) from van body (26).
6. Remove grommet (4) from van body (26).
7. Remove nut (19), two screws (17), and connector halves (18) from harness (23) and cover (22).
8. Remove nut (3), two screws (25), and connector halves (1) from harness (23) and ceiling truss (2).
9. Pull harness (23) through hole in ceiling truss (2) and remove harness (23) from load center (15).

b. Repair

For van wiring harness repair, refer to para. 3-131.

c. Installation

1. Push harness (23) through hole in ceiling truss (2) and position harness (23) on load center (15).

NOTE

Perform steps 2 through 5 for left and right side of van body.

2. Install two connector halves (1) on harness (23) and ceiling truss (2) with two screws (25) and nut (3).
3. Install cover (22) on van body (26) with six screws (24).
4. Install two connector halves (18) on harness (23) and cover (22) with two screws (17) and nut (19).
5. Install grommet (4) on van body (26).
6. Install wires (7) and (8) on rear door blackout switch (6) with screws (5) and (9).
7. Install wire (11) on relay (12) with screw (10).
8. Install wire (13) on neutral buss bar (16) with screw (14).
9. Install cover (20) on load center (15) with six screws (21).

4-157. RIGHT AND LEFT SIDE BLACKOUT HARNESS MAINTENANCE

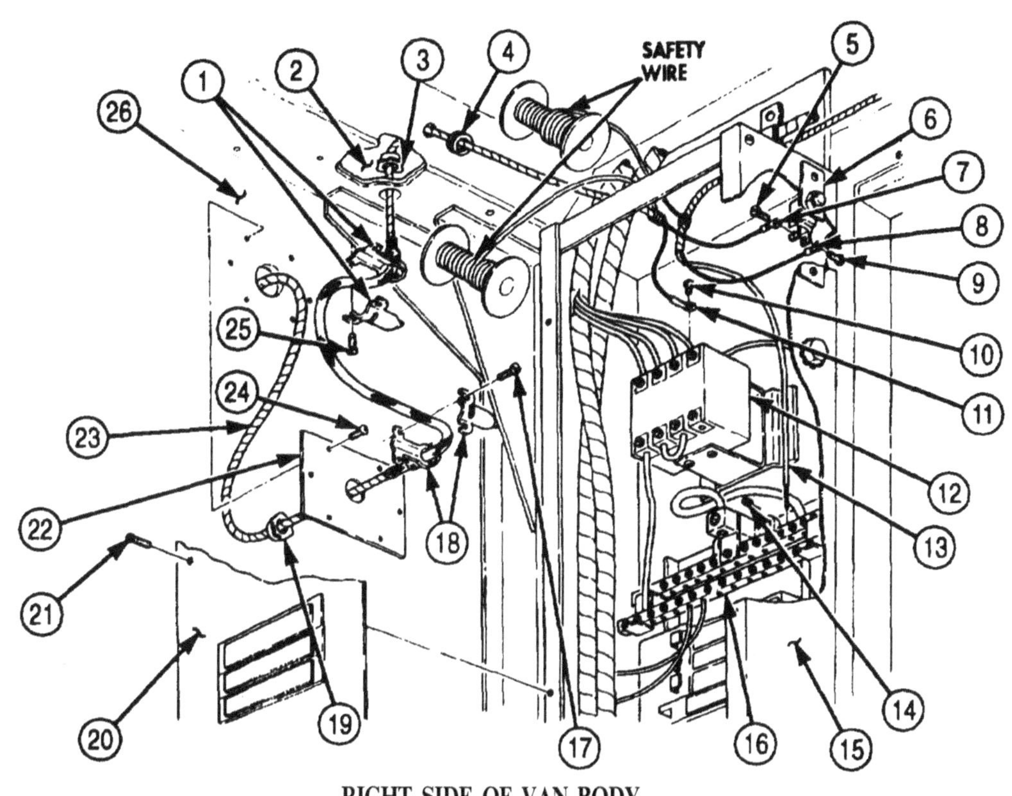

RIGHT SIDE OF VAN BODY

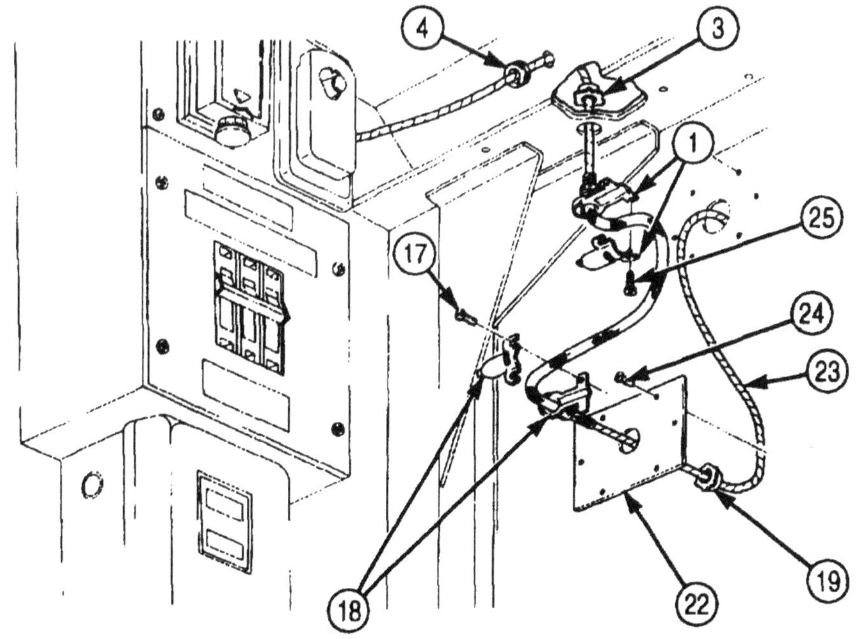

LEFT SIDE OF VAN BODY

FOLLOW-ON TASKS:
- Install hinged roof-operated blackout circuit plungers (para. 3-377).
- Retract van body sides (TM 9-2320-272-10).

4-158. CIRCUIT BREAKER HARNESS AND CIRCUIT BREAKER MAINTENANCE

THIS TASK COVERS:

a. Circuit Breaker Harness Removal
b. Circuit Breaker Removal
c. Repair
d. Circuit Breaker Installation
e. Circuit Breaker Harness Installation

INITIAL SETUP:

APPLICABLE MODELS
M934/A1/A2

TOOLS
General mechanic's tool kit (Appendix E, Item 1)

MATERIALS/PARTS
Two locknuts (Appendix D, Item 291)
Lockwasher (Appendix D, Item 381)

MANUAL REFERENCES (TM)
TM 9-2320-272-10
TM 9-2320-272-24P

EQUIPMENT CONDITION
- Parking brake set (TM 9-2320-272-10).
- Battery ground cables disconnected (para. 3-126).

a. Circuit Breaker Harness Removal

NOTE

Tag wires for installation.

1. Disconnect wire (2) from circuit breaker (1).
2. Remove two screws (4) and clamps (3) from wiring harness (15) and van body (16).
3. Remove locknut (10), screw (13), and clamp (14) from spare tire carrier mounting bracket (5). Discard locknut (10).
4. Remove locknut (6), screw (12), washer (11), wires (8) and (9), and lockwasher (7) from spare tire carrier mounting bracket (5). Discard locknut (6) and lockwasher (7).
5. Remove wiring harness (15) from van body (16).

b. Circuit Breaker Removal

1. Disconnect emergency lamp wiring harness (18) from circuit breaker (1).
2. Remove two screws (17) and circuit breaker (1) from van body (16).

c. Repair

For van wiring harness repair, refer to para. 3-131.

d. Circuit Breaker Installation

1. Install circuit breaker (1) on van body (16) with two screws (17).
2. Connect emergency lamp wiring harness (18) to circuit breaker (1).

e. Circuit Breaker Harness Installation

1. Install new lockwasher (7) and wires (8) and (9) on spare tire carrier mounting bracket (5) with washer (11), screw (12), new locknut (6).
2. Install wiring harness (15) on spare tire carrier mounting bracket (5) with screw (13), clamp (14), and new locknut (10).
3. Install two clamps (3) and wiring harness (15) and van body (16) with two screws (4).
4. Connect wire (2) to circuit breaker (1).

4-158. CIRCUIT BREAKER HARNESS AND CIRCUIT BREAKER MAINTENANCE

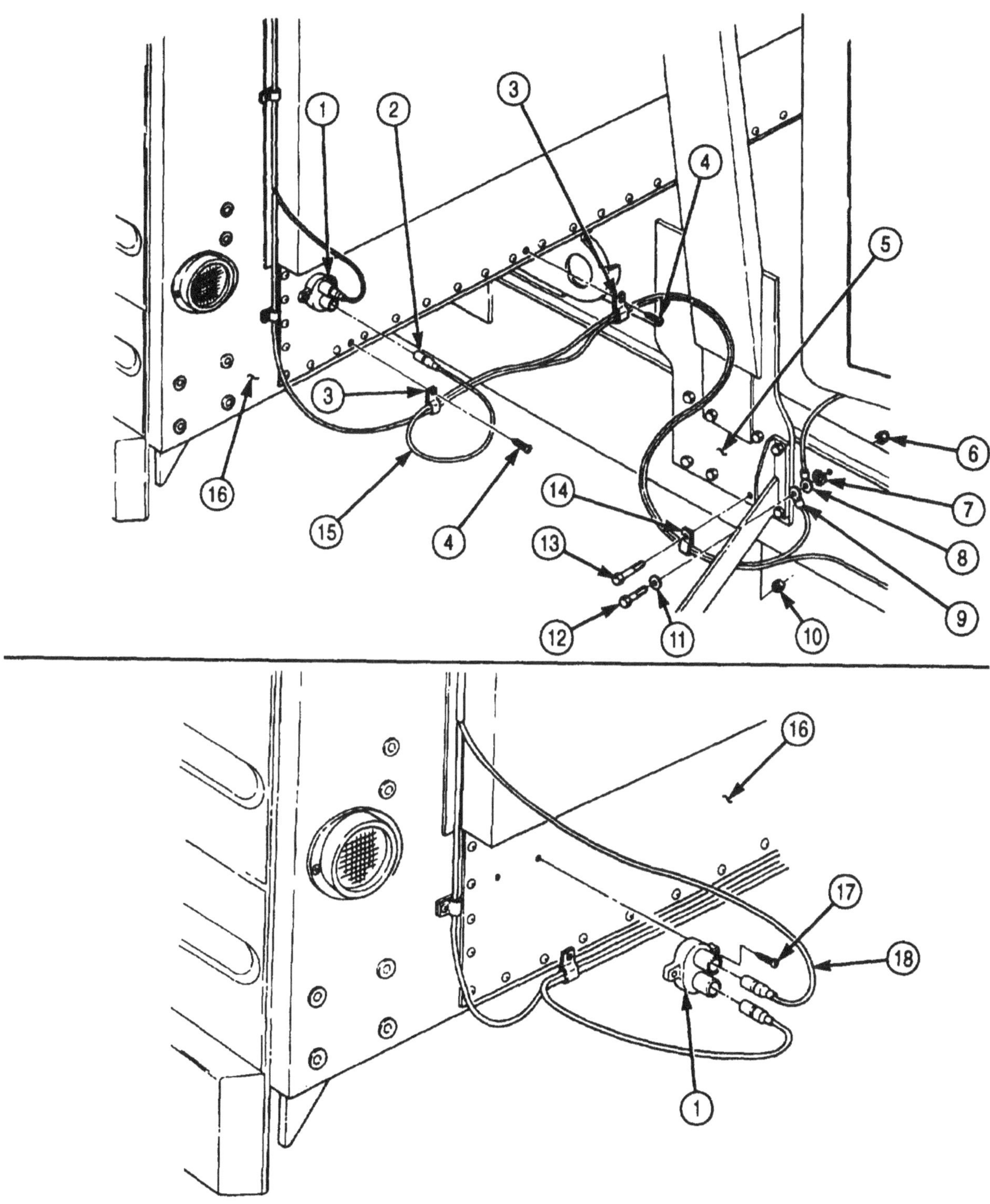

FOLLOW-ON TASK: Connect battery ground cables (para. 3-126).

4-159. ENTRANCE RECEPTACLE 220V 3-PHASE WIRING HARNESS MAINTENANCE

THIS TASK COVERS:

a. Removal
b. Repair
c. Installation

INITIAL SETUP:

APPLICABLE MODELS
M934/A1/A2

TOOLS
General mechanic's tool kit (Appendix E, Item 1)

MATERIALS/PARTS
Four locknuts (Appendix D, Item 276)
Safety wire (Appendix C, Item 80)

PERSONNEL REQUIRED
TWO

MANUAL REFERENCES (TM)
TM 9-2320-272-10
TM 9-2320-272-24P

EQUIPMENT CONDITION
- Parking brake set (TM 9-2320-272-10).
- External power source removed (TM 9-2320-272-10).
- Battery ground cables disconnected (para. 3-126).

a. Removal

NOTE

Tag wires for installation.

1. Remove six screws (14) and cover (1) from load center (5).
2. Remove three screws (16) and wires (17) from circuit breakers (6).
3. Remove setscrew (3) and wire (2) from neutral buss bar (4).
4. Remove three screws (8) and plate (7) from electrical box (13).
5. Remove four locknuts (15), screws (9), and receptacle (10) from van body (12). Discard locknuts (15).
6. Disconnect connector (11) from receptacle (10).

NOTE

Safety wire will be used to route wires through van body.

7. Remove wires (2) and (17) from electrical box (13), load center (5), and van body (12).

b. Repair

For van wiring harness repair, refer to para. 3-131.

c. Installation

NOTE

Safety wire will be used to route wires through van body.

1. Feed wires (2) and (17) through electrical box (13), load center (5), and van body (12).
2. Connect connector (11) to receptacle (10).
3. Install receptacle (10) on van body (12) with four screws (9) and new locknuts (15).
4. Install plate (7) on electrical box (13) with three screws (8).
5. Install wire (2) on neutral buss bar (4) with setscrew (3).
6. Install three wires (17) on circuit breakers (6) with three screws (16).
7. Install cover (1) on load center (5) with six screws (14).

4-159. ENTRANCE RECEPTACLE 220V 3-PHASE WIRING HARNESS MAINTENANCE (Contd)

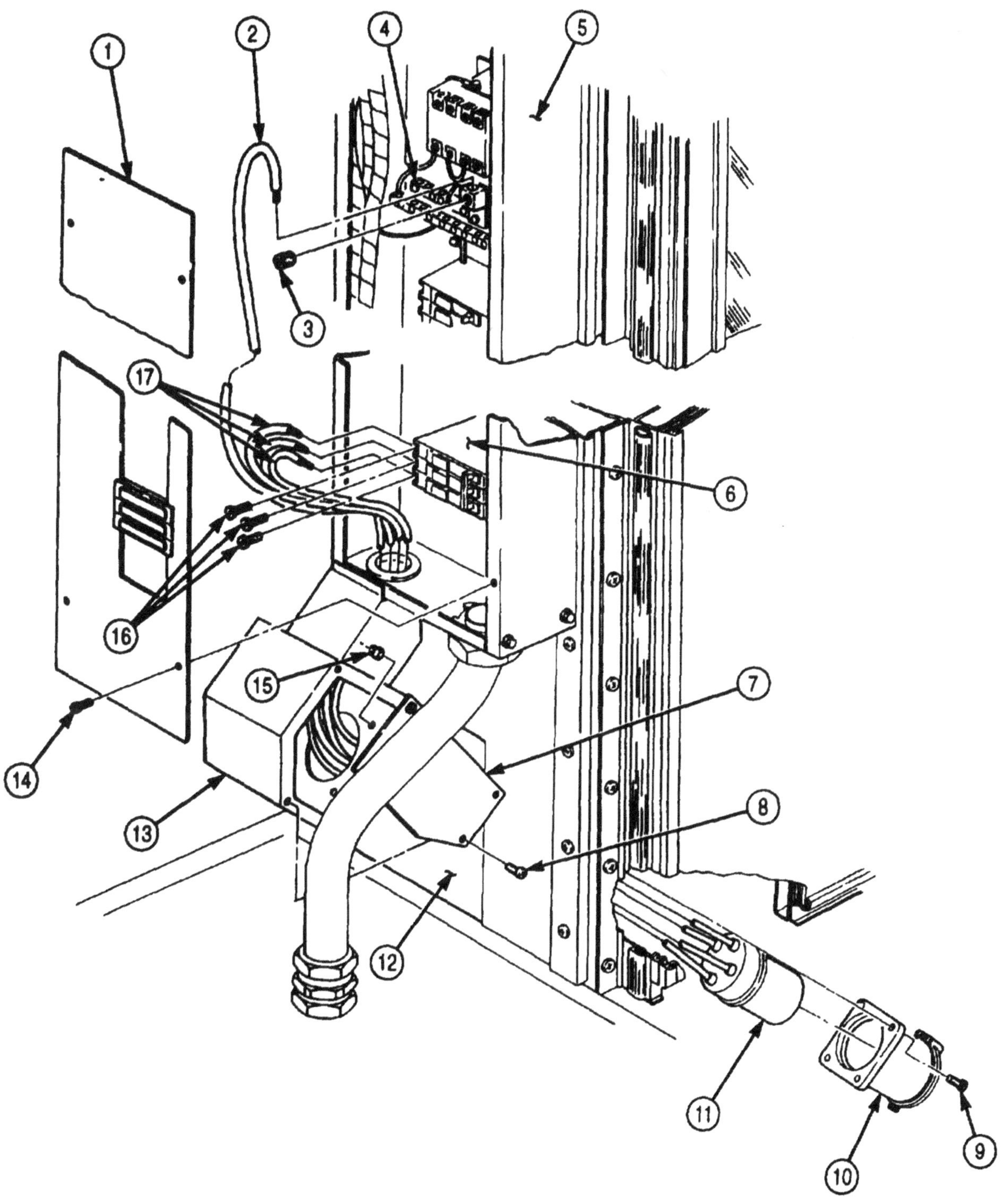

FOLLOW-ON TASKS:
- Connect battery ground cables (para. 3-126).
- Install external power source (TM 9-2320-272-10).

4-160. FLEXIBLE CONVERTER WIRING HARNESS MAINTENANCE

THIS TASK COVERS:

a. Removal
b. Repair
c. Installation

INITIAL SETUP:

APPLICABLE MODELS
M934/A1/A2

TOOLS
General mechanic's tool kit (Appendix E, Item 1)

MATERIALS/PARTS
Gasket (Appendix D, Item 211)
Lockwasher (Appendix D, Item 390)

MANUAL REFERENCES (TM)
TM 9-2320-272-10
TM 9-2320-272-24P

EQUIPMENT CONDITION
Parking brake set (TM 9-2320-272-10).

a. Removal

NOTE

Tag wires for installation.

1. Remove connector (10) from junction box (11).
2. Remove four screws (1), cover (2), and gasket (3) from converter (5). Discard gasket (3).
3. Remove three nuts (7), wires (8), and lockwasher (6) from converter (5). Discard lockwasher (6).
4. Remove wiring harness (9) from van body (4).

b. Repair

For van wiring harness repair, refer to para. 3-131.

c. Installation

1. Install wiring harness (9) on van body (4).
2. Install three wires (8) and new lockwasher (6) on converter (5) with three nuts (7).
3. Install new gasket (3) and cover (2) on converter (5) with four screws (1).
4. Install connector (10) on junction box (11).

4-160. FLEXIBLE CONVERTER WIRING HARNESS MAINTENANCE (Contd)

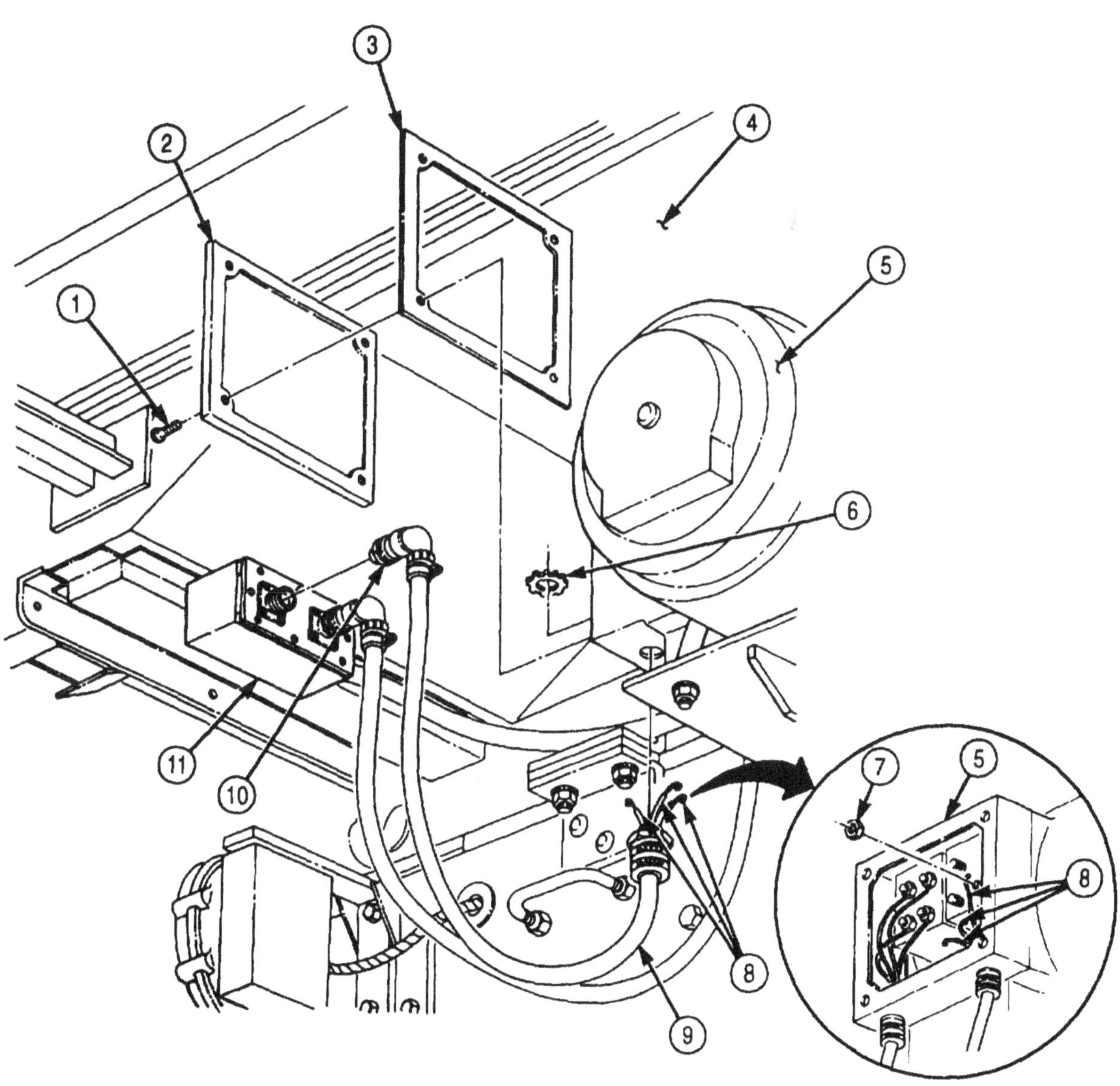

4-161. ELECTRICAL LOAD CENTER BOX MAINTENANCE

THIS TASK COVERS:

a. Removal
b. Disassembly
c. Assembly
d. Installation

INITIAL SETUP:

APPLICABLE MODELS
M934/A1/A2

TOOLS
General mechanic's tool kit (Appendix E, Item 1)

MATERIALS/PARTS
Three lockwashers (Appendix D, Item 391)
Sealing compound (Appendix C, Item 62)

MANUAL REFERENCES (TM)
TM 9-2320-272-10
TM 9-2320-272-24P
TM 9-237

EQUIPMENT CONDITION
- Parking brake set (TM 9-2320-272-10).
- Battery ground cables disconnected (para. 3-126).
- External power source disconnected (TM 9-2320-272-10)

a. Removal

NOTE

Tag wires for installation.

1. Remove six screws (1) and cover (2) from load center (7).
2. Remove three screws (20) and wires (3) from relay (6).
3. Remove screw (14) and wire (19) from neutral buss bar (8).
4. Remove screws (13) and (14) and two wires (16) and (17) from circuit breakers (10) and (9).
5. Remove screw (12) and wire (15) from circuit breaker (11).
6. Remove screw (4), clamp (5), and right main wiring harness (21) from load center (7).
7. Remove three screws (34) and wires (22) from relay (6).
8. Remove screw (32) and wire (33) from neutral buss bar (8).
9. Remove screw (28) and wires (31) from circuit breaker (23).
10. Remove screw (27) and wires (30) from circuit breaker (24).
11. Remove screw (26) and wire (29) from circuit breaker (25).
12. Remove left main wiring harness (35) from load center (7).

4-161. ELECTRICAL LOAD CENTER BOX MAINTENANCE (Contd)

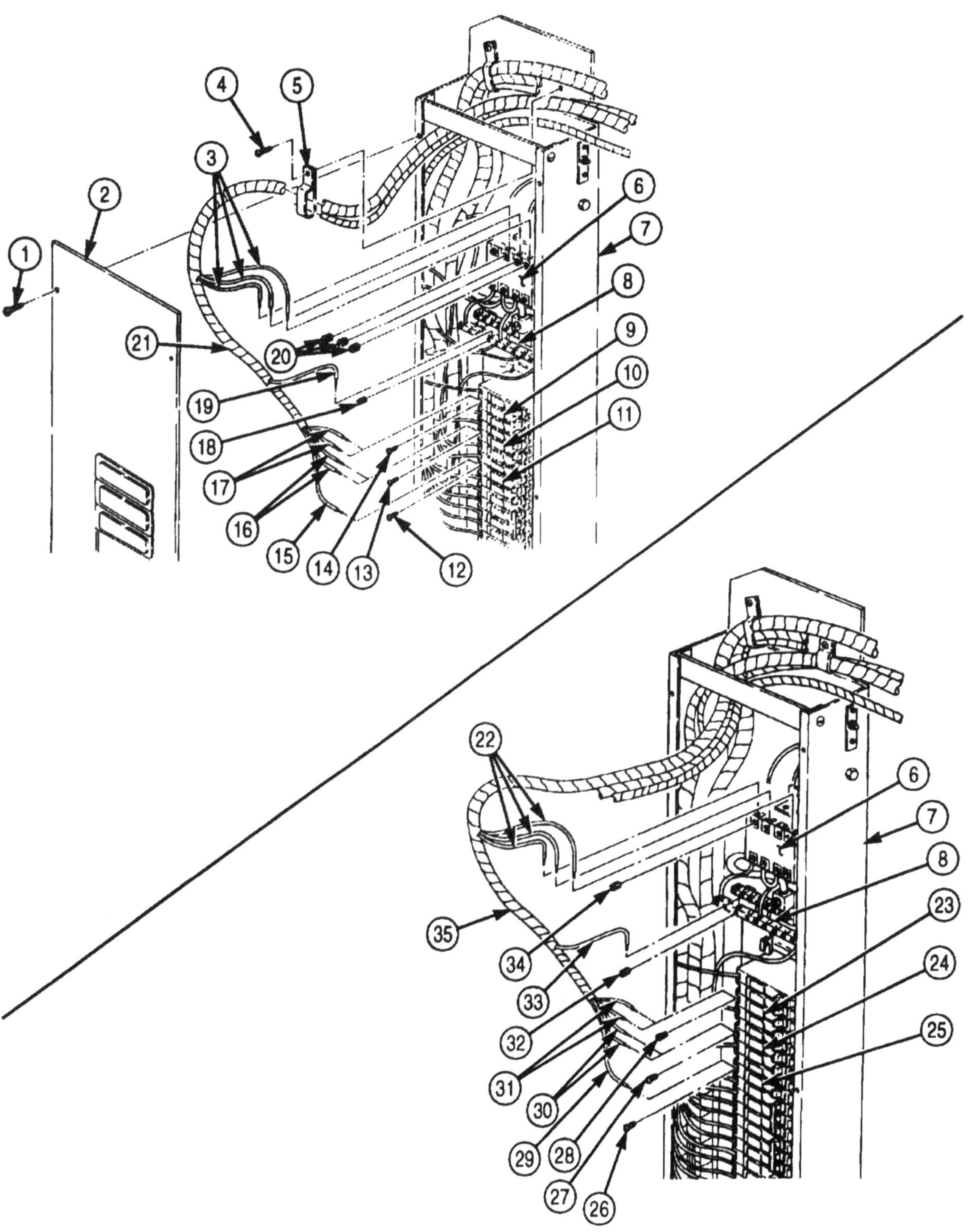

4-161. ELECTRICAL LOAD CENTER BOX MAINTENANCE (Contd)

13. Remove screw (1) and wires (2) and (9) from relay (4).
14. Remove screw (7) and wire (8) from circuit breaker (6).
15. Remove blackout bypass wiring harness (3) from load center (5).
16. Remove nut (12). screw (15). two washers (11). cable assembly (14), ground cable (13), and cable (10) from load centers (5).
17. Remove three screws (17) and wires (18) from three circuit breakers (16).
18. Remove air conditioner wiring harness (19) from load center (5).
19. Remove screw (26) and wire (25) from neutral buss bar (21).
20. Remove three screws (23) and wires (24) from three circuit breakers (22).
21. Remove right 10 kW electric heater wiring harness (20) from load center (5).
22. Remove screw (32) and wire (31) from neutral buss bar (21).
23. Remove screw (29) and wire (30) from circuit breaker (28).
24. Remove right electric heater wiring harness (27) from load center (5).
25. Disconnect cables (37) and (38) from connectors (36) and (40).
26. Remove eight screws (39) and cover (41) from junction box (33).
27. Remove wires (34) and (35) from connectors (36) and (40).

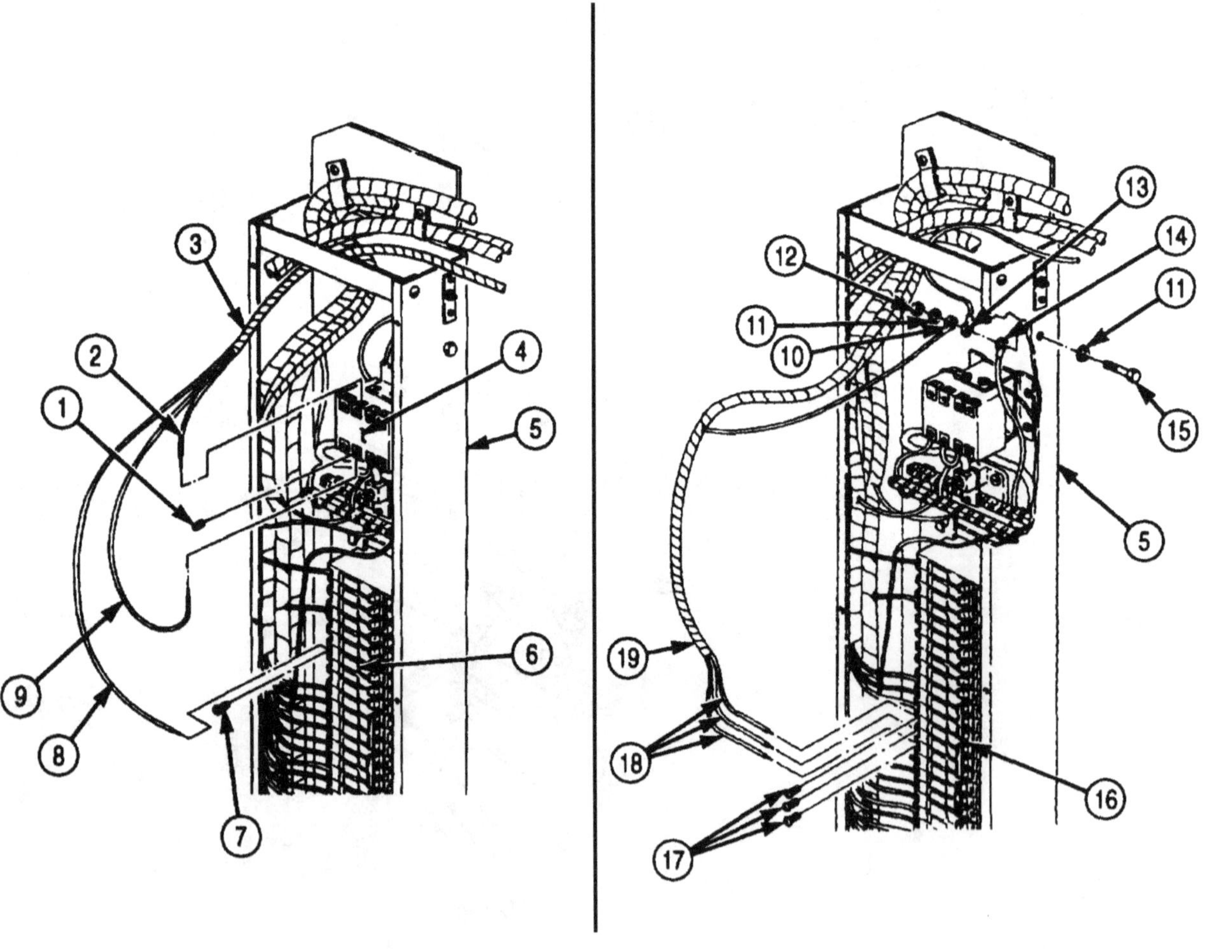

4-161. ELECTRICAL LOAD CENTER BOX MAINTENANCE (Contd)

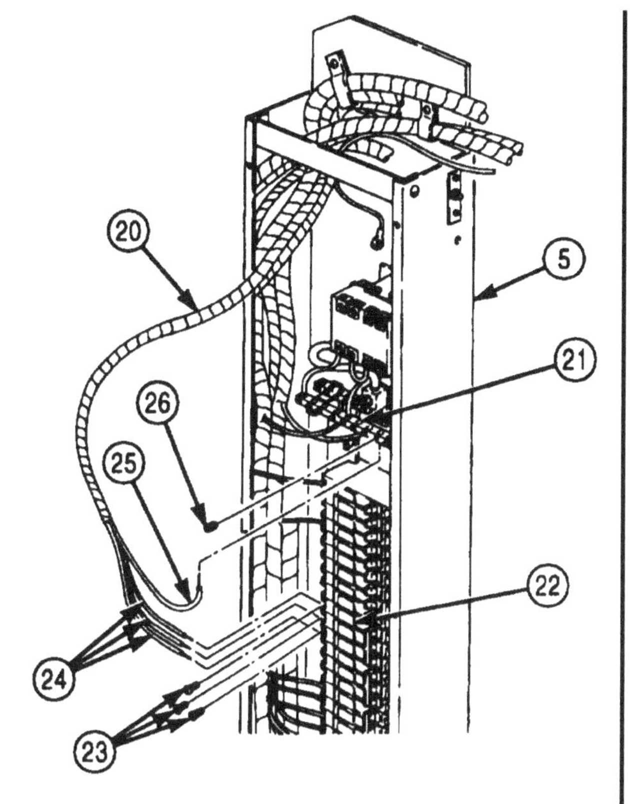

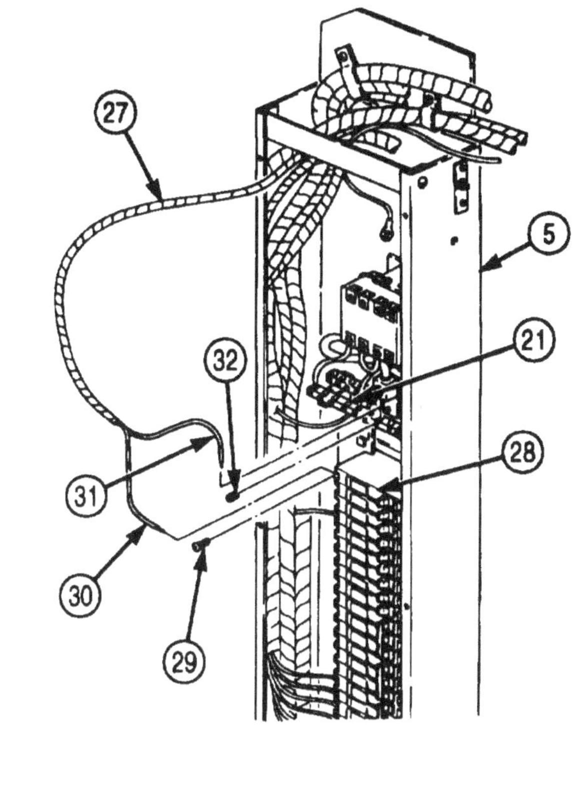

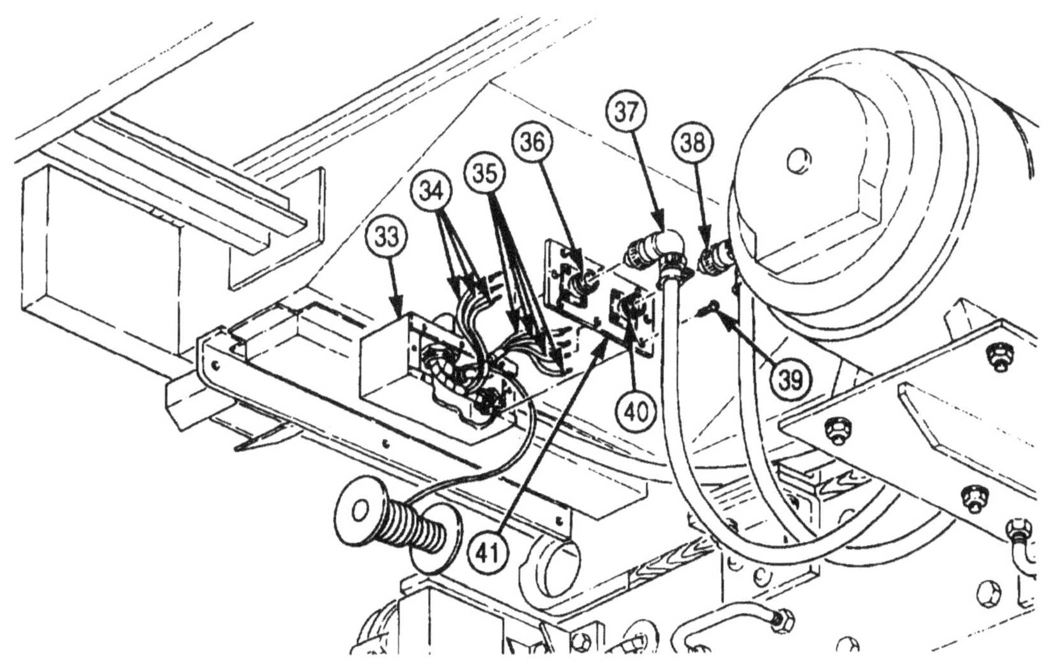

4-161. ELECTRICAL LOAD CENTER BOX MAINTENANCE (Contd)

28. Remove five screws (5) and wires (6) from five circuit breakers (3).
29. Remove converter wiring harness (1) from load center (2).
30. Remove screw (13) and wire (8) from neutral buss bar (9).
31. Remove three screws (11) and wires (12) from three circuit breakers (10).
32. Remove wire (7) and entrance receptacle wiring harness (8) from load center (2).
33. Remove screw (14) and wire (25) from relay (24).
34. Remove two screws (17) and wires (18) and (20) from switch (19).
35. Remove screw (23) and wire (21) from neutral buss bar (22).
36. Remove right and left blackout wiring harnesses (15) and (16) from load center (2).
37. Remove screw (31) and wire (27) from neutral buss bar (22).
38. Remove three screws (29) and wires (30) from three circuit breakers (28).
39. Remove right 3-phase receptacle wiring harness (26) from load center (2).
40. Remove screw (32) and wire (34) from neutral buss bar (22).
41. Remove three screws (36) and wires (37) from three circuit breakers (35).
42. Remove left 3-phase receptacle wiring harness (33) from load center (2).
43. Remove four screws (38) and washers (39) from load center (2).
44. Remove special nut (42) from conduit nut (41).
45. Remove load center (2) from van body (40).

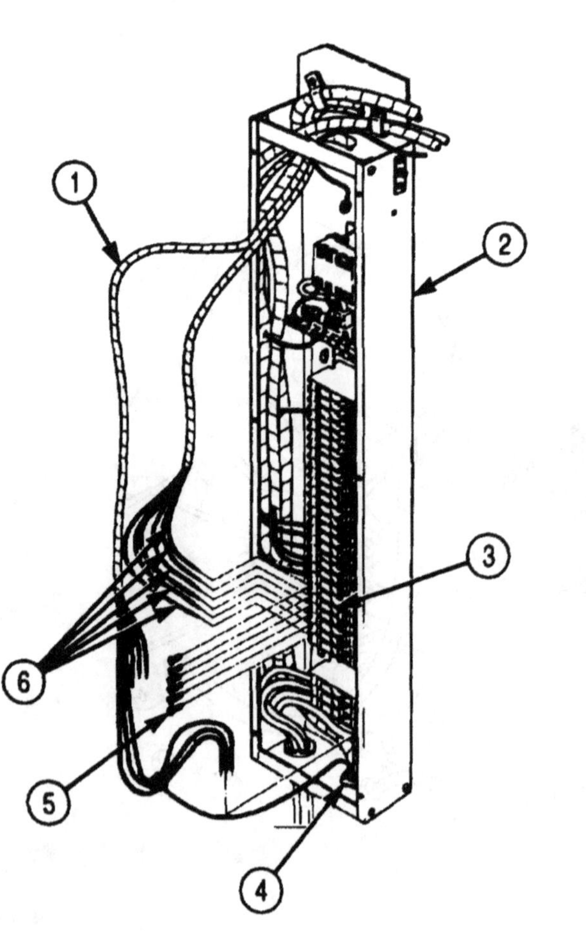

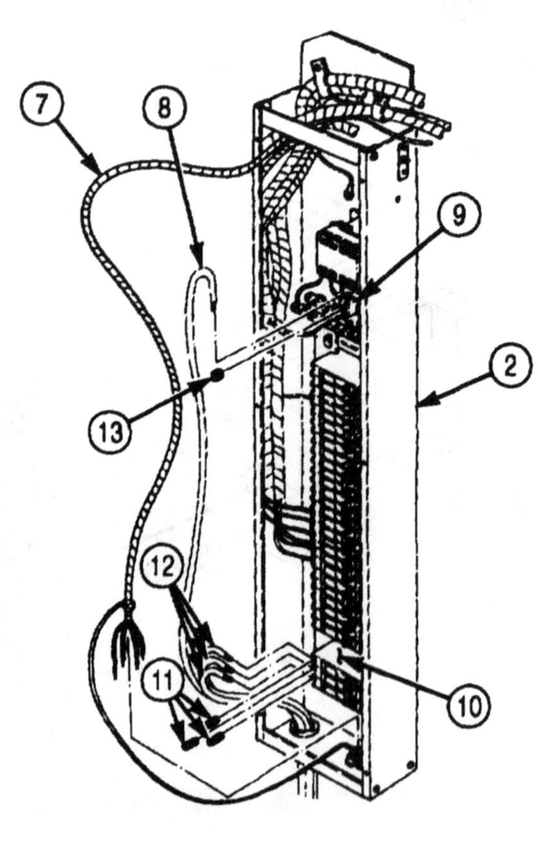

4-161. ELECTRICAL LOAD CENTER BOX MAINTENANCE (Contd)

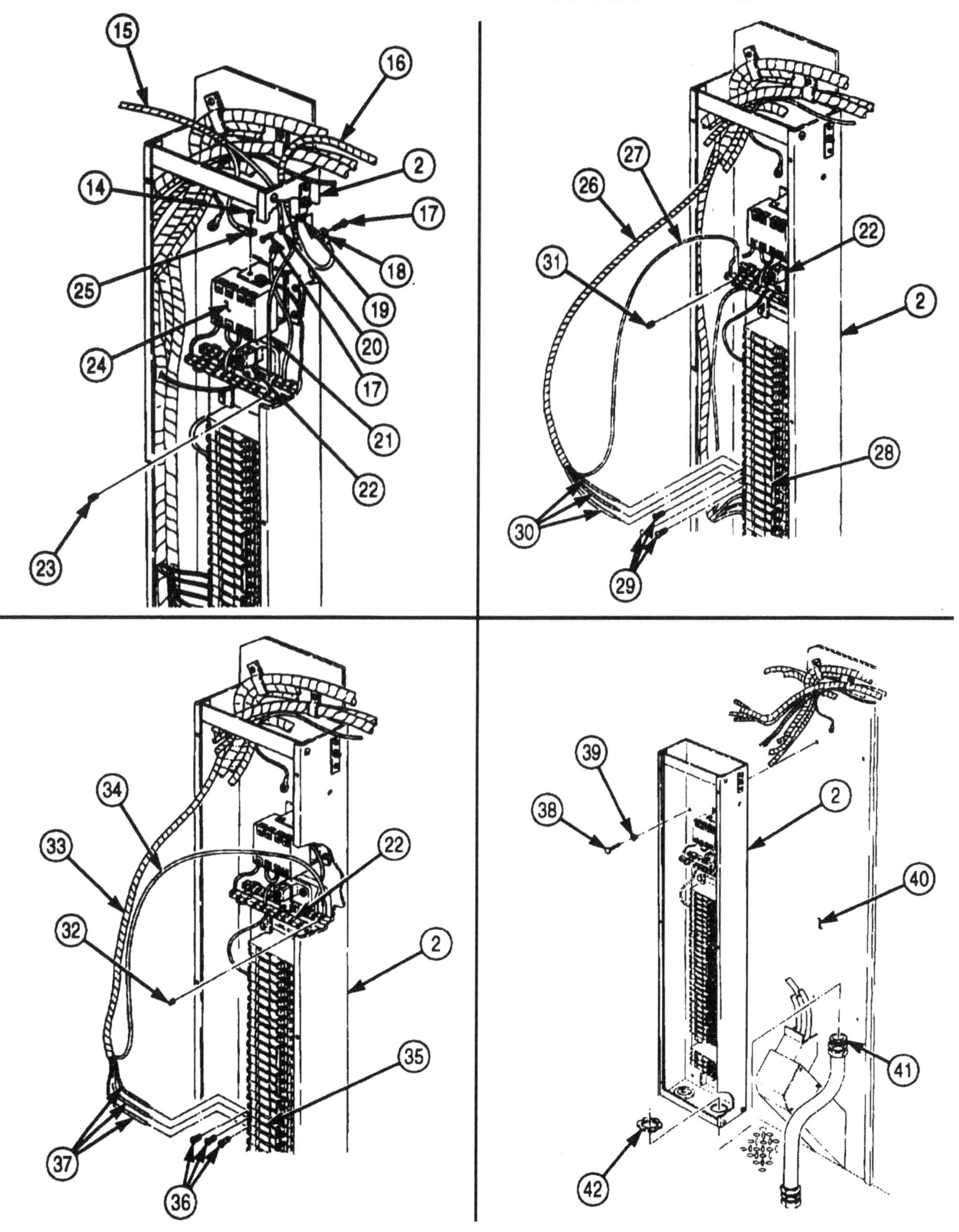

4-161. ELECTRICAL LOAD CENTER BOX MAINTENANCE (Contd)

b. Disassembly

1. Remove six plates (1) from load center cover (2).
2. Remove two screws (32), jumper wire (31), and ground B cable (29) from relay (4).
3. Remove screws (33) and (28) and ground N cable (30) from relay (4) and neutral buss bar (11).
4. Remove screw (34) and ground B cable (29) from neutral buss bar (11).
5. Remove screws (25) and (5) and relay cable (3) from 20-amp circuit breaker (26) and relay (4).
6. Remove three screws (6), lockwashers (7), and relay (4) from load center box (12). Discard lockwashers (7).
7. Remove screw (27) and cable (8) from neutral buss bar (11).
8. Remove six screws (16) and insulators (17) from load center (12).
9. Remove nine screws (24) and nine 20-amp circuit breakers (26) from load center (12).
10. Remove three screws (22) and 30-amp circuit breaker (23) from load center (12).
11. Remove three screws (20) and 40-amp circuit breaker (21) from load center (12).
12. Remove twelve screws (19) and four 20-amp circuit breakers (18) from load center (12).
13. Remove three screws (15) and 100-amp circuit breaker (14) from load center (12).
14. Remove grommet (13) from load center (12).
15. Remove two screws (10) and blackout door switch (9) from load center (12).

c. Assembly

1. Install 100-amp circuit breaker (14) on load center (12) with three screws (15).
2. Install four 20-amp circuit breakers (18) on load center (12) with twelve screws (19).
3. Install 40-amp circuit breaker (21) on load center (12) with three screws (20).
4. Install 30-amp circuit breaker (23) on load center (12) with three screws (22).
5. Install nine 20-amp circuit breakers (26) on load center (12) with nine screws (24).
6. Install six insulators (17) on load center (12) with six screws (16).
7. Install relay (4) on load center (12) with three new lockwashers (7) and screws (6).
8. Install cable (8) on neutral buss bar (11) with screw (27).
9. Install relay cable (3) on 20-amp circuit breaker (26) and relay (4) with two screws (25) and (5).
10. Install ground B cable (29) on neutral buss bar (11) with screw (34).
11. Install ground N cable (30) on relay (4) and neutral buss bar (11) With two screws (33) and (28).
12. Install jumper wire (31) and ground B cable (29) on relay (4) with two screws (32).
13. Install six plates (1) on load center cover (2).
14. Install blackout door switch (9) on load center (12) with two screws (10).
15. Install grommet (13) on load center (12).

4-161. ELECTRICAL LOAD CENTER BOX MAINTENANCE (Contd)

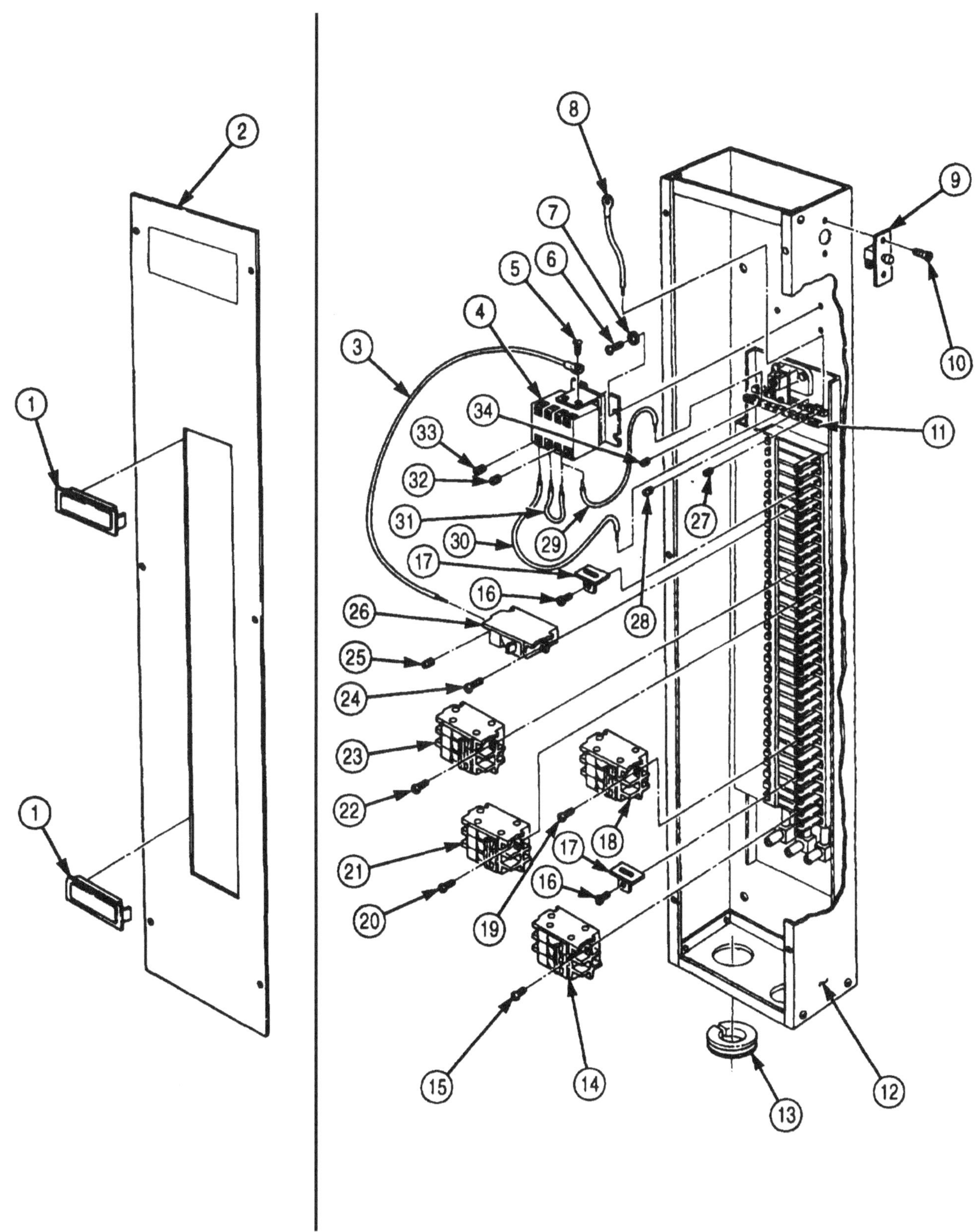

4-161. ELECTRICAL LOAD CENTER BOX MAINTENANCE (Contd)

d. Installation

1. Install load center (3) on van body (4) and conduit (5) with four washers (2), screws (1), and special nut (6).
2. Install left 3-phase receptacle wiring harness (8) and three wires (13) in load center (3) and on triple circuit breaker (11) with three screws (12).
3. Install wire (9) on neutral buss bar (10) with screw (7).

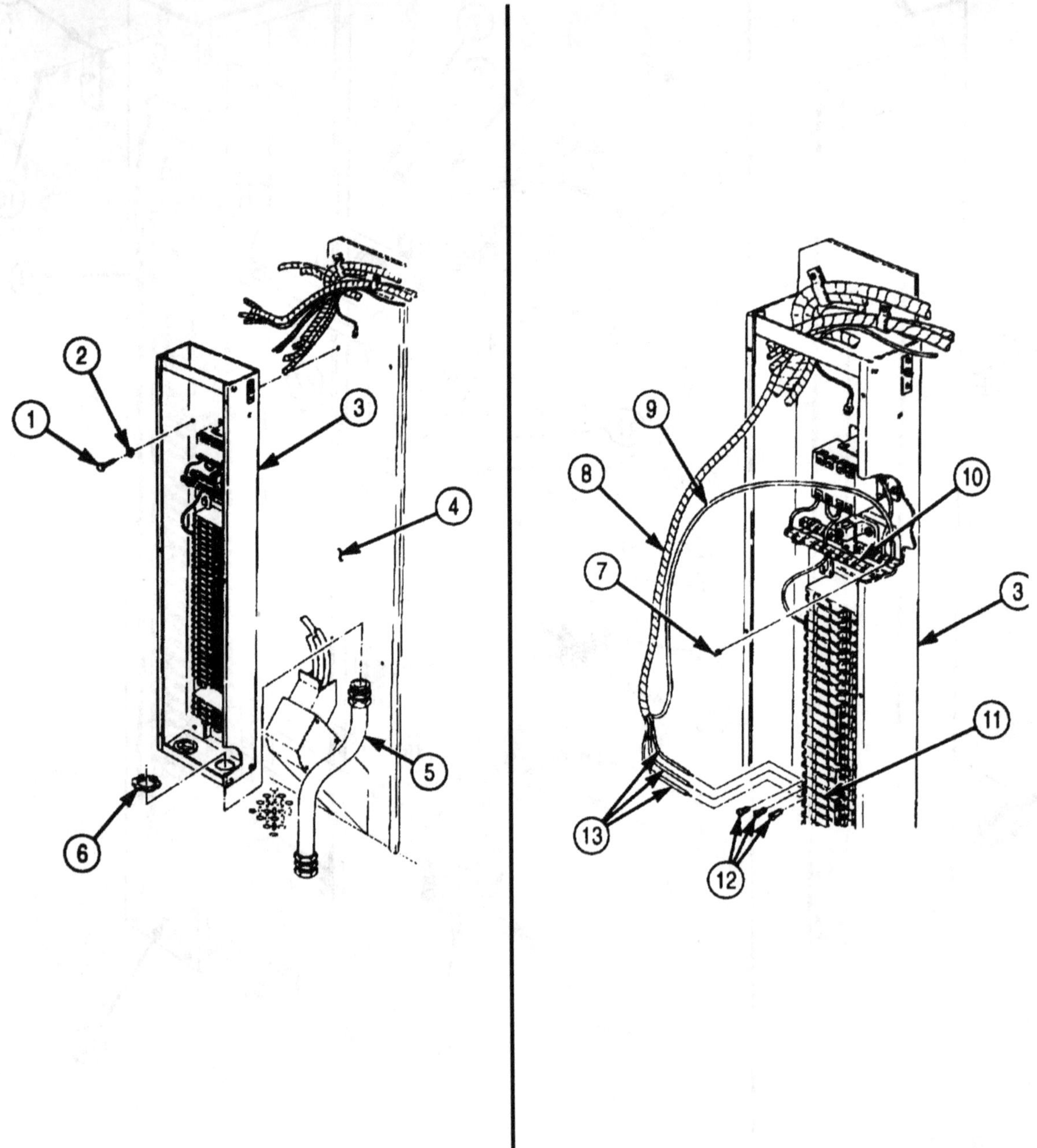

4-161. ELECTRICAL LOAD CENTER BOX MAINTENANCE (Contd)

4. Install right 3-phase receptacle wiring harness (14) and three wires (18) in load center (3) and on three circuit breakers (16) with three screws (17).
5. Install wire (15) on neutral buss bar (10) with screw (19).
6. Install right and left wiring harnesses (21) and (22) and wire (28) in load center (3) and on neutral buss bar (10) with screw (29).
7. Install two wires (24) and (26) on switch (25) with screws (23) and (27).
8. Install wire (31) on relay (30) with screw (20).

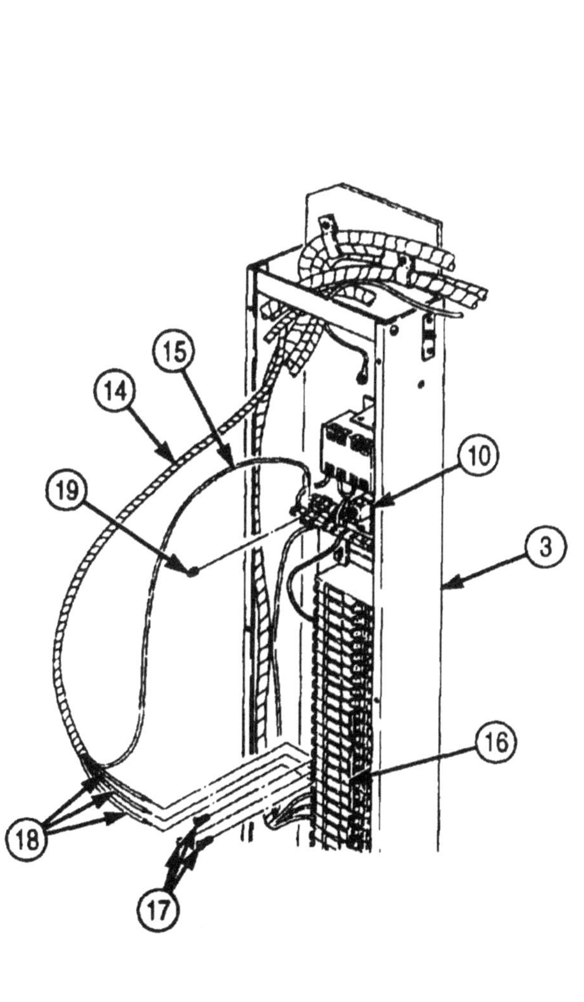

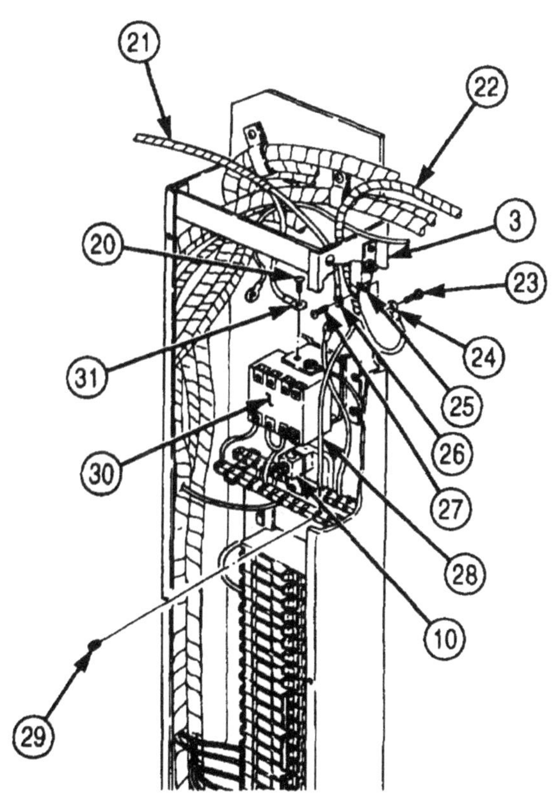

4-161. ELECTRICAL LOAD CENTER BOX MAINTENANCE (Contd)

9. Install three wires (7) and entrance receptacle harness (1) on circuit breakers (5) in load center (4) with three screws (6).
10. Install wire (2) on neutral buss bar (3) with screw (8).
11. Install five wires (13) and converter wiring harness (14) on circuit breakers (10) in load center (4) with five screws (12). Route auxiliary pump wiring harness (9) through coupling (11).
12. Install wires (16) and (17) on connectors (18) and (22) (TM 9-237).

NOTE

Apply sealing compound before performing step 13.

13. Install cover (23) on junction box (15) with eight screws (21).
14. Connect cables (19) and (20) to connectors (18) and (22).
15. Install wires (27) and (28) and right 10 kW electric heater wiring harness (24) on circuit breaker (25) in load center (4) and neutral buss bar (3) with screws (26) and (29).
16. Install wire (33), three wires (32), and left electric heater wiring harness (35) on neutral buss bar (3) and-circuit breakers (30) with screw (34) and three screws (31).

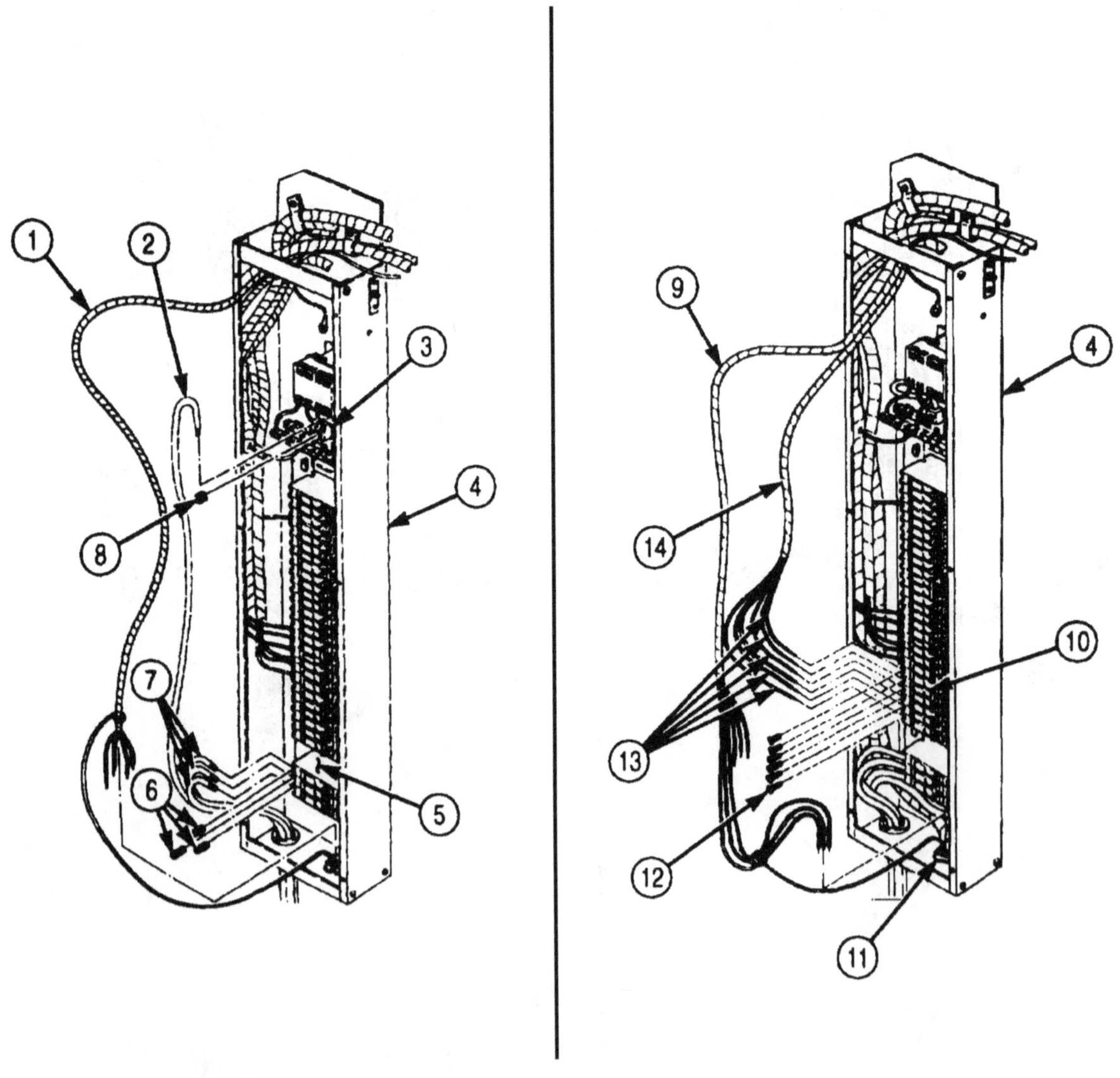

4-161. ELECTRICAL LOAD CENTER BOX MAINTENANCE (Contd)

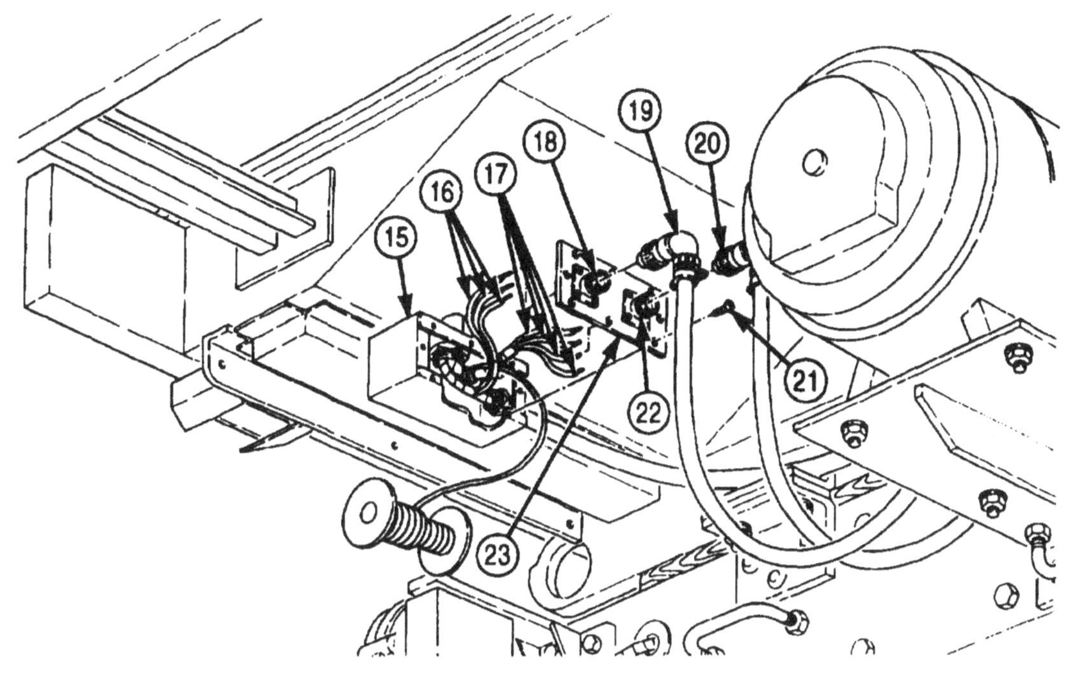

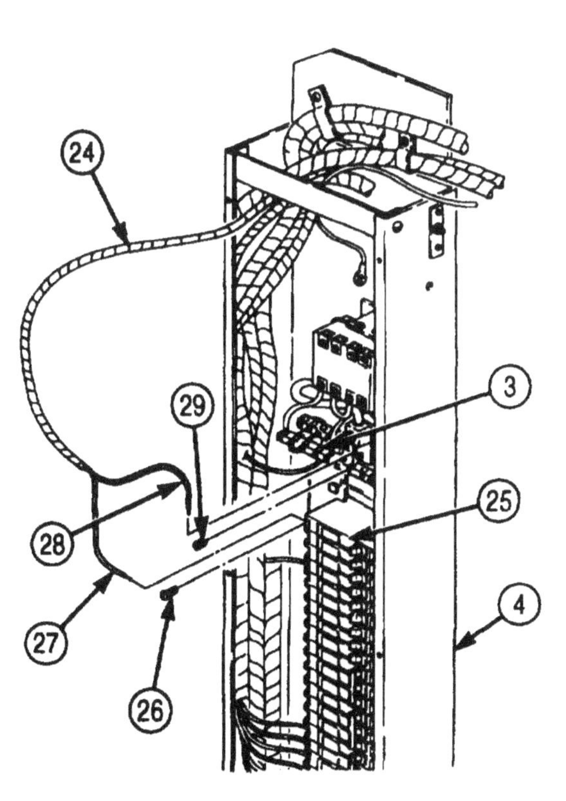

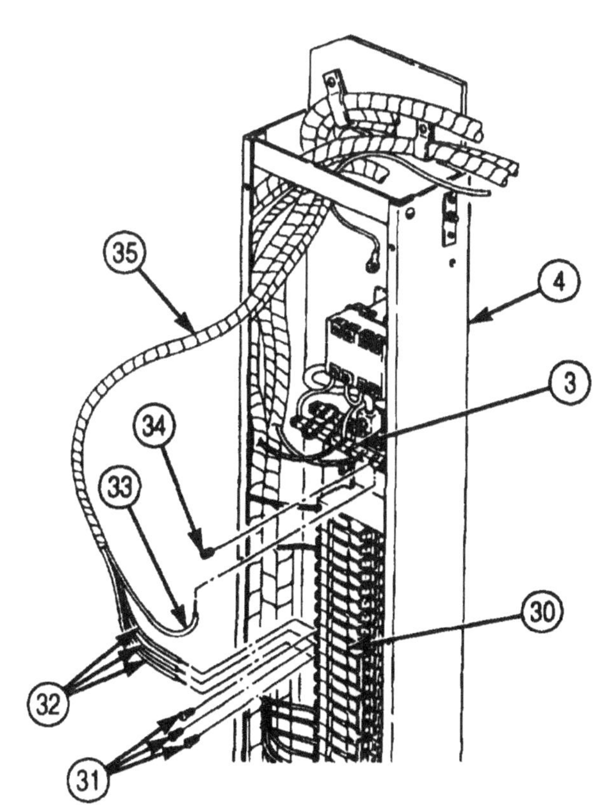

4-161. ELECTRICAL LOAD CENTER BOX MAINTENANCE (Contd)

17. Install three wires (11) and air conditioner wiring harness (12) in load center (8) and on circuit breaker (9) with three screws (10).
18. Install cable assembly (5), ground cable (4), and cable (1) on load center (8) with screw (7), washers (6) and (2), and nut (3).
19. Install wires (14), (20), and (19), and blackout bypass wiring harness (15) in load center (8) and on relay (16) and circuit breaker (17) with two screws (13) and screw (18).
20. Position left main wiring harness (22) on load center (8).
21. Install two wires (32) on circuit breaker (24) with screw (29).
22. Install two wires (31) on circuit breaker (25) with screw (28).
23. Install wire (30) on circuit breaker (26) with screw (27).
24. Install wire (34) on neutral buss bar (23) with screw (33).
25. Install three wires (21) on relay (16) with three screws (35).
26. Install two wires (49) and (48), wire (47), right main wiring harness (53), clamp (40), and screw (39) in load center (8) and on three circuit breakers (41), (42), and (43) with three screws (44), (45), and (46).
27. Install wire (51) on neutral buss bar (23) with screw (50).
28. Install three wires (38) on relay (16) with three screws (52).
29. Install cover (37) on load center (8) with six screws (36).

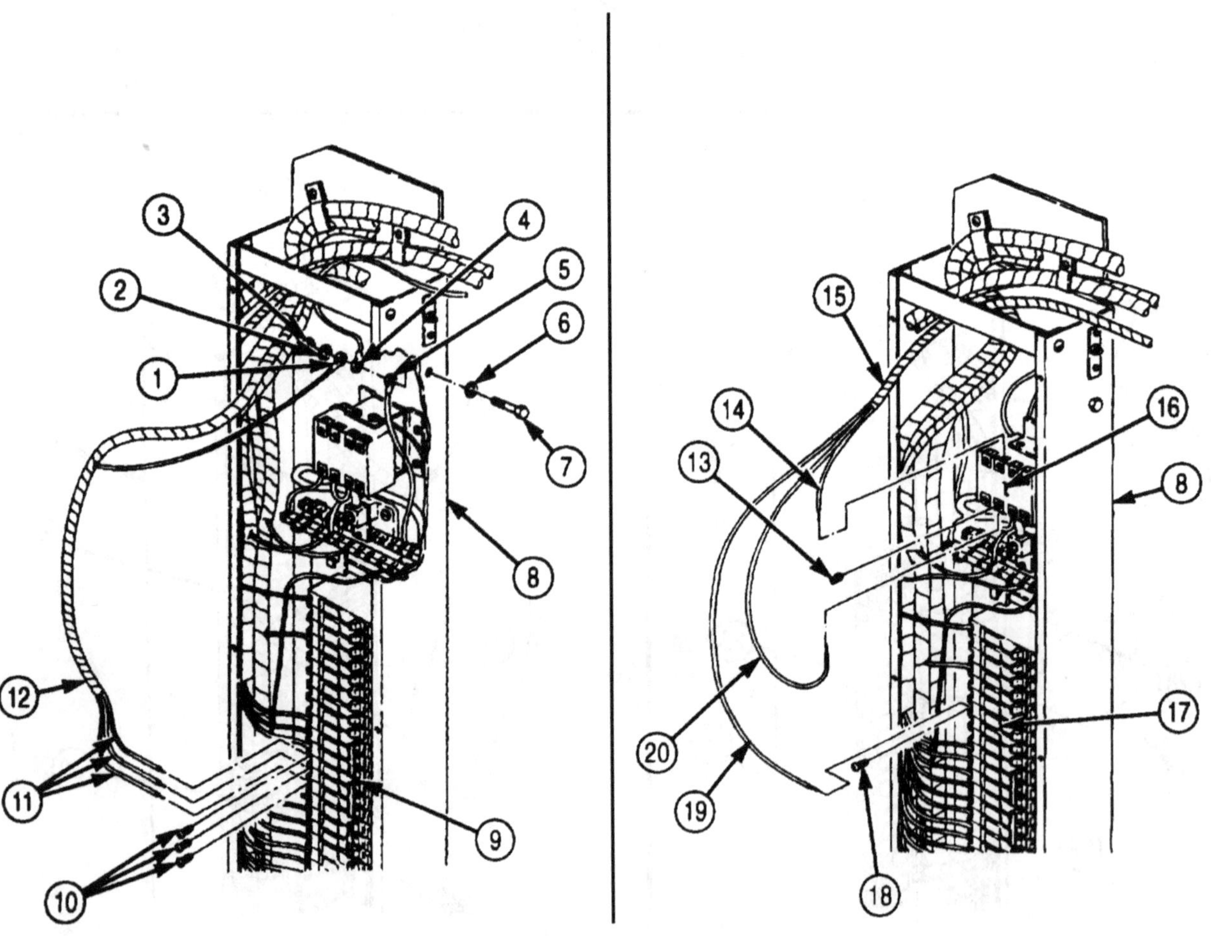

4-161. ELECTRICAL LOAD CENTER BOX MAINTENANCE (Contd)

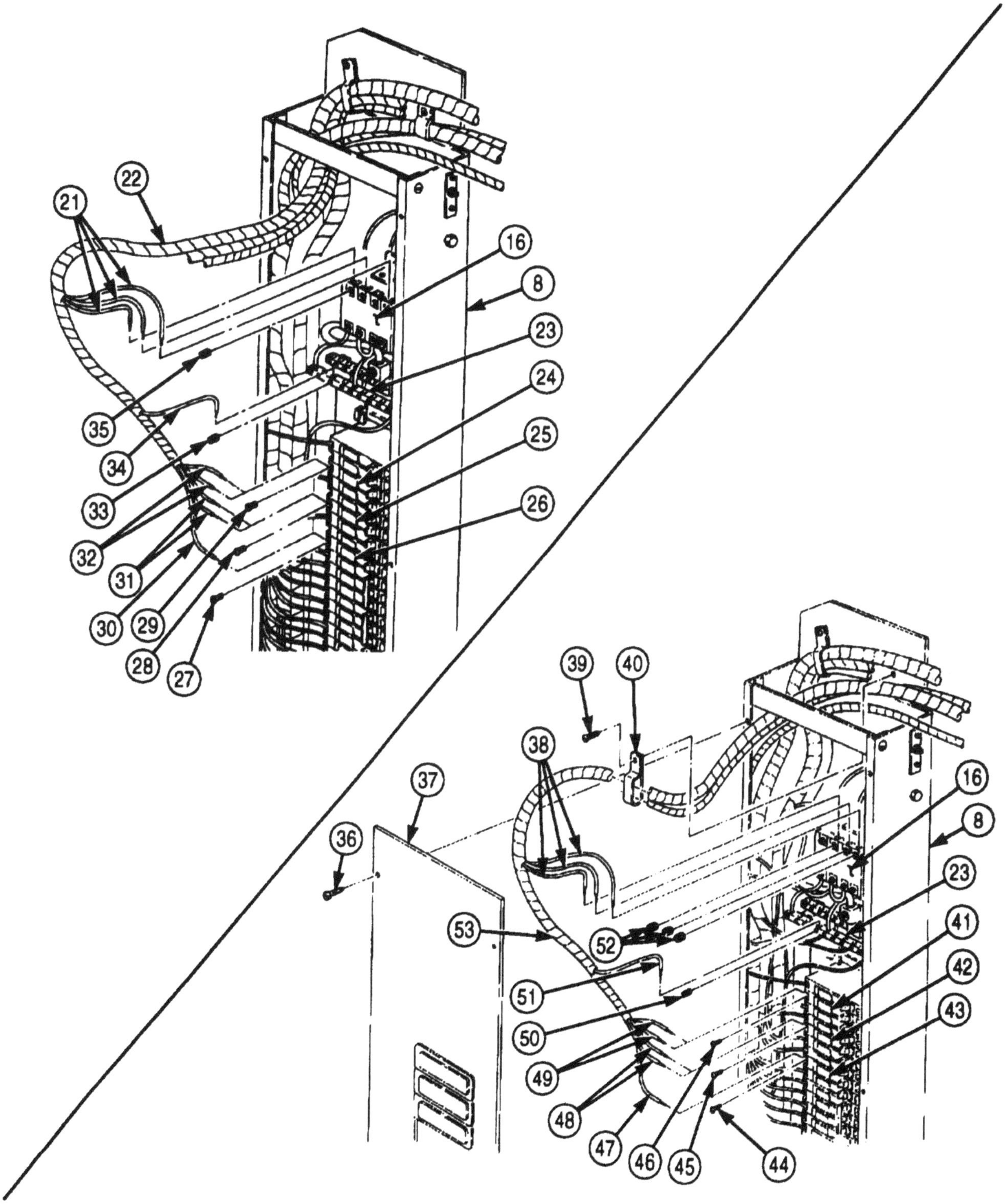

21 22 16 8 23 24 25 26 35 34 33 32 31 30 29 28 27

39 40 38 37 36 16 8 23 41 42 43 53 52 51 50 49 48 47 46 45 44

FOLLOW-ON TASKS: Connect battery ground cables (para. 3-126).
• Connect external power source (TM 9-2320-272-10).

4-162. ELECTRICAL BOX MAINTENANCE

THIS TASK COVERS:

a. Removal
b. Inspection
c. Installation

INITIAL SETUP:

APPLICABLE MODELS
M934/A1/A2

TOOLS
General mechanic's tool kit (Appendix E, Item 1)

MANUAL REFERENCES (TM)
TM 9-2320-272-10
TM 9-2320-272-24P

EQUIPMENT CONDITION
- Parking brake set (TM 9-2320-272-10).
- Battery ground cables disconnected (para. 3-126).
- Entrance receptacle 220V 3-phase harness removed (para. 4-159).

a. Removal

1. Remove electrical box (7) from van body end panel (4).
2. Remove six screws (2) and bracket (3) from electrical box (7).
3. Remove grommet (1) from electrical box (7).
4. Remove screw (5) and plate (6) from electrical box (7).

b. Inspection

Inspect all parts for damage. Replace damaged parts.

c. Installation

1. Install plate (6) on electrical box (7) with screw (5).
2. Install grommet (1) on electrical box (7).
3. Install bracket (3) on electrical box (7) with six screws (2).
4. Position electrical box (7) on van body end panel (4).

4-162. ELECTRICAL BOX MAINTENANCE (Contd)

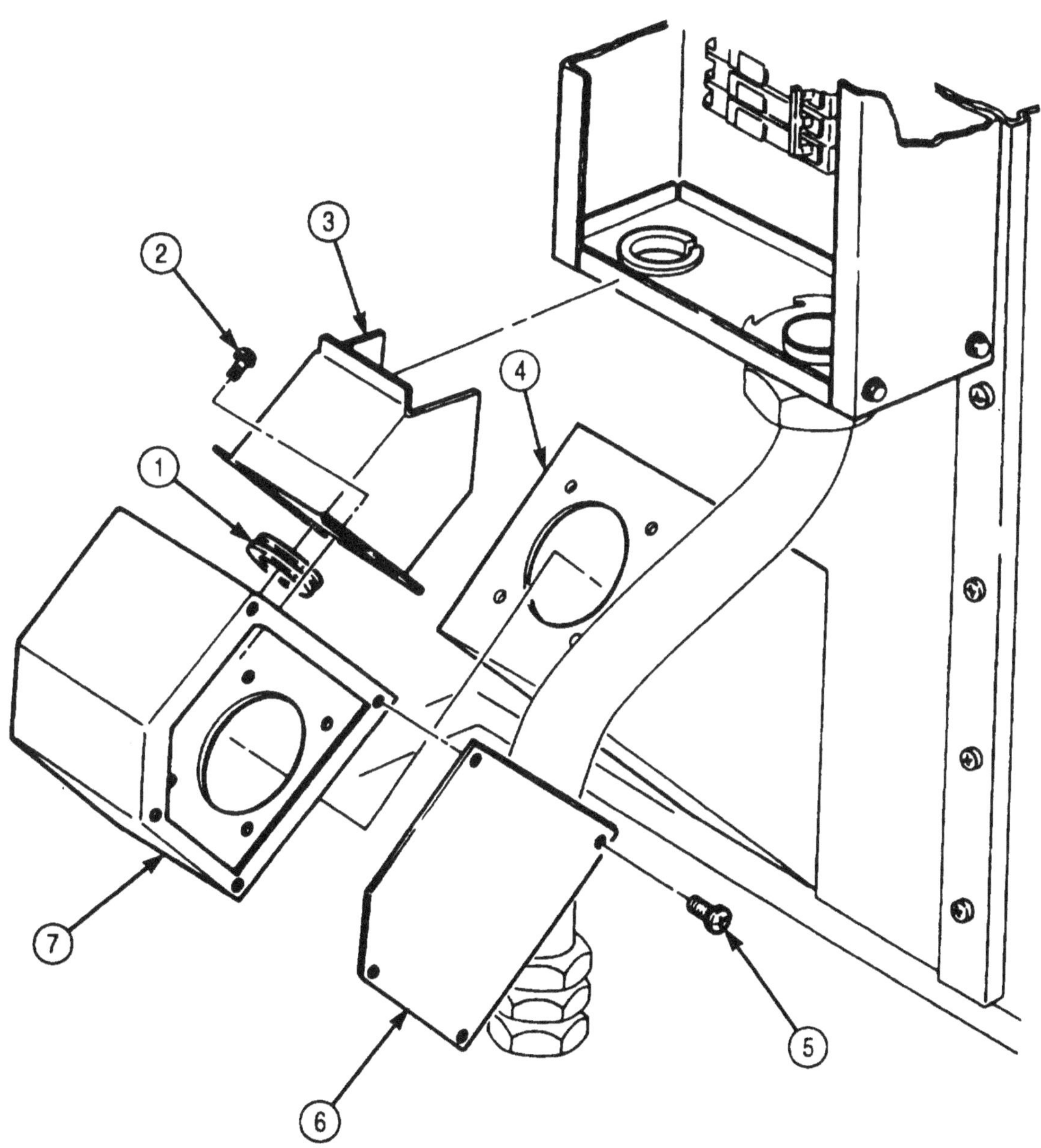

FOLLOW-ON TASKS: Install entrance receptacle 220V 3-phase harness (para. 4-159).
- Connect battery ground cables (para. 3-126).

4-163. ELECTRICAL JUNCTION BOX REPLACEMENT

THIS TASK COVERS:

a. Removal **b. Installation**

INITIAL SETUP:

APPLICABLE MODELS
M934/A1/A2

TOOLS
General mechanic's tool kit (Appendix E, Item 1)

MATERIALS/PARTS
Eight locknuts (Appendix D, Item 318)
Sealing compound (Appendix C, Item 65)

MANUAL REFERENCES (TM)
TM 9-2320-272-10
TM 9-2320-272-24P
TM 9-237

EQUIPMENT CONDITION
- Parking brake set (TM 9-2320-272-10).
- Battery ground cables disconnected (para. 3-126).

a. Removal

1. Remove two cables (3) from connectors (1).
2. Remove eight screws (4) and cover (5) from junction box (7).

NOTE

Tag wires for installation.

3. Remove harness leads (6) and (8) from two connectors (1) (refer to TM 9-237).
4. Remove eight locknuts (9), screws (2), and two connectors (1) from cover (5). Discard locknuts (9).
5. Remove special nut (14), nut (15), adapter (13), and nut (12) from junction box (7).
6. Remove junction box (7) from conduits (10) and (11) and slide off wires (6) and (8).

b. Installation

1. Place nut (15) on conduit (10).
2. Feed harness leads (6) and (8) through holes in junction box (7).
3. Position junction box (7) on conduits (10) and (11).
4. Install nut (15) on junction box (7)
5. Feed conduit (11) through nut (12), adapter (13), junction box (7), and special nut (14).
6. Install special nut (14) on junction box (7).
7. Install two connectors (1) on cover (5) with eight screws (2) and locknuts (9).
8. Connect harness leads (6) and (8) to two connectors (1) (TM 9-237).
9. Coat edge of cover (5) with sealing compound and install cover (5) on junction box (7) with eight screws (4).
10. Connect two cables (3) to connectors (1).

4-163. ELECTRICAL JUNCTION BOX MAINTENANCE (Contd)

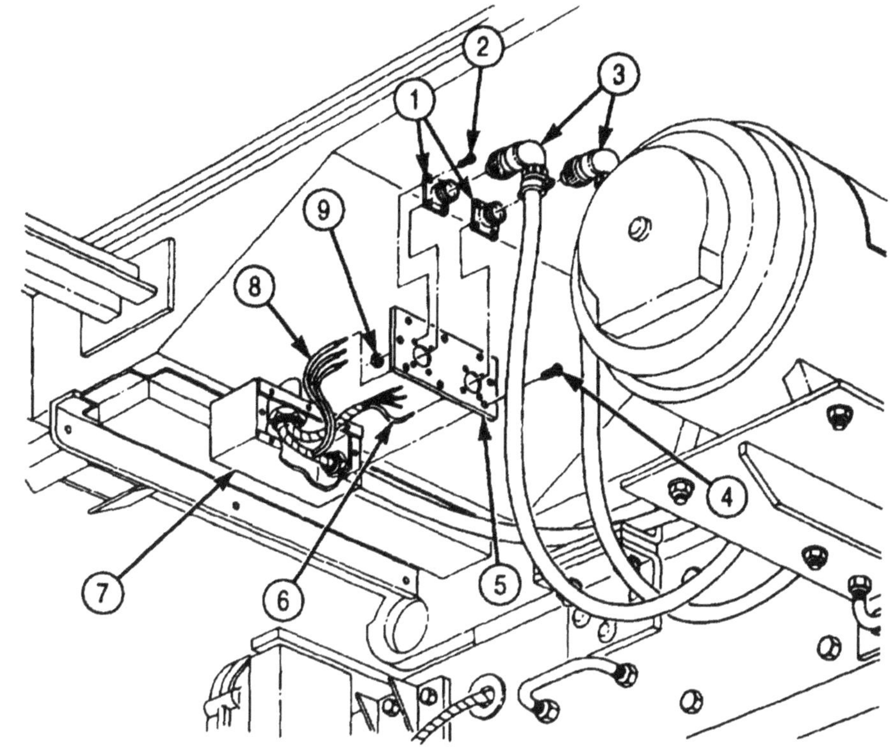

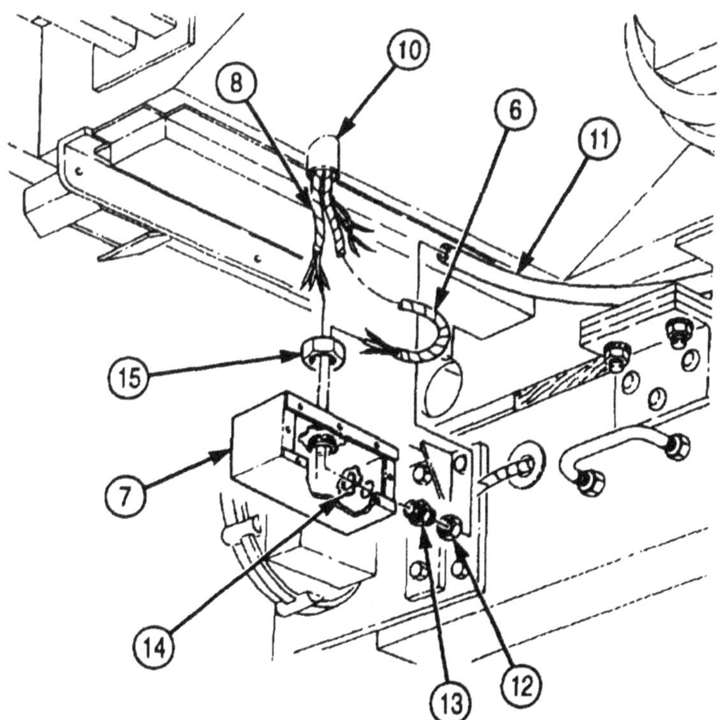

FOLLOW-ON TASK: Connect battery ground cables (para. 3-126).

4-164. 400 HZ CONVERTER HARNESS CONDUIT REPLACEMENT

THIS TASK COVERS:

a. Removal **b. Installation**

INITIAL SETUP:

APPLICABLE MODELS
M934/A1/A2

TOOK
General mechanic's tool kit (Appendix E, Item 1)

MANUAL REFERENCES (TM)
TM 9-2320-272-10
TM 9-2320-272-24P

EQUIPMENT CONDITION
- Parking brake set (TM 9-2320-272-10).
- Battery ground cables disconnected (para. 3-126).
- Electrical junction box removed (para. 4-163).

a. Removal

1. Remove six screws (4) and clamps (5) from conduits (1) and (6).
2. Remove nuts (2) and (7) and coupling (3) from conduits (1) and (6).
3. Remove conduits (1) and (6) from van body (8).

b. Installation

1. Install nuts (2) and (7) and coupling (3) on conduits (1) and (6).
2. Install conduits (1) and (6) on van body (8) with six clamps (5) and screws (4).

4-164. 400 HZ CONVERTER HARNESS CONDUIT REPLACEMENT (Contd)

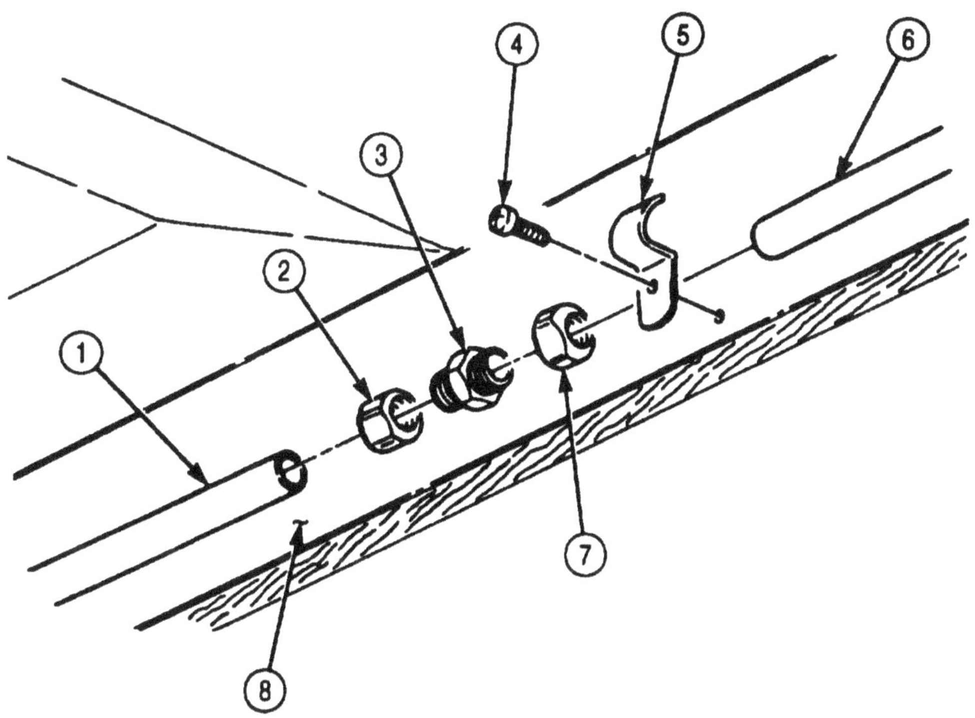

FOLLOW-ON TASKS:
- Connect battery ground cable (para. 3-126).
- Install electrical junction box (para. 4-163).

4-165. HEATER THERMOSTAT AND 10 kW HEATER THERMOSTAT REPLACEMENT

THIS TASK COVERS:

a. Heater Thermostat Removal
b. 10 kW Heater Thermostat Removal
c. Heater Thermostat Installation
d. 10 kW Heater Thermostat Installation

INITIAL SETUP:

APPLICABLE MODELS
M934/A1/A2

TOOLS
General mechanic's tool kit (Appendix E, Item 1)
Electrical tool kit (Appendix D, Item 40)

MANUAL REFERENCES (TM)
TM 9-2320-272-10
TM 9-2320-272-24P

EQUIPMENT CONDITION
- Parking brake set (TM 9-2320-272-10).
- Battery ground cables disconnected (para. 3-126).

a. Heater Thermostat Removal

1. Remove cover (3) from thermostat (5).
2. Remove two screws (4) and thermostat (5) from control center box cover (1).
3. Remove two screws (6) and wires (2) from thermostat (5).
4. Remove thermostat (5) from control center box cover (1).

b. 10 kW Heater Thermostat Removal

1. Remove cover (10) from thermostat (8).
2. Remove two screws (9) and thermostat (8) from control center box cover (12).
3. Disconnect three insulated wire splices (11) from three wires (7) and (13). Discard insulated wire splices (11).
4. Remove thermostat (8) from control center box cover (12).

c. Heater Thermostat Installation

1. Connect three wires (7) and (13) with three new insulated wire splices (11).
2. Position thermostat (8) on control center box cover (12) and install two screws (9).
3. Install cover (10) on thermostat (8).

d. 10 kW Heater Thermostat Installation

1. Install two wires (2) on thermostat (5) with two screws (6).
2. Position thermostat (5) on control center box cover (1) and install two screws (4).
3. Install cover (3) on thermostat (5).

4-165. HEATER THERMOSTAT AND 10 kW HEATER THERMOSTAT REPLACEMENT (Cont)

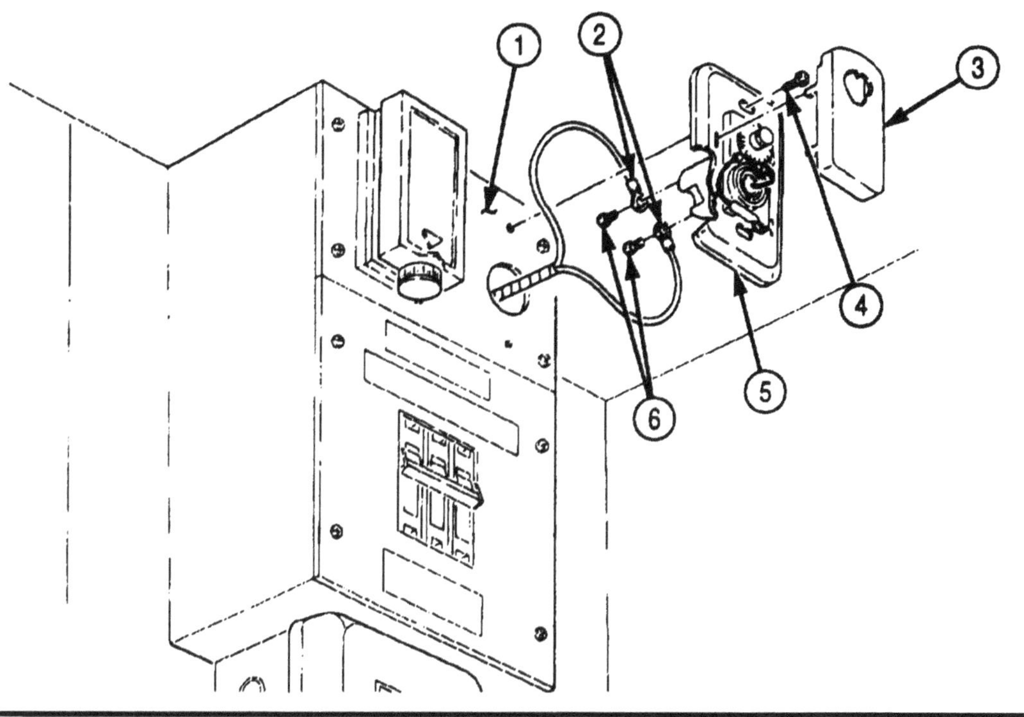

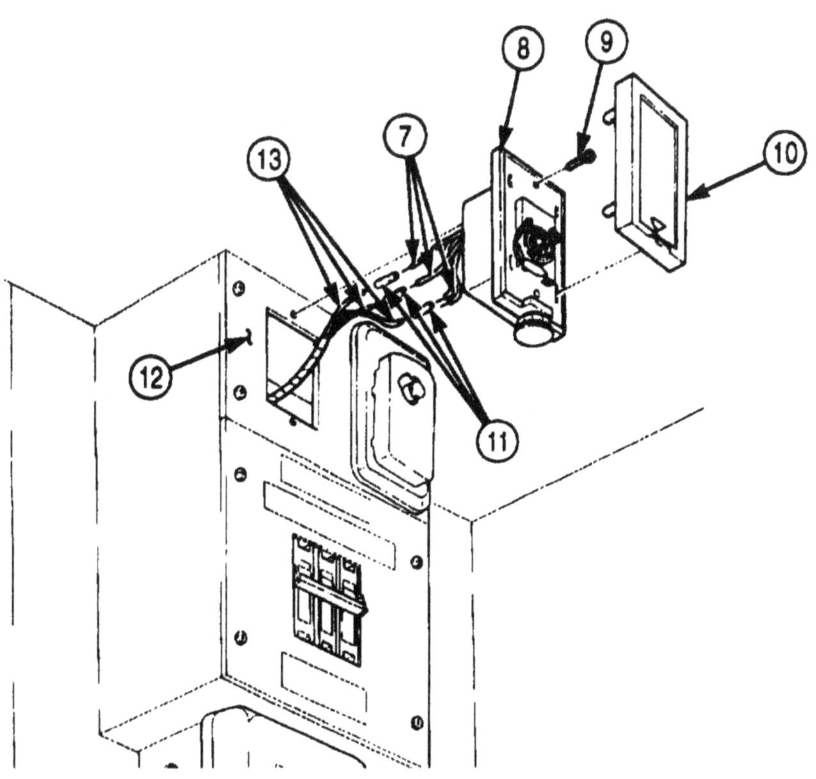

FOLLOW-ON TASK: Connect battery ground cables (para. 3-126).

4-166. BLACKOUT AND EMERGENCY LIGHT FIXTURES REPLACEMENT

THIS TASK COVERS:

a. Removal **b. Installation**

INITIAL SETUP:

APPLICABLE MODELS
M934/A1/A2

TOOLS
General mechanic's tool kit (Appendix E, Item 1)

MANUAL REFERENCES (TM)
TM 9-2320-272-10
TM 9-2320-272-24P

EQUIPMENT CONDITION
- Parking brake set (TM 9-2320-272-10).
- Battery ground cables disconnected (para. 3-126).
- Emergency and blackout lamps removed (para. 3-372).

a. Removal

1. Remove four screws (5) and housing (4) from van ceiling (6).
2. Remove four screws (2), two jumper cables (3), and wiring harness (1) from housing (4).

b. Installation

1. Install two jumper cables (3) and wiring harness (1) on housing (4) with four screws (2).
2. Install housing (4) on van ceiling (6) with four screws (5).

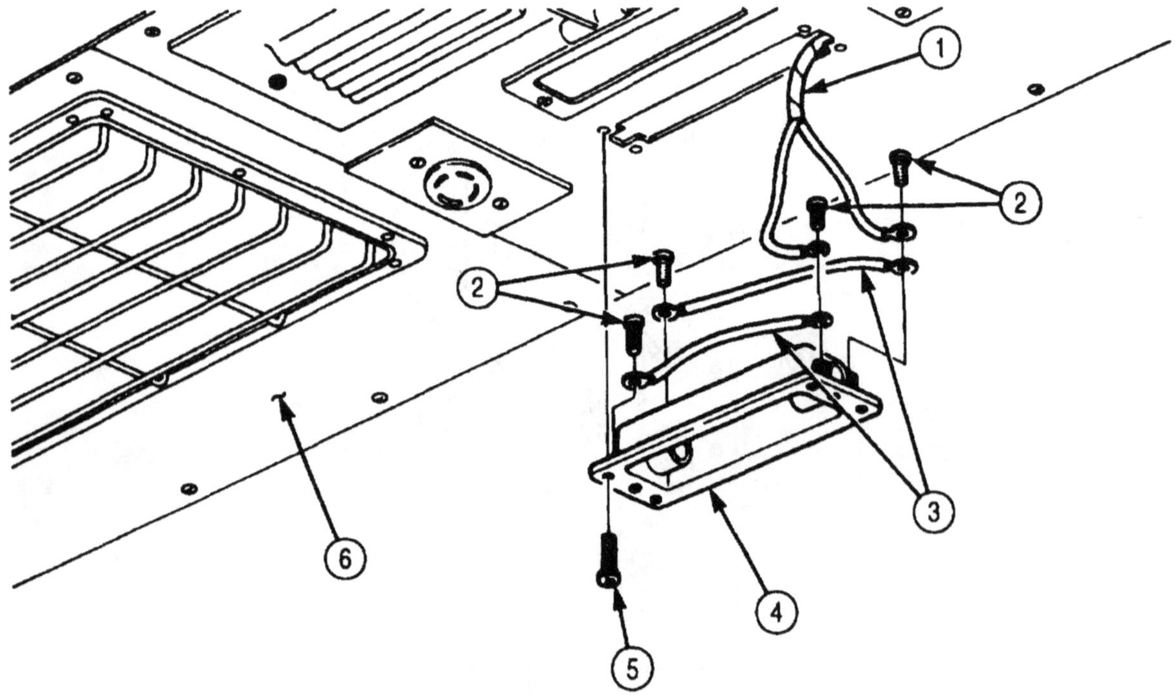

FOLLOW-ON TASKS:
- Connect battery ground cables (para. 3-126).
- Install emergency and blackout lamps (para. 3-372).

4-167. AIR CONDITIONER MANUAL STARTER SWITCHES REPLACEMENT

THIS TASK COVERS:

a. Removal **b. Installation**

INITIAL SETUP:

APPLICABLE MODELS
M934/A1/A2

TOOLS
General mechanic's tool kit (Appendix E, Item 1)
Electrical tool kit (Appendix E, Item 40)

MANUAL REFERENCES (TM)
TM 9-2320-272-10
TM 9-2320-272-24P

EQUIPMENT CONDITION
- Parking brake set (TM 9-2320-272-10).
- Battery ground cables disconnected (para. 3-126).

a. Removal

NOTE

Tag wires for installation.

1. Remove covers (5) and (7) from air conditioner starter switch boxes (1) and (8).
2. Remove six wires (3) from six insulated wire splices (2). Discard insulated wire splices (2).
3. Remove six wires (4) from six insulated wire splices (6). Discard insulated wire splices (6).

4-167. AIR CONDITIONER MANUAL STARTER SWITCHES REPLACEMENT (Contd)

4. Remove two nuts (2) from two connectors (10).
5. Remove two nuts (3) from two connectors (5).
6. Remove two nuts (6) from two connectors (5).
7. Remove four screws (8), manual starter switch box (7), and two connectors (5) from van body (9).
8. Remove four screws (1) and manual starter switch box (4) from van body (9).

b. Installation

1. Feed three wires (22), (19), (20), and (21) into manual starter switch box (4).
2. Install manual starter switch box (4) on two connectors (10) and van body (9) with four screws (1) and two nuts (2).
3. Install two connectors (5) and manual starter switch box (7) on van body (9) and manual starter switch box (4) with four screws (8) and two nuts (3) and (6).
4. Feed three wires (19) and (20) through two connectors (5) and manual starter switch box (7).
5. Connect six wires (22) and (21) to six switch leads (11) and (12) with six insulated wire splices (14).
6. Connect six wires (20) and (19) to six switch leads (16) and (17) with six insulated wire splices (18).
7. Install covers (13) and (15) on air conditioner starter switch boxes (4) and (7).

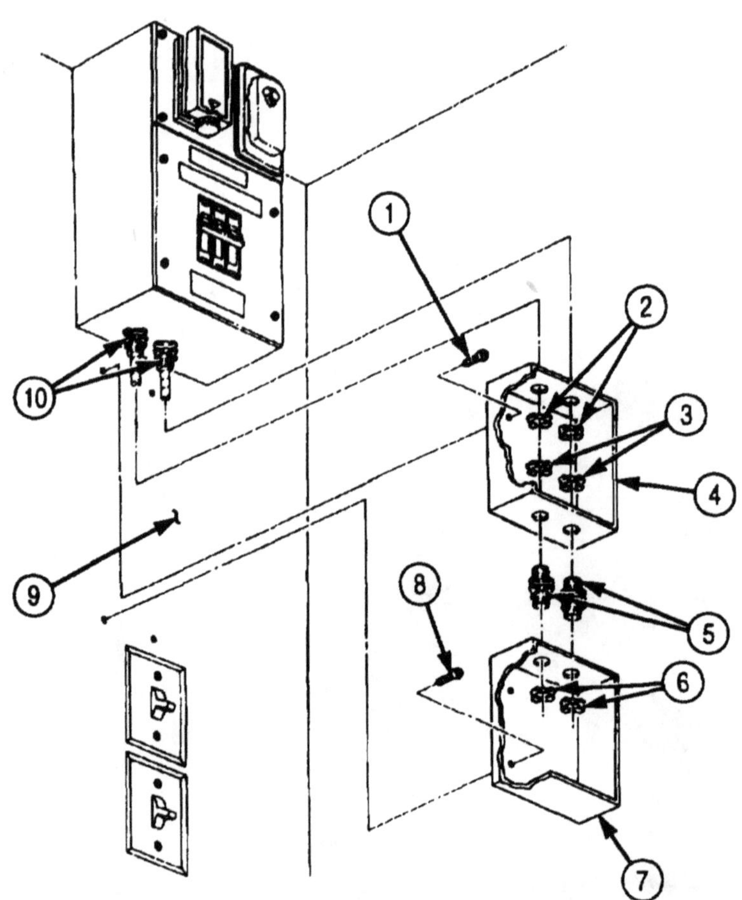

4-167. AIR CONDITIONER MANUAL STARTER SWITCHES REPLACEMENT (Contd)

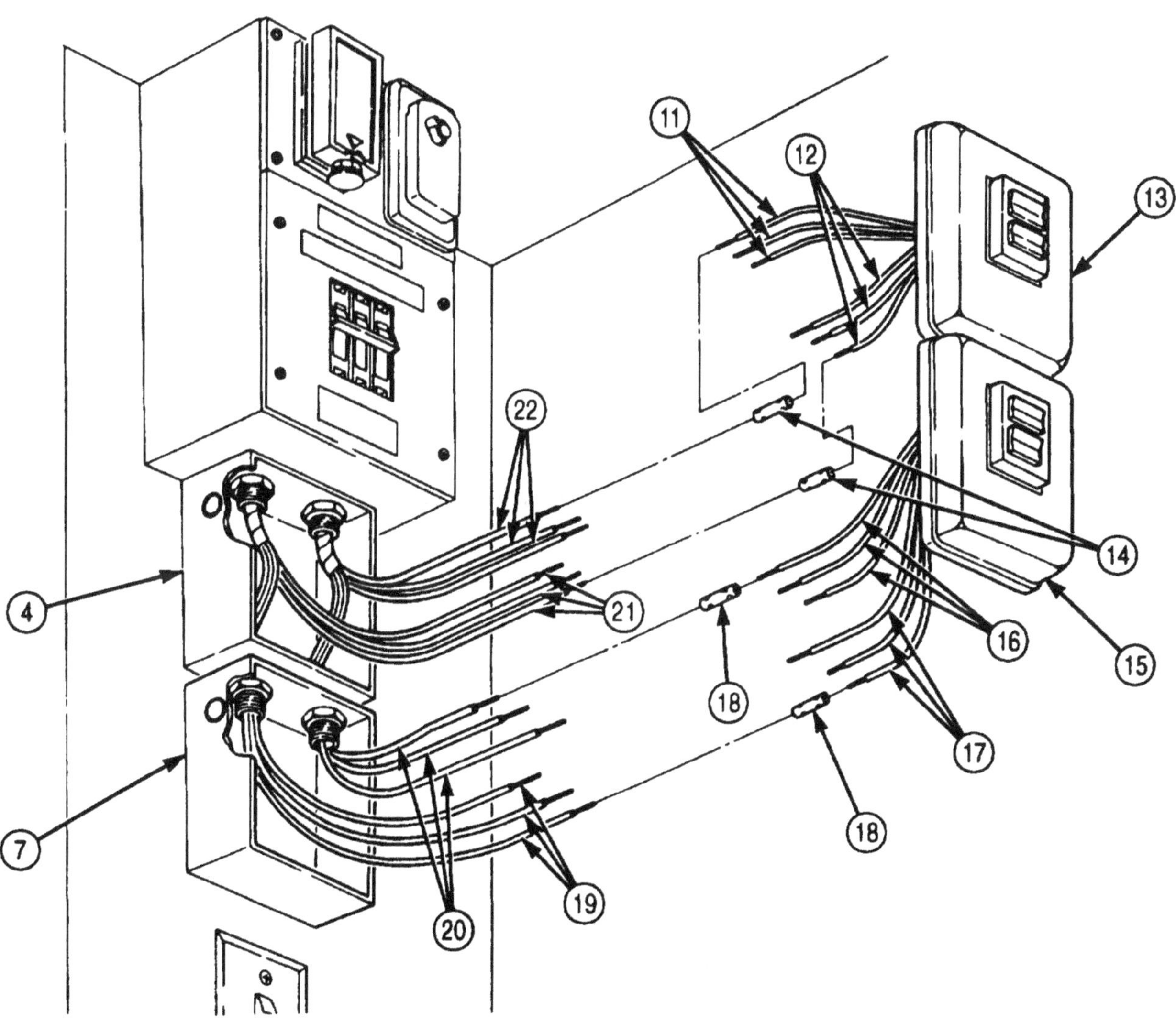

FOLLOW-ON TASK: Connect battery ground cables (para. 3-126).

4-168. FLUORESCENT LIGHT FIXTURES MAINTENANCE

THIS TASK COVERS:

a. Removal
b. Disassembly
c. Inspection
d. Assembly
e. Installation

INITIAL SETUP:

APPLICABLE MODELS
M934/A1/A2

TOOLS
General mechanic's tool kit (Appendix E, Item 1)
Electrical tool kit (Appendix E, Item 40)

MATERIALS/PARTS
Lockwasher (Appendix D, Item 381)

MANUAL REFERENCES (TM)
TM 9-2320-272-10
TM 9-2320-272-24P

EQUIPMENT CONDITION
- Parking brake set (TM 9-2320-272-10).
- Battery ground cables disconnected (para. 3-126).
- Fluorescent light tubes removed (para. 3-371).

NOTE

All eight fluorescent light fixtures are removed and installed the same way This procedure covers one fluorescent light fixture.

a. Removal

1. Remove twenty screws (7) and light fixture body (8) from van ceiling (1).
2. Remove nut (6), lockwasher (5), screw (9), and wire (4) from light fixture body (8). Discard lockwasher (5).
3. Remove two wires (3) and (10) and insulated wire splices (2) from light fixture body (8). Discard insulated wire splices (2).

b. Disassembly

1. Remove four screws (19) and light fixture (20) from light fixture body (8).
2. Remove six screws (18), nuts (17), and sockets (14) from light fixture (20).
3. Remove twelve screws (16) and wires (13) from six sockets (14).
4. Remove six starters (15) from sockets (14).
5. Remove grommet (11) from harness (12).
6. Remove harness (12) from light fixture (20).

c. Inspection

Inspect six starters (15) for burns or damage. Replace if burnt or damaged.

d. Assembly

1. Install six starters (15) on six sockets (14).
2. Install twelve wires (13) on six sockets (14) with twelve screws (16).
3. Install six sockets (14) on light fixture (20) with six screws (18) and nuts (17).
4. Install grommet (11) on harness (12).
5. Install light fixture (20) on light fixture body (8) with four screws (19).

4-168. FLUORESCENT LIGHT FIXTURES MAINTENANCE (Contd)

e. Installation

1. Install wire (4) on light fixture body (8) with nut (6), new lockwasher (5), and screw (9).
2. Install two wires (3) and (10) with insulated wire splices (2).
3. Install light fixture body (8) on van ceiling (1) with twenty screws (7).

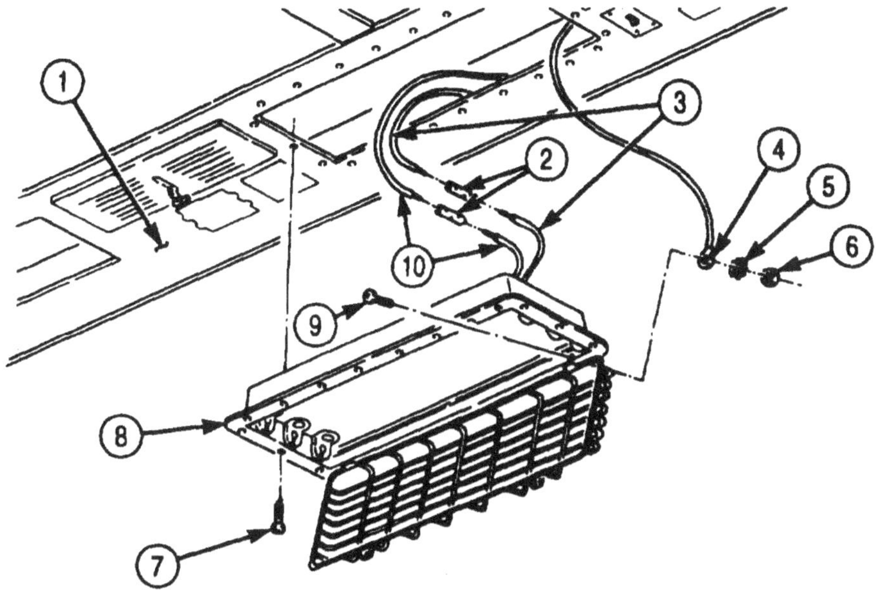

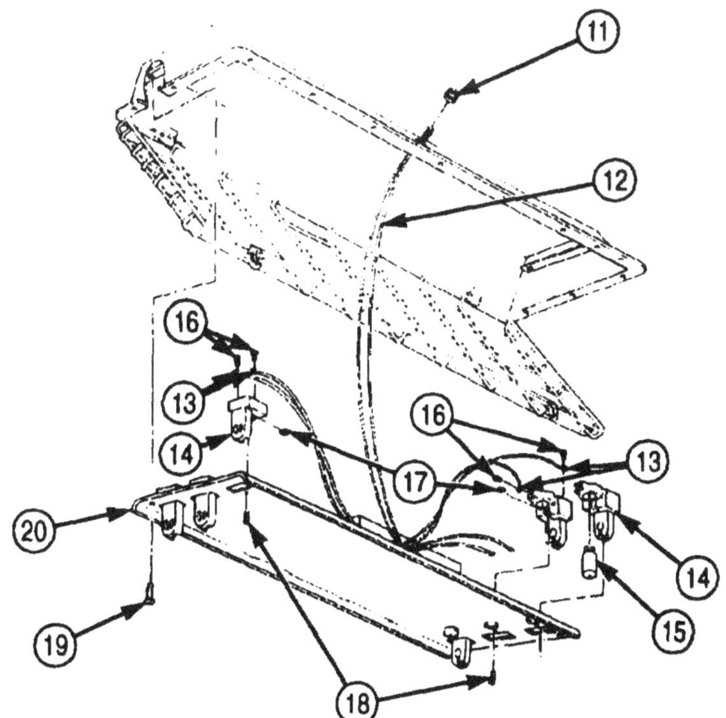

FOLLOW-ON TASKS: Connect battery ground cables (para. 3-126).
- Install fluorescent light tubes (para. 3-371).

4-169. 400 HZ CONVERTER REPLACEMENT

THIS TASK COVERS:

a. Removal **b. Installation**

INITIAL SETUP:

APPLICABLE MODELS
M934/A1/A2

TOOLS
General mechanic's tool kit (Appendix E, Item 1)

MATERIALS/PARTS
Two lockwashers (Appendix D, Item 390)
Two locknuts (Appendix D, Item 312)
Gasket (Appendix D, Item 211)

MANUAL REFERENCES (TM)
TM 9-2320-272-10
TM 9-2320-272-24P

EQUIPMENT CONDITION
- Parking brake set (TM 9-2320-272-10).
- Battery ground cables disconnected (para. 3-126).

a. Removal

1. Remove four screws (1), cover (2), and gasket (3) from converter (4). Discard gasket (3).
2. Remove six nuts (9) and wires (10) from converter (4).
3. Remove two lockwashers (8) and wiring harnesses (6) and (7) from converter (4).
4. Remove four nuts (17), washers (18), and screws (11) from support bracket (5).
5. Remove converter (4) from support bracket (5).
6. Remove two locknuts (19), U-bolt (15), and nuts (16) from support bracket (5). Discard locknuts (19).
7. Remove four nuts (13), screws (14), and support bracket (5) from van body (12).

b. Installation

1. Install support bracket (5) on van body (12) with four screws (14) and nuts (13).
2. Install two nuts (16) and U-bolt (15) on support bracket (5) with and new locknuts (19).
3. Install converter (4) on support bracket (5) with four screws (11), washers (18), and nuts (17).
4. Install two wiring harnesses (6) and (7) on converter (4) with two new lockwashers (8).
5. Install six wires (10) on converter (4) with six nuts (9).
6. Install gasket (3) and cover (2) on converter (4) with four screws (1).

4-169. 400 HZ CONVERTER REPLACEMENT (Contd)

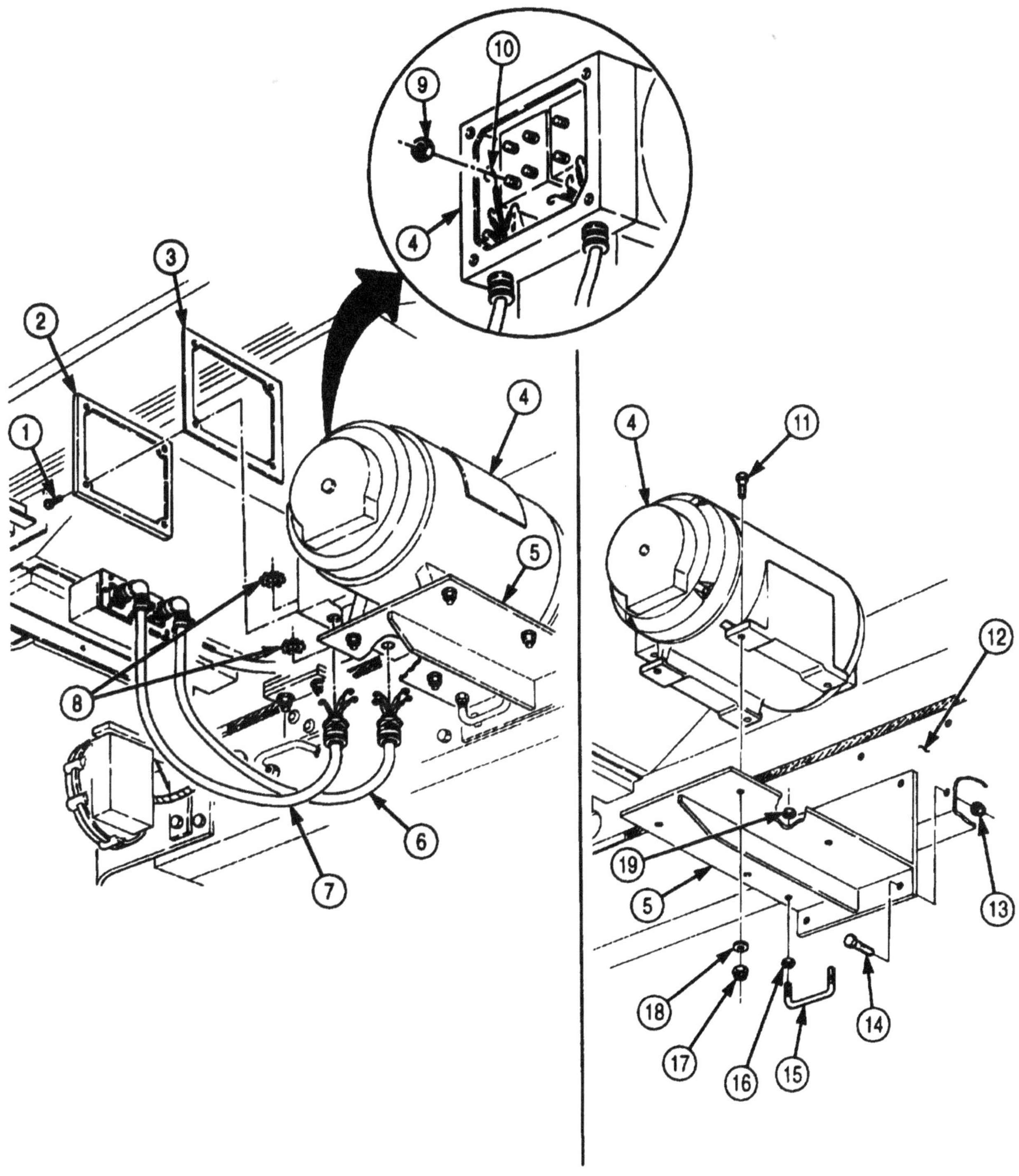

FOLLOW-ON TASK: Connect battery ground cables (para. 3-126).

4-170. ELECTRICAL LOAD CENTER CONDUIT REPLACEMENT

THIS TASK COVERS:

a. Removal **b. Installation**

INITIAL SETUP:

APPLICABLE MODELS
M934/A1/A2

TOOLS
General mechanic's tool kit (Appendix E, Item 1)

MATERIALS/PARTS
Sealing compound (Appendix C, Item 66)

PERSONNEL REQUIRED
TWO

MANUAL REFERENCES (TM)
TM 9-2320-272-10
TM 9-2320-272-24P
TM 9-237

EQUIPMENT CONDITION
Parking brake set (TM 9-2320-272-10).

a. Removal

1. Remove cables (5) and (6) from connectors (4) and (8).
2. Remove eight screws (7) and cover (9) from junction box (1).
3. Disconnect wiring harness (2) and (3) from connectors (4) and (8) (TM 9-237).
4. Remove six screws (10) and cover (11) from load center (14).
5. Pull wiring harnesses (2) and (3) from conduit (20) and load center (14).
6. Remove nut (15), coupling (13), and nut (12) from conduit (20) and load center (14).

NOTE

Assistant will help with step 7.

7. Remove two nuts (18) and coupling (19) from conduits (20) and (16). Remove conduit (20) from van body (17).

b. Installation

1. Position conduit (20) on van body (17).
2. Install coupling (19) on conduits (20) and (16) with two nuts (18).
3. Install coupling (13) on conduit (20) and load center (14) with nut (12) and nut (15).
4. Route wiring harnesses (2) and (3) through conduit (20) and load center (14).
5. Install cover (11) on load center (14) with six screws (10).
6. Connect wiring harnesses (2) and (3) to connectors (4) and (8) (TM 9-237).

NOTE

Apply sealing compound before performing step 7.

7. Install cover (9) on junction box (1) with eight screws (7).
8. Install cables (5) and (6) on connectors (4) and (8).

4-170. ELECTRICAL LOAD CENTER CONDUIT REPLACEMENT (Contd)

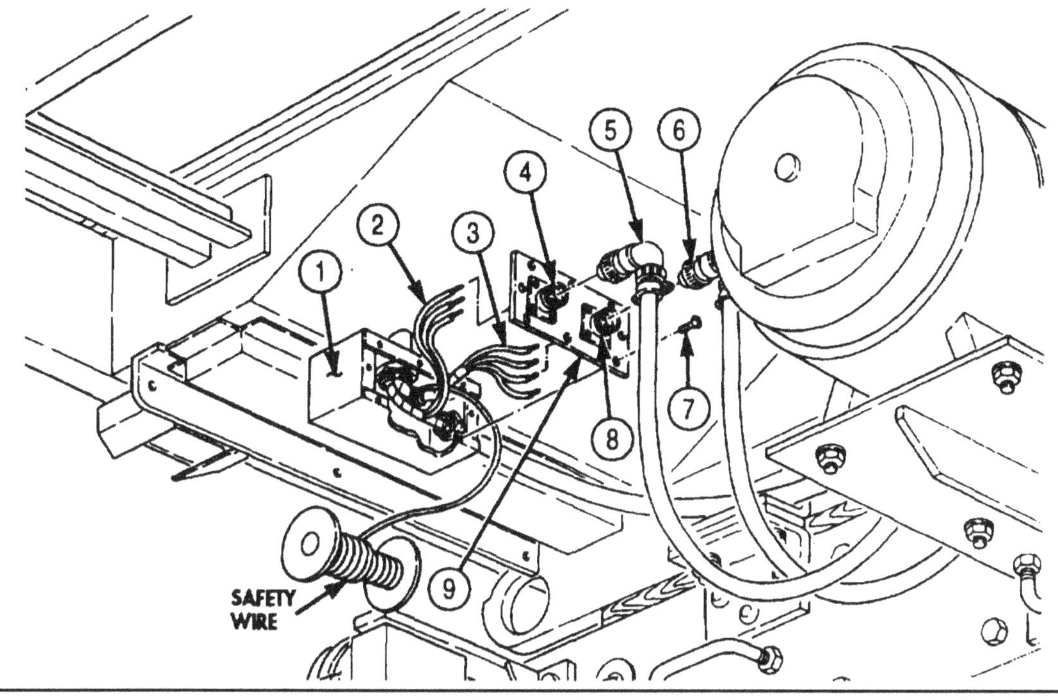

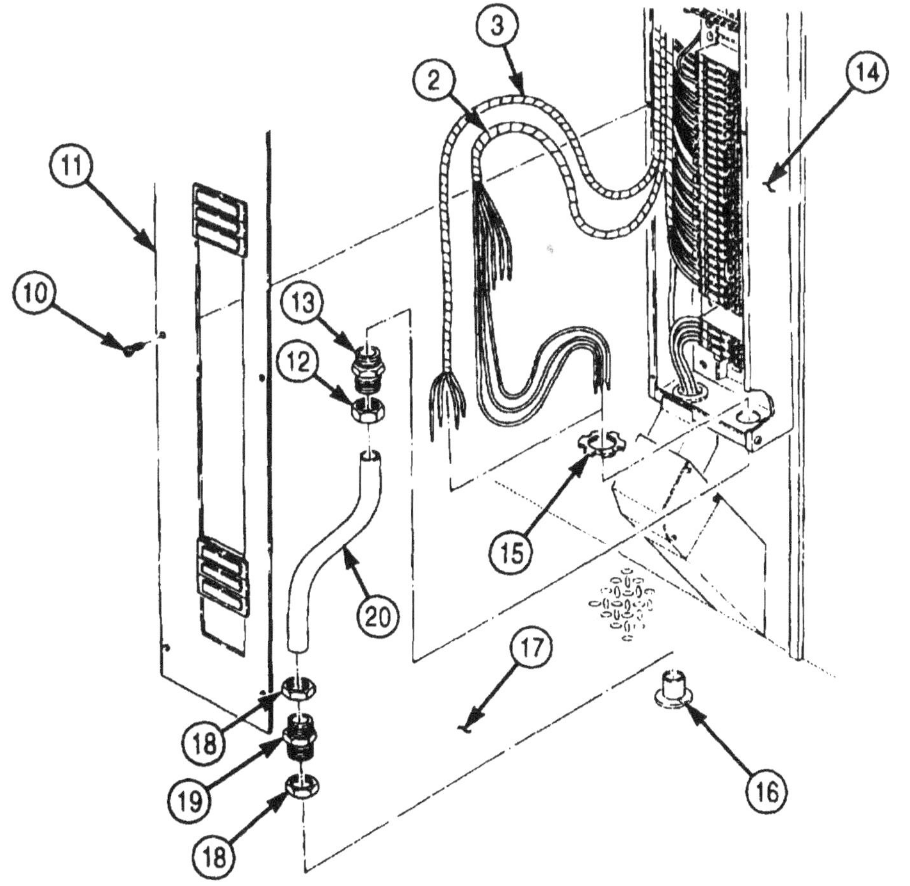

Section XV. WINCH, HOIST, AND POWER TAKEOFF MAINTENANCE

4-171. WINCH, HOIST, AND POWER TAKEOFF MAINTENANCE INDEX

4-172. FRONT WINCH REPAIR

THIS TASK COVERS:

a. Disassembly into Subassemblies
b. Disassembly of Subassemblies
c. Cleaning, Inspection, and Repair
d. Assembly of Subassemblies
e. Assembly of Winch

INITIAL SETUP:

APPLICABLE MODELS

M925/A1/A2, M928/A1/A2, M930/A1/A2, M932/A1/A2, M936/A1/A2

TOOLS

General mechanic's tool kit (Appendix E, Item 1)
Spring tester (Appendix E. Item 132)
Outside micrometer (Appendix E, Item 80)
Inside micrometer (Appendix E, Item 83)
Dial indicator (Appendix E, Item 36)
Puller kit (Appendix E, Item 101)
Mallet
Arbor press
Soft-faced hammer
Chain
Lifting device

MATERIALS/PARTS

Cotter pin (Appendix D, Item 46)
Two gaskets (Appendix D, Item 214)
Two seals (Appendix D, Item 607)
Nine lockwashers (Appendix D, Item 350)
Two keys (Appendix D, Item 265)
Four lockwashers (Appendix D, Item 393)
Seal (Appendix D, Item 608)
Gasket (Appendix D, Item 212)

MATERIALS/PARTS (Contd)

Woodruff key (Appendix D, Item 731)
Woodruff key (Appendix D, Item 730)
Gasket (Appendix D, Item 214)
Gasket (Appendix D, Item 213)
Four lockwashers (Appendix D, Item 377)
GAA grease (Appendix C, Item 28)
Sealing compound (Appendix C, Item 65)
Rag (Appendix C, Item 58)

REFERENCES (TM)

TM 9-2320-272-10
TM 9-2320-272-24P

EQUIPMENT CONDITION

- Parking brake set (TM 9-2320-272-10).
- Front winch removed (para. 3-329).
- Winch oil drained (LO 9-2320-272-12).
- Winch roller removed (para. 4-175).
- Winch cable tensioner removed (para. 4-173).
- Winch level wind removed (if equipped) (para. 4-174).

GENERAL SAFETY INSTRUCTIONS

Use hoist during front winch repair to prevent injury to personnel.

WARNING

Vehicle front winch weighs 353 lb (160 kg), except the M936/A1/A2 winch, which weighs 472 lbs (214 kg). Use hoist during repair to prevent injury to personnel.

a. Disassembly into Subassemblies

1. Remove four screws (3), lockwashers (4), and top channel (2) from winch (1). Discard lockwashers (4).
2. Remove four screws (6), lockwashers (5), and rear channel (7) from winch (1). Discard lockwashers (5).
3. Remove drag brake adjusting screw (12) and spring (13) from end frame (8).
4. Remove nut (11) from end of tie rod (10).

NOTE

Support drum with overhead hoist and chain.

5. Separate end frame (8) from drum (15) and drum shaft (14).

4-172. FRONT WINCH REPAIR (Contd)

6. Remove clutch (17) and thrust washer (16) from drum shaft (14).
7. Remove tie rod (10) from gearcase (9).
8. Remove two keys (18) from drum shaft (14). Discard keys (18).
9. Remove thrust ring (20), drum (15), and seal (19) from drum shaft (14). Discard seal (19).

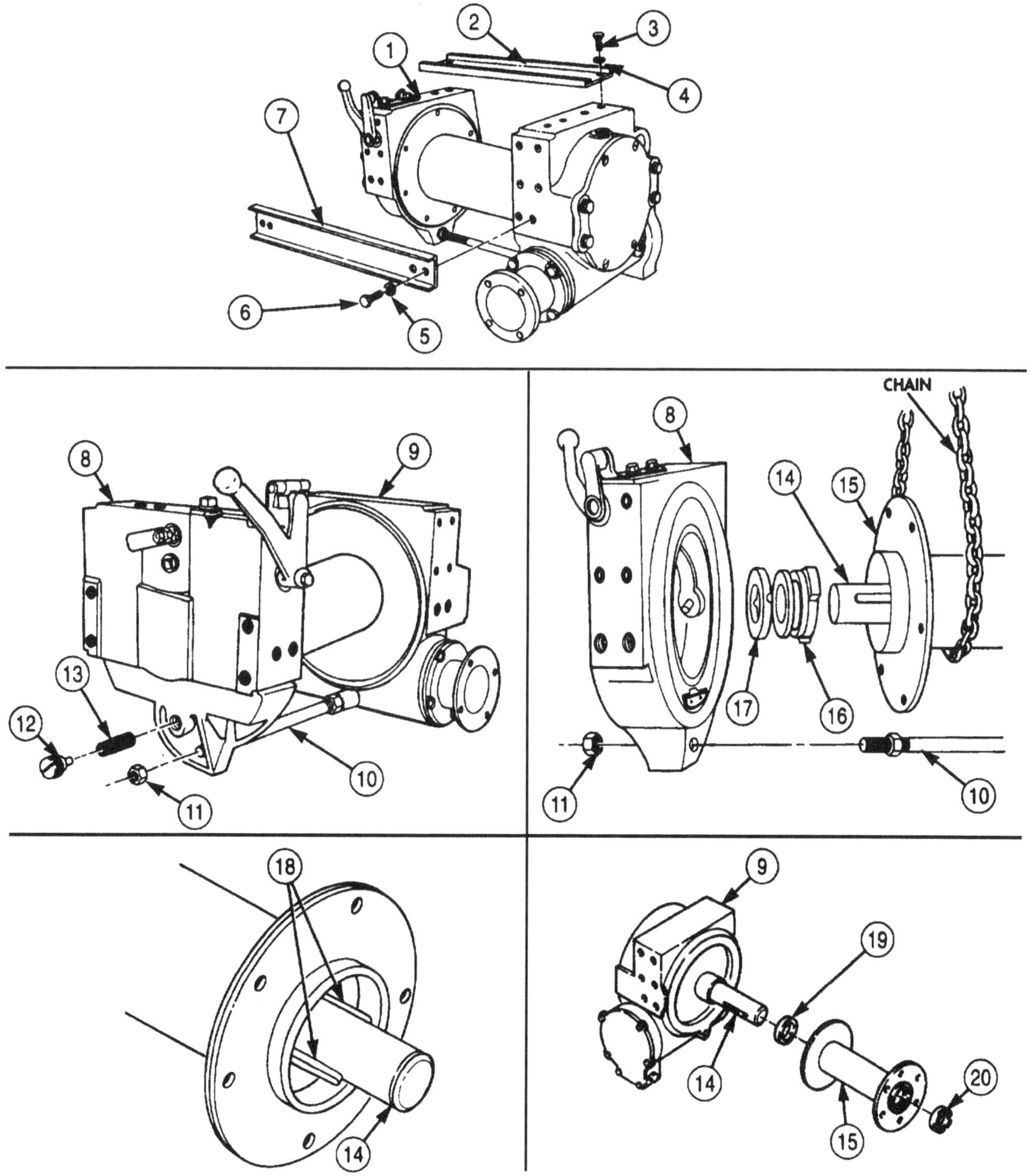

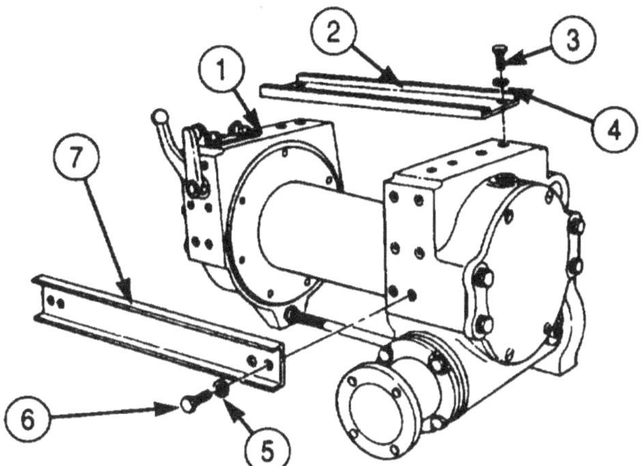

4-172. FRONT WINCH REPAIR (Contd)

b. Disassembly of Subassemblies

1. Remove four screws (7), screws (8), lockwashers (9), end cover (6), and gasket (5) from gearcase (4). Discard lockwashers (9) and gasket (5). Clean gasket remains from mating surfaces.
2. Remove four screws (15), lockwashers (16), and winch motor adapter (14) from gearcase (4). Discard lockwashers (16).
3. Remove adapter gasket (13) from adapter (14). Discard gasket (13). Clean gasket remains from mating surfacea.
4. Remove cotter pin (2), pin (1), and coupling (17) from worm gear shaft (10). Discard cotter pin (2).
5. Remove seal retainer (12) and gasket (11) from worm gear shaft (10). Discard gasket (11). Clean gasket remains from mating surfaces.
6. Remove seal (3) from worm gear shaft (10). Discard seal (3).
7. Remove six screw-assembled washers (21), automatic brake cover (18), and gasket (19) from automatic brake housing (20). Discard gasket (19). Clean gasket remains from mating surfaces.
8. Remove adjusting screw (24), washer (25), and O-ring (26) from brake housing (20).
9. Remove spring (27) and brake band (22) from housing (20).
10. Remove screw (29), lockwasher (30), and washer (28) from automatic brake drum (23). Discard lockwasher (30).

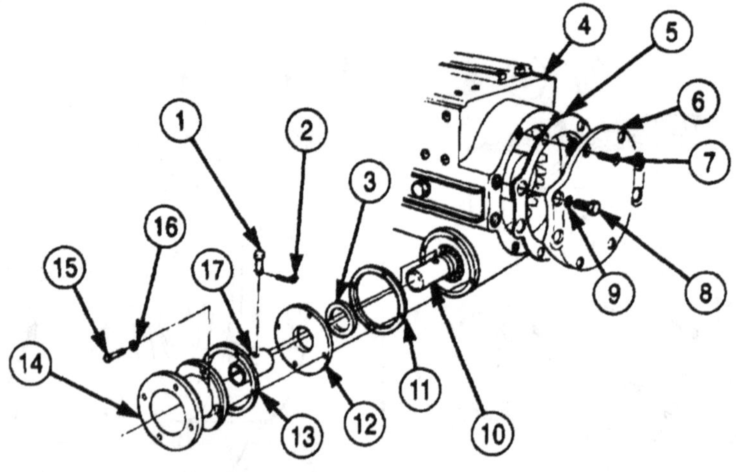

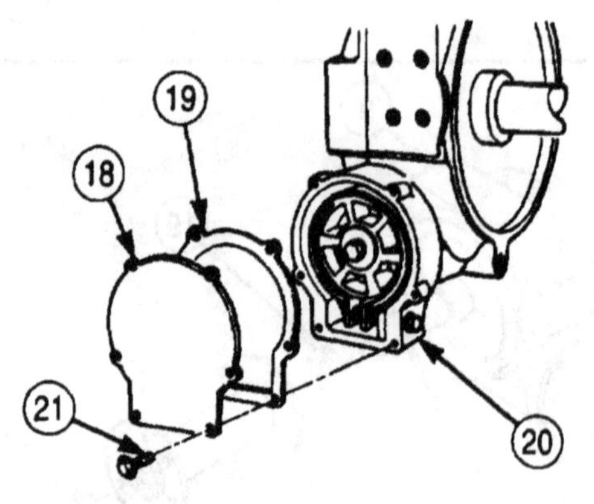

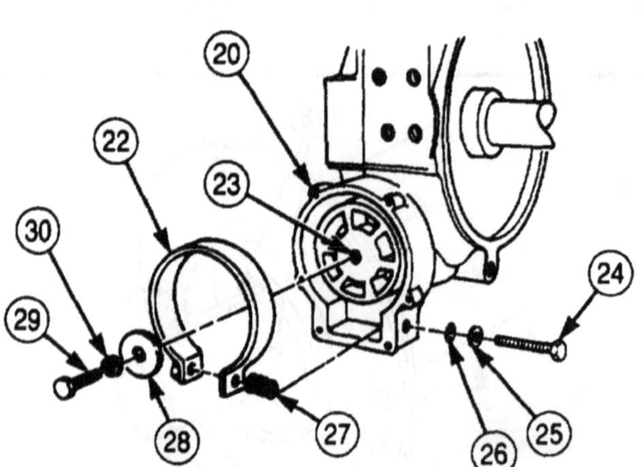

4-172. FRONT WINCH REPAIR (Contd)

11. Using a puller, remove brake drum (23) and woodruff key (31) from worm gear shaft (10). Discard woodruff key (31).

NOTE

Ensure gearcase and brake housing are marked for assembly.

12. Remove four screws (35), lockwashers (32), and brake housing (18) from gearcase (4). Discard lockwashers (32).
13. Remove brake housing gasket (34) from gearcase (4). Discard gasket (34). Clean gasket remains from mating surfaces.
14. Remove brake housing seal (33) from gearcase (4). Discard seal (33).

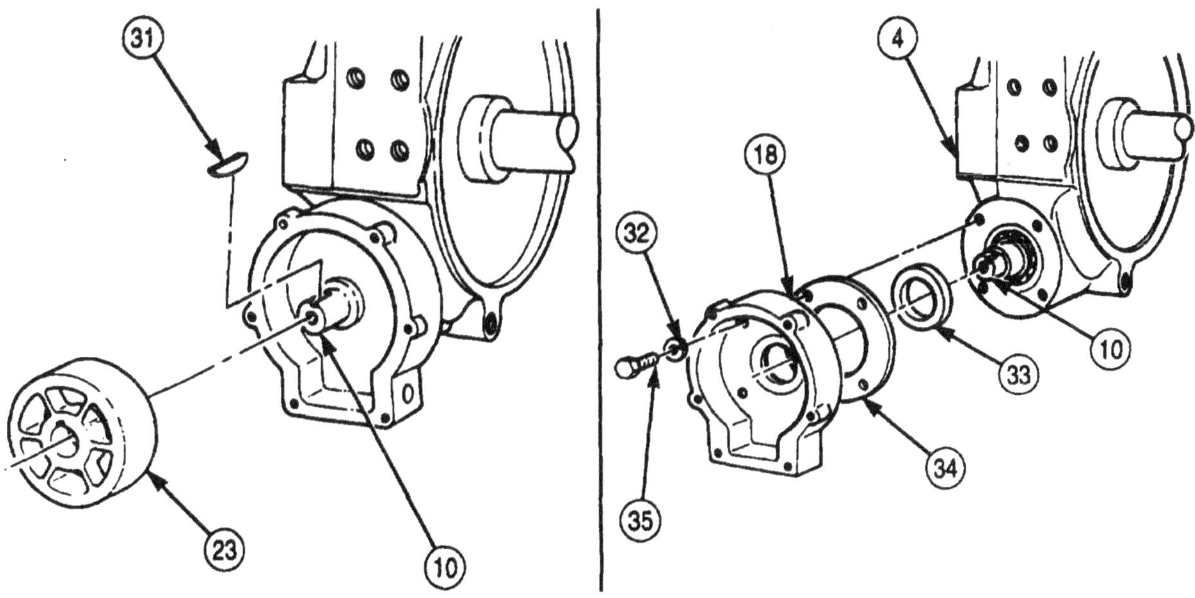

4-172. FRONT WINCH REPAIR (Contd)

NOTE

Ensure a rag is placed in bottom of gearcase under worm gear to protect worm gear during removal.

16. Tap brake end of worm gear shaft (2) with soft-faced hammer and allow worm gear shaft (2) to drop on rag in gearcase (1).
16. Press front bearing (5) off worm shaft (2).
17. Remove drum shaft (4) and gear (3) from gearcase (1).
18. Using hammer and drift pin, remove worm gear shaft (2) from gearcase (1).
19. Remove rear bearing (6) from gearcsse (1).
20. Remove drag brake (8) from end frame (7).
21. Remove bearing sleeve (9) from end frame (7).

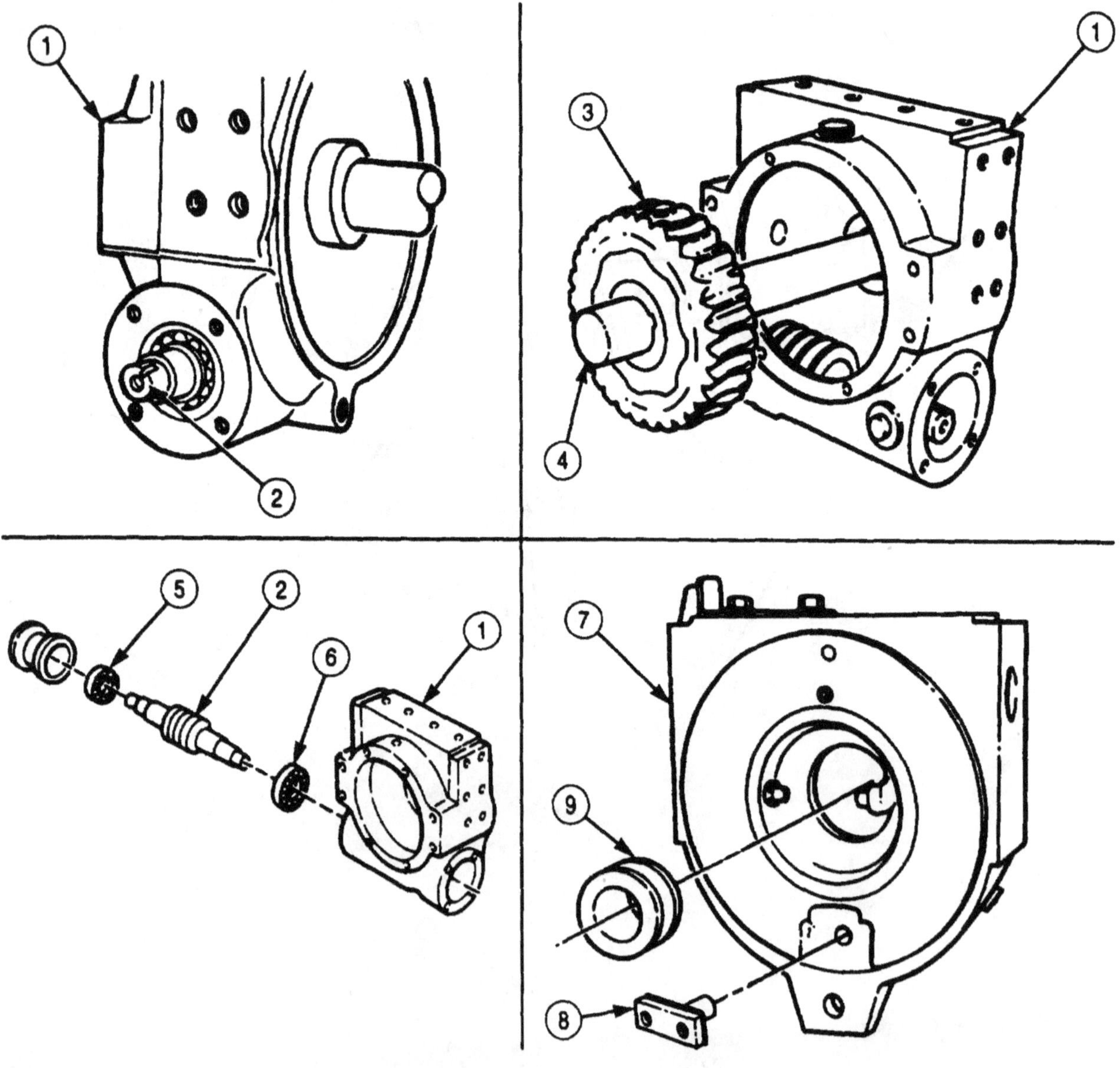

4-172. FRONT WINCH REPAIR (Contd)

22. Remove pipe plug (12) from end frame (7).
23. Remove setscrew (10) from shifter fork (11).
24. Remove two nuts (13), poppet latch (14), nut (16), and poppet assembly (15) from end frame (7).
25. Remove woodruff key (22) from shifter shaft (17). Discard woodruff key (22).
26. Remove shifter shaft (17) from end frame (7). Ball (19) and spring (18) will fall out of shifter shaft (17) as it is removed.
27. Remove oil seals (21) and (20) from end frame (7).

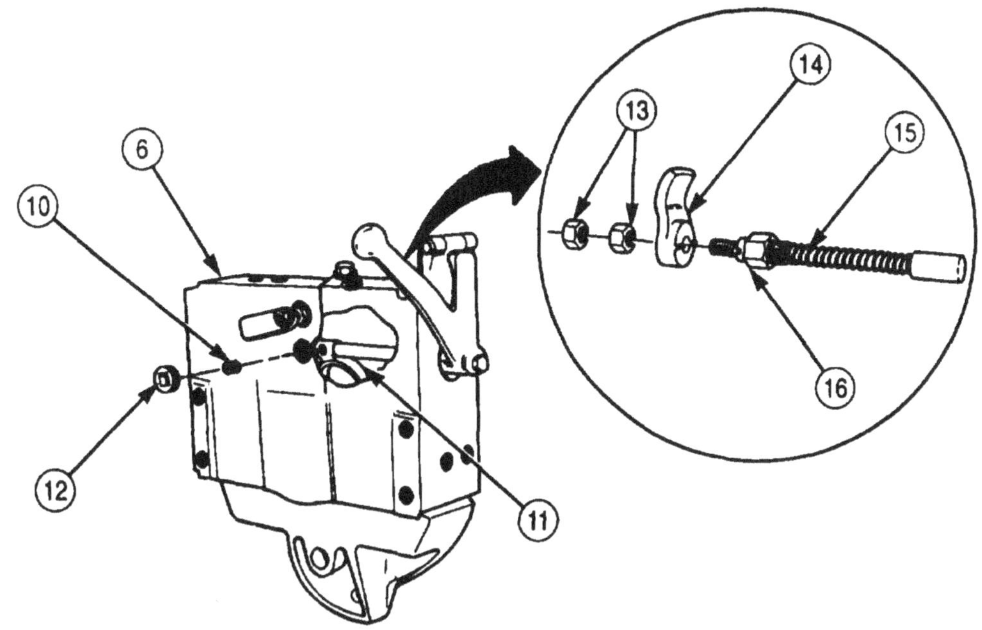

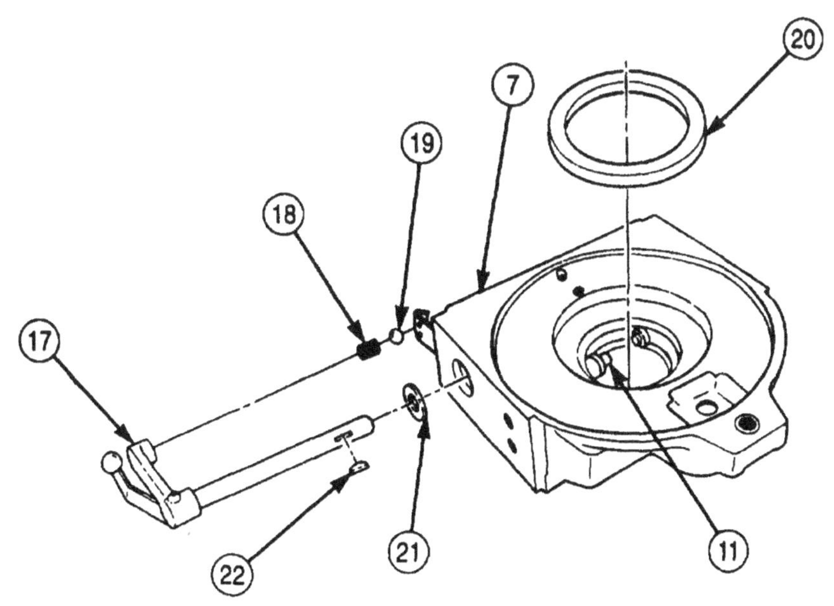

4-172. FRONT WINCH REPAIR (Contd)

c. Cleaning, Inspection, and Repair

1. Wipe thrust ring (1) clean and inspect for cracks and wear. Measure thickness with micrometer. Discard thrust ring (1) if cracked or if thickness measurement is less than 0.049 in. (12.5 mm).
2. Wipe clutch (2) clean and inspect where clutch (2) engages drum (4). Replace if cracked or chipped.
3. Clean thrust ring (3) and inspect for cracks and wear. Measure thickness with micrometer. Discard if cracked or if thickness is less than 0.486 in. (12.3 mm).
4. Clean drum (4) and inspect for breaks, cracks, and elongated screw holes. Refer to para. 2-14 for cleaning instructions. Replace if broken, cracked, or if holes are elongated more than 0.063 in. (1.59 mm).
5. Clean two bushings (5) and measure thickness for wear. Use a micrometer and replace if inner diameter exceeds 2.136 in. (54.25 mm).

NOTE

Perform step 6 only if bushings are to be replaced.

6. Remove bushings (5) from drum (4) and replace with new bushings (5).
7. Clean drum shaft (6) and measure diameter of all five bearing surfaces with micrometer. Replace if outer diameter is less than 2.123 in. (53.92 mm) on any bearing surface.
8. Clean drum shaft gear (7) and inspect for broken, chipped, or scored teeth. Replace if teeth are broken, chipped, or scored.

NOTE

Perform steps 9 and 10 only when shaft gear or drum shaft are to be replaced.

9. Place in arbor press and, using a mandrel, press drum shaft (6) out of drum shaft gear (7).

NOTE

Remove drum shaft gear keys only if they are damaged. If damaged, discard keys and install new keys.

10. Using arbor press, press drum shaft gear (7) on drum shaft (6) until keys (8) are centered in gear (7).
11. Clean winch motor adapter (9) and inspect worm gear shaft (11) and worm gear bearings (10) for breaks and cracks. Replace if broken or cracked.
12. Clean and inspect worm gear shaft (11) and worm gear bearings (10) for chips and breaks on worm gear shaft (7), rough or chipped ball bearings, and races. Replace if ball bearings or races are rough or chipped.
13. Inspect gearcase (12) for breaks and cracks. Replaced if broken or cracked.
14. Inspect gearcase cover (13) for breaks or cracks. Replace if broken or cracked.
15. Inspect automatic brake housing (14) for cracks. Replace if cracked.
16. Inspect brakedrum (15) for cracks and grooves. Replace if cracked or grooved.
17. Inspect spring (16) for cracked or collapsed coils. Discard if coils are collapsed or cracked.
18. Measure free length of spring (16). Discard spring (16) if free length is less than 11.5 in. (38 mm).
19. Test spring (16) with spring tester and torque wrench. Discard if test reading is less than 52 lb-ft (71 N•m) when compressed to 1 in. (25.4 mm).

4-172. FRONT WINCH REPAIR (Contd)

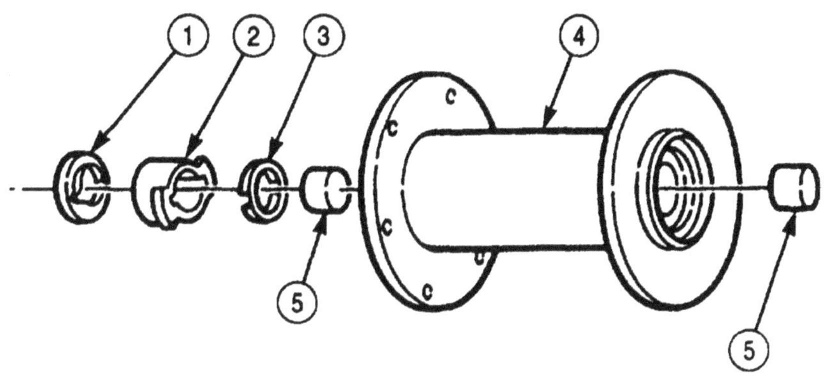

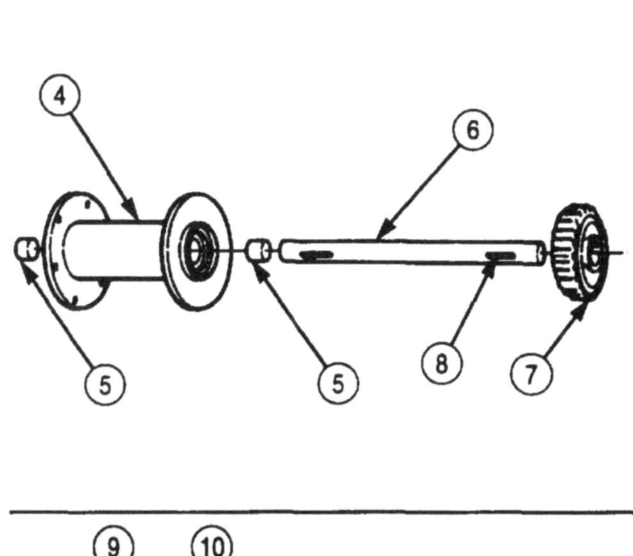

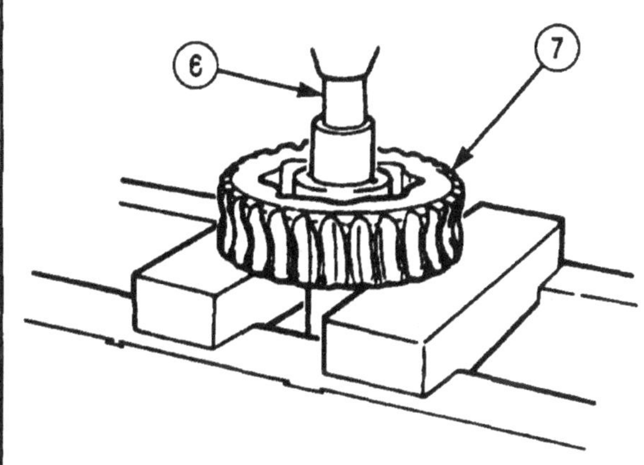

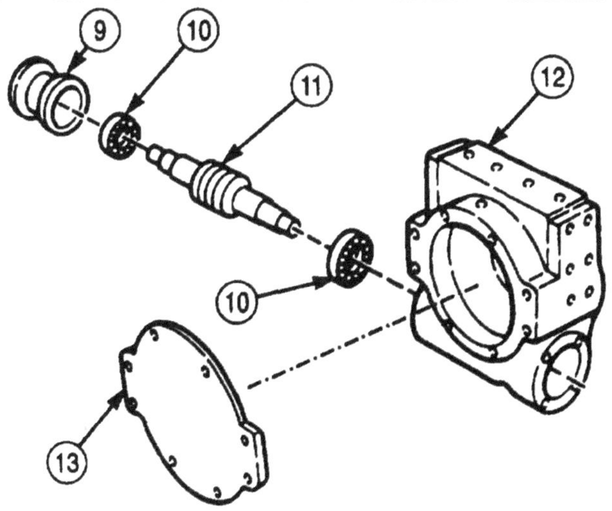

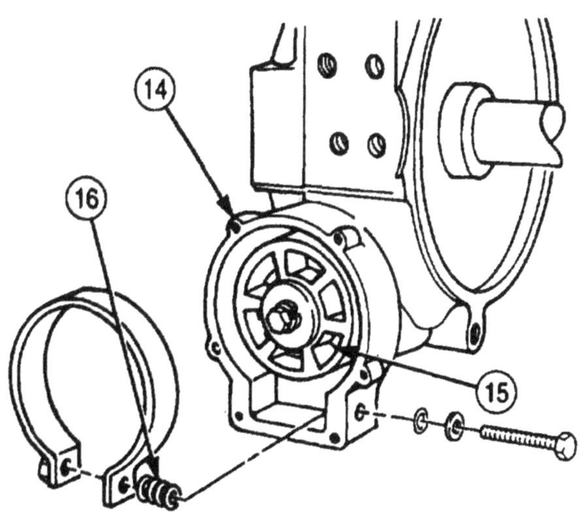

4-172. FRONT WINCH REPAIR (Contd)

20. Inspect end frame sleeve pin (5) for wear and for sufficient height to hold end frame sleeve (2) away from end frame (4).
21. Inspect end frame (4) for breaks and cracks. Replace if cracked or broken.
22. Inspect poppet spring assembly (7) for cracks and bends. Replace if cracked or bent.
23. Inspect hinge (6) for breaks. Replace if broken.
24. Wipe clean and inspect clutch shifter shaft (9) and fit of new woodruff key (8) for bends, snug key fit, and scoring on seal surface. Replace shaft (9) if bent, loose key fit, or scored.
25. Inspect clutch fork (3) for bends or cracks. Replace if bent or cracked.
26. Clean end frame sleeve (2) and inspect for cracks. Replace if cracked.
27. Clean bushing (1) and, using a micrometer, measure inner diameter for wear. Replace if inner diameter exceeds 2.1 in. (54 mm).
28. Inspect drag brake (10) for breaks and cracks. Replace if broken or cracked.

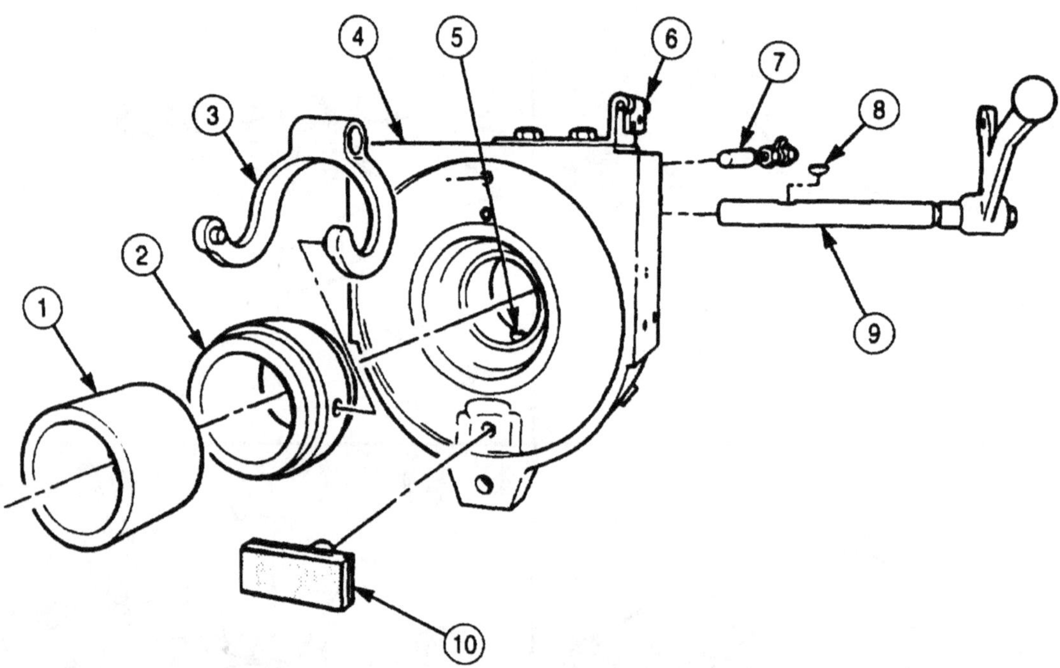

4-172. FRONT WINCH REPAIR (Contd)

d. Assembly of Subassemblies

NOTE

Coat outside diameter of seals with sealing compound before assembly.

1. Install new oil seals (13) and (15) in end frame (4).
2. Insert shifter shaft (9) through seal (15) and then install new woodruff key (8) in shifter shaft (9).
3. Position shifter fork (14) in end frame (4) on shifter shaft (9) and install shifter lever ball (12) and spring (11) in handle of shifter shaft (9).
4. Install setscrew (19) into shifter fork (14) and shifter shaft (9).
5. Install pipe plug (18) on end frame (4).
6. Coat threads of nut (22) with sealing compound and install poppet spring assembly (7) on end frame (4) with nut (22), latch (21), and two nuts (20).
7. Install bearing sleeve (2) in end frame (4).
8. Install drag brake (10) and spring (16) in end frame (4) with screw (17). Install screw (17) all the way in (adjustment of drag brake (10) is made after winch is installed on vehicle). Refer to para. 3-324.

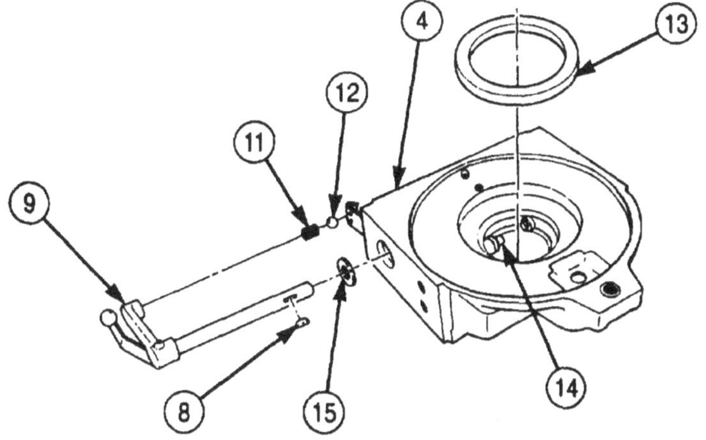

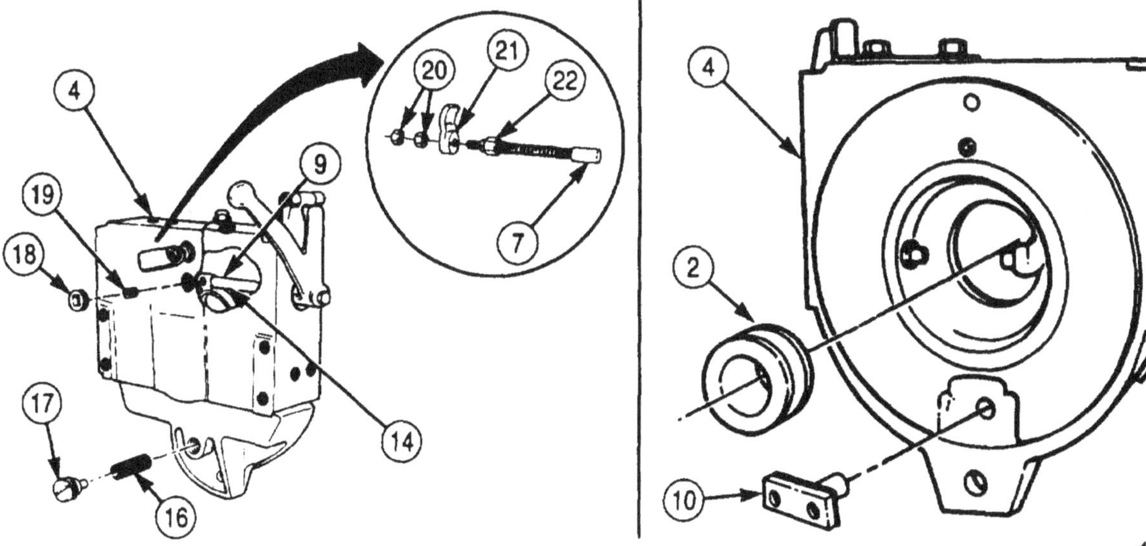

4-172. FRONT WINCH REPAIR (Contd)

9. Using arbor press, install two bearings (6) on worm gear shaft (3).
10. Place worm gear shaft (3) and bearings (5) in bottom of gearcase (1). Do not install in gearcase (1) at this time.

NOTE

Keyway end of worm gear shaft must extend from the marked side of gearcase.

11. Position drum shaft and gear (4) in gearcase (1).
12. Install worm gear shaft (3) and bearings (5) in gearcase (1). Tap shaft (3) with soft-faced hammer until bearings (6) are flush with gearcase (1) and worm gear (2) meshes with gear (4).
13. Using sealing compound, coat outside diameter of new brake housing seal (6) and install in gearcase (1).
14. Coat new gasket (7) with light film of GM grease and position on gearcase (1).
15. Matching alignment marks, install brake housing (8) on gearcase (1) with four new lockwashers (10) and screws (9).
16. Place woodruff key (11) in worm gear shaft (3).
17. Align brake drum (12) with key (13) on worm gear shaft (3) and install with washer (14), new lockwasher (16), and screw (16).

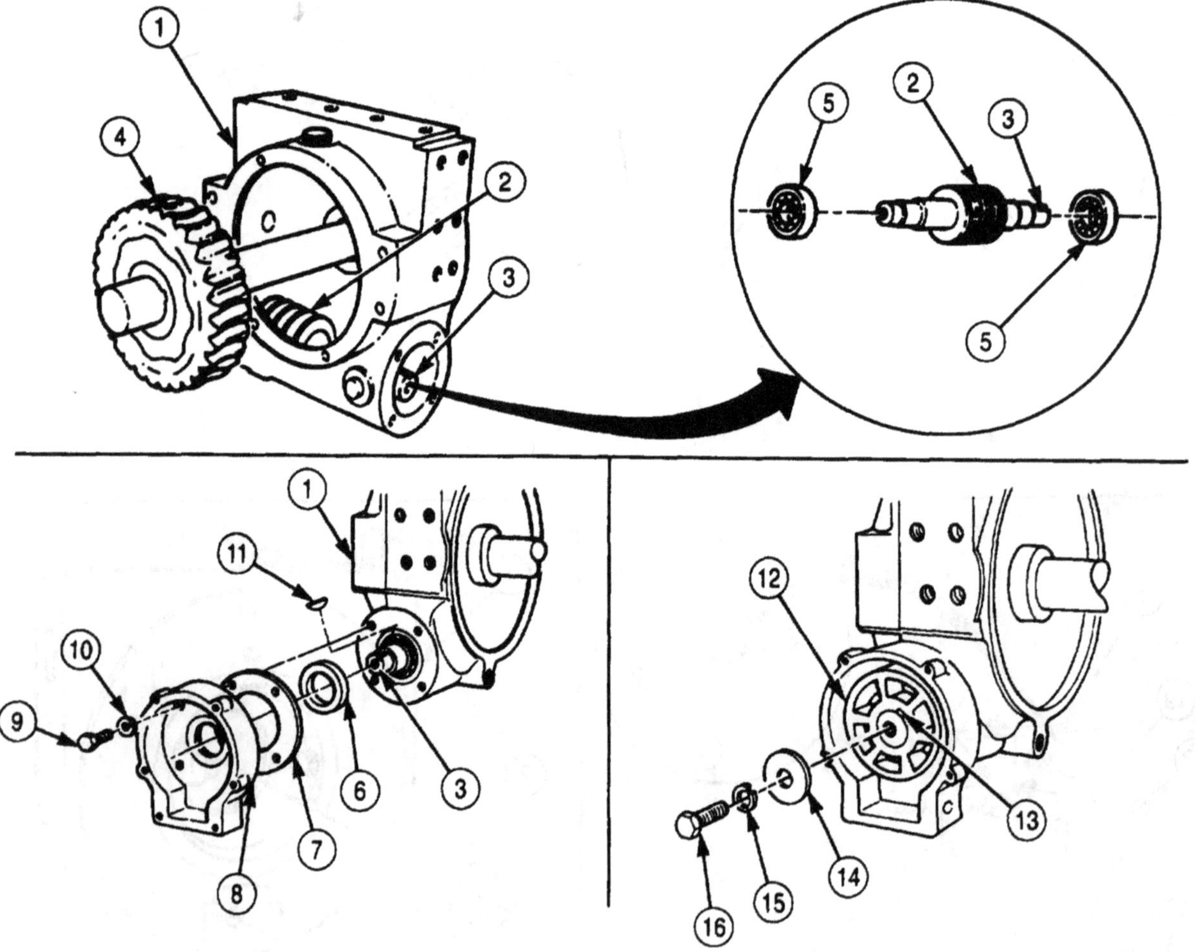

4-172. FRONT WINCH REPAIR (Contd)

18. Coat new gasket (22) with light film of GM grease and position on gearcase (1). Measurements in step 24 may require more gaskets (22).
19. Coat outside diameter of new seal (21) with sealing compound and install on worm gear shaft (3).
20. Position seal retainer (23) over worm gear shaft (3)
21. Align holes in coupling (18) with hole in worm gear shaft (3) and install coupling (18) on shaft (3) with pin (19) and new cotter pin (20).
22. Apply a light film of GAA grease to coat gasket (24) and position on seal retainer (23).
23. Align screw holes in adapter (25) with holes in gearcase (1) and install adapter (25) on gearcase (1) with four new lockwashers (17) and screws (26).
24. Place dial indicator on gearcase (1) and measure end play of worm gear shaft (3). If end play is less than 0.005 in. (0.127 mm), add gaskets (22) as indicated in step 18 until end play reads greater than 0.006 in. (0.127 mm), but less than 0.015 in.(0.37 mm). Remove dial indicator when proper adjustment has been obtained.

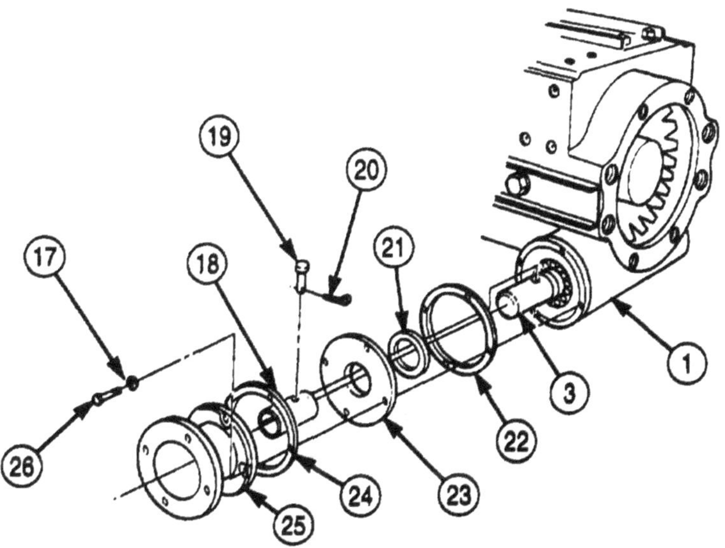

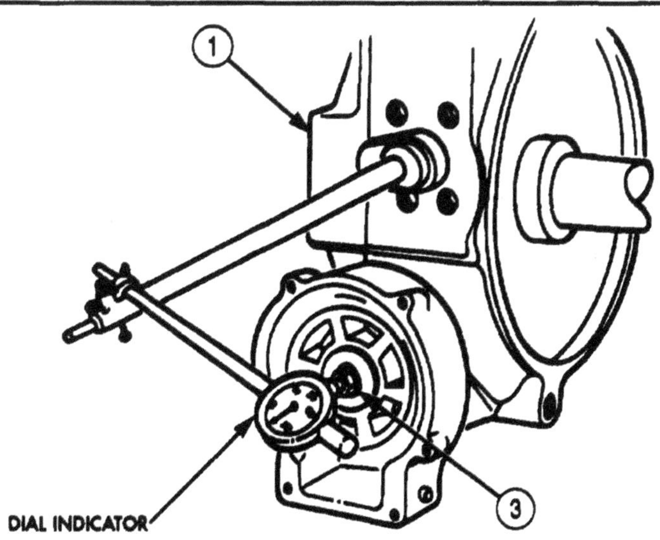

4-172. FRONT WINCH REPAIR (Contd)

25. Install brake band (11) around automatic brake drum (9). Position cage nut (10) to left.
26. Position spring (12) between housing (5) and brake band (11) and install with lockwasher (8), washer (7), and screw (6) through one end of brake band (11) into nut (10) until spring (12) is compressed 1.125-1.187 in. (28-30 mm) in length.
27. Coat new gasket (15) with light film of GM grease and position on housing (5).
28. Position cover (14) over gasket (15) and install with six screw-assembled washers (13).
29. Coat outer diameter of seal (4) with sealing compound and install in drum (2).
30. Install drum (2) on drum shaft (1).

NOTE

Support drum with overhead hoist and chain.

31. Install thrust washer (3) on drum shaft (1). Ensure flat side of thrust washer (3) is against drum (2).
32. Align two new keys (17) so that end of keys (17) fit into thrust washer (3) and install two keys (17) in drum shaft (16). Tap keys (17) lightly with mallet to seat them.

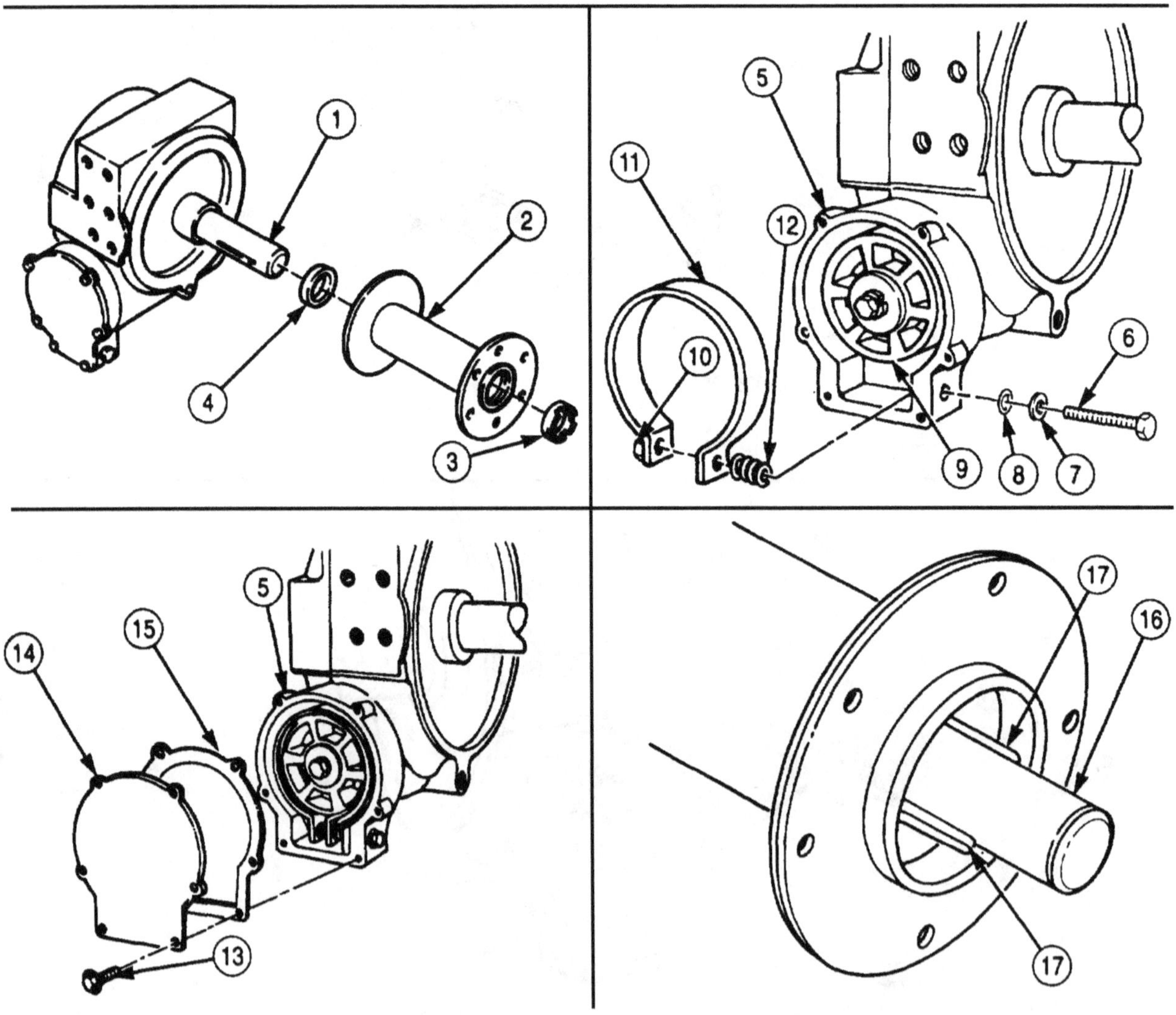

4-172. FRONT WINCH REPAIR (Contd)

e. Assembly of Winch

1. Install clutch (20) and thrust washer (19) on drum shaft (16). Ensure clutch (20) engages drum (9).
2. Install tie rod (21) into gearcase (27) as far as it will go.
3. Ensuring that clutch shifter fork (22) engages in slot of clutch (20), install end frame (18) on drum shaft (16) with tie rod nut (23) on tie rod (21). Do not tighten nut (23).
4. Position rear channel (30) on winch (31) and install with four new lockwashers (28) and screws (29). Do not tighten screws (29).
5. Position top channel (24) on winch (31) and install with four new lockwashers (25) and screws (26). Do not tighten screws (26).
6. Mount dial indicator on gearcase (27) and measure end play of drum shaft (16).
7. Using two nuts (23) on tie rod (21), adjust drum shaft (16) end play to 0.005-0.015 in. (0.127-0.381 mm).
8. When end play of drum shaft (16) is correct, tighten tie rod nuts (23) and screws (29) and (26).

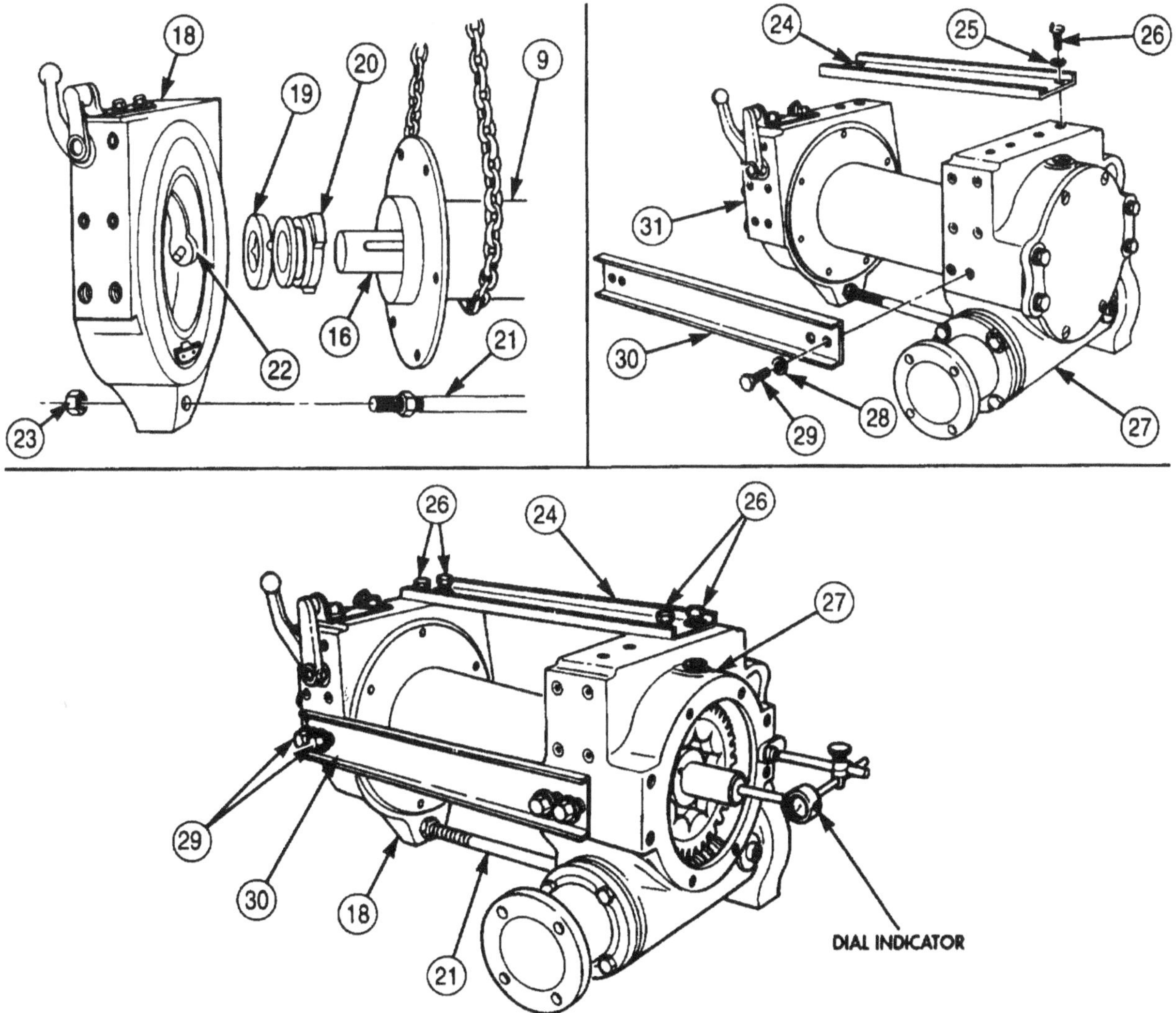

4-172. FRONT WINCH REPAIR (Contd)

9. Coat both sides of new gasket (2) with sealing compound and position gasket (2) on gearcase (1).
10. Position cover (3) on gearcase (1) and install with two new lockwashers (6), four screws (4), and screws (5).

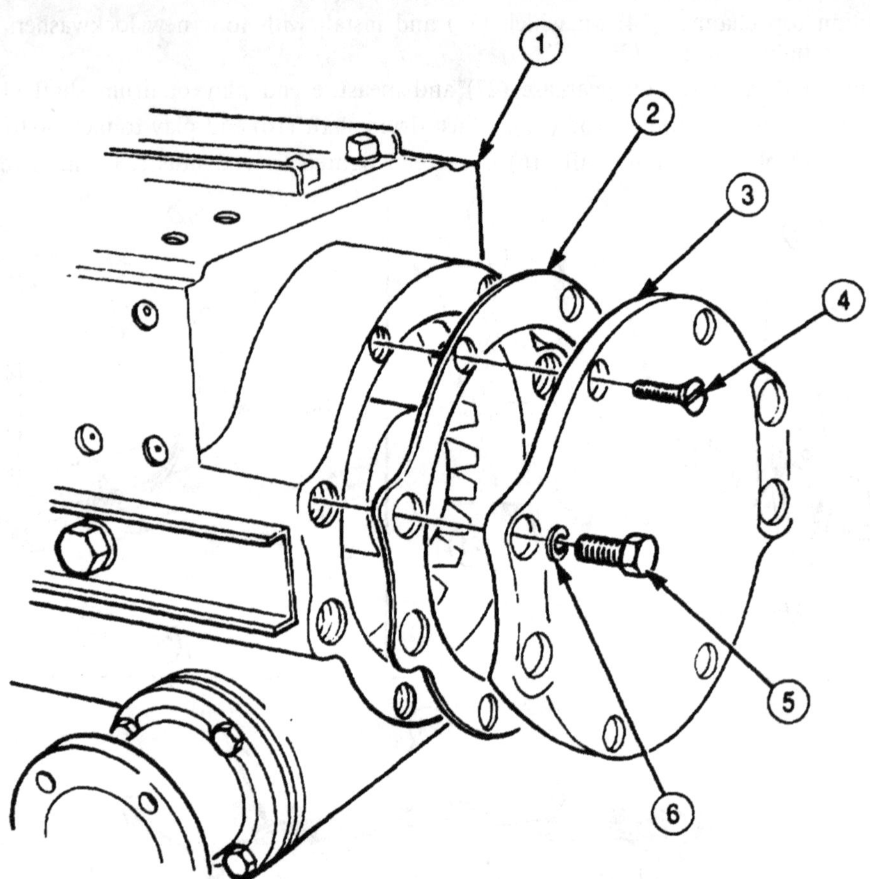

FOLLOW-ON TASKS:
- Install front winch level wind (if removed) (para. 4-174).
- Install winch cable tensioner (para. 4-173).
- Install winch roller (para. 4-175).
- Install front winch (para. 3-329).
- Fill winch with lubricant (LO 9-2320-272-12).

4-173. FRONT WINCH CABLE TENSIONER MAINTENANCE

THIS TASK COVERS:

a. Removal
b. Disassembly
c. Cleaning and Inspection
d. Assembly
e. Installation

INITIAL SETUP:

APPLICABLE MODELS
M936/A1/A2

TOOLS
General mechanic's tool kit (Appendix E, Item 1)
Arbor press
Soft-faced hammer

MATERIALS/PARTS
Four lockwashers (Appendix D, Item 350)
Four lockwashers (Appendix D, Item 354)
Four felt washers (Appendix D, Item 113)
Woodruff key (Appendix D, Item 732)
GAA grease (Appendix C, Item 28)
Cleaning rags (Appendix C, Item 58)

REFERENCES (TM)
TM 9-2320-272-10
TM 9-2320-272-24P

EQUIPMENT CONDITION
- Parking brake set (TM 9-2320-272-10).
- Winch cable removed (para. 3-325).

a. Removal

Remove four nuts (2). lockwashers (3), screws (4), and cable tensioner (8) from brackets of roller assembly (1). Discard lockwashers (3).

b. Disassembly

1. Remove two grease fittings (9) from cable tensioner (8).
2. Remove setscrew (6), lever (5), and poppet assembly (7) from tensioner (8).
3. Remove latch nut (14), latch (10), and winch poppet nut (13) from poppet (12).
4. Remove spring (11) and poppet (12) from poppet nut (13).

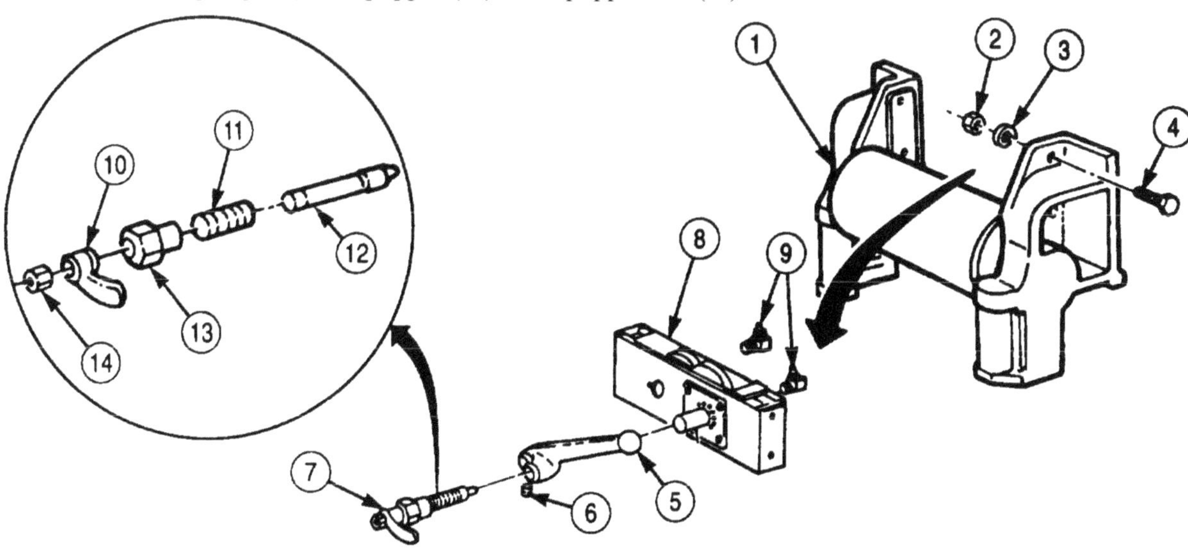

4-173. FRONT WINCH CABLE TENSIONER MAINTENANCE (Contd)

5. Remove four screws (l), lockwashers (2), and block assembly (13) from frame (3). Discard lockwashers (2).
6. Remove shaft (11) from frame (3).
7. Remove bushing (14), woodruff key (12), and bushing (10) from shaft (11). Discard woodruff key (12) and tag parts for installation.
8. Remove sheave (7), two felt washers (8) and (5), and thrust washers (4) and (9) from frame (3). Discard felt washers (5) and (8).
9. Remove bearing (6) from sheave (7).
10. Remove two snaprings (19) from tensioner sheave pin (20).
11. Remove tensioner sheave pin (20) and pin (21) from frame (3).
12. Remove sheave (18), two felt washers (16), and thrust washers (15) from frame (3). Discard felt washers (16).
13. Remove bearing (17) from sheave (18).

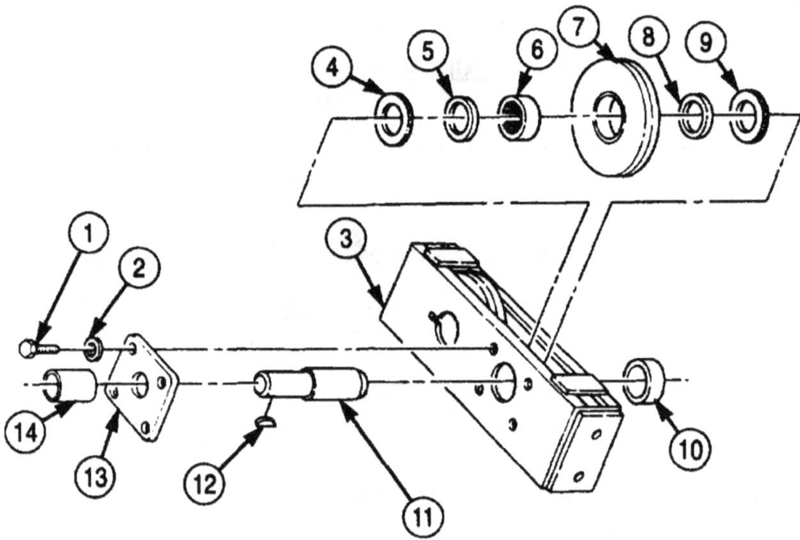

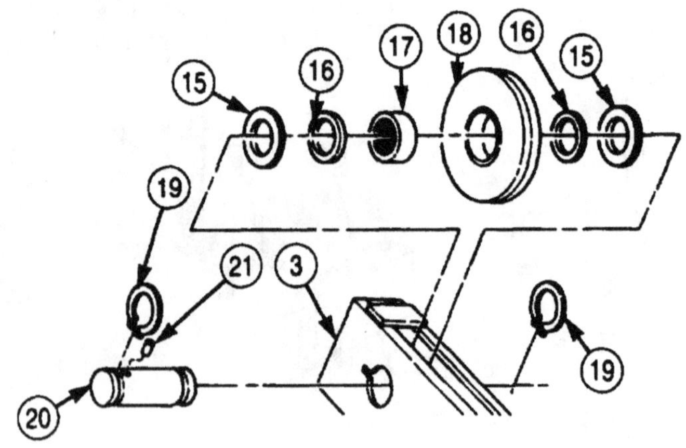

4-173. FRONT WINCH CABLE TENSIONER MAINTENANCE (Contd)

c. Cleaning and Inspection

1. Inspect frame (3) for breaks and cracks. Replace if broken or cracked.
2. Clean tensioner sheave pin (20) and inspect for pits, scoring, and wear. Replace if scored, pitted, or if outside diameter is less than 0.9995 in. (25.39 mm).
3. Clean sheave (18) and inspect for breaks and cracks. Replace if broken or cracked.
4. Clean bearing (17) and inspect for cracks, chips, and broken cage. Replace if cracked, chipped, broken, or if inner diameter exceeds 1.000 in. (25.4 mm).
5. Clean two thrust washers (15) and inspect for cracks, chips, and wear. Replace if cracked, chipped, or if thickness is less than 0.0615 in. (1.56 mm).

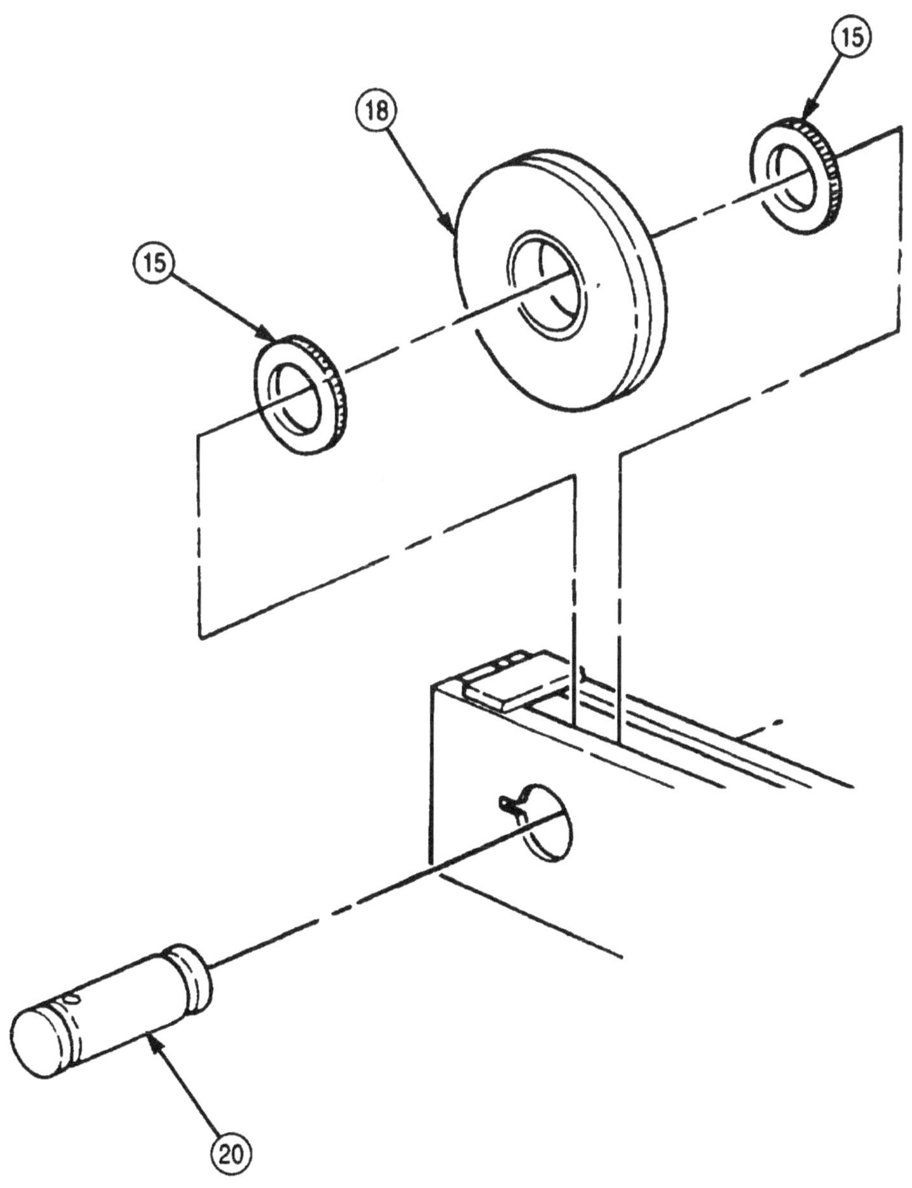

4-173. FRONT WINCH CABLE TENSIONER MAINTENANCE (Contd)

6. Clean block (1) and inspect for cracks and elongated holes. Replace if cracked or holes are elongated.
7. Clean shaft (4) and inspect for cracks. Replace if cracked.
8. Measure outer cam diameter of shaft (4). Replace if cam outer diameter is less than 0.995 in. (26.36 mm).
9. Measure outer diameter of bearing (11). Replace if outer diameter is less than 0.748 in. (18.99 mm).
10. Clean sheave (12) and inspect for cracks and wear. Replace if cracked or broken.
11. Clean two thrust washers (9) and inspect for cracks and wear. Replace if cracked or if thickness is less than 0.062 in. (1.56 mm).
12. Clean bearing (11) and inspect for cracks, chips, and broken cage. Replace if cracked, chipped, or broken.
13. Measure inner diameter of bearing (11) and replace if inner diameter is greater than 1.000 in. (25.4 mm).
14. Clean two bushings (3) and inspect for cracks, chips, and wear. Replace if cracked, chipped, or if inner diameter exceeds 0.754 in. (19.15 mm).
15. Clean frame (2) and replace if damaged.
16. Clean latch (8) and lever (7) and inspect for breaks and cracks. Replace if broken or cracked.
17. Clean poppet (5) and inspect for bends and breaks. Replace if bent or broken.
18. Clean spring (6) and inspect for broken and collapsed coils. Replace if coils are broken or collapsed.

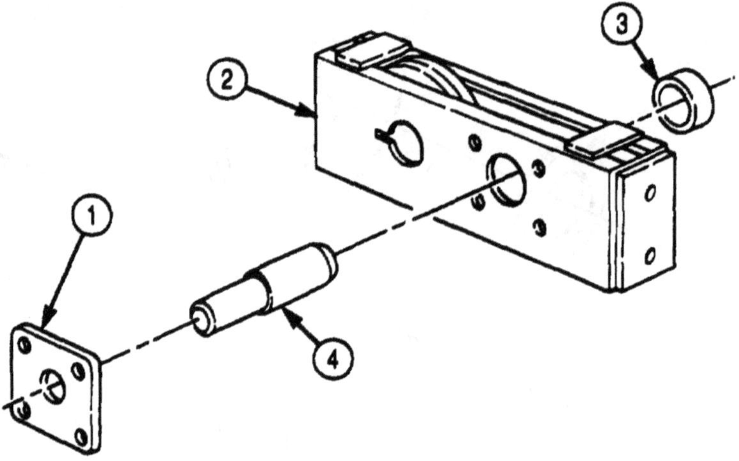

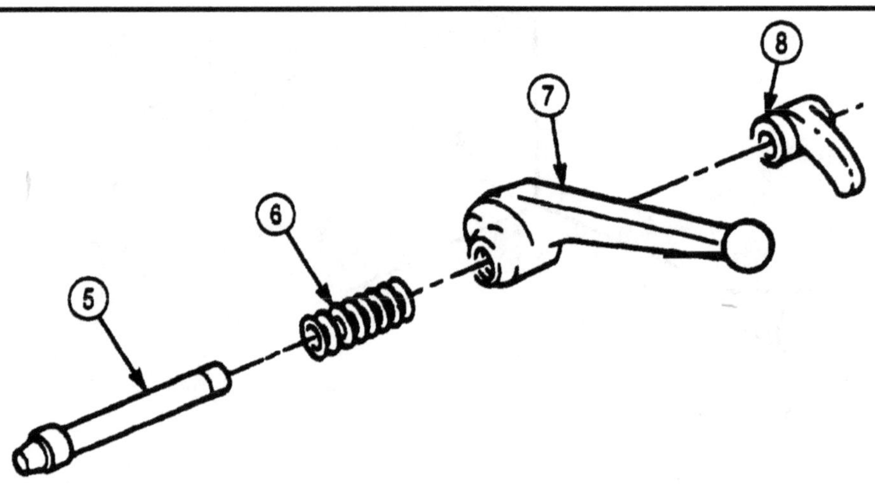

4-173. FRONT WINCH CABLE TENSIONER MAINTENANCE (Contd)

d. Assembly

NOTE

Ensure all bearings are packed with GAA grease before installation.

1. Using arbor press, install bearing (11) in sheave (12).
2. Using soft-faced hammer, drive pin (16) into tensioner sheave pin (14) until seated.
3. Install snapring (15) on pin (14).
4. Position felt side of two new felt washers (10) against bearing (11).
5. Position sheave (12) and two thrust washers (9) in frame (2) and install with tensioner sheave pin (14).
6. Secure tensioner sheave pin (14) on frame (2) with snapring (15). Ensure tensioner sheave pin (14) engages slot (17) in frame (2).

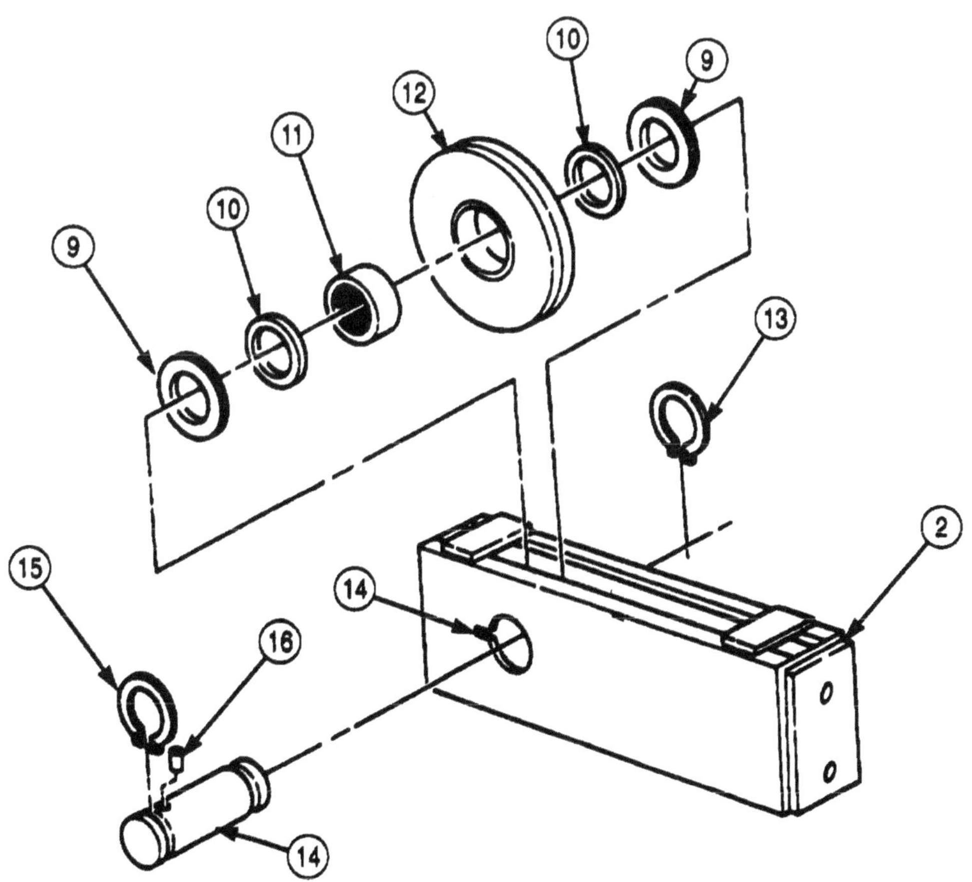

4-173. FRONT WINCH CABLE TENSIONER MAINTENANCE (Contd)

7. Using arbor press, install two bushings (10) and (6) in tensioner (13).
8. Using arbor press, install bearing (3) in sheave (4).
9. Position two new felt washers (2) and (5), with felt sides against ends of bearing (3), in sheave (4).
10. Install woodruff key (8) in shaft (7).
11. Position sheave (4) and two thrust washers (1) in tensioner (13) and install shaft (7).
12. Install block (9) on shaft (7) with four new lockwashers (12) and screws (11).
13. Install lever (18) on shaft (7) with setscrew (19).
14. Install spring (23) and poppet (24) in poppet nut (25).
15. Install latch (22) on poppet (24) with latch nut (26).
16. Install poppet assembly (20) in lever (18) and install with poppet nut (25).
17. Install two grease fittings (21) on cable tensioner (13).

e. Installation

Install cable tensioner (13) on roller assembly (14) with four new screws (17), lockwashers (16), and nuts (15).

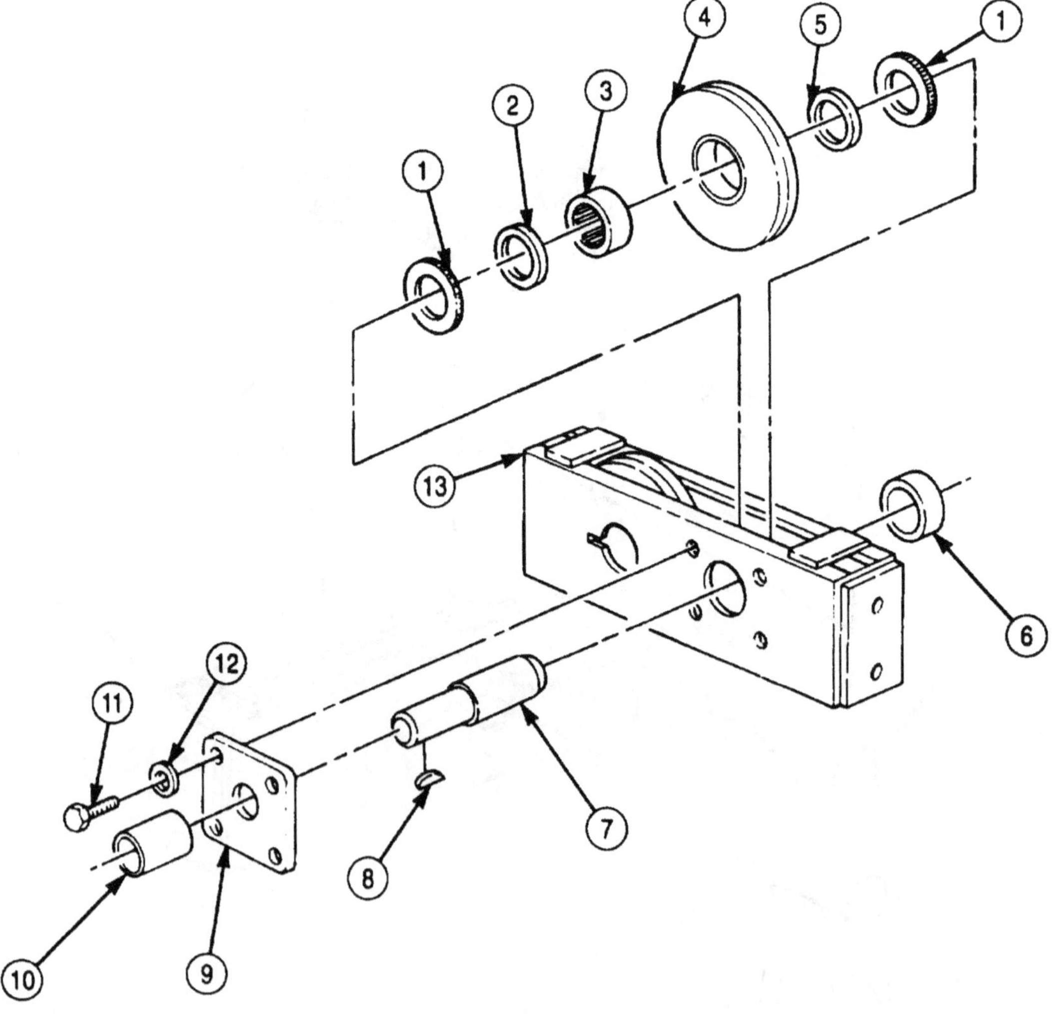

4-173. FRONT WINCH CABLE TENSIONER MAINTENANCE (Contd)

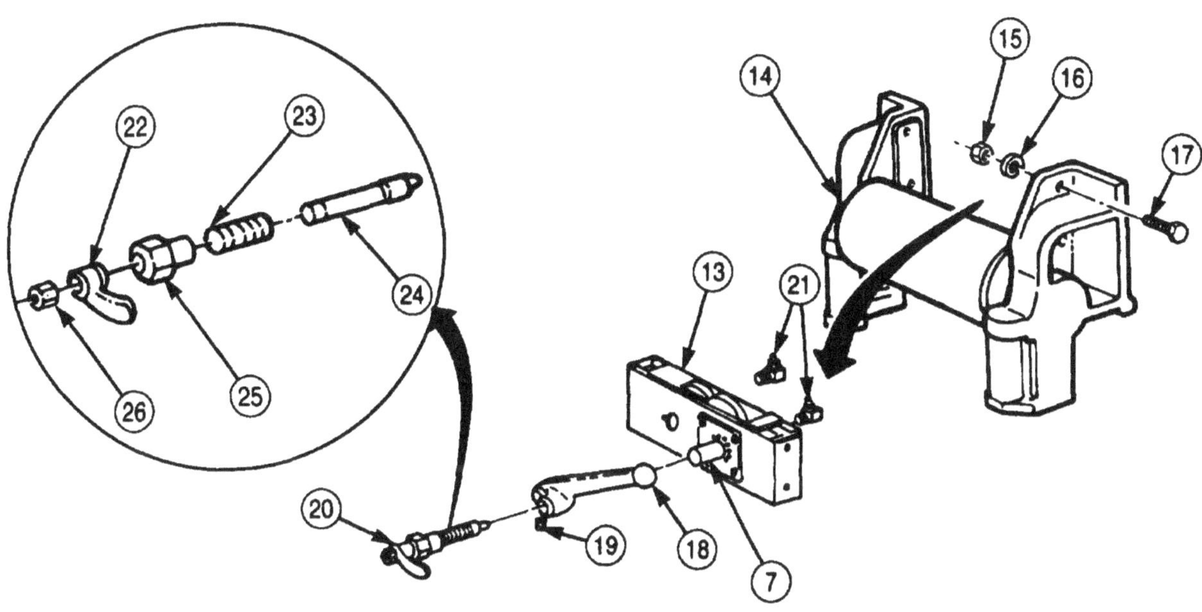

FOLLOW-ON TASKS:
- Install winch cable (para. 3-325).
- Lubricate tensioner assembly (LO 9-2320-272-12).

4-174. FRONT AND REAR WINCH LEVEL WIND MAINTENANCE

THIS TASK COVERS:

a. Removal
b. Disassembly
c. Cleaning, Inspection, and Repair
d. Assembly
e. Installation

INITIAL SETUP:

APPLICABLE MODELS
M936/A1/A2

TOOLS
General mechanic's tool kit (Appendix E, Item 1)
Arbor press

MATERIALS/PARTS
Bushing (Appendix D, Item 24)
Cotter pin (Appendix D, Item 49)
Felt seal (Appendix D, Item 110)
Two felt washers (Appendix D, Item 114)
Five lockwashers (Appendix D, Item 377)
Two felt washers (Appendix D, Item 115)
'Iwo locknuts (Appendix D, Item 314)
Eight lockwashers (Appendix D, Item 354)
Felt washer (Appendix D, Item 116)
GAA grease (Appendix C, Item 28)
Lubricating oil (Appendix C, Item 50)

REFERENCES (TM)
LO 9-23220-272-12
TM 9-2320-272-10
TM 9-2320-272-24P

EQUIPMENT CONDITION
- Parking brake set (TM 9-2320-272-10).
- Winch cable removed (para. 3-325).

NOTE

Front and rear wench level wind are maintained the same way. This procedure covers front level wind.

a. Removal

1. Remove four screws (8) and lockwashers (9) from winch track (3). Discard lockwashers (9).
2. Remove winch track (3) and level wind (4) from winch (10).

b. Disassembly

NOTE

Rectangular holes are provided for removing screws. Slide the level wind backward and forward until screws are visible.

1. Remove four screws (1), lockwashers (2), and level wind (4) from winch track (3). Discard lockwashers (2).
2. Remove two nuts (14), drum lock latch (13), nut (12), spring (11), and poppet (7) from track (3).
3. Remove two outer nuts (6), screws (5), and inner nuts (6) from track (3).

NOTE

Both sets of trolley wheels are removed the same.

4. Remove grease fitting (24) from axle (16).
5. Remove screw (23) and lockwasher (22) from axle (16). Discard lockwasher (22).
6. Remove three washers (18), two felt washers (19), wheel (21), and thrust washer (17) from axle (16). Discard felt washers (19).

4-174. FRONT AND REAR WINCH LEVEL WIND MAINTENANCE (Contd)

7. Using arbor press, remove axle (16) from level wind (15).
8. Using arbor press, remove bearing (20) from wheel (21).

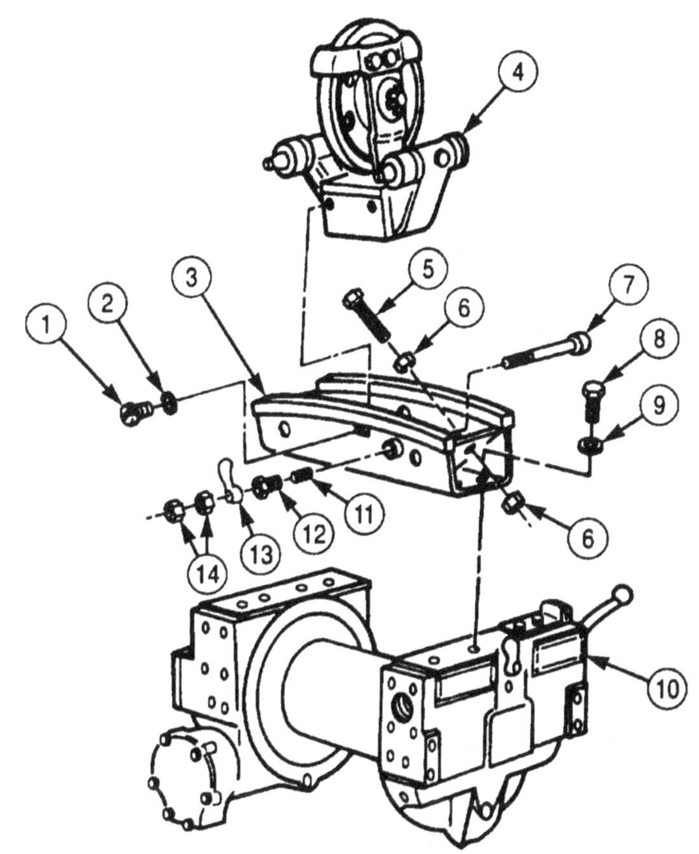

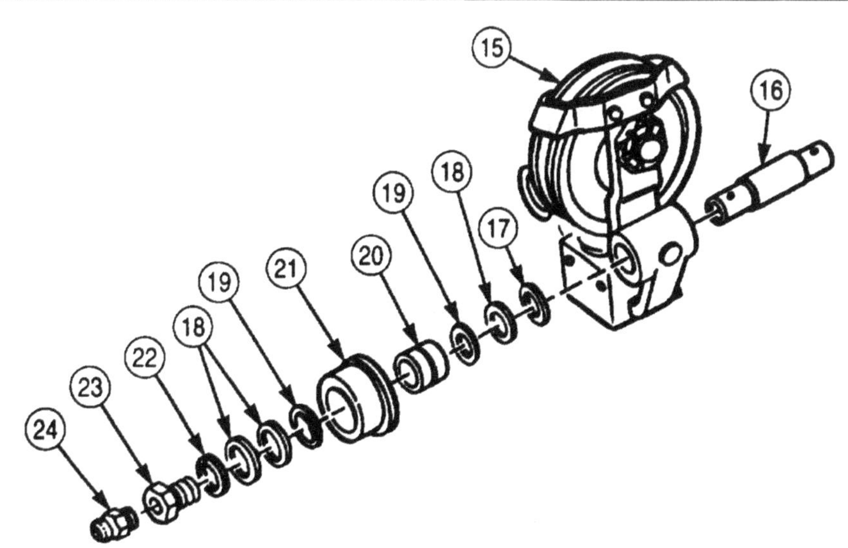

4-174. FRONT AND REAR WINCH LEVEL WIND MAINTENANCE (Contd)

9. Remove four screws (5), lockwashers (4), and sheave guard (3) from frame (11). Discard lockwashers (4).
10. Remove grease fitting (13), cotter pin (8), slotted nut (9), and washer (10) from screw (12). Discard cotter pin (8).
11. Remove screw (12) and sheave (14) from frame (11). Thrust washers (1) will fall out.
12. Remove two felt washers (2), sleeve (7), and bearing (6) from sheave (14). Discard felt washers (2).
13. Remove grease fitting (15), locknut (16), washer (17), and felt washer (18) from end of swivel (19) in frame (11). Discard locknut (16) and felt washer (18).
14. Using arbor press, remove swivel (19) from frame (11).
15. Remove bushing (22), felt seal (25), inner race (26), forty-five ball bearings (21), outer race (27), and bearing (20) from frame (11). Discard felt seal (25) and bushing (22).

NOTE

An arbor press may be required to perform step 16.

16. Remove locknut (24) and extension (23) from swivel (19). Discard locknut (24).

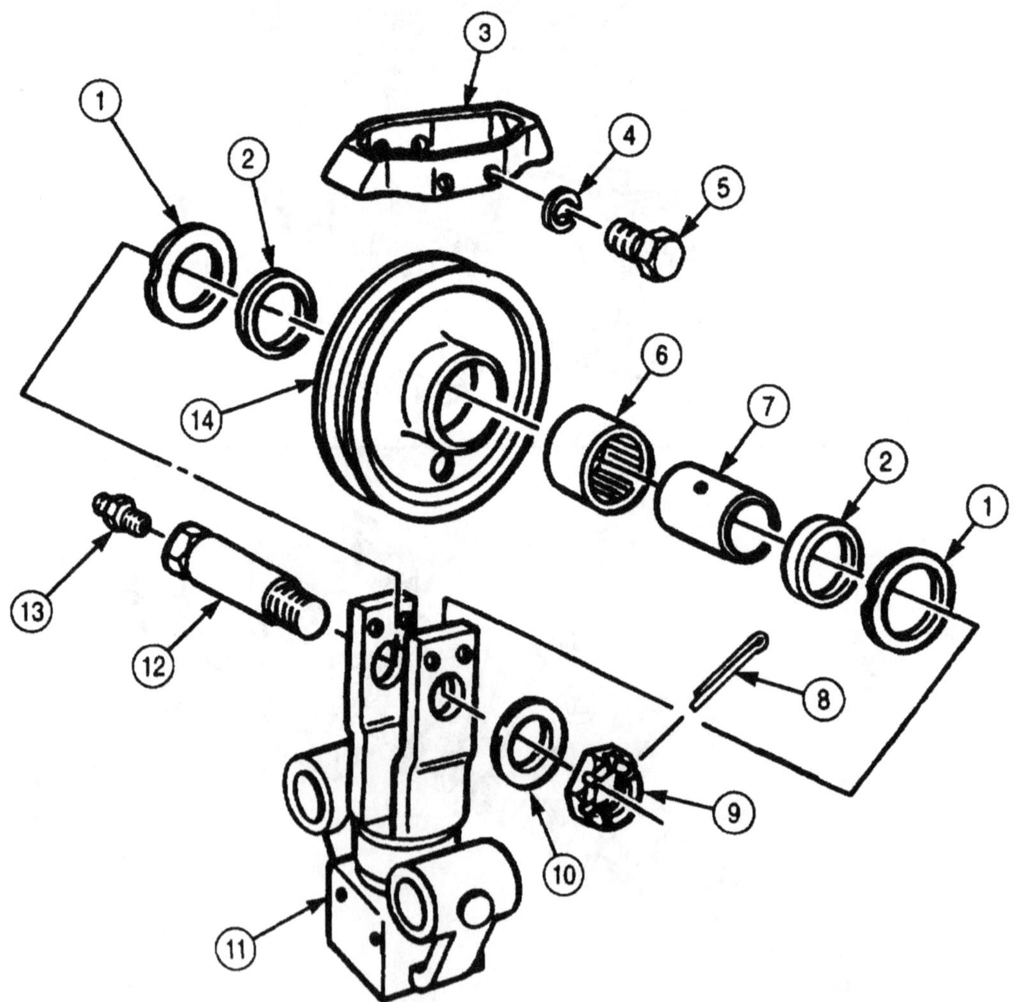

4-174. FRONT AND REAR WINCH LEVEL WIND MAINTENANCE (Contd)

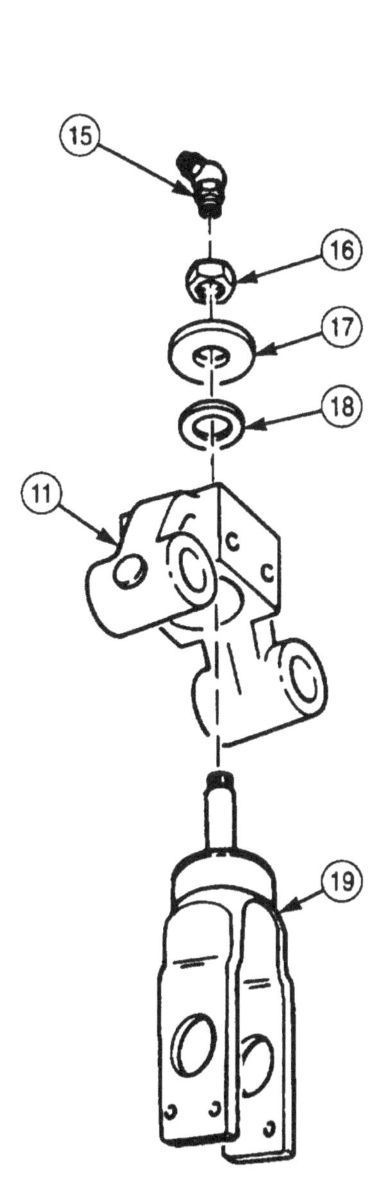

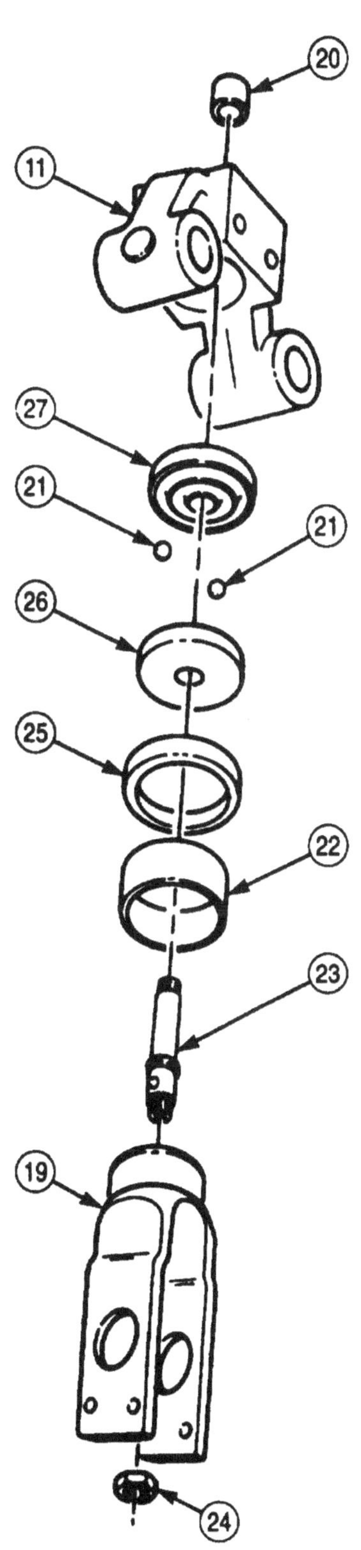

4-174. FRONT AND REAR WINCH LEVEL WIND MAINTENANCE (Contd)

c. Cleaning, Inspection, and Repair

1. Clean track (1) and inspect for cracks and chipped or nicked wheel surface. Discard if cracked. Use file to smooth wheel surface if chipped or nicked.
2. Clean poppet (8) and inspect for cracks, bends, and damaged threads. Discard if cracked, bent, or threads are damaged.
3. Clean spring (7) and inspect for broken or collapsed coils. Discard if coils are broken or collapsed.
4. Clean nut (9) and inspect for damaged threads. Discard if threads are damaged.
5. Clean latch (6) and inspect for cracks. Discard if cracked.
6. Clean nuts (5) and inspect for damaged threads. Discard if threads are damaged.
7. Clean stopscrews (2) and nuts (3) and (4) and inspect for bends, cracks, or damaged threads. Discard if bent, cracked, or threads are damaged.

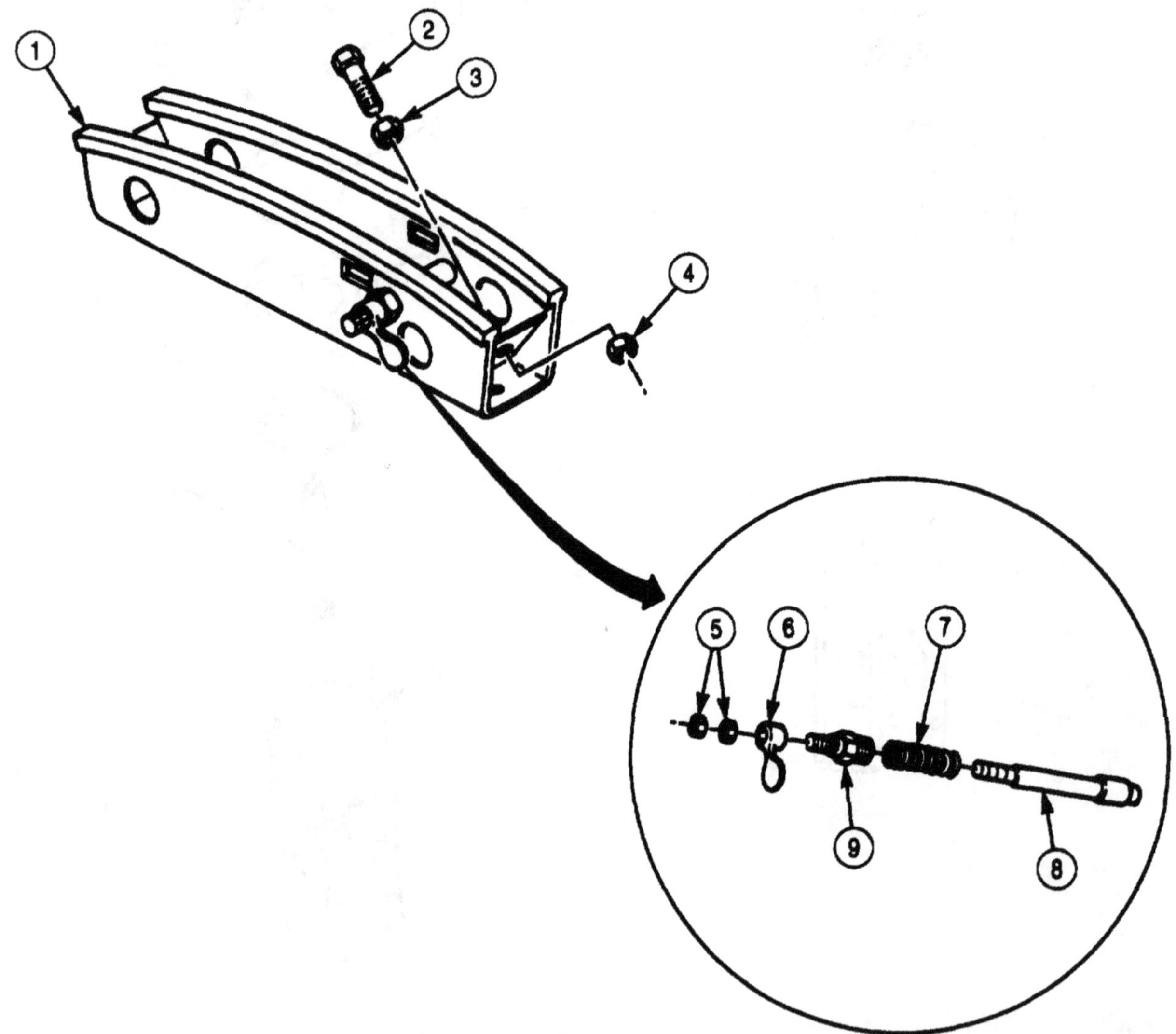

4-174. FRONT AND REAR WINCH LEVEL WIND MAINTENANCE (Contd)

8. Clean wheel (10) and inspect for cracks and damaged roller surface, Discard if cracked or roller surface is damaged.
9. Inspect bearing (11) for chips, cracks, and damaged cage. Discard if chipped, cracked, or cage is damaged.
10. Clean thrust washer (12) and inspect for cracks and wear. Measure thickness. Discard if cracked or thickness is less than 0.058 in. (1.47 mm).
11. Clean axle (13) and inspect for cracks and scoring. Discard if cracked or scored.
12. Clean screw (18) and inspect for cracks, scoring, and damaged threads. Discard if cracked, scored, or threads are damaged.
13. Clean two thrust washers (14) and inspect for cracks and wear. Measure thickness. Discard if cracked or thickness is less than 0.058 in. (1.47 mm).
14. Clean sheave (17) and inspect for cracks. Discard if cracked.
15. Clean sleeve (16) and inspect for cracks and scoring. Discard if cracked or scored.
16. Inspect bearing (15) for chips, cracks, and damaged cage. Discard if chipped, cracked, or cage is damaged.

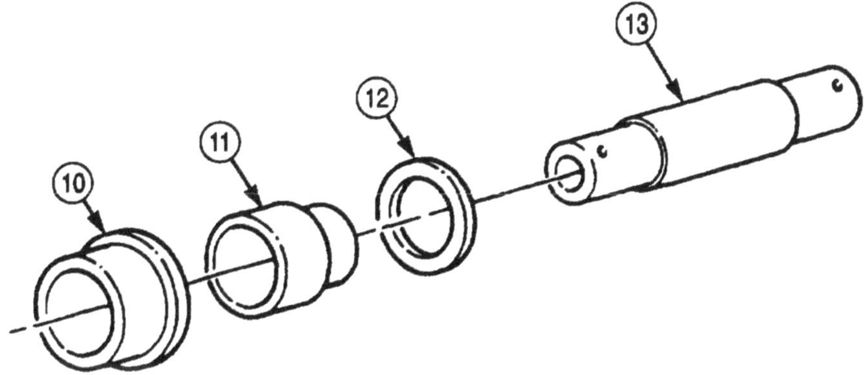

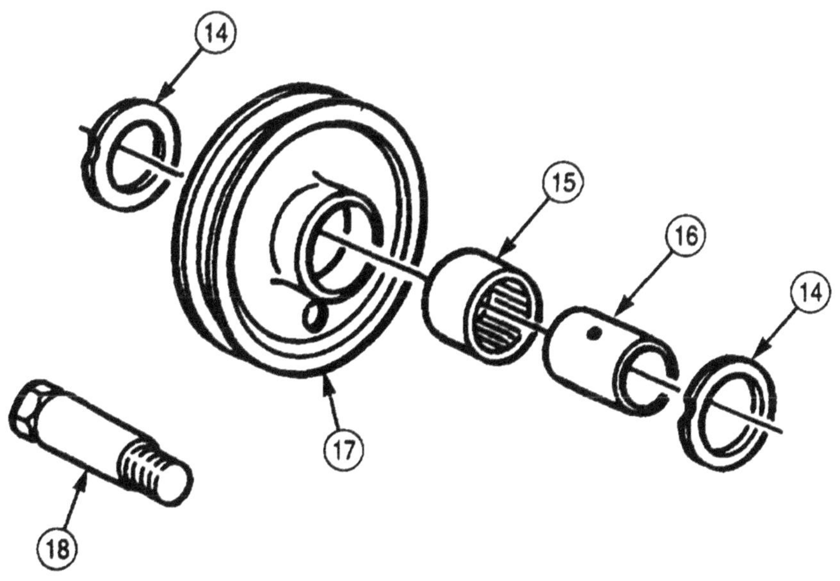

4-174. FRONT AND REAR WINCH LEVEL WIND MAINTENANCE (Contd)

17. Inspect bearing (5) for chips, cracks, and damaged cage. Discard if chipped, cracked, or cage is damaged.
18. Clean inner race (1) and outer race (3) and inspect for pits, chips, and cracks. Discard if pitted, chipped, or cracked.
19. Inspect forty-five bearing balls (2) for cracks, chips, and out-of-round condition. Discard any bearing balls (2) that are cracked, chipped, or out-of-round.
20. Clean frame (4) and inspect for cracks and damaged threads. Discard if cracked or threads are damaged.

4-174. FRONT AND REAR WINCH LEVEL WIND MAINTENANCE (Contd)

d. Assembly

NOTE

- Ensure all felt seals are soaked in lubricating oil before installation.
- Ensure all bearings are packed with GAA grease before installation.

1. Using arbor press, install extension (10) in swivel (6) with new locknut (11).
2. Using arbor press, install outer race (3) into frame (4).
3. Position forty-five ball bearings (2) on outer race (3).
4. Using arbor press, install inner race (1) in frame (4).
5. Install new felt seal (9) over inner race (1).
6. Using arbor press, install new bushing (7) in frame (4).
7. Install bearing (8) in bottom of frame (4).

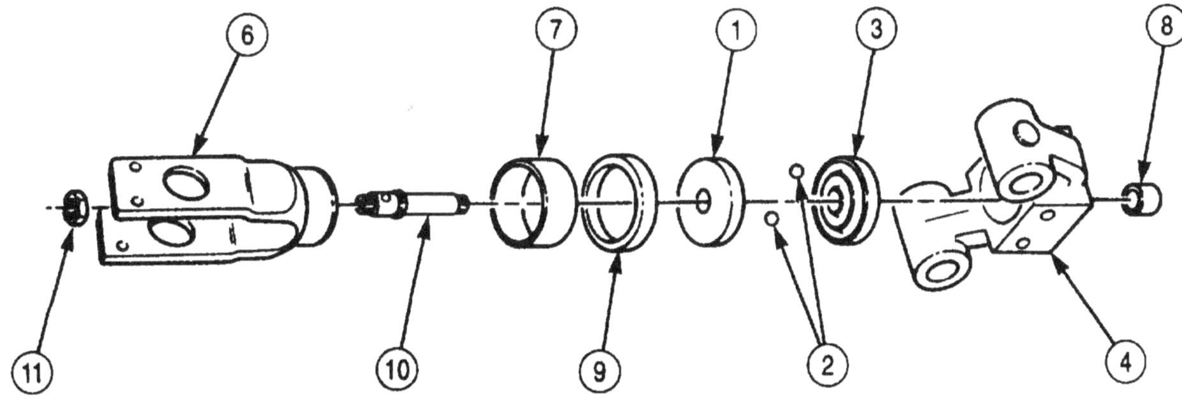

4-174. FRONT AND REAR WINCH LEVEL WIND MAINTENANCE (Contd)

8. Using arbor press, install swivel (1) with extension (2) in frame (7).
9. With swivel (1) and extension (2) seated in frame (7), install new felt washer (3), washer (6), and new locknut (4) on end of extension (2).
10. Install grease fitting (5) on end of extension (2).
11. Install bearing (14), sleeve (15), and two new felt washers (9) in sheave (13).
12. Coat one side of each of two thrust washers (8) with grease and install on sheave (13) with grease side next to bearing (14).
13. Carefully place sheave (13) and thrust washers (8) in swivel (1) and install with screw (19), washer (18), and slotted nut (17). Ensure that slots in nut (17) align with hole in screw (19).
14. Install new cotter pin (16) in slots of nut (17) and hole in screw (19).
15. Install grease fitting (20) in head end of screw (19).
16. Position sheave guard (10) over sheave (13) and install on swivel (1) with four new lockwashers (11) and screws (12).

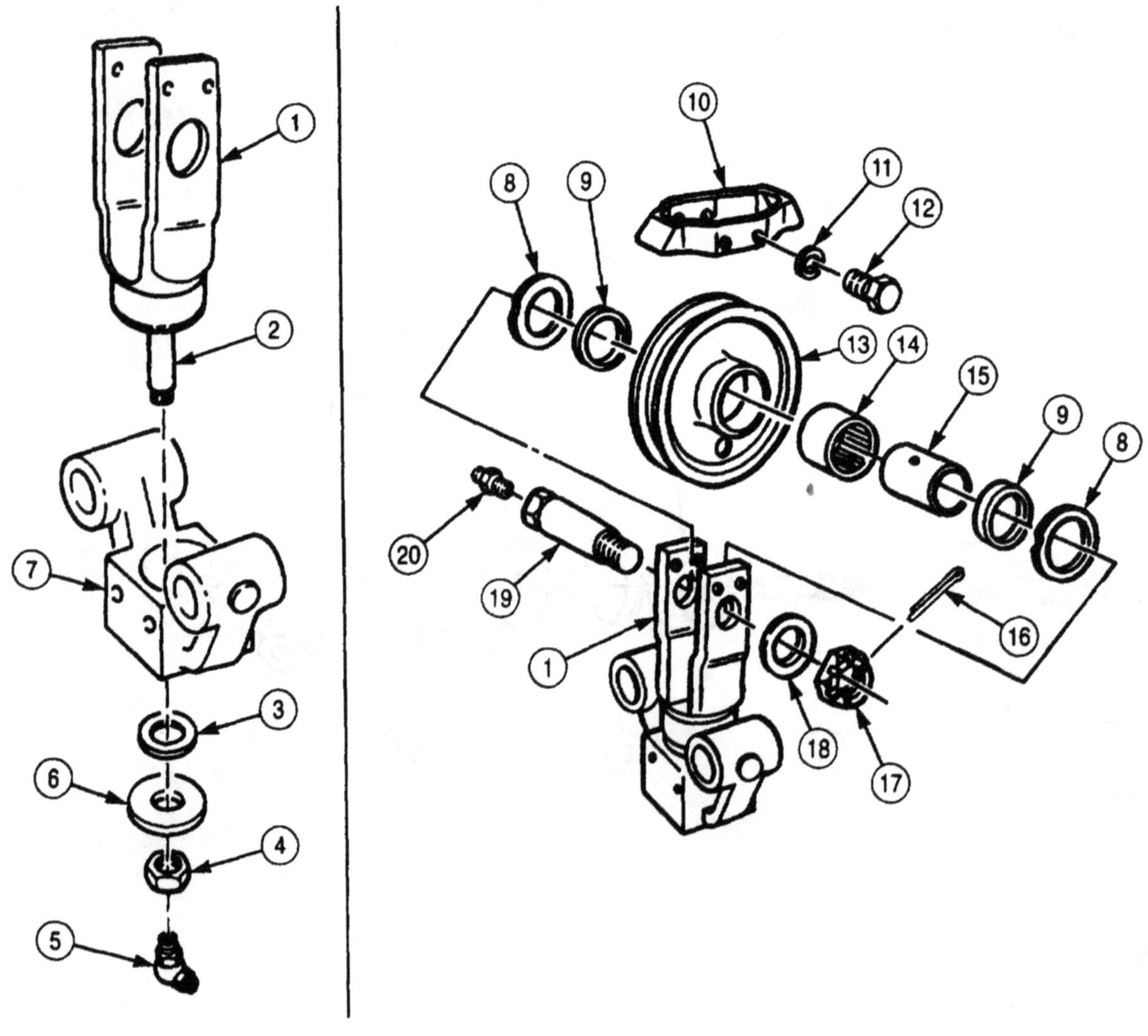

4-174. FRONT AND REAR WINCH LEVEL WIND MAINTENANCE (Contd)

NOTE

- Both sets of trolley wheels are installed the same.
- An arbor press may be required to perform step 17.

17. Install axle (9) in frame (10).
18. Install thrust washer (8) on axle (9).
19. Install bearing (7) in wheel (6).
20. Install two washers (4), new felt washer (5), wheel (6), new felt washer (5), two washers (4), new lockwasher (3), and screw (2) on axle (9).
21. Install grease fitting (1) on screw (2).
22. Place spring (18) over poppet (14) and install with nut (19).
23. Install poppet (14) and latch (20) in track (11) with two nuts (21).
24. Place level wind assembly (10) on track (11) and install with four new lockwashers (23) and screws (22).
25. Install two stopscrews (12) on track (11) with each screw (12) positioned with two nuts (13). Level wind (10) receives final adjustment when winch (17) is mounted on vehicle.

e. Installation

Install level wind assembly (10) and track (11) on winch (17) with four new lockwashers (16) and screws (15).

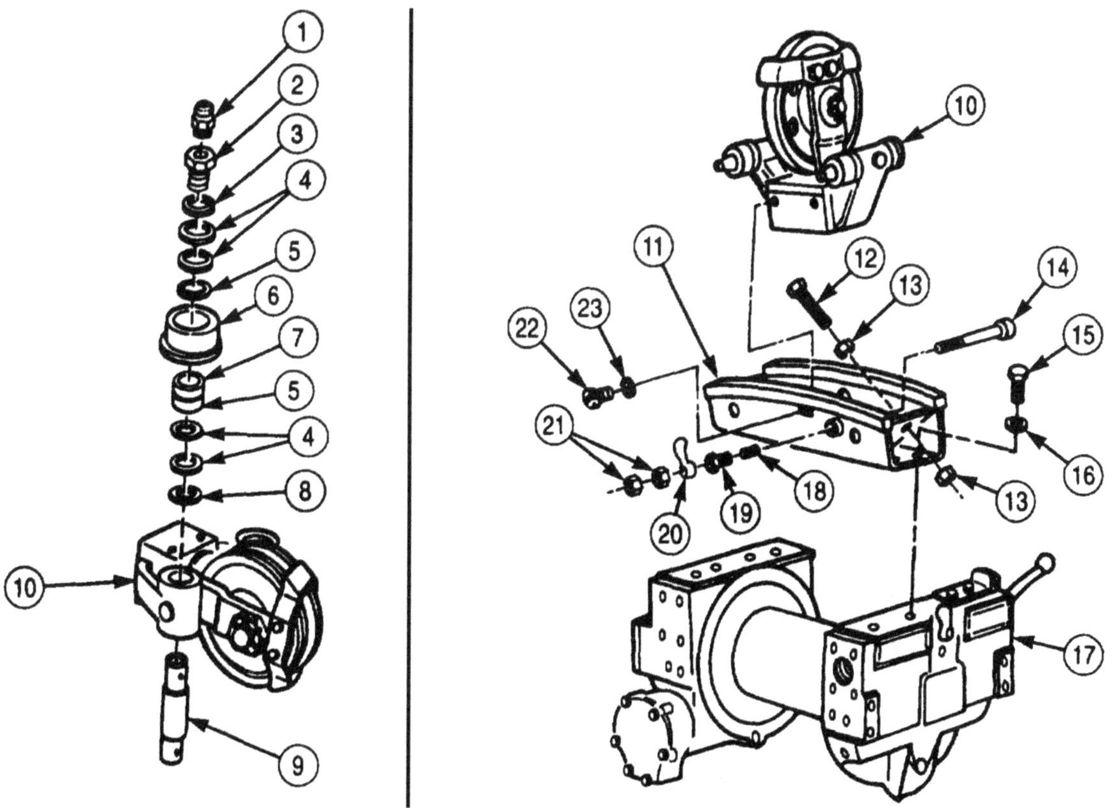

FOLLOW-ON TASKS:
- Lubricate level wind (LO 9-2320-272-12).
- Install winch cable (para. 3-325).

4-175. FRONT WINCH ROLLER ASSEMBLY MAINTENANCE

THIS TASK COVERS:

a. Removal
b. Disassembly
c. Cleaning, Inspection, and Repair
d. Assembly
e. Installation

INITIAL SETUP:

APPLICABLE MODELS
All

TOOLS
General mechanic's tool kit (Appendix E, Item 1)
Arbor press
Mandrel

MATERIALS/PARTS
Two lockwashers (Appendix D, Item 410)
Two felt washers (Appendix D, Item 117)
Two lockwashers (Appendix D, Item 350)
Four lockwashers (Appendix D, Item 393)
Eight lockwashers (Appendix D, Item 377)
GAA grease (Appendix C, Item 28)
Lubricating oil (Appendix C, Item 50)
Drycleaning solvent (Appendix C, Item 71)

REFERENCES (TM)
LO 9-2320-272-12
TM 9-2320-272-10
TM 9-2320-272-24P

EQUIPMENT CONDITION
- Parking brake set (TM 9-2320-272-10).
- Front winch removed (para. 3-329).

GENERAL SAFETY INSTRUCTIONS
- Drycleaning solvent is flammable and toxic. Do not use near open flame.
- Keep fire extinguisher nearby when using drycleaning solvent.

a. Removal

1. Remove eight screws (5) and lockwashers (3) from winch roller assembly (1). Discard lockwashers (3).
2. Remove four screws (4), lockwashers (3), and roller assembly (1) from winch (2). Discard lockwashers (3).

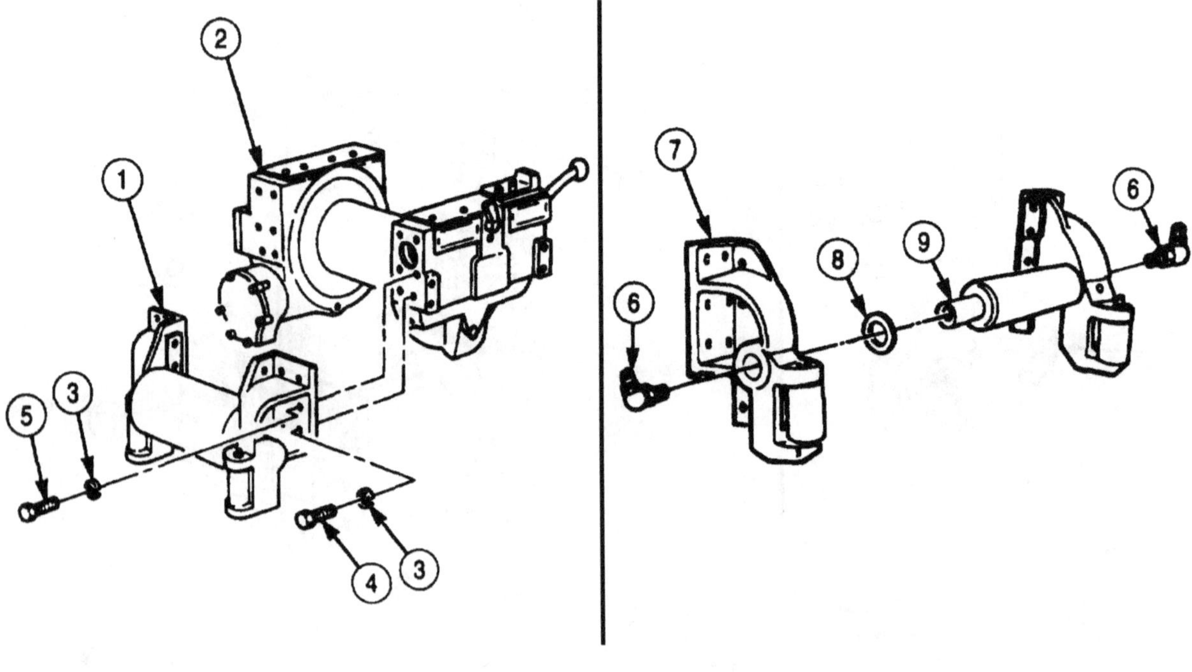

4-175. FRONT WINCH ROLLER ASSEMBLY MAINTENANCE (Contd)

b. Disassembly

1. Remove two grease fittings (6), right roller bracket (7), and thrust washer (8) from roller shaft (9).
2. Remove two felt washers (10), bearings (11), and roller (12) from shaft (14). Discard felt washers (10).
3. Remove thrust washer (13) from shaft (14).

NOTE

Both setscrews are in the same hole.

4. Remove two setscrews (16) from left roller bracket (15).
5. Remove shaft (14) from bracket (15).

NOTE

Both side rollers are disassembled the same.

6. Remove grease fitting (23) from side roller shaft (24).
7. Remove pin (27) from roller bracket (15) and shaft (24).
8. Remove two thrust washers (25) and roller (26) from bracket (15).
9. Remove two screws (17), lockwashers (18), and small side bracket (28) from large bracket (21). Discard lockwashers (18).
10. Remove two screws (22), nuts (19,) lockwasher (20), and large bracket (21) from roller bracket (15). Discard lockwashers (20).

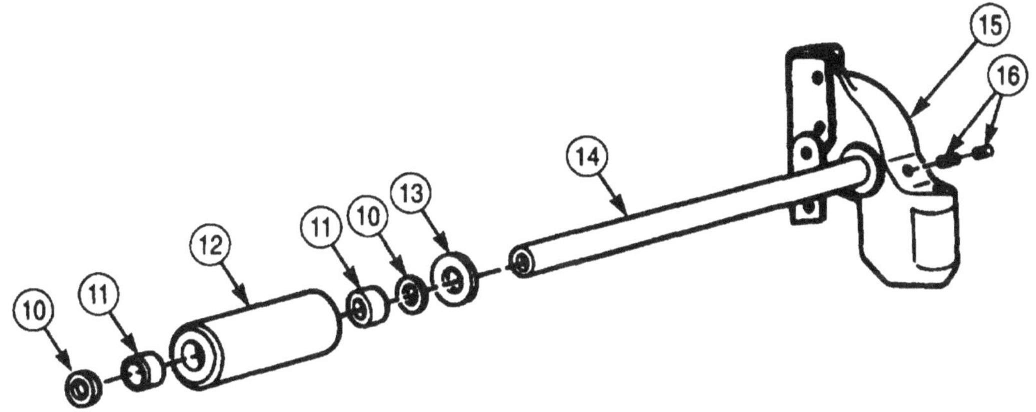

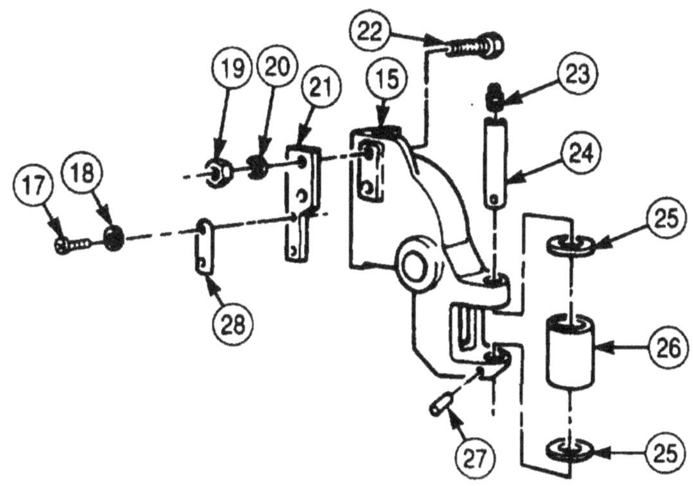

4-175. FRONT WINCH ROLLER ASSEMBLY MAINTENANCE (Contd)

c. Cleaning, Inspection, and Repair

WARNING

Drycleaning solvent is flammable and toxic. Do not use near an open flame and always have a tire extinguisher nearby when solvents are used. Use only in well-ventilated places, wear protective clothing, and dispose of cleaning rags in approved container. Failure to do this may result in injury to personnel and/or damage to equipment.

1. Clean all items with drycleaning solvent before inspection (para. 2-14).
2. Inspect side roller shaft (4) for cracks and wear. Measure outer diameter of shaft (4) at bushing (6) contact points. Discard if cracked or if any outer diameter is less than 0.994 in. (25.25 mm).
3. Inspect bushings (6) for scoring and wear. Measure inner diameter of bushings (6). Discard if scored or inner diameter is more than 1.010 in. (25.65 mm).

NOTE

Perform steps 4 and S only if bushings are to be replaced.

4. Using arbor press and mandrel, remove two bushings (6) from roller (7). Discard bushings (6).
5. Install two new bushings (6) on roller (7).
6. Inspect thrust washers (5) for cracks and wear. Measure thickness of thrust washers (5). Discard if cracked or if thickness is less than 0.040 in. (1.02 mm).
7. Inspect roller shaft (3) for cracks, scoring, and wear. Measure diameter where bearings (1) contact shaft (3). Discard if cracked, scored, or any measured diameter is less than 1.495 in. (37.97 mm).
8. Inspect thrust washers (2) for chips, cracks, scoring, and wear. Measure thickness of thrust washers (2). Discard if chipped, cracked, scored, or thickness is less than 0.052 in. (1.30 mm).
9. Inspect two bearings (1) for chips, pitting, cracks, and damaged cage. Discard if chipped, pitted, cracked, or cage is damaged.

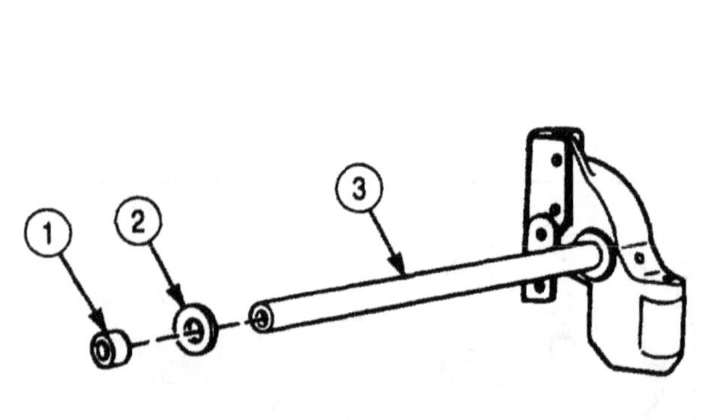

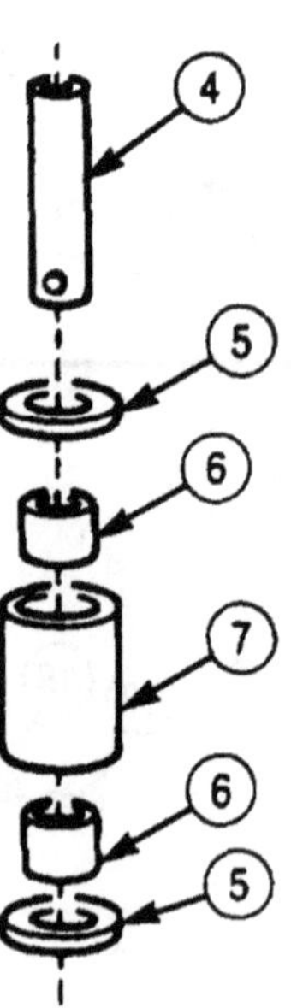

4-175. FRONT WINCH ROLLER ASSEMBLY MAINTENANCE (Contd)

d. Assembly

1. Install large side bracket (10) on roller bracket (14) with two screws (11), new lockwashers (9), and nuts (8).
2. Install small side bracket (15) on large bracket (14) with two new lockwashers (16) and screws (17).
3. Lightly coat two thrust washers (5) with grease and position on ends of roller (7).
4. Install roller (7) and thrust washers (5) on roller bracket (14). Align holes in washers (5) and roller (7) with holes in bracket (14).
5. Align pin (13) holes in shaft (4) and bracket (14), and install shaft (4) on bracket (14) and through roller (7).
6. Install pin (13) in bracket (14) and shaft (4).
7. Install grease fitting (12) in shaft (4).

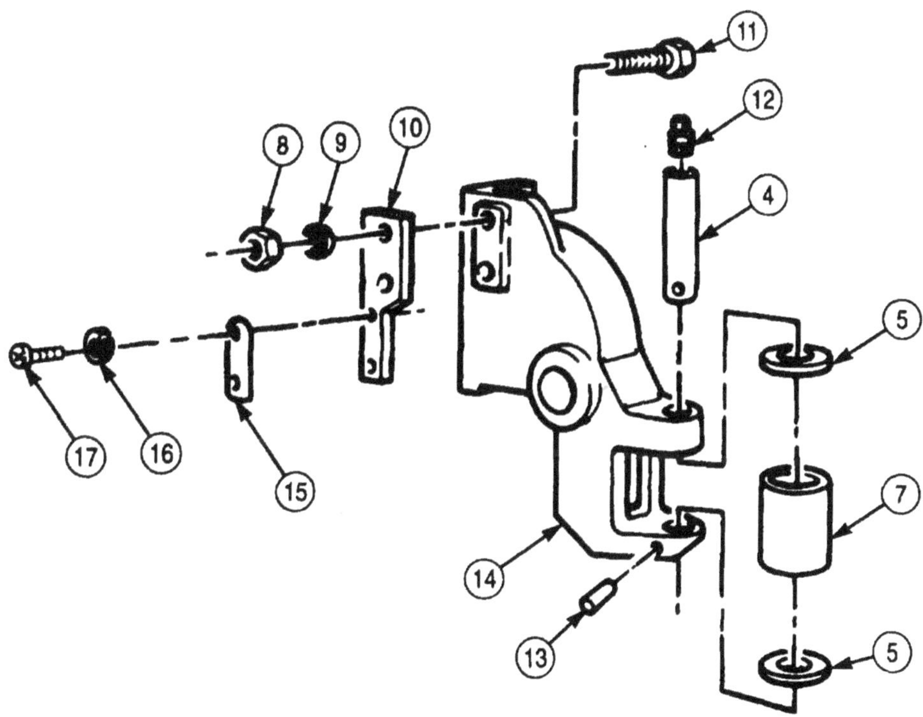

4-175. FRONT WINCH ROLLER ASSEMBLY MAINTENANCE (Contd)

8. Align screw holes in roller shaft (6) and roller bracket (7), and install roller shaft (6) in roller bracket (7).
9. Install two setscrews (8) in roller bracket (7) and shaft (6).

NOTE

- Soak felt washers in oil and coat bearings with GAA grease before installation.
- Steel side of felt washers must be on bearing side on shaft.

10. Install bearings (2) and (4) in roller (3).
11. Install thrust washer (5), new felt washer (1), roller (3), and new felt washer (1) on roller shaft (6).
12. Install thrust washer (10) and right roller bracket (7) on shaft (11).
13. Install two grease fittings (9) on ends of shaft (11).

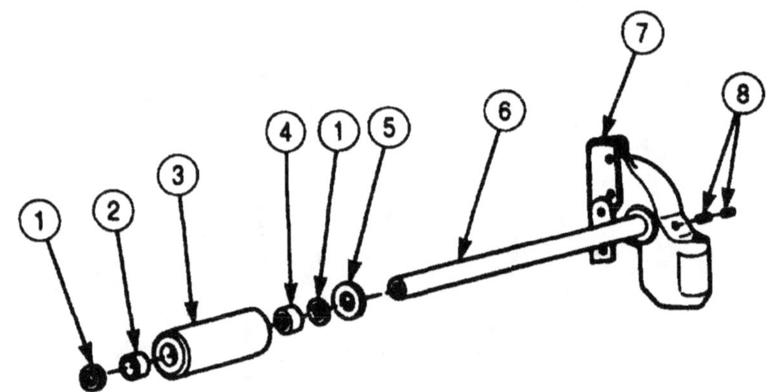

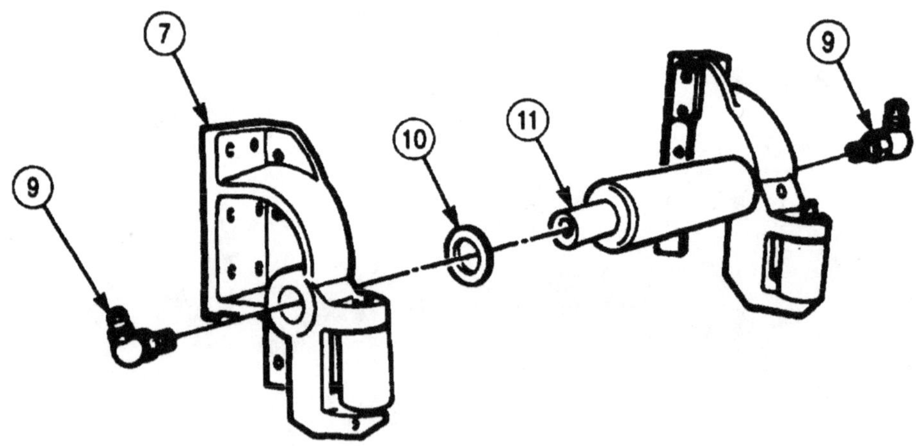

4-175. FRONT WINCH ROLLER ASSEMBLY MAINTENANCE (Contd)

e. Installation

Install winch roller assembly (16) on winch (12) with twelve new lockwashers (13), eight screws (15), and four screws (14).

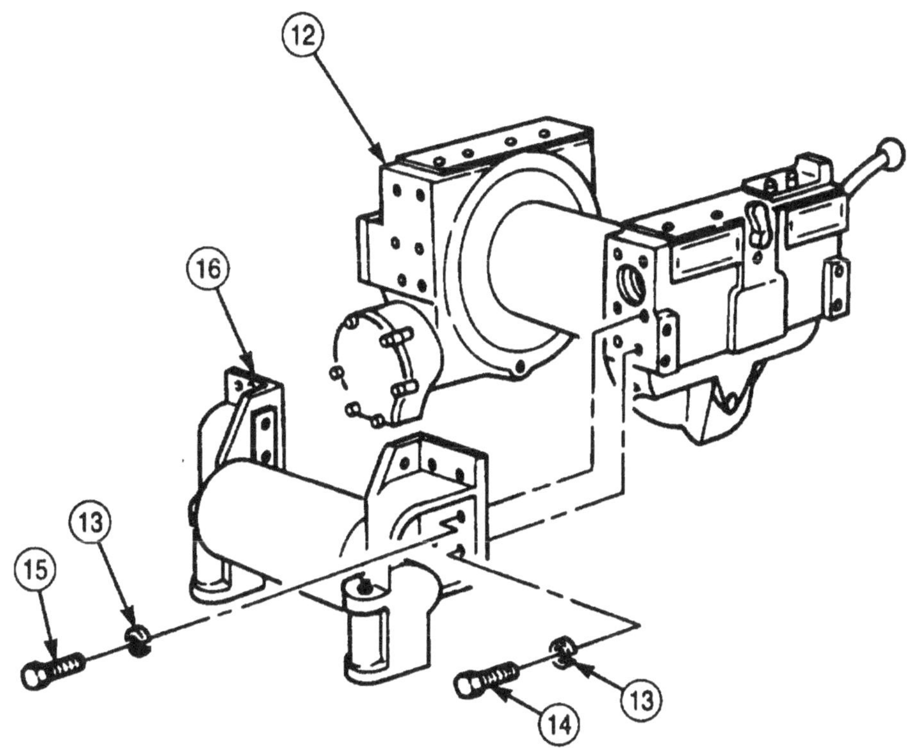

FOLLOW-ON TASKS: • Lubricate roller assembly (LO 9-2320-272-12).
• Install front winch (para. 3-329).

4-176. FRONT WINCH HYDRAULIC PUMP REPLACEMENT

THIS TASK COVERS:

a. Removal **b. Installation**

INITIAL SETUP:

APPLICABLE MODELS

M92/A1/A2, M928/Al/A2, M929/A1/A2, M93/A1/A2, M932/A1/A2, M936/A1/A2

TOOLS

General mechanic's tool kit (Appendix E, Item 1)

MATERIALS/PARTS

O-ring (Appendix D, Item 455)
Four lockwashers (Appendix D, Item 354)
Two locknuts (Appendix D, Item 294)
O-rings (Appendix D, Item 456)
Four lockwashers (Appendix D, Item 350)

REFERENCES (TM)

LO 9-2320-272-12
TM 9-2320-272-10
TM 9-2320-272-24P

EQUIPMENT CONDITION

- Parking brake set (TM 9-2320-272-10).
- Hydraulic oil reservoir drained (para. 4-188).

a. Removal

1. Remove setscrew (2) and driveshaft (1) from pump shaft (17).
2. Remove four screws (8), lockwashers (9), two pipe flange swivels (6), tube (7), and O-ring (10) from hydraulic pump (5). Discard O-ring (10) and lockwashers (9).
3. Remove four screws (13), lockwashers (14), two pipe flange swivels (11), tube (12), and O-ring (15) from hydraulic pump (5). Discard O-ring (15) and lockwashers (14).
4. Remove two screws (3), locknuts (4), and hydraulic pump (5) from pump bracket (16). Discard locknuts (4).

b. Installation

1. Install pump (5) on bracket (16) with two screws (3) and new locknuts (4).
2. Install new O-ring (10), two pipe flange swivels (6), and tube (7) on pump (5) with four new lockwashers (9) and screws (8).
3. Install new O-ring (15), two pipe flange swivels (11), and tube (12) on pump (5) with four new lockwashers (14) and screws (13).
4. Ensure key is in keyway and connect driveshaft (1) to pump shaft (17).
5. Install setscrew (2) in driveshaft (1) and tighten.

4-176. FRONT WINCH HYDRAULIC PUMP REPLACEMENT (Contd)

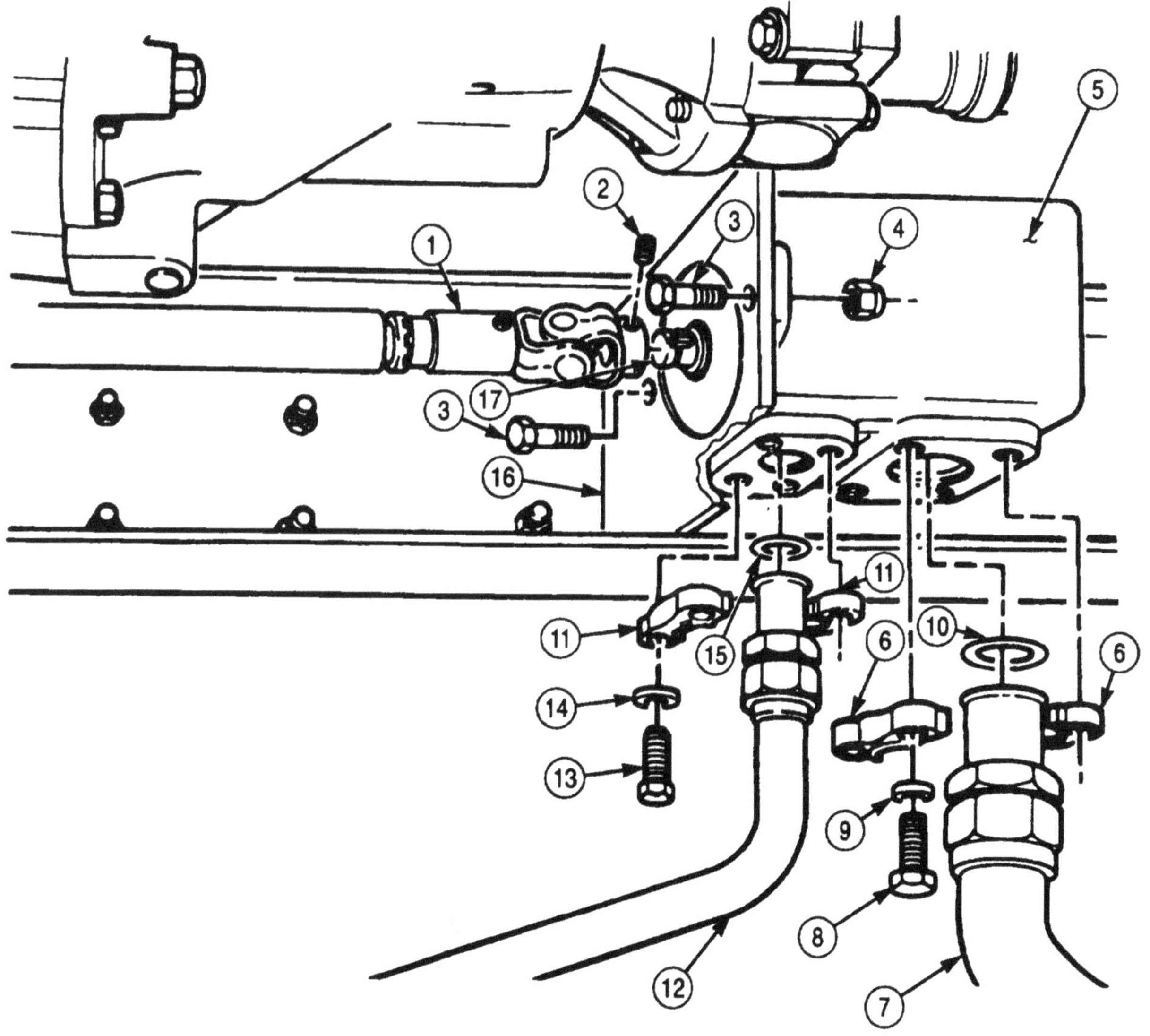

FOLLOW-ON TASKS:
- Fill hydraulic tank (LO 9-2320-272-12).
- Operate winch system and check for leaks (TM 9-2320-272-10).

4-177. FRONT WINCH CONTROL VALVE CABLE REPLACEMENT

THIS TASK COVERS:

a. Removal **b. Installation**

INITIAL SETUP:

APPLICABLE MODELS
M939/A1/A2, M928/A1/A2, M930/A1/A2, M932/A1/A2, M936/A1/A2

TOOLS
General mechanic's tool kit (Appendix E, Item 1)

MATERIALS/PARTS
Cotter pin (Appendix D, Item 84)
Cotter pin (Appendix D, Item 83)
Cotter pin (Appendix D, Item 62)
Tiedown straps (Appendix D, Item 685)

REFERENCES (TM)
TM 9-2320-272-10
TM 9-2320-272-24P

EQUIPMENT CONDITION
Parking brake set (TM 9-2320-2722-10).

a. Removal

NOTE
Remove tiedown straps as needed.

1. Remove cotter pin (6), washer (5), and clevis pin (4) from clevis (3). Discard cotter pin (6).
2. Remove clevis (3) and nut (7) from cable (11).
3. Remove nut (8), washer (9), and cable (11) from bracket (10).
4. Remove washer (2) and nut (1) from cable (11).

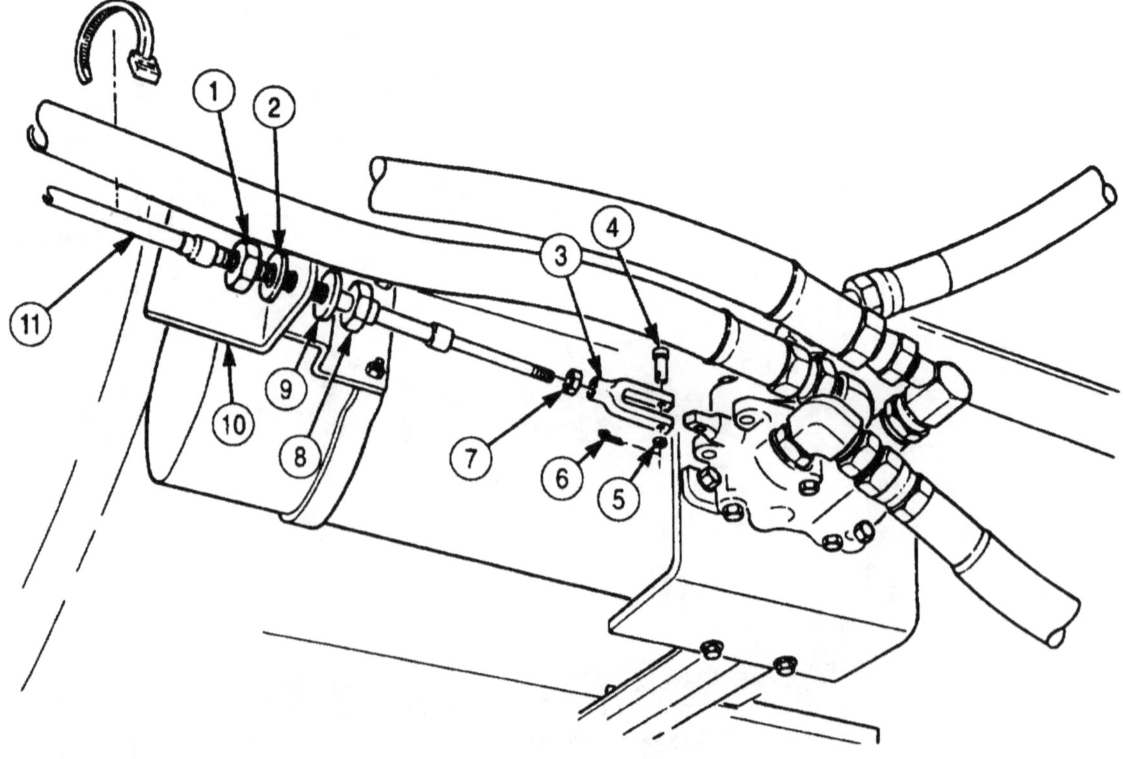

4-177. FRONT WINCH CONTROL VALVE CABLE REPLACEMENT (Contd)

5. Remove six screws (21) and shift panel cover (22) from shift panel (14).
6. Remove cotter pin (26), clevis pin (23), washer (24), and cable clevis (25) from lever (13). Discard cotter pin (26).
7. Remove cotter pin (15), pin (12), washer (27), and lever (13) from shift panel (14). Discard cotter pin (15).
8. Remove two nuts (16), screws (20), clamp (19), shim (18), and cable (11) from vehicle.

NOTE

Grommet may remain on cable during removal. Transfer grommet to new cable.

b. Installation

NOTE

Install tiedown straps as required.

1. Install lever (13) in shift panel (14) with washer (27), pin (12), and new cotter pin (15). Bend ends of cotter pin (15).
2. Ensure grommet (17) is positioned in firewall, and install cable clevis (25) on lever (13) with washer (24), clevis pin (23), and new cotter pin (26). Bend ends of cotter pin (26).
3. Install shim (18) and cable (11) on shift panel (14) with two screws (20) and nuts (16).
4. Install shift panel cover (22) on shift panel (14) with six screws (21).

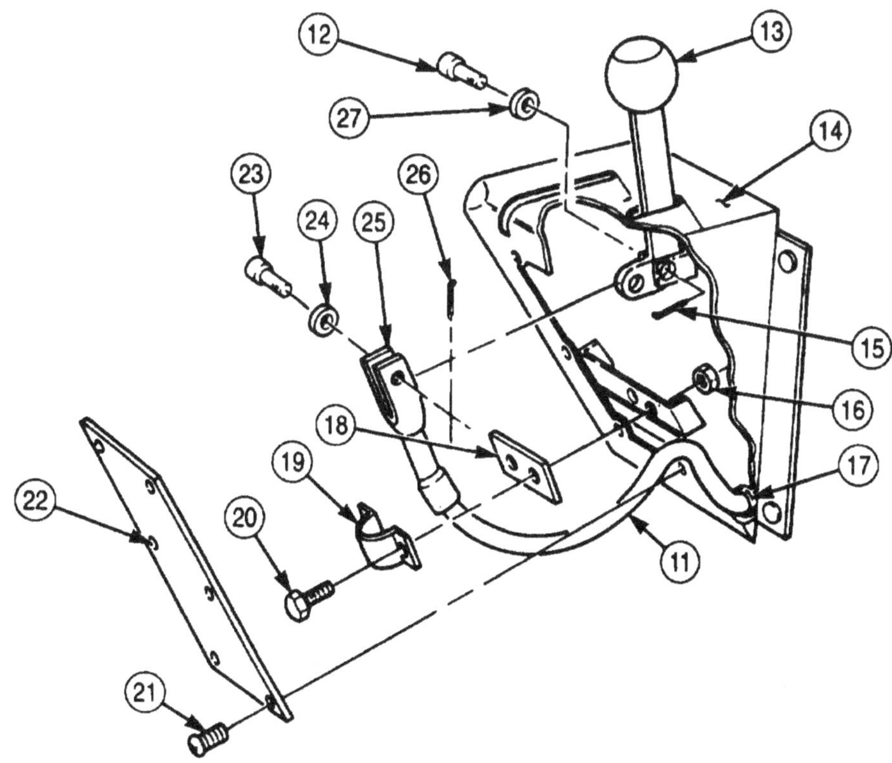

4-177. FRONT WINCH CONTROL VALVE CABLE REPLACEMENT (Contd)

NOTE

Place shift lever in NEUTRAL position.

5. Install nut (1) and washer (2) on cable (12) and feed end of cable (12) through bracket (11).
6. Install washer (10), nuts (9) and (8), and clevis (3) on end of cable (12).
7. Position clevis (3) on valve shaft (5) and adjust nuts (9) and (8) until holes in end of cable (12) and shaft (5) align.
8. Install clevis pin (4), washer (6), and new cotter pin (7) in clevis (3) and shaft (5). Bend ends of cotter pin (7).
9. Tighten nuts (8), (9), and (11).

4-177. FRONT WINCH CONTROL VALVE CABLE REPLACEMENT (Contd)

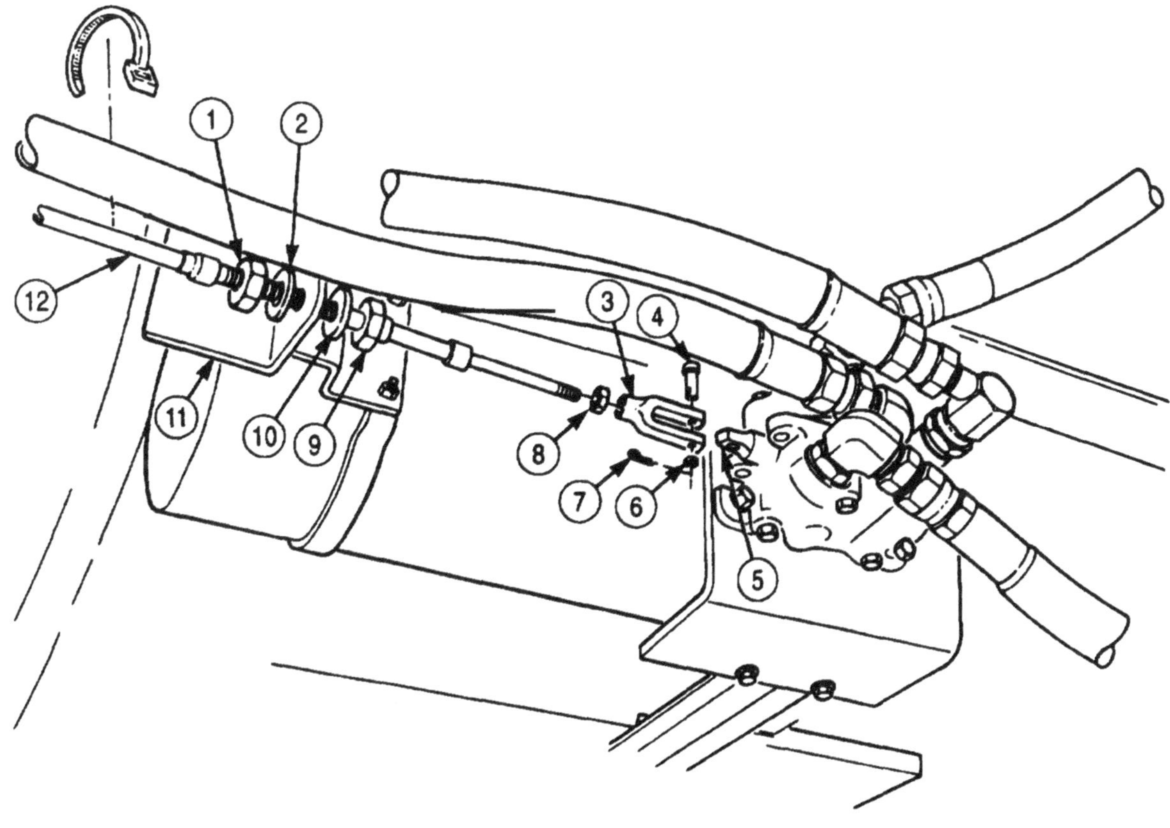

FOLLOW-ON TASK: Start engine (TM 9-2320-2722-10) and check winch control valve operation.

4-178. FRONT AND REAR WINCH CONTROL VALVE REPLACEMENT

THIS TASK COVERS:

a. Removal | b. Installation

INITIAL SETUP:

APPLICABLE MODELS
M925/A1/A2, M928/A1/A2, M932/A1/A2, M936/A1/A2

TOOLS
General mechanic's tool kit (Appendix E, Item 1)

MATERIALS/PARTS
Cotter pin (Appendix D, Item 62)
O-ring (Appendix D, Item 458)
Two O-rings (Appendix D, Item 457)
Two O-rings (Appendix D, Item 430)
Four locknuts (Appendix D, Item 315)
Three locknuts (Appendix D, Item 291)
Cap and plug set (Appendix C, Item 14)

REFERENCES (TM)
LO 9-2320-272- 12
TM 9-2320-272-10
TM 9-2320-272-24P

EQUIPMENT CONDITION
•Parking brake set (TM 9-2320-272-10).
•Hydraulic oil reservoir drained (para. 4-188).

GENERAL SAFETY INSTRUCTIONS
Do not cross hoses when installing.

CAUTION

Plug hydraulic hoses to prevent dirt from entering.

NOTE

All hoses and fittings must be tagged for installation.

a. Removal

1. Disconnect hoses (1), (4), (5), and (6) from control valve (7). Tag for installation.
2. Remove cotter pin (14), washer (15), and clevis pin (3) from cable clevis (2). Discard cotter pin (14).
3. Remove three locknuts (13), screws (10), and control valve (7) from bracket (12). Discard locknuts (13).
4. Remove four locknuts (8), screws (11), and bracket (12) from frame (9). Discard locknuts (8).
5. Remove elbow (20) and adapter (25) from fitting (19) and elbow (26).
6. Remove fittings (16) and (19) and elbows (18) and (26) from control valve (7).
7. Remove four O-rings (17) from fittings (16) and (19) and elbows (18) and (26). Discard O-rings (17).
8. Remove plug (21), spring (23), and relief valve (24).
9. Remove O-ring (22) from plug (21). Discard O-ring (22).

b. Installation

1. Install new O-ring (22) on plug (21) and install relief valve (24), spring (23), and plug (21) in control valve (7).
2. Install four new O-rings (17) and fittings (16), (18), (19), and (26) on control valve (7).
3. Install fittings (20) and (25) on fittings (19) and (26).
4. Install bracket (12) on frame (9) with four screws (11) and new locknuts (8).
5. Install control valve (7) on bracket (12) with three screws (10) and new locknuts (13).

WARNING

Do not cross hoses during installation. Crossing hoses may result in injury to personnel.

4-178. FRONT AND REAR WINCH CONTROL VALVE REPLACEMENT (Contd)

6. Install hoses (1), (4), (5), and (6) on control valve (7).
7. Install cable clevis (2) on control valve (7) with clevis pin (3), washer (15), and new cotter pin (14). Bend ends of cotterpin (14).

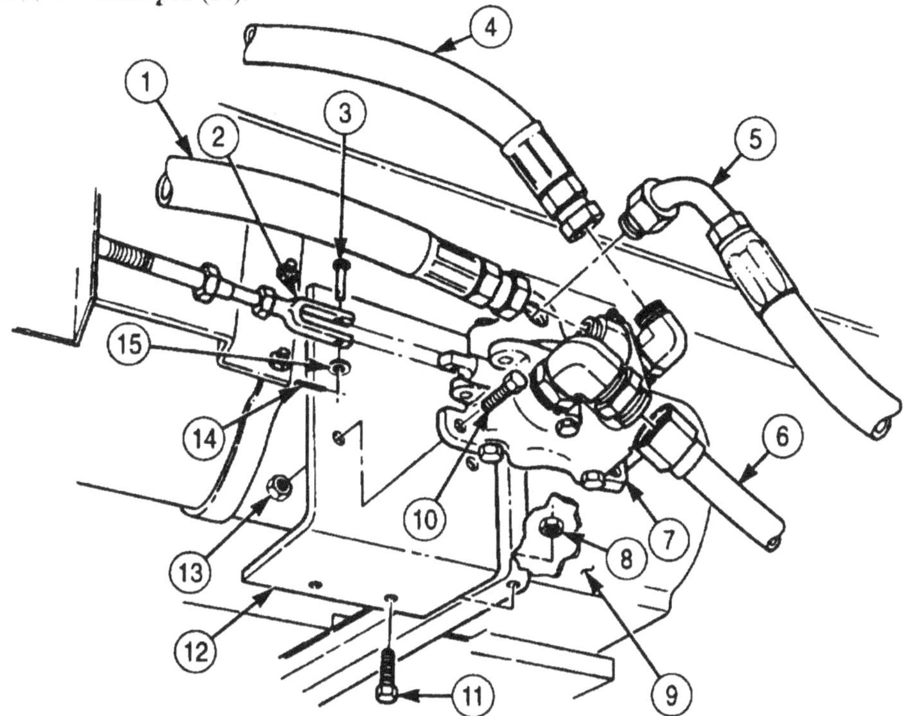

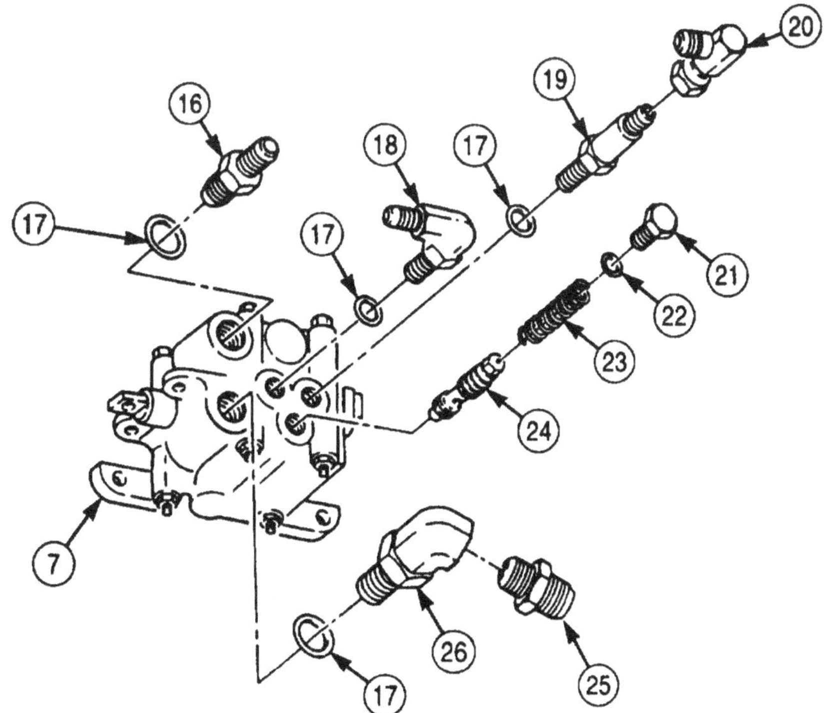

FOLLOW-ON TASKS:
- Fill hydraulic oil reservoir (LO 9-2320-272-12).
- Start engine (TM 9-2320-272-10) and check hydraulic pump operation.

4-179. FRONT AND REAR WINCH CONTROL VALVE (M930/A1/A2) REPLACEMENT

THIS TASK COVERS:

a. Removal **b. Installation**

INITIAL SETUP:

APPLICABLE MODELS
M930/A1/A2

TOOLS
General mechanic's tool kit (Appendix E, Item 1)

MATERIAL/PARTS
O-ring (Appendix D, Item 458)
Two O-rings (Appendix D, Item 457)
Four O-rings (Appendix D, Item 430)
Three locknuts (Appendix D, Item 291)
Two cotter pins (Appendix D, Item 62)
Cap and plug set (Appendix C, Item 14)

REFERENCES (TM)
LO 9-2320-272-12
TM 9-2320-272-10
TM 9-2320-272-24P

EQUIPMENT CONDITION
- Parking brake set (TM 9-2320-272-10).
- Hydraulic oil reservoir drained (para. 4-188).

GENERAL SAFETY INSTRUCTIONS
Do not cross hoses when installing.

CAUTION

Plug all hydraulic hoses to prevent dirt from entering.

NOTE

Front and rear winch control valves are replaced basically the same. This procedure covers the front winch control valve.

a. Removal

NOTE

Tag all hoses and fittings for installation.

1. Disconnect hoses (1), (3), (5), (6), (7), and (8) from control valve (9).

NOTE

Both upper and lower cable clevises are removed the same way.

2. Remove cotter pin (16), washer (15), and clevis pin (4) from two cable clevises (2). Discard cotter pin (16).
3. Remove three locknuts (14), screws (13), and control valve (9) from bracket (11). Discard locknuts (14).
4. Remove four nuts (10), screws (12), and bracket (11) from frame (17).
5. Remove elbows (19) and (24) and adapter (29) from valve fittings (18) and (23) and elbow (30).
6. Remove fittings (31), (18), and (23) and elbows (21), (22), and (30) from control valve (9).
7. Remove six O-rings (20) from fittings (31), (18), and (23) and elbows (21) and (30). Discard O-rings (20).
8. Remove plug (25), O-ring (26), spring (27), and relief valve (28) from control valve (9). Discard O-ring (26).

b. Installation

1. Install six new O-rings (20), fittings (31), (18), and (23), and elbows (21) and (30) on control valve (9). Position as noted in removal.
2. Install elbows (19) and (24) and adapter (29) on valve fittings (18) and (23) and elbow (30).
3. Install relief valve (28), spring (27), new O-ring (26), and plug (25) in control valve (9).
4. Install bracket (11) on frame (17) with four screws (12) and nuts (10).
5. Install control valve (9) on bracket (11) with three screws (13) and new locknuts (14).

4-179. FRONT AND REAR WINCH CONTROL VALVE (M930/A1/A2) REPLACEMENT (Contd)

Do not cross hoses during installation. Crossing hoses may result in injury to personnel.

6. Install hoses (1), (3), (5), (6), (7), and (8) on control valve (9).

NOTE

Cable clevises are installed the same.

7. Install two cable clevises (2) on control valve (9) rod with clevis pin (4), washer (15), and new cotter pin (16). Bend over end of cotter pin (16).

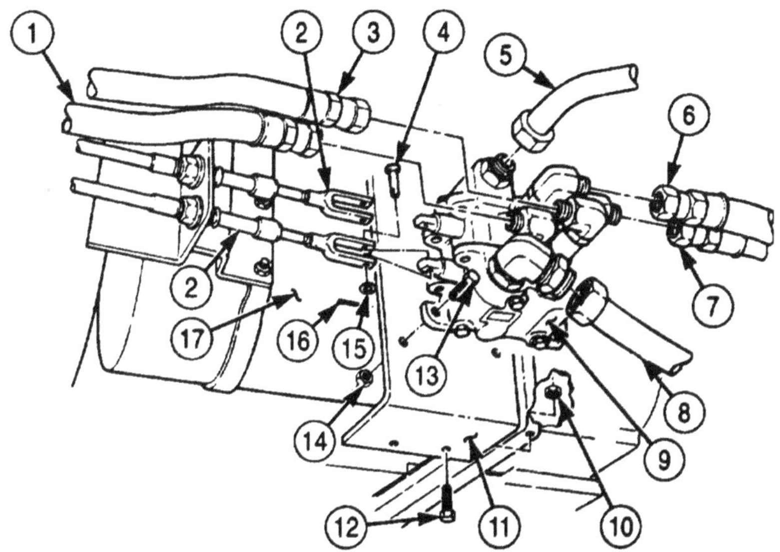

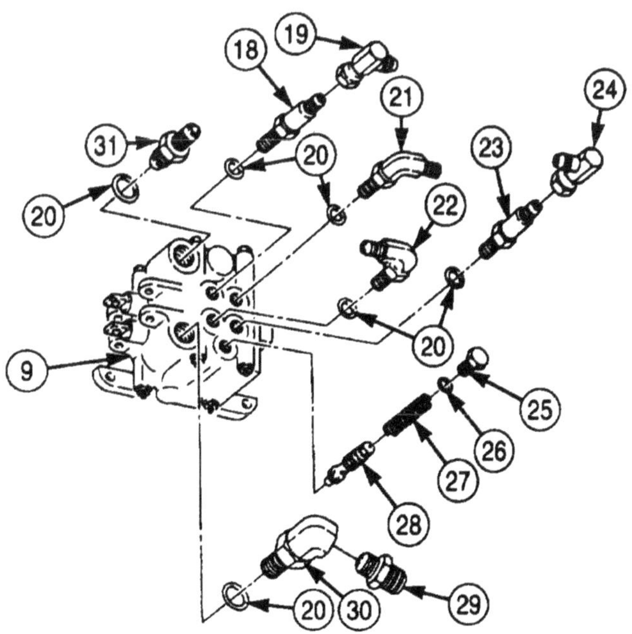

FOLLOW-ON TASKS:
- Fill hydraulic oil reservoir (LO 9-2320-272-12).
- Start engine (TM 9-2320-272-10) and check hydraulic pump operation.

4-180. FRONT WINCH HYDRAULIC HOSE AND TUBE REPLACEMENT

THIS TASK COVERS:

a. Removal

b. Installation

INITIAL SETUP:

APPLICABLE MODELS

M925/A1/A2, M928/A1/A2, M929/A1/A2, M930/A1/A2, M932/A1/A2, M936/A1/A2

TOOLS

General mechanic's tool kit (Appendix E, Item 1)

MATERIALS/PARTS

Two locknuts (Appendix D, Item 306)
Six tiedown straps (Appendix D, Item 690)
Cap and plug set (Appendix C, Item 14)

REFERENCES (TM)

LO 9-2320-272-12
TM 9-232-272-10
TM 9-2320-272-24P

EQUIPMENT CONDITION

- Parking brake set (TM 9-2320-272-10).
- Hydraulic oil reservoir drained (para. 4-188).

GENERAL SAFETY INSTRUCTIONS

Do not cross hoses when installing.

a. Removal

CAUTION

Plug all openings to prevent dirt from entering and damaging components.

NOTE

- Remove and discard tiedown straps holding hoses to vehicle.
- Have drainage container ready to catch oil.
- Perform steps 1 through 4 for all vehicles except M936/A1/A2.

1. Holding filter adapter (10), loosen hose nut (9) and disconnect hose (7) from hydraulic oil reservoir (1).
2. Remove nut (8), screw (4)`, and clamp (5) from mounting bracket (6).
3. Loosen hose nut (3) and disconnect hose (7) from reservoir elbow (2).
4. Remove hose (7) from reservoir (1).

NOTE

Clamp brackets remain attached to frame rail.

5. Remove two locknuts (14), screws (12), and two pairs of hose clamps (13) from right frame rail (11) and two hydraulic hoses (15). Discard locknuts (14).

4-180. FRONT WINCH HYDRAULIC HOSE AND TUBE REPLACEMENT (Contd)

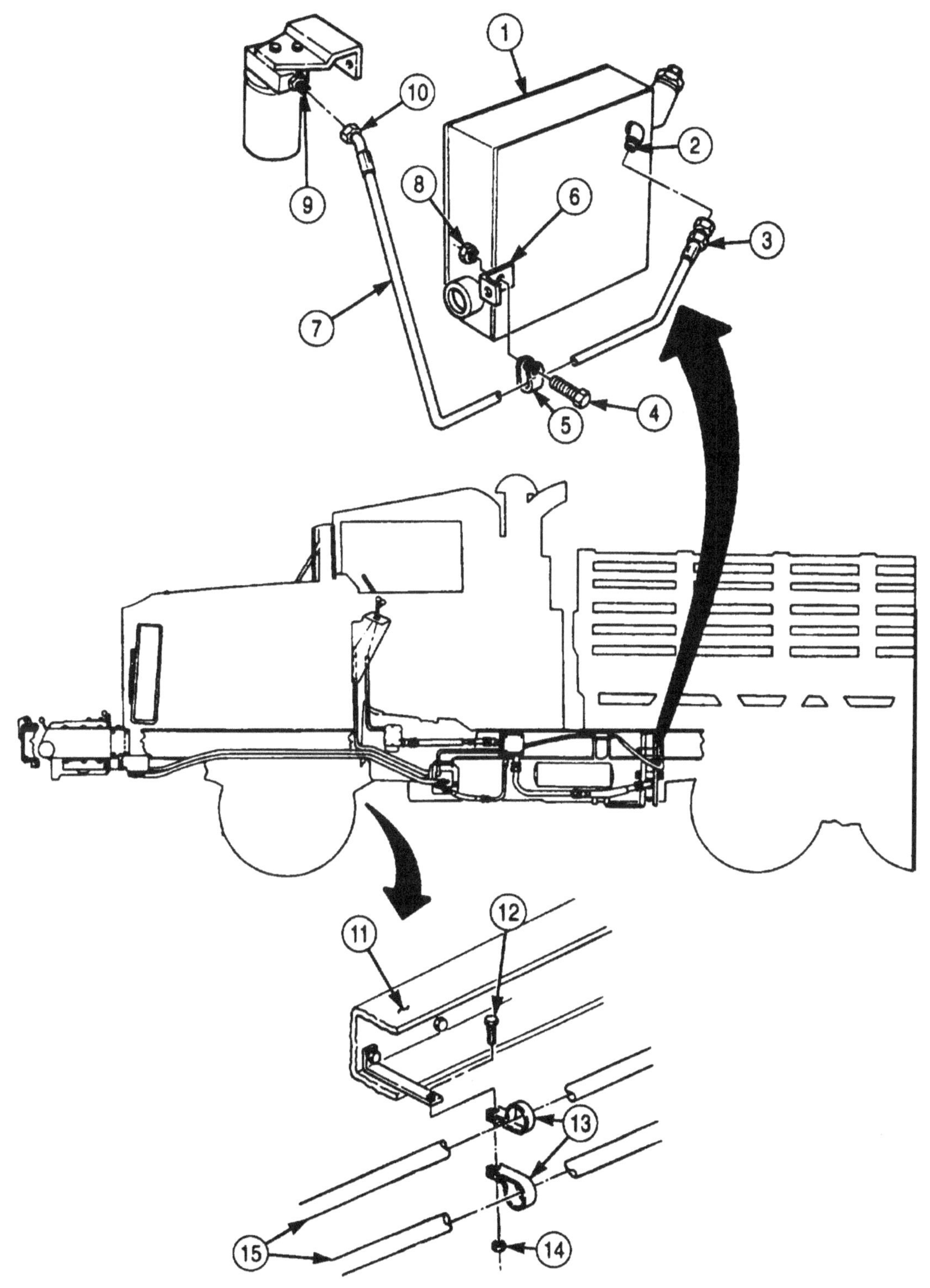

4-180. FRONT WINCH HYDRAULIC HOSE AND TUBE REPLACEMENT (Contd)

NOTE

Tag hoses for installation.

6. Using two wrenches, disconnect swivel flare nut (1) and hose end (2), and remove four hoses (3) and two tubes (4) and (5).

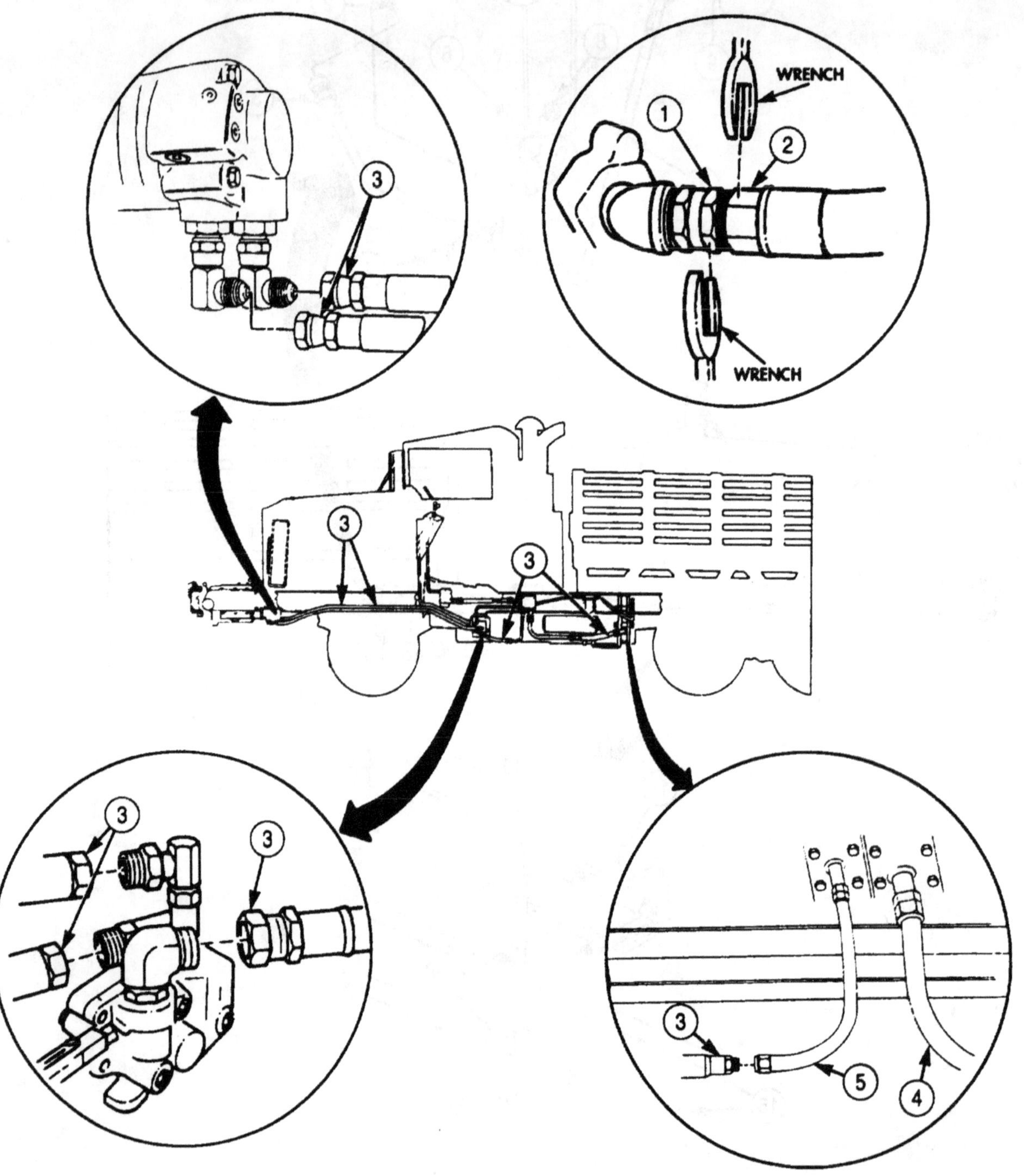

4-180. FRONT WINCH HYDRAULIC HOSE AND TUBE REPLACEMENT (Contd)

NOTE

Perform steps 7 and 8 for all vehicles except M936/A1/A2.

7. Using wrench to prevent adapter (7) from turning, disconnect hose (6) from adapter (7) on filter (8) and from control valve (9).
8. Disconnect two tubes (4) and (5) from adapters (11) and (12) on hydraulic pump (10).

b. Installation

1. Connect tubes (4) and (5) to adapters (11) and (12) on hydraulic pump (10).

NOTE

Perform steps 2 for all vehicles except M936/A1/A2.

2. Connect hose (6) to control valve (9) and adapter (7) on filter (8).

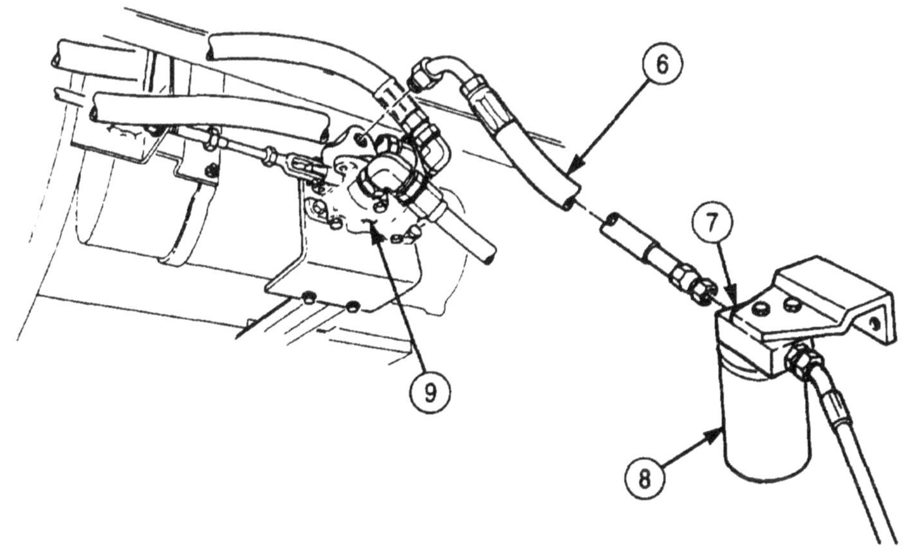

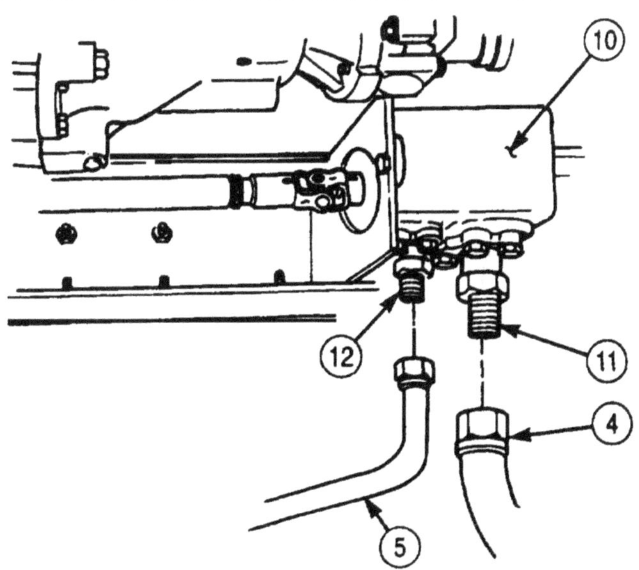

4-180. FRONT WINCH HYDRAULIC HOSE AND TUBE REPLACEMENT (Contd)

NOTE

Perform steps 3 and 4 for all vehicles except M936/A1/A2.

3. Connect hydraulic hose (6) to elbow (2) and adapter (1).
4. Install hose (6) on mounting bracket (5) with clamp (4), screw (3), and nut (7).
5. Using care not to bend tube (10), connect hose (11) to elbow (12) and tube (10).
6. Using care not to bend tube (9), connect hose (8) to tube (9) and control valve elbow (16).
7. Connect two hose ends (23) as tagged to two control valve elbows (21).

WARNING

Do not cross hoses during installation. Crossing hoses may result in injury to personnel.

8. Connect two hose ends (13) to two elbows (14).
9. Position two pairs of clamps (16) on hoses (19) and install clamps (16) on two brackets (20) with two screws (15) and new locknuts (17).
10. Install six tiedown straps (18) as necessary over two hoses (19).

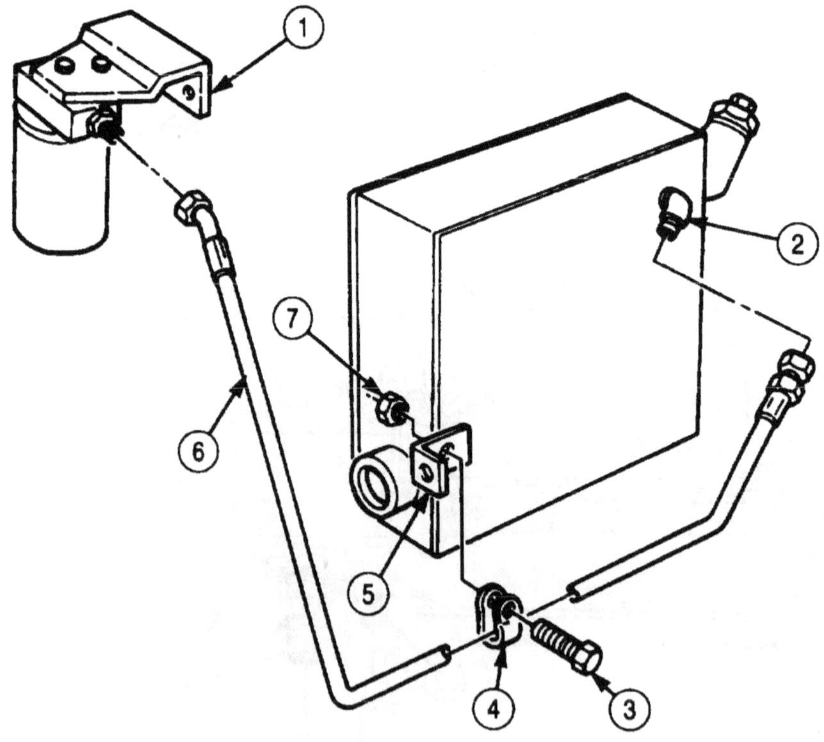

4-180. FRONT WINCH HYDRAULIC HOSE AND TUBE REPLACEMENT (Contd)

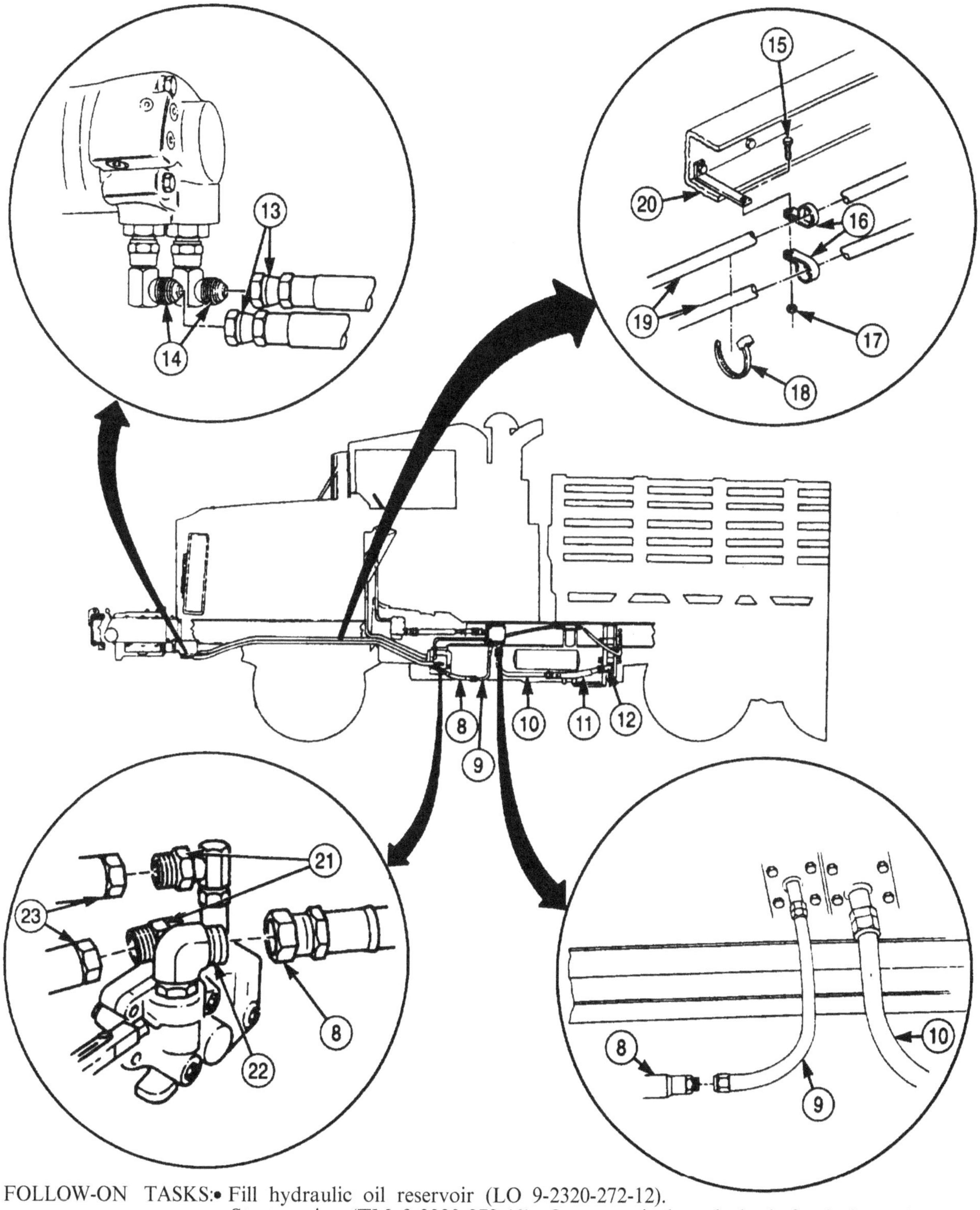

13
14
15
20
16
19
17
18
8
9
10
11
12
21
23
8
22
8
9
10

FOLLOW-ON TASKS:
- Fill hydraulic oil reservoir (LO 9-2320-272-12).
- Start engine (TM 9-2320-272-10). Operate winch and check for leaks and proper operation.

4-181. REAR WINCH REPAIR

THIS TASK COVERS:

a. Disassembly
b. Cleaning and Inspection
c. Assembly

INITIAL SETUP:

APPLICABLE MODELS
M936/A1/A2

TOOLS
General mechanic's tool kit (Appendix E, Item 1)
Puller (Appendix E, Item 110)
Arbor press
Feeler gauge
Mandrel

TEST EQUIPMENT
Spring tester (Appendix E, Item 132)
Dial indicator (Appendix E, Item 36)
Micrometer (Appendix E, Item 80)

MATERIALS/PARTS
Brake band (Appendix D, Item 15)
Four cotter pins (Appendix D, Item 57)
Two gaskets (Appendix D, Item 217)
Two seals (Appendix D, Item 610)
O-ring (Appendix D, Item 459)
Bushing (Appendix D, Item 26)
Gasket (Appendix D, Item 216)
Seal (Appendix D, Item 609)
Seal (Appendix D, Item 611)
Five lockwashers (Appendix D, 350)

MATERIALS/PARTS (Contd)
Eighteen lockwashers (Appendix D, Item 393)
Gasket (Appendix D, Item 218)
Six lockwashers (Appendix D, Item 392)
Two bushings (Appendix D, Item 25)
Two woodruff keys (Appendix D, Item 737)
GAA grease (Appendix C, Item 28)
Lubricating oil (Appendix C, Item 50)
Rags (Appendix C, Item 58)
Sealing compound (Appendix C, Item 65)

PERSONNEL REQUIRED
Two

REFERENCES (TM)
LO 9-2320-272-12
TM 9-2320-272-10
TM 9-2320-272-24P

EQUIPMENT CONDITION
- Rear winch removed (para. 3-332).
- Rear winch level wind removed (para. 4-174).

GENERAL SAFETY INSTRUCTIONS
The rear winch is a heavy component. Use hoist during repair to prevent injury.

a. Disassembly

1. Remove four cotter pins (10), pin (91, pin (11), and lever (7) from bracket (8). Discard cotter pins (10).
2. Remove four screws (16) and lockwashers (17) from bracket (15). Discard lockwashers (17).
3. Remove two screws (18), nuts (20), lockwashers (19), and bracket (15) from tensioner frame (14). Discard lockwashers (19).
4. Remove two nuts (l), lockwashers (2), screws (6), and tensioner frame (14) from bracket (8). Discard lockwashers (2).
5. Remove two nuts (4), lockwashers (5), four screws (13), four washers (12), and bracket (8) from rear winch (3). Discard lockwashers (5).

4-181. REAR WINCH REPAIR (Contd)

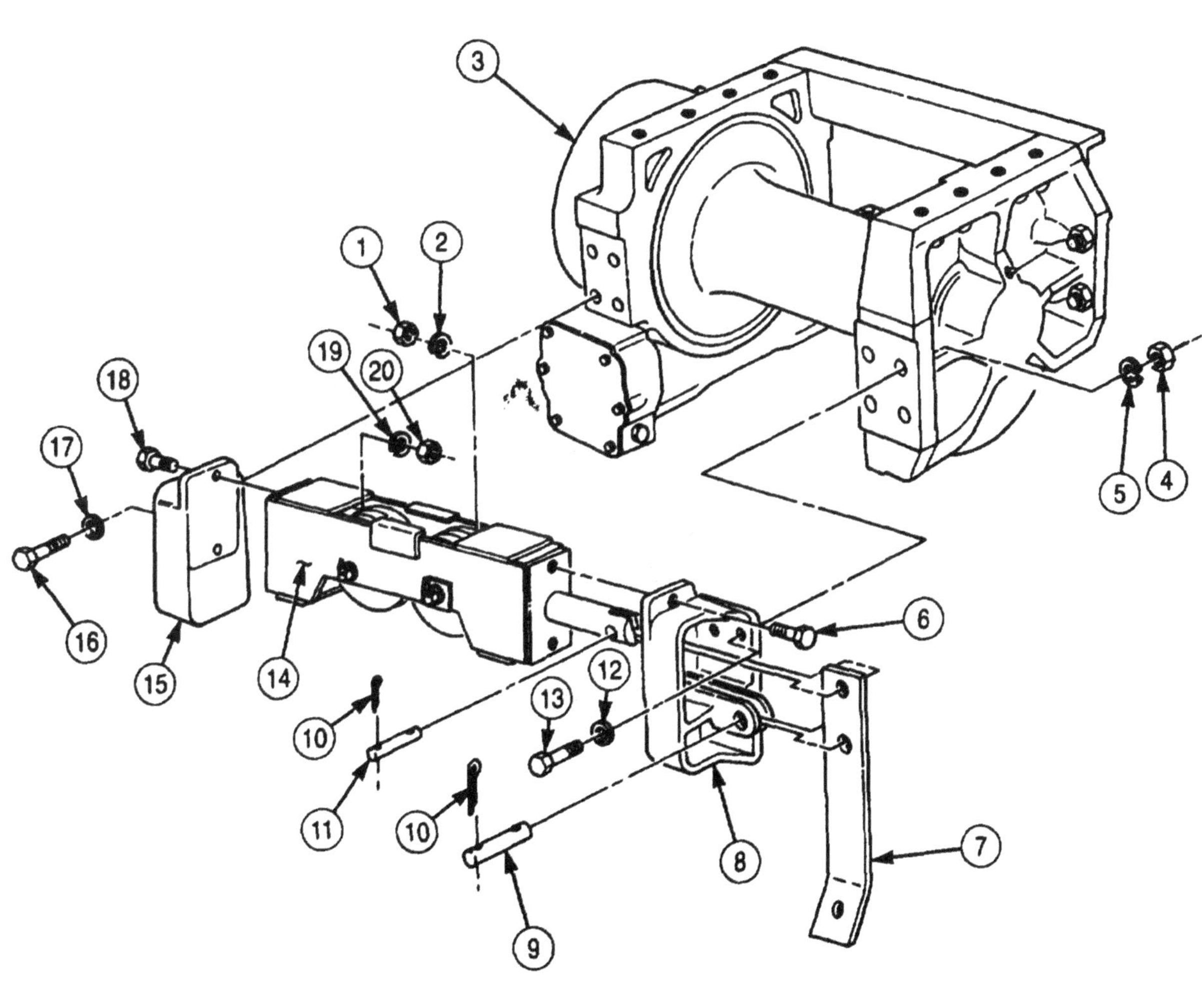

4-181. REAR WINCH REPAIR (Contd)

6. Remove three screws (3) and lockwashers (2) from rear winch (1) and channel (5). Discard lockwashers (2).
7. Remove nut (7), lockwasher (6), screw (4), and front channel (5) from rear winch (1). Discard lockwasher (6).

NOTE

Assistant will help with step 8.

8. Remove two screws (16), six screws (17), lockwashers (18), end cover (19), and gasket (14) from gearcase (11). Discard lockwashers (18) and gasket (14).
9. Remove bushing (15) from end cover (19). Discard bushing (15) if inner diameter is more than 3.150 in. (80.01 mm).
10. Remove thrust washer (13) from drum shaft (12).
11. Remove four screws (8), lockwashers (9). input shaft bearing cap (21), and gasket (10) from gearcase (11). Discard lockwashers (9) and gasket (10).
12. Remove seal (20) from bearing cap (21). Discard seal (20).

NOTE

Support drum with overhead hoist.

13. Remove end frame (26) from drum shaft (12) and gearcase (11).
14. Remove drum (25), seal (24), thrust washer (23), two woodruff keys (22), and grease fitting (27) from shaft (12). Discard seal (24) and two woodruff keys (22).

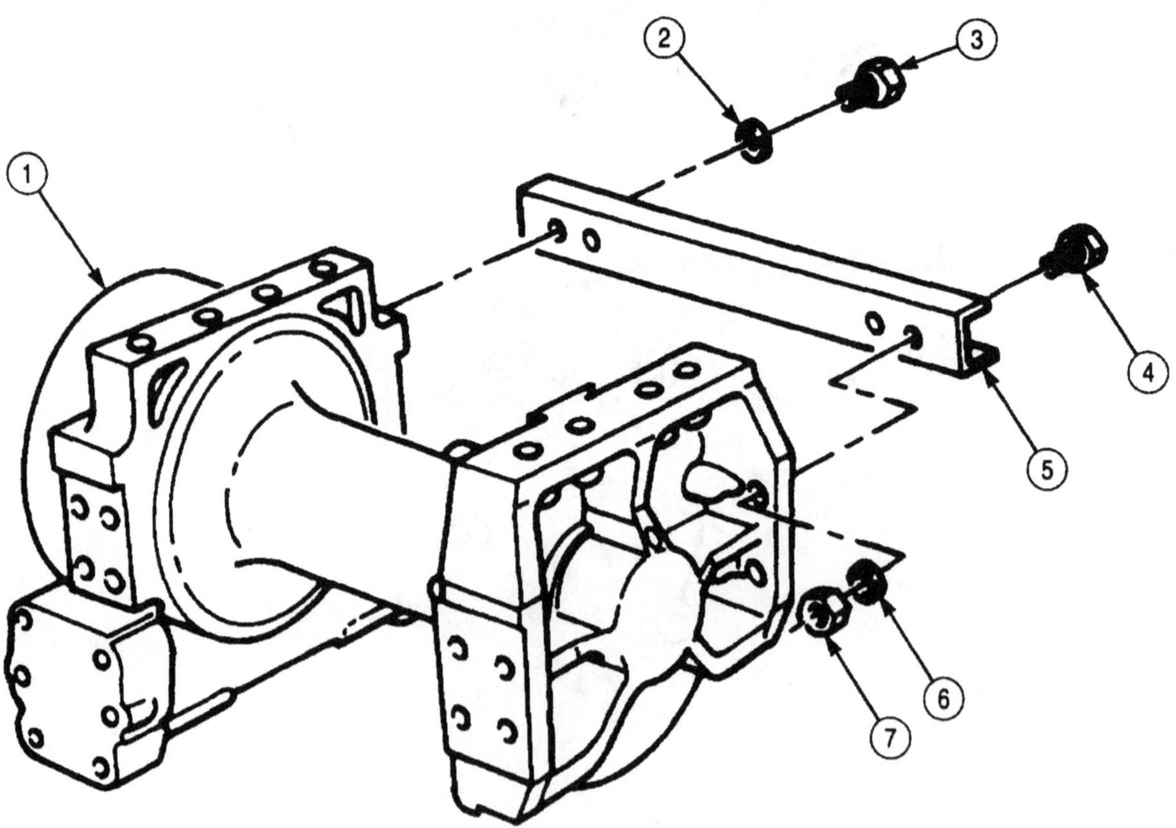

4-181. REAR WINCH REPAIR (Contd)

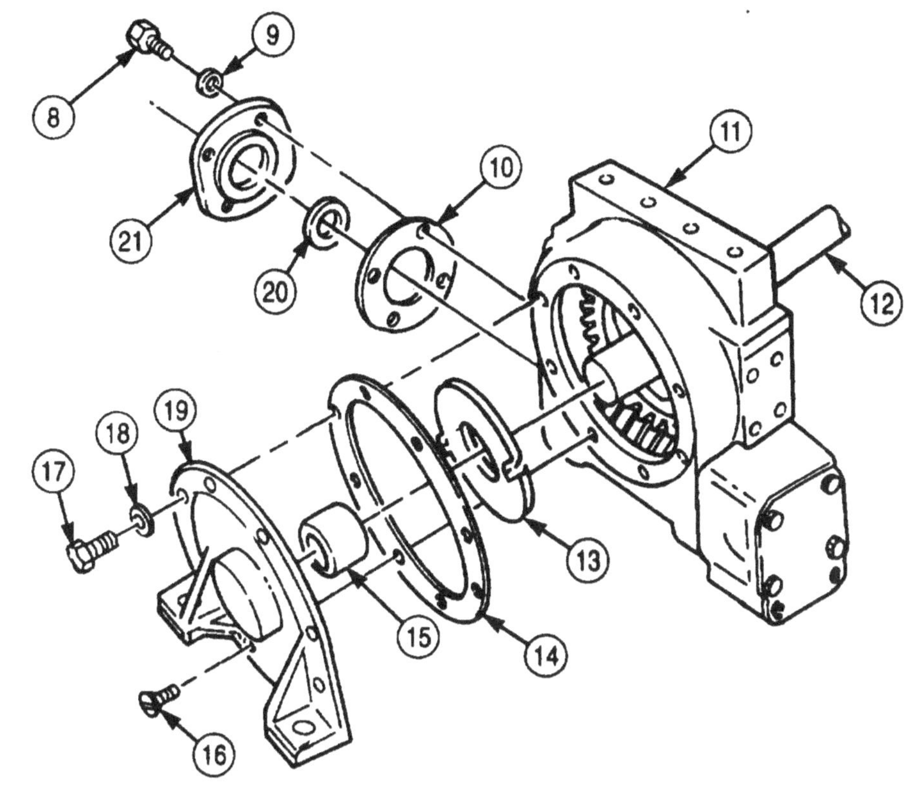

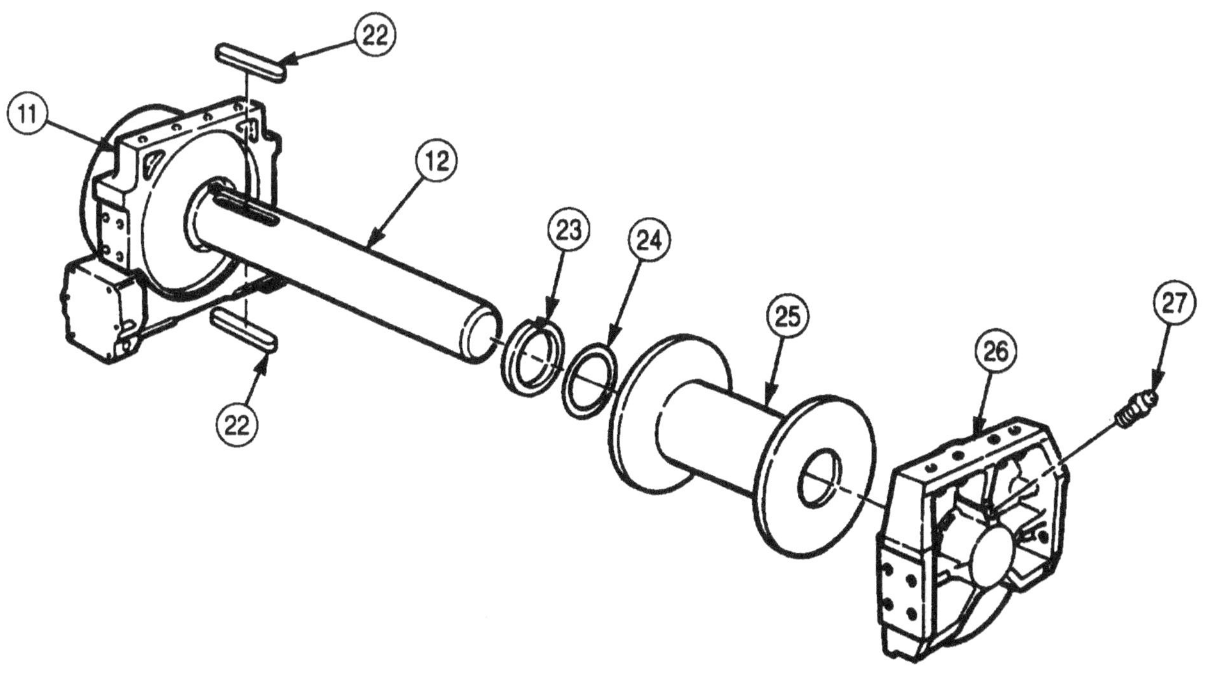

4-181. REAR WINCH REPAIR (Contd)

15. Remove six screw-assembled washers (8), housing cover (9), and gasket (10) from automatic brake housing (2). Discard gasket (10). Clean gasket remains from mating surfaces.
16. Remove adjusting screw (16), washer (15), and O-ring (14) from housing (2). Discard O-ring (14).
17. Remove brake band (7) and spring (11) from brake housing (2). Discard brake band (7).

NOTE

Assistant will help with step 18.

18. Remove screw (6), lockwasher (12), washer (5) and, using a puller, remove brake drum (13) from input shaft (20). Discard lockwasher (12).
19. Scribe alignment marks (21) at mating surfaces of gearcase (1) and brake housing (2) for installation.
20. Remove four screws (4), lockwashers (3), brake housing (2), and gasket (18) from gearcase (1). Discard lockwashers (3) and gasket (18). Clean gasket (18) remains from mating surfaces.
21. Remove seal (17) from brake housing (2). Discard seal (17).
22. Remove woodruff key (19) from worm gear shaft (20). Discard woodruff key (19).

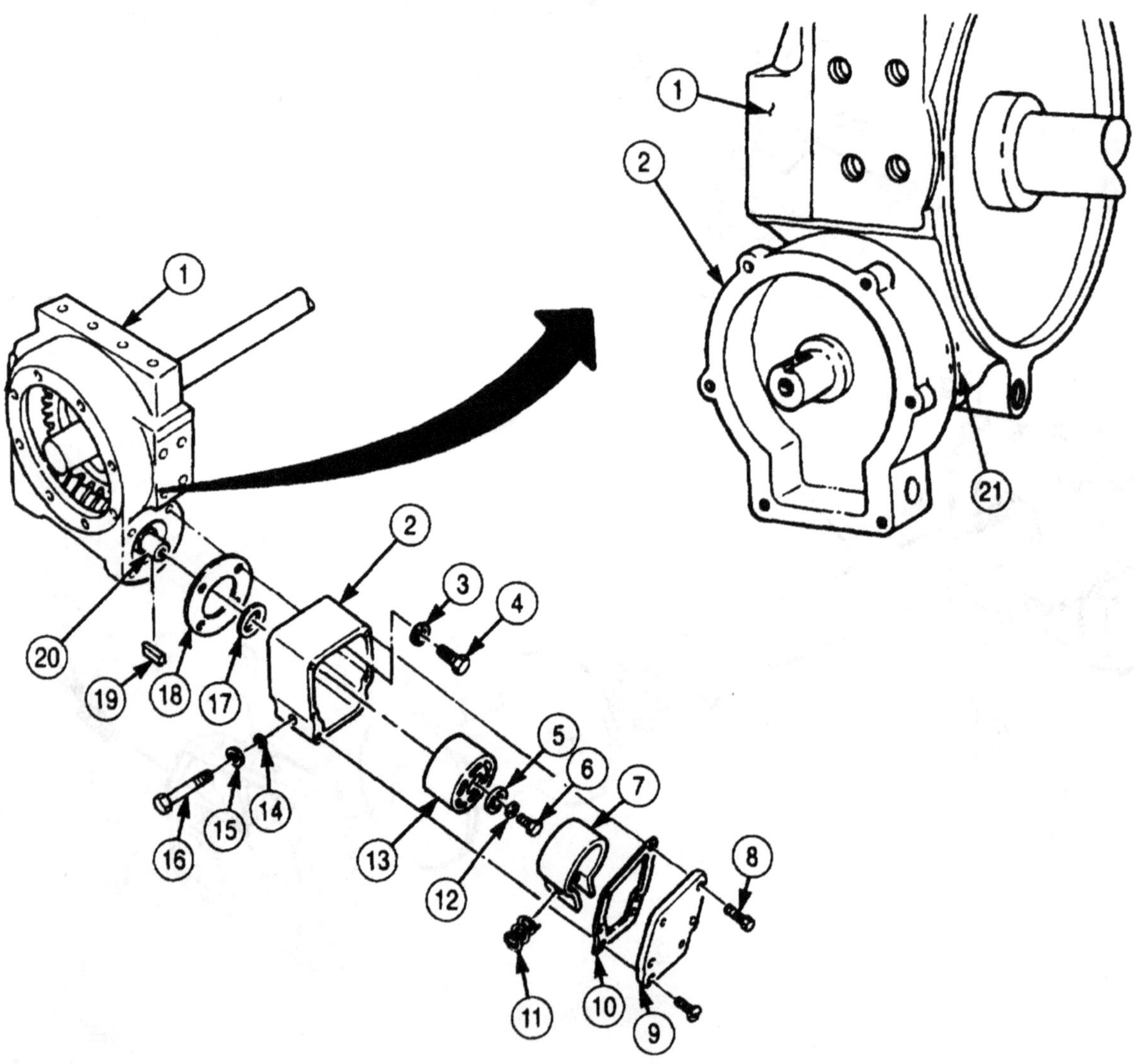

4-181. REAR WINCH REPAIR (Contd)

23. Using a soft-faced hammer, tap on worm gear shaft (20) at brake end and allow worm gear (23) to drop to bottom of gearcase (1).
24. Remove gear (21), drum shaft (24), and thrust washer (22) from gearcase (1).
25. Remove worm gear shaft (20) from gearcase (1).
26. Remove seal (25), thrust washer (26), and sleeve (28) from end frame (29). Discard seal (25).
27. Remove bushing (27) from sleeve (28). Discard bushing (27).

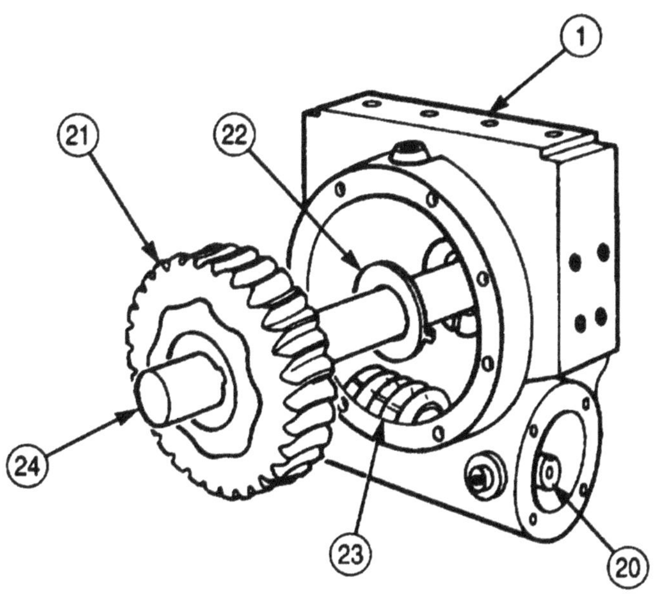

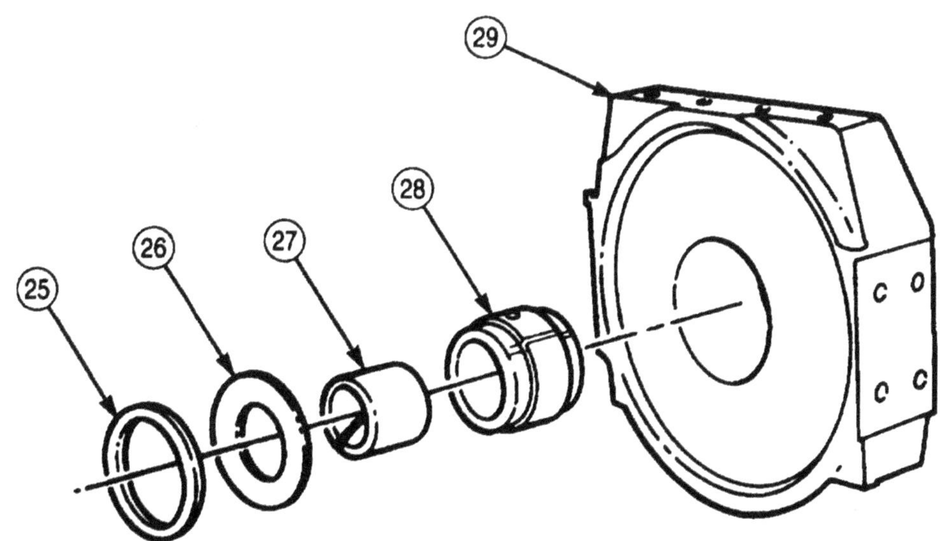

4-181. REAR WINCH REPAIR (Contd)

b. Cleaning and Inspection

1. Clean and inspect all thrust washers (1) for cracks and measure thickness. Replace if cracked or if thickness is less than 0.059 in. (1.50 mm).
2. Clean drum shaft (3) for cracks, chips, straightness, and wear. Roll drum shaft (2) on a flat surface to check wear. Measure bearing surfaces for wear with a micrometer. Replace if cracked, chipped, bent, or any bearing surface is less than 2.995 in. (76.07 mm) in diameter.
3. Clean and inspect three bushings (3), one in end frame and two in drum, for cracks, chips, scoring, and wear. Measure wear with an inside micrometer. Replace bushings (3) if cracked, chipped, scored, or inner diameter is more than 3.010 in. (76.45 mm).
4. Clean and inspect worm shaft (4) and worm gear (5) for cracks, chips, scoring, or damaged bearings. Replace if cracked, chipped, scored or if balls, rollers, or cage is damaged.

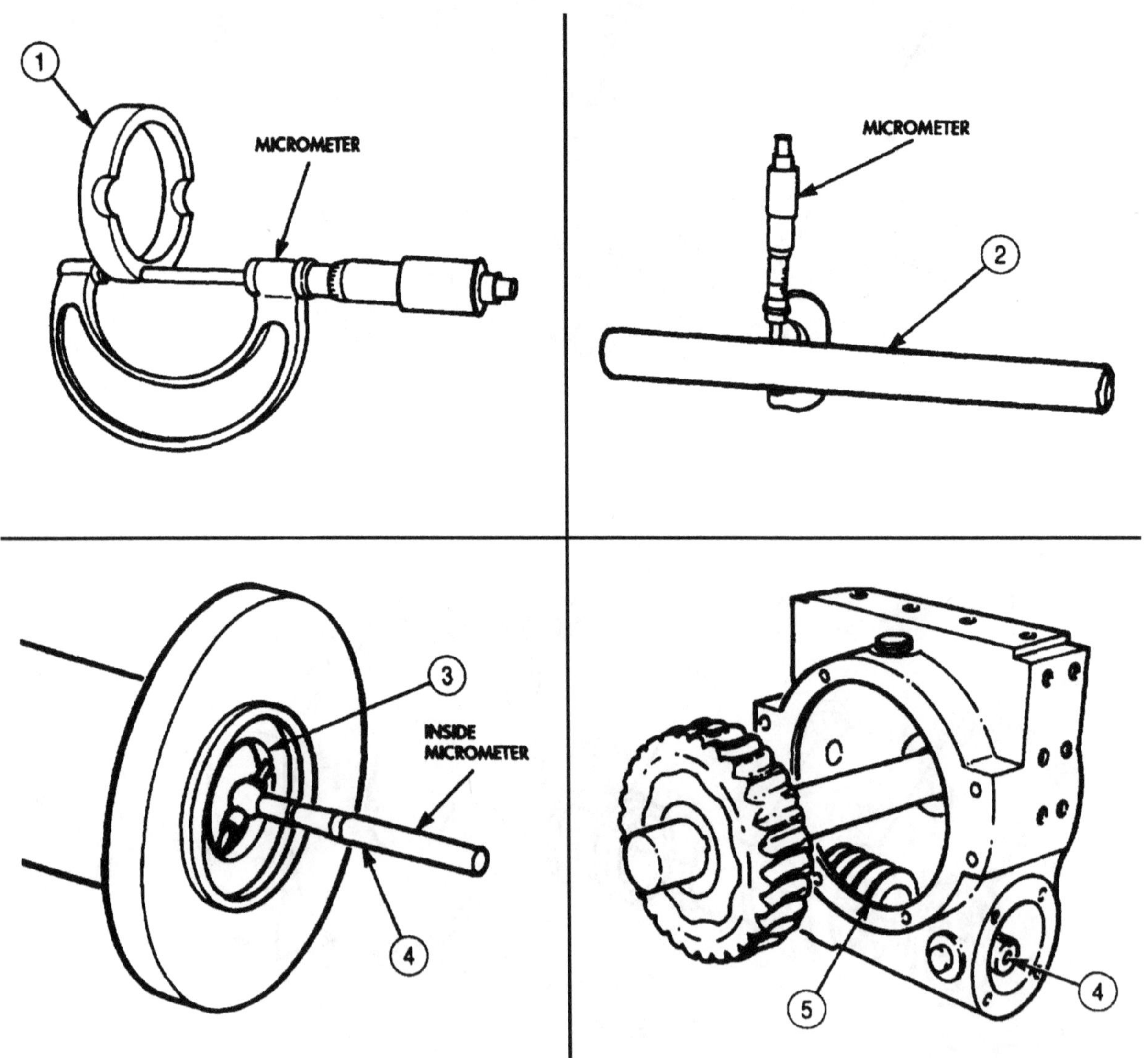

4-181. REAR WINCH REPAIR (Contd)

5. Clean spring (6) and inspect for distortion or cracked or collapsed coils. Replace if distorted or cracked, or coil is collapsed.
6. Using spring tester, test spring (6) strength. Replace if force to compress spring (6) to 2 in. (50.8 mm) is less than 97 lb-ft (131.5 N•m).
7. Clean and inspect all housings, covers, and cases for cracks, broken parts, or damaged threads. Replace if cracked, broken, or threads are damaged.
8. Clean and inspect drum shaft gear (7) for broken, chipped, or scored teeth. Replace if teeth are broken, chipped, or scored.

NOTE

If drum shaft gear does not require replacement, go to task c.

9. Using arbor press and mandrel, remove drum shaft (2) from drum shaft gear (7).

NOTE

Do not remove drum shaft keys if they are not damaged. If removed, install new woodruff keys.

10. Press shaft (2) into gear (7) until woodruff keys are centered in gear (7) hub. Use arbor press and mandrel.

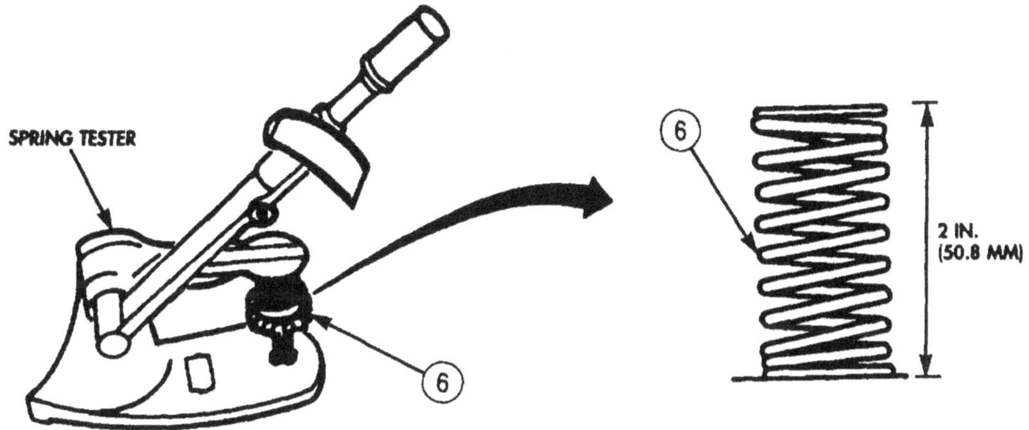

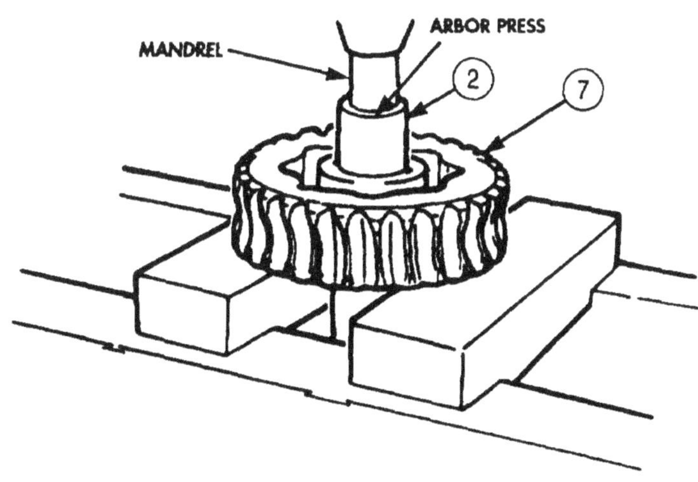

4-181. REAR WINCH REPAIR (Contd)

c. Assembly

NOTE

Apply a light film of oil to all parts during assembly. Do not lubricate brake band or brake drum.

1. Place worm shaft (4) and worm gear (5) in gearcase (3) with keyway end of worm shaft (4) at mating mark side of gearcase (3). Let worm gear (5) rest on rag in gearcase (3).
2. Coat thrust washer (2) with light coat of GAA grease and slide over long end of drum shaft (6). Ensure locking tab on thrust washer (2) points toward long end of drum shaft (6).
3. Install drum shaft (6) and gear (7) part way through bushing (13) in gearcase (3) and position tab of thrust washer (2) into notch in gearcase (3)
4. Complete installation of drum shaft (6) and gear (7) into gearcase (3) until gear (7) is seated on thrust washer (2).
5. Position worm gear shaft (4) in gearcase (3), ensuring worm gear (5) is engaging drum shaft gear (7).
6. Coat thrust washer (1) with coat of GAA grease and place over shaft (6) with tabs pointing away from drum shaft gear (7).
7. Install bushing (13) in end cover (10).

NOTE

Do not install gasket at this time.

8. Position end cover (10) on shaft (6), ensuring tabs of thrust washer (11) engage slots in end cover (10).
9. Measure clearance between gearcase (3) and cover (10) with feeler gauge.
   a. If there is zero clearance, remove cover (10) and install gasket (12) and cover (10) on gearcase (3) (step c.).
   b. If a measurable gap, add one to four gaskets (12) and remeasure gap. Allowable gap is 0.005-0.015 in. (0.13-0.38 mm).
   c. When proper clearance is obtained, remove cover (10) and gasket(s) (12), coat gasket(s) (12) with sealing compound, and install gasket(s) (12) and cover (10) on gearcase (3) with two screws (14) and six new lockwashers (9) and screws (8).

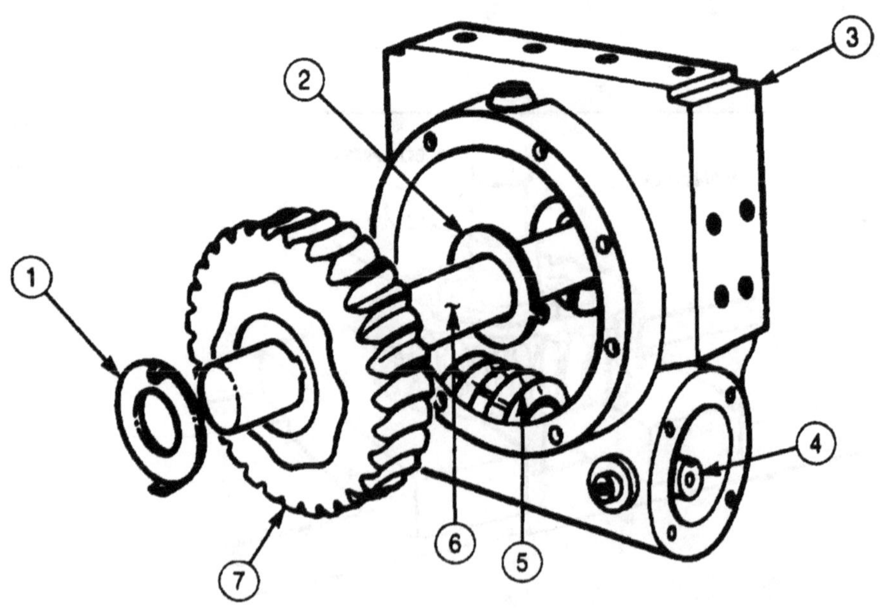

4-181. REAR WINCH REPAIR (Contd)

10. Install new seal (27) in brake housing (20).
11. Position brake housing (20) and new gasket (19) on gearcase (3) and install with four new lockwashers (21) and screws (22).
12. Install new seal (29) in input shaft bearing cap (17).
13. Install new gasket (18) and input shaft bearing cap (17) on gearcase (3) with four new lockwashers (16) and screws (15).
14. Install woodruff key (28) on input shaft (4).
15. Ensure woodruff key (25) slot in brakedrum (26) engages woodruff key CW, position brakedrum (26) on input shaft (4), and install with washer (23), new lockwasher (25), and screw (24).

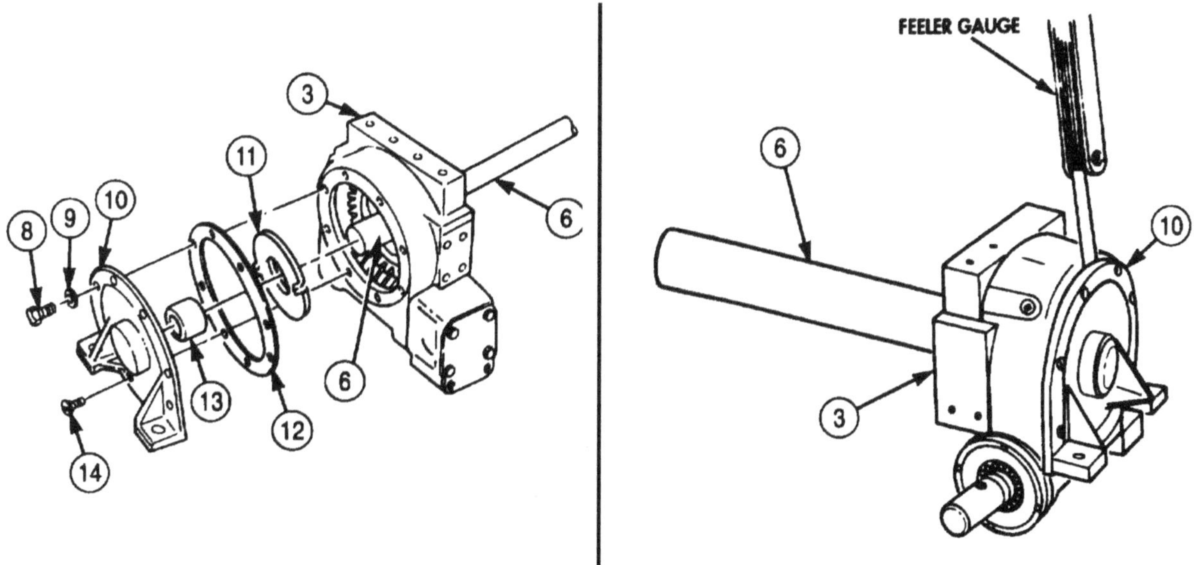

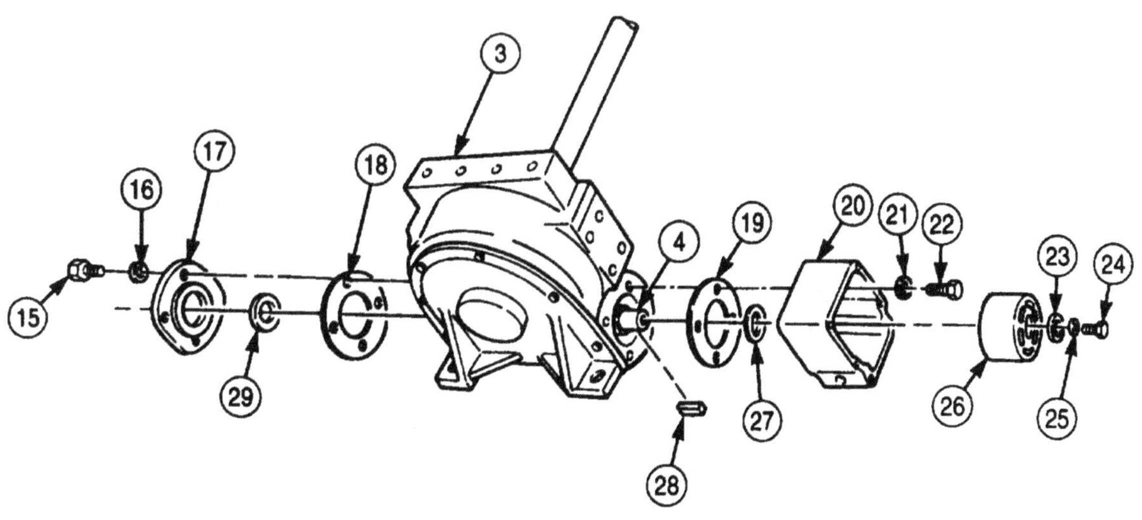

4-181. REAR WINCH REPAIR (Contd)

16. Install dial indicator and extension on gearcase (1) and measure end play of worm gear shaft (2).
    a. Correct end play is 0.005 in. (0.13 mm).
    b. If end play is less than correct, remove bearing cap (item 17 of step 13) and add gaskets (item 18 of step 13) one at a time and remeasure until end play is correct.
17. Remove dial indicator.
18. Position new brake band (5) around drum (4) with caged nut (14) on the right side in housing (3).
19. Position spring (10) between brake band (5) and left side of housing (3), and install new O-ring (13), washer (12), and screw (11). Ensure that screw (11) goes through spring (10), brake band (5), and threads into caged nut (14). Tighten screw (11) until spring (10) length is 2.125-2.187 in. (51.0-52.5 mm).
20. Position new gasket (6) and cover (7) on brake housing (4) and install with two screw-assembled washers (8) in top holes and four screw-assembled washers (9) in bottom four holes of cover (7) and brake housing (4).

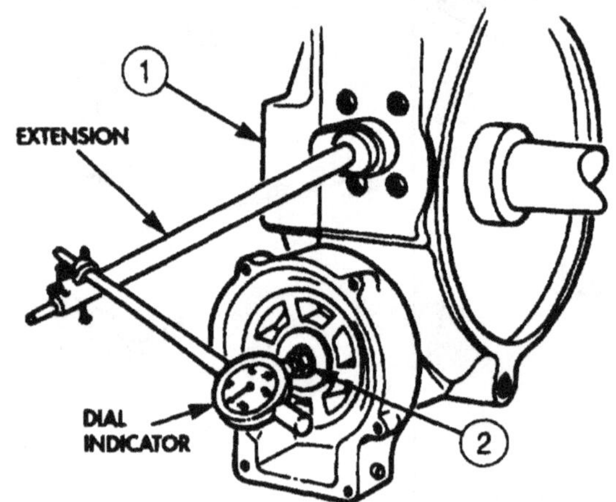

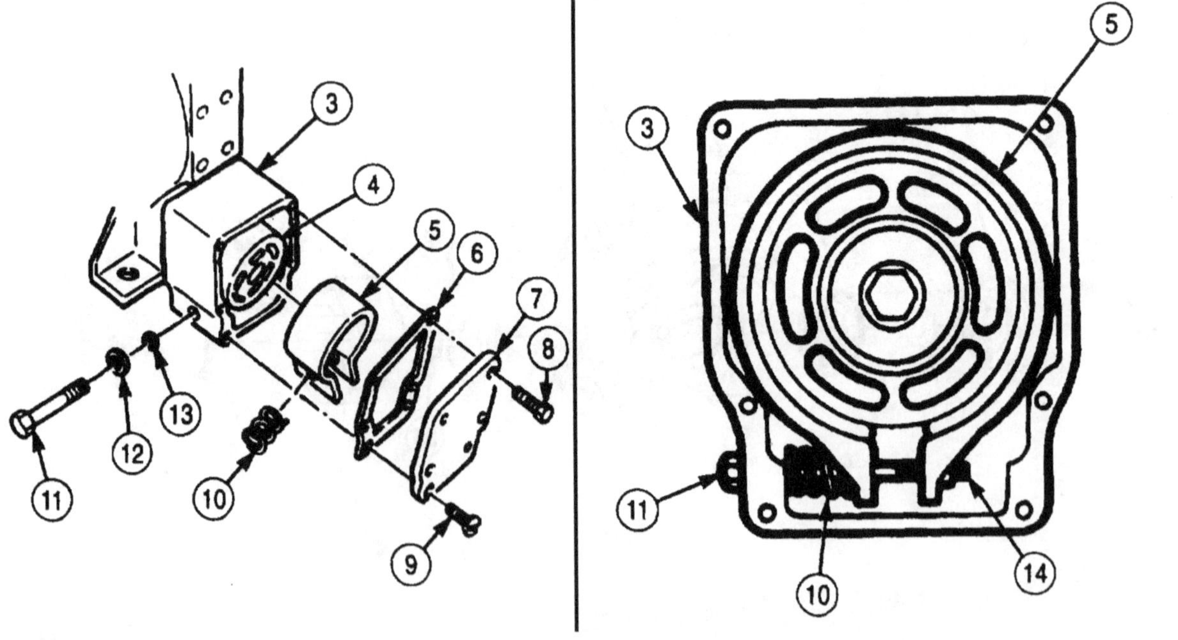

4-181. REAR WINCH REPAIR (Contd)

21. Install bushing (17) in sleeve (18) and position sleeve (18) over pin in end frame (19).
22. Place thrust washer (16) in end frame (19). Ensure that tab on thrust washer (16) engages slot in end frame (19).
23. Install new seal (15) in end frame (19).
24. Install new seal (23) in drum (24).
25. Goat thrust washer (22) with GAA grease, install on shaft (21), and position against gearcase (1). Ensure that thrust washer (22) engages slot in gearcase (1).
26. Using a soft-faced hammer as necessary, install two new keys (20) in shaft (21).
27. Align keyways in drum (24) to keys (20) on shaft (21) and drive drum (24) on shaft (21) until seated.
28. Install grease fitting (25) on end frame (19).
29. Install end frame (19) on shaft (21).

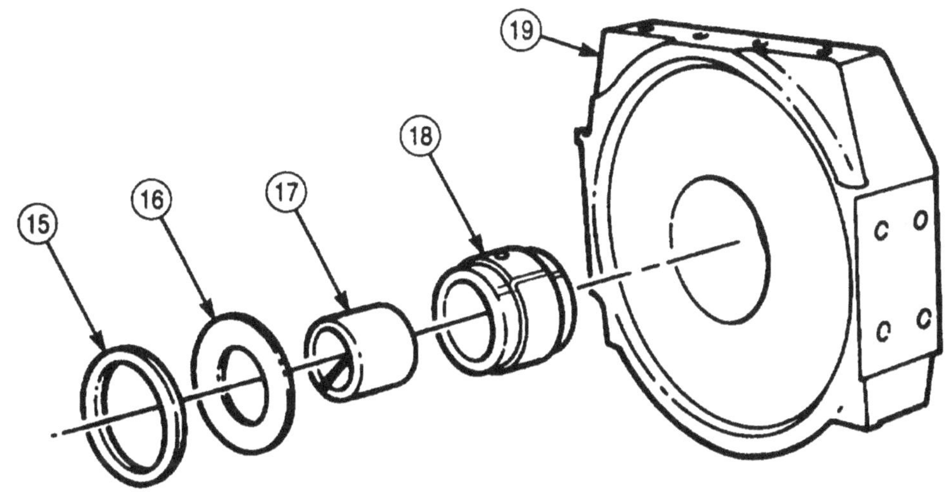

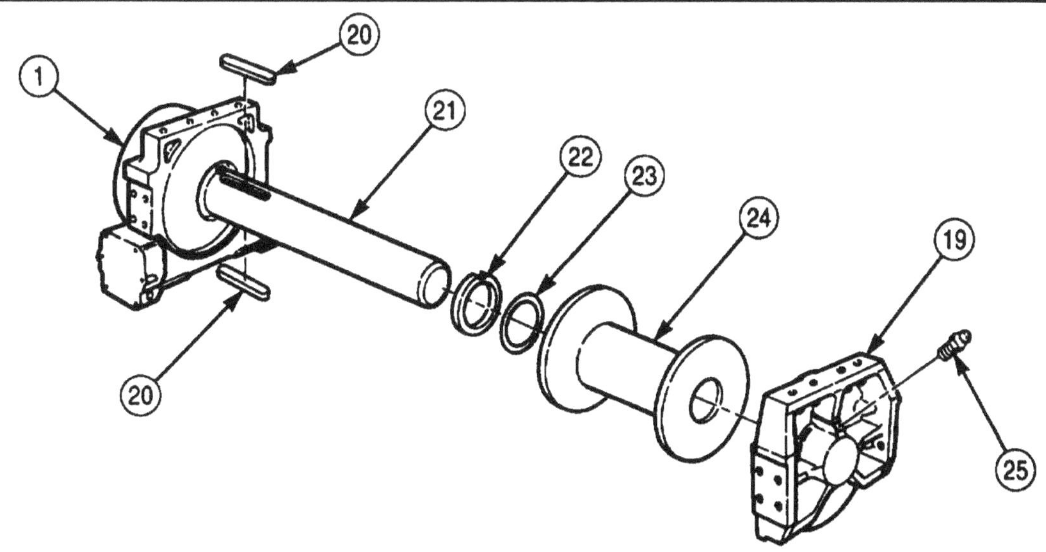

4-181. REAR WINCH REPAIR (Contd)

30. Install front channel (5) on gearcase (1) with two new lockwashers (2) and screws (3).
31. Install front channel (5) on end frame (8) with two screws (4), new lockwashers (6), and nuts (7).
32. Install bracket (20) on end frame (8) with two washers (23), screws (24), new lockwashers (17), and nuts (16).
33. Install tensioner frame (28) on bracket (20) with two screws (18), new lockwashers (15), and nuts (14).
34. Install bracket (29) on tensioner frame (28) with two screws (ll), new lockwashers (12), and nuts (13).
35. Install bracket (29) on gearcase (1) with four new lockwashers (10) and screws (9).
36. Install lever (19) on bracket (20) and rod (25) with pin (21), pin (26), and two new cotter pins (22), and two new cotter pins (27).

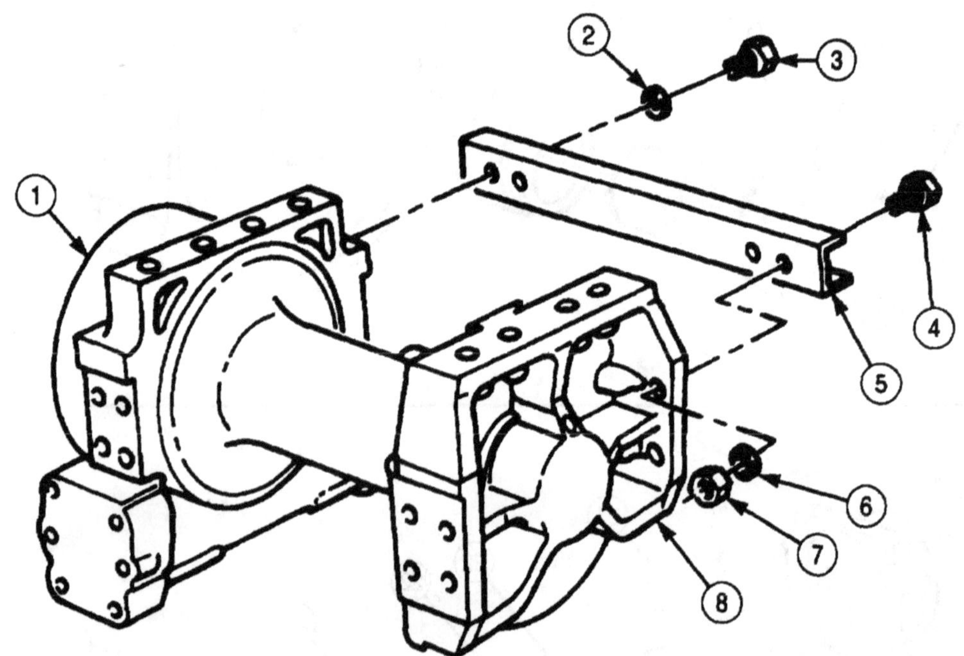

4-181. REAR WINCH REPAIR (Contd)

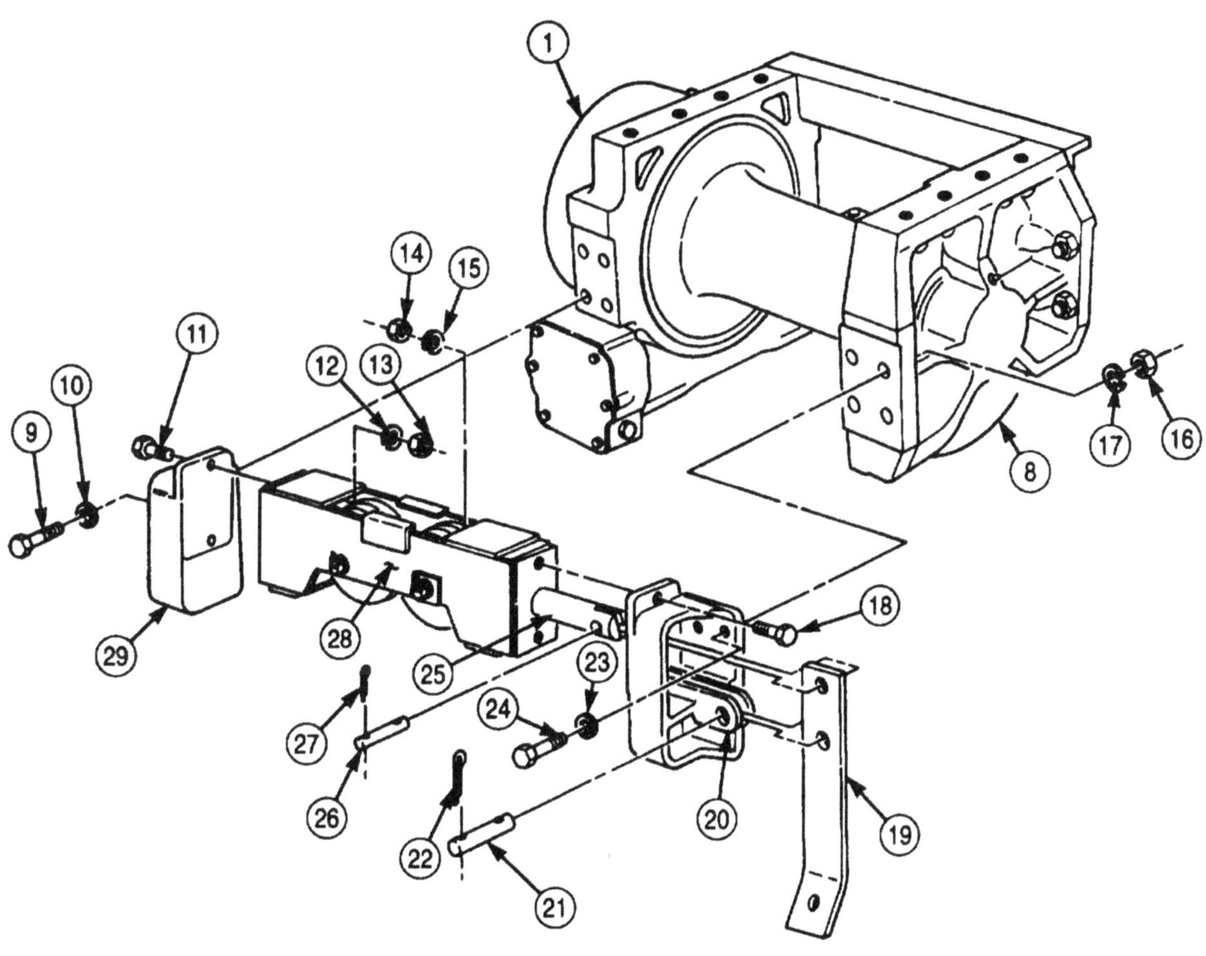

FOLLOW-ON TASKS:
- Install rear winch level wind (para. 4-174).
- Install rear winch (para. 3-332).

THIS TASK COVERS:

a. Disassembly
b. Cleaning, Inspection, and Repair
c. Assembly

INITIAL SETUP:

APPLICABLE MODELS
All

TOOLS
General mechanic's tool kit (Appendix E, Item 1)
Micrometer (Appendix E, Item 80)

MATERIALS/PARTS
Four felt washers (Appendix D, Item 114)
Four cotter pins (Appendix D, Item 57)
Two cotter pins (Appendix D, Item 67)
Four lockwashers (Appendix D, Item 350)
Eight lockwashers (Appendix D, Item 393)
GAA grease (Appendix C, Item 28)

REFERENCES (TM)
LO 9-2320-272-12
TM 9-237
TM 9-2320-272-10
TM 9-2320-272-24P

EQUIPMENT CONDITION
Rear winch cable removed (para. 3-331).

---

1. Remove four cotter pins (10), pins (9) and (ll), and lever (7) from bracket (8). Discard cotter pins (10).
2. Remove four screws (16) and lockwashers (17) from bracket (15). Discard lockwashers (17).
3. Remove two nuts (20), lockwashers (19), screws (18), and bracket (15) from tensioner frame (14). Discard lockwashers (19).
4. Remove two nuts (1). lockwashers (2). screws (6), and tensioner frame (14) from frame bracket (8). Discard lockwashers (2).
5. Remove four nuts (4), lockwashers (5), screws (13), washers (12), and bracket (8) from rear winch (3). Discard lockwashers (5).

4-182. REAR WINCH CABLE TENSIONER REPAIR (Contd)

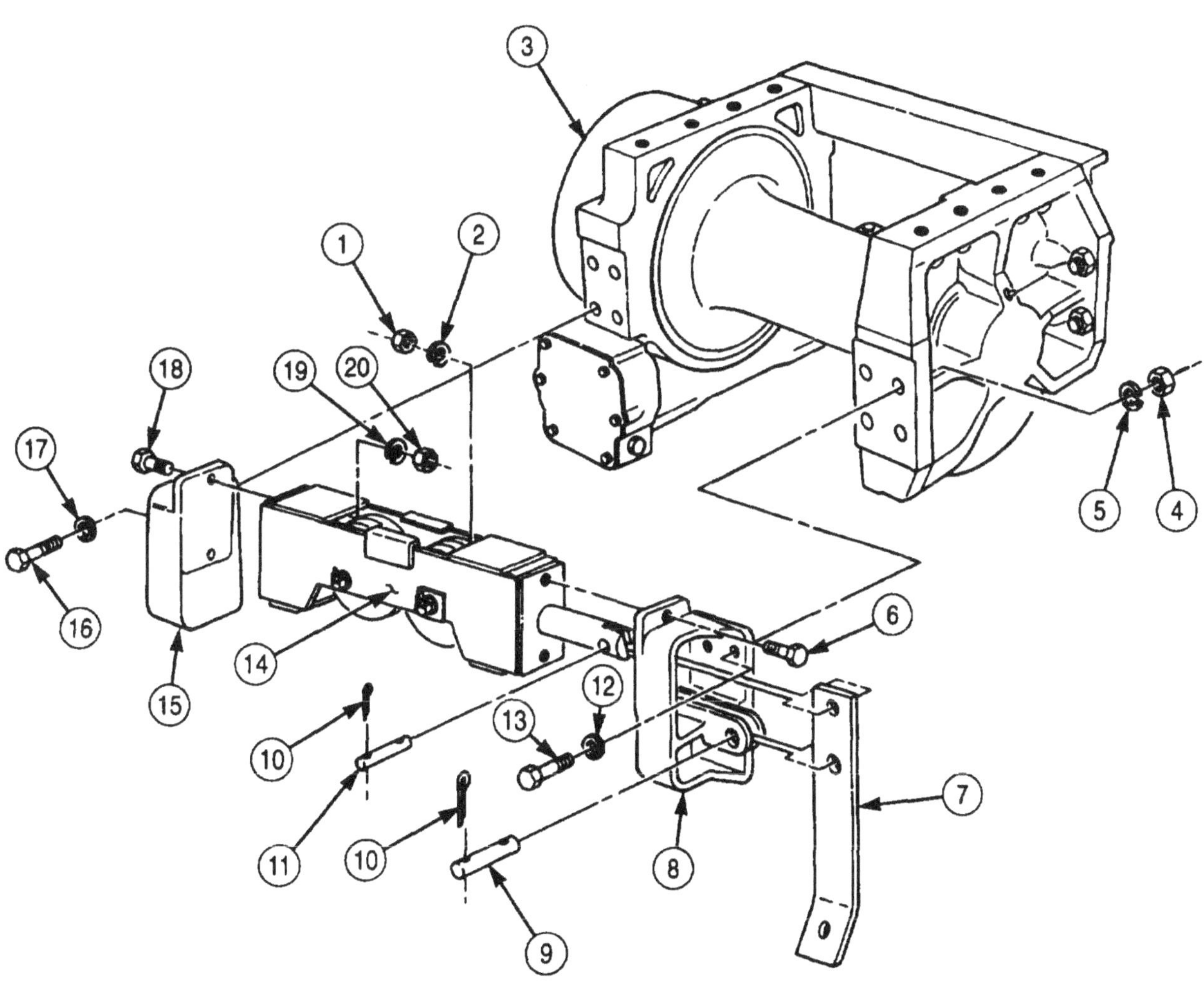

4-182. REAR WINCH CABLE TENSIONER REPAIR (Contd)

6. Remove cotter pin (1) and pin (8) from tensioner frame (2). Discard cotter pin (1).
7. Remove grease fitting (7) from pin (8).
8. Remove sheave (5), two thrust washers (3), and two felt washers (4) from tensioner frame (2). Discard felt washers (4).
9. Remove bearing (6) from sheave (5).
10. Remove cotter pin (9) and pin (16) from tensioner frame (2). Discard cotter pin (9).
11. Remove grease fitting (15) from pin (16).
12. Remove sheave (13), two thrust washers (11), and two felt washers (12) from frame (2). Discard felt washers (12).
13. Remove bearing (14) from sheave (13).
14. Remove frame (10) from tensioner frame (2).

b. Cleaning, Inspection, and Repair

NOTE

Parts for both sheaves are inspected the same way.

1. Clean and inspect thrust washers (3) and (11) for cracks or wear. Measure thickness with a micrometer and discard thrust washers (3) and (11) if thickness is less than 0.055 in. (1.40 mm) or if cracked.
2. Clean and inspect bearings (6) and (14) for chips, cracks, scoring, or damaged cage. Discard if chipped, cracked, scored, or cage is damaged.
3. Clean and inspect pins (8) and (16) for cracks or wear. Measure with a micrometer and discard if outer diameter is less than 1.245 in. (31.63 mm) or if cracked.

NOTE

Refer to TM 9-237 for welding and straightening techniques.

4. Inspect sheave frame (10) for bends and cracks. Repair minor bends or cracks.
5. Clean and inspect tensioner frame (2) for bends, cracks, and broken welds. Repair minor bends, cracks, and broken welds. Replace if not repairable.

4-182. REAR WINCH CABLE TENSIONER REPAIR (Contd)

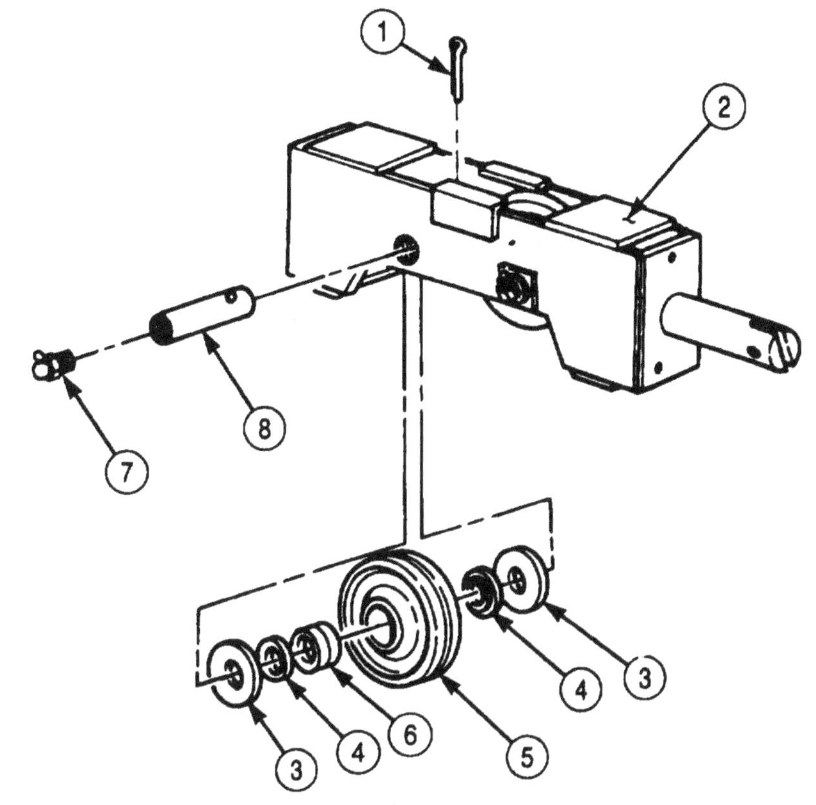

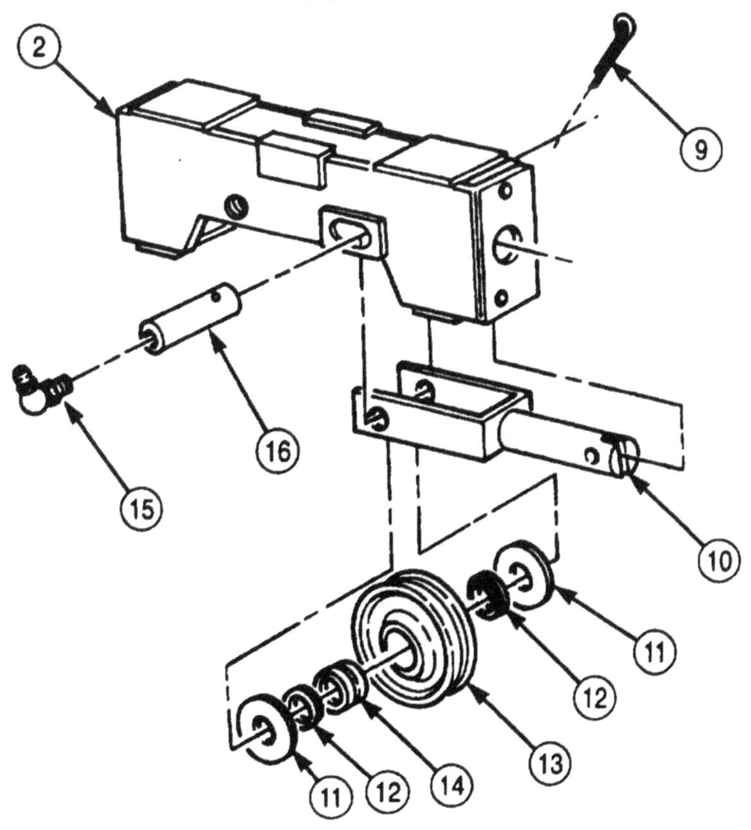

4-182. REAR WINCH CABLE TENSIONER REPAIR (Contd)

c. Assembly

1. Position sheave frame (3) in tensioner frame (1).
2. Coat bearing (7) with GAA grease and install in sheave (6).
3. Install grease fitting (9) in pin (8).
4. Place steel sides of two new felt washers (5) on each side of bearing (7) in sheave (6) and place sheave (6) and two thrust washers (4) in frame (3).
5. Install sheave (6) in tensioner frame (1) and frame (3) with pin (8) and new cotter pin (2).
6. Coat bearing (14) with GAA grease and install in sheave (13).
7. Install grease fitting (16) in pin (15).
8. Place steel sides of two new felt washers (12) on each side of bearing (14) in sheave (13) and place sheave (13) and two thrust washers (11) in tensioner frame (1).
9. Install sheave (13) in tensioner frame (1) with pin (15) and new cotter pin (10).

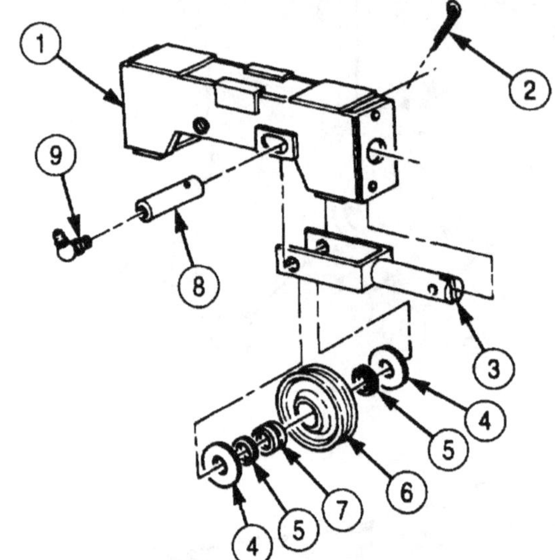

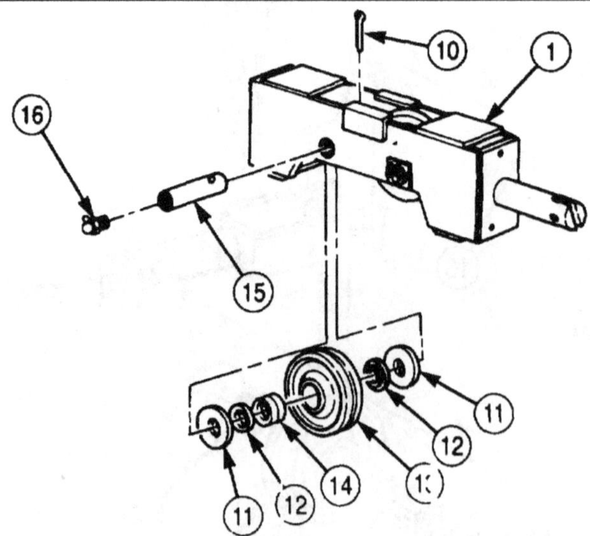

4-182. REAR WINCH CABLE TENSIONER REPAIR (Contd)

10. Install bracket (24) on rear winch (19) with four washers (27), screws (28), new lockwashers (21), and nuts (20).
11. Install tensioner frame (1) on bracket (24) with two screws (22), new lockwashers (18), and nuts (17).
12. Install bracket (30) on tensioner frame (1) with two screws (33), new lockwashers (34), and nuts (35).
13. Install bracket (30) on rear winch (19) with four new lockwashers (32) and screws (31).
14. Install lever (23) on bracket (24) and rod of frame (1) with two pins (25) and (29) and four new cotter pins (26).

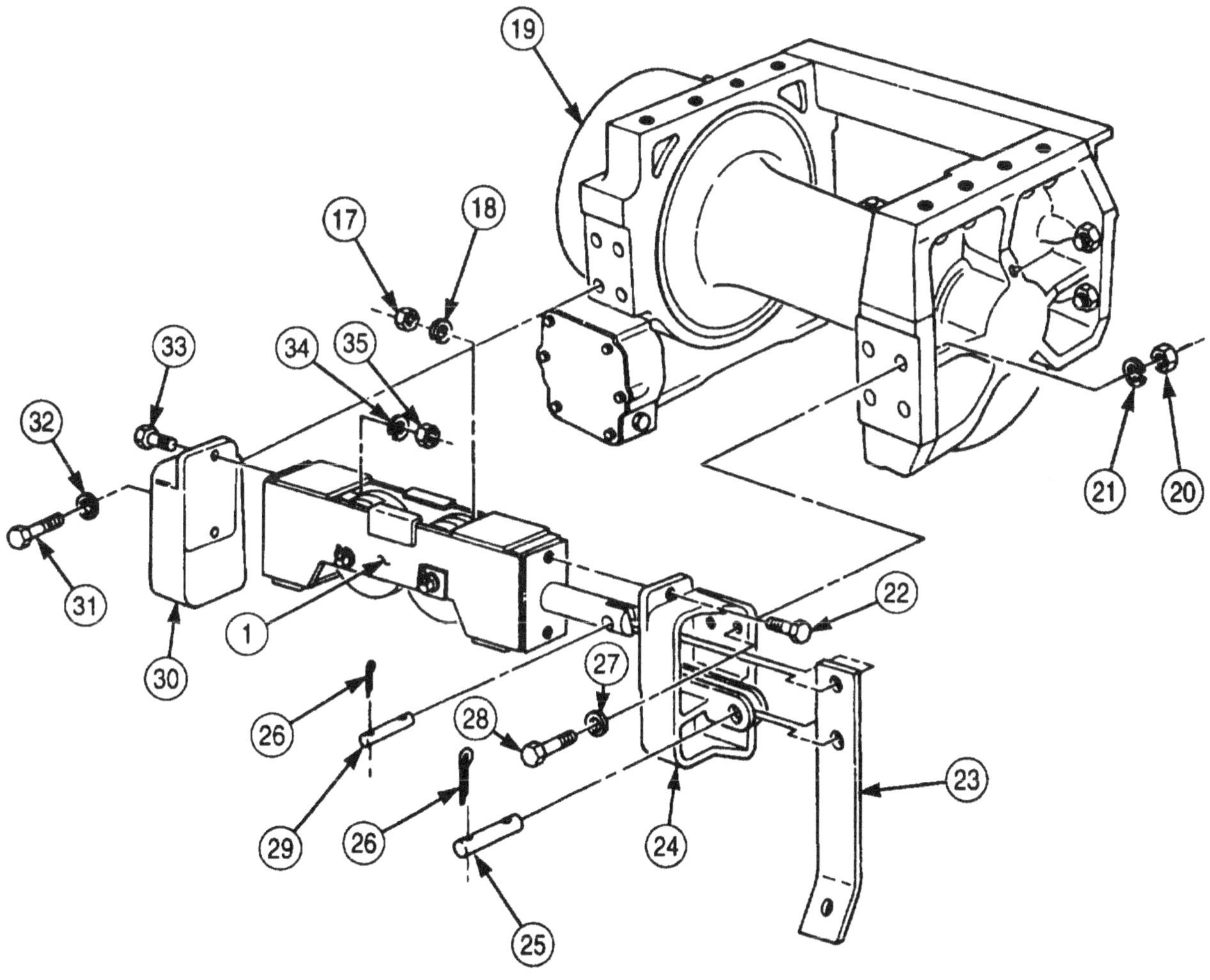

FOLLOW-ON TASKS: • Lubricate rear winch tensioner (LO 9-2320-272-12).
• Install rear winch cable (para. 3-331).

4-183. REAR WINCH SIDE ROLLER REPAIR

THIS TASK COVERS:

a. Disassembly
b. Cleaning and Inspection
c. Assembly

INITIAL SETUP:

APPLICABLE MODELS
M936/A1/A2

TOOLS
General mechanic's tool kit (Appendix E, Item 1)

MATERIALS/PARTS
Two lockwashers (Appendix D, Item 377)
Two lockwashers (Appendix D, Item 354)
Two felt washers (Appendix D, Item 112)
GAA grease (Appendix C, Item 28)
Lubricating oil (Appendix C, Item 50)
Drycleaning Solvent (Appendix C, Item 71)

REFERENCES (TM)
TM 9-214
TM 9-2320-272-10
TM 9-2320-272-24P
LO 9-2320-272-12

EQUIPMENT CONDITION
Parking brake set (TM 9-2320-272-10).

GENERAL SAFETY INSTRUCTIONS
- Keep fire extinguisher nearby when using drycleaning solvent.
- Drycleaning solvent is flammable and toxic. Do not use near open flame.

NOTE

Repair procedures for both side rollers are the same.

a. Disassembly

1. Remove two screws (6), lockwashers (7), and lockplate (8) from bracket (3). Discard lockwashers (7).

NOTE

Assistant will help with steps 2 and 3.

2. Remove shaft (5) from bracket (3) and roller (12).
3. Remove roller (12), two thrust washers (9), and felt washers (10) from bracket (3). Discard felt washers (10).
4. Remove two bearings (11) from roller (12).
5. Remove two screws (l), lockwashers (14), plate (2), and spacer (13) from bracket (3). Discard lockwashers (14).
6. Remove two grease fittings (4) from shaft (5).

b. Cleaning and Inspection

WARNING

Drycleaning solvent is flammable and toxic. Do not use near an open flame and always have a fire extinguisher nearby when solvents are used. Use only in well-ventilated places, wear protective clothing, and dispose of cleaning rags in approved container. Failure to do this may result in injury to personnel and/or damage to equipment.

1. Clean all roller (12) and shaft (5) components in drycleaning solvent and allow to air-dry.
2. Inspect shaft (5) for cracks and wear. Measure outer diameter at bearing (11) surfaces. If outer diameter is less than 2.245 in. (57.02 mm) or cracked, replace shaft (5).
3. Inspect two thrust washers (9) for cracks and wear. Measure thickness. If cracked, or thickness is less than 0.055 in. (1.40 mm), replace thrust washer(s) (9).

4-183. REAR WINCH SIDE ROLLER REPAIR (Contd)

4. Inspect bearings (11) (TM 9-214). Replace if damaged.

c. Assembly

1. Install two grease fittings (4) in shaft (5).
2. Install spacer (13) on bracket (3) with plate (2), two new lockwashers (14), and screws (1).
3. Pack two bearings (11) with GAA grease and install in roller (12) (TM 9-214).

NOTE

New felt washers must be soaked in lubricating oil before installation.

4. Position a new felt washer (10) and thrust washer (9) on each end of roller (12) with felt washer (10) next to bearing (11).
5. Slide roller (12), two bearings (11), felt washers (11), and thrust washers (9) into bracket (3).
6. Ensure locking slot of shaft (5) is on bottom and facing away from winch and install shaft (5) through bracket (3) and roller (12).
7. Install lockplate (8) in slot in shaft (5) and on bracket (3) with two new lockwashers (7) and screws (6).

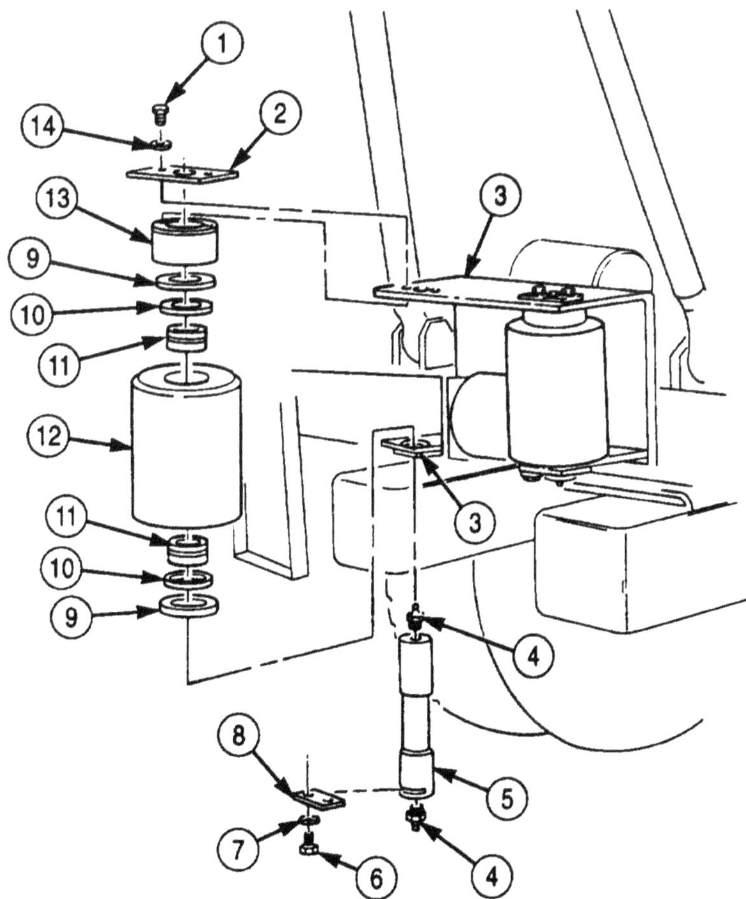

FOLLOW-ON TASK: Lubricate roller assembly (LO 9-2320-272-12).

4-184. REAR WINCH TOP AND BOTTOM ROLLER REPAIR

THIS TASK COVERS:

a. Disassembly
b. Cleaning and Inspection
c. Assembly

INITIAL SETUP:

APPLICABLE MODELS
M936/A1/A2

TOOLS
General mechanic's tool kit (Appendix E, Item 1)

MATERIALS/PARTS
Two felt washers (Appendix D, Item 112)
Two lockwashers (Appendix D, Item 354)
GAA grease (Appendix C, Item 28)
Lubricating oil (Appendix C, Item 50)
Drycleaning solvent (Appendix C, Item 71)

REFERENCES (TM)
TM 9-214
TM 9-2320-272-10
TM 9-2320-272-24P
LO 9-2320-272-12

EQUIPMENT CONDITION
- Parking brake set (TM 9-2320-272-10).
- Left side roller removed (for bottom roller repair only) (para. 4183).

GENERAL SAFETY INSTRUCTIONS
Keep fire extinguisher nearby when using drycleaning solvent.

NOTE
Repair procedures for top and bottom rollers are the same.

a. Disassembly

1. Remove two grease fittings (7) from shaft (8).
2. Remove two screws (10), lockwashers (11), and lockplate (9) from bracket (12). Discard lockwashers (11).
3. Remove shaft (8) from roller (6). Thrust washers (3) may fall out.
4. Remove two grease fittings (2) from tensioner sheave shafts (1).

NOTE
Assistant will help with step 5.

5. Remove roller (6) and, if present, two thrust washers (3) from bracket (12).
6. Remove two felt washers (4) and bearings (5) from roller (6). Discard felt washers (4).

b. Cleaning and Inspection

WARNING

Drycleaning solvent is flammable and toxic. Do not use near open flame and always have a fire extinguisher nearby when solvents are used. Use only in well-ventilated places, wear protective clothing, and dispose of cleaning rags in approved container. Failure to do this may result in injury to personnel and/or damage to equipment.

1. Clean all roller (6) and shaft (8) components in drycleaning solvent and allow to air-dry.
2. Inspect shaft (8) for cracks and wear. Measure outer diameter of bearing surfaces. Replace shaft (8) if cracked or outer diameter is less than 2.245 in. (57.02 mm).
3. Inspect thrust washers (3) for cracks and wear. Measure thickness. Replace any thrust washers (3) if cracked or thickness is less than 0.055 in. (1.397 mm).
4. Inspect bearings (5) (TM 9-214). Replace if damaged.

4-184. REAR WINCH TOP AND BOTTOM ROLLER REPAIR (Contd)

c. Assembly

1. Pack two bearings (5) with GAA grease and install in roller (6) (TM 9-214).
2. Soak two new felt washers (4) in lubricating oil and place on bearings (5) at each end of roller (6).
3. Place thrust washer (3) over felt washers (4) on each end of roller (6).

NOTE

Assistant will help with steps 4 and 5.

4. Slide roller (6) with bearings (5), felt washers (4), and thrust washers (3) into bracket (12).
5. Ensure slot in shaft (8) faces out, and install shaft (8) through bracket (12) and roller (6).
6. Align lockplate (9) with slot in shaft (8), and install with two new lockwashers (11) and screws (16).
7. Install two grease fittings (7), one in each end of shaft (8).
8. Install two grease fittings (2) in tensioner sheave shafts (1).

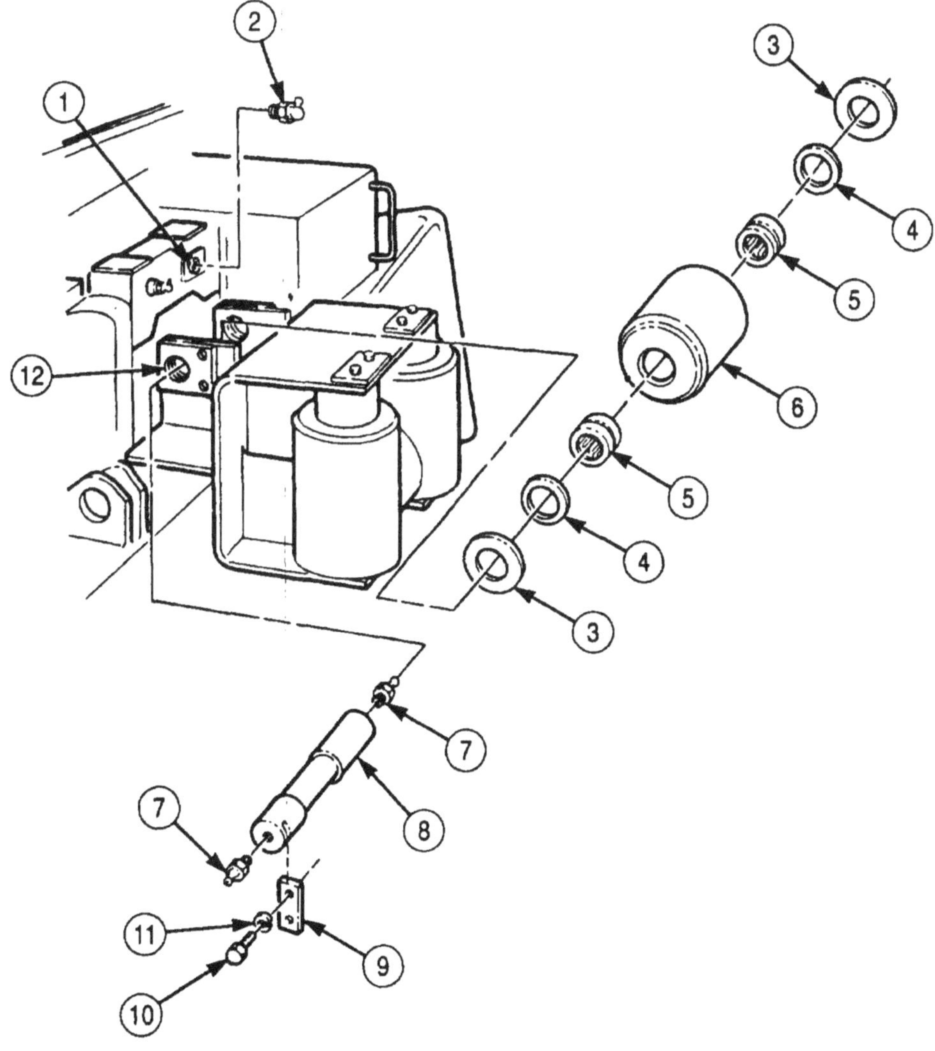

FOLLOW-ON TASKS:
- Lubricate roller assembly (LO 9-2320-272-12).
- Install left side roller, if removed (para. 4-183).

4-185. REAR WINCH HYDRAULIC MOTOR REPLACEMENT

THIS TASK COVERS:

a. Removal

b. Installation

INITIAL SETUP:

APPLICABLE MODELS

M936/A1/A2

TOOLS

General mechanic's tool kit (Appendix E, Item 1)
Torque wrench (Appendix E, Item 144)

MATERIALS/PARTS

Cotter pin (Appendix D, Item 81)
Four locknuts (Appendix D, Item 294)
Six lockwashers (Appendix D, Item 392)
Two O-rings (Appendix D, Item 560)
O-ring (Appendix D, Item 477)
O-ring (Appendix D, Item 456)
Five lockwashers (Appendix D, Item 354)
Adhesive sealant (Appendix C, Item 4)
Antiseize tape (Appendix C, Item 72)

PERSONNEL REQUIRED

TWO

REFERENCES (TM)

LO 9-2320-272-12
TM 9-2320-272-10
TM 9-2320-272-24P

EQUIPMENT CONDITION

- Parking brake set (TM 9-2320-272-10).
- Hydraulic oil reservoir drained (para. 4-188).

a. Removal

1. Remove five screws (4), lockwashers (5), washers (6), and cover (3) from floor plate (15). Discard lockwashers (5).
2. Remove four screws (2), washers (1), and floor plate (15) from frame (7).
3. Remove two locknuts (8), screws (11), and washers (10) from front bracket (9) and frame (7). Discard locknuts (8).

NOTE

Have drainage container ready to catch oil.

4. Disconnect hose connector (16) from elbow (21).
5. Remove elbow (21), adapter (20), and valve (19) from elbow (18). Tag adapter (20) and valve (19) for installation.
6. Rotate elbow (18) upward on pipe nipple (17) to allow removal of front bracket (9).
7. Remove three screws (12), lockwashers (13), and front bracket (9) from winch motor (14).

4-185. REAR WINCH HYDRAULIC MOTOR REPLACEMENT (Contd)

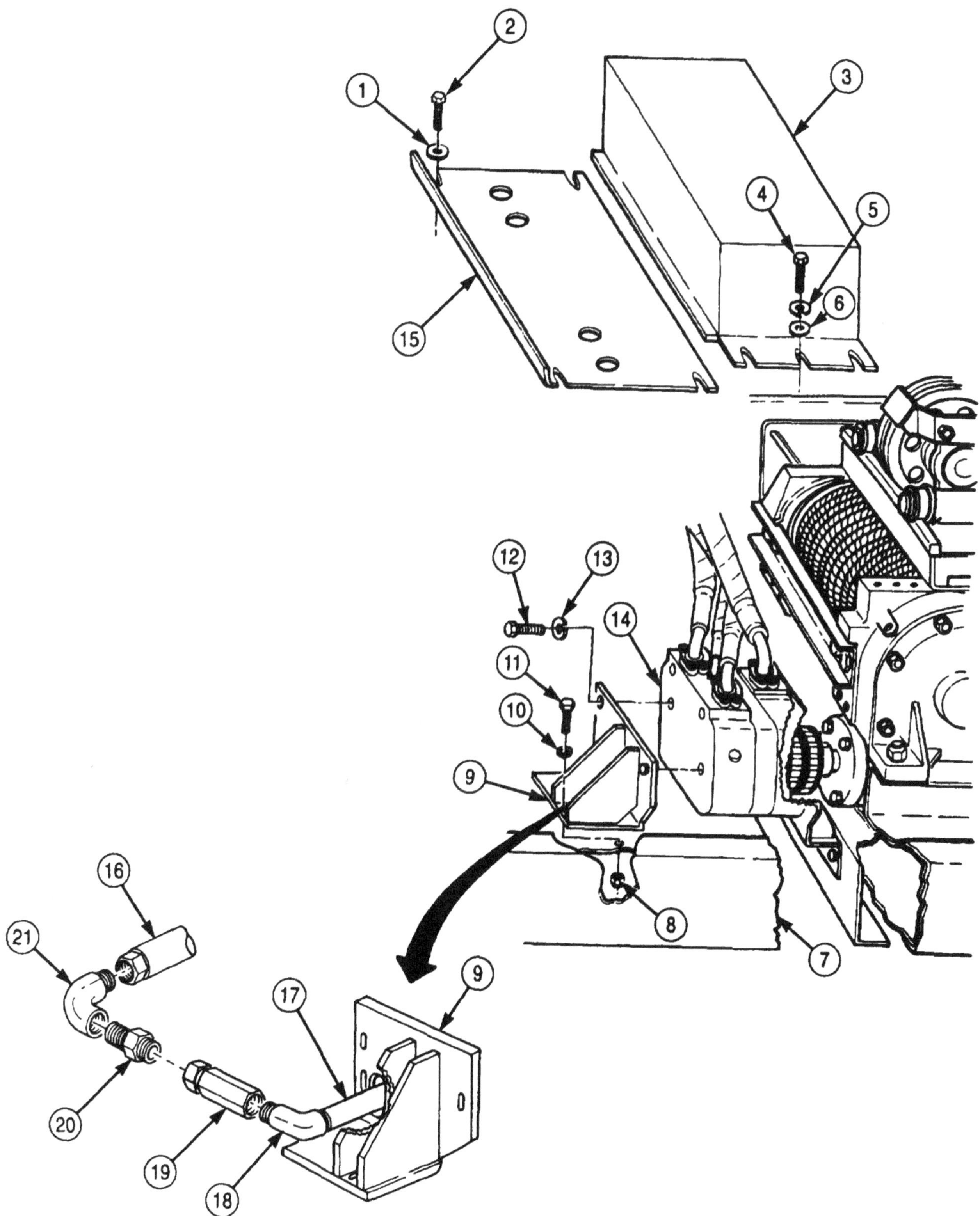

4-185. REAR WINCH HYDRAULIC MOTOR REPLACEMENT (Contd)

8. Remove elbow (7) and pipe nipple (6) from winch motor (8).
9. Remove twelve screws (2), washers (3), and six clamps (4) from hydraulic motor (8).

NOTE

Tag hydraulic hoses for installation.

10. Remove three hydraulic hoses (1) and O-rings (5) from hydraulic motor (8). Discard O-rings (5).
11. Disconnect hydraulic hose (11) from elbow (10) on winch motor (8).
12. Remove elbow (10) and O-ring (9) from winch motor (8). Discard O-ring (9).
13. Disconnect two hose connectors (31) and (32) from tee (33).
14. Loosen setscrew (26) until flush with front coupling hub (14).
15. Remove tee (33), adapter (30), valve (29), and elbow (7) from pipe nipple (6). Tag valve (29) for installation.
16. Remove two locknuts (271, screws (20), and washers (21) from frame crossmember (22) and rear bracket (28). Discard locknuts (27).

NOTE

Assistant will help with step 17.

17. Remove winch motor (8) and rear bracket (28) from frame crossmember (22).
18. Remove key (13) from winch motor shaft (12).
19. Remove six screws (23), lockwashers (24), and rear bracket (28) from winch motor (8). Discard lockwashers (24).
20. Remove pipe nipple (6) from winch motor (8). Tag pipe nipple (6) for installation.
21. Remove cotter pin (18) and shear pin (19) from rear coupling hub (16). Discard cotter pin (18).
22. Remove coupling (15) from winch driveshaft (17).

b. Installation

1. Align holes in rear coupling hub (16) and driveshaft (17) and install coupling (15) on winch driveshaft (17) with shear pin (19) and new cotter pin (18).

NOTE

Wrap all male pipe threads with antiseize tape before installation.

2. Install rear bracket (28) on winch motor (8) with six new lockwashers (24) and screws (23). Tighten screws (23) 60-70 lb-ft (81-95 N•m).

NOTE

Assistant will help with step 3.

3. Place key (13) in slot of winch motor shaft (12) and install winch motor (8) on coupling (15). Ensure key (13) is visible in setscrew hole (25).
4. Install rear bracket (28) on frame crossmember (22) with four washers (21), two screws (20), and two new locknuts (27). Tighten locknuts (27) 60-70 lb-ft (81-95 N•m).
5. Install pipe nipple (6), elbow (7), valve (29), adapter (30), and tee (33) on winch motor (8).
6. Connect two hose connectors (31) and (32) to tee (33).
7. Apply adhesive sealant to threads of setscrew (26) and install in hole (25) in front coupling hub (14).
8. Install elbow (10) on winch motor (8) with new O-ring (9).
9. Connect hydraulic hose (11) on elbow (10).
10. Connect three hydraulic hoses (1) to winch motor (8) with new O-rings (5), six clamps (4), twelve washers (3), and screws (2).
11. Install pipe nipple (17) and elbow (18) on winch motor (14) with elbow (18) pointed up.

4-185. REAR WINCH HYDRAULIC MOTOR REPLACEMENT (Contd)

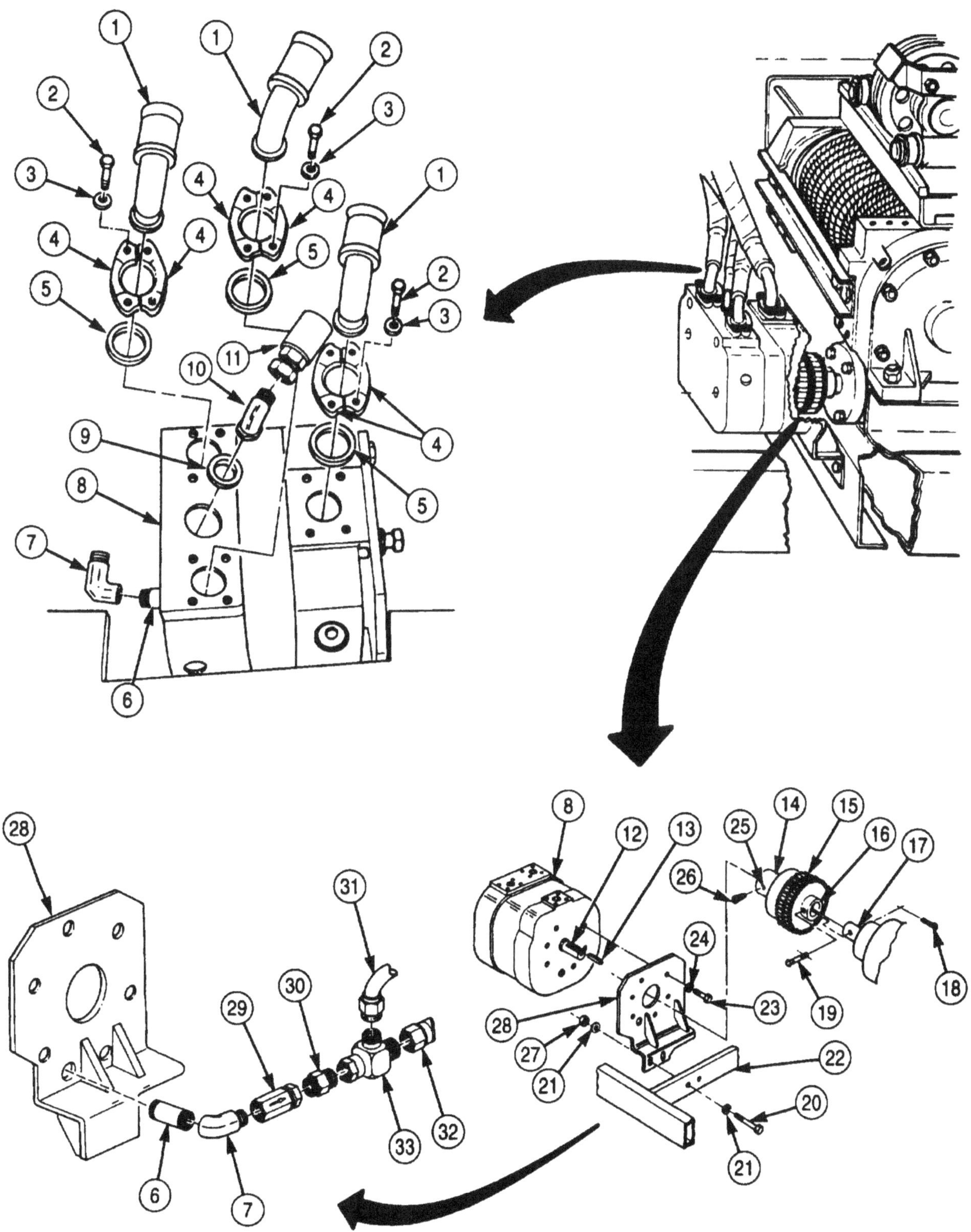

4-185. REAR WINCH HYDRAULIC MOTOR REPLACEMENT (Contd)

12. Install front bracket (9) on winch motor (14) with three new lockwashers (13) and screws (12). Tighten screws (12) 60-70 lb-ft (81-95 N•m). Turn elbow (18) to point left.
13. Install valve (19), adapter (20), and elbow (21) on elbow (18).
14. Connect hose connector (16) to elbow (21).
15. Install front bracket (9) on frame (7) with two washers (10), screws (11), and new locknuts (8). Tighten locknuts (8) 60-70 lb-ft (81-95 N•m).
16. Install floor plate (15) on frame (7) with four washers (1) and screws (2). Tighten screws (2) 60-70 lb-ft (81-95 N•m).
17. Install cover (3) on frame (7) with five washers (6), new lockwashers (5), and screws (4). Tighten screws (4) 60-70 lb-ft (81-95 N-m).

4-185. REAR WINCH HYDRAULIC MOTOR REPLACEMENT (Contd)

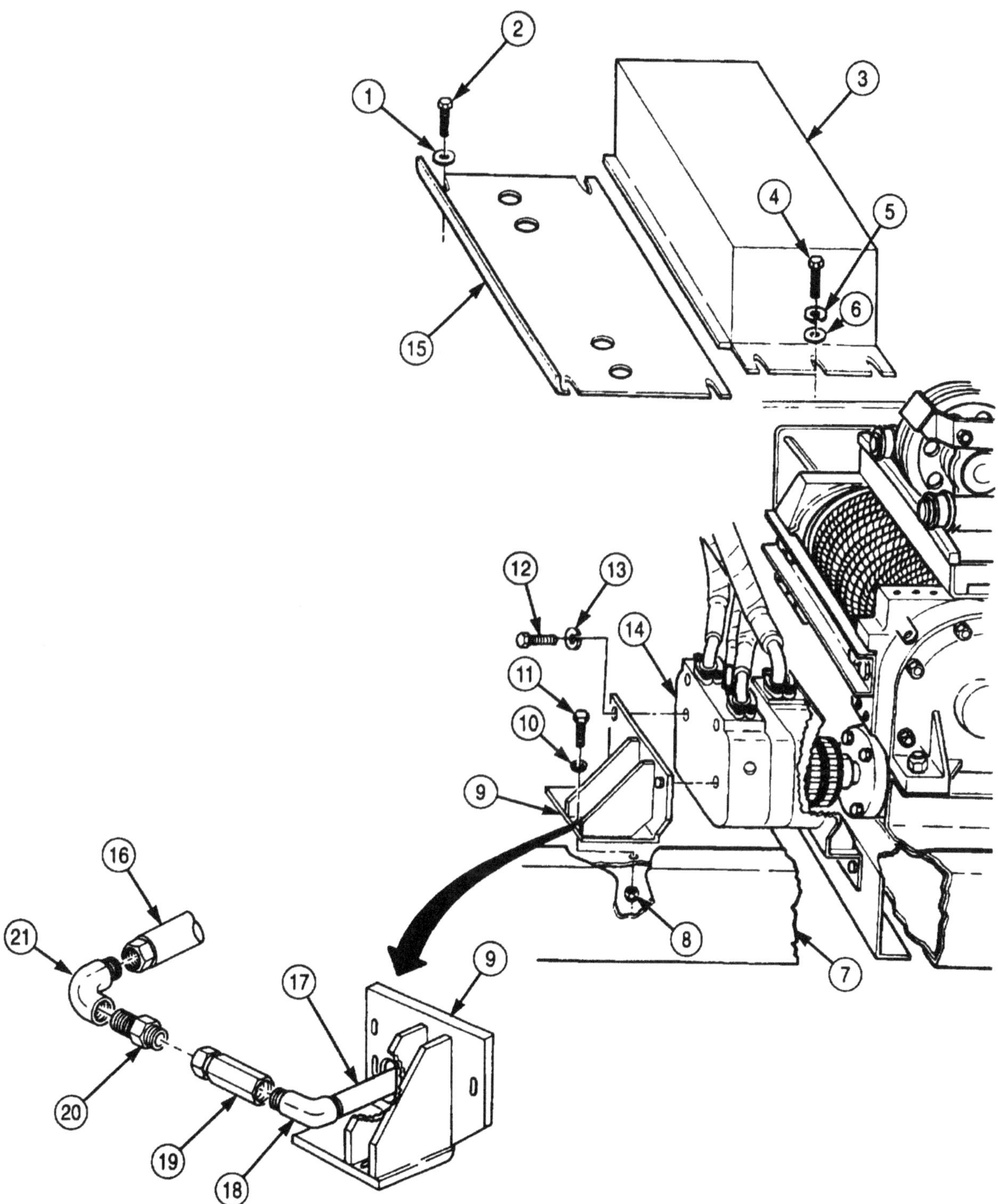

FOLLOW-ON TASKS
- Fill hydraulic oil reservoir to proper level (LO 9-2320-272-12).
- Check rear winch hydraulic motor for proper operation (TM 9-2320-272-10).

4-186. HYDRAULIC OIL FILTER HOUSING REPLACEMENT

THIS TASK COVERS:

a. Removal

b. Installation

INITIAL SETUP:

APPLICABLE MODELS

M936/A1/A2

TOOLS

General mechanic's tool kit (Appendix E, Item 1)

MATERIALS/PARTS

Filter gasket (Appendix D, Item 237)
Cap and plug set (Appendix C, Item 14)
Antiseize tape (Appendix C. Item 72)

REFERENCES (TM)

LO 9-2320-272-12
TM 9-2320-272-10
TM 9-2320-272-24P

EQUIPMENT CONDITION

- Parking brake set (TM 9-2320-272-10).
- Hydraulic oil reservoir drained (para. 4-188).
- Fuel can bracket removed (para. 4-187).

CAUTION

After disconnecting hydraulic lines and hoses, plug all openings to prevent dirt from entering and causing internal parts damage.

a. Removal

NOTE

Have drainage container ready to catch oil.

1. Disconnect flex hose (5) from hydraulic oil filter elbow (4).
2. Remove hydraulic oil filter elbow (4) from filter housing (3).
3. Remove four screws (7) and washers (6) from filter housing (3) and mounting flange (1).
4. Remove filter housing (3) and gasket (2) from mounting flange (1). Discard gasket (2) and clean gasket remains from mating surfaces.

b. Installation

1. Install new gasket (2) and filter housing (3) on mounting flange (1) with four washers (6) and screws (7).
2. Wrap hydraulic oil filter elbow (4) with antiseize tape and install on hydraulic oil filter housing (3).
3. Connect flex hose (5) on hydraulic oil filter elbow (4).

4-186. HYDRAULIC OIL FILTER HOUSING REPLACEMENT (Contd)

FOLLOW-ON TASKS:
- Install fuel can bracket (para. 4-187).
- Fill hydraulic oil reservoir to proper oil level (LO 9-2320-272-12).
- Check hydraulic oil filter for proper operation (TM 9-2320-272-10).

4-187. FUEL CAN BRACKET REPLACEMENT

THIS TASK COVERS:

a. Removal

b. Installation

INITIAL SETUP:

APPLICABLE MODELS
M9361A1/A2

TOOLS
General mechanic's tool kit (Appendix E, Item 1)

MATERIALS/PARTS
Four locknuts (Appendix D, Item 291)
Seven locknuts (Appendix D, Item 299)
Two lockwashers (Appendix D. Item 354)

REFERENCES TM
TM 9-2320-272-10
TM 9-2320-272-24P

EQUIPMENT CONDITION
Parking brake set (TM 9-2320-272-10).

a. Removal

1. Remove four screws (l), locknuts (7), and fuel can holder (2) from upper bracket (6). Discard locknuts (7).
2. Remove two screws (3), washers (5), lockwashers (4), and lower bracket (8) from hydraulic oil filter housing (10). Discard lockwashers (4).
3. Remove seven locknuts (11) and screws (12) from upper bracket (6) and hydraulic oil reservoir (9). Discard locknuts (11).
4. Remove brackets (6) and (8) from oil reservoir (9).

b. Installation

1. Position lower bracket (8) and upper bracket (6) on hydraulic oil filter housing (10) and install with two new lockwashers (4), washers (5), and screws (3).
2. Install bracket (6) on oil reservoir (9) with seven screws (12) and new locknuts (11).
3. Install fuel can holder (2) on upper bracket (6) with four screws (1) and new locknuts (7).

4-187. FUEL CAN BRACKET REPLACEMENT (Contd)

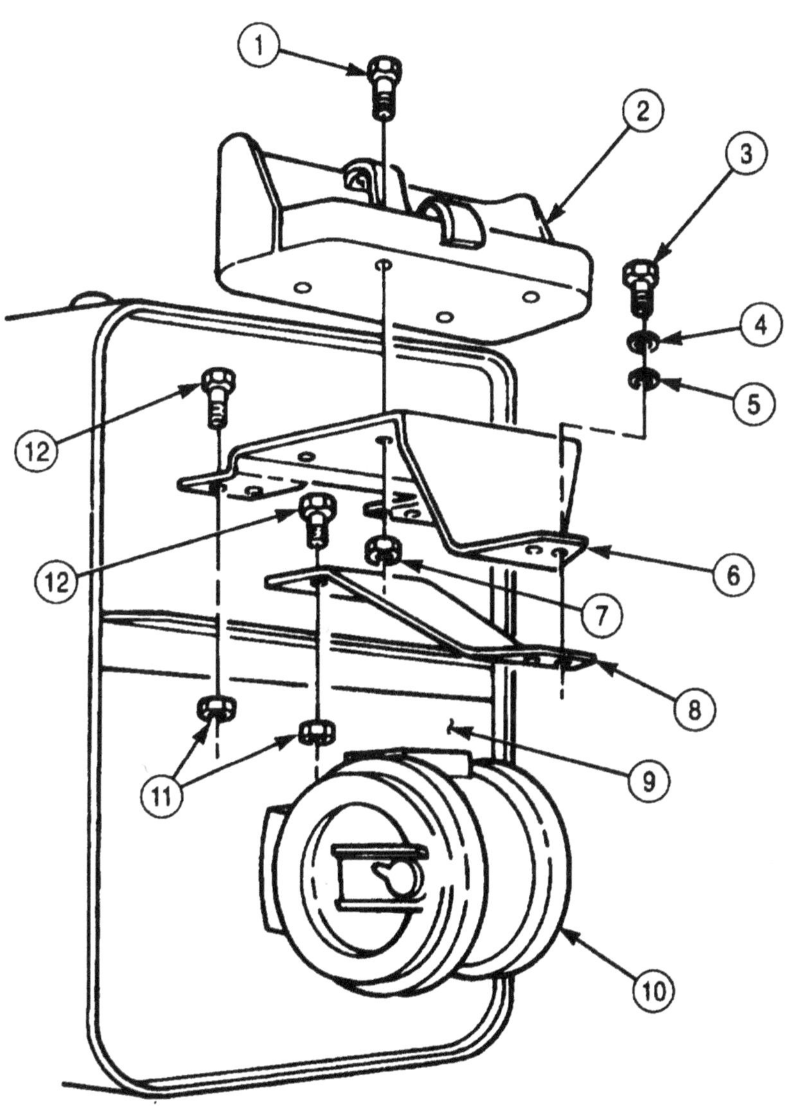

4-188. HYDRAULIC OIL RESERVOIR MAINTENANCE

THIS TASK COVERS:

a. Removal
b. Disassembly
c. Cleaning, Inspection, and Repair
d. Assembly
e. Installation

INITIAL SETUP:

APPLICABLE MODELS
M936/A1/A2

TOOLS
General mechanic's tool kit (Appendix E, Item 1)

MATERIALS/PARTS
Two locknuts (Appendix D, Item 291)
Three cap and plug sets (Appendix C, Item 14)
Antiseize tape (Appendix C, Item 72)
Drycleaning solvent (Appendix C, Item 71)

PERSONNEL REQUIRED
Two

REFERENCES (TM)
LO 9-2320-272-12
TM 9-2320-272-10
TM 9-2320-272-24P

EQUIPMENT CONDITION
- Parking brake set (TM 9-2320-272-10).
- Floodlight assembly removed (TM 9-2320-272-102
- Fuel can bracket removed (para. 4-187).
- Hydraulic oil filter housing removed (para. 4-186).
- Hydraulic oil reservoir drained (para. 4-188).

CAUTION

When disconnecting hydraulic lines and hoses, plug all openings to prevent dirt from entering and causing damage to parts.

NOTE

Have drainage container ready to catch oil.

1. Disconnect hydraulic oil return hose (6) from adapter elbow (7).
2. Remove two locknuts (1) and screws (2) from two oil reservoir retaining straps (4). Discard locknuts (1) and separate straps (4) from hydraulic oil reservoir (3).

NOTE

Assistant will help with step 3.

3. Remove hydraulic oil reservoir (3) from two oil reservoir brackets (5).

NOTE

If a new hydraulic oil reservoir is to be installed, perform step 4.

4. Remove oil gauge (8), filter screen (9), two pipe plugs (10) and (13), adapter elbow (7), and drainvalve (11) with pipe plug (12) from oil reservoir (3).

b. Disassembly

1. Remove oil gauge (8) and filter screen (9) from top of oil reservoir (3).
2. Remove adapter elbow (7), drainvalve (11), and pipe plugs (10) and (13) from oil reservoir (3).
3. Remove pipe plug (12) from drainvalve (11).

4-188. HYDRAULIC OIL RESERVOIR MAINTENANCE (Contd)

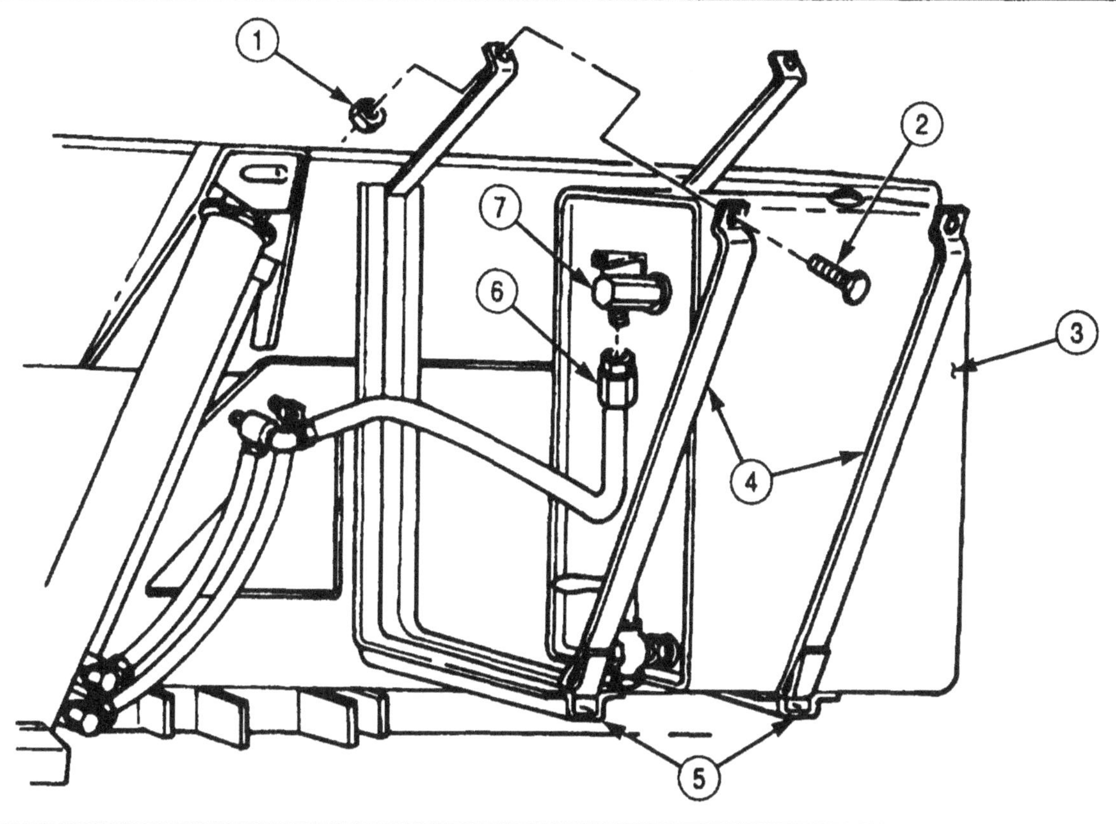

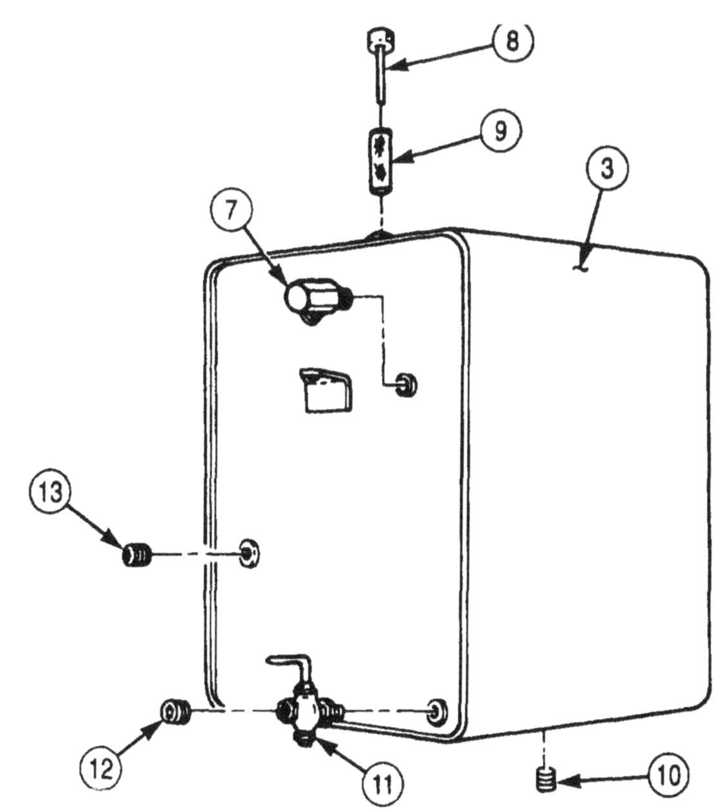

4-188. HYDRAULIC OIL RESERVOIR MAINTENANCE (Contd)

c. Cleaning, Inspection, and Repair

WARNING

Drycleaning solvent is flammable and toxic. Do not use near open flame and always have a fire extinguisher nearby when solvents are used. Use only in well-ventilated places, wear protective clothing, and dispose of cleaning rags in approved container. Failure to do this may result in injury to personnel and/or damage to equipment.

1. Steam clean exterior of reservoir (3).
2. Flush interior of reservoir (3) with drycleaning solvent,
3. Inspect reservoir (3) for leaks:
   a. Plug all openings in reservoir (3) except for oil gauge (1) and filter screen (2) opening.
   b. Insert air hose in filter (2) inlet and cover opening around air hose.
   C. Coat exterior of reservoir (3) with soapy water.
   d. Apply 6 psi (41 kPa) air pressure for a minimum of two minutes and check for air bubbles indicating leaks.
   e. Repair defective reservoir (3) if any leaks are found (TM 9-237). Replace reservoir (3) if leaks cannot be repaired.

d. Assembly

1. Install two pipe plugs (4) and (7) on reservoir (3).
2. Install adapter elbow (8) on reservoir (3).
3. Install drainvalve (5) on reservoir (3).
4. Install pipe plug (6) in drainvalve (5).
5. Install filter screen (2) and oil gauge (1) in reservoir (3).

e. Installation

NOTE

- If installing a new hydraulic oil reservoir, perform step 1.
- Wrap all male pipe threads with antiseize tape before installation.

1. Install drainvalve (5) with pipe plug (6), two pipe plugs (4) and (7), adapter elbow (8), reservoir oil filter (2), and oil gauge (1) on hydraulic oil reservoir (3).

NOTE

Assistant will help with step 2.

2. Position hydraulic oil reservoir (3) on two reservoir brackets (12).
3. Position two retaining straps (11) around hydraulic oil reservoir (3) and install with two screws (10) and new locknuts (9).
4. Connect hydraulic oil return hose (13) to adapter elbow (8).

4-188. HYDRAULIC OIL RESERVOIR MAINTENANCE (Contd)

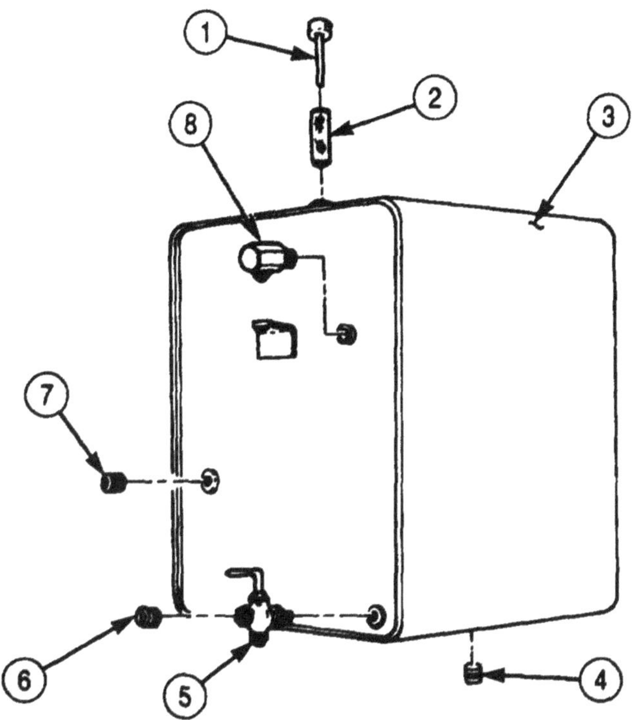

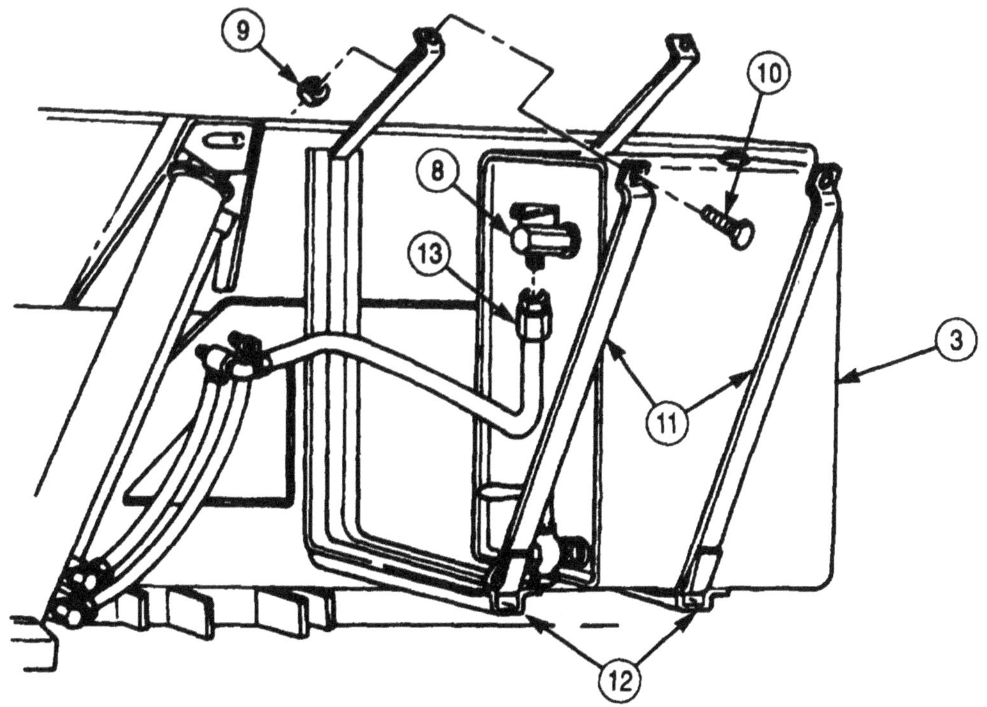

FOLLOW-ON TASKS:
- Install hydraulic oil filter housing (para. 4-186).
- Install fuel can bracket (para. 4-187).
- Install floodlight assembly (TM 9-2320-272-10).
- Fill hydraulic oil reservoir (LO 9-2320-272-12).
- Start engine (TM 9-2320-272-10).

4-189. HYDRAULIC OIL RESERVOIR BRACKETS AND STRAPS REPLACEMENT

THIS TASK COVERS:

a Removal b. Installation

INITIAL SETUP:

APPLICABLE MODELS
M936/A1/A2

TOOLS
General mechanic's tool kit (Appendix E, Item 1)

MATERIALS/PARTS
Eight lockwashers (Appendix D, Item 350)
Four locknuts (Appendix D, Item 291)
Adhesive sealant (Appendix C, Item 4)

REFERENCES (TM)
TM 9-2320-272-10
TM 9-2320-272 24P

EQUIPMENT CONDITION
- Parking brake set (TM 9-2320-272-10).
- Hydraulic oil reservoir removed (para. 4-188).

NOTE
Both hydraulic oil reservoir brackets are replaced the same way.

a. Removal

1. Remove two locknuts (2), screws (9), straps (3) and (4) and insulator strips (5) from reservoir bracket (1). Discard locknuts (2).
2. Remove two insulator strips (6) from bracket (1).
3. Remove four screws (7), lockwashers (8), and bracket (1) from vehicle. Discard lockwashers (8).

b. Installation

1. Install bracket (1) on vehicle with four new lockwashers (8) and screws (7).
2. Install two insulator strips (6) on bracket (1).
3. Install two straps (3) and (4) on bracket (1) with two new locknuts (2) and screws (9).

NOTE
Use adhesive sealant to install old or new insulator strips to brackets and/or straps.

4. Install two insulator strips (5) under straps (3) and (4).

4-189. HYDRAULIC OIL RESERVOIR BRACKETS AND STRAPS REPLACEMENT (Contd)

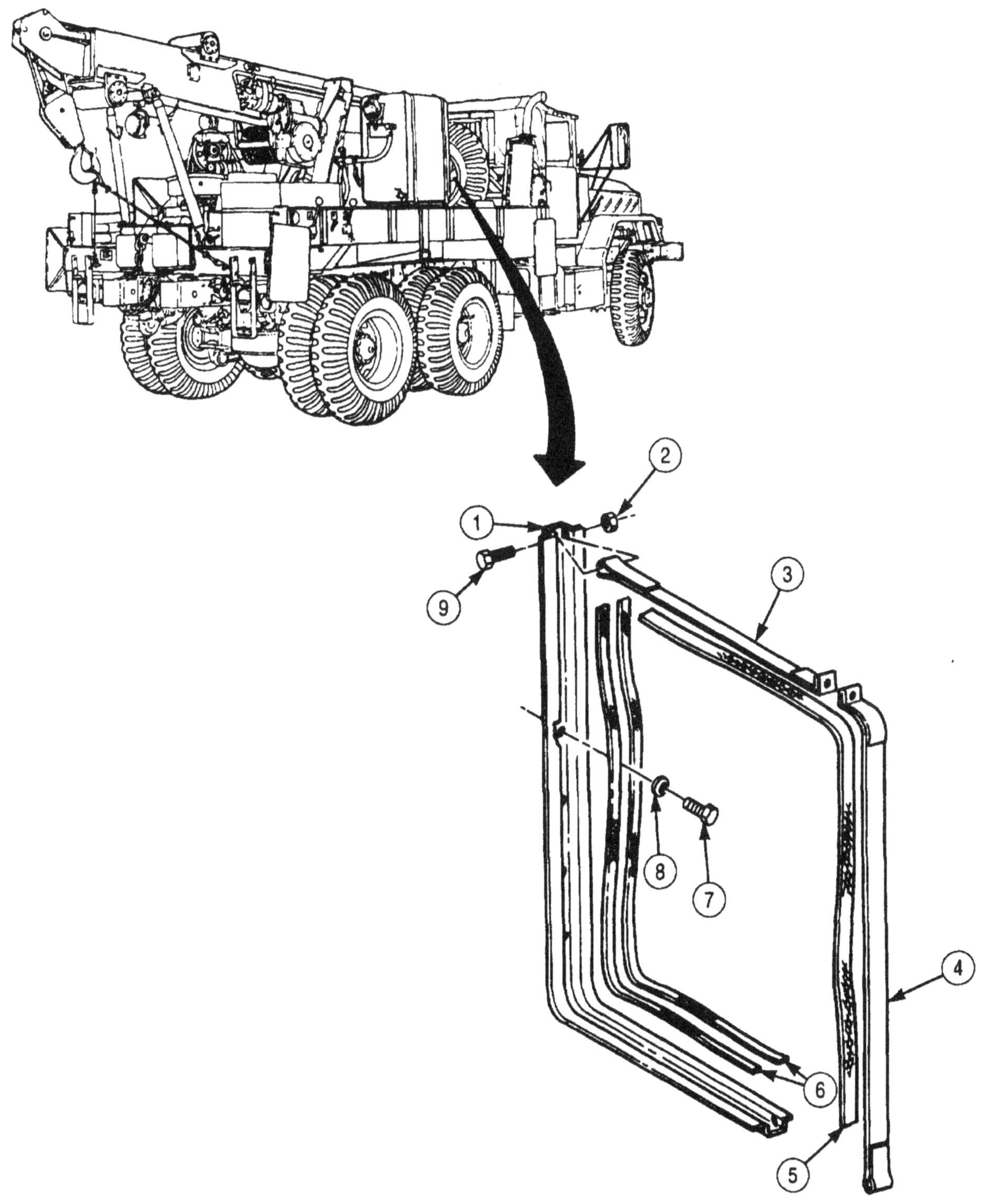

FOLLOW-ON TASK: Install hydraulic oil reservoir (para. 4-188).

4-190. HYDRAULIC HOIST WINCH MOTOR AND LINES REPLACEMENT

THIS TASK COVERS:

a. Removal b. Installation

INITIAL SETUP:

APPLICABLE MODELS

M936/A1/A2

TOOLS

General mechanic's tool kit (Appendix E, Item 1)
Torque wrench (Appendix E, Item 144)

MATERIALS/PARTS

Four locknuts (Appendix D, Item 284)
Four lockwashers (Appendix D, Item 377)
Gasket (Appendix D, Item 166)
Woodruff key (Appendix D, Item 735)
Cap and plug set (Appendix C, Item 14)

REFERENCES (TM)

LO 9-2320-272-12
TM 9-2320-272-10
TM 9-2320-272-24P

EQUIPMENT CONDITION

- Parking brake set (TM 9-2320-272-10).
- Hydraulic oil reservoir removed (para. 4-188).

a. Removal

CAUTION

When disconnecting hydraulic lines and hoses, plug all openings to prevent dirt from entering and causing internal damage to parts.

NOTE

- Have drainage container ready to catch oil.
- Tag lines and hoses for installation.

1. Disconnect two steel hydraulic lines (5) from elbows (1).
2. Disconnect two hydraulic flex hoses (9) from two steel hydraulic lines (5).
3. Remove four locknuts (8), washers (7), and clamps (6) from boom (4). Discard locknuts (8).
4. Turn two steel hydraulic lines (5) until clear of hoist motor (2).
5. Remove four screws (10), lockwashers (11), hoist motor (2), and gasket (3) from winch gearbox adapter (12). Discard lockwashers (11) and gasket (3). Clean gasket remains from mating surface.
6. Remove woodruff key (13) from gear shaft (14).

b. Installation

1. Install woodruff key (13) in slot of gear shaft (14) on hoist motor (2).
2. Install hoist motor (2) on winch gearbox adapter (12) with four new lockwashers (11) and screws (10). Tighten screws (10) 60-70 lb-ft (81-95 N-m).
3. Connect two steel hydraulic lines (5) to elbows (1).
4. Install two steel hydraulic lines (5) on boom (4) with four clamps (6), washers (7), and new locknuts (8).
5. Connect two hydraulic flex hoses (9) to two steel hydraulic lines (5).

4-190. HYDRAULIC HOIST WINCH MOTOR AND LINES REPLACEMENT (Contd)

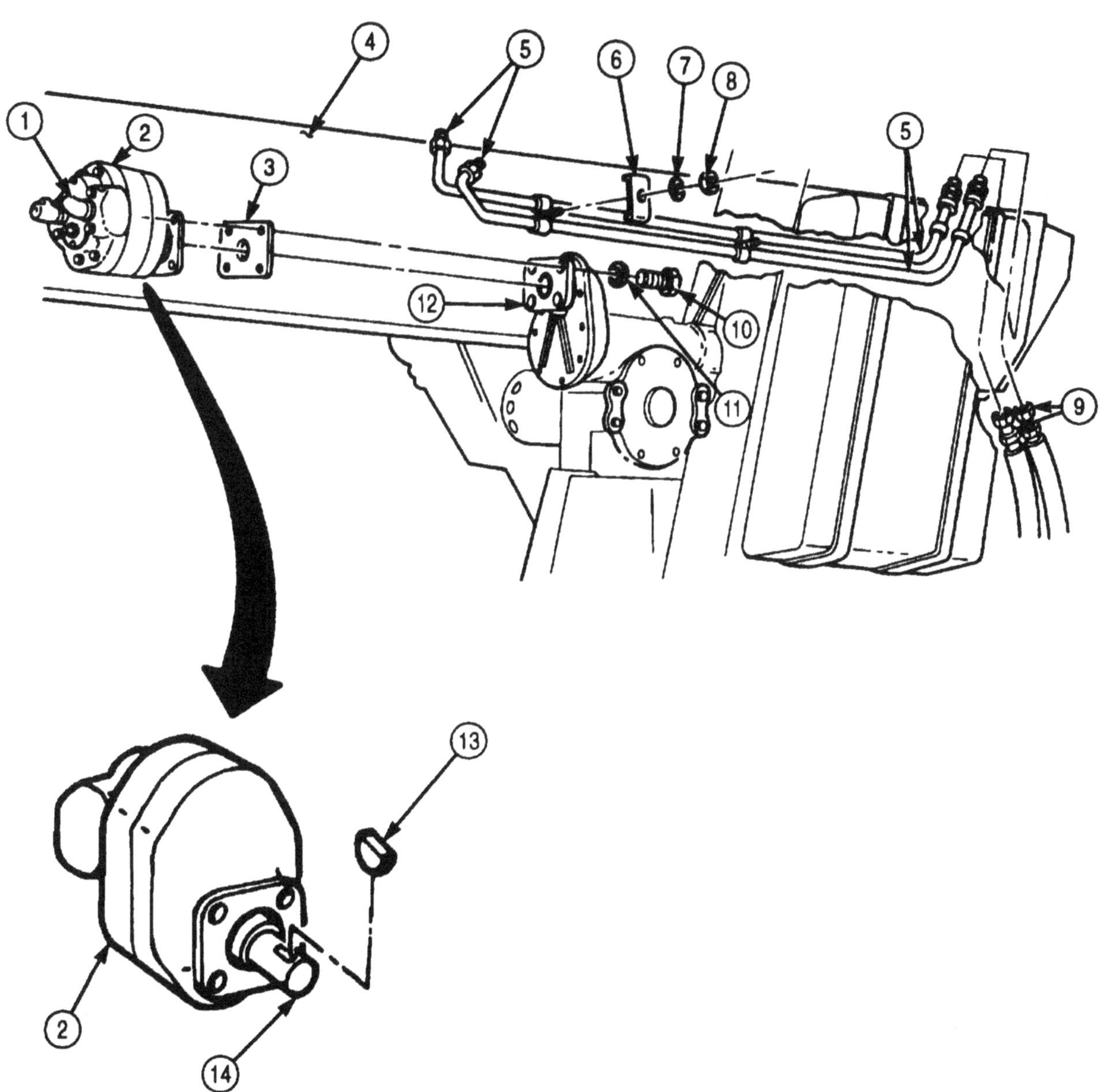

FOLLOW-ON TASKS:
- Fill hydraulic oil reservoir to proper level (LO 9-2320-272-12).
- Check hoist motor for proper operation (TM 9-2320-272-10).

4-191. HOIST LEVEL WIND MAINTENANCE

THIS TASK COVERS:

a. Removal
b. Disassembly
c. Assembly
d. Installation

INITIAL SETUP:

APPLICABLE MODELS
M936/A1/A2

TOOLS
General mechanic's tool kit (Appendix E, Item 1)

MATERIALS/PARTS
Eight locknuts (Appendix D, Item 285)
Two cotter pins (Appendix D, Item 76)

REFERENCES (TM)
TM 9-2320-272-10
TM 9-2320-272-24P

EQUIMENT CONDITION
Parking brake set (TM 9-2320-272-10).

NOTE

Front and rear roller braces are maintained the same way.

a. Removal

Remove four locknuts (9), screws (2), and two front roller braces (4) from boom (1). Discard locknuts (9).

b. Disassembly

1. Remove two cotter pins (3) from shaft (7). Discard cotter pins (3).
2. Remove shaft (7) and level wind roller (6) from two roller braces (4).
3. Remove shaft (7) from roller (6).
4. Remove two bearings (5) from roller (6).
5. Remove grease fitting (8) from shaft (7).

c. Assembly

1. Install two bearings (5) in roller (6).
2. Install grease fitting (8) in shaft (7).
3. Install shaft (7) in roller (6).
4. Install two braces (4) on shaft (7) with two new cotter pins (3).

d. Installation

Install two roller braces (4) on boom (1) with four screws (2) and new locknuts (9).

4-191. HOIST LEVEL WIND MAINTENANCE (Contd)

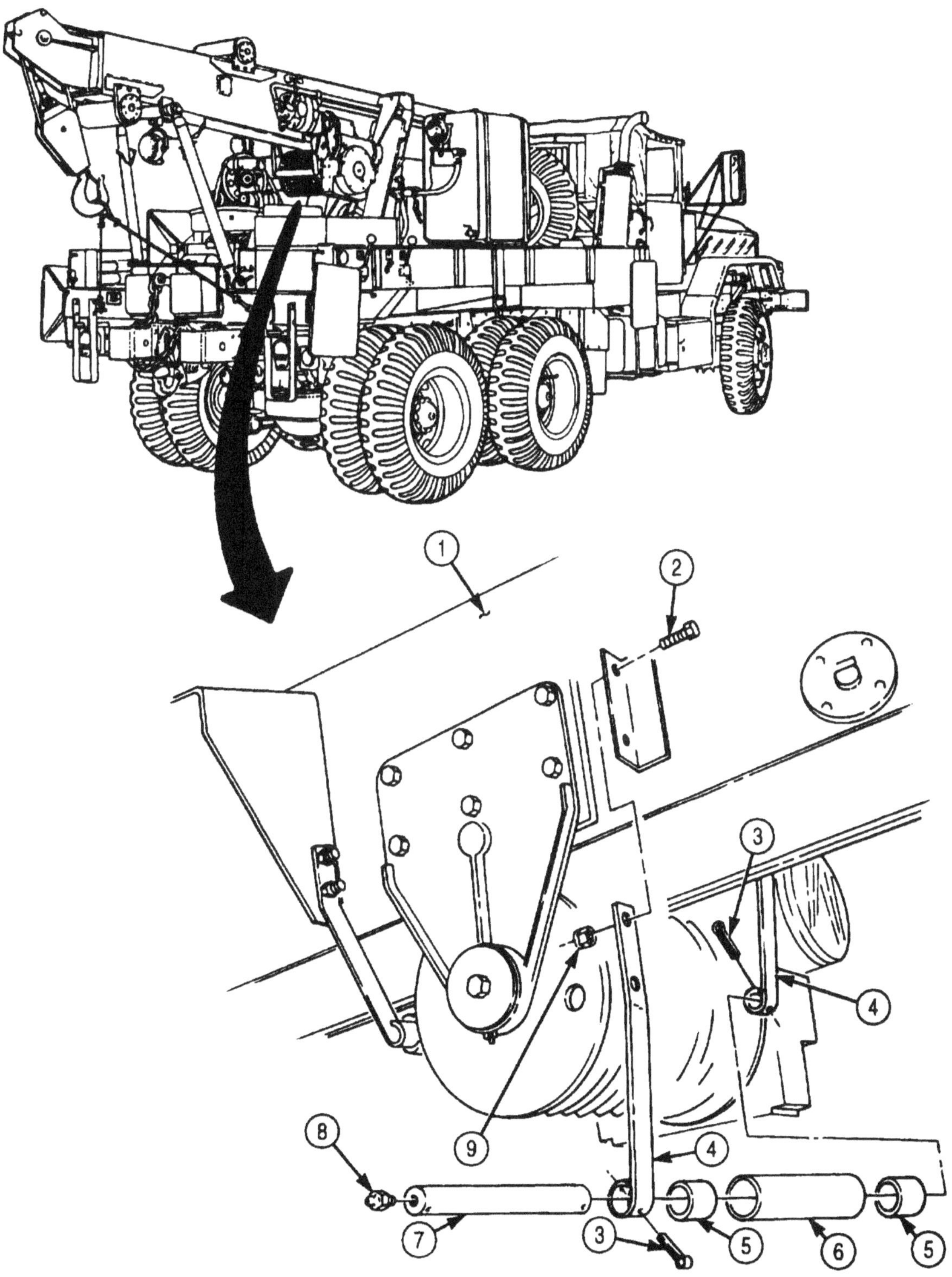

4-192. HOIST WINCH REPLACEMENT

THIS TASK COVERS:

a. Removal b. Installation

INITIAL SETUP:

APPLICABLE MODELS
M936/A1/A2

TOOLS
General mechanic's tool kit (Appendix E, Item 1)
Lifting device
Chain

MATERIAL/PARTS
Twelve lockwashers (Appendix D, Item 393)

PERSONNEL REQUIRED
Two

REFERENCES (TM)
LO 9-2320-272-12
TM 9-2320-272-10
TM 9-2320-272-24P

EQUIPMENT CONDITION
- Parking brake set (TM 9-2320-272-10).
- Boom floodlight wire removed (para. 3-385).
- Hoist winch cable removed (para. 3-384).
- Hydraulic hoist winch motor and lines removed (para. 4-190).
- Hoist level wind removed (para. 4-191).

GENERAL SAFETY INSTRUCTIONS
All personnel must stand clear during hoisting operations.

WARNING

All personnel must stand clear during hoisting operations. A snapped cable, or a shifting or swinging load, may cause injury to personnel.

a. Removal

NOTE

Have drainage container ready to catch oil.

1. Remove drainplug (5) from bottom of hoist winch (4).
2. Install drainplug (5) in hoist winch (4) when oil drainage is completed.
3. Attach utility chain to hoist winch (4).
4. Attach utility chain to lifting device and remove slack from utility chain.

NOTE

Assistant will help with step 5.

5. Remove twelve screws (3) and lockwashers (2) from boom (1) and hoist winch (4). Discard lockwashers (2).
6. Carefully lower hoist winch (4) to wrecker bed.
7. Remove utility chain from hoist winch (4).

b. Installation

NOTE

Assistant will help with steps 1, 2, and 3.

1. Attach utility chain to hoist winch (4).
2. Attach of lifting device to utility chain and lift hoist winch (4) into position on boom (1).
3. Align screw holes in hoist winch (4) with screw holes in boom (1) and install hoist winch (4) on boom (1) with twelve new lockwashers (2) and screws (3).
4. Release lifting device hook from utility chain and remove utility chain from hoist winch (4).

4-192. HOIST WINCH REPLACEMENT (Contd)

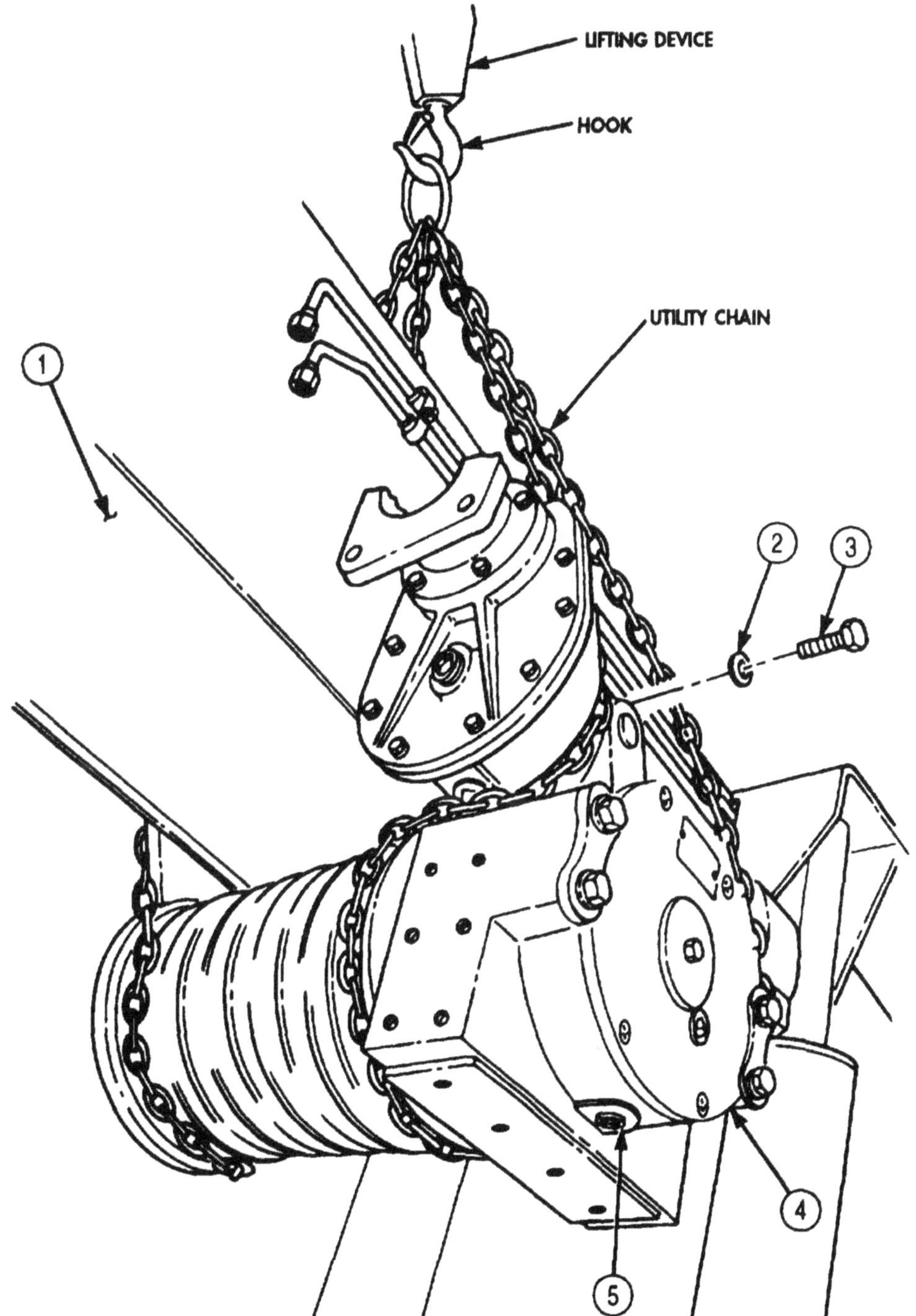

FOLLOW-ON TASKS:
- Install hoist level wind (para. 4-191).
- Install hydraulic hoist winch motor and lines (para. 4-190).
- Install hoist winch cable (para. 3-384).
- Install boom floodlight wire (para. 3-385).
- Fill hoist winch to proper oil level (LO 9-2320-272-12).

4-193. BOOM ELEVATING CYLINDER REPLACEMENT

THIS TASK COVERS:

a. Removal b. Installation

INITIAL SETUP:

APPLICABLE MODEL
M936/A1/A2

TOOLS
General mechanic's tool kit (Appendix E, Item 1)
Torque wrench (Appendix E, Item 144)
Lifting device
Chain

MATERIALS/PARTS
Cotter pin (Appendix D, Item 59)
Lockwasher (Appendix D, Item 350)
Cap and plug set (Appendix D, Item 14)

PERSONNEL REQUIRED
Two

REFERENCES (TM)
LO 9-2320-272-12
TM 9-2320-272-10
TM 9-2320-272-24P

EQUIPMENT CONDITION
- Parking brake set (TM 9-2320-272-10).
- Boom lowered (TM 9-2320-272-10).
- Hydraulic oil reservoir drained (para 4-188).

GENERAL SAFETY INSTRUCTIONS
- All personnel must stand clear during hoisting operations.
- Elevating cylinder is heavy. Use caution when removing or installing elevating cylinder.

CAUTION

When disconnecting hydraulic lines and hoses, plug all openings to prevent dirt from entering and causing internal parts damage.

NOTE

Left and right elevating cylinders are removed and installed in the same way. Bight elevating cylinder is covered in this procedure.

a. Removal

NOTE

- Have drainage container ready to catch oil.
- Tag lines for installation.

1. Disconnect cylinder oil supply line (6) and oil return line (5) from cylinder adapter fitting (10) and snubber valve (9) at bottom of elevating cylinder (11).

WARNING

All personnel must stand clear during hoisting operations. A snapped cable, or shifting or swinging load, may cause injury to personnel.

2. Using chain and lifting device, lift boom (1) until boom (1) weight is fully supported and upper retaining pin (2) is free.
3. Remove screw (4), lockwasher (3), and upper retaining pin (2) from elevating cylinder (11) and boom (1). Discard lockwasher (3).

WARNING

Elevating cylinder is heavy. Remove with the aid of assistant and a lifting device or injury to personnel may result.

4. Remove cotter pin (7), lower retaining pin (8), and elevating cylinder (11) from turntable (12). Discard cotter pin (7).

4-193. BOOM ELEVATING CYLINDER REPLACEMENT (Contd)

5. Using chain and lifting device, lift boom (1) and install crane shipper braces (TM 9-2320-272-10).

b. Installation

1. Using utility chain and lifting device, raise boom (1) and remove shipper braces (TM 9-2320-272-10).
2. Install elevating cylinder (11) on turntable (12) with lower retaining pin (8) and new cotter pin (7).
3. Install elevating cylinder (11) on boom (1) with upper retaining pin (2), new lockwasher (3), and screw (4). Tighten screw (4) 44-61 lb-ft (60-83 N•m).
4. Connect cylinder oil supply line (5) to snubber valve (9).
5. Connect oil return (5) line to cylinder adapter fitting (10).
6. Remove lifting device and chain from boom (1).

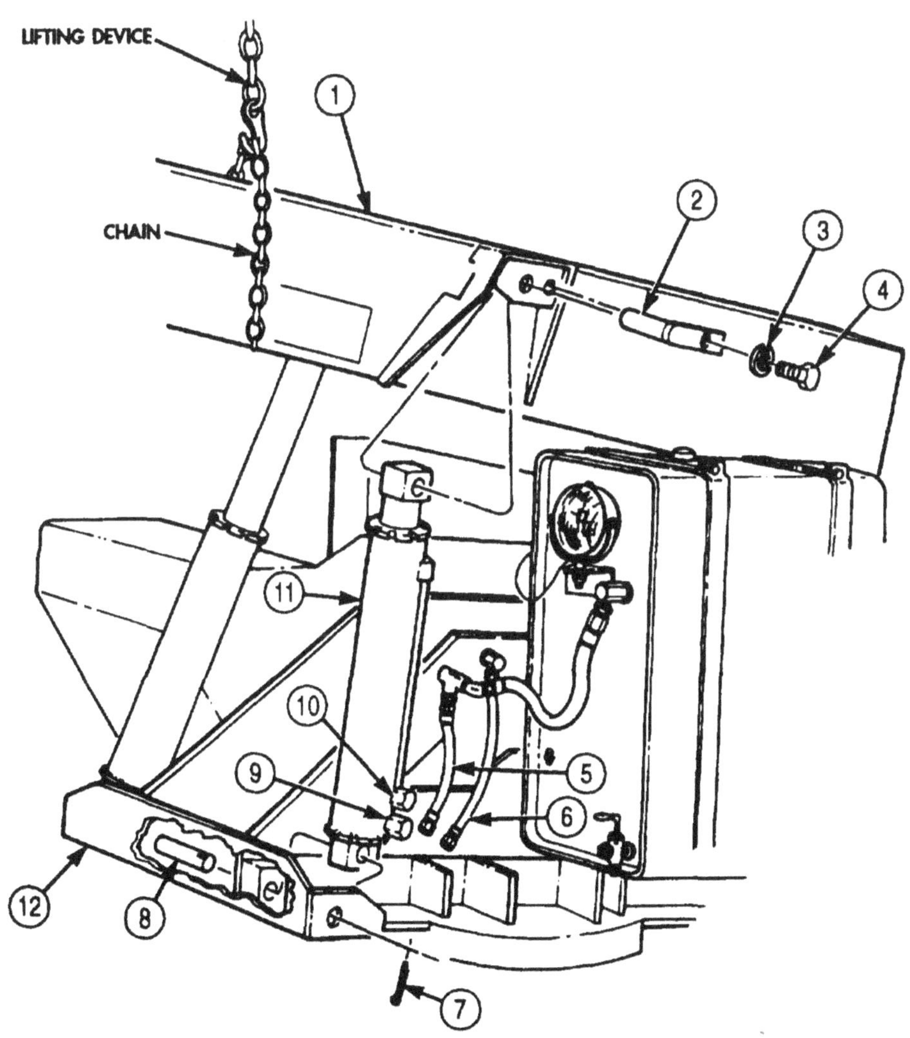

FOLLOW-ON TASKS: • Fill hydraulic oil reservoir to proper level (LO 9-2320-272-12).
• Operate crane through full elevating range (TM 9-2320-272-10) and check for leaks at oil lines and cylinder.

4-194. BOOM REPLACEMENT

THIS TASK COVERS:

a. Removal b. Installation

INITIAL SETUP:

APPLICABLE MODELS
M936/A1/A2

TOOLS
General mechanic's tool kit (Appendix E, Item 1)
Torque wrench (Appendix E, Item 145)
Lifting device
Chains

MATERIALS/PARTS
Locknut (Appendix D, Item 285)
Two lockwashers (Appendix D, Item 350)
Cap and plug set (Appendix C, Item 14)

REFERENCES (TM)
LO 9-2320-272-12
TM 9-2320-272-10
TM 9-2320-272-24P

EQUIPMENT CONDITION
- Parking brake set (TM 9-2320-272-10)
- Boom floodlight wire removed (para. 3-385).
- Hydraulic oil reservoir drained (para. 4-188).
- Hoist winch removed (para. 4-192).

GENERAL SAFETY INSTRUCTIONS
- All personnel must stand clear during hoisting operations.
- Ensure inner boom is secured to outer boom before hoisting.

WARNING

All personnel must stand clear during hoisting operations. A snapped cable, or shifting or swinging load, may cause injury to personnel.

a. Removal

CAUTION

When disconnecting hydraulic lines and hoses, plug all openings to prevent dirt from entering and causing internal damage to parts.

NOTE

- Have drainage container ready to catch oil.
- Tag hoses for installation.

1. Disconnect cylinder extension hose (5) and retracting hose (4) from two boom adapter elbows (3).
2. Secure inner boom (1) to outer boom (2) with utility chain to prevent inner boom (1) movement during hoisting operation.
3. Attach two utility chains to outer boom (2).
4. Raise outer boom (2) until weight of boom (2) is fully supported and elevating cylinder retaining pins (16) are free.
5. Remove two screws (14), lockwashers (15), and retaining pins (16) from outer boom (2) and two elevating cylinders (13). Discard lockwashers (15).

WARNING

Before hoisting outer boom away from wrecker, ensure inner boom is properly secured to outer boom as outlined in step 2 to prevent injury to personnel.

6. Remove grease fittings (6) and (12) from pivot pin (7).
7. Remove locknut (11), pivot pin (7), two sleeve bearings (9), and crane sheave (10) from turntable (8). Discard locknut (11).

4-194. BOOM REPLACEMENT (Contd)

8. Using lifting device and utility chains and, with assistant guiding movement, move outer boom (2) off vehicle and onto jack stands.
9. Remove utility chains from outer boom (2).

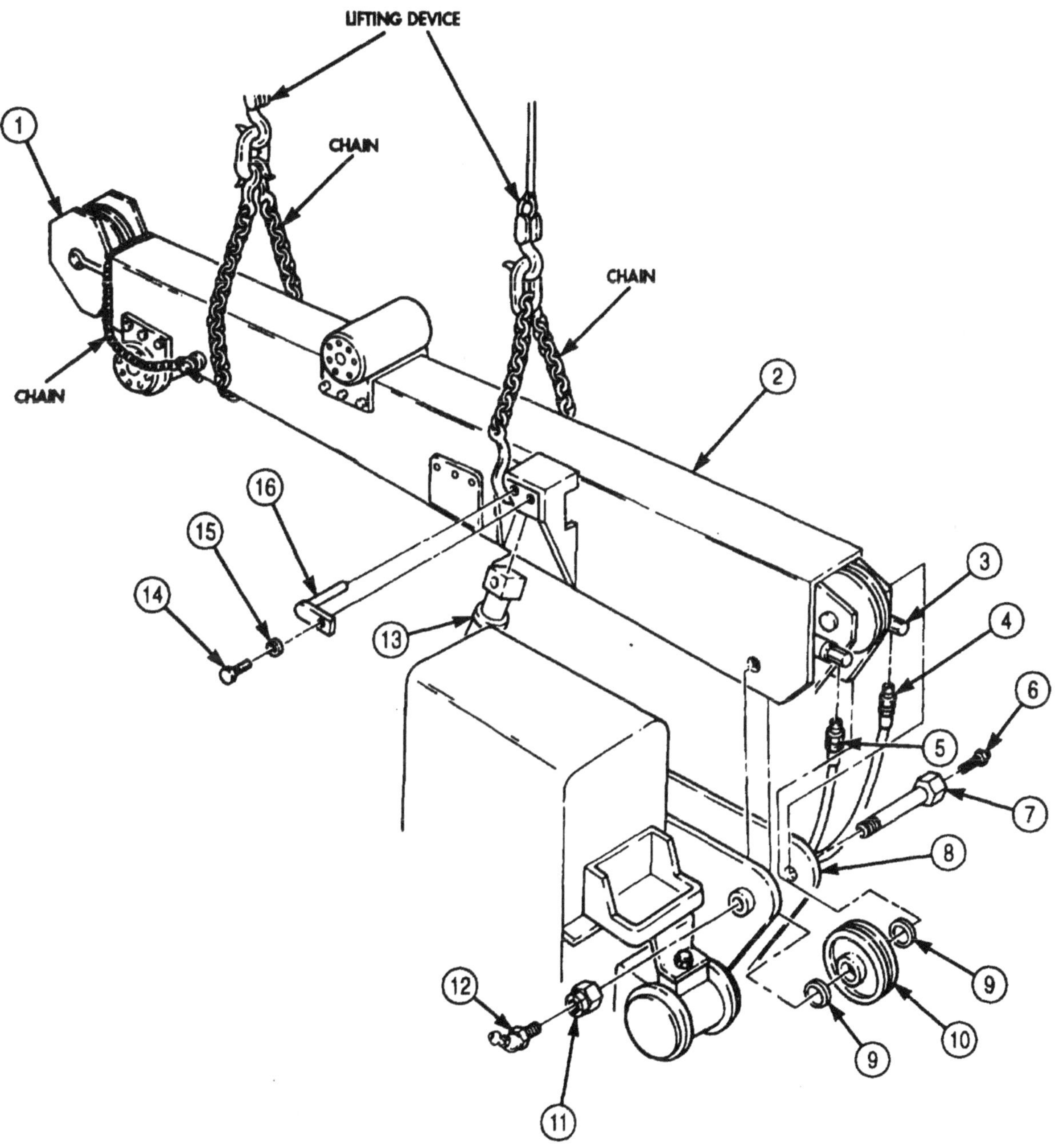

4-194. BOOM REPLACEMENT (Contd)

b. Installation

WARNING

Before hoisting outer boom to wrecker, ensure that inner boom is secured to outer boom as outlined in steps of removal procedure.

NOTE

If installing new boom assembly, use hardware from old boom.

1. Install utility chain to secure inner boom (1) to outer boom (2).
2. Attach two utility chains on outer boom (2).
3. Hoist outer boom (2) over wrecker and position on turntable (8).
4. Assemble two sleeve bearings (9) and sheave (10), align sheave (10) with holes in turntable (8) and outer boom (2), and install boom (2) on turntable (8) with pivot pin (7) and new locknut (11). Tighten locknut (11) 800-1,000 lb-ft(1,185-1,356 N•m).
5. Install grease fittings (6) and (12) on pivot pin (7).
6. Connect two elevating cylinders (13) to boom (2) with two retaining pins (16), new lockwashers (15), and screws (14). Tighten screws (14) 44-61 lb-ft (60-83 N•m).
7. Using lifting device and utility chains, lift outer boom (2) and disconnect shipper braces from boom (2) (TM 9-2320-272-10).
8. Connect cylinder extension hose (5) and retracting hose (4) to two boom adapter elbows (3).
9. Remove utility chains from outer boom (2).

4-194. BOOM REPLACEMENT (Contd)

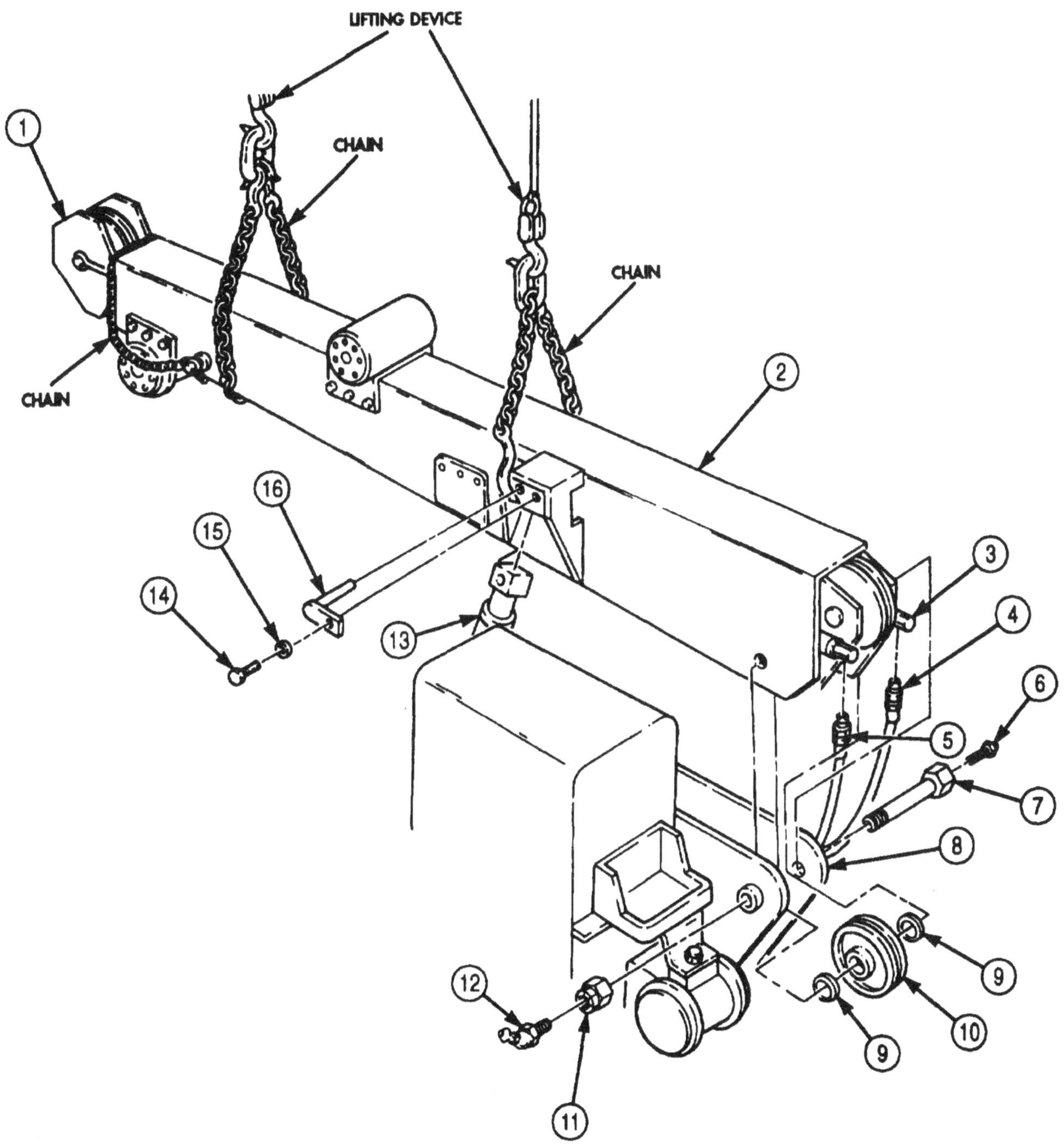

FOLLOW-ON TASKS:
- Install boom floodlight wire (para. 3-385).
- Fill hydraulic oil reservoir to proper level (LO 9-2320-272-12).
- Lubricate boom (LO 9-2320-272-12).
- Install hoist winch (para. 4-192).
- 0perate boom through full range of motion (TM 9-2320-272-10) and check for leaks and proper movement.

4-195. INNER BOOM REPLACEMENT

THIS TASK COVERS:

a. Removal **b. Installation**

INITIAL SETUP:

APPLICABLE MODELS
M963/A1/A2

TOOLS
General mechanic's tool kit (Appendix E, Item 1)
Lifting device
Chains

MATERRALS/PARTS
Cotter pin (Appendix D, Item 67)
Eight lockwashers (Appendix D, Item 350)

PERSONNEL REQUIRED
TWO

REFERENCES (TM)
TM 9-2320-272-10
TM 9-2320-272-24P

EQUIPMENT CONDITION
Parking brake set (TM 9-2320-272-10).
Boom removed (para. 4-194).

GENERAL SAFETY INSTRUCTION
All personnel must stand clear during hoisting operations.
Inner boom must be properly supported.

WARNING

All personnel must stand clear during hoisting operations. A snapped cable, or shifting or swinging load, may cause injury to personnel.

NOTE

Extension cylinder will be removed and installed with the inner boom during this procedure.

a. Removal

1. Remove cotter pin (5) and pin (4) from extension cylinder (6) and outer boom (3). Discard cotter pin (5).
2. Remove eight screws (7), lockwashers (8), and two boom stops (9) from outer boom (3). Discard lockwashers (8).

WARNING

Inner boom must he supported at sheave to prevent tilting until hoist chain can be properly positioned around inner boom or injury to personnel may result.

3. Attach lifting device to sheave (1) of inner boom (2).
4. While assistant monitors and adjusts lifting device (1), slowly pull inner boom (2) out of outer boom (3) until utility chain can be positioned around inner boom (2).

NOTE

If only one lifting device is available, support must be provided under sheave end of inner boom while transferring lifting hook to utility chain on inner boom.

5. Attach lifting device to utility chain on inner boom (2) and remove inner boom (2) from outer boom (3).
6. Place inner boom (2) on jack stands and remove lifting device and utility chain from inner boom (2).

4-195. INNER BOOM REPLACEMENT (Contd)

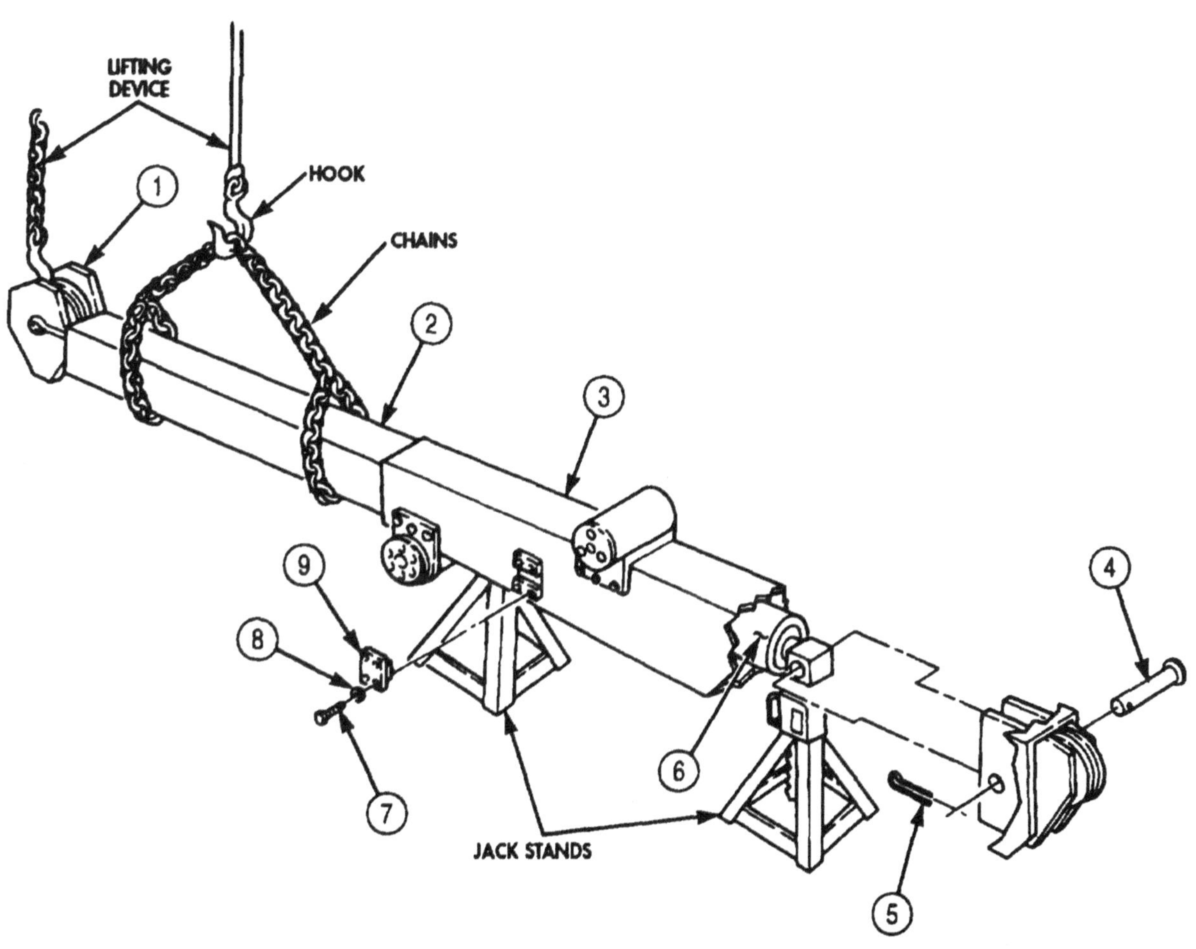

4-195. INNER BOOM REPLACEMENT (Contd)

b. Installation

1. Attach utility chain to straddle balance point of inner boom (2).

NOTE

- Assistant will help with steps 2 through 5.
- Ensure pin hole in extension cylinder is parallel to top surface of inner boom.

2. Attach lifting device to chain, align inner boom (2) with outer boom (3) and start inner boom (2) into outer boom (3).

NOTE

If a second lifting device is not available, use sheave end of inner boom to permit transfer of chain and lifting device.

3. Remove chain from inner boom (2).
4. Attach utility chain and lifting device to sheave (1) of inner boom (2).
5. While adjusting position of sheave (1) end of inner boom (2), slide inner boom (2) into outer boom (3) until holes in front of outer boom (3) and extension cylinder (6) align.
6. Install extension cylinder (6) on outer boom (3) with pin (4) and new cotter pin (5).
7. Remove lifting device and chain from sheave (1).
8. Install two boom stops (9) on outer boom (3) with eight new lockwashers (8) and screws (7).

4-195. INNER BOOM REPLACEMENT (Contd)

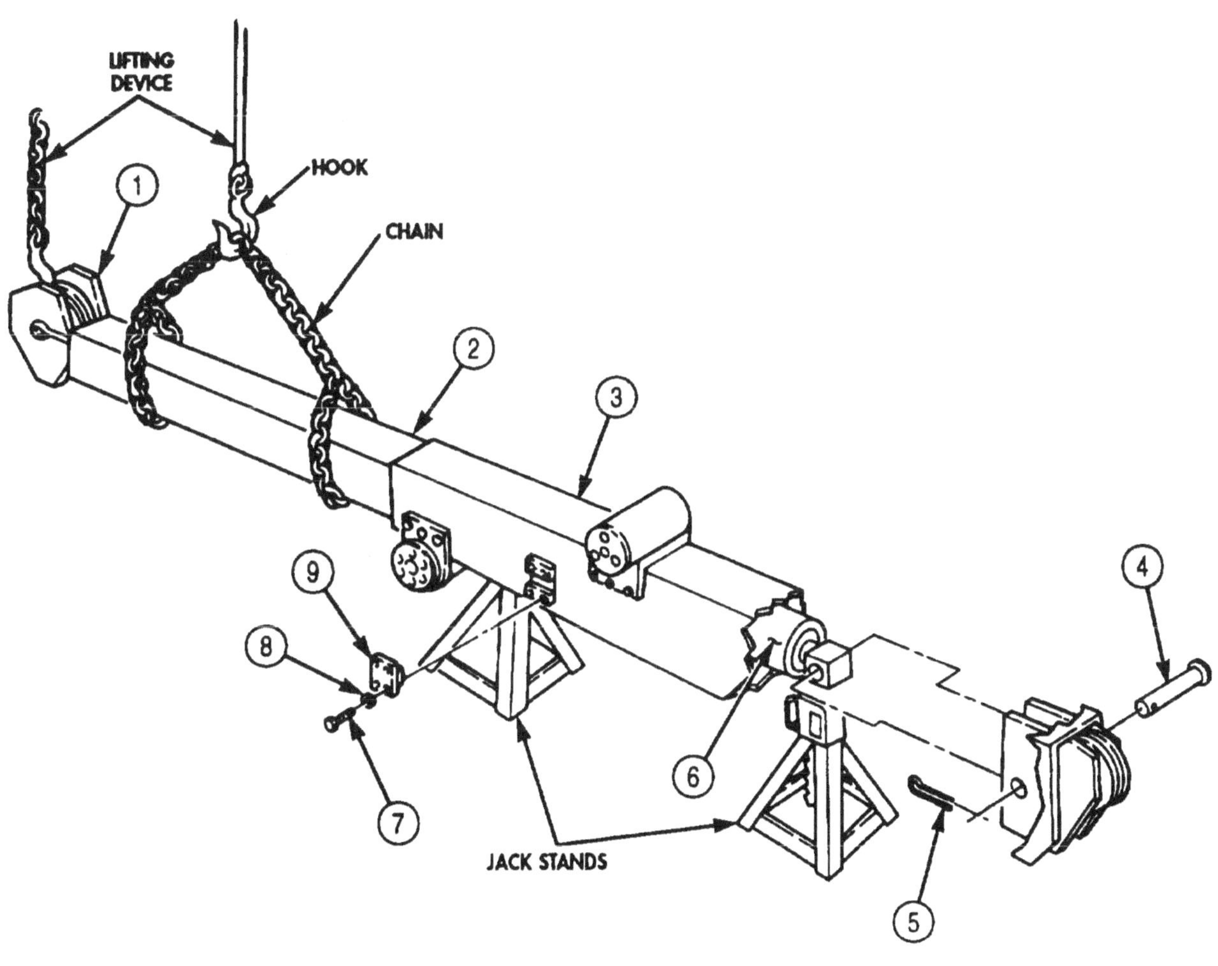

FOLLOW-ON TASKS:
- Install boom (para. 4-194).
- Check boom roller lateral and vertical adjustment (para. 4-197).

4-196. BOOM EXTENSION CYLINDER REPLACEMENT

THIS TASK COVERS:

a. Removal **b. Installation**

INITIAL SETUP:

APPLICABLE MODELS
M936/A1/A2

TOOLS
General mechanic's tool kit (Appendix E, Item 1)
Lifting device
Chains

MATERIALS/PARTS
Cotter pin (Appendix D, Item 69)

REFERENCES (TM)
TM 9-2320-272-10
TM 9-2320-272-24P

EQUIPMENT CONDITION
- Parking brake set (TM 9-2320-272-10).
- Inner boom removed (para. 4-195).

GENERAL SAFETY INSTRUCTIONS
All personnel must stand clear during lifting operations.

a. Removal

1. Remove cotter pin (3) from extension cylinder (1) and inner boom (2). Drive out retaining pin (4) from inner boom (2) and extension cylinder (1). Discard cotter pin (3).
2. Attach chains around extension cylinder (1) and attach to lifting device.

WARNING

All personnel must stand clear during lifting operations. A snapped cable, or shifting or swinging load, may cause injury to personnel.

3. Slowly pull extension cylinder (1) out of inner boom (2) until chains can be positioned. Place. extension cylinder (1) on shop horses or jack stands.
4. Remove chains and lifting device from extension cylinder (1).

b. Installation

1. Attach chains around extension cylinder (1) and attach to lifting device.
2. Lift extension cylinder (1) to inner boom (2) and slide into inner boom (2).
3. Install extension cylinder (1) on inner boom (2) with retaining pin (4) and new cotter pin (3).
4. Remove chains from extension cylinder (1).

4-196. BOOM EXTENSION CYLINDER REPLACEMENT (Contd)

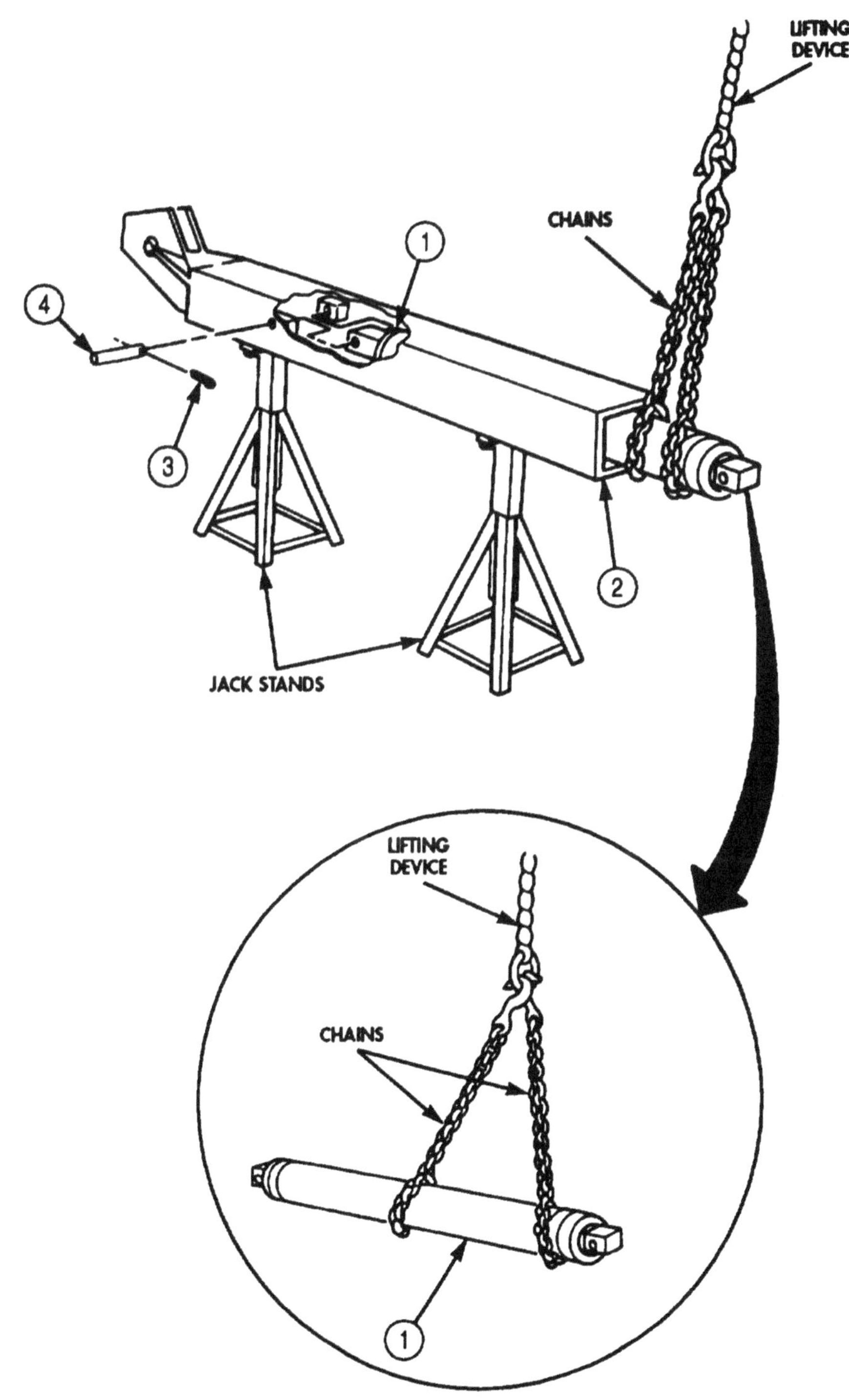

FOLLOW-ON TASK: Install inner boom (para. 4-195).

4-197. BOOM ROLLER MAINTENANCE

THIS TASK COVERS:

a. Removal
b. Disassembly
c. Cleaning and Inspection
d. Assembly
e. Installation

INITIAL SETUP:

APPLICABLE MODELS
M936/A1/A2

TOOLS
General mechanic's tool kit (Appendix E, Item
Arbor press

MATERIALS/PARTS
Eight lockwashers (Appendix D, Item 350)
Two woodruff keys (Appendix D, Item 729)
Six lockwashers (Appendix D, Item 377)
Drycleaning solvent (Appendix C, Item 71)

REFERENCES (TM)
TM 9-2320-272-10
TM 9-2320-272-2413

EQUIPMENT CONDITION
- Parking brake set (TM 9-2320-272-10).
- Inner boom removed (para. 4-195).

GENERAL SAFETY INSTRUCTIONS
- Keep fire extinguisher nearby when using drycleaning solvent.
- Drycleaning solvent is flammable and toxic. Do not use near open flame.

NOTE

Both boom rollers are removed in the same way with boom on or off the vehicle. This procedure applies to the bottom boom roller with inner boom removed.

a. Removal

Remove six screws (3), lockwashers (4), and roller (1) from outer boom (2). Discard lockwashers (4).

b. Disassembly

1. Thoroughly clean exterior of boom roller (1).
2. Remove eight screws (5), lockwashers (6), left and right mounting caps (7), and four shims (8) from boom roller frame (9). Discard lockwashers (6).
3. Remove two rollers (10) and roller shaft (13) from frame (9).
4. Remove two rollers (10) from roller shaft (13).
5. Using an arbor press, remove roller bearings (11) from two rollers (10).
6. Remove two woodruff keys (12) from roller shaft (13). Discard woodruff keys (12).

4-197. BOOM ROLLER MAINTENANCE (Contd)

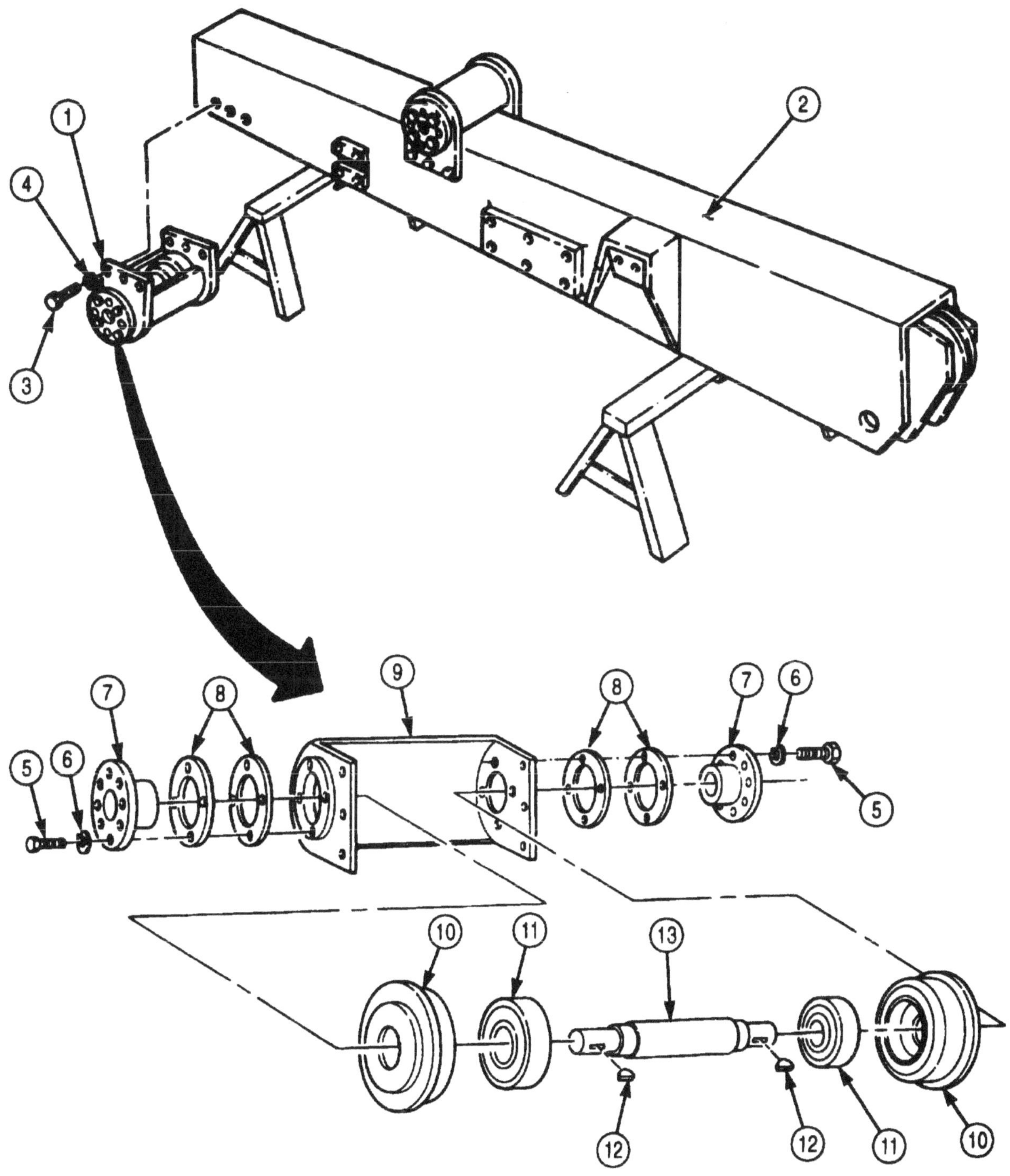

4-197. BOOM ROLLER MAINTENANCE (Contd)

c. Cleaning and Inspection

1. Clean all boom roller parts with drycleaning solvent (para. 2-14).
2. Inspect roller bearings (7) and rollers (6) for pitting and scoring. Replace bearings (7) or rollers (6) if pitted or scored.
3. Inspect boom roller frame (5) and mounting caps (3) for cracks or damaged threads. Replace roller frame (5) or mounting caps (3) if cracked or threads are damaged
4. Inspect roller shaft (9) for scoring and breaks. Replace roller shaft (9) if scored or broken.

d. Assembly

WARNING

Drycleaning solvent is flammable and toxic. Do not use near an open flame and always have a fire extinguisher nearby when solvents are used. Use only in well-ventilated places, wear protective clothing, and dispose of cleaning rags in approved container. Failure to do this may result in injury to personnel and/or damage to equipment.

1. Using arbor press, install two roller bearings (7) on rollers (6).
2. Install two new woodruff keys (8) in roller shaft (9).
3. Install two rollers (6) with bearings (7) on roller shaft (9).
4. Position rollers (6) and shaft (9) as an assembly in frame (5).
5. Position left and right mounting caps (3) over shims (4) on frame (5).
6. Measure clearance between mounting caps (3) and rollers (6). Clearance should be 0.062-0.125 in. (1.6-3.2 mm). If clearance is more than above, remove one or more shims (4). If clearance is less than allowable, add one or more shims (4).
7. Install left and right mounting caps (3) on frame (5) with eight new lockwashers (2) and screws (1).

e. Installation

Install boom roller (11) on outer boom (10) with six new lockwashers (13) and screws (12).

4-197. BOOM ROLLER MAINTENANCE (Contd)

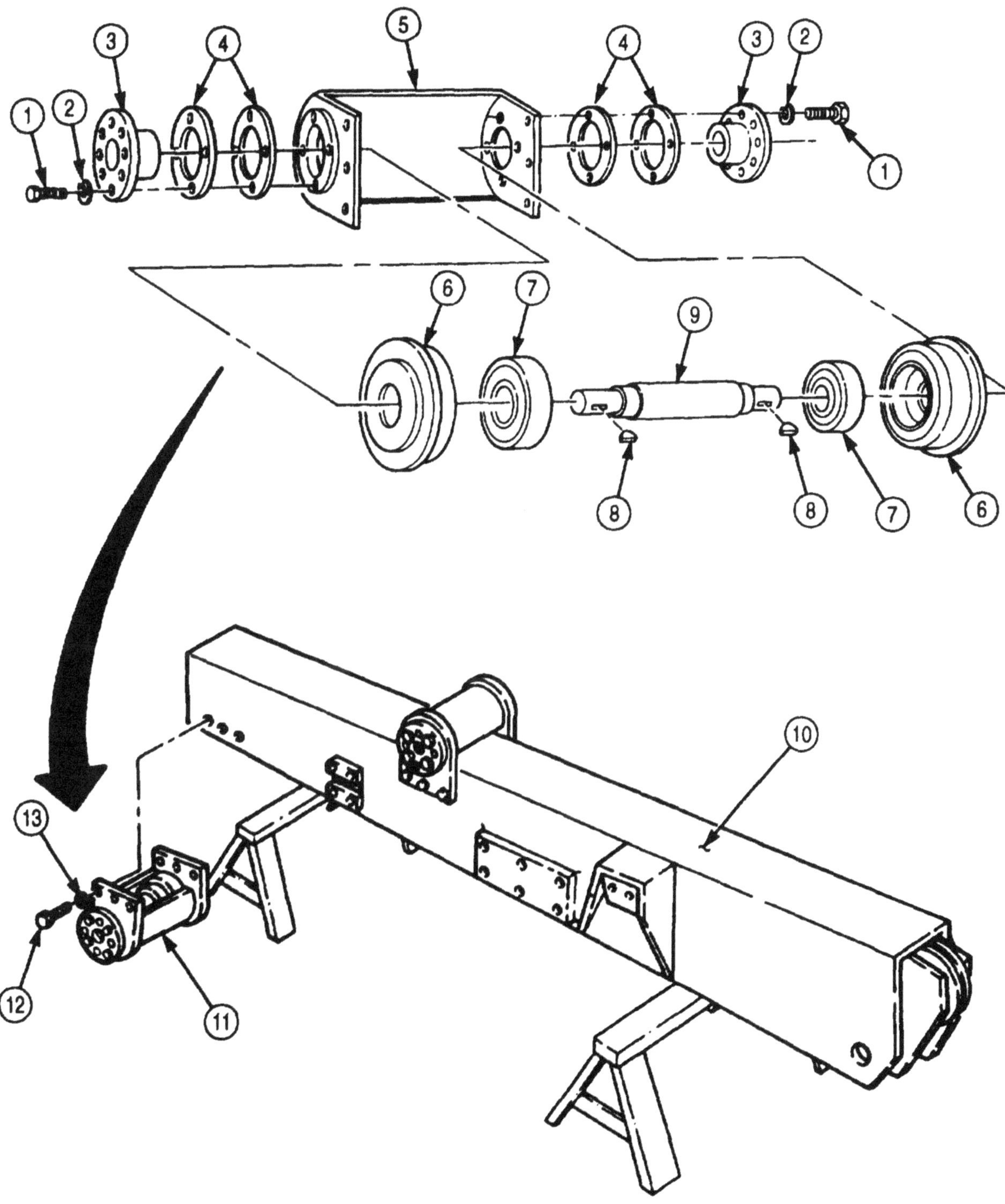

FOLLOW-ON TASK: Install inner boom (para. 4-195).

4-198. CRANE SWIVEL VALVE MAINTENANCE

THIS TASK COVERS:

a. Wiring Harness Disconnection
b. Removal
c. Installation
d. Wiring Harness Connection

INITIAL SETUP:

APPLICABLE MODELS
M936/A1/A2

TOOLS
General mechanic's tool kit (Appendix E, Item 1)
Torque wrench (Appendix E, Item 144)

MATERIALS/PARTS
Six lockwashers (Appendix D, Item 354)
Lockwasher (Appendix D, Item 350)
Lockwasher (Appendix D, Item 379)
Cap and plug set (Appendix C, Item 14)

REFERENCES (TM)
TM 9-2320-272-10
TM 9-2320-272-24P

EQUIPMENT CONDITION
- Parking brake set (TM 9-2320-272-10)
- Boom removed (para. 4-194).
- Battery ground cables disconnected (para. 3-126).

GENERAL SAFETY INSTRUCTIONS
All personnel must stand clear during lifting operations.

a. Wiring Harness Disconnection

NOTE

Tag all parts for installation.

1. Disconnect wires (5) and (6) of electrical swivel harness (4) from crane wiring harness (7).
2. Remove nut (8), ground wire (9), nut (8), and lockwasher (10) from frame (11) through underside of vehicle. Discard lockwasher (10).
3. Disconnect swivel wire harness (13), wires (14) and (17), and ground wire (15) from floodlight harness (16).
4. Remove screw (3), lockwasher (2), and washer (1) from electrical swivel (12). Discard lockwasher (2).

4-198. CRANE SWIVEL VALVE MAINTENANCE (Contd)

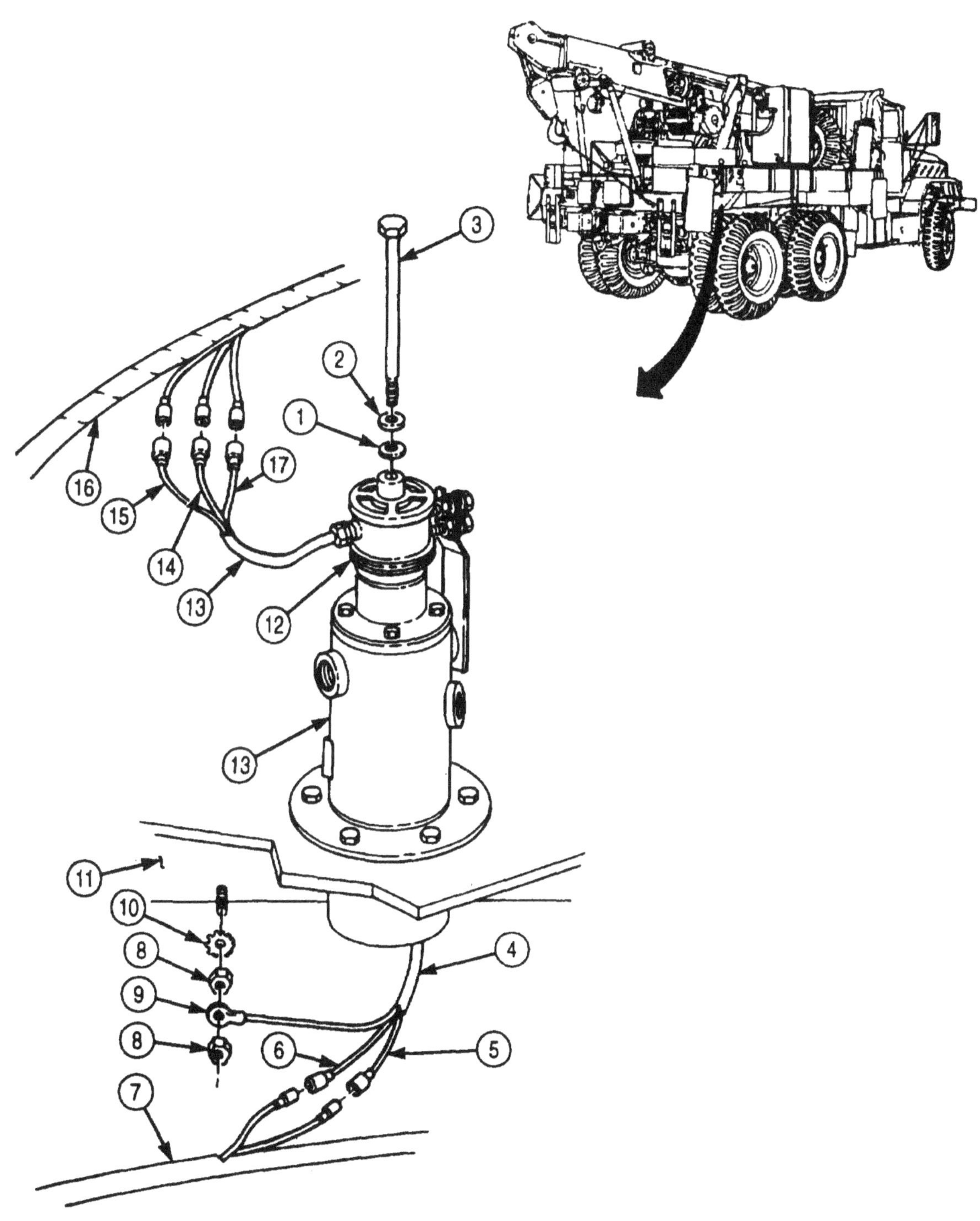

4-198. CRANE SWIVEL VALVE MAINTENANCE (Contd)

b. Removal

WARNING

All personnel must stand clear during lifting operations. A snapped cable, or shifting or swinging load, may cause injury to personnel.

CAUTION

When disconnecting hydraulic lines and hoses, plug all openings to prevent dirt from entering and causing internal damage to parts.

NOTE

- Have drainage container ready to catch oil.
- Tag all hydraulic lines and fittings for installation.

1. Disconnect hydraulic lines (7), (10), and (13) from lower end of hydraulic swivel valve (3).
2. Remove adapter fittings (6), (8), (9), (11), (12), (14), (16), and (17) from lower end of hydraulic swivel valve (3).
3. Disconnect hydraulic lines (1) and (18) from hydraulic swivel valve (3).
4. Disconnect hydraulic tube (5) from elevating cylinder tee (4) and swivel valve (3).
5. Remove adapter fittings (2) and (19) from hydraulic swivel valve (3).
6. Remove six screws (22) and lockwashers (23) from hydraulic swivel valve (3) and turntable (24). Discard lockwashers (23).
7. Install utility chain on hydraulic swivel valve (3).

NOTE

Assistant will assist with step 8.

8. Attach hoist hook to chain and lift valve (3) from turntable (24).
9. Remove utility chain and hoist hook from hydraulic swivel valve (3).

4-198. CRANE SWIVEL VALVE MAINTENANCE (Contd)

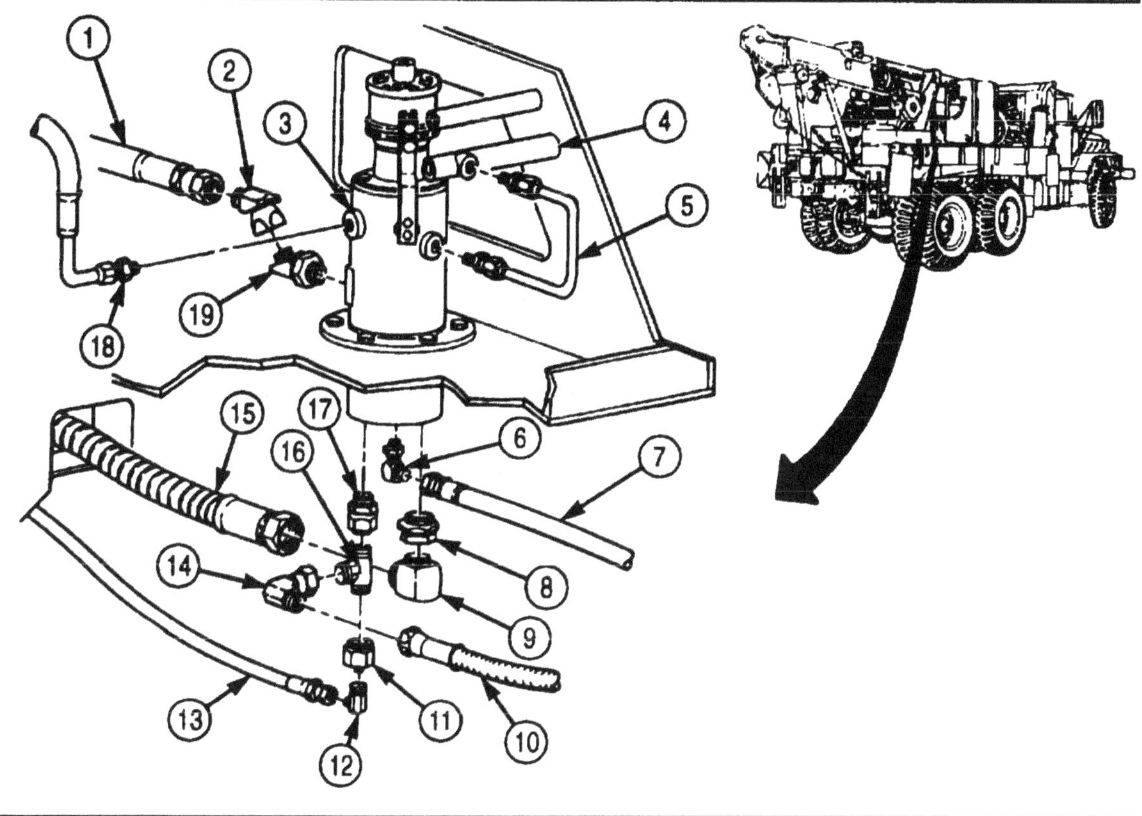

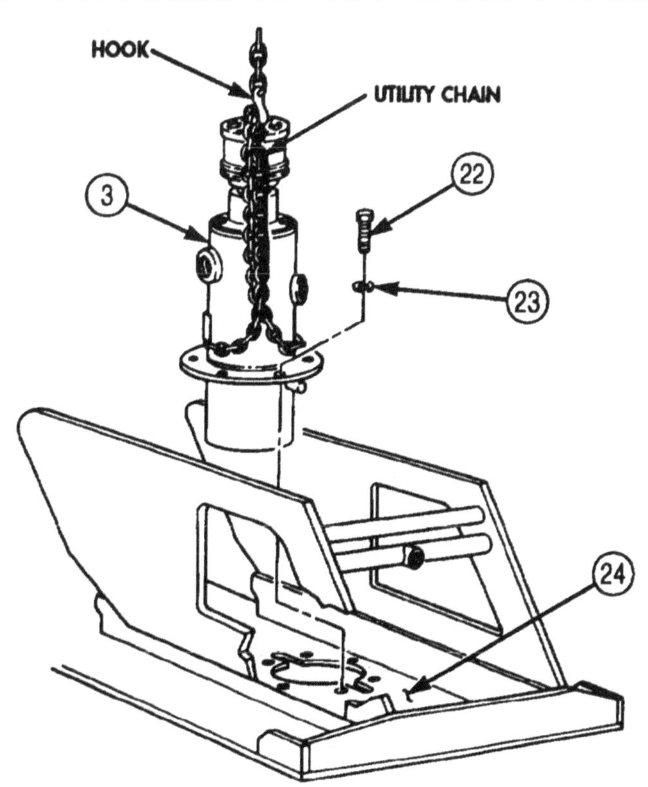

4-198. CRANE SWIVEL VALVE MAINTENANCE (Contd)

c. Installation

1. Install utility chain and hoist hook on hydraulic swivel valve (6).

WARNING

All personnel must stand clear during lifting operations. A snapped cable, or shifting or swinging load, may cause injury to personnel.

NOTE

Assistant will help with step 2.

2. Align valve guide pins (3) on swivel valve (6) with slots (5) in turntable (4) and install with six new lockwashers (2) and screws (1). Tighten screws (1) 44-61 lb-ft (60-83 N•m).
3. Remove hoist hook and utility chain from hydraulic swivel valve (6).
4. Install adapter fitting (23) on swivel valve (6) and adapter fitting (8) on adapter fitting (23).
5. Connect hydraulic tube (10) to swivel valve (6) and elevating cylinder tee (9).
6. Connect hydraulic lines (7) and (24) to hydraulic swivel valve (6).
7. Install adapter fittings (11), (13), (14), (22), (21), (19), (16), and (17) on lower end ofhydraulic swivel valve (6).
8. Connect hydraulic lines (20), (12), (15), and (18) to fittings on lower end of hydraulic swivel valve (6).

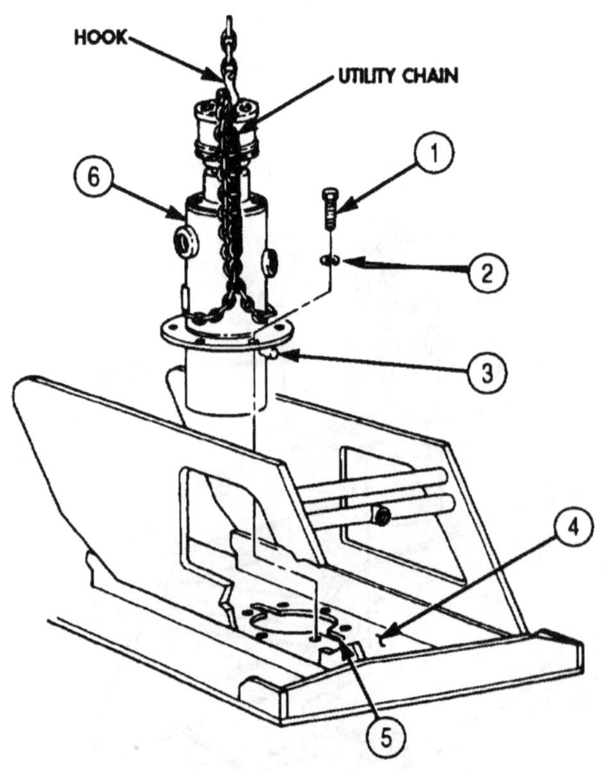

4-198. CRANE SWIVEL VALVE MAINTENANCE (Contd)

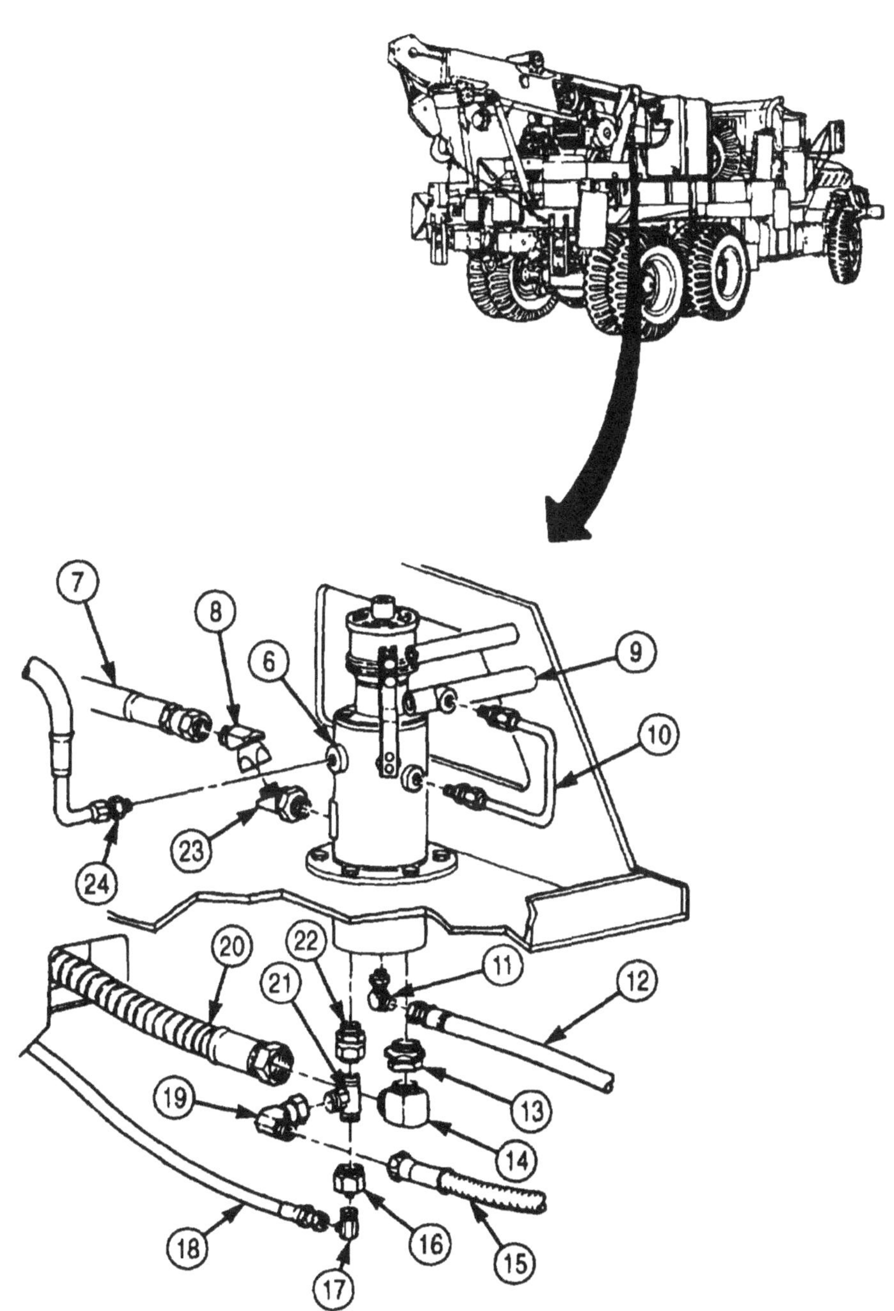

4-198. CRANE SWIVEL VALVE MAINTENANCE (Contd)

d. Wiring Harness Connection

1. Install washer (l), new lockwasher (2), and screw (3) on electrical swivel (12).
2. Connect wires (14), (17), ground wire (15) and swivel sire harness (13) to floodlight harness (16).
3. Connect wires (5) and (6) of swivel harness (4) to crane wiring harness (7) at underside of vehicle.
4. At underside of vehicle, install new lockwasher (10), nut (8), g-round wire (9), and nut (8) on frame (11).

4-198. CRANE SWIVEL VALVE MAINTENANCE (Contd)

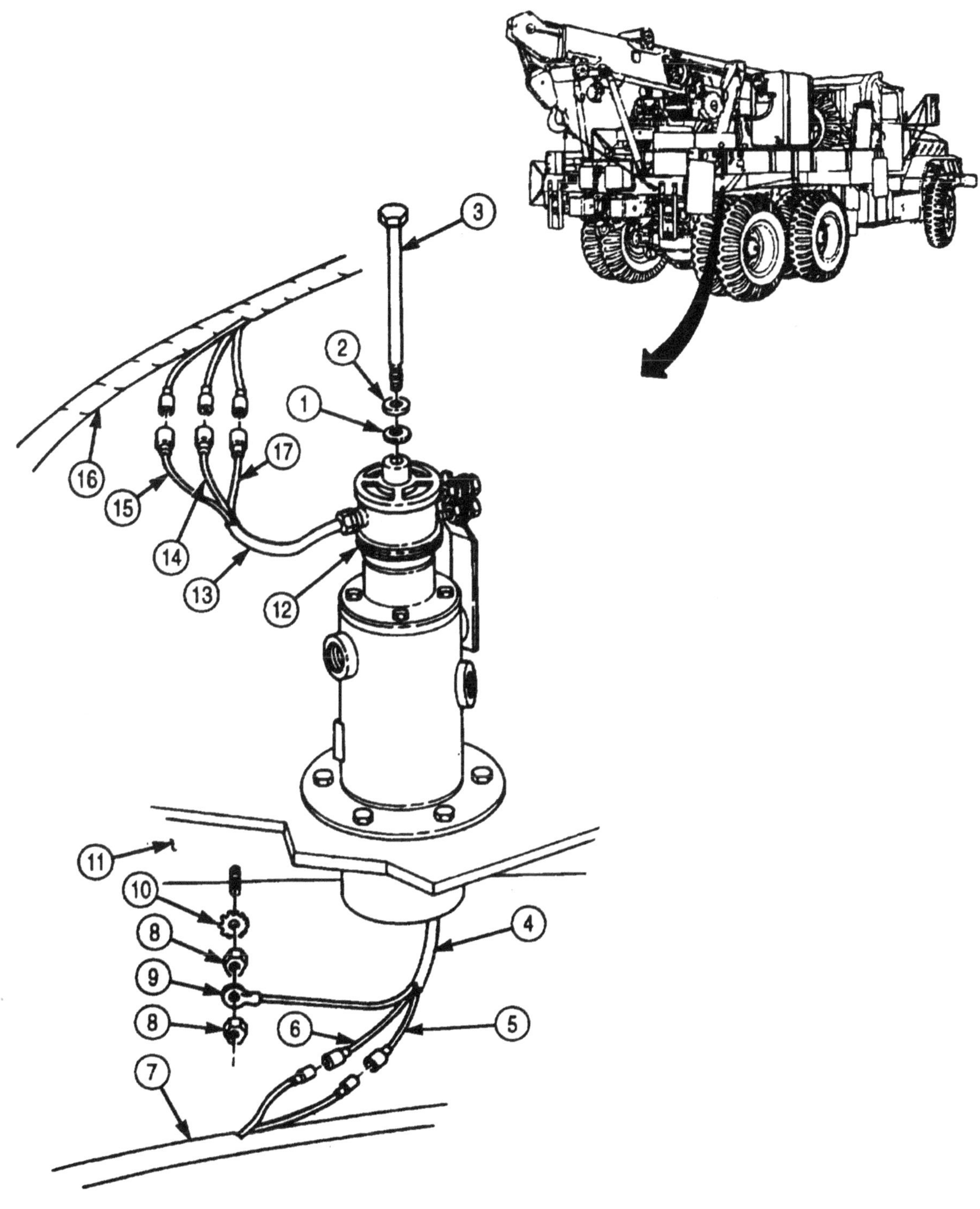

FOLLOW-ON TASKS:
- Install boom (para. 4-194) and check for leaks.
- Connect battery ground cables (3-126).

4-199. CRANE HYDRAULIC SWINGER MOTOR REPLACEMENT

THIS TASK COVERS:

a Removal **b. Installation**

INITIAL SETUP:

APPLICABLE MODELS

M936/A1/A2

TOOLS

General mechanic's tool kit (Appendix E Item 1)
Torque wrench (Appendix E, Item 144)

MATERIALS/PARTS

Four lockwashers (Appendix D, Item 377)
Cap and plug set (Appendix C, Item 14)

REFERENCES (TM)

LO 9-2320-272-12
TM 9-2320-272-10
TM 9-2320-272-24P

EQUIPMENT CONDITION

Parking brake set (TM 9-2320-272-10).

a. Removal

CAUTION

When disconnecting hydraulic lines, plug all openings to prevent dirt from entering and causing internal damage to parts.

NOTE

- Have drainage container ready to catch oil.
- Tag all hydraulic flex lines and adapter elbows for installation.

1. Disconnect hydraulic flex lines (1) and (2) from elbows (3) and (4).
2. Remove four screws (7) and lockwashers (6) from swinger motor (5) and gearcase (8). Discard lockwashers (6).
3. Remove swinger motor (5) from gearcase (8).
4. Remove elbows (3) and (4) from swinger motor (5).

1. Install adapter elbows (3) and (4) on swinger motor (5).
2. Install swinger motor (5) on gearcase (8) with four new lockwashers (6) and screws (7). Tighten mews (7) 44-61 lb-ft (60-83 N•m).
3. Connect hydraulic flex lines (1) and (2) to elbows (3) and (4).

4-199. CRANE HYDRAULIC SWINGER MOTOR REPLACEMENT (Contd)

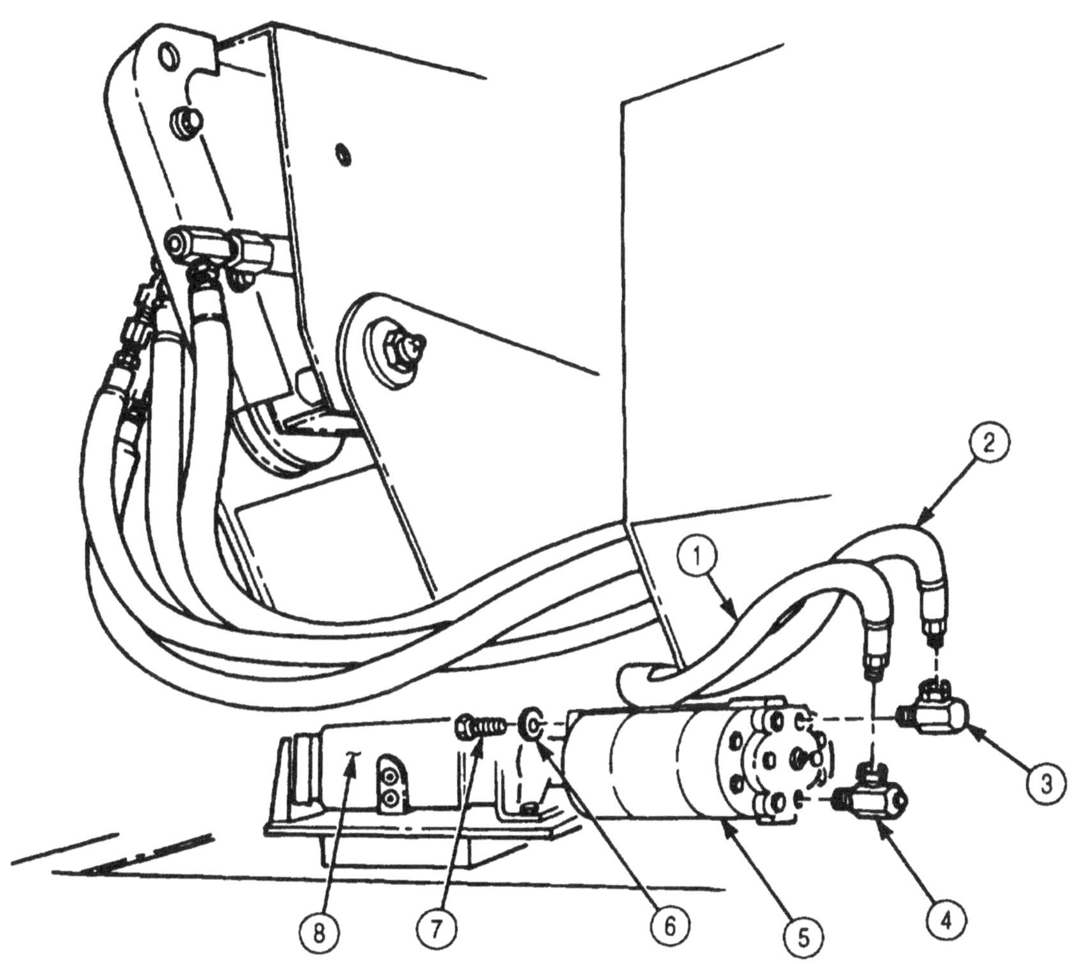

FOLLOW-ON TASKS:
- Fill hydraulic oil reservoir to proper oil level (LO 9-2320-272-12).
- Operate swing control lever to check swinger motor for proper operation (TM 9-2320-272-10).

4-200. CRANE SWINGER GEARCASE REPLACEMENT

THIS TASK COVERS:

a. Removal **b. Installation**

INITIAL SETUP:

APPLICABLE MODELS
M936/A1/A2

TOOLS
General mechanic's tool kit (Appendix E, Item 1)

MATERIALS/PARTS
Six lockwashers (Appendix D, Item 377)

REFERENCES (TM)
TM 9-2320-272-10
TM 9-2320-272-24P

EQUIPMENT CONDITION
- Parking brake set (TM 9-2320-272-10).
- Crane hydraulic swinger motor removed (para. 4-199).

a. Removal

1. Remove six screws (4) and lockwashers (3) from crane swinger gearcase (1) and turntable (2). Discard lockwashers (3).
2. Remove gearcase (1) from turntable (2).

b. Installation

Install gearcase (1) on turntable (2) with six new lockwashers (3) and screws (4).

4-200. CRANE SWINGER GEARCASE REPLACEMENT (Contd)

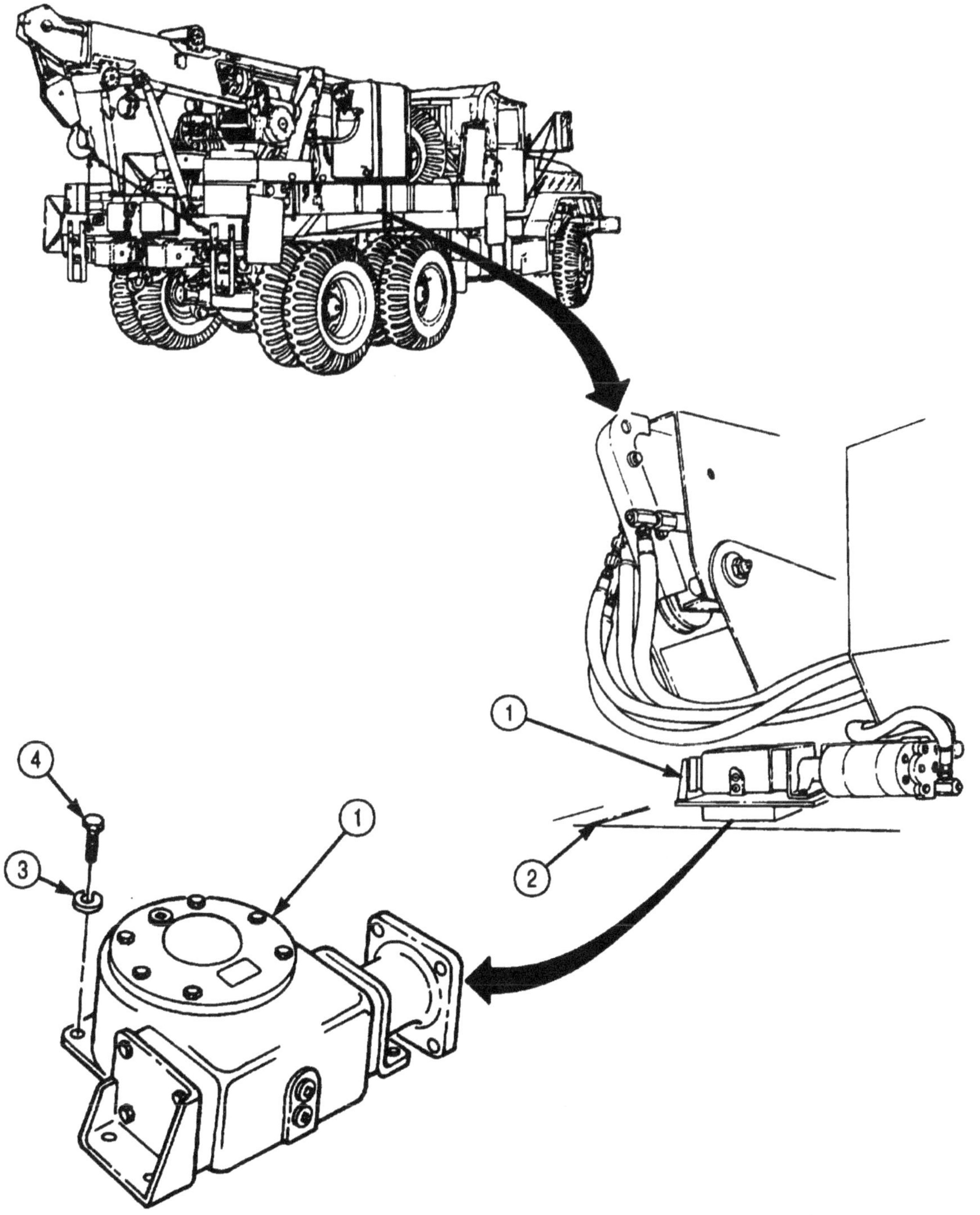

FOLLOW-ON TASK: Install crane hydraulic swinger motor (para. 4-199).

4-201. CRANE CONTROL VALVE REPLACEMENT

THIS TASK COVERS:

a. Removal

b. Installation

INITIAL SETUP:

APPLICABLE MODELS
M936/A1/A2

TOOLS
General mechanic's tool kit (Appendix E, Item 1)
Torque wrench (Appendix E, Item 144)

MATERIALS/PARTS
Four locknuts (Appendix D, Item 285)
Cap and plugs (Appendix C, Item 14)

REFERENCES (TM)
LO 9-2320-272-12
TM 9-2320-272-10
TM 9-2320-272-24P

EQUIPMENT CONDITION
- Parking brake set (TM 9-2320-272-10).
- Hydraulic oil reservoir drained (LO 9-2320-272-12).

a. Removal

1. Remove five screws (1) and control valve cover (2) from gondola (3).

CAUTION

When disconnecting hydraulic lines, plug all openings to prevent dirt from entering and causing internal damage to parts.

NOTE

- Have drainage container ready to catch oil.
- Tag hydraulic lines for installation.

2. Disconnect hydraulic oil reservoir crossover tube (10) from control valve elbow (11).
3. Remove control valve elbow (11) from control valve adapter (4).
4. Disconnect two hydraulic swinger motor flex lines (17) from control valve adapter fittings (16).
5. Disconnect two hydraulic crowd cylinder flex lines (18) from control valve adapter fittings (15).

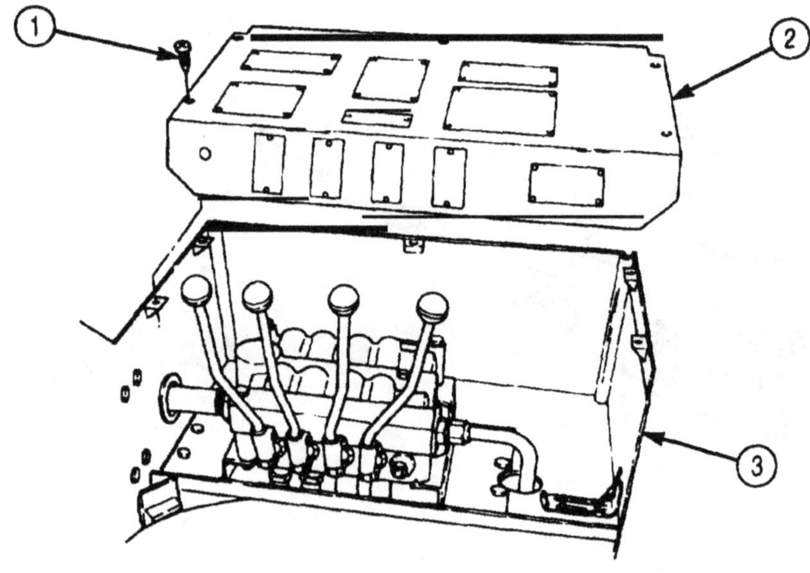

4-201. CRANE CONTROL VALVE REPLACEMENT (Contd)

6. Disconnect two hydraulic hoist motor flex lines (19) from control valve adapter fittings (14).
7. Disconnect hydraulic elevating cylinder crossover tube (9) from control valve adapter fitting (5).
8. Remove control valve tube (7) from hydraulic swivel valve flex line (8) and hydraulic control valve adapter (6).
9. Remove four locknuts (20), screws (12), and control valve (13) from gondola (3). Discard locknuts (20).

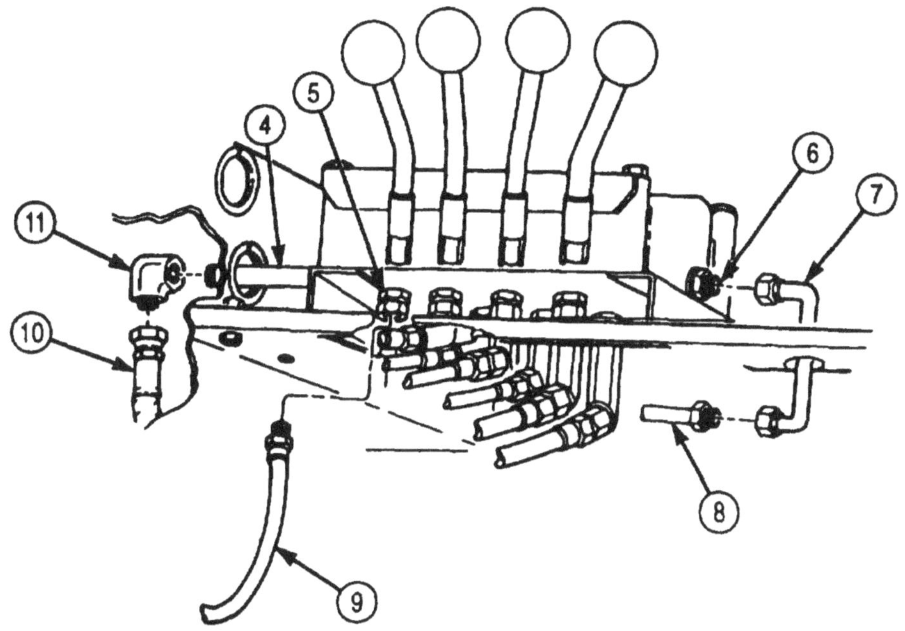

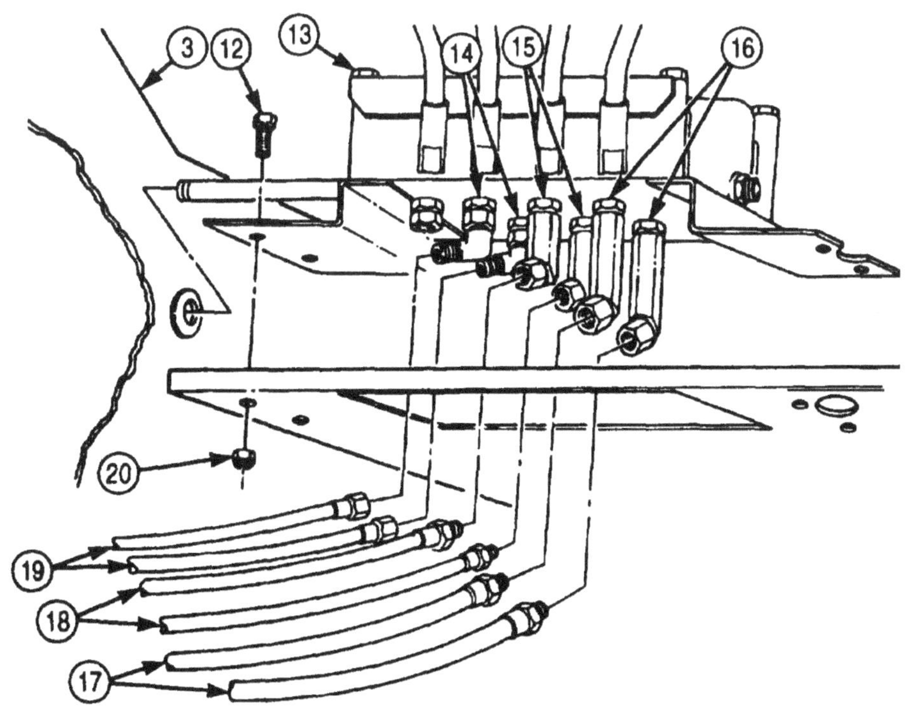

4-201. CRANE CONTROL VALVE REPLACEMENT (Contd)

b. Installation

NOTE

- If new crane control valve is being installed, fittings from old crane control valve may be used. Two side control plates and base may be used. Fittings must be cleaned and inspected for cracks and stripped threads.
- During installation, make sure all hydraulic lines are connected at their marked locations.

1. Install control valve (1) on gondola (10) with four screws (9) and new locknuts (8). Tighten screws (9) 44-61 lb-ft (60-83 N•m).
2. Install two hydraulic hoist motor flex lines (7) on control valve adapter fittings (2).
3. Install two hydraulic crowd cylinder flex lines (6) on control valve adapter fittings (3).
4. Install two hydraulic swinger motor flex lines (5) on control valve adapter fittings (4).
5. Install control valve tube (14) on hydraulic swivel valve flex line (15) and hydraulic control valve adapter (13).
6. Install hydraulic elevating cylinder crossover tube (16) on control valve adapter (12).
7. Install control valve elbow (18) on hydraulic control valve adapter (11).
8. Install hydraulic oil reservoir crossover tube (17) on control valve elbow (18).
9. Install control valve cover (20) on gondola (10) with five screws (19).

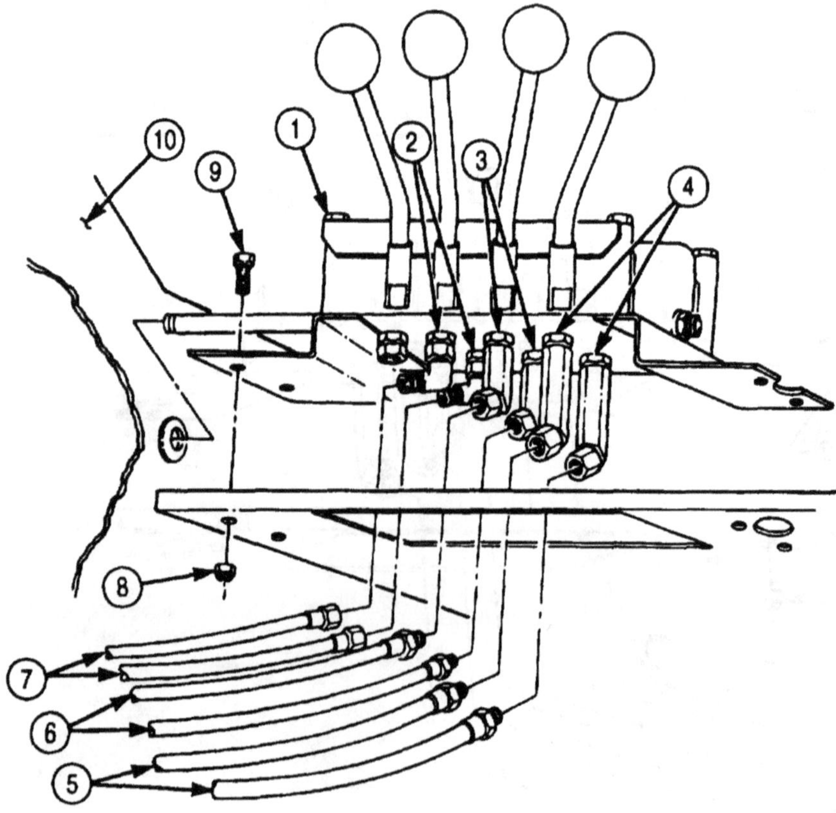

4-201. CRANE CONTROL VALVE REPLACEMENT (Contd)

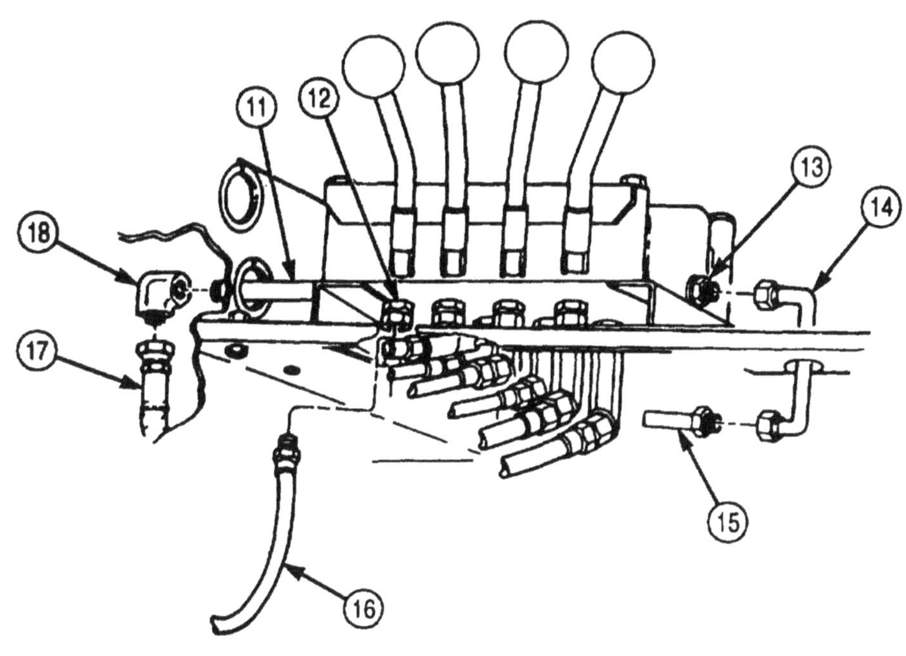

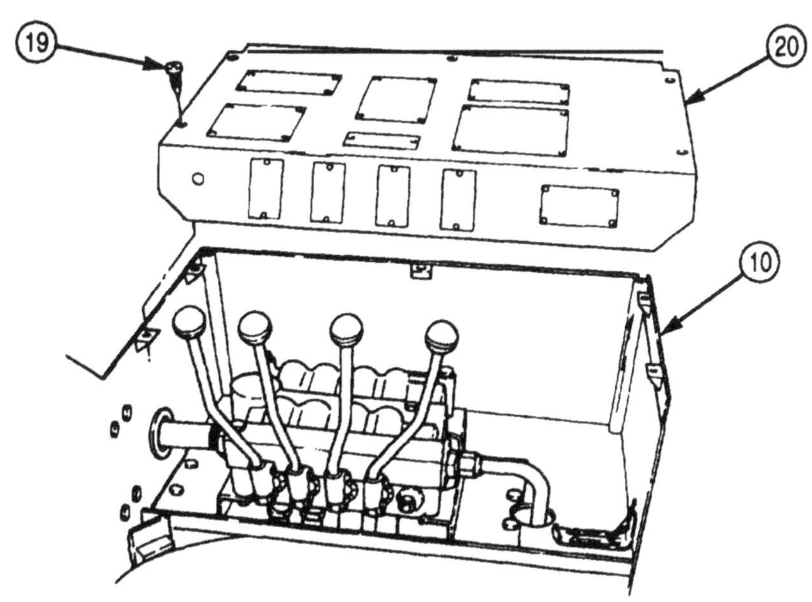

FOLLOW-ON TASKS:
- Fill hydraulic oil reservoir to proper oil level (LO 9-2320-272-12).
- Check pressure relief valve adjustment (para. 3-389).

4-202. GONDOLA REPLACEMENT

THIS TASK COVERS:

a Removal

b. Installation

INITIAL SETUP:

APPLICABLE MODELS

M936/A1/A2

TOOLS

General mechanic's tool kit (Appendix E, Item 1)
Torque wrench (Appendix E, Item 144)
Lifting device
Chains

MATERIALS/PARTS

Three locknuts (Appendix D, Item 291)
Twelve locknuts (Appendix D, Item 285)

REFERENCES (TM)

TM 9-2320-272-10
TM 9-2320-272-24P

EQUIPMENT CONDITION

- Parking brake set (TM 9-2320-272-10)
- Crane control valve removed (para. 4-201).
- Floodlight assembly removed (para. 3-121).

GENERAL SAFETY INSTRUCTIONS

All personnel must stand clear during lifting operations.

WARNING

All personnel must stand clear during lifting operations. A snapped cable, or shifting or swinging load, may cause injury to personnel.

a. Removal

1. Remove hydraulic tube (2) from elevating cylinder crossover tee (8) and control valve flex line (3).

NOTE

Tag all lines for installation.

2. Remove three locknuts (6), clamps (5), and six hydraulic flex lines (7) from gondola (1). Discard locknuts (6).
3. Remove hydraulic swivel valve flex line (4) from gondola (1).
4. Remove gondola guard (9) from gondola (1).
5. Attach two utility chains to gondola (1).
6. Attach hoist hook to utility chains and remove slack.

CAUTION

Ensure all hydraulic lines and wires are fastened clear of gondola to avoid snagging during removal.

NOTE

Assistant will help with step 7.

7. Remove twelve locknuts (11) and screws (12) from turntable side plate (10) and lift gondola (1) away from side plate (10). Discard locknuts (11).
8. Remove lifting device and two utility chains from gondola (1).

4-202. GONDOLA REPLACEMENT (Contd)

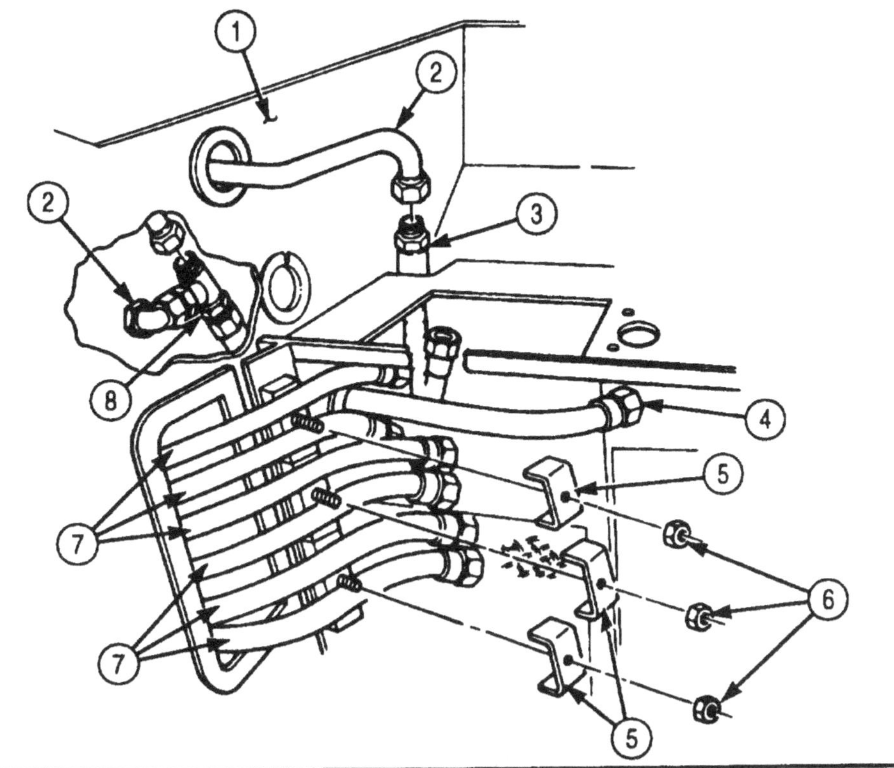

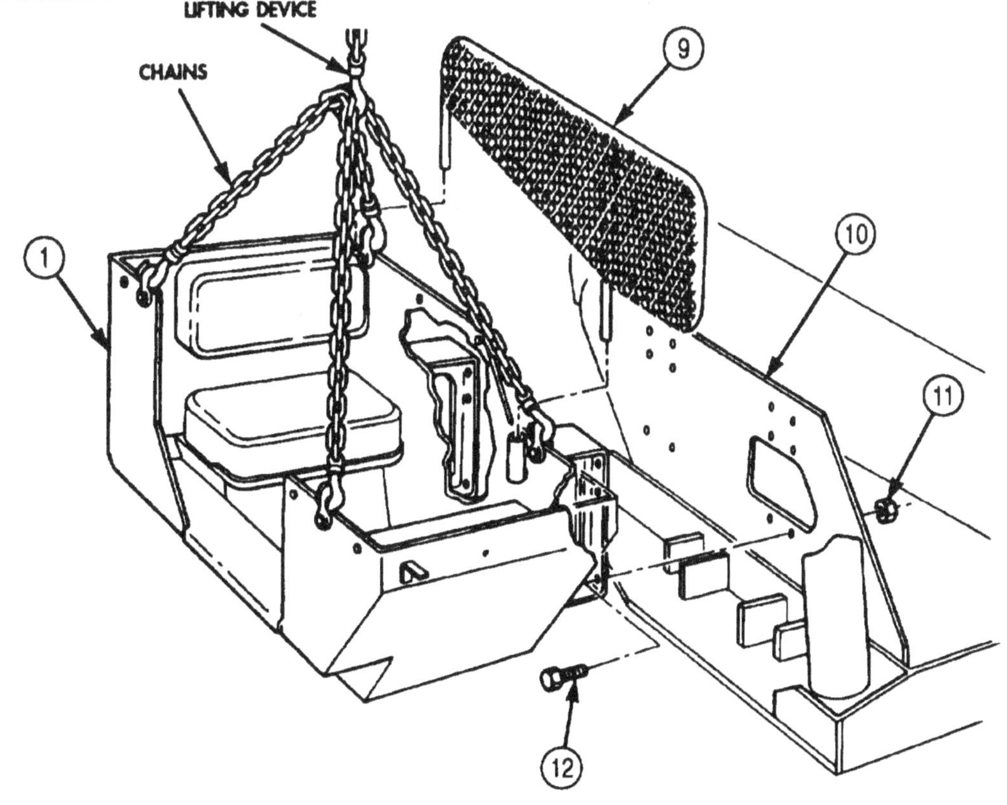

4-202. GONDOLA REPLACEMENT (Contd)

b. Installation

1. Attach two utility chains to gondola (5).
2. Attach lifting device to two utility chains.

NOTE

- Assistant will help with step 3.
- When positioning gondola, guide all hydraulic lines through opening.

3. Lift gondola (5) to turntable side plate (2) and install with twelve screws (4) and new locknuts (3). Tighten screws (4) 44-61 lb-ft (60-83 N•m).
4. Remove two utility chains and hoist hook from gondola (5).
5. Install gondola guard (1) on gondola (5).
6. Guide hydraulic swivel valve flex line (8) through hole in gondola (5).
7. Install six hydraulic flex lines (11) on gondola (5) with three clamps (9) and new locknuts (10).
8. Install hydraulic tube (6) on elevating cylinder crossover tee (12) and control valve flex line (7).

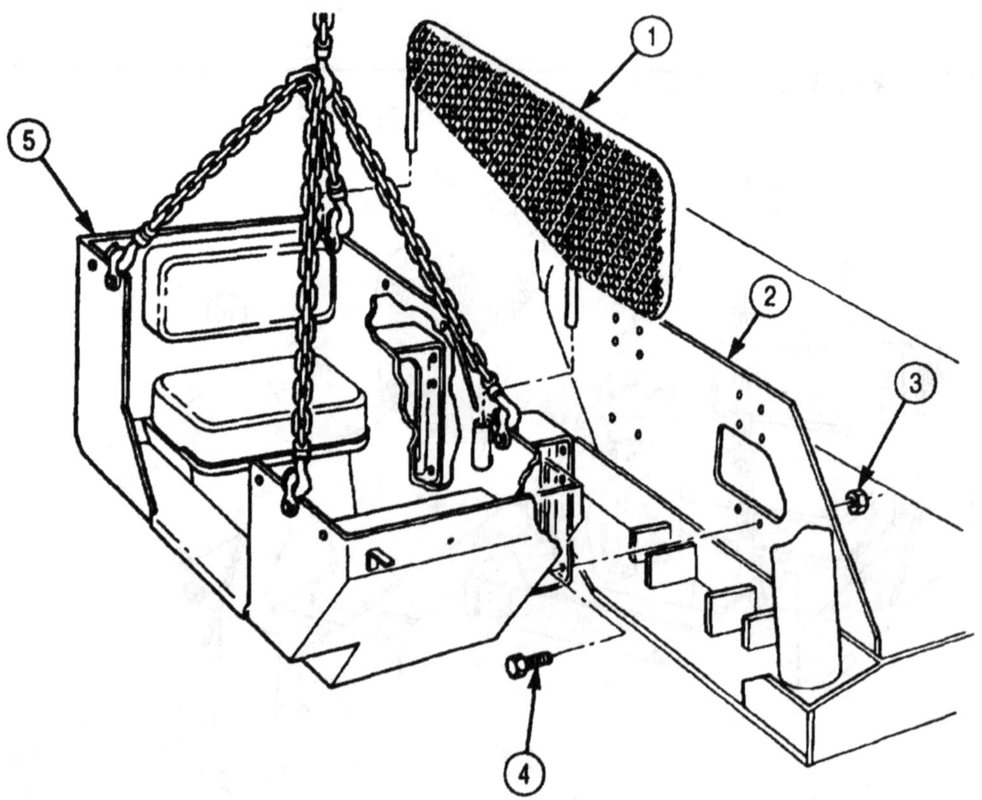

4-202. GONDOLA REPLACEMENT (Contd)

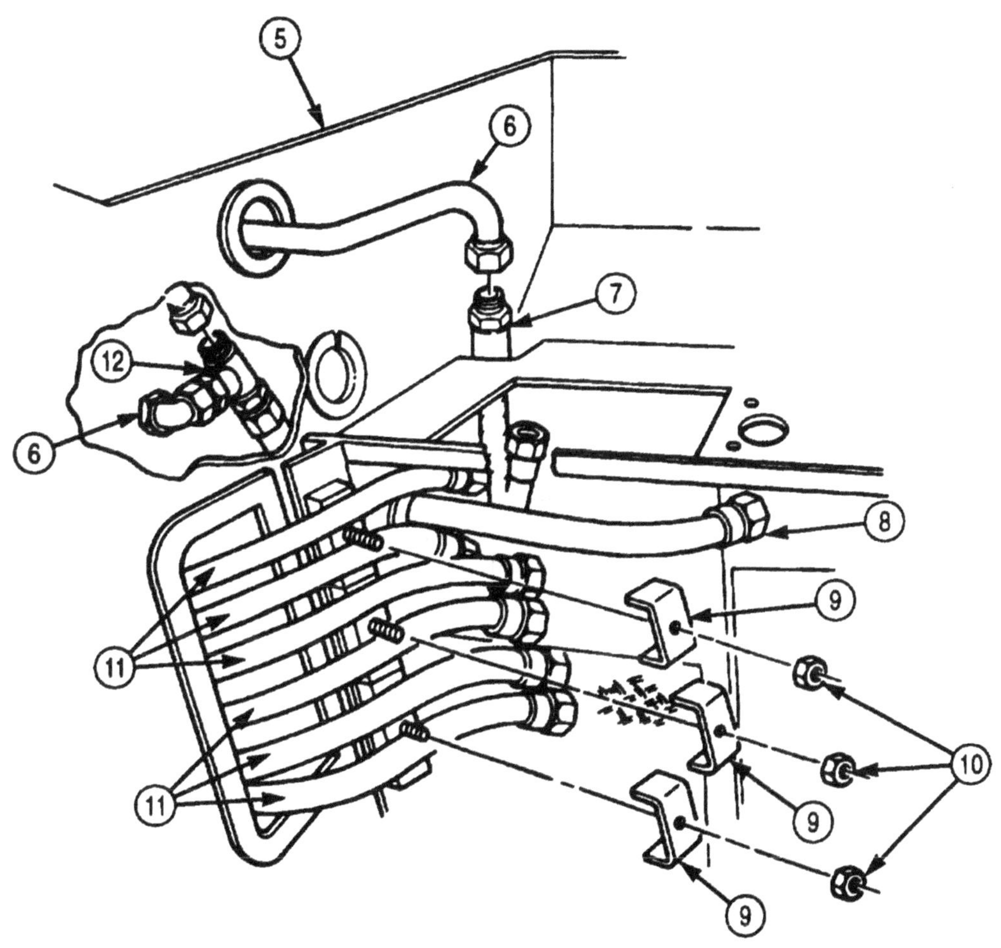

FOLLOW-ON TASKS: • Install crane control valve (para. 4-201).
• Install floodlight assembly (para. 3-121).

4-203. TURNTABLE REPAIR

THIS TASK COVERS:

a. Removal
b. Disassembly
c. Cleaning, Inspection, and Repair
d. Assembly
e. Installation

INITIAL SETUP:

APPLICABLE MODELS
M936/A1/A2

Tools
General mechanic's tool kit (Appendix E, Item 1)
Torque wrench (Appendix E, Item 145)
Lifting device
Chains

MATERIALS/PARTS
Eighteen locknuts (Appendix D, Item 316)
Drycleaning solvent (Appendix C, Item 71)

REFERENCES (TM)
LO 9-2320-272-12
TM 9-2320-272-10
TM 9-2320-272-24P
TM 9-237

EQUIPMENT CONDITION
- Parking brake set (TM 9-2320-272-10).
- Hydraulic oil reservoir removed (para. 4-188).
- Boom elevating cylinders removed (para. 4-1931.
- Hydraulic swivel valve removed (para. 4-198).
- Crane swinger gearcase removed (para. 4-200).
- Gondola removed (para. 4-202).

GENERAL SAFETY INSTRUCTIONS
- All personnel must stand clear during lifting operations.
- Keep fire extinguisher nearby when using drycleaning solvent.
- Drycleaning solvent is flammable and toxic. Do not use near an open flame.

WARNING

All personnel must stand clear during lifting operations. A snapped cable, or shifting or swinging load, may cause injury to personnel.

a. Removal

1. Remove eighteen locknuts (3), screws (5), two turntable side plates (1), and crossover tubes (2) from turntable base plate (4). Discard locknuts (3).
2. Remove eighteen screws (6) from turntable gear and bearing (7) and crane body (8) by rotating turntable base plate (4) so access hole exposes each screw (6).

NOTE

Assistant will help with steps 3 and 4.

3. Attach two utility chains to turntable base plate (4).
4. Attach lifting device to chains and lift turntable base plate (4) away from crane body (8).
5. Remove two utility chains and lifting device from turntable base plate (4).

b. Disassembly

Remove eighteen screws (9) and turntable gear and bearing assembly (7) from underside of turntable base plate (4).

4-203. TURNTABLE REPAIR (Contd)

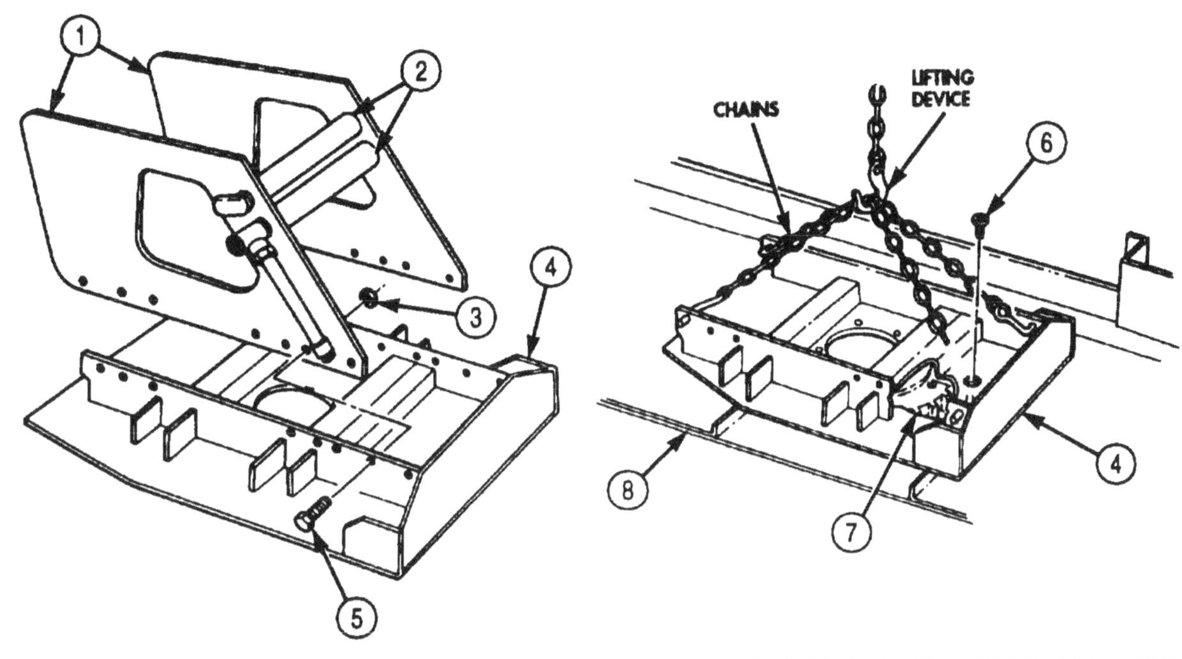

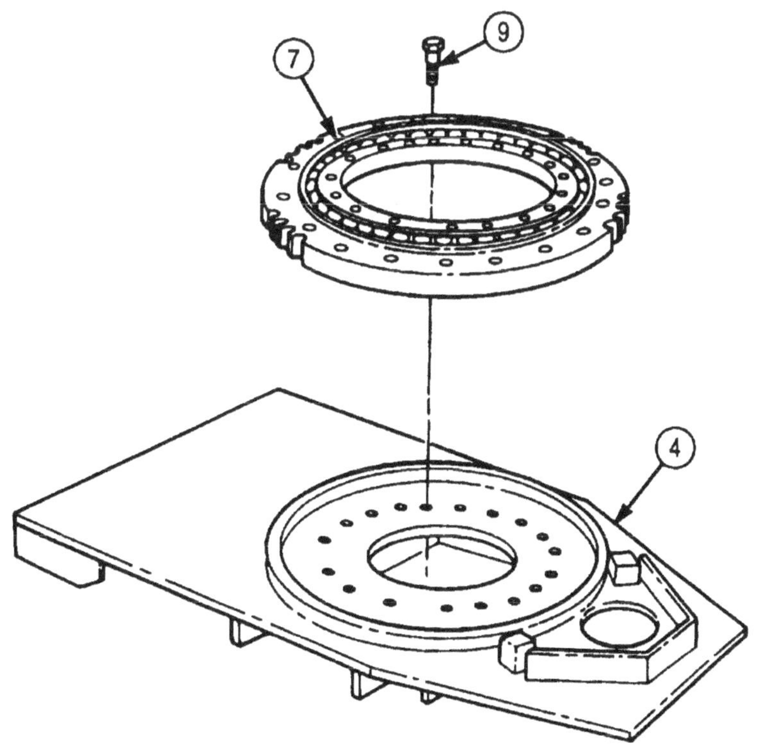

4-203. TURNTABLE REPAIR (Contd)

c. Cleaning, Inspection, and Repair

WARNING

Drycleaning solvent is flammable and toxic. Do not use near open flame and always have a fire extinguisher nearby when solvents are used. Use only in well-ventilated places, wear protective clothing, and dispose of cleaning rags in approved container. Failure to do this may result in injury to personnel and/or damage to equipment.

1. Clean turntable base plate (3) and gear and bearing assembly (1) with drycleaning solvent.
2. Inspect turntable base plate (3) for cracks and breaks. If cracked or broken, refer to TM 9-237.
3. Inspect turntable gear and bearing assembly (1) for cracks, pits, scores, and broken teeth. If cracked, pitted, scored, or broken teeth are evident, replace turntable gear and bearing assembly (1).

d. Assembly

Install turntable gear and bearing assembly (1) on underside of turntable base plate (3) with eighteen screws (2). Tighten screws (2) 170-200 lb-ft (231-271 N•m).

e. Installation

WARNING

All personnel must stand clear during lifting operations. A snapped cable, or shifting or swinging load, may cause injury to personnel.

NOTE

Assistant will help with steps 1 and 2.

1. Attach two utility chains to turntable base plate (3).
2. Attach lifting device to two utility chains and lift turntable base plate (3) onto crane body (5).
3. Remove two utility chains and hoist hook from turntable base plate (3).
4. Rotate turntable base plate (3) on gear bearing assembly (1) to expose each screw hole in turntable base plate (3) and install on crane body (5) with eighteen screws (4). Tighten screws (4) 170-200 lb-ft (231-271 N•m).
5. Install turntable side plates (6) and crossover tubes (7) on turntable base plate (3) with eighteen screws (9) and new locknuts (8). Tighten screws (9) 170-200 lb-ft (231-271 N•m).

4-203. TURNTABLE REPAIR (Contd)

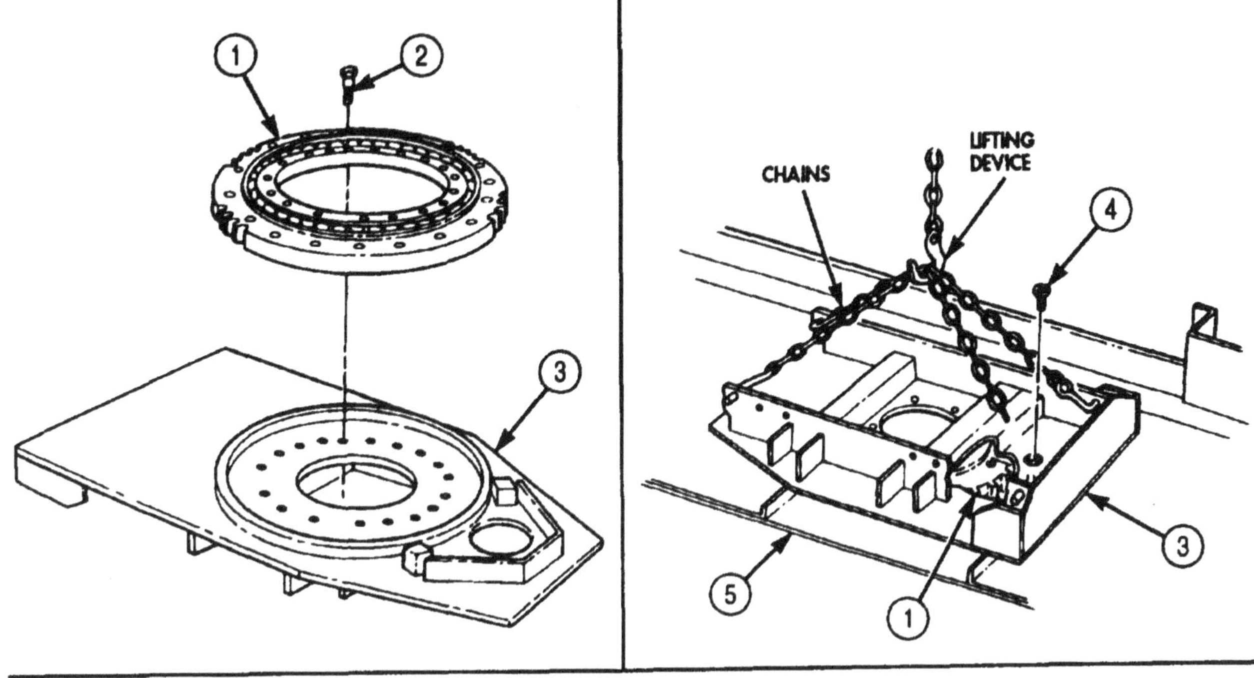

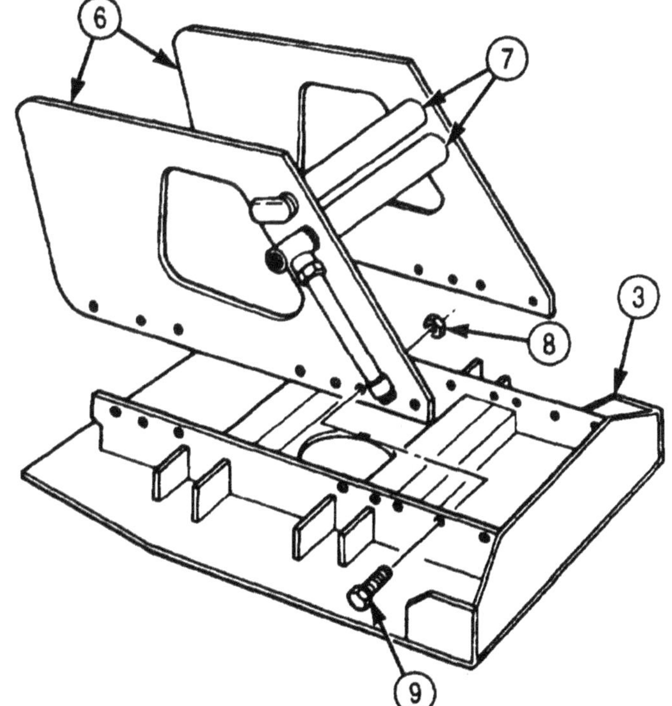

FOLLOW-ON TASKS: • Install gondola (para. 4-202).
- Install crane swinger gearcase (para. 4-200).
- Install hydraulic swivel valve (para. 4-198).
- Install boom elevating cylinders (para. 4-193).
- Lubricate gear bearing (LO 9-2320-272-12).
- Install hydraulic oil reservoir (para. 4-188).

4-204. BOOM REPAIR AND ADJUSTMENT

THIS TASK COVERS:

a. Disassembly
b. Cleaning, Inspection, and Repair
c. Assembly
d. Vertical Adjustment

INITIAL SETUP:

APPLICABLE MODELS
M936/A1/A2

TOOLS
General mechanic's tool kit (Appendix E, Item 1)
Arbor press
Mandrel

MATERIALS/PARTS
Cotter pins (Appendix D, Item 68)
Cotter pin (Appendix D, Item 77)
Two locknuts (Appendix D, Item 311)
Twenty-four lockwashers (Appendix D, Item 377)
GAA grease (Appendix C, Item 28)
Drycleaning solvent (Appendix C, Item 71)

REFERENCES (TM)
LO 9-2320-272-12
TM 9-2320-272-24P
TM 9-237

EQUIPMENT CONDITION
- Hoist winch removed (para. 4-192).
- Inner boom removed (para. 4-195).
- Extension cylinder removed (para. 4-196).
- Boom rollers removed (para. 4-197).

GENERAL SAFETY INSTRUCTIONS
- Keep fire extinguisher nearby when using drycleaning solvent.
- Drycleaning solvent is flammable and toxic. Do not use near an open flame.

a. Disassembly

1. Remove cotter pin (10), crane sheave pin (13), and crane sheave (12) from outer boom (14). Discard cotter pin (10).
2. With arbor press and mandrel, remove two bearings (11) from crane sheave (12)
3. Remove two locknuts (5), screws (17), and spacers (4) from inner boom (9). Discard locknuts (5).
4. Remove cotter pin (6), cable sheave pin (15), and two cable sheaves (1) from inner boom (9). Discard cotter pin (6).
5. Using arbor press and mandrel, remove two sheave bushings (2), thrust washers (3), and spacer (18) from two cable sheaves (1).
6. Remove one hundred and four screws (7) and four boom tracks (8) from inner boom (9).
7. Remove grease fitting (16) from cable sheave pin (15).

b. Cleaning, Inspection, and Repair

WARNING

Drycleaning solvent is flammable and toxic. Do not use near open flame and always have a fire extinguisher nearby when solvents are used. Use only in well-ventilated places, wear protective clothing, and dispose of cleaning rags in approved container. Failure to do this may result in injury or death to personnel and/or damage to equipment.

1. Clean all boom components with drycleaning solvent.
2. Inspect inner boom (9) and outer boom (14) for bends and cracked or broken welds. If welds are cracked or broken, refer to TM 9-237. If bent, replace.
3. Inspect four boom tracks (8) for cracks, breaks, and bends. Replace if cracked, bent, or broken.
4. Inspect cable sheave pins (13) and (15) for breaks or out-of-round condition. Replace if broken or out-of-round.

4-204. BOOM REPAIR AND ADJUSTMENT (Contd)

5. Inspect two cable sheaves (l), cable sheave bushings (2), crane sheave (12), and crane sheave bearings (11) for breaks and out-of-round condition. Replace if broken or out-of-round.

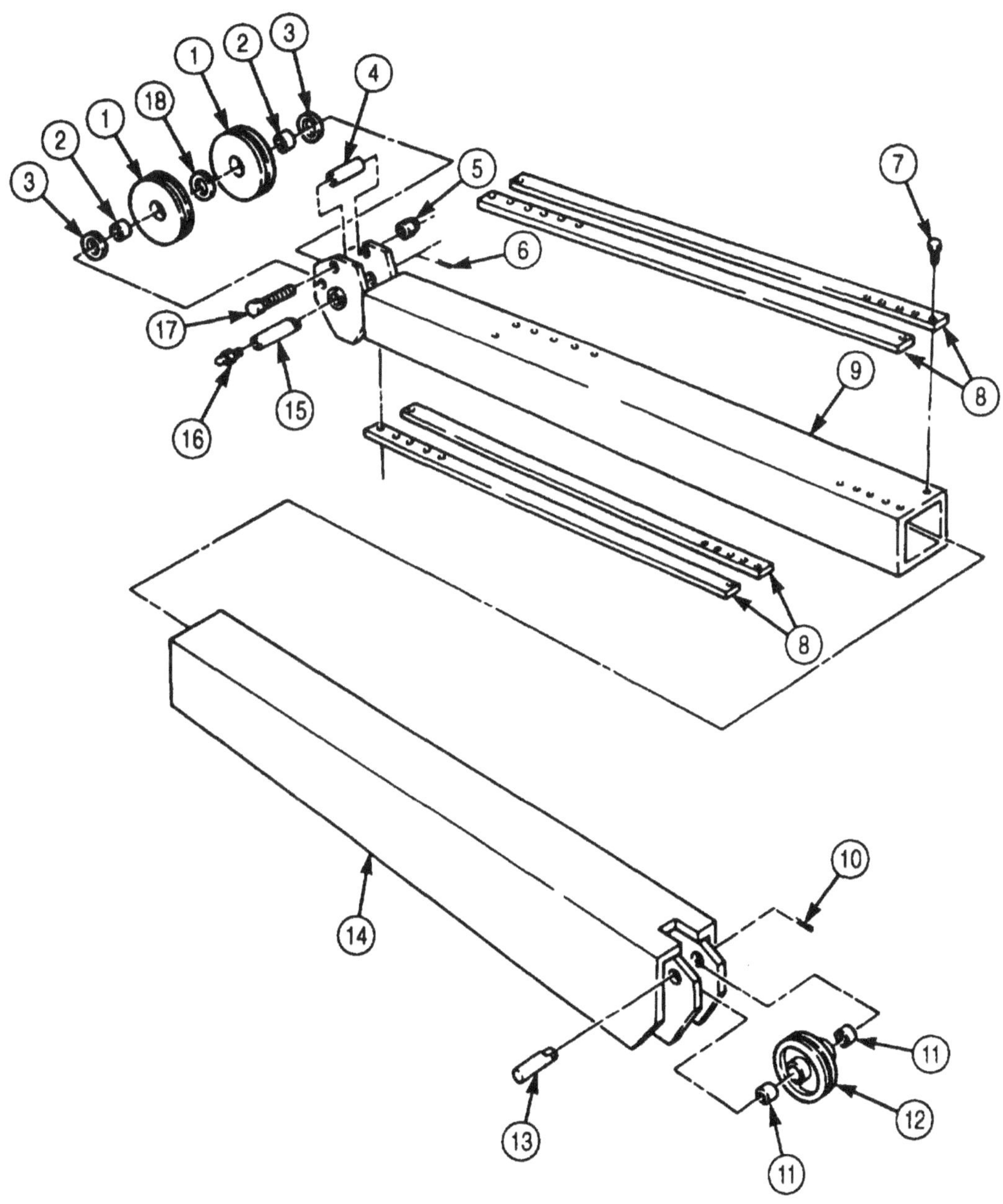

4-204. BOOM REPAIR AND ADJUSTMENT (Contd)

c. Assembly

1. Install four boom tracks (8) to inner boom (9) with one hundred and four screws (7) and apply coat of GAA grease to four boom tracks (8).
2. Using arbor press and mandrel, install two cable sheave bushings (2) into two cable sheaves (1).
3. Align two cable sheave thrust washers (3) and spacer (18) with holes in inner boom (9) and install with cable sheave pin (15) and new cotter pin (6).
4. Install grease fitting (16) in cable sheave pin (15).
5. Position two spacers (4) to inner boom (9) and install with two screws (17) and new locknuts (6).
6. Using arbor press and mandrel, install two crane sheave bearings (11) into crane sheave (12).
7. Align crane sheave (12) with holes in outer boom (14) and install with crane sheave pin (13) and new cotter pin (10).
8. Install boom rollers (para. 4-197).

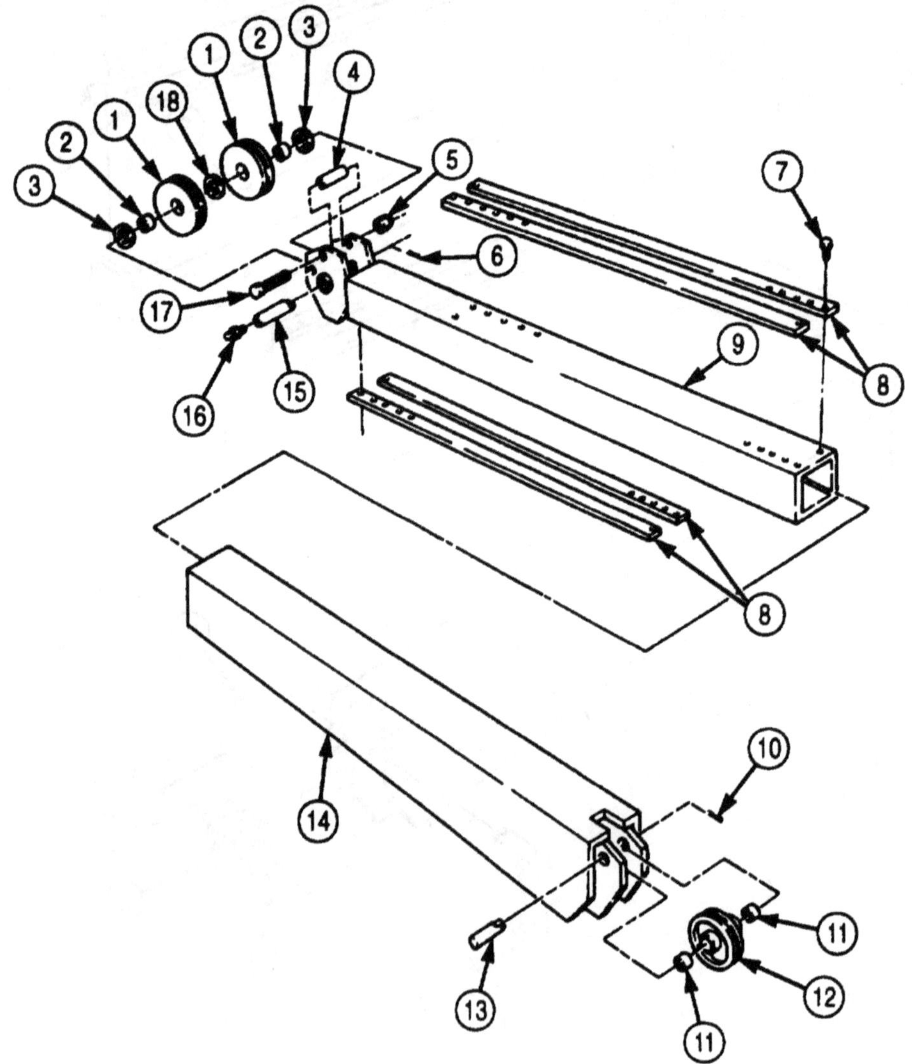

4-204. BOOM REPAIR AND ADJUSTMENT (Contd)

9. Install extension cylinder (para. 4-196).
10. Install inner boom (para. 4-195).
11. Install hoist winch (para. 4-192).

d. Vertical Adjustment

NOTE

Vertical adjustment can be performed with boom assembly either installed or removed from wrecker body.

1. Remove sixteen screws (21) and lockwashers (22) connecting four mounting caps (23) to lower boom roller assembly (20) and upper boom roller assembly (19). Discard lockwashers (22).
2. Turn two mounting caps (23) so top edge of inner boom (9) and top edge of outer boom (14) are approximately parallel.
3. Install two mounting caps (23) on lower boom roller assembly (20) with eight new lockwashers (22) and screws (21).
4. Manually extend and retract inner boom (9) to verify that lower edge clears outer boom (14).

NOTE

Perform steps 5 and 6 only if inner boom catches on upper boom roller assembly after adjustment is checked.

5. Remove eight screws (21) and lockwashers (22) from two mounting caps (23) and upper roller assembly (19) and turn mounting caps (23) one additional screw hole. Discard lockwashers (22).
6. Install two mounting caps (23) on upper roller (19) with eight new lockwashers (22) and screws (21).

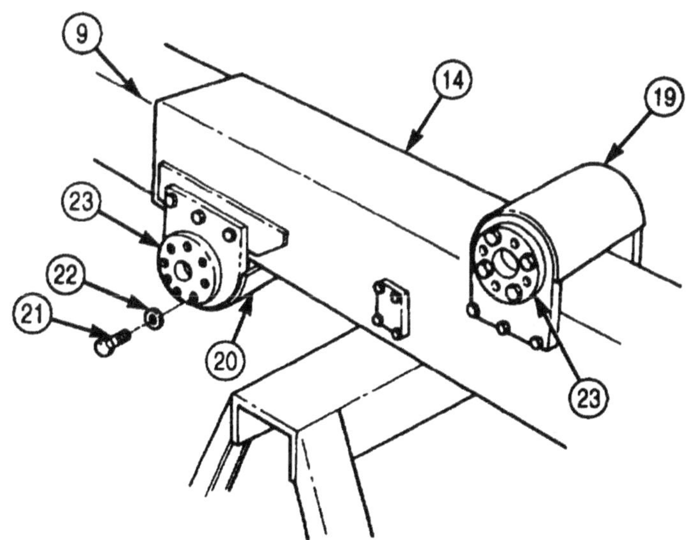

FOLLOW-ON TASK: Lubricate inner and outer booms (LO 9-2320-272-12).

4-205. TRANSFER CASE POWER TAKEOFF (PTO) CONTROL CABLE REPLACEMENT

THIS TASK COVERS:

a. Removal **b. Installation**

INITIAL SETUP:

APPLICABLE MODELS
M936/A1

TOOK
General mechanic's tool kit (Appendix E, Item 1)

MATERIALS/PARTS
Two locknuts (Appendix D, Item 291)
Locknut (Appendix D, Item 299)
Two locknuts (Appendix D, Item 313)
Cotter pin (Appendix D, Item 70)
Tiedown straps (Appendix D, Item 689)

REFERENCES (TM)
TM 9-2320-272-10
TM 9-2320-272-24P

EQUIPMENT CONDITION
Parking brake set (TM 9-2320-272-10).

---

a. Removal

1. Remove tiedown straps from Power Takeoff (PTO) cable (4) as required. Discard tiedown straps.
2. Remove cotter pin (10), clevis pin (2), and PTO cable clevis (3) from PTO select lever (1). Discard cotter pin (10).
3. Remove two locknuts (9), screws (6), clamp (5), PTO cable (4), and shim (7) from PTO cable bracket (8). Discard locknuts (9).
4. Remove PTC clevis (3) from PTO cable (4).
5. Remove two locknuts (14), screws (11), clamp (15), PTO cable (4), and shim (13) from PTO cable bracket (12). Discard locknuts (14).
6. Remove locknut (19) and PTO cable swivel bolt (17) from select lever (18). Discard locknut (19).
7. Remove swivel bolt (17) and cable clevis (16) from PTO cable (4).

b. Installation

1. Install swivel bolt (17) and cable end (16) on PTO cable (4).
2. Install PTO cable swivel bolt (17) on select lever (18) with new locknut (19).
3. Install shim (13), PTO cable (4), and clamp (15) on PTO cable bracket (12) with two screws (11) and new locknuts (14).
4. Install PTO clevis (3) on PTO cable (4).
5. Install shim (7), PTO cable (4), and clamp (5) on PTO cable bracket (8) with two screws (6) and new locknuts (9).

NOTE

Transfer case PTO cable can be adjusted at two points. Adjustments can be made at transfer case select lever cable clevis or transfer case PTO select lever cable collar.

6. Adjust PTO cable clevis (16) forward until aligned with transfer PTO select lever (18). If necessary, reposition transfer PTO cable (4) at PTO cable collar at cross-shaft end of cable (4).
7. Install PTO clevis (3) on PTO select lever (1) with clevis pin (2) and new cotter pin (10).
8. Install tiedown straps on PTO cable (4) as required.

4-205. TRANSFER CASE POWER TAKEOFF (PTO) CONTROL CABLE REPLACEMENT (Contd)

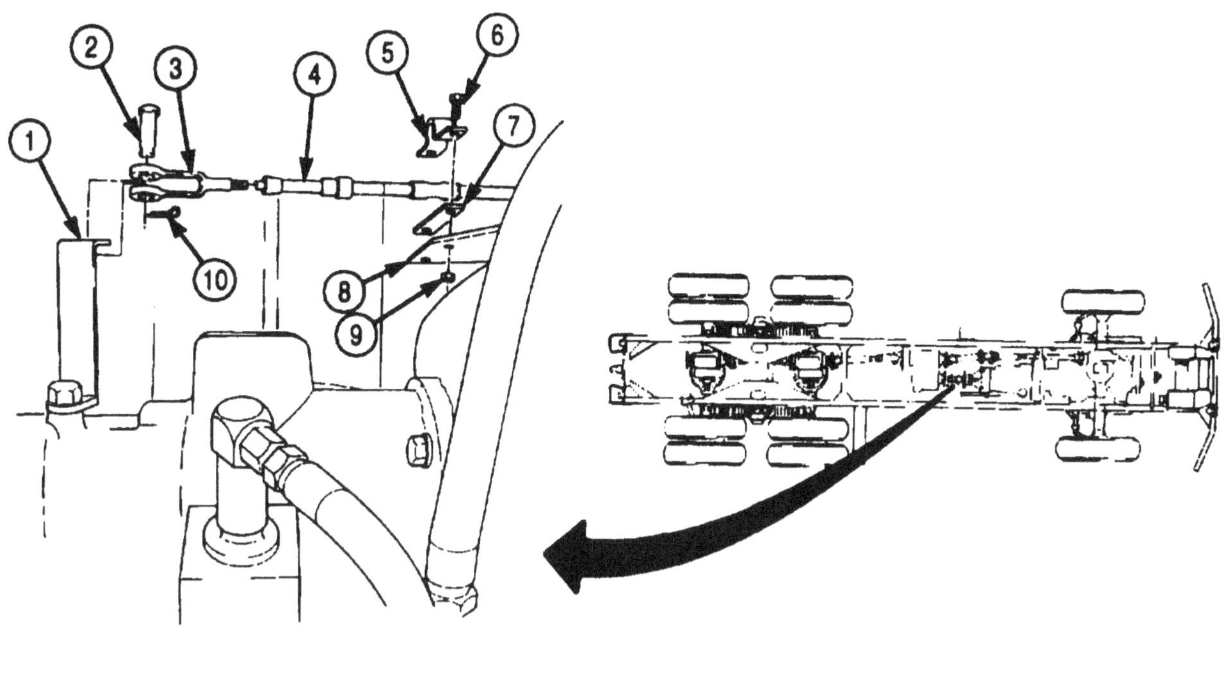

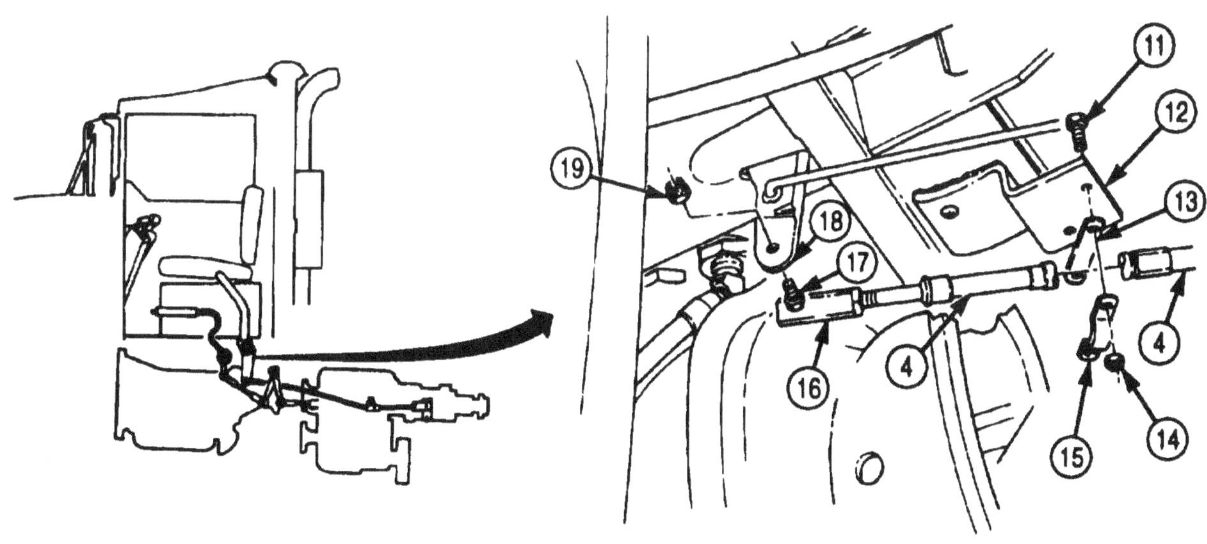

FOLLOW-ON TASK: Check transfer case PTO for proper operation (TM 9-2320-272-10).

4-206. TRANSFER CASE POWER TAKEOFF (PTO) CROSS SHAFT (CONTROL LEVER) REPLACEMENT

THIS TASK COVERS:

a. Removal

b. Installation

INITIAL SETUP:

APPLICABLE MODELS

M936/A1/A2

TOOLS

General mechanic's tool kit (Appendix E, Item 1)

MATERIALS/PARTS

Five locknuts (Appendix D, Item 291)
Six locknuts (Appendix D, Item 307)
Locknut (Appendix D, Item 294)
Cotter pin (Appendix D, Item 70)
Woodruff key (Appendix D, Item 728)
Two locknuts (Appendix D, Item 313)
Locknut (Appendix D, Item 299)
GAA grease (Appendix C, Item 28)

REFERENCES (TM)

TM 9-2320-272-10
TM 9-2320-272-24P

EQUIPMENT CONDITION

- Parking brake set (TM 9-2320-272-10).
- Driver's seat removed (para. 3-283).

a. Removal

1. Remove locknut (21) and PTO cable swivel bolt (22) from cross shaft lever (2). Discard locknut (21).
2. Remove two locknuts (19), screws (8), clamp (20), PTO control cable (23), and shim (10) from bracket (9). Discard locknuts (19).

NOTE

Assistant will help with step 3.

3. Remove four locknuts (18), screws (5), two cross shaft retaining clamps (6), and shims (7) from cab floor (26). Discard locknuts (18).
4. Remove grease fittings (4) from cross shaft retaining clamps (6).
5. Remove cotter pin (29), washer (31), and brake lock control valve rod (24) from cross shaft lever (2). Discard cotter pin (29).
6. Remove locknut (3) and screw (1) from cross shaft lever (2). Discard locknut (3).

NOTE

Assistant will help with step 7.

7. Remove six locknuts (25), screws (30), ring seal (28), rubber seal (27), and cross shaft lever (2) from cab floor (26). Discard locknuts (25).
8. Remove woodruff key (17) and two cross shaft retaining clamps (6) from cross shaft (11). Discard woodruff key (17).
9. Remove locknut (13), washer (14), helical spring (15), and PTO lever pin (16) from PTO lever (12). Discard locknut (13).

b. Installation

1. Install PIO lever pin (16), helical spring (15), and washer (14) on PTO lever (12) with new locknut (13).
2. Install two cross shaft retaining clamps (6) on cross shaft (11).
3. Install grease fittings (4) on two cross shaft retaining clamps (6).

4-206. TRANSFER CASE POWER TAKEOFF (PTO) CROSS SHAFT (CONTROL LEVER) REPLACEMENT (Contd)

4. Apply a small amount of GAA grease to rounded edge of new woodruff key (17) and install on cross shaft (11). Grease will hold woodruff key (17) in place.
5. Insert cross shaft (11) into cross shaft lever (2) and install with screw (1) and new locknut (3).

NOTE

Assistant will help with step 6 through 8.

6. Install rubber seal (27) and ring seal (28) on cab floor (26) with six screws (30) and new locknuts (25).
7. Position cross shaft (11), cross shaft lever (2), and cross shaft retaining clamps (6) on cab floor (26).
8. Install two shims (7) and cross shaft retaining clamps (6) on cab floor (26) with four screws (5) and new locknuts (18).
9. Install brake lock control valve rod (24) on cross shaft lever (2) with washer (31) and new cotter pin (29).
10. Install PTO cable swivel bolt (22) on cross shaft lever (2) with new locknut (21).
11. Install shim (10) and PTO control cable (23) on bracket (9) with clamp (20), two screws (8), and new locknuts (19).

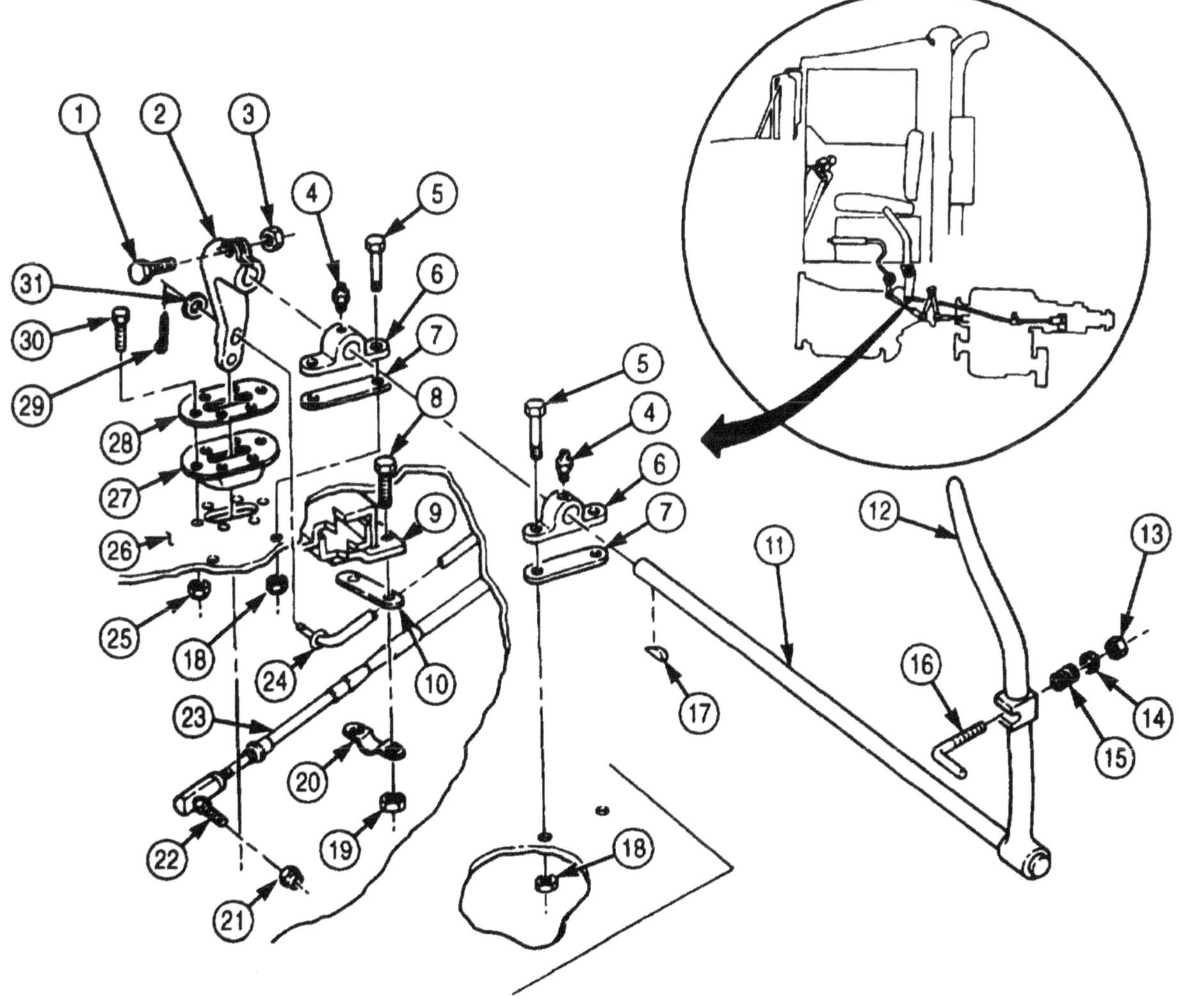

FOLLOW-ON TASKS: • Install driver's seat (para. 3-283).
• Check transfer case PTO for proper operation (TM 9-2320-272-10).

4-207. TRANSFER CASE POWER TAKEOFF (PTO) REPLACEMENT

THIS TASK COVERS:

a. Removal **b. Installation**

INITIAL SETUP

APPLICABLE MODELS

M936/A1/A2

TOOLS

General mechanic's tool hit (Appendix E, Item 1)
Torque wrench (Appendix E, Item 144)

MATERIALS/PARTS

Oil seal (Appendix D. Item 627)
Cotter pin (Appendix D, Item 70)
Adhesive sealant (Appendix C, Item 4)
Cap and plug set (Appendix C, Item 14)
Antiseize tape (Appendix C, Item 72)

REFERENCES (TM)

LO 9-2320-272-12
TM 9-2320-272-10
TM 9-2320-272-24P

EQUIPMENT CONDITION

Transfer case PTO-to-hydraulic pump propeller shaft removed (para. 3-334).

---

a. Removal

1. Remove cotter pin (1), washer (2), clevis pin (4), and clevis (3) from transfer case Power Takeoff (PTO) select lever (5). Discard cotter pin (1).

NOTE

Have drainage container ready to catch oil.

2. Disconnect transfer case oil return line (9) from adapter (10) and plug opening of transfer case oil return line (9).
3. Remove adapter (10) from PTO unit (11).
4. Remove six screws (6) and washers (7) from transfer case (8).
5. Remove PTO unit (11) from transfer case (8).
6. Remove setscrew (12) from PTO oil pump drive gear assembly (15).
7. Slide PTO oil pump drive gear assembly (15) off transfer case main input shaft (13).
8. Remove oil seal (14) from PTO oil pump drive gear assembly (15). Discard oil seal (14).

4-207. TRANSFER CASE POWER TAKEOFF (PTO) REPLACEMENT (Contd)

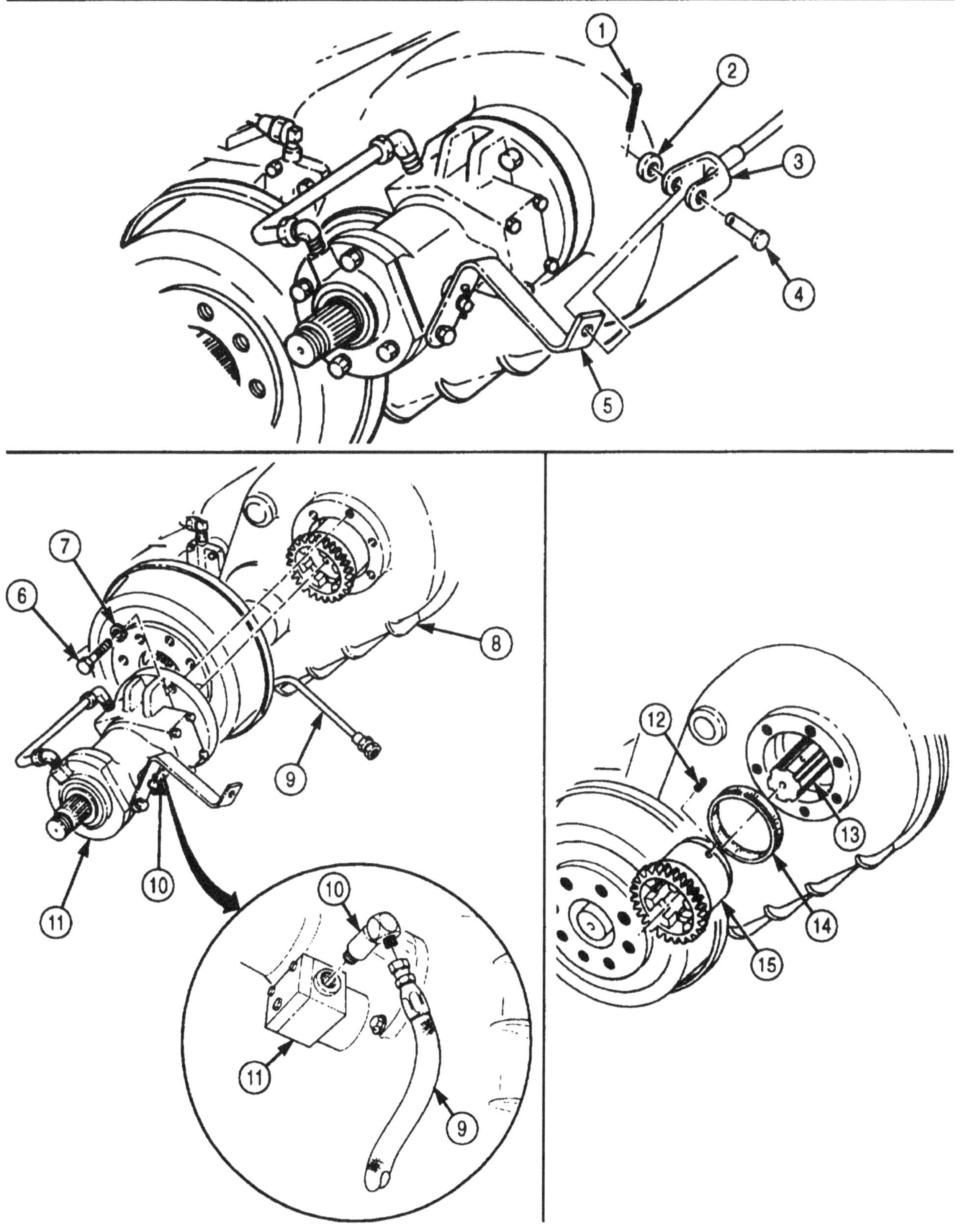

4-207. TRANSFER CASE POWER TAKEOFF (PTO) REPLACEMENT (Contd)

b. Installation

1. Install new oil seal (3) on PTO oil pump drive gear assembly (4).
2. Install PTO oil pump drive gear assembly (4) on transfer case main input shaft (2). Align set-screw (1) hole in gear assembly (4) with recess in main input shaft (2).
3. Install setscrew (1) in PTO oil pump drive gear assembly (4) and apply adhesive sealant to threads of setscrew (1). Tighten setscrew (1) 84-108 lb-in. (10-12 N•m).
4. Install PTO (10) on transfer case (7) with six washers (6) and screws (5). Tighten screws (5) 40-65 lb ft, (54-88 N•m).

NOTE

Wrap male pipe threads with antiseize tape prior to installation.

5. Install adapter (9) on PTO (10).
6. Connect transfer case oil return line (8) to adapter (9).
7. Install PTO clevis (13) on PTO select lever (15) with clevis pin (14), washer (12), and new cotter pin (11).

4-207. TRANSFER CASE POWER TAKEOFF (PTO) REPLACEMENT (Contd)

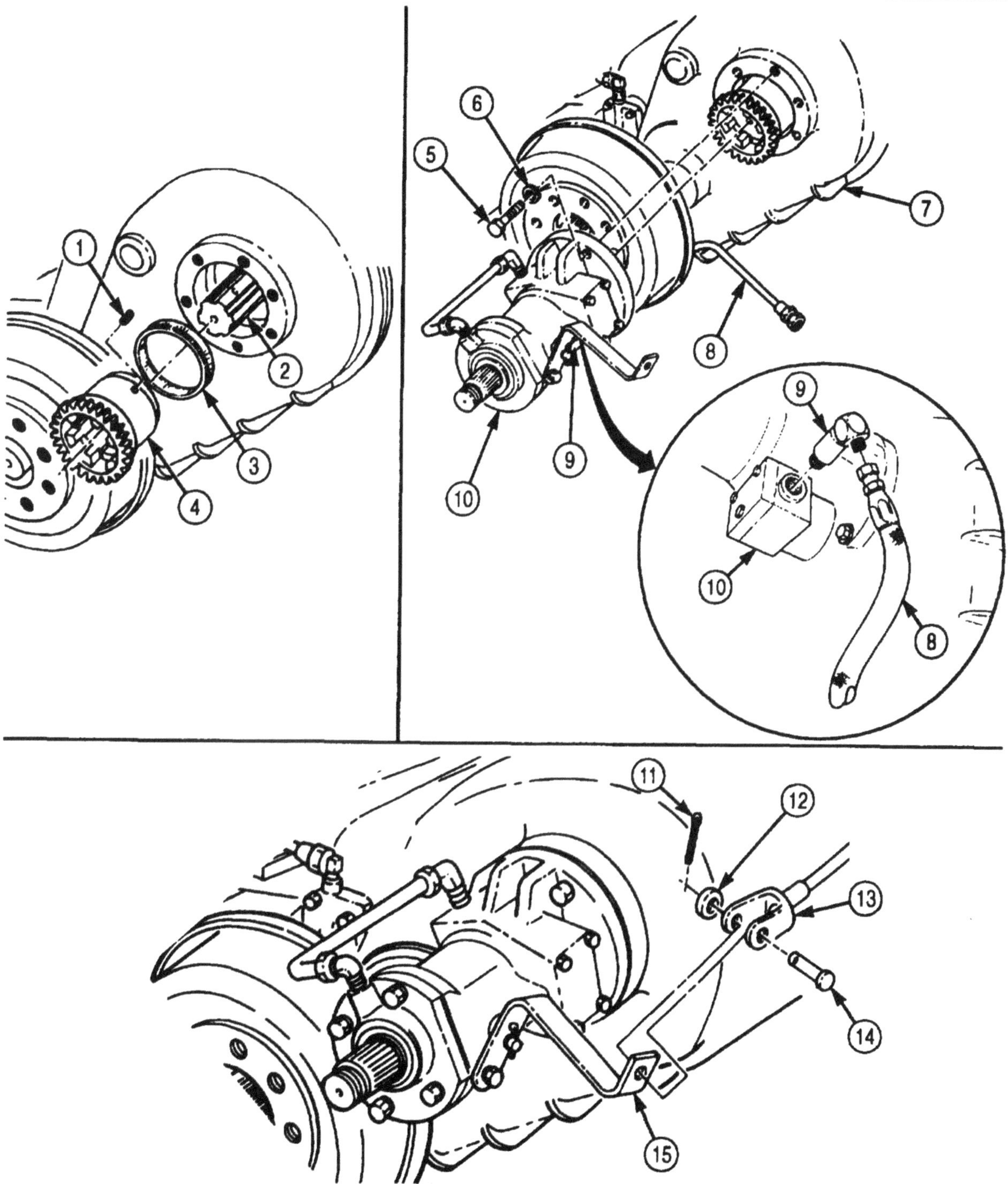

FOLLOW-ON TASKS: • Adjust transfer case PTO cable (para. 4-205).
• Install transfer case PTO-to-hydraulic pump propeller shaft (para. 3-334).
• Fill transfer case to proper oil level (LO 9-2320-272-12).
• Check transfer case PTO for proper operation (TM 9-2320-272-10).

4-208. TRANSFER CASE POWER TAKEOFF (PTO) MAINTENANCE

THIS TASK COVERS:

a. Disassembly
b. Cleaning and Inspection
c. Assembly

INITIAL SETUP:

APPLICABLE MODELS
M936/A1/A2

TOOLS
General mechanic's tool kit (Appendix E, Item 1)
Torque wrench (Appendix E, Item 144)
Dial indicator (Appendix E, Item 36)
Soft-faced hammer
Arbor press
Feeler gauge

MATERIALS/PARTS
Adapter plug gasket (Appendix D, Item 220)
Shifter shaft O-ring (Appendix D, Item 486)
Oil seal (Appendix D, Item 504)
Cotter pin (Appendix D, Item 82)
Snapring (Appendix D, Item 661)
Woodruff key (Appendix D, Item 732)
Oil seal sleeve (Appendix D, Item 507)
Lint-free cloth (Appendix C, Item 21)
Drycleaning solvent (Appendix C, Item 71)

REFERENCES (TM)
TM 9-2320-272-24P
TM 9-214

EQUIPMENT CONDITION
Transfer case PTO removed (para. 4-207).

GENERAL SAFETY INSTRUCTIONS
- Keep fire extinguisher nearby when using drycleaning solvent.
- Drycleaning solvent is flammable and toxic. Do not use near an open flame.

a. Disassembly

1. Remove oil line (2) and adapter elbows (1) and (5) from oil pump housing (3) and bearing cap (4).
2. Remove two nuts (10), washers (7), screws (6), washers (7), and PTO housing (9) from oil pump adapter (8).
3. Remove four screws (15) and washers (14) from bearing cap (4) and PTO housing (9).

NOTE

Tag shims for installation

4. Remove bearing cap (4), shims (13), outer bearing race (12), and PTO shaft (11) from PTO housing (9).

4-208. TRANSFER CASE POWER TAKEOFF (PTO) MAINTENANCE (Contd)

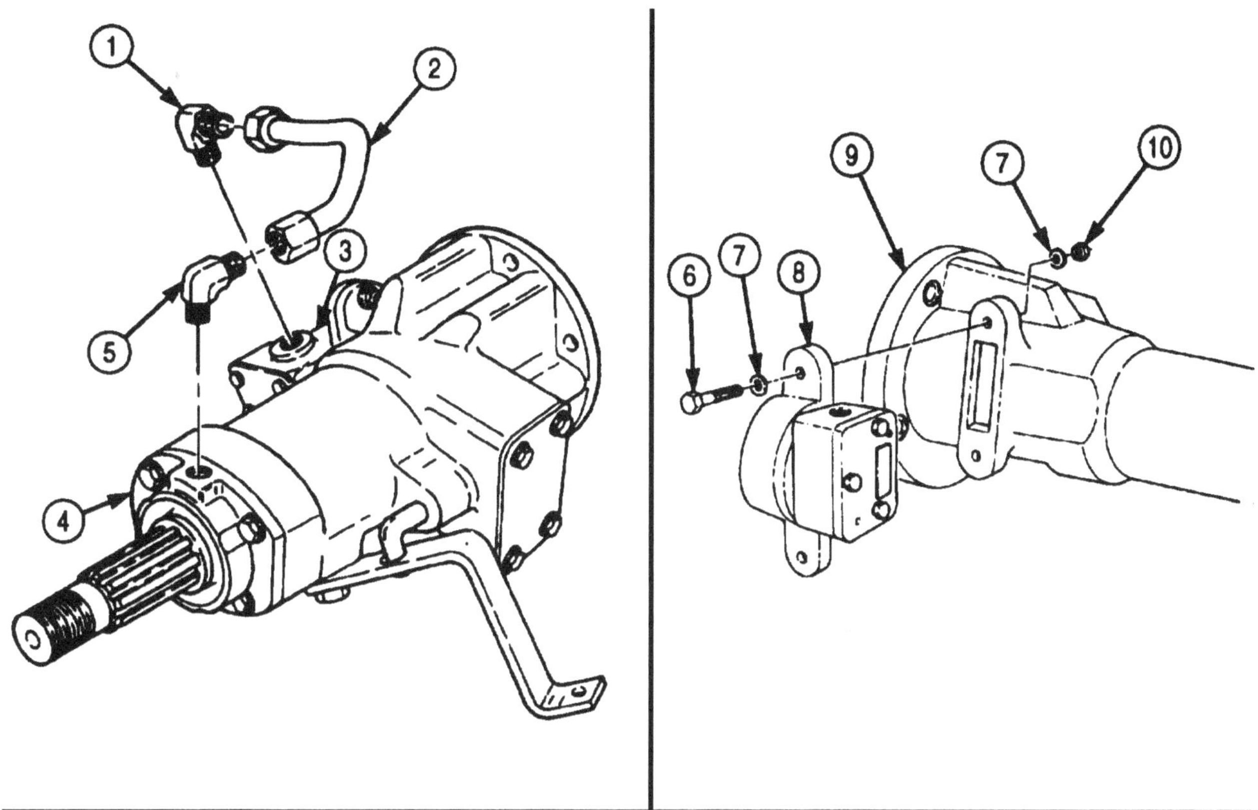

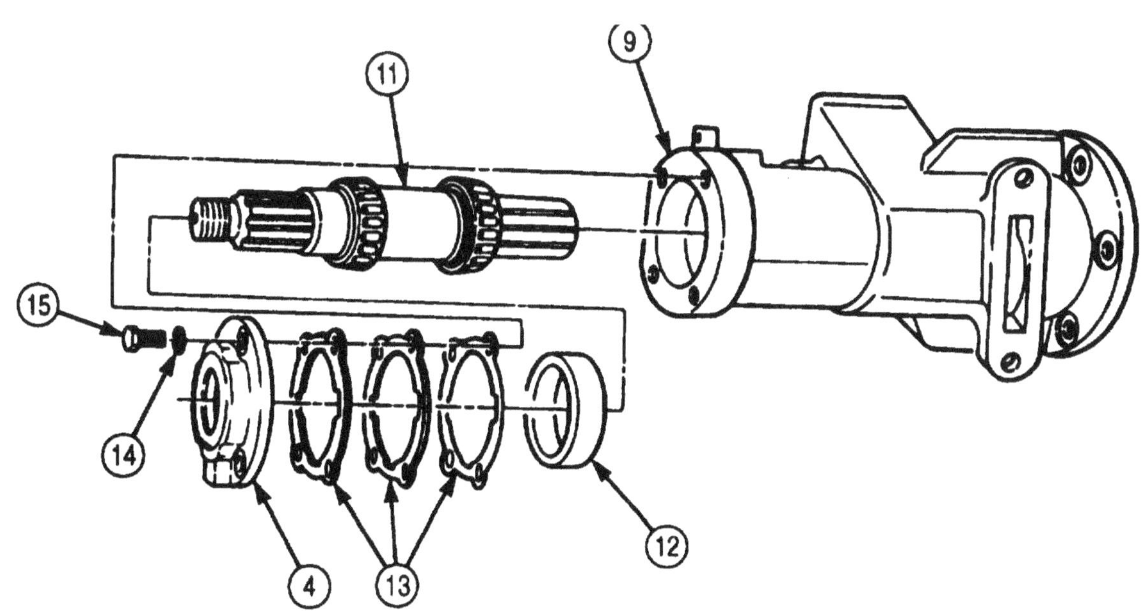

4-208. TRANSFER CASE POWER TAKEOFF (PTO) MAINTENANCE (Contd)

5. Remove cotter pin (7) and washer (6) from PTO shift lever (5) and shifter shaft (10). Discard cotter pin (7).
6. Remove screw (8), washer (9), and PTO shift lever (5) from PTO housing (1).
7. Remove four screws (4), washers (3), and plate (2) from PTO housing (1).

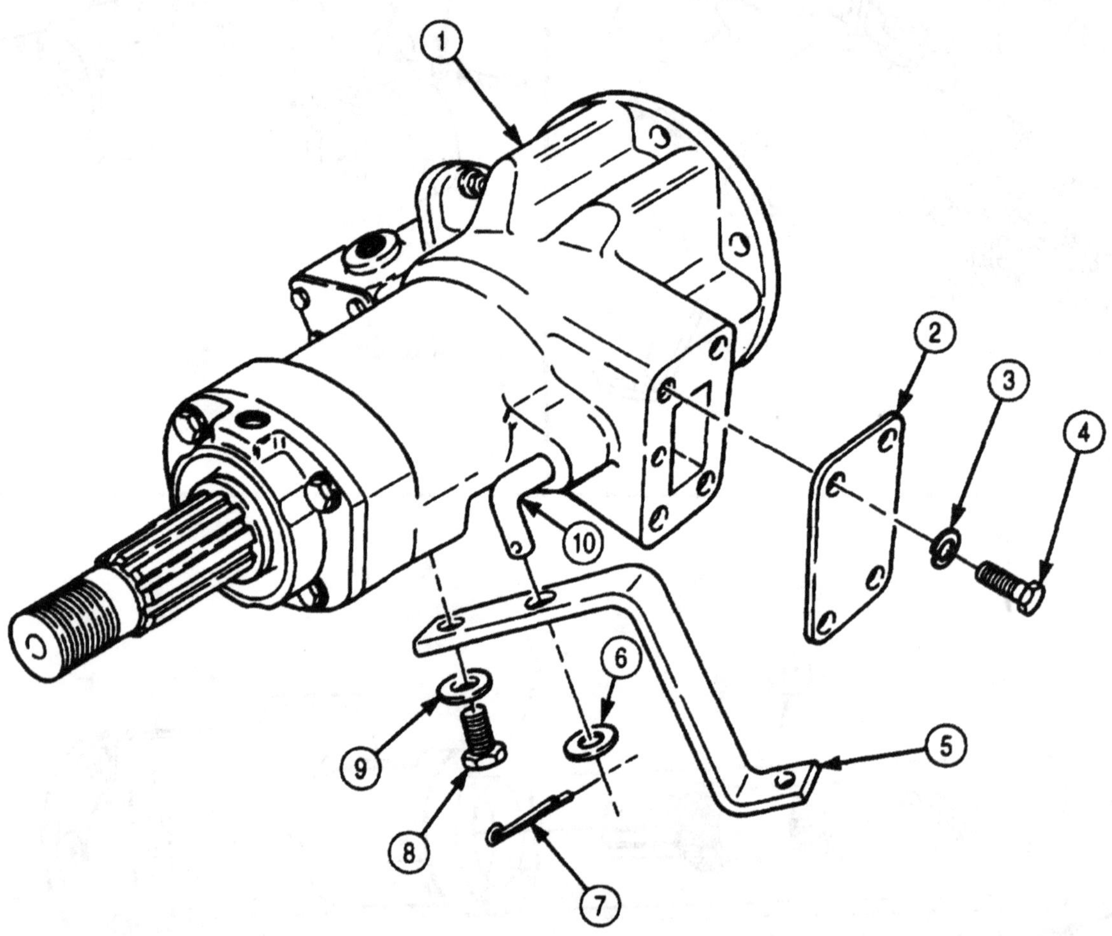

4-208. TRANSFER CASE POWER TAKEOFF (PTO) MAINTENANCE (Contd)

8. Tilt PTO housing (1) until dowel (14), spring (15), and detent ball (16) fall out.
9. Remove setscrew (13) and shifter shaft (10) from shifter fork (12).
10. Remove O-ring (11) from shifter shaft (10). Discard O-ring (11).

NOTE

Mark position of shifter fork for installation.

11. Remove shifter fork (12) from PTO housing (1).
12. Tilt PTO housing (1) until clutch collar (17) falls out.

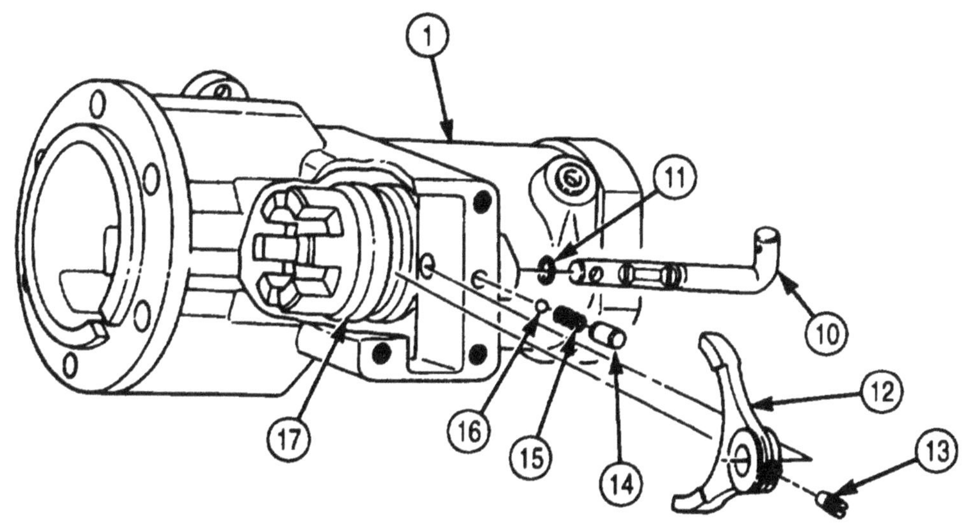

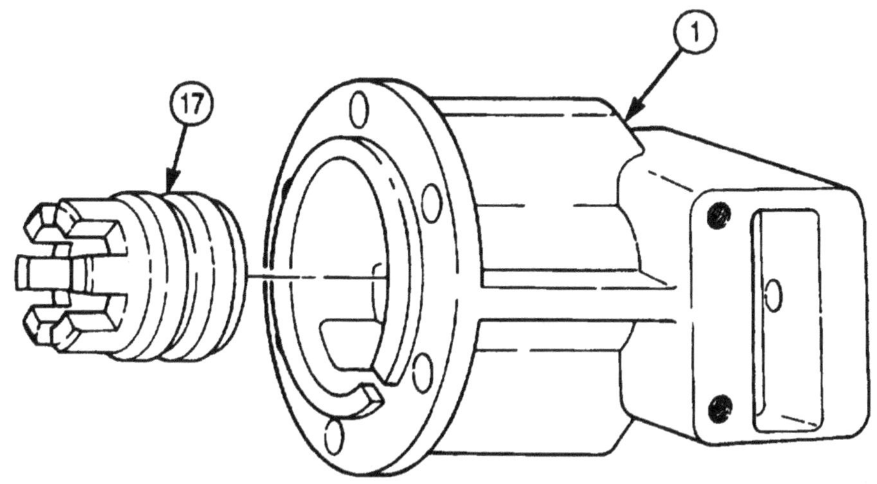

4-208. TRANSFER CASE POWER TAKEOFF (PTO) MAINTENANCE (Contd)

13. Remove inner bearing race (5) from PTO housing (6).
14. Remove access plug (15) and gasket (16) from oil pump adapter (9). Discard gasket (16).
16. Remove outer snapring (17) from pump shaft (13). Discard snapring (17).
16. Remove three screws (12), oil pump adapter (9), and drive gear (14) from oil pump (11).
17. Remove woodruff key (10) from oil pump shaft (13). Discard woodruff key (10).
18. Remove outer (1) and inner (4) output shaft bearings from PTO shaft (3).
19. Remove oil seal sleeve (2) from PTO shaft (3). Discard oil sleeve (2).
20. Remove PTO oil seal (8) from bearing cap (7). Discard PTO oil seal (8).

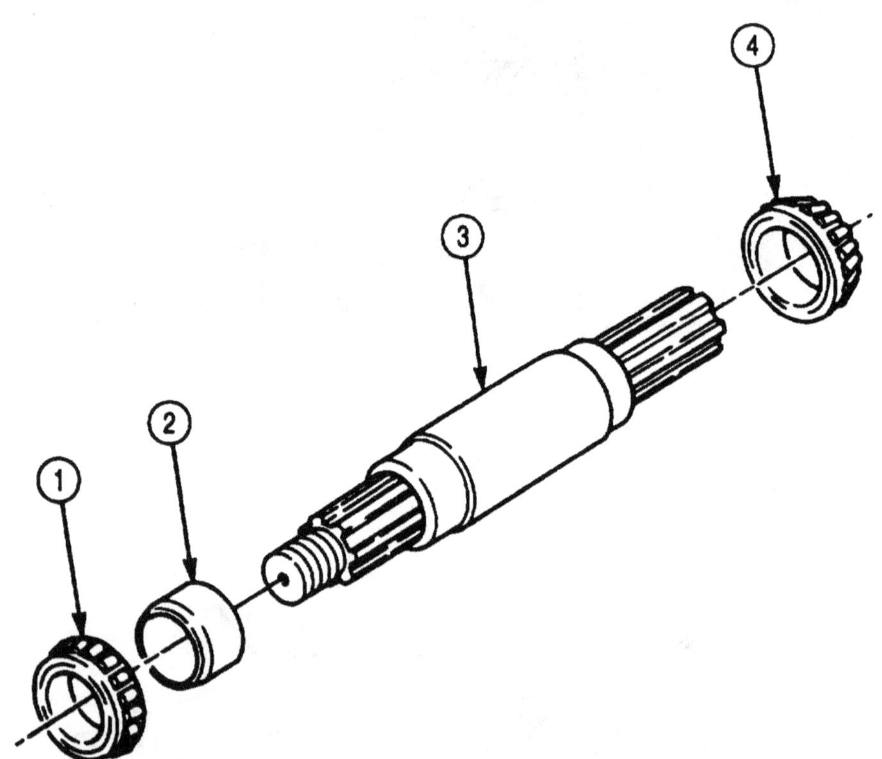

4-208. TRANSFER CASE POWER TAKEOFF (PTO) MAINTENANCE (Contd)

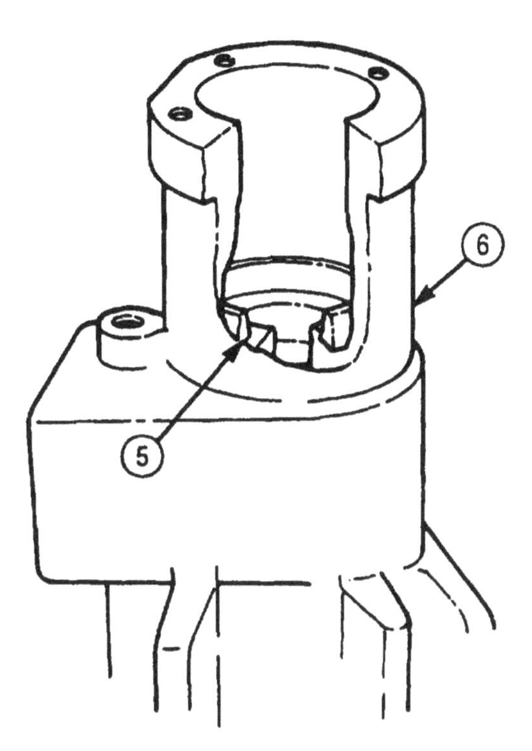

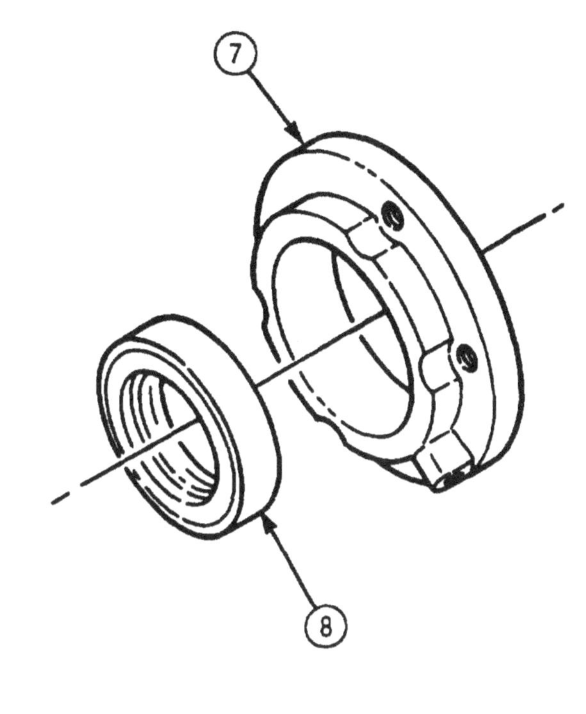

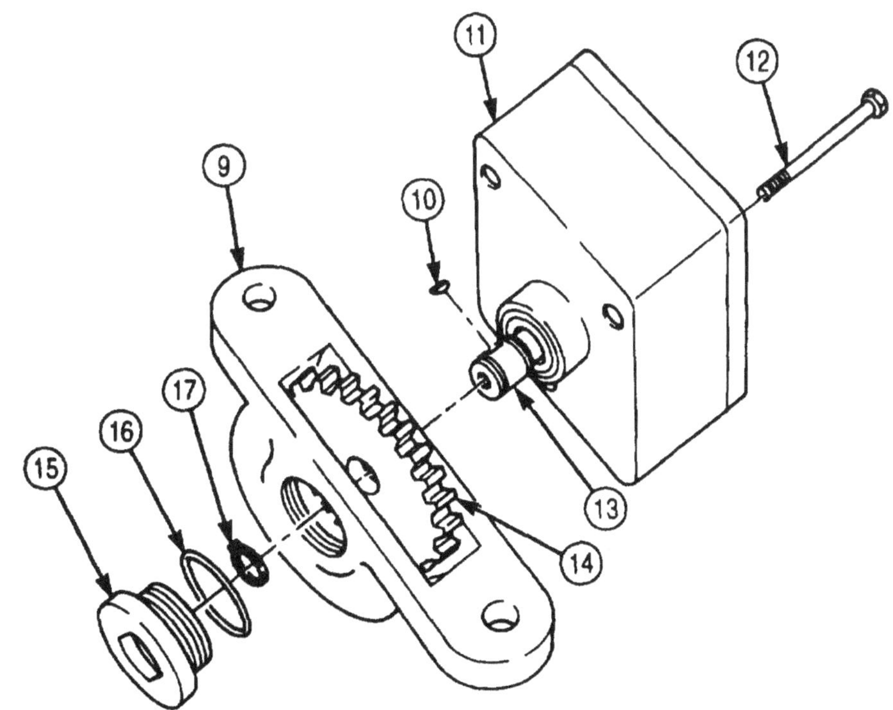

4-208. TRANSFER CASE POWER TAKEOFF (PTO) MAINTENANCE (Contd)

b. Cleaning and Inspection

WARNING

Drycleaning solvent is flammable and toxic. Do not use near open flame and always have a fire extinguisher nearby when solvents are used. Use only in well-ventilated places, wear protective clothing, and dispose of cleaning rags in approved container. Failure to do this may result in injury or death to personnel and/or damage to equipment.

NOTE

Cleaning and inspection instructions include PTO oil pump drive gear and collar assembly removed from transfer case input shaft in para. 4-207.

1. Clean all transfer case PTO components with drycleaning solvent and dry with lint-free cloth.
2. Inspect PTO shaft (3) for pitting, galling, and scoring, If pitted, galled, or scored, replace PTO shaft (3).
3. Inspect PTO housing (13) for cracks, or pitting in bores and on machined surfaces. If PTO housing (13) is cracked or pitted in bores or on machined surfaces, replace transfer case PTO.
4. Inspect inner (4) and outer (2) shaft bearings and two bearing races (1) and (5) for cracking, pitting, scoring, and discoloration. If any matched bearing (4) or (2) or races (1) or (5) are pitted, cracked, scored, or discolored, discard and replace all four (TM 9-214).
5. Inspect PTO oil pump drive gear (8) and collar (7) for cracks, pitting, scoring, or missing teeth. If cracked, pitted, scored, or teeth are missing, replace PTO oil pump drive gear (8) and collar (7).
6. Inspect oil pump adapter (12) for cracks. If cracked, replace pump adapter (12).
7. Inspect oil pump (9) and shaft (10) for cracks and scoring. If cracked or scored, replace oil pump (9) and shaft (10).
8. Inspect bearing cap (6) for cracks. If cracked, replace bearing cap (6).

4-208. TRANSFER CASE POWER TAKEOFF (PTO) MAINTENANCE (Contd)

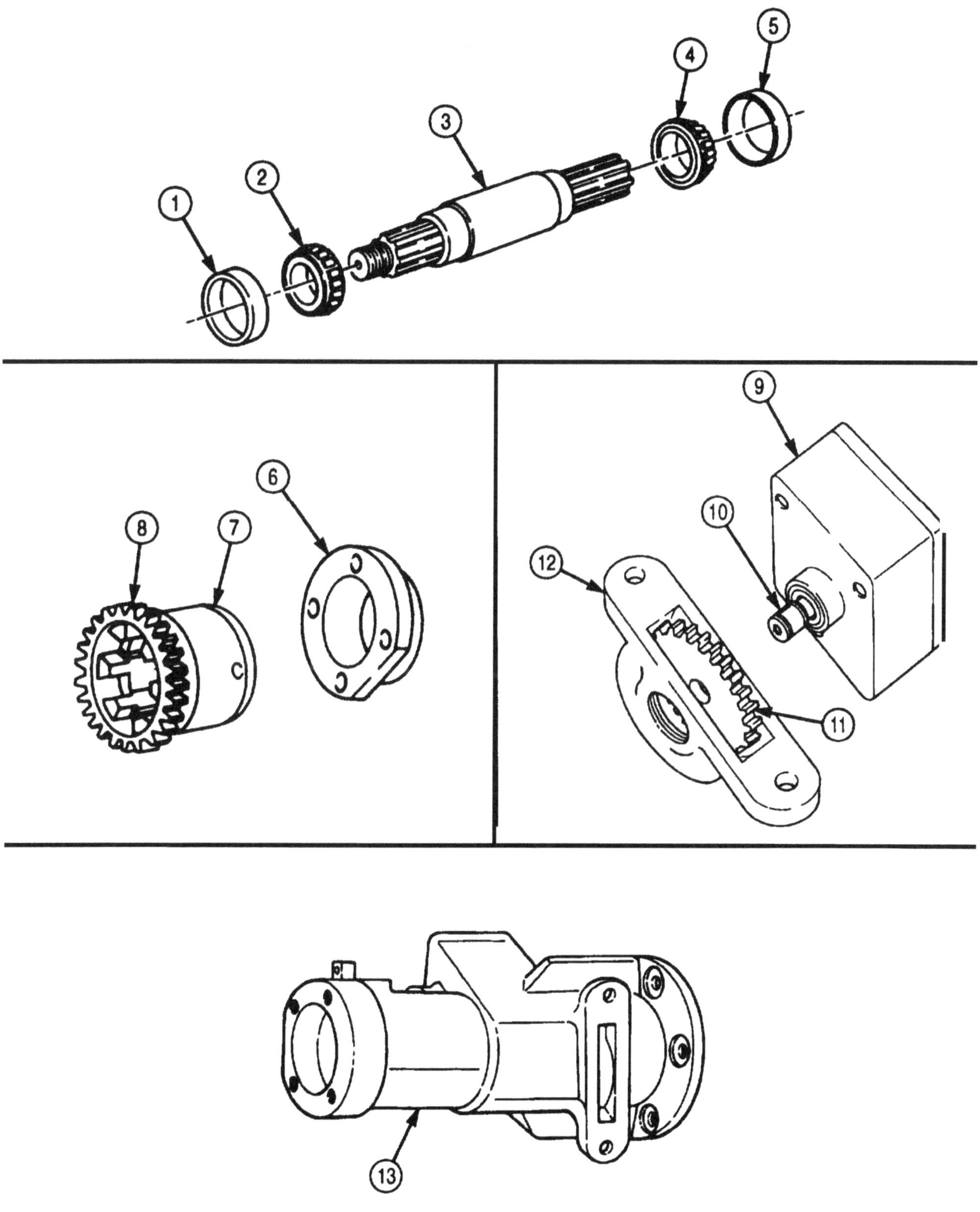

1
2
3
4
5

6
7
8

9
10
11
12

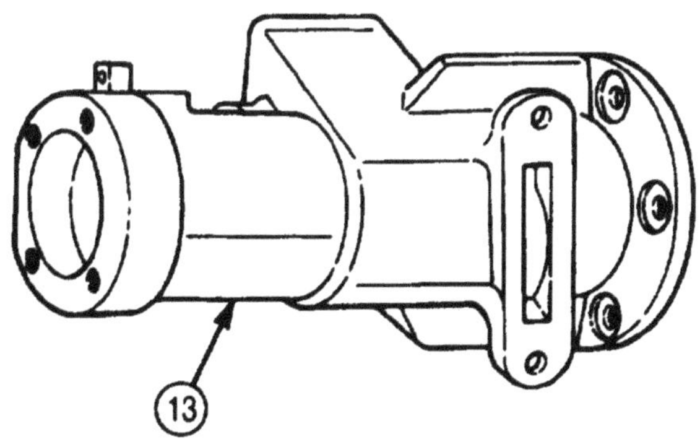

4-208. TRANSFER CASE POWER TAKEOFF (PTO) MAINTENANCE (Contd)

c. Assembly

1. Install new woodruff key (12) on oil pump shaft (15).
2. Position oil pump adapter (10) against oil pump (13), align oil pump gear (11) on oil pump shaft (15), and install three screws (14). Tighten screws (14) 9-11 lb-ft (12-15 N•m).
3. Install new snapring (18) on oil pump shaft (15).
4. Install new adapter plug gasket (17) and adapter plug (16) on oil pump adapter (10). Tighten adapter plug (16) 80 lbft (108 N•m).
5. Install inner bearing (5), outer bearing (2), and new oil seal sleeve (3) on PTO shaft (4) with arbor press and mandrel.
6. Position inner bearing race (6) over inner bearing (5) on PTO shaft (4).
7. Position PTO shaft (4) in PTO housing (9) through bearing cap end (8) of PTO housing (9). Ensure inner bearing race (6) seats squarely in bore of PTO housing (9) on shoulder (7).
8. Position outer bearing race (1) in PTO housing (9). Ensure outer bearing race (1) seats squarely in bore of PTO housing (9).

NOTE

Ensure OUTSIDE stamping is up.

9. Using arbor press and mandrel, install new oil seal (19) in bearing cap (20).

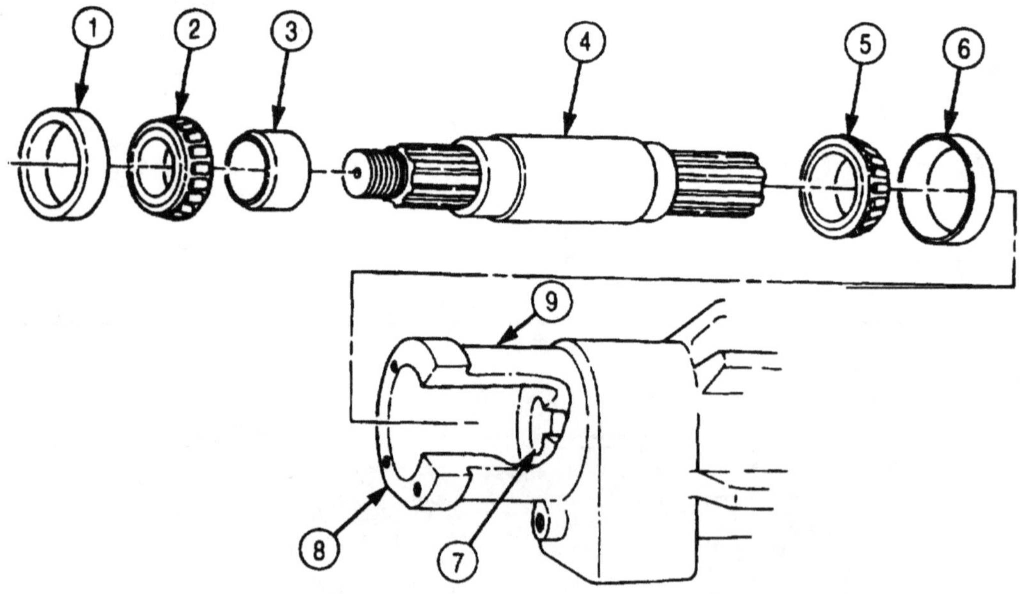

4-208. TRANSFER CASE POWER TAKEOFF (PTO) MAINTENANCE (Contd)

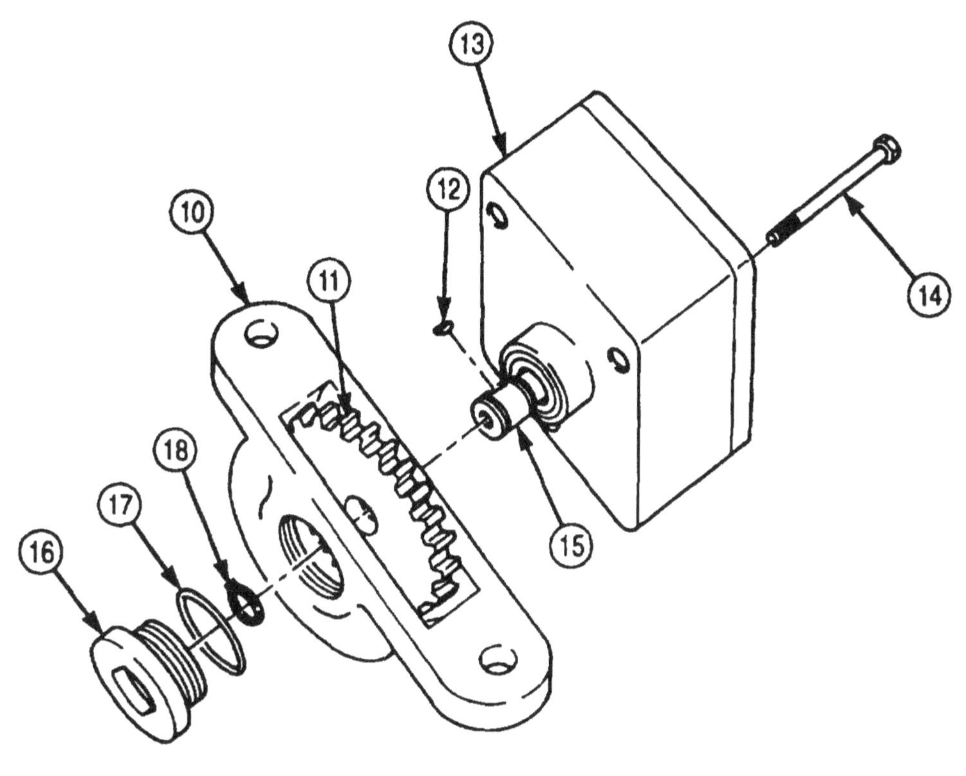

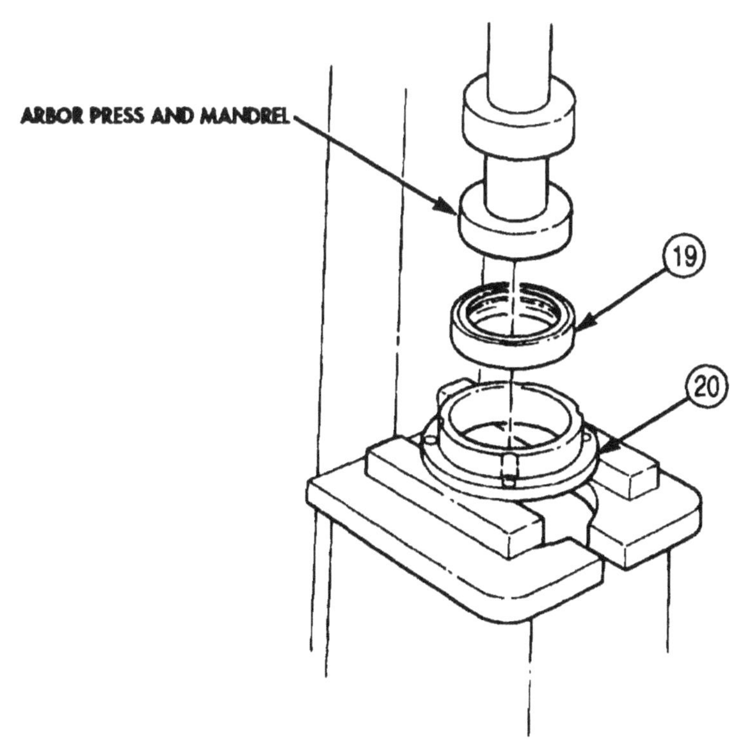

4-208. TRANSFER CASE POWER TAKEOFF (PTO) MAINTENANCE (Contd)

NOTE

Steps 10 and 11 establish starting shim pack thickness.

10. Install bearing cap (3) on transfer case PTO (4) andtap alternately around outer edge with soft-faced hammer.
11. Using feeler gauge, measure clearance between cap (3) and PTO gear (4). Record measurement.
12. Remove bearing cap (3) from transfer case PTO (4).

CAUTION

Do not apply sealer on mating surfaces.

NOTE

Use measurement obtained in step 11 plus a .003 in. (0.0762 mm) shim for starting shim pack thickness.

13. Install starting shim pack (5) and bearing cap (3) on transfer case PTO (4) with four washers (2) and screws (1). Tighten screws (1) 18-24 lb-R (24-32 N•m).
14. Position dial indicator plunger against PTO shaft (6).

NOTE

Steps 15 through 17 check transfer PTO shaft end play.

15. Force transfer PTO shaft (6) to rear of transfer case PTO (4).
16. Set dial indicator to zero.
17. Force transfer PTO shaft (6) to front of transfer case PTO (4) and record reading shown on dial indicator. End play should be .001-.005 in. (.0254-0.127 mm).
18. Remove four screws (1), washers (2), and bearing cap (3) from transfer case PTO (4).

CAUTION

Do not apply sealer to mating surfaces.

NOTE

Use reading obtained in step 17 plus a .003 in. (.0762 mm) shim for starting shim pack thickness.

19. Install starting shim pack (5) and bearing cap (3) on transfer case PTO (4) with four washers (2) and screws (1). Tighten screws (1) 18-24 lb-R (24-32 N•m).
20. Position dial indicator plunger against PTO shaft (6).

NOTE

Steps 21 through 23 check transfer PTO shaft end play.

21. Force transferPTO shaft (6) to rear of transfer case PTO (4).
22. Set dial indicator to zero and force transfer PTO shaft (6) to front of transfer case PTO (4).
23. Record reading shown on dial indicator. End play should be .001-.005 in. (.00254-0.127 mm).
24. Remove four screws (1), washers (2), and bearing cap (3) from transfer case PTO (4).

NOTE

Use reading obtained in step 23 for number or thickness of shims to be removed or added for transfer case PTO shaft adjustment.

25. Remove or add shims (5) as necessary.
26. Remove shim pack (5).

NOTE

Do not apply sealer to shims.

27. Apply a light coating of sealer to mating surfaces of transfer case PTO (4) and bearing cap (3).
28. Install shim pack (5) and bearing cap (3) on transfer case PTO (4) with four washers (2) and screws (1). Tighten screws (1) 18-24 lb-R (24-32 N•m).

4-208. TRANSFER CASE POWER TAKEOFF (PTO) MAINTENANCE (Contd)

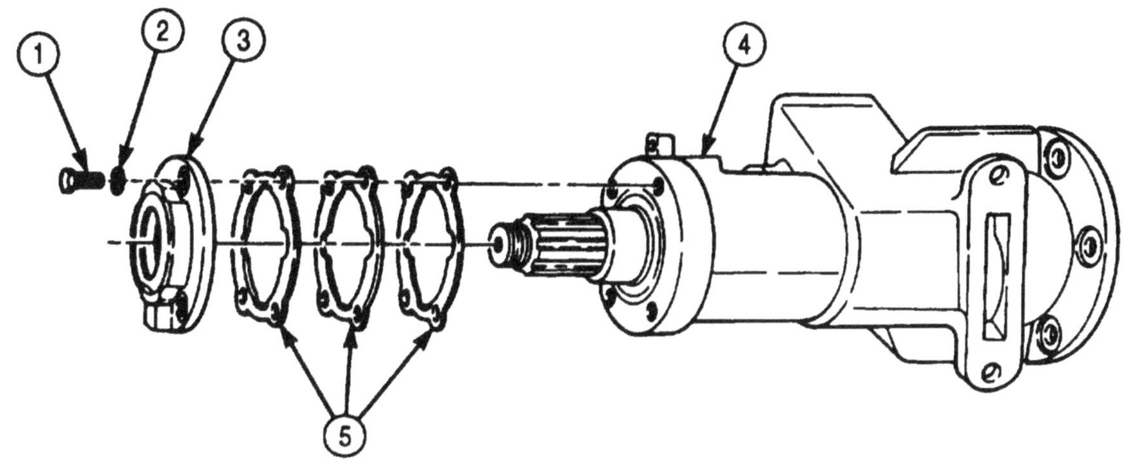

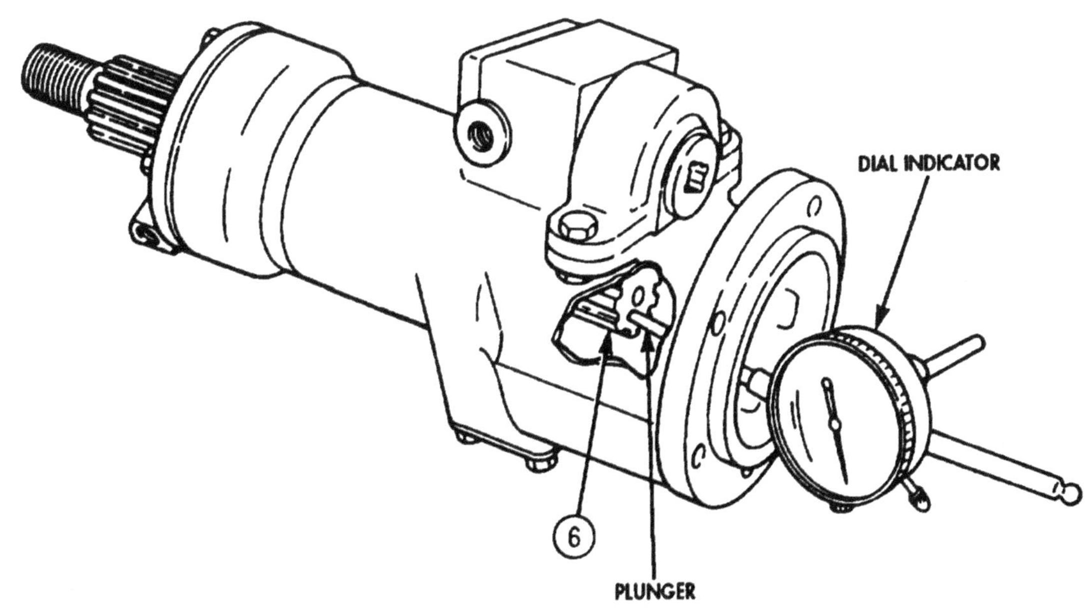

4-208. TRANSFER CASE POWER TAKEOFF (PTO) MAINTENANCE (Contd)

CAUTION

Clutch collar and shifter must be properly installed to ensure proper PTO operation. Ensure clutch collar internal splines face away from PTO mounting flange and thick edge of shifter fork faces shifter shaft.

29. Position splined end (19) of clutch collar (13) over PTO shaft (18).
30. Position finger ends (6) of shifter fork (6) in center groove (12) of clutch collar (13).
31. Install new O-ring (2) on shifter shaft (4).
32. Place shift shaft (4) through PTO housing bore (11) and fork (6) and install with setscrew (7) into hole (3). Tighten setscrew (7) 7-9 lb-ft (10-12 N•m).
33. Place detent ball (10), spring (9), and dowel (8) into PTO housing bore (1).
34. Install cover plate (15) on PTO housing (14) with four washers (16) and screws (17). Tighten screws (17) 18-24 lb-ft (24-32 N•m).

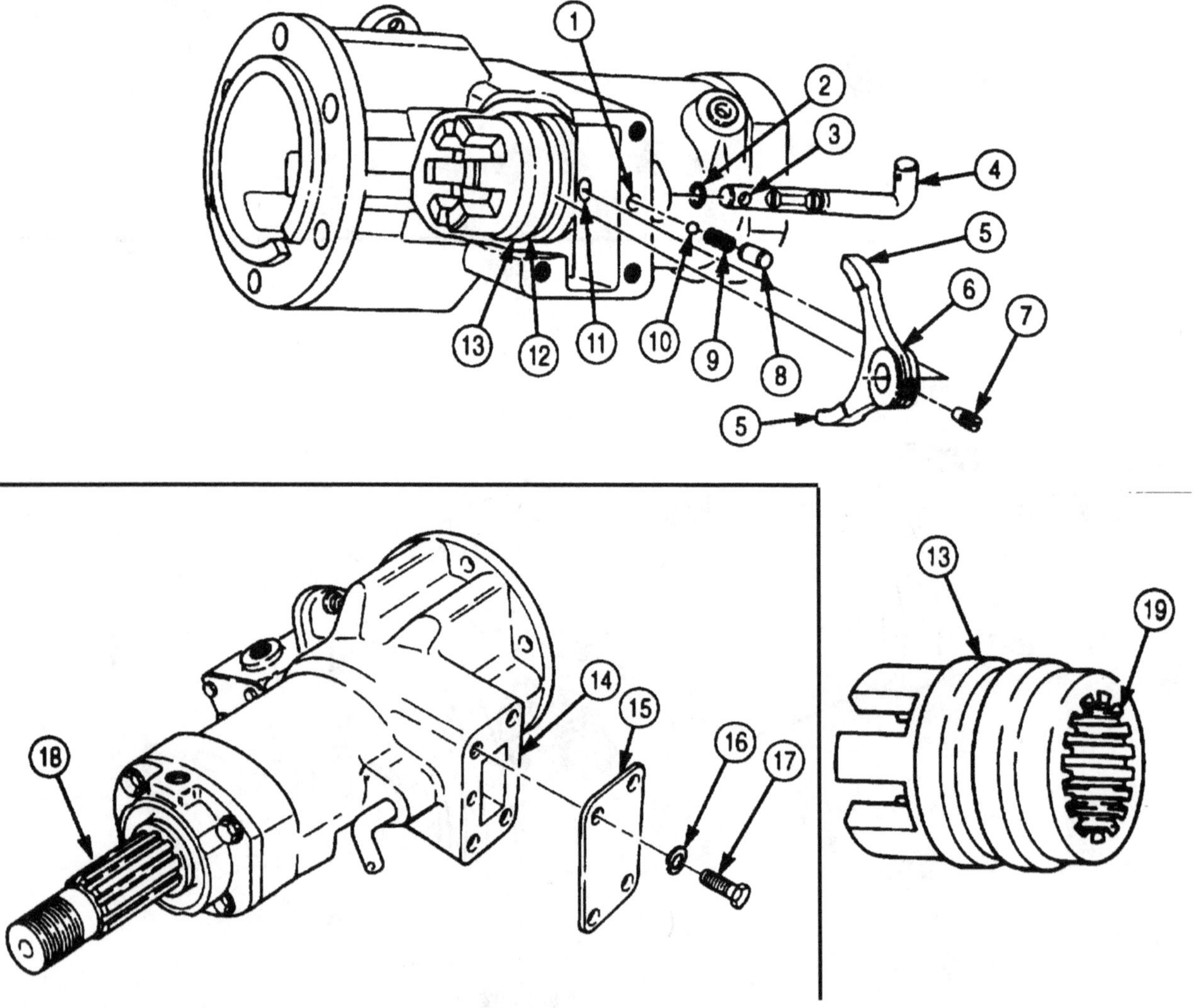

4-208. TRANSFER CASE POWER TAKEOFF (PTO) MAINTENANCE (Contd)

NOTE

At this point, check operation of PTO shifter shaft. Shifter shaft must be able to be moved into two positions.

35. Install PTO shift lever (20) on shifter shaft (4) with washer (21) and new cotter pin (22).
36. Install PTO shift lever (20) on PTO housing end (14) with washer (24) and screw (23).
37. Install oil pump adapter (27) on PTO housing (28) with four washers (26), two screws (25), and nuts (29). Tighten screws (25) 15-20 lb-ft (20-27 N•m).
38. Install adapter elbows (31) and (30) on PTO housing (14).
39. Install oil line (32) to PTO adapter elbows (30) and (31).

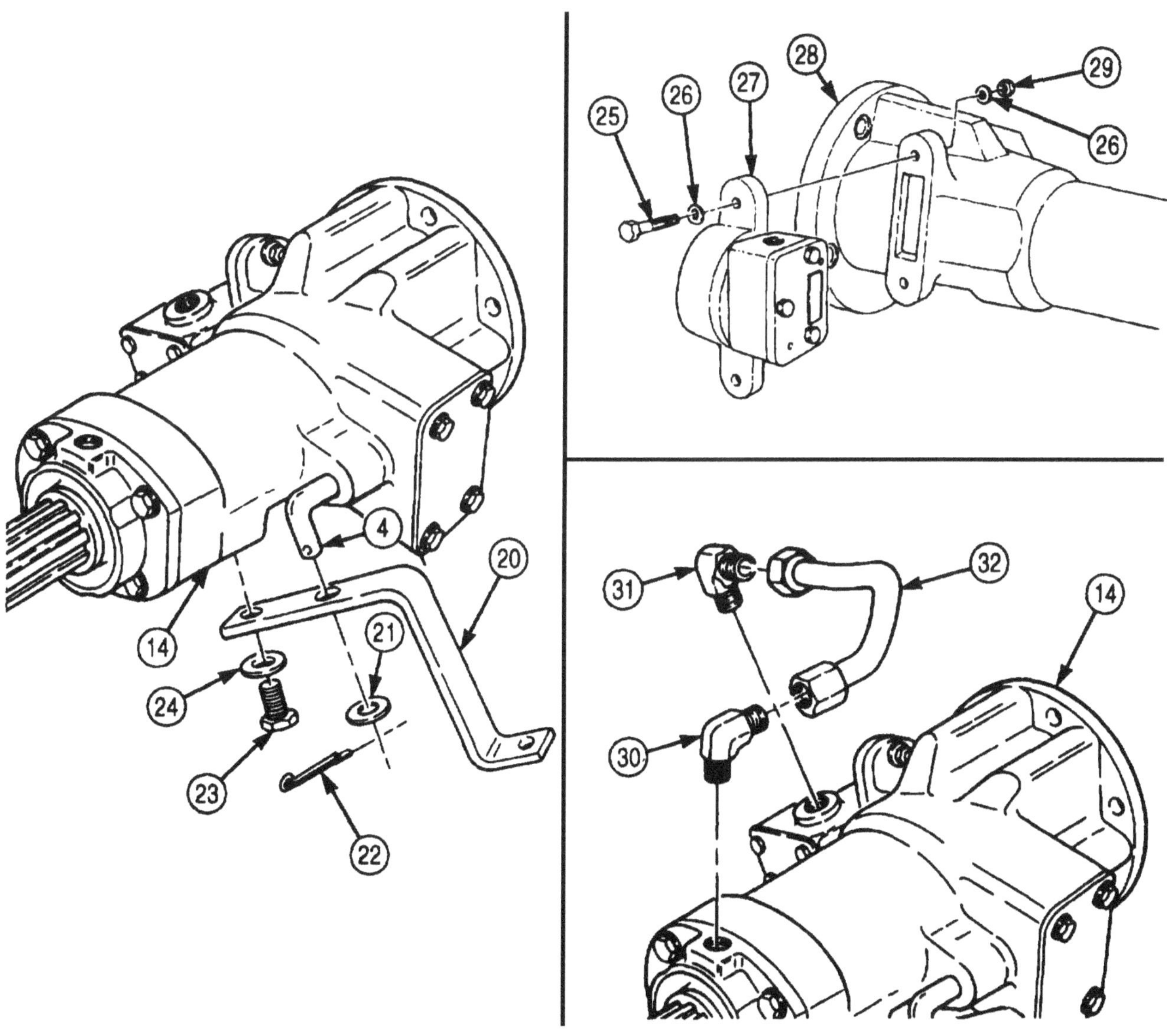

FOLLOW-ON TASK: Install transfer case PTO (para. 4-207).

4-209. TRANSMISSION POWER TAKEOFF (PTO) CONTROL CABLE REPLACEMENT

THIS TASK COVERS:

a. Removal **b. Installation**

INITIAL SETUP:

APPLICABLE MODELS
M925, M928, M929, M930, M932, M936

TOOLS
General mechanic's tool kit (Appendix E, Item 1)

MATERIALS/PARTS
Cotter pin. (Appendix D, Item 84)
Four locknuts (Appendix D, Item 307)
Cotter pin (Appendix D, Item 66)

REFERENCES (TM)
TM 9-2320-272-10
TM 9-2320-272-24P

EQUIPMENT CONDITION
Parking brake set (TM 9-2320-272- 10).

a. Removal

1. Open transmission dipstick access door (1) inside of cab (TM 9-2320-272-10)
2. Remove two locknuts (2), screws (5), clamp (6), and spacer plate (4) from PTO cable bracket (3). Discard locknuts (2).
3. Remove cotter pin (11) and washer (10) and lift PTO cable pin (8) free of select lever (9). Discard cotter pin (11).
4. Remove PTO cable pin (8)from PTO cable (7).

NOTE

Tag all ends of PTO cable for installation.

5. Remove six screws (20) and access cover (19) from PTO control panel (12).
6. Remove two locknuts (18), screws (14), clamp (17), and spacer (16) from PTO control panel (12). Discard locknuts (18).
7. Remove cotter pin (22), washer (23), and clevis pin (25) from PTO control lever (13), and pull end of control cable (7) clear of PTO control panel (12). Discard cotter pin (22).
8. Loosen nut (21) on PTO control cable (7) and remove cable clevis (24).
9. Pull PTO control cable (7) on PTO control panel (12) through grommet (15) and into cab.

b. Installation

1. Feed transmission select lever end of PTO control cable (7) through rear of PTO control panel (12) and into grommet (15).
2. Attach PTO cable clevis (24) to PTO control panel end of cable (7) with nut (21).
3. Install PTO cable clevis (24) on PTO control lever (13) with clevis pin (25), washer (23), and new cotter pin (22).
4. Install PTO control cable (7) on PTO control panel (12) with two screws (14), spacer (16), clamp (17), and two new locknuts (18).
5. Install PTO control panel access cover (19) on PTO control panel (12) with six screws (20).
6. Pull PTO control cable (7) down to PTO from engine compartment and install PTO cable pin (8) on end of PTO control cable (7).
7. Connect PTO cable pin (8) to PTO select lever (9) with washer (10) and new cotter pin (11).
8. Attach PTO control cable (7) to PTO cable bracket (3) with clamp (6), spacer plate (4), two screws (5), and new locknuts (2).

4-209. TRANSMISSION POWER TAKEOFF (PTO) CONTROL CABLE REPLACEMENT (Contd)

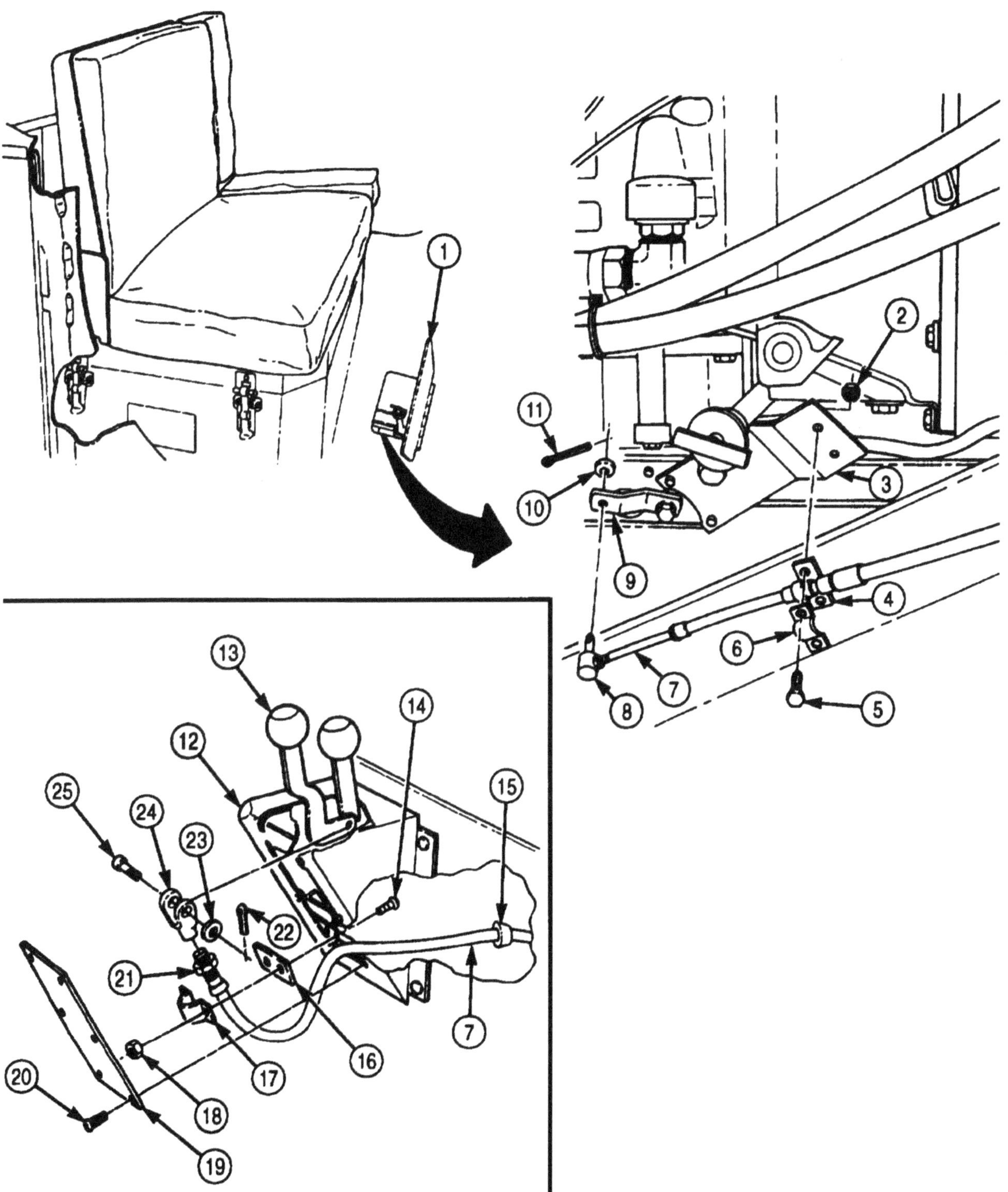

FOLLOW-ON TASK: Start engine (TM 9-2320-272-10) and check transmission PTO for proper operation.

4-210. TRANSMISSION POWER TAKEOFF (PTO) CONTROL CABLE (M939A2) REPLACEMENT

THIS TASK COVERS:

a. Removal **b. Installation**

INITIAL SETUP:

APPLICABLE/MODELS
M939A2

TOOLS
General mechanic's tool kit (Appendix E, Item 1)

MATERIALS/PARTS
Cotter pin (Appendix D, Item 84)
Two locknuts (Appendix D, Item 307)
Cotter pin (Appendix D, Item 66)

REFERENCES (TM)
TM 9-2320-272-10
TM 9-2320-272-24P

EQUIPMENT CONDITION
- Parking brake set (TM 9-2320-272-10).
- Exhaust pipe removed (para. 3-50).

---

a. Removal

1. Remove cotter pin (9), washer (10), and PTO control cable (5) from PTO select lever (11). Discard cotter pin (9).
2. Loosen nut (13) and remove cable pin (12) from PTO control cable (5).
3. Remove two locknuts (7), screws (4), clamp (3), PTO control cable (5), and spacer (2) from bracket (6). Discard locknuts (7).

NOTE

Perform step 4 only if bracket requires replacement.

4. Remove two screws (1) and bracket (6) from PTO (8).
5. Remove six screws (18) and access cover (19) from shift panel (15).
6. Remove two nuts (24), screws (28), clamp (27), PTO control cable (5), and spacer (26) from shift panel (15).
7. Remove cotter pin (17), washer (20), clevis pin (23), and control cable (5) from control lever (16). Discard cotter pin (17).
8. Loosen nut (22) and remove cable clevis (21) from PTO control cable (5).
9. Remove grommet (25) and PTO control cable (5) from firewall (14).

4-2 10. TRANSMISSION POWER TAKEOFF (PTO) CONTROL CABLE (M939A2) REPLACEMENT (Contd)

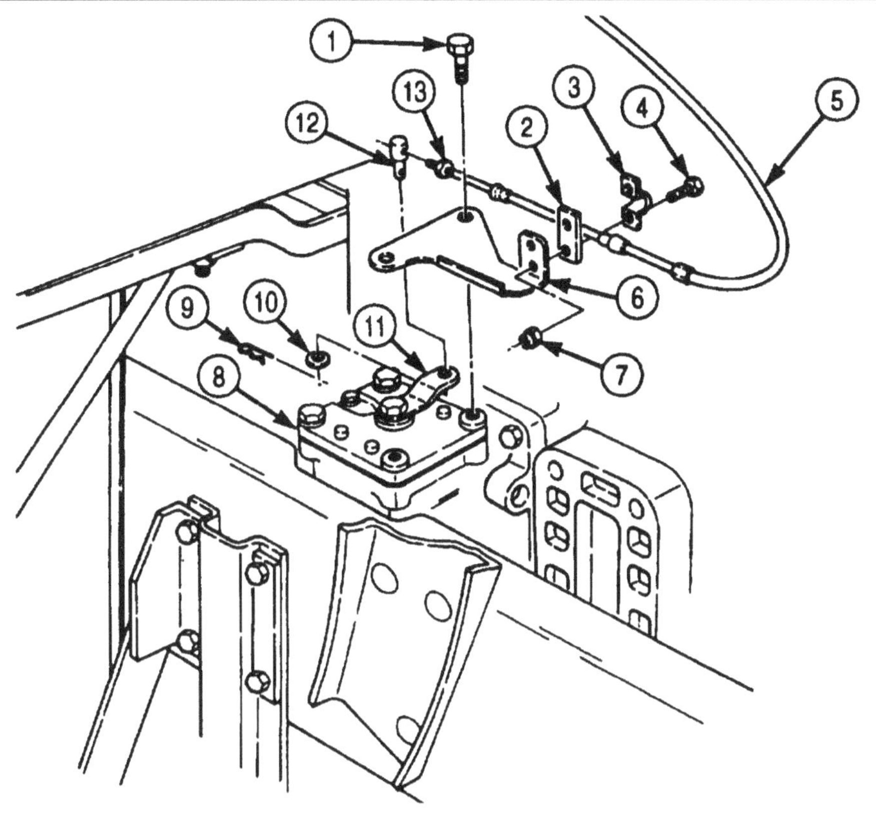

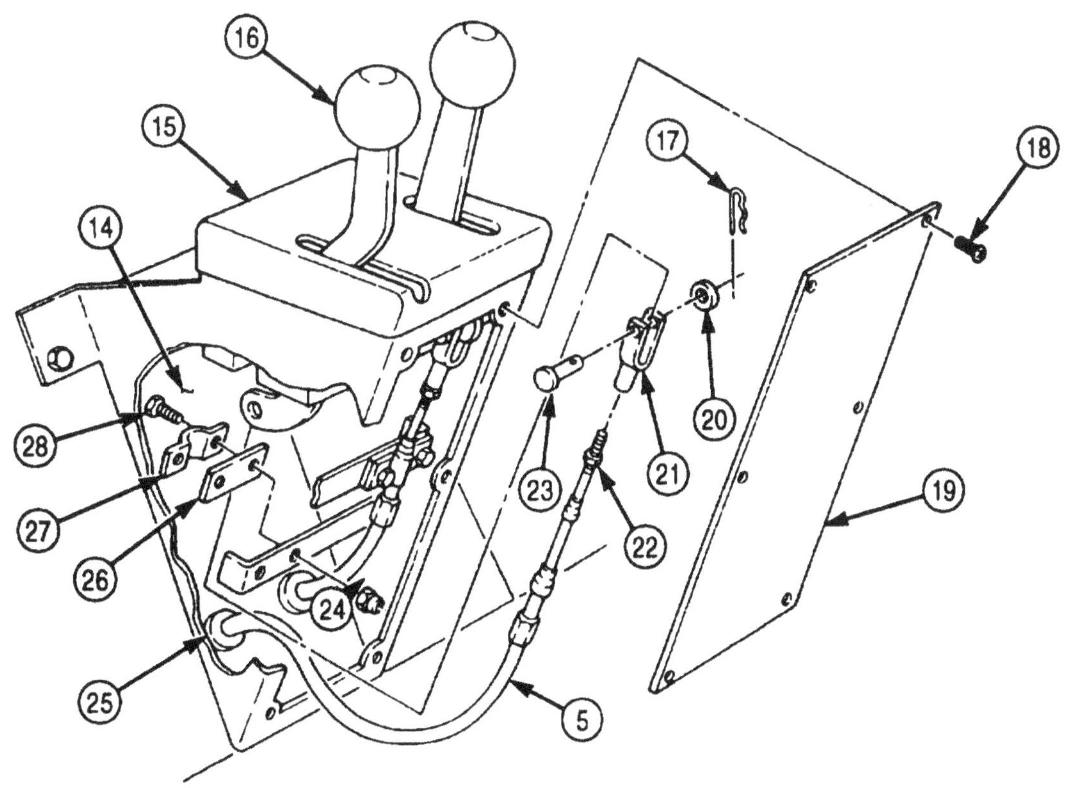

4-210. TRANSMISSION POWER TAKEOFF (PTO) CONTROL CABLE (M939A2) REPLACEMENT (Contd)

b. Installation

1. Insert PTO control cable (11) through hole in firewall (1).
2. Place grommet (13) on PTO control cable (11) and install on firewall (1).
3. Install cable clevis (8) on control cable (11). Tighten nut (10) against cable clevis (8).
4. Install PTO control cable (11) on control lever (3) with clevis pin (9), washer (7), and new cotter pin (4).
5. Install spacer (14) and PTO control cable (11) on shift panel (2) with clamp (15), two screws (16), and nuts (12).
6. Install access cover (6) on shift panel (2) with six screws (5).

NOTE

Perform step 7 if bracket was removed.

7. Install bracket (23) on PTO (28) with two screws (19).
8. Install cable pin (17) on PTO control cable (11). Tighten nut (18) against cable pin (17).
9. Install PTO control cable (11) on select lever (25) with washer (26) and new cotter pin (27).
10. Install spacer (20) and PTO control cable (11) on bracket (23) with clamp (21), two screws (22), and new locknuts (24).

4-210. TRANSMISSION POWER TAKEOFF (PTO) CONTROL CABLE (M939A2) REPLACEMENT (Contd)

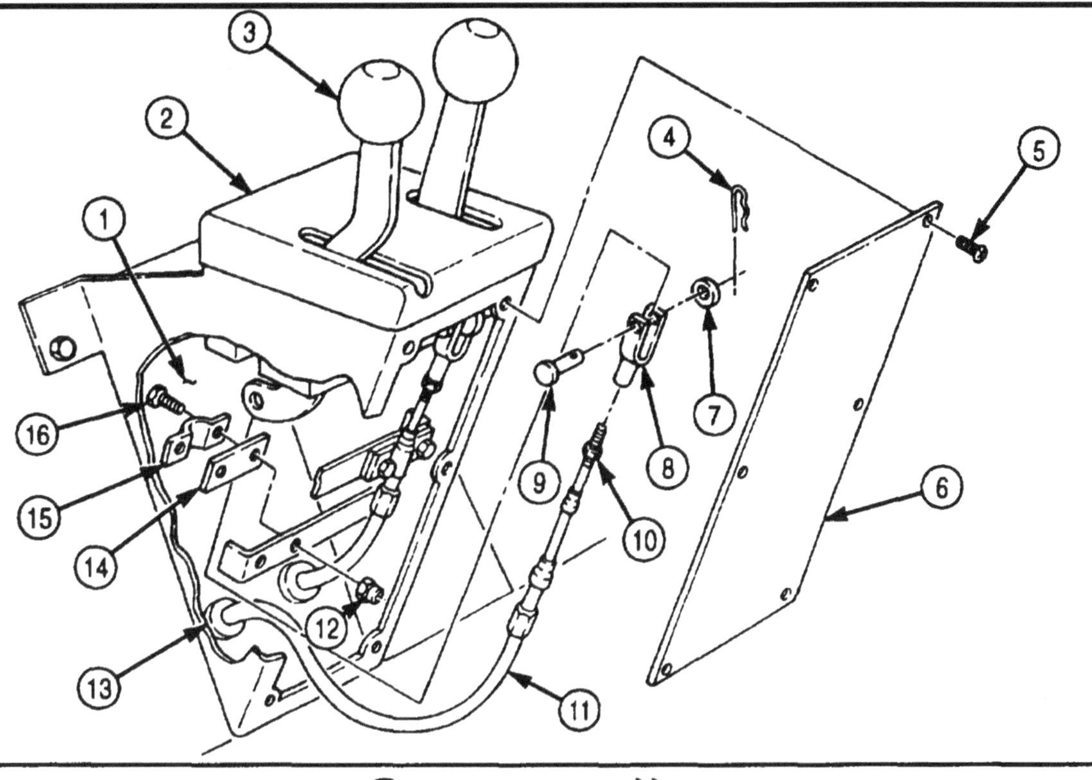

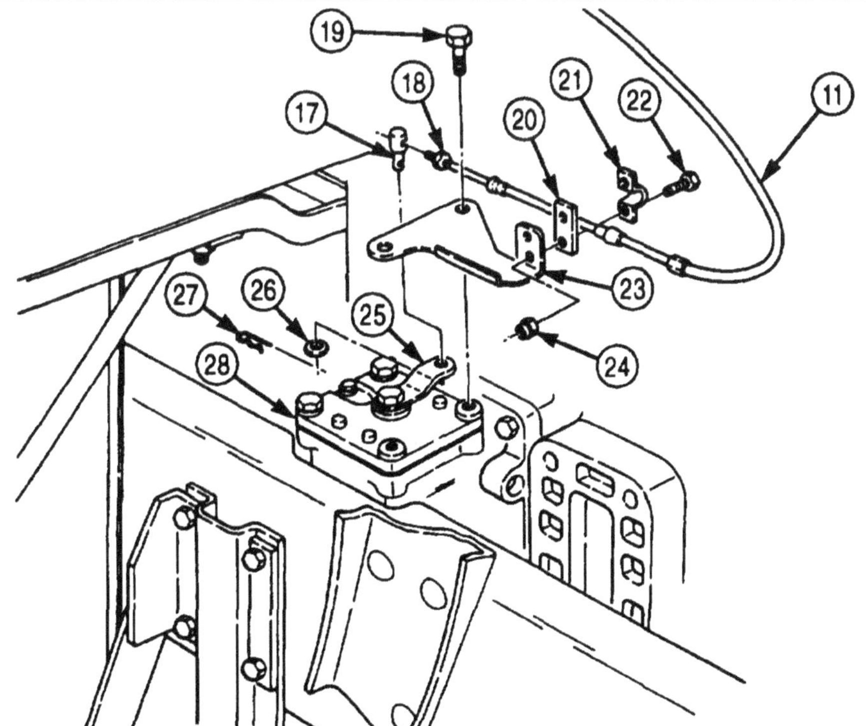

FOLLOW-ON TASKS:
- Install exhaust pipe (para. 3-50).
- Start engine (TM 9-2320-272-10) and check transmission PTO for proper operation.

4-211. TRANSMISSION POWER TAKEOFF (PTO) REPLACEMENT

THIS TASK COVERS:

a. Removal **b. Installation**

INITIAL SETUP:

APPLICABLE MODELS
M925, M928, M929, M930, M932, M936

TOOLS
General mechanic's tool kit (Appendix E, Item 1)

MATERIALS/PARTS
Cotter pin (Appendix D, Item 84)
Locknut (Appendix D, Item 288)
Gasket (Appendix D, Item 238)
Cap and plug set (Appendix C, Item 14)
Lubricating oil (Appendix C, Item 48)
Antiseize tape (Appendix C, Item 72)

REFERENCES (TM)
LO 9-2320-272-12
TM 9-2320-272-10
TM 9-2320-272-24P

EQUIPMENT CONDITION
- Parking brake set (TM 9-2320-272-10).
- Transmission PTO-to-hydraulic pump propeller shaft removed (para. 3-334).
- Transmission oil dipstick tube removed (para. 3-3).

a. Removal

1. Open transmission dipstick access door (1) on inside of cab (TM 9-2320-272-10).
2. Remove two nuts (3), screws (8), clamp (7), and spacer (5) from PTO cable bracket (4).
3. Remove cotter pin (12) and washer (11) from PTO cable pin (9), and lift PTO cable pin (9) away from select lever (10). Discard cotter pin (12).
4. Tie PTO cable (6) clear of work area.
5. Remove screw (16), washer (15), and bracket (4) from PTO (19).
6. Remove screw (17), washer (18), two screws (14), and washers (13) from PTO (19) and transmission (2).

4-211. TRANSMISSION POWER TAKEOFF (PTO) REPLACEMENT (Contd)

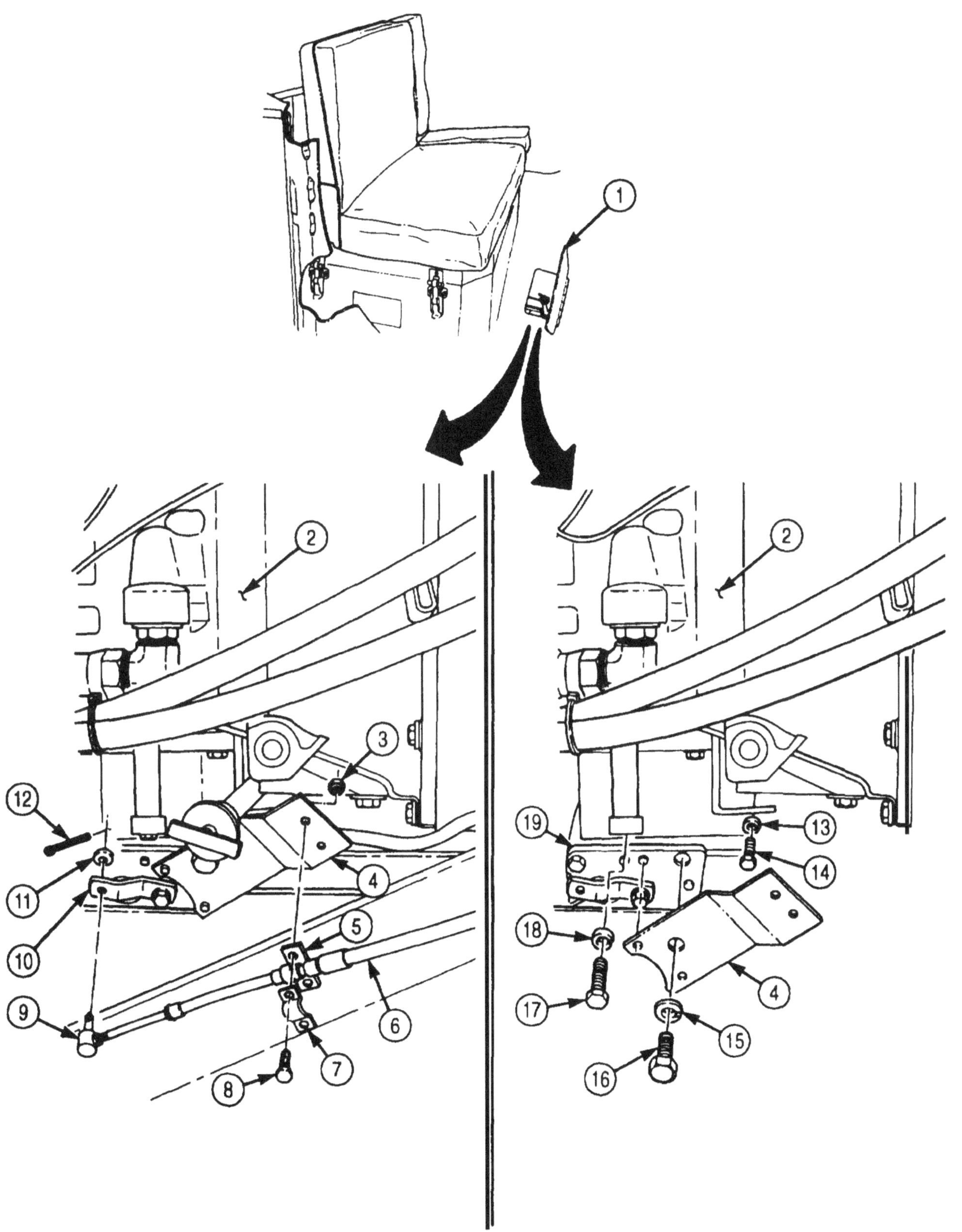

4-211. TRANSMISSION POWER TAKEOFF (PTO) REPLACEMENT (Contd)

7. Disconnect oil hose (8) from PTO oil hose adapter (7) and position oil hose (8) clear of work area.
8. Plug end of oil hose (8).
9. Remove two nuts (5), washers (6), screw (3), and washer (4) from PTO (2) and transmission (1).
10. Remove locknut (12) and screw (10) from right frame rail (13) and bracket (11). Discard locknut (12).
11. Remove PTO (2) and gasket (9) from transmission (1). Discard gasket (9).
12. Clean gasket remains from mating surfaces of transmission (1) and PTO (2).
13. Remove PTO oil hose adapter (7) from PTO (2) and plug orifice of PTO (2).

4-211. TRANSMISSION POWER TAKEOFF (PTO) REPLACEMENT (Contd)

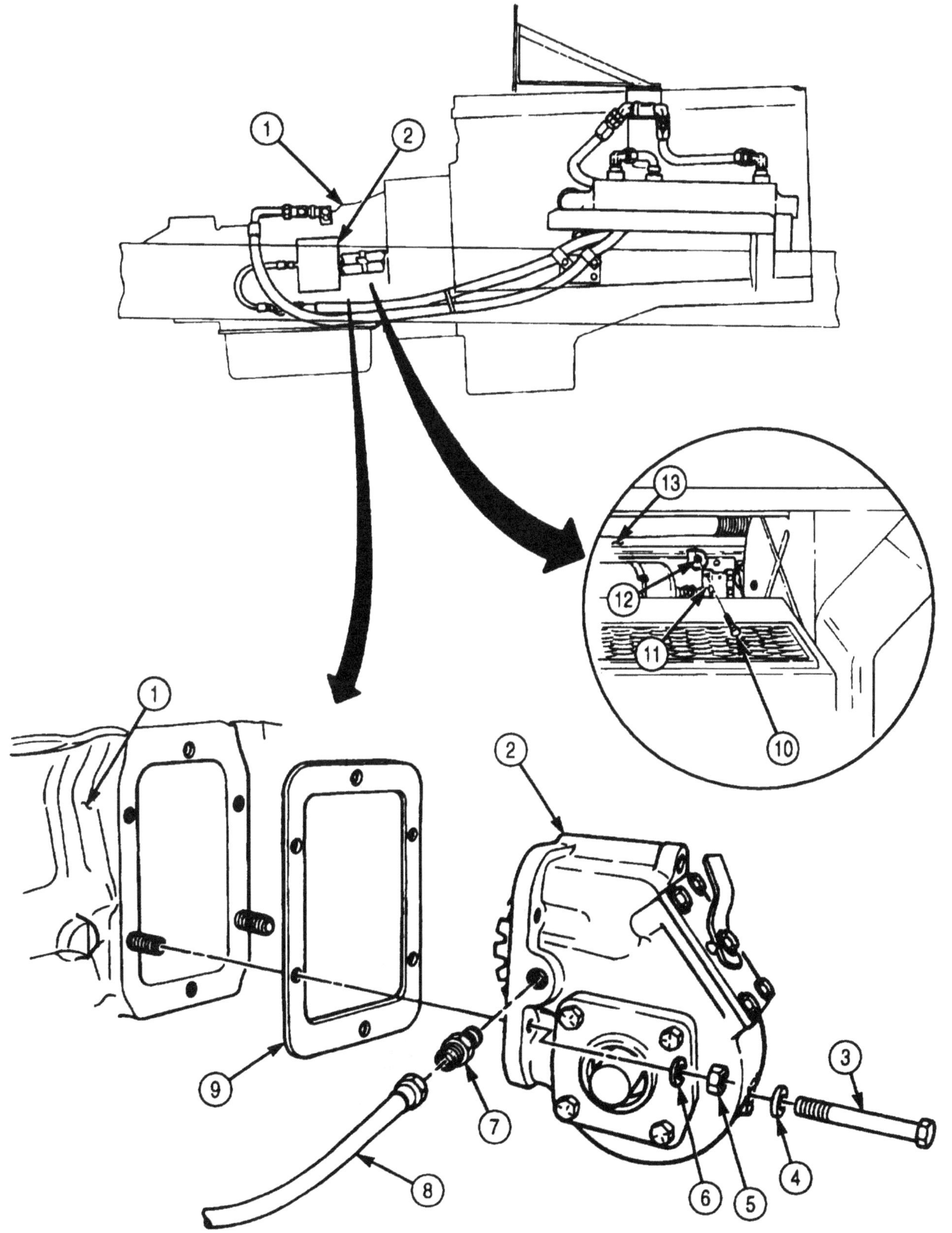

4-211. TRANSMISSION POWER TAKEOFF (PTO) REPLACEMENT

b. Installation

NOTE

- Wrap male pipe threads with antiseize tape before installation.
- Remove plug from PTO (if present) before installing PTO oil hose adapter.

1. Install PTO oil hose adapter (7) on PTO (2).
2. Loosely install washer (4) and screw (3) in PTO (2).
3. Install new gasket (9) and PTO (2) on transmission (1) with two washers (6) and nuts (5).
4. Tighten screw (3).
5. Remove plug from PTO oil hose (8) and connect PTO oil hose (8) to adapter (7).
6. Install screw (10) and new locknut (12) on frame rail (13) and bracket (11).

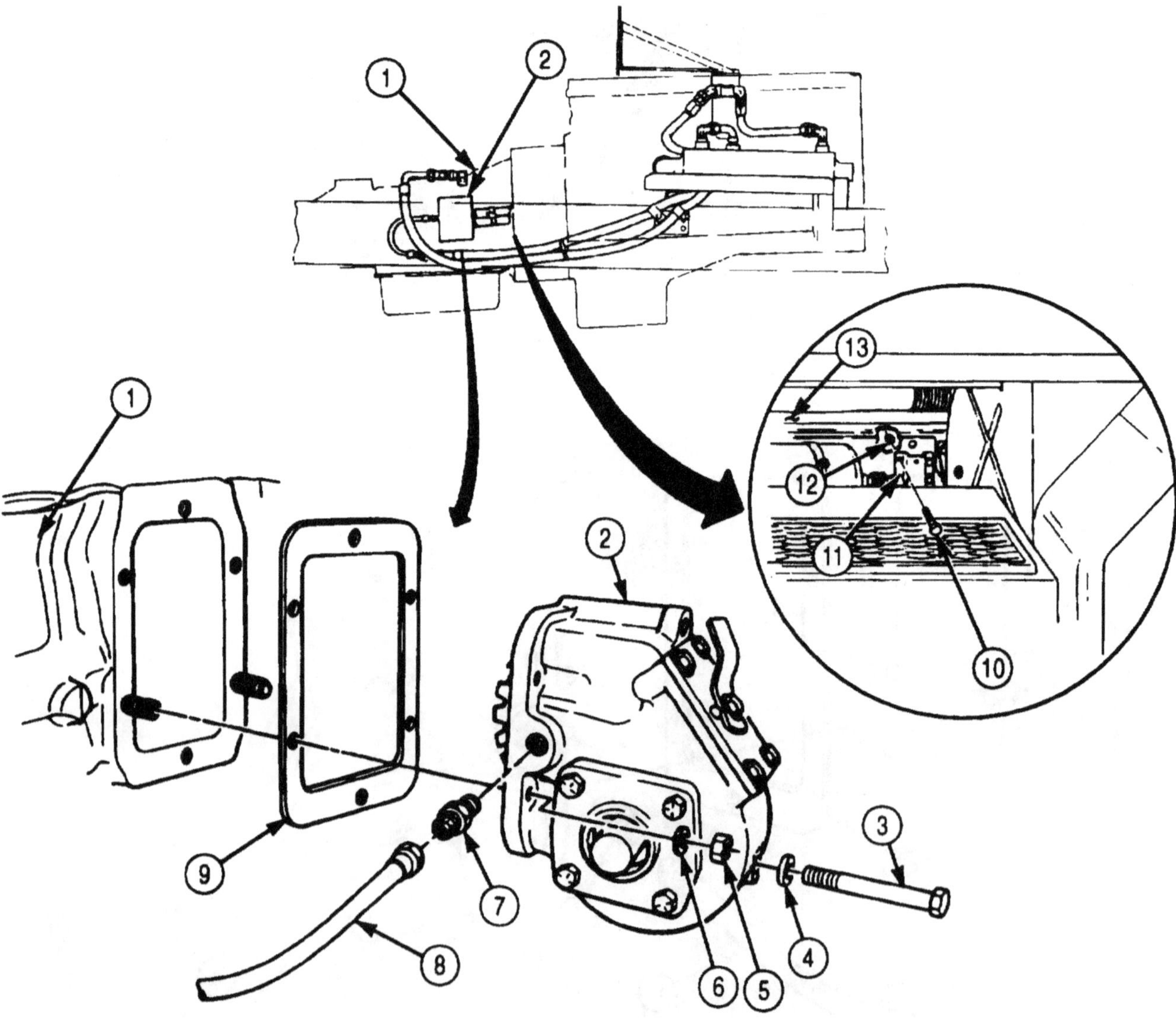

4-211. TRANSMISSION POWER TAKEOFF (PTO) REPLACEMENT (Contd)

7. Install two washers (24), screw (25), washer (29), and screw (28) on PTO (30) and transmission (1).
8. Install PTO cable bracket (15) on PTO (30) with washer (26) and screw (27).
9. Install PTO cable (19) and PTO cable pin (20) on PTO select lever (21) with washer (22) and new cotter pin (23).
10. Install spacer (16) and clamp (17) on PTO cable bracket (15) with two screws (18) and nuts (14).

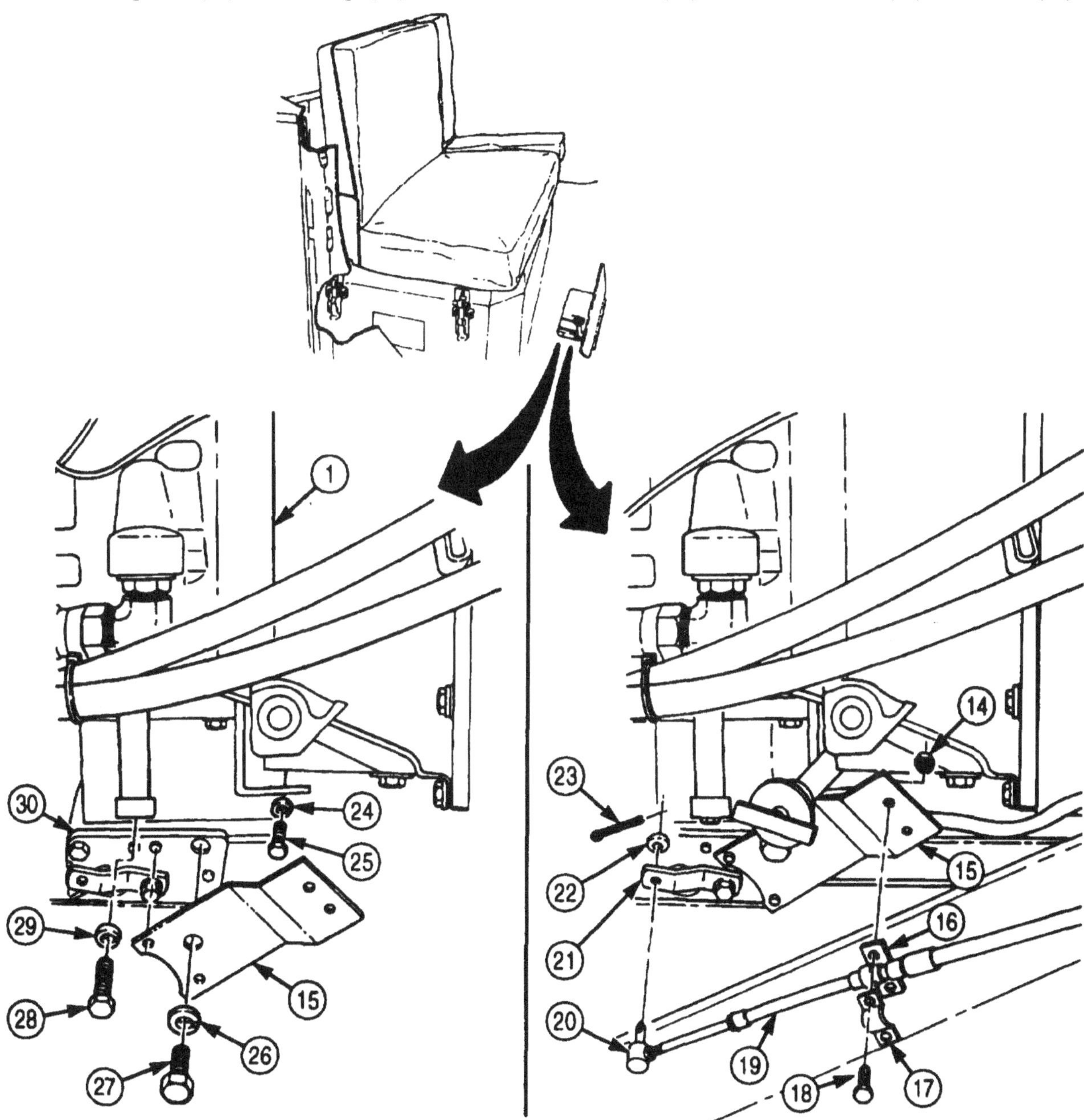

FOLLOW-ON TASKS:
- Install transmission PTO-to-hydraulic pump propeller shaft (para. 3-334).
- Install transmission oil dipstick tube (para. 3-3).
- Fill transmission to proper oil level (LO 9-2320-272-12).
- Start engine, check for leaks, and proper operation of transmission PTO (TM 9-2320-272-10).

4-212. TRANSMISSION POWER TAKEOFF (PTO) MAINTENANCE

THIS TASK COVERS:

a. Disassembly
b. Cleaning and Inspection
c. Assembly

INITIAL SETUP:

APPLICABLE MODELS

M925, M928, M929, M930, M932, M936

TOOLS

General mechanic's tool kit (Appendix E, Item 1)
Torque wrench (Appendix E, Item 144)
Dial indicator (Appendix E, Item 36)
Arbor press

MATERIAL/PARTS

Two shift cover gaskets (Appendix D, Item 221)
Two bearing cap gaskets (Appendix D, Item 222)
Two thrust washers (Appendix D, Item 683)
Output shaft oil seal (Appendix D, Item 501)
Snapring (Appendix D, Item 664)
O-ring (Appendix D, Item 459)
Lockwasher (Appendix D, Item 394)
Woodruff key (Appendix D, Item 733)
Oil-soluble grease (Appendix C, Item 53)
Drycleaning solvent (Appendix C, Item 71)
Lubricating oil (Appendix C, Item 48)

REFERENCES (TM)

TM 9-2320-272-24P

EQUIPMENT CONDITION

Transmission PTO removed (para. 4-211).

GENERAL SAFETY INSTRUCTIONS

- Keep tire extinguisher nearby when using drycleaning solvent.
- Drycleaning solvent is flammable and toxic. Do not use near an open flame.
- When cleaning with compressed air, wear eyeshields and ensure source pressure does not exceed 30 psi (207 kPa).

a. Disassembly

1. Remove woodruff key (7) from PTO output driveshaft (8). Discard woodruff key (7).
2. Remove four screws (3), shifter cover (2), gaskets (4) and (6), and spacer (5) from PTO housing (1). Discard gaskets (4) and (6).
3. Remove setscrew (11) and pipe plug (13) from PTO mating flange (12).
4. Using arbor press, press idler pin (10) completely through PTO housing (1).
5. Remove lube adapter (9) (if present) with idler pin (10).

4-212. TRANSMISSION POWER TAKEOFF (PTO) MAINTENANCE (Contd)

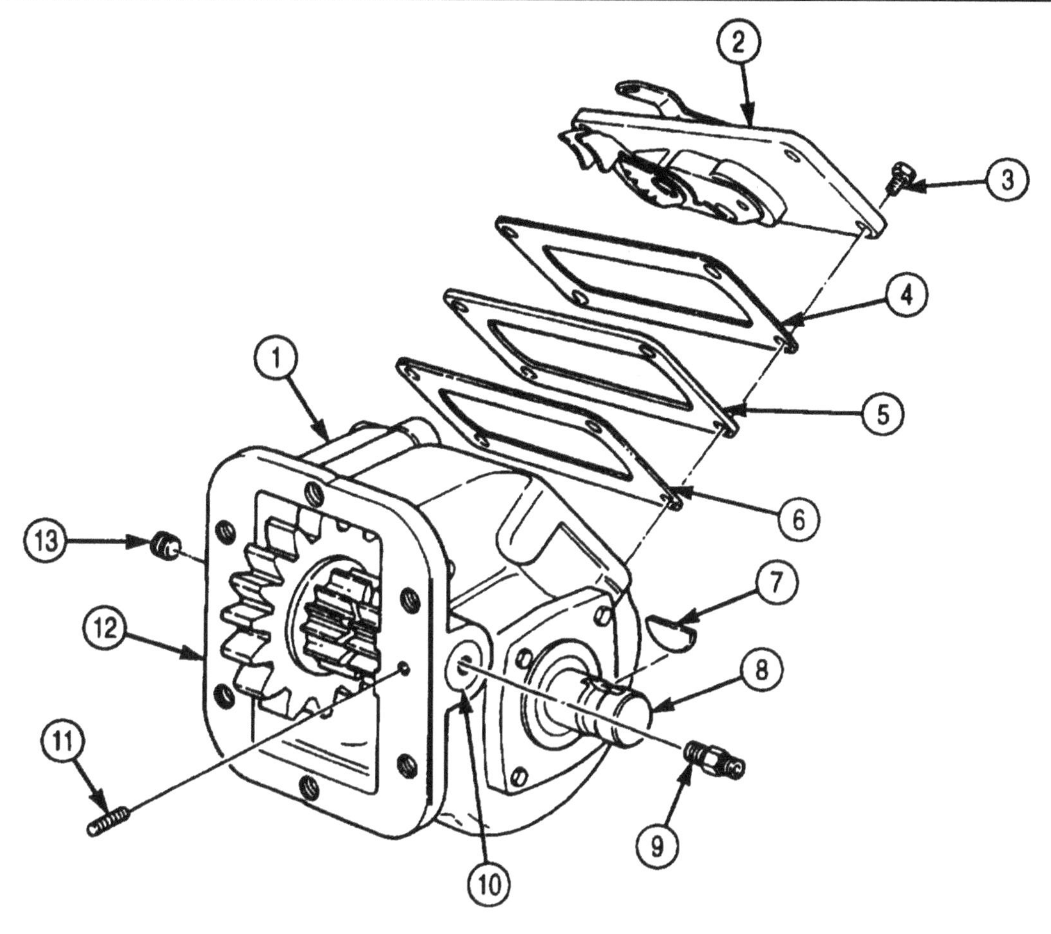

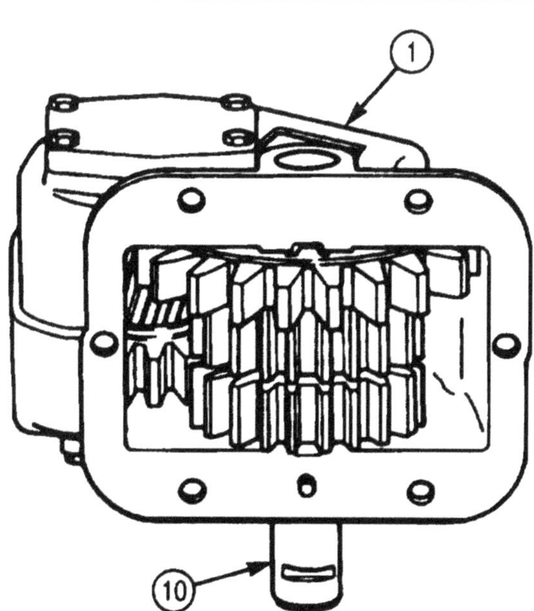

4-212. TRANSMISSION POWER TAKEOFF (PTO) MAINTENANCE (Contd)

6. Remove input gear (6), drive gear (5), and two thrust washers (2) from PTO housing (1). Discard thrust washers (2).
7. Slide input gear (6) off drive gear (5).
8. Remove nineteen needle bearings (3) from each end of drive gear (5).
9. Remove spacer (4) from drive gear (5).

NOTE

- Before removing input bearing cap from PTO housing, mark cap and housing for installation.
- Exact number of gaskets found when removing bearing caps may vary.

10. Remove four screws (22), output bearing cap (21), and gasket (20) from PTO housing (1). Discard gasket (20).
11. Remove four screws (17), input bearing cap (18), and gasket (19) from PTO housing (1). Discard gasket (19).

NOTE

Mark lever, shaft, shifter cover, and hole containing poppet and spring for installation.

12. Remove screw (7), lockwasher (8), lever (9), washer (10), gear controller (14), poppet (13), and spring (12) from shifter cover (11). Discard lockwasher (8).
13. Remove O-ring (16) from gear controller shaft (15). Discard O-ring (16).

4-212. TRANSMISSION POWER TAKEOFF (PTO) MAINTENANCE (Contd)

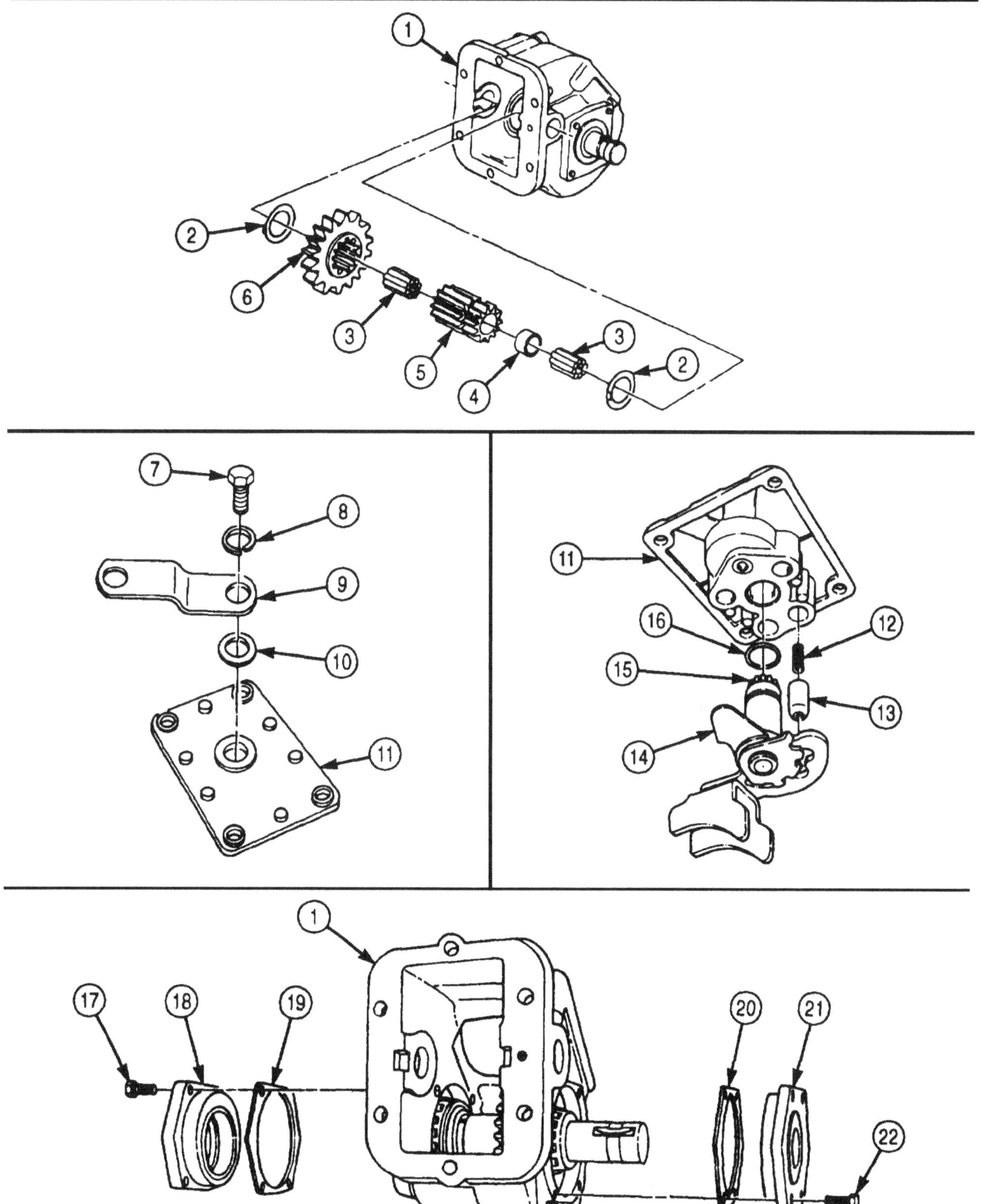

4-212. TRANSMISSION POWER TAKEOFF (PTO) MAINTENANCE (Contd)

NOTE

Disassembly of output and input bearing caps is the same. This procedure covers output bearing cap.

14. Remove bearing cup (2) from bearing cap (1).

NOTE

Step 15 is performed on the output bearing cap only.

15. Remove output shaft oil seal (3) from bearing cap (1). Discard output shaft oil seal (3).

NOTE

Tag input bearing and output bearing for installation.

16. Remove input bearing (4) and snapring (6) from output shaft (5). Discard snapring (6).
17. Remove output shaft (5) from PTO housing (8). Spacer (7) and output gear (10) will slide off output shaft (5) when removed.
18. Remove output bearing (9) from output shaft (5).

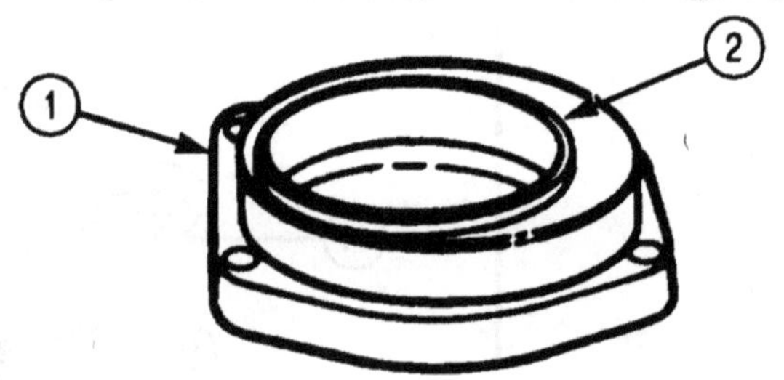

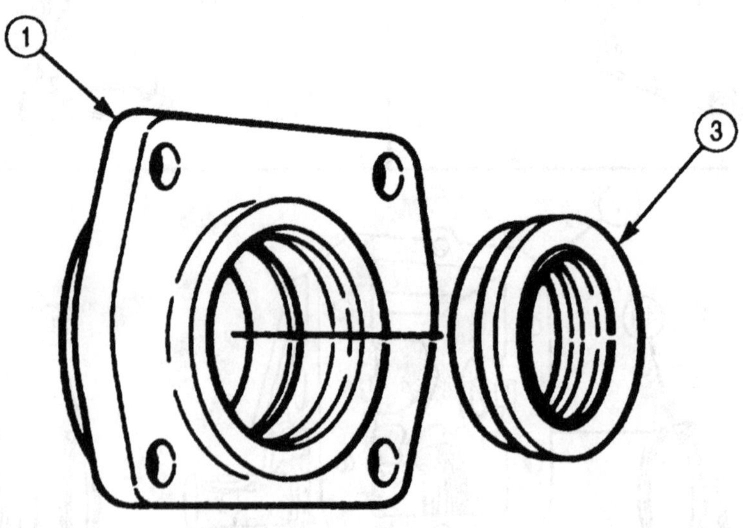

4-212. TRANSMISSION POWER TAKEOFF (PTO) MAINTENANCE (Contd)

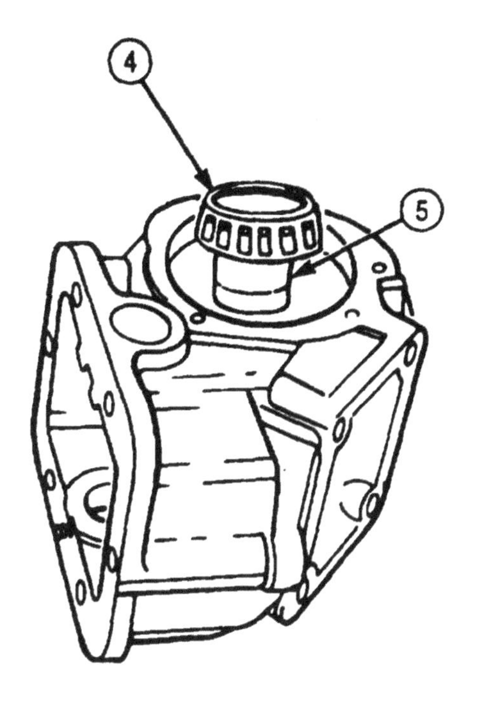

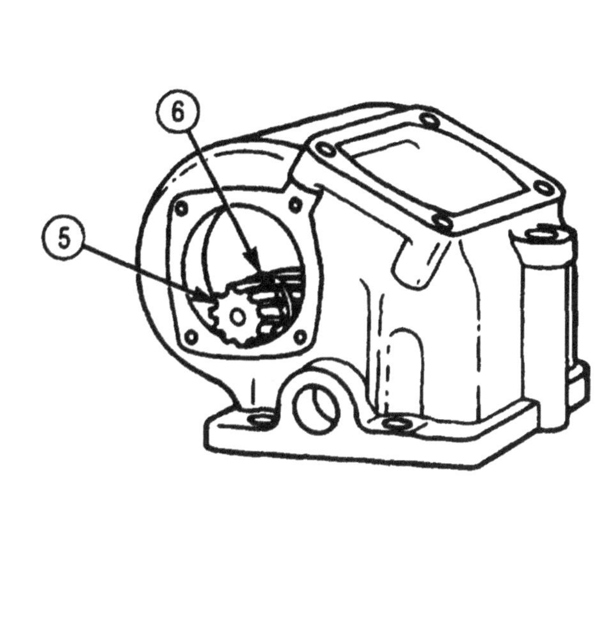

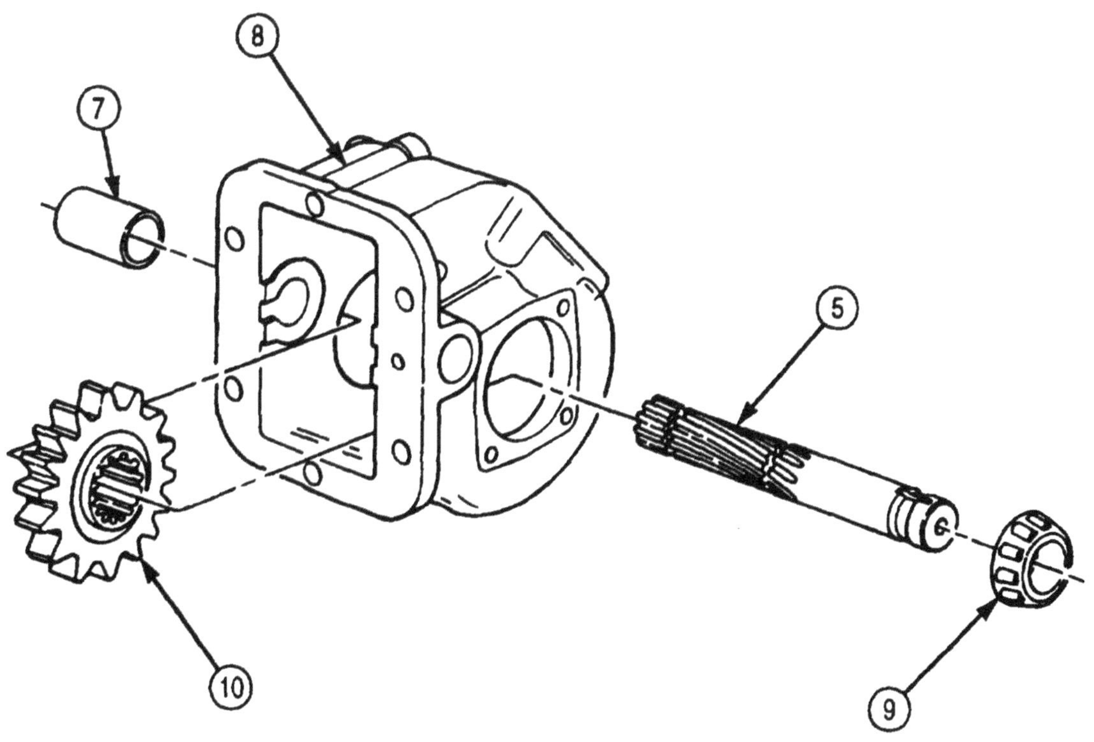

4-212. TRANSMISSION POWER TAKEOFF (PTO) MAINTENANCE (Contd)

b. Cleaning and Inspection

WARNING

- Drycleaning solvent is flammable and toxic. Do not use near open flame and always have a fire extinguisher nearby when solvents are used. Use only in well-ventilated places, wear protective clothing, and dispose of cleaning rags in approved container. Failure to do this may result in injury or death to personnel and/or damage to equipment.
- Eyeshields must be worn when cleaning with compressed air. Compressed air source will not exceed 30 psi (207 kPa). Failure to do so may result in injury to personnel.

NOTE

For general cleaning instructions, refer to para. 2-14.

1. Clean idler pin (9), input gear (11), drive gear (10), output gear (4), and output shaft (5) with drycleaning solvent and inspect for breaks, cracks, chips, pitting, and wear. If broken, cracked, chipped, pitted, or worn, replace.
2. Clean thirty-eight needle bearings (8), input bearing (2), and output bearing (6) with drycleaning solvent. Do not dry bearings (2) and (6) with compressed air.
3. Inspect thirty-eight needle bearings (8), input bearing (2), and output bearing (6) for pits, scores, cracks, and chips. If pitted, scored, cracked, or chipped, replace.
4. Clean PTO housing (3) with drycleaning solvent and remove gasket remains from mating surfaces.
5. Inspect PTO housing (3) for cracks and chips.
6. Clean screw (18), lever (17), washer (16), shifter cover (15), spring (14), poppet (13), gear controller (12), and bearing caps (1) and (7) with drycleaning solvent. Clean gasket remains from mating surfaces.
7. Inspect screw (18), lever (17), washer (16), shifter cover (15), spring (14), poppet (13), gear controller (12), and bearing caps (1) and (7) for pits, breaks, cracks, chips, and wear. If pitted, broken, cracked, chipped, or worn, replace.

4-212. TRANSMISSION POWER TAKEOFF (PTO) MAINTENANCE (Contd)

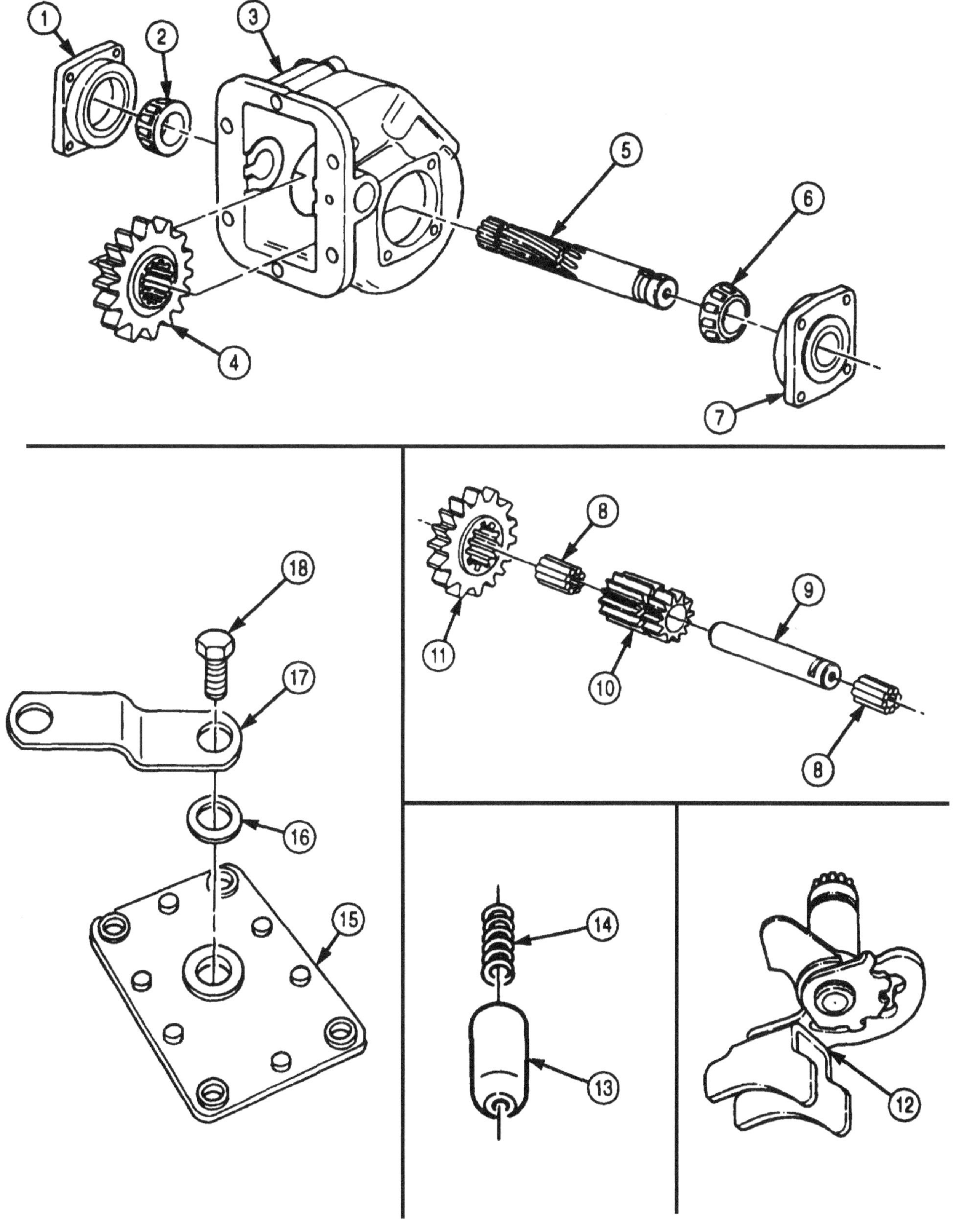

4-212. TRANSMISSION POWER TAKEOFF (PTO) MAINTENANCE (Contd)

c. Assembly

NOTE

Apply lubricating oil to all PTO components during assembly.

1. Using wood block and hammer, install one bearing cup (2) into output bearing cap (3) and the other bearing cup (2) into input bearing cap (1).
2. Place PTO housing (7) on workbench with PTO-to-transmission flange (12) facing you. Ensure mark on output bearing cap side (8) is to your right.
3. Place output gear (13) in right side of PTO housing (7). Ensure flat edge of output gear (13) faces output shaft keyway (11).
4. Place output shaft (10) through right (output) side (8) of PTO housing (7) and output gear (13).
5. Slide spacer (6) and new snapring (5) over input side of output shaft (10). Ensure snapring seats in groove (9).
6. Install input bearing (4) to input side of output shaft (10) until bottomed at snapring (5).

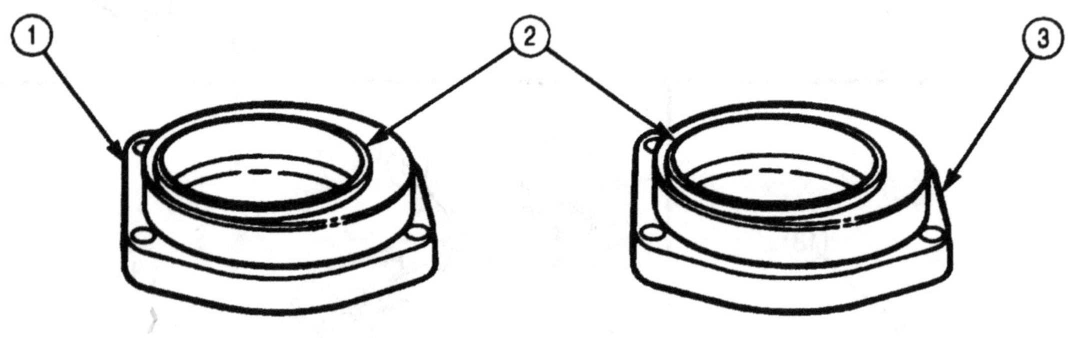

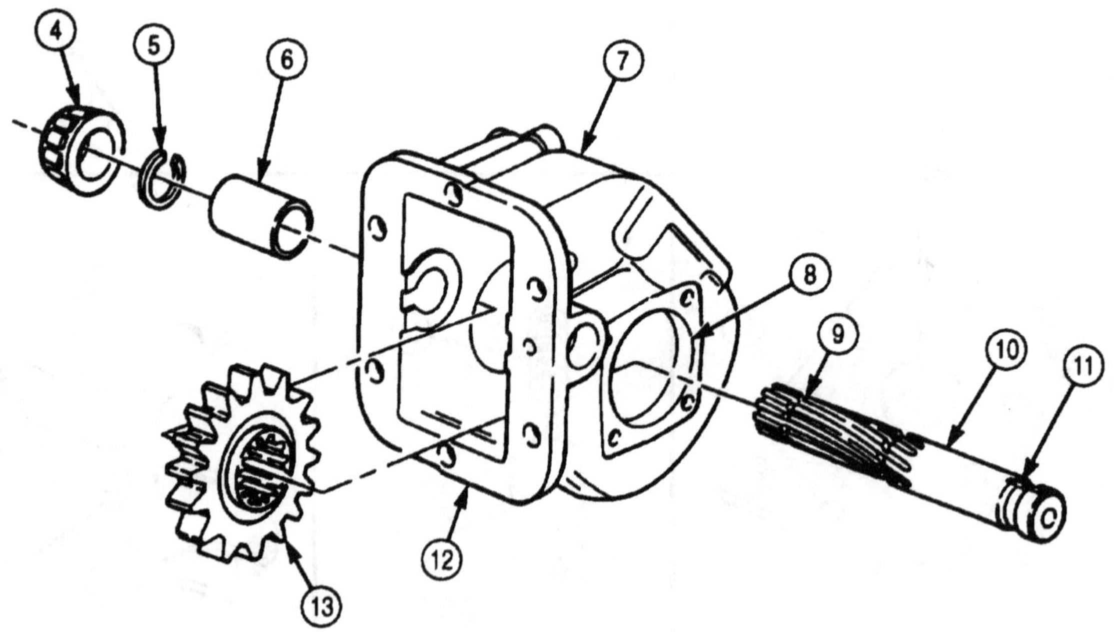

4-212. TRANSMISSION POWER TAKEOFF (PTO) MAINTENANCE (Contd)

7. Using arbor press, install output bearing (4) on output side of output shaft (10) until seated on output gear (13).

NOTE

For correct bearing cap installation, ensure arrow on input bearing cap points toward shifter cover opening. Once positioned, the output bearing cap can be correctly installed.

8. Install new gasket(s) (15) and input bearing cap (1) on unmarked side of PTO housing (7) with four screws (14). Tighten screws (14) 8-10 lb-ft (11-14 N•m).
9. Install new gasket(s) (16) and output bearing cap (3) on marked side of PTO housing (7) with four screws (17). Tighten screws (17) 8-10 lb-ft (11-14 N•m).

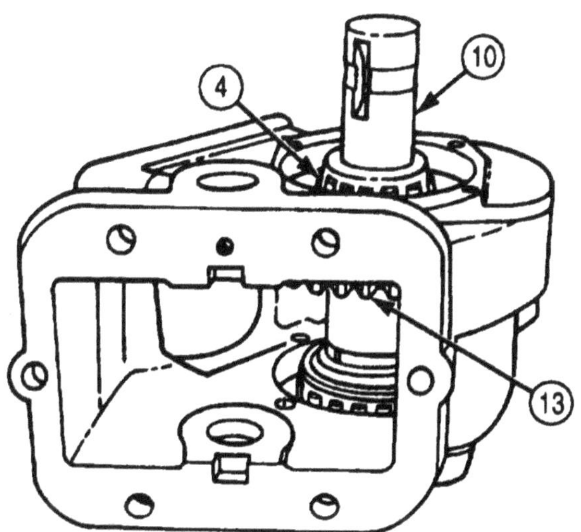

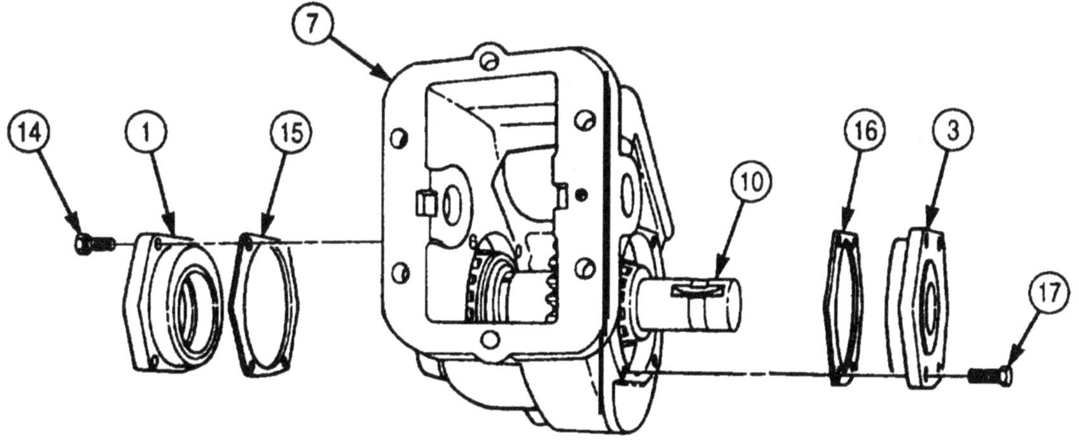

4-212. TRANSMISSION POWER TAKEOFF (PTO) MAINTENANCE (Contd)

10. Check output shaft (1) end play with dial indicator. Output shaft (1) should turn freely and end play must not exceed .006 in. (0.152 mm).

NOTE

If end play exceeds .006 in. (0.152 mm), remove gasket(s). If end play is too small, add gasket(s) to increase end play.

11. Adjust output shaft (1) end play by using prybar through hole (2) and forcing shaft (1) up and down.
12. Install new output shaft oil seal (3) on output shaft (1) until flush with outside edge of output bearing cap (4).

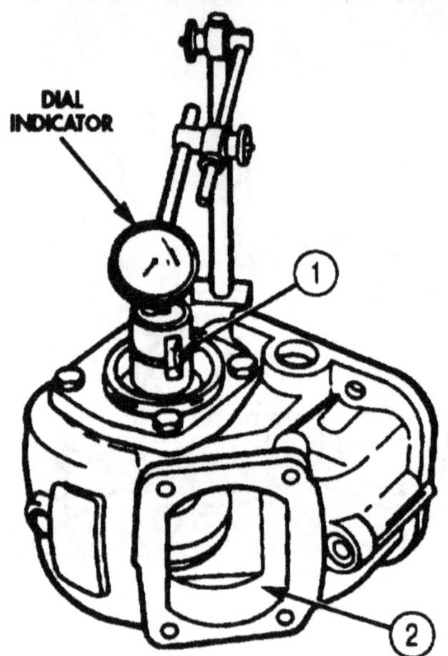

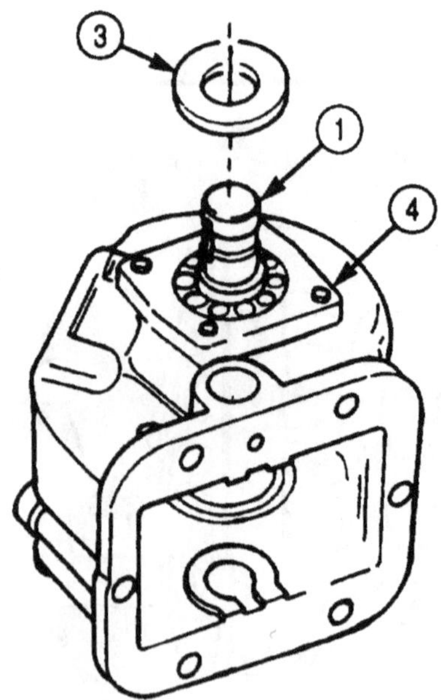

4-212. TRANSMISSION POWER TAKEOFF (PTO) MAINTENANCE (Contd)

13. Apply coat of oil-soluble grease to inside of drive gear (8) and set drive gear (8) flat on workbench.

NOTE

Use oil-soluble grease to hold needle bearings in place in steps 14 and 15.

14. Insert first nineteen needle bearings (7) in drive gear (8).
15. Position spacer (6) in drive gear (8) next to first nineteen needle bearings (7) and insert second nineteen needle bearings (5) in drive gear (8) until bottomed against spacer (6).
16. Using arbor press, press idler pin (9) through idler pin hole (13) from output bearing cap (4) side of housing (11) until flush with inside wall of housing (11). Ensure idler pin slot (10) will align with screw hole (12) when idler pin (9) is pressed through housing (11).

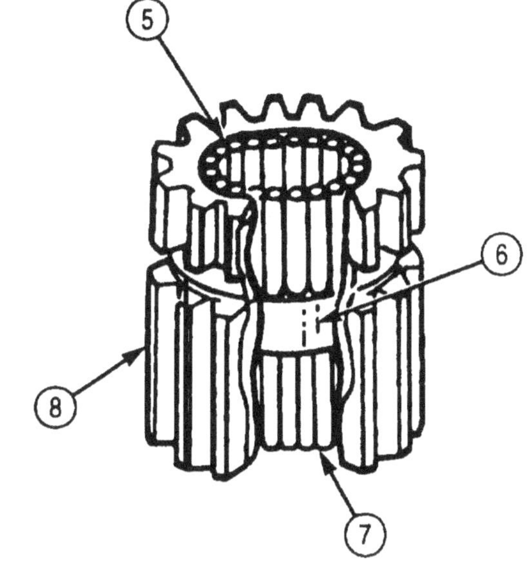

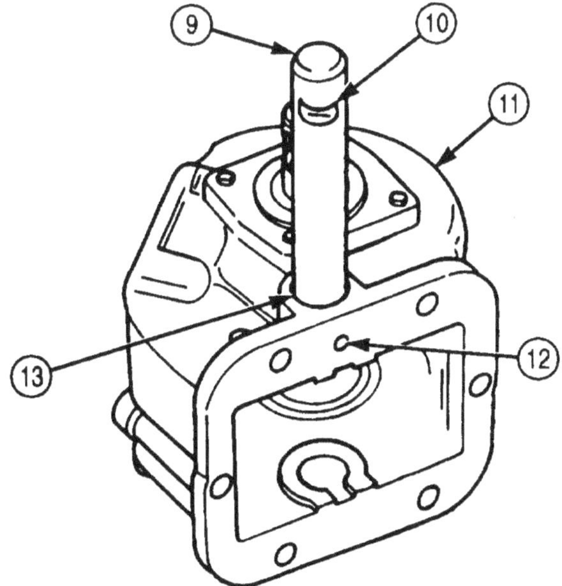

4-212. TRANSMISSION POWER TAKEOFF (PTO) MAINTENANCE (Contd)

17. Slide input gear (3) onto long side of drive gear (2) with large flange side first.
18. Position drive gear (2) in PTO housing (1) with short part of drive gear (2) meshing with output gear (9).

NOTE

Use oil-soluble grease to hold thrust washer in place.

19. Place new thrust washer (7) between input gear (3) and idler pin hole (4) so tab (8) sits in housing slot (6).
20. Place new thrust washer (7) between drive gear (2) and idler pin hole (4) so tab (8) sits in housing slot (6).
21. Using arbor press, press idler pin (5) completely through PTO housing (1) until both ends of idler pm (5) are flush with PTO housing (1).
22. Install lube adapter (15) and plug (19) in idler pin (5).
23. Install setscrew (17) in PTO-to-transmission flange (18). Ensure screw (17) is aligned with groove in idler pin (5).
24. Install new O-ring (23) on shaft of gear controller (12).
25. Install spring (21) and poppet (22) in hole scribed for installation in shifter cover (10).
26. Install gear controller (12) on shifter cover (10) with new lockwasher (27), screw (26), lever (29), and washer (28).
27. Align lever (29) to scribe marks on shifter cover (10). Ensure point (24) on gear controller (12) engages dimple (25) in poppet (22).
28. Install two new gaskets (13), spacer (20), and shifter cover (10) on PTO housing (1) with four screws (11). Tighten screws (11) 8-10 lb-ft (11-14 N•m). Ensure gear controller (12) engages input gear (3).
29. Install new woodruff key (14) on PTO output driveshaft (16).

4-212. TRANSMISSION POWER TAKEOFF (PTO) MAINTENANCE (Contd)

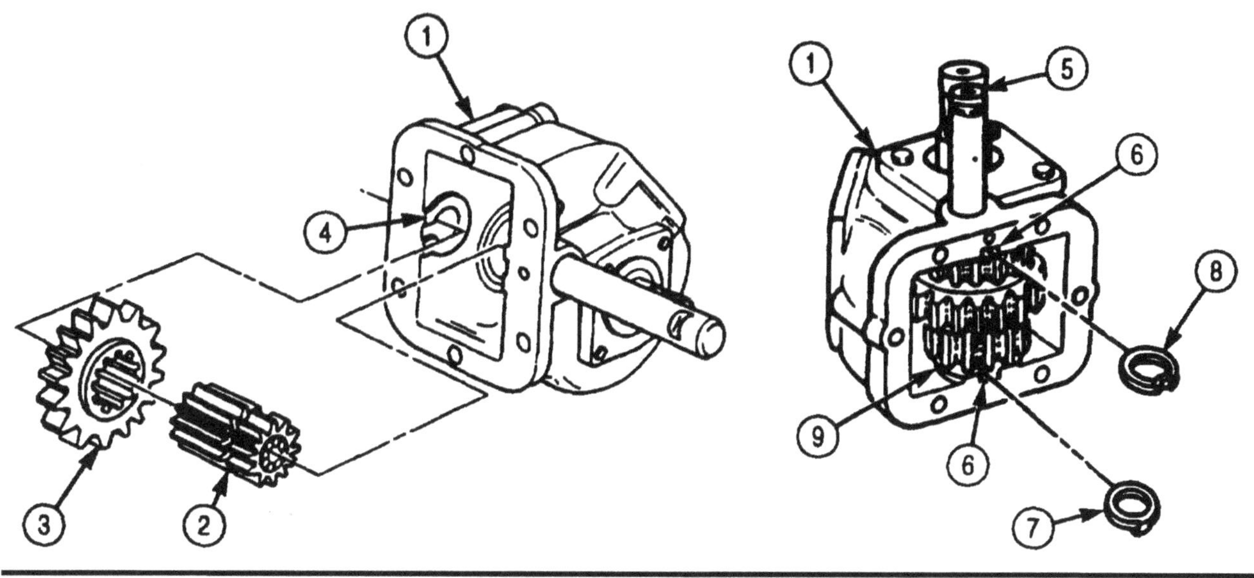

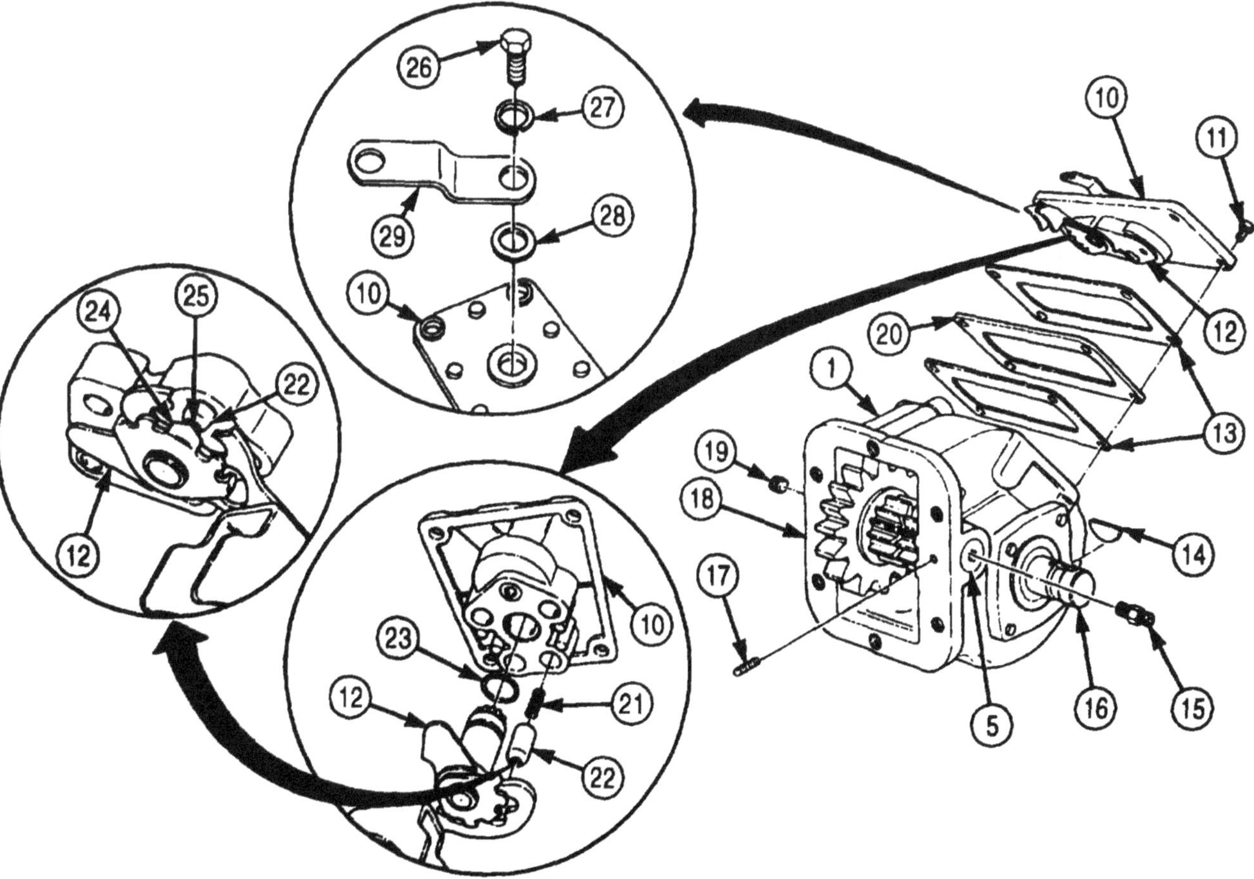

FOLLOW-ON TASK: Install transmission PTO (para. 4-211).

Section XVI. SPECIAL PURPOSE KITS MAINTENANCE

4-213. SPECIAL PURPOSE KITS MAINTENANCE INDEX

4-214. PERSONNEL FUEL BURNING HEATER CONTROL BOX REPLACEMENT

THIS TASK COVERS:

a. Removal **b. Installation**

INITIAL SETUP:

APPLICABLE MODELS
All

TOOLS
General mechanic's tool kit (Appendix E, Item 1)

MATERIALS/PARTS
Two lockwashers (Appendix D, Item 345)
Two locknuts (Appendix D, Item 313)

REFERENCES (TM)
TM 9-2320-272-10
TM 9-2320-272-24P

EQUIPMENT CONDITION
- Parking brake set (TM 9-2320-272-10).
- Battery ground cables disconnected (para. 3-126).

a. Removal

1. Disconnect coolant heater harness (4) from coolant heater control box (7).
2. Disconnect connector (5) from control box wire (6) on coolant heater box (7).

4-214. PERSONNEL FUEL BURNING HEATER CONTROL BOX REPLACEMENT (Contd)

3. Remove two nuts (2) and lockwashers (3) from studs (8) and remove coolant heater box (7) from mounting bracket (10). Discard lockwashers (1).
4. Remove two locknuts (1), screws (9), and mounting bracket (10) from underside of dash (11). Discard locknuts (1).

b. Installation

1. Install mounting bracket (10) on underside of dash (11) with two screws (9) and new locknuts (1).
2. Install coolant heater box (7) on mounting bracket (10) with two new lockwashers (3) and nuts (2).
3. Connect connector (5) to control box wire (6) on coolant heater box (7).
4. Connect coolant heater harness (4) to coolant heater box (7).

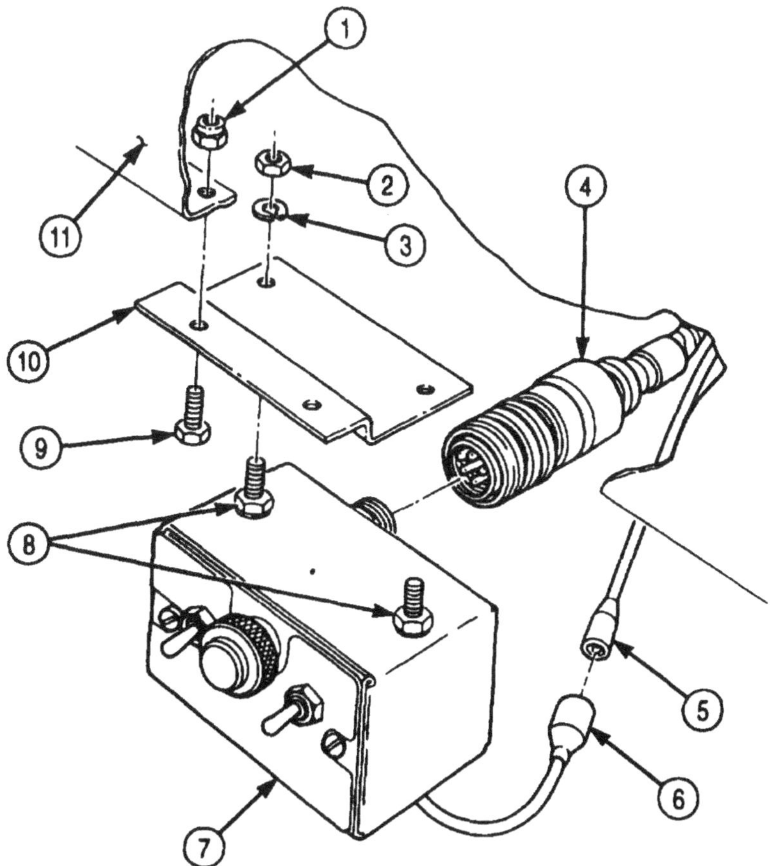

FOLLOW-ON TASKS:
- Connect battery ground cables (para. 3-126).
- Check personnel fuel burning heater for proper operation (TM 9-2320-272-10).

4-215. PERSONNEL FUEL BURNING HEATER (M939/A1) REPLACEMENT

THIS TASK COVERS:

a. Removal **b. Installation**

INITIAL SETUP:

APPLICABLE MODELS
M939/A1

TOOLS
General mechanic's tool kit (Appendix E, Item 1)

MATERIALS/PARTS
Cotter pin (Appendix D, Item 59)

REFERENCES (TM)
TM 9-2320-272-10
TM 9-2320-272-24P

EQUIPMENT CONDITION
- Parking brake set (TM 9-2320-272-10).
- Hood raised and secured (TM 9-2320-272-10).

GENERAL SAFETY INSTRUCTIONS
Diesel fuel is flammable. Do not perform this procedure near open flames.

a. Removal

WARNING

Diesel fuel is flammable. Do not perform this procedure near open flames. Injury to personnel may result.

1. Disconnect heater harness (2) from fuel burning heater (1).
2. Disconnect fuel line (7) from heater elbow (6).
3. Remove cotter pin (4) from exhaust tube (5) and fuel burning heater (1). Discard cotter pin (4).
4. Loosen two clamps (3) and remove fuel burning heater (1) from clamps (3).

b. Installation

1. Install fuel burning heater (1) into clamps (3) and tighten clamps (3).
2. Install exhaust tube (5) on fuel burning heater (1) with new cotter pin (4).
3. Connect fuel line (7) to heater elbow (6).
4. Connect heater harness (2) to fuel burning heater (1).

4-215. PERSONNEL FUEL BURNING HEATER (M939/A1) REPLACEMENT (Contd)

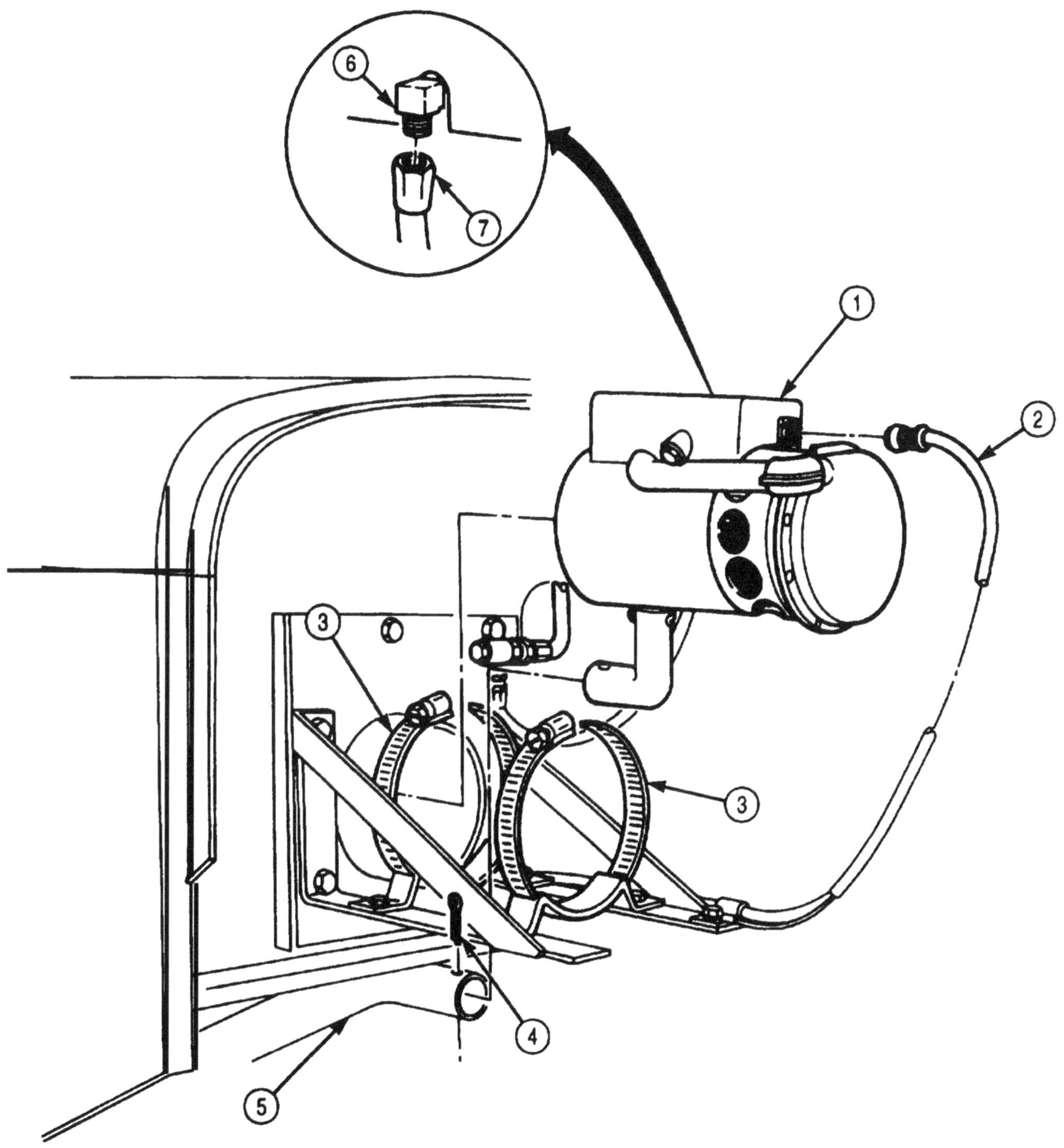

FOLLOW-ON TASK: Check personnel fuel burning heater for proper operation (TM 9-2320-272-10).

4-216. PERSONNEL FUEL BURNING HEATER MOUNTING BRACKET (M939/A1) REPLACEMENT

THIS TASK COVERS:

a. Removal **b. Installation**

INITIAL SETUP:

APPLICATION MODELS
M939/A1

TOOLS
General mechanic's tool kit (Appendix E, Item 1)

REFERENCES (TM)
TM 9-2320-272-10
TM 9-2320-272-24P

EQUIPMENT CONDITION
- Parking brake set (TM 9-2320-272-10).
- Hood raised and secured (TM 9-2320-272-10).
- Personnel fuel burning heater removed (para. 4-215).

a. Removal

Remove four screws (4) and mounting brackets (2) and (3) from firewall (1).

b. Installation

Install mounting brackets (2) and (3) on firewall (1) with four screws (4).

4-216. PERSONNEL FUEL BURNING HEATER MOUNTING BRACKET (M939/A1) REPLACEMENT (Contd)

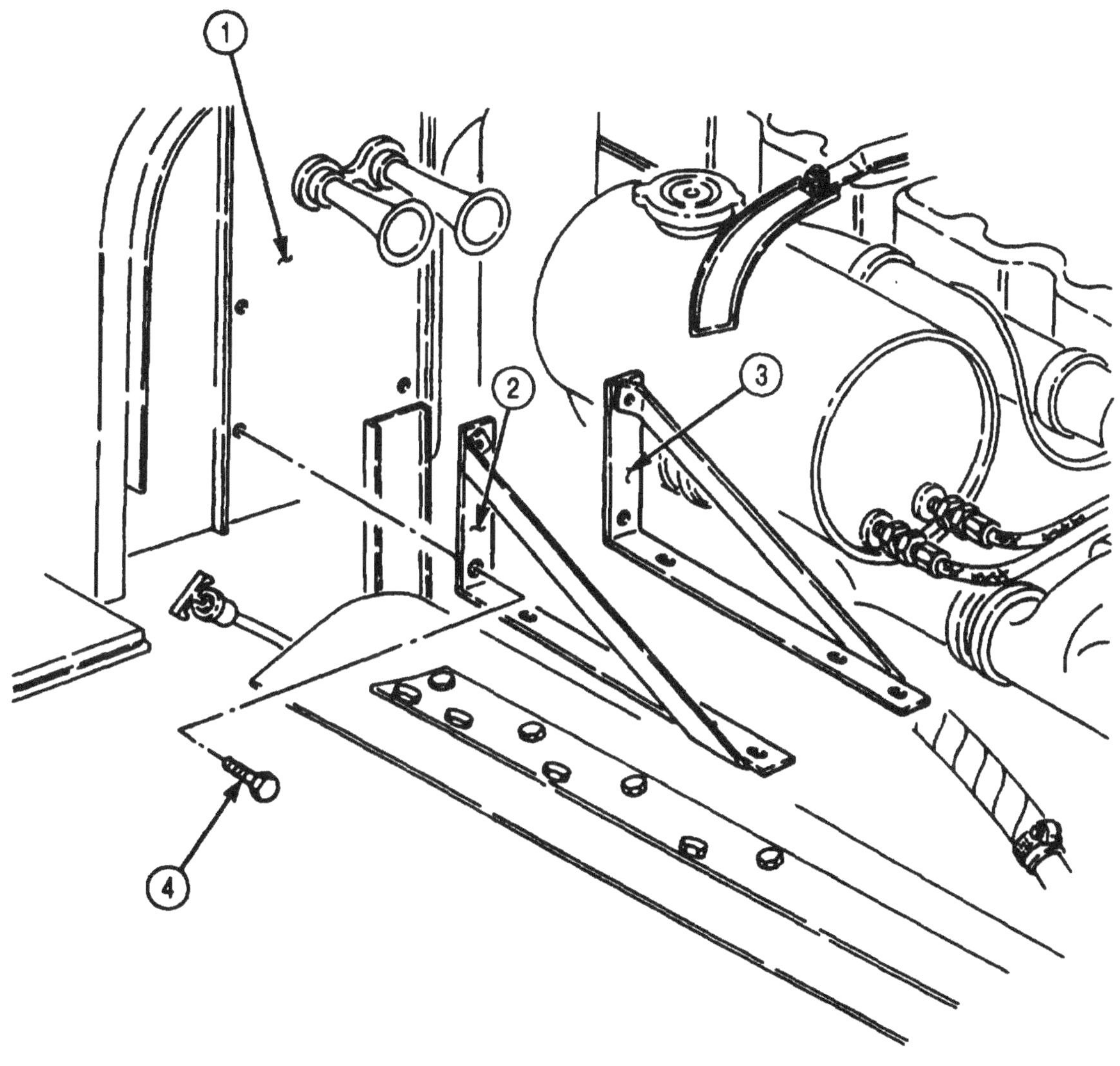

FOLLOW-ON TASK: Install personnel fuel burning heater (para. 4-215).

4-217. PERSONNEL FUEL BURNING HEATER AND MOUNTING BRACKET (M939A2) REPLACEMENT

THIS TASK COVERS:

a. Removal **b. Installation**

INITIAL SETUP:

APPLICABLE MODELS
M939A2

TOOLS
General mechanic's tool kit (Appendix E, Item 1)

MATERIALS/PARTS
Lockwasher (Appendix D, Item 353)
Four locknuts (Appendix D, Item 299)
Cotter pin (Appendix D, Item 71)
Antiseize tape (Appendix C, Item 72)

PERSONNEL REQUIRED
Two

REFERENCES (TM)
TM 9-2320-272-10
TM 9-2320-272-24P

EQUIPMENT CONDITION
- Parking brake set (TM 9-2320-272-10).
- Hood raised and secured (TM 9-2320-272-10).
- Fuel shutoff valve closed (TM 9-2320-272-10).
- Battery ground cables disconnected (para. 3-126).

GENERAL SAFETY INSTRUCTIONS
Diesel fuel is flammable. Do not perform this procedure near open flames.

a. Removal

WARNING

Diesel fuel is flammable. Do not perform this procedure near open flames. Injury to personnel may result.

1. Disconnect heater harness (13) from heater (1).
2. Disconnect hose (4) from adapter (3).
3. Remove adapter (3) from heater (1).
4. Remove cotter pin (10) from elbow (11) and exhaust port (14). Discard cotter pin (10).
6. Bemove two clamps (2) and heater (1) from brackets (9) and (21).
6. Remove two locknuts (8), washers (7), screws (5), washers (6), and bracket (9) from two mounting brackets (12). Discard locknuts (8).
7. Remove two locknuts (19), washers (20), screws (15), bracket (21), two washers (18), lockwasher (17), and ground wire (16) from two mounting brackets (12). Discard lockwasher (17) and locknuts (19).

b. Installation

NOTE

Clean all male pipe and hose threads and wrap with antiseize tape before installation.

1. Install bracket (21) on two mounting brackets (12) with two washers (18), new lockwasher (17), ground wire (16), two screws (15), washers (20), and new locknuts (19).
2. Install bracket (9) on two mounting brackets (12) with two washers (6), screws (5), washers (7), and new locknuts (8).
3. Install heater (1) on brackets (9) and (21) with two clamps (2).
4. Install new cotter pin (10) on elbow (11) and exhaust port (14).
5. Install adapter (3) on heater (1).
6. Connect hose (4) to adapter (3).
7. Connect heater harness (13) to heater (1).

4-217. PERSONNEL FUEL BURNING HEATER AND MOUNTING BRACKET (M939A2) REPLACEMENT (Contd)

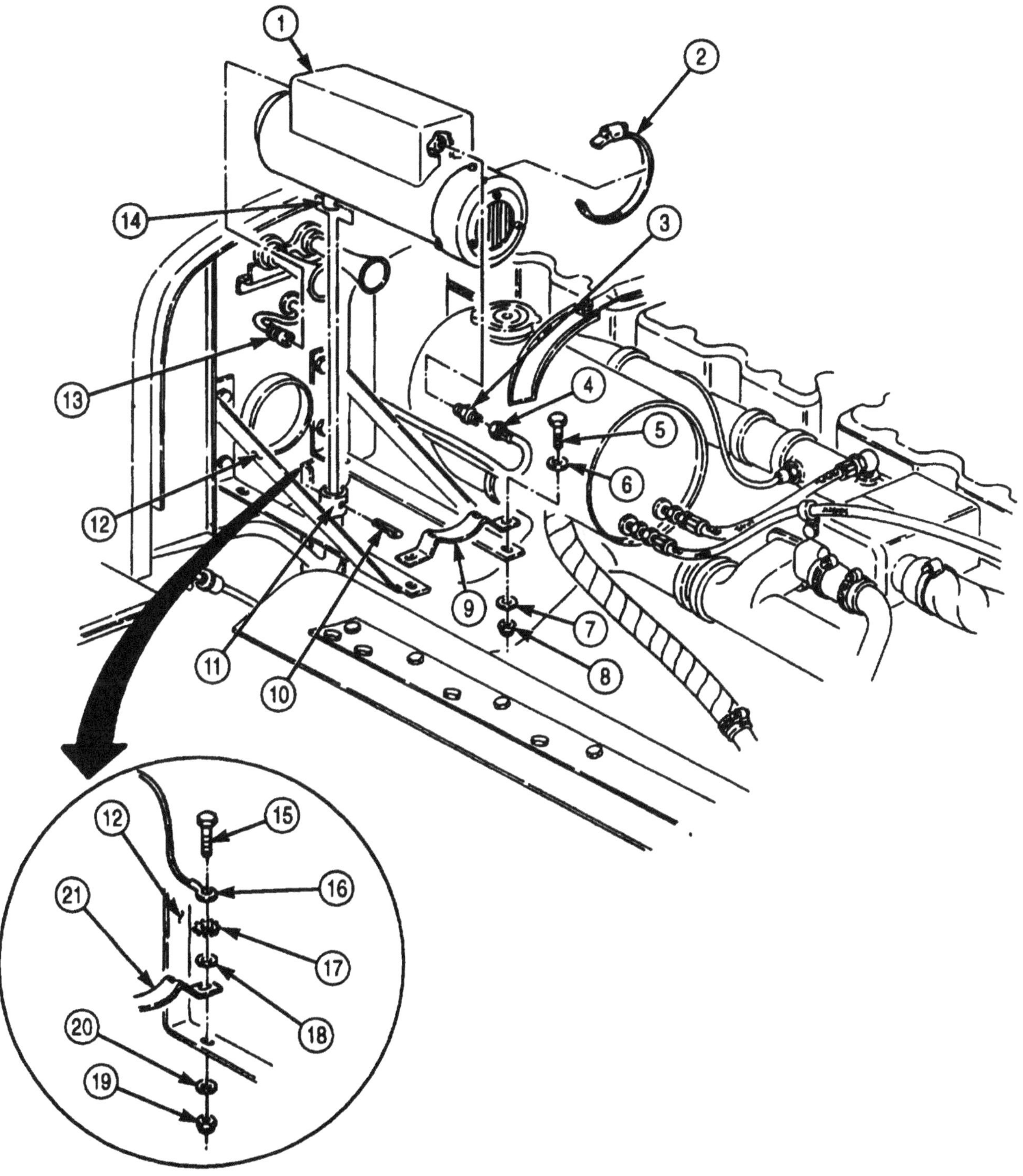

FOLLOW-ON TASKS:
- Connect battery ground cables (para. 3-126).
- Open fuel shutoff valve (TM 9-2320-272-10).
- Check personnel fuel burning heater for proper operation (TM 9-2320-272-10).

4-218. PERSONNEL FUEL BURNING HEATER FUEL PUMP (M939/A1) REPLACEMENT

THIS TASK COVERS:

a. Removal **b. Installation**

INITIAL SETUP:

APPLICABLE MODELS
M939/A1

TOOLS
General mechanic's tool kit (Appendix E, Item 1)

MATERIALS/PARTS
Two locknuts (Appendix D, Item 313)
Antiseize tape (Appendix C, Item 72)
Cap and plug set (Appendix C, Item 14)

PERSONNEL REQUIRED
Two

REFERENCES (TM)
TM 9-2320-272-10
TM 9-2320-272-24P

EQUIPMENT CONNECTION
- Parking brake set (TM 9-2320-272-10).
- Battery ground cables disconnected (para. 3-126).

GENERAL SAFETY INSTRUCTIONS
Diesel fuel is flammable. Do not perform this procedure near open flames.

a. Removal

WARNING

Diesel fuel is flammable. Do not perform this procedure near open flames. Injury to personnel may result.

1. Disconnect wire (17) from wire (16) on electrical wire clip (15).

CAUTION

Cover or plug all openings after disconnecting lines and hoses to prevent contamination. Failure to do this may result in fuel pump damage.

2. Disconnect hose (4) from adapter (3).
3. Disconnect hose (5) from adapter (6).
4. Remove locknut (10) and screw (12) from left-hand splash panel (11) and fuel pump mounting bracket (9). Discard locknut (10).
5. Remove locknut (10), screw (12), fuel pump (8), washer (13), wire clip (15), and washer (14) from left-hand splash panel (11). Discard locknut (10).
6. Remove adapter (6) from elbow (7).
7. Remove adapter (3) from fuel shutoff valve (2).
8. Remove fuel shutoff valve (2) from elbow (1).
9. Remove elbows (1) and (7) from fuel pump (8).

b. Installation

NOTE

Clean all male pipe threads and wrap with antiseize tape before installation.

1. Install elbows (7) and (1) on fuel pump (8).
2. Install fuel shutoff valve (2) on elbow (1).
3. Install adapter (3) on fuel shutoff valve (2).
4. Install adapter (6) on elbow (7).

4-218. PERSONNEL FUEL BURNING HEATER FUEL PUMP (M939/A1) REPLACEMENT (Contd)

5. Install fuel pump (8) on left-hand splash panel (11) with washer (14), electrical wire clip (15), washer (13), screw (12), and new locknut (10).
6. Install screw (12) and new locknut (10) on left-hand splash panel (11) and fuel pump mounting bracket (9).
7. Connect hose (5) to adapter (6).
8. Connect hose (4) to adapter (3).
9. Connect, wire (17) to wire (16) on wire clip (15).

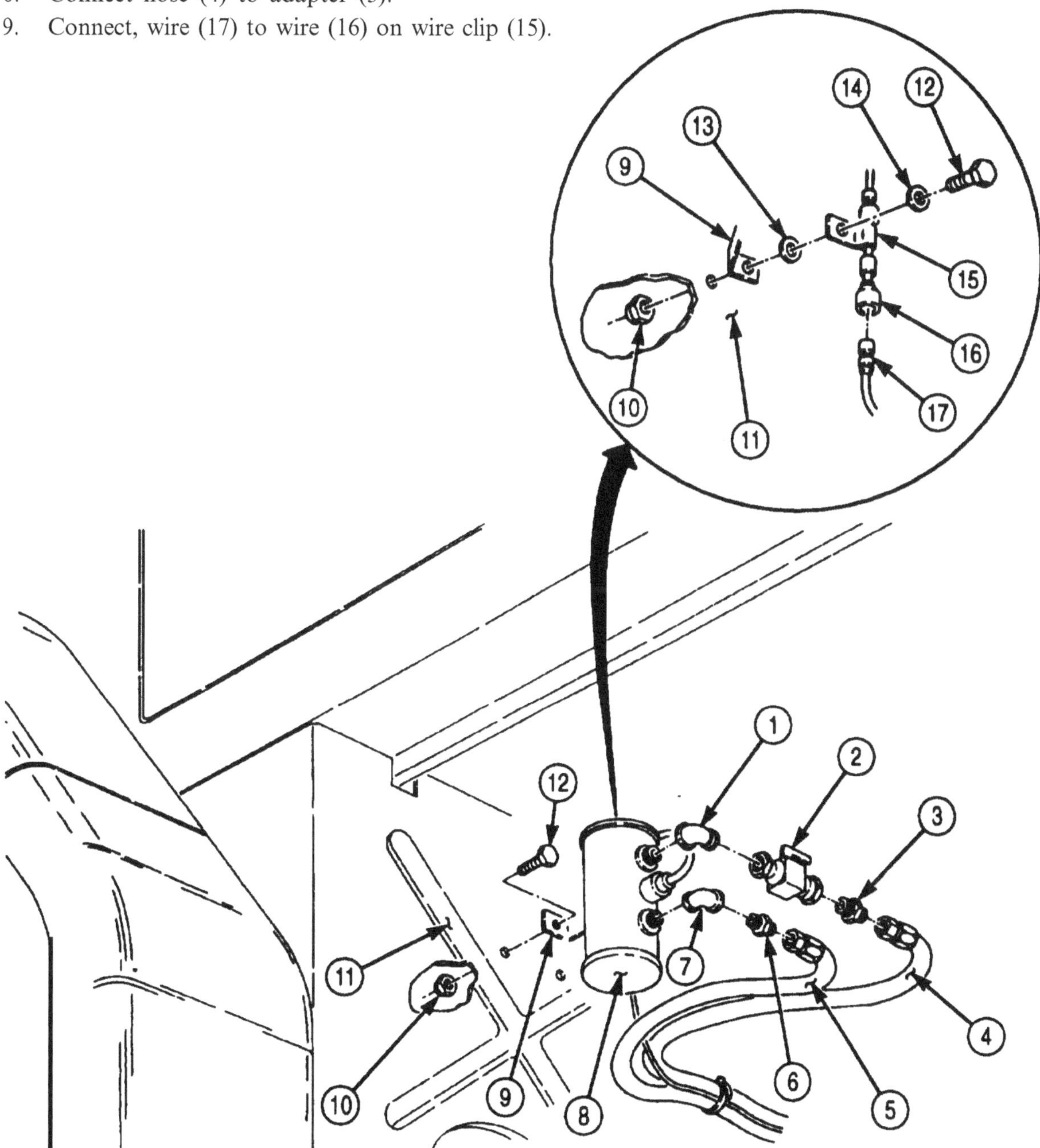

FOLLOW-ON TASKS:
- Connect battery ground cables (para. 3-126).
- Check personnel fuel burning heater for proper operation (TM 9-2320-272-10).

4-219. PERSONNEL FUEL BURNING HEATER ELECTRIC FUEL PUMP (M939A2) REPLACEMENT

THIS TASK COVERS:

a. Removal **b. Installation**

INITIAL SETUP:

APPLICABLE MODELS
M939A2

TOOLS
General mechanic's tool kit (Appendix E, Item 1)

MATERIALS/PARTS
Two locknuts (Appendix D, Item 313)
Cap and plug set (Appendix C, Item 14)

REFERENCES (TM)
TM 9-2320-272-10
TM 9-2320-272-24P

EQUIPMENT CONDITION
- Parking brake set (TM 9-2320-272-10).
- Battery ground cables disconnected (para. 3-126).

GENERAL SAFETY INSTRUCTIONS
Diesel fuel is flammable. Do not perform this procedure near open flame.

WARNING

Diesel fuel is flammable. Do not perform this procedure near open flame. Injury to personnel may result.

NOTE

Procedures for changing personnel fuel burning and engine coolant heater electric fuel pumps are similar, this procedure is for personnel fuel burning heater electric fuel pump.

a. Removal

1. Disconnect connector (3) from electric fuel pump wire (6).

CAUTION

When disconnecting lines and hoses from pump, immediately plug all open ports. Failure to do so can cause damage to pump.

NOTE

Tag all lines and ports for installation.

2. Disconnect fuel inlet line (8) and outlet line (2) from adapter fittings (4).
3. Remove two screws (5), locknuts (1), and electric fuel pump (7) from left-hand splash panel (9). Discard locknuts (1).

b. Installation

1. Install electric fuel pump (7) on left-hand splash panel (9) with two screws (5) and new locknuts (1).
2. Connect fuel inlet line (8) and outlet line (2) to adapter fittings (4).
3. Connect connector (3) to electric fuel pump wire (6).

4-219. PERSONNEL FUEL BURNING HEATER ELECTRIC FUEL PUMP (M939A2) REPLACEMENT (Contd)

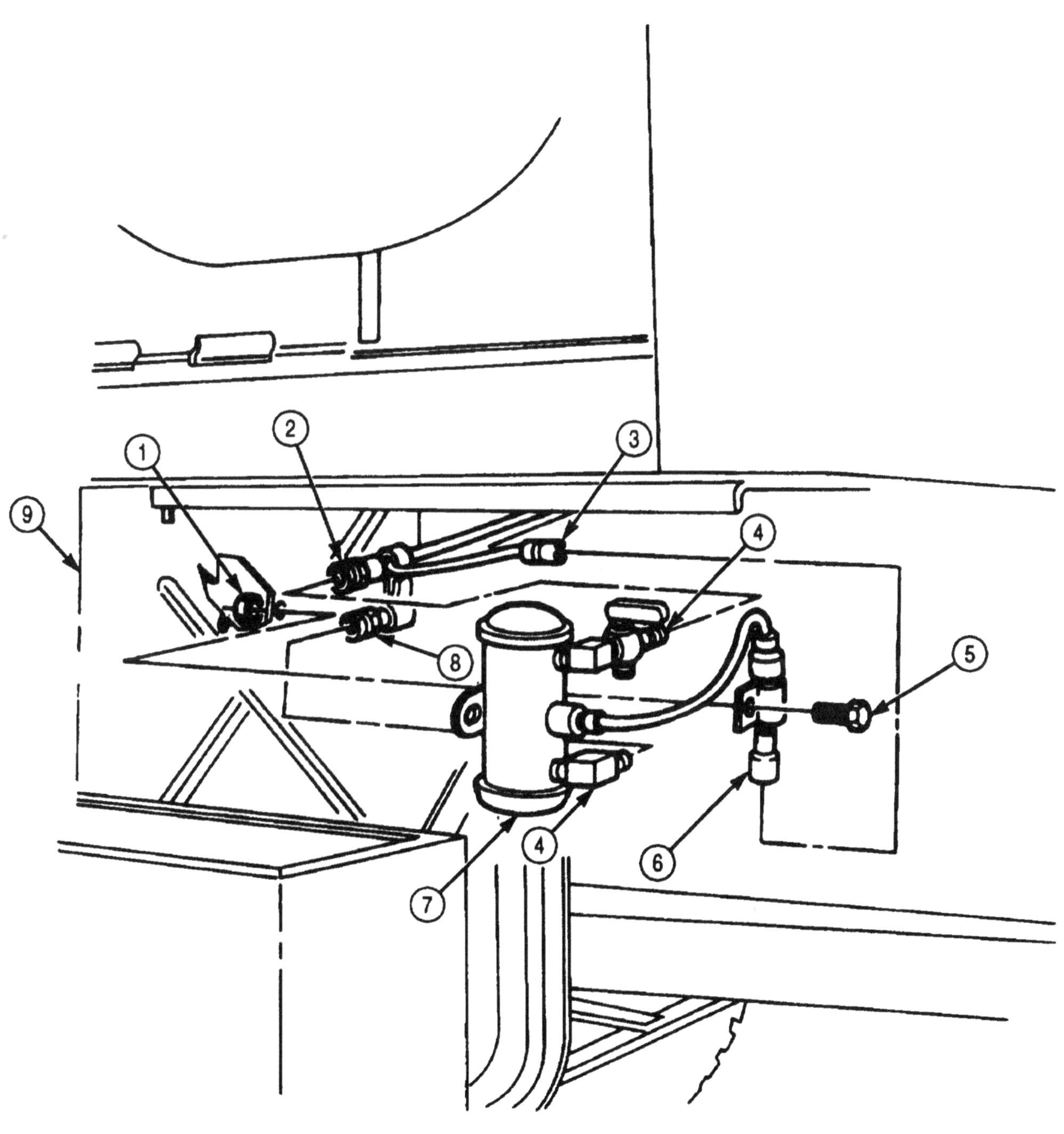

FOLLOW-ON TASKS: • Connect battery ground cables (para. 3-126).
• Check personal fuel burning heater for proper operation (TM 9-2320-272-10).

4-220. PERSONNEL FUEL BURNING HEATER FUEL SHUTOFF VALVE REPLACEMENT

THIS TASK COVERS:

a. Removal **b. Installation**

INITIAL SETUP:

APPLICABLE MODELS
All

TOOLS
General mechanic's tool kit (Appendix E, Item 1)

MATERIALS/PARTS
Antiseize tape (Appendix C, Item 72)
Cap and plug set (Appendix C, Item 14)

REFERENCES (TM)
TM 9-2320-272-10
TM 9-2320-272-24P

EQUIPMENT CONDITION
Parking brake set (TM 9-2320-272-10).

GENERAL SAFETY INSTRUCTIONS
Diesel fuel is flammable. Do not perform this procedure near open flames.

a. Removal

WARNING

Diesel fuel is flammable. Do not perform this procedure near open flames. Injury to personnel may result.

CAUTION

Cover or plug all openings after disconnecting lines and hoses to prevent contamination. Failure to do this may result in fuel pump damage.

1. Disconnect hose (4) from adapter (3).
2. Remove adapter (3) from fuel shutoff valve (2).
3. Remove fuel shutoff valve (2) from elbow (1).

b. Installation

NOTE

Clean all male pipe threads and wrap with antiseize tape before installation.

1. Install fuel shutoff valve (2) on elbow (1).
2. Install adapter (3) on fuel shutoff valve (2).
3. Connect hose (4) to adapter (3).

4-220. PERSONNEL FUEL BURNING HEATER FUEL SHUTOFF VALVE REPLACEMENT (Contd)

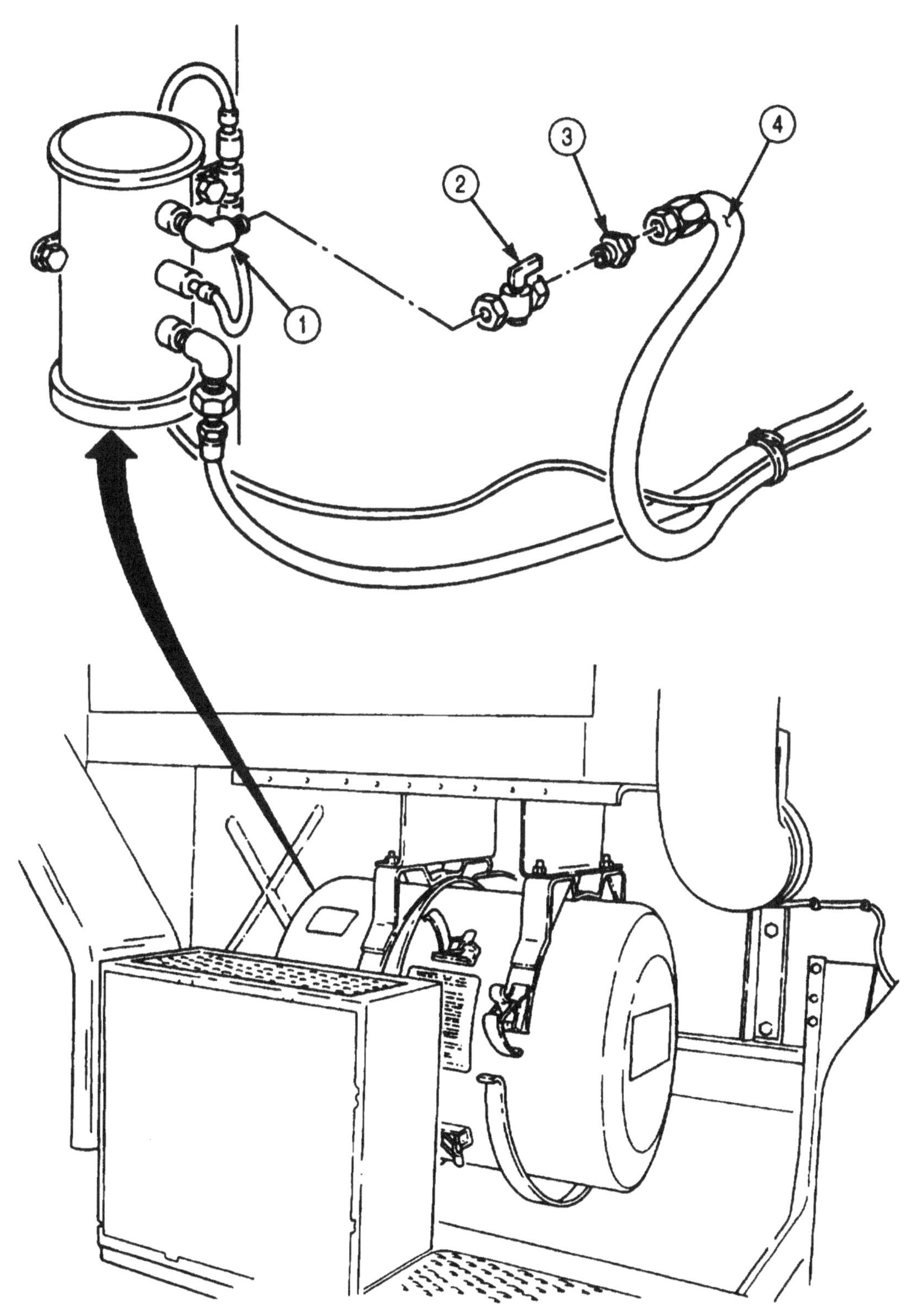

4-221. PERSONNEL FUEL BURNING HEATER WIRING HARNESS REPLACEMENT

THIS TASK COVERS:

a. Removal **b. Installation**

INITIAL SETUP:

APPLICABLE MODELS
All

TOOLS
General mechanic's tool kit (Appendix E, Item 1)

MATERIALS/PARTS
Lockwasher (Appendix D, Item 353)
Locknut (Appendix D, Item 299)
Eleven tiedown straps (Appendix D, Item 685)

PERSONNEL REQUIRED
TWO

REFERENCES (TM)
TM 9-2320-272-10
TM 9-2320-272-24P

EQUIPMENT CONDITION
- Parking brake set (TM 9-2320-272-10).
- Hood raised and secured (TM 9-2320-272-10).
- Right and left splash shield removed (TM 9-2320-272-10).
- Battery ground cables disconnected (para. 3-126).

---

a. Removal

1. Remove locknut (28), washer (27), screw (25), washer (27), lockwasher (26), and ground wire (22) from mounting bracket (19) and angle bracket (17). Discard lockwasher (26) and locknut (28).

NOTE

Note routing of wiring harness and leads for installation.

2. Disconnect heater harness connector (1) from personnel heater (18) and push harness connector (1) into cab through grommet (2).
3. Remove grommet (2) from firewall knockout (20) and fuel pump lead (13).
4. Remove four tiedown straps (14) from fuel pump lead (13) and front harness (16) on engine compartment firewall (3). Discard tiedown straps (14).
5. Disconnect leads (13) and (24) from connector (23) on cab (21) and push lead (13) through firewall (3) into engine compartment.
6. Remove three tiedown straps (30) from fuel pump fuel lines (31). Discard tiedown straps (30).
7. Disconnect leads (29) and (13) from fuel pump (32).
8. Disconnect heater harness (9) and lead (11) from heater control box (8) and lead (10).
9. Disconnect lead (12) from lead (15) of wiring harness (5).
10. Remove four tiedown straps (4) from heater harness (9) on cab (21). Discard tiedown straps (4).
11. Disconnect wire (7) from pin A of battery switch (6) and remove wiring harness (5) from cab (21).
12. Remove heater harness (9) from cab (21).

4-221. PERSONNEL FUEL BURNING HEATER WIRING HARNESS REPLACEMENT (Contd)

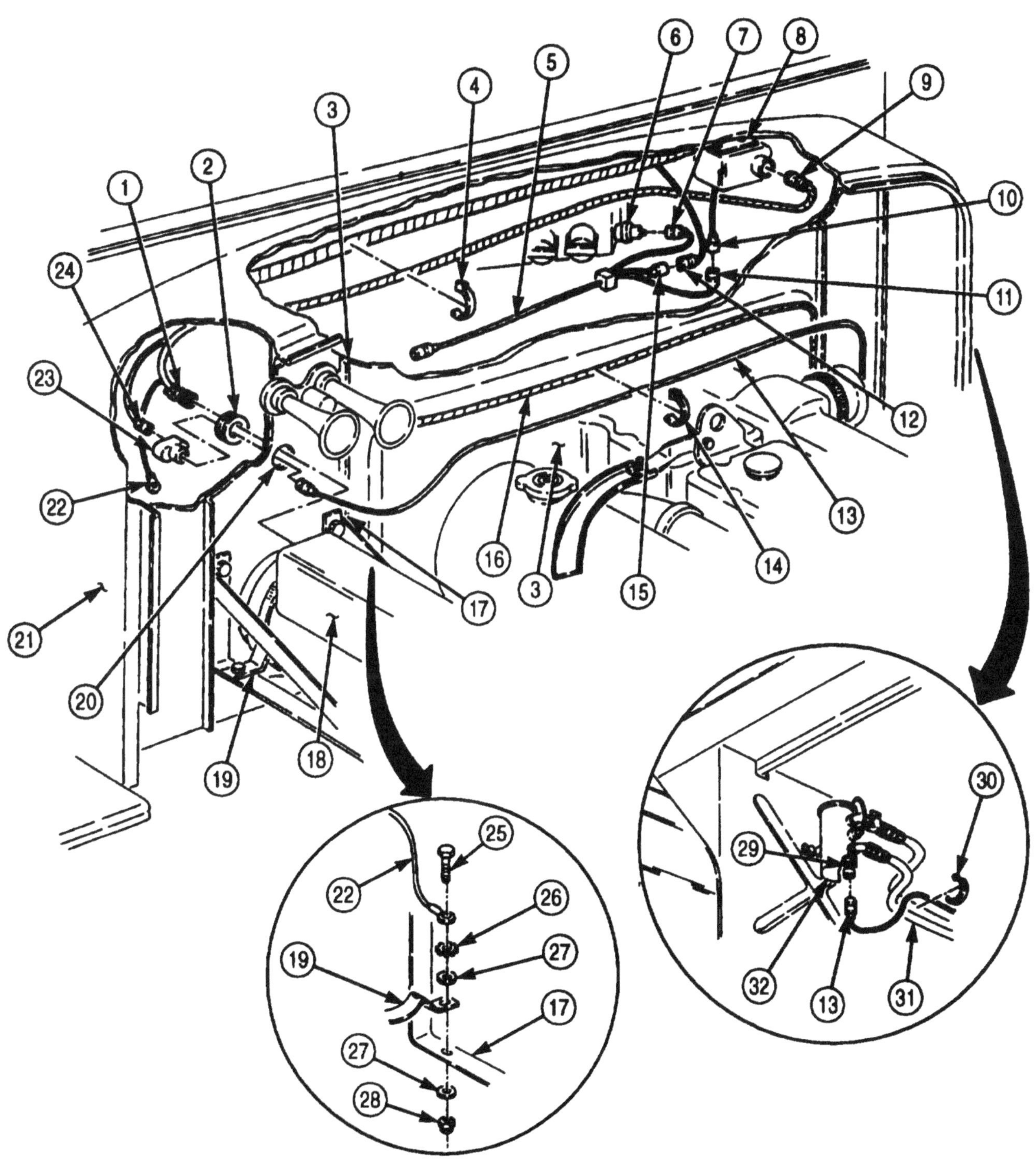

4-221. PERSONNEL FUEL BURNING HEATER WIRING HARNESS REPLACEMENT (Contd)

b. Installation

1. Position heater harness (9) on cab (21).
2. Connect wire (7) to pin A of battery switch (6) and install wiring harness (5) on cab (21).
3. Install heater harness (9) on cab (21) with four new tiedown straps (4).
4. Connect lead (12) to lead (15) of wiring harness (5).
5. Connect heater harness (9) and lead (11) to heater control box (8) and lead (10).
6. Connect leads (29) and (13) to fuel pump (32).
7. Install three new tiedown straps (30) on fuel pump fuel lines (31).
8. Push lead (13) through firewall (3) into cab (21) and connect leads (13) and (24) to connector (23) on cab (21).
9. Install four new tiedown straps (14) on fuel pump lead (13) and front harness (16) on engine compartment firewall (3).
10. Install grommet (2) on heater harness (9) and firewall knockout (20).
11. Push harness connector (1) into engine compartment through firewall knockout (20) and connect heater harness connector (1) to personnel heater (18).
12. Install washer (27), new lockwasher (26), ground wire (22), screw (25), washer (27), and new locknut (28) on mounting bracket (19) and angle bracket (17).

4-221. PERSONNEL FUEL BURNING HEATER WIRING HARNESS REPLACEMENT (Contd)

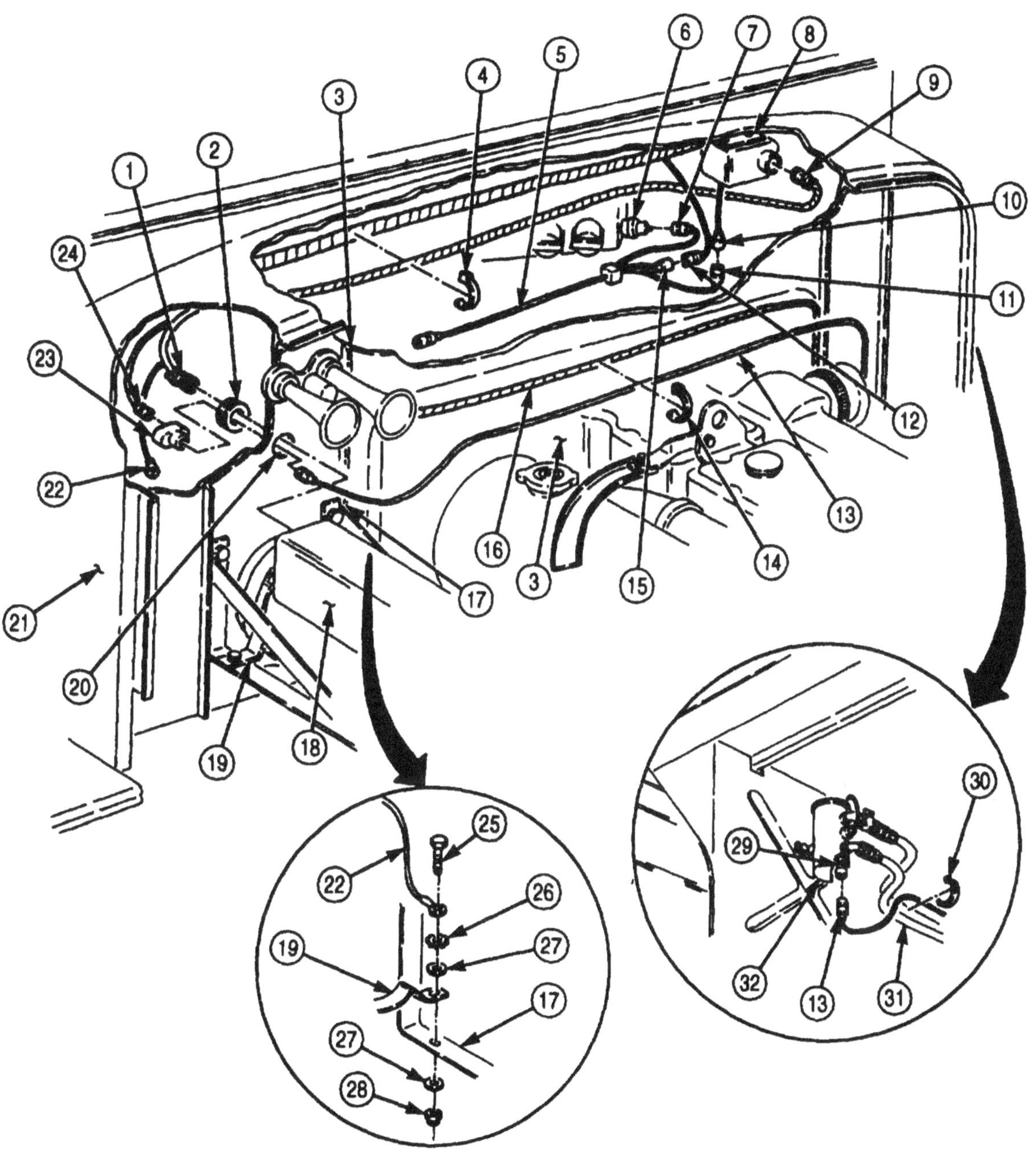

FOLLOW-ON TASKS: • Connect battery ground cables (para. 3-126).
• Install right and left splash shield (TM 9-2320-272-10).

4-222. PERSONNEL FUEL BURNING HEATER EXHAUST TUBE (M939/A1) REPLACEMENT

THIS TASK COVERS:

a. Removal

b. Installation

INITIAL SETTUP:

APPLICABLE MODELS
M939/A1

TOOLS
General mechanic's tool kit (AppendiE, Item 1)

MATERIALS/PARTS
Two cotter pins (Appendix D, Item 59)
Two locknuts (Appendix D, Item 291)

REFERENCES (TM)
TM 9-2320-272-10
TM 9-2320-272-24P

EQUIPMENT CONDITION
Parking brake set (TM 9-2320-272-10).

GENERAL SAFETY INSTRUCTIONS
Do not touch hot exhaust system componentswith bare hands.

WARNING

Do not handle hot exhaust tube with bare hands. Injury to personnel may result.

a. Removal

1. Remove cotter pin (3) from exhaust elbow (1) and fuel burning heater (2). Discard cotter pin (3)
2. Remove cotter pin (4) from exhaust elbow (1) and exhaust tube (7). Discard cotter pm (4).
3. Remove two locknuts (6), screws (9), clamps (5), tube (7), and exhaust elbow (1) from fuel burning heater (2) and splash panel (8).

b. Installation

1. Install exhaust tube (7) and elbow (1) on splash panel (8) with two clamps (5), screws (9), and new locknuts (6).
2. Install exhaust elbow (1) on fuel burning heater (2) with new cotter pin (3).
3. Install exhaust elbow (1) on exhaust tube (7) with cotter pin (4).

4-222. PERSONNEL FUEL BURNING HEATER EXHAUST TUBE (M939/Al) REPLACEMENT (Contd)

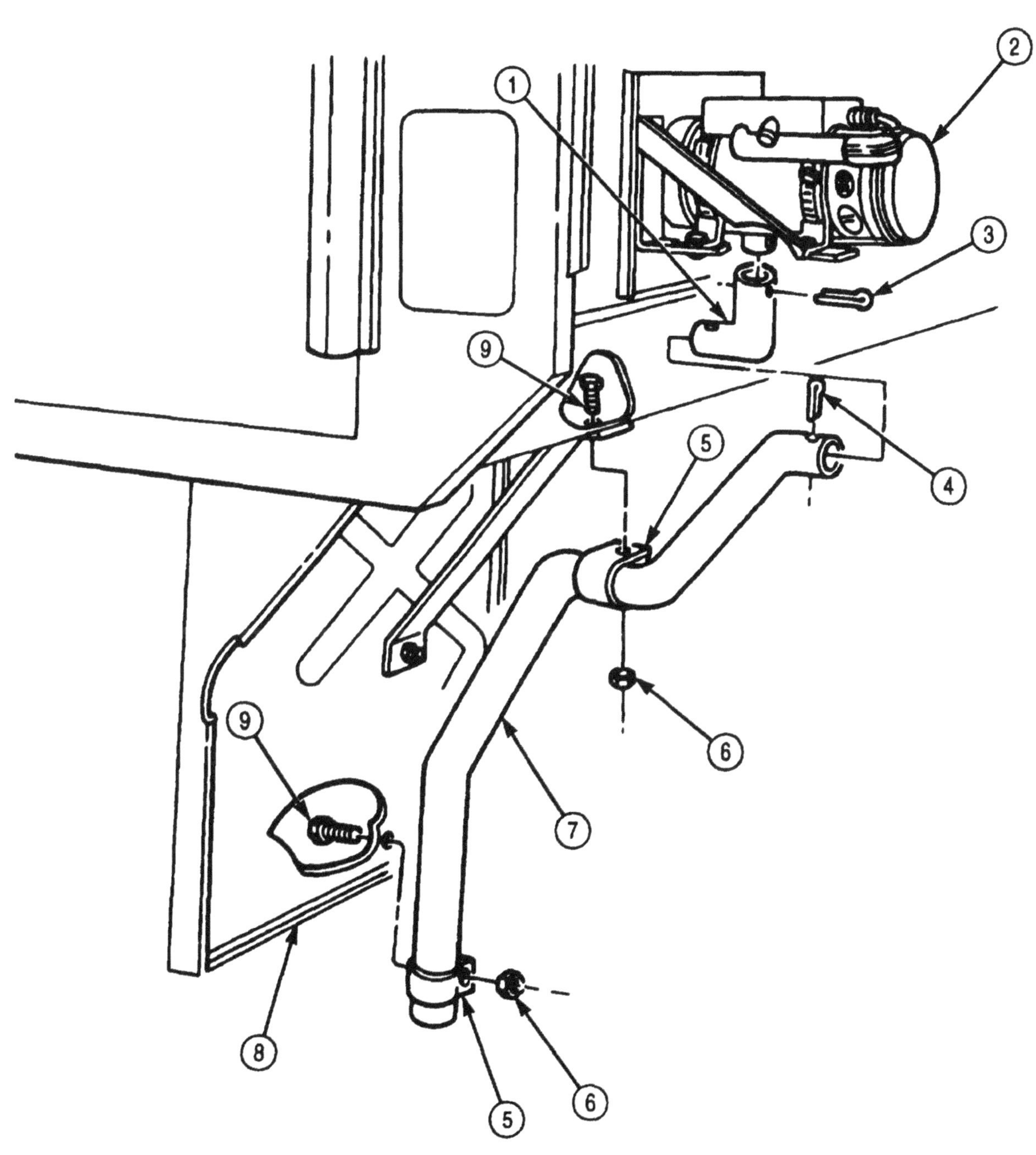

4-223. PERSONNEL FUEL BURNING HEATER EXHAUST TUBE (M939A2) REPLACEMENT

THIS TASK COVERS:

a. Removal **b. Installation**

INITIAL SETUP:

APPLICABLE MODELS
M939A2

TOOLS
General mechanic's tool kit (Appendix E, Item 1)

MATERIALS/PARTS
Three cotter pins (Appendix D, Item 71)
Locknut (Appendix D,Item 334)

REFERENCES (TM)
TM 9-2320-272-10
TM 9-2320-272-24P

EQUIPMENT CONDITION
- Parking brake set (TM 9-2320-272-10).
- Hood raised and secured (TM 9-2320-272-10).

GENERAL SAFETY INSTRUCTIONS
Do not touch hot exhaust system components with bare hands.

a. Removal

WARNING

Do not handle hot exhaust tube with bare hands. Injury to personnel may result.

1. Remove locknut (3), washer (2), screw (l), and clamp (12) from firewall (6) and exhaust tube (10). Discard locknut (3).
2. Remove cotter pin (9) and exhaust tube (10) from elbow (8). Discard cotter pin (9).
3. Remove cotter pins (7) and (4) and elbows (8) and (11) from heater exhaust outlet (5). Discard cotter pins (4) and (7).

b. Installation

1. Install elbows (11) and (8) on heater exhaust outlet (5) with new cotter pins (4) and (7).
2. Install exhaust tube (10) on elbow (8) with new cotter pin (9).
3. Install clamp (12) on exhaust tube (10) and firewall (6) with screw (l), washer (2), and new locknut (3).

4-223. PERSONNEL FUEL BURNING HEATER EXHAUST TUBE (M939A2) REPLACEMENT(Contd)

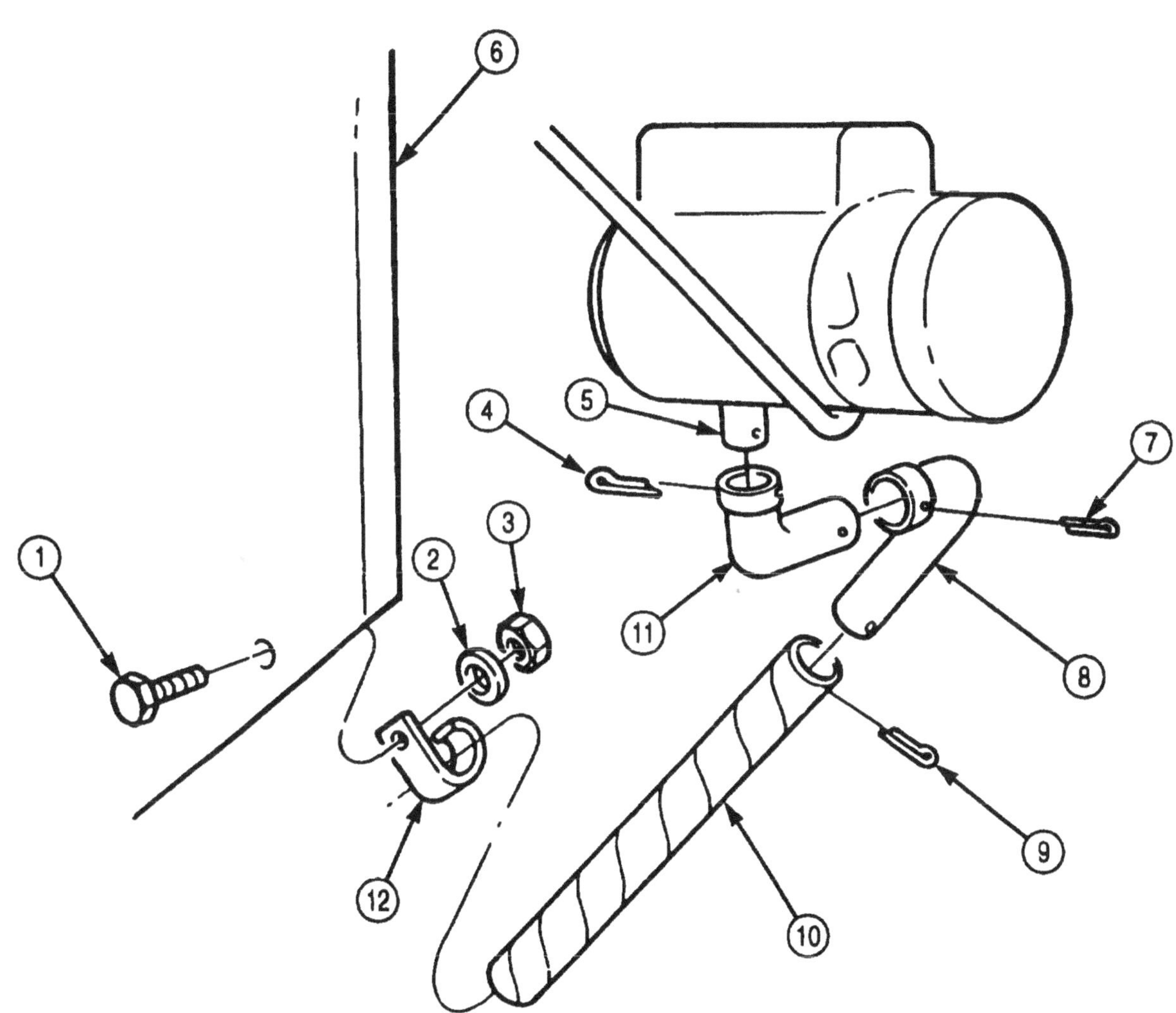

4-224. PERSONNEL FUEL BURNING HEATER CIRCUIT BREAKER REPLACEMENT

THIS TASK COVERS:

a. Removal **b. Installation**

INITIAL SETUP:

APPLICABLE MODELS
All

TOOLS
General mechanic's tool kit (Appendix E, Item 1)

REFERENCES (TM)
TM 9-2320-272-10
TM 9-2320-272-24P

EQUIPMENT CONDITION
- Parking brake set (TM 9-2320-272-10).
- Battery ground cables disconnected (para. 3-126).

a. Removal

1. Disconnect connectors (3) and (4) from heater circuit breaker (1).
2. Remove two screws (2) and heater circuit breaker (1) from left underside of dash (5).

b. Installation

1. Install heater circuit breaker (1) on left underside of dash (5) with two screws (2).
2. Connect connectors (3) and (4) to heater circuit breaker (1).

4-224. PERSONNEL FUEL BURNING HEATER CIRCUIT BREAKER REPLACEMENT (Contd)

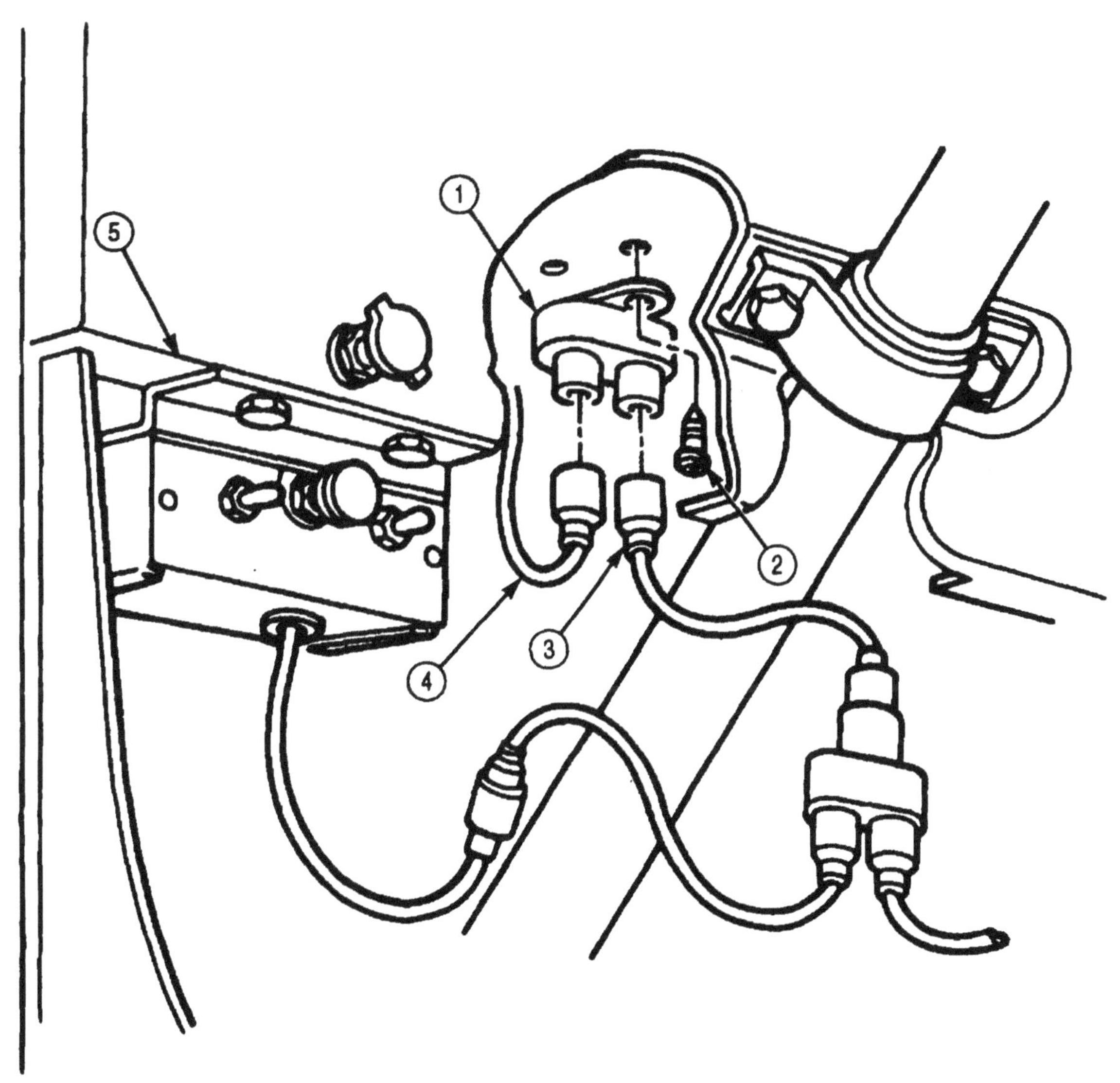

FOLLOW-ON TASK: Connect battery ground cables (para. 3-126).

4-225. DEEPWATER FORDING CONTROL REPLACEMENT

THIS TASK COVERS:

a. Removal **b. Installation**

INITIAL SETUP:

APPLICABLE MODELS
All

TOOLS
General mechanic's tool kit (Appendix E, Item 1)

MATERIALS/PARTS
Two locknuts (Appendix D, Item 313)
Two locknuts (Appendix D, Item 283)
Lockwasher (Appendix D, Item 368)

REFERENCES (TM)
TM 9-2320-272-10
TM 9-2320-272-24P

EQUIPMENT CONDITION
- Parking brake set (TM 9-2320-272-10).
- Hood raised and secured (TM 9-2320-272-10).

a. Removal

1. Loosen screw (1) and remove control cable (10) from nut (2) on pressurization valve (3).
2. Remove two locknuts (11), screws (4), clamp (5), and control cable (10) from fording cable bracket (6). Discard locknuts (11).
3. Remove nut (16) and lockwasher (17), and slide down control cable (10).
4. Pull hand control (14) and control cable (10) from instrument panel (15) and grommet (12) in firewall (13), and remove nut (16) and lockwasher (17) from control cable (10). Discard lockwasher (17).
5. Remove two locknuts (9), screws (7), and fording cable bracket (6) from surge tank mounting bracket (8). Discard locknuts (9).

b. Installation

1. Install fording cable bracket (6) on surge tank mounting bracket (8) with two screws (7) and new locknuts (9).
2. Insert control cable (10) and hand control (14) through instrument panel (15), new lockwasher (17), nut (16), and grommet (12) in firewall (13), and tighten nut (16).
3. Install control cable (10) on fording cable bracket (6) with clamp (5), two screws (4), and new locknuts (11).
4. Install control cable (10) in nut (2) on pressurization valve (3), and tighten screw (1).

4-225. DEEPWATER FORDING CONTROL REPLACEMENT (Contd)

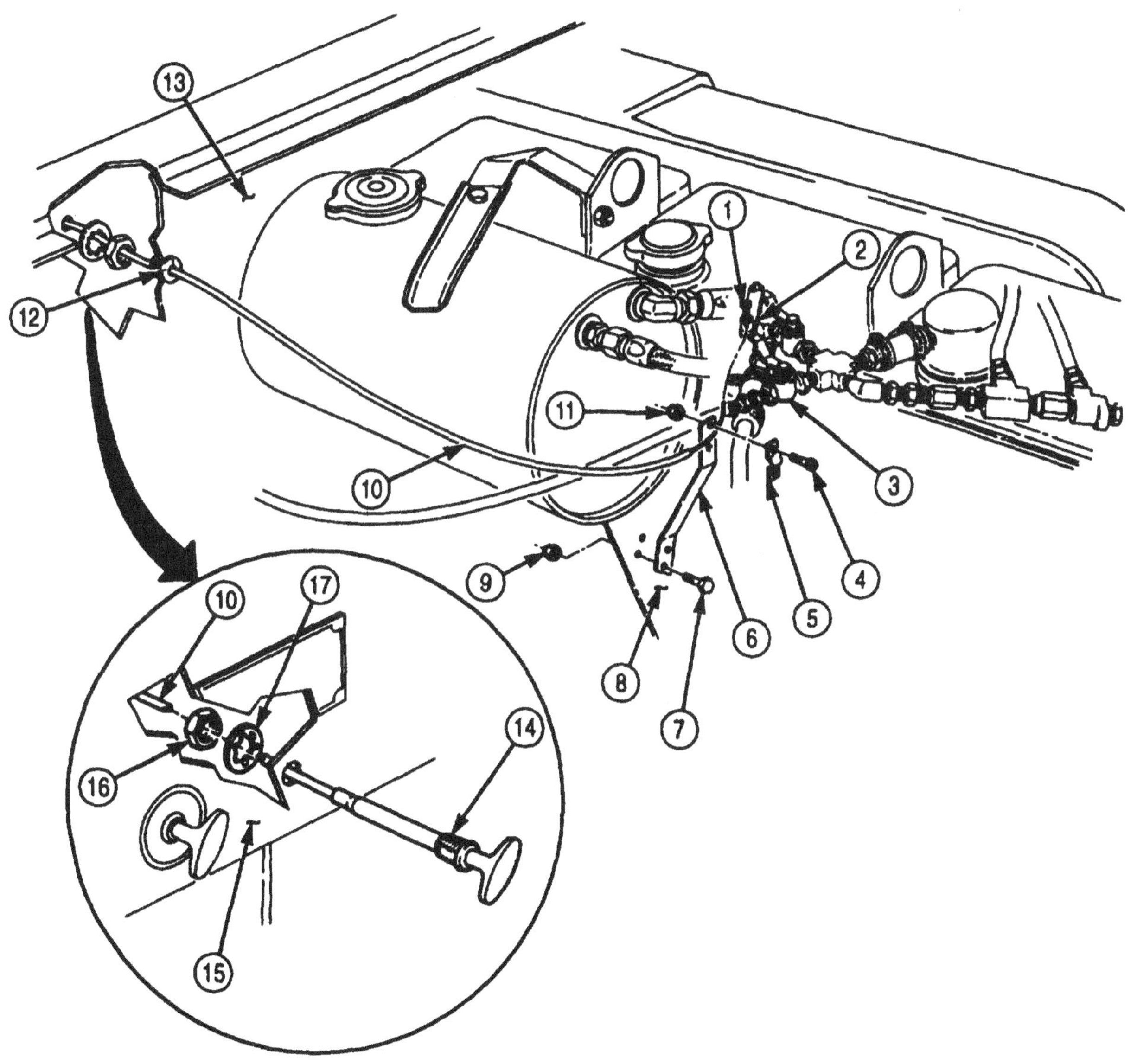

4-226. DEEPWATER FORDING PRESSURIZATION VALVE REPLACEMENT

THIS TASK COVERS

a. Removal (M9389/A1)
b. Removal (M939A2)
c. Installation (M939A2)
b. Installation (M939/A1)

INITIAL SETUP:

APPLICABLE MODELS
All

TOOLS
General mechanic's tool kit (Appendix E, Item 1)

MATERIALS/PARTS
Two locknuts (M939/Al) (Appendix D. Item 283)
Locknut (M939/Al) Appendix D, Item 318)
Lockwasher (M939/Al) (Appendix D, Item 354)
Antiseize tape (Appendix C, Item 72)

REFERENCES (TM)
TM 9-2320-272-10
TM 9-2320-272-24P

EQUIPMENT CONDITION
- Parking brake set (TM 9-2320-272-10).
- Hood raised and secured (TM 9-2320-272-10).
- Air reservoirs drained (TM 9-2320-272-10).
- Air pressure hose removed from valve (M939A2) (para. 4-227).

GENERAL SAFETY INSTRUCTIONS
Do not disconnect air lines before draining air reservoir.

WARNING

Do not disconnect air lines before draining air reservoirs. Small parts under pressure may shoot out with high velocity causing injury to personnel.

a. Removal (M939/A1)

NOTE

Tag all hoses and lines for installation.

1. Remove two locknuts (25), screws (18), and clamp (19) from fording cable bracket (24). Discard locknuts (25).
2. Loosen screw (29) and remove control cable (26) from nut (28).
3. Disconnect supply hose (20) from pressurized valve adaptor (21).

NOTE

Perform step 4 for vehicles equipped with positive crankcase ventilation system.

4. Remove two clamps (33), hose (35), and crankcase vent tube (32) from pressurization valve elbow (34).
5. Remove two clamps (16), hose (17), and crankcase breather tube (15) from pressurization valve elbow (27).
6. Disconnect pressurization hose (6) and vent hose (5) from pressurization valve connectors (7).
7. Remove locknut (a), screw (12), clamp (ll), and power steering pump hose (13) from bracket (9) on radiator support bracket (10). Discard locknut (8).
8. Disconnect power steering pump hose (13) from pressurization valve connector (14).
9. Remove two clamps (1) and hose (2) from crankcase breather (3) and pressurization valve (22).
10. Remove screw (30), lockwasher (31), and pressurization valve (22) with bracket (23) from rocker lever cover (4). Discard lockwasher (3 1).

4-226. DEEPWATER FORDING PRESSURIZATION VALVE REPLACEMENT (Contd)

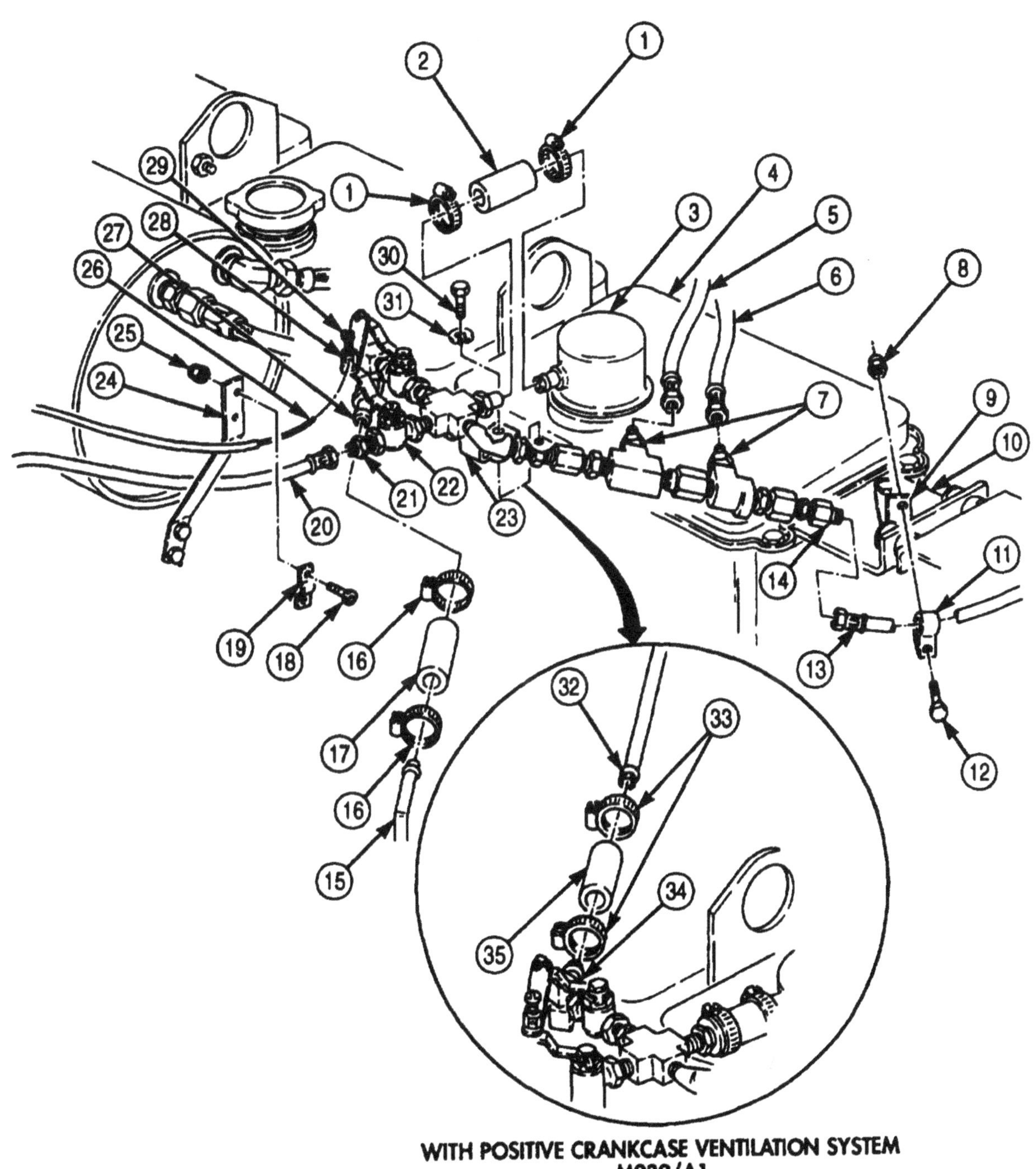

WITH POSITIVE CRANKCASE VENTILATION SYSTEM
M939/A1

4-226. DEEPWATER FORDING PRESSURIZATION VALVE REPLACEMENT (Contd)

b. Removal (M939A2)

1. Remove two Screws (15), washers (14), and pressurization valve (5) from mounting bracket (13) and bracket (12).
2. Remove screw (7), washer (8), spacer (9), and arm (6) from pressurization valve (5).
3. Remove connector (3) from pressurization valve (5).
4. Remove tee (4) from pressurization valve (5).
5. Remove adapter (19) from tee (1).
6. Remove tee (1) and connector (2) from pressurization valve (5).
7. Remove regulator valve (18) from elbow (17).
8. Remove elbow (17) and connector (16) from pressurization valve (5).
9. Remove elbow (10) from pressurization valve (5).
10. Remove elbow (11) from pressurization valve (5).

c. Installation (M939A2)

NOTE

All male threads must be wrapped with antiseize tape before installation.

1. Install elbow (11) on pressurization valve (5).
2. Install elbow (10) on pressurization valve (5).
3. Install connector (16) and elbow (17) on pressurization valve (5).
4. Install regulator valve (18) on elbow (17).
5. Install connector (2) and tee (1) on pressurization valve (5).
6. Install adapter (19) on tee (1).
7. Install tee (4) on pressurization valve (5).
8. Install connector (3) on pressurization valve (5).
9. Install arm (6) on pressurization valve (5) with spacer (9), washer (8), and screw (7).
10. Install pressurization valve (5) on mounting bracket (13) and bracket (12) with two washers (14) and screws (15).

4-226. DEEPWATER FORDING PRESSURIZATION VALVE REPLACEMENT (Contd)

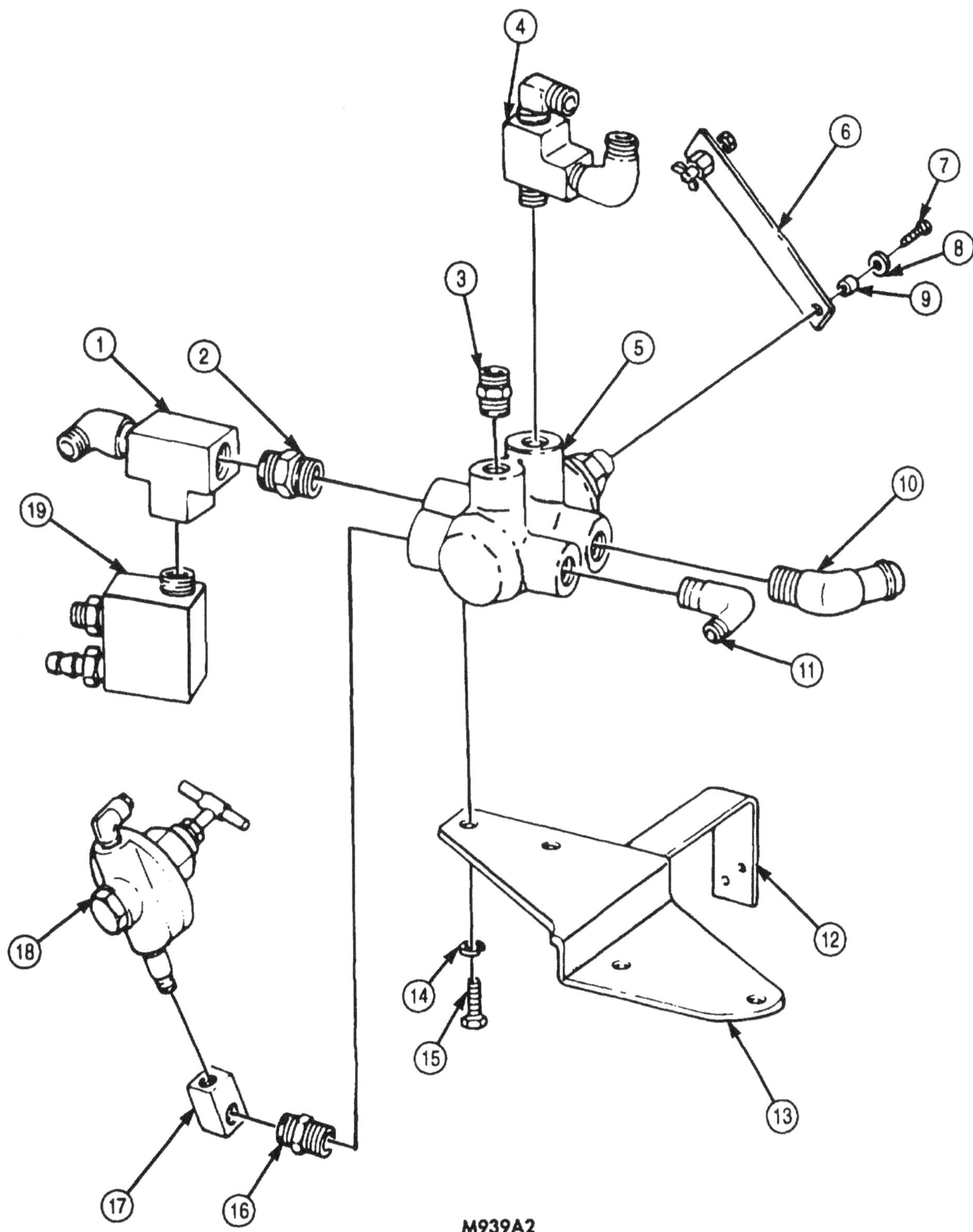

4-226. DEEPWATER FORDING PRESSURIZATION VALVE REPLACEMENT (Contd)

d. Installation (M939/A1)

NOTE

All male threads must be wrapped with antiseize tape before installation,

1. Connect pressurization valve (22) on crankcase breather (3) with hose (2) and two clamps (1).
2. Install pressurization valve bracket (23) on rocker lever cover (4) with new lockwasher (31) and screw (30).
3. Connect power steering pump hose (13) to pressurization valve connector (14).
4. Install power steering pump hose (13) on radiator support bracket (10) and bracket (9) with clamp (ll), screw (12), and new locknut (8).
5. Connect pressurization hose (6) and vent hose (5) to pressurization valve connectors (7).
6. Install crankcase breather tube (15) on pressurization valve elbow (27) with hose (17) and two clamps (16).

NOTE

Perform step 7 for vehicles equipped with positive crankcase ventilation system.

7. Install crankcase vent tube (32) on pressurization valve elbow (34) with hose (35) and two clamps (33).
8. Connect supply hose (20) to pressurized valve adapter (21).
9. Install control cable (26) on nut (28) and tighten screw (29).
10. Install control cable (26) on fording cable bracket (24) with clamp (19), two screws (18), and new locknuts (25).

4-226. DEEPWATER FORDING PRESSURIZATION VALVE REPLACEMENT (Contd)

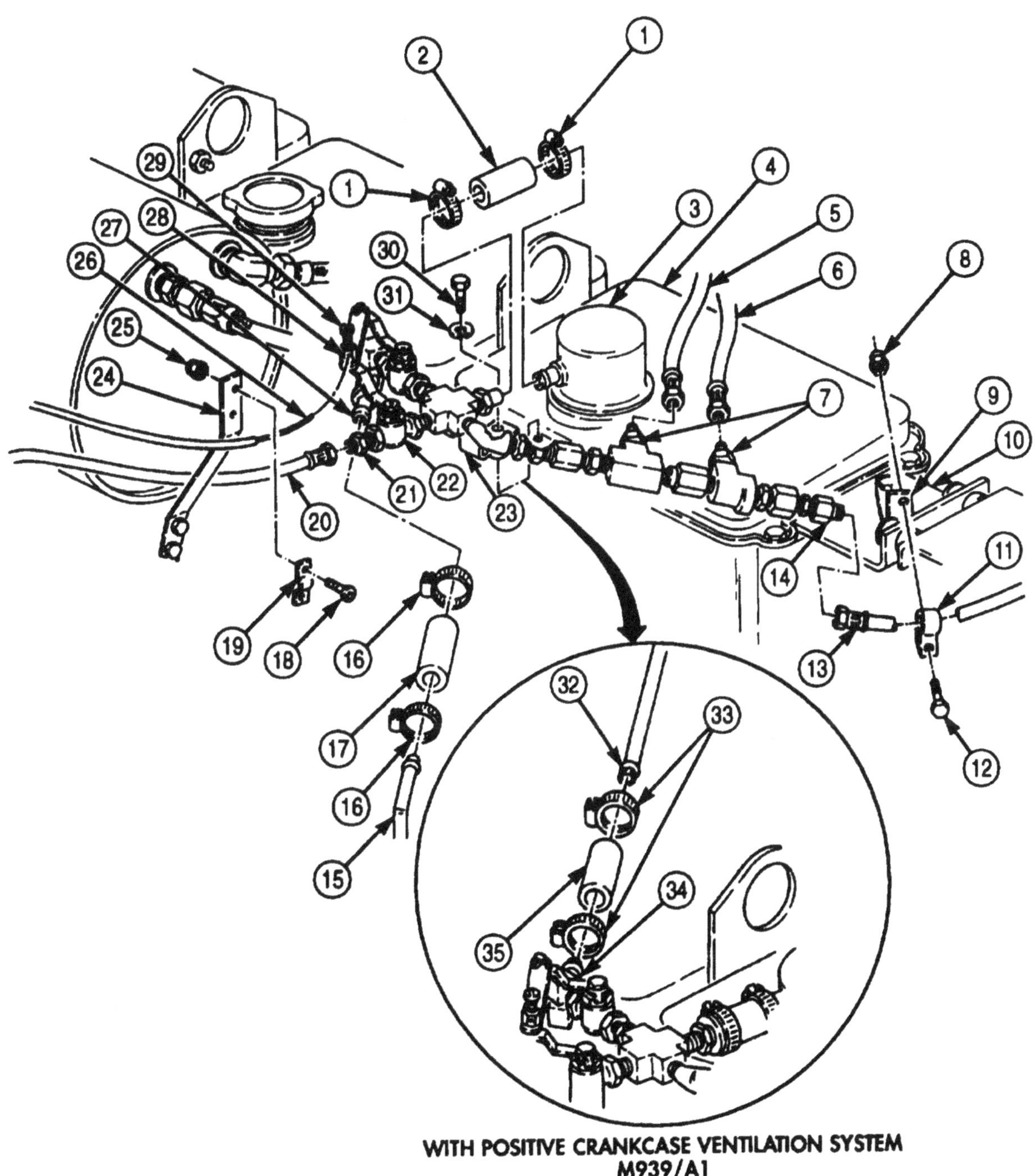

WITH POSITIVE CRANKCASE VENTILATION SYSTEM
M939/A1

FOLLOW-ON TASKS:
- Connect air pressure hose to valve (M939A2) (para. 4-227).
- Start engine (TM 9-2320-272-10) and allow air pressure to build up to normal operating range. Check for air leaks. Road test vehicle.

4-227. DEEPWATER FORDING AIR PRESSURE HOSE REPLACEMENT

THIS TASK COVERS:

a. Removal (M939/A1)
b. Removal (M939A2)
c. Installation (M939A2)
d. Installation (M939/A1)

INITIAL SETUP:

APPLICABLE MODELS
All

TOOLS
General mechanic's tool kit (Appendix E, Item 1)

MATERIALS/PARTS
Locknut (M939/Al) (Appendix D, Item 318)
Five tiedown straps (M939/A1)
(Appendix C, Item 684)
Antiseize tape (Appendix C, Item 72)

REFERENCES (TM)
TM 9-2320-272-10
TM 9-2320-272-24P

EQUIPMENT CONDITION
- Parking brake set (TM 9-2320-272-10).
- Hood raised and secured (TM 9-2320-272-10).
- Air reservoirs drained (TM 9-2320-272-10).

GENERAL SAFETY INSTRUCTIONS
Do not disconnect air lines before draining air reservoir.

WARNING

Do not disconnect air lines before draining air reservoirs. Small parts under pressure may shoot out with high velocity causing injury to personnel.

a. Removal (M939/A1)

1. Remove nut (5), screw (2), and clamp (3) from supply hose (4) and remove supply hose (4) from pressurized valve (6) and pressurized valve adapter (1).
2. Remove locknut (26), screw (24), clamp (25), and power steering pump hose (28) from bracket (22). Discard locknut (26).
3. Remove power steering pump hose (28) from pressurized valve connector (27) and adapter (20).
4. Remove five tiedown straps (18) from pressure hose (7) and vent hose (8). Discard tiedown straps (18).
5. Remove pressure hose (7) and bushing (17) from toe (16) and pressurized valve connector (31).
6. Remove vent hose (8) and bushing (15) from toe (14) and pressurized valve connector (30).
7. Remove air line (9) from shutoff valve (10) and alcohol evaporator (13).
8. Remove drain valve (11) and elbow (12) from alcohol evaporator (13).
9. Remove adapter (20) from power steering pump (21).
10. Remove shutoff valve (10) from air compressor (19).
11. Remove screw (29) and bracket (22) from radiator support bracket (23).

4-227. DEEPWATER FORDING AIR PRESSURE HOSE REPLACEMENT (Contd)

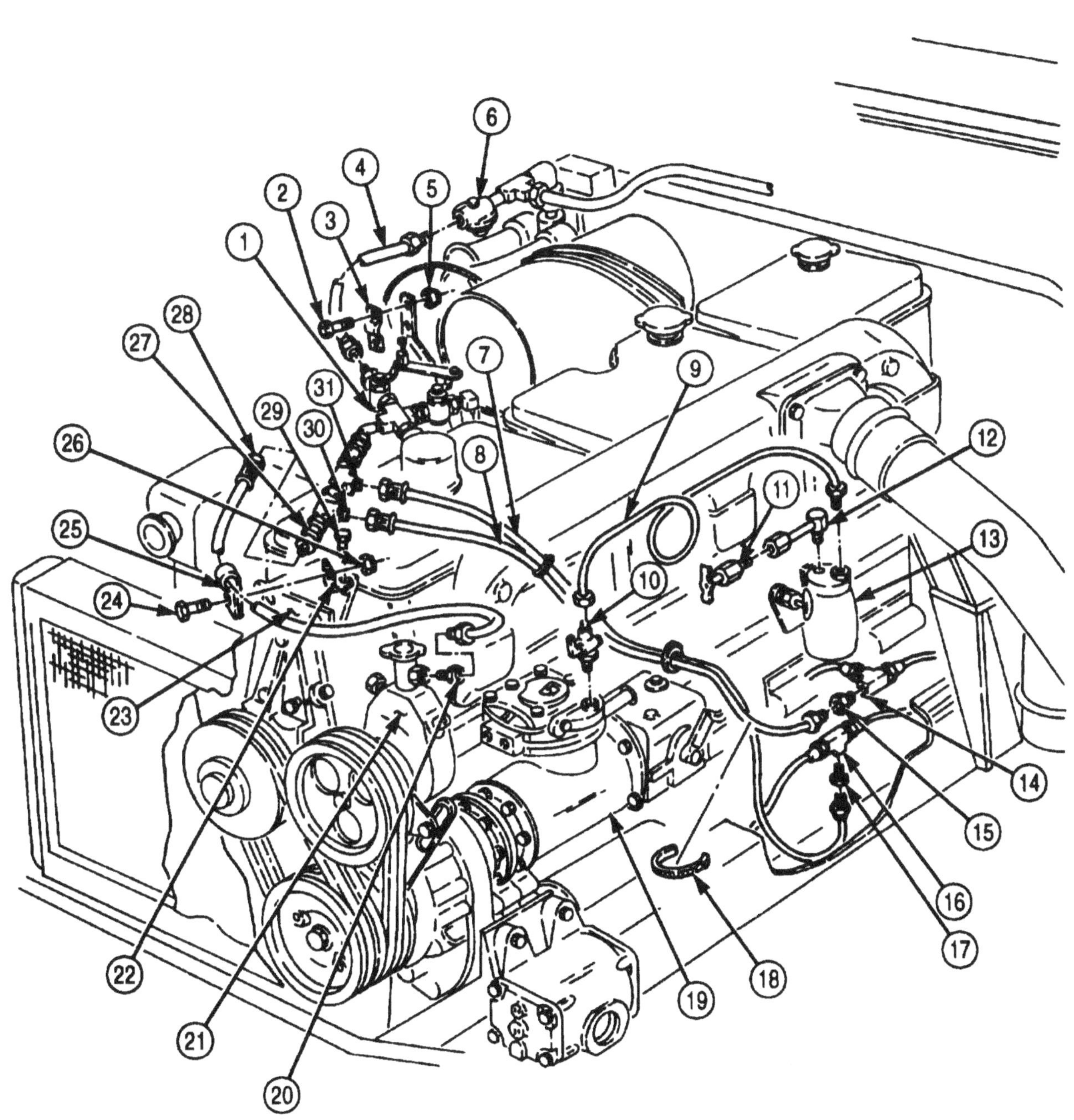

4-227. DEEPWATER FORDING AIR PRESSURE HOSE REPLACEMENT (Contd)

b. Removal (M939A2)

1. Remove hose (31) from elbows (32) and (29) on air governor (1) and pressurization valve (21).
2. Remove hose (18) from elbow (22) and connector (19) on pneumatic controller (23) and pressurization valve (21).
3. Remove hose (17) from connectors (16) and (20) on air intake tube (15) and pressurization valve (21).
4. Remove hose (2) from connectors (3) and (30) on tee (4) and pressurization valve (21).
5. Remove hose (5) from elbows (9) and (6) on power steering reservoir (13) and pressurization valve (21).
6. Remove hose (14) from elbows (12) and (28) on pressurization valve (21).
7. Remove clamp (10) and hose (8) from elbow (11) on pressurization valve (21).
8. Remove two clamps (26) and hose (27) from elbows (25) and (7) on rocker cover (24) and pressurization valve (21).

c. Installation (M939A2)

NOTE

All male threads must be wrapped with antiseize tape before installation.

1. Install hose (27) on elbows (7) and (25) on rocker cover (24) and pressurization valve (21) with two clamps (26).
2. Install hose (8) on elbow (11) on pressurization valve (21) with clamp (10).
3. Install hose (14) on elbows (12) and (28) on pressurization valve (21).
4. Install hose (5) on elbows (9) and (6) on power steering reservoir (13) and pressurization valve (21).
5. Install hose (2) on connectors (3) and (30) on tee (4) and pressurization valve (21).
6. Install hose (17) on connectors (16) and (20) on air intake tube (15) and pressurization valve (21).
7. Install hose (18) on elbow (22) and connector (19) on pneumatic controller (23) and pressurization valve (21).
8. Install hose (31) on elbows (32) and (29) on air governor (1) and pressurization valve (21).

4-227. DEEPWATER FORDING AIR PRESSURE HOSE REPLACEMENT (Contd)

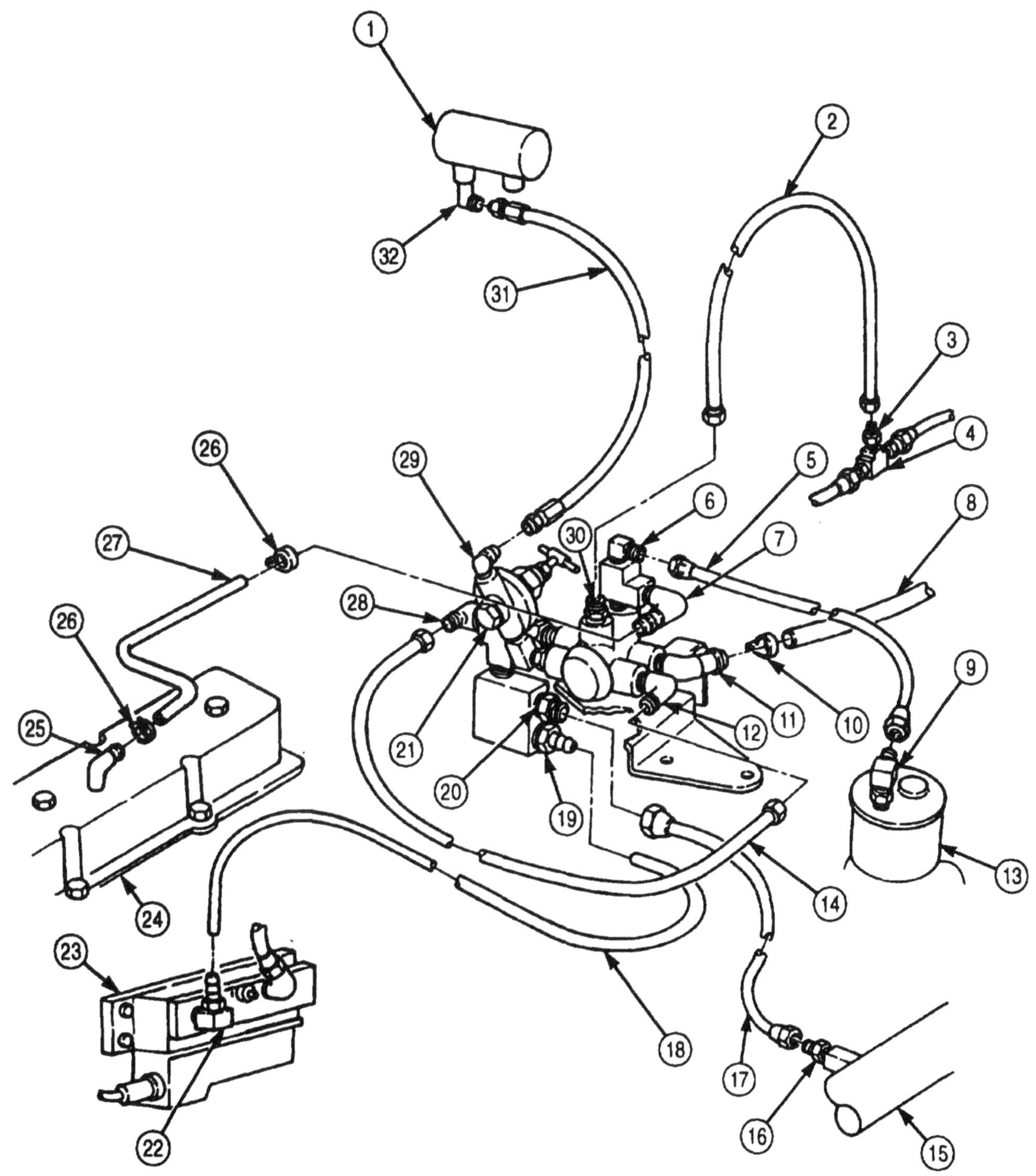

4-227. DEEPWATER FORDING AIR PRESSURE HOSE REPLACEMENT (Contd)

d. Installation (M939/A1)

NOTE

All male threads must be wrapped with antiseize tape before installation.

1. Install bracket (22) on radiator support bracket (23) with screw (29).
2. Install shutoff valve (10) on air compressor (19).
3. Install elbow (12) and drain valve (11) on alcohol evaporator (13).
4. Install air line (91 on shutoff valve (10) and alcohol evaporator (13).
5. Install bushing (16) on tee (14)
6. Install vent hose (8) on bushing (15) and pressurized valve connector (30).
7. Install bushing (17) on tee (16).
8. Install pressure hose (7) on bushing (17) and pressurized valve connector (31).
9. Install five new tiedown straps (18) on pressure hose (71 and vent hose (8).
10. Install adapter (20) on power steering pump (21).
11. Install power steering pump hose (28) on pressurized valve connector (27) and adapter (20).
12. Install power steering pump hose (28) on bracket (22) with clamp (25), screw (24), and new locknut (26).
13. Install supply hose (4) on pressurized valve adapter (1) and regulator valve (6).
14. Install clamp (3) on supply hose (4) with screw (2) and nut (5).

4-227. DEEPWATER FORDING AIR PRESSURE HOSE REPLACEMENT (Contd)

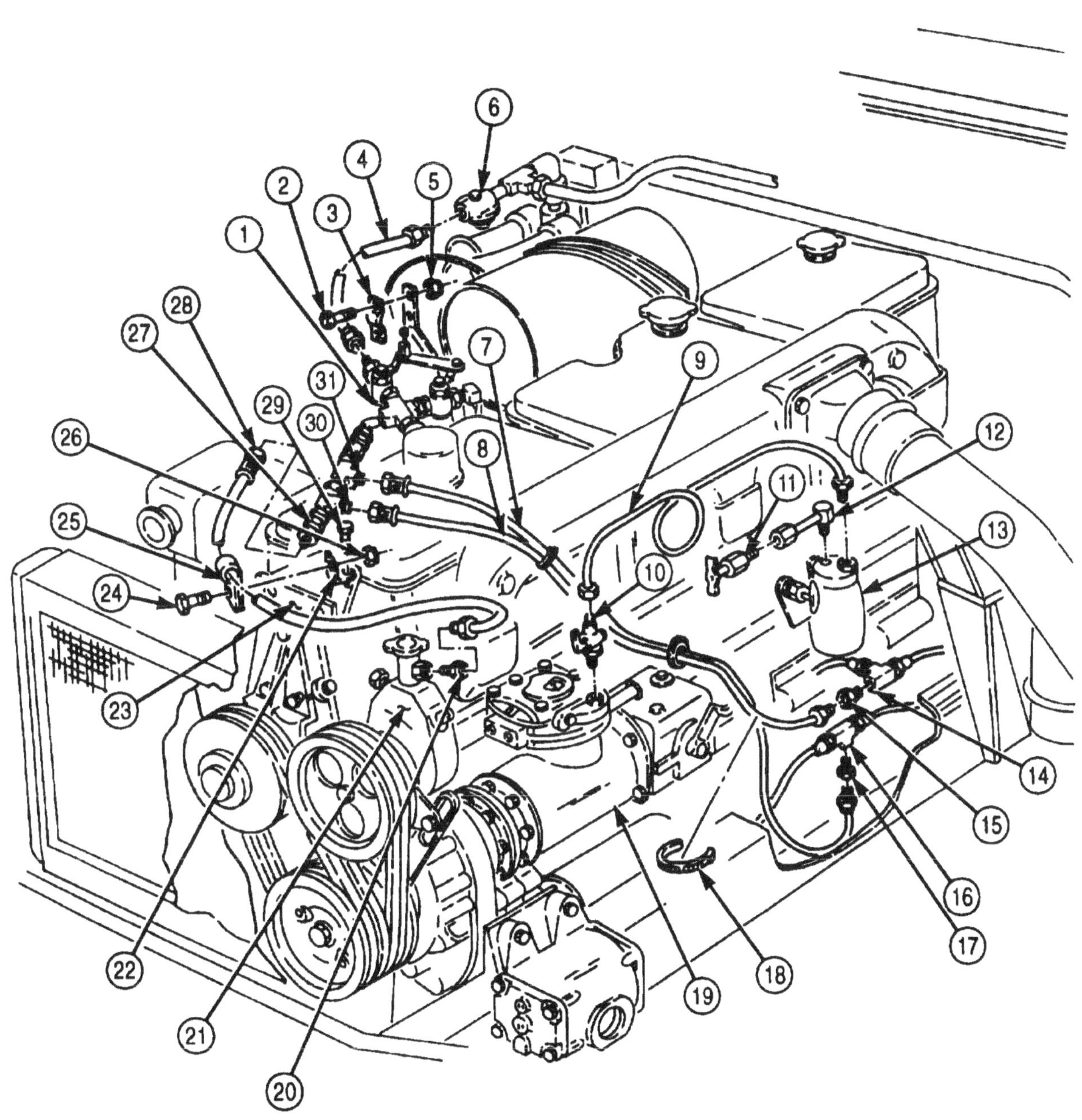

FOLLOW-ON TASK: Start engine (TM 9-2320-272-10) and allow air pressure to build up to normal operating range. Check for air leaks. Road test vehicle.

4-228. DEEPWATER FORDING REGULATOR VALVE REPLACEMENT

THIS TASK COVERS:

a. Removal **b. Installation**

INITIAL SETUP:

APPLICABLE/MODELS
All

TOOLS
General mechanic's tool kit (Appendix E, Item 1)

MATERIALS/PARTS
Antiseize tape (Appendix C, Item 72)

REFERENCES (TM)
TM 9-2320-272-10
TM 9-2320-272-24P

EQUIPMENT CONDITION
- Parking brake set (TM 9-2320-272-10).
- Hood raised and secured (TM 9-2320-272-10).
- Air reservoirs drained (TM 9-2320-272-10).

GENERAL SAFETY INSTRUCTIONS
Do not disconnect air linesbefore draining air reservoir.

WARNING

Do not disconnect air lines before draining air reservoirs. Small parts under pressure may shoot out with high velocity causing injury to personnel.

a. Removal

1. Disconnect supply hose (7) from regulator valve (6).
2. Remove air line (4) and adapter (3) from tee (2).
3. Remove regulator valve (6), nipple (5), tee (2), elbow (l), and reducer (9) from air horn solenoid (8).

b. Installation

NOTE

All male threads must be wrapped with antiseize tape before installation.

1. Install reducer (9), elbow (1), tee (2). nipple (5), and regulator valve (6) on air horn solenoid (8).
2. Install air line (4) and adapter (3) on tee (2).
3. Connect supply hose (7) to regulator valve (6).

INDEX

INDEX (Contd)

INDEX (Contd)

INDEX (Contd)

INDEX (Contd)

INDEX (Contd)

INDEX (Contd)

INDEX (Contd)

INDEX (Contd)

INDEX (Contd)

INDEX (Contd)

INDEX (Contd)

INDEX (Contd)

INDEX (Contd)

INDEX (Contd)

INDEX (Contd)

INDEX (Contd)

www.ingramcontent.com/pod-product-compliance
Lightning Source LLC
Chambersburg PA
CBHW080413030426
42335CB00020B/2436
9781954285651